Springer Proceedings in Complexity

Springer Proceedings in Complexity publishes proceedings from scholarly meetings on all topics relating to the interdisciplinary studies of complex systems science. Springer welcomes book ideas from authors. The series is indexed in Scopus.

Proposals must include the following:

- name, place and date of the scientific meeting
- a link to the committees (local organization, international advisors etc.)
- scientific description of the meeting
- list of invited/plenary speakers
- an estimate of the planned proceedings book parameters (number of pages/articles, requested number of bulk copies, submission deadline)

Submit your proposals to: christoph.baumann@springer.com

More information about this series at http://www.springer.com/series/11637

Christos H. Skiadas · Yiannis Dimotikalis
Editors

13th Chaotic Modeling and Simulation International Conference

Springer

Editors
Christos H. Skiadas
ManLab
Technical University of Crete
Chania, Greece

Yiannis Dimotikalis
Department of Management Science
and Technology
Hellenic Mediterranean University
Agios Nikolaos, Crete, Lasithi, Greece

ISSN 2213-8684 ISSN 2213-8692 (electronic)
Springer Proceedings in Complexity
ISBN 978-3-030-70794-1 ISBN 978-3-030-70795-8 (eBook)
https://doi.org/10.1007/978-3-030-70795-8

This Springer imprint is published by the registered company Springer Nature Switzerland AG
The registered company address is: Gewerbestrasse 11, 6330 Cham, Switzerland

Contents

Contributors

Olga P. Abramova Donetsk National University, Donetsk, Ukraine

Andrii V. Abramov Donetsk National University, Donetsk, Ukraine

Valeriy S. Abramov Donetsk Institute for Physics and Engineering Named After A.A. Galkin, Donetsk, Ukraine

Maricel Agop Department of Physics, Gheorghe Asachi Technical University from Iaşi, Iaşi, Romania

Elena S. Alekseeva Tashkent, UzbekistanVuzgorodok,

Aija Anisimova Department of Mathematics, University of Latvia, Riga, Latvia

Juan F. R. Archilla Group of Nonlinear Physics, Universidad de Sevilla, ETSII, Avda. Reina Mercedes s/n, Sevilla, Spain

R. Atlas University of Maryland, Baltimore County, Baltimore, USA

M. Axenides Institute of Nuclear and Particle Physics, N.C.S.R. Demokritos, Agia Paraskevi, Greece

I. Kusbeyzi Aybar Faculty of Education, Department of Computer Education and Instructional Technology, Yeditepe University, Istanbul, Turkey

O. O. Aybar Piri Reis University, Tuzla, Istanbul, Turkey

Nilufar A. Azamova Tashkent, UzbekistanVuzgorodok,

Svetlana Azarenka Almaty Academy of the Ministry of Internal Affairs of the Republic of Kazakhstan, Nur-Sultan, Kazakhstan

J.-J. Baik Seoul National University, Seoul, South Korea

C. Barone INFN, Gruppo Collegato Salerno and Department of Physics, University of Salerno, Fisciano (SA), Italy

Szymon Bauch Institute of Physics, Polish Academy of Sciences, Warsaw, Poland

Philippe Beltrame UMR1114 EMMAH INRAE - Avignon Université, Avignon, France

Małgorzata Białous Institute of Physics, Polish Academy of Sciences, Warsaw, Poland

Jean-Marie Bilbault ImViA, Université de Bourgogne, Dijon Cedex, France

Stéphane Binczak ImViA, Université de Bourgogne, Dijon Cedex, France

Farzaneh Boroumand Department of Biostatistics, Health School, Mashhad University of Medical Sciences, Mashhad, Iran;
Department of Mathematics and Statistics, Faculty of Science and Engineering, Macquarie University, NSW, Australia

Andre Botha Department of Physics, University of South Africa, Florida Park, South Africa

Belqassim Bouteghrine LCOMS, Université de Lorraine, Metz, France

Inese Bula Department of Mathematics, University of Latvia, Riga, Latvia

S. Carvalho CEAFEL. ADM, ISEL-Engineering Superior Institute of Lisbon, Polytechnic Institute of Lisbon, Lisboa, Portugal

Pengxin Cheng Collaborative Innovation Center of Advanced Nuclear Energy Technology, Key Laboratory of Advanced Reactor Engineering and Safety of Ministry of Education, Institute of Nuclear and New Energy Technology, Tsinghua University, Beijing, People's Republic of China

Gabriel Crumpei Faculty of Psychology and Education Sciences, "Alexandru Ioan Cuza" of Iaşi, Iaşi, Romania

J. Cui North Carolina State University, Raleigh, USA;
San Diego State University, San Diego, USA

Cristina P. de Campos Department of Earth and Environmental Sciences, University of Munich—DEES/LMU, Munich, Germany

Dan Dimitriu Faculty of Physics, Alexandru Ioan Cuza University from Iaşi, Iaşi, Romania

Donald B. Dingwell Department of Earth and Environmental Sciences, University of Munich—DEES/LMU, Munich, Germany

Serhii Donetskyi National Technical University of Ukraine "Igor Sikorsky Kyiv Polytechnic Institue", Kyiv, Ukraine

Hassan Doosti Department of Mathematics and Statistics, Faculty of Science and Engineering, Macquarie University, NSW, Australia

D. P. Dowling School of Mechanical and Materials Engineering, University College Dublin, Dublin, Belfield, Ireland

Vasileios Drakopoulos Department of Computer Science and Biomedical Informatics, University of Thessaly, Lamia, Greece

Mengmeng Du State Key Laboratory for Strength and Vibration of Mechanical Structures, School of Aerospace Engineering, Xi'an Jiaotong University, Xi'an, China

M. Dukarić Center for Applied Mathematics and Theoretical Physics, University of Maribor, Maribor, Slovenia;
SSI Schaefer, Maribor, Slovenia

Elena Efremova Kotelnikov Institute of Radio Engineering and Electronics, Moscow, Russia

Safwan El Assad Université de Nantes/Polytech Nantes, IETR UMR CNRS, Nantes, France

Werner Ertel-Ingrisch Department of Earth and Environmental Sciences, University of Munich—DEES/LMU, Munich, Germany

Elena A. Eseva Russian University of Transport (MIIT), Moscow, Russia

Lucian Eva N. Oblu Clinical Emergency Hospital Iaşi, Iaşi, Romania

S. Faghih-Naini San Diego State University, San Diego, USA;
University of Bayreuth, Bayreuth, Germany;
Friedrich-Alexander University Erlangen-Nuremberg, Erlangen, Germany

B. Ferčec Faculty of Energy Technology, University of Maribor, Krško, Slovenia;
Center for Applied Mathematics and Theoretical Physics, University of Maribor, Maribor, Slovenia

G. Filatrella INFN Gruppo Collegato Salerno and Department of Sciences and Technologies, University of Sannio, Benevento, Italy

Giovanni Filatrella Department of Sciences and Technologies and Salerno Unit of CNISM, University of Sannio, Benevento, Italy

Tatiana F. Filippova Krasovskii Institute of Mathematics and Mechanics, Russian Academy of Sciences, Ural Federal University, Ekaterinburg, Russian Federation;
Department of Optimal Control, N.N. Krasovskii Institute of Mathematics and Mechanics, Ural Branch of Russian Academy of Sciences, Ekaterinburg, Russian Federation

E. Floratos Institute of Nuclear and Particle Physics, N.C.S.R. Demokritos, Agia Paraskevi, Greece;
Department of Nuclear and Particle Physics, National and Kapodistrian University of Athens, Athens, Greece

Konstantina Founta Department of Economics, University of Thessaly, Volos, Greece

Alina Gavriluţ Faculty of Mathematics, Alexandru Ioan Cuza University from Iaşi, Iaşi, Romania

Jean-Marc Ginoux Aix Marseille University, Université de Toulon, CNRS, CPT, Marseille, France

Claudio Guarcello Department of Physics, University of Salerno, Fisciano (SA), Italy;
Dipartimento di Fisica "E.R. Caianiello", Università di Salerno, Fisciano, SA, Italy;
INFN, Sezione di Napoli Gruppo Collegato di Salerno, Complesso Universitario di Monte S. Angelo, Naples, Italy

Adrian-Josue Guel-Cortez Centre for Fluid and Complex Systems, Coventry University, Coventry, UK

Nan Gui Collaborative Innovation Center of Advanced Nuclear Energy Technology, Key Laboratory of Advanced Reactor Engineering and Safety of Ministry of Education, Institute of Nuclear and New Energy Technology, Tsinghua University, Beijing, People's Republic of China

Boris Gutkin Centre for Cognition and Decision Making, National Research University Higher School of Economics, Moscow, Russia;
Group of Neural Theory, LNC INSERM U960, École Normale Supérieure PSL*, University, Paris, France

Harold M. Hastings Bard College At Simon's Rock, Barrington, MA, USA

Max-Olivier Hongler Ecole polytechnique fédérale de Lausanne, STI-DO, Lausanne, Switzerland

Ştefan Irimiciuc Plasma and Radiation, National Institute for Lasers, Bucharest, Romania

Aliyu Isah ImViA, Université de Bourgogne, Dijon Cedex, France;
Kano University of Science and Technology, Kano, Nigeria

Aliasghar Janalizade Department of Physics, Institute for Advanced Studies in Basic Sciences, Zanjan, Iran

Haijun Jia Collaborative Innovation Center of Advanced Nuclear Energy Technology, Key Laboratory of Advanced Reactor Engineering and Safety of Ministry of Education, Institute of Nuclear and New Energy Technology, Tsinghua University, Beijing, People's Republic of China

Bin Jiang College of Power and Energy Engineering, Harbin Engineering University, Heilongjiang, China

Shengyao Jiang Collaborative Innovation Center of Advanced Nuclear Energy Technology, Key Laboratory of Advanced Reactor Engineering and Safety of Ministry of Education, Institute of Nuclear and New Energy Technology, Tsinghua University, Beijing, People's Republic of China

Marian Jurčišin Institute of Experimental Physics, Slovak Academy of Sciences, Košice, Slovakia

Eva Jurčišinová Institute of Experimental Physics, Slovak Academy of Sciences, Košice, Slovakia

D. Katsinis Institute of Nuclear and Particle Physics, N.C.S.R. Demokritos, Agia Paraskevi, Greece;
Department of Nuclear and Particle Physics, National and Kapodistrian University of Athens, Athens, Greece

Shunji Kawamoto Osaka Prefecture University, Sakai, Osaka, Japan

Andrii Khrinenko Department of Applied Mathematics and Information Technologies, Odessa National Polytechnic University, Odessa, Ukraine

Eun-jin Kim Centre for Fluid and Complex Systems, Coventry University, Coventry, UK

O. M. Kiselev Institute of Mathematics, Ufa, Russia;
Innopolis University, Innopolis, Russia

Dmitriy S. Klyuev Povolzhskiy State University of Telecommunications and Informatics, Samara, Russia

Mohammad R. Kolahchi Department of Physics, Institute for Advanced Studies in Basic Sciences, Zanjan, Iran

Nino Kordzakhia Department of Mathematics and Statistics, Faculty of Science and Engineering, Macquarie University, NSW, Australia

Maria V. Korovina Lomonosov MSU, Moscow, Russia

Oksana I. Kos Moscow Aviation Institute (National Research University), Moscow, Russia

Alexander M. Krot Laboratory of Self-Organization System Modeling, United Institute of Informatics Problems, National Academy of Sciences of Belarus, Minsk, Belarus

Pavel Kurasov Department of Mathematics, Stockholm University, Stockholm, Sweden

Ilknur Kusbeyzi Aybar Faculty of Education, Department of Computer Education and Instructional Technology, Yeditepe University, Istanbul, Turkey

Lev Kuzmin Kotelnikov Institute of Radio Engineering and Electronics, Moscow, Russia

V. J. Law School of Mechanical and Materials Engineering, University College Dublin, Dublin, Belfield, Ireland

Michał Ławniczak Institute of Physics, Polish Academy of Sciences, Warsaw, Poland

J. Leonel Rocha CEAUL. ADM, ISEL-Engineering Superior Institute of Lisbon, Polytechnic Institute of Lisbon, Lisboa, Portugal

Dong Liang State Key Laboratory for Strength and Vibration of Mechanical Structures, School of Aerospace Engineering, Xi'an Jiaotong University, Xi'an, China

Linyao Li School of Mathematics and Statistics, Xi'an Jiaotong University, Xi'an, China

Xiang Li College of Power and Energy Engineering, Harbin Engineering University, Heilongjiang, China

G. Linardopoulos Institute of Nuclear and Particle Physics, N.C.S.R. Demokritos, Agia Paraskevi, Greece;
Department of Nuclear and Particle Physics, National and Kapodistrian University of Athens, Athens, Greece

Rodica Luca Department of Mathematics, Gh. Asachi Technical University, Iaşi, Romania

Zheng Luo State Key Laboratory for Strength and Vibration of Mechanical Structures, School of Aerospace Engineering, Xi'an Jiaotong University, Xi'an, China

Nikolai Magnitskii Federal Research Center "Computer Science and Control" of RAS, Lomonosov Moscow State University, Moscow, Russia

Petar Mali Department of Physics, Faculty of Science, University of Novi Sad, Novi Sad, Serbia

Leila S. Marques Instituto de Astronomia, Geofísica e Ciências Atmosféricas—IAG/USP, São Paulo, Brazil

Julio J. Martinell Instituto de Ciencias Nucleares, UNAM, México D.F., Mexico

Hovik A. Matevossian Federal Research Centre "Computer Science & Control" & Steklov Mathematical Institute, Russian Academy of Sciences, Moscow, Russia;
Moscow Avaiation Institute (National Research University), Moscow, Russia

Ashutosh Maurya Department of Mathematics, Institute of Science, Banaras Hindu University, Varanasi, India

Santiago Medina-Carrasco X-Ray Laboratory, CITIUS, Universidad de Sevilla, Avda Reina Mercedes 4B, Sevilla, Spain

Giovanni Migliaccio Università di Pisa, Pisa, Italy

Anatoly M. Neshcheret Povolzhskiy State University of Telecommunications and Informatics, Samara, Russia

Matteo Nicolini Polytechnic Department of Engineering and Architecture, University of Udine, Udine, Italy

Giorgio Nordo MIFT - Università di Messina, Messina, Italy

Roman Opiata Department of Applied Mathematics and Information Technologies, Odessa National Polytechnic University, Odessa, Ukraine

Oleg V. Osipov Povolzhskiy State University of Telecommunications and Informatics, Samara, Russia

S. Pagano INFN, Gruppo Collegato Salerno and Department of Physics, University of Salerno, Fisciano (SA), Italy

Du Yong Pak Department of Mathematics, University of Science, Pyongyang, Democratic People's Republic of Korea

Kosmas Papadopoulos Department of Economics, Democritus University of Thrace, University Campus, Komotini, Greece

Sabrina Francesca Pellegrino Department of Mathematics, Università degli Studi di Bari Aldo Moro, Bari, Italy

Diego Perugini Department of Physics and Geology, University of Perugia, Perugia, Italy

A. S. Piedjou Komnang Department of Physics, University of Salerno, Fisciano (SA), Italy

R. A. Pielke Sr. CIRES and ATOC, University of Colorado At Boulder, Boulder, USA

Vincenzo Pierro INFN, Sezione di Napoli Gruppo Collegato di Salerno, Complesso Universitario di Monte S. Angelo, Naples, Italy;
INFN Gruppo Collegato Salerno and Department of Engineering, University of Sannio, Benevento, Italy

Alexander A. Potapov V. A. Kotelnikov Institute of Radio Engineering and Electronics, Russian Academy of Sciences, Moscow, Russia;
JNU-IREE RAS Joint Lab. of Information Technology and Fractal Processing of Signals, JiNan University, Guangzhou, China

Anupam Priyadarshi Department of Mathematics, Institute of Science, Banaras Hindu University, Varanasi, India

E. Prozorova Mathematical-Mechanical Department St. Petersburg State University, Peterhof, Russia

Natalia Puzanov Ariel University, Ariel, Israel;
Princeton University, Princeton, NJ, USA

Zongchao Qiao Ecole Centrale de Nantes, LS2N UMR CNRS, Nantes, France

Alexander A. Rassadin Nizhnij Novgorod, Russia

Richard Remecky Institute of Experimental Physics, Slovak Academy of Sciences, Košice, Slovakia;
Bogoliubov Laboratory of Theoretical Physics, Joint Institute for Nuclear Research, Dubna, Moscow Region, Russian Federation

A. Rettaroli INFN, Laboratori Nazionali di Frascati, Frascati (Roma) Italy and Department of Mathematics and Physics, University of Roma Tre, Roma, Italy

T. A. L. Reyes San Diego State University, San Diego, USA

SongIl Ri Department of Mathematics, University of Science, Pyongyang, Democratic People's Republic of Korea

Julio Rodriguez Swissgrid AG, Aarau, Switzerland

F. Michael Russell School of Computing and Engineering, University of Huddersfield, Huddersfield, UK

Cristina Marcela Rusu Faculty of Physics, Alexandru Ioan Cuza University from Iaşi, Iaşi, Romania

Volodymyr Rusyn Department of Radio Engineering and Information Security, Yuriy Fedkovych Chernivtsi National University, Chernivtsi, Ukraine

Mazen Saad Ecole Centrale de Nantes, Laboratoire de Mathématiques Jean Leray, UMR CNRS, Nantes, France

Said Sadoudi Ecole Militaire Polytechnique, Algiers, Algeria

Masato Sakai Graduate School of Natural Science and Technology, Kanazawa University, Kakumamachi, Kanazawa, Japan

Mahdi Salehi Department of Mathematics and Statistics, University of Neyshabur, Neyshabur, Iran

Aceng Sambas Department of Mechanical Engineering, Universitas Muhammadiyah Tasikmalaya, Tasikmalaya, Indonesia

Georges Sarafopoulos Department of Economics, Democritus University of Thrace, University Campus, Komotini, Greece

Masayuki Sato Graduate School of Natural Science and Technology, Kanazawa University, Kakumamachi, Kanazawa, Japan

Mohammad T. Shakeri Department of Biostatistics, Health School, Mashhad University of Medical Sciences, Mashhad, Iran

Bo-Wen Shen Department of Mathematics and Statistics, San Diego State University, San Diego, CA, USA

Yuri M. Shukrinov BLTP, JINR, Dubna, Moscow Region, Russia;
Dubna State University, Dubna, Russia

Aleksandr Shvets National Technical University of Ukraine "Igor Sikorsky Kyiv Polytechnic Institute", Kyiv, Ukraine

A. J. Sievers Laboratory of Atomic and Solid State Physics, Cornell University, Ithaca, NY, USA

Leszek Sirko Institute of Physics, Polish Academy of Sciences, Warsaw, Poland

Christos H. Skiadas ManLab, Technical University of Crete, Chania, Crete, Greece

Evangelos N. Skoubris University of West Attica, Egaleo, Athens, Greece

Ilya N. Smirnov Lomonosov MSU, Moscow, Russia

Vladimir Yu. Smirnov Moscow Aviation Institute (National Research University), Moscow, Russia

Michael Snyder Technical Space Science Center, Morehead State University, Morehead, KY, USA

Siavash H. Sohrab Department of Mechanical Engineering, Robert R. McCormick School of Engineering and AppliedScience, Northwestern University, Evanston, IL, USA

Vyacheslav Somsikov Al-Farabi Kazakh National University, Almaty, Kazakhstan

Bernardo Spagnolo Dipartimento di Fisica e Chimica "Emilio Segrè", Group of Interdisciplinary Theoretical Physics, Università di Palermo and CNISM, Unità di Palermo, Palermo, Italy;
Radiophysics Department, Lobachevsky State University, Nizhny Novgorod, Russia

Carolina A. Tafoya Instituto de Ciencias Nucleares, UNAM, México D.F., Mexico

Camel Tanougast LCOMS, Université de Lorraine, Metz, France

Ina Taralova Ecole Centrale de Nantes, LS2N UMR CNRS, Nantes, France

A. S. Tchakoutio Nguetcho LISSAS, Faculté des Sciences, Université de Maroua, Maroua, Cameroon

Jasmina Tekić Laboratory for Theoretical and Condensed Matter Physics - 020, "Vinča" Institute of Nuclear Sciences, National Institute of the Republic of Serbia, University of Belgrade, Belgrade, Serbia

Frangiskos V. Topalis National Technical University of Athens, Zografou, Athens, Greece

Jorge Torres Instituto de Ciencias Nucleares, UNAM, México D.F., Mexico

Elias D. Tsirbas National Technical University of Athens, Zografou, Athens, Greece

Jiyuan Tu Collaborative Innovation Center of Advanced Nuclear Energy Technology, Key Laboratory of Advanced Reactor Engineering and Safety of Ministry of Education, Institute of Nuclear and New Energy Technology, Tsinghua University, Beijing, People's Republic of China;
School of Engineering, RMIT University, Melbourne, VIC, Australia

Adriana Pedrosa Biscaia Tufaile Soft Matter Lab, Escola de Artes, Ciências E Humanidades, University of São Paulo, São Paulo, Brazil

Alberto Tufaile Soft Matter Lab, Escola de Artes, Ciências E Humanidades, University of São Paulo, São Paulo, Brazil

Davide Valenti Dipartimento di Fisica e Chimica "Emilio Segrè", Group of Interdisciplinary Theoretical Physics, Università di Palermo and CNISM, Unità di Palermo, Palermo, Italy;
CNR-IRIB, Consiglio Nazionale delle Ricerche - Istituto per la Ricerca e l'Innovazione Biomedica, Palermo, Italy

Timm A. Vanderelli Ferrocell USA, Ligonier, PA, USA

Sergey Varbanets Department of Computer Algebra and Discrete Mathematics, Odessa I.I. Mechnikov National University, Odessa, Ukraine

Ferdinand Verhulst Department of Mathematics, University of Utrecht, TA Utrecht, The Netherlands

Caio M. Vicentini Department of Earth and Environmental Sciences, University of Munich—DEES/LMU, Munich, Germany;
Instituto de Astronomia, Geofísica e Ciências Atmosféricas—IAG/USP, São Paulo, Brazil

Yakov Vorobyov Department of Mathematics, Informatics and Information Activities, Izmail State University of Humanities, Odessa, Ukraine

George Vostrov Department of Applied Mathematics and Information Technologies, Odessa National Polytechnic University, Odessa, Ukraine

Ying Wu State Key Laboratory for Strength and Vibration of Mechanical Structures, School of Aerospace Engineering, Xi'an Jiaotong University, Xi'an, China

Asher Yahalom Ariel University, Ariel, Israel

Xingtuan Yang Collaborative Innovation Center of Advanced Nuclear Energy Technology, Key Laboratory of Advanced Reactor Engineering and Safety of Ministry of Education, Institute of Nuclear and New Energy Technology, Tsinghua University, Beijing, People's Republic of China

Tai Young-Taft Bard College At Simon's Rock, Barrington, MA, USA

Loukas Zachilas Department of Economics, University of Thessaly, Volos, Greece

Denis Zakharov Centre for Cognition and Decision Making, National Research University Higher School of Economics, Moscow, Russia;
Institute of Biology and Biomedicine, Lobachevsky State University of Nizhni Novgorod, Nizhny Novogorod, Russia

Andrei Zală Electrical Engineering, Energetics and Applied Informatics Faculty, Gheorghe Asachi Technical University from Iaşi, Iaşi, Romania

X. Zeng The University of Arizona, Tucson, USA

Qun Zheng College of Power and Energy Engineering, Harbin Engineering University, Heilongjiang, China

Higgs Boson and Higgs Field in Fractal Models of the Universe: Active Femtoobjects, New Hubble Constants, Solar Wind, Heliopause

Valeriy S. Abramov

Abstract Theoretically the relationship between the main parameters of active femtoobjects and the Higgs boson in fractal models of the Universe was investigated. To describe the structure of the solar wind, heliopause, new Hubble constants are proposed. Estimates of the main parameters are conformed with the experimental data obtained by the Planck space observatory (based on Fermi-LAT and Cerenkov telescopes), UTR-2 and URAN-2 radio telescopes, Parker Solar Probe, Voyager 2 and Voyager 1. Within the framework of the anisotropic model, a description of the main characteristics of the model femtoobject and its relationships with the parameters of the Higgs boson and the Higgs field was performed. To take into account the stochastic behavior of the parameters of a model femtoobject (an active object with dimensions of the order of the classical electron radius), random variables are introduced. Using the example of a hydrogen atom, we estimated the radius of a proton, its mean square deviation, and compared it with an experiment. Estimates of the anomalous contributions to the magnetic moments of leptons based on the lepton quantum number are obtained.

Keywords Model femtoobject · Higgs boson and Higgs field · Fractal models of the Universe · Hubble constants · Structure of the solar wind · Heliopause · Hydrogen atom · Proton and electron radii · Magnetic moments of leptons

1 Introduction

To describe fractal cosmological objects (using binary black holes and neutron stars as an example), the model was proposed in [1, 2] that takes into account the relation between the parameters of the Higgs boson and relict photons, gravitons. Within the framework of this model, the possibility of radiation of gravitational waves from such cosmological objects in the superradiation regime is shown [2]. Higgs field accounting made it possible to propose an anisotropic model of fractal cosmology,

V. S. Abramov (✉)
Donetsk Institute for Physics and Engineering Named After A.A. Galkin, Donetsk, Ukraine
e-mail: vsabramov@mail.ru

C. H. Skiadas et al. (eds.), *13th Chaotic Modeling and Simulation International Conference*, Springer Proceedings in Complexity,
https://doi.org/10.1007/978-3-030-70795-8_1

within the framework of which it is possible to describe the effect of accelerated expansion of the Universe [3]. In this case, a transition to the description of atomic defects, active nanoobjects, and neutrinos is possible [4, 5]. Active objects in fractal quantum systems have their own characteristic features of behavior [6–8]. In this case, superradiative states of active objects may appear [7]. When describing various physical fields (gravitational, electromagnetic, neutrino, deformation, stress) in fractal quantum systems, it is necessary to take into account the ordering effect of the corresponding operators [8]. Coherent laser spectroscopy methods and the modern development of nanotechnology make it possible to study active femtoobjects (protons, neutrons, atomic and muon hydrogens, leptons) in fractal quantum systems. Estimates of the characteristic sizes for the proton radius and Rydberg constant in atomic and muon hydrogens were obtained in [9–11]. Note that active femtoobjects such as leptons have anomalies in magnetic properties [12–14]. For neutrinos, the effect of oscillations (mutual transformations of the electron, muon neutrino and τ-neutrino into each other) is observed [13].

The relationships between the Higgs boson parameters and active nanoelements in fractal systems were studied in [15–17]. Features of the behavior of coupled states of a vortex–antivortex pair were considered in [16]. In [17], the description of the relations of the Higgs boson parameters with cosmological objects in the Universe was proposed. For the accelerated expansion of the Universe, within the framework of this model [17], the relationships of the Hubble constant (old value) with the parameters of the Higgs boson and relict radiation were obtained. The experimental data on the attenuation of gamma rays against an intergalactic background, obtained by the Planck space observatory (based on Fermi-LAT and Cerenkov telescopes), made it possible to determine new values of the Hubble constant and the density of matter in the Universe [18]. The authors explain these new values by the interaction of γ rays with relic photons. In this case, it becomes necessary to agreement the old and new values of the Hubble constants both within the framework of our model and with the cosmological model ΛCDM (plane cosmology). On the other hand, experimental data on the compound, structure, and behavior of the solar wind (flows of various particles) near the Sun [19–24], Earth [25] and in interstellar space (near the heliopause) [26–30] should also be associated with new values of the Hubble constant, the expansion rate, and the density of matter in the Universe.

The aim of this work is to describe the main characteristics of active femtoobjects, the solar wind, heliopause and their relationships with the parameters of the Higgs boson and the Higgs field in fractal models of the Universe.

2 Description of Model Femtoobject

The compound of the solar wind may include active nanoobjects [4–7] and femtoobjects. Based on the results of [1, 2, 4–7], we introduce the main parameters ξ_{2p}, Ω_{A0}, r_p of a model femtoobject

$$\xi_{2p} = \chi_0/n'_F = 1/(N'_p - N); \; \Omega_{A0} = n_{A0}E_e/E_{H0}; \; r_p = 2r_e/(z'_\mu n_F), \quad (1)$$

which are related with the known parameters of quantum electrodynamics

$$r_e = e^2/(m_e c_0^2); \hbar c_0 = e \cdot e_{\alpha 0}; e_{\alpha 0} = e \cdot \alpha_0; \alpha_0 = \hbar c_0/e^2;$$
$$E_e = m_e c_0^2 = e^2/r_e; r_{0p} = m_e r_e/m_p = e^2/E_p; E_p = m_p c_0^2 = e^2/r_{0p};$$
$$\mu_B = e\hbar/2m_e; \mu_N = e\hbar/2m_p. \quad (2)$$

Here r_e and r_{0p}, m_e and m_p, E_e and E_p are classical radii, rest masses, rest energies for electron and proton, respectively; c_0 is limited speed of light in vacuum; $\hbar$ is Planck's constant; e is electron charge; α_0 is fine structure constant; $e_{\alpha 0}$ is renormalized electron charge; μ_B is Bohr magneton; μ_N is nuclear magneton. Next we will use the numerical values $E_e = 0.51099907$ eV, $m_p/m_e = 1836.152701$, $E_p = 938.2723226$ eV, $r_e = 2.81794092$ fm, $r_{0p} = 1.534698568$ am. Note that in this work, model femtoobjects are active objects with sizes of the order of the classical electron radius r_e. Model attoobjects with sizes of the order of the classical proton radius r_{0p} describe the internal structure of nucleons (the presence of a core and scalar, vector clouds [12]). In fractal quantum systems (such as atomic and muon hydrogen), model attoobjects can lead to a change in the main parameters (1), anomalies in magnetic properties (2) and stochastic behavior [8] of model femtoobjects and leptons. In our model, the main parameters of the model femtoobject are related to the resting energy of the Higgs boson E_{H0}, the main parameter n_{A0} for black holes [1, 2], the number of quanta n_F, n'_F of the fermionic field ($n_F + n'_F = 1$) from the anisotropic model (taking into account the presence of the Higgs field) [3], and the cosmological redshift z'_μ [1, 2], the effective susceptibility χ_0 in the absence of the Higgs field [4–7] and the effective number N in the Dicke superradiation model [2]. The numerical values of these parameters are: $E_{H0} = 125.03238$ GeV, $n_{A0} = 58.04663887$, $n_F = 0.945780069$, $n'_F = 0.054219931$, $z'_\mu = 7.18418108$, $\chi_0 = 0.257104198$, $N = 17.0073101$. Using formulas (1), we find the numerical values of the main parameters of the model femtoobject $\xi_{2p} = 4.741876161$, $\Omega_{A0} = 237.232775 \cdot 10^{-6}$, $r_p = 0.829458098$ fm and $N'_p = 17.21819709$.

To take into account the stochastic behavior of the parameters of the model femtoobject, we introduce a random variable $\hat{\xi}_{rp}$ with two possible values ξ_{1p}, ξ_{2p} and their corresponding probabilities P_{1p}, P_{2p}, and expected value $M(\hat{\xi}_{rp}) = 1$. Based on the parameters ξ_{2p}, Ω_{A0} from (1) we find the probabilities P_{1p}, P_{2p}, possible value ξ_{1p}, variance $D(\hat{\xi}_{rp})$, standard deviation $\sigma(\hat{\xi}_{rp})$

$$P_{1p} = \xi_{2p}/(\xi_{2p} + \Omega_{A0}); \; P_{2p} = \Omega_{A0}/(\xi_{2p} + \Omega_{A0}); \; P_{1p} + P_{2p} = 1$$

$$\xi_{1p} = (1 - \xi_{2p}P_{2p})/P_{1p}; \; D(\hat{\xi}_{rp}) = (\xi_{2p} - \xi_{1p})^2 P_{1p}P_{2p}; \; \sigma(\hat{\xi}_{rp}) = D^{1/2}(\hat{\xi}_{rp}). \quad (3)$$

The values of these parameters from (3) are equal: $P_{1p} = 0.999949973$, $P_{2p} = 50.027 \cdot 10^{-6}$, $\xi_{1p} = 0.999812796$, $D(\hat{\xi}_{rp}) = 700.495 \cdot 10^{-6}$, $\sigma(\hat{\xi}_{rp}) = 0.026466865$.

Next, we introduce a random variable $\hat{r}_p = r_p \cdot \hat{\xi}_{rp}$ with two possible values r_p^*, r_e^* and their corresponding probabilities P_{1p}, P_{2p}. If r_p is a constant value, then the possible values r_p^*, r_e^*, expected value $M(\hat{r}_p)$, variance $D(\hat{r}_p)$, standard deviation $\sigma(\hat{r}_p)$ are found by the formulas

$$r_p^* = \xi_{1p} r_p; \; r_e^* = \xi_{2p} r_p; \; M(\hat{r}_p) = r_p^* P_{1p} + r_e^* P_{2p} = r_p;$$
$$D(\hat{r}_p) = (r_e^* - r_p^*)^2 P_{1p} P_{2p}; \; \sigma(\hat{r}_p) = D^{1/2}(\hat{r}_p) \tag{4}$$

The numerical values are equal: $r_p^* = 0.82930282$ fm, $r_e^* = 3.933187582$ fm, $D(\hat{r}_p) = 481.936 \cdot 10^{-6} (fm)^2$, $\sigma(\hat{r}_p) = 0.021953046\, fm$. Our calculated value of the proton radius r_p^* almost coincides with the new experimental value of 0.8293 fm for the proton radius in the hydrogen atom, obtained by 2S-4P spectroscopy (based on quantum interference) [11].

Based on the anisotropic model [1, 2, 4], we find the relationship of the radii r_p, r_p^* with other characteristic parameters r'_p, x_p, y_p, $r_{p\tau}$, $r_{p\tau}^*$, $r_{p\mu}$

$$r'_p - r_p = x_p; \; x_p = r_p \,\mathrm{sn}(u_\mu; k_\mu); \; y_p = r_p \,\mathrm{cn}(u_\mu; k_\mu); \; x_p^2 + y_p^2 = r_p^2;$$
$$r_p - y_p = 3(r'_p - r_{p\tau}); \; r_{p\tau}^* = r_{p\tau}\, n_{F\tau}; \; 2r_{p\mu} = r'_p(1 - S_{1u} - S_{2u}) - 4(r_p - r_p^*). \tag{5}$$

The parameter $\mathrm{sn}(u_\mu; k_\mu) = \sin\varphi_\mu = 0.057234291$ is related to the angle φ_μ [1, 2]; quantum numbers $n_{F\tau} = 0.950987889$, $n'_{F\tau} = 1 - n_{F\tau}$ are related with the lepton quantum number $\Omega_{\tau L} = (n'_{F\tau})^2 = 0.002402187$ from [5]; parameters $|S_{1u}| = 0.046741575$, $S_{2u} = 0.033051284$ defined in [4]. Further, based on expressions (5), we find the numerical values of the characteristic parameters: $r'_p = 0.876931544$ fm, $x_p = 0.047473446$ fm, $y_p = 0.828098429$ fm, $r_{p\tau} = 0.876478321$ fm, $r_{p\tau}^* = 0.833520268$ fm, $r_{p\mu} = 0.841841587$ fm. Our calculated values r'_p and $r_{p\mu}$ practically coincide with the values of 0.8768 fm (the CODATA value) and 0.84184 fm (determined on the basis of fine and ultrafine splitting in the framework of quantum electrodynamics) [9], respectively. Our calculated value $r_{p\tau}^*$ practically coincides with the value of 0.8335 fm for muonic hydrogen [10]. Our anisotropic model [1, 2, 4] also makes it possible to estimate the measurement error δr_p, $\delta r'_p$ using the formulas

$$\delta r_p = \chi_{32}\, r'_p = r_{p\chi}\, \mathrm{sn}(u_\mu; k_\mu)[1 + \mathrm{sn}(u_\mu; k_\mu)]; \; r_{p\chi} = 2r_e\, \chi_{11}/(z'_\mu\, n_F);$$
$$\delta r'_p = r_{d\tau} S_{2u}; \; r_{d\tau} = |\chi_{ef}| r_{F\tau}; \; r_{F\tau} = n_{F\tau}\, r'_p. \tag{6}$$

Taking into account $\chi_{11} = 0.181800122$, $\chi_{32} = 0.010405201$, $|\chi_{ef}| = 0.250425279$ from [1, 2] and expressions (6) we find estimates of measurement errors $\delta r_p = 0.009124649$ fm, $\delta r'_p = 0.006902512$ fm, which do not disagree the experimental estimates of 0.0091 fm from [11], 0.0069 fm from [9], respectively. In this case, the calculated value of the radius $r_{d\tau} = 0.208842481$ fm in our model is near the mean square radius of the electric charge distribution in the core of nucleons equal to 0.21 fm [12]. The radius $r_{F\tau} = 0.833951278$ fm is related with the characteristic radii $r'_{F\tau}$, $r_{\tau L}$ and the value $\Omega'_{\tau L} = 0.97597813$ by the expressions

$$r'_{F\tau} = n'_{F\tau} r'_p;\ (r'_{F\tau})^2 + (r_{\tau L})^2 = (r'_p)^2;$$
$$r^2_{\tau L} = \Omega'_{\tau L}(r'_p)^2;\ \Omega'_{\tau L} = 1 - \Omega_{\tau L} = n_{F\tau}(1 + n'_{F\tau}). \quad (7)$$

The values of these radii are equal: $r'_{F\tau} = 0.042980266$ fm, $r_{\tau L} = 0.866334751$ fm. Anomalies in the magnetic moments of leptons can be determined by the influence of CMB radiation. In this case, relict radiation can lead to effects of renormalization of the initial parameters: fine structure constant α_0, electron charge e, limiting speed of photon propagation in vacuum c_0; rest masses m_e, m_μ, m_τ and magnetons μ_B, $\mu_\mu = e\hbar/2m_\mu$, $\mu_\tau = e\hbar/2m_\tau$ for electron, muon, τ-lepton, respectively. The magnetic moments of leptons $< \hat{\mu}_e >$, $< \hat{\mu}_\mu >$, $< \hat{\mu}_\tau >$ for an electron, muon, τ-lepton, respectively, are determined by the expressions

$$2 < \hat{\mu}_e >= (2 + \Omega_{\mu e})\mu_B;\ 2 < \hat{\mu}_\mu >= (2 + \Omega_{\mu\mu})\mu_\mu;\ 2 < \hat{\mu}_\tau >= (2 + \Omega_{\mu\tau})\mu_\tau. \quad (8)$$

Anomalous contributions to magnetic moments and renormalization effects are described by parameters $\Omega_{\mu e}$, $\Omega_{\mu\mu}$, $\Omega_{\mu\tau}$ for electron, muon, τ-lepton, respectively, based on the lepton number $\Omega_{\tau L}$

$$\Omega_{\mu e} = \Omega_{\tau L} - \Omega_{HL};\ \Omega_{HL} = E_{HL}/E_{H0};\ E_{HL} = n'_{H3}E_e;\ N' = 17.21088699; \quad (9)$$

$$\Omega_{\mu\mu} = \Omega_{\tau L} - \Omega'_{NL};\ \Omega'_{NL} = E'_{NL}/E_{H0};\ E'_{NL} = N'E_e;\ (N' - N) \cdot \chi_0 = n'_{\mu F}; \quad (10)$$

$$\Omega_{\mu\tau} = \Omega_{\tau L} - 0.5(\Omega_{HL} + \Omega_{GL});\ \Omega_{GL} = E_{GL}/E_{H0};\ E_{GL} = n_G E_e. \quad (11)$$

Additional contributions Ω_{HL}, Ω'_{NL}, Ω_{GL} are determined based on the energies E_{HL}, E'_{NL}, E_{GL} and the resting energy of the Higgs boson E_{H0}. From (9–11) it follows that these additional energies are determined by the numbers of quanta n'_{H3}, N', n_G and the rest energy of the electron E_e. Wherein

$$n'_{H3} = n_{H3}/(1 + \Omega_{0\nu});\ 1 + \Omega_{0\nu} = 1 + (n'_F)^2 = 1 + (N'_p - N)^2 \cdot \chi_0^2; \quad (12)$$

$$n_{H3} = Q_{H3}n_{h2} = 0.5Q_{H3}n_{A0};\ n_{A0} = z'_\mu(z'_\mu + 1) - n_Q/n_g;\ n_Q = 2n_G. \quad (13)$$

Here $n_g = 8$, $n_Q = 6$, $n_G =< \hat{c}_G \hat{c}^+_G >= 3$ and $n'_G =< \hat{c}^+_G \hat{c}_G >= 2$ can be interpreted as the numbers of quanta of the gluon, quark, excited, and ground states of the gravitational fields, respectively; neutrino density $\Omega_{0\nu} = 0.002939801$ [4]. Based on (13) we find $n_{H3} = 20.33926863$. Further, taking into account (12, 10), we obtain $n'_{H3} = 20.27965049$, $n'_{\mu F} = 0.052340473$. Based on Eqs. (9–11) we find the energies $E_{HL} = 10.36288254\,\text{MeV}$, $E'_{NL} = 8.794747246\,\text{MeV}$, $E_{GL} = 1.53299721\,\text{MeV}$; additional contributions $\Omega_{HL} = 82.88159067 \cdot 10^{-6}$, $\Omega'_{NL} = 70.33975716 \cdot 10^{-6}$, $\Omega_{GL} = 12.26080164 \cdot 10^{-6}$. The found parameters $\Omega_{\mu e}/2 = 1159.652705 \cdot 10^{-6}$, $\Omega_{\mu\mu}/2 = 1165.923621 \cdot 10^{-6}$, $\Omega_{\mu\tau}/2 = 1177.307902 \cdot 10^{-6}$ coincide with the data [14] for anomalies of the magnetic moments of leptons.

3 New Hubble Constants

The parameters of active nanoobjects and femtoobjects are related with cosmological parameters. To describe accelerated expansion of the Universe in model I [17] and the anisotropic model [1, 2, 4], the Hubble constants H_{01}, H_{02}, H_0, characteristic distances L_{01}, L_{02}, L_0, speeds υ_{01}, υ_{02}, υ_0 were introduced

$$H_{01} = c_0/L_{01} = \upsilon_{01}/L_0; \;\; H_{02} = c_0/L_{02} = \upsilon_{02}/L_0; \;\; H_0 = \upsilon_0/L_0; \;\; H_0 = \upsilon_0/L_0. \tag{14}$$

The values $L_0 = 1\text{Mpc}$, $H_{01} = 73.2\,\text{km} \cdot \text{s}^{-1} \cdot \text{Mpc}^{-1}$, $L_{01} = 4.0954948\,\text{Gpc}$ (distance to supernova type 1a), $\upsilon_{01} = 73.2\,\text{km} \cdot \text{s}^{-1}$ and $L_{02} = 4.2574359\,\text{Gpc}$ (event horizon), $H_{02} = 70.415674\,\text{km} \cdot \text{s}^{-1} \cdot \text{Mpc}^{-1}$, $\upsilon_{02} = 70.415674\,\text{km} \cdot \text{s}^{-1}$ were obtained on the basis of the analysis of supernova type 1a [3] and measurements by Cepheids, respectively. The Hubble constant $H_0 = 67.83540245\,\text{km} \cdot \text{s}^{-1} \cdot \text{Mpc}^{-1}$, velocity $\upsilon_0 = 67.83540245\,\text{km} \cdot \text{s}^{-1}$ were introduced in [1, 2, 4] to describe the radiation of gravitational waves, relict photons from binary black holes, neutron stars based on the expression

$$\upsilon_0 = \upsilon_{01}/\Omega_{tH}; \;\; \Omega_{tH} = Q_{H0} + |S'_{01}|; \;\; Q_{H0} = \upsilon_{01}/\upsilon_{02} = H_{01}/H_{02} = L_{02}/L_{01}. \tag{15}$$

Here are $Q_{H0} = 1.039541282$, $|S'_{01}| = 0.039541282$. New experimental data on the attenuation of γ-rays against an intergalactic background [18] make it possible to introduce a new Hubble constant H_0^*, velocity υ_0^*, and matter density Ω_m based on expressions

$$H_0^* = \upsilon_0^*/L_0; \;\; \upsilon_0^* = \upsilon_{01}/\Omega_{tH}^*; \;\; \Omega_{tH}^* - \Omega_{tH} = S_{012}; \;\; S_{012} = |S'_{01}| - S'_{02}. \tag{16}$$

Here is $S'_{02} = 0.03409$. The numerical values of $H_0^* = 67.49443576\,\text{km} \cdot \text{s}^{-1} \cdot \text{Mpc}^{-1}$, $\upsilon_0^* = 67.49443576\,\text{km} \cdot \text{s}^{-1}$, $\Omega_m = (n'_F + \Omega'_{c1})/2 = 0.141145722$ (the parameter $\Omega'_{c1} = 0.228071512$ is related to the gap in the energy spectrum of relict photons) are close to the experimental data from [18]. From [16] it follows the connection of parameters H_{01}, υ_{01} for the accelerated expansion of the Universe with new parameters H_0^*, υ_0^*. Our parameters H_0, υ_0 and new parameters H_0^*, υ_0^* are close to the main parameters H_0', υ_0' of the model ΛCDM (plane cosmology). In our model H_0', υ_0' are defined by expressions

$$H_0' = \upsilon_0'/L_0;\ \ \upsilon_0' = \upsilon_{01}/\Omega_{tH}';\ \ \Omega_{tH}' = \Omega_{tH}^* + \Omega_{0\nu} + n_g\Omega_{A0}/n_{A0}. \quad (17)$$

Values $H_0' = 67.30995226\,\text{km}\cdot\text{s}^{-1}\cdot\text{Mpc}^{-1}$, $\upsilon_0' = 67.30995226\,\text{km}\cdot\text{s}^{-1}$ are close to the parameters of the planar cosmology model.

4 Solar Wind and Heliopause

The Sun is the source of solar wind (flows of photons and various particles) [19]. Photons achieve the Earth after 8 min, and high-energy particles arrive with a delay of 100 min [20]. To estimate the characteristic distances and times, we use

$$L_{ES}' = L_{ES}/Q_{H0} = c_0 t_{ES} = \upsilon_{H0}\, t_{ES}';\ \ n_{H0} = Q_{H0}^2 = (1 + |S_{01}'|)^2 = \upsilon_{01}^2/\upsilon_{02}^2, \quad (18)$$

where $\upsilon_{H0}^2 = c_0^2/n_{H0}$. Taking into account the numerical values of the distance from the Earth to the Sun $L_{ES} = 1\,\text{au} = 1.495995288\cdot 10^8\text{km}$, the limiting speed of light in a vacuum $c_0 = 2.99792458\cdot 10^5\text{km}\,\text{s}^{-1}$, we find estimates of the refractive index of the medium $n_{H0} = 1.080646077$, the speed of photon propagation in the medium $\upsilon_{H0} = 2.883891801\cdot 10^5\text{km}\,\text{s}^{-1}$, the distance $L_{ES}' = 0.961962759\,\text{au}$, and the times of arrival of photons to the Earth from the Sun in vacuum $t_{ES} = 480.0293392\,\text{s}$ and in the medium $t_{ES}' = 499.0103147\,\text{s}$.

To estimate the delay time t_{0m} of particles, arriving on the Earth from the Sun, we use the expressions

$$2\,t_{0m} = \tau_{0\gamma}\ln N_{0m};\ \ \tau_{0\gamma} = \tau_{0\alpha}/n_{0\alpha};\ \ \tau_{0\alpha} = \nu_{0\alpha}^{-1};\ \ n_{0\alpha} = 1.5 + |\xi_{0H}|^2;$$

$$\ln N_{0m} = 2n_{0\alpha}\ln N_{0\alpha};\ \ Q_{H2}N_{0\alpha} = 0.5 + \Omega_{c1}' + n_{F\tau}';\ \ \nu_{0\alpha} = \nu_{H0}/N_{0A}. \quad (19)$$

Expressions (19) were obtained in the framework of the Dicke theory of superradiance and describe the main parameters $\tau_{0\gamma}$, t_{0m} of the superradiance pulse in a medium from a state with the number of particles N_{0m}.

Based on the numerical values $N_{0A} = 3.557716045\cdot 10^5$, $\nu_{H0} = 50.182731\,\text{Hz}$, $|\xi_{0H}|^2 = 0.181800122$, $Q_{H2} = 1/3$, $n_{F\tau}' = 0.049012111$ we find estimates of the frequency $\nu_{0\alpha} = 141.0532217\mu Hz$, relaxation time $\tau_{0\alpha} = 118.1587096\,\text{min}$, fractal parameter $n_{0\alpha} = 1.681800122$, coherent spontaneous relaxation time $\tau_{0\gamma} =$

70.25728449 min, effective numbers of active particles $N_{0\alpha} = 2.331250869$ and $N_{0m} = 17.23047995$, delay time $t_{0m} = 100.0101199$ min.

To estimate the characteristic parameters for the region near the boundary of the heliopause, we first find the relationships between the rest energies E_{0E} and E_{H0}, rest masses M_E and m_{H0}, the gravitational radii of Schwarzschild R_{GE} and R_{H0} for the Earth and the Higgs boson, respectively, by the formulas

$$E_{0E}/N_a E_{H0} = M_E/N_a m_{H0} = R_{GE}/N_a R_{H0} = n_{0E};\ E_{H0} = c_0^2 m_{H0};$$

$$m_{H0} = c_0^2 R_{H0}/2GN_a;\ R_{GE} = A_G E_{0E};\ A_G = R_{H0}/E_{H0} = 2GN_a/c_0^4;$$

$$R_{GE} = N_{GE} L_{ES} = n_{0E} N_a R_{H0};\ M_E = 5.977 \cdot 10^{27}\text{g}. \tag{20}$$

Based on (20) we find the parameters of the theory $A_G = 0.960836162$ fm (eV)$^{-1}$, $n_{0E} = 73.87419814$, $R_{GE} = 5.347530124 \cdot 10^{18}$km, $N_{GE} = 3.574563481 \cdot 10^{10}$.

Taking into account (18) in the framework of the anisotropic model [4] we find the characteristic velocities υ_{hS}, υ'_{hS}, distances L_{hS}, L'_{hS}, time of arrival of the signal from the heliopause to the Earth t_{hS} from the expressions

$$\upsilon'_{hS} = Q_{H0}\,\upsilon_{hS} = |\chi_{ef}|\upsilon_{01};\ L_{hS} = N_{hS} L_{ES};\ L'_{hS} = N'_{hS} L_{ES};\ L^*_{hS} = L_{hS}/Q_{H0};$$

$$N'_{hS} = n_{H0} N_{hS};\ L_{hS}/R_{GE} = \upsilon^2_{hS}/c_0^2;\ L_{hS}/L_{ES} = t_{hS}/t_{ES}. \tag{21}$$

Based on (18–21), the values $|\chi_{ef}| = 0.250425279$ from [4], we find the estimates $\upsilon_{hS} = |\chi_{ef}|\,\upsilon_{02} = 17.63386481$ km s^{-1}, $N_{hS} = 123.6734916$, $N'_{hS} = 133.6472735$, $L_{hS} = 1.850149607 \cdot 10^{10}$km, $t_{hS} = t_{ES} N_{hS} = 16.49080679$ hour. The speed υ_{hS} is close to the speed $\upsilon_{V2} = 17.5$ km s^{-1} of the V2 probe; the distance $L^*_{hS} = 118.9692932\,au$ is near the distance to the heliopause boundary $L_{V2} = 119$ au from [26].

To describe the transition region near the boundary of the heliopause, we introduce the times t_1, t_2, t_3, distances L_1, L_2, L_3. Next, we find the characteristic time intervals t_{31}, t_{21}, t_{32} by the formulas

$$t_{31} = t_3 - t_1 = 1/\nu_{31};\ \nu_{31} = (1 - \psi_{02})\nu_{H0} S_{2u}/N_{0A};\ t_{21} = t_2 - t_1 = t_{31} P_\tau;$$

$$t_{32} = t_3 - t_2 = t_{31} P'_\tau;\ P_\tau + P'_\tau = 1;\ P'_\tau = 1/(2 + S'_{03}). \tag{22}$$

Using the parameters $\psi_{02} = 0.984494334$, $S'_{03} = 0.460458718$ from [4], we obtain numerical values: frequency $\nu_{31} = 0.072287263\mu Hz$; probabilities $P_\tau = 0.593571722$, $P'_\tau = 0.406428278$; time intervals $t_{31} = 160.1122188$ day, $t_{21} = 95.03808539$ day, $t_{32} = 65.07413336$ day. The obtained values of the intervals t_{21}

and t_{32} practically coincide with the time intervals of 95 days and 65 days for the transition region near the heliopause boundary from [26, Fig. 1a].

The characteristic distance L_3 for interstellar space (outside the heliopause at $L_3 > L_2$) is determined from the expressions

$$L_3 = N_{L3} L_{ES};\ N_{L3} = (1 - \Omega_{hL} - S_{2u}) N_{hS}. \tag{23}$$

Using the parameters $\Omega_{hL} = 0.000118617$ from [4, 5], N_{hS} from (21), we find the value $N_{L3} = 119.5712542$ and the estimate of the distance $L_3 = 119.5712542$ au. To estimate the distance L_1 (inside the heliosphere for $L_1 < L_2$), we use the characteristic distances $L_{\mu e}$, $L_{\mu\mu}$, $L_{\mu\tau}$ for e, μ, τ-leptons, respectively, determined by the expressions

$$L_{\mu e} = N_{\mu e} L_{ES};\ N_{\mu e} = n_{\mu e} N_{hS};\ n_{\mu e} = (2 + \Omega_{\mu e}) - (1 + S_{1u});$$

$$L_{\mu\mu} = N_{\mu\mu} L_{ES};\ N_{\mu\mu} = n_{\mu\mu} N_{hS};\ n_{\mu\mu} = (2 + \Omega_{\mu\mu}) - (1 + S_{1u});$$

$$L_{\mu\tau} = N_{\mu\tau} L_{ES};\ N_{\mu\tau} = n_{\mu\tau} N_{hS};\ n_{\mu\tau} = (2 + \Omega_{\mu\tau}) - (1 + S_{1u}). \tag{24}$$

Using the parameters $\Omega_{\mu e}$, $\Omega_{\mu\mu}$, $\Omega_{\mu\tau}$ from (9–11), based on (24) we find the estimates of distances $L_{\mu e} = 118.1796344$ au, $L_{\mu\mu} = 118.1811855$ au, $L_{\mu\tau} = 118.1840014$ au. For search of the characteristic distance L_2 (as the heliopause boundary), we consider a random variable $\hat{L}_2$ with two possible values L_3 from (23), $L_1 = L_{\mu e}$ from (24) and their corresponding probabilities $P_{\psi 01}$, $P'_{\psi 01}$. For expected value $M(\hat{L}_2)$, variance $D(\hat{L}_2)$, deviation $\sigma(\hat{L}_2)$, we have

$$M(\hat{L}_2) = P_{\psi 01} L_3 + P'_{\psi 01} L_{\mu e} = L_2;\ D(\hat{L}_2) = (L_3 - L_{\mu e})^2 P_{\psi 01} P'_{\psi 01};\ \sigma(\hat{L}_2) = D^{1/2}(\hat{L}_2);$$

$$P_{\psi 01} + P'_{\psi 01} = 1;\ P'_{\psi 01} = \psi_{01}/(1 + S'_{03} + \psi_{01});\ \psi_{01} = 1.015268884. \tag{25}$$

The numerical values of the distance $L_2 = 119.0005661$ au and space intervals $L_{32} = L_3 - L_2 = 0.57068813$ au, $L_{21} = L_2 - L_{\mu e} = 0.8209317$ au practically coincide with the characteristic values of 119 au, 0.57 au, 0.82 au, respectively, from [26, Fig. 1a]. Based on (22), (25), we find the average values of the velocities υ_{21} (inside the heliosphere), υ_{32} (outside the heliopause), the jump in velocities $\delta\upsilon_{21}$ (at the heliopause) and the ratio of velocities $\upsilon_{32}/\upsilon_{21}$

$$\upsilon_{21} = L_{21}/t_{21} = L_{31} P_{\psi 01}/t_{31} P_\tau;\ \upsilon_{32} = L_{32}/t_{32} = L_{31} P'_{\psi 01}/t_{31} P'_\tau;\ L_{31} = L_3 - L_1;$$

$$\delta\upsilon_{21} = \upsilon_{32} - \upsilon_{21};\ \upsilon_{32}/\upsilon_{21} = \psi_{01} = \varepsilon_{01}/E_{H0} = \nu_{01}/\nu_{H0}. \tag{26}$$

The numerical values are equal: $\upsilon_{21} = 14.95635805\,\text{km s}^{-1}$, $\upsilon_{32} = 15.18472495\,\text{km s}^{-1}$, $\delta\upsilon_{21} = 228.366896\,\text{m s}^{-1}$. We note, that the probabilities $P_{\psi 01}$ and P_τ are coupled through a conditional probability $P_{\psi\tau}$, and the ratio of the velocities and the jump in velocities allow us to introduce probabilities P_ψ, P'_ψ using expressions of the type

$$P_{\psi 01} = P_\tau P_{\psi\tau};\;\; P_{\psi\tau} = (2 + S'_{03})/(1 + S'_{03} + \psi_{01}) = 1/(1 + n_{01});\;\; P_\psi + P'_\psi = 1;$$

$$P_\psi = 1/\psi_{01} = \upsilon_{21}/\upsilon_{32};\;\; P'_\psi = \delta\upsilon_{21}/\upsilon_{32};\;\; n_{01} = (\psi_{01} - 1)/(2 + S'_{03}). \quad (27)$$

From (27) it follows, that n_{01} is a function of two arguments ψ_{01} and S'_{03}. If the Higgs field is absent ($\psi_{01} = 1$), then from (27) we obtain: $n_{01} = 0$; probabilities $P_{\psi\tau} = 1$, $P_{\psi 01} = P_\tau$, $P_\psi = 1$, $P'_\psi = 0$; jump in speed $\delta\upsilon_{21} = 0$ and equality of speeds $\upsilon_{21} = \upsilon_{32}$. The presence of the Higgs field ($\psi_{01} \neq 1$) leads to the appearance of a velocity jump, when crossing the heliopause boundary. Replacing the parameter S'_{03} in (27) with other parameters S'_{0x}, S_{xu} ($x = 1, 2, 3, 4$) of the energy (frequency) spectra leads to a change in the probabilities and stochastic behavior of the velocities υ_{21}, υ_{32}.

The anisotropic model [4] and expressions (1, 4) allow us to obtain relationships of velocities υ_{32}, υ_{21} with characteristic velocities $\upsilon_{\psi u}$, υ_{eu} (active nanoobjects, femtoobjects that are part of the solar and galactic wind) of the type

$$\upsilon_{32} = n'_F\,\upsilon_{\psi u} = \chi_0\,\upsilon_{eu} = \psi_{01}\,\upsilon_{21};\;\; \upsilon_{\psi u} = \xi_{2p}\,\upsilon_{eu};\;\; \xi_{2p} = r_e^*/r_p. \quad (28)$$

Based on (28) we find the velocity estimates $\upsilon_{eu} = 59.04358906\,\text{km s}^{-1}$, $\upsilon_{\psi u} = 279.9773874\,\text{km s}^{-1}$. On the other hand, the characteristic solar wind velocity $\upsilon_{\psi u}$ is related to the Hubble constants H_{01} and H_{02}, H_0, H_0^*, H'_0, velocities υ_{01} and υ_{02}, υ_0, υ_0^*, υ'_0 for models from (14, 15, 16, 17), respectively, by expressions of the type

$$0.5\upsilon_{\psi u} = 2\upsilon_{02} - \upsilon_{0A} = \upsilon_W - \upsilon_q - \upsilon_{0A};\;\; \upsilon_q = \upsilon_{01} - \upsilon_{02} = \upsilon_W - 2\upsilon_{02};$$

$$\upsilon_W = \upsilon_{01} + \upsilon_{02} = \upsilon_0\,\Omega_{tH} + \upsilon_{02} = \upsilon_0^*\,\Omega_{tH}^* + \upsilon_{02} = \upsilon'_0\,\Omega'_{tH} + \upsilon_{02};\;\; \upsilon_{0A} = c_0/N_{0A}. \quad (29)$$

Values of speeds are equal: $\upsilon_{0A} = 0.84265426\,\text{km s}^{-1}$, $\upsilon_W = 143.615674\,\text{km s}^{-1}$, $\upsilon_q = 2.784326\,\text{km s}^{-1}$.

The velocity υ_{hS} from (21) is related to the characteristic velocities of relict photons υ_{ra}, υ_{ra}^* and the velocities υ_{02}, υ_0^*, $\upsilon_{0\rho}$, υ_W, $\upsilon_{h\rho}$ by expressions of the type

$$2\,\upsilon_{hS}\,\upsilon_{ra} = \upsilon_{ra}^*\,\upsilon_{02};\;\; \upsilon_{ra} = c_0/N_{ra};\;\; \upsilon_{ra}^* = 2|\chi_{ef}|\upsilon_{ra};$$

$$\upsilon_{ra}^*\,\upsilon_{0\rho} = \upsilon_{ra}\,\upsilon_0^*;\;\; \upsilon_W^2 = \upsilon_{0\rho}^2 + \upsilon_{h\rho}^2;\;\; N_{ra} = 1041.293475; \quad (30)$$

Values of speeds are equal: $\upsilon_{ra} = 287.9039053\,\text{km s}^{-1}$, $\upsilon^*_{ra} = 144.1968316\,\text{km s}^{-1}$, $\upsilon_{0\rho} = 134.7596298\,\text{km s}^{-1}$, $\upsilon_{h\rho} = 49.65182785\,\text{km s}^{-1}$.

The experimental data obtained by the Wind probe (the interval of solar wind speed changes of 600–300 km s^{-1}, Fig. 6 from [25]), on the UTR-2, URAN-2 radio telescopes (Fig. 5 from [25]) showed, that the solar wind in orbit and beyond the Earth's orbit consists of a set of particle flows with different velocities and densities. The structure of these flows depends on time and solar activity [19, 20]. An analysis [25] of intermode (intramode) interactions of particles of different flows was performed by the interplanetary scintillation method based on the behavior of space and time correlation functions for radiation intensity. The velocities $2\,\upsilon_{0\rho}$, $\upsilon_{\psi u}$ and υ_{ra} are close to the characteristic velocities of 270, 280 and 290 km s^{-1} of separate solar wind modes from [25]. The detailed analysis of the multimode structure of the solar wind in our model is possible based on spectra of type $\upsilon_{\psi ux} = 2\,\upsilon_{\psi u} S_{xu}$ and $\upsilon_{rax} = 2\,\upsilon_{ra} S_{xu}$. From (30) it follows that the velocities $\upsilon_{0\rho}$ and $\upsilon_{h\rho}$ can be interpreted as both the radial and transverse components of the total velocity υ_W. The presence of transverse components $\pm\,\upsilon_{h\rho}$ of the solar wind near the Sun is confirmed by experimental data collected by the Parker Solar Probe [21–24]. The behavior of the transverse component (Fig. 2 from [22]) is stochastic and varies in the range from 50 to –50 km s^{-1}. In [24], such a behavior of the slow solar wind is associated with the presence of equatorial coronal holes in the Sun. A fast solar wind with speeds $2\,\upsilon_{0\rho}$ occurs near the poles of the Sun.

In our model, it is also possible to describe the multimode structure of the solar and galactic winds at the crossing of the heliopause based on the velocities υ_{eu} from (28), υ_W from (29), υ^*_{ra} from (30) and the corresponding velocity spectra. The experimental data (Fig. 4d from [27], Fig. 2 from [29]) confirm the stochastic behavior and change in the velocity of solar wind particles when the heliopause crosses from 150 km s^{-1} to 100 km s^{-1}. The complex dynamic behavior of the plasma components (Figs. 3, 4 from [29]) with velocities near υ_{eu}, $2\,\upsilon_{eu}$ inside the heliosphere indicates the presence of a boundary layer near the heliopause.

To estimate the characteristic energies ε_{0A}, E_{0A}, $\varepsilon_{\lambda A}$, effective wavelength λ_A, effective number N_{0n} of particles, we use expressions of the type

$$E_{H0}/\varepsilon_{0A} = E_{0A}/E_G = N_{0A};\;\; E_{H0}/E_{0A} = \varepsilon_{0A}/E_G = N_{0n};$$

$$E_{H0}/E_G = N_{HG} = N_{0n} N_{0A};\;\; \varepsilon^2_{\lambda A} = \varepsilon_{0A} E_{0A} = E_{H0} E_G;\;\; \lambda_A = a_\lambda/\varepsilon_{\lambda A}. \quad (31)$$

Taking into account $N_{0A} = 3.557716045 \cdot 10^5$, $N_{HG} = 1.031830522 \cdot 10^{16}$, a_λ from [6] we find the estimates: $\varepsilon_{0A} = 351.4400206\,\text{keV}$, $E_{0A} = 4.311073329\,\text{eV}$, $\varepsilon_{\lambda A} = 1.230887363\,\text{keV}$, $\lambda_A = 1.007114093\,\text{nm}$, $N_{0n} = 2.900261036 \cdot 10^{10}$.

The presence of a multimode structure of the solar and galactic wind, the Higgs field leads to the replacement $\varepsilon_{\lambda A}$, λ_A by $\varepsilon^*_{\lambda A}$, λ^*_A by the formulas

$$\varepsilon^*_{\lambda A} = \psi_{rc}\varepsilon_{bb};\;\; \lambda^*_A = a_\lambda/\varepsilon^*_{\lambda A} = 2R_{\lambda A};\;\; E_{\lambda A} = R_{\lambda A}/A_G;$$

$$\varepsilon_{bb} = \varepsilon_{0A}(| S_{1u} | + S_{2u}); \ \psi_{rc} = 2\Delta_{rc}/E_{0A} = (\varepsilon_{01} - \varepsilon_{02})S_{1u}/\varepsilon_{02}S_{2u}. \quad (32)$$

The values are equal: $\varepsilon_{bb} = 28.042404\,\text{keV}$, $\psi_{rc} = 0.04420725$, $\Delta_{rc} = 95.290347\,\text{meV}$, $\varepsilon^*_{\lambda A} = 1.239677565$ keV, $\lambda^*_A = 0.999972933\,\text{nm}$, $E_{\lambda A} = 0.520365996\,\text{MeV}$. The energy $E_{\lambda A}$ (for solar wind particles inside the heliosphere) is associated with the energy $E_{\lambda L}$ (for galactic wind particles behind the heliopause)

$$E_{\lambda A} = (\Omega_{\tau L} + n_g\Omega_{0G})E_{\lambda L}; \ \Omega_{0G}N_{0A} = 1.5 + \Omega'_{c1} + n'_{F\tau};$$

$$E^2_{rc} = E^2_{0A} - 4\Delta^2_{rc}; \ (E'_{rc})^2 = E^2_{0A} + 4\Delta^2_{rc}. \quad (33)$$

The numerical values are equal: $\Omega_{0G} = 4.99501253 \cdot 10^{-6}$, $E_{\lambda L} = 213.0772532\,\text{MeV}$, $E_{rc} = 4.306858745\,\text{eV}$, $E'_{rc} = 4.315283797\,\text{eV}$. The energy estimates ε_{bb}, $E_{\lambda L}$ obtained in our model are consistent with the energies of 28 keV, 213 MeV from [26], and the energy $E_{\lambda A}$ is consistent with the energy of 0.5 MeV from [28].

The magnetic characteristics of solar and galactic wind particles have features of the behavior at the intersection of the heliopause: a jump in the magnetic field from 0.42 to 0.68 nT is observed (Fig. 1a from [27]); components of the magnetic field can have different signs (Fig. 3 from [27]); the presence of a magnetic barrier (Fig. 4a from [27]); a change in the direction of the magnetic field components (Fig. 6b, c from [27]). In our model, to estimate the components of magnetic fields $B_{y\beta x}$, $B^*_{y\beta x}$ we use frequency spectra of the type

$$\nu_{y\beta x} = \gamma_n B_{y\beta x}/2\pi = 2\nu_{y\beta}S'_{0x}; \ \nu^*_{y\beta x} = \gamma_n B^*_{y\beta x}/2\pi = 2\nu_{y\beta}S_{ux}; \ y = 0, 1, 2;$$

$$\nu_{y\beta} = \nu_{0y}/N_{ra}; \ B'_{2\beta 1} = B^*_{2\beta 1}/(1.5 + n'_{zg} + S_{012}); \ \nu_{00} = \nu_{H0}; \ \nu_{02} = \psi_{02}\nu_{H0}. \quad (34)$$

Here we use the well-known nuclear gyromagnetic ratio $\gamma_n/2\pi = 0.6535\,\text{MHz/kO}$ for the deuteron (^{2}H) [4, 12]. Based on (34) we find estimates: frequencies $\nu^*_{2\beta 1} = 4.4353480\,\text{mHz}$; jump of magnetic fields from $B'_{2\beta 1} = 0.4190147\,\text{nT}$ to $B^*_{2\beta 1} = 0.6787067\,\text{nT}$ at the intersection of heliopause. The numerical values of the fields deviations of the type $\delta B = B_{0\beta 1} - B_{0\beta 2} = 0.0804015\,\text{nT}$, $\delta B^* = B^*_{0\beta 1} - B^*_{0\beta 2} = 0.2019195\,\text{nT}$ and the sum of the deviations $\delta B + \delta B^* = 0.282321\,\text{nT}$ are characteristic of the stochastic behavior of the magnetic field on time inside the heliosphere (consistent with data Fig. 6 from [27]).

5 Conclusions

In fractal quantum systems the model femtoobjects, as active objects with sizes of the order of the classical electron radius, are considered. The main parameters of the model femtoobject, which are coupled with the known parameters from quantum electrodynamics and the Higgs boson, are introduced. To take into account the stochastic behavior of the parameters, random variables with two possible values and the corresponding probabilities are introduced. It was shown, that the obtained estimates of the proton radius, measurement errors using the example of the hydrogen atom, and estimates of the anomalies in the magnetic moments of leptons are consistent with the experimental data.

The parameters of active nanoobjects and femtoobjects are coupled with cosmological parameters, with new values of the Hubble constants. These active objects can determine the compound, structure and behavior of the solar wind (flows of various particles) near the Sun, Earth and in interstellar space (near the heliopause). The relationships of such active objects with the parameters of the Higgs boson and the Higgs field are determined. Estimates of the main parameters are conformed with the experimental data, obtained by the Planck space observatory (based on Fermi-LAT and Cerenkov telescopes), UTR-2 and URAN-2 radio telescopes, Parker Solar Probe, Voyager 2 and Voyager 1.

The results can be used to find a solution to the problem associated with the Covid-2019 virus (based on active femtoobjects and nanoobjects), in cosmic medicine.

References

1. V.S. Abramov, Gravitational Waves, Relic Photons and Higgs Boson in a Fractal Models of the Universe, in *11th Chaotic Modeling and Simulation International Conference, Springer Proceedings in Complexity*, ed. by C.H. Skiadas, I. Lubashevsky (Springer Nature Switzerland AG 2019), pp. 1–14
2. V.S. Abramov, Superradiance of gravitational waves and relic photons from binary black holes and neutron stars. Bull. Russian Acad. Sci. Phys. **83**(3), 364–369 (2019)
3. A.G. Riess, A.V. Filippenko, P. Challis et al., Observational evidence from supernovae for an accelerating Universe and a cosmological constant. Astron. J. **116**(3), 1009–1038 (1998)
4. V.S. Abramov, Active nanoobjects, neutrinos and higgs fields in anisotropic models of fractal cosmology. Bull. Russian Acad. Sci. Phys. **83**(12), 1516–1520 (2019)
5. V.S. Abramov, Supernonradiative states, neutrino and Higgs Boson in fractal quantum systems. Bull. Russ. Acad. Sci. Phys. **84**(3), 284–288 (2020)
6. V.S. Abramov, Active Nanoobjects, Neutrino and Higgs Boson in a Fractal Models of the Universe, in *12th Chaotic Modeling and Simulation International Conference, Springer Proceedings in Complexity*, ed. by C.H. Skiadas, Y. Dimotikalis (Springer Nature Switzerland AG, 2020), pp. 1–14
7. V. Abramov, Super-nonradiative states in fractal quantum systems, XIII International Workshop on Quantum Optics (IWQO-2019). EPJ Web of Conferences 2019, 220, p. 2
8. O.P. Abramova, A.V. Abramov, Effect of Ordering of Displacement Fields Operators of Separate Quantum Dots, Elliptical Cylinders on the Deformation Field of Coupled Fractal Structures, in *11th Chaotic Modeling and Simulation International Conference*, ed. by C.H. Skiadas, I.

Lubashevsky (Springer Proceedings in Complexity, Springer Nature Switzerland AG, 2019), pp. 15–27
9. R. Pohl, A. Antognini, F. Nez et al., The size of the proton. Nature **466**, 213–217 (2010)
10. A. Beyer, L. Maisenbacher, A. Matveev et al., The Rydberg constant and proton size from atomic hydrogen. Science **358**, 79–85 (2017)
11. N. Kolachevsky, 2S–4P spectroscopy in hydrogen atom: the new value for the Rydberg constant and the proton charge radius. IWQO—2019: Collection of abstracts. Vladimir, September 9–14, 2019 (Moscow, Trovant, 2019), p. 32
12. S.V. Vonsovsky, *Magnetism of Microparticles* (Nauka, Moscow, 1973).
13. N. Agafonova, A. Alexandrov, A. Anokhina et al., Final results of the OPERA experiment on $\nu\tau$ appearance in the CNGS neutrino beam. Phys. Rev. Lett. **120**(211801), 1–7 (2018)
14. R.M. Barnett, C.D. Carone, D.E. Groom et al., Review of particle physics. Phys. Rev. D **54**(1), 1–22 (1996)
15. V.S. Abramov, The Higgs boson in fractal quantum systems with active nanoelements. Bull. Russ. Acad. Sci. Phys. **80**(7), 859–865 (2016)
16. V.S. Abramov, Pairs of vortex-antivortex and Higgs boson in a fractal quantum system. CMSIM J. **1**, 69–83 (2017)
17. V.S. Abramov, Cosmological parameters and Higgs boson in a fractal quantum system. CMSIM J. **4**, 441–455 (2017)
18. A. Domínguez, R. Wojtak, J. Finke et al., A new measurement of the Hubble constant and matter content of the Universe using extragalactic background light γ-ray attenuation. Astrophys. J. **885**(2), 137 (2019)
19. N. Schukina, The Sun is the source of life and the cause of catastrophes. Part I. A star named the Sun. Universe, Space, Time **8**(145), 4–11 (2016)
20. N. Schukina, The Sun is the source of life and the cause of catastrophes. Part II. Light giving life. Universe, Space, Time **9**(146), 16–23 (2016)
21. D.J. McComas, E.R. Christian, C.M.S. Cohen et al., Probing the energetic particle environment near the Sun. Nature **576**, 223–227 (2019)
22. J.C. Kasper, S.D. Bale, J.W. Belcher et al., Alfvénic velocity spikes and rotational flows in the near-Sun solar wind. Nature **576**, 228–231 (2019)
23. R.A. Howard, A. Vourlidas, V. Bothmer et al., Near-Sun observations of an F-corona decrease and K-corona fine structure. Nature **576**, 232–236 (2019)
24. S.D. Bale, S.T. Badman, J.W. Bonnell et al., Highly structured slow solar wind emerging from an equatorial coronal hole. Nature **576**, 237–242 (2019)
25. N.N. Kalinichenko, M.R. Olyak, A.A. Konovalenko et al., Large-scale structure of solar wind beyond Earth's orbit reconstructed by using data of two-site interplanetary scintillation observations at decameter radiowaves. Kinemat. Phys. Celest. Bodies **35, 1**(205), 27–41 (2019)
26. S.M. Krimigis, R.B. Decker, E.C. Roelof et al., Energetic charged particle measurements from Voyager 2 at the heliopause and beyond. Nat. Astron. **3**, 997–1006 (2019)
27. L.F. Burlaga, N.F. Ness, D.E. Berdichevsky et al., Magnetic field and particle measurements made by Voyager 2 at and near the heliopause. Nat. Astron. **3**, 1007–1012 (2019)
28. E.C. Stone, A.C. Cummings, B.C. Heikkila et al., Cosmic ray measurements from Voyager 2 as it crossed into interstellar space. Nat. Astron. **3**, 1013–1018 (2019)
29. J.D. Richardson, J.W. Belcher, P. Garcia-Galindo et al., Voyager 2 plasma observations of the heliopause and interstellar medium. Nat. Astron. **3**, 1019–1023 (2019)
30. D.A. Gurnett, W.S. Kurth, Plasma densities near and beyond the heliopause from the Voyager 1 and 2 plasma wave instruments. Nat. Astron. **3**, 1024–1028 (2019)

Qubits and Fractal Structures with Elements of the Cylindrical Type

Olga P. Abramova and Andrii V. Abramov

Abstract By the method of numerical simulation, the behavior of the deformation field of both separated and related model fractal structures of a cylindrical type was investigated. It is shown, that for the considered structures, the behavior of the deformation field essentially depends on the choice of stochastic processes (realized during iterations) and on the states of the qubit in the perpendicular plane to the axis of the cylinder. It is shown that the structure of the complex deformation field for a circular (elliptical) cylinder essentially depends on the initial basic, superposition states of the qubit. Due to the presence of various qubit states for coupled (using the example of circular and elliptic cylinders) fractal structures, the appearance of random matrices during iterations is characteristic. There is a need to use commutators and anti-commutators, products of separate deformation field operators. At this, the structure of the complex deformation field has own characteristic features of behavior.

Keywords Fractal structure · Qubits · Random matrices · Complex deformation field · Ordering of operators · Quantum chaos

1 Introduction

Earlier in [1–3], to describe the total deformation field of coupled fractal structures in an iterative process, the sum of the displacement field operators of separate fractal structures was used. The deformation field of the coupled structure essentially depends on the sequence of separate operators of displacement fields in the iterative process. On the examples of quantum dots [4], elliptic [1, 2] and circular [3, 5] cylinders the influence of the ordering of separate operators of displacement fields on the total deformation field of the coupled structure was shown. The presence of variable semiaxes and variable moduli leads to stochastic behavior of the

O. P. Abramova (✉) · A. V. Abramov
Donetsk National University, Donetsk, Ukraine
e-mail: oabramova@ua.fm

C. H. Skiadas et al. (eds.), *13th Chaotic Modeling and Simulation International Conference*, Springer Proceedings in Complexity,
https://doi.org/10.1007/978-3-030-70795-8_2

complex deformation field of such structures. Based on pairs of same fractal structures with opposite orientations of the deformation fields, complex zero operators were introduced [3, 5]. It is shown that changes in the order of the sequence of separate operators in the zero operator for a coupled structure leads to the appearance of a nonzero complex deformation field. At the same time, noise tracks appear on the background of stochastic peaks. The noise track is a stochastic ring, the inside region of which is regular region.

For describe quantum chaos random matrices are used [6]. Elements of random matrices can be formed as a result of an iterative process. In this case, the need arises for the use of commutators and anti-commutators, products of separate operators, qubit states [7, 8] of the deformation field. Quantum computers [9–12] encode information in qubits. The physical systems that realise qubits can be any objects having two quantum states. Different nanostructures and metamaterials [13] can be chosen as active objects. These active objects can be in superposition qubit states and exhibit stochastic properties, quantum entanglement.

The aim of this work is to describe the deformation fields of fractal coupled structures consisting of two separate structures (circular and elliptical cylinders) with different qubit states. In this case, the deformation fields of coupled structures are considered as the sum and product (scalar and matrix) of the deformation fields of separate structures.

2 Description of the Deformation Field of Separate Fractal Structures in Various Qubit States

We consider a model fractal structure (circular or elliptical cylinder), located in a bulk discrete lattice $N_1 \times N_2 \times N_3$, whose nodes are given by integers n, m, j. By analogy with [5, 11–13] nonlinear equations for the dimensionless displacement function u of the lattice node are

$$u = k_u^2(1 - 2\mathrm{sn}^2(u - u_0, k_u'));\tag{1}$$

$$k_u^2 = (1-\alpha)/Q;\; k_u' = \left(1 - k_u^2\right)^{1/2};\; p = p_0 + p_1 n + p_2 m + p_3 j\tag{2}$$

$$Q = p - b_1(n-n_0)^2/n_c^2 - b_2(m-m_0)^2/m_c^2 - b_3(j-j_0)^2/j_c^2.\tag{3}$$

Here u_0 is the constant (critical) displacement; α is the fractal dimension of the deformation field u along the axis Oz ($\alpha \in [0, 1]$); variable modules k_u, k_u' are functions of indices n, m, j nodes of the bulk discrete lattice. The choice of the positive sign of the module k_u' is associated with the choice of the second branch of the displacement function u [14]. Function Q determines the form of the fractal structure, the type of attractors and take into account the interaction of the nodes of both in the main plane of the discrete rectangular lattice $N_1 \times N_2$ as well as interplane interactions. The parameters b_1, b_2, b_3, n_0, n_c, m_0, m_c, j_0, j_c characterize different fractal structures. The choice of function p depends on the choice of parameters p_i, $i = \overline{0, 3}$. In this paper, we are limited to consideration of qubit states with $p_1 \neq 0$, $p_2 \neq 0$, $p_3 = 0$ and shift $p_0 \neq 0$.

The iterative procedure on index n for Eqs. (1)–(3) simulates stochastic processes on a rectangular discrete lattice with dimensions $N_1 \times N_2$.

By numerical modelling, it was assumed that $N_1 = 240$, $N_2 = 240$, $\alpha = 0.5$, $u_0 = 29.537$, $p_0 = 1.0423$, $b_1 = b_2 = 1$, $n_0 = 121.1471$, $m_0 = 120.3267$, $j_0 = 31.5279$, $j_c = 11.8247$, $b_3 = 0$. For a circular cylinder, the semiaxes were equal $n_c = m_c = 57.4327$, and for an elliptical cylinder they were as follows $n_c = 55.2537$, $m_c = 14.9245$.

To describe the deformation field of an separate circular (elliptical) cylinder, the basic states of the qubit and their superposition in the plane nOm are introduced. The various states of the qubit in the plane nOm are described by nonzero coefficients for linear terms in the functions p, Q.

The initial state of an separate circular (elliptical) cylinder is the state (0, 0) with the coefficients $p_1 = 0$, $p_2 = 0$.

The basic states of a qubit are states (1, 0), (0, 1), (−1, 0), (0, −1) with the coefficients $p_1 \neq 0$ or $p_2 \neq 0$. So for state (1, 0) are $p_1 = 0.00423$, $p_2 = 0$; for state (0, 1) are $p_1 = 0$, $p_2 = 0.00572$; for state (−1, 0) are $p_1 = -0.00423$, $p_2 = 0$; for state (0, −1) are $p_1 = 0$, $p_2 = -0.00572$.

For superpositional states of qubits, we have, respectively:

$$\begin{aligned}
&\text{state } (1, 1) \text{ are } p_1 = 0.00423,\ p_2 = 0.00572;\\
&\text{state } (1, -1) \text{ are } p_1 = 0.00423,\ p_2 = -0.00572;\\
&\text{state } (-1, 1) \text{ are } p_1 = -0.00423,\ p_2 = 0.00572;\\
&\text{state } (-1, -1) \text{ are } p_1 = -0.00423,\ p_2 = -0.00572.
\end{aligned} \tag{4}$$

Figure 1 shows the behavior of the cross sections of the deformation field u for elliptical (Fig. 1a, d) and circular (Fig. 1b, e) cylinders in the initial state (0, 0) and basis states of qubits.

Figure 1c, f, g, h, i, j, k, l shows the behavior of the cross sections of the deformation field u for a circular cylinder in the basic states of qubits: Reu (Fig. 1c, g, h, i), Imu (Fig. 1f, j, k, l). In this case, the peak amplitudes and the variation range Reu, Imu for the elliptical cylinder are smaller than for the circular one.

For the initial (0, 0) and basic states (1, 0), (0, 1) of qubits the regular behavior Reu in the inner region is characterized (Fig. 1b, c, g).

By changing qubit states, the features of behavior of the deformation field is observed: the concave part of the inner region (Fig. 1b) changes to the convex (Fig. 1c) and then to the convex-concave (Fig. 1g). Such behavior allows the interpretation of the inner region as a membrane with the possible alteration of its states due to a change in the states of qubits.

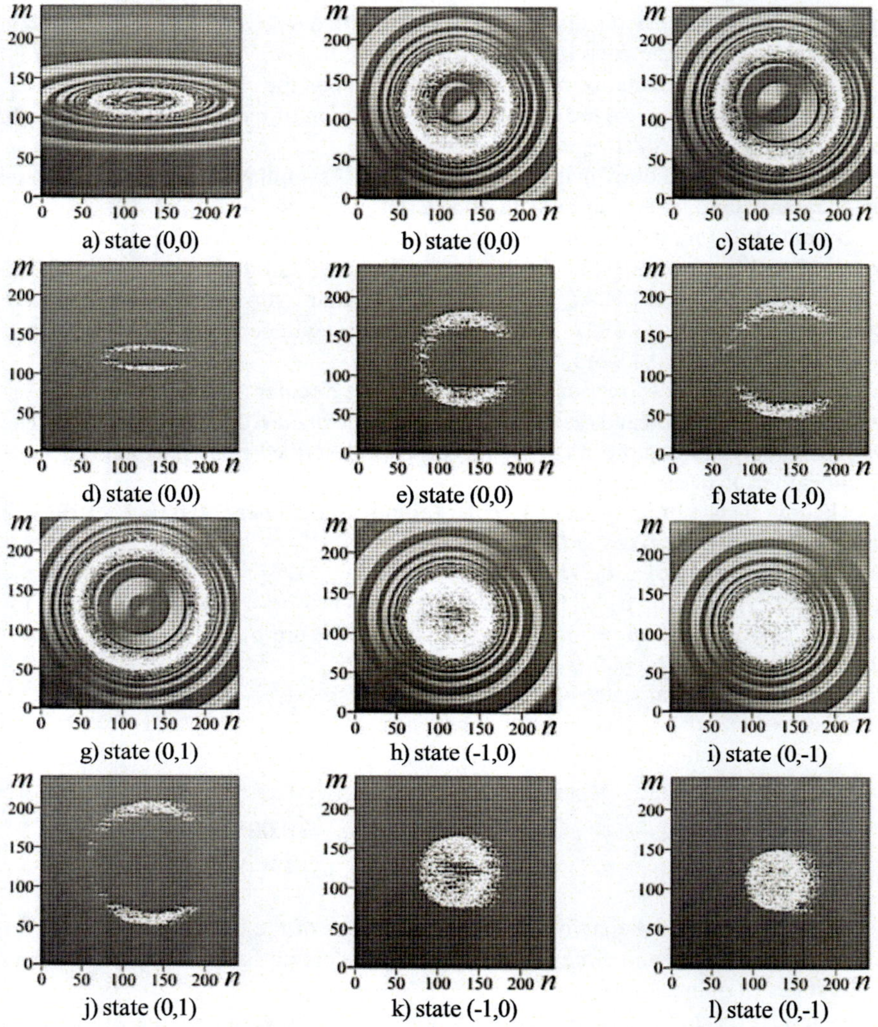

Fig. 1 The behavior of the cross sections u (top view) depending on the states of qubits of separate structures: Re$u \in [-1; 1]$—(a, b, c, g h, i); Im$u \in [-1; 1]$—(d, e, f, j, k, l). The initial states of qubits (0, 0) for elliptical (a, d) and circular (b, e) cylinders. The basic states of qubits for a circular cylinder (c, f, g–l)

The regular behavior of the inner region Reu is limited by the stochastic boundary (stochastic ring). The outer region Reu is characterized by wave-like behavior, which is explained by the presence of variable modules k_{ui}, k'_{ui} (2) in expressions (1). Localized in the region of boundary rings with discontinuities (Fig. 1e, f, j) the stochastic behavior is characteristic for Im*u*, while Im*u* = 0 is characteristic in the inner and outer regions of the rings.

For the other basic states (−1, 0), (0, −1) of qubits characteristic stochastic behavior Re*u* in the inner region and wave-like behavior in the outer region (Fig. 1h, i), that indicates a significant alteration of the structures.

For these states Im*u* has a stochastic structure, localized in the inner region of the cylinder (Fig. 1k, l), and outside the region Im*u* = 0. The imaginary part Im*u* indicates the presence of an effective damping. By changing these basis states, the character of damping changes.

The presence of superpositional states of qubits in separate structures leads to a change in the behavior of the complex deformation field. As an example, Fig. 2 shows the behavior Re*u* (Fig. 2a, d) and cross sections (Fig. 2b, c, e, f) of an separate structure (circular cylinder) in superposition states of qubits (1, 1) and (−1, −1). The characteristic features of the behavior of the deformation field for state (1, 1) (Fig. 2b, c) are close to state (0, 0) (Fig. 1b, e). The characteristic features the cross sections behavior of deformation field for the state (−1, −1) (Fig. 2e, f) are close to the states (−1, 0) (Fig. 1h, k), (0, −1) (Fig. 1i, l).

However, the behavior Re*u* for the superposition state (−1, −1) (Fig. 2d) differs significantly from the characteristic behavior Re*u* of all other superposition states of qubits (1, 1) (Fig. 2a), (1, −1), (−1, 1). Instead of a structure such as a circular stochastic dislocation (Fig. 2a), a structure like a stochastic funnel (Fig. 2d) arises.

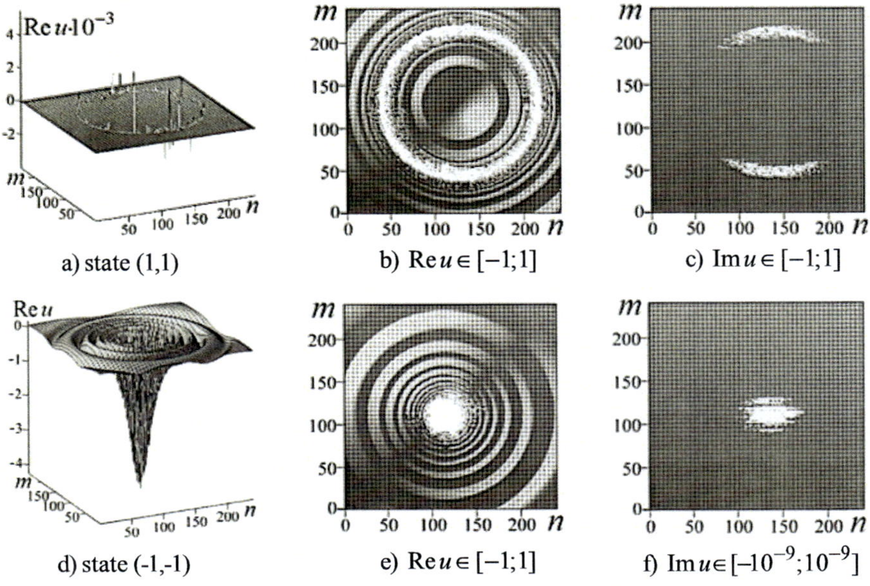

Fig. 2 Superpositional states of qubits of a separate structure (circular cylinder). Behavior Re*u* (a, d) and cross sections (top view) (b, c, e, f) in the states: (1, 1)—(a, b, c); (−1, −1)—(d, e, f)

In this case, the amplitudes Reu and Imu for the state $(-1, -1)$ are significantly smaller than the amplitudes for other states of qubits.

3 Fractal Coupled Structures with Initial States of Qubits of Separate Structures

Consider the model fractal coupled structures (I,II), (II,I), consisting of two separate structures (I) and (II) with the same initial qubit states (0, 0). By analogy with (1)–(3) nonlinear equations for the dimensionless complex displacement function u of the lattice node are

$$u = \sum_{i=1}^{2} u_{Ri};\; u_{Ri} = R_i k_{ui}^2 (1 - 2\mathrm{sn}^2(u - u_{0i}, k_{ui}')); \tag{5}$$

$$k_{ui}^2 = (1 - \alpha_i) \Big/ Q_i;\; k_{ui}' = \left(1 - k_{ui}^2\right)^{1/2};\; p_i = p_{0i} + p_{1i}n + p_{2i}m + p_{3i}j; \tag{6}$$

$$Q_i = p_i - b_{1i}(n - n_{0i})^2/n_{ci}^2 - b_{2i}(m - m_{0i})^2/m_{ci}^2 - b_{3i}(j - j_{0i})^2/j_{ci}^2. \tag{7}$$

Here, all parameters have the same meaning as for expressions (1)–(3). Parameters R_i ($i= 1, 2$) determine the orientation of the deformation fields of separate structures in a coupled system. For separate structures (I) and (II), the deformation fields $u = u_{R1}$ and $u = u_{R2}$ correspond to the matrices $\mathbf{M}_{R1}$ and $\mathbf{M}_{R2}$, whose elements are found independently from each other by the iteration method. In this case, the iterative procedure on index n for Eqs. (5)–(7) simulates two independent stochastic processes on a rectangular discrete lattice with dimensions $N_1 \times N_2$. Earlier in [5], ordered operators of displacement fields of a coupled structure were introduced as the sum of the operators of separate structures. Here, for the sum of the matrices $\mathbf{M}_{R1}$, $\mathbf{M}_{R2}$ the relation is fulfilled.

$$\mathbf{M}_{R1} + \mathbf{M}_{R2} = \mathbf{M}_{R2} + \mathbf{M}_{R1}. \tag{8}$$

The deformation fields for the coupled structures (I,II), (II,I) correspond to the ordered operators

$$u = u_{\,(\mathrm{I,II})} = u_{R1} + u_{R2},\; u = u_{\,(\mathrm{II,I})} = u_{R2} + u_{R1} \tag{9}$$

and matrices $\mathbf{M}_{(\mathrm{I,II})}$, $\mathbf{M}_{(\mathrm{II,I})}$, whose elements are found by the iteration method. The iterative procedure on index n for Eqs. (5)–(7) simulates two other independent stochastic processes for matrices $\mathbf{M}_{(\mathrm{I,II})}$, $\mathbf{M}_{(\mathrm{II,I})}$. In this case, the relations are fulfilled

$$\mathbf{M}_{(\mathrm{I,II})} = \mathbf{M}_{R1} + \mathbf{M}_{R2} = \mathbf{M}_{R2} + \mathbf{M}_{R1} = \mathbf{M}_{(\mathrm{II,I})};\; \mathbf{M}_{(\mathrm{I,II})} - \mathbf{M}_{(\mathrm{II,I})} = 0 \tag{10}$$

To describe the deviation of the deformation field of the coupled structures (I,II) and (II,I), we introduce the ordered operator

$$u_\delta = (u_{R1} + f_2(u_{R2})) - (u_{R2} + f_1(u_{R1})), \tag{11}$$

which corresponds to the matrix $\mathbf{M}_\delta$. An iterative procedure on index n simulate stochastic process for a matrix $\mathbf{M}_\delta$, which does not coincide with stochastic processes for matrices $\mathbf{M}_{(\mathrm{I,II})}$, $\mathbf{M}_{(\mathrm{II,I})}$, $\mathbf{M}_{R1}$, $\mathbf{M}_{R2}$. In this case

$$\mathbf{M}_\delta \neq \mathbf{M}_{(\mathrm{I,II})} - \mathbf{M}_{(\mathrm{II,I})};\, \mathbf{M}_\delta \neq 0. \tag{12}$$

From (12) follows, that stochastic processes for matrices $\mathbf{M}_{(\mathrm{I,II})}$, $\mathbf{M}_{(\mathrm{II,I})}$ become dependent. If in (11) assume $f_2(u_{R2}) = u_{R2}$, $f_1(u_{R1}) = u_{R1}$, then $\mathbf{M}_{(\mathrm{I,II})} = \mathbf{M}_{(\mathrm{II,I})}$, what confirms the independence conditions for stochastic processes (10). Attractors of the deformation field of the coupled fractal structure are located on the surface, the core of which is determined from the condition

$$Q_1 \cdot Q_2 = 0. \tag{13}$$

By numerical modeling, it was assumed, that:$\alpha_i = 0.5$, $u_{0i} = 29.537$, $p_{0i} = 1.0423$, $b_{1i} = b_{2i} = 1$, $n_{0i} = 121.1471$, $m_{0i} = 120.3267$, $n_{c1} = m_{c1} = 57.4327$, $j_{0i} = 31.5279$, $j_{ci} = 11.8247$, $p_{1i} = p_{2i} = p_{3i} = 0$, $b_{3i} = 0$. In this case, in fractal coupled structures (I, II) and (II, I), the structure (I) is a circular cylinder and the structure (II) is an elliptical cylinder with variable semi-axes n_{c2}, m_{c2}. The variable semiaxes were chosen so that the cross-sectional area of the ellipse $S = \pi\, n_{c2} m_{c2}$ did not change and was equal to the cross-sectional area of the circular cylinder $S = 824.6316\pi$ from [2, 3]. For an elliptical cylinder (II), the semiaxes n_{c2}, m_{c2} were defined as follows:

variant 1 are $n_{c2} = 43.0746$, $m_{c2} = 19.1443$ (the elliptical cylinder is inside the circular cylinder);

variant 2 are $n_{c2} = 55.2537$, $m_{c2} = 14.9245$ (the elliptical cylinder approaches to the circular cylinder along the axis On);

variant 3 are $n_{c2} = 119.9327$, $m_{c2} = 6.8758$ (the elliptical cylinder extends beyond the boundaries of the circular cylinder along the axis On).

Figure 3 shows the behavior of attractors for all three variants of fractal structures (I, II) and the complex deformation field $u_{(\mathrm{I,II})}$. The different behavior of attractors (Fig. 3a, b, c) and cross sections of the complex deformation field (Fig. 3d–i) confirm the stochastic nature of the deformation field of the structure (I, II) and its dependence on the semiaxes of the elliptic cylinder (II).

The behavior of the deformation field of the structure (II, I) in this paper is not given. However, completed researches performed make it possible to estimate the deviations (12) $\mathrm{Re}\mathbf{M}_\delta \approx 10^{-9}$, $\mathrm{Im}\mathbf{M}_\delta \approx 10^{-25}$, which indicates to the dependence of stochastic processes in (11).

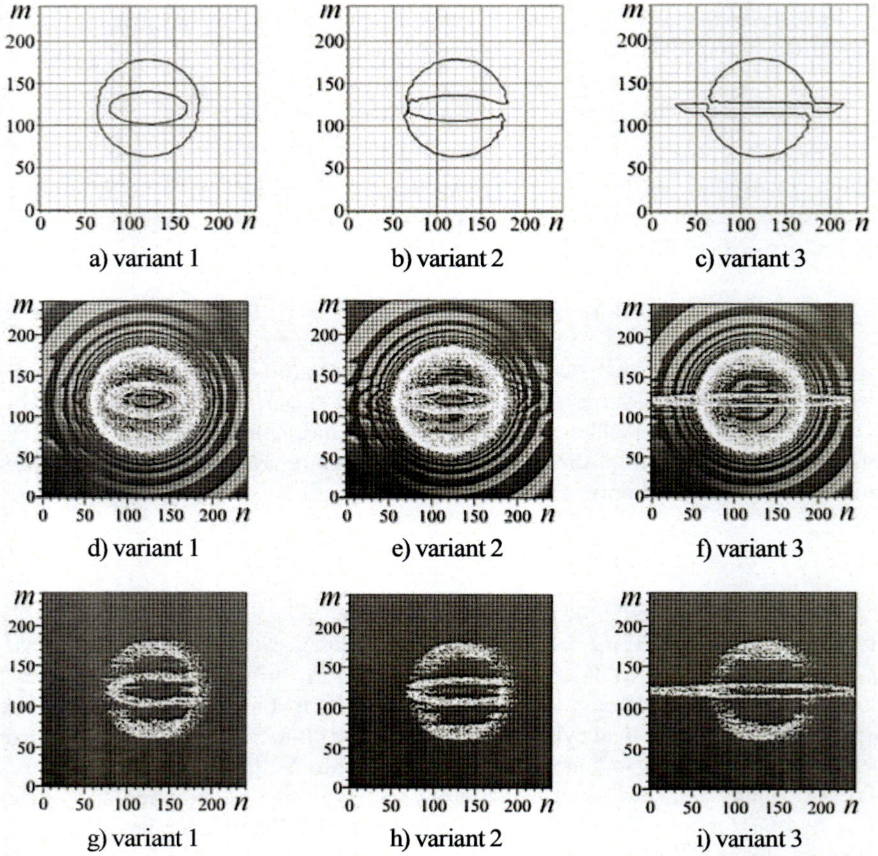

Fig. 3 The behavior of attractors (a, b, c) and the deformation field u of coupled structure (I,II): (d, e, f)—Re$u_{(\mathrm{I,II})} \in [-1; 1]$, (g, h, i)—Im$u_{(\mathrm{I,II})} \in [-1; 1]$ cross sections (top view)

As a result of the iterative process, elements of random matrices are formed, which depend on various qubit states of separate structures in a coupled structure. Random matrices are used to describe quantum chaos [6]. In this case, there is a need to use commutators and anti-commutators, products of separate operators of the deformation field. Next, we consider fractal coupled structures (III) and (IV), the deformation fields of which u_3 and u_4 are described by the product of the deformation fields of separate structures (I) and (II) with the same initial qubit states (0, 0). The deformation fields of structures (III) and (IV) correspond to the matrices $\mathbf{M}_3 = \mathbf{M}_{R1} \cdot \mathbf{M}_{R2}$ and $\mathbf{M}_4 = \mathbf{M}_{R2} \cdot \mathbf{M}_{R1}$. Here, the dot symbol describes the operation of ordinary matrix multiplication. Figure 4 shows the behavior of the complex deformation field for structures (III) and (IV). In this case, structure (II) parameters were chosen corresponding to variant 2. The attractors of structures (III) and (IV) coincide with the

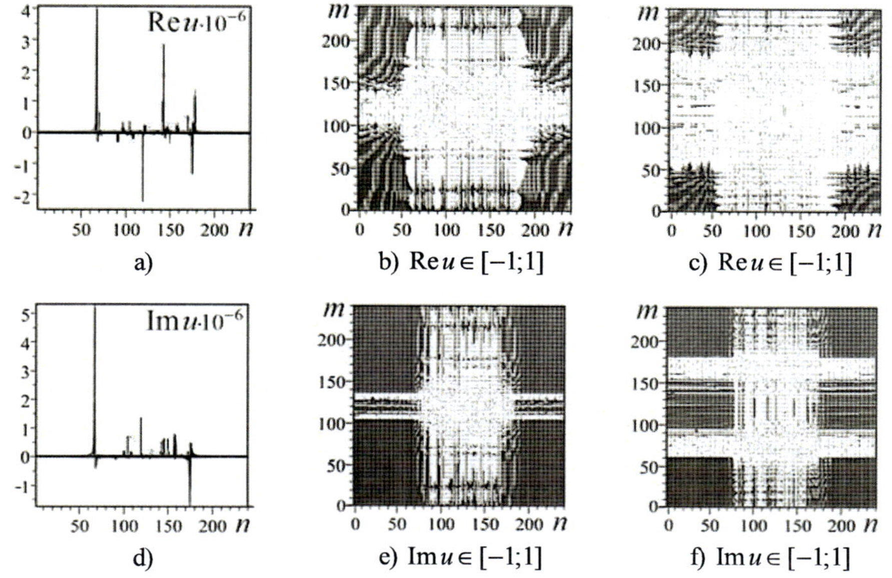

Fig. 4 Deformation fields of structures (III), (IV): $u = u_3$ (a, d)—projections onto the plane nOu, (b, e)—cross sections (top view); $u = u_4$ (c, f)—cross sections (top view)

attractor from Fig. 3b. Cross sections (Fig. 4b, e), projections onto the plane nOu (Fig. 4a, d) confirm the stochastic and fractal behavior of the deformation field of structure (III), which differs significantly from the behavior of the deformation field of structure (IV) (Fig. 4c, f). This confirms the non-commutativity the operation of ordinary matrix multiplication $\mathbf{M}_3 - \mathbf{M}_4 = \mathbf{M}_{R1} \cdot \mathbf{M}_{R2} - \mathbf{M}_{R2} \cdot \mathbf{M}_{R1} \neq 0$.

Changing the operation of ordinary matrix multiplication on the scalar multiplication of complex deformation fields leads to the replacement of the coupled structures (III) and (IV) on structures (V) and (VI). In this case, the iterative procedure on index n simulates the coupled (dependent) stochastic processes of the initial independent stochastic processes for separate structures (I) and (II) with the same initial qubit states (0, 0).

The deformation fields of structures (V) and (VI) are described by the functions $u_5 = u_{R1} f_5(u_{R2})$ and $u_6 = u_{R2} f_6(u_{R1})$, to which the matrices $\mathbf{M}_5$ and $\mathbf{M}_6$ correspond. If by modeling we use independent iterative processes for structures (I) and (II), then

$$f_5(u_{R2}) = u_{R2};\ f_6(u_{R1}) = u_{R1};\ u_5 = u_{R1}u_{R2} = u_{R2}u_{R1} = u_6;\ \mathbf{M}_5 = \mathbf{M}_6. \quad (14)$$

Matrix equality confirms the independence of iterative processes.

Figure 5 shows the behavior of the complex deformation field for structures (V) and (VI). In this case, structure (II) parameters were chosen corresponding to

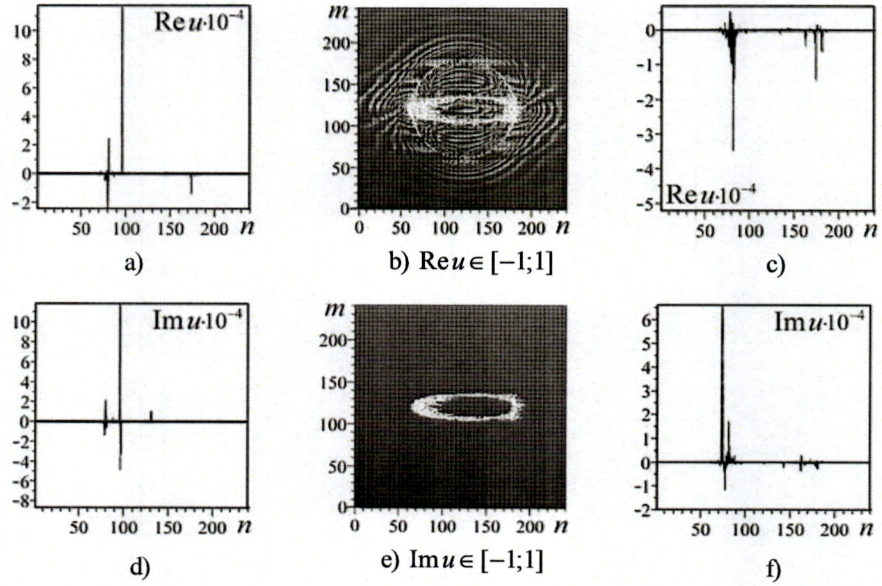

Fig. 5 Deformation fields of structures (V), (VI): $u = u_5$ (a, d)—projections onto the plane nOu, (b, e)—cross sections (top view); $u = u_6$ (c, f)—projections onto the plane nOu

variant 2. The attractors of structures (V) and (VI) coincide with the attractor from Fig. 3b. Cross sections (Fig. 5b, e), projections onto the plane nOu (Fig. 5a, d) confirm another (compared to Fig. 4) stochastic and fractal deformation field behavior of the structure (V), which also differs significantly from the deformation field behavior of the structure (VI) (Fig. 5c, f). This is due to the dependence of the stochastic processes ($\mathbf{M}_5 \neq \mathbf{M}_6$).

4 Fractal Coupled Structures with Various Superpositional Qubits States of Separate Structures

Next, we consider the superpositional qubits states of fractal coupled structures (V) and (VI). The deformation fields of these structures are described by functions $u_5 = u_{R1} f_5(u_{R2})$ and $u_6 = u_{R2} f_6(u_{R1})$ with the corresponding matrices $\mathbf{M}_5$ and $\mathbf{M}_6$, where the scalar multiplication of complex deformation fields of separate structures (I), (II) is realized. In this case, the iterative procedure on index n simulates coupled (dependent) stochastic processes for the initial independent stochastic processes for structures (I) and (II), the deformation fields of which are described by the functions $u = u_{R1}$ and $u = u_{R2}$. As an example, Fig. 6 shows the behavior of the complex deformation field for structure (V). In this case, the separate structure (I) is a circular

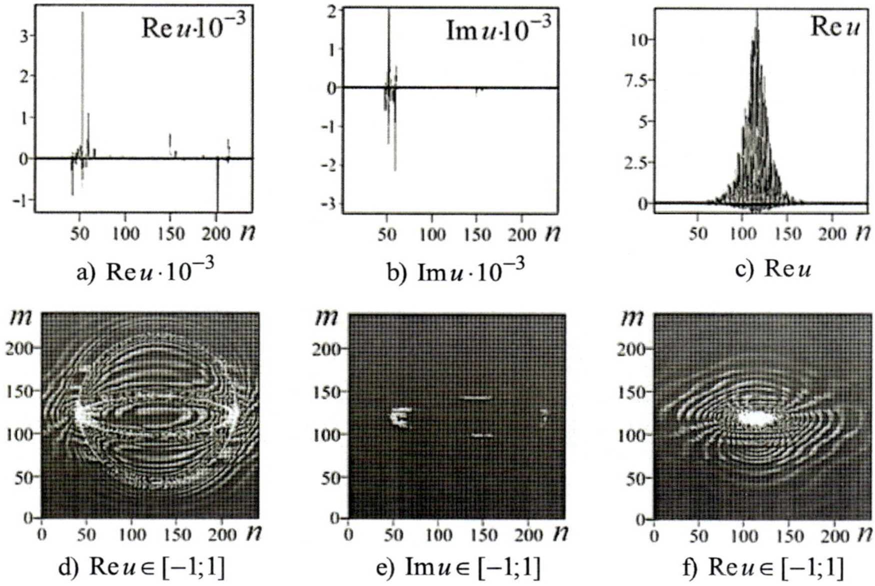

Fig. 6 The behavior of the displacement u of the fractal coupled structure (V): separate structures (I) and (II) have the same superposition qubit states: (1, 1)—(a, b, d, e); (−1, −1)—(c, f)

cylinder with parameters as for Fig. 1, and the parameters of a separate structure (II) (elliptical cylinder) correspond to variant 2 (the elliptical cylinder approaches the circular cylinder along the axis On). In the coupled structure (V), the separate structures (I), (II) have the same superposition qubit states (1, 1) (Fig. 6a, b, d, e) and (−1, −1) (Fig. 6c, f). The behavior of the deformation field of the coupled structure (V) with the same initial qubit states (0, 0) is given on Fig. 5a, b, d, e. The presence of same superpositional qubit states (1, 1) of separate structures in a coupled system (Fig. 6a, b, d, e) leads to a change in the complex deformation field compared to Fig. 5a, b, d, e: the decrease amplitudes of peaks, the shift of peaks (Fig. 6a, b), the change of structure (Fig. 6d, e) are observed. An original feature of the deformation field behavior of the coupled structure (V) with the same superpositional states (−1, −1) of separate structures is the absence of the imaginary part of the displacement function in all region (Imu_5 = 0), that indicates the absence of effective attenuation. This makes it possible to interpret the coupled structure (V) with the same superpositional states (−1, −1) of the separate structures (I), (II) as a memory cell. For Reu_5 the presence of a broadened stochastic peak up is characteristic (Fig. 6c). In this case the cross-sectional structure (Fig. 6f) for state (−1, −1) differs from the cross-sectional structure (Fig. 6d) for state (1, 1).

By changing the superposition qubit states of separate structures (I), (II), one can change and control the behavior of the complex deformation field of the coupled structure (V). As an example, Fig. 7 shows the behavior of cross sections Reu_5

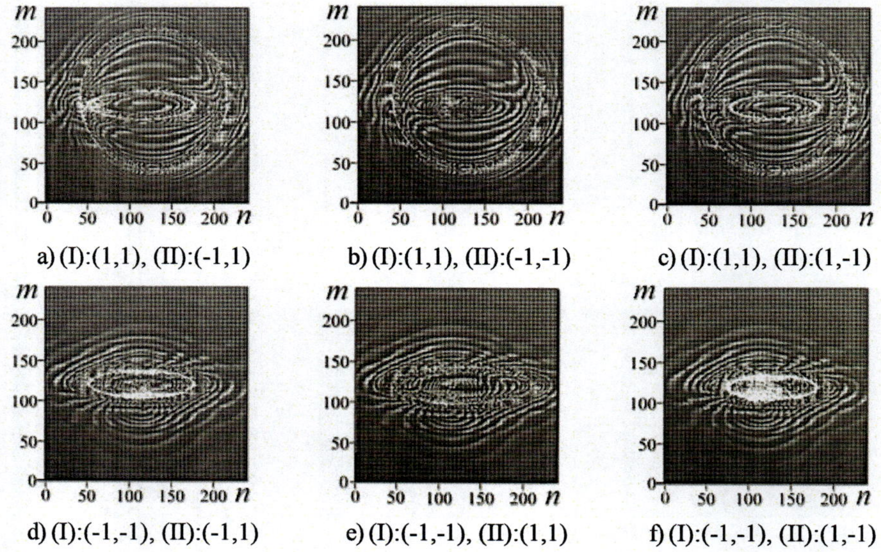

Fig. 7 The behavior of the cross sections Re$u \in [-1; 1]$ (top view) for fractal coupled structure (V). Separate structures (I) and (II) have different superpositional states of qubits

of the fractal coupled structure (V), when changing superposition qubit states of separate structures (I), (II). If structure (I) is in state (1, 1), and the qubit states of structure (II) change (Fig. 7a–c), then the complex deformation field of structure (V) changes significantly compared to Fig. 6d, e: for sections Reu_5, the effect of mixing of separate trajectories in the inner region, a change in behavior Imu_5 are observed. If structure (I) is in the state $(-1, -1)$, and the qubit states of structure (II) change (Fig. 7d–f), then the complex deformation field of structure (V) in comparison with Fig. 6c, f (where Im$u_5 = 0$) arises. In this case, an alteration of the structure of the inner region with the formation of stochastic boundary rings, the effect of mixing of individual trajectories for the cross sections Reu are observed. Using additional (external or internal) action the transitions of separate structures from one qubit state to another can be realized.

Similarly, the behavior of the deformation field of the coupled structure (VI), depending on the qubit states of separete structures (II), (I) was studied. In the general case, the deformation field of the coupled structure (VI) is complex. In this case, the conditions

$$u_6 - u_5 = u_{R2} f_6(u_{R1}) - u_{R1} f_5(u_{R2}) \neq 0, \mathbf{M}_6 - \mathbf{M}_5 \neq 0, \tag{15}$$

are satisfied, that is connected with the dependence of this stochastic processes. This indicates, that the displacement field operators of the separate structures (II), (I) and (I), (II) do not commute in the coupled structures (VI) and (V). As for structure (V), a feature of the deformation field behavior of the coupled structure (VI) with the same superposition states $(-1, -1)$ of separate structures is the absence of effective attenuation in all region ($\mathrm{Im}u_6 = 0$). For $\mathrm{Re}u_6$ the presence of the broadened stochastic peak with a structure close to the peak $\mathrm{Re}u_5$ (Fig. 6c) is also characteristic, but $\mathrm{Re}u_6 - \mathrm{Re}u_5 \neq 0$.

5 Conclusions

By the numerical modelling method the behavior of the deformation field of the coupled fractal structures (circular and elliptical cylinders) in various (initial, basic, superpositional) qubit states was investigated. It is shown, that when the qubit states change, features of the behavior of the complex deformation field of a separate structure are observed. The regular behavior of the inner region $\mathrm{Re}u$ is limited by the stochastic boundary (stochastic ring), wherein the concave part of the inner region changes to convex and then to convex-concave. The wave-like behavior for outer region $\mathrm{Re}u$ is characteristic. Such behavior allows the interpretation of the inner region as a membrane with the possible alteration of its states due to the change of qubit states. The stochastic behavior for $\mathrm{Im}u$, localized in the region of boundary rings with discontinuities is characteristic, wherein in the inner and outer regions of the rings $\mathrm{Im}u = 0$.

For fractal coupled structures with initial states of qubits of separate structures, the behavior of attractors and the complex deformation field is considered. It is shown, that the behavior of the deformation field essentially depends on the choice of stochastic processes realized during iterations. As examples, the features of the behavior of the deformation fields resulting from the sum, scalar and matrix products of independent and dependent stochastic processes are investigated.

Fractal coupled structures with various superpositional states of qubits of separate structures are considered. It is shown, that the presence of same superpositional qubit states of separate structures in the coupled system leads to the change in the complex deformation field: there is the decrease in peak amplitudes, peak displacement, and the change in structure. The original feature of the behavior of the deformation field of the coupled structure with the same superpositional states $(-1, -1)$ of separate structures is the absence of effective attenuation ($\mathrm{Im}u = 0$), which allows one to interpret the such structure as the memory cell.

By changing the superpositional qubit states of separate structures, one can change and control the behavior of the complex deformation field of the coupled structure. In this case, for the cross sections $\mathrm{Re}u$, the alteration of the inner region structure with the formation of stochastic boundary rings, the effect of mixing of separate trajectories is observed. Using additional (external or internal) action transitions of separate structures from one qubit state to another can be realized.

In the general case, the operators of the displacement field of coupled structures depend on the qubit states of separate structures and do not commute.

The results can be used to describe neural networks with variable parameters, in medicine when modeling blood vessels, for quantum information processing.

References

1. O.P. Abramova, Ordering of the displacement field operators of separate quantum dots, elliptical cylinders in coupled fractal structures. Bull. Donetsk Natl. Univer., A, **1**, 3–14 (2018)
2. O.P. Abramova, A.V. Abramov, Effect of ordering of displacement fields operators of separate quantum dots, elliptical cylinders on the deformation field of coupled fractal structures, in *11th Chaotic Modeling and Simulation International Conference, Springer Proceedings in Complexity,* ed. by C.H. Skiadas and I. Lubashevsky (Springer Nature Switzerland, AG, 2019), pp. 15–26
3. O.P. Abramova, Complex zero operators in coupled fractal structure with elements of cylindrical type. Bull. Donetsk Natl. Univer., A, **1**, 25–35 (2019)
4. V.S. Abramov, Quantum dots in a fractal multilayer system. Bull. Russian Acad. Sci. Phys. **81**(5), 625–632 (2017)
5. O.P. Abramova, A.V. Abramov, Coupled fractal structures with elements of cylindrical type, in *12th Chaotic Modeling and Simulation International Conference, Springer Proceedings in Complexity*, ed. by C.H. Skiadas and Y. Dimotikalis (Springer Nature Switzerland, AG, 2020), pp. 15–26
6. H.-J. Stöckmann, *Quantum Chaos* (An Introduction. Philipps-Universität Marburg, Germany, 2007).
7. A.N. Omelyanchuk, E.V. Ilyichev, S.N. Shevchenko, *Quantum Coherent Phenomena in Josephson Qubits* (Naukova Dumka, Kiev, 2013).
8. M.N. Fedorov, I.A. Volkov, Y.M. Mikhailova, Coutrites and Kukwarts in spontaneous parametric scattering of light, correlation and entanglement of states. JETP **142**(1,7), 20–43 (2012)
9. M. Nielsen, I. Chuang, *Quantum Computation and Quantum Information* (Cambridge University Press, New York, 2010).
10. D. Boumeister, A. Eckert, A. Zeilinger, *Physics of Quantum Information* (Springer, New York, 2001).
11. Y. Ozhigov, Quantum computers speed up classical with probability zero. Chaos Solitons Fractals **10**, 1707–1714 (1999)
12. D. Castelvecchi, Quantum computers ready to leap out of the lab. Nature **541**, 9–10 (2017)
13. Y.S. Kivshar and N.N. Rozanov (ed.), *Nonlinearities in Periodic Structures and Metamaterials* (Moscow, Fizmatlit, 2014)
14. V.S. Abramov, Alteration of the stochastic state of the deformation field in the model multilayer nanosystem. Bull Donetsk Natl. Univer., A, **2**, 81–89 (2014)

Fractal Atomicity, a Fundamental Concept in the Dynamics of Complex Systems

Maricel Agop, Alina Gavriluţ, Lucian Eva, and Gabriel Crumpei

Abstract Applying a fractal method of analyzing the dynamics of the structural units of any complex system, a mathematical concept is built, namely that of fractal atomicity. The construction of such a concept involves defining dynamic variables in the form of fractal functions, defining scale resolutions, defining a principle of scale covariance as a fundamental principle of motion, equations of evolution, etc. Finally, some specific mathematical properties of the fractal atom are also established.

Keywords Fractal atomicity · Complex systems dynamics · Fractal functions · Holographic atom

1 Introduction

The notion of non-atomicity for set functions plays a key role in Measure Theory and its applications and extensions. For classical measures taking values in finite dimensional Banach spaces, it guarantees the connectedness of range. Even just replacing σ-additivity with finite additivity for measures requires some stronger non-atomicity property for the same conclusion to hold.

Because of their multiple applications in game theory or mathematical economics, the study concerning atoms and non-atomicity for additive, respectively, non-additive set functions has developed. Particularly, (non)atomic (purely) measures have been studied in different forms due to their special form and their special properties

M. Agop
Department of Physics, "Gheorghe Asachi" Technical University from Iaşi, Iaşi, Romania

A. Gavriluţ (✉)
Faculty of Mathematics, "Alexandru Ioan Cuza" University of Iaşi, Iaşi, Romania
e-mail: gavrilut@uaic.ro

L. Eva
N. Oblu Clinical Emergency Hospital Iaşi, Iaşi, Romania

G. Crumpei
Faculty of Psychology and Education Sciences, "Alexandru Ioan Cuza" of Iaşi, Iaşi, Romania

C. H. Skiadas et al. (eds.), *13th Chaotic Modeling and Simulation International Conference*, Springer Proceedings in Complexity,
https://doi.org/10.1007/978-3-030-70795-8_3

(Chiţescu [4, 5], Cavaliere and Ventriglia [3], Gavriluţ and Agop [13], Gavriluţ and Croitoru [10–12], Gavriluţ [7–9], Gavriluţ et al. [14], Khare and Singh [23], Li et al. [24, 25], Pap [28–30], Pap et al. [31], Rao and Rao [32], Suzuki [40], Wu and Bo [41] and many others). Modifications of non-additive Measure Theory (Pap [29, 30]) led to Quantum Measure Theory (Gudder [15–19], Salgado [33], Sorkin [35–38], Surya and Waldlden [39]. In these papers, an extended notion of a measure has been introduced and certain applications to interference, probability have been highlighted (Schweizer and Sklar [34]). In fact, Quantum Measure Theory (introduced by Sorkin [35–38]) represents a generalization of Quantum Theory, where physical predictions are computed from a matrix known as decoherence functional. Quantum measures are an useful tool to describe Quantum Mechanics and its applications to Quantum Gravity and Cosmology (Hartle [20, 21], Phillips [31]).

Despite the continuous efforts of numerous scientists, reconciling General Relativity with Quantum Theory remains one of the most important open problems in Physics. The framework of General Relativity suggests that one promising approach to such a unification will be by means of a reformulation of Quantum Theory in terms of histories rather than states. Following this idea, Sorkin [35–38], has proposed a history-based framework, which can accommodate both standard Quantum Mechanics as well as physical theories beyond the quantum formalism.

As we shall prove in this paper, in such framework, Schrödinger's equation from Quantum Mechanics can be identified with a particular type of geodesic of the fractal space. In consequence, fundamental concepts of Quantum Mechanics can be extended to similar concepts, but on fractal manifolds. In this paper, we extend the concept of atoms/pseudo-atoms to the concept of fractal minimal atom/fractal pseudo-atom, respectively. We also give characterizations from a mathematical viewpoint to these new concepts and we make explicit certain physical implications. The notion of a fractal minimal atom as a particular case of fractal atom is also discussed. In this framework, we are looking for certain physical correspondences in the Quantum Mechanics context.

The present paper is organized as follows. After an Introductory part, Sect. 2 contains some results concerning the properties of different types of atoms, introduced from the Quantum Measure Theory mathematical perspective. Certain physical implications and interpretations are provided. In Sect. 3, elements of Fractal Mechanics are provided in order to build the background for extending the notions that are specific to atomicity, to those involving fractal atomicity. From this perspective, new concepts as that of a fractal minimal atom is introduced and some of its properties are discussed in Sect. 4.

2 Types of Atoms in the Mathematical Approach

In what follows, T denotes an abstract nonvoid set and $\mathcal{C}$ a ring of subsets of T. Suppose $(V, +, \cdot)$ is a real linear space, with the origin 0.

If $p \in \mathbb{N}^*$, then by $i = \overline{1, p}$ we usually mean $i \in \{1, 2, \ldots, p\}$.

Definition 2.1 (*Gavriluţ and Agop* [13]) Let $m : \mathcal{C} \to V$ be a set function, with $m(\emptyset) = 0$. m is said to be:

(i) finitely additive if $m(\bigcup_{i=1}^{n} E_i) = \sum_{i=1}^{n} m(E_i)$, for any arbitrary pairwise disjoint sets $(E_i)_{i\in\{1,2,\ldots,n\}} \subset \mathcal{C}$, $n \in \mathbb{N}^*$;

(ii) null-additive[1] if $m(E \cup F) = m(E)$, for every disjoint $E, F \in \mathcal{C}$, with $m(F) = 0$;

(iii) null-additive[2] if $m(E \cup F) = m(E)$, for every $E, F \in \mathcal{C}$, with $m(F) = 0$;

(iv) null-null-additive if $m(E \cup F) = 0$, for every $E, F \in \mathcal{C}$, with $m(E) = m(F) = 0$;

(v) null-equal if $m(E) = m(F)$, for every $E, F \in \mathcal{C}$, with $m(E \cup F) = 0$;

(vi) diffused if $m(\{t\}) = 0$, whenever $\{t\} \in \mathcal{C}$.

Definition 2.2 (*Gavriluţ et al.* [14]) If V is, moreover, a Banach lattice, a set function $m : \mathcal{C} \to V$, with $m(\emptyset) = 0$, is said to be:

(i) null-monotone if for every $E, F \in \mathcal{C}$, with $E \subseteq F$, if $m(F) = 0$, then $m(E) = 0$;

(ii) monotone (or, fuzzy) if $m(E) \leq m(F)$, for every $E, F \in \mathcal{C}$, with $E \subseteq F$;

(iii) a submeasure (in the sense of Drewnowski [6]) if m is monotone and subadditive, i.e., $m(E \cup F) \leq m(E) + \nu(F)$, for every (disjoint) $E, F \in \mathcal{C}$;

(iv) σ-additive (or, a (vector) measure) if $m(\bigcup_{n=1}^{\infty} E_n) = \lim_{n\to\infty} \sum_{k=1}^{n} m(E_k)$, for every pairwise disjoint sets $(E_n)_{n\in\mathbb{N}^*} \subset \mathcal{C}$, with $\bigcup_{n=1}^{\infty} E_n \in \mathcal{C}$.

Definition 2.3 If $\mathcal{A}$ is an arbitrary σ-algebra of T and if $m : \mathcal{A} \to \mathbb{R}_+$ is a measure on $\mathcal{A}$, with $m(T) = 1$, then:

(i) The space $(T, \mathcal{A}, m)$ is said to be a sample space and m is said to be a probability measure;

(ii) The elements of T are called sample points or outcomes and the elements of $\mathcal{A}$ are called events.

In this case, for every $E \in \mathcal{A}$, $m(E)$ is interpreted as the probability of the event E to occur.

Remark 2.4 (i) The notion of a null-equal-measure has the following physical interpretation (Gavriluţ and Agop [13]): in the situation involving destructive interference, in order for two waves to produce complete destructive interference, thereby "cancelling out" each other, their original amplitudes must have been equal.

(ii) If $m(T) > 0$, then one can immediately generate a probability measure by means of a normalization process.

Remark 2.5 (I) (i) One observes that a set function $m : \mathcal{C} \to V$ is diffused if the measure of any singleton of the space is null. This means in the construction of a physical theory, the vacuum condition of the matter should be considered as its complement.

(ii) Shannon's entropy is a subadditive real-valued set function [2].

(iii) If V is a Banach lattice, $T = \{t_1, t_2, \ldots, t_n\}$, $n \in \mathbb{N}^*$, is an arbitrary finite metric space and $m : \mathcal{P}(T) \to V$ (or, more general, if T is a T_1-separated topological space, $\mathcal{B}$ is the Borel σ-algebra of T generated by the lattice of all compact subsets of T and $m : \mathcal{B} \to V$) is null-additive and diffused, then $m(T) = 0$ (i.e., the space T is composed of particles which annihilate one each other).

(II) If $m : \mathcal{C} \to V$ is null-monotone, then:

(i) m is null-additive^1 if and only if it is null-additive^2. In this case, m will be simply called null-additive.

(ii) If m is null-null-additive, then it is null-equal.

Definition 2.6 (*Gavriluţ and Croitoru* [10–12]) Let $m : \mathcal{C} \to \mathbb{R}_+$ be a set function, with $m(\emptyset) = 0$.

(i) A set $E \in \mathcal{C}$ is said to be an atom of ν if $m(E) > 0$ and for every $F \in \mathcal{C}$, with $F \subseteq E$, we have $m(F) = 0$ or $m(E \backslash F) = 0$ (in a certain sense, an atom can be interpreted as being a black hole);

In consequence, an atom is a measurable set which has positive “measure” and contains no set of smaller positive “measure”. In fact, the main examples of atomsare singletons that have positive “measure”.

(ii) *m is said to be non-atomic (or, atomless) if it has no atoms (i.e., for every $E \in \mathcal{C}$ with $m(E) > 0$, there exists $F \in \mathcal{C}$, $F \subseteq E$, such that $m(F) > 0$ and $m(E \backslash F) > 0$);*

(iii) A set $E \in \mathcal{C}$ is called a pseudo-atom of ν if $m(E) > 0$ and $F \in \mathcal{C}$, $F \subseteq E$ implies $m(F) = 0$ or $m(F) = m(E)$;

(iv) m is said to be non-pseudo-atomic if it has no pseudo-atoms (i.e., for every $E \in \mathcal{C}$ with $m(E) > 0$, there exists $F \in \mathcal{C}$, $F \subseteq E$, such that $m(F) > 0$ and $m(E) \neq m(F)$);

(v) m is said to be finitely purely atomic if there is a finite family $(E_i)_{i \in \{1,2,\ldots,n\}}$ of pairwise disjoint atoms of m so that $T = \bigcup_{i=1}^{n} E_i$ (in this case, the space T is a finite collection of pairwise disjoint atoms).

I is well-known that the Lebesgue measure on the real line has no atoms. It is also that the entropy of a non-atomic measure must be infinite, while the entropy of an atomic measure vanishes. The following statements easily follow:

Proposition 2.7 *Suppose $m : \mathcal{C} \to \mathbb{R}_+$ is so that $m(\emptyset) = 0$.*

(i) If m is finitely additive, then $E \in \mathcal{C}$ is an atom of m if and only if E is a pseudo-atom of m.

(ii) Any $\{t\} \subseteq T$, provided $\{t\} \in \mathcal{C}$ and $m(\{t\}) > 0$, is an atom of m.

(ii) If m is null-additive^1, then every atom of m is also a pseudo-atom. The converse is not generally valid.

Example 2.8 Let $T = \{t_1, t_2\}$ be a finite abstract space composed of two elements.

(i) We consider the set function $m : \mathcal{P}(T) \to \mathbb{R}_+$ defined for every $E \subset T$ by

$$m(E) = \begin{cases} 2, E = T \\ 1, E = \{t_1\} \\ 0, E = \{t_2\} \text{ or } E = \emptyset. \end{cases}$$

Then T is an atom and it is not a pseudo-atom of m.

(ii) We define $m : \mathcal{P}(T) \to \mathbb{R}_+$ by $m(E) = \begin{cases} 1, E \neq \emptyset \\ 0, E = \emptyset \end{cases}$, for every $E \subset T$.

Then m is null-additive and $T = \{t_1, t_2\}$ is a pseudo-atom of m, but it is not an atom.

(iii) Let be $T = \{1, 2, \ldots 10\}$ and $m : \mathcal{P}(T) \to \mathbb{R}_+$, $m(A) = \text{card}A$, $\forall A \subseteq T$. Then each of the singletons $\{i\}$, $i \in \{1, \ldots, 10\}$ is an atom.

(iv) Let T be a countable set, $\mathcal{A} = \{A \subseteq T;\ A \text{ is finite or } T \setminus A \text{ is finite }\}$ and $m : \mathcal{A} \to \mathbb{R}_+$,

$$m(A) = \begin{cases} 0, & A \text{ is finite} \\ 1, & T \setminus A \text{ is finite} \end{cases}, \forall A \in \mathcal{A}.$$

Then every set $A \in \mathcal{A}$ so that $T \setminus A$ is finite is an atom of m.

(v) On $T = \mathbb{R}$, let be the algebra $\mathcal{A} = \{\emptyset, T\}$ and $m : \mathcal{A} \to \mathbb{R}_+$, $m(\emptyset) = 0$, $m(\mathbb{R}) = 1$. Then $\mathbb{R}$ is an atom and obviously it is not a singleton set.

Proposition 2.9 *If $m : \mathcal{C} \to \mathbb{R}_+$ is null-monotone and null-additive and if $E \cup F$ is an atom of m, then E, F are m-compatible that is, the following property holds: $m(E \cup F) + m(E \cap F) = m(E) + m(F)$ (thus, any two components of an atom must be compatible).*

Proof 1. If $m(E) = 0$, then $m(E \cap F) = 0$ and $m(E \cup F) = m(F)$, so the conclusion follows.

2. If $m(E) > 0$, then by Proposition 2.7 - (i) E is an atom, too and $m((E \cup F)\setminus E) = m(F\setminus E) = 0$. Since $F = (F\setminus E) \cup (F \cap E)$, then $m(F) = m(E \cap F)$ and since $E \cup F = E \cup (F\setminus E)$, we get $m(E \cup F) = m(E)$.

In what follows, let $\mathcal{K}$ be the lattice of all compact subsets of a locally compact Hausdorff space T and $\mathcal{B}$ be the Borel σ-algebra generated by $\mathcal{K}$. In such framework, the following definition is consistent:

Definition 2.10 (*Pap* [28–30]) $m : \mathcal{B} \to \mathbb{R}_+$ is said to be regular if for every $E \in \mathcal{B}$ and every $\varepsilon > 0$, there exist $K \in \mathcal{K}$ and an open set $D \in \mathcal{B}$ such that $K \subset E \subset D$ and $m(D\setminus K) < \varepsilon$.

Theorem 2.11 (Pap *[28–30]*) *Suppose $m : \mathcal{B} \to \mathbb{R}_+$ is a monotone null-additive regular set function. If $E \in \mathcal{B}$ is an atom of m, there exists only one point $e \in E$ so that $m(E\setminus\{e\}) = 0$ (and so, $m(E) = m(\{e\})$.*

Remark 2.12 The previous theorem has the following physical interpretation: in an atom, the entire "information" is concentrated in each of its points.

3 Fractal Theory of Motion

As a rule, the classical models used in the description of the dynamics of the systems are based on the assumption, otherwise unjustified, of the differentiability of the variables that describe them. The success of these models must be understood gradually/sequentially, on "domains" in which the differentiability and integrability are still valid. But the differentiable and integrable mathematical procedures "suffer" when we want to describe the dynamics of the systems, because only these ones "support" and "operate" with nonlinearities and chaoticities.

In order to describe dynamics, while still being dependent on the differentiable and integrable mathematical procedures, it is necessary to explicitly introduce the scale resolution in the expressions of the physical variables that describe them and implicitly in the expressions of the fundamental equations that govern these dynamics. This means that any variable, used in describing the dynamics of the systems, dependent in the classical sense, both on the spatial and time coordinates, depends, in the context mentioned above, on the scale resolution. In other words, instead of "operating" for example with a single variable described by a strictly non-differentiable mathematical function, we will "operate" only with approximations of this mathematical function obtained by mediating it at different scale resolutions. Consequently, any variable designed to describe system dynamics will "function" as the limit of a family of mathematical functions, this being non-differentiable for a zero-scale resolution and differentiable for a non-zero scale resolution.

This way of describing the dynamics of the systems, where any "explicitation" of dynamics is done at finite scale resolution, obviously implies both the development of a new geometrical structure and of models conforming to these geometrical structures for which the laws of motion, invariant to spatial and temporal transformations, are" integrated" with scale laws invariant to transformations of scale resolution. In our opinion, such a geometric structure can be based on the concept of fractal, and the corresponding physical model, the Multifractal Theory of Motion in the form of Scale Relativity Theory with arbitrary and constant fractal dimension [27].

Let us suppose that on a fractal space-time manifold the motions of systems take place on fractal curves. Then:

Remark 3.1 Any fractal curve is explicitly scale resolution dependent (which will be referred as $\delta\tau$). Its length tends to infinity when its proper time interval, $\Delta\tau$, tends to zero (an extension of the Lebesgue theorem on a fractal space-time manifold).

Remark 3.2 In the limit $\Delta\tau \to 0$, a curve in a fractal space-time manifold is a zigzagged as one can imagine. Thus, it exhibits the property of self-similarity in all its points of a fractal space-time manifold, which can be translated into an extension property of holography (every part of a fractal space-time manifold reflects the whole of the same space-time manifold).

Remark 3.3 The differential proper time reflection invariance of any variable is broken. Then every variable $Q(\tau)$ is replaced by the fractal variable $Q(\tau, d\tau)$ explicitly dependent on the proper time resolution interval whose derivative is undefined only in the limit, $\Delta\tau \to 0$. As a consequence, two derivatives of every fractal variable as explicit functions of τ and $d\tau$ will be defined. For example, the two derivatives of the 4-coordinate $X^{\mu}(\tau, \Delta\tau)$ takes the form:

$$\frac{d_+X^\mu}{d\tau} = \lim_{\Delta\tau\to 0_+} \frac{X^\mu(\tau+\Delta\tau,\Delta\tau) - X^\mu(\tau,\Delta\tau)}{\Delta\tau}$$

$$\frac{d_-X^\mu}{d\tau} = \lim_{\Delta\tau\to 0_-} \frac{X^\mu(\tau,\Delta\tau) - X^\mu(\tau-\Delta\tau,\Delta\tau)}{\Delta\tau} \tag{1}$$

The sign $+$ corresponds to the forward process and the sign $-$ to the backwards one.

Remark 3.4 The differential of 4-coordinate $dX^\mu(\tau,\Delta\tau)$ can be expressed as the sum of two differentials, one not scale dependent, $d_\pm x^\mu(\tau))$, and other scale dependent, $d_\pm\xi^\mu(\tau,d\tau))$, i.e.,

$$d_\pm X^\mu(\tau,\Delta\tau) = d_\pm x^\mu(\tau) + d_\pm\xi^\mu(\tau,d\tau); \tag{2}$$

Remark 3.5 $d_\pm\xi^\mu$ satisfies the equation

$$d_\pm\xi^\mu(\tau,d\tau) = \lambda^\mu_\pm(d\tau)^{1/f(\alpha)} \tag{3}$$

where $\lambda^\mu_\pm$ are constant coefficients whose statistical significance will be given in what follows, $f(\alpha)$ is the singularity spectrum of order α, α is the singularity index and D_F is the fractal dimension of the motion curves from the fractal space-time manifold.

Remark 3.6 There are many modes and thus various selection of definitions of fractal dimensions: fractal dimension in the Kolmogorov sense, fractal dimension in the Hausdorff–Besikovici sense etc. Selecting one of these definitions and operating with it in the dynamics of systems, the value of the fractal dimension must be constant and arbitrary: $D_F < 2$ for correlative processes, $D_F > 2$ for non-correlative processes etc. In such conjecture we can identify not only the "areas" of the dynamics of the system that are characterized by a certain fractal dimension, but also the number of "areas" whose fractal dimensions are situated in an interval values. Moreover, through the singularity spectrum we can identify "classes of universality" in the dynamics of the system, even when strange attractors have various aspects.

Remark 3.7 The differential proper time reflection invariance is recovered by combining the derivatives $d_+/d\tau$ and $d_-/d\tau$ in the fractal operator (Cresson procedure):

$$\frac{\hat{d}}{d\tau} = \frac{1}{2}\left(\frac{d_+ + d_-}{d\tau}\right) - \frac{i}{2}\left(\frac{d_+ - d_-}{d\tau}\right) \tag{4}$$

Applying, for example, the fractal operator to the 4-coordinate X^μ yields the 4-complex velocity:

$$\hat{V}^\mu = \frac{\hat{d}X^\mu}{d\tau} = \frac{1}{2}\left(\frac{d_+X^\mu + d_-X^\mu}{d\tau}\right) - \frac{i}{2}\left(\frac{d_+X^\mu - d_-X^\mu}{d\tau}\right) = V^\mu - iU^\mu \tag{5}$$

with

$$V^{\mu}=\frac{1}{2}(v_{+}^{\mu}+v_{-}^{\mu}),\ U^{\mu}=\frac{1}{2}(v_{+}^{\mu}-v_{-}^{\mu}),\ v_{+}^{\mu}=\frac{d_{+}x^{\mu}+d_{+}\xi^{\mu}}{d\tau},\ v_{-}^{\mu}=\frac{d_{-}x^{\mu}+d_{-}\xi^{\mu}}{d\tau}$$

The real part V^{μ} is scale resolution independent, while the imaginary one U^{μ} is scale resolution dependent.

Remark 3.8 An infinite number of geodesics can be found relating any pair of points of a fractal space-time manifold, and this is true on all scale resolutions of the physical system dynamics. Then, in the fractal space-time manifold, all the entities of the system are substituted with the geodesics themselves so that any external constraint can be interpreted as a selection of geodesics in the same fractal space-time manifold.

Remark 3.9 The infinity of geodesics in the bundle, their multifractality, the two values of the derivative etc., imply a generalized statistical fluid-like description (fractal fluid). Thus, one provides the multifractalisation type through stochastic processes. For example, we can choose the average of $d_{\pm}X^{i}$ in the form

$$< d_{\pm}X^{i} >\equiv d_{\pm}x^{i} \tag{6}$$

which by (2) implies

$$< d_{\pm}\xi^{i} >= 0$$

Theorem 3.10 *The transition from the dynamics of the Special Relativity to the dynamics of Multifractal Theory of Motion in the form of Scale Relativity Theory on a fractal space-time manifold can be described through the scale covariant derivative:*

$$\frac{\hat{d}}{d\tau}=\partial_{\tau}+\hat{V}^{\mu}\partial_{\mu}+\frac{1}{4}(d\tau)^{(2/f(\alpha))-1}D^{\mu\nu}\partial_{\mu}\partial_{\nu} \tag{7}$$

where

$$\begin{aligned} D^{\mu\nu} &= d^{\mu\nu}-i\bar{d}^{\mu\nu}\\ d^{\mu\nu} &= \lambda_{+}^{\mu}\lambda_{+}^{\nu}-\lambda_{-}^{\mu}\lambda_{-}^{\nu},\ \bar{d}^{\mu\nu}=\lambda_{+}^{\mu}\lambda_{+}^{\nu}+\lambda_{-}^{\mu}\lambda_{-}^{\nu} \end{aligned} \tag{8}$$

Proof The proof of the above statements is given in [1].

Remark 3.11 If the multifractality of the motion curve is achieved through Markov type stochastic process

$$\lambda_+^\mu \lambda_+^\nu = \lambda_-^\mu \lambda_-^\nu = -\lambda \eta^{\mu\nu} \tag{9}$$

where $\eta^{\mu\nu}$ is the Minkowski metric, then the scale covariant derivative (8) takes the form

$$\frac{\hat{d}}{d\tau} = \partial_\tau + \hat{V}^\mu \partial_\mu + i\frac{\lambda}{2}(d\tau)^{(2/D_F)-1}\partial_\mu \partial^\mu \tag{10}$$

Remark 3.12 Applying the scale covariant operator to the complex velocity (5), the geodesics equation on a fractal space-time is obtained:

$$\frac{\hat{d}\hat{V}^\mu}{d\tau} = \partial_\tau \hat{V}^\mu + \hat{V}^\nu \partial_\nu \hat{V}^\mu + \frac{1}{4}(d\tau)^{(2/f(\alpha))-1} D^{\alpha\beta}\partial_\alpha \partial_\beta \hat{V}^\mu \equiv 0 \tag{11}$$

Remark 3.13 Through separation of motions on scale resolutions (the real part from the imaginary one) on a fractal space-time, the geodesics equation in real terms becomes:

$$\begin{aligned}
&\tfrac{\hat{d}V^\mu}{d\tau} = \partial_\tau V^\mu + V^\nu \partial_\nu V^\mu - U^\nu \partial_\nu U^\mu + \tfrac{1}{4}(d\tau)^{(2/f(\alpha))-1} d^{\alpha\beta}\partial_\alpha \partial_\beta V^\mu - \\
&-\tfrac{1}{4}(d\tau)^{(2/f(\alpha))-1}\overline{d}^{\alpha\beta}\partial_\alpha \partial_\beta U^\mu = 0 \\
&\tfrac{\hat{d}U^\mu}{d\tau} = \partial_\tau U^\mu + V^\nu \partial_\nu U^\mu + U^\nu \partial_\nu V^\mu + \tfrac{1}{4}(d\tau)^{(2/f(\alpha))-1} d^{\alpha\beta}\partial_\alpha \partial_\beta U^\mu + \\
&+\tfrac{1}{4}(d\tau)^{(2/f(\alpha))-1}\overline{d}^{\alpha\beta}\partial_\alpha \partial_\beta V^\mu = 0
\end{aligned} \tag{12}$$

Theorem 3.14 *Choosing the 4-complex velocity from (5) in terms of a scalar complex field* Ψ,

$$\hat{V}^\mu = i\lambda (d\tau)^{(2/f(\alpha))-1}\partial_\mu \ln \Psi \tag{13}$$

the geodesics of a fractal space-time manifold in the case of multifractalisation by means of Markovian stochastic processes, for a null value of the integration constant takes the form of the generalized Schrödinger type equation

$$\lambda^2 (d\tau)^{(4/f(\alpha))-2}\partial_\mu \partial^\mu \Psi + i\lambda \partial_\tau \Psi = 0 \tag{14}$$

Remark 3.15 For non-relativistic dynamics in 3D space on Peano type curve, $f(\alpha) \equiv D_F = 2$, at Compton scale resolution, $\lambda = \hbar/2m_0$, with $\hbar$ the Planck reduced constant and m_0 the rest mass of the particle, from (13) it results the standard Schrödinger equation

$$\frac{\hbar^2}{2m_0}\partial_i \partial^i \Psi + i\hbar \partial_\tau \Psi = 0 \tag{15}$$

4 From the Standard Mathematical Atom to the Fractal Atom by Means of a Physical Procedure

Let T be an abstract nonvoid set, $\mathcal{G}$ a lattice of subsets of T and $m : \mathcal{G} \to \mathbb{R}_+$ an arbitrary set function with $m(\emptyset) = 0$. Evidently, one can immediately generalize the notions of a pseudo-atom / minimal atom, respectively, to this context when $\mathcal{G}$ is only a lattice and not necessarily a ring.

Example 4.1 (i) It T is a nonempty metric space, then the Hausdorff dimension $\dim_{Haus} : \mathcal{P}(T) \to \mathbb{R}$ (Mandelbrot [26]) is a monotone real function. Evidently, $\dim_{Haus}(\emptyset) = 0$.

(ii) For every $d \geq 0$, the Hausdorff measure $H^d : \mathcal{P}(T) \to \mathbb{R}$ is a submeasure.

Remark 4.2 (i) The union of two sets E and F having the fractal dimensions D_E, respectively, D_F, has the fractal dimension $D_{E\cup F} = \max\{D_E, D_F\}$;

(ii) The intersection of two sets E and F having the fractal dimensions D_E, respectively, D_F has the fractal dimension $D_{E\cap F} = D_E + D_F - d$, where d is the embedding Euclidean dimension (Iannaccone and Khokha [22]).

The following definition is then consistent:

Definition 4.3 A pseudo-atom/minimal atom, respectively, $E \in \mathcal{G}$ of m having the fractal dimension D_E is said to be a fractal pseudo-atom/fractal minimal atom, respectively.

Therefore, we can give:

Proposition 4.4 *If $E, F \in \mathcal{G}$ are fractal pseudo-atoms of m and if $m(E \cap F) > 0$, then $E \cap F$ is a fractal pseudo-atom of m and $m(E \cap F) = m(E) = m(F)$.*

5 Conclusions

The main conclusions of the present paper are the following:

(i) (Pseudo)-atomicity is treated from the Quantum Measure theory mathematical perspective and several physical applications are given;

(ii) Minimal atomicity in correspondence with Quantum Measure Theory is also discussed. In such context, some physical applications are provided;

(iii) The concept of atomicity (and, particularly, that of minimal atomicity) is extended in the form of fractal atomicity, respectively, fractal minimal atomicity. Some mathematical properties of fractal minimal atomicity are given. In such approach, an inverse method with respect to the common developments concerning the atomicity concept has been used, observing that Quantum Mechanics identifies as a particular case of Fractal Mechanics for a given scale resolution. Precisely, we talk about a fractality through Markov type stochastic processes, in which case the standard Schrödinger equation identifies with the geodesics of a fractal space for motions of a complex system structural units on Peano type curves at Compton scale resolution.

The concept of fractal atomicity introduces a minimal code which could correspond to the DNA that is specific to each person. In this paper, we introduced the basis for a multivalent logic, which could open new perspectives in genomics for instance, in the decipher of the intimate mechanisms at biostructures level.

References

1. M. Agop, A. Gavriluţ, G. Ştefan, B. Doroftei, Implications of non-differentiable entropy on a space-time manifold. Entropy **17**, 2184–2197 (2015)
2. K.M.R. Audenaert, Subadditivity of q-entropies for $q > 1$. J. Math. Phys. **48** (2007)
3. P. Cavaliere, F. Ventriglia, On nonatomicity for non-additive functions. J. Math. Anal. Appl. **415**(1), 358–372 (2014)
4. I. Chiţescu, Finitely purely atomic measures and $\mathcal{L}^p$-spaces. An. Univ. Bucureşti Şt. Natur. **24**, 23–29 (1975)
5. I. Chiţescu, Finitely purely atomic measures: coincidence and rigidity properties. Rend. Circ. Mat. Palermo **50**(3), 455–476 (2001)
6. L. Drewnowski, Topological rings of sets, continuous set functions, integration, I, II, III. Bull. Acad. Polon. Sci. **20**, 269–276, 277–286, 439–445 (1972)
7. A. Gavriluţ, Non-atomicity and the Darboux property for fuzzy and non-fuzzy Borel, Baire multivalued set functions. Fuzzy Sets Syst. **160**, 1308–1317 (2009). Erratum in Fuzzy Sets Syst. **161**, 2612–2613 (2010)
8. A. Gavriluţ, Fuzzy Gould integrability on atoms. Iran. J. Fuzzy Syst. **8**(3), 113–124 (2011)
9. A. Gavriluţ, *Regular Set Multifunctions* (Pim Publishing House, Iaşi, 2012)
10. A. Gavriluţ, A. Croitoru, On the Darboux property in the multivalued case. Annals of the University of Craiova. Math. Comput. Sci. Ser. **35**, 130–138 (2008)
11. A. Gavriluţ, A. Croitoru, Non-atomicity for fuzzy and non-fuzzy multivalued set functions. Fuzzy Sets Syst. **160**, 2106–2116 (2009)
12. A. Gavriluţ, A. Croitoru, Pseudo-atoms and Darboux property for set multifunctions. Fuzzy Sets Syst. **161**(22), 2897–2908 (2010)
13. A. Gavriluţ, M. Agop, *An Introduction to the Mathematical World of Atomicity Through a Physical Approach* (ArsLonga Publishing House, Iaşi, 2016)
14. A. Gavriluţ, A. Iosif, A. Croitoru, The Gould integral in Banach lattices. Positivity **19**(1), 65–82 (2015)
15. S. Gudder, Quantum measure and integration theory. J. Math. Phys. **50** (2009)
16. S. Gudder, Quantum integrals and anhomomorphic logics (2009), arXiv:quant-ph (0911.1572)
17. S. Gudder, Quantum measure theory. Math. Slovaca **60**, 681–700 (2010)
18. S. Gudder, Quantum measures and the coevent interpretation. Rep. Math. Phys. **67**, 137–156 (2011)
19. S. Gudder, Quantum measures and integrals
20. J.B. Hartle, *The Quantum Mechanics of Cosmology*. Lectures at Winter School on Quantum Cosmology and Baby Universes, Jerusalem, Israel, Dec 27, 1989–Jan 4, 1990 (1989)
21. J.B. Hartle, Spacetime quantum mechanics and the quantum mechanics of spacetime, in *Proceedings of the Les Houches Summer School on Gravitation and Quantizations*, ed. by J. Zinn-Justin, B. Julia, Les Houches, France, 6 Jul–1 Aug 1992 (North-Holland, 1995), arXiv:gr-qc/9304006
22. P.M. Iannaccone, M. Khokha, *Fractal Geometry in Biological Systems: An Analitical Approach* (1995)

23. M. Khare, A.K. Singh, Atoms and Dobrakov submeasures in effect algebras. Fuzzy Sets Syst. **159**(9), 1123–1128 (2008)
24. J. Li, R. Mesiar, E. Pap, Atoms of weakly null-additive monotone measures and integrals. Inf. Sci. 134–139 (2014)
25. J. Li, R. Mesiar, E. Pap, E.P. Klement, Convergence theorems for monotone measures. Fuzzy Sets Syst. **281**, 103–127 (2015)
26. B.B. Mandelbrot, *The Fractal Geometry of Nature*, Updated and augm. edn. (W.H. Freeman, New York, 1983)
27. I. Mercheş, M. Agop, *Differentiability and Fractality in Dynamics of Physical Systems* (World Scientific, 2015)
28. E. Pap, The range of null-additive fuzzy and non-fuzzy measures. Fuzzy Sets Syst. **65**(1), 105–115 (1994)
29. E. Pap, *Null-Additive Set Functions*. Mathematics and Its Applications, vol. 337 (Springer, 1995)
30. E. Pap, Handbook of measure theory, in *Some Elements of the Classical Measure Theory* (2002), pp. 27–82
31. E. Pap, A. Gavriluţ, M. Agop, Atomicity via regularity for non-additive set multifunctions. Soft Comput. (Found.) 1–6 (2016). https://doi.org/10.1007/s00500-015-2021-x
32. K.P.S.B. Rao, M.B. Rao, *Theory of Charges* (Academic Press Inc., New York, 1983)
33. R. Salgado, Some identities for the q-measure and its generalizations. Mod. Phys. Lett. A **17**, 711–728 (2002)
34. B. Schweizer, A. Sklar, *Probabilistic Metric Spaces* (Elsevier Science Publishing Co., Inc., 1983). Republished in 2005 by Dover Publications, Inc., with a new preface, errata, notes, and supplementary references
35. R.D. Sorkin, Quantum mechanics as quantum measure theory. Mod. Phys. Lett. A **9**, 3119–3128 (1994)
36. R.D. Sorkin, Quantum measure theory and its interpretation, in *Quantum Classical Correspondence: Proceedings of the* 4th *Drexel Symposium on Quantum Non-integrability*, ed. by D.H. Feng, B.-L. Hu (International Press, Cambridge Mass, 1997), pp. 229–251
37. R. Sorkin, Quantum dynamics without the wave function. J. Phys. A: Math. Theory **40**, 3207–3231 (2007)
38. R. Sorkin, *Quantum mechanics as quantum measure theory*, arXiv:gr-qc/9401003
39. S. Surya, P. Waldlden, Quantum covers in q-measure theory (2008), ArXiv: quant-ph 0809.1951
40. H. Suzuki, Atoms of fuzzy measures and fuzzy integrals. Fuzzy Sets Syst. **41**, 329–342 (1991)
41. C. Wu, S. Bo, Pseudo-atoms of fuzzy and non-fuzzy measures. Fuzzy Sets Syst. **158**, 1258–1272 (2007)

Statistical Methods and Nonlinear Dynamics for Analyzing Brain Activity. Theoretical and Experimental Aspects

Maricel Agop, Andrei Ză, Dan Dimitriu, Ştefan Irimiciuc, Alina Gavriluţ, Cristina Marcela Rusu, Gabriel Crumpei, and Lucian Eva

Abstract In this paper, we propose statistical methods and nonlinear dynamics for analyzing brain activity in epileptic patients, using the PhysioNet database. Thus, the analysis by statistical methods (the time variation of the standard deviation of the component signals of the electroencephalogram, the time variation of the signal variance, the time variation of the skewness, the time variation of the kurtosis, the construction of the recurrence maps corresponding to both normal functioning of the brain, as well as of the pre-crisis period, respectively of the crisis, the evolution in time of the spatial–temporal entropy, the variations of the Lyapunov coefficients, etc.) allows us to determine not only the epilepsy time based on a specific strange attractor but also that the entry into the epileptic seizure can be determined at least twenty minutes in advance. Finally, utilyzing elements of nonlinear dynamics and chaos, one builds in the states space certain attractors corresponding to a wide "class" of signals of encephalographic type. These classes dictate the normal or the abnormal

M. Agop
Department of Physics, Gheorghe Asachi Technical University from Iaşi, Iaşi, Romania

A. Ză
Electrical Engineering, Energetics and Applied Informatics Faculty, Gheorghe Asachi Technical University from Iaşi, Iaşi, Romania

D. Dimitriu · C. M. Rusu
Faculty of Physics, Alexandru Ioan Cuza University from Iaşi, Iaşi, Romania
e-mail: dimitriu@uaic.ro

Ş. Irimiciuc
Plasma and Radiation, National Institute for Lasers, Bucharest, Romania
e-mail: stefan.irimiciuc@inflpr.ro

A. Gavriluţ (✉)
Faculty of Mathematics, Alexandru Ioan Cuza University from Iaşi, Iaşi, Romania
e-mail: gavrilut@uaic.ro

G. Crumpei
Faculty of Psychology and Education Sciences, Alexandru Ioan Cuza from Iaşi, Iaşi, Romania

L. Eva
Oblu Clinical Emergency Hospital Iaşi, Iaşi, Romania

C. H. Skiadas et al. (eds.), *13th Chaotic Modeling and Simulation International Conference*, Springer Proceedings in Complexity,
https://doi.org/10.1007/978-3-030-70795-8_4

functioning (the epileptic one) of the brain so that a possible classification of the types of epilepsy can be given.

Keywords Brain activity · Epilepsy · Electroencephalogram · Fractal model · Signal · Skewness · Kurtosis

1 Introduction

Epilepsy is a group of long-term neurological disorders characterized by one or more epileptic seizures. Epileptic seizures are the result of excessive or abnormal activity of nerve cells in the cortex in the brain. Epilepsy cannot be cured, but seizures can be controlled with medication in about 70% of cases. In cases where seizures do not respond to medication, surgery, neurostimulation, or dietary changes may be considered. Normally, the electrical activity of the brain is not synchronous. In the case of epileptic seizures, due to structural or functional problems in the brain, a group of neurons discharges in an abnormal, excessive and synchronized manner. Crisis prediction can play a particularly important role and refers to the attempt to predict epilepsy seizures based on EEG. Our work is in this direction.

The scope of the present paper is to develop a ''procedure" through which not only the prediction but also the type of epilepsy can be established.

Although the first electroencephalograms (EEG) were recorded 143 years ago, progress in interpreting them is extremely slow. So far, there is no classification of the structures that appear in the EEG, so that there is a correspondence between them and the activity of the brain. The clinical interpretation of electroencephalograms is mainly performed by visual recognition of certain structures and by associations made by the specialist physician [1]. The Fouriér analysis cannot be applied because the signals associated with the electroencephalograms are not stationary. The signals are extremely weak, in the domain of microvolts, "submerged in high noise" [2]. For this reason, special attention must be paid to the quality of the electrodes used and their positioning. Also, the identification and analysis of artifacts should not be underestimated, as they may occur due to slight movements of the electrodes, or contraction of the muscles below the electrodes. The analyzed electroencephalograms were downloaded from the PhysioNet database (https://physionet.org/physiobank/database/chbmit/), (Fig. 1). This allows all researchers to access a free collection of physiological signals (PhysioBank), recorded from a wide range of patients, as well as specialized software for viewing and analyzing them. It is supported by the National Institute of General Medical Science (NIGMS) and the National Institute of Biomedical Imaging and Bioengineering (NIBIB), and free access is made in accordance with ODC Public Domain Dedication and License v1.0. Existing resources are made available to stimulate current research in the domain of studying complex biomedical and physiological signals.

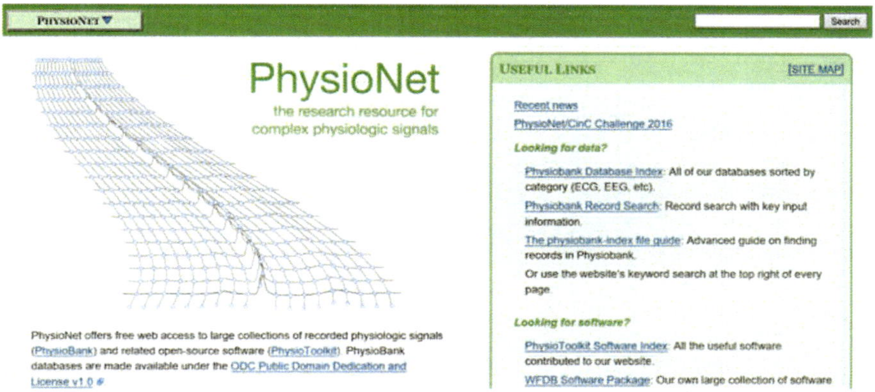

Fig. 1 Interface of PhysioNet database

2 Statistical and Nonlinear Procedures

In the present paper we analyzed an EEG recorded from an epileptic patient aged 11 years using the statistical and nonlinear procedures (standard deviation and variance, spatianl-temporal entropy, Lyapunov exponents etc.). The characteristics of this EEG are as follows:

- the signals were collected on 23 channels;
- the resolution of each signal was 16 bit;
- the sampling time of 4 ms;
- the duration of the signal was 60 min;
- the duration of the epileptic crisis was of 40 s.

Figure 2 graphically shows the signal recorded on channel FP1-F7. It can be observed that neuronal activity does not have regular dynamics. The brain's operating period can be divided into four areas of interest:

- the normal activity area of the brain (range 0–1800s), which is characterized by a chaotic dynamic, with a relatively high signal amplitude;
- the pre-crisis area (range 1800–3000 s), characterized by a decrease in signal amplitude;
- the epileptic crisis zone (range 3000–3040 s), in which the amplitude of the signal reaches its maximum value in a very short period of time, having a more regular behavior due to the synchronization of the neurons activity;
- the post-crisis zone (range 3040–3600 s), where the signal amplitude decreases to a relatively small value, but increases to the value corresponding to the area of normal neuronal activity.

In Figs. 3, 4, 5 and 6 the EEG corresponding to the four areas described above are represented. The corresponding signals were analyzed with a series of statistical

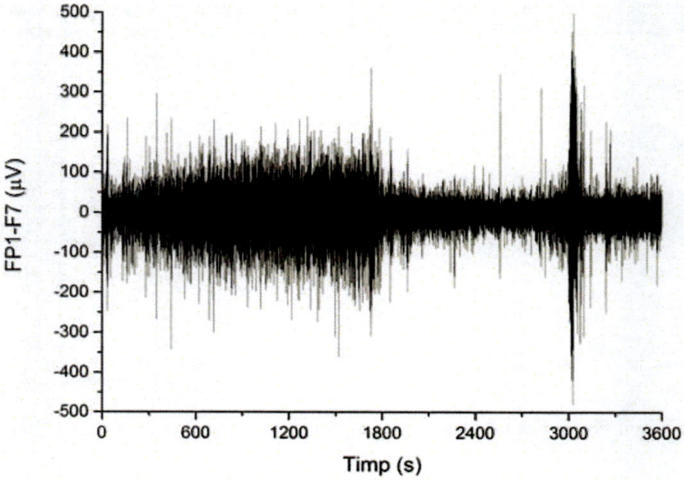

Fig. 2 Graphical representation of the signal recorded on channel FP1-F7

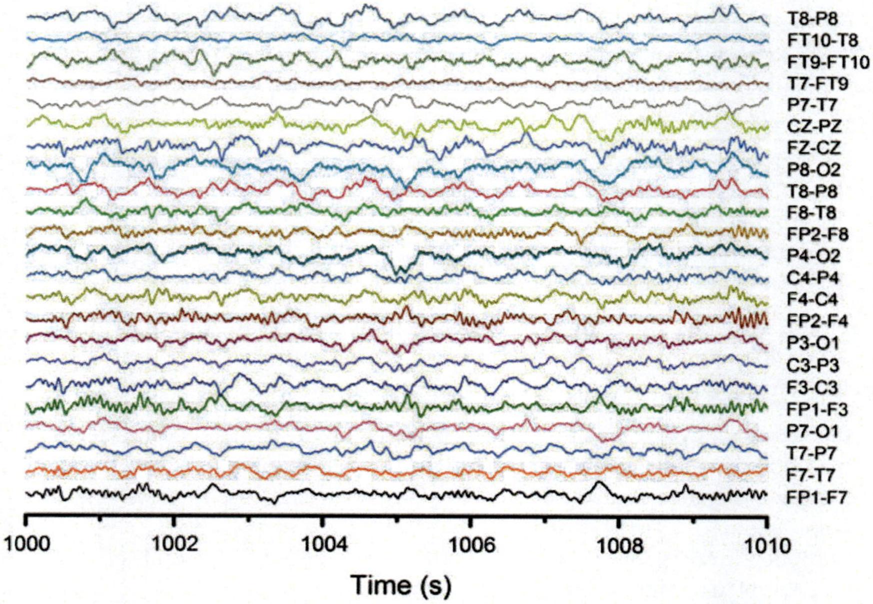

Fig. 3 Electroencephalogram corresponding to the normal functioning of the brain

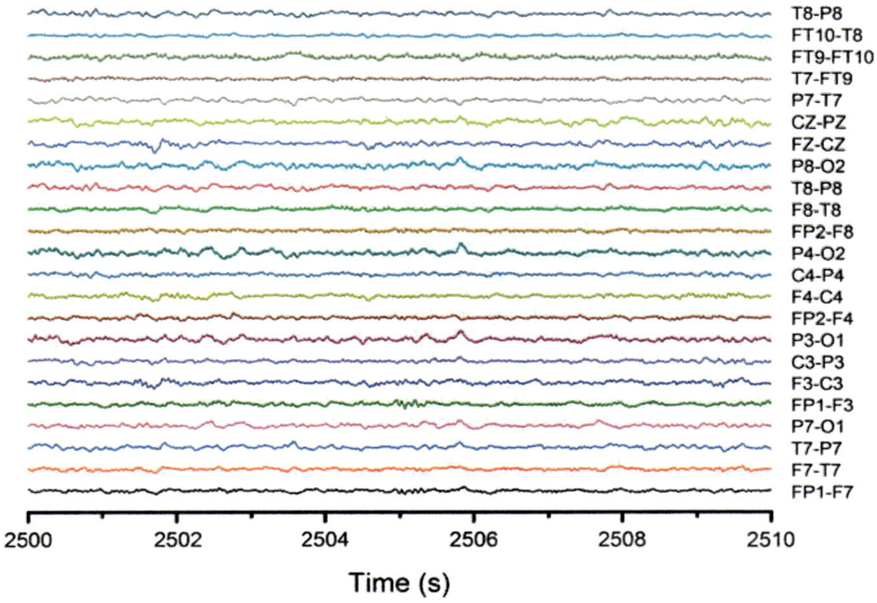

Fig. 4 Electroencephalogram corresponding to the pre-crisis period

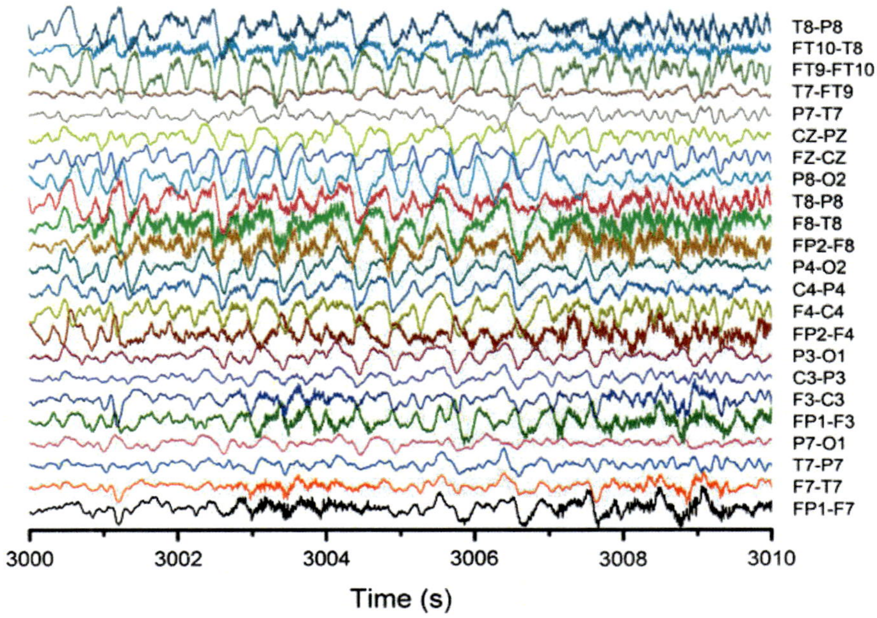

Fig. 5 Electroencephalogram corresponding to the period of epileptic crisis

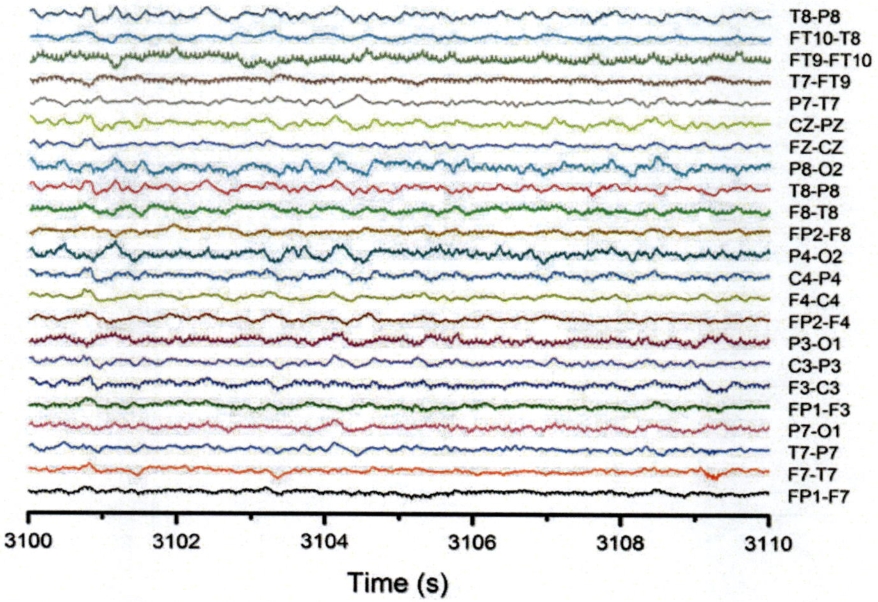

Fig. 6 Electroencephalogram corresponding to the post-crisis period

methods and nonlinear dynamics and only those results that allowed to extract some information of interest are described. The graphical representation of the standard deviation (Fig. 7) shows that, before the pre-crisis, its value drops sharply (approximately until second 1800) and then remains approximately constant until near the crisis (second 3000). During the epileptic crisis, the standard deviation presents an accentuated maximum. Since the standard deviation is an indicator of data dispersion, the fact that it remains at a small, approximately constant value, during the pre-crisis period, denotes that the recorded potentials have small, relatively equal values, so the nerve impulses at the neuron level are of small amplitude and with a "quiet" dynamic. During the crisis the values of the potentials deviate strongly from the average value.

The same result, but much better outlined, with smaller errors, is obtained from the graphical representation of the variance in time (Fig. 8).

Figures 9 and 10 show the time variations of skewness and kurtosis, parameters that indicate the deviation from a normal Gaussian distribution. Figure 9 it can observed that skewness has an average value close to zero, with the exception of pronounced positive maxima that appear in the pre-crisis and crisis regions, but only on a few channels (FP1-F7 and FP1-F3), which it is an indication that the epileptic crisis is most likely a focal one, located in the part of the brain that is in the immediate vicinity of the FP1 electrode.

Regarding kurtosis, it has positive average values, but lower than 3, except for high maximum of high values on channels FP1-F7, FP1-F3 and FP2-F4, correlated with

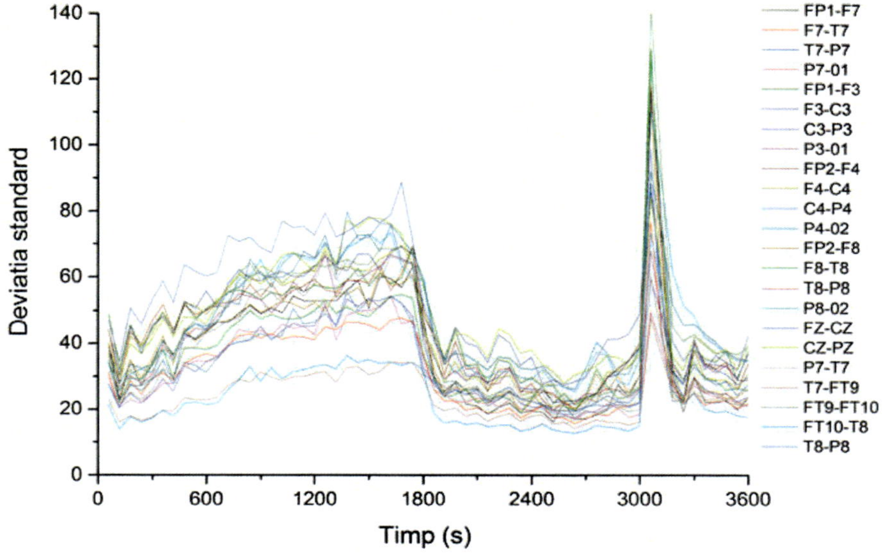

Fig. 7 Variation in time of the standard deviation of the component signals of the electroencephalogram

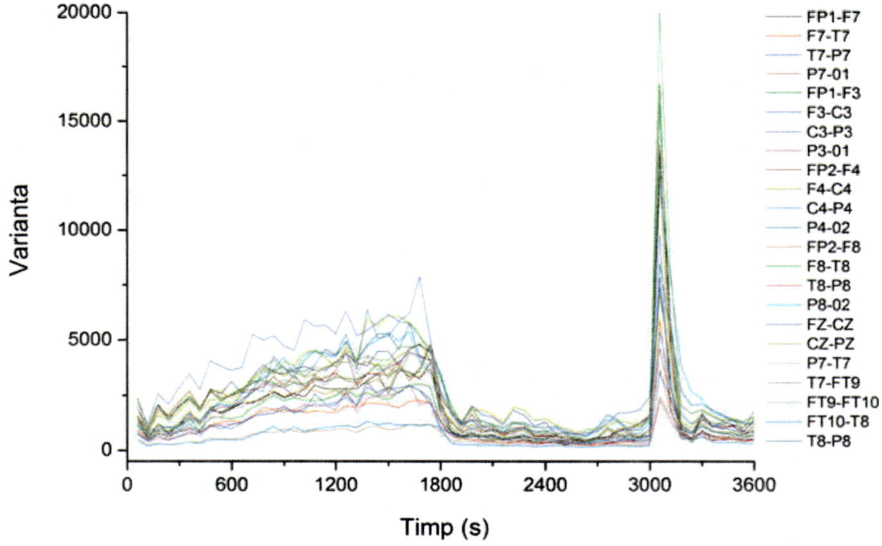

Fig. 8 The variation in time of the variance of the component signals of the electroencephalogram

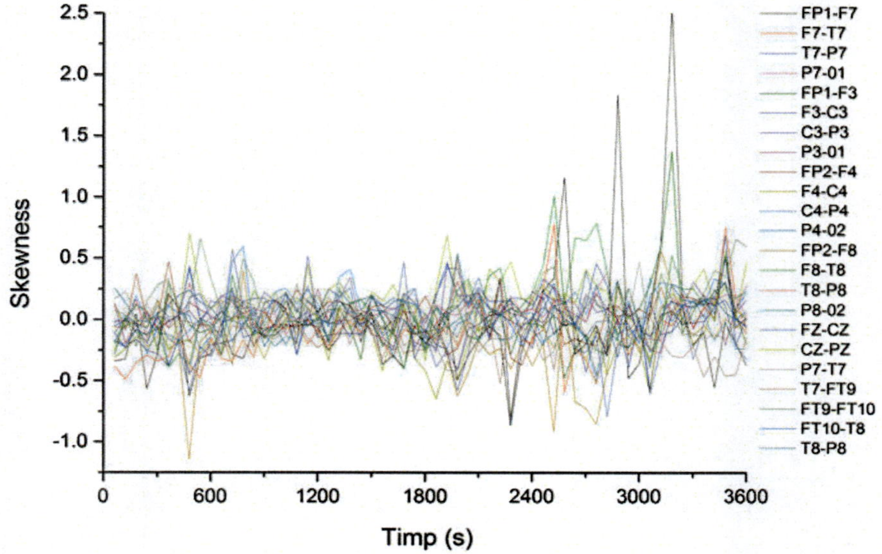

Fig. 9 The variation in time of the skewness, calculated for the component signals of the electroencephalogram

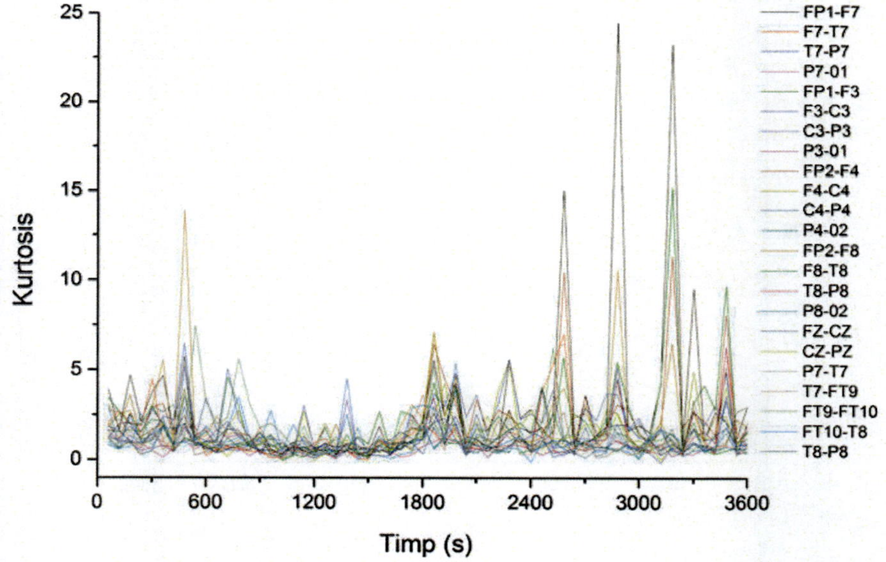

Fig. 10 Variation in time of kurtosis, calculated for the component signals of the electroencephalogram

the maximum observed for skewness. The behavior of this parameter confirms that, most likely, we are dealing with a focal epileptic crisis. The recurrence map will give us global information about the dynamics of the brain and, for this reason, we will not get information about the focal or global character of the epileptic crisis. The recurrence maps for the signal recorded on channel FP1-F7, corresponding to the normal functioning of the brain, the pre-crisis period and, respectively, the crisis period in Figs. 11, 12 and 13 are represented. These were obtained with the Visual Recurrence Analysis v.4.7 free application, developed by Eugene Kononov, (https://web.archive.org/web/20070131023353), https://www.myjavaserver.com/~nonlinear/vra/download.html.

The lack of homogeneity of the maps indicates the existence of a non-stationary signal, and the single points, isolated, indicate strong fluctuations in the system. During the epileptic crisis, the regular component of the system dynamics is much more evident, in agreement with previous observations.

For a more detailed quantitative analysis, the variation in time of the spatio-temporal entropy for 5 channels in Fig. 14 is represented. There is a decrease of this until the beginning of the pre-crisis period, when it shows a rapid growth, remaining at a high value throughout the pre-crisis and crisis period. On some channels (FP1-F7,

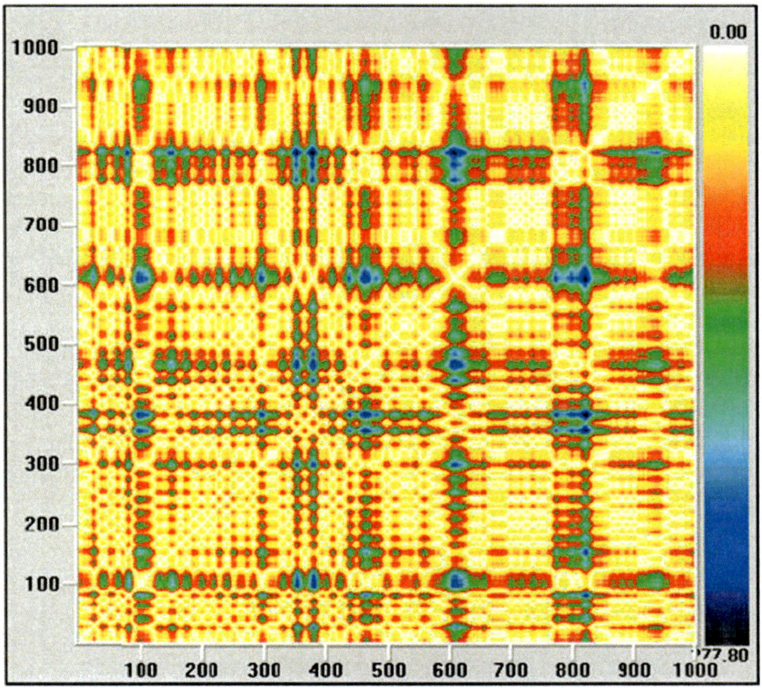

Fig. 11 The recurrence map corresponding to the normal functioning of the brain

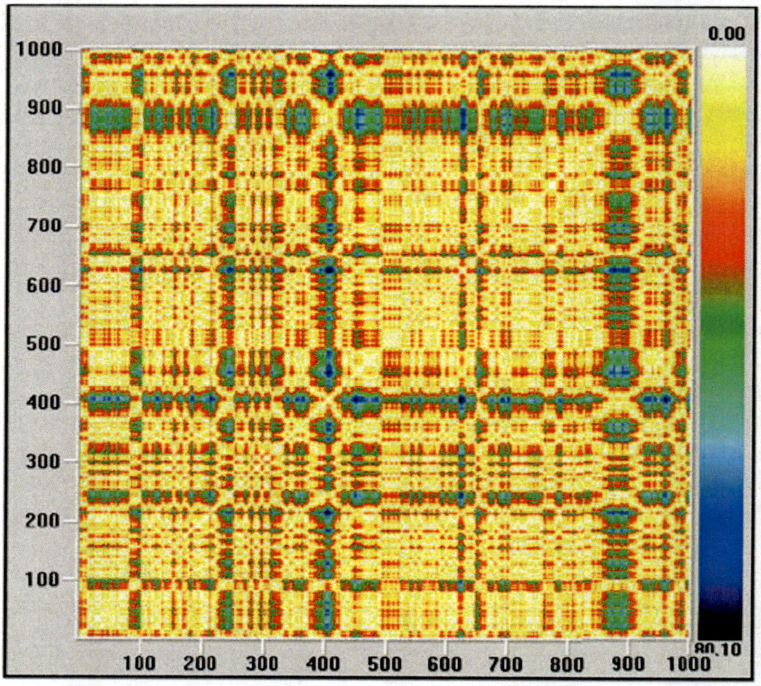

Fig. 12 The recurrence map corresponding to the pre-crisis period

FP1-F3 and F7-T7) the existence of several minimums is observed, the spatial—temporal value of entropy decreasing to values close to the regularity limit. This is best evidenced by the evolution of the signal corresponding to channel FP1-F7, represented in Fig. 15. In this case, the decrease in the entropy value occurs exactly during the epileptic crisis.

Figure 16 shows the time variation of the largest Lyapunov exponent, calculated for 10 channels of the electroencephalogram using the subroutine "Largest Lyapunov exponent" from the Santis application. It is found that the largest Lyapunov exponent is positive, with an average value of about 0.09. This means that the brain dynamics are chaotic. During the crisis and the pre-crisis, the largest Lyapunov exponent shows some sharp decreases to values close to zero, i.e. to the regularity limit.

3 Nonlinear Dynamics Procedures

In this section, we analyzed wide ''classes" of signals corresponding both to the functioning of the normal brain but also to the brain affected by epilepsy, based on the EEGs generated by the same database PhysioNet. Then the analysis of these

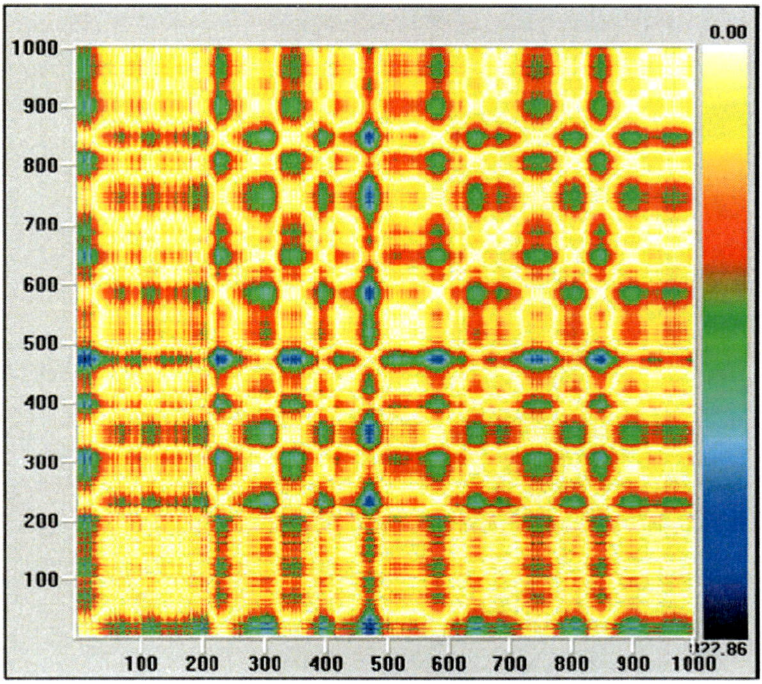

Fig. 13 Recurrence map corresponding to the epileptic crisis

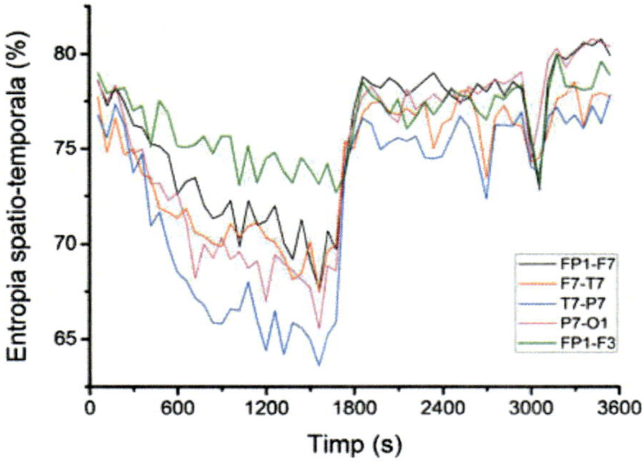

Fig. 14 Evolution in time of the spatial–temporal entropy for 5 component signals of the electroencephalogram

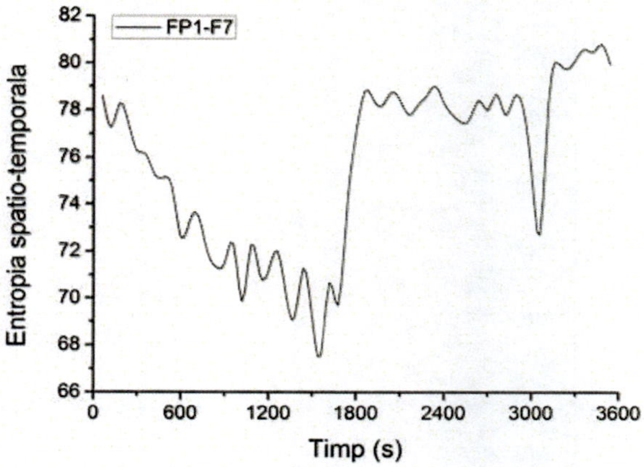

Fig. 15 Spatial–temporal entropy variation for the signal corresponding to channel FP1-F7

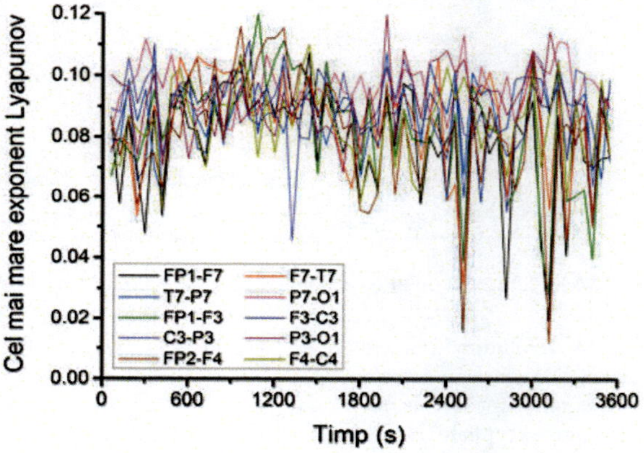

Fig. 16 Variation of the largest Lyapunov exponent corresponding to the 10-channel signals of the electroencephalogram

signals allowed us the construction of the attractors in the states space, see Figs. 17a–n. One observes that to each class of signals, it corresponds a specific attractor and this enables us to conclude that to each type of epilepsy, in the states space it could correspond a specific attractor. Moreover, one could give a classification of the types of epilepsy based on the type of the specific attractor.

4 Conclusions

The analyzes performed on the signals corresponding to the electroencephalogram of an epileptic patient show that some statistical parameters, such as standard deviation or variance, as well as the spatial–temporal entropy, can be used to predict in advance (about 20 min before the electroencephalogram investigated here) the onset of the epileptic crisis. To do this, these parameters (or at least one of them) need to be monitored permanently, and the warning system must be coupled to a system of automatic intervention on the patient, by drug or electrophysiology, so that the onset of the crisis is prevented. Thus, the basis of a functional electronic device, which can be carried and controlled permanently by the epileptic patient, can be laid (as soon as a sensor notices the occurrence of a dynamic behavior of a pre-crisis type, a treatment that avoids the onset of the epileptic crisis).

In the future, statistical analysis should be extended to other types of electroencephalograms, in which multiple epileptic crises occur at short intervals. Also, methods for analyzing more complex signals, specific to non-stationary signals, such as wavelet transform or Hilbert-Huang transform, must be tried.

Concerning the analysis of nonlinear dynamics and chaos, they showed to us that the type of the strange attractor form the states space could allow us to distinguish between the diverse types of epilepsy. Such result could be of very much help for the doctor in establishing the diagnosis accurately.

References

1. B.J. West, *Fractal Physiology and Chaos in Medicine*, 2nd edn. (World Scientific, Singapore, 2013).
2. S.P. Layne, G. Mayer-Kress, J. Holzfuss, Problems associated with dimensional analyis of electroencephalogram data, în *Dimensions and Entropies in Chaotic Systems*, ed. by G. Mayer-Kress (Springer, Berlin, 1986), pp. 246–256
3. https://physionet.org/physiobank/database/chbmit/. Accessed 30 Apr 2018
4. https://web.archive.org/web/20070131023353), https://www.myjavaserver.com/~nonlinear/vra/download.html. Accessed 16 Feb 2018

Behaviour of Solutions of a Neuron Model

Aija Anisimova and Inese Bula

Abstract We consider a discrete-time network of a single neuron as the discrete dynamical system

$$x_{n+1} = \beta x_n - g(x_n), \quad n = 0, 1, \ldots, \tag{1}$$

where $\beta > 1$ and an internal decay rate, g is a step signal function given by a piecewise constant function which consists of five steps in the form

$$g(x) = \begin{cases} b, & x \geq \alpha \\ a, & \alpha > x > 0 \\ 0, & x = 0 \\ -a, & 0 > x > -\alpha \\ -b, & -\alpha \geq x \end{cases}, \quad b > a > 0, \alpha > 0. \tag{2}$$

The considered model is quite simple as a mathematical expression, but with complex dynamics of its solutions. The model is highly sensitive to initial conditions and parameters. Small differences in an initial value and parameters yield widely diverging outcomes for the model, giving a great amount of different periodic orbits. Periodic orbits have been discussed according to the different rage of β. We can find some values of parameters such our considered model has the chaotic behaviour.

Keywords Nonlinear difference equation · Discrete dynamical system · Periodic orbits · Chaotic maps

A. Anisimova (✉) · I. Bula
Department of Mathematics, University of Latvia, Jelgavas 3, Riga 1004, Latvia
e-mail: aija-anisimova@inbox.lv

I. Bula
e-mail: ibula@lanet.lv

C. H. Skiadas et al. (eds.), *13th Chaotic Modeling and Simulation International Conference*, Springer Proceedings in Complexity,
https://doi.org/10.1007/978-3-030-70795-8_5

1 Introduction and Preliminaries

We are inspired by works of *May*, *Wu*, *Zhou*, *Huang*, *Zhu* [14, 20–23], where had been considered a discrete-time model of neurons with different signal functions. Typical signal functions are step functions, piecewise linear functions and sigmoid functions. Models involving a step signal function are referred as McCulloch-Pitts models. A sigmoid function is the most common form of a signal function. It is defined as a strictly increasing smooth bounded function satisfying certain concavity and asymptotic properties. Our major question is: whether we are able to describe the behaviour of these simple single neuron models with different kinds of signal functions, would it be possible to do similar research with more complex models—describing two or more neurons network with more complicated characteristics of function f is the fundamental motivation of the present work. We expect that our investigation in the future bring us to more general results in neuron science, perhaps giving the importance of modelling more realistic neuron models describing by discrete dynamical systems. In [2] we have already obtained some results about the periodicity of a neuron model (1) with parameter $0 < \beta \leq 1$ and a signal function (2). Also in [3] we had analysed model (1) with a different step signal function—a step function with two thresholds. In [4, 5] is considered model (1) with periodic internal decay rate.

We conclude that the model (1) with the signal function (2) describes more general situation as considered in [21] (also [6, 18, 19, 22, 23]).

For the general theory of difference equations, one can refer to the monographs of *Elaydi* [9], *Holmgren* [11], *Kulenovic* [12], as well as *Zhou* [21] article, where have been mentioned basic concepts which are necessary for studies of periodic points, periodic orbits and their stability. Before the discussion of our results we give an overview of theory recalling some definitions which will be used in our research.

Consider a first-order difference equation

$$x_{n+1} = f(x_n), \quad n = 0, 1, \ldots, \tag{3}$$

where $f : \mathbf{R} \to \mathbf{R}$ is a given function. A solution of (3) is a sequence $(x_n)_{n\in\mathbf{N}}$ satisfying equation (3) for all $n = 0, 1, \ldots$. If an initial condition $x_0 \in \mathbf{R}$ is given, then the *orbit* $O(x_0)$ of a point x_0 is defined as a set of points

$$O(x_0) = \{x_0,\ x_1 = f(x_0),\ x_2 = f(x_1) = f^2(x_0),\ x_3 = f(x_2) = f^3(x_0), \ldots\}.$$

Definition 1 A point x_s in the domain of f is said to be a *fixed point* of the map f defined by (3) (or an *equilibrium* or a *stationary state*) if $f(x_s) = x_s$.

Note that for a stationary state x_s the orbit consists only of the point x_s.

Definition 2 A *stationary state* x_s of (3) is *stable* if

$$\forall \varepsilon > 0\ \exists \delta > 0\ \forall x_0 \in \mathbf{R}\ \forall n \in \mathbf{N}\ |x_0 - x_s| < \delta \Rightarrow |f^n(x_0) - x_s| < \varepsilon.$$

Otherwise, the stationary state x_s is called *unstable*.

Definition 3 A *stationary state* x_s of (3) is *asymptotically stable* if it is stable and attracting, i.e., it is stable and if there exists $\nu > 0$ such that $|x_0 - x_s| < \nu$ implies $\lim_{n\to\infty} f^n(x_0) = x_s$.

Definition 4 An orbit $O(x_0)$ of the initial point $x_0 \in \mathbf{R}$, is said to be *eventually stationary state* to x_s if

$$\exists N\, \forall n \geq N \quad x_{n+1} = x_n = x_s.$$

Definition 5 An orbit $O(x_0)$, $x_0 \in \mathbf{R}$, is said to be *asymptotically stationary state* to x_s if $\lim_{n\to\infty} f^n(x_0) = x_s$.

Definition 6 An orbit $O(x_0)$ of the initial point x_0 of (3) is said to be *periodic of period* $p \geq 2$ if

$$x_p = x_0 \text{ and } x_i \neq x_0, \quad 1 \leq i \leq p-1.$$

So, we say that (3) has a p-periodic orbit.

Definition 7 A *periodic orbit* $O(x_0) = \{x_0, x_1, x_2, \ldots, x_{p-1}, \ldots\}$ of period p is *stable* if each point x_i, $i = 0, 1, \ldots, p-1$, is a stable stationary state of the difference equation $x_{n+1} = f^p(x_n)$. A *periodic orbit* of period p which is not stable is said to be *unstable*.

Definition 8 A point z is said to be a *limit point* of $O(x_0)$ if there exists a subsequence $(x_{n_k})_{k=0,1,2,\ldots}$ of $O(x_0)$ such that $\lim_{k\to+\infty} |x_{n_k} - z| = 0$. The *limit set* $L(x_0)$ of the orbit $O(x_0)$ is a set of all limit points of the orbit.

Definition 9 An *orbit* $O(x_0)$ is said to be *asymptotically periodic* if its limit set is a periodic orbit. An *orbit* $O(x_0)$ such that $x_{n+p} = x_n$ for some $n \geq 1$ and some $p \geq 2$ is said to be *eventually periodic*.

2 Stationary Points

We consider (1) with $\beta > 1$. We find periodic orbits of period 2, 4 and others and discuss equilibrium points for the model (1) with the signal function (2), which arises from iterating the function

$$h(x) = \begin{cases} \beta x - b, & x \geq \alpha \\ \beta x - a, & \alpha > x > 0 \\ 0, & x = 0 \\ \beta x + a, & 0 > x > -\alpha \\ \beta x + b, & -\alpha \geq x \end{cases}, \quad b > a > 0, \alpha > 0. \tag{4}$$

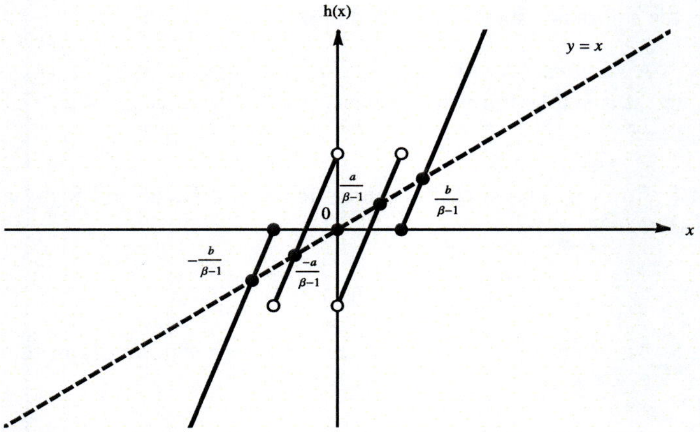

Fig. 1 Stationary states

To find stationary states (fixed points of function (4)) of (1) we solve the equation $h(x^*) = x^*$ (or graphically find intersections with $y = x$ line). It is possible that there are 5 or only 3 stationary states depends on parameters α, β, a, b. Zero is always a stationary point of (1). If $\frac{a}{\beta-1} < \alpha < \frac{b}{\beta-1}$, then there are 5 stationary states of (1): $0, \pm\frac{a}{\beta-1}, \pm\frac{b}{\beta-1}$ (see Fig. 1), otherwise for every difference equation (1) exists at least 3 stationary states: $0, \pm\frac{b}{\beta-1}$ or $0, \pm\frac{a}{\beta-1}$ (see Fig. 2).

If all points of orbit belong to external parts of the graphic of the function (4) we can get a set $\{\frac{b}{\beta}, \frac{b(1+\beta)}{\beta^2}, \ldots, \frac{b(1+\beta+\beta^2+\cdots+\beta^n)}{\beta^{n+1}}, \ldots\}$, which consists of points, which are eventually stationary state to 0 and in this case in the previous iteration we always get $x_n = \frac{b}{\beta}, n = 0, \ldots, n$ and inequalities $\frac{b}{\beta}\frac{1+\beta+\beta^2+\cdots+\beta^n}{\beta^n} > \frac{b}{\beta} \geq \alpha$ hold. Analogy is in a situation when the points lie on the negative external line of function (4), namely, the points that are obtained from the equation $h(x) = \beta x + b$. In such way we can obtain those points which are eventually stationary state to $zero$, for example, considering situations when all points lie on internal segments and other combinations.

3 Periodic Orbits of Period 2

Let us consider (1) with the signal function (2). Here we find all periodic orbits of given initial point of period 2. Depends on parameters α, β and a, b the following inequalities are satisfied. Six different cases are possible

(1) If $\frac{a+b\beta}{\beta^2-1} \geq \alpha$ and $\frac{\beta a+b}{\beta^2-1} < \alpha$, then $\{\frac{a+b\beta}{\beta^2-1}, \frac{\beta a+b}{\beta^2-1}\}$ is a periodic orbit of period 2.
(2) If $\frac{-a+b\beta}{\beta^2-1} \geq \alpha$ and $-\alpha < \frac{b-\beta a}{\beta^2-1} < 0$, then $\{\frac{-a+b\beta}{\beta^2-1}, \frac{b-\beta a}{\beta^2-1}\}$ is a periodic orbit of period 2.
(3) If $\alpha \leq \frac{b}{\beta+1}$, then $\{\frac{b}{\beta+1}, -\frac{b}{\beta+1}\}$ is a periodic orbit of period 2.

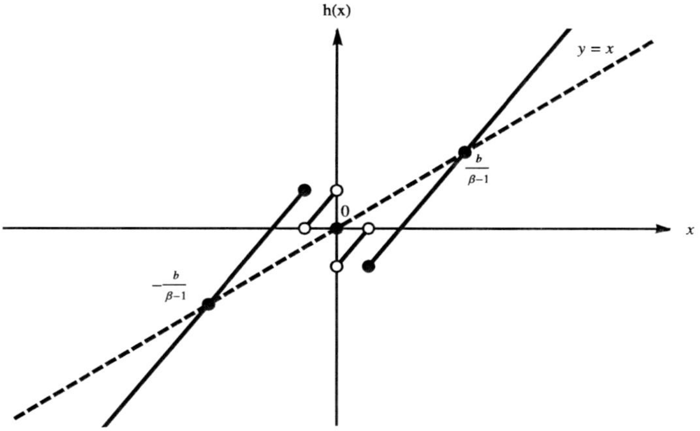

Fig. 2 Existence of at least 3 stationary states

(4) If $\frac{a}{\beta-1} < \alpha$, then $\{\frac{a}{\beta+1}, -\frac{a}{\beta+1}\}$ is a periodic orbit of period 2.
(5) If $0 < \frac{\beta a-b}{\beta^2-1} < \alpha$ and $\frac{a-\beta b}{\beta^2-1} \leq -\alpha$, then $\{\frac{\beta a-b}{\beta^2-1}, \frac{a-\beta b}{\beta^2-1}\}$ is a periodic orbit of period 2.
(6) If $-\alpha < \frac{-\beta a-b}{\beta^2-1} < 0$ and $\frac{-a-\beta b}{\beta^2-1} \leq -\alpha$, then $\{\frac{-\beta a-b}{\beta^2-1}, \frac{-a-\beta b}{\beta^2-1}\}$ is a periodic orbit of period 2.

There are possible various combinations of these periodic orbits with the same values of parameters α, β, a, b. These combinations depend on fulfillment of following inequalities:

$$\frac{a+b\beta}{\beta^2-1} > \frac{-a+b\beta}{\beta^2-1} > \frac{b}{\beta+1} \geq \alpha > \frac{a}{\beta+1} > \frac{-b+a\beta}{\beta^2-1} > 0 > \frac{-b-a\beta}{\beta^2-1} > -\alpha$$

$$\frac{-\beta b-a}{\beta^2-1} < \frac{a-\beta b}{\beta^2-1} < -\frac{b}{\beta+1} \leq -\alpha < -\frac{a}{\beta+1} < \frac{b-\beta a}{\beta^2-1} < 0 < \frac{b+\beta a}{\beta^2-1}.$$

For example, if $\beta a > b$ and $\frac{a}{\beta+1} < \alpha \leq \frac{b}{\beta+1}$, then there exists 4 periodic orbits of period 2, i.e., $\{\frac{-a+b\beta}{\beta^2-1}, \frac{b-\beta a}{\beta^2-1}\}$, $\{\frac{b}{\beta+1}, -\frac{b}{\beta+1}\}$, $\{\frac{a}{\beta+1}, -\frac{a}{\beta+1}\}$, $\{\frac{\beta a-b}{\beta^2-1}, \frac{a-\beta b}{\beta^2-1}\}$ and we have an arrangement

$$\frac{-a+b\beta}{\beta^2-1} > \frac{b}{\beta+1} \geq \alpha > \frac{a}{\beta+1} > \frac{-b+a\beta}{\beta^2-1} > 0$$

$$\frac{a-\beta b}{\beta^2-1} < -\frac{b}{\beta+1} \leq -\alpha < -\frac{a}{\beta+1} < \frac{b-\beta a}{\beta^2-1} < 0.$$

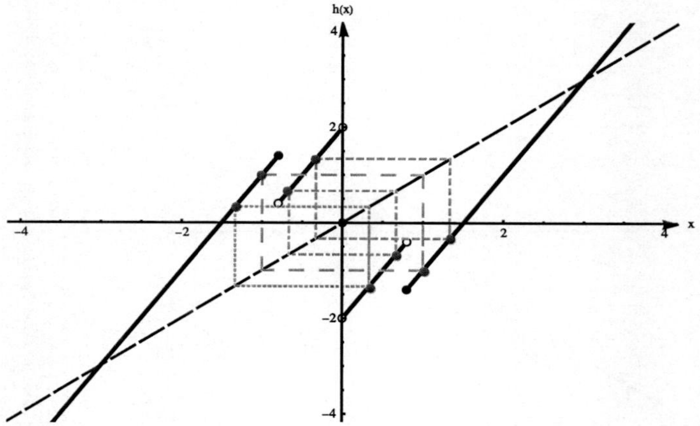

Fig. 3 Periodic orbits of period 2

As an example, if $\beta = 2$, $a = 2$, $b = 3$, $\alpha = 0.8$, then we have 4 periodic orbits of period 2: $\{\frac{4}{3}, -\frac{1}{3}\}$, $\{1, -1\}$, $\{\frac{2}{3}, -\frac{2}{3}\}$, $\{\frac{1}{3}, -\frac{4}{3}\}$ (see Fig. 3).

If additionally a condition $\beta > \frac{2b}{b-a}$ is assumed it is possible to obtain six different periodic orbits of period 2 together, e.g., choosing $\beta = 8, a = 2, b = 3, \alpha = 0.32$, we obtain periodic orbits of period 2: $\{\frac{26}{63}, \frac{19}{63}\}$, $\{\frac{22}{63}, -\frac{13}{63}\}$, $\{\frac{1}{3}, -\frac{1}{3}\}$, $\{\frac{2}{9}, -\frac{2}{9}\}$, $\{\frac{13}{63}, -\frac{22}{63}\}$, $\{-\frac{26}{63}, -\frac{19}{63}\}$.

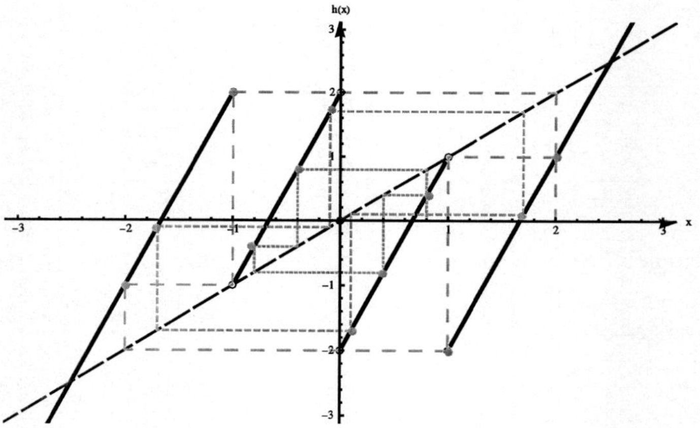

Fig. 4 Periodic orbits of period 4 of cases (1), (3), (4)

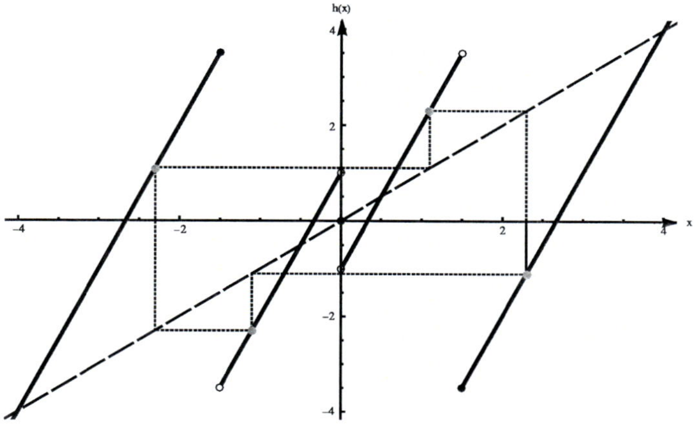

Fig. 5 Periodic orbits of period 4 of cases (2)

4 Periodic Orbits of Period 4

Now we demonstrate some periodic orbits of period 4 of the difference equation (1). There are many different options of orbits which can be obtained depending on parameters α, β and a, b. We consider some of them

(1) $\{x_0 = \frac{a\beta+a}{\beta^2+1}, x_1 = \frac{a\beta-a}{\beta^2+1}, x_2 = \frac{-a\beta-a}{\beta^2+1}, x_3 = \frac{-a\beta+a}{\beta^2+1}\}$ is a periodic orbit of period 4. In this case it is necessary that $\alpha > x_0 > x_1 > 0 > x_3 > x_2 > -\alpha$. For example, if we choose parameters $\beta = 3, a = 2, b = 5, \alpha = 1$, we obtain a periodic orbit $\{\frac{4}{5}, \frac{2}{5}, -\frac{4}{5}, -\frac{2}{5}\}$ (see Fig. 4).

(2) $\{x_0 = \frac{b\beta-a}{\beta^2+1}, x_1 = \frac{-a\beta-b}{\beta^2+1}, x_2 = \frac{-b\beta+a}{\beta^2+1}, x_3 = \frac{a\beta+b}{\beta^2+1}\}$ is a periodic orbit of period 4. In this case it is necessary that $x_0 \geq \alpha > x_3 > 0 > x_1 > -\alpha \geq x_2$. For example, if we choose parameters $\beta = 3, a = 1, b = 8, \alpha = 1.5$, we obtain a periodic orbit $\{\frac{23}{10}, -\frac{11}{10}, -\frac{23}{10}, -\frac{11}{10}\}$ (see Fig. 5).

(3) $\{x_0 = \frac{b\beta+a}{\beta^2+1}, x_1 = \frac{a\beta-b}{\beta^2+1}, x_2 = \frac{-b\beta-a}{\beta^2+1}, x_3 = \frac{-a\beta+b}{\beta^2+1}\}$ is a periodic orbit of period 4. In this case is necessary that $x_0 \geq \alpha > x_1 > 0 > x_3 > -\alpha \geq x_2$. For example, if we choose parameters $\beta = 3, a = 2, b = 5, \alpha = 1$, we obtain a periodic orbit $\{\frac{17}{10}, \frac{1}{10}, -\frac{17}{10}, -\frac{1}{10}\}$ (see Fig. 4).

(4) $\{x_0 = \frac{b\beta+b}{\beta^2+1}, x_1 = \frac{b\beta-b}{\beta^2+1}, x_2 = \frac{-b\beta-b}{\beta^2+1}, x_3 = \frac{-b\beta+b}{\beta^2+1}\}$ is a periodic orbit of period 4. In this case is necessary that $x_0 > x_1 \geq \alpha > 0 > -\alpha \geq x_3 > x_2$. For example, if we choose parameters $\beta = 3, a = 2, b = 5, \alpha = 1$, we obtain a periodic orbit $\{2, 1, -2, -1\}$ (see Fig. 4).

(5) $\{x_0 = \frac{b\beta^3+b\beta^2+b\beta-b}{\beta^4-1}, x_1 = \frac{b\beta^3+b\beta^2-b\beta+b}{\beta^4-1}, x_2 = \frac{b\beta^3-b\beta^2+b\beta+b}{\beta^4-1}, x_3 = \frac{-b\beta^3+b\beta^2+b\beta+b}{\beta^4-1}\}$ is a periodic orbit of period 4. In this case is necessary that $x_0 > x_1 > x_2 \geq \alpha > 0 > -\alpha \geq x_3$. If we take $\beta = 3$, $a = 4$, $b = 6$ and $\alpha = 1$, we get a periodic orbit $\{\frac{57}{20}, \frac{51}{20}, \frac{33}{20}, -\frac{21}{20}\}$ (see Fig. 6).

(6) $\{x_0 = \frac{b\beta^3+b\beta^2+b\beta-a}{\beta^4-1}, x_1 = \frac{b\beta^3+b\beta^2-a\beta+b}{\beta^4-1}, x_2 = \frac{b\beta^3-a\beta^2+b\beta+b}{\beta^4-1},$
$x_3 = \frac{-a\beta^3+b\beta^2+b\beta+b}{\beta^4-1}\}$ is a periodic orbit of period 4. In this case is necessary that $x_0 > x_1 > x_2 \geq \alpha > 0 > x_3 > -\alpha$. If we take $\beta = 3$, $a = 4$, $b = 6$ and $\alpha = 1$, we get a periodic orbit $\{\frac{23}{8}, \frac{21}{8}, \frac{15}{8}, -\frac{3}{8}\}$ (see Fig. 6).

(7) $\{x_0 = \frac{-b\beta^3-b\beta^2-b\beta+a}{\beta^4-1}, x_1 = \frac{-b\beta^3-b\beta^2+a\beta-b}{\beta^4-1}, x_2 = \frac{-b\beta^3+a\beta^2-b\beta-b}{\beta^4-1},$
$x_3 = \frac{a\beta^3-b\beta^2-b\beta-b}{\beta^4-1}\}$ is a periodic orbit of period 4. In this case is necessary that $\alpha > x_3 > 0 > -\alpha \geq x_2 > x_1 > x_0$. If we take $\beta = 3$, $a = 4$, $b = 6$ and $\alpha = 1$, we get a periodic orbit $\{-\frac{23}{8}, -\frac{21}{8}, -\frac{15}{8}, \frac{3}{8}\}$ (see Fig. 6).

(8) $\{x_0 = \frac{-b\beta^3-b\beta^2-b\beta+b}{\beta^4-1}, x_1 = \frac{-b\beta^3-b\beta^2+b\beta-b}{\beta^4-1}, x_2 = \frac{-b\beta^3+b\beta^2-b\beta-b}{\beta^4-1},$
$x_3 = \frac{b\beta^3-b\beta^2-b\beta-b}{\beta^4-1}\}$ is a periodic orbit of period 4. In this case is necessary that $x_3 \geq \alpha > 0 > -\alpha \geq x_2 > x_1 > x_0$. If we take $\beta = 3$, $a = 4$, $b = 6$ and $\alpha = 1$, we get a periodic orbit $\{-\frac{57}{20}, -\frac{51}{20}, -\frac{33}{20}, \frac{21}{20}\}$ (see Fig. 6).

(9) $\{x_0 = \frac{a\beta^3+a\beta^2+a\beta-a}{\beta^4-1}, x_1 = \frac{a\beta^3+a\beta^2-a\beta+a}{\beta^4-1}, x_2 = \frac{a\beta^3-a\beta^2+a\beta+a}{\beta^4-1},$
$x_3 = \frac{-a\beta^3+a\beta^2+a\beta+a}{\beta^4-1}\}$ is a periodic orbit of period 4. In this case is necessary that $\alpha > x_0 > x_1 > x_2 > 0 > x_3 > -\alpha$. For example, if we choose parameters $\beta = 3$, $a = 2$, $b = 5$, $\alpha = 1$, we obtain a periodic orbit $\{\frac{19}{20}, \frac{17}{20}, \frac{11}{20}, -\frac{7}{20}\}$ (see Fig. 7).

(10) $\{x_0 = \frac{a\beta^3+a\beta^2+a\beta-b}{\beta^4-1}, x_1 = \frac{a\beta^3+a\beta^2-b\beta+a}{\beta^4-1}, x_2 = \frac{a\beta^3-b\beta^2+a\beta+a}{\beta^4-1},$
$x_3 = \frac{-b\beta^3+a\beta^2+a\beta+a}{\beta^4-1}\}$ is a periodic orbit of period 4. In this case is necessary that $\alpha > x_0 > x_1 > x_2 > 0 > -\alpha \geq x_3$. For example, if we choose parameters $\beta = 3$, $a = 2$, $b = 5$, $\alpha = 1$, we obtain a periodic orbit $\{\frac{73}{80}, \frac{59}{80}, \frac{17}{80}, -\frac{109}{80}\}$ (see Fig. 7).

(11) $\{x_0 = \frac{-a\beta^3-a\beta^2+a\beta+a}{\beta^4-1}, x_1 = \frac{-a\beta^3-a\beta^2+a\beta-a}{\beta^4-1}, x_2 = \frac{a\beta^3-a\beta^2-a\beta-a}{\beta^4-1},$
$x_3 = \frac{a\beta^3-a\beta^2-a\beta-a}{\beta^4-1}\}$ is a periodic orbit of period 4. In this case is necessary that $\alpha > x_3 > 0 > x_2 > x_1 > x_0 > -\alpha$. For example, if we choose parameters $\beta = 3$, $a = 2$, $b = 5$, $\alpha = 1$, we obtain a periodic orbit $\{-\frac{19}{20}, -\frac{17}{20}, -\frac{11}{20}, \frac{7}{20}\}$ (see Fig. 7).

(12) $\{x_0 = \frac{-a\beta^3-a\beta^2-a\beta+b}{\beta^4-1}, x_1 = \frac{-a\beta^3-a\beta^2+b\beta-a}{\beta^4-1}, x_2 = \frac{-a\beta^3+b\beta^2-a\beta-a}{\beta^4-1},$
$x_3 = \frac{b\beta^3-a\beta^2-a\beta-a}{\beta^4-1}\}$ is a periodic orbit of period 4. In this case is necessary that $x_3 \geq \alpha > 0 > x_2 > x_1 > x_0 > -\alpha$. For example, if we choose parameters $\beta = 3$, $a = 2$, $b = 5$, $\alpha = 1$, we obtain a periodic orbit $\{-\frac{73}{80}, -\frac{59}{80}, -\frac{17}{80}, \frac{109}{80}\}$ (see Fig. 7).

5 Periodic Orbits of Period 2*k*

In this section the sufficient conditions for existence or periodic $2k$-periodic orbit are obtained. But we can construct periodic orbits of an arbitrary period. We show in the next two theorems this construction when all points of the orbit belong to external

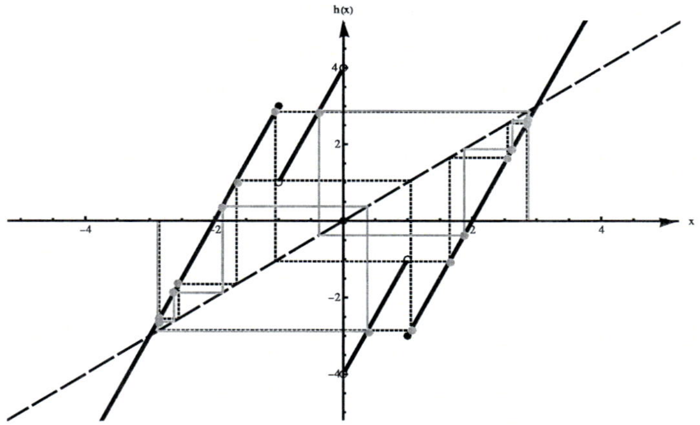

Fig. 6 Periodic orbits of period 4 of cases (5), (6), (7), (8)

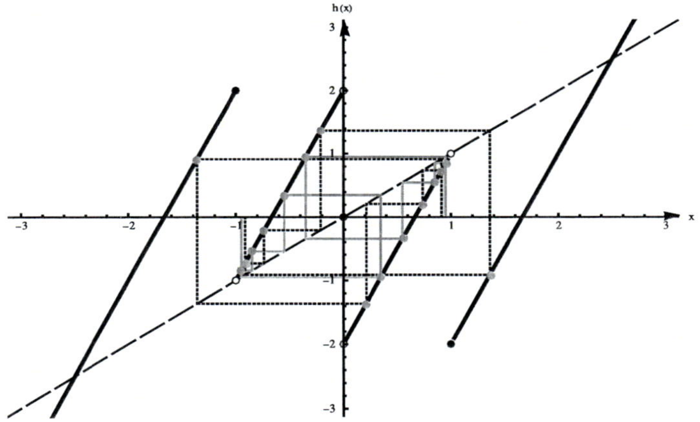

Fig. 7 Periodic orbits of period 4 of cases (9), (10), (11), (12)

lines of the function (4), i.e., points of the orbit are greater or equal to α or less or equal to $-\alpha$.

Theorem 1 *If there exists a positive integer k such that*

$$\frac{b(\beta^{2k} - 2\beta^{2k-2} + 1)}{(\beta^{2k} - 1)(\beta + 1)} \geq \alpha \tag{5}$$

then the difference equation (1) has a periodic orbit of period $2k$*. So more each periodic orbit is unstable.*

Proof We construct a periodic orbit $O(x_0)$ of period $2k$. Let us take

$$\begin{aligned} &x_0 > \alpha, \quad x_1 \geq \alpha, \quad x_2 < -\alpha, \quad x_3 < -\alpha, \\ &(-1)^i x_i > \alpha \text{ for } i = 4, \ldots, 2k-1, \\ &x_{2k} = x_0. \end{aligned} \tag{6}$$

Then

$$\begin{aligned} &x_0 > \alpha, \\ &x_1 = \beta x_0 - b \geq \alpha, \\ &x_2 = \beta^2 x_0 - b\beta - b < -\alpha, \\ &x_3 = \beta^3 x_0 - b\beta^2 - b\beta + b < -\alpha, \\ &x_4 = \beta^4 x_0 - b\beta^3 - b\beta^2 + b\beta + b > \alpha, \\ &x_5 = \beta^5 x_0 - b\beta^4 - b\beta^3 + b\beta^2 + b\beta - b < -\alpha, \\ &x_6 = \beta^6 x_0 - b\beta^5 - b\beta^4 + b\beta^3 + b\beta^2 - b\beta + b > \alpha, \\ &\ldots \\ &x_{2k-1} = \beta^{2k-1} x_0 - b\beta^{2k-2} - b\beta^{2k-3} + b\beta^{2k-4} + b\beta^{2k-5} - b\beta^{2k-6} + \\ &\qquad + b\beta^{2k-7} - \cdots + b\beta - b < -\alpha, \\ &x_{2k} = \beta^{2k} x_0 - b\beta^{2k-1} - b\beta^{2k-2} + b\beta^{2k-3} + b\beta^{2k-4} - b\beta^{2k-5} + \\ &\qquad + b\beta^{2k-6} - b\beta^{2k-7} + \cdots + b\beta^2 - b\beta + b = x_0 > \alpha. \end{aligned}$$

Since $x_{2k} = x_0$, then

$$x_0 = \frac{b(\beta^{2k-1} + \beta^{2k-2} - \beta^{2k-3} - \beta^{2k-4} + \beta^{2k-5} - \beta^{2k-6} + \cdots - \beta^2 + \beta - 1)}{\beta^{2k} - 1}, \tag{7}$$

therefore

$$x_0 = \frac{b(\beta^{2k} + 2\beta^{2k-1} - 2\beta^{2k-3} - 1)}{(\beta^{2k} - 1)(\beta + 1)}. \tag{8}$$

Then by (5) and (8)

$$\begin{aligned} x_1 = \beta x_0 - b &= \frac{b(\beta^{2k+1} + 2\beta^{2k} - 2\beta^{2k-2} - \beta)}{(\beta^{2k} - 1)(\beta + 1)} - b \\ &= \frac{b(\beta^{2k+1} + 2\beta^{2k} - 2\beta^{2k-2} - \beta - \beta^{2k+1} - \beta^{2k} + \beta + 1)}{(\beta^{2k} - 1)(\beta + 1)} \\ &= \frac{b(\beta^{2k} - 2\beta^{2k-2} + 1)}{(\beta^{2k} - 1)(\beta + 1)} \geq \alpha. \end{aligned}$$

We show that

$$\min\{x_0, x_1, x_4, x_6, \ldots, x_{2k-2}\} = x_1 \geq \alpha$$

or

$$\alpha \leq x_1 < x_0 \quad \text{and} \quad \alpha \leq x_1 < x_4 < x_6 < \cdots < x_{2k-2}.$$

At first we notice that

$$x_1 = \frac{b(\beta^{2k} - 2\beta^{2k-2} + 1)}{(\beta^{2k} - 1)(\beta + 1)} < \frac{b(\beta^{2k} + 2\beta^{2k-1} - 2\beta^{2k-3} - 1)}{(\beta^{2k} - 1)(\beta + 1)} = x_0.$$

It is true because for all $k = 2, 3, \ldots$ and $\beta > 1$

$$2(\beta^{2k-1} - \beta^{2k-3} + \beta^{2k-2} - 1) > 0.$$

Further we show that $x_1 < x_4$. We notice that $x_4 = \beta^3 x_1 - \beta^2 b + \beta b + b$. Then

$$\begin{aligned}
& x_1 < x_4 = \beta^3 x_1 - \beta^2 b + \beta b + b \\
& \Leftrightarrow 0 < (\beta^3 - 1)x_1 - \beta^2 b + \beta b + b \\
& \Leftrightarrow 0 < \tfrac{b(\beta^{2k} - 2\beta^{2k-2} + 1)(\beta^3 - 1)}{(\beta^{2k} - 1)(\beta + 1)} - \beta^2 b + \beta b + b \\
& \Leftrightarrow 0 < \beta^{2k+3} - 2\beta^{2k+1} + \beta^3 - \beta^{2k} + 2\beta^{2k-2} - 1 - \beta^{2k+3} + \beta^{2k+2} + \beta^{2k+1} - \\
& \qquad - \beta^{2k+2} + \beta^{2k+1} + \beta^{2k} + \beta^3 - \beta^2 - \beta + \beta^2 - \beta - 1 \\
& = 2\beta^{2k-2} + 2\beta^3 - 2\beta - 2 = 2\beta(\beta^{2k-3} - 1) + 2(\beta^3 - 1)
\end{aligned}$$

because $\beta > 1$ and $k \geq 2$. Finally we show that $x_{2m-2} < x_{2m}$, $3 \leq m < k$. Indeed

$$\begin{aligned}
& x_{2m-2} = \beta^{2m-2} x_0 - b\beta^{2m-3} - b\beta^{2m-4} + b\beta^{2m-5} + b\beta^{2m-6} - b\beta^{2m-7} + \\
& \qquad + b\beta^{2m-8} - \cdots - b\beta + b \\
& < x_{2m} = \beta^{2m} x_0 - b\beta^{2m-1} - b\beta^{2m-2} + b\beta^{2m-3} + b\beta^{2m-4} - b\beta^{2m-5} + \\
& \qquad + b\beta^{2m-6} - \cdots - b\beta + b \\
& \Leftrightarrow 0 < (\beta^{2m} - \beta^{2m-2}) x_0 - b\beta^{2m-1} - b\beta^{2m-2} + 2b\beta^{2m-3} + 2b\beta^{2m-4} - 2b\beta^{2m-5} \qquad (9) \\
& \Leftrightarrow 0 < \tfrac{\beta^{2m-2}(\beta^2 - 1) b(\beta^{2k} + 2\beta^{2k-1} - 2\beta^{2k-3} - 1)}{(\beta^{2k} - 1)(\beta + 1)} - \\
& \qquad - b\beta^{2m-1} - b\beta^{2m-2} + 2b\beta^{2m-3} + 2b\beta^{2m-4} - 2b\beta^{2m-5} \\
& = \tfrac{2b\beta^{2m-5}}{\beta^{2k} - 1}(\beta^3 - \beta^2 - \beta + 1).
\end{aligned}$$

The function $\beta^3 - \beta^2 - \beta + 1$ is strictly increasing in the interval $]1, +\infty[$. If $\beta = 1$, the function $\beta^3 - \beta^2 - \beta + 1$ is equal to 0 but if $\beta > 1$ it is greater than 0. Hence we get

$$\frac{2b\beta^{2m-5}}{\beta^{2k} - 1}(\beta^3 - \beta^2 - \beta + 1) > 0.$$

We show that

$$x_2 < -\alpha \text{ and } \max\{x_3, x_5, x_7, \ldots, x_{2k-1}\} = x_{2k-1} < -\alpha.$$

Since (5) holds, then $\frac{b(2\beta^{2k-2} - \beta^{2k} - 1)}{(\beta^{2k} - 1)(\beta + 1)} \leq -\alpha$. At first we show that $x_2 < -\alpha$:

$$
\begin{aligned}
x_2 &= \beta^2 x_0 - b\beta - b = \tfrac{\beta^2 b(\beta^{2k}+2\beta^{2k-1}-2\beta^{2k-3}-1)}{(\beta^{2k}-1)(\beta+1)} - b(\beta+1) \\
&= \tfrac{b(-\beta^{2k}-2\beta^{2k-1}+2\beta+1)}{(\beta^{2k}-1)(\beta+1)} = \tfrac{b(2\beta^{2k-2}-\beta^{2k}-1-2\beta^{2k-2}-2\beta^{2k-1}+2\beta+2)}{(\beta^{2k}-1)(\beta+1)} \\
&\le -\alpha - \tfrac{2b(\beta^{2k-2}+\beta^{2k-1}-\beta-1)}{(\beta^{2k}-1)(\beta+1)} < -\alpha.
\end{aligned}
$$

By the construction $x_{2k-1} = \frac{x_0-b}{\beta}$, then

$$
\begin{aligned}
x_{2k-1} &= \tfrac{b(\beta^{2k}+2\beta^{2k-1}-2\beta^{2k-3}-1)}{(\beta^{2k}-1)(\beta+1)\beta} - \tfrac{b}{\beta} \\
&= \tfrac{b(2\beta^{2k-1}-2\beta^{2k-3}-\beta^{2k+1})}{(\beta^{2k}-1)(\beta+1)\beta} = \tfrac{b(2\beta^{2k-2}-2\beta^{2k-4}-\beta^{2}k)}{(\beta^{2k}-1)(\beta+1)} \\
&= \tfrac{b(2\beta^{2k-2}-\beta^{2k}-1+1-2\beta^{2k-4})}{(\beta^{2k}-1)(\beta+1)} \le -\alpha - \tfrac{b(2\beta^{2k-4}-1)}{(\beta^{2k}-1)(\beta+1)} < -\alpha,
\end{aligned}
$$

because $\beta > 1$ and $k \ge 2$. Similar as (9) it is possible to show that $x_{2m-3} < x_{2m-1}$ for all $3 \le m \le k$.

If assumptions of Theorem 1 holds, the orbit $O(x_0)$ where x_0 is defined by (8), satisfies (6). Thus (1) has $2k$-periodic orbit $O(x_0)$.

6 Periodic Orbits of Period $2k - 1$

The following theorem give the sufficient conditions for existence of an arbitrary odd periodic orbit of (1).

Theorem 2 *If there exists a positive integer k such that*

$$
\frac{b(\beta^{2k+1} - 2\beta^{2k-1} - 1)}{\beta^{2k+1} - 1} \ge \alpha, \tag{10}
$$

then the difference equation (1) has a periodic orbit of period $2k + 1$*. This periodic orbits are unstable.*

Proof We will construct a periodic orbit $O(x_0)$ of period $2k + 1$ such that

$$
\begin{aligned}
&x_0 \ge \alpha, \quad x_1 < -\alpha, \quad x_2 < -\alpha, \\
&(-1)^i x_i < -\alpha \text{ for } i = 3, \ldots, 2k, \\
&x_{2k+1} = x_0.
\end{aligned} \tag{11}
$$

Then

$$
\begin{aligned}
&x_0 \geq \alpha,\\
&x_1 = \beta x_0 - b < -\alpha,\\
&x_2 = \beta^2 x_0 - b\beta + b < -\alpha,\\
&x_3 = \beta^3 x_0 - b\beta^2 + b\beta + b > \alpha,\\
&x_4 = \beta^4 x_0 - b\beta^3 + b\beta^2 + b\beta - b < -\alpha,\\
&\dots\\
&x_{2k} = \beta^{2k} x_0 - b\beta^{2k-1} + b\beta^{2k-2} + b\beta^{2k-3} - b\beta^{2k-4} + b\beta^{2k-5} -\\
&\qquad\qquad -b\beta^{2k-6} + \dots + b\beta - b < -\alpha,\\
&x_{2k+1} = \beta^{2k+1} x_0 - b\beta^{2k} + b\beta^{2k-1} + b\beta^{2k-2} - b\beta^{2k-3} + b\beta^{2k-4} -\\
&\qquad\qquad -b\beta^{2k-5} + b\beta^{2k-6} + \dots + b\beta^2 - b\beta + b = x_0 > \alpha.
\end{aligned}
$$

Since $x_{2k+1} = x_0$, then

$$
x_0 = \frac{b(\beta^{2k} - \beta^{2k-1} - \beta^{2k-2} + \beta^{2k-3} - \beta^{2k-4} + \beta^{2k-5} - \dots - \beta^2 + \beta - 1)}{\beta^{2k+1} - 1},
$$

therefore

$$
x_0 = \frac{b(\beta^{2k+1} - 2\beta^{2k-1} - 1)}{(\beta^{2k+1} - 1)(\beta + 1)}. \tag{12}
$$

Since (10) holds, then $x_0 \geq \alpha$.

We show that

$$
\alpha \leq x_0 < x_{2k-1} < x_{2k-3} < \dots < x_5 < x_3.
$$

At first we show that

$$
x_0 < x_{2k-1} = \beta^{2k-1} x_0 - b\beta^{2k-2} + b\beta^{2k-3} + b\beta^{2k-4} - b\beta^{2k-5} + b\beta^{2k-6} - \dots - b\beta + b.
$$

Because (12) holds we prove that

$$
0 < \frac{(\beta^{2k-1} - 1) b(\beta^{2k+1} - 2\beta^{2k-1} - 1)}{(\beta^{2k+1} - 1)(\beta + 1)} - b\beta^{2k-2} + b\beta^{2k-3} + b\beta^{2k-4} - \dots - b\beta + b.
$$

The right side of the last inequality by the algebraic transformations is equal to

$$
\frac{b(2\beta^{2k-1} - 2\beta^{2k-3})}{(\beta^{2k+1} - 1)(\beta + 1)} = \frac{2b\beta^{2k-3}(\beta - 1)}{\beta^{2k+1} - 1},
$$

which is greater than 0 since $\beta > 1$ and $k \geq 2$.

Secondly we show that $x_{2m-1} < x_{2m-3}$, $3 \leq m < k$. Indeed

$$
\begin{aligned}
&(x_{2m-1} = \beta^{2m-1}x_0 - b\beta^{2m-2} + b\beta^{2m-3} + b\beta^{2m-4} - b\beta^{2m-5} + b\beta^{2m-6} - \\
&\qquad\qquad -b\beta^{2m-7} - \cdots - b\beta + b \\
&< x_{2m-3} = \beta^{2m-3}x_0 - b\beta^{2m-4} + b\beta^{2m-5} + b\beta^{2m-6} - b\beta^{2m-7} + b\beta^{2m-8} - \\
&\qquad\qquad -b\beta^{2m-9} - \cdots - b\beta + b) \\
&\Leftrightarrow 0 < (\beta^{2m-3} - \beta^{2m-1})x_0 + b\beta^{2m-2} - b\beta^{2m-3} - 2b\beta^{2m-4} + 2b\beta^{2m-5} \\
&\Leftrightarrow 0 < \tfrac{\beta^{2m-3}(1-\beta^2)b(\beta^{2k+1}-2\beta^{2k-1}-1)}{(\beta^{2k+1}-1)(\beta+1)} + \\
&\qquad\qquad + b\beta^{2m-2} - b\beta^{2m-3} - 2b\beta^{2m-4} + 2b\beta^{2m-5} \\
&\Leftrightarrow 0 < \tfrac{2b\beta^{2m-5}(\beta-1)}{\beta^{2k+1}-1}.
\end{aligned}
\tag{13}
$$

Since $\beta > 1$ and $m \geq 3$, the last inequality holds.

Now we need to show that

$$\max\{x_1, x_2, x_4, \ldots, x_{2k}\} < -\alpha.$$

In this case it is possible to prove that $x_1 < x_{2k} < x_{2k-2} < \cdots < x_2 < -\alpha$.

From (12) follows that

$$x_1 = \beta x_0 - b = \frac{b\beta(\beta^{2k+1} - 2\beta^{2k-1} - 1)}{(\beta^{2k+1} - 1)(\beta + 1)} - b. \tag{14}$$

By the construction

$$x_{2k} = \frac{x_0}{\beta} - \frac{b}{\beta} = \frac{b\beta(\beta^{2k+1} - 2\beta^{2k-1} - 1)}{\beta(\beta^{2k+1} - 1)(\beta + 1)} - \frac{b}{\beta}.$$

Inequality $x_1 < x_{2k}$ is fulfilled, if the following inequalities holds

$$
\begin{aligned}
&\tfrac{\beta^2(\beta^{2k+1}-2\beta^{2k-1}-1)}{(\beta^{2k+1}-1)(\beta+1)} - \beta < \tfrac{\beta(\beta^{2k+1}-2\beta^{2k-1}-1)}{(\beta^{2k+1}-1)(\beta+1)} - 1 \\
&\Leftrightarrow \tfrac{\beta^{2k+3}-2\beta^{2k+1}-\beta^2-\beta^{2k+3}-\beta^{2k+2}+\beta^2+\beta}{(\beta^{2k+1}-1)(\beta+1)} < \tfrac{\beta^{2k+1}-2\beta^{2k-1}-1-\beta^{2k+2}-\beta^{2k+1}+\beta+1}{(\beta^{2k+1}-1)(\beta+1)} \\
&\Leftrightarrow 0 < \tfrac{2\beta^{2k_1}(\beta^2-1)}{(\beta^{2k+1}-1)(\beta+1)}.
\end{aligned}
$$

The last inequality holds since $\beta > 1$ and therefore $x_1 < x_{2k}$.

Now we show that $x_2 < -\alpha$. We note that from (10) follows that

$$\frac{b(2\beta^{2k-1} - \beta^{2k+1} + 1)}{(\beta^{2k+1} - 1)(\beta + 1)} \leq -\alpha. \tag{15}$$

Then (considering (14) and (15))

$$
\begin{aligned}
x_2 &= \beta x_1 + b = \tfrac{b\beta^2(\beta^{2k+1}-2\beta^{2k-1}-1)}{(\beta^{2k+1}-1)(\beta+1)} - b\beta + b \\
&= \tfrac{b(\beta^{2k+3}-2\beta^{2k+1}-\beta^2-\beta^{2k+3}+\beta^{2k+1}+\beta^2-1)}{(\beta^{2k+1}-1)(\beta+1)} = \tfrac{b(-\beta^{2k+1}-1)}{(\beta^{2k+1}-1)(\beta+1)} \\
&= \tfrac{b(2\beta^{2k-1}-\beta^{2k+1}+1-2\beta^{2k-1}-2)}{(\beta^{2k+1}-1)(\beta+1)} \\
&\leq -\alpha - \tfrac{2b(\beta^{2k-1}+1)}{(\beta^{2k+1}-1)(\beta+1)} < -\alpha.
\end{aligned}
$$

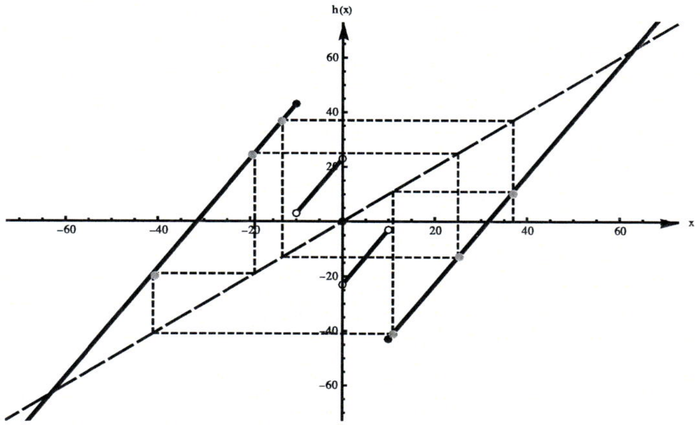

Fig. 8 Periodic orbits of period 6 with $\alpha = 10$

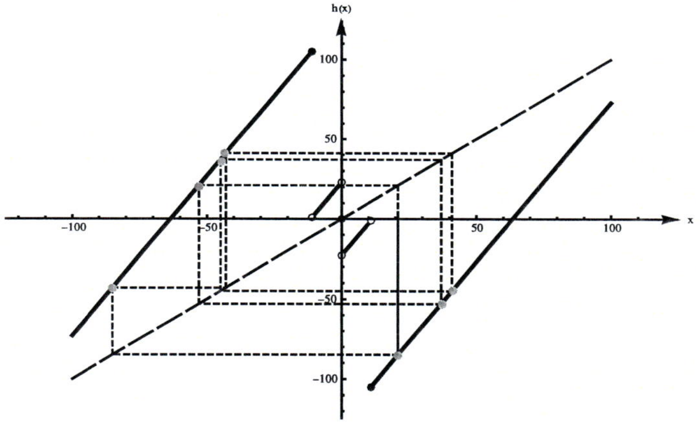

Fig. 9 Periodic orbits of period 7 with $\alpha = 11$

Similar as (13) it is possible to show that $x_{2m} < x_{2m-2}$ for all $2 \leq m \leq k$.

If assumptions of Theorem 2 hold, the orbit $O(x_0)$ where x_0 is defined by (12), satisfies (11) and is periodic orbit of period $2k + 1$ for (1).

We illustrate Theorems 1 and 2 with the following example. Let us take $k = 3$. If $\beta = 2$, $a = 23$, $b = 63$, $\alpha = 10$ then by Theorem 1 the point $x_0 = 37$ gives a periodic orbit of period 6 $\{37, 11, -41, -19, 25, -13\}$ (see Fig. 8) and if $\beta = 2$, $a = 23$, $b = 127$, $\alpha = 11$ by Theorem 2 the point $x_0 = 21$ gives a periodic orbit of period 7 $\{21, -85, -43, 41, -45, 37, -53\}$ (see Fig. 9). We can observe that the values of points of these orbits are greater or equal to α or less or equal to $-\alpha$, that is, the points belong to external lines of the function h.

We remark that these theorems find just some periodic orbits but not all. It is possible to construct periodic orbit if all points of orbit belong to internal lines of the function (4), i.e., points of orbits are in the set $]-\alpha; \alpha[\setminus \{0\}$. Furthermore we remark that in the case $\beta > 1$ there exist infinitely many eventually periodic orbits.

7 Chaotic Behaviour

The fact that the difference equation (1) has periodic orbits of any period is one of the indicators of its complicated dynamics. One more important reflection of this complexity is sensitive dependence on initial conditions, which is the hallmark of chaos. In this part we discuss chaotic behaviour of (4).

Sharkovsky's Theorem (see for example, [8, 9, 11, 17]) is the basis of the *Li* and *Yorke* proof ([13], (1975)). This Theorem shows that any one-dimensional system which exhibits a regular cycle of period three will also display regular cycles of every other length under assumption that used function is continuous. By Theorem 2 we can find some values of parameters such that our model has a periodic orbit of period 3 for $\beta > 1$. But the function in our model (1) is not continuous. There are known the another definitions of chaotic snapping in the case that function f is not a continuous one. Frequently in literature is used the following definition given by *Devaney* ([7]).

Let (X, ρ) be a metric space.

Definition 10 The function $f : X \to X$ is *chaotic* if

(a) the periodic points of f are dense in X,
(b) f is topologically transitive,
(c) f exhibits sensitive dependence on initial conditions.

Also mappings with one property—sensitive dependence on initial conditions—frequently are considered as chaotic (see [10]).

Definition 11 The function $f : X \to X$ *exhibits sensitive dependence on initial conditions* if

$$\exists \delta > 0 \; \forall x \in X \; \forall \varepsilon > 0 \; \exists y \in X \; \exists n \in \mathbf{N} : \quad \rho(x, y) < \varepsilon \;\&\; \rho(f^n(x), f^n(y)) > \delta.$$

Definition 12 The function $f : X \to X$ is *topologically transitive* on X if

$$\forall x, y \in X \; \forall \varepsilon > 0 \; \exists z \in X \; \exists n \in \mathbf{N} : \quad \rho(x, z) < \varepsilon \;\&\; \rho(f^n(z), y) < \varepsilon.$$

Definition 13 Let $A, B \subseteq X$ and $A \subseteq B$. Then A is *dense* in B if for each point $x \in B$ and each $\varepsilon > 0$, there exists $y \in A$ such that $\rho(x, y) < \varepsilon$.

In particular case if $a = 0$, then, for example, with parameters $\beta = 2$, $b = 6$, $\alpha = 3$ in interval [0; 6] our function is a doubling map which is chaotic [1, 15–17]. This particular situation is shown in Fig. 10a.

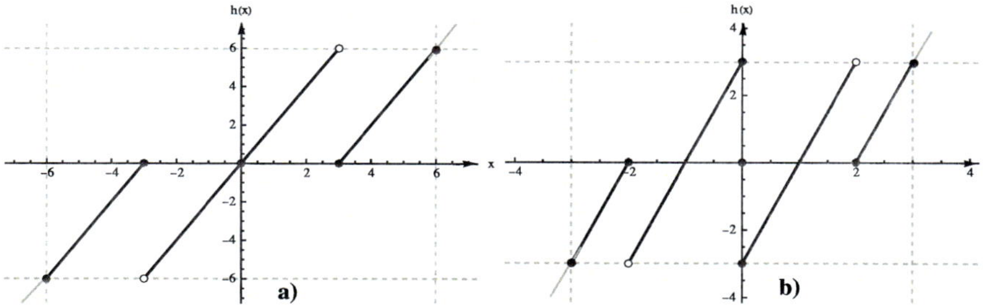

Fig. 10 Chaotic functions

In our case we use theorem from the book by *Robinson* [17].

Theorem 3 *Assume that*

(1) f is a function defined on a closed interval $[a, b] \subset \mathbf{R}$ with k discontinuities,
(2) $f([a, b]) \subset [a, b]$,
(3) the map f is differentiable at all points x different from the discontinuities points, with a derivative that satisfies $|f'(x)| > 1$ at all these points, with x different from the discontinuities points.

Then f is a chaotic mapping.

Therefore we can formulate a following theorem.

Theorem 4 *There exists parameters β, α, b and a such that the function*

$$h(x) = \begin{cases} \beta x - b, & \alpha \leq x, \\ \beta x - a, & 0 < x < \alpha, \\ 0, & x = 0, \\ \beta x + a, & -\alpha < x < 0, \\ \beta x + b, & x \leq -\alpha \end{cases}$$

is a chaotic mapping in some interval.

For example in case if $\frac{b}{\beta-1} = a = \beta\alpha - a$ the interval is $[-\frac{b}{\beta-1}, \frac{b}{\beta-1}]$. If $\beta = 3$, $a = 3$, $b = 6$, $\alpha = 2$ and the function $h(x)$ is chaotic in the interval $[-3, 3]$ (see Fig. 10b).

References

1. K.T. Alligood, T.D. Sauer, J.A. Yorke, *CHAOS: An Introduction to Dynamical Systems* (Springer-Verlag, New York Inc., 1997)
2. A. Anisimova, M. Avotina, I. Bula, Periodic orbits of single neuron models with internal decay rate $0<\beta \leq 1$. Math. Model. Anal. **18**, 325–345 (2013)

3. A. Anisimova, M. Avotina, I. Bula, Periodic and chaotic orbits of a neuron model. Math. Model. Anal. **20**(1), 30–52 (2015)
4. I. Bula, M.A. Radin, Periodic orbits of a neuron model with periodic internal decay rate. Appl. Math. Comput. **266**, 293–303 (2015)
5. I. Bula, M.A. Radin, N. Wilkins, Neuron model with a period three internal decay rate. Electron. J. Qual. Theory Differ. Equ. **46**, 1–19 (2017)
6. Y. Chen, All solutions of a class of difference equations are truncated periodic. Appl. Math. Lett. **15**, 975–979 (2002)
7. R. Devaney, *An Introduction to Chaotic Dynamical Systems*, 2nd edn. (Addison-Wesley, 1989)
8. S.N. Elaydi, *An Introduction to Difference Equations*, 2nd edn. (Springer-Verlag, New York Inc., 1999)
9. S.N. Elaydi. *Discrete Chaos. With Applications in Science and Engineering*, 2nd edn. (Chapman & Hall, CRC, 2008)
10. D. Gulick, *Encounters with Chaos* (McGraw-Hill, 1992)
11. R.A. Holmgren, *A First Course in Discrete Dynamical Systems*, 2nd edn. (Springer-Verlag, New York, 1996)
12. M.R.S. Kulenovic, O. Merino, *Discrete Dynamical Systems and Difference Equations with Mathematica* (Chapman & Hall, CRC, 2002)
13. T.Y. Li, J.A. Yorke, Period three implies chaos. Am. Math. Mon. **82**, 985–992 (1975)
14. R.M. May, Simple mathematical models with very complicated dynamics. Nature **261**, 459–467 (1976)
15. H.O. Peitgen, H. Juergen, D. Saupe, *Chaos and Fractals: New Frontiers of Science*, 2nd edn. (Springer-Verlag, New York Inc., 2004)
16. H.O. Peitgen, H. Juergen, D. Saupe, *Chaos - Bausteine der Ordnung* (Springer-Verlag Klett-Cotta, 1994)
17. R.C. Robinson, *An Introduction to Dynamical Systems: Continuous and Discrete* (Pearson Education, 2004)
18. Z. Yuan, L. Huang, Y. Chen, Convergence and periodicity of solutions for a discrete-time network model of two neurons. Math. Comput. Model. **45**, 941–950 (2002)
19. Z. Wei, L. Huang, Y. Meng, Unboundedness and periodicity of solutions for a discrete-time network model of three neurons. Appl. Math. Model. **32**, 1463–1474 (2008)
20. J. Wu, *Introduction to Neural Dynamics and Signal Transmission Delay* (De Gruyter, Berlin, 2001)
21. Z. Zhou, Periodic orbits on discrete dynamical systems. Comput. Math. Appl. **45**, 1155–1161 (2003)
22. Z. Zhou, J. Wu, Stable periodic orbits in nonlinear discrete-time neural networks with delayed feedback. Comput. Math. Appl. **45**, 935–942 (2003)
23. H. Zhu, L. Huang, Dynamics of a class of nonlinear discrete-time neural networks. Comput. Math. Appl. **48**, 85–94 (2004)

M-Theory as a Dynamical System Generator

M. Axenides, E. Floratos, D. Katsinis, and G. Linardopoulos

Abstract We review our recent work on ellipsoidal M2-brane solutions in the large-N limit of the BMN matrix model. These bosonic finite-energy membranes live inside $\mathfrak{so}(3) \times \mathfrak{so}(6)$ symmetric plane-wave spacetimes and correspond to local extrema of the energy functional. They are static in $\mathfrak{so}(3)$ and stationary in $\mathfrak{so}(6)$. Chaos appears at the level of radial stability analysis through the explicitly derived spectrum of eigenvalues. The angular perturbation analysis is suggestive of the presence of weak turbulence instabilities that propagate from low to high orders in perturbation theory.

Keywords Dynamical systems · Chaos · M-theory · BMN matrix model · Relativistic membranes

1 Introduction

M-theory By the end of the first superstring revolution (1984–1994), five seemingly different 10-dimensional superstring theories had emerged:

$$\text{Types I, II (IIA, IIB), Heterotic } (\mathfrak{so}(32), \mathrm{E}_8 \times \mathrm{E}_8).$$

M. Axenides · E. Floratos · D. Katsinis · G. Linardopoulos (✉)
Institute of Nuclear and Particle Physics, N.C.S.R. Demokritos, 153 10 Agia Paraskevi, Greece
e-mail: glinard@inp.demokritos.gr

M. Axenides
e-mail: axenides@inp.demokritos.gr

E. Floratos
e-mail: mflorato@phys.uoa.gr

D. Katsinis
e-mail: dkatsinis@phys.uoa.gr

E. Floratos · D. Katsinis · G. Linardopoulos
Department of Nuclear and Particle Physics, National and Kapodistrian University of Athens, 157 84 Athens, Greece

C. H. Skiadas et al. (eds.), *13th Chaotic Modeling and Simulation International Conference*, Springer Proceedings in Complexity,
https://doi.org/10.1007/978-3-030-70795-8_6

During the subsequent second superstring revolution (1994–2003), it was found that the 5 superstring theories are connected via a web of dualities (T-duality, S-duality, U-duality, mirror symmetry). What is more, it was realized that the five 10-dimensional superstring theories were just limiting cases of an 11-dimensional theory. This theory was called "M-theory"; it is obtained in the strong-coupling limit ($g_s \to \infty$) of IIA superstring theory. The letter "M" stands for "magic, mystery and matrix" according to one of its founders, Witten [17]. Others have associated the letter "M" with "membranes" [2].

Relativistic membranes The idea behind the theory of relativistic membranes is simple: replace 1-dimensional lines (strings) with 2-dimensional surfaces (membranes), much like lines/strings replace 0-dimensional points/particles in the passage from quantum field theory to string theory. Like point particles and strings, membranes are Poincaré invariant objects that can be supersymmetrized. It has been proven that supermembranes can only be defined consistently in 11 spacetime dimensions. Higher-dimensional extended supersymmetric objects (Mp-branes) can be defined in an analogous fashion. Nonetheless, there are reasons to believe that supermembranes (or "M2-branes") are the fundamental objects of the 11-dimensional M-theory, just like strings are the fundamental objects of 10-dimensional string theory.

Matrix models According to the *matrix theory conjecture* of Banks, Fischler, Shenker and Susskind (BFSS) [3], a theory of matrix-discretized supermembranes provides a realization of M-theory in flat spacetime. In the language of matrix models, membranes are fuzzy objects that are represented by $N \times N$ matrices. In the limit of very large matrix dimensions ($N \to \infty$), these matrix models are known to reduce to supermembrane theories.

In 2002, Berenstein, Maldacena and Nastase (BMN) [4] proposed a reformulation of the BFSS matrix model on a particular type of a background that consists of a weakly curved spacetime that is known as a plane-wave, supported by a constant (4-form) field strength:

$$ds^2 = -2dx^+dx^- - \left[\frac{\mu^2}{9}\sum_{i=1}^{3} x_i x_i + \frac{\mu^2}{36}\sum_{j=1}^{6} y_j y_j\right] dx^+dx^+ + \sum_{i=1}^{3} dx_i dx_i + \\ + \sum_{j=1}^{6} dy_j dy_j, \qquad F_{123+} = \mu. \tag{1}$$

Briefly, the BMN matrix model is a deformation of the BFSS matrix model by mass terms and a flux (aka Myers) term. In the large-N limit it is again known [5] that the BMN matrix model reduces to a theory of supermembranes in the 11-dimensional plane-wave background (1). Interestingly, M(atrix) theory has quite recently been applied to the study of chaotic phenomena that take place on the horizons of black holes.

Black holes Black holes (BHs) are regions of spacetime where the force of gravity is so strong that nothing (not even light) can escape. The 2-dimensional surface beyond which it is (classically) impossible for matter or information to escape the gravitational pull of a BH is known as the BH's event horizon. In 1974 Stephen Hawking predicted that it is (quantum-mechanically) possible for BHs to emit thermal radiation and thus slowly evaporate. Because Hawking's radiation is purely thermal, all the information that is stored in BHs seems to get lost.

To resolve the ensuing BH *information paradox* we ultimately need to understand the mechanisms with which information is being stored and processed in BHs. One such mechanism is known as *fast scrambling* or ultra-fast thermalization [6]. More generally, it is widely believed that chaotic phenomena are a dominant feature of BH horizons. Because it is inherently nonlocal, M(atrix) theory turns out to be a valuable tool in the study of information processing by BHs. More precisely, M(atrix) theory can be used to model the dynamics of the microscopic degrees of freedom that are present on BH horizons [7, 8].

2 General Setup

Our starting point is the Hamiltonian of a bosonic relativistic membrane in the 11-dimensional maximally supersymmetric plane-wave background (1). The Hamiltonian reads, in the so-called light-cone gauge $x^+ = \tau$ [5]:

$$H = \frac{T}{2}\int d^2\sigma\Big[\pi_i^2 + \frac{1}{2}\left\{x_i, x_j\right\}^2 + \frac{1}{2}\left\{y_i, y_j\right\}^2 + \left\{x_i, y_j\right\}^2 + \frac{\mu^2 x^2}{9} + \frac{\mu^2 y^2}{36} - \\ - \frac{\mu}{3}\epsilon_{ijk}\left\{x_i, x_j\right\}x_k\Big]. \tag{2}$$

From now on the indices of the coordinates x_i will implicitly be taken to run from 1 to 3, while those of the coordinates y_j will run from 1 to 6.[1] In (2) T stands for the membrane tension and

$$\pi_i^2 \equiv \sum_{i=1}^{3}\dot{x}_i\dot{x}_i + \sum_{j=1}^{6}\dot{y}_j\dot{y}_j, \qquad x^2 \equiv \sum_{i=1}^{3}x_i x_i, \qquad y^2 \equiv \sum_{j=1}^{6}y_j y_j. \tag{3}$$

The definition of the Poisson bracket $\{f, g\}$ that we will be using is

$$\{f\,,\,g\} \equiv \frac{\epsilon_{rs}}{\sqrt{w\left(\sigma\right)}}\,\partial_r f\,\partial_s g = \frac{1}{\sqrt{w\left(\sigma\right)}}\left(\partial_1 f\,\partial_2 g - \partial_2 f\,\partial_1 g\right), \tag{4}$$

[1] Note also that there's no distinction between upper and lower indices, so that these will be henceforth used interchangeably.

where $d^2\sigma = \sqrt{w(\sigma)}\, d\sigma_1\, d\sigma_2$ is the spatial volume element of the worldvolume and ϵ_{rs} is the 2-dimensional Levi-Civita symbol. In a flat worldvolume it's $w(\sigma) = 1$ and the usual definition of the Poisson bracket is retrieved.

The Lagrangian equations of motion for the spatial coordinates x and y corresponding to the Hamiltonian (2) are:

$$\ddot{x}_i = \{\{x_i, x_j\}, x_j\} + \{\{x_i, y_j\}, y_j\} - \frac{\mu^2}{9} x_i + \frac{\mu}{2}\epsilon_{ijk}\{x_j, x_k\} \tag{5}$$

$$\ddot{y}_i = \{\{y_i, y_j\}, y_j\} + \{\{y_i, x_j\}, x_j\} - \frac{\mu^2}{36} y_i. \tag{6}$$

The coordinates x and y can also be shown to obey the Gauss law constraint:

$$\sum_{i=1}^{3}\{\dot{x}_i, x_i\} + \sum_{j=1}^{6}\{\dot{y}_j, y_j\} = 0. \tag{7}$$

3 The Spherical Ansatz

Let us make the following ansatz for the spatial coordinates x and y [9, 10]:

$$x_i \equiv x_{1i} = \tilde{x}_{1i}(\tau)\, e_1(\sigma), \quad i = 1, \ldots, q_1 \tag{8}$$

$$x_{q_1+j} \equiv x_{2j} = \tilde{x}_{2j}(\tau)\, e_2(\sigma), \quad j = 1, \ldots, q_2 \qquad \& \quad q_1 + q_2 + q_3 = 3 \tag{9}$$

$$x_{q_1+q_2+k} \equiv x_{3k} = \tilde{x}_{3k}(\tau)\, e_3(\sigma), \quad k = 1, \ldots, q_3 \tag{10}$$

and

$$y_i \equiv y_{1i} = \tilde{y}_{1i}(\tau)\, e_1(\sigma), \quad i = 1, \ldots, s_1 \tag{11}$$

$$y_{s_1+j} \equiv y_{2j} = \tilde{y}_{2j}(\tau)\, e_2(\sigma), \quad j = 1, \ldots, s_2 \qquad \& \quad s_1 + s_2 + s_3 = 6 \tag{12}$$

$$y_{s_1+s_2+k} \equiv y_{3k} = \tilde{y}_{3k}(\tau)\, e_3(\sigma), \quad k = 1, \ldots, s_3. \tag{13}$$

The ansatz (8)–(13) splits the coordinates x and y into three groups

$$x_{ai} = \tilde{x}_{ai}(\tau)\, e_a \qquad \& \qquad y_{bj} = \tilde{y}_{bj}(\tau)\, e_b, \tag{14}$$

where $i = 1, \ldots, q_a$, $j = 1, \ldots, s_b$, $a, b = 1, 2, 3$. Going over to spherical coordinates, $(\sigma_1, \sigma_2) \to (\theta, \phi)$, we define[2]:

---

[2] We use the volume element in (θ, ϕ) space which implies that $\sqrt{w(\sigma)} = \sin\theta$ should be used in the definition (4) of the Poisson bracket. For alternative parametrizations such as

$$(e_1, e_2, e_3) = (cn(\phi|m)\, sn(\theta|n), sn(\phi|m)\, sn(\theta|n), sn(\theta|n)), \tag{15}$$

$$(e_1, e_2, e_3) = (\cos\phi\sin\theta, \sin\phi\sin\theta, \cos\theta), \qquad \phi \in [0, 2\pi), \quad \theta \in [0, \pi] \tag{16}$$

$$\{e_i, e_j\} = \epsilon_{ijk}\, e_k, \qquad \int e_i\, e_j\, d^2\sigma = \frac{4\pi}{3}\,\delta_{ij}. \tag{17}$$

Note that the Gauss law constraint (7) is automatically satisfied by the ansatz (8)–(13). Now consider the following solutions:

$$\tilde{\mathbf{x}}_1(\tau) = e^{\Omega_{x1}\tau}\cdot\tilde{\mathbf{x}}_{10}, \qquad \tilde{\mathbf{x}}_2(\tau) = e^{\Omega_{x2}\tau}\cdot\tilde{\mathbf{x}}_{20}, \qquad \tilde{\mathbf{x}}_3(\tau) = e^{\Omega_{x3}\tau}\cdot\tilde{\mathbf{x}}_{30} \tag{18}$$

$$\tilde{\mathbf{y}}_1(\tau) = e^{\Omega_{y1}\tau}\cdot\tilde{\mathbf{y}}_{10}, \qquad \tilde{\mathbf{y}}_2(\tau) = e^{\Omega_{y2}\tau}\cdot\tilde{\mathbf{y}}_{20}, \qquad \tilde{\mathbf{y}}_3(\tau) = e^{\Omega_{y3}\tau}\cdot\tilde{\mathbf{y}}_{30}. \tag{19}$$

As in the case of flat space (worked out in [11]) it can be shown that the radii

$$r_{x1}^2 \equiv \tilde{x}_1^2 = \sum_{i=1}^{q_1}\tilde{x}_{10i}\tilde{x}_{10i}, \; r_{x2}^2 \equiv \tilde{x}_2^2 = \sum_{j=1}^{q_2}\tilde{x}_{20j}\tilde{x}_{20j}, \; r_{x3}^2 \equiv \tilde{x}_3^2 = \sum_{k=1}^{q_3}\tilde{x}_{30k}\tilde{x}_{30k} \tag{20}$$

$$r_{y1}^2 \equiv \tilde{y}_1^2 = \sum_{i=1}^{s_1}\tilde{y}_{10i}\tilde{y}_{10i}, \; r_{y2}^2 \equiv \tilde{y}_2^2 = \sum_{j=1}^{s_2}\tilde{y}_{20j}\tilde{y}_{20j}, \; r_{y3}^2 \equiv \tilde{y}_3^2 = \sum_{k=1}^{z_3}\tilde{y}_{30k}\tilde{y}_{30k} \tag{21}$$

of the ansatz (18)–(19) can be determined (for all the antisymmetric matrices Ω_{x1}, Ω_{x2}, Ω_{x3}, Ω_{y1}, Ω_{y2}, Ω_{y3}) in terms of the conserved angular momenta

$$(\ell_{x1})_{ij} \equiv \dot{\tilde{x}}_{1i}\tilde{x}_{1j} - \tilde{x}_{1i}\dot{\tilde{x}}_{1j}, \qquad (\ell_{y1})_{ij} \equiv \dot{\tilde{y}}_{1i}\tilde{y}_{1j} - \tilde{y}_{1i}\dot{\tilde{y}}_{1j} \tag{22}$$

$$(\ell_{x2})_{ij} \equiv \dot{\tilde{x}}_{2i}\tilde{x}_{2j} - \tilde{x}_{2i}\dot{\tilde{x}}_{2j}, \qquad (\ell_{y2})_{ij} \equiv \dot{\tilde{y}}_{2i}\tilde{y}_{2j} - \tilde{y}_{2i}\dot{\tilde{y}}_{2j} \tag{23}$$

$$(\ell_{x3})_{ij} \equiv \dot{\tilde{x}}_{3i}\tilde{x}_{3j} - \tilde{x}_{3i}\dot{\tilde{x}}_{3j}, \qquad (\ell_{y3})_{ij} \equiv \dot{\tilde{y}}_{3i}\tilde{y}_{3j} - \tilde{y}_{3i}\dot{\tilde{y}}_{3j}, \tag{24}$$

by minimizing the corresponding effective potential of the membrane. This is completely equivalent to plugging the ansatz (18)–(19) into the equations of motion (5)–(6) and determining the relation between the radii $r_{x1}, r_{x2}, r_{x3}, r_{y1}, r_{y2}, r_{y3}$ and the components of the matrices $\Omega_{x1}, \Omega_{x2}, \Omega_{x3}, \Omega_{y1}, \Omega_{y2}, \Omega_{y3}$ (which in turn always combine to form the conserved angular momenta $\ell_{x1}, \ell_{x2}, \ell_{x3}, \ell_{y1}, \ell_{y2}, \ell_{y3}$).

---

where $\phi \in [0, 4\mathbb{K}(m))$ and $\theta \in [0, 2\mathbb{K}(n)]$, the corresponding volume element is $\sqrt{w(\sigma)} = sn(\theta|n)\, dn(\theta|n)\, dn(\phi|m)$.

4 Effective Potentials

The energy of the membrane (2) becomes:

$$E = \frac{2\pi T}{3}\Big[\dot{\tilde{x}}_1^2 + \dot{\tilde{x}}_2^2 + \dot{\tilde{x}}_3^2 + \dot{\tilde{y}}_1^2 + \dot{\tilde{y}}_2^2 + \dot{\tilde{y}}_3^2 + \tilde{x}_1^2\tilde{x}_2^2 + \tilde{x}_2^2\tilde{x}_3^2 + \tilde{x}_3^2\tilde{x}_1^2 + \tilde{y}_1^2\tilde{y}_2^2 + \tilde{y}_2^2\tilde{y}_3^2 + \\ + \tilde{y}_3^2\tilde{y}_1^2 + \tilde{x}_1^2\left(\tilde{y}_2^2 + \tilde{y}_3^2\right) + \tilde{x}_2^2\left(\tilde{y}_3^2 + \tilde{y}_1^2\right) + \tilde{x}_3^2\left(\tilde{y}_1^2 + \tilde{y}_2^2\right) + \frac{\mu^2}{9}\tilde{x}^2 + \\ + \frac{\mu^2}{36}\tilde{y}^2 - 2\mu\,\epsilon_{ijk}\,\tilde{x}_{1i}\tilde{x}_{2j}\tilde{x}_{3k}\Big]. \tag{25}$$

We now proceed to the following decomposition of the coordinates:

$$\dot{\tilde{x}}_1^2 \equiv \dot{\tilde{x}}_{1i}\dot{\tilde{x}}_{1i} = \dot{r}_{x1}^2 + \frac{\ell_{x1}^2}{r_{x1}^2}, \qquad \dot{\tilde{y}}_1^2 \equiv \dot{\tilde{y}}_{1j}\dot{\tilde{y}}_{1j} = \dot{r}_{y1}^2 + \frac{\ell_{y1}^2}{r_{y1}^2} \tag{26}$$

$$\dot{\tilde{x}}_2^2 \equiv \dot{\tilde{x}}_{2i}\dot{\tilde{x}}_{2i} = \dot{r}_{x2}^2 + \frac{\ell_{x2}^2}{r_{x2}^2}, \qquad \dot{\tilde{y}}_2^2 \equiv \dot{\tilde{y}}_{2j}\dot{\tilde{y}}_{2j} = \dot{r}_{y2}^2 + \frac{\ell_{y2}^2}{r_{y2}^2} \tag{27}$$

$$\dot{\tilde{x}}_3^2 \equiv \dot{\tilde{x}}_{3i}\dot{\tilde{x}}_{3i} = \dot{r}_{x3}^2 + \frac{\ell_{x3}^2}{r_{x3}^2}, \qquad \dot{\tilde{y}}_3^2 \equiv \dot{\tilde{y}}_{3j}\dot{\tilde{y}}_{3j} = \dot{r}_{y3}^2 + \frac{\ell_{y3}^2}{r_{y3}^2}. \tag{28}$$

Plugging (20)–(21) and (26)–(28) into (25), we find that the energy of the membrane becomes

$$E = \frac{2\pi T}{3}\Big[\dot{r}_{x1}^2 + \dot{r}_{x2}^2 + \dot{r}_{x3}^2 + \dot{r}_{y1}^2 + \dot{r}_{y2}^2 + \dot{r}_{y3}^2 + \frac{\ell_{x1}^2}{r_{x1}^2} + \frac{\ell_{x2}^2}{r_{x2}^2} + \frac{\ell_{x3}^2}{r_{x3}^2} + \frac{\ell_{y1}^2}{r_{y1}^2} + \frac{\ell_{y2}^2}{r_{y2}^2} + \\ + \frac{\ell_{y3}^2}{r_{y3}^2} + r_{x1}^2 r_{x2}^2 + r_{x2}^2 r_{x3}^2 + r_{x3}^2 r_{x1}^2 + r_{y1}^2 r_{y2}^2 + r_{y2}^2 r_{y3}^2 + r_{y3}^2 r_{y1}^2 + \\ + r_{x1}^2\left(r_{y2}^2 + r_{y3}^2\right) + r_{x2}^2\left(r_{y3}^2 + r_{y1}^2\right) + r_{x3}^2\left(r_{y1}^2 + r_{y2}^2\right) + \frac{\mu^2}{9}(r_{x1}^2 + \\ + r_{x2}^2 + r_{x3}^2) + \frac{\mu^2}{36}\left(r_{y1}^2 + r_{y2}^2 + r_{y3}^2\right) - 2\mu\,\epsilon_{ijk}\tilde{x}_{1i}\tilde{x}_{2j}\tilde{x}_{3k}\Big], \tag{29}$$

so that the corresponding effective potential reads

$$V_{\text{eff}} = \frac{2\pi T}{3}\Bigg[\frac{\ell_{x1}^2}{r_{x1}^2} + \frac{\ell_{x2}^2}{r_{x2}^2} + \frac{\ell_{x3}^2}{r_{x3}^2} + \frac{\ell_{y1}^2}{r_{y1}^2} + \frac{\ell_{y2}^2}{r_{y2}^2} + \frac{\ell_{y3}^2}{r_{y3}^2} + r_{x1}^2 r_{x2}^2 + r_{x2}^2 r_{x3}^2 + r_{x3}^2 r_{x1}^2 +$$
$$+ r_{y1}^2 r_{y2}^2 + r_{y2}^2 r_{y3}^2 + r_{y3}^2 r_{y1}^2 + r_{x1}^2\left(r_{y2}^2 + r_{y3}^2\right) + r_{x2}^2\left(r_{y3}^2 + r_{y1}^2\right) +$$
$$+ r_{x3}^2\left(r_{y1}^2 + r_{y2}^2\right) + \frac{\mu^2}{9}\left(r_{x1}^2 + r_{x2}^2 + r_{x3}^2\right) + \frac{\mu^2}{36}\left(r_{y1}^2 + r_{y2}^2 + r_{y3}^2\right) -$$
$$- 2\mu\,\epsilon_{ijk}\tilde{x}_{1i}\tilde{x}_{2j}\tilde{x}_{3k}\Bigg]. \tag{30}$$

The above potential (30) contains four different kinds of terms, either repulsive or attractive: (1) kinetic/angular momentum terms (repulsive), (2) quartic interaction terms (attractive), (3) mass terms (attractive), and (4) a cubic Myers flux term (repulsive). The last two kinds of terms (i.e. the mass terms (3) and the Myers term (4)) are μ-dependent and so they drop out in the $\mu \to 0$ limit (flat space) that was studied in [11]. In both cases (either $\mu = 0$ or $\mu \neq 0$), it is the equilibration of attractive and repulsive forces that determines the extrema of the potential. The two extra repulsive/attractive terms for $\mu \neq 0$ (induced by the plane-wave background) increase the complexity of the resulting dynamical system, as it will become apparent below.

There are three ways to distribute the $\mathfrak{so}(3)$ coordinates x_i $(i = 1, 2, 3)$ into the three groups that are specified by the units e_i in (16), so that we can generally distinguish three main types of membrane configurations. The first two of them (labelled types I and II below) describe rotating membranes (tops) that are point-like (collapsed) in one or two $\mathfrak{so}(3)$ directions and have a vanishing Myers flux term. The third type (III) is probably the most interesting one as it contains all four kinds of repulsive and attractive terms that we described above and extends into the full geometric background of $\mathfrak{so}(3) \times \mathfrak{so}(6)$. Let us now introduce these three types of configurations.

4.1 Type I: $q_1 = 3$, $q_2 = q_3 = 0$

For $q_1 = 3$, $q_2 = q_3 = 0$ we have

$$r_x \equiv r_{x1}, \quad r_{x2} = r_{x3} = 0 \quad \& \quad \ell_x \equiv \ell_{x1}, \quad \ell_{x2} = \ell_{x3} = 0 \tag{31}$$

and the flux term vanishes. The effective potential (30) of the membrane becomes:

$$V_{\text{eff}} = \frac{2\pi T}{3}\Bigg[\frac{\ell_x^2}{r_x^2} + \frac{\ell_{y1}^2}{r_{y1}^2} + \frac{\ell_{y2}^2}{r_{y2}^2} + \frac{\ell_{y3}^2}{r_{y3}^2} + r_{y1}^2 r_{y2}^2 + r_{y2}^2 r_{y3}^2 + r_{y3}^2 r_{y1}^2 + r_x^2\left(r_{y2}^2 + r_{y3}^2\right)$$
$$+ \frac{\mu^2 r_x^2}{9} + \frac{\mu^2}{36}\left(r_{y1}^2 + r_{y2}^2 + r_{y3}^2\right)\Bigg]. \tag{32}$$

Apart from the completely symmetric (single-radius) configuration $r = r_x = r_{y1} = r_{y2} = r_{y3}$, $\ell = \ell_x = \ell_{y1} = \ell_{y2} = \ell_{y3}$, the radii and the momenta of the effective potential (32) may be grouped into 5 different axially symmetric (2-radii) configurations and 4 more configurations with 3 different radii. Each of these potentials possesses a local minimum that corresponds to a stationary top solution with time-independent radius and nonzero total angular momentum. There are no static solutions (i.e. having constant radius and zero angular momentum) in this case.

4.2 Type II: $q_1 = 2$, $q_2 = 1$, $q_3 = 0$

For $q_1 = 2$, $q_2 = 1$ and $q_3 = 0$,

$$r_{x3} = 0 \quad \& \quad \ell_{x2} = \ell_{x3} = 0 \tag{33}$$

and the flux term vanishes again. The effective potential (30) becomes:

$$\begin{aligned} V_{\text{eff}} = \frac{2\pi T}{3}\Bigg[&\frac{\ell_{x1}^2}{r_{x1}^2} + \frac{\ell_{y1}^2}{r_{y1}^2} + \frac{\ell_{y2}^2}{r_{y2}^2} + \frac{\ell_{y3}^2}{r_{y3}^2} + r_{x1}^2 r_{x2}^2 + r_{y1}^2 r_{y2}^2 + r_{y2}^2 r_{y3}^2 + r_{y3}^2 r_{y1}^2 + \\ &+ r_{x1}^2\left(r_{y2}^2 + r_{y3}^2\right) + r_{x2}^2\left(r_{y3}^2 + r_{y1}^2\right) + \frac{\mu^2}{9}\left(r_{x1}^2 + r_{x2}^2\right) + \\ &+ \frac{\mu^2}{36}\left(r_{y1}^2 + r_{y2}^2 + r_{y3}^2\right)\Bigg]. \end{aligned} \tag{34}$$

Although again this case does not lead to any static configuration (with constant radius and zero angular momentum), we may construct one single-radius ($r = r_{x1} = r_{x2} = r_{y1} = r_{y2} = r_{y3}$, $\ell = \ell_{x1} = \ell_{y1} = \ell_{y2} = \ell_{y3}$) solution, 13 axially symmetric (2-radii) tops and 21 tops with 3 different radii.

For example let us consider a type II configuration with all the $\mathfrak{so}(6)$ variables set equal to zero:

$$x_1 = x(\tau) \cdot e_1, \quad x_2 = y(\tau) \cdot e_1, \quad x_3 = z(\tau) \cdot e_2, \quad y_i = 0, \quad i = 1, \ldots, 6, \tag{35}$$

where the time-dependent part has the form (18). In this case the effective potential (34) becomes:

$$V_{\text{eff}} = \frac{2\pi T}{3}\left[\frac{\ell^2}{x^2 + y^2} + \left(x^2 + y^2\right) z^2 + \frac{\mu^2}{9}\left(x^2 + y^2 + z^2\right)\right], \tag{36}$$

after setting $\ell_{x1} = \ell$ for simplicity. The corresponding extremisation condition $\nabla V_{\text{eff}} = 0$ implies

$$x\,z^2+\frac{\mu^2 x}{9}-\frac{x\,\ell^2}{\left(x^2+y^2\right)^2}=y\,z^2+\frac{\mu^2 y}{9}-\frac{y\,\ell^2}{\left(x^2+y^2\right)^2}=z\left(x^2+y^2\right)+\frac{\mu^2 z}{9}=0,$$

which is solved by

$$x^2+y^2=\frac{3\ell}{\mu} \quad \& \quad z=0. \tag{37}$$

Complying with (18), we can choose e.g.:

$$x(\tau)=\sqrt{\frac{3\ell}{\mu}}\cos\frac{\mu\,\tau}{3}, \qquad y(\tau)=\sqrt{\frac{3\ell}{\mu}}\sin\frac{\mu\,\tau}{3}, \qquad z(\tau)=0. \tag{38}$$

Equivalently we could have directly plugged (35) into the equations of motion (5)–(6):

$$\ddot{x}\cdot e_1=-x\,z^2\cdot e_1-\frac{\mu^2 x}{9}\cdot e_1+\mu\,y\,z\cdot e_3 \tag{39}$$

$$\ddot{y}\cdot e_1=-y\,z^2\cdot e_1-\frac{\mu^2 y}{9}\cdot e_1+\mu\,x\,z\cdot e_3 \tag{40}$$

$$\ddot{z}\cdot e_2=-z\left(x^2+y^2\right)\cdot e_2-\frac{\mu^2 z}{9}\cdot e_2. \tag{41}$$

It is easily seen that any solution of the type (18) will again satisfy (37).

4.3 Type III: $q_1=q_2=q_3=1$

For $q_1=q_2=q_3=1$, we write:

$$x_1=r_{x1}e_1,\ x_2=r_{x2}e_2,\ x_3=r_{x3}e_3 \quad \& \quad \ell_{x1}=\ell_{x2}=\ell_{x3}=0. \tag{42}$$

Note that r_{x1}, r_{x2}, r_{x3} are not radii anymore, but coordinates. The effective potential (30) of the membrane can be written as:

$$\begin{aligned} V_{\text{eff}}=\frac{2\pi T}{3}\Bigg[&\frac{\ell_{y1}^2}{r_{y1}^2}+\frac{\ell_{y2}^2}{r_{y2}^2}+\frac{\ell_{y3}^2}{r_{y3}^2}+r_{x1}^2r_{x2}^2+r_{x2}^2r_{x3}^2+r_{x3}^2r_{x1}^2+r_{y1}^2r_{y2}^2+r_{y2}^2r_{y3}^2+\\ &+r_{y3}^2r_{y1}^2+r_{x1}^2\left(r_{y2}^2+r_{y3}^2\right)+r_{x2}^2\left(r_{y3}^2+r_{y1}^2\right)+r_{x3}^2\left(r_{y1}^2+r_{y2}^2\right)+\\ &+\frac{\mu^2}{9}\left(r_{x1}^2+r_{x2}^2+r_{x3}^2\right)+\frac{\mu^2}{36}\left(r_{y1}^2+r_{y2}^2+r_{y3}^2\right)-2\mu r_{x1}r_{x2}r_{x3}\Bigg]. \end{aligned} \tag{43}$$

By combining the various radii (along with the corresponding angular momenta) into groups of one, two or three, we obtain 30 different top configurations, one of which corresponds to a completely symmetric top, 9 to axially symmetric (2-radii) tops and 10 to tops that have 3 different radii.

5 Simple Type III Solutions

The $\mathfrak{so}(3) \times \mathfrak{so}(3) \times \mathfrak{so}(3) \subset \mathfrak{so}(3) \times \mathfrak{so}(6)$ invariant ansatz

$$x_i = \tilde{u}_i(\tau)\, e_i, \qquad y_j = \tilde{v}_j(\tau)\, e_j, \qquad y_{j+3} = \tilde{w}_j(\tau)\, e_j, \qquad i, j = 1, 2, 3 \tag{44}$$

was studied in [12]. The ansatz (44) is obviously of the form (42) (type III) and it describes rotating and pulsating membranes of spherical topology. The corresponding Hamiltonian

$$H = \frac{2\pi T}{3}\left(\tilde{p}_u^2 + \tilde{p}_v^2 + \tilde{p}_w^2\right) + U, \tag{45}$$

is obtained by integrating out the worldvolume coordinates θ and ϕ. The potential energy U reads

$$U = \frac{2\pi T}{3}\left[\tilde{u}_1^2\tilde{u}_2^2 + \tilde{u}_2^2\tilde{u}_3^2 + \tilde{u}_3^2\tilde{u}_1^2 + \tilde{r}_1^2\tilde{r}_2^2 + \tilde{r}_2^2\tilde{r}_3^2 + \tilde{r}_3^2\tilde{r}_1^2 + \tilde{u}_1^2\left(\tilde{r}_2^2 + \tilde{r}_3^2\right) + \right.$$
$$+ \tilde{u}_2^2\left(\tilde{r}_3^2 + \tilde{r}_1^2\right) + \tilde{u}_3^2\left(\tilde{r}_1^2 + \tilde{r}_2^2\right) + \frac{\mu^2}{9}\left(\tilde{u}_1^2 + \tilde{u}_2^2 + \tilde{u}_3^2\right) +$$
$$\left. + \frac{\mu^2}{36}\left(\tilde{r}_1^2 + \tilde{r}_2^2 + \tilde{r}_3^2\right) - 2\mu\tilde{u}_1\tilde{u}_2\tilde{u}_3\right], \quad \tilde{r}_j^2 \equiv \tilde{v}_j^2 + \tilde{w}_j^2, \; j = 1, 2, 3. \tag{46}$$

The manifest $\mathfrak{so}(2) \times \mathfrak{so}(2) \times \mathfrak{so}(2)$ symmetry of the Hamiltonian (45)–(46) with respect to the $\mathfrak{so}(6)$ coordinates $\tilde{v}_i$ and $\tilde{w}_i$ implies that any solution of the equations of motion preserves three $\mathfrak{so}(2)$ angular momenta ℓ_i $(i = 1, 2, 3)$. The kinetic terms of the Hamiltonian (45) can be expressed in terms of the conserved angular momenta ℓ_i as

$$\tilde{p}_v^2 + \tilde{p}_w^2 = \sum_{i=1}^{3}\left(\dot{\tilde{r}}_i^2 + \frac{\ell_i^2}{\tilde{r}_i^2}\right), \tag{47}$$

leading to the effective potential

$$V_{\text{eff}} = U + \frac{2\pi T}{3}\left(\frac{\ell_1^2}{\tilde{r}_1^2} + \frac{\ell_2^2}{\tilde{r}_2^2} + \frac{\ell_3^2}{\tilde{r}_3^2}\right). \tag{48}$$

5.1 The $\mathfrak{so}(3)$ Symmetric Membrane

Let us now consider the simplest possible subsystem of (44) where the $\mathfrak{so}(6)$ variables $\tilde{v}_i$ and $\tilde{w}_i$ are set to zero [12]:

$$\tilde{v}_i = \tilde{w}_i = 0, \qquad i = 1, 2, 3. \tag{49}$$

Scaling out the mass parameter μ by setting

$$t = \mu\tau, \qquad \tilde{u}_i = \mu u_i \tag{50}$$

leads to the form

$$V_{\text{eff}} = \frac{2\pi T \mu^4}{3}\left[u_1^2 u_2^2 + u_2^2 u_3^2 + u_1^2 u_3^2 + \frac{1}{9}\left(u_1^2 + u_2^2 + u_3^2\right) - 2u_1 u_2 u_3\right] \tag{51}$$

of the membrane effective potential (48) and the Hamilton equations of motion,

$$\dot{u}_1 = p_1, \qquad \dot{p}_1 = -\left[u_1\left(u_2^2 + u_3^2\right) + \tfrac{u_1}{9} - u_2 u_3\right] \tag{52}$$

$$\dot{u}_2 = p_2, \qquad \dot{p}_2 = -\left[u_2\left(u_3^2 + u_1^2\right) + \tfrac{u_2}{9} - u_3 u_1\right] \tag{53}$$

$$\dot{u}_3 = p_3, \qquad \dot{p}_3 = -\left[u_3\left(u_1^2 + u_2^2\right) + \tfrac{u_3}{9} - u_1 u_2\right]. \tag{54}$$

The effective potential (51) is a particular instance of the generalized 3-dimensional Hénon–Heiles potential that was introduced in [13],

$$V_{\text{HH}} = \frac{1}{2}\left(u_1^2 + u_2^2 + u_3^2\right) + K_3\, u_1 u_2 u_3 + K_0\left(u_1^2 + u_2^2 + u_3^2\right)^2 + \\ + K_4\left(u_1^4 + u_2^4 + u_3^4\right), \tag{55}$$

with $K_3 = -9$, $K_0 = -K_4 = 9/4$. The critical points of the effective potential (51) are:

$$\mathbf{u}_0 = 0, \qquad \mathbf{u}_{1/6} = \frac{1}{6}\cdot(1, 1, 1), \qquad \mathbf{u}_{1/3} = \frac{1}{3}\cdot(1, 1, 1). \tag{56}$$

6 more critical points can be obtained by flipping the sign of exactly two u_i's. This is consistent with the manifest tetrahedral (T_d) symmetry of the potential (51). The extrema $\mathbf{u}_0$ (point-like membrane) and $\mathbf{u}_{1/3}$ (Myers dielectric sphere) are global degenerate minima of the potential while $\mathbf{u}_{1/6}$ is a saddle point:

$$V_{\text{eff}}(0) = V_{\text{eff}}\left(\frac{1}{3}\right) = 0, \qquad V_{\text{eff}}\left(\frac{1}{6}\right) = \frac{2\pi T\mu^4}{6^4}. \tag{57}$$

Table 1 Radial spectrum of the $\mathfrak{so}(3)$ symmetric membrane

Critical point	Eigenvalues λ^2(#)	Stability
$\mathbf{u}_0$	$\frac{1}{9}$ (3), $\frac{1}{36}$ (6)	Center (S)
$\mathbf{u}_{1/6}$	$-\frac{1}{18}$ (1), $\frac{5}{18}$ (2), $\frac{1}{12}$ (6)	Saddle point
$\mathbf{u}_{1/3}$	$\frac{1}{9}$ (1), $\frac{4}{9}$ (2), $\frac{1}{4}$ (6)	Center (S)

Radial spectrum [12] By radially perturbing the 9 critical points u_i^0 in (56) as

$$u_i = u_i^0 + \delta u_i\,(t)\,, \qquad \delta u_i\,(t) = \sum_{k=1}^{6} c_k e^{i\lambda t}\xi_{ik}, \tag{58}$$

we may confirm the above conclusion by examining the corresponding Hessian matrix. It turns out that $\mathbf{u}_0$ and $\mathbf{u}_{1/3}$ are global minima (positive-definite Hessian) and $\mathbf{u}_{1/6}$ is a saddle point (indefinite Hessian). These results are summarized in the following Table 1.

Angular spectrum [14] We may also perform more general (angular/multipole) perturbations of the following form:

$$x_i\,(t) = x_i^0 + \delta x_i\,(t)\,, \qquad i = 1, 2, 3, \tag{59}$$

where δx_i is expanded in spherical harmonics $Y_{jm}\,(\theta, \phi)$ as

$$x_i\,(t) = \mu u_i\,(t)\,e_i, \quad x_i^0 \equiv \mu u_i^0 e_i, \quad \delta x_i\,(t) = \mu \cdot \sum_{j=1}^{\infty} \sum_{m=-j}^{j} \eta_i^{jm}\,(t)\,Y_{jm}\,(\theta, \phi)\,. \tag{60}$$

For the critical points $\mathbf{u}_0$, $\mathbf{u}_{1/6}$, $\mathbf{u}_{1/3}$ we find the eigenvalues [12]:

$$\mathbf{u}_0 : \lambda_P^2 = \lambda_\pm^2 = \frac{1}{9}, \; \lambda_\theta^2 = \frac{1}{36} \tag{61}$$

$$\mathbf{u}_{1/6} : \lambda_P^2 = 0, \; \lambda_+^2 = \frac{1}{36}\,(j+1)\,(j+4)\,, \; \lambda_-^2 = \frac{j\,(j-3)}{36},$$
$$\lambda_\theta^2 = \frac{1}{36}\,(j^2 + j + 1) \tag{62}$$

$$\mathbf{u}_{1/3} : \lambda_P^2 = 0, \; \lambda_+^2 = \frac{1}{36}\,(j+1)^2\,, \; \lambda_-^2 = \frac{j^2}{9}, \; \lambda_\theta^2 = \frac{1}{36}\,(2j+1)^2\,, \tag{63}$$

with multiplicities $d_P = 2j+1$, $d_+ = 2j+3$, $d_- = 2j-1$ and $d_\theta = 6\,(2j+1)$, respectively.

The critical point $\mathbf{u}_0$ (point-like membrane) is obviously stable. $\mathbf{u}_{1/3}$ has a zero mode of degeneracy $2d_P$ while all its other eigenvalues are stable for $j = 1, 2, \ldots$ $\mathbf{u}_{1/6}$

has one $2d_P$-degenerate zero mode for every j and a 10-fold degenerate zero mode for $j = 3$. It is unstable for $j = 1$ (2-fold degenerate) and $j = 2$ (6-fold degenerate). The above results were first obtained by [5] from the matrix model. In the flat-space limit ($\mu \to 0$), we recover the results of [15, 16].

5.2 The $\mathfrak{so}(3) \times \mathfrak{so}(3) \times \mathfrak{so}(3)$ Symmetric Membrane

Similar perturbative analyses can be carried out in the $\mathfrak{so}(3) \times \mathfrak{so}(6)$ sector. A solution of the corresponding equations of motion is given by

$$u_i^0 = u_0, \quad v_j^0(t) = v_0 \cos\left(\omega t + \varphi_j\right), \quad w_j^0(t) \equiv v_{j+3}^0(t) = v_0 \sin\left(\omega t + \varphi_k\right), \tag{64}$$

where (u_0, v_0) are the critical points of the axially symmetric potential

$$V \equiv \frac{V_{\text{eff}}}{2\pi T \mu^4} = u^4 + 2u^2v^2 + v^4 + \frac{u^2}{9} + \frac{v^2}{36} - \frac{2u^3}{3} + \frac{\ell^2}{v^2} \tag{65}$$

and $\ell\mu^3 \equiv \ell_1 = \ell_2 = \ell_3$. It can be proven that the critical points (u_0, v_0) always lie within the interval:

$$\frac{1}{6} \leq u_0 \leq \frac{1}{3} \qquad \& \qquad 0 \leq v_0 \leq \frac{1}{12}. \tag{66}$$

Radial spectrum [12] To obtain the radial spectrum we set

$$u_i = u_i^0 + \delta u_i(t), \qquad v_i = v_i^0(t) + \delta v_i'(t), \qquad w_i = w_i^0(t) + \delta w_i'(t), \tag{67}$$

finding six zero eigenvalues and four nonzero ones (quadruply and doubly degenerate):

$$\lambda_{1\pm}^2 = \frac{5u_0}{2} - \frac{1}{9} \pm \sqrt{\frac{1}{9^2} - \frac{u_0}{9} - \frac{5u_0^2}{12} + 4u_0^3} \tag{68}$$

$$\lambda_{2\pm}^2 = \frac{5u_0}{2} - \frac{5}{18} \pm \sqrt{\frac{5^2}{18^2} - \frac{35u_0}{18} + \frac{163u_0^2}{12} - 20u_0^3}. \tag{69}$$

The plots of these eigenvalues can be found in the following Fig. 1.

Angular spectrum [14] Going further, we again set out to perform angular/multipole perturbations of the form:

$$x_i = x_i^0 + \delta x_i, \quad i = 1, 2, 3 \qquad \& \qquad y_k = y_k^0 + \delta y_k, \quad k = 1, \ldots, 6, \tag{70}$$

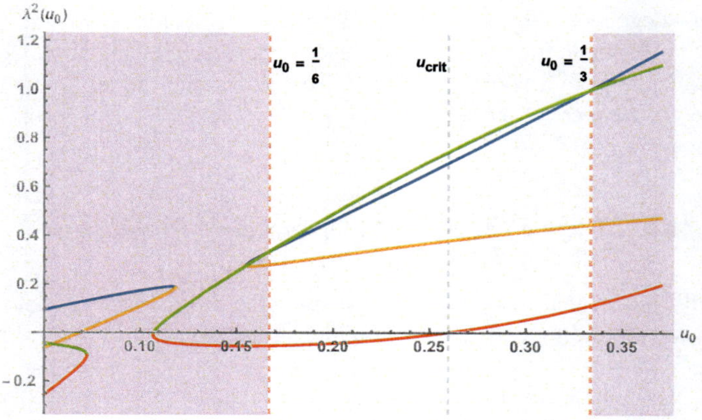

Fig. 1 Radial spectrum of the $\mathfrak{so}(3) \times \mathfrak{so}(6)$ symmetric membrane

where the δx_i, δy_k are expanded around the classical solution,

$$x_i^0 = \mu u_0 e_i, \quad i = 1, 2, 3, \qquad y_i^0 = \mu v_i^0(t)\, e_1, \quad i = 1, 2 \tag{71}$$

$$y_k^0 = \mu v_k^0(t)\, e_2, \quad k = 3, 4 \tag{72}$$

$$y_l^0 = \mu v_l^0(t)\, e_3, \quad l = 5, 6, \tag{73}$$

in spherical harmonics $Y_{jm}(\theta, \phi)$:

$$\delta x_i = \mu \cdot \sum_{j,m} \eta_i^{jm}(\tau)\, Y_{jm}(\theta, \phi), \qquad \delta y_k = \mu \cdot \sum_{j,m} \epsilon_k^{jm}(\tau)\, Y_{jm}(\theta, \phi) \tag{74}$$

$$\delta y_l = \mu \cdot \sum_{j,m} \zeta_l^{jm}(\tau)\, Y_{jm}(\theta, \phi), \tag{75}$$

for $i = 1, 2, 3$, $k = 1, 3, 5$, $l = 2, 4, 6$. We find that one of the eigenvalues always vanishes, two others are given by the following analytic expression

$$\lambda_P^2 = \frac{1}{2}\left(j^2 + j + 2\right) u_0 - \frac{1}{18}\Bigg(1 + j\,(j+1) \pm \pm 3\sqrt{144\left(j^2 + j - 2\right) u_0^3 - 12\left(j^2 + j - 14\right) u_0^2 - 24 u_0 + 1}\Bigg), \tag{76}$$

while 6 more eigenvalues $\lambda_\pm$ are also known in closed forms but are too complicated to be included here. The corresponding multiplicities of the eigenvalues are $d_P = 2j + 1$, $d_+ = 2j + 3$, $d_- = 2j - 1$. For $j = 1$ four eigenvalues vanish, while two others coincide with the eigenvalues (68)–(69) that were found from radial perturbations:

$$\lambda_P^2 = 4u_0 + \frac{1}{3}, \qquad \lambda_+^2 = \frac{5u_0}{2} - \frac{1}{9} \pm \sqrt{\frac{1}{9^2} - \frac{u_0}{9} - \frac{5u_0^2}{12} + 4u_0^3} \tag{77}$$

$$\lambda_-^2 = \frac{5u_0}{2} - \frac{5}{18} \pm \sqrt{\frac{5^2}{18^2} - \frac{35u_0}{18} + \frac{163u_0^2}{12} - 20u_0^3}. \tag{78}$$

For $j = 2$ there's one zero eigenvalue while $\lambda_P > 0$. We can also plot the $j = 2$ eigenvalues of $\lambda_\pm$ (Fig. 2):

- The squared nonzero $j = 1$ eigenvalues are all positive/stable in the interval (66), except $\lambda^2_{-(-)}$ which is positive/stable only for $u_{\text{crit}} < u_0 < 1/3$, where $u_{\text{crit}} \equiv \frac{1}{60}\left(11 + \sqrt{21}\right)$.
- For $j = 2$, the λ_P, λ_+ and one of the λ_- squared eigenvalues are positive/stable in the interval (66). The remaining λ_-^2 eigenvalue is negative/unstable in the interval $\frac{1}{6} \le u_0 \le 0.207245 < u_{\text{crit}}$.
- For $j \geq 3$ all the squared eigenvalues are non-negative inside the interval (66) and so the system is stable .

Here's a summary of the angular/multipole spectrum (Table 2):

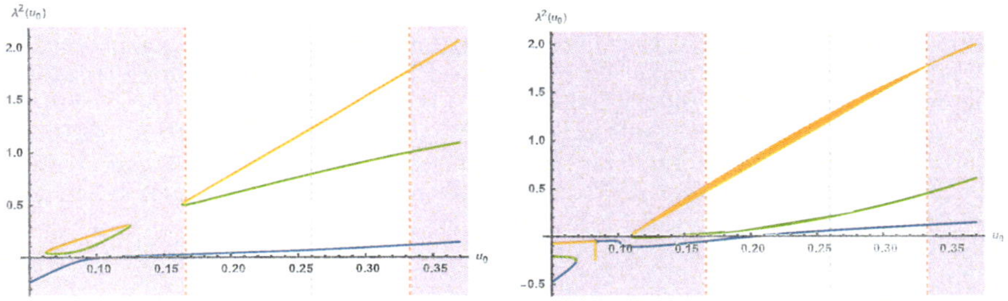

Fig. 2 $\lambda_\pm^2$ for $j = 2$ as a function of u_0

Table 2 Angular spectrum of the $\mathfrak{so}(3) \times \mathfrak{so}(6)$ symmetric membrane

Eigenvalues	$j = 1$	$j = 2$	$j \geq 3$	Degeneracy
λ_P^2	0, 0, +	0, +, +	0, +, +	$d_P = 2j + 1$
λ_+^2	0, +, +	+, +, +	+, +, +	$d_+ = 2j + 3$
λ_-^2	0, +, {0, ±} (positive for $u_0 > u_{\text{crit}}$)	+, +, {0, ±} (positive for $u_0 > 0.207245$)	+, +, +	$d_- = 2j - 1$

Higher-order perturbations [17] Beyond linearized perturbation theory (always inside the interval (66)), we anticipate a cascade of instabilities that originates from the $j = 1, 2$ multipoles and propagates towards all higher modes ($j = 3, 4, \ldots$). The perturbative expansion becomes

$$x_i = \sum_{n=0}^{\infty} \varepsilon^n \delta x_i^n = x_i^0 + \sum_{n=1}^{\infty} \varepsilon^n \delta x_i^n, \quad i = 1, 2, 3 \tag{79}$$

$$y_i = \sum_{n=0}^{\infty} \varepsilon^n \delta y_i^n = y_i^0 + \sum_{n=1}^{\infty} \varepsilon^n \delta y_i^n, \quad i = 1, \ldots, 6. \tag{80}$$

It follows that any given mode j at any given order n in perturbation theory couples to all the modes of the previous orders $1, \ldots, n-1$ through an effective forcing term that emerges in the corresponding system of fluctuation equations. The perturbations are expanded in spherical harmonics as

$$\delta x_i^n = \mu \cdot \sum_{j,m} \eta_i^{njm}(\tau)\, Y_{jm}(\theta, \phi), \qquad \eta_i^{njm}(0) = 0, \qquad i = 1, 2, 3 \tag{81}$$

$$\delta y_i^n = \mu \cdot \sum_{j,m} \theta_i^{njm}(\tau)\, Y_{jm}(\theta, \phi), \qquad \theta_i^{njm}(0) = 0, \qquad i = 1, \ldots, 6. \tag{82}$$

For example it can be shown that the ($n = 1$, $j = 1, 2$) instabilities we found above couple to every mode ($j = 1, 2, \ldots$) of the second order ($n = 2$) in perturbation theory.

Acknowledgements The authors would like to thank Christos Skiadas for the invitation to participate to the 13th CHAOS conference. The present research is funded in the context of the project "Chaotic dynamics and black holes in BMN theory" E-12386 (MIS 5047794) under the call for proposals "Supporting researchers with an emphasis on young researchers—Cycle B" (EDULLL 103). The project is co-financed by Greece and the European Union (European Social Fund—ESF) by the Operational Programme Human Resources Development, Education and Lifelong Learning 2014–2020.

References

1. E. Witten, Magic, mystery, and matrix. Not. Am. Math. Soc. **45**, 1124 (1998)
2. M.J. Duff, Benchmarks on the brane [hep-th/0407175]
3. T. Banks, W. Fischler, S.H. Shenker, L. Susskind, M-theory as a matrix model: a conjecture. Phys. Rev. **D55**, 5112 (1997) [hep-th/9610043]
4. D. Berenstein, J. Maldacena, H. Nastase, Strings in flat space and pp waves from $\mathcal{N} = 4$ super Yang-Mills. JHEP **04**, 013 (2002) [hep-th/0202021]
5. K. Dasgupta, M.M. Sheikh-Jabbari, M. Van Raamsdonk, Matrix perturbation theory for M-theory on a pp-wave. JHEP **05**, 56 (2002) [hep-th/0205185]

6. Y. Sekino, L. Susskind, Fast scramblers. JHEP **10**, 065 (2008), arXiv:0808.2096
7. G. Gur-Ari, M. Hanada, S.H. Shenker, Chaos in classical D0-brane mechanics. *JHEP* **02**, 091 (2016), arXiv:1512.00019
8. Y. Asano, D. Kawai, K. Yoshida, Chaos in the BMN matrix model. JHEP **06**, 191 (2015), arXiv:1503.04594
9. P.A. Collins, R.W. Tucker, Classical and quantum mechanics of free relativistic membranes. Nucl. Phys. B **112**, 150 (1976)
10. T. Harmark, K.G. Savvidy, Ramond-Ramond field radiation from rotating ellipsoidal membranes. Nucl. Phys. **B585**, 567 (2000) [hep-th/0002157]
11. M. Axenides, E. Floratos, Euler top dynamics of Nambu-Goto p-branes. JHEP **03**, 093 (2007) [hep-th/0608017]
12. M. Axenides, E. Floratos, G. Linardopoulos, M2-brane dynamics in the classical limit of the BMN matrix model. Phys. Lett. **B773**, 265 (2017), arXiv:1707.02878
13. K. Efstathiou, D. Sadovskií, Perturbations of the 1:1:1 resonance with tetrahedral symmetry: a three degree of freedom analogue of the two degree of freedom Hénon-Heiles Hamiltonian. Nonlinearity **17**, 415 (2004)
14. M. Axenides, E. Floratos, G. Linardopoulos, Multipole stability of spinning M2 branes in the classical limit of the BMN matrix model. Phys. Rev. **D97**, 126019 (2018), arXiv:1712.06544
15. M. Axenides, E.G. Floratos, L. Perivolaropoulos, Metastability of spherical membranes in supermembrane and matrix theory. JHEP **11**, 020 (2000) [hep-th/0007198]
16. M. Axenides, E.G. Floratos, L. Perivolaropoulos, Quadrupole instabilities of relativistic rotating membranes. Phys. Rev. **D64**, 107901 (2001) [hep-th/0105292]
17. M. Axenides, E. Floratos, D. Katsinis, G. Linardopoulos (to appear)

Hopf Bifurcation Analysis for the Fitzhugh-Nagumo Model of a Spiking Neuron

Ilknur Kusbeyzi Aybar

Abstract The Fitzhugh-Nagumo model, which describes a pulse transmission activity in a neuron, is first called the Bonhoeffer-van der Pol model since it is originally transformed from the well-known van der Pol model. The complexity of the neural dynamical models consist of multi-parameter nonlinear systems often allow studying only a particular case for some given values of parameters and prevent obtaining general results. In this study, we present general parameter regions for the existence and the stability of Hopf bifurcation for the Fitzhugh-Nagumo model.

Keywords Fitzhugh-Nagumo model · Limit cycle · Stability · Periodic solutions

1 Introduction

Neurons, the smallest members of the brain, transmit information between each other through electrical activities. The electrical activities of a single neuron can be modeled and analyzed by dynamical systems. The communication among neurons observed as firing or spikes occurs as an oscillation formation or loss around a singular point of the dynamical system, i.e., Hopf bifurcation, when a parameter exceeds a threshold value.

The Fitzhugh-Nagumo model, which was proposed by R. Fitzhugh in 1961 and simulated by J. Nagumo et al. in 1962, is governed by the ODE system

$$
\begin{aligned}
\frac{dx}{dt} &= x(x-a)(1-x) - y = F(x, y),\\
\frac{dy}{dt} &= \epsilon(x - \gamma y) = G(x, y),
\end{aligned}
\tag{1}
$$

I. Kusbeyzi Aybar (✉)
Faculty of Education, Department of Computer Education and Instructional Technology, Yeditepe University, 34755 Istanbul, Turkey
e-mail: ikusbeyzi@yeditepe.edu.tr

C. H. Skiadas et al. (eds.), *13th Chaotic Modeling and Simulation International Conference*, Springer Proceedings in Complexity,
https://doi.org/10.1007/978-3-030-70795-8_7

where state variables $x(t)$ and $y(t)$ represent the change in the membrane voltage (action potential) and the change in the number of open potassium channels on the membrane of a single neuron, respectively, over time. In this model, the voltage passes the threshold value a; spiking occurs, then it starts to decrease and stabilizes at the neuron membrane's resting potential. The parameter ϵ denotes the speed of the change in the number of open ion channels. The parameter γ states how strongly y depends on x. All system parameters are assumed to be positive, except for the parameter a, which can be negative.

2 Dynamics of the Fitzhugh-Nagumo Model

The eigenvalues of the Jacobian matrix can be calculated at the singular points to determine whether the system is stable or unstable at that point. The investigation of the stability for a dynamical system is important because by this analysis it is then possible to know if all nearby trajectories approach that point of singularity. If the real parts of the eigenvalues of the system's Jacobian matrix are all negative at that point, the system is stable at this point of singularity; otherwise, it is unstable at that point [8]. If the real parts of the eigenvalues of the Jacobian matrix are a pair of pure imaginary values, then the phenomenon called Hopf bifurcation occurs at that point, which implies that at that point, the system is oscillating, that is, this singularity point is an oscillatory solution of the system. The limit cycle can be either stable or unstable, depending on the behaviour of the system. If all nearby trajectories approach the limit cycle inward and outward, it is a stable limit cycle. In order to determine the parameter regions for the existence of a stable limit cycle, we calculate the first Lyapunov coefficient [4]. This study contributes to the comprehension of the electrical activities of neurons by using the dynamical analysis methods since the spikes in the communication of neurons are mathematically stable oscillatory solutions.

Proposition 1 *System* (1) *has three singular points* $E_0 = (0, 0)$,

$$E_- = \left(\frac{1}{2}\left(1 + a - \sqrt{(a-1)^2 - \frac{4}{\gamma}}\right), \frac{1}{2\gamma}\left(1 + a - \sqrt{(a-1)^2 - \frac{4}{\gamma}}\right)\right),$$

and

$$E_+ = \left(\frac{1}{2}\left(1 + a + \sqrt{(a-1)^2 - \frac{4}{\gamma}}\right), \frac{1}{2\gamma}\left(1 + a + \sqrt{(a-1)^2 - \frac{4}{\gamma}}\right)\right).$$

2.1 System Dynamics at E_0

At E_0, the membrane voltage is zero, and the ion channels are all closed. The eigenvalues of the Jacobian matrix at E_0, which is at the resting potential, are

$$\lambda_{1,2} = \frac{1}{2}(-a - \epsilon\gamma \pm \sqrt{(a - \epsilon\gamma)^2 - 4\epsilon}). \tag{2}$$

Proposition 2 *The singular point E_0 is a stable singular point when the system parameters a, ϵ, and γ are all positive or*

$$-\gamma\epsilon < a \leq 0.$$

Theorem 1 *According to the eigenvalues given in (2), system (1) exhibits Hopf bifurcation when*

$$a < 0, \epsilon > a^2, a = -\epsilon\gamma. \tag{3}$$

Proof To investigate Hopf bifurcation at E_0, we accept conditions (3) and calculate the first Lyapunov coefficient at this point. We look for a Lyapunov function of the form

$$h(x, y) = \alpha x^2 + \beta xy + \sigma y^2 + h.o.t.$$

that satisfies

$$\frac{\partial h}{\partial x}F(x, y) + \frac{\partial h}{\partial y}G(x, y) = g_1(x^2 + y^2)^2 + \cdots . \tag{4}$$

By solving (4) together with (1), we obtain

$$\alpha = \frac{\beta\epsilon}{2a}$$

and

$$\sigma = -\frac{\beta}{2\epsilon\gamma}.$$

The quadratic form

$$4\alpha\sigma - \beta^2 = -4 + \frac{4}{\epsilon\gamma^2}$$

is positive-definite when we choose $\beta = 2$ because of condition (3) which reduces to

$$\epsilon > 0, 0 < \gamma < \sqrt{\frac{1}{\epsilon}}$$

in this case. We calculate the first Lyapunov coefficient as

$$g_1 = \frac{6 + 2\gamma(\epsilon\gamma(1 - 2\epsilon\gamma) - 2)}{\gamma(3 + \epsilon(2 + \epsilon(3 + 4\gamma^2)))}. \tag{5}$$

Stable oscillations are indicated by a negative first Lyapunov coefficient. Therefore, we study the case $g_1 < 0$ together with the conditions given in (3) and we conclude in the following Theorem.

Theorem 2 *System* (1) *exhibits supercritical Hopf bifurcation when one of the following cases are satisfied:*

$$\begin{aligned}
&(i)\ 0 < \gamma < 1, \frac{1}{4\gamma} + \frac{1}{4}\sqrt{-3(\frac{(5\gamma - 8)}{\gamma^3})} < \epsilon < \frac{1}{\gamma^2},\\
&(ii)\ 1 < \gamma \leq \frac{3}{2}, \frac{1}{4\gamma} + \frac{1}{4}\sqrt{-3(\frac{5\gamma - 8}{\gamma^3})} < \epsilon < \frac{1}{\gamma^2},\\
&(iii)\ \frac{3}{2} < \gamma \leq \frac{8}{5}, 0 < \epsilon < \frac{1}{4\gamma} - \frac{1}{4}\sqrt{-3(\frac{5\gamma - 8}{\gamma^3})},\\
&(iv)\ \frac{3}{2} < \gamma \leq \frac{8}{5}, \frac{1}{4\gamma} + \frac{1}{4}\sqrt{-3(\frac{5\gamma - 8}{\gamma^3})} < \epsilon < \frac{1}{\gamma^2},\\
&(v)\ \gamma > \frac{8}{5}, 0 < \epsilon < \frac{1}{\gamma^2}.
\end{aligned} \tag{6}$$

Example 1 To demonstrate supercritical Hopf bifurcation at the origin we choose the parameter set $(a, \epsilon, \gamma) = (-1.75, 0.5, 3.5)$. The eigenvalues of the Jacobian matrix are $\lambda_{1,2} = \pm 0.661438i$. In Fig. 1, we illustrate that system (1) exhibits stable oscillatory regime, i.e. limit cycle. The trajectories outside and inside approach to the limit cycle. In Fig. 1a we have two initial points, one at the outside, and one at the inside of the limit cycle. This is clearly obvious since the first Lyapunov coefficient at this point is $g_1 = -0.0127119$, as we obtain by (5). In Fig. 1b, we choose ten different initial points.

2.2 System Dynamics at E_- and E_+

In this section, we investigate the local dynamics of system (1) at the singular point E_-. However, same results are obtained for the singular point E_+. First, we make assumption that the state variables and parameter values to be real numbers to represent realistic values, which means that the expressions in the root must be nonnegative. For this reason, we assume the square root terms in the singular points are positive or at least zero. Hence, we begin our investigation for Hopf bifurcation at $E_\pm$ by accepting the condition

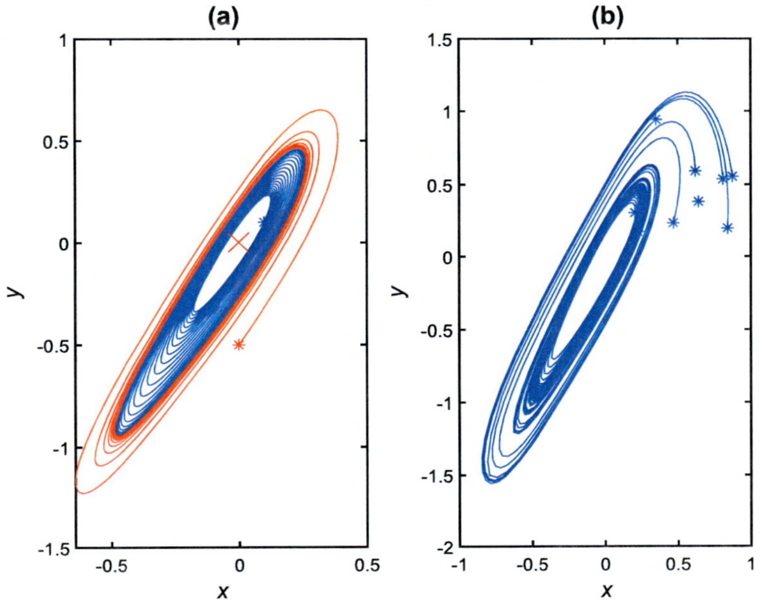

Fig. 1 **a** The supercritical Hopf bifurcation of system (1) when $(a, \epsilon, \gamma) = (-1.75, 0.5, 3.5)$ around the origin. **b** The trajectories moving towards the limit cycle from ten different initial points outside

$$\Delta = (a-1)^2\gamma - 4 \geq 0. \tag{7}$$

The eigenvalues of the Jacobian matrix at E_- are

$$\begin{aligned}\lambda_{1,2} = &\frac{1}{4\gamma}(2 - \Delta - 2\epsilon\gamma^2 + (1+a)\sqrt{\gamma\Delta} \\ &\pm \frac{1}{2}\sqrt{-32\epsilon\gamma^2(\Delta - (1+a)\sqrt{\gamma\Delta}) + 4(2 - \Delta - 2\epsilon\gamma^2 + (1+a)\sqrt{\gamma\Delta})^2}).\end{aligned} \tag{8}$$

Proposition 3 *E_- is a stable singular point under one of the following cases:*

$$\begin{aligned}&i.\ a < 1, \Delta = 0, \epsilon > \frac{2-\Delta}{2\gamma^2},\\ &ii.\ a < 1, \Delta > 0, \epsilon > \frac{2 - \Delta + \sqrt{\gamma\Delta}(1+a)}{2\gamma^2},\\ &iii.\ a > 1, \Delta = 0, \epsilon > \frac{2-\Delta}{2\gamma^2},\\ &iv.\ a > 1, \Delta > 0, \epsilon > \frac{2 - \Delta + \sqrt{\gamma\Delta}(1+a)}{2\gamma^2}.\end{aligned} \tag{9}$$

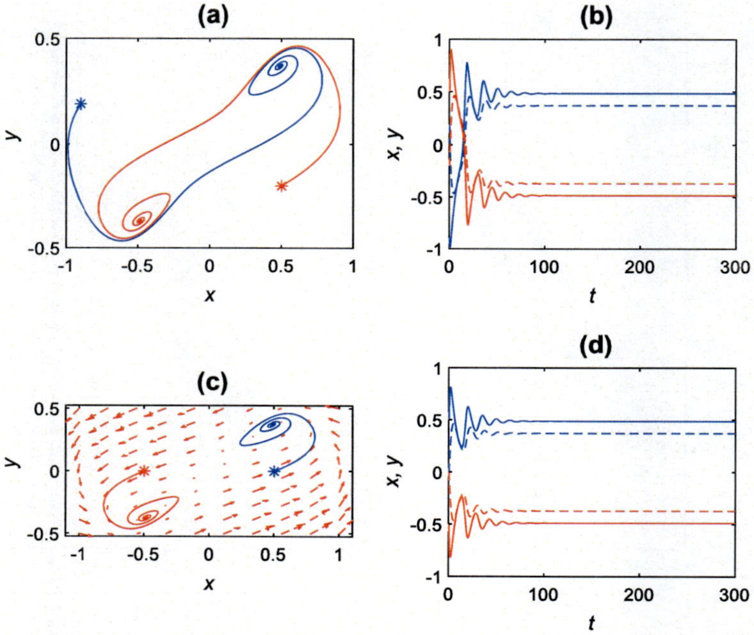

Fig. 2 **a** Stable spiral foci at $E_{-} = (-0.486458, -0.371342)$, and $E_{+} = (0.486458, 0.371342)$ for system (10). **b** Time series plots for the case given in (**a**). **c** The case given in (**a**) for different initial points. **d** Time series plots of (**c**)

Example 2 As an example of Proposition 3, we consider the parameters $(a, \epsilon, \gamma) = (-1, 0.32, 1.31)$. For this parameter set, system (1) is rewritten as

$$\begin{aligned} \frac{dx}{dt} &= x - x^3 - y, \\ \frac{dy}{dt} &= 0.32x - 0.4192y. \end{aligned} \tag{10}$$

The singular points of system (10) are $E_{-} = (-0.486458, -0.371342)$, $E_0 = (0, 0)$, and $E_{+} = (0.486458, 0.371342)$. The eigenvalues of the Jacobian matrix at E_{-} and E_{+} are $\lambda_{1,2} = -0.0645618 \pm 0.440717i$. Since the real parts of the eigenvalues are negative, E_{-} and E_{+} are stable singular points. The origin is a saddle point with a positive, $\lambda_1 = 0.718807$, and a negative, $\lambda_2 = -0.138007$, real eigenvalue. The phase portrait for this parameter set is given in Fig. 2. The solid lines indicate the voltage, $x(t)$, and the dashed lines indicate the number of open channels, $y(t)$. We observe that the state variables end up at E_{-} and E_{+}, after little fluctuations.

Hopf bifurcation indicates the birth or death of a periodic solution at a singular point under a small perturbation of a parameter. Hopf bifurcation occurs when a complex conjugate pair of eigenvalues of the Jacobian matrix at a singular point becomes pure imaginary. In this case, a limit cycle, which is a stable (unstable) isolated periodic orbit, exists, and the bifurcation is called supercritical (subcritical) Hopf bifurcation.

Considering assumption (7), we find the conditions for the real part of (8) to be zero and the expression inside the root to be negative so that the eigenvalues (8) are pure imaginary.

Theorem 3 *System* (1) *undergoes Hopf bifurcation at* E_- *when* $a = -1$ *if*

$$1 < \gamma < \frac{3}{2}, \epsilon = \frac{3-2\gamma}{\gamma^2}. \tag{11}$$

Moreover, the Hopf bifurcation at E_- *is always supercritical (i.e. there exists a stable limit cycle.).*

Proof To investigate the existence and the stability of Hopf bifurcation, we apply a change of variables to move E_- to the origin by the linear transformation

$$x = u + \frac{1}{2}\left(1 + a - \sqrt{\frac{\Delta}{\gamma}}\right),$$

$$y = v + \frac{1}{2\gamma}\left(1 + a - \sqrt{\frac{\Delta}{\gamma}}\right),$$

and obtain the following system:

$$\begin{aligned}\frac{dx}{dt} &= \frac{1}{2\gamma}((2 - \Delta + (1+a)\sqrt{\gamma\Delta})x + (-(1+a)\gamma + 3\sqrt{\gamma\Delta})x^2 - 2\gamma x^3 - 2\gamma y),\\ \frac{dy}{dt} &= \epsilon(x - \gamma y).\end{aligned} \tag{12}$$

System (12) has Hopf bifurcation at the origin when $a = -1$. The eigenvalues of the Jacobian matrix at the singular point at the origin are

$$\lambda_{1,2} = \frac{1}{2\gamma}(3 - \gamma(2 + \epsilon\gamma) + \sqrt{9 + \gamma(-12 + \gamma(4 + \epsilon(2 + \gamma(-4 + \epsilon\gamma))))}). \tag{13}$$

The eigenvalues given in (13) are pure imaginary if

$$1 < \gamma < \frac{3}{2}, \epsilon = \frac{3-2\gamma}{\gamma^2}. \tag{14}$$

Hence, system (12) exhibits Hopf bifurcation. Under these conditions, we have positive definite quadratic form.

We observe that when $\Delta = 0$, there is no Hopf bifurcation at the system. Hence, we consider $\Delta > 0$. We also consider $a \neq 1$ for the system to be defined.

Calculating the first Lyapunov coefficient at this point we have

$$\alpha = \frac{\beta\epsilon\gamma}{-2 - \Delta - (1+a)\sqrt{\gamma\Delta}}$$

$$\epsilon = \frac{2 - \Delta + (1+a)\sqrt{\gamma\Delta}}{2\gamma^2}$$

$$\sigma = -\frac{be}{2\epsilon\gamma}$$

We assume the coefficient as $\beta = 2$. Then, we obtain

$$g_1 = -\frac{24(-1+\gamma)\gamma^3}{27 + \gamma(-36 + \gamma(54 + \gamma(-52 + 19\gamma)))} \quad (15)$$

for the first Lyapunov coefficient of system (1) at E_-. When g_1 given in (15) is negative, then there exists a supercritical Hopf bifurcation for system (1). When we solve

$$g_1 < 0$$

and

$$1 < \gamma < \frac{3}{2}, \epsilon = \frac{3 - 2\gamma}{\gamma^2}$$

together, we obtain again condition (14). Therefore, if there is Hopf bifurcation at E_- and E_+, it is a supercritical Hopf bifurcation. Hence, there exists a stable limit cycle.

We explain Theorem (3) with the following numerical example.

Example 3 As an example for Hopf bifurcation, we choose parameter values $(a, \epsilon, \gamma) = (-1, 0.32, 1.25)$, satisfying (3). In this case, system (1) can be written as

$$\begin{aligned} \frac{dx}{dt} &= x - x^3 - y, \\ \frac{dy}{dt} &= 0.32x - 0.4y. \end{aligned} \quad (16)$$

System (16) has the singular points $E_- = (-0.447214, -0.357771)$, $E_0 = (0, 0)$, and $E_+ = (0.447214, 0.357771)$. System (16) possesses Hopf bifurcation at the singular points E_- and E_+, which are located symmetrically on the trajectory plot. This result is due to the eigenvalues of the Jacobian matrix, which are $\lambda_{1,2} = \pm 0.4i$. The stability of Hopf bifurcation is determined by the first Lyapunov coefficient is

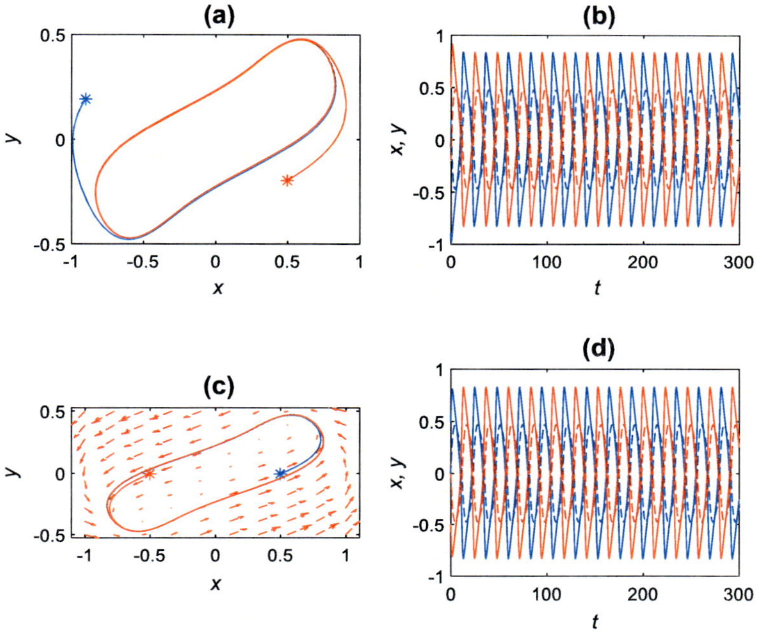

Fig. 3 **a** The stable limit cycle for Example 3 when $(a, \epsilon, \gamma) = (-1, 0.32, 1.25)$ around $E_+ = (0.447214, 0.357771)$ and $E_- = (-0.447214, -0.357771)$. **b** The oscillations in the state variables, i.e. the voltage and the number of open channels given in (**a**). **c** The trajectories moving towards the limit cycle. **d** The oscillations in the state variables, i.e. the voltage and the number of open channels given in (**c**)

$g_1 = -1.04639$ as calculated from (15). At the origin, E_0 is a saddle point represented by a positive ($\lambda_1 = 0.712311$) and a negative ($\lambda_2 = -0.112311$) real eigenvalues. The stable limit cycle for this example is illustrated in Fig. 3. In Fig. 3a, the blue star is at $(-0.9, 0.19)$ and the red star is at $(0.5, -0.2)$. The trajectories initiating from the blue and the red star approach towards the limit cycle. In Fig. 3b, the voltage, $x(t)$, is represented by a solid line, and the number of open ion channels, $y(t)$ is represented by a dashed line. For the two initial points, the blue and the red star, we see the oscillations in $x(t)$ and $y(t)$ for the phase portrait given in Fig. 3a. In Fig. 3c, two points move along the limit cycle. In Fig. 3d, we observe the oscillations of the state variables for the phase portrait given in Fig. 3c.

References

1. I.K. Aybar, O.O. Aybar, B. Fercec, V.G. Romanovski, S. Swarup Samal, A. Weber, Investigation of invariants of a chemical reaction system with algorithms of computer algebra. MATCH Commun. Math. Comput. Chem. **74**, 465–480 (2015)
2. L.N.M. Duysens, J. Amesz, Fluorescence spectrophotometry of reduced phosphopyridine nucleotide in intact cells in the near-ultraviolet and visible region. Biochim. Biophys. Acta. **24**(1), 19–26 (1957)
3. B. Ferčec, Integrability and local bifurcations in polynomial systems of ordinary differential equations. Ph.D. Thesis, 2013
4. Y.A. Kuznetsov, *Elements of Applied Bifurcation Theory* (Springer, New York, 1995)
5. A.M. Liapunov, *Stability of Motion*, with a contribution by V. Pliss. Translated by F. Abramovici, M. Shimshoni (Academic Press, New York, 1996)
6. Mathematica 9.0 Wolfram Research Inc., https://www.wolfram.com
7. MATLAB 9.7 MathWorks 2019, https://www.mathworks.com
8. L. Perko, *Differential Equations and Dynamical Systems, Texts in Applied Mathematics*, 7th edn. (Springer-Verlag, New York, 2001)
9. V.G. Romanovski, D.S. Shafer, *The Center and Cyclicity Problems, A Computational Algebra Approach* (Birkhauser, Boston-Basel-Berlin, 2009)
10. R. Fitzhugh, Impulses and physiological states in theoretical models of nerve membrane. Biophys. J. **1**, 445–466 (1961)
11. J. Nagumo, S. Arimoto, S. Yoshizawa, An active pulse transmission line simulating nerve axon. Proc. IRE. **50**, 2061–2070 (1962)
12. Y. Xia, M. Grašič, W. Huang, V.G. Romanovski, Limit cycles in a model of olfactory sensory neurons. Int. J. Bifurcat. Chaos **29**(3), 1950038 (2019)
13. L. Zhang, J. Yu, Invariant algebraic surfaces of the FitzHugh-Nagumo system. J. Math. Anal. Appl. **483**(2) (2020)
14. A.G. Korotkov, A.O. Kazakov, T.A. Levanova, Effects of memristor-based coupling in the ensemble of FitzHugh-Nagumo elements. Eur. Phys. J. Spec. Top. **228**, 2325–2337 (2019)

Limit Cycles of the Schnakenberg Chemical Reaction Model

I. Kusbeyzi Aybar, B. Ferčec, O. O. Aybar, and M. Dukarić

Abstract Dynamical systems contribute to the mathematical modeling of chemical reactions of biological or ecological phenomena producing limit cycle oscillations. In this study, we present a computational approach to examine the bifurcations of limit cycles of the two-dimensional simple chemical reaction model known as the Schnakenberg model. With our approach, we obtain conditions on parameters of the system of the chemical reaction model which gives Hopf bifurcation. Using the Lyapunov function we show the stability of Hopf bifurcation. We illustrate the results with a numerical example.

Keywords Schnakenberg · Limit cycle · Chemical reaction · Lyapunov function

1 Introduction

During biochemical reactions, which are the transformations of molecules to other molecules inside the cell, enzymes play roles of biological catalysts and change concentration rates [10]. Biochemical reactions enable cell functions such as digestion

I. K. Aybar (✉)
Faculty of Education, Department of Computer Education and Instructional Technology, Yeditepe University, 34755 Istanbul, Turkey
e-mail: ikusbeyzi@yeditepe.edu.tr

B. Ferčec
Faculty of Energy Technology, University of Maribor, Hočevarjev trg 1, 8270 Krško, Slovenia

B. Ferčec · M. Dukarić
Center for Applied Mathematics and Theoretical Physics, University of Maribor, Mladinska 3, 2000 Maribor, Slovenia

O. O. Aybar
Piri Reis University, 34940 Tuzla, Istanbul, Turkey

M. Dukarić
SSI Schaefer, 2000 Maribor, Slovenia

C. H. Skiadas et al. (eds.), *13th Chaotic Modeling and Simulation International Conference*, Springer Proceedings in Complexity,
https://doi.org/10.1007/978-3-030-70795-8_8

and respiration to reproduction and contribute to maintaining the life processes of living organisms. The biochemical reaction, named glycolysis, which is the destruction of glucose to enzymatic acid with enzymes to generate energy, happens in all living organisms. Moreover, in all living organisms, the same enzymes act in this reaction. The fluctuations in the concentrations of the substances lead to glycolytic oscillations, which depend on the concentration rates. The glycolytic oscillations were first observed experimentally by Duysens and Amesz [2]. This phenomenon is known as a fascinating biochemical reaction represented mathematically by a generalized version of the Schnakenberg model [13].

2 The Schnakenberg Model

In 1978, J. Schnakenberg introduced the simple chemical reaction system with a limit cycle behaviour, the so called Schnakenberg model, an autocatalytic chemical reaction model with oscillatory behaviour. This model is characterized by the following three chemical reactions that involve two chemical components and two chemical resources:

$$A \rightleftharpoons X,\, B \to Y,\, 2X + Y \to 3X.$$

The dynamical system for this chemical reaction is obtained by considering the state variables x and y as the concentrations at a given time of the chemical substances X and Y which leads to the following system of ordinary differential equations:

$$\begin{aligned} \frac{dx}{dt} &= x^2y - x + b, \\ \frac{dy}{dt} &= -x^2y + a. \end{aligned} \tag{1}$$

Here, the parameters $a > 0$ and $b > 0$ denote the concentration rates of the chemical resources A and B, respectively.

To demonstrate the oscillations in this model, Hwang et al. [4] showed that the dimensionless system of the Schnakenberg model possesses at most one limit cycle in $\mathbb{R}^2_+$. The nonexistence and existence of a positive non-constant steady-state solution to the Schnakenberg system are studied by Li, in 2011 [6]. In this paper, we investigate Hopf bifurcation for the Schnakenberg model by computing the Lyapunov function [12].

System (1) has one unique singular point at $E = (a + b, \frac{a}{(a+b)^2})$, with the eigenvalues of Jacobian matrix given as

$$\lambda_{1,2} = \frac{-1}{2}(1 + (a + b)^2 \pm \sqrt{k(a, b)}), \tag{2}$$

where

$$k(a, b) = (-a + b + (a + b)^3)^2 - 4(a + b)^4. \tag{3}$$

It is obvious that these eigenvalues are pure imaginary if

$$1 + (a + b)^2 = 0, \; k(a, b) < 0, \; a, b > 0,$$

which simplifies to the condition $0 < a < 1$ and $(a + b)^3 = a - b$. In this case the singularity at E can either be a center (all trajectories are closed), or a focus (all trajectories are spirals) [11]. When the condition

$$(a + b)(1 + (a + b)^2) \neq 0$$

is satisfied, the singular point at E is a focus, if $(a + b)(1 + (a + b)^2) < 0$ all trajectories in a neighborhood of E are moving towards the singularity E (stable focus) and if $(a + b)(1 + (a + b)^2) > 0$ all trajectories in a neighborhood of E are moving away the singularity at E. For instance, Fig. 1a shows stable limit cycle of the system (1) with $a = 0.6$ and $b = 0.17037459017229974$ and eigenvalues are $\lambda_{1,2} = -0.0178966 \pm 0.77016668i$ and Fig. 1b shows stable limit cycle of the system (1) with $a = 0.6$ and $b = 0.15037459017229974$ and eigenvalues are $\lambda_{1,2} = 0.01806962 \pm 0.75015699i$ [9].

3 The Existence of Hopf Bifurcation

To study Hopf bifurcation for system (1), we first move the singular point E to origin by the linear transformation

$$x \rightarrow X + (a + b) \text{ and } y \rightarrow Y + \frac{a}{(a + b)^2}, \tag{4}$$

and obtain

$$\begin{aligned} \dot{x} &= \frac{1}{a+b}((a + b)x^2y + ax^2 + +2(a + b)^2xy + (a - b)x + (a + b)^3y) = F(x, y) \\ \dot{y} &= \frac{-1}{a+b}((a + b)x^2y + ax^2 + +2(a + b)^2xy + 2ax + (a + b)^3y) = G(x, y), \end{aligned} \tag{5}$$

where X is rewritten as x, and Y as y. The necessary condition for the existence of Hopf bifurcation at the origin for system (5) is when the trace of the linear approximation of system (5)

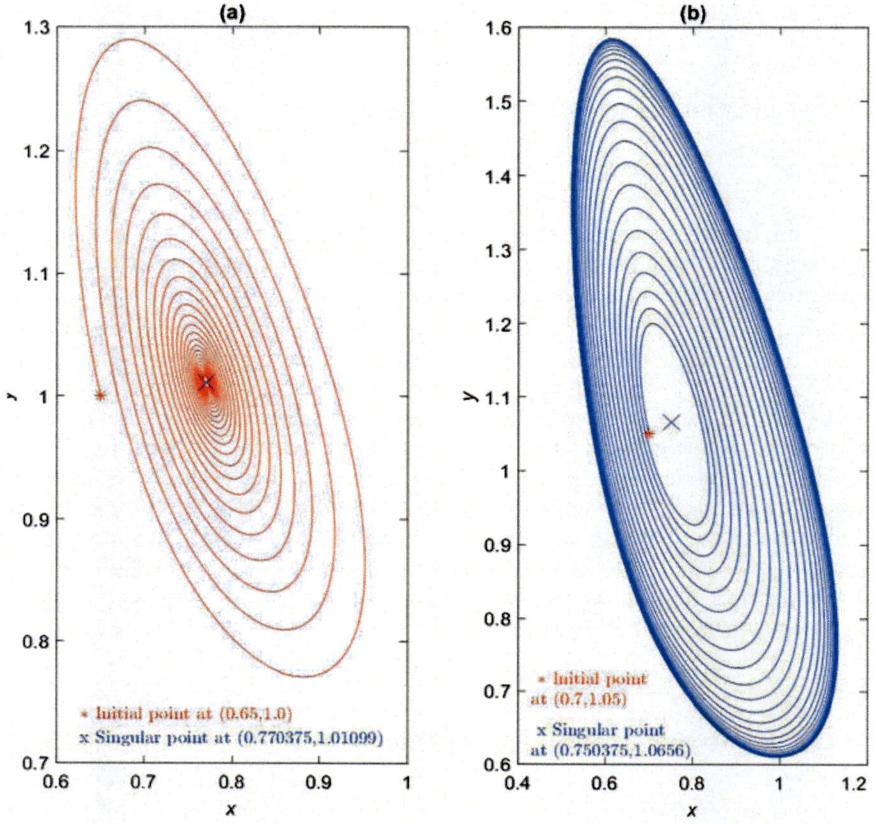

Fig. 1 **a** Stable and **b** unstable foci of system (1)

$$\frac{a-b}{a+b} - (a+b)^2,$$

is zero. This condition also satisfies that the real part of the eigenvalues is zero. The second necessary condition is for $k(a, b)$ given in (3) to be negative. These two conditions, with additional two on parameters a and b for the chemical system, $a, b > 0$, form the following system of semi-algebraic equations:

$$\begin{aligned}&(a > 0) \wedge (b > 0) \wedge ((a+b)^3 - a + b = 0) \wedge \\ &(((a+b)^3 - a + b)^2 - 4(a+b)^4 < 0).\end{aligned} \tag{6}$$

System (16) can be solved using `Mathematica` routine `Reduce` [8] and we obtain

$$0 < a < 1 \text{ and } (a+b)^3 = a - b. \tag{7}$$

A commonly used approach for the determination of Hopf bifurcation is the computation of normal forms. However, in this work, we adopt an approach employing the Lyapunov function which we describe now. For a system

$$\begin{aligned} \dot{x} &= -y + P(x, y) = P_1(x, y), \\ \dot{y} &= x + Q(x, y) = Q_1(x, y), \end{aligned} \tag{8}$$

we can always find a function of the form

$$\Psi(x, y) = x^2 + y^2 + \sum_{j+k=3} \psi_{jk} x^j y^k,$$

such that

$$\frac{\partial \Psi}{\partial x} P_1(x, y) + \frac{\partial \Psi}{\partial y} Q_1(x, y) = g_1(x^2 + y^2)^2 + g_2(x^2 + y^2)^3 + \cdots . \tag{9}$$

Based on the Lyapunov Theorem of asymptotic stability [7], we determine the type of the focus stability by using the first nonzero coefficient g_i of the extension of

$$\frac{\partial \Psi}{\partial x} P_1(x, y) + \frac{\partial \Psi}{\partial y} Q_1(x, y).$$

Then, a focus is stable if g_i is negative, and unstable if g_i is positive [1].

If the system is of the form

$$\begin{aligned} \dot{x} &= a_1 x + b_1 y + P(x, y) = P_1(x, y), \\ \dot{y} &= c_1 x - a_1 y + Q(x, y) = Q_1(x, y), \end{aligned} \tag{10}$$

for which the trace of the linear approximation matrix is zero, the resulting expressions involve radicals. To avoid this, we search for a positive-definite Lyapunov function of the form

$$\Psi(x, y) = \alpha x^2 + \beta xy + \gamma y^2 + \sum_{j+k=3} \psi_{jk} x^j y^k$$

which satisfies (9). This is the case if the conditions

$$\alpha = -\frac{c_1 \beta}{2a_1}, \ \gamma = \frac{b_1 \beta}{2a_1} \tag{11}$$

hold, and it is known that the quadratic form

$$\alpha x^2 + \beta xy + \gamma y^2 \tag{12}$$

is positive-definite if $\alpha > 0$ and $4\alpha\gamma - \beta^2 > 0$. Inserting (11) into the expression

$$4\alpha\gamma - \beta^2, \tag{13}$$

we obtain $4\alpha\gamma - \beta^2 = \frac{-\beta^2(a_1^2+b_1c_1)}{a_1^2}$, and when the origin is a center or a focus for system (10), the quadratic form (12) is positive-definite [14].

Theorem 1 *The singular point at the origin (or at E) of system (5) (or (1)) satisfying conditions (7) is a stable focus.*

Proof When (16) are satisfied, the eigenvalues of the linear approximation matrix of system (1) are

$$\lambda_{1,2} = \pm\frac{\sqrt{-4(a+b)^4 + (-a+b+(a+b)^3)^2}}{2(a+b)}.$$

We look for a Lyapunov function up to degree 8,

$$\Psi_8(x, y) = \sum_{k+s=2}^{8} \psi_{ks} x^k y^s \tag{14}$$

satisfying the equation

$$\frac{\partial \Phi_8}{\partial x} F(x, y) + \frac{\partial \Phi_8}{\partial y} G(x, y) = g_1(x^2+y^2)^2 + g_2(x^2+y^2)^3 + g_3(x^2+y^2)^4. \tag{15}$$

One can see that

$$\alpha = \frac{a\beta}{a-b} \text{ and } \gamma = \frac{\beta}{2}.$$

This condition can be determined by equating the coefficients of the same monomials on both sides of equation (15).

Let $\beta = 2$, then $\alpha = \frac{2a}{a-b}$, $\gamma = 1$, and $4\alpha\gamma - \beta^2 = \frac{4(a+b)}{a-b} > 0$. For $a > b$, the quadratic form (12) of (14) is positive-definite.

The first nonzero coefficient g_i is

$$
\begin{aligned}
g_1 = {} & 4a(a+b)^2(12a^{14} + 168a^{13}b + b^6(1+b^2)(2b^2-5)(1+2b^2)(1+3b^2)+ \\
& + 2a^{12}(546b^2-1) + 6a^{11}b(728b^2-5) + 7a^9b(5-110b^2+3432b^4)+ \\
& + 2a^{10}(1-99b^2+6006b^4) + a^7(-57b+300b^3-3564b^5+41184b^7)+ \\
& + a^5b(-3+167b^2-966b^4-4356b^6+24024b^8) + ab^3(3+5b^2-123b^4- \\
& + 325b^6-90b^8+168b^{10}) + a^3b(9-2b^2+13b^4-1860b^6-1430b^8+ \\
& + 4368b^{10}) + a^8(-17+33b^2(5-60b^2+1092b^4))+ \\
& + a^6(6+b^2(-13+924b^4(-5+39b^2)))+ \\
& + a^4(-3+b^2(-25+3b^2(75-630b^2-990b^4+4004b^6)))+ \\
& + a^2b^2(-9+b^2(24+b^2(-167+42b^2(-25-11b^2+26b^4)))))/ \\
& ((a-b)(2a^6+12a^5b+b^2(2+b^2)(1+2b^2)+a^4(30b^2-3)+ \\
& + a^3(-4b+40b^3)+a^2(2+6b^2+30b^4)+4ab(-1+3(b^2+b^4))) \\
& (3a^{12}+36a^{11}b+b^6(1+b^2)(3+b^2)(1+3b^2)+ \\
& + 10a^9b(13+66b^2)+a^{10}(13+198b^2)+12a^7b(9+130b^2+198b^4)+ \\
& + a^8(17+585b^2+1485b^4)+2ab^5(3+42b^2+65b^4+18b^6)+ \\
& + 4a^3b^3(5+113b^2+390b^4+165b^6)+a^2b^4(5+248b^2+585b^4+198b^6)+ \\
& + a^6(11+304b^2+2730b^4+2772b^6)+a^5(38b+508b^3+3276b^5+ \\
& + 2376b^7)+3a^4(16+5b^2(3+38b^2+182b^4+99b^6)))).
\end{aligned}
$$

The semi-algebraic system

$$
\begin{aligned}
& (g_1 \geq 0) \wedge (a > 0) \wedge (b > 0) \wedge ((a+b)^3 - a + b = 0) \wedge \\
& (((a+b)^3 - a + b)^2 - 4(a+b)^4 < 0).
\end{aligned} \tag{16}
$$

is an unsolvable system (checked with `Reduce` of `Mathematica`). Since $g_1 < 0$, the derivative with respect to a vector field is negative-definite. Hence the focus is stable [5].

Next theorem summarizes the conditions for the existence of Hopf bifurcation in system (1).

Theorem 2 *For the parameters, a and b, that satisfy the conditions given in (7), Hopf bifurcation can occur at the singular point* $E = (a+b, \frac{a}{(a+b)^2})$ *of system (1). The bifurcation is always supercritical, i.e., a stable limit cycle is born from E.*

Proof As demonstrated in the proof of the Theorem 1, the singular point that satisfies conditions (7) is a stable focus. By slightly varying the parameter b, we slightly perturb the system (1), changing the real parts of the eigenvalues (2) to positive. Hence, point E becomes an unstable focus, and the results is a stable limit cycle [3].

3.1 Numerical Example

To show the existence of stable limit cycles, we choose parameters a and b as $a = 0.6$ and $b = 0.16037459017229974$. The corresponding system (1) has singular point at $E = (0.760375, 1.03776)$, the eigenvalues given in (2) are $\lambda_{1,2} = \pm 0.760375i$ and the coefficient g_1 of (15) is negative, $g_1 = -0.17187 < 0$. As we can see in Fig. 2a, the trajectories move towards the singular point E, i.e. E is stable focus. Now, we perturb b slightly as $b = 0.16037459017229974 - \frac{1}{100} = 0.150375$. Then, eigenvalues become $\lambda_{1,2} = 0.0180689 \pm 0.750157i$ with positive real parts. When we choose initial point $(0.7, 1.05)$, we see that the trajectory moves away from singular point E, on the other hand, the second trajectory plotted from another initial point $(0.6, 0.6)$ moves towards the singular point E. Both trajectories approach to the limit cycle as seen in Fig. 2b.

Note that if the eigenvalues are pure imaginary, the local phase portrait in the neighbourhood of the singularity can not be a center, since $g_1 \neq 0$ for positive values of a and b for system (1).

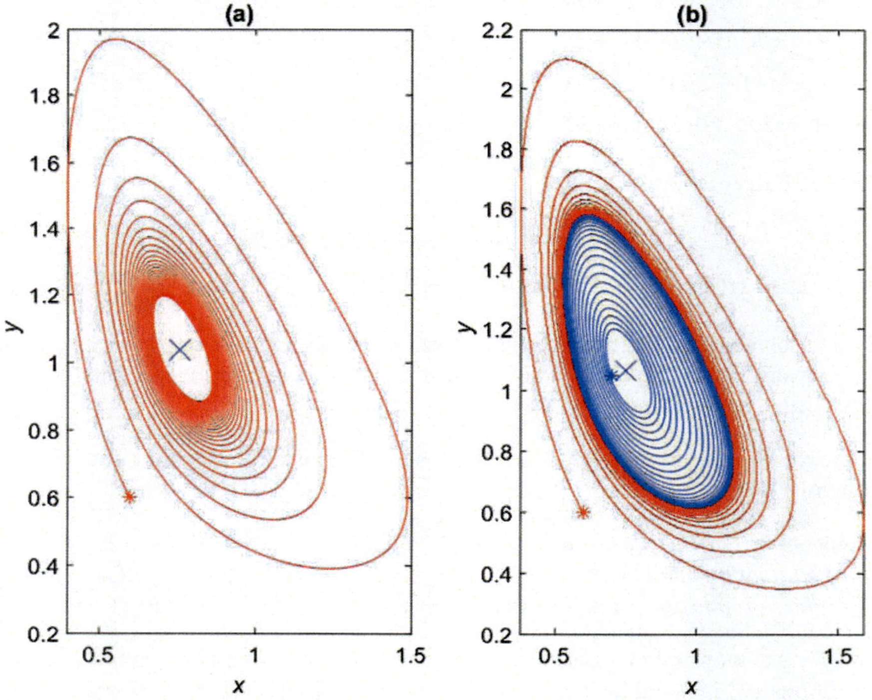

Fig. 2 **a** Stable focus for parameter values $a = 0.6$ and $b = 0.16037459017229974$. **b** A supercritical Hopf bifurcation appearing for system (1) with $a = 0.6$ and $b = 0.150375$

Acknowledgements This work is supported by the Slovenian Research Agency (research core no. “P1-0306” and the project no. “BI-TR/19-22-003”) and the Scientific and Technological Research Council of Turkey (TUBITAK) under project no. 119F017.

References

1. I.K. Aybar, O.O. Aybar, B. Fercec, V.G. Romanovski, S. Swarup Samal, A. Weber, Investigation of invariants of a chemical reaction system with algorithms of computer algebra. MATCH Commun. Math. Comput. Chem. **74**, 465–480 (2015)
2. L.N.M. Duysens, J. Amesz, Fluorescence spectrophotometry of reduced phosphopyridine nucleotide in intact cells in the near-ultraviolet and visible region. Biochim. Biophys. Acta **24**(1), 19–26 (1957)
3. B. Ferčec, Integrability and local bifurcations in polynomial systems of ordinary differential equations. Ph.D. Thesis, 2013
4. T.-W. Hwang, H.-J. Tsai, Uniqueness of limit cycles in theoretical models of certain oscillating chemical reactions. J. Phys. A **38**(38), 8211–8223 (2005)
5. Y.A. Kuznetsov, *Elements of Applied Bifurcation Theory* (Springer, New York, 1995)
6. Y. Li, Steady-state solution for a general Schnakenberg model nonlinear anal. Real World Appl. **12**, 1985–1990 (2011)
7. A.M. Liapunov, *Stability of Motion*, with a contribution by V. Pliss. Translated by F. Abramovici, M. Shimshoni (New York, Academic Press, 1996)
8. Mathematica 9.0 Wolfram Research Inc., https://www.wolfram.com
9. MATLAB 9.7 MathWorks 2019, https://www.mathworks.com
10. J.D. Murray. *Mathematical Biology: I. An Introduction*, 3rd edn. (Springer, New York, 2002)
11. L. Perko, *Differential Equations and Dynamical Systems, Texts in Applied Mathematics*, 7th edn. (Springer-Verlag, New York, 2001)
12. V.G. Romanovski, D.S. Shafer, *The Center and Cyclicity Problems, A Computational Algebra Approach* (Birkhauser, Boston-Basel-Berlin, 2009)
13. J. Schnakenberg, Simple chemical reaction systems with limit cycle behaviour. J. Theor. Biol. **81**(3), 389–400 (1979)
14. Y. Xia, M. Grašič, W. Huang, V.G. Romanovski, Limit cycles in a model of olfactory sensory neurons. Int. J. Bifurcat. Chaos **29**(3), 1950038 (2019)

Fractality and the Internal Dirichlet Problem

Nilufar A. Azamova, Elena S. Alekseeva, Alexander A. Potapov, and Alexander A. Rassadin

Abstract In this chapter influence of fractality on solution of the two-dimensional internal Dirichlet problem is analyzed. Two different situations are considered namely the first of them deals with fractal boundary condition on the unit disk. In this case exact solution of the Laplace equation proves to obey to some analog of the de Rham functional equation. Also norm and the Dirichlet integral for this solution has been estimated. In the second situation boundary condition is supposed to be regular but boundary of the domain is fractally perturbed. For clarification of this case both approximate conformal mapping technique and the Potapov concept of physical fractals has been applied.

Abbreviations

IDP	The internal Dirichlet Problem
WF	The Weierstrass Function
TWF	The truncated Weierstrass Function

N. A. Azamova · E. S. Alekseeva
Universitetskaya str., 4, Tashkent, Uzbekistan, Vuzgorodok 100174
e-mail: kometarella@mail.ru

A. A. Potapov (✉)
V. A. Kotelnikov Institute of Radio Engineering and Electronics, Russian Academy of Sciences, Moscow, Russia
e-mail: potapov@cplire.ru

JiNan University, Guangzhou 510632, China

Mokhovaya str., 11-7, 125009 Moscow, Russia

A. A. Rassadin
Gagarin ave., 23, 603950 Nizhnij Novgorod, Russia
e-mail: brat_ras@list.ru

C. H. Skiadas et al. (eds.), *13th Chaotic Modeling and Simulation International Conference*, Springer Proceedings in Complexity,
https://doi.org/10.1007/978-3-030-70795-8_9

1 The internal Dirichlet problem and fractality of boundary condition

In this section we discuss the internal Dirichlet problem (IDP) on the unit disk Ω_0.

First of all let us consider the two-dimensional Laplace equation:

$$\frac{\partial^2 u}{\partial r^2} + \frac{1}{r}\frac{\partial u}{\partial r} + \frac{1}{r^2}\frac{\partial^2 u}{\partial \varphi^2} = 0, \quad r < 1, \tag{1}$$

where r and φ are polar coordinates on this disk;

$u(r, \varphi)$ is unknown function.

Further Eq. (1) ought to be provided by boundary condition:

$$u(1, \varphi) = W(\varphi). \tag{2}$$

A peculiarity of our point of view on IDP (1–2) is in consideration of fractality of its boundary condition.

As a model of fractal boundary condition we take the well-known Weierstrass function (WF):

$$W(\varphi) = \sum_{n=1}^{\infty} a^n \cdot \cos(b^n \cdot \varphi), \tag{3}$$

where a and b are its parameters.

To provide 2π-periodicity of this function over polar angle φ one is obliged to choose function parameter $b = 2, 3, 4, \ldots$.

If these parameters obey to inequalities $0 < a < 1$ and $a \cdot b \geq 1$ then WF (3) is a continuous but nowhere differentiable function (see [1] and references there in). Moreover in this case WF possesses by the following fractal dimension:

$$D_F = 2 + \frac{\ln a}{\ln b}. \tag{4}$$

Formula (4) demonstrates that one can vary fractal dimension of WF under fixed b continuously from $D_F = 1$ to $D_F = 2$ by means of changing of its parameter a from $a = 1/b$ to $a = 1$. That is why we choose WF as the model of fractal boundary condition for IDP (1–2).

General solution of the Laplace Eq. (1) without singulatity in the center of the unit disk is equal to [2]:

$$u(r, \varphi) = \sum_{m=0}^{\infty} r^m \cdot (A_m \cdot \cos(m \cdot \varphi) + B_m \cdot \sin(m \cdot \varphi)). \tag{5}$$

Substituting $r = 1$ into series (5) and comparing the result with WF (3) it is easy to find that exact solution of IDP (1–3) is equal to:

$$u(r, \varphi) = \sum_{n=1}^{\infty} a^n \cdot r^{b^n} \cdot \cos(b^n \cdot \varphi). \tag{6}$$

In particular in the center of the disk $u(0, \varphi) = 0$.

Moreover one can check that function (6) obeys to the next functional equation:

$$u(r, \varphi) = a \cdot r^b \cdot \cos(b \cdot \varphi) + a \cdot u(r^b, b \cdot \varphi), \tag{7}$$

expressing its self-similarity.

Substituting $r = 1$ into functional Eq. (7) and using formula (2) one can easily obtain the de Rham functional equation for WF [1]:

$$W(\varphi) = a \cdot \cos(b \cdot \varphi) + a \cdot W(b \cdot \varphi). \tag{8}$$

Further let us calculate on the solution (6) the well-known Dirichlet integral:

$$I[u] = \iint_{\Omega_0} (\nabla u)^2 r dr d\varphi, \tag{9}$$

where ∇ is the operator of two-dimensional gradient.

It is easy to see from functional series (6) that vector $\nabla u(1, \varphi)$ consists from nowhere differentiable functions therefore it is convenient to calculate functional (9) as the following limit:

$$I[u] = \lim_{r \to 1-0} \int_0^{2\pi} u(r, \varphi) \cdot \frac{\partial u}{\partial r}(r, \varphi) \cdot r \cdot d\varphi. \tag{10}$$

Under $b = 2, 3, 4, \ldots$ functions $\{\cos(b^n \varphi)\}_{n=1}^{\infty}$ are orthogonal on the interval $[0, 2\pi]$:

$$\int_0^{2\pi} \cos(b^n \varphi) \cdot \cos(b^m \varphi) \cdot d\varphi = \pi \cdot \delta_{nm}, \tag{11}$$

hence using this relation one can obtain from (10) that the Dirichlet integral is equal to:

$$I[u] = \pi \cdot \sum_{n=1}^{\infty} (a^2 \cdot b)^n. \tag{12}$$

It is obvious that numerical series (12) converges if $a^2b < 1$ and diverges otherwise. We underline that under growing value of parameter b length of segment of convergence for a: $a \in (1/b, 1/\sqrt{b})$ tends to zero. One can rewrite condition of convergence for sum (12) via fractal dimension (4) of WF:

$$I[u] = \begin{cases} \frac{\pi \cdot a^2 \cdot b}{1 - a^2 \cdot b}, & 1 \le D_F < \frac{3}{2} \\ \infty, & \frac{3}{2} \le D_F < 2 \end{cases}. \tag{13}$$

Formula (13) means that if boundary condition (2) of IDP in some sense is "weakly" nondifferentiable then series (12) is convergent and if WF (3) is "strongly" nondifferentiable then series (12) is divergent.

Let us now consider norm of function (6) in Hilbert space $L^2(\Omega_0)$:

$$||u|| = \sqrt{\iint\limits_{\Omega_0} u^2(r, \varphi) \cdot r dr d\varphi}. \tag{14}$$

Using formula (11) it is not difficult to find that value (14) is equal to:

$$||u|| = \sqrt{\frac{\pi}{2} \cdot \sum_{n=1}^{\infty} \frac{a^{2n}}{b^n + 1}}. \tag{15}$$

Under $b \ge 2$ numerical series contained in the right-hand side of the formula (15) can be majorized by sum $\sum\limits_{n=1}^{\infty} (a^2 b^{-1})^n$ of convergent geometrical progression. Thus norm (14) can be estimated as follows:

$$||u|| \le \sqrt{\frac{\pi}{2} \cdot \frac{a^2}{b - a^2}}. \tag{16}$$

Upper bound (16) is valid under all admissible values of parameter a.

In practice to deal with solution (6) of IDP (1–2) with boundary condition (3) possessing by fractal dimension (4) it is convenient to transfer from mathematical fractals to physical ones [1]. The procedure of this transfer is presented on Fig. 1. Of course this approach destroys relations (7) and (8) but it brings new advantages namely let one take only a limited number N of terms of the series (3) then the resulting function:

$$W_N(\varphi) = \sum_{n=1}^{N} a^n \cdot \cos(b^n \cdot \varphi) \tag{17}$$

will be differentiable as many times as required.

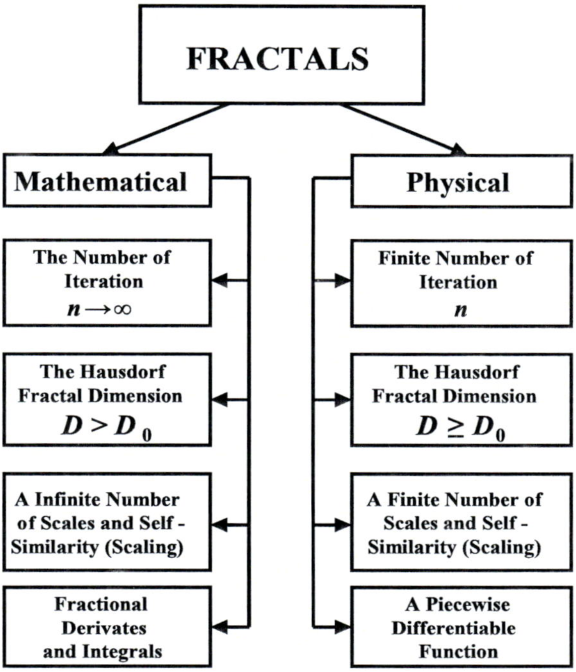

Fig. 1 The Potapov scheme of division of fractals on mathematical and physical

Using inequality $|\cos\varphi| \leq 1$ which is valid for all real φ it is not difficult to establish that for any fixed $\delta > 0$ inequality

$$|W(\varphi) - W_N(\varphi)| < \delta \tag{18}$$

is true under

$$N > \left[\frac{\ln \delta(1-a)}{\ln a}\right] + 1. \tag{19}$$

In this case we shall call function (17) as the truncated Weierstrass function (TWF). In other words if one takes number N for TWF obeying to condition (19) then such TWF approximates WF in accordance with inequality (18).

Graph of the TWF under $a = 0.75$ and $b = 2$ with $\delta = 0.01$ is presented on Fig. In this case $D_F \approx 1.58$ and $N = 21$.

At last let us apply procedure of truncation for series (6) with the same number of terms N as in condition (19):

$$u_N(r, \varphi) = \sum_{n=1}^{N} a^n \cdot r^{b^n} \cdot \cos(b^n \cdot \varphi). \tag{20}$$

Because of $r < 1$ it is easy to estimate that:

$$|u(r, \varphi) - u_N(r, \varphi)| \leq \frac{a^{N+1}}{1-a} \cdot r^{b^{N+1}}. \tag{21}$$

Inequality (21) means that in this case function (20) approximates function (6) even better then TWF approximates WF. From this inequality one can see that in fact fractality of exact solution (6) really exists only in narrow ring in the vicinity of the unit circle $\Gamma_0 : |z| = 1$ which is the boundary of the domain Ω_0. Moreover it is clear that under $r << 1$ it is enough only the first term of the series (6) to describe behaviour of exact solution of IDP under consideration quite precisely.

Graph of function (20) with parameters corresponding to the TWF presented on Fig. 2 is shown on Fig. 3.

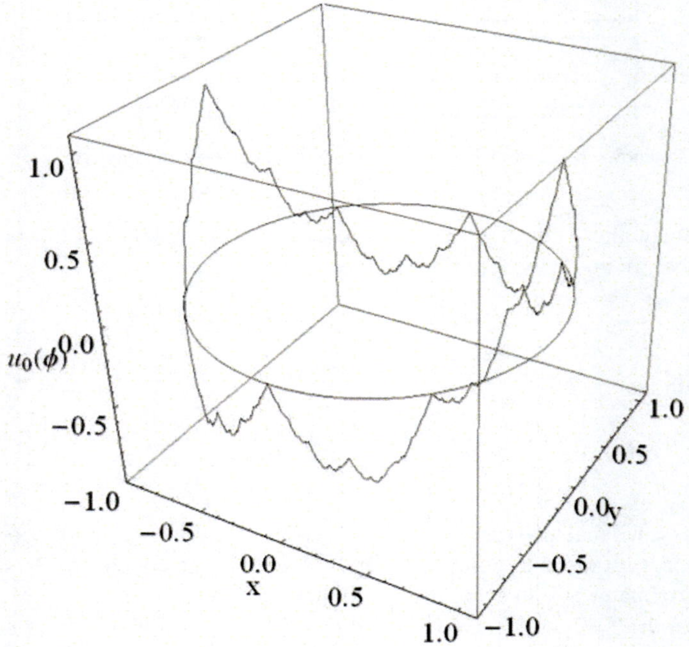

Fig. 2 Graph of the boundary condition

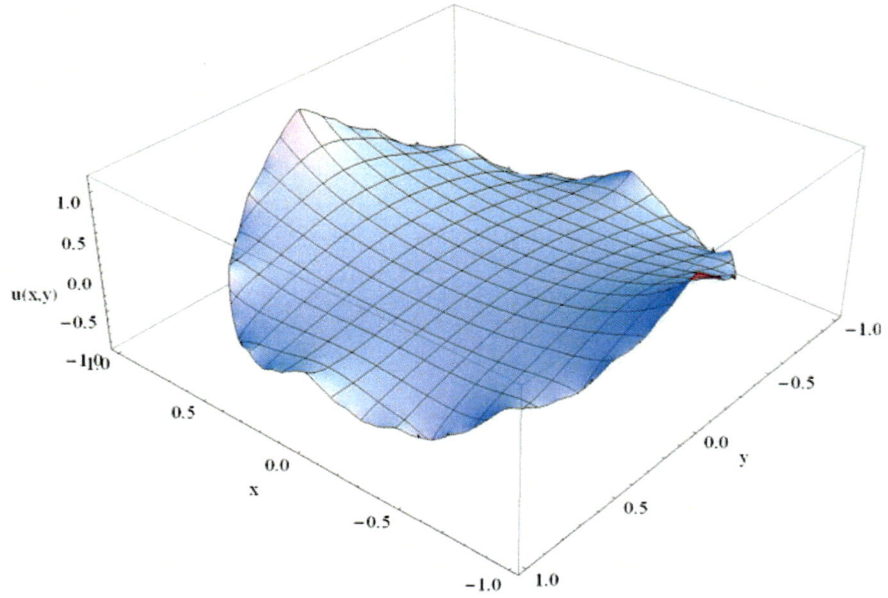

Fig. 3 Graph of exact solution

2 The Internal Dirichlet Problem on the Nearly Circular Domain Bounded by Fractal Curve

In this section we consider IDP on the nearly circular domain:

$$\frac{\partial^2 u}{\partial x^2} + \frac{\partial^2 u}{\partial y^2} = 0, \quad z = x + iy \in \Omega_\varepsilon, \quad u|_{z \in \Gamma_\varepsilon} = u_0(z), \tag{22}$$

where Ω_ε is star-shaped domain on complex plane C containing the origin of coordinates;

$\Gamma_\varepsilon = \partial\Omega_\varepsilon$ is boundary of domain Ω_ε;

ε is a small parameter ($0 < \varepsilon << 1$) characterizing proximity of domain Ω_ε to the unit disk Ω_0;

$u(z, \varepsilon)$ is unknown function;

$u_0(z)$ is boundary condition.

Let one choose polar equation of Γ_ε in the next form:

$$r(\varphi) = 1 - \varepsilon \cdot \chi(\varphi), \quad \varphi \in [0, 2\pi], \tag{23}$$

where $\chi(\varphi)$ is 2π-periodic function over polar angle φ and let one suppose that $\chi \in C^2([0, 2\pi])$ hence closed curve (23) differs slightly from the unit circle Γ_0 both its location and its curvature.

Further the Riemann theorem [3] claims that in this case there is a holomorphic function $w = f(z, \varepsilon)$ realizing conformal mapping of domain Ω_ε on the unit disk Ω_0.

Moreover exact solution of IDP (22) can be written via this function $f(z, \varepsilon)$ as follows [4]:

$$u(z, \varepsilon) = \mathrm{Re}\left[\frac{1}{2 \cdot \pi \cdot i} \oint\limits_{\Gamma_\varepsilon} u_0(\zeta) \cdot \frac{f(\zeta, \varepsilon) + f(z, \varepsilon)}{f(\zeta, \varepsilon) - f(z, \varepsilon)} \cdot \frac{f'_\zeta(\zeta, \varepsilon)}{f(\zeta, \varepsilon)} \cdot d\zeta\right]. \tag{24}$$

Generally speaking one may derive this mapping $f(z, \varepsilon)$ explicitly in the framework of formalism of the harmonic moments of exterior domain $\overline{C} \backslash \Omega_\varepsilon$ [5] but this way is too hard. On the other hand due to representation (23) of Γ_ε we can restrict ourselves by construction of approximate conformal mapping of the nearly circular domain on unit disk.

In this case for function $w = f(z, \varepsilon)$ realizing conformal mapping of domain Ω_ε on the unit disk Ω_0 and obeying to conditions $f(0, \varepsilon) = 0$ and $f'_z(0, \varepsilon) > 0$ the following asymptotic formula is known to be valid [3, 6]:

$$w = f(z, \varepsilon) = z + \varepsilon \cdot f_1(z) + O(\varepsilon^2), \tag{25}$$

where

$$f_1(z) = z \cdot \int\limits_0^{2\pi} \chi(\varphi) \cdot \frac{\exp(i\varphi) + z}{\exp(i\varphi) - z} \cdot \frac{d\varphi}{2\pi}. \tag{26}$$

A feature of our approach is taking into account a fractality of domain's boundary Γ_ε. On the other hand in accordance with methods developed in [3, 6] for derivation of the desired mapping the boundary Γ_ε must be quite smooth. In order to overcome this obstacle we apply concept of physical fractals (see Fig. 1) namely we use as function $\chi(\varphi)$ TWF (17). Of course this TWF must contain enough terms in correspondence with inequality (19) to approximate the input WF with fixed accuracy (18). Graph of typical shape of such boundary is shown on Fig. 4, dotted line corresponding to the unit circle Γ_0.

Calculating integral (26) with function $W_N(\varphi)$ we obtain the next lacunary polynomial of complex variable z:

$$f_1(z) = \sum_{n=1}^{N} a^n \cdot z^{b^n+1}. \tag{27}$$

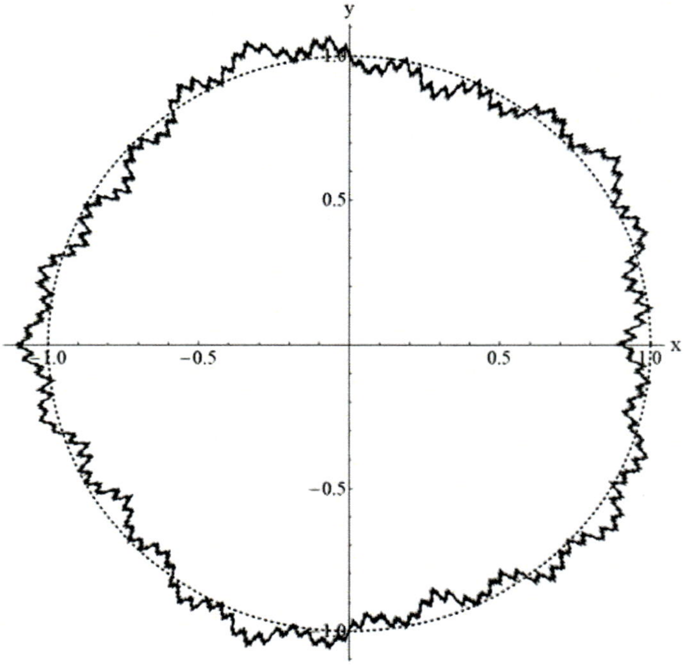

Fig. 4 Graph of the domain with fractal boundary

Substituting expression (25) with function (27) into formula (24) and expanding its integrand on ε one can find approximate solution of IDP (22–23) in the following form:

$$u(z,\varepsilon) = u^{(0)}(z) + \varepsilon \cdot u^{(1)}(z) + O(\varepsilon^2), \quad z \in \Omega_\varepsilon. \tag{28}$$

In formula (28)

$$u^{(0)}(z) = \mathrm{Re}\left[\frac{1}{2\cdot\pi\cdot i}\oint\limits_{\Gamma_\varepsilon} u_0(\zeta)\cdot\frac{\zeta+z}{\zeta-z}\cdot\frac{d\zeta}{\zeta}\right] \tag{29}$$

and

$$u^{(1)}(z) = u_1^{(1)}(z) + u_2^{(1)}(z) + u_3^{(1)}(z), \tag{30}$$

where

$$u_1^{(1)}(z) = \mathrm{Re}\left[\frac{1}{2\cdot\pi\cdot i}\oint\limits_{\Gamma_\varepsilon} u_0(\zeta)\cdot\left(f_1'(\zeta) - \frac{f_1(\zeta)}{\zeta}\right)\cdot\frac{\zeta+z}{\zeta-z}\cdot\frac{d\zeta}{\zeta}\right], \tag{31}$$

$$u_2^{(1)}(z) = -\mathrm{Re}\left[2\cdot z\cdot\frac{1}{2\cdot\pi\cdot i}\oint\limits_{\Gamma_\varepsilon}\frac{u_0(\zeta)\cdot f_1(\zeta)\cdot d\zeta}{\zeta\cdot(\zeta-z)^2}\right] \tag{32}$$

and

$$u_3^{(1)}(z) = \mathrm{Re}\left[2\cdot f_1(z)\cdot\frac{1}{2\cdot\pi\cdot i}\oint\limits_{\Gamma_\varepsilon}\frac{u_0(\zeta)\cdot d\zeta}{(\zeta-z)^2}\right]. \tag{33}$$

Zero-order term in asymptotic expansion (28) looks like the well-known Poisson integral for the unit disk Ω_0 [3, 4], but we stress that integration in expression (29) is performed along the curve Γ_ε with polar Eq. (23) hence to calculate integral in formula (29) one ought to substitute into this formula $\zeta(\varphi) = (1-\varepsilon\cdot\chi(\varphi))\cdot\exp(i\cdot\varphi)$.

Integrals (31), (32) and (33) representing contribution of boundary roughness into the first order term (30) in asymptotic expansion (28) must be estimated in the same manner.

This calculation seems to be very awkward but in practice contour integrals in formulae (31), (32) and (33) can be found as sums over residues of its integrands in domain Ω_ε.

To demonstrate this technique let us derive the influence of fractal roughness of boundary of round cylindrical hole in conductor on distribution in it of electrostatic potential and electric field strength. And let us remind that approximate conformal mapping (25) of domain with fractal boundary is determined by function (27) corresponding to TWF (17).

According to general principles of electrostatics let us set:

$$u_0(z) = 1, \quad z \in \Gamma_\varepsilon. \tag{34}$$

It is easy to check that in this case calculation in correspondence with formula (29) gives us that:

$$u^{(0)}(z) = 1, \quad z \in \Omega_\varepsilon. \tag{35}$$

Further let one consider the following function:

$$f_1'(z) - \frac{f_1(z)}{z} = \sum_{n=1}^{N}(a\cdot b)^n\cdot z^{b^n}. \tag{36}$$

Lacunary polynomial (36) is a holomorphic function of complex variable z on Ω_ε therefore the result of application of the theory of residues to integral in expression (31) is equal to:

$$u_1^{(1)}(z) = 2 \cdot \mathrm{Re}\left[f_1'(z) - \frac{f_1(z)}{z} \right]. \tag{37}$$

Function $f_1(z)/z$ also has no singularities in domain Ω_ε hence using the Cauchy formula for derivative one can obtain from formula (32) that:

$$u_2^{(1)}(z) = -2 \cdot \mathrm{Re}\left[z \cdot \frac{d}{dz} \frac{f_1(z)}{z} \right]. \tag{38}$$

At last in accordance with the Cauchy theorem after homotopy of curve Γ_ε into the unit circle Γ_0 in integral (33) it is not difficult to see that

$$u_3^{(1)}(z) = 0. \tag{39}$$

Thus combining formulae (37), (38) and (39) in correspondence with expression (30) we obtain the following unexpected result:

$$u^{(1)}(z) = 0, \quad z \in \Omega_\varepsilon \tag{40}$$

Formula (40) means that electrostatic potential of this domain varies only in the second order on ε:

$$u(z, \varepsilon) = 1 + O(\varepsilon^2), \quad z \in \Omega_\varepsilon. \tag{41}$$

It is immediately succeed from expression (41) that in any point $z \in \Omega_\varepsilon$ electric field strength is equal to:

$$\vec{E}(z, \varepsilon) = -\nabla u(z, \varepsilon) = O(\varepsilon^2). \tag{42}$$

Formula (42) gives us useful consequence for technique of electric measurements namely under electrostatic screening using round cylindrical hole in conductor one may not take care about precision of boundary processing.

After some generalization on three spatial dimensions the ideology of calculations developed in this section may be useful for description of electrically charged fractal core–shell nanoparticles and elastic tensions around such structures [7].

Acknowledgements AAP and AER are grateful to the Russian Foundation for Basic Research grant No. 18-08-01356-a. AAP is grateful to the China grant "Leading Talent Program in Guangdong Province" (No. 00201502, 2016-2020) JiNan University (China, Guangzhou).

References

1. A.A. Potapov, V. Gulyaev Yu, S.A. Nikitov, A.A. Pakhomov, V.A. German, The Modern Methods of Image Processing, ed. by A.A. Potapov (FIZMATLIT, 2008) [in Russian]
2. P.M. Morse, H. Feshbach, Methods of theoretical physics. Part II. McGraw-Hill (1953)
3. M.A. Lavrentyev, B.V. Shabat, Methods of the Theory of Functions of a Complex Variable—Nauka (1973) [in Russian]
4. V.S. Vladimirov, Equations of Mathematical Physics—Marcel Dekker Inc. (1971)
5. S. Natanson, Towards an effectivisation of the Riemann theorem. Ann Global Anal. Geom. **28**(3), 233–255 (2005)
6. G.V. Siryk, On a conformal mapping of near domains. Uspekhi Matem Nauk. **9**(5), 57–60 (1956). ([in Russian])
7. A.V. Shishulin, A.A. Potapov, V.B. Fedoseev, Phase equilibria in fractal core-shell nanoparticles of Pb5(VO4)3Cl–Pb5(PO4)3Cl system: The influence of size and shape, in *Advances in Artificial Systems for Medicine and Education II*, ed. by Z. Hu, S. Petoukhov, M. He (Part of the Advances in Intelligent Systems and Computing book series—AISC, v. 902). (Springer Int. Publ., Cham, Switzerland, 2019), pp. 405–413

Selective Transport of Suspending Micro-Particles in an Oscillating Fluid Through Micro-Channels

Philippe Beltrame

Abstract This paper analyzes the possibility to obtain selective transport of microparticles depending their size. The particles are suspended in a fluid confined in modulated channels and a periodic pumping moves back and forth the fluid without net displacement. Using numerical simulation and bifurcation analysis tools, we show the existence of particle drift under the Stokes assumption of the fluid flow. For specific parameter ranges, the particle transport can be selective. The transport solution and the selectivity are related to (de)synchronization transitions in forced non-linear oscillators. We reveal that chaotic transitions are a key factor to drop from a bounded dynamics to a net transport. This transport phenomenon can be relevant for heavy particles in suspended in the air in microgravity environnement.

Keywords Particle transport · Synchronization · Chaotic dynamics · Bifurcation analysis · Microfluidic

1 Introduction

Sorting suspended particles in a fluid is an issue in many domains such as the food industry, medical analyses or wastewater treatment. Many processes are based on the microfiltration using a membrane. However, at high permeation rates, this method suffers from the accumulation of non-permeating particles above the membrane surface, thereby blocking the pores [1]. In recent decades, alternative techniques using flow in a periodic and asymmetric structure of micro-channels have been developed. In these systems, the particles are driven mainly by the viscous force. Micro-particles may drift from the streamline mainly due to the lift force. In a confined geometry, the lift effect is strongly dependent on the particle size and induces a selective trajectory. Based on this principle, passive micro-fluidic devices to sort the particles have been

P. Beltrame (✉)
UMR1114 EMMAH INRAE - Avignon Université, Avignon, France
e-mail: philippe.beltrame@univ-avignon.fr

C. H. Skiadas et al. (eds.), *13th Chaotic Modeling and Simulation International Conference*, Springer Proceedings in Complexity,
https://doi.org/10.1007/978-3-030-70795-8_10

developed such as branching channels, pinched flow fractionation, spiral channels or media with a periodic pattern of micro-posts [2–6]. The particle sorting is done using a continuous flow.

Such a method does not apply to the removal of specific particles from a basin. Particle selection using an oscillating flow, i.e. without net displacement of the fluid, is possible as shown in [7]. The particle transport occurs through a periodic structure of triangular columns. For a range of particle sizes, the lift force acts asymmetrically during the back and forth cycle of the fluid flow. This results in a drift that is orthogonal to the oscillating flow for a parameter range. Therefore in all systems the lift force is the key phenomenon of particle drift and sorting.

In contrast, the present paper focuses on longitudinal transport, i.e. the drift takes place along the axis of fluid oscillations. Indeed few studies are devoted to the onset of transport and to the kind of transition. These questions are useful for the design optimization of these devices.

We consider a micro-device similar to [8] where two basins are connected via modulated channels filled with a liquid. A periodic pumping confers a back and forth fluid motion dragging the particles in suspension. For oscillating Stokes flow, the 1D transport of particles is usually explained by the Stokes drift [9]. The particle follows the flow drift due to the traveling wave. In our context, the fluid is, however, driven by a standing wave which does not lead to fluid drift. The flow therefore needs to be ratchet like. In the early 2000s, the transport of overdamped particles in ratchets in many fields in physics was interpreted as a Brownian motor in which transport results from the action of noise in an asymmetric potential [10, 11]. Such a drift ratchet phenomenon may occur in the microfluidic context considered here [12] and the experiment in [8] corroborated this theory. Nevertheless, further experiments revealed that the thermal fluctuations are negligible and the experiment in [8] does not evidence transport due to a Brownian ratchet.

Recently, we highlighted different 1D transport mechanisms in a Stokes flow, called *ratchet flow*, for a simple model of inertial particle [13]. We showed that the spatial variations of the fluid flow induce a 'ratchet effect'. For instance, for moderate damping chaotic dynamics are a key component of the transport. However, the parameter domains of the transport require that the particle radius is not negligible compared to the channel radius [14]. Therefore, the drag coefficient depends on the channel walls and hence on the particle position. Such a variation may induce a friction ratchet [10, 15].

The goal of this paper was to determine whether such 1D transport mechanisms as in [13] exist for a particle radius comparable to the channel radius, and then to determine a possible dependence of the transport direction on the particle size. To answer these questions, we computed the friction for the 3D axisymmetric problem and we used bifurcation analysis tools and continuation of periodic orbit to provide a comprehensive overview of the dynamics in the phase space.

2 Modeling

Let us consider a L-periodically modulated channel infinitely extended along the line (Ox) through which a Newtonian fluid with the viscosity μ is T-periodically pumped. We call 'cell' the channel portion of length L (Fig. 1). The cell is axisymmetric and its radius $r(x)$ varies sinusoidally:

$$R(x) = R_m\,(1 + c_r \cos(2\pi x)) \tag{1}$$

where R_m is the mean radius and c_r is the channel camber. In this study, we fix the channel geometry such as $R_m = 0.14 \cdot L$ and $c_r = 0.56$ which corresponds to the shape shown in Fig. 1. We assume that the flow is a quasi-static Stokes flow. Then, a periodic pumping implies a periodic fluid velocity field in space and time. Moreover, the time dependence of the velocity field is governed by the periodic pumping $\mathbf{v}_0(\mathbf{r}, t) = \mathbf{u}_0(\mathbf{r})A(t)$, where $A(t)$ is the pumping amplitude and $\mathbf{u}_0(\mathbf{r})$ is L-periodic. We note $[p]$ the amplitude of the pressure difference between the cell inlet and outlet. Pressure, length and time are scaled by $[p]$, L and T respectively. We consider spherical particle of mass m with the adimensional radius r_p. If, in addition, we assume that the particle moves only along the axis and the particle does not rotate then the particle position $x(t)$ is governed by the dimensionless ODE:

$$\ddot{x} + \mathcal{P}_\gamma \hat{\gamma}(x)\dot{x} = \mathcal{P}_\gamma \mathcal{P}_v \hat{\gamma}(x)\hat{u}_{eq}(x)A(t). \tag{2}$$

In the latter equation, we have introduced two bifurcation dimensionless parameters:

$$\mathcal{P}_v = \frac{[p]T}{\mu} \tag{3a}$$

$$\mathcal{P}_\gamma = \frac{LT\mu}{m}. \tag{3b}$$

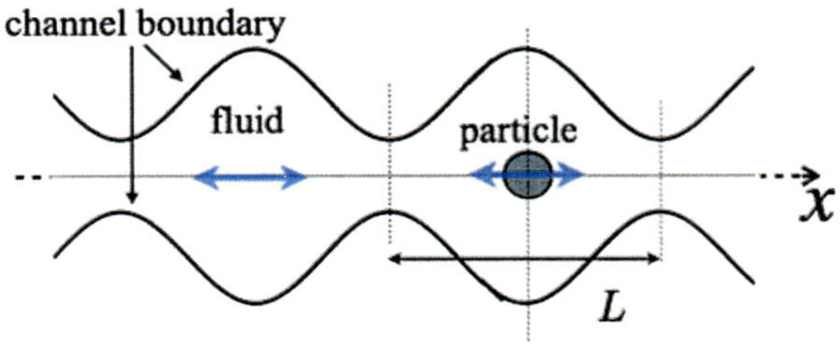

Fig. 1 Sketch of the problem: the particle translates along the x-axis of a periodic modulated channel. It is dragged by the periodic motion of a viscous fluid

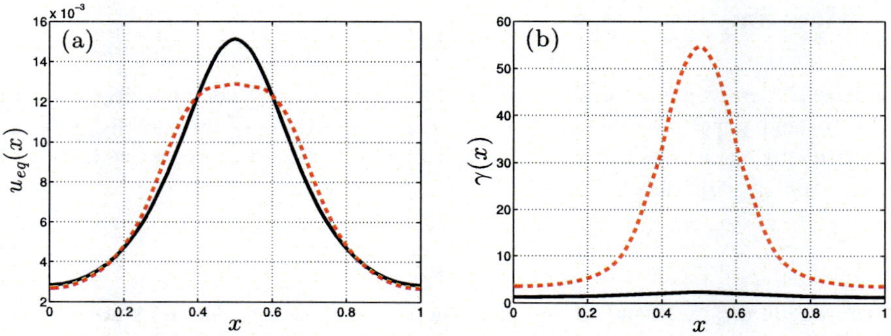

Fig. 2 Profile of (left panel) $\hat{u}_{eq}(x)$ and (right panel) $\hat{\gamma}(x)$ functions for two particle radii: (black line) $r_p = 0.05$ and (red dashed line) $r_p = 0.1$

The field $\hat{\gamma}(x) > 0$ is the normalized drag coefficient of the particle. It depends on the channel boundary and on the particle size [14, 16]. The field $\hat{u}_{eq}(x)$ is related to the flow field velocity without particle and also to the particle radius r_p. Figure 2 displays the fields $\hat{u}_{eq}(x)$ and $\hat{\gamma}(x)$ for a sinusoidal cell profile such as $c_r = 0.56\, r_{min} = 0.14$. The computation of these coefficients is detailed in [17, 18]. The particle size does not notably affect the veolcity field $\hat{u}_{eq}$ (Fig. 2a) whereas the friction is very sensitive to the particle radius (Fig. 2b). The friction is maximal in the narrow region of the cell and minimal in the larger region. For $r_m = 0.1$, the ratio between the maximum and minimum value is large in contrast to $r_m = 0.05$, for which $\hat{\gamma}(x)$ is almost constant. In the latter case, Eq. 2 is similar to the ratchet flow model studied in [13, 19, 20] and we expect similar transport dynamics. If $r_p = 0.1$, the large friction contrast may induce, in addition, a friction ratchet [10].

Note that the functions $u_{eq}(x)$ and $\gamma(x)$ are 1-periodic and have the parity-symmetry like the geometry of the problem. If, in addition, the pumping $A(t)$ varies sinusoidally, then the problem is invariant by the parity symmetry $x \to -x$. More precisely, if $x(t)$ is a trajectory given by Eq. 2 then $-x(t + 1/2)$ is also the solution for a symmetric initial condition. In order to break of the parity-symmetry, the back and forth phases of the pumping should be different, i.e. this means that $A(t + 1/2) \neq -A(t)$. Let us introduce the parameter α such as $0 \leq \alpha < 1$ and define the function $A(t)$:

$$A(t) = \left| \begin{array}{ll} 1 - \alpha & \text{, if } 0 \leq (t \text{ modulo } 1) < \alpha \\ \cos\left(2\pi \frac{t-\alpha}{1-\alpha}\right) - \alpha & \text{, if } \alpha \leq (t \text{ modulo } 1) < 1 \end{array} \right. \tag{4}$$

If $\alpha = 0$ is zero, $A(t) = \cos(2\pi t)$ and the problem is symmetric. Otherwise, the pressure difference is constant during the first step in the interval $[t_0, t_0 + \alpha[$ followed by a sinusoidal pumping in the interval $[t_0 + \alpha, t_0 + 1[$. In this case $A(t + 1/2) \neq -A(t)$. Note that the mean value of the pumping is still zero if $\alpha \neq 0$.

In this paper, we consider two particle sizes: $r_p = 0.05$ or $r_p = 0.1$. The transport dynamics are explored in the parameter space using the time integration and the path-following method of the periodic solutions. The bifurcation parameters are $\mathcal{P}_\gamma$, $\mathcal{P}_v$ and α. Branches of T-periodic solution on bifurcation diagrams are represented by the norm $||.||$ such as: $||s|| = [\frac{1}{T}\int_0^T(\dot{x}(t))^2 dt]^{1/2}$.

3 Drift at Large Drag $\mathcal{P}_\gamma$

In this section we analyze the transport solutions when the drag $\mathcal{P}_\gamma$ is large. According to [13], if $\mathcal{P}_v \ll 1$ we can prove that there are two periodic solutions of the particle motion which are centered at the extrema of the velocity field u_{eq} noted s_0 for the maximum and s_m for the minimum. In the symmetric case ($\alpha = 0$), by increasing $\mathcal{P}_v$, we find out a third solution noted s_a bifurcated from either s_0 or s_m via a spontenaeous symmetry breaking. Figure 3 shows the time evolution of the three solutions s_0, s_m and s_a. As in [13], one of the three solutions are stable and attracts all of the dynamics if $\alpha = 0$. Therefore, the transport solutions can occur only if $\alpha \neq 0$.

3.1 Transitions to Transport Solutions

We follow the periodic branches of solutions s_0 and s_m by varying α and we fix $\mathcal{P}_v = 1350$ (Fig. 4). The branches s_0 and s_m annihilate in a saddle-node bifurcation for $\alpha \simeq 0.2807$. A similar scenario arises starting from $\alpha = 1$ for which the pumping is zero: a pair of saddle orbits annihilate in a saddle-node for $\alpha \simeq 0.8143$. In the large range [0.3; 0.8], no solution is found. The saddle-node bifurcations correspond to the intermittent bifurcation type-I [21] as explained in [13]. The stroboscopic time evolution of the particle position at every period (Fig. 5) displays a regular descending

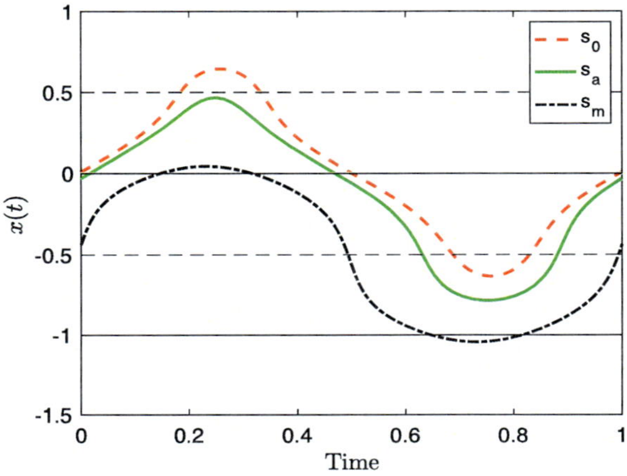

Fig. 3 Time evolution of the periodic solutions s_0, s_m and s_a for $r_p = 0.05$, $\alpha = 0$, $\mathcal{P}_\gamma = 79$ and $\mathcal{P}_v = 676$

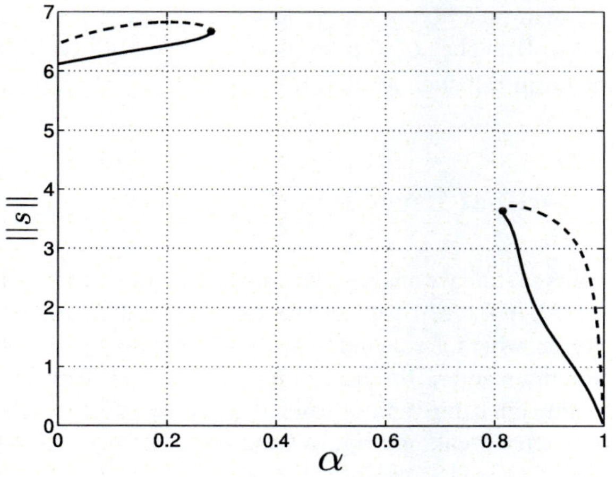

Fig. 4 Continuation of 1-periodic solutions for $r_p = 0.05$, $\mathcal{P}_v = 1350$, $\mathcal{P}_\gamma = 79$ by varying α. Plain [dashed] line indicates stable [unstable] orbit. Black dots indicate pitchfork bifurcations

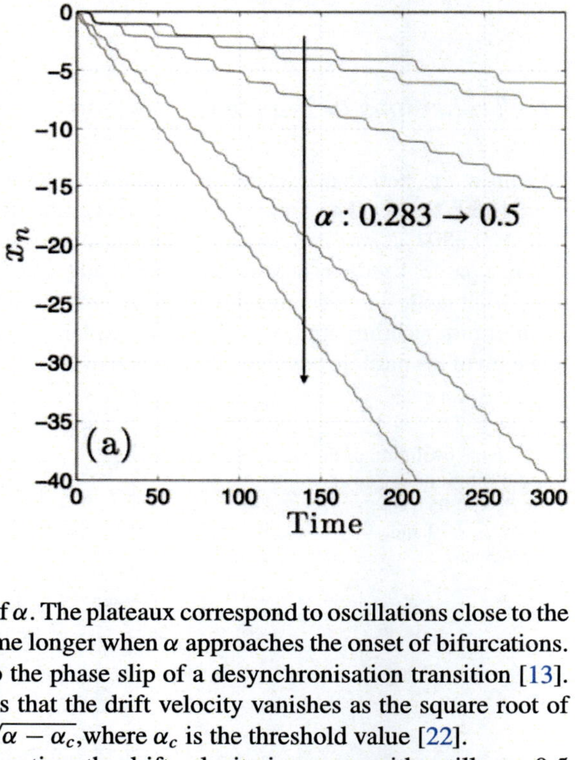

Fig. 5 Discrete particle positions $x_n = x(n)$, $n \in \mathbb{N}$ near the onset. Different values of $\alpha =$ 0.283; 0.285; 0.3; 0.4; 0.5, other parameters as in Fig. 4

staircase for different values of α. The plateaux correspond to oscillations close to the threshold. The plateaux become longer when α approaches the onset of bifurcations. Such a dynamics is similar to the phase slip of a desynchronisation transition [13]. A well-known consequence is that the drift velocity vanishes as the square root of the threshold distance: $c \propto \sqrt{\alpha - \alpha_c}$,where α_c is the threshold value [22].

According to the time integration, the drift velocity increases with α till $\alpha \simeq 0.5$ and remains almost constant till 0.6. By further increasing α the velocity decreases to zero when α approaches the critical value of the second saddle node. Consequently, the optimal transport is about $\alpha = 0.5$. In the next section, we fix α at 0.5 and we seek the parameter domains of particle transport.

Note that according to the discussion in [13], the particle drift phenomenon is part of a class of dissipative rocking ratchets for which the transport direction is determined by the asymmetry [23, 24]. In the current problem, it means that the direction of transport depends only on the sign of the α parameter, i.e. the kind of pumping. The particle drifts to negative values of x if $\alpha > 0$, otherwise the particle drifts to positive values.

3.2 Domain of Intermittent Drift

We explore the transport domain by varying the parameters $\mathcal{P}_v$ and $\mathcal{P}_\gamma$. We trace the saddle-node loci of the periodic solutions, in the $(\mathcal{P}_\gamma, \mathcal{P}_v)$ plane (Fig. 6) which represents the possible onset of transport. By varying $\mathcal{P}_\gamma$ the two saddle-nodes form a vertical band which ends at a minimal value of $\mathcal{P}_\gamma$ except if $\mathcal{P}_v$ is about 1000 (see Fig. 6). The transport arises in the region outside these bands and when the bands do not overlap (gray region in Fig. 6). Therefore, the transport domain is roughly a sector in the $(\mathcal{P}_v, \mathcal{P}_\gamma)$ plane. Then, the transport occurs if $\mathcal{P}_v$ is large enough, in other words, if the pumping amplitude is large. There is additional tapered vertical spaces for specific values of $\mathcal{P}_v$ for which transport may occur for large $\mathcal{P}_\gamma$ values. The specific $\mathcal{P}_v$ values are only slightly affected by the particle size: the tapered region occurs for $\mathcal{P}_v$ about 1500, 2200 and 2900 regardless of r_p.

In general, the domains of intermittent drift are qualitatively similar by varying r_p. However, the existence domain differs quantitatively, which allows to find specific parameters for which the drift arises while for the other particle size the dynamics is still periodic and so bounded. Therefore, the transport only occurs for specific particle sizes.

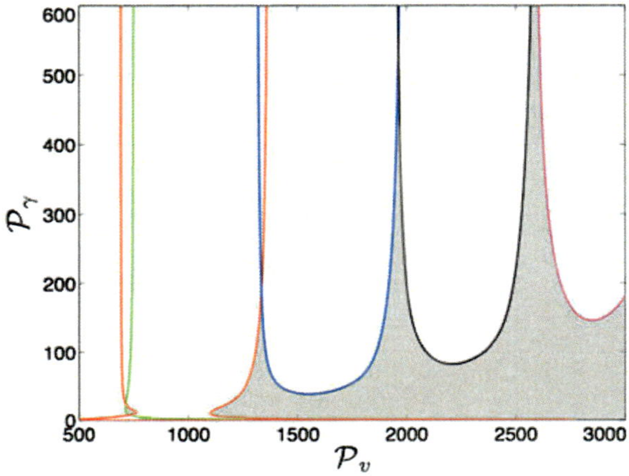

Fig. 6 Saddle-node loci of 1-periodic solutions in the $(\mathcal{P}_\gamma, \mathcal{P}_v)$ plane for $r_p = 0.1$. The gray region displays the domain of intermittent drift

4 Transport for Moderate $\mathcal{P}_\gamma$

In this section, we consider smaller values of $\mathcal{P}_\gamma$. Thus, the particle damping is smaller implying a larger desynchronization between fluid movement and the particle trajectory. This leads to more complex dynamics especially chaotic dynamics.

4.1 Synchronized Transport

We study the behavior of the intermittent drift when $\mathcal{P}_\gamma$ is decreasing. We observe that the drift velocity c increases and the discrete dynamics presents shorter and shorter plateaus as in Fig. 5. According to Fig. 7, the plateaus may disappear and the transport occurs almost with a constant velocity. The representation of the dynamics in the co-moving frame ($c = -1$) points out an intermittent and a regular behavior. Indeed, the dynamics in the co-moving frame is due to the phase slip at a synchronization transition for weakly nonlinear oscillator. It is a similar scenario as described in Sect. 3 but in the co-moving frame. The threshold is a saddle-node from which a pair of periodic solutions emerge. In the laboratory frame: these periodic solutions correspond to a synchronized transport solution: after a entire number n of time periods the particle moves by an entire number of spatial periods m, then the velocity is a rational $c = m/n$. This result is typical of phase locking (here is the c velocity) of a forced non-linear oscillator [22]. In the example of Fig. 7, we have $c = -1$, but we found also other velocities for other parameters: $c = -1/2$ and $c = -2$.

Another difference with the intermittent drift is that the synchronized transport can be not an attractor especially if $\mathcal{P}_\gamma$ is not large. For instance, for $\mathcal{P}_\gamma = 7.46$, we plotted the bifurcation diagram of the synchronized transport by varying the $\mathcal{P}_v$ parameter (Fig. 8). The transport emerges via a saddle-node bifurcation related to a

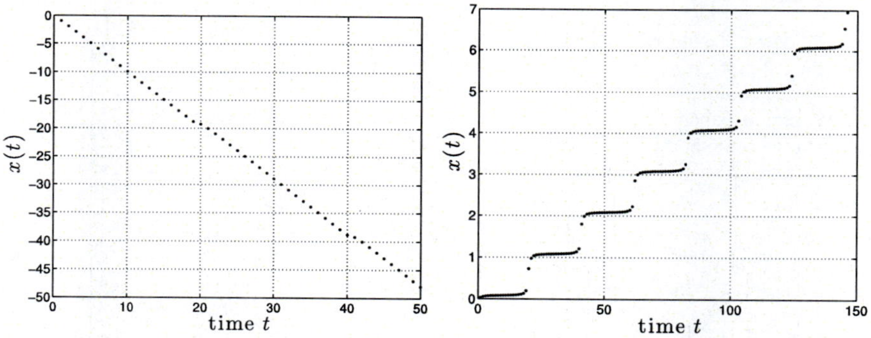

Fig. 7 [left] Discrete particle trajectory x_n near the onset of the synchronized transport $c = -1$. [right] The same discrete particle trajectory represented in the comoving frame $x_n + n$. Parameters are: $r_p = 0.1, \alpha = 0.5, \mathcal{P}_\gamma = 24.2, \mathcal{P}_v = 2250$

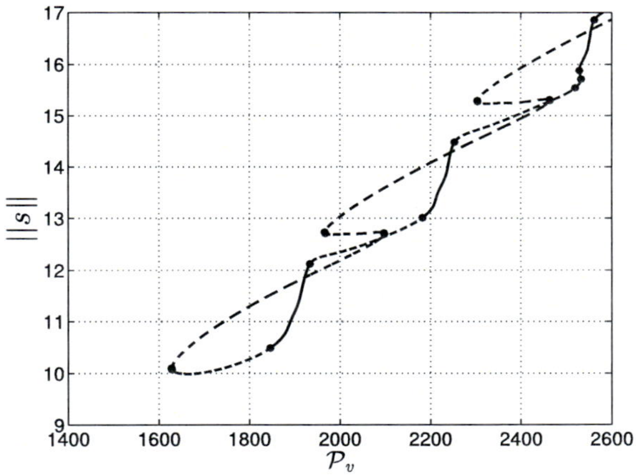

Fig. 8 Continuation of synchronized transport $c = -1$ with $\mathcal{P}_\gamma = 7.46$ and $r_p = 0.1$. Plain [dashed] lines indicate stable [unstable] solutions. Black dots indicate period-doubling bifurcations

synchronization phenomenon at about $\mathcal{P}_v = 1628$. From this saddle, two branches display zigzags with further saddle-nodes and period-doubling bifurcations. Due to the period doubling, the branch changes its stability. According to Fig. 8, there exist $\mathcal{P}_v$ ranges for which the transport $c = -1$ is a stable solution. From the period-doubling bifurcations a cascade of period-doubling bifurcations occurs leading to a chaotic dynamics. This dynamics being bounded in the co-moving frame the transport velocity remains equal to $c = -1$. The scenario is similar to the one described in [13] in this context and it involves universal results of (de)synchronization of periodic oscillators by periodic external action [22]. As a result, in the range delimited by period-doubling, there may exist chaotic transport solutions without locked velocity. The transport velocity is then lower than in the synchronized transport case but the transport does not vanish.

4.2 *Chaotic Transitions*

The synchronization transition at the saddle-node of Fig. 8 ($\mathcal{P}_v = 1628$, $\mathcal{P}_\gamma = 7.46$) differs from the synchronization at $\mathcal{P}_\gamma = 24.1$. The dynamics is no longer quasi-periodic but chaotic. In [13], we found intermittency occuring at chaotic time intervals: this behavior is typical of large forcing of an oscillator [22]. However, the particle dynamics displayed in Fig. 10b does not corroborate the existence of intermittency. We guess that the coexistence of attractors hides the intermittency.

Now, we detail the transitions from the periodic solutions to the synchronized transport by increasing $\mathcal{P}_v$ when $\mathcal{P}_\gamma = 7.46$. Indeed, we retrieve all the bifurcations scenario explained in detail in [13]. The route to synchronized transport is

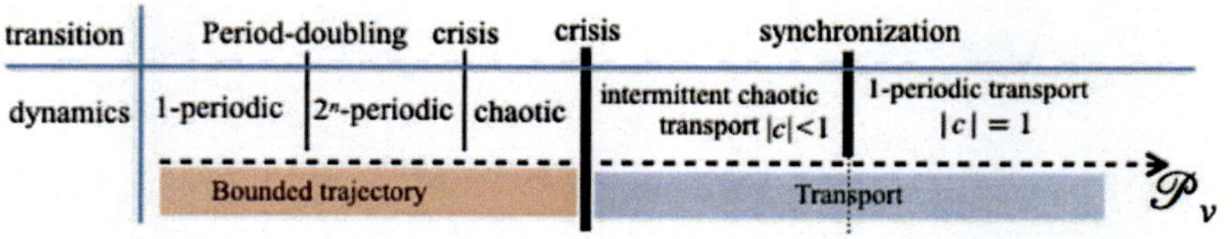

Fig. 9 Route to synchronized transport: dynamics and transitions by increasing $\mathcal{P}_v$ for moderate $\mathcal{P}_\gamma$

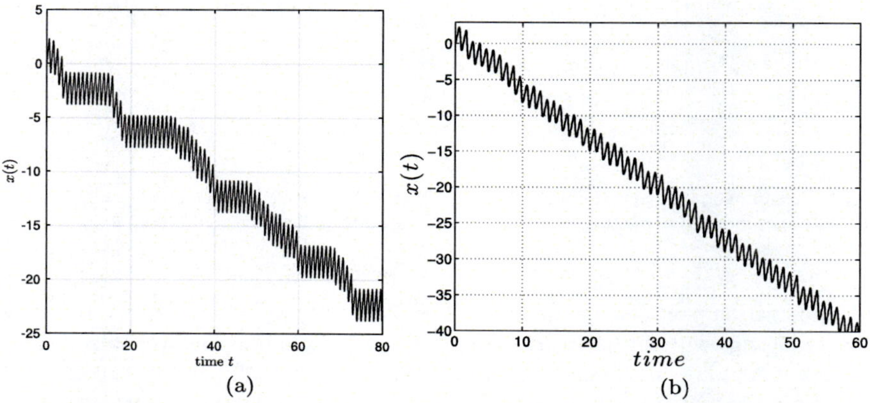

Fig. 10 Chaotic dynamics $r_p = 0.1, \mathcal{P}_\gamma = 7.46, \alpha = 0.5$ for two different values of $\mathcal{P}_v$. **a** near the crisis of the unbounded dynamics ($\mathcal{P}_v = 1458.3$). **b** near the syncrhonization transition ($\mathcal{P}_v = 1467.5$)

sketched in Fig. 9. The periodic solutions s_0 or s_m (Fig. 2), by increasing $\mathcal{P}_v$, involve a period-doubling cascade leading to a chaotic unbounded dynamics. A merging crisis may appears and because of the spatial periodicity the strange attractor is no longer bounded. If the problem does not have the parity symmetry ($\alpha \neq 0$), we expect a preferential direction. In Fig. 10a, the dynamics of the particle near the onset displays a drift to negative values in a intermittent manner. However, the intermittency is not quasi-periodic but chaotic. By increasing further $\mathcal{P}_v$, the drift velocity increases and the dynamics is still chaotic (Fig. 10b).

4.3 Chaotic Drift

The previous transport solutions are obtained for $\alpha = 0.5$ and we find only transport to negative direction. A non-zero value of α is required for the existence of intermittent and quasi-periodic drift. However, if $\mathcal{P}_\gamma$ is not large, unbounded dynamics and also synchronized transport may exist for $\alpha = 0$. The scenario from periodic solu-

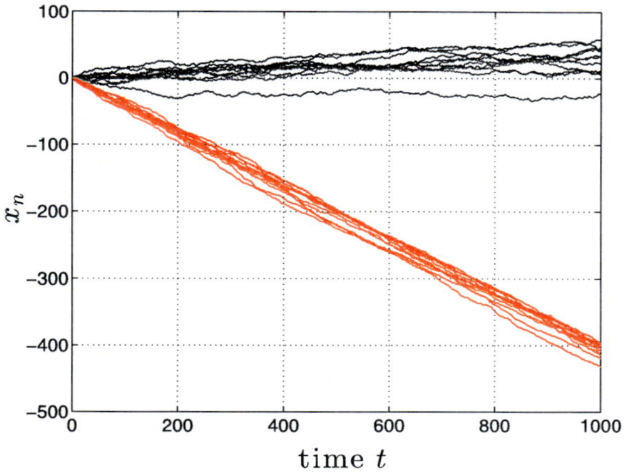

Fig. 11 Stroboscopic particles positions x_n at entire times for radii $r_p = 0.05$ (black lines) and $r_p = 0.1$ (red lines) for eleven initial conditions $(x_0, v_0) = (0, i/10)$ with $i = 0, 1, \ldots, 10$. Other parameters are $\alpha = 0.1, \mathcal{P}_\gamma = 6.7, \mathcal{P}_v = 2000$

tion to the unbounded dynamics in Fig. 9 involves, this time, spontaneous symmetry breaking. The chaotic dynamics has diffusion like behavior: no prefential direction is observed. When there is a synchronized transport then the synchronized transport solution in the opposite direction exists too. To have an effective transport, we need to slightly breaks the parity symmetry. As explained in [19], the transport solutions still exist but for slightly different parameter ranges when α is small. Therefore, we expect there is parameter domains for which the transport direction depends on the particle size.

We found for $\alpha = 0.1$, $\mathcal{P}_\gamma = 6.7$ and $\mathcal{P}_v = 2000$ different drift directions depending on the particle size: a net drift appears for $r_p = 0.1$ and while, for $r_p = 0.05$, a slight drift to positive direction occurs (Fig. 11). The dynamics is chaotic and notably it depends on the initial conditions. For $r_p = 0.05$, the mean value over the initial conditions of the velocity transport is slightly positive ($c \simeq 0.04$). In addition, there is a diffusive-like behavior too: the trajectory deviation increases with the time. Indeed, the chaotic dynamics results from the competition between opposite transport solutions $c = \pm 1$ which are unstable. For $r_p = 0.1$ the diffuse behavior is weak and the mean velocity remains close to $c = -2/5$. A possible explanation is that we are in the vicinity of the onset of the synchronized transport $c = -2/5$. The multiplicity of solutions makes it difficult to find this synchronized transport.

5 Concluding Remarks

We have shown using time-integration and bifurcation analysis that a selective transport of micro-particles is possible depending on the particle size.

For large drag $\mathcal{P}_\gamma$, the possible slow drift is quasi-periodic and determined by the pumping asymmetry, i.e. the parameter α. Then, the drift requires a non zero value of

α typically a value about 0.5. By decreasing the drag $\mathcal{P}_\gamma$ we found out synchronized transport solutions. The velocity is then locked to a rational value. If α is about 0.5, the transport velocity is negative. Therefore, for α about 0.5, we found either bounded periodic solutions or transport in the negative direction. However, if the selectivity is possible, we do not find a change of transport direction depending on the particle size.

Such a property was found for moderate value of $\mathcal{P}_\gamma < 10$ and α about 0.1. Indeed, for this parameters values, the asymmetry of the problem is no longer required. We retrieved the scenarios of the ratchet problem of a point-like particle. In particular, the dynamics can be chaotic. We found parameters such as the direction of the parameter drift depends on the particle size.

This theoretical study may have application of heavy particles in the air in micro-gravity environment [25]. Indeed, the small value of $\mathcal{P}_\gamma$ requires a low density fluid and the gravity could break the phenomena transport. Because of the chaotic dynamics could be strongly influenced by the noise [20], a further work would be to study the influence of noise on the transport selectivity.

References

1. T. Kulrattanarak, R. van der Sma, C. Schroënn, R. Boom, *Advances in Colloid and Interface Science* 142200853
2. N. Pamme, Lab Chip200771644-1659
3. R. Huang, T.A. Barber, M.A. Schmidt, R.G. Tompkinsand, M. Toner, D.W. Bianchiand, R. Kapur, W.L Flejter, Prenatal DIAGNOSIS282008892
4. A.E. Hasni, Göbbels K., A. Thiebes, Bräunig P., W. Mokwa and U. Schnakenberg, Procedia Eng. 2520111197 eurosensorsXXV
5. J. Autebert, B. Coudert, F.-C. Bidard, J.-Y. Pierga, S. Descroix, L. Malaquin, J.-L. Viovy, Flow cytometry and cell sorting: the next generation. Methods 5720122970
6. G.-H. Lee, S.-H. Kim, K. Ahn, S.-H. Lee, J.Y. Park, J. Micromech. Microeng. 262016013003
7. K. Loutherback, J. Puchalla, R.H. Austin, J.C. Sturm Phys. Rev. Lett. 1022009045301
8. S. Matthias, F. Müller, Nature 424200353
9. F. Santamaria, G. Boffetta, M.M. Afonso, A. Mazzino, M. Onorato, D. Pugliese, EPL (Europhy. Lett.) 102201314003
10. P. Reimann, Phys. Rep. 361200257
11. P., Hänggi, F. Marchesoni, Rev. Mod. Phys. 812009387
12. C. Kettner, P. Reimann, P. Hänggi, F. Müller, Phys. Rev. E 612000312
13. P. Beltrame, M. Makhoul, M. Joelson, Phys. Rev. E 932016012208
14. M. Makhoul, P. Beltrame, M. Joelson, Int. J. Mech. 92015260
15. R.H. Luchsinger, Phys. Rev. E 622000272
16. H. Brenner, J. Fluid Mech. **18**, 144 (1964)
17. M. Makhoul, P. Beltrame, M. Joelson, Particle drag force in a periodic channel: wall effects, in Proceedings of topical problems of fluid mechanics: proceedings, prague, pp. 141–148 (2015)
18. M. Makhoul, Modélisation du transport de particule dans un écoulement de stokes à effet cliquet Ph.D. thesis Université d'Avignon (juillet 2016)
19. P. Beltrame, Absolute negative mobility in a ratchet flow, *The Foundations of Chaos Revisited: From Poincaré to Recent Advancements Understanding Complex Systems (Springer)* (2016), pp. 249–261
20. P. Beltrame, J. Chaotic Model. Simul. 12018101

21. Y. Pomeau, P. Manneville, Commun. Math. Phys. 741980189
22. A. Pikovsky, M. Rosenblum, J. Kurths, Int. J. Bifurc. Chaos 1020002291
23. A. Wickenbrock, D. Cubero, N.A.A. Wahab, P. Phoonthong, F. Renzoni, Phys. Rev. E 842011021127
24. J.A. Cuesta, N.R. Quintero, R. Alvarez-Nodarse, Phys. Rev. X 32013041014
25. Makhoul, M., Beltrame, P. Selective Transport of airborne microparticles through micro-channels under microgravity. Microgravity Sci. Technol. **33**, 4 (2021). https://doi.org/10.1007/s12217-020-09855-3

Application of New 4-D Chaotic Map for Secure IP-Communications

Belqassim Bouteghrine, Camel Tanougast, and Said Sadoudi

Abstract Chaos systems have been studied for decades due to their applications in several domains such as: economy, communications, cryptography, etc. In recent years, designing and proposing new and higher dimensional chaotic systems become an increased tendency in particular for chaotic systems applied in network security domain. In this paper, we propose a new 4-dimension chaotic map with four (04) control parameters and five (05) non-linear terms. Then, we investigate the chaotic behaviors of the proposed system by considering the bifurcation and the Lyapunov exponents (LE) theories. The proposed map is applied for generating cipher keys to perform data encryption and secure an Internet Protocol (IP) communication.

Keywords Chaotic · Dimension · Bifurcation · Lyapunov exponents · Network security · Encryption · IP-communication

1 Introduction

Chaos systems have been studied for decades due to their applications in several domains such as: electronic circuits [1, 2], network security [3, 4], encryption domain [5, 6] and in power control [7]. In one hand, chaotic systems' properties have been studied and investigated using the bifurcation and the Lyapunov exponents theories [8, 9]. In the other hand, among the existing chaotic systems, researchers have investigated for new and more complex systems. By combining two coexisting attractors, in [10] the evolution of a new 4-dimension chaotic system is presented and analyzed by using bifurcation diagrams and Lyapunov exponents' spectrum. In [11] authors

B. Bouteghrine (✉) · C. Tanougast
LCOMS, Université de Lorraine, 57070 Metz, France
e-mail: belqassim.bouteghrine@univ-lorraine.fr

C. Tanougast
e-mail: Camel.Tanougast@univ-lorraine.fr

S. Sadoudi
Ecole Militaire Polytechnique, 16000 Algiers, Algeria

C. H. Skiadas et al. (eds.), *13th Chaotic Modeling and Simulation International Conference*, Springer Proceedings in Complexity,
https://doi.org/10.1007/978-3-030-70795-8_11

proved that using only sine or cosine functions and modifying two (02) variables in the function, n-scroll attractors can be generated. Moreover looking for higher dimension, a new 4-D hyperchaotic continuous time system is introduced in [12] and its main specifications are analyzed by means of equilibrium points, stabilities and power spectrum. In [13] authors showed that, using 4-D Lorenz system for secure TCP communicationonl, consume a huge amount of ressources and computations.

Consequently, we present through this paper an optimized new 4-dimensional discrete time system for encryption purpose. Then, the chaotic behavior of the proposed map is investegated based on the bifurcation and Lyapunov exponent theories. The rest of this paper is structured as follows. The 4-D map is introduced and the chaotic behavior is investigated in Sect. 2. In Sect. 3, software implementation of secure communication is illustrated using the proposed map. Finally, a conclusion is given in Sect. 4.

2 The Proposed 4-D Map

2.1 *System Description*

The proposed 4-D map with (04) bifurcation parameters and (05) non-linear terms is given as follows:

$$\begin{aligned} X(n+1) &= 1 - a * X(n)^2 + (Y(n) * Z(n) * P(n)) \\ Y(n+1) &= 1 - b * Y(n)^2 + (X(n) * Z(n) * P(n)) \\ Z(n+1) &= c * (X(n) * Y(n) * P(n)) \\ P(n+1) &= d * X(n) \end{aligned} \tag{1}$$

where X, Y, Z and P are the state variables and a, b, c, d are the control parameters or the bifurcation parameters.

2.2 *Chaos Behavior Investigation*

Following the lines given in [12, 13], the chaotic behavior of the proposed system (1) is investigated by considering mainly the bifurcation and the Lyapunov exponents (LE) theories.

2.2.1 Bifurcation Process

Bifurcation theory is concerned with changes in the solutions' behavior of the proposed system (1) as the parameters a, b, c and d are varying.

Figure 1 shows that chaotic behavior appears in several intervals of the parameters $a \in [0.22, 1.00]$, $b \in [0.90, 2.00]$, $c \in [-0.95, 0.95]$ and for the values of the parameter $d \in [-1.05, 2.5]$.

2.2.2 LE Process

Computing the LE values gives a possibility not only to detect all resonances in the response function, but also to detect the presence of chaos [14]. To search for strange attractors or for chaos in the proposed model (1), we proceed as follows:

First, we select the values of the parameters a, b, c and d calculated in the previous step (bifurcation process). Then we iterate equations (1) repeatedly until the Lyapunov exponent becomes small or negative, in which case the solution is probably a fixed point or limit cycle. In either event, we choose a different combination of a, b, c, d and start over the process.

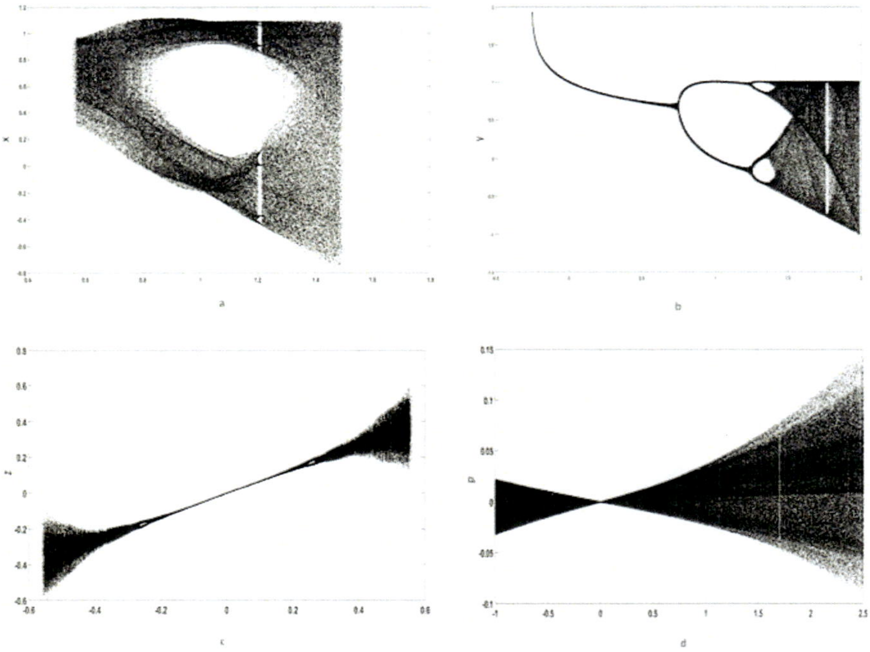

Fig. 1 The bifurcation graphs of the proposed 4-D map (a-b-c-d)

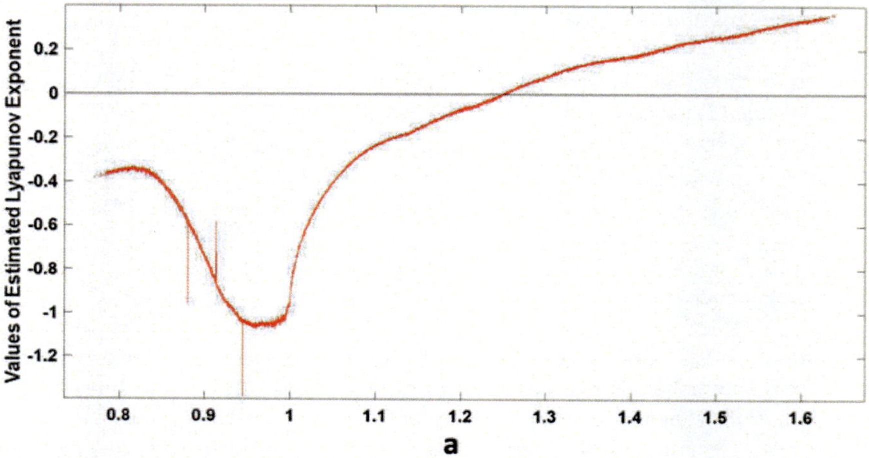

Fig. 2 The LE graph of the proposed 4-D map

If, after a few thousand iterations, the solution is bounded (not enormous) and the Lyapunov exponent is positive, then it is likely that you have found a strange attractor that corresponds to the chosen values of a, b, c and d (Fig. 2).

As a final result, the chaotic behavior is obtained and illustrated by the trajectory graphs and the signal graphs in Figs. 3 and 4, respectively.

3 Software Implementation of Secure Communication

Chaotic systems have been introduced in the network communication security domain since that they are characterized by their sensitivity, unpredictability and their widespread spectrum. Hence; we propose to introduce the 4-D chaos map for securing Client-Server Communication. In the first order, we prepare the Client-Server platform composed of two (02) client stations connected through an Ethernet network to one (01) server machine (Fig. 5).

The proposed platform works considering the following steps:

- The server listens for client connections;
- The client initiates the connection;
- The connection is established;
- The key-Generator load the first random sequence and the next sample;
- The client uses the first key to encrypt and send data to the server;
- The server use the same key to decrypt and send confirmation of receiving data.

Finally, the client terminates the communication by closing the channel, and the key-Generator module saves samples for next use with the server.

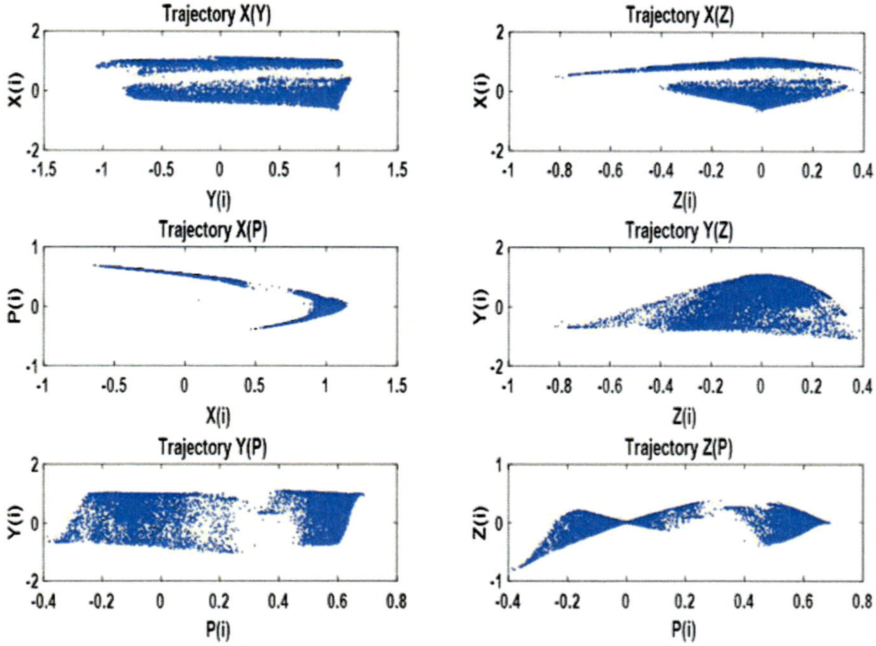

Fig. 3 The trajectory graphs of the proposed 4-D map

In the second order, we develop a C# software application for exchanging securely messages between connected computers (Fig. 6). The developed application includes a chaos-based cryptosystem using the proposed 4-D map which is described by the system (1); with $a = 1.55$; $b = 1.7$; $c = 1.6$; $d = 0.40$.

Considering that we are targeting to generate a 32-bit key using the chosen map, then we have the key space of the cryptosystem $2^{32*(4+4)} = 2^{256}$ which is considered very good value in for encryption since that the required value should be more than 2^{100} to satisfy the encryption requirements for key space [16].

4 Conclusion

In recent years; researching and developping new and higher dimensional chaotic systems become a rising trend in perticular for secure communication purposes.

In this paper, we propose new 4-Dimensional chaotic map in first order. Secondly, the chaotic behavior of the proposed system is investigated using an algorithm based on the bifurcation and the Lyapunov exponents properties. Finally, we apply the proposed system to generate random keys in order to perform a secure IP-communication.

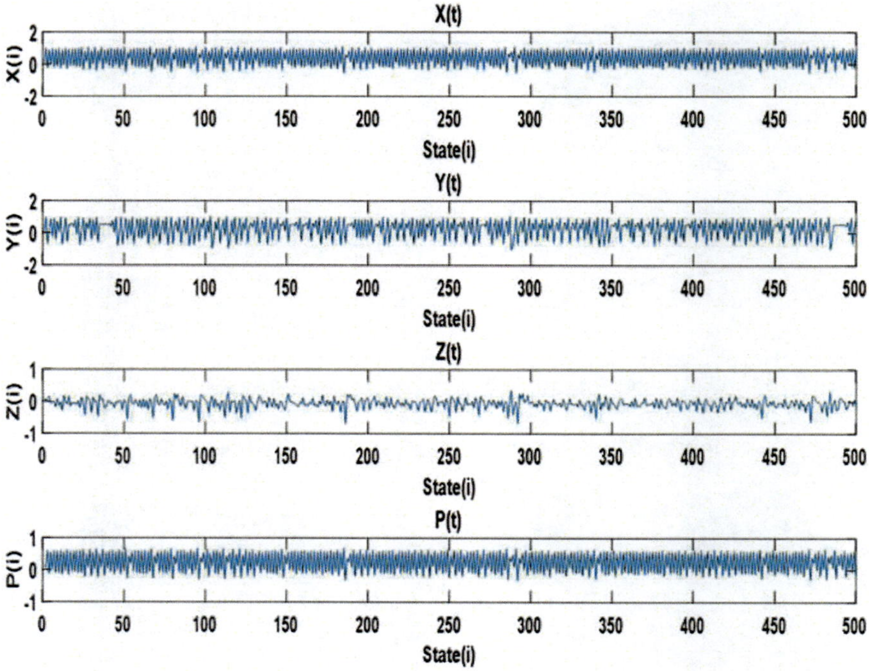

Fig. 4 The signal graphs of the proposed 4-D map

Fig. 5 The platform test bench

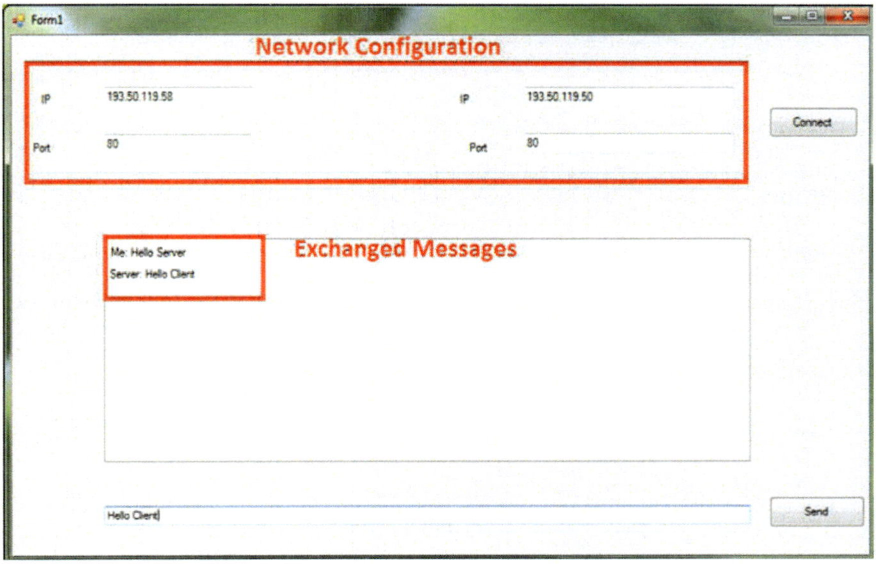

Fig. 6 The developped application's interface

As future work, statistical and security tests of the proposed system will be considered as well as the hardware implementation of the proposed scheme.

References

1. Y. Zhou, Z. Hua, C.M. Pun, C.L. Philip Chen, Cascade chaotic system with applications. IEEE Trans. Cybern. **45**(9), 2001–2012 (2015)
2. G. Mykolaitis, Temperature-stable hyperchaotic oscillator. IEEE Trans. Circuits Syst. Fundam. Theory Appl. **45**(11), 1198–1200 (1998)
3. A. Chonka, J. Singh, W. Zhou, Chaos theory based detection against network mimicking DDoS attacks. IEEE Commun. Lett. **13**(9), 717–719 (2009)
4. Y. Chen, X. Ma, X. Wu, CDDoS detection algorithm based on preprocessing network traffic predicted method and chaos theory. IEEE Commun. Lett. **17**(5), 1052–1054 (2013)
5. X. Zhang, L. Wang, Z. Zhou, Y. Niu, A chaos-based image encryption technique utilizing Hilbert curves and H-fractals. IEEE Access **7**, 74734–74746 (2019)
6. U. Cavusoglu, A. Akgul, S. Kacar, I. Pehlivan, A. Zengin, A novel chaos-based encryption algorithm over TCP data packet for secure communication. Secur. Commun. Netw. **9**(11), 1285–1296 (2016)
7. A. Al Talabani, A. Nallanathan, H.X. Nguyen, A novel chaos based cost function for power control of cognitive radio networks. IEEE Commun. Lett. **19**(4), 657–660 (2015)
8. J.C. Sprott. *Chaos and Time-Series Analysis*, vol. 69 (Oxford University Press, Oxford, 2003)
9. Z. Sandor, B. Erdi, A. Szell, B. Funk, The relative Lyapunov indicator: an efficient method of chaos detection. Celest. Mech. Dyn. Astron. **90**(1), 127–138 (2004)
10. Q. Lai, C. Chen, X.W. Zhao, J. Kengne, C. Volos, Constructing chaotic system with multiple coexisting attractors. IEEE Access **7**, 24051–24056 (2019)

11. W.K.S. Tang, G.Q. Zhong, G. Chen, K.F. Man, Generation of n-scroll attractors via sine function. IEEE Trans. Circuits Syst. Fundam. Theory Appl. **48**(11), 1369–1372 (2001)
12. C. Shen, S. Yu, J. Lu, G. Chen, A systematic methodology for constructing hyperchaotic systems with multiple positive Lyapunov exponents and circuit implementation. IEEE Trans. Circuits Syst. Regul. Pap. **61**(3), 854–864 (2014)
13. C.J. Feng, X.Z. Li, L.Y. Liu, Bifurcation and global dynamical behavior of the f(T) theory. Mod. Phys. Lett. A **29**(07), 1450033 (2014)
14. J.K. Hunter, *Introduction to Dynamical Systems* (U. C. Davis, Mathematics, MAT A, 207, 2011)
15. M.A. Ababneh, New four-dimensional chaotic attractor. Ain Shams Eng. J. **9**(4), 1849–1854 (2018)
16. B. Bouteghrine, M. Rabiai, C. Tanougast, S. Sadoudi, Hardware implementation of secured socket communication based on chaotic cryptosystem, in *2019 International Conference on Cyber Security and Protection of Digital Services (Cyber Security)* (Oxford, United Kingdom, 2019), pp. 1–4. https://doi.org/10.1109/CyberSecPODS.2019.8884879

Liutex-Based Investigation of Vortex in Multiphase Flow Past 2-D Cylinder Using GPU-Accelerated LBM

Pengxin Cheng, Nan Gui, Xingtuan Yang, Jiyuan Tu, Shengyao Jiang, and Haijun Jia

Abstract Multiphase flow past obstacles is extensively applied in industries, under which condition the interplay of different phases associated with impact of solid renders the flow characteristics and vortex field sophisticated. In this paper, we employ the GPU-accelerated Lattice Boltzmann Method to study the process of multiphase flow past a 2-D cylinder. The drag force components induced by continuous and dispersed phase respectively as well as the total force are illustrated and the underlying mechanisms are interpreted. The vortex-identification methods based on traditional approaches and Liutex is compared. The relationship between the extremums of disparate vortex identification variables and bubble deformation process is investigated.

Keywords Flow past cylinder · Two-phase flow · Vortex identification · Liutex · Lattice Boltzmann Method (LBM)

1 Introduction

Flow past obstacles is a prevailing phenomenon widely encountered in various fields of engineering. In the past decades, a multitude of experiments and simulation were conducted concerning lift and drag force, flow-induced vibration, vortex separation, etc. The multiphase flow past obstacles is of significant importance in practice, for example the heat exchangers in power engineering.

P. Cheng (✉) · N. Gui · X. Yang · J. Tu · S. Jiang · H. Jia
Collaborative Innovation Center of Advanced Nuclear Energy Technology, Key Laboratory of Advanced Reactor Engineering and Safety of Ministry of Education, Institute of Nuclear and New Energy Technology, Tsinghua University, Beijing 100084, People's Republic of China
e-mail: chengpx15@mails.tsinghua.edu.cn

N. Gui
e-mail: guinan@mail.tsinghua.edu.cn

J. Tu
School of Engineering, RMIT University, Melbourne, VIC 3083, Australia

C. H. Skiadas et al. (eds.), *13th Chaotic Modeling and Simulation International Conference*, Springer Proceedings in Complexity,
https://doi.org/10.1007/978-3-030-70795-8_12

In view of the influence of bubbles on the solid, the involvement of bubbles into the vortex as well as the impact of bubble migration on vortex shedding, the interaction in two-phase cross flow between three phases is complexly coupled, where the conventional computational fluid dynamics approaches may probably face difficulties.

The Lattice Boltzmann method (LBM) has enjoyed steady development in the past years, especially in the domain of multiphase flow as well as flow in complicated geometry. LBM provides a powerful new method with efficient algorithm and solid physics background from the perspective of mesoscopic scale. Since computation is mostly restricted to the local nodes, LBM is inherently suitable for implementation on parallelized hardware such as GPU, which manifests an obvious advantage over CPU in High Performance Computing. The LBM running on GPU could remarkably accelerate calculation with reliable performance.

As a fundamental research target, the vortex is of crucial significance for an intuitive understanding of turbulence. Up to now, there still exist plentiful ambiguous issues waiting to be settled, for instance with no agreement on a widely-adopted definition of vortex. Over the past decades, the vortex identification method has evolved from the first generation intuitive method based on vorticity that fails to distinguish rotational motion from a shear layer in viscous turbulent flow, to the second generation that depend on other parameters based on eigenvalue including $\Delta, \lambda_2, \lambda_{ci}$, etc., which unfortunately rely on uncertain case-related threshold and unable to evaluate the vortical strength to some degree.

Lately Liu and Gao et al. proposed the third generation Liutex method [1, 2], which presents a precise perspective of mathematical vortex definition. In essence, the vorticity could break down into the non-rotational portion, mainly shear, as well as the rotational portion named Liutex, i.e. $\vec{\omega} = \vec{S} + \vec{R}$. The magnitude of R is utilized to capture the local rotational strength. Wang et al. [3] put forward an explicit formula to determine it in a simple way. The direction of Liutex vector could be determined by real eigenvector of velocity gradient tensor, which characterizes the local rotational axis. With the aid of Liutex, ongoing research has been conducted on various applications including vortex in swirling jets [4, 5] and the hairpin vortex [6].

In this paper we will present a detailed investigation on the vortex and flow field of two-phase flow past cylinder. Then we will analyze the alteration of multiphase drag force corresponding to the movement of fluid. Afterwards we will discuss the evolution of the parameters during the bubble deformation for a deep understanding of the relationship with vortex development. Finally, we will investigate the relationship between bubble deformation process and extremums of different variables including Liutex.

2 Models and Methods

2.1 Fundamentals of Lattice Boltzmann Method

Lattice Boltzmann Method could be derived from the Boltzmann equation that is based on kinetic molecular dynamics.

$$\frac{\partial f}{\partial t} + \vec{\xi} \cdot \nabla_{\vec{x}} f + \vec{a} \cdot \nabla_{\vec{\xi}} f = \Omega(f),$$

where $f(\vec{x}, \vec{\xi}, t)$ represents the particle distribution function indicating the density of particles at position $\vec{x}$ and time t with velocity $\vec{\xi}$. According to single-relaxation-time approximation put forward by Bhatnagar–Gross–Krook (BGK) [7], the collision operator Ω_f could be simplified as follows.

$$\Omega(f) = -\frac{f - f_{eq}}{\tau}$$

By discretizing in the physical space, time and specially velocity space, one could derive the discrete Lattice Boltzmann Equation (LBE).

$$f_i\left(\vec{x} + \vec{c_i}\,\delta t, t + \delta t\right) - f_i(\vec{x}, t) = (\Omega(\vec{x}, t) + F_i(\vec{x}, t))\delta t$$

where $\vec{c_i}$ denotes the discrete velocity direction, $f_i(\vec{x}, t)$ is the particle distribution function along $\vec{c_i}$. DdQm is the widely adopted discretization model, where d and m represent number of dimensions and velocities respectively. For 2-D simulation D2Q9 is usually employed.

We could calculate the macroscopic parameters including density and velocity via moments of discrete distribution function. under the limitation of small Mach number, the LBGK model could recover continuity equation as well as Navier–Stokes equation at macroscopic scale using the multi-scale Chapman-Enskog expansion.

2.2 Multi-Component Multiphase Lattice Boltzmann Methods

In terms of multiphase flow simulation, Shan and Chen [8] proposed a concise model incorporating the Shan-Chen force, which does not require explicit interface tracking. Among the multiphase model, the Shan-Chen model has desirable performance in accuracy and efficiency, which is adopted in this paper.

Considering a multiphase fluid system consisting of n components, the multi-component LBE is as follows,

$$f_i^{\sigma}(\vec{x}+\vec{c}_i\delta t, t+\delta t) - f_i^{\sigma}(\vec{x}, t) = \frac{f_i^{eq(\sigma)} - f_i^{\sigma}}{\tau^{\sigma}}\delta t + \left(1 - \frac{\delta t}{2\tau^{\sigma}}\right) F_i^{\sigma}(\vec{x}, t)$$

One could calculate the total interaction force exerting on the σ component can by adding the force inflicted by all the components.

$$F_{SC}^{\sigma}(\vec{x}) = -\psi^{\sigma}(\vec{x}) \sum_{\sigma'} G_{\sigma\sigma'} \sum_i w_i \psi^{\sigma'}(\vec{x} + \vec{c_i}\,\delta t)\vec{c_i}\,\delta t$$

where $\psi(\rho)$ denotes pseudo-potential and $G(\vec{x}, \vec{x}')$ represents Green function used to evaluate the range and strength of interaction.

In this paper the two-component multiphase case is studied with no phase change, i.e. $n = 2$ and $G_{\sigma\sigma} = 0$. For repulsive force between disparate components $G_{\sigma\sigma'}$ is positive. One could compute the overall density, average velocity as well as pressure as follows.

$$\rho = \sum_{\sigma} \rho_{\sigma}, \vec{u} = \frac{\sum_{\sigma} (\rho_{\sigma}\vec{u}_{\sigma}/\tau_{\sigma})}{\sum_{\sigma} (\rho_{\sigma}/\tau_{\sigma})}, p = c_s^2(\rho_{\sigma} + \rho_{\sigma'}) + c_s^2 G \rho_{\sigma}\rho_{\sigma'}$$

2.3 Bounce-Back Method and Momentum Exchange Algorithm

In LBM the Bounce-Back method is the prevailing method to implement the no-slip wall boundary condition, with advantage of simplicity in implementation and guaranteed mass conservation. Traditional Bounce-Back method in the regular lattice could merely approximate arbitrary boundaries with staircase shapes, which might introduce large errors at cells nearby boundaries. While by increasing the mesh resolution the error could be reduced and the results approaches precise solution.

In the context of Bounce-Back scheme, in this paper we adopt the Momentum Exchange Algorithm (MEA) to calculate the force. The fundamental idea of MEA is to firstly recognize the populations across boundary and then compute the net momentum transfer including two items, i.e. the momentum transferred from fluid to wall f_i^{in} as well as the momentum transported in the opposite direction $f_{\bar{i}}^{out}$ [9].

The drag forces, which are opposed to direction of flow, are evaluated using MEA by the components of total force in the vertical direction. Because the force is calculated based on the populations f, considering the multi-component case, we could attain the force of the σ component from the populations of corresponding component f^{σ}.

2.4 Parallel Implementation of LBM with GPU

In practice, a widespread option is to decompose the main procedure of computation of LBE into two steps, i.e. collision and streaming. With the features of 'non-linearity is local' in collision step and 'non-locality is linear' in streaming step, LBM is ideal for computation on parallel architecture such as GPU. GPU is capable of executing a large number of threads in parallel, which could make full use of thousands of cores in GPU and notably accelerate the computation. The implementation of LBM on GPU has aroused worldwide interest and delivered satisfactory acceleration performance. In the present work, we adopt the open-source LBM software Sailfish for simulation [10] and run calculation on GPU of NVIDIA Tesla K20.

2.5 Vortex Identification Method and Liutex

Several widely adopted identification parameters are chosen including the vorticity ω [11], Liutex R [4], Q [12], Omega Ω [13, 14], Omega-Liutex Ω_R [15],which are defined by:

$$\begin{aligned}
&\omega = \|\boldsymbol{\omega}\| = \|\nabla \times \boldsymbol{u}\|,\\
&\boldsymbol{R} = R\boldsymbol{r} = \left(\langle \boldsymbol{\omega}, \boldsymbol{r}\rangle - \sqrt{\langle \boldsymbol{\omega}, \boldsymbol{r}\rangle^2 - 4\lambda_{ci}^2}\right)\boldsymbol{r},\\
&Q = 0.5\left(\|\boldsymbol{B}\|_F^2 - \|\boldsymbol{A}\|_F^2\right),\\
&\Omega = \frac{\|\boldsymbol{B}\|_F^2}{\|\boldsymbol{A}\|_F^2 + \|\boldsymbol{B}\|_F^2} = \frac{b}{a+b},\\
&\Omega_R = \frac{\beta^2}{\alpha^2 + \beta^2 + \varepsilon}
\end{aligned} \tag{1}$$

The details of definiation could refer to prior work [16]. In this paper, we will adopt these criteria for analysis of vortex in comparison.

3 Validation

In the present section, we firstly validate the LBM model against 2-D single-phase flow past a cylinder to verify the precision of force evaluation. Then we conduct the multi-component Laplace test to inspect the multiphase model.

For two-dimensional case, the drag and lift coefficient and the Strouhal number are calculated as follows. The characteristic length equals the obstacle's diameter.

$$C_D = \frac{F_D}{\frac{1}{2}\rho D\overline{u}^2}, C_L = \frac{F_L}{\frac{1}{2}\rho D\overline{u}^2}, \mathrm{St} = \frac{fD}{\overline{u}}$$

Table 1 Validation of flow past a 2-D cylinder

Re	20		100		
	C_D	C_L	C_D^{max}	C_L^{max}	St
Present result	5.5499	0.0118	3.2877	0.9842	0.3062
Benchmark lower bound	5.57	0.0104	3.22	0.99	0.295
Benchmark upper bound	5.59	0.0110	3.24	1.01	0.305

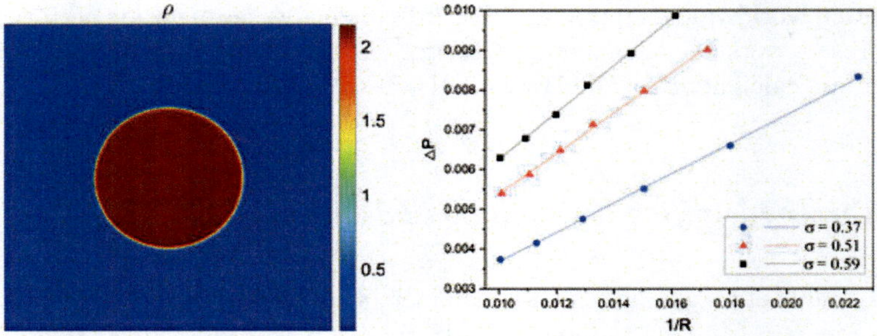

(a) Density contour (b) Data fitting

Fig. 1 The validation of multi-component Laplace test

The comparison of simulation results with validation reference is displayed in Table 1. As a whole, the results show in good agreement with benchmark in Schaefer's paper [17].

As for the Laplace test, we initialize a circular droplet with radius R of one phase that is surrounded by the another phase. As shown in Fig. 1, the relationship between Δp and $1/R$ is linear, in desirable conformity with Laplace Law, i.e. $\Delta p = \sigma/R$.

We completed the validation of two-phase model in previous paper [18] and tested the performance of LBM on GPU in another prior paper [19]. On the whole, we have comprehensively validated the LBM model, GPU acceleration performance as well as two-phase scheme.

4 Results and Discussion

4.1 Numerical Setup

The discretized flow domain consists of 256 × 1024 mesh grids. A two-dimensional cylinder with the diameter of 100 nodes is set in the center of channel, which is called *cylinder* later. At the upper side of channel, a bubble of the dispersed phase is initialized with the diameter equal to 100 nodes, which is referred to as *bubble* in the subsequent sections. The remaining field is occupied by continuous phase. It is noteworthy that the density contour represents that of continuous phase. Referring to prior LBM research [20], G is determined via $G\rho = 1.8$ and same viscosities are set for the two phases. The initial velocity of the bubble and the force is downward.

4.2 *The Evolution of Two-Phase Flow*

As shown in Fig. 2, the typical cases during evolution of density contour are selected. The bubble moves downwards to the cylinder as a consequence of initial velocity and imposed force. As the bubble approaches the cylinder, the hindrance from cylinder flattens the bottom of bubble (at $T = 6300$). Subsequently it is split by the solid (at $T = 7600$) and stretched a great deal (at $T = 10{,}400$) Subsequently it decomposes into several daughter bubbles when the filament turns unstable (at $T = 10{,}800$). The four small bubbles move along the periphery of cylinder and successively detach from the solid (at $T = 12{,}300$). Influenced by the vortex field behind the obstacle, bubbles move towards center-line and become closer (at $T = 14{,}000$). They coalesce in pairs for the first time (at $T = 15{,}600$) and eventually merge into one large bubble (at $T = 18{,}100$).

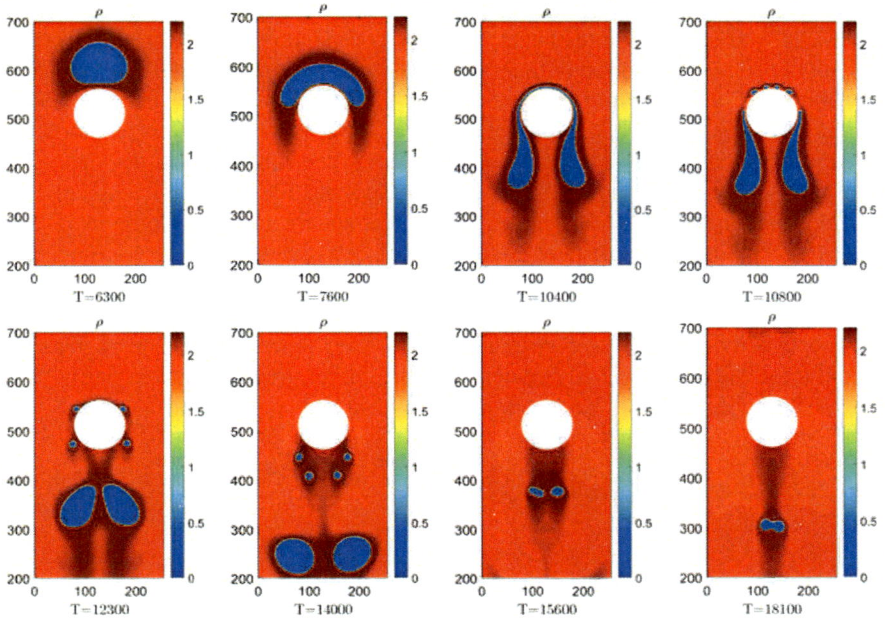

Fig. 2 The contour of density

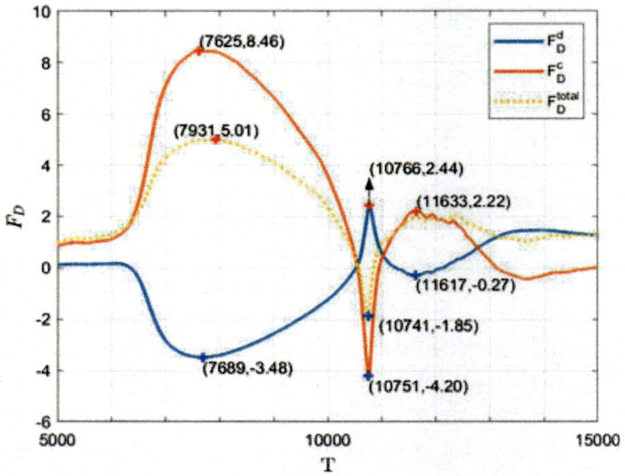

Fig. 3 The evolution of component and total drag force

4.3 *The Evolution of Drag Force*

The evolution of drag force is shown in Fig. 3. With the bubble approaching the cylinder, the drag force component of continuous phase F_D^c increment steadily while that of the disperse phase F_D^d declines inversely, leading to the increase of total drag force F_D^{total}. This variation results from the compression of the fluid between solid and bubble due to the motion of the bubble. This trend continues until F_D^c attains its maximum at $T = 7625$ and F_D^d arrives at the minimum at $T = 7689$, when the bubble surrounds nearly the upper half part of cylinder. Then the bubble covers the majority of cylinder periphery and the tendency of the curves converts into the opposite, namely F_D^c declines till its minimum at $T = 10{,}751$ and F_D^d increases conversely till it attains maximum at $T = 10{,}766$. F_D^{total} plunges and attained its minimum at $T =$ 10,741. With the portion of bubble coverage of cylinder periphery increasing, F_D^d consequently enhances and F_D^c reduces, until the bubble is over stretched and break up into pieces. Therewith the continuous phase regains the contact with cylinder and F_D^c boosts, arriving at a local maximum at $T = 11{,}633$. In contrast F_D^d drops towards a local minimum at $T = 11{,}617$. Afterwards the curves gradually flatten out.

4.4 *The Evolution of R, S, ω*

According to the relation of $\vec{R} + \vec{S} = \vec{\omega}$, the evolution of vorticity ω and R, S are shown in Figs. 4 and 5.

The evolution of breakup process is shown in Fig. 4. As the bubble brings into contact with solid and deformation influences the distribution of R, there exist zones with negative R in the center and zones with positive R on the lateral at $T = 7600$. When the break-up of bubble occurs at $T = 10{,}800$, magnitude of Liutex near the

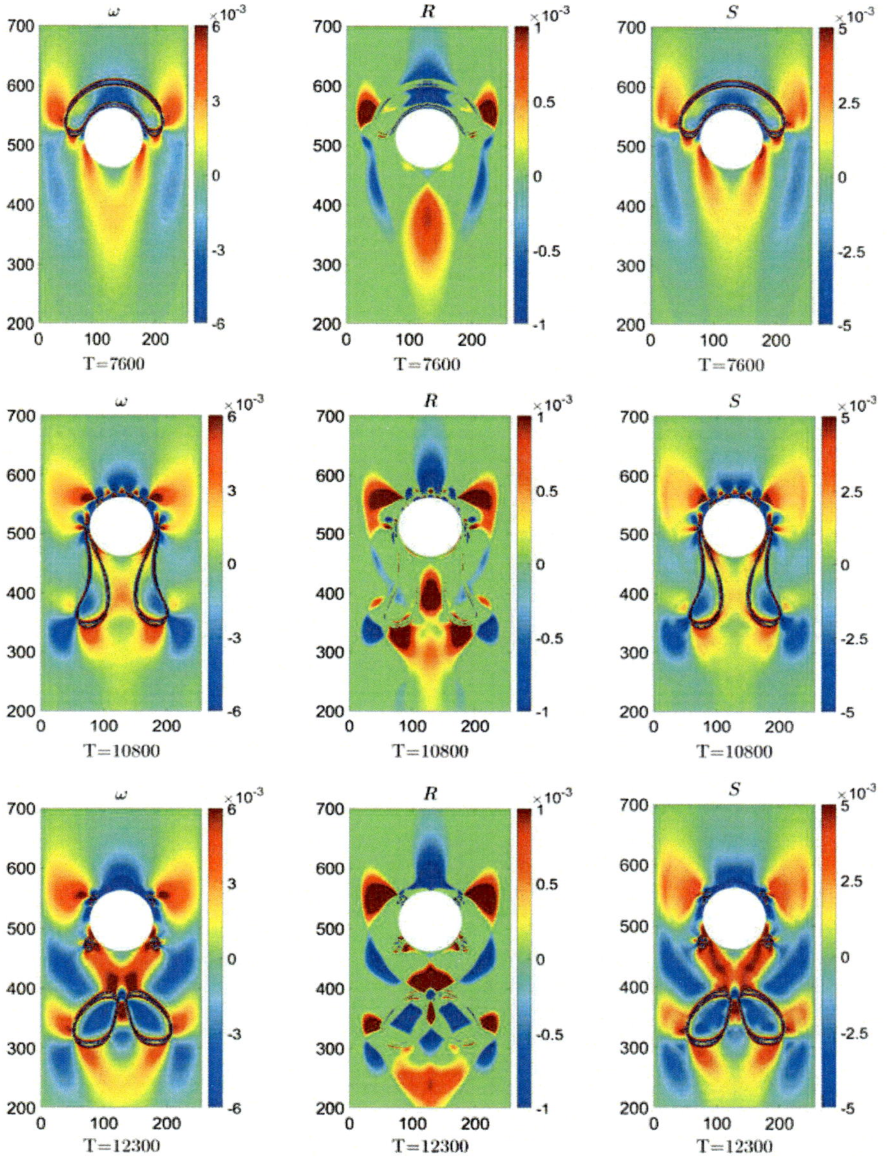

Fig. 4 The evolution of ω, R, S of breakup process

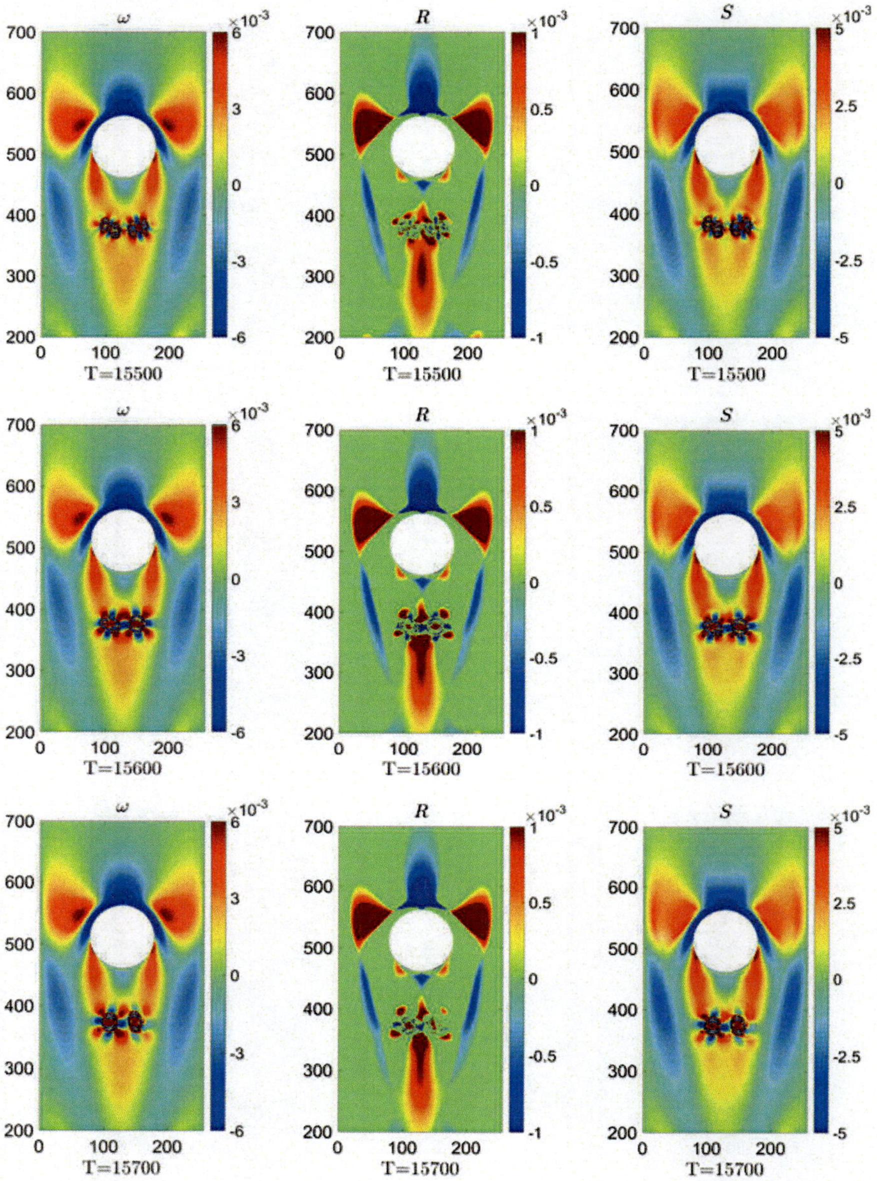

Fig. 5 The evolution of ω, R, S of coalescence process

position of fracture is relatively high in positive or negative value. There exist zones with positive R between and ahead of the stretched bubbles, as well as zones with negative R laterally adjacent to them. As the bubbles evolve into a nearly round shape due to surface tension at $T = 12{,}300$, there exist zones with negative R inside the bubbles.

As for the evolution of coalescence process in Fig. 5, when the small bubbles merge together, the distortion results in zones with high level of R both around and inside the bubble. Taking the coalescence of two bubbles in the right as an example, at $T = 15{,}500$ there appear negative Liutex zone near the line of bubble contact. At $T = 15{,}600$, there are positive Liutex zone inside the bubble and alternatively positive and negative Liutex outside the bubble. At $T = 15{,}700$, the negative Liutex zones have disappeared and the positive zones also shrink in area and decrease in magnitude.

On the whole, S and Ω show similarity with one another in magnitudes and distributions. Shear deformation plays the dominating role on the vortex development characteristics owing to the relatively low velocity. The influence of variations of R is fairly secondary. However, the sheer rotational motion indicated by R also shows conspicuous features dominated by the evolvement of bubbles during deformation, breakup as well as coalescence.

4.5 *The Evolution of* R^{max} *and* S^{max}

The evolution of statistics of primary variables of the dispersed and continuous phases will be examined. The blue and red marks in the figures denote the local minimums and maximums.

Firstly, the maximum value of R (namely R^{max}) is displayed in Fig. 6. Note that data of R_d^{max} is merely selected from dispersed phase, i.e. the interior of the bubble. There exist three distinct peaks at $T = 10{,}830$, $T = 15{,}480$ and $T = 18{,}050$. Data of R_c^{max} is sampled from continuous phase, i.e. the exterior of the bubble. The remarkable peaks locate at $T = 10{,}770$, $T = 15{,}460$ and $T = 18{,}030$.

For the continuous phase, the distinct increase of R_c^{max} at about $T = 10770$ corresponds to the very moment the bubble starts to split into daughter bubbles, whereas the peak of R_c^{max} at $T = 15460$ is close to the moment when daughter bubbles merge together for the first time. Similarly, the peak of R_c^{max} at $T = 18030$ is adjacent to the moment when bubbles eventually merge into a large one. For the dispersed phase, the evident peaks of R_d^{max} coincide with that of R_c^{max} correspondingly. The slight differences of peak times between R_c^{max} and R_d^{max} results from the small differences between rotational movement of vortices in the two phases.

Moreover, as shown in the figures, the peaks of R_d^{max} are later than that of R_c^{max}. One may deduce that the vortical motion could possibly be spread from the exterior towards the interior of the bubble, that is to say the external rotational or shear motions is likely to induce the vortex inside the bubble.

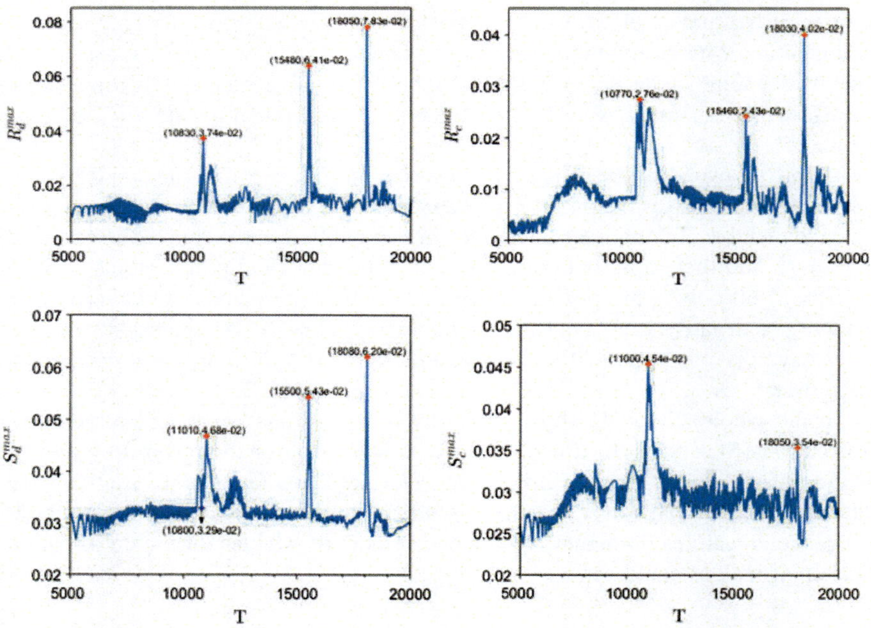

Fig. 6 The evolution of R^{max} and S^{max}

Regarding the evolution of S_d^{max} for the dispersed phase, the curve manifests three remarkable peaks at $T = 11{,}010$, $T = 15{,}500$ and $T = 18{,}080$, with the location approximate to that in figure of R_d^{max}. As to S_c^{max} for the continuous phase, a distinct peak locates at $T = 11{,}000$ and a minor peak at $T = 18{,}050$.

Comparing subfigures in Fig. 6, it is also clear that the peaks of R^{max} are close to that of S^{max}. In terms of traditional applications, we always use vorticity to identify the vortex. In the present case it is mainly dominated by shear instead of pure rotation. However, like in Fig. 6, since the peaks of real vortex indicated by R^{max} is adjacent to the peaks of shear S^{max}, using conventional vorticity identification methods could be acceptable in some degree when one doesn't need the most precise description of vortex.

However, the first peak of R^{max} is closer to the moment of bubble break-up than that of S^{max}. Besides there is no peak near the first coalescence of bubbles in Figs. (6d). Therefore, it is mostly encouraged to adopt R to evaluate the pure rotational vortex in two-phase flows, no matter for vortices in continuous phase or in dispersed phase.

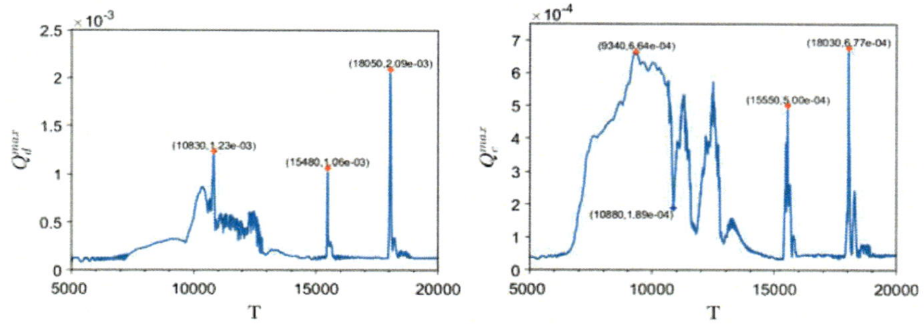

Fig. 7 The evolution of Q^{max}

4.6 The Evolution of Q^{max}

Figure 7 shows the evolution of Q^{max}. Respecting the maximum of Q_d^{max}, there exist three distinct peaks located at $T = 10{,}830$ and the other two at $T = 15{,}480$ and $T = 18{,}050$. Comparing with prior figures, these peaks are on the location close to the bubble coalescence or break-up. In subfigure of Q_c^{max}, two peaks lie at $T = 15{,}550$ and $T = 18{,}030$, while the curve reaches minimum at $T = 10{,}880$ and experiences drastic fluctuations.

In summary, Q^{max} could manifest the multiphase vortical motion, which is a usable indicator for rotational motion of fluids as well as shape deformation of bubbles to some extent, yet not as good as R^{max}.

4.7 The Evolution of Ω^{ave} and $\Omega^{ave}{}_R$

The evolution of Ω^{ave} and Ω_R^{ave} are displayed in Fig. 8. Concerning Ω_d^{ave} of dispersed phase, it first climbs to a maximum at $T = 6180$. Subsequently it falls back yet rises again and exhibits dramatic fluctuations afterwards. The variation of Ω_d^{ave} apparently signifies the variation of vortical motion in the bubbles, which is more smoothly and varied more slowly than that of Ω_c^{ave}. As a whole, neither Ω^{ave} of the dispersed phase nor that of the continuous phase could well indicate the instantaneous change of vortex motion of two-phases, inferior to using R^{max}.

The subfigures show the variations of Ω_R^{ave} for respective phases. Compared with Ω^{ave}, the variation of Ω_R^{ave} shows much distinct characteristics. In other words, Ω_R^{ave} is always fluctuating more intermittently and obviously than Ω^{ave}. In addition, the peaks of Ω_R^{ave} is adjacent to that of previous vortex identification parameters, e.g. R^{max} or Q^{max}, such as the peaks at $T = 11{,}050$, $T = 15{,}920$ and $T = 18{,}470$.

Referring to Eq. (1), the mechanism may result from the definition of parameters. Generally speaking, in terms of defining mere rotational motion, α and β are preferable to **A** and **B**. α and β are adopted to define the absolute rotational vortical

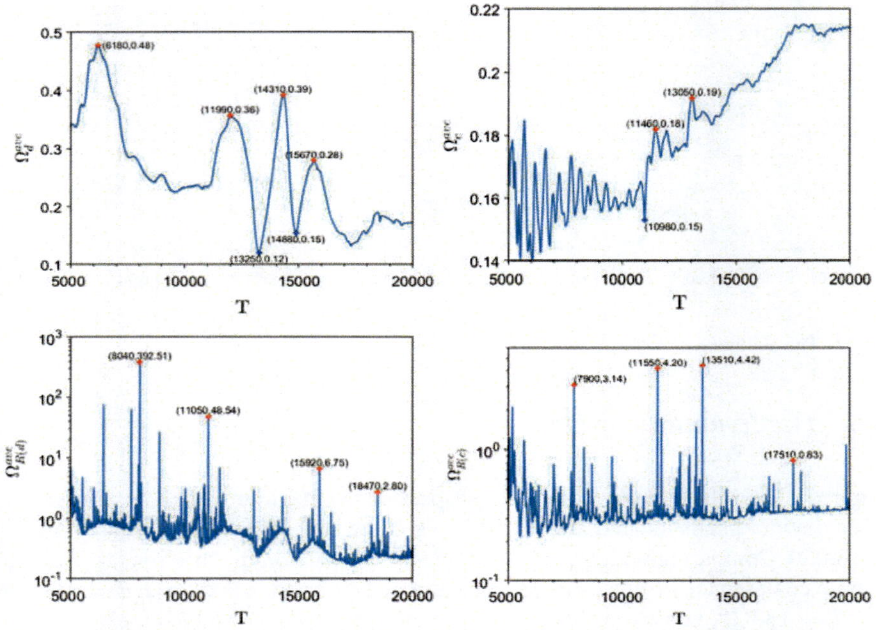

Fig. 8 The evolution of Ω^{ave} and Ω_R^{ave}

motion, namely *R*. In contrast **A** and **B** are unable to represent the pure rotational motion characteristics. Consequently, using α and β, thus Ω_R^{ave}, is more suitable to evaluate the magnitude of rotational motion.

5 Conclusions

In this work, we presented the features of drag force and vortex of two-phase flow past a two-dimensional cylinder utilizing Liutex-based analysis via multi-component multiphase LBM model on GPU computation. The following are the main findings in brief:

(1) Regarding the evolution feature, the bubble could be stretched and deformed till the final breakup during the flow past the cylinder. Subsequently, the daughter bubbles might gradually merge in the downstream to constitute bigger ones successively.
(2) Affected by the bubble behavior, the components of drag force induced by the continuous phase and dispersed phase show the opposite trend. The mechanism of variation in force components results from the change of density and velocity of the two phases.

(3) The vortex identification method Liutex R successfully characterizes sheer rotation of fluids in two-phase flow, though compared to the shear its magnitude is relatively small.
(4) Via averaging over each phase fields, Ω_R manifests evident superiority compared with Ω, because Ω_R is not obscured by averaging procedure, while Ω^{ave} fails to show such type of characteristics.

Acknowledgements The authors are grateful for the support of this research by the National Natural Science Foundations of China (Grant No. 51576211), the National High Technology Research and Development Program of China (863) (2014AA052701), the Science Fund for Creative Research Groups of National Natural Science Foundation of China (Grant No. 51621062), and the Foundation for the Author of National Excellent Doctoral Dissertation of PR China (FANEDD, Grant No. 201438).

References

1. C. Liu, Y. Gao, S. Tian, X. Dong, Rortex a new vortex vector definition and vorticity tensor and vector decompositions. Phys. Fluids. **30**(3) (2018)
2. Y. Gao, C. Liu, Rortex and comparison with eigenvalue-based vortex identification criteria. Phys. Fluids **30**(8), 085107 (2018)
3. Y.-Q. Wang, Y.-S. Gao, J.-M. Liu, C. Liu, Explicit formula for the liutex vector and physical meaning of vorticity based on the liutex-shear decomposition. J. Hydrodyn. **31**(3), 464–474 (2019)
4. N. Gui, H.-B. Qi, L. Ge, P.-X. Cheng, H. Wu, X.-T. Yang, J.-Y. Tu, S.-Y. Jiang, Analysis and correlation of fluid acceleration with vorticity and liutex (rortex) in swirling jets. J. Hydrodyn. **31**(5), 864–872 (2019)
5. N. Gui, L. Ge, P.X. Cheng, X.T. Yang, J.Y. Tu, S.Y. Jiang, Comparative assessment and analysis of rortex vortex in swirling jets. J. Hydrodyn. **31**(3), 495–503 (2019)
6. Y.S. Gao, J.M. Liu, Y.F. Yu, C. Liu, A liutex based definition and identification of vortex core center lines. J. Hydrodyn. **31**(3), 445–454 (2019)
7. P.L. Bhatnagar, E.P. Gross, M. Krook, A model for collision processes in gases. i. Small amplitude processes in charged and neutral one-component systems. Phys. Rev. **94**(3), 511–525 (1954)
8. X. Shan, H. Chen, Lattice boltzmann model for simulating flows with multiple phases and components. Phys. Rev. E **47**(3), 1815–1819 (1993)
9. K. Timm, H. Kusumaatmaja, A. Kuzmin, O. Shardt, G. Silva, E. Viggen, *The Lattice Boltzmann Method: Principles and Practice* (Germany, Springer, Berlin, 2016).
10. M. Januszewski, M. Kostur, Sailfish: A flexible multi-GPU implementation of the lattice Boltzmann method. Comput. Phys. Commun. **185**(9), 2350–2368 (2014)
11. J.-M. Liu, Y.-Q. Wang, Y.-S. Gao, C. Liu, Galilean invariance of omega vortex identification method. J. Hydrodyn. **31**(2), 249–255 (2019)
12. Y. Wang, Y. Gao, H. Xu, X. Dong, J. Liu, W. Xu, M. Chen, C. Liu, Liutex theoretical system and six core elements of vortex identification. J. Hydrodyn. (2020)
13. X. Dong, Y. Wang, X. Chen, Y. Dong, Y. Zhang, C. Liu, Determination of epsilon for omega vortex identification method. J. Hydrodyn. **30**, 541–548 (2018)
14. J. Liu, Y. Gao, Y. Wang, C. Liu, Objective omega vortex identification method. J. Hydrodyn. **31**, 455–463 (2019)

15. C. Liu, Y. S. Gao, X.R. Dong, Y.Q. Wang, J.M. Liu, Y.N. Zhang, X.S. Cai, N. Gui, Third generation of vortex identification methods: Omega and liutex/rortex based systems. J. Hydrodyn. **31**, 205–223 (2019)
16. P. Cheng, N. Gui, X. Yang, J. Tu, S. Jiang, H. Jia, Liutex-based analysis of drag force and vortex in two-phase flow past 2-D square obstacle using LBM on GPU. J. Hydrodyn. **32**, 820–833 (2020)
17. M. Schaefer, S. Turek, Benchmark computations of laminar flow around a cylinder **48**, 547–566 (1996)
18. T. Sun, N. Gui, X. Yang, J. Tu, S. Jiang, Numerical study of patterns and influencing factors on flow boiling in vertical tubes by thermal lbm simulation. Int. Commun. Heat Mass Transfer **86**, 32–41 (2017)
19. P. Cheng, N. Gui, X. Yang, J. Tu, S. Jiang, Application of lattice boltzmann methods for the multiphase fluid pipe flow on graphical processing unit. J. Comput. Multiph. Flow. **10**(3), 109–118 (2018)
20. H. Huang, D.T. Thorne, M.G. Schaap, M.C. Sukop, Proposed approximation for contact angles in shan-and-chen-type multicomponent multiphase lattice boltzmann models. Phys. Rev. E **76**(6), 066701 (2007)

Generalised Univariable Fractal Interpolation Functions

Vasileios Drakopoulos, Du Yong Pak, and SongIl Ri

Abstract We show how to construct a generalised iterated function system whose graph is the attractor, a fractal set, of some continuous function which interpolates a given set of data. Moreover, Rakotch contractions and vertical scaling factors as (continuous) 'contraction functions' are used in order to obtain generalised fractal interpolation functions with extensive practical applications, including data fitting and approximation of functions. A special generalised fractal interpolation function is introduced as an explicit illustrative example to show the effectiveness of the proposed method as compared to other existing methods. In particular, fractal interpolation functions which are widely presented in the literature can be obtained as particular cases of our construction.

Keywords Iterated function system (IFS) · Fractal interpolation function (FIF) · Rakotch contraction · Function vertical scaling factors

Mathematics Subject Classification: 37C45 · 28A80 · 37L30

1 Introduction

The concept of fractal interpolation functions (FIFs) was introduced by Barnsley [1] on the basis of the theory of iterated function systems (IFSs). Barnsley defined a fractal interpolation function (FIF) in 1986 and presented a construction of fractal functions by fractal interpolation. In the developments of theory of FIFs, many researchers have generalized the notion of FIFs in different ways. The fractal interpolation functions have been discussed in detail in the literature (see [1–4, 6–11]).

V. Drakopoulos (✉)
Department of Computer Science and Biomedical Informatics, University of Thessaly, Lamia, Greece
e-mail: vdrakop@uth.gr

D. Y. Pak · S. Ri
Department of Mathematics, University of Science,Pyongyang, Democratic People's Republic of Korea

C. H. Skiadas et al. (eds.), *13th Chaotic Modeling and Simulation International Conference*, Springer Proceedings in Complexity,
https://doi.org/10.1007/978-3-030-70795-8_13

Fractal interpolation functions have become a powerful tool for modeling many natural objects and have wide applications in mathematics and several other areas of applied sciences. For example, the fractal interpolation functions have been widely used in approximation theory, image compression, computer graphics and modeling of natural functions (or surfaces) such as rocks, metals, terrains and so on. As we know, a fractal interpolation function is generated by an IFS that consists of a finite set of some continuous functions on a complete metric space. Vertical scaling factors in the continuous functions have a decisive influence on the shape of the corresponding FIF because the vertical scaling factors uniquely determine the corresponding FIF provided that the interpolation points are prescribed in advance (see [14]).

How to construct fractal functions (rough functions) and analyse their complexity has become one of the most important topics in fractals (see [11]). The graph of a fractal interpolation function is an attractor of some iterated function system (see [1]). The concept of iterated function system was introduced as a natural generalization of the well-known Banach contraction principle (see [1], see [14], cf. [11]). Iterated function systems have become powerful tools for construction and analysis of new fractal interpolation functions. In order to ensure more flexibility in modeling natural shapes and phenomena or in image processing, researchers proposed many types of fractal interpolation functions by using iterated function systems (see [1, 2, 11, 14]).

The connectivity of attractors of iterated function systems is very important in the construction of fractal interpolation functions. The graphs of linear one variable fractal interpolation functions are always continuous functions. In usual approaches, the existence of linear FIFs follows from Banach's fixed point theorem (see [1]). Furthermore, in almost all the papers, the various types of FIFs are limited within the cases of constant vertical scaling factors and Banach's fixed point theorem. In 2011, to get the FIFs with more flexibility and diversity in a more general sense, Wang and Fan introduced a natural generalization of Barnsley's affine fractal interpolation function by using special function vertical scaling factors and Banach's fixed point theorem (see [5]). In order to construct new iterated function systems and fractal interpolation functions, one can use the well-known fixed point results obtained in the fixed point theory (see [4, 10, 11, 13]). As far as we know, the first significant generalization of Banach's principle was obtained by Rakotch in 1962 (see [6], p. 124).

In 2017, Ri presented a method to generate generalized fractal interpolation functions by using the Rakotch fixed point theorem [8] instead of the Banach fixed point theorem in certain concrete case (see [11]). However, results of [11] do not directly apply to the general case which often occurs in practical applications. In fact, in general case, the methods of proof of results in [11] fail because generalized transformations can involve special function vertical scaling factors that are not constant vertical scaling factors. The results of [5] and [11] inspire us to find possible vertical scaling factors (not necessarily constant vertical scaling factors) and contractions (not necessarily Banach contractions) for the existence of generalized FIFs. In this paper, in order to obtain generalized fractal interpolation functions, we use Rakotch contractions and special function vertical scaling factors. Dealing with generalized fractal interpolation functions is better than the one provided by [11]. In particular, we give an explicit illustrative example to demonstrate the effectiveness of obtained results.

This paper is organized as follows. In Sect. 2 we recall some results needed in constructing general FIFs. In Sect. 3 we introduce a new type of iterated function systems that will be used in our discussion for a special class of FIFs with function vertical scaling factors. In Sect. 4 we give a generalized FIF with special function vertical scaling factors as the fixed point of certain Read-Bajraktarević operator. In Sect. 5 we ensure that a generalized iterated function systems with special function vertical scaling factors has a unique invariant set. In Sect. 6 we give an explicit illustrative example to demonstrate the effectiveness of the preceding theory. Finally, in Sect. 7 we draw our conclusions.

2 Preparatory Facts

In this section, we describe some basic notions and theorems on fixed point theory. The following results will be the key in the proof of our main results.

Definition 1 (*see* [13], *p. 100, see* [8], *see* [6], *p. 144*) (1) If for some function $\varphi : (0, +\infty) \to (0, +\infty)$ and a self-map f of a metric space (X, d), we have

$$\forall_{x,y\in X}\ d(f(x), f(y)) \leq \varphi(d(x, y)),$$

then we say that f is a φ-contraction. (2) If f is a φ-contraction for some function $\varphi : (0, +\infty) \to (0, +\infty)$ such that for any $t > 0$, $\alpha(t) := \frac{\varphi(t)}{t} < 1$ and the function $(0, +\infty) \ni t \to \frac{\varphi(t)}{t}$ is non-increasing, then we call such a function a Rakotch contraction.

Remark 1 (see [6], p. 144, diagram) Each Banach contraction is a Rakotch contraction, since a map $f : X \to X$ is a Banach contraction if it is a φ-contraction for a function $\varphi(t) = \alpha t$, for some $0 \leq \alpha < 1$.

Theorem 1 (see [8], cf. [13], cf. [4], cf. [6]) *(1) Let X be a complete metric space and $f : X \to X$ be a Rakotch contraction. Then there is a unique fixed point $k \in X$ of f, and for each $x \in X$,*

$$\lim_{n\to+\infty} f^n(x) = k.$$

(2) Let X be a complete metric space and $\{X; f_1, \ldots, f_N\}$ be an iterated function system consisting of Rakotch contractions. Then there is a unique non-empty compact set $K \subset X$ such that

$$K = \bigcup_{i=1}^{N} f_i(K).$$

Now we describe some basic results on fractal interpolation theory. Let N be a positive integer greater than one and $I := [x_0, x_N] \subset \mathbb{R}$. Let a set of data points

$\{(x_i, y_i) \in I \times \mathbb{R} : i = 0, 1, 2, \ldots, N\}$ be given, where $\{x_0, x_1, \ldots, x_N\}$ is a partition of I (i.e., $x_0 < x_1 < x_2 < \cdots < x_N$) and $y_0, y_1, \ldots, y_N$ are given real numbers. Set $I_i := [x_{i-1}, x_i] \subset I$ and let $l_i : I \to I_i$ for $i = 1, 2, \ldots, N$ be contractive homeomorphisms such that

$$l_i(x_0) = x_{i-1}, l_i(x_N) = x_i,$$
$$|l_i(x') - l_i(x'')| \leq \lambda |x' - x''| \quad whenever \quad x', x'' \in I$$

for some $0 \leq \lambda < 1$. Furthermore, let mappings $F_i : I \times \mathbb{R} \to \mathbb{R}$ be continuous with, for some $k \geq 0$ and $0 \leq \alpha < 1$,

$$F_i(x_0, y_0) = y_{i-1}, F_i(x_N, y_N) = y_i.$$
$$|F_i(x', y') - F_i(x'', y'')| \leq k|x' - x''| + \alpha|y' - y''|$$

for all $x', x'' \in I$, $y', y'' \in \mathbb{R}$, and $i = 1, 2, \ldots, N$.

Now define functions $w_i : I \times \mathbb{R} \to I \times \mathbb{R}$ for $i = 1, 2, \ldots, N$ by

$$w_i := (l_i(x), F_i(x, y)).$$

Barnsley presented the following famous result.

Theorem 2 (cf. [1], p. 306) *The IFS*$\{I \times \mathbb{R}, w_i : i = 1, 2, \ldots, N\}$ *defined above has a unique nonempty compact set* $G \subset \mathbb{R}^2$ *such that*

$$G = \bigcup_{i=1}^{N} w_i(G).$$

Then G *is the graph of a continuous function* $f : I \to \mathbb{R}$ *which obeys*

$$f(x_i) = y_i \quad for \quad i = 0, 1, \ldots, N.$$

The function f whose graph is the attractor of an IFS is called a fractal interpolation function , or FIF for short, corresponding to the data $\{(x_i, y_i) : i = 0, 1, \ldots, N\}$ (cf. [1], p. 306).

Remark 2 In accordance with the idea of Barnsley, researchers proposed many types of FIFs. In [1, 11, 14],

$$l_i(x) := \frac{x_i - x_{i-1}}{x_N - x_0} x + \frac{x_N x_{i-1} - x_0 x_i}{x_N - x_0}.$$

(1) In [1, 2, 14], the maps $w_i(x, y)$ are chosen so that functions $F_i(x, y)$ are Banach contractions with respect to the second variable.
In the affine fractal interpolation function (cf. [1], p. 308),

$$F_i(x, y) := c_i x + d_i y + f_i,$$

where $|d_i| < 1$, and in the fractal interpolation function with function vertical scaling factors (see [5], see [14], p. 3–4, cf. [2]),

$$F_i(x, y) := d_i(x)y + q_i(x),$$

where $\sup_{x\in I} |d_i(x)| < 1$. (2) In [11], one type of fractal interpolation functions is considered, where the maps $w_i(x, y)$ are chosen so that functions $F_i(x, y)$ are Rakotch contractions with respect to the second variable. In the nonlinear fractal interpolation function (see [11]),

$$F_i(x, y) := c_i x + s_i(y) + f_i,$$

where s_i is a Rakotch contraction.

3 A Certain Generalised IFS

In this section, we introduce a new type of IFSs that will be used in our discussion for a special class of FIFs with function vertical scaling factors. Barnsley's functional condition for existence of a fractal interpolation function can be replaced by another functional conditions (see [11]). In order to obtain a new generalized fractal interpolation function, we use Rakotch contractions and special function vertical scaling factors in the construction of a generalized IFS (cf. [11], cf. [14]).

Let N be a positive integer greater than one and $I := [x_0, x_N] \subset \mathbb{R}$. We will work in the complete metric space $I \times \mathbb{R}$, with the Euclidean metric d_0. Let a set of data points $\{(x_i, y_i) \in I \times \mathbb{R} : i = 0, 1, 2, \ldots, N\}$ be given, where $x_0 < x_1 < x_2 < \cdots < x_N$ and $y_0, y_1, y_2, \ldots, y_N \in \mathbb{R}$. Set $I_i := [x_{i-1}, x_i] \subset I$ and define contractive homeomorphisms $l_i : I \to I_i$ by

$$l_i(x) := a_i x + e_i,$$

where for all $i = 1, 2, \ldots, N$, the real numbers a_i, b_i are chosen to ensure that $l_i(I) = I_i$. Let $\varphi : (0, +\infty) \to (0, +\infty)$ be a non-decreasing continuous function such that for any $t > 0$, $\alpha(t) := \frac{\varphi(t)}{t} < 1$ and the function $(0, +\infty) \ni t \to \frac{\varphi(t)}{t}$ is non-increasing. Let $d_i : I \to \mathbb{R}$ be a continuously differentiable function such that

$$\max_{x\in I} |d_i(x)| \leq 1.$$

Then by the differential mean value theorem and the existence theorem of maximum value and minimum value of continuous function, we can see that for some $L_{d_i} > 0$,

$$|d_i(x') - d_i(x'')| \leq L_{d_i}|x' - x''|,$$

where $x', x'' \in I$. Hence d_i is Lipschitz function defined on I satisfying $\max_{x\in I} |d_i(x)| \leq 1$. Consider an IFS of the form $\{I \times \mathbb{R}; w_i, i = 1, 2, \ldots, N\}$ in which the maps are generalized transformations of the special structure

$$w_i \begin{pmatrix} x \\ y \end{pmatrix} = \begin{pmatrix} l_i(x) \\ F_i(x, y) \end{pmatrix} = \begin{pmatrix} a_i x + e_i \\ c_i x + d_i(x)s_i(y) + f_i \end{pmatrix},$$

where the transformations are constrained by the data according to

$$w_i \begin{pmatrix} x_0 \\ y_0 \end{pmatrix} = \begin{pmatrix} x_{i-1} \\ y_{i-1} \end{pmatrix}, w_i \begin{pmatrix} x_N \\ y_N \end{pmatrix} = \begin{pmatrix} x_i \\ y_i \end{pmatrix}$$

for $i = 1, 2, \ldots, N$, and s_i are some Rakotch contractions (with the same function φ). Then for all $(x, y'), (x, y'') \in I \times \mathbb{R}$,

$$\begin{aligned} |F_i(x, y') - F_i(x, y'')| &= |d_i(x)||s_i(y') - s_i(y'')| \\ &\leq |s_i(y') - s_i(y'')| \leq \varphi(|y' - y''|). \end{aligned}$$

That is, each $w_i(x, y)$ is chosen so that function $F_i(x, y)$ is Rakotch contraction with respect to the second variable. Also, analytically, we obtain (compare with a_i, e_i, c_i, f_i of [11]).

$$\begin{aligned} a_i &= \frac{x_i - x_{i-1}}{x_N - x_0}, \\ e_i &= \frac{x_N x_{i-1} - x_0 x_i}{x_N - x_0} \\ c_i &= \frac{y_i - y_{i-1}}{x_N - x_0} - \frac{d_i(x_N)s_i(y_N) - d_i(x_0)s_i(y_0)}{x_N - x_0}, \\ f_i &= \frac{x_N y_{i-1} - x_0 y_i}{x_N - x_0} - \frac{x_N d_i(x_0)s_i(y_0) - x_0 d_i(x_N)s_i(y_N)}{x_N - x_0}. \end{aligned}$$

Remark 3 (1) Our bivariable function $d_i(x)s_i(y)$ is a generalization of bivariable function $d_i(x)y$ in the fractal interpolation function with function vertical scaling factors (see [5], see [14], p. 3–4, cf. [2]). In fact, in the case when $0 < \max_{x\in I} |d_i(x)| < 1$ (see [14], p. 3), obviously,

$$d_i(x)y = \frac{d_i(x)}{\max_{x\in I} |d_i(x)|} \max_{x\in I} |d_i(x)|y.$$

Let $s_i(y) := \max_{x\in I} |d_i(x)|y$ and $d_i^*(x) := \frac{d_i(x)}{\max_{x\in I} |d_i(x)|}$. Then $d_i(x)y = d_i^*(x)s_i(y)$, $\max_{x\in I} |d_i^*(x)| = 1$ and s_i is a Banach (or Rakotch) contraction.

(2) Our functional condition $\max_{x\in I} |d_i(x)| \leq 1$ is the essential condition to show the difference between Banach contractibility of $F_i(\cdot, y)$ and Rakotch contractibility of $F_i(\cdot, y)$ (compare with [14]). In fact, since $\varphi(t) < t$ for any $t > 0$,

$$\begin{aligned} |F_i(x, y') - F_i(x, y'')| &= |d_i(x)||s_i(y') - s_i(y'')| \\ &\leq \max_{x\in I} |d_i(x)||s_i(y') - s_i(y'')| \\ &\leq \max_{x\in I} |d_i(x)|\varphi(|y' - y''|) \\ &\leq \max_{x\in I} |d_i(x)||y' - y''|, \end{aligned}$$

where $(x, y'), (x, y'') \in \mathbb{R}^2$.
Hence, if $\max_{x\in I} |d_i(x)| < 1$, as can be seen, notwithstanding each s_i is a Rakotch contraction that is not Banach contraction, each F_i is Banach contraction with respect to the second variable because

$$|F_i(x, y') - F_i(x, y'')| \leq \max_{x\in I} |d_i(x)||y' - y''|.$$

On the other hand, if $\max_{x\in I} |d_i(x)| = 1$, then we can conclude that each F_i is Rakotch contraction (that is not Banach contraction) with respect to the second variable whenever each s_i is a Rakotch contraction (that is not Banach contraction) because

$$|F_i(x, y') - F_i(x, y'')| \leq \max_{x\in I} |d_i(x)|\varphi(|y' - y''|).$$

4 Fixed Point of a Certain Operator

In this section, we introduce a generalized FIF with as continuous 'contraction function' vertical scaling factors as the fixed point of certain Read-Bajraktarević operator (see [7]). By using Rakotch fixed point theorem, we show that the graph of a generalized FIF with as continuous 'contraction function' vertical scaling factors is the invariant set of some generalized IFS.

Denote by $C(I)$ the set of continuous functions $f : I = [x_0, x_N] \to \mathbb{R}$. Let $C^*(I) \subset C(I)$ denote the set of continuous functions $f : I \to \mathbb{R}$ such that $f(x_0) = y_0$ and $f(x_N) = y_N$, that is,

$$C^*(I) := \{f \in C(I) : f(x_0) = y_0, f(x_N) = y_N\}.$$

Let $C^{**}(I) \subset C^*(I) \subset C(I)$ be the set of continuous functions that pass through the given data points $\{(x_i, y_i) \in I \times \mathbb{R} : i = 0, 1, 2, \ldots, N\}$, that is,

$$C^{**}(I) := \{f \in C^*(I) : f(x_i) = y_i, i = 0, 1, \ldots, N\}.$$

Define a metric $d_{C(I)}$ on $C(I)$ by

$$d_{C(I)}(g,h) := \max_{x\in[x_0,x_N]} |g(x)-h(x)|$$

for all $g, h \in C(I)$. Then $(C(I), d_{C(I)})$, $(C^*(I), d_{C(I)})$ and $(C^{**}(I), d_{C(I)})$ are complete metric spaces. For all $f \in C^*(I)$, define a mapping $T : C^*(I) \to C(I)$ by

$$\begin{aligned} Tf(x) :&= F_i(l_i^{-1}(x), f(l_i^{-1}(x))) \\ &= c_i l_i^{-1}(x) + d_i(l_i^{-1}(x)) s_i(f(l_i^{-1}(x))) + f_i \end{aligned}$$

for $x \in [x_{i-1}, x_i]$ and $i = 1, 2, \ldots, N$. Obviously, T is a form of Read-Bajraktarević operator as defined in [7].

Lemma 1 *$Tf \in C^{**}(I)$ for all $f \in C^*(I)$. That is, $T : C^*(I) \to C^{**}(I)$ and $T^n : C^{**}(I) \to C^{**}(I)$ for all $n \geq 2$.*

Proof Since

$$w_i\begin{pmatrix} x_0 \\ y_0 \end{pmatrix} = \begin{pmatrix} x_{i-1} \\ y_{i-1} \end{pmatrix}, \; w_i\begin{pmatrix} x_N \\ y_N \end{pmatrix} = \begin{pmatrix} x_i \\ y_i \end{pmatrix}$$

for $i = 1, 2, \ldots, N$, we obtain $l_i(x_0) = x_{i-1}, l_i(x_N) = x_i, l_i^{-1}(x_{i-1}) = x_0, l_i^{-1}(x_i) = x_N$, $F_i(x_0, y_0) = y_{i-1}$ and $F_i(x_N, y_N) = y_i$ for $i = 1, 2, \ldots, N$. Hence if $x_i \in [x_{i-1}, x_i]$ for $i = 1, 2, \ldots, N$, then since $f \in C^*(I)$, we obtain

$$\begin{aligned} Tf(x_i) &= F_i(l_i^{-1}(x_i), f(l_i^{-1}(x_i))) \\ &= c_i l_i^{-1}(x_i) + d_i(l_i^{-1}(x_i)) s_i(f(l_i^{-1}(x_i))) + f_i \\ &= c_i x_N + d_i(x_N) s_i(f(x_N)) + f_i \\ &= F_i(x_N, f(x_N)) = F_i(x_N, y_N) = y_i \end{aligned}$$

and if $x_i \in [x_i, x_{i+1}]$ for $i = 0, 1, 2, \ldots, N-1$, then since $f \in C^*(I)$, we obtain

$$\begin{aligned} Tf(x_i) &= F_{i+1}(l_{i+1}^{-1}(x_i), f(l_{i+1}^{-1}(x_i))) \\ &= c_{i+1} l_{i+1}^{-1}(x_i) + d_{i+1}(l_{i+1}^{-1}(x_i)) s_{i+1}(f(l_{i+1}^{-1}(x_i))) + f_{i+1} \\ &= c_{i+1} x_0 + d_{i+1}(x_0) s_{i+1}(f(x_0)) + f_{i+1} \\ &= F_{i+1}(x_0, f(x_0)) = F_{i+1}(x_0, y_0) = y_i. \end{aligned}$$

So $f(x_i) = y_i$ for all $i = 0, 1, 2, \ldots, N$ and $Tf(x)$ is continuous at each of the points $x_1, x_2, \ldots, x_{N-1}$. By definition of the mapping T, $Tf(x)$ is continuous on the interval $[x_{i-1}, x_i]$ for all $i = 1, 2, \ldots, N$. Hence $Tf \in C^{**}(I)$ and $T^n : C^{**}(I) \to C^{**}(I)$ for all $n \geq 2$. □

Using Lemma 1 and the technique introduced in [11], we can obtain the following Theorem that will be used in our discussion for a special class of FIFs with function vertical scaling factors.

Theorem 3 *Let N be a positive integer greater than one. Let $\{I \times \mathbb{R}; w_i, i = 1, 2, \ldots, N\}$ denote the IFS defined above, associated with the set of data*

$$\{(x_i, y_i) : i = 0, 1, \ldots, N\}.$$

Then the operator T is a Rakotch contraction (considered as a map $T: C^*(I) \to C^*(I)$). Hence there is a unique continuous function $f: I \to \mathbb{R}$ which is a fixed point of T. In particular, $f(x_i) = y_i$ for $i = 0, 1, \ldots, N$. Moreover, the graph G of f is invariant with respect to $\{I \times \mathbb{R}; w_1, \ldots, w_N\}$, i.e.,

$$G = \bigcup_{i=1}^{N} w_i(G).$$

Proof Since $\max_{x\in I} |d_i(x)| \leq 1$, we obtain that for all $g, h \in C^*(I) \subset C(I)$,

$$\begin{aligned}
d_{C(I)}(Tg, Th) &= \max_{x\in[x_0,x_N]} |Tg(x) - Th(x)| \\
&= \max_{i=1,2,\ldots,N} \max_{x\in[x_{i-1},x_i]} |Tg(x) - Th(x)| \\
&= \max_{i=1,2,\ldots,N} \max_{x\in[x_{i-1},x_i]} |c_i l_i^{-1}(x) + d_i(l_i^{-1}(x)) s_i(g(l_i^{-1}(x))) + f_i \\
&\quad - c_i l_i^{-1}(x) + d_i(l_i^{-1}(x)) s_i(h(l_i^{-1}(x))) + f_i| \\
&= \max_{i=1,2,\ldots,N} \max_{x\in[x_{i-1},x_i]} |d_i(l_i^{-1}(x)) s_i(g(l_i^{-1}(x))) - d_i(l_i^{-1}(x)) s_i(h(l_i^{-1}(x)))| \\
&\leq \max_{i=1,2,\ldots,N} \max_{x\in[x_{i-1},x_i]} |s_i(g(l_i^{-1}(x))) - s_i(h(l_i^{-1}(x)))| \\
&\leq \max_{i=1,2,\ldots,N} \sup_{x\in[x_{i-1},x_i]} \varphi(|g(l_i^{-1}(x)) - h(l_i^{-1}(x))|),
\end{aligned}$$

where $\varphi: (0, +\infty) \to (0, +\infty)$ is some non-decreasing function such that $\varphi(t) < t$ for $t > 0$ and $t \to \frac{\varphi(t)}{t}$ is non-increasing. Since $\varphi: (0, +\infty) \to (0, +\infty)$ is non-decreasing continuous function and $l_i^{-1} : [x_{i-1}, x_i] \to [x_0, x_N]$ for all $i = 1, 2, \ldots, N$, we obtain that for $i_0 \in \{1, 2, \ldots, N\}$ and $x_0 \in [x_{i_0-1}, x_{i_0}]$,

$$\begin{aligned}
\varphi(|g(l_{i_0}^{-1}(x_0)) - h(l_{i_0}^{-1}(x_0))|) &\leq \varphi(\max_{x\in[x_{i_0-1},x_{i_0}]} |g(l_{i_0}^{-1}(x)) - h(l_{i_0}^{-1}(x))|) \\
&\leq \varphi(\max_{x\in[x_0,x_N]} |g(x) - h(x)|) \\
&= \varphi(d_{C(I)}(g, h)).
\end{aligned}$$

Since x_0 was arbitrary,

$$\sup_{x\in[x_{i_0-1},x_{i_0}]} \varphi(|g(l_{i_0}^{-1}(x) - h(l_{i_0}^{-1}(x))|) \leq \varphi(d_{C(I)}(g, h))$$

and since i_0 was arbitrary,

$$\max_{i=1,2,\ldots,N} \sup_{x\in[x_{i-1},x_i]} \varphi(|g(l_i^{-1}(x) - h(l_i^{-1}(x))|) \leq \varphi(d_{C(I)}(g,h)).$$

Hence we obtain

$$\begin{aligned} d_{C(I)}(Tg, Th) \leq & \max_{i=1,2,\ldots,N} \sup_{x\in[x_{i-1},x_i]} \varphi(|g(l_i^{-1}(x) - h(l_i^{-1}(x))|) \\ & \leq \varphi(d_{C(I)}(g,h)) = \varphi(d_{C(I)}(g,h)). \end{aligned}$$

So we conclude that $T\colon C^*(I) \to C^{**}(I) \subset C^*(I)$ is a same function φ) on the complete metric space $(C^*(I), d_{C(I)})$. Theorem 1 (1) implies that T possesses a unique fixed point in $C^*(I)$. That is, there exists a continuous function $f \in C^*(I)$ such that for all $x \in [x_0, x_N]$,

$$Tf(x) = f(x).$$

Since $T : C^*(I) \to C^{**}(I)$ (by Lemma 1), we have $f = Tf \in C^{**}(I)$. That is, there is a continuous function f that passes through the given data points $\{(x_i, y_i) \in I \times \mathbb{R} : i = 0, 1, 2, \ldots, N\}$. Let G denote the graph of $f \in C^{**}(I)$, that is, $G := \{(x, f(x)) : x \in [x_0, x_N]\}$. Since f is a fixed point of the operator T and if $x \in [x_{i-1}, x_i]$, then

$$Tf(x) = F_i(l_i^{-1}(x), f(l_i^{-1}(x))),$$

we obtain that for all $x \in [x_0, x_N]$,

$$\begin{aligned} f(l_i(x)) =& Tf(l_i(x)) \\ =& F_i(l_i^{-1}(l_i(x)), f(l_i^{-1}(l_i(x)))) \\ =& F_i(x, f(x)). \end{aligned}$$

Since $w_i(x, y) = (l_i(x), F_i(x, y))$ for all for $i = 1, 2, \ldots, N$, we obtain that

$$\begin{aligned} w_i(G) =& w_i(\{(x, f(x)) : x \in [x_0, x_N]\}) \\ =& \{w_i(x, f(x)) : x \in [x_0, x_N]\} \\ =& \{(l_i(x), F_i(x, f(x))) : x \in [x_0, x_N]\} \\ =& \{(l_i(x), f(l_i(x))) : x \in [x_0, x_N]\} \\ =& \{(x, f(x)) : x \in [x_{i-1}, x_i]\}. \end{aligned}$$

Hence

$$\begin{aligned} G &= \{(x, f(x)) : x \in [x_0, x_N]\} \\ &= \bigcup_{i=1}^{N} \{(x, f(x)) : x \in [x_{i-1}, x_i]\} \\ &= \bigcup_{i=1}^{N} w_i(G). \end{aligned}$$

This completes the proof. □

Remark 4 In the case where the vertical scaling factor parameters are constants, Barnsley investigated the existence of affine FIFs by using the Banach fixed point theorem (see [1]), and Wang and Fan introduced a natural generalisation of Barnsley's affine FIFs by using special function vertical scaling factors and Banach's fixed point theorem (see [5]). Here, we study the existence of generalized FIFs with function vertical scaling factors by using the Rakotch fixed point theorem, and the techniques used in Theorem 3 is completely different from those used in [1, 2]. But Theorem 3 does not ensure that the IFS$\{I \times \mathbb{R}; w_i, i = 1, 2, \ldots, N\}$ has a unique invariant set. The uniqueness of invariant set is determined explicitly in Theorem 4.

5 Attractor of a Certain IFS

In this section, we ensure that a generalized IFS$\{I \times \mathbb{R}; w_i, i = 1, 2, \ldots, N\}$ with special function vertical scaling factors has a unique invariant set (attractor). Theorem 4 that is our main theorem in this paper improves upon a result proved by [11], and the proof of Theorem 4 is based on arguments first applied in [11]. If we combine the both of Theorem 3 and Theorem 4 into a party, we can easily know that the graph of a generalized FIF with special function vertical scaling factors is a unique attractor of a certain IFS.

Theorem 4 *Let N be a positive integer greater than one. Let each s_i be a bounded Rakotch contraction. Let $\{I \times \mathbb{R}; w_i, i = 1, 2, \ldots, N\}$ denote the IFS defined above, associated with the set of data*

$$\{(x_i, y_i) : i = 0, 1, \ldots, N\}.$$

Then there is a metric d_θ on $I \times \mathbb{R}$, equivalent to the Euclidean metric d_0, such that for all $i = 1, \ldots, N$, w_i are Rakotch contractions with respect to d_θ. In particular, there exists a unique nonempty compact set $G \subset I \times \mathbb{R}$ such that

$$G = \bigcup_{i=1}^{N} w_i(G).$$

Proof We define a metric d_θ on $I \times \mathbb{R}$ by

$$d_\theta((x', y'), (x'', y'')) := |x' - x''| + \theta|y' - y''|,$$

where θ is a positive real number which is specified below. Since $|d_i(x') - d_i(x'')| \leq L_{d_i}|x' - x''|$ and $F_i(x, y) := c_i x + d_i(x)s_i(y) + f_i$,

$$\begin{aligned}
|F_i(x', y') - F_i(x'', y'')| &= \\
&= |c_i x' + d_i(x')s_i(y') + f_i - (c_i x'' + d_i(x'')s_i(y'') + f_i)| \\
&\leq |c_i||x' - x''| + |d_i(x')s_i(y') - d_i(x'')s_i(y'')| \\
&\leq |c_i||x' - x''| + |d_i(x')||s_i(y') - s_i(y'')| + |s_i(y'')||d_i(x') - d_i(x'')| \\
&\leq |c_i||x' - x''| + |s_i(y') - s_i(y'')| + \sup_{y'' \in D(s_i)} |s_i(y'')||d_i(x') - d_i(x'')| \\
&\leq (|c_i| + \sup_{y'' \in D(s_i)} |s_i(y'')|L_{d_i})|x' - x''| + |s_i(y') - s_i(y'')|,
\end{aligned}$$

where $D(s_i) \subset \mathbb{R}$ is the domain of definition of s_i. Let

$$k := \max_{i=1,2,\ldots,N} (|c_i| + \sup_{y'' \in D(s_i)} |s_i(y'')|L_{d_i}),$$

Then for all $(x', y'), (x'', y'') \in I \times \mathbb{R}$,

$$|F_i(x', y') - F_i(x'', y'')| \leq k|x' - x''| + \varphi(|y' - y''|),$$

where $\varphi : (0, +\infty) \to (0, +\infty)$ is some non-decreasing function such that $\varphi(t) < t$ for $t > 0$ and $t \to \frac{\varphi(t)}{t}$ is nonincreasing. That is, each $F_{i,j}$ is a Rakotch contraction (with the same function φ) with respect to the second variable, and Lipschitz with respect to the first variable. Hence we obtain for all $(x', y'), (x'', y'') \in I \times \mathbb{R}$,

$$\begin{aligned}
d_\theta(w_i(x', y'), w_i(x'', y'')) &= d_\theta((l_i(x''), F_i(x', y')), (l_i(x''), F_i(x'', y''))) \\
&= |l_i(x') - l_i(x'')| + \theta|F_i(x', y')) - F_i(x'', y'')| \\
&\leq |a_i||x' - x''| + \theta(k|x' - x''| + \varphi(|y' - y''|)) \\
&= |a_i||x' - x''| + \theta k|x' - x''| + \theta\varphi(|y' - y''|) \\
&\leq (|a_i| + \theta k)|x' - x''| + \theta\varphi(|y' - y''|).
\end{aligned}$$

Let $(x', y'), (x'', y'') \in I \times \mathbb{R}$ and $(x', y') \neq (x'', y'')$. Since $\varphi : (0, +\infty) \to (0, +\infty)$ is non-decreasing continuous function and $\varphi(t) < t$ for all $t > 0$, we obtain that

$$\begin{aligned}
d_\theta(w_i(x', y'), w_i(x'', y'')) &\leq (|a_i| + \theta k)|x' - x''| + \theta\varphi(|y' - y''|) \\
&= (|a_i| + \theta k)|x' - x''| + \theta \frac{\varphi(|y' - y''|)}{|x' - x''| + |y' - y''|}(|x' - x''| + |y' - y''|) \\
&= (|a_i| + \theta k + \theta \frac{\varphi(|y' - y''|)}{|x' - x''| + |y' - y''|})|x' - x''| \\
&\quad + \theta \frac{\varphi(|y' - y''|)}{|x' - x''| + |y' - y''|}(|y' - y''|) \\
&\leq (|a_i| + \theta k + \theta \frac{\varphi(|x' - x''| + |y' - y''|)}{|x' - x''| + |y' - y''|})|x' - x''|
\end{aligned}$$

$$
\begin{aligned}
&+\theta \tfrac{\varphi(|x'-x''|+|y'-y''|)}{|x'-x''|+|y'-y''|}(|y'-y''|)\\
\le\ & (|a_i|+\theta k+\theta)|x'-x''| + \theta \tfrac{\varphi(|x'-x''|+|y'-y''|)}{|x'-x''|+|y'-y''|}|y'-y''|)\\
\le\ & \max\{|a_i|+\theta k+\theta, \tfrac{\varphi(|x'-x''|+|y'-y''|)}{|x'-x''|+|y'-y''|}\}(|x'-x''|+\theta|y'-y''|)\\
=\ & \max\{|a_i|+\theta k+\theta, \tfrac{\varphi(|x'-x''|+|y'-y''|)}{|x'-x''|+|y'-y''|}\}d_\theta((x',y'),(x'',y''))\\
\le\ & \max\{\max_{i=1,2,\ldots,N}|a_i|+\theta k+\theta, \tfrac{\varphi(|x'-x''|+|y'-y''|)}{|x'-x''|+|y'-y''|}\}d_\theta((x',y'),(x'',y'')).
\end{aligned}
$$

Since $N > 1$, we obtain $0 < a_i := \frac{x_i - x_{i-1}}{x_N - x_0} < 1$ for all $i = 1, 2, \ldots, N$.

Let

$$\theta := \frac{1 - n = 1, 2, \ldots, N|a_i|}{2(k+1)}.$$

Then $0 < n = 1, 2, \ldots, N|a_i| + \theta k + \theta < 1$ and since $k \ge 0$, we obtain $0 < \theta < 1$.

Let for all $t > 0$,

$$\beta(t) := \max\{\max_{i=1,2,\ldots,N} |a_i| + \theta p + \theta, \frac{\varphi(t)}{t}\}.$$

Then because $\alpha(t) := \frac{\varphi(t)}{t}$ and $\alpha : (0, +\infty) \to [0, 1)$ is a non-increasing, we can see that $\beta : (0, +\infty) \to [0, 1)$ is a non-increasing and for each $(x', y'), (x'', y'') \in I \times \mathbb{R}$, $(x', y') \neq (x'', y'')$,

$$d_\theta(w_i(x', y'), w_i(x'', y'')) \le \beta(d((x', y'), (x'', y'')))d_\theta((x', y'), (x'', y'')),$$

where $d((x', y'), (x'', y'')) := |x' - x''| + |y' - y''|$. Since $0 < \theta < 1$, for all $(x', y'), (x'', y'') \in I \times \mathbb{R}$, $(x', y') \neq (x'', y'')$,

$$|x' - x''| + \theta|y' - y''| \le |x' - x''| + |y' - y''|.$$

That is,

$$d_\theta((x', y'), (x'', y'')) \le d((x', y'), (x'', y'')).$$

Since $\beta : (0, +\infty) \to [0, 1)$ is a non-increasing, we can see that

$$
\begin{aligned}
d_\theta(w_i(x', y'), w_i(x'', y'')) &\le \beta(d((x', y'), (x'', y'')))d_\theta((x', y'), (x'', y''))\\
&\le \beta(d_\theta((x', y'), (x'', y'')))d_\theta((x', y'), (x'', y'')).
\end{aligned}
$$

Hence w_i are Rakotch contractions in $(I \times \mathbb{R}, d_\theta)$. On the other hand, metric d_θ is equivalent to the Euclidean metric d_0 on $I \times \mathbb{R}$ (see [11]). So $(I \times \mathbb{R}, d_\theta)$ is a complete metric space. Hence $w_i : I \times \mathbb{R} \to I \times \mathbb{R}$ is a Rakotch contraction in $(I \times \mathbb{R}, d_\theta)$ and by Theorem 1 (1), there exists an unique fixed point in $I \times \mathbb{R}$. By

Theorem 1 (2), for the complete metric space $(I \times \mathbb{R}, d_\theta)$, there is a unique nonempty compact set $G \subset I \times \mathbb{R}$ such that

$$G = \bigcup_{i=1}^{N} w_i(G).$$

By the definition of Hausdorff metric, equivalence of two metrics implies the equivalence of Hausdorff metrics generated by them (see [10], p.91, Lemma 3.6). Hence for $(I \times \mathbb{R}, d_0)$, there is a unique nonempty compact set $G \subset I \times \mathbb{R}$ such that

$$G = \bigcup_{i=1}^{N} w_i(G).$$

This completes the proof. □

Remark 5 The boundedness of s_i is the essential condition to establish a unique invariant set of an iterated function system.

In the fractal interpolation function with function vertical scaling factors, $0 < \max_{x\in I} |d_i(x)| < 1$ (see [5], see [14], p. 3–4, cf. [2]). Let $M := \max_{x\in I} |c_i x + f_i|$ and $h \geq \frac{M}{1-\max_{x\in I} |d_i(x)|}$. Then for all $y \in [-h, h]$,

$$\begin{aligned} |F_i(x, y)| &= |c_i x + d_i(x)y + f_i| \\ &\leq M + \max_{x\in I} |d_i(x)||y| \\ &\leq M + \max_{x\in I} |d_i(x)|h \leq h. \end{aligned}$$

So for all $(x, y) \in I \times [-h, h]$, we can see that $F_i(x, y) \in [-h, h]$. That is, an iterated function system,

$$\{I \times [-h, h]; w_i : i = 1, 2, \ldots, N\},$$

has been constructed (cf. [3], p. 1897). Thus $D(s_i) = [-h, h]$ and $s_i(y) := \max_{x\in I} |d_i(x)|y$ is bounded in $D(s_i)$ (see Remark 3). Hence the boundedness of s_i in $D(s_i)$ is the essential condition to establish a unique invariant set of an iterated function system (see the proof of Theorem 4, cf. [3], p. 1897, the proof of Theorem 2.1).

Remark 6 Our result is a substantial generalization of [1, 11, 14]. The function whose graph is the attractor of an IFS as described in Theorem 3 and Theorem 4 generalizes the affine fractal interpolation function (see [1]), the fractal interpolation function with function vertical scaling factors (see [5], see [14]) and the nonlinear fractal interpolation function (see. [11]). (1) In the affine fractal interpolation function (cf. [1], p.308, Example 1), for all $t > 0$,

$$\varphi(t) := \max_{i=1,2,\ldots,N} |d_i| t,$$

where $|d_i| < 1$ for all $i = 1, 2, \ldots, N$.
(2) In the fractal interpolation function with function vertical scaling factors (cf. [5], cf. [14], p. 3), for all $t > 0$,

$$\varphi(t) := \max_{i=1,2,\ldots,N} \max_{x \in I} |d_i(x)| t,$$

where $d_i(x)$ is Lipschitz function defined on I satisfying $\sup_{x\in I} |d_i(x)| < 1$ for all $i = 1, 2, \ldots, N$. (3) In the nonlinear fractal interpolation function (cf. [11]), $d_i(x) \equiv 1$ and $\varphi : (0, +\infty) \to (0, +\infty)$ is a non-decreasing continuous function such that for any $t > 0$, $\alpha(t) := \frac{\varphi(t)}{t} < 1$ and the function $(0, +\infty) \ni t \to \frac{\varphi(t)}{t}$ is non-increasing. Thus, we improve upon results proved by [1, 11, 14].

6 A Certain Generalized Fractal Interpolation Function

In this section we focus on generalized FIFs with a special structure by means of results obtained in the previous sections. For this purpose, we give an extremely explicit simple example to demonstrate the effectiveness of the preceding theory. We may assume, without loss of generality, that $[x_0, x_N] = [0, 1]$. This special case can always be achieved by means of an affine transformation (which does not change the existence of FIF) (see [2]).

Let $\varphi(t) := \frac{t}{1+t}$ for $t \in (0, +\infty)$. Then $\varphi : (0, +\infty) \to (0, +\infty)$ is a non-decreasing continuous function and $t \to \frac{\varphi(t)}{t}$ is non-increasing continuous function.

Let a set of data $\{(x_i, y_i) : i = 0, 1, \ldots, N\}$ be given, where $x_i, y_i \in [0, 1]$ for all $i = 0, 1, \ldots, N$. Let for all $i = 1, 2, \ldots, N$,

$$d_i(x) := 2^{2i} x^i (1 - x)^i.$$

Then

$$\max_{x\in[x_0,x_N]} |d_i(x)| = 1$$

and by differential mean value theorem, for all $x', x'' \in [0, 1]$, there is $L_{d_i} > 0$ such that

$$|d_i(x') - d_i(x'')| \le L_{d_i} |x' - x''|.$$

Let for $y \in [0, +\infty)$ and $i = 1, 2, \ldots, N$,

$$s_i(y) := \frac{y}{1 + iy}.$$

Then, for $y', y'' \in [0, +\infty)$,

$$|s_i(y') - s_i(y'')| = |\frac{y'}{1+iy'} - \frac{y''}{1+iy''}| \leq \frac{|y'-y''|}{1+i|y'-y''|}$$
$$\leq \frac{|y'-y''|}{1+|y'-y''|} = \varphi(|y'-y''|).$$

That is, each s_i is Rakotch contraction (with the same function φ) that is not Banach contraction on $[0, +\infty)$ (see [11], cf. [12], p. 848, cf. [9], p. 262).
Let for all $i = 1, 2, \ldots, N$,

$$w_i(x, y) := (a_i x + e_i, c_i x + d_i(x)s_i(y) + f_i),$$

where

$$a_i = x_i - x_{i-1}, \quad e_i = x_{i-1},$$
$$c_i = y_i - y_{i-1}, \quad f_i = y_{i-1}.$$

Then, by Theorems 3 and 4, there exists a continuous function $f : [0, 1] \to \mathbb{R}$ that interpolates the given data $\{(x_i, y_i) : i = 0, 1, \ldots, N\}$. Moreover, the graph G of f is invariant with respect to $\{[0, 1] \times \mathbb{R}; w_1, w_2, \ldots, w_N\}$, i.e.,

$$G = \bigcup_{i=1}^{N} w_i(G).$$

This clearly shows that our method is much more effective than the method due to [1, 11].

Remark 7 We refer to f as a generalized fractal interpolation function with function vertical scaling factors. The reason for this name is that the functions F_i take the form

$$F_i(x, y) = c_i x + d_i(x)s_i(y) + f_i,$$

where $\max_{x \in I} |d_i(x)| \leq 1$ and each s_i is Rakotch contraction.

That is, each F_i, in general, is generalized with respect to the second variable y (cf. [11]).

In fact, in [1, 14], since $0 < |d_i(x)| \equiv |d_i| < 1$ or $0 < \max_{x\in I} |d_i(x)| < 1$ and

$$d_i(x)y = \frac{d_i(x)}{\max_{x\in I} |d_i(x)|} \max_{x\in I} |d_i(x)|y,$$

we can see that

$$\begin{aligned} F_i(x, y) &= c_i x + d_i(x)y + f_i \\ &= c_i x + d_i^*(x)s_i(y) + f_i, \end{aligned}$$

where $d_i^*(x) := \frac{d_i(x)}{\max_{x\in I} |d_i(x)|}$ and $s_i(y) := \max_{x\in I} |d_i(x)|y$, and thus each s_i is a special Banach contraction and linear. That is, each $F_i(x, y)$ is a special Banach contraction and linear with respect to the second variable y. Then the corresponding FIF is an affine FIF introduced by Barnsley (see [1]) or a FIF with function vertical scaling factors (see [5], see [14], cf. [2]). Obviously, we can say that the generalized FIFs with function vertical scaling factors may have more flexibility and applicability.

In below, we give the graph of an affine FIF, the graph of a nonlinear FIF of [11] that is not an affine FIF, and the graph of a generalized FIF that is not a FIF of [11] (see Figs. 1, 2, 3). Here we omit their details to avoid the repetition.

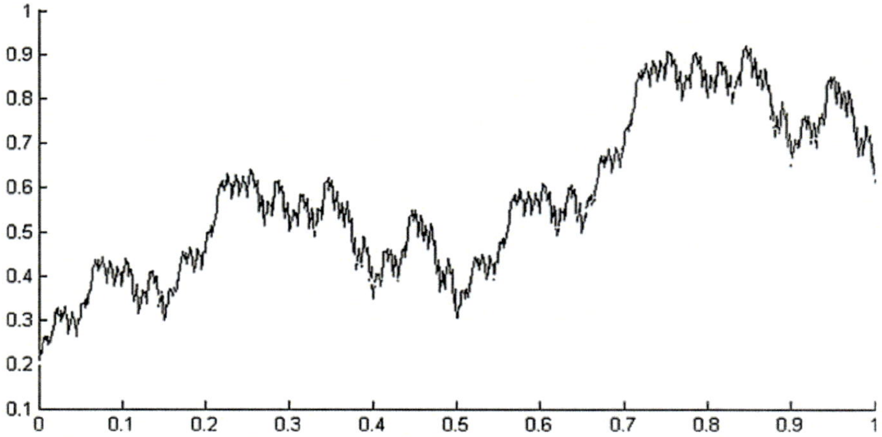

Fig. 1 An affine FIF

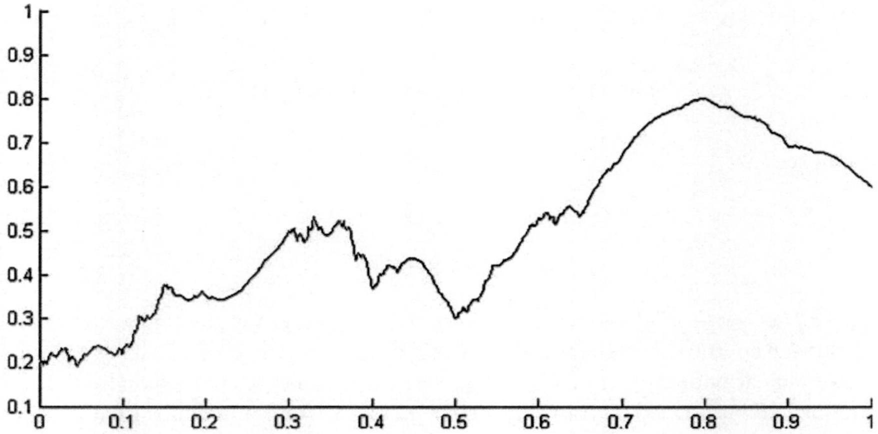

Fig. 2 A nonlinear FIF of [11] that is not an affine FIF

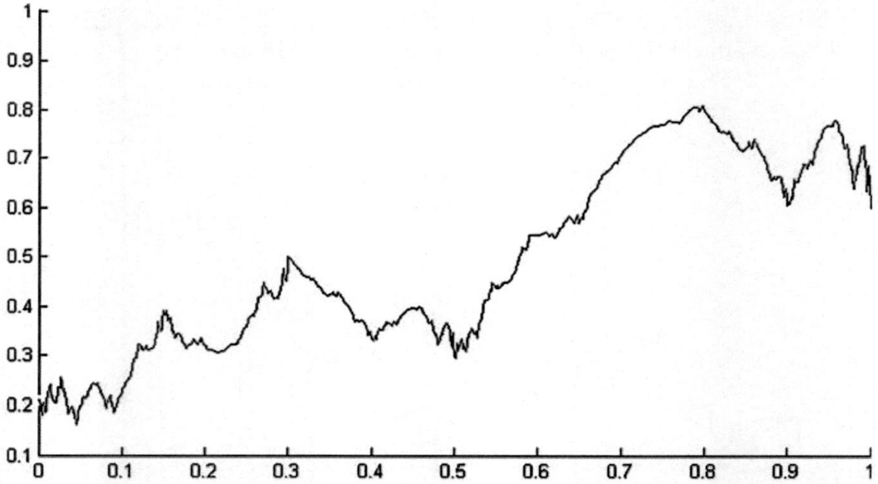

Fig. 3 A generalized FIF that is not a FIF of [11]

7 Conclusions

The FIFs have been widely used in approximation theory, image compression, computer graphics and modeling of natural surfaces such as rocks, metals, terrains and so on. In order to get more flexibility and diversity in modeling natural shapes and phenomena or in image processing, we introduced new generalized FIFs which generalize widely used linear FIFs. In order to obtain new generalized FIFs, we use Rakotch contractions and special function vertical scaling factors, and we have presented the principle and the method of generalized fractal interpolation in detail. Dealing with generalized fractal interpolation functions is better than the one provided by [11]. Theorems 3 and 4 ensure that an attractor of constructed generalized IFS is a graph of some continuous function which interpolates the given data. In particular, an explicit illustrative example shows that our result remains still true under essentially weaker conditions on the maps of IFS. Comparing linear FIFs with function vertical scaling factors and generalized FIFs with function vertical scaling factors, we can know that the FIFs considered in this paper have more flexibility and diversity and are more suitable to the fitting and approximation of many complicated functions.

References

1. M.F. Barnsley, Fractal functions and interpolation. Constr. Approx. **2**, 303–329 (1986)
2. M.F. Barnsley, P.R. Massopust, Bilinear fractal interpolation and box dimension. J. Approx. Theory **192**, 362–378 (2015)
3. Z. Feng, Y. Feng, Z. Yuan, Fractal interpolation surfaces with function vertical scaling factors. Appl. Math. Lett. **25**(11), 1896–1900 (2012)
4. G. Gwóźdź-Łukawska, J. Jachymski, The Hutchinson-Barnsley theory for infinite iterated function systems. Bull. Aust. Math. Soc. **72**, 441–454 (2005)
5. W. Hong-Yong, Z.L. Fan, Analytical characteristics of fractal interpolation functions with function vertical scaling factors. Acta Math. Sinica (Chin. Ser.) **54**(1), 147–158 (2011)
6. J. Jachymski, I. Jóźwik, Nonlinear contractive conditions: a comparison and related problems. Banach Center Publ. **77**, 123–146 (2007)
7. P.R. Massopust, *Fractal Functions, Fractal Surfaces and Wavelets* (Academic Press, San Diego, 1994)
8. E. Rakotch, A note on contractive mappings. Proc. Am. Math. Soc. **13**, 459–465 (1962)
9. B.E. Rhoades, A comparison of various definitions of contractive mappings. Trans. Am. Math. Soc. **226**, 257–290 (1977)
10. S. Ri, A new fixed point theorem in the fractal space. Indagationes Mathematicae **27**(1), 85–93 (2016)
11. S. Ri, A new nonlinear fractal interpolation function. Fractals **25**(6), 1750063 (12 pages) (2017)
12. N.-A. Secelean, Generalized iterated function systems on the space $l^{\infty}(X)$. J. Math. Anal. Appl. **410**(2), 847–858 (2014)
13. F. Strobin, Attractors of generalized IFSs that are not attractors of IFSs. J. Math. Anal. Appl. **422**(1), 99–108 (2015)
14. H.-Y. Wang, Yu. Jia-Shan, Fractal interpolation functions with variable parameters and their analytical properties. J. Approx. Theory **175**, 1–18 (2013)

Analysis of Thermal and Quantum Escape Times of Josephson Junctions for Signal Detection

G. Filatrella, C. Barone, Claudio Guarcello, A. S. Piedjou Komnang, Vincenzo Pierro, A. Rettaroli, and S. Pagano

Abstract In this work we investigate the limits to the possibility to reveal the existence of weak microwave signals through Josephson junctions. Even if the Josephson element is capable to reveal the electromagnetic field, thermal noise is to be quantified by means of signal theory, as a confounding factor that limits the detection. We show how the decision problem can be embedded in the frame of signal detection. As a consequence, the optimization of the detection probability and the minimization of the false alarm probability give a guide to select the Josephson junction parameters that best suit the purpose.

Keywords Josephson junctions · Escape time · Signal detection · Particle detection

G. Filatrella (✉)
INFN Gruppo Collegato Salerno and Department of Sciences and Technologies, University of Sannio, Via F. De Sanctis, 82100 Benevento, Italy
e-mail: giovanni.filatrella@unisannio.it

C. Barone · S. Pagano
INFN, Gruppo Collegato Salerno and Department of Physics, University of Salerno, Via Giovanni Paolo II, 84084 Fisciano (SA), Italy

C. Guarcello · A. S. Piedjou Komnang
Department of Physics, University of Salerno, Via Giovanni Paolo II, 84084 Fisciano (SA), Italy

V. Pierro
INFN Gruppo Collegato Salerno and Department of Engineering, University of Sannio, Corso Garibaldi 107, 82100 Benevento, Italy

A. Rettaroli
INFN, Laboratori Nazionali di Frascati, Frascati (Roma) Italy and Department of Mathematics and Physics, University of Roma Tre, 00100 Roma, Italy

C. H. Skiadas et al. (eds.), *13th Chaotic Modeling and Simulation International Conference*, Springer Proceedings in Complexity,
https://doi.org/10.1007/978-3-030-70795-8_14

1 Introduction

When Josephson junctions (JJ) are employed as detectors of microwave signals, they can reach a very high response (of the order of kV/W [32]), close to the quantum sensitive limit [14, 15, 26]. Moreover, as superconducting elements, the devices can work no matter how the temperature is lowered, thus allowing to minimize thermal noise, at least until the contribution to the escapes through quantum tunneling processes becomes statistically dominant. These premises are, in a nutshell, the basis for the intensity of the efforts devoted to the development of highly-sensitive detectors based on JJ [3], as well as Josephson calorimeters [16, 35]. To reveal even very weak electromagnetic signals, the energy to induce the transition between two states of the JJ should be close to the single photon energy. As the interaction between the JJ and the signal is generally mediated by a resonant cavity, also the latter excitation should be on the same energy scale [20]. Thus, for a more detailed description of the device potentialities and limits, it is necessary to embed the problem in the context of microwave photon manipulation through superconducting electronics [1, 13]. In doing so, the energy landscape of the device, as well as the antenna interaction with the microwave photon field, are essential for a device description [4]. It is in fact the JJ potential that dictates the disturbing switches due to thermal escapes and quantum tunneling [4, 33]. A model for the detector consists of a current drive that perturbs the JJ dynamics and favors the switches to a finite voltage state; the statistics of the escape times are analyzed to highlight the presence of the perturbation.

Several remarkable achievements have been reached so far. The minimum photon content that can be revealed through JJ has been estimated in the order of 10^2, a limit that can be possibly lowered [36]. It has also been demonstrated that it is possible to resolve the number of photons in a propagating mode [11]. Quite naturally, to decide about the presence of a weak signal, it is necessary to achieve a silent enough state, that is a device capable to stay quite if no signal is applied. Some schemes that exploit the phase diffusion regime have been recently proposed with such purpose [31]. However, for the purpose to exploit the high sensitivity of JJ in the search for photons resulting from elusive particles as axions [3, 7–10, 22, 29, 37], it is necessary to set up a different detection scheme to ascertain the existence of extra photons, above and beside the background [5]. The purpose of this work is to analyze a scheme for the detection of such excess photons through the analysis of the JJ *switching currents* [2, 30], embedded in the frame of signal detection [12]. In fact, as the Josephson phase is of quantum nature it is not directly accessible, detection is possible only if the photons cause a *switch*, a passage from the superconducting state to a finite voltage state (mathematically, the problem amounts to determine the first passage time across the separatrix of a potential well [17]). As the passage also occurs because of thermal fluctuations, a careful analysis is necessary to prove that the passages are a consequence of some external field. The framework of signal detection allows to make simple estimates of the temperature constraints and of the experimental set-up. As a consequence, the optimization of the detection probability (and the minimization of the false alarm probability) gives a guide to select the JJ parameters that best suit to reveal weak microwave signals. In brief, detection amounts to the following question: how is it possible to infer the presence of microwave extra photons, apart the thermal ones, from the analysis of the switches to the finite voltage of a JJ? A tentative answer to this question is the subject of the present paper, organized as follows: in Sect. 2 the problem will be extensively formulated, alongside with the description of an electrical model for the dynamics of a JJ and of a perturbation. In Sect. 3 the signal analysis indications for the performances of the detection will be recounted. Last Sect. 4 concludes.

2 The Problem

We shall consider a small tunnel JJ coupled to the environment through the bias current:

$$C\frac{\hbar}{2e}\frac{d^2\varphi}{dt'^2} + \frac{1}{R}\frac{\hbar}{2e}\frac{d\varphi}{dt'} + I_0 \sin\varphi = I_b + I_N(t') + I_S(t') \tag{1}$$

The right hand side of this equation summarizes the current supplied to the junction. In this formulation, therefore, one assumes that the impedance of the Josephson element is much lower than the impedance of any source, and consequently the external world is seen as a current source (although other configurations are possible, in which a voltage bias is more appropriated [24]). Equation (1) includes inertia (determined by the capacitance C), dissipation (as governed by the dissipative element R) and fluctuations (the random current $I_N(t')$ supplied by the resistance), the nonlinear periodic term (the oscillating Josephson current of amplitude I_0), a constant bias current I_b, and the signal current $I_S(t)$ (that summarizes the effects of the photon field). Fluctuations are assumed to be Gaussian with:

$$\langle I_N(t')\rangle = 0, \tag{2}$$

$$\langle I_N(t')I_N(t'-s')\rangle = \frac{2k_BT}{R}\delta(t'-s'). \tag{3}$$

where k_B is the Boltzmann constant, T is the absolute temperature, δ the Dirac function, $\langle\cdot\rangle$ is the expectation operator.

The usual normalized units [6, 34] are as follows (here, as usual, $\hbar$ is the reduced Planck constant, and e is the elementary charge):

- The current is normalized to the critical current I_0:

$$\gamma = \frac{I_b}{I_0}. \tag{4}$$

- Time is normalized to ω_J^{-1}, where $\omega_J = \sqrt{\hbar C/2eI_0}$ is the frequency of the linear oscillator:

$$t = t'\omega_J. \tag{5}$$

Introducing the normalized temperature $D = k_B T \omega_J / R I_0^2$ and the normalized conductance $1/\beta_c = (1/R)\sqrt{\hbar/(2eI_0C)}$, leads to the normalized versions of (1), (2), (3):

$$\frac{d^2\varphi}{dt^2} + \frac{1}{\beta_c}\frac{d\varphi}{dt} + \sin\varphi = \gamma + \gamma_N(t) + \gamma_S(t), \tag{6}$$

$$\langle \gamma_N(t) \rangle = 0, \tag{7}$$

$$\langle \gamma_N(t)\gamma_N(t-s) \rangle = 2D\delta(t-s), \tag{8}$$

where γ's indicate the normalized current terms. The dynamics of the JJ is characterized by the bias-dependent small oscillation frequency:

$$\omega_0(\gamma) = \left(1-\gamma^2\right)^{1/4} \tag{9}$$

and the energy barrier that cages the dynamics is:

$$\Delta U(\gamma) = 2\left[\sqrt{1-\gamma^2} - \gamma\cos^{-1}(\gamma)\right]. \tag{10}$$

Finally, the photon field γ_S is supposed to be modeled as a succession of normalized impulses of amplitude A and duration δ_T that arrive regularly with a period T. The response of a JJ to such a field, and how the response can be exploited to infer the existence of the field, is the problem dealt with in this paper. Some indications are presented in the next Section.

3 Results

A JJ described by (6), (7), (8) can undergo switches between the zero voltage and the finite voltage, as shown in Fig. 1. When the system switches, the phase difference increases and, according to the a.c. Josephson relation $V = (h/2e)d\varphi/dt$ [18, 19], a voltage drop across the junction can be measured. The switches can occur either because of thermal current (7), (8), or because of the current pulse train. The starting point is therefore to collect the sequence of times at which a passage has occurred in the absence of the photon field drive:

$$t_1^0,\ t_2^0, \ldots, t_N^0. \tag{11}$$

For this case, one expects that, on average, the escapes (the passages across the separatrix between the localized oscillations and unbound runaways) occur with a rate given by the Kramers approximation [17]:

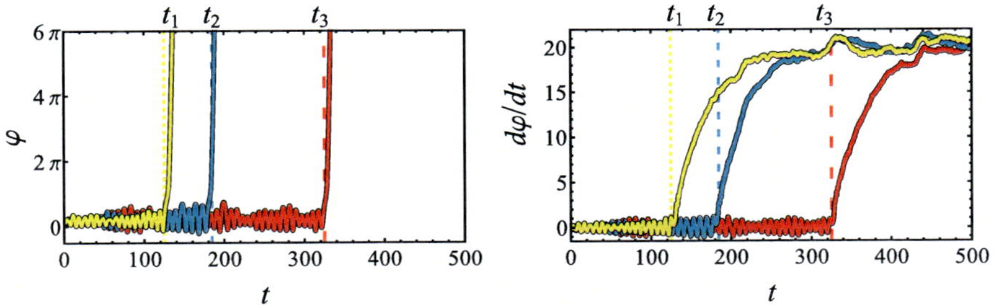

Fig. 1 Time dependent dynamics of the phase φ and the voltage $d\varphi/dt$. The dashed lines denote the time at which a passage to the finite voltage is detected

$$r_0(\gamma, D) = \frac{\omega_0}{2\pi} e^{-\frac{\Delta U(\gamma)}{D}} \tag{12}$$

(for the Gaussian noise; other kinds of noise give pretty different results [25]).

In the presence of a signal, which mimics the absorption of some photons that is capable to induce a current into the JJ, the sequence of switches is presumably altered, and the photons can be possibly revealed through deviations from the purely thermal sequence (11); let us call the sequence of the escapes in the presence of the extra-photons:

$$t_1^1, t_2^1, \ldots, t_N^1. \tag{13}$$

This sequence is expected to have a larger number of events in the same observation time respect to the unperturbed case (11). An example of two histograms for $\{t^0\}$ (no signal) and $\{t^1\}$ (with signal) are shown in Fig. 2. We underline that, at variance with the cases in which one is interested in counting the number of photons (e.g., [20]), to reveal a source of photons (e.g., the presence of an axion field [3]) it suffices to statistically determine if an observed collection of switches is more likely drawn from sequence (11) or from sequence (13).

In the simplest approximation, one expects that the rate r_1 associated to the escapes (13) in the presence of photons is just the sum of the unperturbed rate r_0 and an additional rate r_A due to the photons:

$$r_1(\gamma, D, \gamma_S) = r_0(\gamma, D) + r_A. \tag{14}$$

In our settings, where the photons are modeled as current pulses with a period T, we have $r_A = 1/T$ for a most efficient detector.

Apart the analytical approximations embodied in (14), for our line of reasoning it is important to underline that the rates r_0 and r_1 are statistical averages, for (12) and (14) predict what happens on average. Thus, in a given measurement time P one expects $1/\langle t_i^0 \rangle = n_0 = P r_0$ escapes in the purely thermal case, and $1/\langle t_i^1 \rangle = n_1 = P r_1$ escapes in the presence of the extra photons, with the obvious inequality

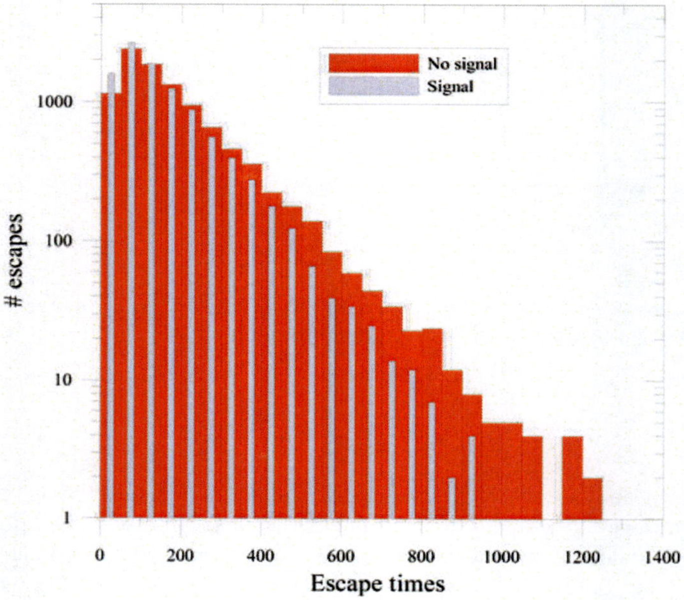

Fig. 2 Histograms of $N = 10^4$ switching times. Parameters of the simulations are: $\gamma = 0.8$, $D = 0.1$, $1/\beta_c = 0.025$. The impulses that mimic the photon fields are of amplitude $A = 0.5$, duration $\delta\tau = 10$, and arrive regularly with a period $T = 100$

$n_1 > n_0$. In an actual measurement, one observes a certain number of switches, say n, and a decision is to be made: in which sense the measured number favors an hypothesis (the switches are just due to thermal activation) or the other (there are extra switches due to the photon field)? Naturally, the more the two rates are apart, the more likely is that the measurement is a clear cut decision. Also, no matter how close they are, with a sufficiently long measurement time P it is always possible to discriminate the two conditions. To make these ideas quantitative is the objective of the next Section.

3.1 *Statistical Analysis of the Switching Times*

To quantify the efficiency of the detection of a photon field one can introduce the Kumar-Caroll (KC) index d_{KC} [21], in analogy with the detection of continuous sinusoidal signals [2, 12]:

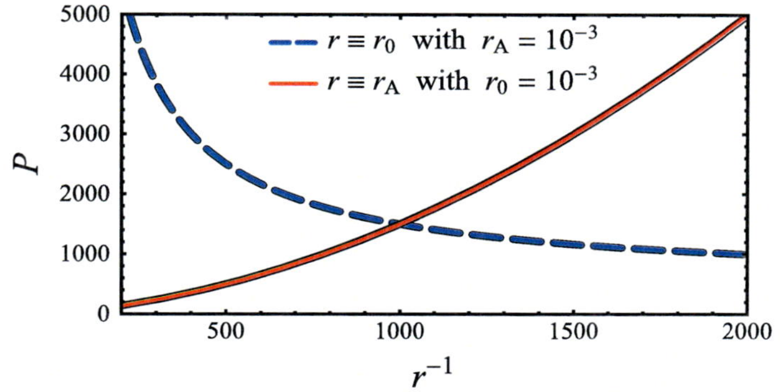

Fig. 3 Observation time P, according to estimate (16), as a function of r_A at a fixed $r_0 = 0.001$ (solid red curve) and as a function of r_0 at a fixed $r_A = 0.001$ (dashed blue curve)

$$d_{\text{KC}} = \frac{\left|\langle t^1\rangle - \langle t^0\rangle\right|}{\sqrt{\frac{1}{2}\left[\sigma^2\left(\langle t^1\rangle\right) + \sigma^2\left(\langle t^0\rangle\right)\right]}} \tag{15}$$

where $\langle t^{0,1}\rangle$ is the average switching time in the absence (presence) of the signal, and $\sigma^2\left(\langle t^{0,1}\rangle\right)$ the corresponding variances of the average. This index is a proxy for the Signal-to-Noise-Ratio (SNR) [21], and as such will be used in this paper.

Assuming a large number of events in the observation time P, and that the escapes due to the photons are an additional rate independent of the thermal rate, and that a reliable detection requires at least $d_{KC} = 1$, one obtains the relation [27]:

$$P\, r_A^2 - \frac{1}{2} r_A - r_0 = 0, \tag{16}$$

between the rate of the photons r_A, the thermal escape rate r_0, and the observation time P. Inspection of (16) reveals that the observation time P is positively related to the thermal rate r_0, see Fig. 3.

Some further elaborations of the estimate (16) are relevant. To begin with, one can define a ratio between the photon arrivals and the thermal spontaneous escapes:

$$r_0 = x\, r_A$$

that allows to explicitly observe that:

$$d_{KC} = \sqrt{\frac{2 P r_A}{2x + 1}}. \tag{17}$$

This equation confirms the intuitive scaling between the observation time P, the photon arrival rate, and the ratio between the thermal and photon rates. In particular,

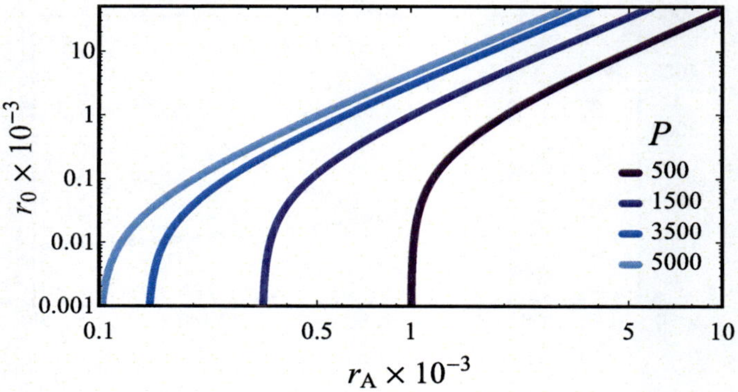

Fig. 4 Relation between the thermal escape rate as a function of the photon arrival time, according to estimate (16), for different values of P

one can insert the matching condition $P = 1/r_A$ in (17) to obtain the relation between the observation time, the temperature and bias point through (9)–(12).

In Fig. 3 it is displayed the behavior of the observation time P, calculated according to (16), as a function of the inverse photon rate, r_A^{-1} (see the red solid curve). If the photon rate is low, e.g. around 10^{-3} Hz, the measurement time for a $d_{KC} = 1$ reads $\sim$1500 s. It is also noticeable that the dependence is quadratic. Should the photon rate be much smaller than the assumed mHz, the detection could prove unfeasible. Conversely, if the photon arrival rate is relatively high, one can considerably increase also the thermal rate, and hence higher temperatures are allowed.

In Fig. 4 it is displayed the behavior of the thermal escape rate, r_0, as a function of the photon arrival time, r_A, for different values of the observation time P. It is evident that the thermal escape rate (that is, the temperature of the system) can be increased as the photon rate increases. However, for any value of P, there is a threshold value of r_A (e.g., $r_A \simeq 10^3$ for $P = 500$) below which thermal escapes to achieve a $d_{KC} = 1$ become vanishingly small. The latter condition entails extremely low temperatures.

In brief, if some SNR is to be reached, the trade-off between the parameters of the experiment can be evaluated. A more detailed analysis of the problem requires to retrieve the index (15) from numerical simulations of the model equations (6), (7), (8), as we shall do below.

3.2 An Example of the Kumar-Carrol Index Usage

To illustrate an application of the signal-to-noise index (15), let us suppose that the index has been computed for several values of the bias current γ to optimize the bias level. To fix the ideas, let us suppose that the highest value of d_{KC} corresponds to $\gamma = 0.8$; the escapes retrieved for such choice are shown in Fig. 2. The index reads

Table 1 Statistics of the escape times of Fig. 2. Here N is the number of switches, $\langle t_i \rangle$ the average escape time, $\sigma(t_i)$ the standard deviation, $\sigma(\langle t_i \rangle_N)$ the standard deviation of the average. The resulting KC-index reads $d_{KC} = 20.8$

Signal	N	$\langle t_i \rangle$	$\sigma(t_i)$	$\sigma(\langle t_i \rangle_N)$
Absent	10^4	182.7	151	1.51
Present	10^4	154.0	124	1.24

$d_{KC} = 20.8$, that indicates a very good SNR—would it be possible to collect 10^4 escapes, the average of the exit times could give a clear cut indication of the presence, or not, of the excess photons. In fact the statistics of the escapes can be summarized in Table 1.

It is evident that for the case under examination the detection performs extremely well. With a KC-index around 20 the SNR is extremely high, and also intuitively there is no doubt that, if the average escape time decreases from ~180 to ~150, something has happened and a signal is present. This is quite reasonable, for the switches occur on average with the same rate as in the incoming pulses period $T = 100$; therefore the efficiency is very high (almost each pulse causes a switch) and the number of data is conspicuous ($N = 10^4$). Under these circumstances, the statistical analysis is just a confirm of the intuition. However, the approach proves useful for the design of an experiment if it is not possible to collect as many as 10^4 switches. Let us suppose, *ceteris paribus*, that the rate of arrival of the pulses is extremely low, say $1/r_A \sim 1h$, and therefore the number of events that can be collected in a day of measurements is around $N = 20$. A principal question would be: how many days of data collection should be planned to decide, with good confidence, about the existence of the pulses?

Let us thus suppose that the actual sample of data to be analyzed consists of M escape times, $M \ll N$, say $M = 20$ to fix ideas. A collection of 500 average escape times (over $M = 20$ events) retrieved binning the 10^4 data of Fig. 2, is shown in Fig. 5. In a single measurement run in which 20 switches are collected, a single average escape time would be obtained, and not always the same value for the statistical fluctuations. Let us summarize the data in Table 2.

In the first place, let us remind of a subtle difference between Figs. 2 and 5, as perhaps better explained in the corresponding Tables 1 and 2. Figure 2 is a collection of all data, any single switch that has occurred. Figure 5 is a collection of hypothetical repetition of the average over 20 switches. The statistical test to decide about the presence of the signal is to be performed on the average over all data in the case of Fig. 2, and hence the very high SNR, compared to the test on a single event of Fig. 5, to which pertains a much smaller SNR.

For the central limit theorem one can assume that the distribution of the averages is approximately Gaussian, centered on the population average (that is estimated through the average of the $N = 10^4$ data) and with a standard deviation which is smaller than the population standard deviation.

Applying the signal analysis means to determine the error of the first type α and of the second type β when a decision on the existence of the photon field is

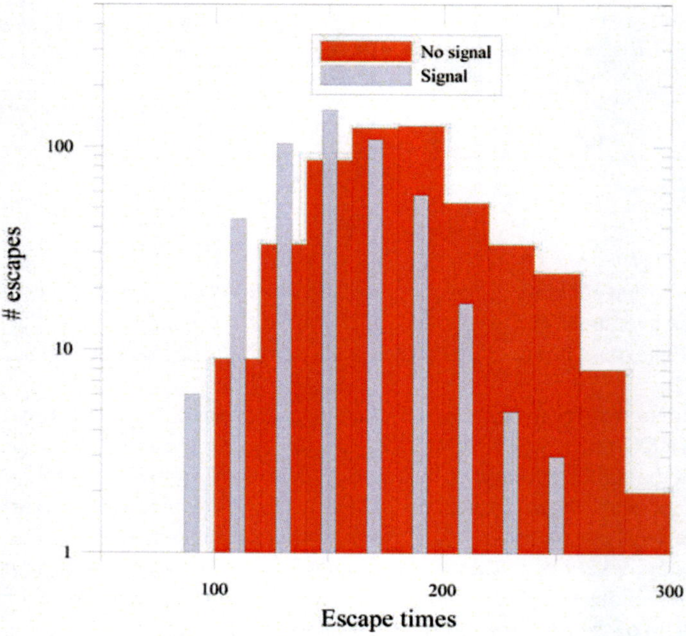

Fig. 5 Histograms of the average over 20 switching times, for the same data of Fig. 2, and thus consisting of $N' = 500$ data. The other parameters are therefore the same as in Fig. 2

Table 2 Statistics of the escape times of Fig. 5, that is the escape times averaged over $M = 20$ events. Here $N' = N/M = 500$ is the number of means, $\langle t_i \rangle$ the average escape time over $N' \times M$ escapes, $\sigma\left(\langle t_i \rangle_{N'}\right)$ the standard deviation of the N' means, $\sigma\left(\langle t_i \rangle_{N' \times M}\right)$ the standard deviation of the overall average. Naturally, $\langle t_i \rangle$ and $\sigma\left(\langle t_i \rangle_{N' \times M}\right)$ coincide with Table 1, as they are computed on the same set of data. The resulting KC-index for a single measurement (that is, an average over 20 data) reads $d_{KC} = 0.93$

Signal	N'	$\langle t_i \rangle$	$\sigma\left(\langle t_i \rangle_{N'}\right)$	$\sigma\left(\langle t_i \rangle_{N' \times M}\right)$
Absent	500	182.7	33.8	1.51
Present	500	154.0	27.7	1.24

to be made, on the basis of the escape time average. Let us do so with the help of the Gaussian approximation, that is to assume that the histogram of the average escape time of Fig. 5 is Gaussian distributed, as schematically illustrated by Fig. 6. To decide between the hypothesis one usually places a threshold for the measured average escape time, $\langle t \rangle_{th}$, see Fig. 6. The decision will favor the hypothesis "1" (the photon field is there) if the actual measured time is below the threshold, and obviously the complementary hypothesis (there is no photon field) will be selected if the actual measured time is above. The choice of the threshold reflects on the decision errors: either the probability of false alarm (type I error α), or the probability of missing

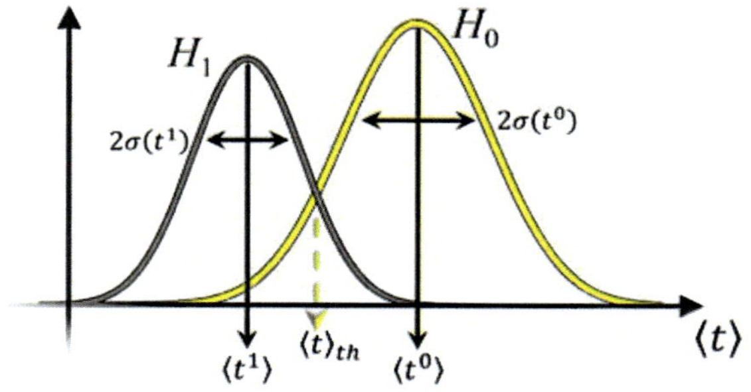

Fig. 6 Application of the detection scheme to the averaged data assumed to be Gaussian distributed

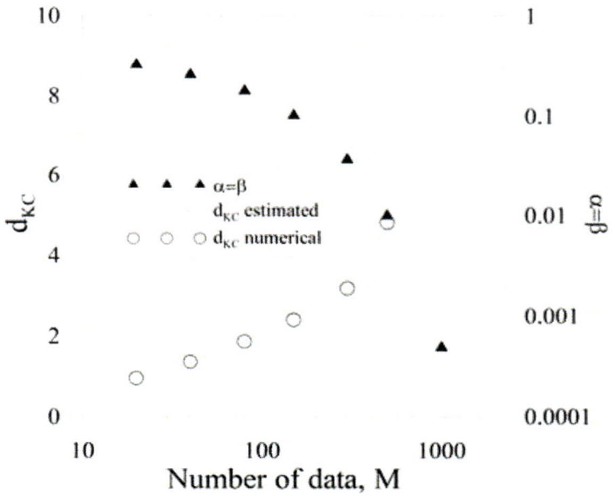

Fig. 7 Application of the detection scheme to the averaged data assumed to be Gaussian distributed. The circles (left axis) indicate the SNR ratio estimated by the index d_{KC}. The triangles (right axis) indicate the $\alpha = \beta$ level of the errors in the detection through the sample average as a function of the number of switches M of the average. The solid line is the estimate obtained assuming that the standard deviations scale as the square root of the number of points N. The other parameters are the same as in Fig. 2

a signal (type II error β). These features are combined in the so-called receiver operator characteristic of the test statistic, that is, of the combination of the errors α and β for each particular choice of the threshold. It is natural, if there are not particular reasons to do otherwise, to choose the case $\alpha = \beta$ that unequivocally individuates a threshold and hence the errors. The features of the detection can thus be obtained by a straightforward application of the central limit theorem with the data of histogram of Fig. 2: it suffices to notice that the standard deviation $\sigma(t_i)$ that appears in the definition of d_{KC} (15) is smaller of a factor $\sim \sqrt{M} \simeq 4.5$, as it is confirmed by Tables 1 and 2. In this manner one can connect the SNR estimated through d_{KC} to the size of type I and II errors, as illustrated in Fig. 7. In the figure, we show the d_{KC} index as a function of the sample size M (open circles and solid line). From each estimated d_{KC} index, according to the condition $\alpha = \beta$, we can uniquely determine the size of the errors of the test (triangles), see Fig. 6. In the same figure, the solid line displays the d_{KC} behavior estimated assuming that the standard deviations in (15) decrease as the square root of the number of data in a sample, $\sqrt{M}$. It is clear that such behavior scales nicely for $M \ll N$. Therefore, if a sufficiently long simulation is available, it is possible to carefully design the experiment to achieve the desired error bound.

Let us summarize how it is possible to put to a good use the definition of the SNR through the KC-index (15). First, it is useful to select the most appropriated parameter values to achieve the best SNR; in the present case we have supposed that the optimization of the SNR has given the best bias point $\gamma = 0.8$. For such parameter, a consistent number of events has been collected, $N = 10^4$. From the data so collected, it has been possible to determine the number of experiments that could suffice to achieve the desired level of error (type I and II, chosen to be identical).

4 Conclusions

We have demonstrated that the application of signal analysis to switching event of a Josephson junction subject to a periodic train of current pulses can be used to carefully plan experiments devised to decide about the existence of the perturbation. If the pulse train is to be interpreted as a photon field that irradiates the junction, this scheme can be applied to decide about the existence of elusive particles [3]. In particular the scheme can be useful if the arrival frequency of the pulses is very low, and therefore it is particularly cumbersome to collect a large number of events to distinguish the signal from the external field from spontaneously, thermally activated, events. Under these circumstances, it is necessary to resort to statistical test. When this is the case, the SNR to be reached for a significant test is the guideline for an accurate experiment design. As the estimate of the SNR calls for extensive simulations, both to optimize the parameters and to collect a statistically relevant number of events, it is probably necessary to resort to parallel simulations, possibly with CUDA architecture [28].

Let us add a word of caution. The analysis here presented is based on sample mean, detection can be improved with maximum likelihood estimators [2], that exploit the full information content of the escape distribution. However, the analytical estimate of the distribution of the escapes is a relatively complicated problem [23] to give a reliable solution for the estimate evaluation.

Acknowledgements The authors wish to acknowledge financial support from Italian National Institute for Nuclear Physics INFN through the Project SIMP and from University of Salerno through projects FARB17PAGAN, FARB19PAGAN.

References

1. B. Abdo, O. Jinka, N.T. Bronn, S. Olivadese, M. Brink, On-chip single-pump interferometric Josephson isolator for quantum measurements (2020), arXiv:2006.01918v1
2. P. Addesso, G. Filatrella, V. Pierro, Characterization of escape times of Josephson junctions for signal detection. Phys. Rev. E **85**, 016708-01–10 (2012)
3. D. Alesini, D. Babusci, C. Barone et al. Status of the SIMP project: toward the single microwave photon detection. J. Low Temp. Phys. **199**, 348–354 (2020); Development of a Josephson junction based single photon microwave detector for axion detection experiments. J. Phys. Conf. Ser. **1559**, 012020-1–7 (2020)
4. D.V. Anghel, K. Kulikov, Y.M. Galperin, L.S. Kuzmin, Electromagnetic radiation detectors based on Josephson junctions: effective Hamiltonian. Phys. Rev. B **101**, 024511-1–10 (2020)
5. E. Aprile et al., Observation of excess electronic recoil events in XENON1T. Phys. Rev. D **102**, 0720041-26 (2020)
6. A. Barone, G. Paternó, *Physics and Applications of the Josephson Effect* (Wiley, New York, 1982)
7. C. Beck, Testing axion physics in a Josephson junction environment. Modern Physics Letters A 26, 2841–2852, 2011
8. C. Beck, Possible resonance effect of axionic dark matter in Josephson junctions. Phys. Rev. Lett. 111, 231801, 2013
9. C. Beck, Axion mass estimates from resonant Josephson junctions. Physics of the Dark Universe **7–8**, 6–11 (2015)
10. C. Beck, Cosmological flux noise and measured noise power spectra in SQUIDs. Scientific Reports 6, 28275, 2016
11. R. Dassonneville, R. Assouly, T. Peronnin, P. Rouchon, B. Huard, Number-resolved photocounter for propagating microwave mode. Phys. Rev. Appl. **14**, 044022-1–18 (2020)
12. G. Filatrella, V. Pierro, Detection of noise-corrupted sinusoidal signals with Josephson junctions. Phys. Rev. E **82**, 046712-1–9 (2010)
13. X. Gu, A.F. Kockum, A. Miranowicz, Y. Liu , F. Nori, Microwave photonics with superconducting quantum circuits. Phys. Rep. **718–719**, 1–102 (2017)
14. C. Guarcello, D. Valenti, B. Spagnolo, V. Pierro, G. Filatrella, Anomalous transport effects on switching currents of graphene-based Josephson junctions. Nanotechnology **28**, 134001 (2017)
15. C. Guarcello, D. Valenti, B. Spagnolo, V. Pierro, G. Filatrella, Josephson-based threshold detector for Lévy-distributed current fluctuations. Phys. Rev. Appl. **11**, 044078-01–09 (2019)
16. C. Guarcello, A. Braggio, P. Solinas, G.P. Pepe, F. Giazotto, Josephson-threshold calorimeter. Phys. Rev. Appl. **11**, 054074-1-16 (2019)
17. P. Hänggi, P. Talkner, M. Borkovec, Reaction-rate theory: fifty years after Kramers. Review of Modern Physics. **62**, 251–341 (1990)
18. B. Josephson. Possible new effects in superconductive tunnelling. Physics Letters, 1, 251, 1962
19. B. D. Josephson. The discovery of tunnelling supercurrents. Rev. Mod. Phys., 46, 251, 1974
20. B. Kubala, J. Ankerhold, A.D. Armour, Electronic and photonic counting statistics as probes of nonequilibrium quantum dynamics. New J. Phys. **22**, 023010-1–10 (2020)
21. B.V.K.V. Kumar, C.W. Carrol, Performances of Wigner distribution function based detection methods. Optical Engineer **23**, 732–737 (1984)
22. L.S. Kuzmin, A.S. Sobolev, C. Gatti, D. Di Gioacchino, N. Crescini, A. Gordeeva, E. Il'ichev, Single Photon Counter Based on a Josephson Junction at 14 GHz for Searching Galactic Axions. IEEE Trans. Appl. Supercond. **28**, 1–5 (2018)
23. B. Lindner, Moments of the First Passage Time under External Driving. J. Stat. Phys. **117**, 703–737 (2004)
24. M. Nashaat, A. E. Botha, and Yu. M. Shukrinov. Devil's staircases in the IV characteristics of superconductor/ferromagnet/superconductor Josephson junctions', Physical Review B, 97, 224514, 2018

25. V.V. Palyulin, G. Blackburn, M.A. Lomholt, N.W. Watkins, R. Metzler, R. Klages, A.V. Chechkin, First passage and first hitting times of Lévy flights and Lévy walks. New J. Phys. **21**, 103028 (2019)
26. F. Paolucci, N. Ligato, V. Buccheri, G. Germanese, P. Virtanen, F. Giazotto, Hypersensitive tunable Josephson escape sensor for gigahertz astronomy. Phys. Rev. Appl. **14**, 034055-1–9 (2020)
27. A.S. Piedjou Komnang, C. Guarcello, C. Barone, C. Gatti, S. Pagano, V. Pierro, A. Rettaroli, G. Filatrella, Analysis of Josephson junctions switching time distributions for the detection of single microwave photons. Chaos Solitons Fract. **142**, 110–496 (2021)
28. V. Pierro, L. Troiano, E. Mejuto and G. Filatrella. Stochastic first passage time accelerated with CUDA. Journal of Computational Physics, 361, 136–149, 2018
29. V.A. Popov, Resonance detection of dark matter axions using a dc SQUID. Journal of Experimental and Theoretical Physics 122, 236–242, 2016
30. O.V. Pountougnigni, R. Yamapi , C. Tchawoua, V. Pierro, G. Filatrella, Detection of signals in presence of noise through Josephson junction switching currents. Phys. Rev. E **101**, 052205-1–10 (2020)
31. L.S. Revin, A.L. Pankratov, A.V. Gordeeva, A.A. Yablokov, I.V. Rakut, V.O. Zbrozhek, L.S. Kuzmin, Microwave photon detection by an Al Josephson junction. Beilstein J. Nanotechnol. **11**, 960–965 (2020)
32. S.J. Rezvani, D. Di Gioacchino, C. Gatti, C. Ligi, M. Cestelli Guidi, S. Cibella, M. Fretto, N. Poccia, S. Lupi, A. Marcelli, Proximity array device: a novel photon detector working in long wavelengths. Condens. Matter **5**, 33-1–9 (2020)
33. A.M. Sokolov, F.K. Wilhelm, A superconducting detector that counts microwave photons up to two. Phys. Rev. Appl. **14**, 064063 (2020)
34. F. Tafuri (ed.), *Fundamentals and Frontiers of the Josephson Effect*. Springer Series in Materials Science, vol. 286 (2019)
35. E.D. Walsh, D.K. Efetov, G. Lee, M. Heuck, J. Crossno, T.A. Ohki, P. Kim, D. Englund, K. Chung, Graphene-based Josephson-junction single-photon detector. Phys. Rev. Appl. **8**, 024022-1–11 (2017)
36. A.A. Yablokov, V.M. Mylnikov, A.L. Pankratov, E.V. Pankratova, A.V. Gordeeva, Suppression of switching errors in weakly damped Josephson junctions. Chaos, Solitons Fractals **136**, 109817-1–6 (2020)
37. J. Yan, C. Beck, Nonlinear dynamics of coupled axion-Josephson junction systems. Physica D **403**, 132294 (2020)

Control for Set-Valued Movements of Dynamical Systems Under Uncertainty with Applications

Tatiana F. Filippova

Abstract The guaranteed control problems for nonlinear dynamical systems with uncertainty in initial states and parameters are studied. The case is investigate when only the bounding sets for initial system states and for system parameters are given without any additional statistical or probabilistic information on these values. Applying the previously developed approaches and new results developed here to evaluating trajectory tubes and reachable sets, we study the properties of optimal control that solves the problem of control for the trajectory tube of a dynamic system with uncertainty and nonlinearity of a quadratic type.

Keywords Nonlinear dynamics · Control · Estimation · Uncertainty · Ellipsoidal calculus · Funnel equations

1 Introduction

The paper investigates the problems associated with the study of reachable sets of a nonlinear control dynamical system (and of a corresponding differential inclusion) with incomplete information on the initial states of the system or on other system parameters, limited by specifying only some special sets containing the unknown elements (Kurzhanski [14], Kurzhanski and Varaiya[16], Allgöwer and Zheng [1], Milanese et al. [18], Scweppe [22], Walter and Pronzato [23]). As indicated in many studies, the geometry of the reachable sets of nonlinear dynamical systems may be very complicated. In these cases, the approximation of reachable sets by domains of a certain canonical form is of interest. As such canonical figures, the most natural

13th CHAOS Conference Proceedings, 9–12 June 2020, Florence, Italy C. H. Skiadas (Ed).

T. F. Filippova (✉)
Krasovskii Institute of Mathematics and Mechanics, Russian Academy of Sciences,
Ural Federal University, Ekaterinburg, Russian Federation
e-mail: ftf@imm.uran.ru

C. H. Skiadas et al. (eds.), *13th Chaotic Modeling and Simulation International Conference*, Springer Proceedings in Complexity,
https://doi.org/10.1007/978-3-030-70795-8_15

are ellipsoids, parallelepipeds, polyhedra and some other canonical figures. A number of important approaches are relevant for assessing the unknown states of control systems and corresponding trajectory tubes of differential inclusions through approximation by canonical sets and tubes of motions with an accurate description of their parameters and dynamic characteristics (Kurzhanski and Valyi [15], Chernousko [5], Kostousova [12], Polyak et al. [21]).

Currently the principal facts and results of the theory of linear differential systems with uncertain parameters are well developed, a number of important and computationally useful algorithms have been constructed for finding the external and internal (with respect to the inclusion of sets) approximations of the set-valued states of dynamical systems in the case of a linear system dynamics. However the presence of nonlinear terms in the state velocities of the control systems causes a loss of the convexity of the reachable sets and, therefore, raises many theoretical questions and therefore requires the development of related mathematical tools and algorithms that are adequate to the indicated problems of nonlinear analysis. Some ideas and approaches to the study of set-valued motions (trajectory tubes) for a number of differential systems with nonlinearity and uncertainty in dynamics were presented earlier in Filippova [7], Filippova and Lisin [8], Filippova and Matviychuk [9] (see also references in the indicated publications).

In this paper we assume that in a dynamic system there are two types of nonlinearity, namely, we have a combination of bilinear and quadratic functions in the state velocities. Earlier, we examined the problems of evaluating the reachable sets of systems under study taking into account all possible controls at once. Knowing the areas of reachability with respect to all parameters of the system under study (for all possible initial states, disturbances, controls) is very useful, since it helps to evaluate the capabilities of the system. However, it seems important to have a description of the trajectory tube generated by a specific choice of a control function, it will allow solving optimization problems for set-valued movements of the considered systems under uncertainty. Note that in this paper we consider a special class of control systems with nonlinearity and uncertainty under other informational assumptions than was done in a recent paper Filippova and Matviychuk [10]. Thus, this research continues and complements developments in the field of mathematical control theory related to the study of the dynamics of multivalued states of nonlinear control systems. The approaches and algorithms presented here may be applied in the study of models with nonlinearity and uncertainty in real systems in robotics, economics, biology and other fields (considered e.g. in Allgöwer and Zheng [1], Bayen and Rapoport [2], Cecarelli et al. [4], Keller et al. [11]).

2 Problem Formulation

2.1 Basic Notations

The main notations used in the paper are basic; however, we define here some additional, most frequently used and important constructions.

We denote by $\mathbb{R}^n$ the n-dimensional vector space and by $\operatorname{comp}\mathbb{R}^n$ the set of all compact subsets of $\mathbb{R}^n$. Also $\mathbb{R}^{n\times m}$ denotes the set of all $n \times m$-matrices.

The usual inner product of $x, y \in \mathbb{R}^n$ is $x'y = (x, y) = \sum_{i=1}^{n} x_i y_i$ with prime as a transpose and also the

$$\|x\| = \|x\|_2 = (x'x)^{1/2}, \quad \|x\|_\infty = \max_{1\le i\le n} |x_i|$$

are corresponding norms for $x \in \mathbb{R}^n$.

For the identity matrix we use the symbol $I \in \mathbb{R}^{n\times n}$. Denote by $\operatorname{Tr}(A)$ a trace (a sum of diagonal elements) of $n \times n$-matrix A. Let $B(a, r) = \{x \in \mathbb{R}^n : \|x - a\| \le r\}$ be a ball in $\mathbb{R}^n$ with a center $a \in \mathbb{R}^n$ and with a radius $r > 0$.

We use here also the notation

$$E(a, Q) = \{x \in \mathbb{R}^n : (Q^{-1}(x - a), (x - a)) \le 1\}$$

for the ellipsoid in $\mathbb{R}^n$, where $a \in \mathbb{R}^n$ is its center and a $n \times n$-matrix Q is symmetric and positive definite.

2.2 Main Problem

We study here the nonlinear control system

$$\dot{x} = A(t)x + f(x)d + u(t), \qquad x_0 \in \mathcal{X}_0, \quad t_0 \le t \le T, \tag{1}$$

here $x, d \in \mathbb{R}^n$, $\|x\| \le K$ $(K > 0)$, the function $f(x)$ is quadratic in x, that is $f(x) = x'Bx$, with a positive definite and symmetric $n \times n$-matrix B.

Functions $u(t)$ (“controls") in (1) are assumed to be Lebesgue measurable on $[t_0, T]$ and

$$u(t) \in \mathcal{U}, \quad for \ \ a.e. \ t \in [t_0, T].$$

We assume that the constraint set $\mathcal{U}$ is given and $\mathcal{U} \in \operatorname{comp}\mathbb{R}^n$. The $n \times n$-matrix function $A(t)$ in (1) has the form

$$A(t) = A^0 + A^1(t), \tag{2}$$

where the $n \times n$-matrix A^0 is given and the measurable $n \times n$-matrix $A^1(t)$ is unknown but bounded, $A^1(t) \in \mathcal{A}^1$ for $t \in [t_0, T]$, namely we have

$$A(t) \in \mathcal{A} = A^0 + \mathcal{A}^1,$$

$$\mathcal{A}^1 = \big\{A{=}\{a_{ij}\}{\in}\mathbb{R}^{n\times n} : |a_{ij}|{\le}c_{ij},\ i, j{=}1, \dots n\big\}, \tag{3}$$

where $c_{ij} \geq 0$ $(i, j = 1, \dots n)$ are given numbers. The latter relations mean that all elements of the matrix $A(t)$ are known only up to certain errors, the values of which are given (this does not exclude the case when some elements of the matrix can be known exactly, this corresponds to the situation when some $c_{ij} = 0$).

Assume that we have the ellipsoid as an initial set $\mathcal{X}_0$ in (1), that is

$$\mathcal{X}_0 = E(a_0, Q_0),$$

with a symmetric and positive definite matrix $Q_0 \in \mathbb{R}^{n\times n}$ and with a center a_0.

If it will be necessary we will use also a notation $x(t; u(\cdot)) = x\big(t; u(\cdot), A(\cdot), x_0\big)$ with indication of additional parameters $A(\cdot)$, x_0 for an absolutely continuous function $x(t)$ which is the solution to (1)–(3) with initial state $x_0 \in \mathcal{X}_0$, with admissible control $u(\cdot)$ and with a matrix $A(\cdot)$ satisfying (2)–(3).

Due to the fact that some quantities are unknown but bounded, we are forced to consider all possible versions of motions compatible with additional data as a generalized solution to the control system, that is, we need to replace a single-valued trajectory by a bundle or tube of motions of the following form $\mathcal{X}(t; u(\cdot))$.

Definition 1 For each admissible control $u(\cdot)$ the *generalized solution tube* $\mathcal{X}(t; u(\cdot))$ (with $t \in [t_0, T]$) of system (1)–(3) is defined as follows,

$$\mathcal{X}(t; u(\cdot)) = \{x \in \mathbb{R}^n : \exists x_0{\in}\mathcal{X}_0,\ \exists A(\cdot){\in}\mathcal{A},$$

$$x = x(t) = x\big(t; u(\cdot), A(\cdot), x_0\big)\big\}.$$

Let us consider the following main problems.

Problem 1 For each feasible control $u(\cdot){\in}\mathcal{U}$, find the optimal external ellipsoidal estimate $E(\hat{a}, \hat{Q}; T, u(\cdot))$ of the reachable set $X(T; u(\cdot))$ of the system (1)–(4), such that

$$\mathcal{X}(T; u(\cdot)) \subset E(\hat{a}, \hat{Q}; T, u(\cdot)).$$

Remark 1 Here we understand the optimality of the desired ellipsoidal estimate, bearing in mind the closest operation with respect to inclusion of related sets.

Problem 2 Given a vector $x^* \in R^n$ find the feasible control $u^*(\cdot) \in \mathcal{U}$ such that the related ellipsoidal estimate is optimal, that is we have

$$d(x^*, E(\hat{a^*}, \hat{Q}^*; T, u^*(\cdot))) = \inf_{u(\cdot)\in\mathcal{U}} d(x^*, E(\hat{a^*}, \hat{Q}^*; T, u(\cdot))) = \epsilon^*.$$

3 Main Results

First, we define an auxiliary parameter k, which is required to formulate the main result (see also Filippova [7]). To do this, consider the matrix $B^{1/2}Q_0B^{1/2}$ and denote its maximal eigenvalue as k^2, that is we have

$$E(a_0, Q_0) \subseteq E(a_0, (k_0^+)^2 B^{-1}), \tag{4}$$

and k_0^+ is the smallest positive number for which this estimate (4) is true.

Theorem 1 *The upper ellipsoidal estimate is true*

$$\mathcal{X}(t_0+\sigma; u(\cdot)) \subseteq E(a^*(t_0+\sigma), Q^*(t_0+\sigma) \mid u(\cdot)) + o(\sigma)B(0, 1) \tag{5}$$

with $\sigma^{-1}o(\sigma) \to 0$ *for* $\sigma \to +0$ *and*

$$a^*(t_0+\sigma) = \tilde{a}(t_0+\sigma) + \sigma(\hat{a} + a_0'Ba_0 \cdot d + k^2 d) + \sigma u(t_0), \tag{6}$$

and with functions $\tilde{a}(t)$, $Q^*(t)$ *satisfying the following equations*

$$\dot{\tilde{a}} = \tilde{A}^0 \tilde{a}, \quad t_0 \le t \le T, \ \tilde{a}(t_0) = a_0, \tag{7}$$

$$\dot{Q}^* = \tilde{A}^0 Q^* + Q^*(\tilde{A}^0)' + qQ^* + q^{-1}G, \ Q^*(t_0) = Q_0, \quad t_0 \le t \le T, \tag{8}$$

where

$$\tilde{A}^0 = A^0 + 2d \cdot a_0' B, \quad q = \big(n^{-1}\,\mathrm{Tr}\,((Q^*)^{-1}G)\big)^{1/2}, \tag{9}$$

$$G = \mathrm{diag}\Big\{(n-v)\big[\sum_{i=1}^{n} c_{ji}|\tilde{a}_i| + \big(\max_{\sigma=\{\sigma_{ij}\}} \sum_{p,q=1}^{n} Q^*_{pq} c_{jp} c_{jq} \sigma_{jp} \sigma_{jq}\big)^{1/2}\big]^2\Big\}, \tag{10}$$

with a maximum in (10) calculated over numbers $\sigma_{ij} = \pm 1$, $i, j = 1, \ldots, n$, *such that we have* $c_{ij} \neq 0$ *and* v *is a number of such indices* i *for which* $c_{ij} = 0$ *for all* $j = 1, \ldots, n$.

Proof The relation (5) is established along the main lines and ideas presented in Filippova [7]. Indeed, from the funnel equation Panasyuk [20] we have

$$X(t_0+\sigma; u(\cdot)) \subseteq \bigcup_{\tilde{x} \in E(0, k_0^{+2} B^{-1})} (a_0 + \tilde{x} + \sigma(A_0 + \mathcal{A}_1)(a_0 +$$

$$\tilde{x}) + \sigma(a_0 + \tilde{x})' B(a_0 + \tilde{x})) + \sigma u(t_0) + o(\sigma)B(0, 1). \tag{11}$$

We remind that we may use here the property that at the boundary points $\tilde{x}$ of the ellipsoid $E(0, (k_0^+)^2 B^{-1})$ we have the equality $\tilde{x}' B \tilde{x} = (k_0^+)^2$ (for a more simple case detailed explanations of the last property may be found also in Filippova [7]). With this property and rearranging the terms in (11), we come to the formulas (5)–(10). ■

Remark 2 We see here that the ellipsoidal estimates of the tube $X(t; u(\cdot))$ for each fixed control $u(\cdot)$ are under investigation here and therefore the parameters of the estimation procedures depend on $u(\cdot)$. We can complicate the problem by additionally assuming the presence of state constraints or by considering a slightly more general class of uncertainty, e.g. in the coefficients of the matrix of linear terms of the state velocities.

Remark 3 It follows from Theorem 1 that we can construct a discrete tube $E(\hat{a}, \hat{Q}; T, u(\cdot))$ with ellipsoidal cross-sections that solves Problem 1 and for which we have the inclusion

$$X(T; u(\cdot)) \subseteq E(\hat{a}^+(T), \hat{Q}^+(T); u(\cdot)) + o(\epsilon)B(0, 1). \tag{12}$$

We emphasize that this discrete construction may be used as a basis for related computational schemes and algorithms allowing to find the trajectory tubes numerically.

Using the results Filippova and Matviychuk [9], we may derive the following result.

Theorem 2 *Let ϵ^*, $u^*(\cdot)$ be the optimal values of the Problem 2. Then we have the relations*

$$\epsilon^* = \min_{u(\cdot)\in\mathcal{U}} \max_{||l||=1} \{r^+(T; u(\cdot))(l'B^{-1}l)^{1/2} +$$

$$l'(a^+(T; u(\cdot)) - x^*)\} = \max_{||l||=1} \{r^+(T; u^*(\cdot))(l'B^{-1}l)^{1/2} + \tag{13}$$

$$l'(a^+(T; u^*(\cdot)) - x^*)\}.$$

Proof First, we find the minimal positive number ϵ such that the following inclusion is true

$$E(a^+(T), Q^+(T); T, u(\cdot)) \subseteq B(x^*, \epsilon),$$

or equivalently

$$\rho(l|E(a^+(T), Q^+(T); T, u(\cdot)) \leq \rho(l|B(x^*, \epsilon)), \quad \forall l \in \mathbb{R}^n.$$

Appling the result of Theorem 1, we get the relation

$$l'a^+(T) + (l'Q^+(T)l)^{1/2} \leq l'x^* + \epsilon||l||,$$

and from the above relations we conclude that

$$\epsilon^* = \min_{u(\cdot)} \max_{||l||=1} ((l' Q^+(T) l)^{1/2} + l'(a^+(T) - x^*)).$$

Taking into account the equality $Q^+(T) = r^+(T) B^{-1}$ we get the equations (13). ■

The proposed results may be used as the basis for the development of computational algorithms for solving applied problems of controlling and estimating the movements of real systems operating in conditions of uncertainty and nonlinearity, in particular, in the fields of robotics, economics and finance, biology and other fields. Related algorithms with computational examples (for lower dimensional systems) that illustrate the approach may be found e.g. in Filippova and Matviychuk [9]. In the next section a more complicated example of a dynamical system in the space $\mathbb{R}^3$ is given and discussed.

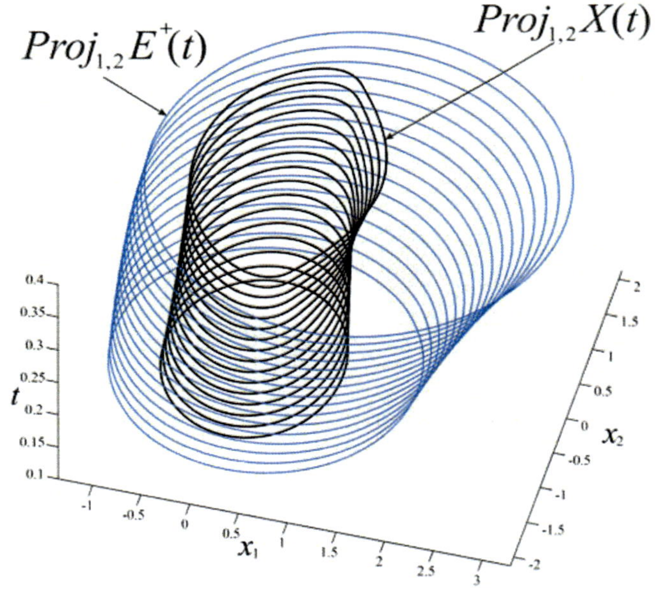

Fig. 1 Projections $Proj_{1,2}E^+(t)$ of ellipsoids $E^+(t) = E(a^+(t), Q^+(t))$ (blue color) and projections $Proj_{1,2}X(t)$ of reachable sets (black color) $X(t)$ at the plane of $\{x_1, x_2, t\}$-coordinates

4 Numerical Simulations

Example. Consider the following control system

$$\begin{cases} \dot{x}_1 = -x_1 + x_1^2 + x_2^2 + 2x_3^2 + u_1(t), \\ \dot{x}_2 = \quad x_2 + u_2(t), \\ \dot{x}_3 = \quad x_3 + u_3(t), \end{cases} \tag{14}$$

Assume that $\mathcal{U} = B(0, 1)$, $x_0 \in X_0 = B(0, 1)$ and $t \in [0, T]$ with $T = 0.4$. The projections of reachable sets $X(t)$ together with related estimating ellipsoids $E^+(t) = E(a^+(t), Q^+(t))$ onto the planes of state coordinates (related planes are (x_1, x_2), (x_1, x_3) and (x_2, x_3), respectively) are shown in Figs. 1, 2, and 3 for time grid $t = 0.1; 0.15; 0.2; 0.25; 0.3; 0.35; 0.4$ (we need to specify here that for simplicity we put $u(t) = 0$ here, in other cases calculations and pictures are similar).

The last Fig. 4 shows the upper estimating ellipsoid $E^+(t) = E(a^+(t), Q^+(t))$ and the reachable set $X(t)$ as they are in the related space $\mathbb{R}^3$ of state variables $\{x_1, x_2, x_3\}$ for $t = 0.4$.

Note that the evaluating ellipsoid touches the reachable set (that is, the external estimate is "tight"), which implies that without changing the structure of parameters

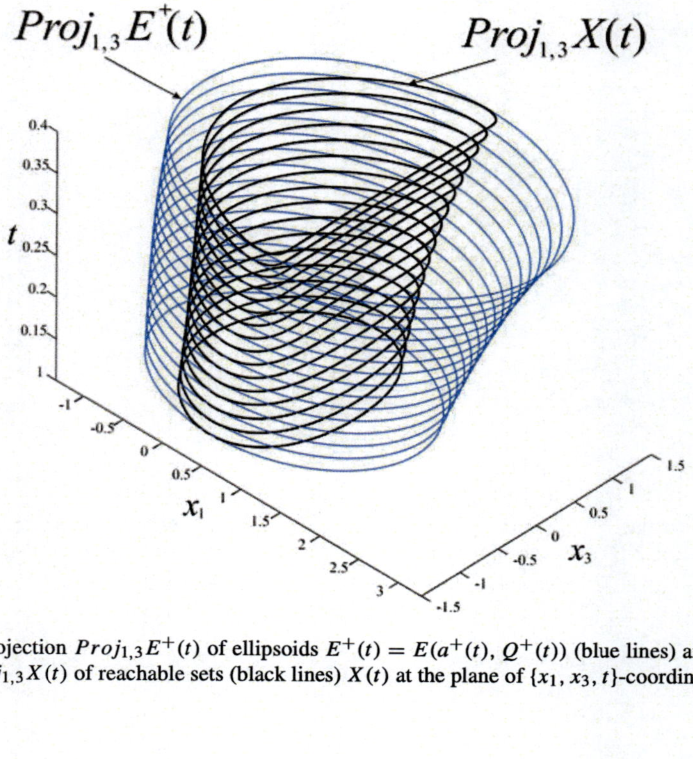

Fig. 2 Projection $Proj_{1,3}E^+(t)$ of ellipsoids $E^+(t) = E(a^+(t), Q^+(t))$ (blue lines) and projections $Proj_{1,3}X(t)$ of reachable sets (black lines) $X(t)$ at the plane of $\{x_1, x_3, t\}$-coordinates

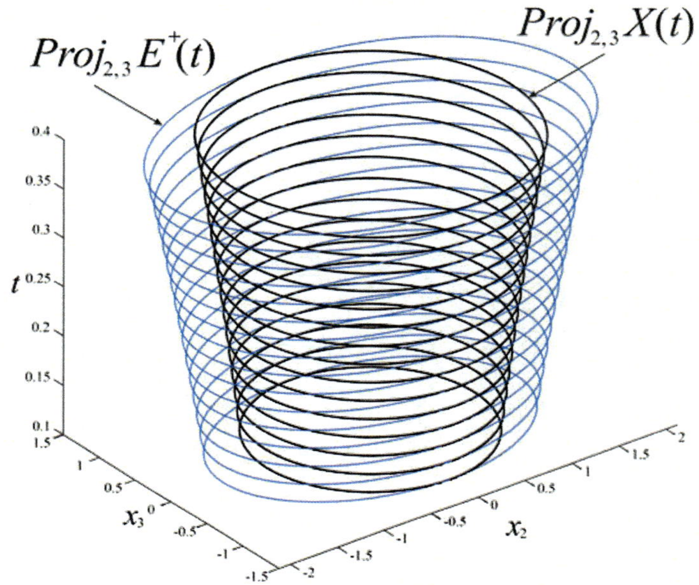

Fig. 3 The projections $Proj_{2,3}E^{+}(t)$ of estimating ellipsoids $E^{+}(t) = E(a^{+}(t), Q^{+}(t))$ (indicated in blue lines) and projections $Proj_{2,3}X(t)$ of reachable sets (indicated in black lines) $X(t)$ at the plane of $\{x_2, x_3, t\}$-coordinates

(for example, without changing the main matrix of coefficients), it cannot be reduced to a smaller ellipsoid.

5 Further Theoretical Directions and Possible Applications

Theoretical schemes and related numerical algorithms for evaluating trajectory tubes and methods for solving control problems for set-valued motions based on Theorems 1–2 can be developed further in many directions, among them we note the following areas:

- studies of optimization and robust stabilization problems for uncertain nonlinear systems with impulsive control functions,
- problems of viability and control for dynamical systems described by nonlinear differential equations and differential inclusions,
- improvement and development of new numerical methods for estimating set-valued motions of nonlinear dynamical systems (ensembles of trajectories) based on the proposed ideas for high-dimensional systems,
- research of new, more complex classes of nonlinearity in the dynamics of controlled systems with uncertain factors,

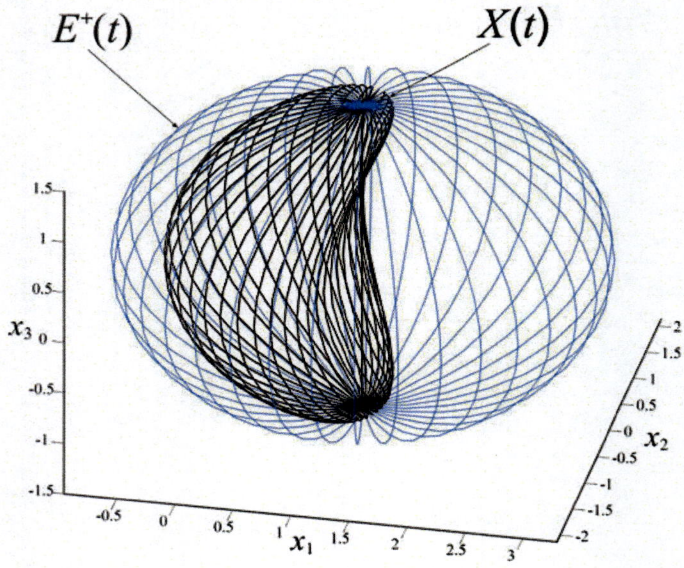

Fig. 4 Reachable set $X(t)$ and its upper ellipsoidal estimate $E^+(t) = E(a^+(t), Q^+(t))$ for $t = 0.4$ (3d-picture in the plane of state variables $\{x_1, x_2, x_3\}$)

- development of theoretical approaches to the estimation of set-valued motions using approximations for set-valued motions based on the use of discrete schemes of the theory of differential inclusions with a large order of accuracy.

The applications of the problems discussed here are in the nonlinear control and estimation theory and related nonlinear models with unknown but bounded errors. Numerous application models can be noted here, in particular, real models in robotics, in transportation systems, in biology, medicine and economics. In these aspects, we would like to highlight, in particular, the studies and results obtained earlier by Bayen and Rapoport [2], Cecarelli et al. [4], Koller et al. [11]), Filippova and Matviychuk [9], Kuntsevich and Volosov [13], Malyshev and Tychinskii [17], Ovsyannikov [19].

6 Conclusion

The paper deals with the state estimation problems for uncertain dynamical control systems for which we assume that the initial state is unknown but bounded with given constraints. We consider here a special case of uncertainty and nonlinearity when the matrix parameters in state velocities are unknown but bounded.

The system nonlinearity under study is generated also by the presence of bilinear terms and quadratic forms in related differential equations. The problem is reformulated as the control problem for the motion of related set-valued states.

Using the ideas developed earlier for some classes of uncertain systems we solve here the control problem with a new class of uncertainty and with a special structure of nonlinearity. So we construct the external ellipsoidal estimates of reachable sets for the system under study and find the solution of the related optimization problem.

Acknowledgements The study was partially supported by the Russian Foundation for Basic Researches (RFBR Project No.18-01-00544).

References

1. F. Allgöwer, A. Zheng (eds.), *Nonlinear Model Predictive Control* (Birkhauser, Basel, 2000)
2. T. Bayen, A. Rapaport, Minimal time crisis versus minimum time to reach a viability kernel: a case study in the prey-predator model. Optim. Control. Appl. Methods (Wiley) **40**(2), 330–350 (2019)
3. R.W. Brockett, On the reachable set for bilinear systems. Lect. Notes Econ. Math. Syst. **111**, 54–63 (1975)
4. N. Ceccarelli, M. Di Marco, A. Garulli, A. Giannitrapani, A set theoretic approach to path planning for mobile robots, in *Proc. 43rd IEEE Conference on Decision and Control, Atlantis, Bahamas, Dec. 2004* (2004), pp. 147–152
5. F.L. Chernousko, *State Estimation for Dynamic Systems* (CRC Press, Boca Raton, 1994)
6. F.L. Chernousko, D.Y. Rokityanskii. Ellipsoidal bounds on reachable sets of dynamical systems with matrices subjected to uncertain perturbations. J. Optim. Theory Appl. **104**, **1**, 1–19 (2000)
7. T.F. Filippova, Differential equations of ellipsoidal estimates for reachable sets of a nonlinear dynamical control system. Proc. Steklov Inst. Math. (Supplementary issues) **271**, suppl.1, S75–S84 (2010)
8. Problems Control Dynam, T.F. Filippova, D.V. Lisin, On the estimation of trajectory tubes of differential inclusions. Proc. Steklov Inst. Math. Systems. Suppl. **2**, S28–S37 (2000)
9. T. F. Filippova and O. G. Matviychuk. Approaches to estimating the dynamics of interacting populations with impulse effects and uncertainty, in *12th Chaotic Modeling and Simulation International Conference, CHAOS 2019*, ed. by C. Skiadas, Y. Dimotikalis, Springer Proceedings in Complexity, Springer, Cham, pp. 85–99 (2020)
10. T.F. Filippova, O.G. Matviychuk, Control problems for set-valued motions of systems with uncertainty and nonlinearity, in *International Conference "Stability, Control, Differential Games", SCDG2019, September 16–20, 2019, Yekaterinburg, Russia*, eds. A. Tarasyev, V. Maksimov, T. Filippova, Lecture Notes in Control and Information Sciences - Proceedings, Springer, Cham, pp. 379–389 (2020)
11. T. Koller, F. Berkenkamp, M. Turchetta and A. Krause. Learning-based model predictive control for safe exploration and reinforcement learning, arXiv:1803.08287v3 [cs.SY], https://arxiv.org/pdf/1803.08287.pdf , pp. 1–9 (2018)
12. E. K. Kostousova. On polyhedral estimates of reachable sets of discrete-time systems with uncertain matrices and integral bounds on additive terms, in *Numerical Computations: Theory and Algorithms*, eds. by Y. Sergeyev, D. Kvasov. NUMTA 2019, Lecture Notes in Computer Science, Springer, Cham, 11974, pp. 124–138 (2000)
13. V.M. Kuntsevich, V.V. Volosov, Ellipsoidal and interval estimation of state vectors for families of linear and nonlinear discrete-time dynamic systems. Cybern. Syst. Anal. **51**(1), 64–73 (2015)
14. A.B. Kurzhanski, *Control and Observation under Conditions of Uncertainty* (Nauka, Moscow, 1977)

15. A.B. Kurzhanski, I. Valyi, *Ellipsoidal Calculus for Estimation and Control* (Birkhauser, Boston, 1997)
16. A.B. Kurzhanski, P. Varaiya, *Dynamics and Control of Trajectory Tubes* (Springer-Verlag, New York, Theory and Computation, 2014)
17. V.V. Malyshev, Y.D. Tychinskii, Construction of attainability sets and optimization of maneuvers of an artificial earth satellite with thrusters in a strong gravitational field. *Proc. of RAS, Theory and Control Systems* **4**, 124–132 (2005)
18. M. Milanese, J.P. Norton, H. Piet-Lahanier, E. Walter (eds.), *Bounding Approaches to System Identification* (Plenum Press, New York, 1996)
19. D.A. Ovsyannikov, Mathematical modeling and optimization of beam dynamics. IFAC Proc. **34**(6), 13–22 (2001)
20. A.I. Panasyuk. On the equation of an integral funnel and its applications. *Differ. Equations***24**(11), 1263–1271 (1988)
21. B. T. Polyak, S. A. Nazin, C. Durieu, E. Walter, Ellipsoidal parameter or state estimation under model uncertainty. *Automatica*, 40, 1171–1179 (2004)
22. F. Schweppe, *Uncertain Dynamic Systems* (Prentice-Hall, Englewood Cliffs, New Jersey, 1973)
23. E. Walter, L. Pronzato, *Identification of Parametric Models from Experimental Data* (Springer-Verlag, Heidelberg, 1997)

Interacting Populations: Dynamics and Viability in Bounded Domains Under Uncertainty

Tatiana F. Filippova

Abstract Nonlinear control systems which describe the dynamics of the interactions of predators and their preys under assumption of uncertainty in related initial conditions are studied. It is assumed that the interaction of populations occurs in limited areas, estimated by corresponding ellipsoids. The possible presence of uncertainty or errors in determining the parameters of these ellipsoids and the uncertainty in the initial conditions of the moving objects and also in some parameters of dynamical systems are taken into account. Procedures and algorithms for evaluating the movements of upper estimates of reachable sets of the system under indicated conditions of uncertainty are proposed. Numerical simulation results related to the proposed techniques and illustrating the results are also included.

Keywords Nonlinear systems · Control · Uncertainty · Ellipsoidal calculus · Viability · State estimation

1 Introduction

In 1920 Alfred Lotka initiated the study of a predator-prey model and showed that the populations could oscillate permanently, he further developed his researches in this direction and published a book (Lotka [20]). In 1926 the Italian mathematician Vito Volterra happened to become interested in the same model (Volterra [31]). Now the classical Lotka–Volterra equations have a long history of use and researches in different branches of theory and applications, in this field it is worth to mention some recent approaches and results (e.g. Skiadas [29], Filippova and Matviychuk [16]) related to this class of problems.

T. F. Filippova (✉)
Department of Optimal Control, N.N. Krasovskii Institute of Mathematics and Mechanics,
Ural Branch of Russian Academy of Sciences, 16 Sofya Kovalevskaya str., 620108 Ekaterinburg, Russian Federation
e-mail: ftf@imm.uran.ru

C. H. Skiadas et al. (eds.), *13th Chaotic Modeling and Simulation International Conference*, Springer Proceedings in Complexity,
https://doi.org/10.1007/978-3-030-70795-8_16

We consider the modification of this classical problem and study it under conditions of uncertainty in initial system states and maybe also under uncertainty in some other parameters. We study the case when there is no precise description of the uncertain data (noises and errors) in the model, but only some bounds on these values are specified. This situation arises when we are trying to create a mathematical model that takes into account the absence of probabilistic data on uncertain parameters of the studied phenomenon, or these data are not reliable enough and therefore cannot be taken into account. To study problems of this class, we use the set-membership estimation approach which deals with a set of all feasible parameters vectors, consistent with the model structure, current measurements and uncertainty features (Bertsekas and Rhodes [3], Kurzhanski and Valyi [18], Kurzhanski and Varaiya [19], Kurzhanski and Filippova [17], Milanese et al. [23], Schweppe [28], Walter and Pronzato [32], Chernousko [7–9], Brockett [6], Dontchev and Lempio [10], Veliov [30], Mazurenko [22], Polyak [26]).

In this paper we continue the study initiated in Filippova [13–15], Filippova and Matviychuk [16] and construct the estimating procedures for reachable sets of nonlinear dynamical control systems of Lotka-Volterra type under conditions of uncertainty and in the complicating assumption of the presence of state constraints on the system trajectories. It should be noted that the study of control systems with state constraints is closely related to the techniques and results of the theory of survival, or, using the already well-established term, viability theory (Aubin and Cellina [1], Bayen and Rapaport [2], Kurzhanski and Filippova [17], Bonneuil and Mullers [5], Filippova [11]).

The viability theory presents additional convenient instruments to design and to develop mathematical and algorithmic techniques for investigating and adapting to viability constraints the complex dynamical systems under uncertainty which may be found in many domains from biology to economics, financial markets, control theory, robotics etc. Basing on above mentioned approaches we formulate here new theoretical schemes and construct new algorithms for determining upper ellipsoidal estimates of reachable sets of the studied control system with uncertainty. Numerical examples and simulation results related to the proposed techniques and to the presented algorithms are also included.

2 Problem Formulation

2.1 *Preliminary Constructions*

We will continue to use the notations defined earlier in Filippova and Matviychuk [16] but for the convenience of the reader we will nevertheless recall some basic concepts. So R^n will be the n–dimensional Euclidean space, $comp R^n$ is the set of all compact subsets of R^n, $R^{n\times n}$ stands for the set of all $n\times n$–matrices and $x'y = (x, y) = \sum_{i=1}^{n} x_i y_i$ be the usual inner product of $x, y \in R^n$ with prime as a transpose, the Euclidean norm is $\|x\| = (x'x)^{1/2}$. We denote as $B(a, r)$ the ball in R^n,

$B(a, r) = \{x \in R^n : \|x - a\| \le r\}$, I is the identity $n \times n$-matrix, $diag\{a_1, \ldots, a_n\}$ is the diagonal $n \times n$–matrix with elements a_i at the main diagonal and with zero entries outside it.

Denote the ellipsoid $E(a, Q) = \{x \in R^n : \ (Q^{-1}(x - a), (x - a)) \le 1\}$ with a center $a \in R^n$ and with a symmetric positive definite $n \times n$–matrix Q. Let $Tr(Y)$ denote the trace of $n \times n$–matrix Y (the sum of its diagonal elements).

Consider the following system

$$\dot{x} = Ax + f(x)d + u(t), \quad x_0 \in X_0, \quad t_0 \le t \le T, \tag{1}$$

where $x, d \in R^n$, $\|x\| \le K$ $(K > 0)$, $f(x)$ is the nonlinear function, which is quadratic in x,

$$f(x) = x'Bx, \tag{2}$$

with a given symmetric and positive definite $n \times n$-matrix B. Control functions $u(t)$ in (1) are assumed to be Lebesgue measurable on $[t_0, T]$ and satisfying the constraint

$$u(t) \in U, \quad \text{for a.e. } t \in [t_0, T], \tag{3}$$

here U is a given set, $U \in compR^m$. The $n \times n$–matrix A in (1) is assumed to be also given.

We will assume that the bounding set X_0 for initial system states x_0 in (1) is an ellipsoid, $X_0 = E(a_0, Q_0)$, with a symmetric and positive definite matrix Q_0 and with a center a_0.

Let the absolutely continuous function $x(t) = x(t;\ u(\cdot),\ x_0)$ be a solution to dynamical system (1)–(17) with initial state $x_0 \in X_0$ and with admissible control $u(\cdot)$.

Definition 1 The reachable set $X(t)$ at time t $(t_0 < t \le T)$ of system (1)–(3) is defined as

$$\begin{aligned} X(t) = \{\, x \in R^n :\ \exists\, x_0 \in X_0,\ \exists\, u(\cdot) \in U,\ \ \text{such that} \\ x = x(t) = x(t;\ u(\cdot),\ x_0)\,\},\ \ t_0 < t \le T. \end{aligned} \tag{4}$$

Remark 1 We can also interpret the control functions $u(\cdot)$ in (1)–(4) as unknown perturbations in the model dynamics. Thus, the tube $X(t)$ can play the role of a set-valued state of the system (1) under conditions of uncertainty in its dynamics.

2.2 Evolution Equations for Dynamical Models with Set-Valued States

The further notions are closely related to the concepts and definitions of set-valued analysis in general and in particular to a solution notion of a differential inclusion (Aubin and Cellina [1])

$$\dot{x} \in \mathcal{F}(t, x) \tag{5}$$

where $\mathcal{F}(t, x)$ is a set-valued function reflecting the variety of models under uncertainty conditions.

Assume that the initial condition to the differential inclusion (5) is unknown but bounded

$$x(t_0) = x_0, \quad x_0 \in X_0 \in comp R^n \tag{6}$$

Denote by $h(A, B)$ the Hausdorff distance between sets $A, B \subseteq R^n$, namely,

$$h(A, B) = \max\{h^+(A, B), h^-(A, B)\},$$

with $h^+(A, B), h^-(A, B)$ being the Hausdorff semidistances between the sets A, B, namely,

$$h^+(A, B) = \sup\{d(x, B) \mid x \in A\},$$

$$h^-(A, B) = h^+(B, A), \quad d(x, A) = \inf\{\| x - y \| \mid y \in A\}.$$

Assuming a set $X_0 \in \text{comp}\, R^n$ to be given, denote as $x[t] = x(t, t_0, x_0)$ $(t \in [t_0, T])$ a solution to (5) (an isolated trajectory) that starts at point $x[t_0] = x_0 \in X_0$.

We take here the Caratheodory–type trajectory $x[\cdot]$, i.e. an absolutely continuous function $x[t]$ that satisfies the inclusion

$$\frac{d}{dt} x[t] = \dot{x}[t] \in \mathcal{F}(t, x[t]) \tag{7}$$

for almost every $t \in [t_0, T]$. We assume that all the solutions $\{x[t] = x(t, t_0, x_0) \mid x_0 \in X_0\}$ are extendable up to T that is possible under some additional assumptions, Filippova [12]. Let us consider the equation

$$\begin{gathered} \lim_{\sigma \to +0} \sigma^{-1} h\Big(\mathcal{X}[t+\sigma], \bigcup_{x \in \mathcal{X}[t]} (x + \sigma \mathcal{F}(t, x))\Big) = 0, \\ t_0 \le t \le T, \ \mathcal{X}[t_0] = X_0. \end{gathered} \tag{8}$$

Theorem 1 (Panasyuk [25], Kurzhanski and Filippova [17]) *The multifunction $\mathcal{X}[t] = \mathcal{X}(t, t_0, X_0)$ is the unique set–valued solution to the equation (8).*

2.3 *State Constraints and Viability*

The main problem of the paper is to find external ellipsoidal estimates of the reachable set $X(t)$ $(t_0 < t \le T)$ and to apply these results to find the upper bounds of reachable sets for nonlinear dynamical control systems of Lotka-Volterra type considered now with the following viability (or state) constraint

$$x[s] \in Y(s), \quad s \in [t_0, t] \tag{9}$$

where $Y(t) \in \text{conv } R^n$ for $t \in [t_0, T]$.

This viability constraint may be induced by state restrictions defined for a given plant model or by the so-called measurement equation

$$y(t) = G(t)x + w, \tag{10}$$

where y is the measurement, $G(t)$ is a given matrix function, w is an unknown but bounded "noise" with a given bound,

$$w \in Q^*(t), \quad Q^*(t) \in \text{comp } R^p,$$

(here $Q^*(t)$ is a given set-valued function).

To start the analysis of the above problem consider the analogy of the funnel (8) but now for the viable trajectory tubes $X[t] = X(t, t_0, X_0)$:

$$\begin{gathered}\lim_{\sigma \to +0} \sigma^{-1} h\Big(X[t+\sigma], \bigcup_{x \in X[t]} (x + \sigma \mathcal{F}(t, x)) \bigcap Y(t+\sigma) \Big) = 0, \\ t_0 \le t \le T, \quad X[t_0] = X_0.\end{gathered} \tag{11}$$

The following result is valid (under some additional assumptions on $\mathcal{F}(t, x)$ and $Y(t)$).

Theorem 2 (Kurzhanski and Filippova [17]) *The set-valued function $X[t] = X(t, t_0, X_0)$ is the unique solution to the equation (11).*

3 Main Results

We apply here the approaches described above to study the classical model of the Lotka-Volterra type, but now we consider these approaches with two complications. Namely, first, we assume that we have uncertain initial states, and second, we assume that we have an additional constraint on the current states of the system. In the recent paper Filippova and Matviychuk [16], we considered only the first problem; a more complicated case of two simultaneous constraints is being investigated here.

3.1 Lotka-Volterra Control Systems Under Uncertainty and State Constraints

Consider the following control system of Lotka-Volterra type which describes the classical ecological predator-prey (or parasite-host) model with additional control functions (Bayen and Rapaport [2], Bonneuil and Mullers [5], Lotka [20], Murray [24], Prostyakov [27]):

$$\begin{cases} \dot{x_1}(t) = & ax_1 - bx_1x_2 + u_1, \\ \dot{x_2}(t) = & -cx_2 + dx_1x_2 + u_2, \end{cases} \quad x(t_0) = x_0, \quad t_0 \le t \le T. \tag{12}$$

Here $a, b, c, d > 0$ are given and we assume that initial state vector x_0 is unknown but bounded, that is we have the inclusion $x_0 \in X_0$, where X_0 is a given compact subset of R^2. This assumption may be interpreted for example in such a way that we do not know exactly the initial states (or amounts) of predators and prey. Control functions $u(t)$ in (12) are assumed to be Lebesgue measurable on $[t_0, T]$, they satisfy the constraint

$$u(t) \in U, \quad \text{for a.e. } t \in [t_0, T], \tag{13}$$

where U is given, $U \in compR^2$. The choice of control can influence, in particular, the rate of change in the number of predators and prey.

Using diagonalization procedures of matrix analysis described in details in Filippova and Matviychuk [16], we transform the system (12) to the following one (here a parameter $\epsilon > 0$)

$$\begin{aligned} \dot{z} = A^* z + f_\varepsilon^{(1)}(z) \cdot d^{(1)} + f_\varepsilon^{(2)}(z) \cdot d^{(2)} + w(t), \\ z_0 \in \mathcal{Z}_0, \quad w \in \mathcal{W}, \quad t_0 \le t \le T, \end{aligned} \tag{14}$$

where

$$A^* = \begin{pmatrix} A & -C \\ -C & A \end{pmatrix},$$

functions $f_\varepsilon^{(1)}(z)$ and $f_\varepsilon^{(2)}(z)$ are positive definite quadratic forms with matrices $B_\varepsilon^{(1)} = diag\{1, \varepsilon^2\}$ and $B_\varepsilon^{(2)} = diag\{\varepsilon^2, 1\}$, respectively. The following result may be used to produce the upper estimate of a reachable set $X(t)$ of the system (12) of Lotka-Volterra type with ellipsoidal constraints on the control function.

Theorem 3 (Matviychuk [21]) *For all $\sigma > 0$ and for reachable set $X(t_0 + \sigma) = X(t_0 + \sigma, t_0, X_0)$ of the system (12) we have the following upper estimate*

$$\begin{aligned} X(t_0 + \sigma) \subseteq & E(a^{(1)}(\sigma), Q^{(1)}(\sigma)) \bigcap E(a^{(2)}(\sigma), Q^{(2)}(\sigma)) + o(\sigma)B(0, 1), \\ & \sigma^{-1} o(\sigma) \to 0 \ \textit{when} \ \sigma \to +0, \end{aligned} \tag{15}$$

where

$$a^{(1)}(\sigma) = Z^{-1}(a(\sigma) + \sigma k_1^2 \lambda_{12}^2 d^{(2)}) + \sigma \hat{a},$$
$$a^{(2)}(\sigma) = Z^{-1}(a(\sigma) + \sigma k_2^2 \lambda_{21}^2 d^{(1)}) + \sigma \hat{a},$$
$$a(\sigma) = (I + \sigma Z A Z') Z a_0 + \sigma (Z a_0)' B_\varepsilon^{(1)} Z a_0 d^{(1)} + \sigma (Z a_0)' B_\varepsilon^{(2)} Z a_0 d^{(2)},$$
$$Q^{(1)}(\sigma) = Z^{-1}\big((p_1^{-1} + 1)(I + \sigma R) k_1^2 (B_\varepsilon^{(1)})^{-1} (I + \sigma R)' +$$
$$+ (p_1 + 1)\sigma^2 (||d^{(2)}||^2 k_1^4 \lambda_{12}^4 \cdot I + Z \hat{Q} Z')\big)(Z^{-1})',$$
$$Q^{(2)}(\sigma) = Z^{-1}\big((p_2^{-1} + 1)(I + \sigma R) k_2^2 (B_\varepsilon^{(2)})^{-1} (I + \sigma R)' +$$
$$+ (p_2 + 1)\sigma^2 (||d^{(1)}||^2 k_2^4 \lambda_{21}^4 \cdot I + Z \hat{Q} Z')\big)(Z^{-1})',$$
$$R = Z A Z' + 2 d^{(1)} (Z a_0)' B_\varepsilon^{(1)} + 2 d^{(2)} (Z a_0)' B_\varepsilon^{(2)},$$

where $B_\varepsilon^{(1)} = diag\{1, \varepsilon^2\}$, $B_\varepsilon^{(2)} = diag\{\varepsilon^2, 1\}$, k_1^2, k_2^2, λ_{12}^2 *and* λ_{21}^2 *are the maximal eigenvalue of the matrices*

$$(B_\varepsilon^{(1)})^{1/2} Z Q_0 Z' (B_\varepsilon^{(1)})^{1/2}, \quad (B_\varepsilon^{(2)})^{1/2} Z Q_0 Z' (B_\varepsilon^{(2)})^{1/2},$$
$$(B_\varepsilon^{(1)})^{-1/2} B_\varepsilon^{(2)} (B_\varepsilon^{(1)})^{-1/2}, \quad (B_\varepsilon^{(2)})^{-1/2} B_\varepsilon^{(1)} (B_\varepsilon^{(2)})^{-1/2},$$

respectively, numbers p_1, p_2 *are the unique positive solutions of related algebraic equations*

$$\sum_{i=1}^{n} \frac{1}{p_1 + \alpha_i} = \frac{n}{p_1(p_1 + 1)}, \quad \sum_{i=1}^{n} \frac{1}{p_2 + \beta_i} = \frac{n}{p_2(p_2 + 1)}$$

with $\alpha_i, \beta_i \geq 0$ $(i = 1, \ldots, n)$ *being the roots of the following equations*

$$\det\big((I + \sigma R) k_1^2 (B_\varepsilon^{(1)})^{-1} (I + \sigma R)' - \alpha \sigma^2 (||d^{(2)}||^2 k_1^4 \lambda_{12}^4 \cdot I + Z \hat{Q} Z')\big) = 0,$$
$$\det\big((I + \sigma R) k_2^2 (B_\varepsilon^{(2)})^{-1} (I + \sigma R)' - \beta \sigma^2 (||d^{(1)}||^2 k_2^4 \lambda_{21}^4 \cdot I + Z \hat{Q} Z')\big) = 0.$$

Because the above formulas look too complicated, before to apply them we slightly modify the system (12) in the same way as it was done in Bratus [4, pp. 44–45]. After this transformation we come to the system

$$\begin{cases} \dot{v_1} = & v_1 - v_1 v_2 + u_1, \\ \dot{v_2} = & -\alpha v_2 + v_1 v_2 + u_2, \end{cases} \quad v(t_0) = v_0, \quad t_0 \leq t \leq T. \tag{16}$$

Here $\alpha > 0$ and initial state vectors v_0 are unknown but bounded,

$$v_0 \in B(0, \mu) = \{v = \{v_1, v_2\} : v_1^2 + v_2^2 \leq 1\}$$

and control functions $u(t)$ satisfy the following constraint (with $r > 0$)

$$u(t) \in B(0, r), \quad \text{for a.e. } t \in [t_0, T]. \tag{17}$$

Replace here ones again the variables as $w_1 = v_1 + v_2$, $w_2 = v_1 - v_2$ and consider the system

$$\begin{gathered} \dot{w} = Dw + d^{(1)} f_1(w) + d^{(2)} f_2(w), \\ w(t_0) = w_0 \in B(0, r), \quad r = 2^{-1/2}, \quad t_0 \le t \le T, \end{gathered} \tag{18}$$

with $f_1(w) = 2w_1^2 + w_2^2$, $f_2(w) = w_1^2 + 2w_2^2$,

$$D = \begin{pmatrix} \frac{1-\alpha}{2} & \frac{1+\alpha}{2} \\ \frac{1+\alpha}{2} & \frac{1-\alpha}{2} \end{pmatrix} \tag{19}$$

and with related trajectory tube denoted as $W(\cdot) = W(\cdot, t_0, X_0)$.

Theorem 4 *For all $\sigma > 0$ and for $W(t_0 + \sigma) = W(t_0 + \sigma, t_0, X_0)$ the following upper estimate is true*

$$W(t_0 + \sigma) \subseteq E(a^{(1)}(\sigma), Q^{(1)}(\sigma)) \bigcap E(a^{(2)}(\sigma), Q^{(2)}(\sigma)) + o(\sigma)B(0, 1), \tag{20}$$

where $\sigma^{-1}o(\sigma) \to 0$ when $\sigma \to +0$ and parameters of the estimating ellipsoids $E(a^{(1)}(\sigma), Q^{(1)}(\sigma))$ and $E(a^{(2)}(\sigma), Q^{(2)}(\sigma))$ are defined in Theorem 3 with the following simplifications

$$k_1^2 = k_2^2 = 2/r^2, \quad \lambda_{12}^2 = \lambda_{21}^2 = 2, \quad a = \hat{a} = 0,$$

$$d^{(1)} = -d^{(2)} = -(0, 1)', \; \hat{Q} = r^2 I,$$

$$B^{(1)} = \begin{pmatrix} 2 & 0 \\ 0 & 1 \end{pmatrix}, \quad B^{(2)} = \begin{pmatrix} 1 & 0 \\ 0 & 2 \end{pmatrix}.$$

Proof The proof of the above upper estimates follows the scheme used in Filippova and Matviychuk [16].

3.2 *Iterative Algorithm and Numerical Simulation*

The following iterative algorithm may produce the external ellipsoidal tube for the reachable set $X(t)$, $t \in [t_0, T]$ in numerical modelling. Note that, in comparison with the algorithm given in Filippova and Matviychuk [16], it is somewhat more complicated in calculations, since additional steps appeared here.

Algorithm. Subdivide the time segment $[t_0, T]$ into subsegments $[t_i, t_{i+1}]$ where $t_i = t_0 + ih$ $(i = 1, \dots, m)$, $\sigma = (T - t_0)/m$, $t_m = T$.

- Given $\mathcal{X}_0 = E(a_0, Q_0)$, find ellipsoids $E(a^{(1)}(\sigma), Q^{(1)}(\sigma))$ and $E(a^{(2)}(\sigma), Q^{(2)}(\sigma))$ from Theorems 3, 4.
- Find the smallest (with respect to some criterion [7, 18]) ellipsoid $E(a_1, Q_1)$ which contains the intersection

$$E(a^{(1)}(\sigma), Q^{(1)}(\sigma)) \bigcap E(a^{(2)}(\sigma), Q^{(2)}(\sigma)) \subseteq E(a_1, Q_1).$$

- Find the new ellipsoid $E(a_2, Q_2)$ such that we have

$$E(a_1, Q_1) \bigcap Y \subseteq E(a_2, Q_2)$$

  (this is a correction taking into account state constraints)
- Consider the system on the next subsegment $[t_1, t_2]$ with $E(a_2, Q_2)$ as the initial ellipsoid at the next instant t_1.
- Further steps repeat above iterations. At the end of the process we will get the external estimate $E(a^+(t), Q^+(t))$ of the tube $X(t)$ with accuracy tending to zero when $m \to \infty$.

Example 1 Consider the following control system of Lotka–Volterra type:

$$\begin{cases} \dot{x_1}(t) = x_1 - x_1 x_2 + u_1, \\ \dot{x_2}(t) = -x_2 + x_1 x_2 + u_2, \end{cases} \quad x_0 \in X_0, \quad t_0 \leq t \leq T. \tag{21}$$

Here we take $t_0 = 0$, $T = 1$, $X_0 = B(0, 1)$ and $U = B(0, 0.1)$. We assume also that we have the following additional state constraint

$$x[s] \in E(d, D), \quad s \in [t_0, t]$$

with $d = (0.1, 0)$ and

$$D = \begin{pmatrix} 2.25 & 0 \\ 0 & 0.56 \end{pmatrix}.$$

The trajectory tube $X(t)$ and some reachable sets are shown in the Figs. 1 and 2. Ellipsoidal estimates for the reachable set $X(0.01)$ which were found on the basis of results of Theorems 3, 4 are shown in Fig. 2. Several steps of the main Algorithm of external ellipsoidal estimating the reachable set $X(t)$ with the resulting ellipsoidal tube $E(a^+(t), Q^+(t))$ are shown in the Fig. 2.

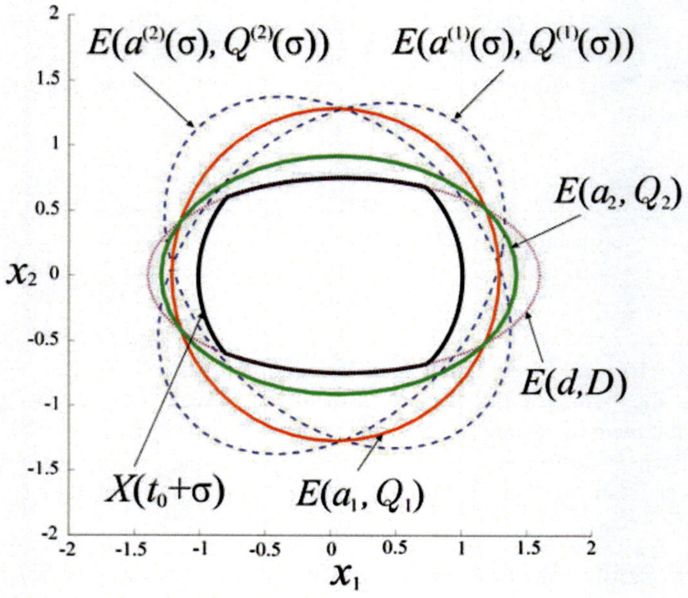

Fig. 1 Estimates of $X(t)$ of the system (21)

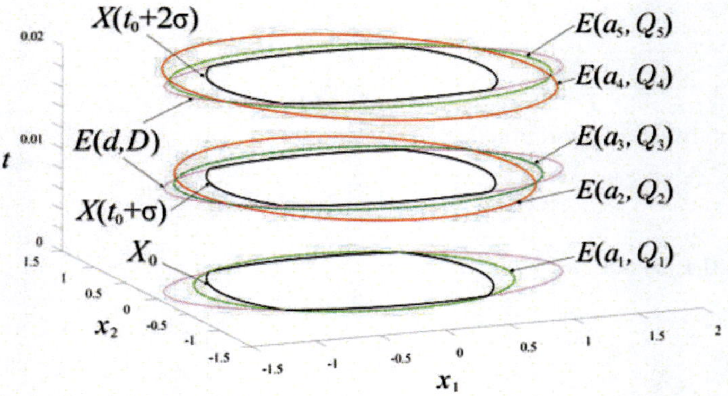

Fig. 2 Steps of the Algorithm of ellipsoidal estimating process for the trajectory tube $X(t)$ ($\sigma = 0.01$)

4 Conclusions

We considered the problems of state estimation for dynamical control systems with unknown but bounded initial state. The solution was studied through the techniques of trajectory tubes of related differential inclusions.

The problem of estimating reachable sets of nonlinear dynamical control systems of Lotka-Volterra type which describe the dynamics of the interaction of predators and their preys under uncertainty conditions and state constraints was studied. Applying results of the theory of trajectory tubes of control systems we found upper ellipsoidal estimates of related reachable sets.

The algorithms of constructing ellipsoidal estimates are given, numerical simulation results related to the proposed techniques and illustrating the algorithms are included.

Acknowledgements The research was supported by the Research and Education Center of Krasovskii Institute of Mathematics and Mechanics (Ural Branch of Russian Academy of Sciences, Ekaterinburg, Russia) in the framework of the Ural Mathematical Center (Project "Set-Valued Dynamics in Control and Estimation Problems for Dynamical Systems with Uncertainty").

References

1. J.-P. Aubin, A. Cellina, *Differential Inclusions: Set-Valued Maps and Viability Theory* (Springer, Berlin, 1984).
2. T. Bayen, A. Rapaport, Minimal time crisis versus minimum time to reach a viability kernel: a case study in the prey-predator model. Optim. Control. Appl. Method **40**(2), 330–350 (2019)
3. D.P. Bertsekas, I.B. Rhodes, Recursive state estimation for a set-membership description of uncertainty. IEEE Trans. Autom. Control. **16**, 117–128 (1971)
4. A.S. Bratus, A.S. Novozhilov, A.P. Platonov, *Dynamical Systems and Models of Biology* (Fizmatlit, Moscow, 2010)
5. N. Bonneuil, K. Mullers, Viable populations in a prey-predator system. J. Math. Biol. **35**, 261–293 (1997)
6. R.W. Brockett, On the reachable set for bilinear systems. Lect. Notes Econ. Math. Syst. **111**, 54–63 (1975)
7. F.L. Chernousko, *State Estimation for Dynamic Systems* (CRC Press, Boca Raton, 1994)
8. F.L. Chernousko, Ellipsoidal approximation of the reachable sets of linear systems with an indeterminate matrix. Appl. Math. Mech. **60**(6), 940–950 (1996)
9. F.L. Chernousko, D.Ya. Rokityanskii. Ellipsoidal bounds on reachable sets of dynamical systems with matrices subjected to uncertain perturbations. J. Optim. Theory Appl. **104**(1), 1–19 (2000)
10. A.L. Dontchev, F. Lempio, Difference methods for differential inclusions: a survey. SIAM Rev. **34**, 263–294 (1992)
11. T.F. Filippova, A note on the evolution property of the assembly of viable solutions to a differential inclusion. Comput. Math. Appl. **25**(2), 115–121 (1993)
12. T.F. Filippova, Trajectory tubes of nonlinear differential inclusions and state estimation problems. J. Concr. Appl. Math. **8**, 454–469 (2010). Eudoxus Press, LLC
13. T.F. Filippova, Set-valued dynamics in problems of mathematical theory of control processes. Int. J. Mod. Phys. Ser. B (IJMPB) **26**(25), 1–8 (2012)

14. T.F. Filippova, Differential equations of ellipsoidal state estimates for bilinear-quadratic control systems under uncertainty. J. Chaotic Model. Simul. (CMSIM) **1**, 85–93 (2017)
15. T.F. Filippova, External estimates for reachable sets of a control system with uncertainty and combined nonlinearity. Proc. Steklov Inst. Math. **301**(suppl.1), 32–43 (2018)
16. T.F. Filippova, O.G. Matviychuk, Approaches to estimating the dynamics of interacting populations with impulse effects and uncertainty, in *Springer Proceedings in Complexity*, ed. by C. Skiadas, Y. Dimotikalis (Springer, Cham, 2020), pp. 85–99
17. A.B. Kurzhanski, T.F. Filippova, On the theory of trajectory tubes – a mathematical formalism for uncertain dynamics, viability and control, in *Advances in Nonlinear Dynamics and Control: a Report from Russia, Progress in Systems and Control Theory*, vol. 17, ed. by A.B. Kurzhanski (Birkhauser, Boston, 1993), pp. 22–188
18. A.B. Kurzhanski, I. Valyi, *Ellipsoidal Calculus for Estimation and Control* (Birkhauser, Boston, 1997).
19. A.B. Kurzhanski, P. Varaiya, *Dynamics and Control of Trajectory Tubes: Theory and Computation* (Springer, New York, 2014).
20. A.J. Lotka, *Elements of Physical Biology* (Williams and Wilkins, Baltimore, 1925).
21. O.G. Matviychuk, Estimates of the reachable set of nonlinear control system under uncertainty. AIP Conf. Proc. **2025**(100005), 1–9 (2018)
22. S.S. Mazurenko, A differential equation for the gauge function of the star-shaped attainability set of a differential inclusion. Doklady Math. **86**(1), 476–479 (2012)
23. M. Milanese, J.P. Norton, H. Piet-Lahanier, E. Walter (eds.), *Bounding Approaches to System Identification* (Plenum Press, New York, 1996)
24. J.D. Murray, *Mathematical Biology* (Springer, New York, 1993).
25. A.I. Panasyuk, Equations of attainable set dynamics, Part 1: integral funnel equations. J. Optim. Theory Appl. **2**, 349–366 (1990)
26. B.T. Polyak, S.A. Nazin, C. Durieu, E. Walter, Ellipsoidal parameter or state estimation under model uncertainty. Automatica **40**, 1171–1179 (2004)
27. P. Prostyakov, Construction of the reachability set of the Lotka-Volterra system. Differ. Equ. **42**(3), 391–399 (2006)
28. F. Schweppe, *Uncertain Dynamic Systems* (Prentice-Hall, Englewood Cliffs, New Jersey, 1973)
29. C.H. Skiadas, A two population model for the stock market problem, in *Chaotic Systems: Theory and Applications*, ed. by C.H. Skiadas, I. Dimotikalis (World Scientific, 2010), pp. 302–308
30. V.M. Veliov, Second order discrete approximations to strongly convex differential inclusion. Syst. Control. Lett. **13**, 263–269 (1989)
31. V. Volterra, Variazioni e fluttuazioni del numero d'individui in specie animali conviventi. Memoria della Reale Accademia Nazionale dei Lincei, Ser. VI **2**, 31–113 (1926)
32. E. Walter, L. Pronzato, *Identification of Parametric Models from Experimental Data* (Springer, Heidelberg, 1997).

Battle of Salamis: Greeks Were Destined to Win

Konstantina Founta and Loukas Zachilas

Abstract In this paper we present a new non–linear, discrete, dynamical system trying to model the historic battle of Salamis (480 BC) between Greeks and Persians. The model describes the most effective strategic behavior between two participants during a battle or in a war. Moreover, we compare the results of the dynamical analysis to Game Theory, considering this conflict as a dynamic game.

Keywords Battle of salamis · Discrete systems · Modeling strategic behavior · Game theory

1 Introduction

The model approaches short–term conflicts between two participants (players), where one is weaker than the other opponent. Also, the parameters (that we use in Eq. 1, see below) are the most crucial factors in order to highlight the optimal way to achieve a decisive victory.

The solution of the equations of the dynamical system (called equilibrium point) shows what kind of behavior each player should adopt, i.e. an aggressive or defensive one. In this way, according to Game Theory, the optimum strategic behavior is called Nash Equilibrium [8].

One of the most representative games of Evolutionary Game Theory is the so-called game "Hawk—Dove", which was originally developed by Smith and Price [9] to describe animal conflicts and is quite similar to our attempt. We will present below the game and its results.

The game "Hawk—Dove" has many applications in everyday life. There are two animals (or two players) fighting for the same resource. Each of them can behave

K. Founta (✉) · L. Zachilas
Department of Economics, University of Thessaly, Volos, Greece
e-mail: kfounta@uth.gr

L. Zachilas
e-mail: zachilas@uth.gr

C. H. Skiadas et al. (eds.), *13th Chaotic Modeling and Simulation International Conference*, Springer Proceedings in Complexity,
https://doi.org/10.1007/978-3-030-70795-8_17

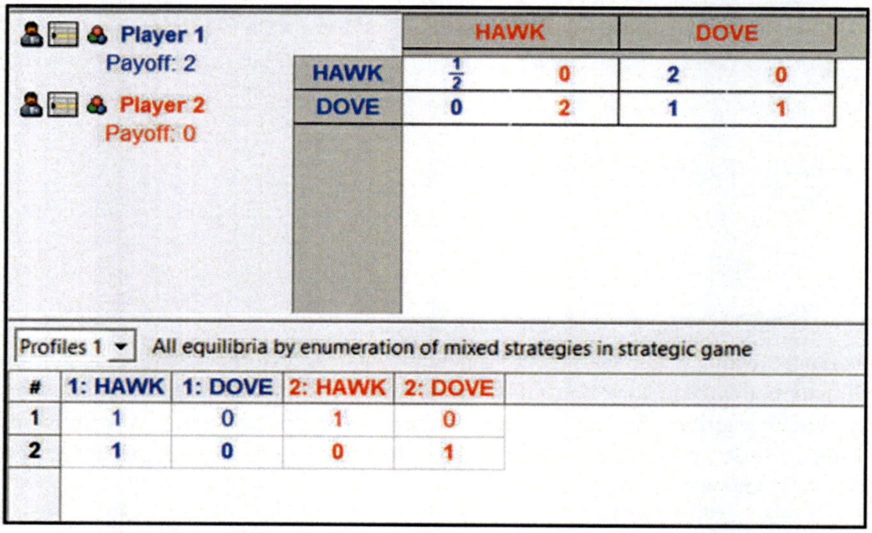

Fig. 1 The results of "Hawk—Dove" game

either as a hawk (i.e. fight for the resource) or as a dove (i.e. abandon the resource before the conflict escalates into a fight). Individuals have a benefit B if they win and a cost C if lose.

If a Hawk meets a Hawk, they will fight and one of them will win the resource; average payoff is *(B-C)/2*. If a Hawk meets a Dove, the Dove immediately withdraws, so the payoff of the Dove is *zero*, while the payoff of the Hawk is *B*. If a Dove meets a Dove, the one who first gets hold of the resource keeps it, while the other does not fight for it; average payoff *B/2*. The strategic form of the game is given by the payoff matrix:

$$Payoff_{H,D} = \begin{pmatrix} (B-C)/2 & B \\ 0 & B/2 \end{pmatrix}$$

2 Solution of the Game "Hawk—Dove"

We set the benefit $B = 2$ if a player wins, and the cost $C = 1$ if a player loses. Using the Gambit[1] software (16.0.1), we find Nash equilibriums and the dominant strategy.

Figure 1 shows us the payoff matrix and the two Nash equilibriums. If both players behave as a Hawk, the one who first injures the other wins. We set the player 1 starts and injures the player 2, thus player 1 wins. If someone behaves as a Hawk and the

[1] McKelvey, Richard D., McLennan, Andrew M., and Turocy, T. L., 2014.

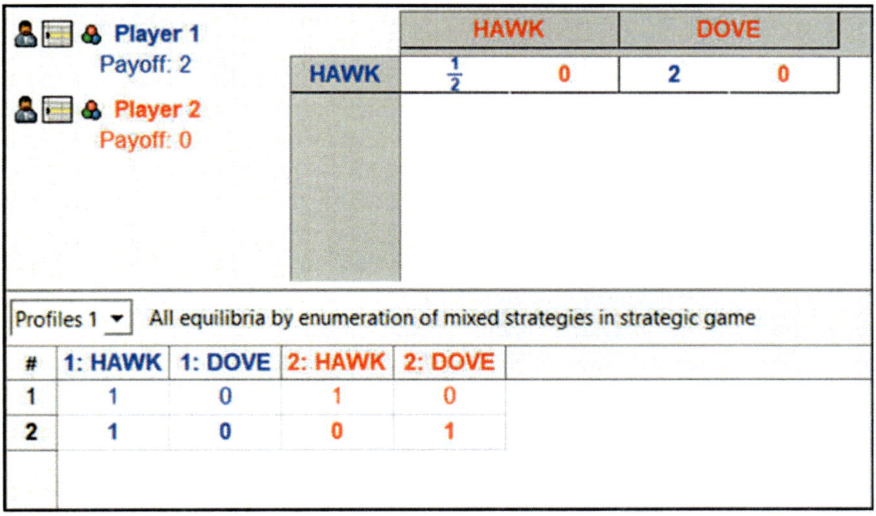

Fig. 2 Dominant strategy

other behaves as a Dove, then the player with the aggressive behavior (Hawk) wins and takes all the resource. If both players behave as a Dove, then they share the resource.

Regarding Nash equilibriums, there are two pure strategies. On the one hand, both players behave as Hawks and on the other hand, player 1 behaves as a Hawk and player 2 as a Dove. Moreover, we can observe that player 1 behaves as a Hawk in both cases and player 2 behaves either as a Hawk or as a Dove, but in each case player 1 wins.

We should note that if player 2 injures first player 1, the Nash equilibriums would be different.

Figure 2 shows the dominant strategy of the game, where player 1 behaves as a Hawk independently of the player's 2 behavior (i.e. Hawk or Dove). Therefore, the first dominant strategy may not be effective, because both players behave as Hawks and player 1 wins the half of the resource and does not maximize his profit. Although, if the player behaves as a Hawk, knowing that the other player behaves as a Dove, then he takes all the resource (maximum profit). Thus, we believe that the second Nash equilibrium is more effective and optimum strategy.

3 The Dynamical Model

It is widely acknowledged that the military strategy is the combination of ends, ways and means [6]. In our attempt to study the strategic behavior of two warring parties, we developed a new non-linear discrete system of two equations based on the above phrase. The main objective of the model is to simulate the way by which the two opponents behave strategically, where the one is weaker than the other.

At the same time, in Game Theory, the war is considered as a dynamic game where the strategies of the players are studied by calculating their optimal strategy (Nash equilibrium). In this research, we tried to compare the results of the Game Theory with those from the analysis of the discrete dynamical system. At the end of the analysis, the optimum and effective strategy for both participants (players) will be suggested.

The model, which is applied in short-term conflicts and describes the strategic behavior of each participant, is given by Eq. 1:

$$\begin{cases} x_{t+1} = P_x + TN_x - G \cdot \left(D_y + E_x\right) \cdot 4y_t \cdot (1 - y_t) \\ y_{t+1} = P_y + TN_y - (1 - G) \cdot \left(D_x + E_y\right) \cdot 4x_t \cdot (1 - x_t) \end{cases} \tag{1}$$

where:

x_t: The strategic behavior of any form of social organization x (state, country, etc.) at the time t.

y_t: The strategic behavior of any form of social organization y (state, country, etc.) at the time t.

x_{t+1}: The optimal strategic behavior of any form of social organization x at the (next moment of) time $t + 1$.

y_{t+1}: The optimal strategic behavior of any form of social organization y at the (next moment of) time $t + 1$.

We consider $x_t, y_t, x_{t+1}, y_{t+1} \in [0, 1]$, because the logistic equation is defined in [0,1], which is derived from the study of biological populations reproduced in discrete time [5]. It's the evolution of the population model of Malthus [7] and shows that the exponential growth cannot tend to infinity, but there is a critical point, i.e. a saturation. In other words, it is not possible for someone to win and the other to lose continuously. Also, each optimal strategic behavior, at the time t, affects the next move—strategic behavior, at the time $t + 1$, of the opponent.

In addition, we can interpret the values of variables (and parameters, as shown below) as percentages or probabilities, which help us to explain the results; these are also explained through the Game Theory.

Moreover, if the value of x_{t+1} (or y_{t+1}, respectively) equals to 0, it indicates the fully defensive strategic behavior of participant x (or y respectively), while if it equals to 1, then it indicates the fully aggressive behavior of participant x (or y respectively).

The parameters of Eq. 1 are the main and most important factors that could affect the strategic behavior of x (or y, respectively). In particular:

The parameter $\boldsymbol{P_x}$ represents the strength (economic, military, population, territorial) of x and $\boldsymbol{P_y}$ is the strength of y, respectively. These two parameters indicate the substance of each form of social organization compared to the other.

$\boldsymbol{TN_x}$ and $\boldsymbol{TN_y}$ represents the Technological Naval capability and evolution of x and y, respectively. These two parameters are also defined in comparison with the technological capability and evolution of the other participant and describe the *means* mentioned by [6].

The parameter $\boldsymbol{G}$ represents the geographical location (geophysical terrain) of the area where the battle or the war is taking place. We believe that this is another part of the military strategy, namely the *ways* [6]. Trying to emphasize the importance of this parameter and how it can be an advantage or disadvantage for each participant, we set in the first equation as $\boldsymbol{G}$ and in the second equation as **1—*G***. The closer to the 1 the value of the parameter, the easier the geophysical terrain of the area is.

The parameter $\boldsymbol{D_x}$ represents the damages caused by x to y and respectively, $\boldsymbol{D_y}$ represents the damages that y brings to x. The damages which we refer to may be economic, territorial, military, etc. or even deception and damaging of the psychological part of the opponent. Moreover, these two parameters complete the last part of the military strategy, namely the *ends* [6].

The parameter $\boldsymbol{E_x}$ represents the expenses of participant x and $\boldsymbol{E_y}$ the expenses of participant y, respectively. In other words, these denote the preparation costs of each participant for a battle (or war), compared to each other.

All the parameters that have been presented above should belong to [0,1]. Namely, $P_x, P_y, TN_x, TN_y, G, D_x, D_y, E_x, E_y \in [0, 1]$.

In the next section, we present the dynamic analysis and the results from the application of Eq. 1 in naval battle of Salamis.

4 The Case of (Naval) Battle of Salamis

The naval battle of Salamis was an important battle of the second Persian invasion in Greece and has been estimated to being held on September 28th, 480 BC in the Salamis straits (in the Saronic Gulf near Athens). The two warring parties were the Greeks (Hellenic alliance) and the Persian Empire [2].

After the fall of Thermopylae, the Persians proceeded to Athens. The Greeks had been advised by the Oracle of Delphi, that only the "wooden walls" would save them and they considered that this referred to a fight in the sea [5].

A few days before the battle, the meeting of the Greek admirals had to decide the geographic location of the battle. On the one hand, the Spartan General Evriviades proposed to fight in the Isthmus of Corinth, under the main argument that in case of failure it would be possible for them to continue to fight into the center of the Peloponnese. On the other hand, the Athenian General Themistocles insisted to fight in Salamis straits. He believed that if he forced the Persians to attack there, the numerous Persian ships couldn't extent highlighting their dominance. Ultimately, the council considered that Themistocles' argument was better and decided to support it [1].

The Greek fleet was estimated by Herodotus in 380 triremes and Aeschylus gave a round 300 triremes, but we can't be certain for the exact number. On contrary, the

Persian fleet was estimated in 500–600 triremes.[2] Herodotus describes the Persian ships as "*better sailing*", when compared to the Greek fleet. This may be attributable to a combination of factors such as lightness of materials and structure of the ship, better seamanship and more extensive naval experience. The triremes of Hellenic alliance were heavier and more durable. However, Herodotus reports that these ships were equipped with an embolism, with which they sank the enemy ships. They used two attacking maneuvers: *diekplous*, (i.e. attack from the rear or sides with a sharp turn) and *periplous*, (flanking or enveloping move, which generally gave an extra benefit against superior numbers in open water). The purpose of both was to ram the enemy in the side. In this way, they achieved serious damages or even the complete destruction of the Persians ships. On the contrary, the Persian tactic was "ramming and boarding" [10].

At dawn (if the date of the battle was indeed 28th Sept.), the two fleets were ready for the naval conflict. Xerxes, sure of his victory, sat on a throne on mountain Aigaleo, to enjoy the war spectacle. The narrowness of the space and the limited extent of the sea did not allow the Persians to use the major of their force in the front line. Thus, the number of ships was approximately equal. In this naval battle, the bravery and dexterity of the Greek fleet played an important role. They fought aggressively to defend their moral values and their freedom [1].

Herodotus reports that "*the Greeks fought with discipline and held their formation, but the Persians did not seem to be following any plan, so things were bound to turn out for them as they did*". Also, Aeschylus mentions that Themistocles must be given the credit for their battle and the winning tactics. The turning point of the battle came as the Persians "*suffered their greatest losses when the ships in their front line were put to fight and those following, pressing forward to impress the King* (i.e. Xerxes) *with their deeds, became entangled with them as they tried to escape*", as Herodotus comments [10].

The naval battle evolved rapidly and by the noon it was visible that the Greeks would win. The Persian fleet had crushed, while the Greek fleet continued to haunt them, killing the helpless, non–swimming soldiers. This brought the battle to an end, leaving the Greek force in full control of the straits [1].

When the battle was over, a Roman source mentions that Greeks lost more than 40 triremes and Persians more than 200 ones [10]. The victory of the Greek force was of major importance, since they managed to cause the collapse of the Persian morale, which is evidenced by the abandonment of the battle. In addition, the right decision of Themistocles for the geographic location of the naval battle was one of the most intelligent movements to bring the Greek victory.

[2] Aeschylus, writing decades earlier, also gives 1,207 triremes, but Herodotus writes, shortly before battle took place, that the Persian fleet wasn't much bigger than Greek. Because of a weather phenomenon (storms) 600 ships sank (400 at the coast of Magnesia, north of Artemisium and 200 in Euboea).

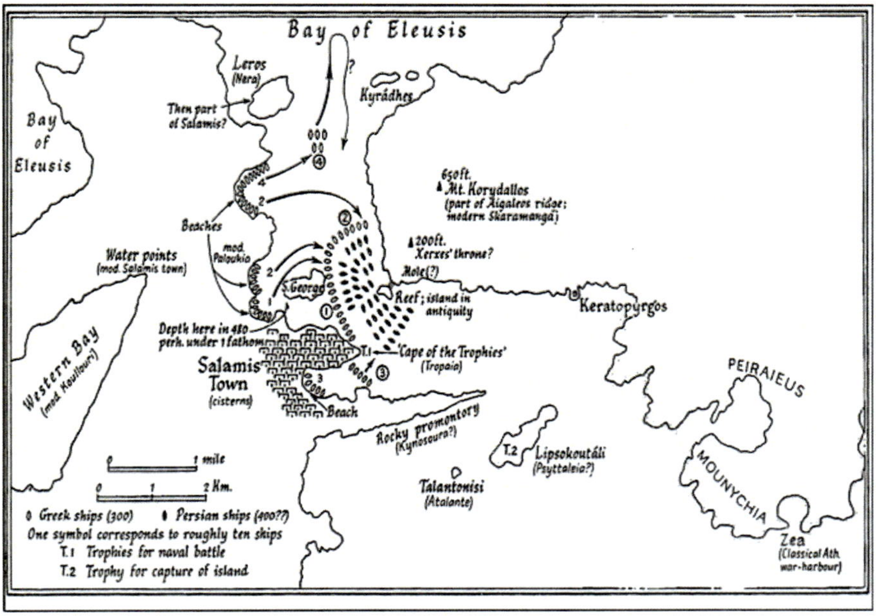

Fig. 3 The battle of Salamis. Source: Burn, A. R. (1962). Persia and the Greeks, New York: Minerva Press

5 Applying the Model in Naval Battle of Salamis—Approaching the Reality

Starting the dynamical analysis of the naval battle of Salamis, we set the initial conditions in Eq. 1, which represent as much as possible the historical events of the battle. Specifically:

(a) We set Greeks as the weak participant—player (x) and Persians as the powerful participant—player (y).
(b) The strength of Hellenic alliance, $P_x = 0.25$ and the strength of Persian empire, $P_y = 0.8$.
(c) The technological naval capability of Greeks, $TN_x = 0.7$ and the technological naval capability of Persians, $TN_y = 0.35$.
(d) The geographic location of the naval battle, $G = 0.4$, i.e. the Salamis straits, which are an advantage point for the Greek fleet.
(e) The damage caused to Persian side was huge, so we set $D_x = 0.8$ and $D_y = 0.2$.
(f) The preparation costs of this battle for each participant: $E_x = 0.3$, $E_y = 0.7$, respectively. According to Kyriazis and Zouboulakis [5], 100 new Athenian triremes were built under the Athenian Naval Law of Themistocles. Each one cost one talent (6000 ancient drachmae), so the total cost was 100 talents (or 600.000 ancient drachmae). In 480 BC, the Athenian fleet was comprised of

200 triremes, equivalent to the two thirds of the total Greek strength. However, the Persian ships were similar in shape, so we assume that the cost of each ship was similar. Thus, it is obvious that the Persians spent more money to support their expedition to the Greek territories than the Greeks.

With these initial conditions, we solve the system (Eq. 1), by using the mathematical software Maxima[3] (5.39.0), calculating the equilibrium points. Then, we study more extensively the behavior of the model and we present bifurcation diagrams and timeseries diagrams using the software E&F Chaos.[4]

Solving the system (Eq. 1), there are two equilibrium points: E_1 ($x^* = 0.75$, $y^* = 0.475$) and E_2 ($x^{**} = 0.96$, $y^{**} = 1.012$). According to Game Theory, these two fixed points are considered as Nash Equilibriums [8]. Below, the stability of the fixed points will be examined.

The Jacobian matrix is:

$$J = \begin{pmatrix} 0 & 0.8y - 0.8(1-y) \\ 3.6x - 3.6(1-x) & 0 \end{pmatrix}$$

We calculate the Jacobian matrix at the equilibrium point E_1:

$$J^* = \begin{pmatrix} 0 & 0.038 \\ 1.803 & 0 \end{pmatrix}$$

The determinant of J^* is det (J^*) = 0.069 > 0.
The trace of J^* is trace(J^*) = 0.
The eigenvalues of J^* is (0.264*i*, -0.264 *i*); two complex roots.
The discriminant = $\Delta = trace(J^*)^2 - 4 \cdot det(J^*)$ 0.2788 < 0.
Therefore, the equilibrium point E_1 is a stable—center.

Studying the second fixed point E_2, the Jacobian matrix at the equilibrium point is:

$$J^{**} = \begin{pmatrix} 0 & 0.82 \\ 3.314 & 0 \end{pmatrix}$$

The determinant of J^{**} is det (J^{**}) = 2.718 < 0.
The trace of J^{**} is trace(J^{**}) = 0.
The eigenvalues of J^{**} is (1.648, 1.6487); two real roots.
The discriminant is $\Delta = trace(J^{**})^2 - 4 \cdot det(J^{**})$ 10.874 > 0.
Therefore, the equilibrium point E_2 is a saddle point.

Consequently, we accept the fixed point E_1 ($x^* = 0.75$, $y^* = 0.475$) and reject E_2 ($x^{**} = 0.96$, $y^{**} = 1.012$), because the value of y^{**} is greater than 1.

---

[3] https://sourceforge.net/projects/maxima/files/Maxima-Windows/5.39.0-Windows/

[4] E & F Chaos: written by Diks, C., Hommes, C., Panchenko, V., van der Weide, R., (2008).

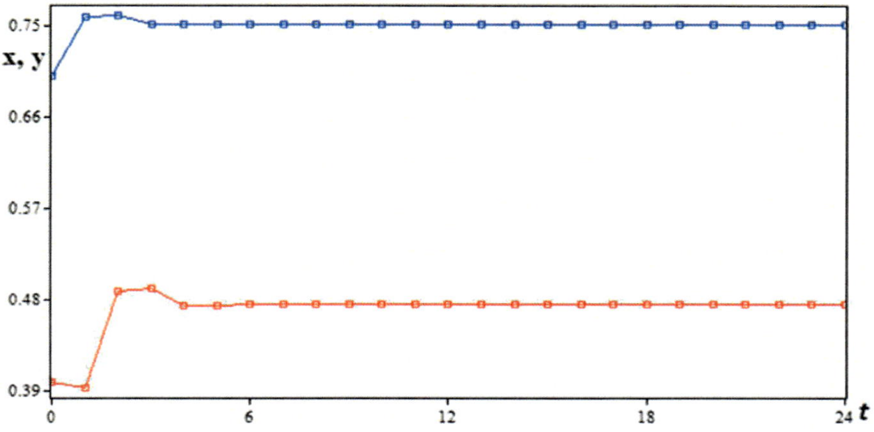

Fig. 4 Time series diagram—x (Greeks; blue) and y (Persians; red)

Thus, we continue the analysis for the fixed point E_1. Interpreting this equilibrium point, we confirm the aggressive (strategic) behavior of Greeks; since the value of x^* is close to 1 and the mild (strategic) behavior of Persians; since they thought it would be an "*easy win*".

Indeed (historically), the courage of the Greeks, their technological naval skills, and the advantageous geographical location contributed in this aggressive behavior. As far as the Persians are concerned, their mild (strategic) behavior is due to the fact that they underestimated their enemy, since they regarded that the Greeks are an easy target, and they would achieve a decisive victory.

Connecting the game "Hawk—Dove" to the naval battle of Salamis, player 1 (red) is "Persians" and player 2 (blue) is "Greeks" (Fig. 4). The Hellenic alliance had an aggressive behavior (Hawk) and the Persians behaved as a Dove. According the Nash equilibriums that have been mentioned above (See 2), the Greeks (player 2) should behave as a Hawk (i.e. aggressive), regardless of Persian's behavior, so as to win this battle.

Figure 4 shows us how the two warring parties behave (strategically). In particular, it represents the optimal strategic behavior of Greeks and Persians in Salamis straits for a time interval of 24 h. We can observe an oscillation, at the beginning, until $t =$ 6 h. (both lines) and then it is normalized and balanced. That means that the duration of the main battle was approximately 6 h. Indeed, according to historical documents, the battle started at dawn (approximately at 06:00 am) and the Greek victory was visible at noon.

Figure 5 presents the strategic behavior of Greeks (blue) and Persians (red) as the parameter G changes. We observe for the positive values of G, the blue line is above the red until $G = 0.64$ (critical value) and for $G > 0.64$ the red line is above the blue. The increase of the value of parameter signifies the change of the geographical location (a more open sea), which becomes more difficult for Greeks and in contrary

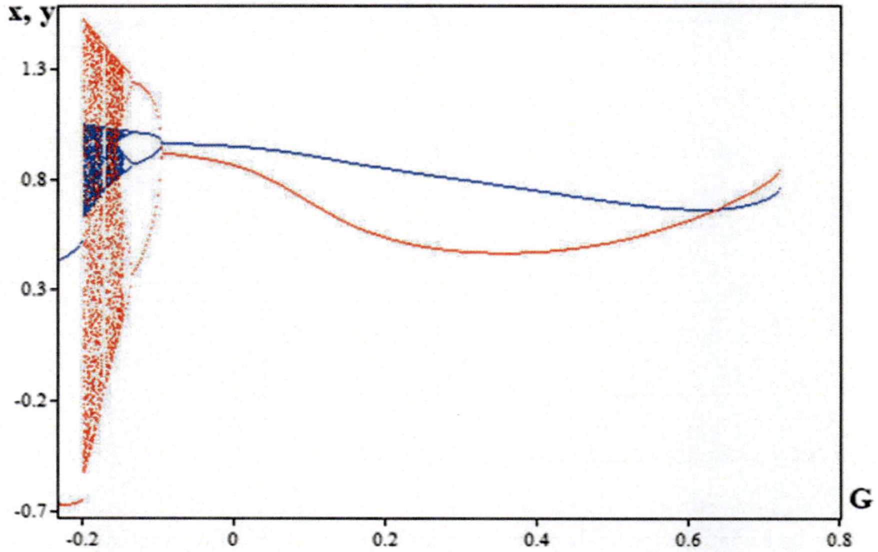

Fig. 5 Bifurcation Diagram for different values of G. x (Greeks; blue) and y (Persians; red)

easier for Persian. Thus, we approve that if the location of the naval battle was in an open sea, the Persians would have a crucial advantage, which would possibly lead to win this conflict.

Although we did not study the negative values of parameter G, we believe that there are some unpredictable geophysical factors (e.g. meteorological phenomena to influence the outcome of the conflict), which are surprisingly interesting. Specifically, we refer to weather conditions, such as air, ripple, etc., which can affect the geophysical terrain of the area. Due to these weather phenomena, period doubling bifurcations and chaos appear and we cannot predict what could happen in the battle for these values of G.

Figure 6 depicts the technological evolution and capability of x (Greeks; blue) and y (Persians; red), respectively. In the left diagram (Fig. 6a), for the negative values of parameter TN_x, we can distinguish a pair of bubble bifurcations, while afterwards we have the well-known period-doubling scenario to chaos. A possible interpretation of this chaotic scenario is the uncertainty of Greeks in technological capability—first attempts to construct ships. The first ships, as Krasanakis [4] mentions, were floating planks and carved tree trucks only with oars. Since the ships were primitive, the situation was unstable (there is chaos in this range of values) because they were not capable to fight in naval battles. Later, the sails were invented, which gave high speed to ships, and they were consisted no more than wood but iron. For this reason, we have bubble bifurcations, which indicate the technological alternatives that existed for the construction of the ships. In the interval of positive values of TN_x, there is stability with two fixed points. Here, it's the beginning of better shipbuilding ability

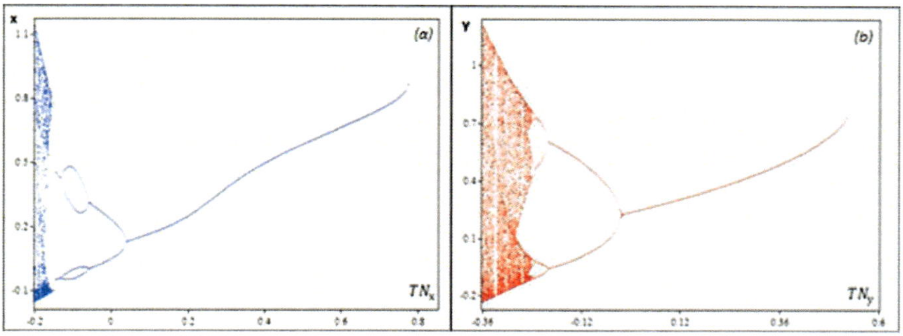

Fig. 6 Bifurcation diagrams for different values of parameters TN_x and TN_y. (**a**) x (Greeks; blue) and (**b**) y (Persians; red)

and new expertise ship construction. Finally, there is one equilibrium point which shows the better version of ships, of that period, namely Triremes. Triremes were wooden warships which move either with sails or oars. Moreover, in the positive values of the parameter TN_x, the increasing of the slope of the curve is visible, which, on the one hand, it means that in 480 BC the triremes were an innovation in shipbuilding and on the other hand, it shows the excellent naval capability of the Greeks.

Persians, through the years, developed technological equipment because of their expansive mania to conquer Greece. Comparing the Figs. 6a and 6b, it seems that Persians had a lower technological development than Greeks, since they focused more on land army than on warships. Their ships were mainly used as troopships rather than battleships [10].

6 Conclusions

In this article, a new non—linear discrete model has been presented, which simulates the optimum strategic behavior of two warring parties for short–term battles. In addition, we try to compare this model with the game “Hawk—Dove”, applying this attempt in the naval battle of Salamis. Based on the results we have extracted, we (mathematically) proved the historical events of this conflict. Specifically, the Greek’s strategic behavior fits with the aggressive behavior of the Hawk and as well as the Persian’s strategic behavior fits with the more defensive behavior of the Dove in the game. Moreover, the estimated duration of the battle was proved as well as the dominance of the Greek fleet in the Salamis straits. Finally, the technological naval capability of the Greek alliance was able to cause serious damages to the opponent and led them to a crashing defeat.

References

1. A.R. Burn, *Persia and the Greeks* (Minerva Press, New York, 1962).
2. W. Goodwin, The battle of Salamis. Harvard Stud. Class. Philol. **17**, 74–101 (1906)
3. P. Green, The Greco—Persian Wars, University of California Press (1998)
4. A. Krasanakis, Naval History of the Greek Nation. Athens Press (2008). ISBN: 9789604841592
5. N. Kyriazis, M. Zouboulakis, Democracy, sea power and institutional change: an economic analysis of the athenian naval law. Eur. J. Law Econ. **17**, 117–132, Kluwer Academic Publishers (2004)
6. E. Lorenz, The problem of deducing the climate from the governing equations. Tellus 16, (1964)
7. A.F. Lykke Jr, Defining military strategy. *Military Rev.* **69**(5) (1989)
8. T. Malthus, An Essay on the Principle of Population. Publisher: J. Johnson, London (1798)
9. M.J. Osborne, A. Rubinstein, A Course in Game Theory (MIT, Cambridge, MA, 1994). ISBN: 9780262150415
10. J. Maynard Smith, G.R. Price, The logic of animal conflict. Nature **246**(5427), 15–18 (1973)
11. W. Shepherd, Salamis 480 BC: The naval campaign that saved Greece. Osprey Publishing (2010). ISBN: 9781846036842

Software

12. C. Diks, C. Hommes, V. Panchenko, R. van der Weide, E&F Chaos: A user-friendly software package for nonlinear economic dynamics. Comput. Econ. **32**, 221–244 (2008)
13. D. McKelvey, Richard, M. McLennan, Andrew, L. Turocy, Theodore, Gambit: Software Tools for Game Theory, Version 16.0.1. (2014). https://www.gambit-project.org
14. https://sourceforge.net/projects/maxima/files/Maxima-Windows/5.39.0-Windows/

Atoms and Pseudo-atoms in Quantum Measure Theory

Alina Gavriluţ

Abstract In this paper, different results concerning (pseudo)-atomicity are obtained from the Quantum Measure Theory mathematical perspective and several physical applications are given. Precisely, the mathematical concept of atomicity (and, particularly, that of minimal atomicity) is extended, based on the non-differentiability of the motion curves associated to the motions of the structural units of a complex system on a fractal manifold.

Keywords Atom · Pseudo-atom · Minimal atom · Non-differentiability · Fractal manifold

1 Introduction

Although classical measure theory imposes strict additivity conditions, in the recent decades, a rich theory of non-additive measures developed. Precisely, modifications of traditional Measure Theory (Pap [12, 13]) led to Quantum Measure Theory (Gudder [7, 8], Salgado [15], Sorkin [17–19] and Surya and Wallden [20]). An extended notion of a measure has been introduced and its applications to the study of interference, probability, and spacetime histories in Quantum Mechanics have been discussed (Schweizer and Sklar [16]). Introduced by Sorkin [17–19], quantum measures are an useful tool which enables us to describe Quantum Mechanics and its applications to Quantum Gravity and Cosmology (Hartle [9, 10], Phillips [14], Salgado [15]).

Quantum Measure Theory indicates a wide variety of applications, as its mathematical structure is used in the standard quantum formalism. In [17–19], Sorkin proposed a history-based framework, which can accommodate both standard Quantum Mechanics and physical theories beyond the quantum formalism. Recently, since Quantum Mechanics can be assimilated with a particular model of Fractal Mechan-

A. Gavriluţ (✉)
Faculty of Mathematics, Alexandru Ioan Cuza University of Iaşi, Iaşi, Romania
e-mail: gavrilut@uaic.ro

C. H. Skiadas et al. (eds.), *13th Chaotic Modeling and Simulation International Conference*, Springer Proceedings in Complexity,
https://doi.org/10.1007/978-3-030-70795-8_18

ics at a given scale resolution in the form of Scale Relativity Theory in a constant fractal dimension and arbitrary (Mercheş and Agop [11]), fundamental concepts of Quantum Mechanics can be extended to similar concepts, but on fractal manifolds.

This paper refers to certain aspects concerning atomicity, which is a very important property in (non)-additive measure theory (Gavriluţ [1], Gavriluţ and Croitoru [3, 4]). This property concerns sets which, from a mathematical "measurement" viewpoint have the property that each subset is either "negligible" or it is almost like the considered set. Precisely, we consider here atomicity, pseudo-atomicity and minimal atomicity from the Quantum Measure Theory mathematical perspective. Several physical applications and considerations are provided and the concept of fractal atomicity is introduced.

2 Atomicity in Quantum Measures Theory

Unless stated otherwise, all over this paper, T denotes an abstract non-empty set and $\mathcal{L}$ a lattice of subsets of T. Suppose $(V, +, \cdot)$ is a real vector space with the origin θ.

In what follows, we give certain key-concepts from Quantum Measure Theory, with slight modifications of the corresponding ones from Gavriluţ et al. [2], Gavriluţ and Agop [5, 6]:

Definition 2.1 Let $m : \mathcal{L} \to V$ be a set function, with $m(\emptyset) = \theta$.

(I) m is said to be:

(i) *finitely additive* (or, *grade-1-additive*) if $m(\bigcup_{i=1}^{p} E_i) = \sum_{i=1}^{p} m(E_i)$ holds, for any arbitrary pairwise disjoint sets $(E_i)_{i\in\{1,2,\ldots,p\}} \subset \mathcal{L}$, $p \in \mathbb{N}^*$;

(ii) a *grade-2-measure* if

$$m(E \cup F \cup G) + m(E) + m(F) + m(G) = m(E \cup F) + m(F \cup G) + m(E \cup G)$$

holds, for any pairwise disjoint sets $E, F, G \in \mathcal{L}$;

(II) Two sets $E, F \in \mathcal{L}$ are called *m-compatible* (denoted by EmF) if

$$m(E \cup F) + m(E \cap F) = m(E) + m(F)$$

holds (i.e., m-compatible sets are those two sets for which the set function m behaves like a grade-1-measure);

(III) An arbitrary fixed set $E \in \mathcal{L}$ which is m-compatible with any set $F \in \mathcal{L}$ is called a *macroscopic set*.

Remark 2.2 (*Gavriluţ* et al. [2], *Gavriluţ and Agop* [5, 6]) (i) Some quantum objects interfere with each other, but others do not. Consequently, one can justify the name of a "macroscopic set" by the fact that it does not interfere with any set and thus it behaves like a non-quantum object in the macroscopic world.

(ii) One can immediately check that m-compatibility generates a relation which is reflexive, symmetric but it is not transitive.

(iii) Evidently, if m is grade-1-additive, then it is also a grade-2-measure, but the converse does not hold.

(iv) If $E \in \mathcal{L}$ is arbitrarily chosen, then E and $\emptyset$ are m-compatible.

(v) Suppose t_i, where $i \in \{1, 2, ..., p\}$, $p \in \mathbb{N}^*$ represent quantum objects or quantum events and let be their collection $T = \{t_1, t_2, ..., t_p\}$. One can need an interpretation of a "measure" on T, in situations when the additivity condition from Definition 2.1.-(ii) is not fulfilled.

For these reasons, in what follows we introduce several notions, that are weaker than classical additivity and also than those from the above Definition 2.1. As before, these notions generalize those from Gavriluţ et al. [2], Gavriluţ and Agop [5, 6]:

Definition 2.3 A set function $m : \mathcal{L} \to V$, with $m(\emptyset) = \theta$, is called:

(i) *disjoint-null-additive* if $m(E \cup F) = m(E)$, for every disjoint $E, F \in \mathcal{L}$, with $m(F) = \theta$;

(ii) *null-additive* if $m(E \cup F) = m(E)$, for every $E, F \in \mathcal{L}$, with $m(F) = \theta$;

(iii) *null-null-additive* if $m(E \cup F) = \theta$, for every $E, F \in \mathcal{L}$, with $m(E) = m(F) = \theta$;

(iv) *null-equal* if $m(E) = m(F)$, for every $E, F \in \mathcal{L}$, with $m(E \cup F) = \theta$;

(v) a *quantum measure* (*q-measure*, for short) if it is a disjoint-null-additive and null-equal grade-2-measure.

Definition 2.4 If, moreover, $(V, \leq)$ is an ordered vector space, then a set function $m : \mathcal{L} \to V$, with $m(\emptyset) = \theta$, is called:

(i) *null-monotone* if for every $E, F \in \mathcal{L}$, with $E \subseteq F$, if $m(F) = \theta$, then $m(E) = \theta$;

(ii) *monotone* if $m(E) \leq m(F)$, for every $E, F \in \mathcal{L}$, with $E \subseteq F$;

(iii) a *submeasure* if m is monotone and subadditive (i.e., $m(E \cup F) \leq m(E) + \nu(F)$, for every (disjoint) $E, F \in \mathcal{L}$).

Example 2.5 (i) It T is a nonempty metric space, then the Hausdorff dimension $\dim_{Haus} : \mathcal{P}(T) \to \mathbb{R}$ is a monotone real function. Evidently, $\dim_{Haus}(\emptyset) = 0$.

(ii) For every $d \geq 0$, the Hausdorff measure $H^d : \mathcal{P}(T) \to \mathbb{R}$ is an outer measure, so, particularly, it is a submeasure.

Definition 2.6 Suppose $(V, \leq)$ is an ordered vector space.

I) If $m : \mathcal{L} \to V$, with $m(\emptyset) = \theta$ is an arbitrary set function, then:

(i) A set $A \in \mathcal{L}$ is called a *pseudo-atom* of m if $m(A) > \theta$ and $B \in \mathcal{L}$, $B \subseteq A$ implies $m(B) = \theta$ or $m(B) = m(A)$;

(ii) m is said to be *non-pseudo-atomic* (NPA, for short) if it has no pseudo-atoms (i.e., for every $A \in \mathcal{L}$ with $m(A) > \theta$, there exists $B \in \mathcal{L}$, $B \subseteq A$, such that $m(B) > \theta$ and $m(A) \neq m(B)$;

(iii) A set $A \in \mathcal{L}$ is called to be *a minimal atom of m* if $m(A) \geq \theta$ and for every $B \in \mathcal{C}$, $B \subseteq A$, it holds either $m(B) = \theta$ or $B = A$;

II) Suppose, moreover, that $\mathcal{C}$ is a ring of subsets of T and let $m : \mathcal{C} \to V$ be a set function, with $m(\emptyset) = \theta$.

(i) A set $A \in \mathcal{C}$ is called an *atom* of ν if $m(A) > \theta$ and for every $B \in \mathcal{C}$, with $B \subseteq A$, we have $m(B) = \theta$ or $m(A \backslash B) = \theta$;

(ii) m is said to be *non-atomic* (NA, for short) if it has no atoms (i.e., for every $A \in \mathcal{C}$ with $m(A) > \theta$, there exists $B \in \mathcal{C}$, $B \subseteq A$, such that $m(B) > \theta$ and $m(A \backslash B) > \theta$);

(iii) m is said to be *finitely purely atomic* if there is a finite family $(A_i)_{i \in \{1,2,\ldots,n\}}$ of pairwise disjoint atoms of m so that $T = \bigcup_{i=1}^{n} A_i$ (in this case, the entire space is assumed to be a finite collection of pairwise disjoint atoms).

In case when $\mathcal{C}$ is a ring of subsets of T and $m : \mathcal{C} \to \mathbb{R}_+$, with $m(\emptyset) = 0$, is an arbitrary set function, then (when they exist), its atoms and pseudo-atoms have certain remarkable properties:

Proposition 2.7 *(Gavriluţ and Croitoru [3, 4]) (i) (self-similarity of atoms) If m is null-monotone, $E \in \mathcal{C}$ is an atom of m and $F \in \mathcal{C}$, $F \subseteq E$ is such that $m(F) > 0$, then F is an atom of m and $m(E \backslash F) = 0$.*

(ii) (self-similarity of pseudo-atoms) If $E \in \mathcal{C}$ is a pseudo-atom of and $F \in \mathcal{C}$, $F \subseteq E$ is such that $m(F) > 0$, then F is a pseudo-atom of m and $m(F) = m(E)$.

(iii) If $E, F \in \mathcal{C}$ are pseudo-atoms of and $m(E \cap F) > 0$, then $E \cap F$ is a pseudo-atom of m and $m(E \cap F) = m(E) = m(F)$.

(iv) Let $m : \mathcal{C} \to \mathbb{R}_+$ be null-additive and let $E, F \in \mathcal{C}$ be pseudo-atoms of m.

1. If $m(E \cap F) = 0$, then $E \backslash F$ and $F \backslash E$ are pseudo-atoms of m and $m(E \backslash F) = m(E)$, $m(F \backslash E) = m(F)$.

2. If $m(E) \neq m(F)$, then $m(E \cap F) = 0$, $m(E \backslash F) = m(E)$ and $m(F \backslash E) = m(F)$.

(v) Let $m : \mathcal{C} \to \mathbb{R}_+$ be null-additive and let $E, F \in \mathcal{C}$ be pseudo-atoms of m. If $m(E \cap F) > 0$ and $m(E \backslash F) = m(F \backslash E) = 0$, then $E \cap F$ is a pseudo-atom of and $m(E \Delta F) = 0$.

Definition 2.8 Suppose $m : \mathcal{C} \to \mathbb{R}_+$, with $m(\emptyset) = 0$, is an arbitrary set function. Let $\overline{m} : \mathcal{P}(T) \to [0, \infty]$, defined for every $A \in \mathcal{P}(T)$ by:

$$\overline{m}(A) = \sup \left\{ \sum_{i=1}^{p} m(A_i);\ A = \bigcup_{i=1}^{p} A_i,\ A_i \in \mathcal{C}, \forall i \in \{1, ..., p\}, A_i \cap A_j = \emptyset,\ i \neq j \right\}.$$

We say that m *is of finite variation* if $\overline{m}(T) < \infty$.

3 Quantum Measures. Decoherence Functions

In Quantum Mechanics, when a wavefunction becomes coupled to its environment, the objects involved interacting with the surroundings, the decoherence phenomenon occurs. It is also known as the "wavefunction collapse" and it allows the classical limit to emerge on the macroscopic scale from a set of quantum events. After decoherence has occurred, the system's components can no longer interfere, so one could assign a well-defined probability to each possible decoherent outcome.

Remark 3.1 In light's classical theory, the intensity of the light in an arbitrary point is determined by the square amplitude of the light. For instance, in Young's two-slit experiment, the intensity of the light on the detector screen is given by the square amplitude of the wave obtained through the overlapping (superposition) of the secondary waves originating from each slit. Of course, this classical wave theory can not be used in this case since it ignores the corpuscular character of the light. However, by analogy, it suggests that in Quantum Mechanics, it can be introduced either a wavefunction which satisfies the Schrödinger equation. The wavefunction is a complex quantity, while the states density is a real one. We expect then, that the states density $\rho(x, y, z, t)$ to find the particle in a given point from the volume V, in a vecinity of the point of coordinates (x, y, z, t) at a momentum t should be proportional with $|\Psi|^2$, that is,

$$\rho(x, y, z, t) \equiv |\Psi(x, y, z, t)|^2 \tag{1}$$

Let Ψ_1 be the wavefunction in a given point from the screen where the interference field is localized, corresponding to the waves propagated through the slit 1. Similarly, let Ψ_2 be the wavefunction in the same point, corresponding to the waves propagated through the slit 2. The two intensity distributions, corresponding to the "experiments" performed with only one open slit are determined by the respective states densities (probability distributions)

$$\rho_1 \equiv |\Psi_1|^2, \rho_2 \equiv |\Psi_2|^2 \tag{2}$$

On the other hand, when both slits are open, the wavefunction is given by the sum of the two contributions Ψ_1 and Ψ_2 :

$$\Psi \equiv \Psi_1 + \Psi_2 \tag{3}$$

The corresponding states density (probability distribution)

$$\rho \equiv |\Psi_1 + \Psi_2|^2 \tag{4}$$

determines then the intensity of the "structure" from the interference field.

Let us explain in the following, Ψ_1 and Ψ_2 in the form

$$\Psi_1 = \sqrt{\rho_1} e^{i\theta_1}, \Psi_2 = \sqrt{\rho_2} e^{i\theta_2} \tag{5}$$

It results

$$\rho \equiv |\Psi|^2 = |\Psi_1|^2 + |\Psi_2|^2 + 2\mathrm{Re}\{|\Psi_1| \cdot |\Psi_2| \exp[i(\theta_2 - \theta_1)]\} \\ \equiv \rho_1 + \rho_2 + 2\sqrt{\rho_1 \rho_2} \cos \Delta\theta, \; \Delta\theta = \theta_2 - \theta_1. \tag{6}$$

Now, if the term $\cos \Delta\theta$ is a time functional

$$\cos \Delta\theta \equiv \cos \Delta\theta(t) \tag{7}$$

then the system is decoherent (the interference field does not exist). If

$$\cos \Delta\theta = \text{const.} \tag{8}$$

then the system is coherent (there exists an interference field).

One can define functions related to interference, as shown in the following. Let $\mathbb{C}$ be the family of complex numbers, "$\overline{\quad}$" the complex conjugate of a complex number and "$|\cdot|$", the modulus of a complex number.

Definition 3.2 Suppose $\mathcal{A}$ is an algebra of subsets of T. A function $D : \mathcal{A} \times \mathcal{A} \to \mathbb{C}$ is said to be a *decoherence function* if the following conditions hold:
(i) $D(E, F) = \overline{D(F, E)}$, for every $E, F \in \mathcal{A}$;
(ii) $D(E, E) \geq 0$, for every $E \in \mathcal{A}$;
(iii) $|D(E, F)| \leq D(E, E) \cdot D(F, F)$, for every $E, F \in \mathcal{A}$;
(iv) $D(E \cup F, G) = D(E, G) + D(F, G)$, for every disjoint $E, F \in \mathcal{A}$ and every $G \in \mathcal{A}$.

Remark 3.3 (i) Since $D(E, E) \in \mathbb{R}$, then the conditions (ii) and (iii) from Definition 3.2. are justified.

(ii) By (i), for arbitrary $E, F \in \mathcal{A}$ representing quantum objects, $Re[D(E, F)]$ can be interpreted as the interference between E and F, as we remark in what follows:

Proposition 3.4 *If $D : \mathcal{A} \times \mathcal{A} \to \mathbb{C}$ is a decoherence function, then the set function $M : \mathcal{A} \to \mathbb{C}$, $M(E) = D(E, E)$ is a q-measure.*

Example 3.5 If V is a pre-Hilbert space and if $m : \mathcal{A} \to V$ is finitely additive, then $D : \mathcal{A} \times \mathcal{A} \to \mathbb{C}$,

$$D(E, F) =< m(E), m(F) >,$$

for every $E, F \in \mathcal{A}$ is a decoherence function.

Moreover, if $m : \mathcal{A} \to \mathbb{C}$ is finitely additive (often interpreted as a quantum amplitude), then one can define the decoherence function defined for every $E, F \in \mathcal{A}$ by

$$D(E, F) = m(E) \cdot \overline{m(F)}.$$

The corresponding q-measure is $M : \mathcal{A} \to \mathbb{C}$,

$$M(E) = D(E, E) = m(E) \cdot \overline{m(E)} = |m(E)|^2,$$

for every $E \in \mathcal{A}$.

Remark 3.6 (i) If $E, F \in \mathcal{A}$ are disjoint, then M is not grade-1-additive. Indeed,

$$\begin{aligned} M(E \cup F) &= |m(E \cup F)|^2 = |m(E) + m(F)|^2 = \\ &= |m(E)|^2 + |m(F)|^2 + 2\mathrm{Re}[m(E)\overline{m(F)}] = \\ &= M(E) + M(F) + 2\mathrm{Re}D(E, F). \end{aligned}$$

Also, $M(E \cup F) = M(E) + M(F)$ if and only if $\mathrm{Re}D(E, F) = 0$ (this means interference is represented by the real part of a decoherence function).

(ii) If $m : \mathcal{A} \to \mathbb{R}$ is a real valued submeasure of finite variation $\overline{m}$, then the set function

$$D : \mathcal{A} \times \mathcal{A} \to \mathbb{R},\ D(E, F) = < \overline{m}(E), \overline{m}(F) >$$

is a decoherence function.

4 From the Standard Mathematical Atom to the Fractal Atom by Means of a Physical Procedure

Let T be an abstract nonvoid set, $\mathcal{L}$ a lattice of subsets of T and $m : \mathcal{L} \to \mathbb{R}_+$ an arbitrary set function with $m(\emptyset) = 0$.

Remark 4.1 (i) The union of two sets A and B having the fractal dimensions D_A, respectively, D_B has the fractal dimension $D_{A\cup B} = \max\{D_A, D_B\}$;

(ii) The intersection of two sets A and B having the fractal dimensions D_A, respectively, D_B has the fractal dimension $D_{A\cap B} = D_A + D_B - d$, where d is the embedding Euclidean dimension.

One can introduce then the following concept:

Definition 4.2 A pseudo-atom, a minimal atom, respectively, $A \in \mathcal{L}$ of m having the fractal dimension D_A is said to be a *fractal pseudo-atom*, a *fractal minimal atom*, respectively.

One easily gets:

Proposition 4.3 *If $A, B \in \mathcal{L}$ are fractal pseudo-atoms of m and if $m(A \cap B) > 0$, then $A \cap B$ is a fractal pseudo-atom of m and $m(A \cap B) = m(A) = m(B)$.*

References

1. A. Gavriluţ, Non-atomicity and the Darboux property for fuzzy and non-fuzzy Borel/Baire multivalued set functions. Fuzzy Sets Syst. **160**, 1308–1317 (2009). Erratum in Fuzzy Sets Syst. **161**, 2612–2613 (2010)
2. A. Gavriluţ, M. Agop, *An Introduction to the Mathematical World of Atomicity through a Physical Approach* (ArsLonga Publishing House, Iaşi, Romania, 2016)
3. A. Gavriluţ, A. Croitoru, Non-atomicity for fuzzy and non-fuzzy multivalued set functions. Fuzzy Sets Syst. **160**(14), 2106–2116 (2009)
4. A. Gavriluţ, A. Croitoru, Pseudo-atoms and Darboux property for set multifunctions. Fuzzy Sets Syst. **161**(22), 2897–2908 (2010)
5. G. Gavriluţ, A. Gavriluţ, M. Agop, Extended Minimal Atomicity through Non-differentiability: A Mathematical-Physical Approach. Adv. Math. Phys. **160**, 1–16 (2019). https://doi.org/10.1155/2019/8298691
6. A. Gavriluţ, I. Mercheş, M. Agop, Atomicity through Fractal Measure Theory: Mathematical and Physical Fundamentals with Applications. Springer (2019). https://doi.org/10.1007/978-3-030-29593-6
7. S. Gudder, Quantum measure theory. Mathematica Slovaca **60**(5), 681–700 (2010)
8. S. Gudder, *Quantum measures and integrals*, arxiv print arXiv:0909.2203
9. J.B. Hartle, The quantum mechanics of cosmology, in *Lectures at Winter School on Quantum Cosmology and Baby Universes*, Jerusalem, Israel, Dec 27, 1989-Jan 4, 1990 (1989)
10. J.B. Hartle, Spacetime quantum mechanics and the quantum mechanics of spacetime, in *Proceedings of the Les Houches Summer School on Gravitation and Quantizations*, Les Houches, France, 6 Jul-1 Aug 1992, J. Zinn-Justin and B. Julia, Eds., North-Holland, Amsterdam (1995)
11. I. Mercheş, M. Agop, *Differentiability and Fractality in Dynamics of Physical Systems* (World Scientific Publishing Co., 2015)
12. E. Pap, Null-additive set functions, in *Vol. 337 of Mathematics and Its Applications* (Springer, 1995)
13. E. Pap, Some elements of the classical measure theory, in *Handbook of MeasureTheory* (2002), pp. 27–82
14. A.C. Phillips, *Introduction to Quantum Mechanics* (Wiley, New York, NY, USA, 2003)
15. R.B. Salgado, Some identities for the q-measure and its generalizations. Mod. Phys. Lett. A **17**(12), 711–728 (2002)
16. B. Schweizer, A. Sklar, *Probabilistic Metric Spaces* (Elsevier Science Publishing Co, 1983)
17. R.D. Sorkin, Quantum mechanics as quantum measure theory. Mod. Phys. Lett. A **9**(33), 3119–3127 (1994)
18. R.D. Sorkin, Quantum dynamics without the wavefunction. J. Phys. A: Math. Gen. **40**(12), 3207–3221 (2007)
19. R.D. Sorkin, Quantum measure theory and its interpretation, in *Quantum Classical Correspondence: Proceedings of the 4th Drexel Symposium on Quantum Non-Integrability*, eds. by D.H. Feng, B.-L.Hu (International Press, Cambridge, Mass, USA, 1997), pp. 229–251
20. S. Surya, P.Wallden, *Quantum covers in q-measure theory*, arxiv print arXiv:0809.1951 (2008)

Albert Einstein and the Doubling of the Deflection of Light

Jean-Marc Ginoux

Abstract One of the three consequences of Einstein's theory of general relativity was the curvature of light passing near a massive body. In 1911, he published a first value of the angle of deflection of light, then a second value in 1915, equal twice the first. In the early 1920s, when he received the Nobel Prize in Physics, a violent controversy broke out over this result. It was then disclosed that the first value he had obtained in 1911 had been calculated more than a century before by a German astronomer named Johann von Soldner. The aim of this article is therefore to compare the methods used by Soldner and then by Einstein leading to this first value and to explain the importance of the doubling of this value in the framework of Einstein's theory of general relativity. Such a consequence of this theory lies at the intersection of several scientific fields such as Mathematics, Physics, Astronomy and Philosophy.

Keywords General relativity · Deflection of light · Johann von Soldner

1 The Genesis of General Relativity and the Curvature of Light

Two years after the publication of his article on special relativity, Albert Einstein considered generalizing his theory. Thus, in 1907, he wrote an article entitled "Relativitätsprinzip und die aus demselben gezogenen Folgerungen" ("On the Principle of Relativity and the Conclusions Drawn from it"), at the request of Johannes Stark, editor of the *Jahrbuch der Radioaktivität*, in which he presented for the first time one of the consequences of his theory. In paragraph V. entitled "The Principle of Relativity and Gravitation", he wrote:

> From this it follows that those light rays that do not propagate along the ξ-axis are bent by the gravitational field [5].

J.-M. Ginoux (✉)
Aix Marseille University, Université de Toulon, CNRS, CPT, Marseille, France
e-mail: jmginoux@orange.fr

C. H. Skiadas et al. (eds.), *13th Chaotic Modeling and Simulation International Conference*, Springer Proceedings in Complexity,
https://doi.org/10.1007/978-3-030-70795-8_19

At that time Einstein thought that the effect of gravitational field on rays of light was too weak to be detected as evidenced by his conclusion:

> Unfortunately, the effect of the terrestrial gravitational field is so small according to our theory (because of the smallness of $\gamma x/c^2$) that there is no prospect of a comparison of the results of the theory with experience [5].

During the autumn 1911, Einstein became full professor of theoretical physics at the German Charles-Ferdinand University in Prague. Einstein lived more than a year at number 7 Leniscka Street with his first wife Mileva Marić and their two children Hans Albert and Eduard. Here he wrote a second article, considered as the starting point of his theory of general relativity: "Einfluss der Schwerkraft auf die Ausbreitung des Lichtes" ("On the influence of Gravitation on the Propagation of Light") in which he gave a first value of the deflection of light passing near a massive body. In paragraph 4 entitled Curvature of light rays in the gravitational field, he wrote:

> By equation (4) a light-ray passing by a heavenly body suffers a deflection to the side of the diminishing gravitational potential, that is, to the side directed toward the heavenly body, of the magnitude
>
> $$\alpha = \frac{1}{c^2}\int_{\nu=-\frac{\pi}{2}}^{\nu=+\frac{\pi}{2}} \frac{kM}{r^2}\cos(\nu)\, ds = \frac{2kM}{c^2\Delta}$$
>
> where k denotes the constant of gravitation, M the mass of the heavenly body, Δ the distance of the ray from the center of the body (and r and ν are as shown in Fig. 3). *A light-ray going past the Sun would accordingly undergo deflection by the amount of* $4.10^{-6} = 0.83$ *seconds of arc*. The angular distance of the star from the center of the Sun appears to be increased by this amount. As the fixed stars in the parts of the sky near the Sun are visible during total eclipses of the Sun, this consequence of the theory may be compared with experimental evidence [6].

Then, he added:

> It would be urgently wished that astronomers take up the question here raised, even though the considerations presented above may seem insufficiently established or even bizarre. For, apart from any theory, there is the question whether it is possible with the equipment at present available to detect an influence of gravitational fields on the propagation of light [6].

In 1912 Einstein was appointed, on the recommendation of Marie Curie and Henri Poincaré, to a chair of mathematical physics at the Swiss Federal Polytechnic in Zürich (later the Eidgenössische Technische Hochschule, ETH) where he had studied a few years earlier. The following year, he joined Berlin to become member of the Prussian Academy. On November 1915, he submitted four papers to the journal of the Prussian Academy of Sciences [7–10]. Then, he published in 1916 in *Annalen der Physik* a 40-page article entitled "Die Grundlage der algemeinen Relativitätstheorie" ("The Foundations of the General Theory of Relativity") considered as the final and complete version of his theory of general relativity. In the last section he presented the three observable physical consequences of his theory and in particular the curvature of the light rays and explained that this latter consequence could be verified by exact astronomical experiments. Then, he provided a second value of the deflection of light passing near a massive body:

> We examine the curvature undergone by a ray of light passing by a masse M at the distance Δ. If we choose the system of co-ordinates in agreement with the accompanying diagram, the total bending of the ray (calculated positively if concave towards the origin) is given in sufficient approximation by[1]
>
> $$B = \frac{\kappa M}{2\pi\Delta}$$
>
> According to this, a ray of light going past the sun undergoes a deflexion of 1.7″ ...[11].

It is very surprising to notice that this second value of the angle of deflection of light provided by Einstein is equal twice the first. It will be learned some years later that the first value provided by Einstein in 1911 was in fact identical to that published a century earlier by the German physicist and astronomer Johann Georg von Soldner. This fact has been reported since by many historians of science such as Thomas Glick [20], Jean Eisenstaedt [14], Jürgen Renn [29] and Milena Wazeck [35] to name but a few. Nevertheless, neither of these references contains any mathematical analysis nor comparison of Soldner's and Einstein's results. According to Abraham Pais:

> An Argentinian eclipse expedition which had gone to Brazil in 1912 and which had the deflection of light on its experimental program was rained out. In the summer of 1914, a German expedition led by Erwin Freundlich and financed by Gustav Krupp, in a less familiar role of benefactor of humanity, headed for the Crimea to observe the eclipse of August 21. (Russian soldiers and peasants were told by their government not to fear evil omens: the forthcoming eclipse was a natural phenomenon.). When the war broke out, the party was warned in time to return and some did so. Those who hesitated were arrested, eventually returned home safely but of course without results. Frustration continued also after November 18, 1915, the day on which Einstein announced the right bending of 1″74. [...].
> An opportunity to observe an eclipse in Venezuela in 1916 had to be passed up because of the war. Early attempts to seek deflection in photographs taken during past eclipses led nowhere. An American effort to measure the effect during the eclipse of June 1918 never gave conclusive results. It was not until May 1919 that two British expeditions obtained the first useful photographs and not until November 1919 that their results were formally announced [28].

When the armistice was signed on November 11, 1918, two expeditions were mounted, one to Sobral in Brazil, led by Andrew Crommelin from the Greenwich Observatory, and one to Principe Island off the coast of Spanish Guinea, led by Eddington. After the return of the expeditions, data analysis began. Einstein could not hide his enthusiasm in the expectation of the results. On September 22, 1919, Hendrik Lorentz sent a telegram to Einstein announcing:

> Eddington found stellar shift at solar limb, tentative value between nine-tenths of a second and twice that [13, p. 95].

On the afternoon of November 6, 1919, at Burlington House in Piccadilly, the Astronomer Royal, Sir Frank Dyson, had the honor of presenting the results of the two expeditions. He described in detail the equipment, the photographs, and the complexities of the computations. His conclusion, however, was simple.

---

[1] The value $\kappa = 8\pi K/c^2$ where $K = 6.67 \times 10^{-11}\ \mathrm{m^3\,kg^{-1}\,s^{-2}}$ is the Cavendish's constant.

> The results of the expeditions to Sobral and Principe leave little doubt that a deflection of light takes place in the neighbourhood of the sun and that it is of the amount demanded by Einstein's generalized theory of relativity.

Thus, it was considered that the astronomical observations had "demonstrated" the curvature of space. The day after, Einstein's name became legendary. *The London Times* published an article entitled "Revolution in Science, New Theory of the Universe, Newtonian ideas overthrown". On November 9, 1919, *The New York Times* published the following article [18, 19]:

> Diversion of light Rays Accepted as Affecting Newton's Principles.

Thus, in 1919 the expedition led by Crommelin and Eddington "confirmed" Einstein's second prediction, that is to say, that the value of the deflection of the light passing near the Sun was equal to 1.7 s of arc. Eddington's results for the solar eclipse observation of the apparent displacement of stars of 1919 have been widely disputed by many historians of science. See for example John Earman and Clark Glymour [4]. See also the works of Daniel Kennefick [23, 24]. Moreover, the question, which was the subject of intense controversy, was whether or not Einstein was aware of Soldner's work when he published his article in 1911.

2 Controversy Around the Curvature of Light

In the early Twenties, Einstein and his theory of relativity were subject to many attacks of various natures [35]. The first, the most odious, had an anti-Semitic character. Then, in 1921, scientists such as Charles Lane Poor, a professor of astronomy, a specialist in Celestial Mechanics at Columbia University, tried to prove that Einstein's theories of relativity were false. To this aim he published an article [27] entitled "Is Einstein wrong?—A debate". When Einstein was invited to the Collège de France in 1922 by Paul Langevin, his theories were challenged by Édouard Guillaume who had worked with him as a patent examiner at the Swiss patent office in Bern and had come purposely from Swiss in order to "destroy relativity". Indeed, he had published a few weeks beforehand an article entitled: "Y a-t-il une erreur dans le premier mémoire d'Einstein ?" which let no doubt concerning his intentions [21]. Unfortunately, this anti-relativistic attitude was shared by a part of the French scientific community and more particularly by many Academicians of Sciences. It has been initially convened that Einstein presents his work at the Academy but some of the members of the Academy had decided as a protest against his presence to rise and leave the hall as soon as he entered, Einstein had to renounce [1]. In August 1922, Einstein, who had received many death threats, decided to leave temporarily Europe. On October 8, 1922, Albert Einstein and his second wife Elsa came aboard the S. S. Kitano Maru in Marseille (south France) and reached then their final destination at Fukuoka in Japan where Einstein gave a lecture at the Daihaku Theater. On November 15, during a stopover in an hotel of Shanghai, Einstein received a telegram from Sweden announcing him that he has been awarded the Nobel Prize of Physics for "his contribution to theoretical physics and more particularly for his discovery of the law of photoelectric effect" and not for his relativity theories [16]. This news almost immediately triggered new reactions. Einstein's most virulent opponent was the German physicist Philipp Lenard, Nobel Prize in 1905 (the year of the relativity theory) and active proponent of the Nazi ideology. Then, in April 1923 Professors Ernst Gehrcke[2] of Berlin, P. Lenard of Heidelberg, O. E. Westin[3] of Stockholm charged Einstein with downright plagiarism, saying:

> From these facts the conclusion seems inevitable that Einstein cannot be regarded as a scientist of real note. He is not an honest investigator." Thus Westin protested to the Directorate of the Nobel Foundation against the reward of Einstein.[4]

What were "theses facts" invoked by the three professors against Einstein? *The New York Times* of April 13 provides the answer:

> ...in 1801 Dr. J. von Soldner, a German physicist of eminence in his day, actually derived the formula recently used by Einstein. This was 122 years ago. Einstein never once mentions Soldner in his writings. This is bad enough, but the worst is yet to come. It has been shown by Professor Dr. E. Gehrcke, Director of the Imperial Physical and Technical Institute of Berlin, a position first filled by Helmohltz and by Professor P. Lenard of Heidelberg, winner of the Nobel Prize in Physics, that Soldner omitted a certain factor in his formula of 1801, which error Einstein also copied when he appropriated the Einstein-Soldner formula in the Einstein paper of 1911. In a subsequent paper to the Berlin Academy of Science, 1915, Einstein camouflaged this fraud as best he could, yet could not prevent its discovery and exposure by Professors Lenard of Heidelberg, Gehrcke of Berlin and Westin or Stockholm.

To these three scientists was added the professor Arvid Reuterdahl, Dean of the Engineering Department of the University of St. Thomas, St. Paul, Minnesota. He sent a letter to the Editor of *The New York Times* published on June 3, 1923 and in which he claimed:

> There are two episodes in the Einsteinian development of the bending of light. In Einstein's 1911 paper the value of the deflection is given as 0.83 of a second. Soldner's value was 0.84 of a second. The two formulae are identical except in the matter of the convenient substitution by Einstein of different letters than those used by Soldner. Compared, letter for letter, the meanings are, however identical.[5] In his 1916 paper Einstein modified his 1911 value to read 1.7 of a second. Einstein has never taken the world into his confidence concerning the reason of this change. He has never admitted that either one or the other of these values must be erroneous.[6]

---

[2] E. Gehrcke (1878–1960) was a German experimental physicist, Director of the Imperial Institute of Berlin. He was a *Privatdozent* at the *Friedrich-Wilhelms-Universität* from 1904 to 1921 and an *außerordentlicher Professor (extraordinarius professor)* from 1921 to 1946.

[3] Oscar Edward Westin (1848–1930) was a Swedish engineer, professor of mechanical engineering at the *Royal Institute of Technology* in Stockholm.

[4] See *The New York Times*, April 13, 1923. This event is also mentioned in Philipp Frank [15, p. 202 and next].

[5] See Sect. 3.

[6] We will see in Sect. 3 that this statement is partially inaccurate.

Indeed, in 1916, in his famous article entitled "The Foundations of the General Theory of Relativity [11]", Einstein realized that his earlier result on the bending of light, he had presented in his previous article of 1911, was too small by a factor of 2. This factor will be proved to be decisive because it enables to reveal a strict separation between Newtonian and Einsteinian theory of gravitation. This was due to the fact that in his 1911 paper Einstein had not taken into account in his computations the curvature of space but, only included the effect of Newtonian gravitational interaction on the four-dimensional space-time that he will express later in a Minkowskian metric. According to Abraham Pais:

> Let us briefly recapitulate Einstein's progress in understanding the bending of light.
> 1907. The clerk at the patent office in Bern discovers the equivalence principle, realizes that this principle by itself implies some bending of light, but believes that the effect is too small to ever be observed.
> 1911. The professor at Prague finds that the effect can be detected for starlight grazing the sun during a total eclipse and finds that the amount of bending in that case is $0''87$. He does not yet know that space is curved and that, therefore, his answer is incorrect. He is still too close to Newton, who believed that space is flat and who could have himself computed the $0''87$ (now called the Newton value) from his law of gravitation and his corpuscular theory of light.
> 1912. The professor at Zurich discovers that space is curved. Several years pass before he understands that the curvature of space modifies the bending of light.
> 1915. The member of the Prussian Academy discovers that general relativity implies a bending of light by the sun equal to $1''74$, the Einstein value, twice the Newton value. This factor of 2 sets the stage for a confrontation between Newton and Einstein [28].

3 Comparison of the Work of Einstein and Soldner

Such a comparison has been subject to many studies and several historians of science have analyzed both works of Soldner and Einstein from a mathematical point of view during these last decades. In 1975, Hans Fuchs published an article entitled "On the history of ideas about the effect of gravity on the light [17]" in which he presented the proofs leading to the value of the deflection of light passing near a massive body obtained by Soldner in 1801 and then by Einstein in 1911 and finally in 1915. After comparing Einstein's paper of 1911 with that of Soldner of 1801, Fuchs [17] wrote:

> Man erhält also trotz Zuhilfenahme des Aquivalenzprinzips wieder den alten falschen klassischen Wert! Wie ist das möglich?.[7]

Then, he explained:

> Obwohl wir von der Erde aus eine Verlangsamung der Vorgänge feststellen, werden wir doch mit gleichen Uhren auf der Sonne wie auf der Erde die gleichen Frequenzen und die gleiche Lichtgeschwindigkeit messen, weil die Uhren in gleichem Masse wie die Naturvorgänge verlangsamt werden! Diese logisch einwandfreie Sicht der Dinge erlangte Einstein erst

[7] "Thus, despite the aid of the principle of equivalence, one gets the old false classic value again! How is that possible?"

durch das tiefere Verständnis, das durch die allgemeinen Relativitätstheorie gebracht wurde (nach 1915)[8].

By using the Fields Equations of Gravitation and the Schwarzchild metric (including the so-called Ricci tensors) Fuchs showed that Einstein was then able to give the "der korrekte relativistische Wert" ("the correct relativistic value"). He also compared the relativistic and classical calculations and stated that the two results differ by a factor of two. However, his proof is different from those presented in this paper. In 1978, Stanley Jaki published an article in which he recalled the historical context of the bending of light [22]. Starting from the seminal works of Newton and Laplace, he presented the controversy triggered out by Lenard, Gehrcke and Westin (see Sect. 2 above). Then, he proposed an English translation of Soldner's article allowing historians of science to study and compare his works with those of Einstein. He gives many details and references concerning Soldner's life but didn't provide any mathematical analysis of his article.[9] In 1980, John Earman and Clark Glymour [4] published a long article in which they compared Einstein's results of 1911 and 1915. As Fuchs, they recalled:

> Einstein had not by 1911 yet absorbed the four-dimensional geometrical way of viewing space-time urged by Minkowski. In certain respects his thinking about space-time was still classical [4].

Then, they explained:

> Einstein gave two arguments for the deflection of light passing near a massive body such as the sun; one argument, given in 1911 before the general theory was in hand, relied on his "principle of equivalence," while the other, given in 1916, used Einstein's own approximate solution to his gravitational field equations together with Huygens' principle from classical optics. The former derivation gave a value for the deflection at the limb of the sun of $0.83''$ of arc, the latter $1.7''$ of arc [4].

They concluded that:

> By 1916 Einstein had obtained two different expressions for the angular deflection of a light ray by a massive gravitational source, both giving the angle as a hyperbolic function of distance of closest approach to the massive body. The two expressions, one from the principle of equivalence and the other from the general theory, differ only by a factor of two [4].

Nevertheless, they didn't compare Einstein's results with Soldner's. In 1981, Hans-Jürgen Treder and Gerhard Jackisch published an article in which they considered that "A factor 2, which had been the occasion for misinterpretation, has to be attributed to the terminology used by German physicists and astronomers of that time [33]." They concluded that:

---

[8] "Although we observe a slowing down of the earth from the earth, we will measure the same frequencies and the same speed of light with the same clocks on the sun as on the earth, because the clocks are slowed down to the same extent as the natural processes! Einstein achieved this logically correct view of things only through the deeper understanding that came with the general theory of relativity (after 1915)."

[9] See also Eisenstaedt [14].

> Soldner did nowhere draw false inferences but fell a victim to the printer's devil, and it is indisputable that Soldner obtained the Newtonian value of the deflection of light, which with respect to the constants of his times amounts to 0″84, and not to Einstein's value [33].

Hence, it is obvious that they compared Soldner's result with that provided by Einstein in 1915 and not with that of 1911. Moreover, their conclusion has been challenged by historians of science such as Ledo Stefanini [32] who wrote: "Some scholars attribute the numerical errors appearing in Soldner's memoir (correctly pointed out by Lenard in 1921 republication) to typographical errors,[10] but this does not suffice to clear the issue." This point will be clarified in the Sect. 3.2.

So, the aim of this work is to mathematically compare the formula obtained by Soldner with the one stated by Einstein in his paper of 1911 and to verify if they are identical or not. By using a simple first-order series expansion, it will thus be proved (to our knowledge for the first time) that both Soldner's and Einstein's formula are perfectly identical. Then, a mathematical analysis of the second formula concerning the bending of light established by Einstein in 1915 will enable to explain the importance of the doubling of this value in the framework of Einstein's theory of gravitation.

3.1 Value of Deflection of Light by Soldner in 1801

Johann Georg von Soldner (1776–1833) was a German physicist, mathematician and astronomer. He first worked in the Berlin Observatory (*Berliner Sternwarte*) and later in 1808 in Munich where he became a member of the Academy of Sciences and the director of the observatory in Bogenhausen. In a paper written in March 1801 and published in 1804, he calculated the amount of deflection of a light ray by a star based on Newton's corpuscular theory of light and wrote:

> It is, of course, true that already through observations and otherwise one was aware of considerable deviations from an assumed law; such as was the case with the aberration of light. There can, however, be deviations which are so small that it is difficult to decide whether they are true deviations or errors of observation. There can also be deviations which are considerable but, being combined with magnitudes one has not yet succeeded in clearly identifying, escape the observer. Of the latter kind may be the deviation of a light ray from straight line when it passes close by a celestial body and is considerably exposed to its attraction [30].

Soldner then presented the following diagram (see Fig. 1).

Then, he indicated that "C (Fig. 1) is the center of the attracting body, A is the location at its surface. From A, a light ray goes into the direction AD or in the horizontal direction, by a velocity with which it traverses the way ν in a second. Yet the light ray, instead of travelling at the straight line AD, will forced by the celestial body to describe a curved line AMQ, whose nature we will investigate (…) g be the

[10] Stefanini quotes Hans-Jüurgen Treder and Gerhard Jackisch [33].

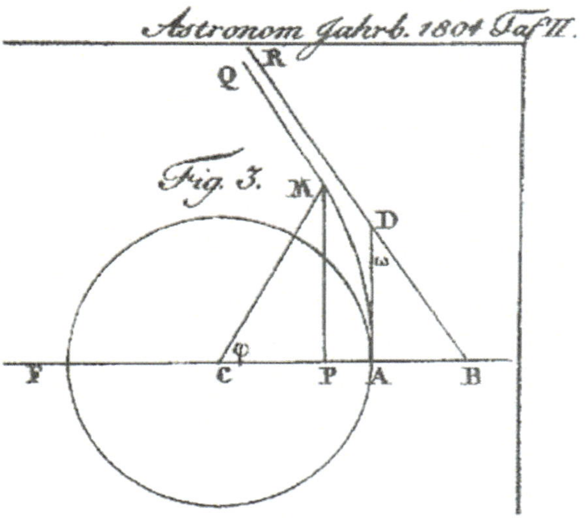

Fig. 1 Soldner's diagram for deflexion of a light ray [30]

gravitational acceleration at the surface of the body. Furthermore $CP = x$, $MP = y$ and the angle φ." Then, he explained:

> The force with which the light ray at M will be pulled by the body in the direction MC, will be $2gr^{-2}$. This force can be decomposed into two others, $\frac{2g}{r^2}\cos(\varphi)$ and $\frac{2g}{r^2}\sin(\varphi)$ according to the directions x and y; and therefore one obtains the following two equations (s. *Traité de mécanique céleste* par Laplace, Tome I, p. 21)
>
> $$\frac{d^2x}{dt^2} = \frac{2g}{r^2}\cos(\varphi)$$
> $$\frac{d^2x}{dt^2} = \frac{2g}{r^2}\sin(\varphi) \tag{1}$$
>
> ... [30].

These two equations correspond to the projection along the x and y directions of Newton's second law according to which the product of the mass by the acceleration (left hand side) is equal to Newton's gravitational force (right hand side). Here g represents the acceleration of gravity on the surface of a celestial body (e.g., of the sun, see Remark below). Soldner's simplification of the "mass of a ray of light" on either side of this equality is consistent with Newton's theory of light then considered as made up of small discrete particles called "corpuscles". Indeed, it is only in 1803, three years after the writing of Soldner's article, that Thomas Young (1773–1829) performed his famous double-slit experiment from which he proposed a wave theory of the light. Thus, starting from the two preceding equations and after a demonstration which does not present any great difficulties, Soldner draws the following conclusion:

> The light ray, however, comes in the direction DA to the eyes of the observer; thus ADB will be the angle of perturbation. If one calls this angle ω then one has, since the triangle ABD

at A is a right triangle

$$\tan\omega = \frac{AB}{AD} \tag{2}$$

If one puts these values for AB and AD in the expression for $\tan\omega$, then one has

$$\tan\omega = \frac{2g}{\nu\sqrt{\nu^2 - 4g}} \tag{3}$$

If one substitutes in the formula for tan ω the acceleration of gravity on the surface of the sun, and one takes the radius of that body for unity, then one finds $\omega = 0''.84$. If one could observe the fixed stars very close to the sun, then one would have to take this very much into account. But since this is not known to happen, the perturbation caused by the sun can also be neglected [30].

At the time, such observations were impossible; Soldner therefore concluded that these effects were minute. He ended his article as follows:

Hopefully, no one would find it objectionable that I treat a light ray as a heavy body. That light rays have all the absolute [basic] properties of matter one can see from the phenomenon of aberration which is possible only because light rays are truly material. And furthermore, one cannot think of a thing which exists and works on our senses that would not have the property of matter [30].

Thus, it appears that Soldner based his computations on the Newton's emission theory, according to which light is made up of particles. As far as Einstein is concerned, he made use of the Huygens principle, that is to say, the variation of the direction of the wavefront as a function of the luminous frequency, as will be seen in the next section.

3.2 *Clarifying the Misinterpretations of Factor 2 in Soldner's Computation*

During these last decades, Soldner's value of deflection of light has been challenged by many historians of science. In the beginning of the eighties, Tredder and Jackisch [33] pointed out that the factor 2 before the acceleration of gravity g on the surface of the considered body (earth, moon and sun) investigated by Soldner was resulting from a different definition of acceleration at Soldner's time. Thus, they explained:

Accordingly SOLDNER's "acceleration of gravity" is only one half of the value required by the analytical definition, $g = d^2s/dt^2$. About the year 1800, whenever LAGRANGE's and LAPLACE's exact analytic mechanics was to be brought together with the usage of German physicists and astronomers this became occasionally a cause of error. In SOLDNER'S theory, however, everthing is correct in the sense of NEWTON in spite of the factor 2, as could also be confirmed by inspecting the first edition of SOLDNER'S treatise of 1801 in BODE's Jahrbuch of 1804 and its reproduction by LENARD (1921) in "Annalen der Physik" of 1921. By no means did SOLDNER give double the Newtonian amount of the deflection of light (which then would be EINSTEIN's). Nevertheless SOLDNER's paper contains two misprints, to which LENARD rightly called attention. In the present exemplar of "Astronomisches Jahrbuch" of the Berliner Sternwarte of 1804 those two misprints have for generations been corrected by hand (it is unknown if by BODE himself). Both misprints consist in that now a factor 2 is missing although the deduction and the context do require it [33].

First of all, let's notice that such a mistake or misprint in Soldner's had already been reported in 1923 by the famous Swiss-American astronomer Robert Julius Trumpler in a paper which has not been quoted by Tredder and Jackisch [33]. Half a century before them, Trumpler clearly explained that:

In setting up the differential equations for the motion of the particle he erroneously used for the gravitational force the expression

$$2gr^{-2}$$

where g = acceleration at the surface of the attracting body, and r = distance from the center of the attracting body (adopting the radius of this body as unit distance). The factor 2 has no justification and should be omitted. Designating by ω the angular deflection of light from a star at infinity until it reaches the surface of the attracting body, Soldner derived the formula

$$\tan \omega = \frac{2g}{\nu\sqrt{\nu^2 - 4g}} \quad (\nu = \text{speed of light})$$

which he applied to the earth and the sun. On account of the mistake mentioned his result for the Sun

$$\text{(half deflection)} \quad \omega = 0.''84$$

is twice too large [34].

Thus, according to Trumpler [34] and Tredder and Jackisch [33], it appears that a factor 2 has been erroneously introduced in the right hand side of Soldner's equation (3) while a factor 2 is missing in the left hand side of his equation which should be correctly rewritten as follows:

$$\tan \frac{\omega}{2} = \frac{g}{\nu\sqrt{\nu^2 - 2g}} \tag{4}$$

In order to verify if this modified formula (4) is correct, let's test it with the "tutorial examples" of deflection of light by massive bodies, i.e., earth and sun used by Soldner. Before, it is important to notice that the translation of Soldner's article by Stanley Jaki [22] also contain a mistake. Soldner [30] wrote:

Unter der Voraussetzung, dass das Licht 564″, 8 DecimalSekunden Zeit brauche, um von der Sonne zur Erde zu kommen, findet man, dass es in einer Decimaleskunde 15,562085 Erdhalbmesser durchlauft.

Jaki [22] translated as:

On the presupposition that light needs 564″.8 decimal seconds of time to come from the sun to the earth, one finds that it traverses in **one-tenth** of a second 15.562085 earth radii.

In fact, Soldner [30] distinguished decimal seconds from angular seconds while Jaki [22] made a confusion in the above sentence which should be translated as:

On the presupposition that light needs 564″.8 decimal seconds of time to come from the sun to the earth, one finds that it traverses in **one-decimal second** 15.562085 earth radii.

Thus, Soldner [30] first applied his formula (3) to the deflection of light by the earth. He wrote:

On the presupposition that light needs 564″.8 decimal seconds of time to come from the sun to the earth, one finds that it traverses in one-decimal second 15.562085 earth radii. Thus $v = 15.562085$. If one takes among the geographical latitudes that whose square of the sine is 1/3 (corresponding to a latitude of 35° 16'), the earth's radius as 6,369,514 meters, and the acceleration of gravity there as 3.66394 meters (see *Traité de mécanique céleste* by Laplace, Vol. I, p. 118), then expressed in earth radii $g = 0.000000575231$. I make use of this set of units so that without special reductions I may take from the *Traité de mécanique céleste* the newest and most available determinations of the magnitude of the earth radius and of the acceleration of gravity. Thereby nothing will change concerning the final results, for here only the relation of the speed of light to the velocity of a body falling to the earth is concerned. The earth's radius and the acceleration of gravity must therefore be taken at the specified degree of latitude, because the earth-spheroid is, with respect to bodily content, similar to a globe which has for its radius the earth's radius, or 6,369,514 meters. When one puts these values for v and g into the equation for tan ω, then one obtains, in sexagesimal seconds, $\omega = 0''.0009798$, or in round figures, $\omega = 0''.001$.

From these paragraph and these data, it is possible to rebuild Soldner's computation. Concerning the time the light needs to come from the sun to the earth, Soldner seems to have used another book from Laplace entitled *Exposition du système du monde* in which he wrote [25, p. 103]: "Il en résulte que la lumière emploie 571″ pour venir du soleil à la terre." Soldner used the value 564″.8 which is very close to that of Laplace. As regards of the values of earth's radius, he effectively used the value coming from the first volume of the *Traité de mécanique céleste* of Pierre Simon de Laplace [26, p. 119]: "ce rayon est égal à 6369514me." Concerning the acceleration of gravity, he recalled that he used the acceleration of gravity taken at the specified degree of latitude whose square of the sine is 1/3 (corresponding to a latitude of 35° 16′) and he expressed it in earth radii following Laplace [26, p. 118 and next]. Soldner also used an estimation of the speed of light. According to Carl Boyer [2, p. 39]: "In German books the velocity generally was expressed in round figures as 40,000 geographic miles—about 184,000 statute miles—per second, although other values up to 42,000 also appeared." Thus, Soldner seems have used the value of 184,800 miles per second which corresponds to 46.68 earth radii and which is approximately equal to 15.56 at the specified degree of latitude λ whose square of the sine is 1/3. So, Soldner's approach seems to have been the following. First, he may have posed:

$$g = \frac{g_\lambda}{R_{earth}}$$
$$\nu = \frac{\nu_\lambda}{R_{earth}} \tag{5}$$

For the earth, Soldner used $g_\lambda = 3.66394 \text{ m s}^{-2}$, $R_{earth} = 6369514$ m and $\nu_\lambda = 46.68 R_{earth}/3 \text{ m s}^{-1}$. So, he obtained: $g = 3.66394/6369514{=}0.000000575231 \text{ s}^{-2}$

and $\nu = \nu_\lambda / R_{earth} = 46.68/3 = 15.562085\ \mathrm{s}^{-1}$ since he has taken "the radius of that body for unity [30]". By replacing in Soldner's formula (3), we find that $\tan\omega \approx 4.75 \times 10^{-9}$, from which we deduce that $\omega \approx 0.0009785$, that is the value obtained by Soldner. By replacing these values in Soldner's corrected formula (4), we obtain for $\omega/2$, of course, half Soldner's value and so, exactly the same value for ω.

Then, Soldner [30] applied his formula (3) to the deflection of light by the sun. He wrote:

> If one substitutes in the formula for $\tan\omega$ the acceleration of gravity on the surface of the sun, and one takes the radius of that body for unity, then one finds $\omega = 0''.84$.

For the sun, Soldner didn't give any value for acceleration of gravity on its surface nor for its radius. Nevertheless, we can make the reasonable assumption that he followed the same approach as those he has used for the earth. Thus, he may have used the various data established by Laplace in his books [25, 26] such as sun's parallax, the ratio of sun's mass to earth's mass, and Kepler's laws to deduce them. Moreover, at that time Soldner was working under the German astronomer Johann Elert Bode, director of the Berlin observatory. So, Bode, who had "accepted his paper for publication in "Astronomisches Jahrbuch" [33]," could have provided to him the necessary data for his computation. However, by reformulating (5) we have:

$$\begin{aligned} g &= \frac{g_S}{R_{sun}} \\ \nu &= \frac{46.68 R_{earth}}{R_{sun}} \end{aligned} \tag{6}$$

By using $g_S = 274\ \mathrm{m\,s^{-2}}$ for the gravity at the sun's surface and $R_{sun} = 6.9634 \times 10^8$ m for its radius, we found that $g = g_S/R_{Sun} \approx 3.93 \times 10^{-7}\ \mathrm{s}^{-2}$ and that the speed of light is $\nu = 46.68 R_{earth}/R_{sun} \approx 0.43\ \mathrm{s}^{-1}$. Then, by replacing these values in Soldner's original formula (3), we obtain that $\tan\omega = 4.25 \times 10^{-6}$ and so, that $\omega \approx 0''.87$ which is consistent with Soldner's value and seems to confirm our assumption. By replacing these values in Soldner's corrected formula (4), we obtain for $\omega/2$, of course, half Soldner's value and so, exactly the same value for $\omega \approx 0''.87$.

Let's notice that according to Trumpler:

> Lenard, it should be said, recognizes the error in Soldner's work to which attention is called in this paper and gives correctly the value for the deflection to which Soldner's theory leads. In these comments (page 603) Lenard transforms Soldner's formula into a notation and form similar to those employed by Einstein [34].

In Soldner's formula (3), Lenard corrected the missing factor 2 in the left hand side of this equation and removed the factor 2 erroneously introduced in the right hand side. He obtained the previous formula (4). Then, he replaced in (6) the acceleration of gravity on the surface of the sun by the well-known formula $g_S = KM/R^2$ where K is the Cavendish's constant (see footnote 1 above), M and R are respectively the mass and radius of the sun. Following Soldner's approach, Lenard obtained the following formula:

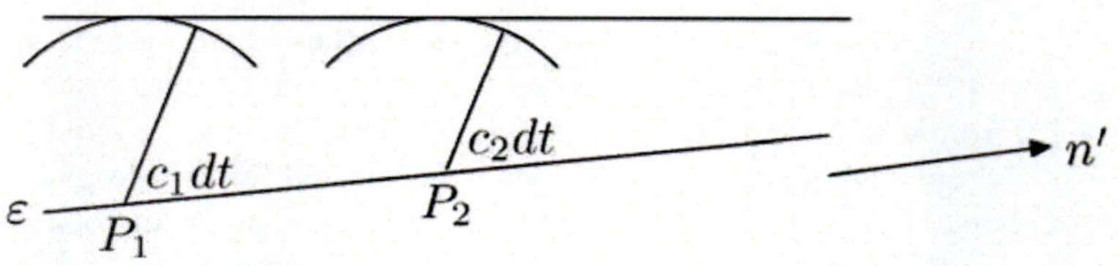

Fig. 2 Einstein's diagram for deflexion of a light ray [6]

$$\tan\frac{\omega}{2} = \frac{KM}{cR\sqrt{c^2 - 2KM/R}} \tag{7}$$

where c is the speed of light in vacuum. Then, while using a first order series expansion, he gave the following approximation of ω:

$$\boxed{\omega = \frac{2KM}{c^2 R}} \tag{8}$$

Finally, he explained [31]:

> Soldner berechnet weiter nach der erhaltenen Formel die Ablenkung für Erde, Mond, Sonne und findet sie sehr klein. Für die Sonne $\omega = 0''.84$. (S. 170) (während in wirkliclilreit nach seiner Formel $\omega/2 = 0''.84$, was mit der Erfahrung, so weit dieselbe heute geht, auch zu stimmen scheint, wie es in der Vorbemerkung erläutert wurde)[11]

Let's notice that according to Tredder and Jackisch:

> So on page 170, for example, of BODE's Jahrbuch a factor 2 should precede the angle [ω in Soldner's formula (3)]; but, as the context proves, this is a mere misprint [33].

But contrary to what they claimed, this is a factor 1/2 and not 2 that should precede the angle ω in Soldner's formula as highlighted in (4) and as it has been stated by Lenard [31], Trumpler [34], Stefanini [32] and previously in this work.

3.3 *Value of Deflection of Light by Einstein in 1911*

In his 1911 paper, Einstein [6] first established that "the velocity of light in the gravitational field is a function of the location." Then, by using Huygens's principle, he stated that "light-rays propagated across a gravitational field undergo deflection."

From Fig. 2, Einstein stated that the direction of the wave front changes by an amount equal to $\partial c/\partial n'$ per unit of distance along the direction of the wave (where

[11] Soldner further calculates the deflection for the earth, moon, and sun using the formula obtained and finds it very small. For the sun $\omega = 0''.84$. (p. 170) (while in fact, according to his formula $\omega/2 = 0''.84$, which seems to be consistent with experience, as far as the same goes today, such as it was explained in the preliminary remark).

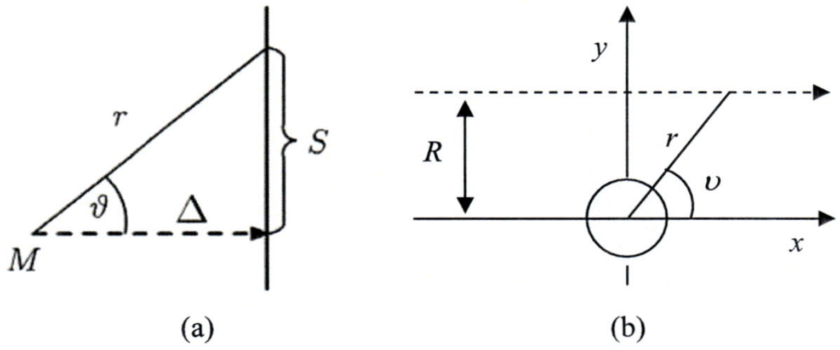

Fig. 3 Deflection of a light ray in polar coordinates [6]

c is the velocity of light) and the "angle of deflection per unit of path of the light-ray is $-\frac{1}{c}\frac{\partial c}{\partial n'}$". Finally, he obtained for the deflection α, which a light-ray experiences toward the side n' on any path (s) the expression

$$\alpha = -\frac{1}{c^2}\int \frac{\partial \Phi}{\partial n'} ds \tag{9}$$

where the integral goes from $-\infty$ to $+\infty$ and $\Phi = kM/r$ is the gravitation potential. Then, Einstein changed variables to polar coordinates as highlighted on the following figure (see Fig. 3a). In order to simplify the understanding of his approach, let's pose in what follows: $n' = y$, $s = x$ and $S = R$ (see Fig. 3b).[12]

According to Fig. 3b, we have $r = \sqrt{x^2 + y^2}$. Thus, the gravitation potential reads:

$$\Phi = \frac{kM}{\sqrt{x^2 + y^2}} \tag{10}$$

It follows that $\frac{\partial \Phi}{\partial n'} = \frac{\partial \Phi}{\partial y} = \frac{\partial}{\partial y}\left(\frac{kM}{\sqrt{x^2 + y^2}}\right) = -\frac{kM}{r^3}y$. Einstein then considers that nearly all of the deflection occurs within some reasonable proximity of the gravitating body. So, we can simply set $y = R$ in the integral which reads:

$$\alpha = \frac{1}{c^2}\int_{-\infty}^{+\infty} \frac{kM}{r^3} y ds = \frac{kMR}{c^2}\int_{-\infty}^{+\infty} \frac{1}{r^3} dx \tag{11}$$

The following variable changes $x = R\tan(\upsilon)$, $y = R$ leads Einstein to:

[12] See also Kevin Brown [3].

$$\begin{aligned}\alpha &= \frac{kMR}{c^2}\int_{-\infty}^{+\infty}\frac{1}{\left(x^2+y^2\right)^{3/2}}dx \\ &= \frac{kMR}{c^2}\int_{-\pi/2}^{+\pi/2}\frac{Rdv/\cos^2(v)}{\left(R^2\tan^2(v)+R^2\right)^{3/2}} \\ &= \frac{kM}{c^2R}\int_{-\pi/2}^{+\pi/2}\cos(v)\,dv = \frac{2kM}{c^2R}\end{aligned} \tag{12}$$

Then, Einstein [6] ended his 1911 paper by this sentence:

> By equation (9) a light-ray passing by a heavenly body suffers a deflection to the side of the diminishing gravitational potential, that is, to the side directed toward the heavenly body, of the magnitude
>
> $$\alpha = \frac{1}{c^2}\int_{\nu=-\frac{\pi}{2}}^{\nu=+\frac{\pi}{2}}\frac{kM}{r^2}\cos(\nu)\,ds = \frac{2kM}{c^2\Delta}, \tag{13}$$
>
> where k denotes the constant of gravitation, M the mass of the heavenly body, Δ the distance of the ray from the center of the body (and r and v are as shown in Fig. 3). *A light-ray going past the Sun would accordingly undergo deflection by the amount of* $4.10^{-6} = 0.83$ *seconds of arc.*

Let's notice that although the reasonings and the computations are different, the result of Einstein is exactly the same as that of Soldner (as shown in the next section). In other words, a light ray passing near the sun will undergo a deflection of nearly $0.83''$.

3.4 *Comparison of Einstein and Soldner Formulas*

This section aims to prove that both formulas established by Soldner in 1801 and Einstein in 1911 are identical except in the choice of letters. Soldner's corrected formula (4) which reads:

$$\tan\frac{\omega}{2} = \frac{g}{\nu\sqrt{\nu^2-2g}}.$$

Thus, by using (6), we pose $g = g_S/\Delta$ and $\nu = c/\Delta$ where Δ is the radius of the considered body (in the case of the Sun $\Delta = R_{sun} = R$ with respect to the notations of Einstein, Soldner and Lenard.). Soldner's formula can be rewritten as

$$\tan\omega = \frac{2g}{\nu\sqrt{\nu^2-4g}} = \frac{2g}{\nu^2}\frac{1}{\sqrt{1-\frac{4g}{\nu^2}}} = \frac{2g}{\nu^2}\left[1-\frac{4g}{\nu^2}\right]^{-\frac{1}{2}} \tag{14}$$

But since, according to Soldner $g \ll \nu$, a first-order[13] series expansion can be made and reads:

$$\tan\omega = \frac{2g}{\nu^2}\left[1 - \frac{4g}{\nu^2}\right]^{-\frac{1}{2}} = \frac{2g}{\nu^2}\left[1 + \frac{2g}{\nu^2} + \ldots\right] \approx \frac{2g}{\nu^2} + O\left(\nu^{-4}\right) \tag{15}$$

The angle ω is considered as infinitely small, so we deduce that $\omega \approx \frac{2g}{\nu^2}$. Then, by posing $g = g_S/\Delta$ and $\nu = c/\Delta$, we have:

$$\boxed{\omega \approx \frac{2g_S\Delta}{c^2}} \tag{16}$$

In 1911, Einstein proposed the following formula: $\alpha = \frac{2kM}{c^2\Delta}$ where

- k is the constant of gravitation (Cavendish's constant),
- M is the mass of the heavenly body,
- Δ is the distance of the ray from the center of the body, i.e., its radius.

In the case of the Sun, we have: $g_S = \frac{kM}{\Delta^2}$. So, it gives

$$\alpha = \frac{kMR}{c^2\Delta} = \frac{2\Delta}{c^2}\left(\frac{kM}{\Delta^2}\right) = \frac{2g_S\Delta}{c^2} \tag{17}$$

Thus, both Soldner and Einstein's formulas are identical.

$$\boxed{\omega = \alpha = \frac{2g_S\Delta}{c^2}} \tag{18}$$

Let's notice that this value is exactly the same as those obtained by Lenard [31] (see (8) in Sect. 3.2).

3.5 *Value of Deflection of Light by Einstein in 1915*

During the autumn 1915, Einstein completed his general theory of relativity. He thus modified his gravitation potential Φ while taking into account as previously the Newton's gravitational interaction (see (10) in Sect. 3.3) but also the curvature of space near a massive body.[14] Then, he expressed the gravitation potential Φ as

[13] In his 1911 and 1915 articles Einstein also made use of first-order approximations as well as Lenard [31] in 1921.

[14] See Kevin Brown [3].

$$\Phi = \frac{kM}{\sqrt{x^2+y^2}} + \frac{kMx^2}{\left(x^2+y^2\right)^{3/2}} \tag{19}$$

In this case $\frac{\partial \Phi}{\partial n'} = \frac{\partial \Phi}{\partial y} = \frac{\partial}{\partial y}\left(\frac{kM}{\sqrt{x^2+y^2}} + \frac{kMx^2}{\left(x^2+y^2\right)^{3/2}}\right) = -\frac{4x^2+y^2}{r^5}kMy.$

The integral reads then:

$$\alpha = \frac{1}{c^2}\int_{-\infty}^{+\infty} \frac{4x^2+y^2}{r^5}kMyds = \frac{kMR}{c^2}\int_{-\infty}^{+\infty} \frac{4x^2+R^2}{r^5}dx \tag{20}$$

By using the same variable changes $x = R\tan(\upsilon)$, $y = R$, Einstein obtained:

$$\begin{aligned} \alpha &= \frac{kMR}{c^2}\int_{-\infty}^{+\infty} \frac{4x^2+R^2}{r^5}dx \\ &= \frac{kMR^3}{c^2}\int_{-\pi/2}^{+\pi/2} \frac{4\tan^2(\upsilon)+1}{(R/\cos(\upsilon))^5}\frac{Rd\upsilon}{\cos^2(\upsilon)} = \frac{4kM}{c^2R} \end{aligned} \tag{21}$$

In his publication of 1916 Einstein [11] provided for the deflection of a light ray the value $\frac{kM}{2\pi\Delta}$ which results of the previous integration. He defined at the page 818 of this same article the constant $\kappa = \frac{8\pi k}{c^2}$ (see (69)). By replacing into the value of B, we have:

$$B = \frac{4kM}{c^2R} = 2\alpha \tag{22}$$

Einstein wrote in his conclusion:

According to this, a ray of light going past the sun undergoes a deflexion of 1.7″…

Thus, it appears that Einstein's computation of the value of deflection of a light ray performed in 1915 led him to twice the amount derived in his 1911 paper.

Where does this doubling come from? How did Einstein justify it?

In fact, contrary to what Arvid Reuterdahl (see Sect. 2) claimed, Einstein has really "taken the world into his confidence concerning the reason of this change". Indeed, as early as 1915, Einstein wrote:

By use of the Huygens principle, one finds through a simple calculation, that a light ray from the Sun at distance Δ undergoes an angular deflection of magnitude $2\alpha/\Delta$, while the earlier calculation had given the value α/Δ. A corresponding light ray from the surface rim of the Sun should give a deviation of 1.7″ (instead of 0.85″) [9].

In 1920, in the Appendix 3 of the third edition of his book written in 1916, he explained:

As a result of this theory, we should expect that a ray of light which is passing close to a heavenly body would be deviated towards the latter. For a ray of light which passes the sun at a distance of Δ sun-radii from its centre, the angle of deflection (α) should amount to

$$\alpha = \frac{1.7 \text{ seconds of arc}}{\Delta} \tag{23}$$

It may be added that, according to the theory, half of this deflection is produced by the Newtonian field of attraction of the sun, and the other half by the geometrical modification ("curvature") of space caused by the sun [12].

4 Conclusion

The author of one of Einstein's most famous biographies, Ronald Clark had written that he was "the man who had bent the light". One would be led to believe that the concept of "curvature of light" which was conceived by Soldner in the early nineteenth century was rediscovered a century later by Einstein. In fact, many authors have shown that this idea was already present in the works of Isaac Newton. Indeed, in the first of the famous Queries of his work entitled *Opticks*, Newton wrote in 1704:

> Do not Bodies act upon Light at a distance, and by their action bend its Rays, and is not this action (*caeteris paribus*) strongest at the least distance?

It thus appears that neither Soldner nor Einstein is the inventor of the concept of "curvature of light". One can then ask whether Einstein had any knowledge of Soldner's work. It is naturally very difficult to answer this question. The elements we have today allow us only to affirm that the values of the deflection of light rays passing near a massive body obtained by Soldner in 1801 and by Einstein in 1911 are perfectly identical, although the computations of Soldner were based on Newton's corpuscular theory, while those of Einstein were based on the Huygens's principle. According to Trumpler:

> From this conclusion the deflection of light could be calculated by using Soldner's method. Einstein, however, follows an entirely different course. For the Sun, Einstein finds $\omega = 0''.83$, but with more accurate data the value
>
> $$\omega = 0''.87$$
>
> is obtained. This formula, which is based on the principles:
>
> 1. Light is subject to gravitation
> 2. Gravitation follows Newton's law, is equivalent to Soldner's formula, but more general.
>
> The increase of this value over that in Einstein's 1911 paper is not due to any mistake in calculation in the earlier paper but is an effect of the difference between Einstein's and Newton's law of gravitation, as the 1916 deflection is essentially based on the principles:

1. Light is subject to gravitation
2. Gravitation follows Einstein's law instead of Newton's [34].

First, it is important to note that the impossibility of measuring the deflection of light during the eclipses of 1912 and 1914 was an extraordinary opportunity for Einstein. Indeed, without this providential rain and without the declaration of war the observations of the astronomers would have absolutely not confirmed the first value that it had provided in 1911 and they would certainly have invalidated his theory. Note then that the method of computing the deflection of light that Einstein used in 1915 is exactly the same one he used in 1911. The only difference is the expression of gravitation potential which took into account the curvature of the space in the vicinity of a massive body. Thus, in his article in 1915, Einstein provided a value ($1.7''$) which was well contained within the range of values observed by the expeditions led by Crommelin and Eddington in 1918 ($0.9''$ to $1.8''$). It is clear from this analysis that the plagiarism accusations against Einstein which are part of an anti-relativist and anti-Semitic movement are absolutely baseless. Even if it has been established that the values of the deflection of a light ray provided by Soldner in 1801 and Einstein in 1911 are identical, they are both wrong because they don't take into account the curvature of space and so, are not consistent with those "obtained" by Crommelin and Eddington and confirmed after by more accurate astronomical observations.

Acknowledgements Author would like to thank Pr. Christos Skiadas for his helpful advices.

Conflict of Interest The author declares that he has no conflict of interest.

References

1. M. Biezunski, *Einstein à Paris* (Presses Universitaires de Vincennes, Saint Denis, Paris, 1991)
2. C.B. Boyer, Estimates of the velocity of light. Isis **33**(1), 24–40 (1941)
3. K. Brown, *Reflections on Relativity* (2017), www.lulu.com
4. J. Earman, C. Glymour, Relativity and eclipses: the British eclipse expeditions of 1919 and their predecessors. Hist. Stud. Phys. Sci. **11**(1), 49–85 (1980)
5. A. Einstein, Relativitätsprinzip und die aus demselben gezogenen Folgerungen (On the principle of relativity and the conclusions drawn from it). Jahrb. der Radioakt. **4**, 411–462 (1907)
6. A. Einstein, Einfluss der Schwerkraft auf die Ausbreitung des Lichtes (On the influence of gravitation on the propagation of light). Ann. der Phys. **4**(35), 898–908 (1911)
7. A. Einstein, Grundgedanken Der Allgemeinen Relativitätstheorie Und Anwendung Dieser Theorie in Der Astronomie (Fundamental ideas of the general theory of relativity and the application of this theory in astronomy). Preuss. Akad. der Wiss. Sitzungsberichte 315 (1915)
8. A. Einstein, Zur allgemeinen Relativitätstheorie (On the general theory of relativity). Preuss. Akad. der Wiss. Sitzungsberichte 778–786 and 799–801 (1915)
9. A. Einstein, Erklärung Der Perihelbewegung Des Merkur Aus Der Allgemeinen Relativitätstheorie (Explanation of the perihelion motion of Mercury from the general theory of relativity). Preuss. Akad. der Wiss. Sitzungsberichte 831–839 (1915)
10. A. Einstein, Feldgleichungen Der Gravitation (The field equations of gravitation). Preuss. Akad. der Wiss. Sitzungsberichte 844–847 (1915)

11. A. Einstein, Die Grundlage der allgemeinen Relativitätstheorie (The foundations of the general theory of relativity). Ann. der Phys. **4**(49), 769–822 (1916)
12. A. Einstein, *Relativity: The Special and General Theory* (Methuen & Co Ltd., London, 1920)
13. A. Einstein, D. Kormos Buchwald, R. Schulmann, J.I.D.J. Kennefick, T. Sauer, *The Collected Papers of Albert Einstein, Vol. 9, The Berlin Years: Correspondence, January 1919–April 1920* (Princeton University Press, Princeton, N.J., 2004)
14. J. Eisenstaedt, De l'influence de la gravitation sur la propagation de la lumière en théorie newtonienne. L'archéologie des trous noirs. Arch. Hist. Exact Sci. **42**(4), 315–386 (1991)
15. P. Frank, *Einstein: His Life and Times* (A. A. Knopf, New York, 1947)
16. R.M. Friedman, *Politics of Excellence: Behind the Nobel Prize in Science* (W. H. Freeman, New York, 2001)
17. H. Fuchs, Zur Geschichte der Ideen über die Wirkung der Schwerkraft auf das Licht. Orion **33**(151), 183–193 (1975)
18. J.M. Ginoux, *Albert Einstein: A Biography through the Time(s)* (Hermann, Paris, 2016)
19. J.M. Ginoux, *Albert Einstein Demystified* (Hermann, Paris, 2020)
20. T. Glick, *The Comparative Reception of Relativity*. Boston Studies in the Philosophy and History of Science, vol. 103 (Springer Netherlands, Dordrecht, D. Reidel, 1987)
21. É. Guillaume, Y a-t-il une erreur dans le premier mémoire d'Einstein? Rev. Générale des Sci. Pures et Appl. **33**, 5–10 (1922)
22. S.L. Jaki, Johann Georg von Soldner and the gravitational bending of light, with an English translation of his essay on it published in 1801. Found. Phys. **8**(11–12), 927–950 (1978)
23. D. Kennefick, Testing relativity from the 1919 eclipse–a question of bias. Phys. Today **62**(3), 37–42 (2009)
24. D. Kennefick, *No Shadow of a Doubt: The 1919 Eclipse That Confirmed Einstein's Theory of Relativity* (Princeton University Press, Princeton, N.J., 2019)
25. P.S. Laplace, *Exposition du sysè du monde* (Imprimerie du Cercle-Social, Paris, 1796)
26. P.S. Laplace, *Traité de mécanique céleste* (Imprimerie de Crapelet, Paris, 1798)
27. C.L. Poor, Is Einstein wrong?–a debate. The Forum 705–715 (1924)
28. A. Pais, *Subtle Is the Lord: The Science and the Life of Albert Einstein* (Oxford University Press, 1982)
29. J. Renn, *The Genesis of Relativity*, Vol. 1–4. Boston Studies in the Philosophy and History of Science, Vol. 250 (Springer Netherlands, Dordrecht, D. Reidel, 2007)
30. J.S. Soldner, Ueber die Ablenkung eines Lichtstrals von seiner geradlinigen Bewegung (On the deviation of a light ray from its motion along a straight line through the attraction of a celestial body which it passes close by). Berl. Astron. Jahrb. 161–172 (1804)
31. J. Soldner, Über die Ablenkung eines Lichtstrahls von seiner geradlinigen Bewegung durch die Attraktion eines Weltkörpers, an welchem er nahe vorbeigeht. Ann. Phys. **370**, 593–604 (1921)
32. L. Stefanini, A misunderstanding in Soldner's interpretation of the gravitational deflection of light. Lett. Mat. **4**(3–4), 167–172 (2017)
33. H.J. Treder, G. Jackisch, On Soldner's value of Newtonian deflection of light. Astron. Nachr. **302**(6), 275–277 (1981)
34. R. Trumpler, Historical note on the problem of light deflection in the Sun's gravitational field. Publ. Astron. Soc. Pac. **35**(206), 185–188 (1923)
35. M. Wazeck, *On Einstein's Opponents: The Public Controversy About the Theory of Relativity in the 1920s* (Cambridge, New York, 2014)

Effects of Thermal and Lévy Noise Sources on the Switching Current Distributions of a Josephson Junction

Claudio Guarcello, Giovanni Filatrella, Bernardo Spagnolo, Vincenzo Pierro, and Davide Valenti

Abstract We discuss the combined effect of Gaussian and α-stable Lévy noise sources on the switching current distribution of a short tunnel Josephson junction, while an external bias current flowing through the junction is linearly swept. At a fixed temperature, if the bias current is repeatedly ramped up from zero, and the value of the bias current at which the system switches to the finite voltage state is recorded, we obtain the distribution of the current values associated to escape events, i.e., the probability distribution of the bias currents at which the junction switches to the finite voltage state from the superconducting zero-voltage state. This information content is promptly available in experiments on Josephson junctions. We demonstrate that a Lévy noise current added to the linearly increased bias current clearly modi-

C. Guarcello (✉)
Dipartimento di Fisica "E.R. Caianiello", Università di Salerno, Via Giovanni Paolo II, 132, 84084 Fisciano, SA, Italy
e-mail: cguarcello@unisa.it

C. Guarcello · V. Pierro
INFN, Sezione di Napoli Gruppo Collegato di Salerno, Complesso Universitario di Monte S. Angelo, 80126 Naples, Italy

G. Filatrella
Department of Sciences and Technologies and Salerno Unit of CNISM, University of Sannio, Via Port'Arsa 11, 82100 Benevento, Italy

B. Spagnolo · D. Valenti
Dipartimento di Fisica e Chimica "Emilio Segrè", Group of Interdisciplinary Theoretical Physics, Università di Palermo and CNISM, Unità di Palermo, Viale delle Scienze, Edificio 18, 90128 Palermo, Italy

B. Spagnolo
Radiophysics Department, Lobachevsky State University, 23 Gagarin Ave., 603950 Nizhny Novgorod, Russia

V. Pierro
Department of Engineering, University of Sannio, Corso Garibaldi 107, 82100 Benevento, Italy

D. Valenti
CNR-IRIB, Consiglio Nazionale delle Ricerche - Istituto per la Ricerca e l'Innovazione Biomedica, Via Ugo La Malfa 153, 90146 Palermo, Italy

C. H. Skiadas et al. (eds.), *13th Chaotic Modeling and Simulation International Conference*, Springer Proceedings in Complexity,
https://doi.org/10.1007/978-3-030-70795-8_20

fies the switching current distribution due to the sole Gaussian-distributed thermal fluctuations. Albeit both Gaussian and Lévy components contribute to the overall noise level, they do not interfere, because they produce switching at different bias levels: the Lévy noise in the lower-current portion of the distribution, the Gaussian noise when the energy barrier becomes comparable to the noise energy. Finally, the analytical expression of the cumulative distribution function of Josephson switching currents can be compared with an analytical estimate.

Keywords Conference · CHAOS · Josephson junction · Lévy noise · Switching current distributions

1 Introduction

Josephson junctions (JJs) have already demonstrated to be a reliable tool to investigate non-Gaussian noise signals [1–6], also embedded in a thermal noise background.

A JJ set-up is a threshold device working on an activation, or switching, mechanism, and this makes Josephson devices nowadays often employed for sensing and detection applications [7–13]. In particular, an additional noise source, added to the thermal fluctuations, can be revealed through the appearance of premature "anomalous" switches from the superconducting to the resistive state, which produce a measurable voltage across the junction. In this regard, the application of JJs for both the detection and the characterization of a specific type of non-Gaussian fluctuations, i.e., the α-stable Lévy noise source, has recently proved particularly effective [14–16]. This type of stochastic processes can drive the degree of freedom associated to the Josephson phase difference φ over a very long "distance" in a single displacement, namely, a *flight*. In fact, the electrical response of a JJ is determined by the behavior of a quantum variable, that is the gauge invariant phase difference φ between the macroscopic phases of the two superconductors forming the device. Its dynamics is described by the celebrated Josephson equations [17, 18] and is not accessible directly, but only indirect electrical measurements, basically of current and voltage, can be indeed monitored. Thus, approaches based on measurements of switching current distributions (SCDs), in the case of a linearly ramping bias current [15] and average voltage drops across a junction biased by a constant current [16], have been recently proposed to investigate the effects of Lévy noise sources. The method of analysis of unknown noise sources through the study of SCDs offers some evident advantages when the noise is characterized by fat tails, i.e., by a finite probability of an infinite-amplitude fluctuation. In principle, this type of noise can pose a serious difficulty to the experimentalists, since it should requires extremely long times to reconstruct the behavior of rare events. Thus, determining the value of the parameter α, which characterizes the noise distribution, demands for long experiments (or simulations) to explore extreme values. In contrast, sweeping the bias current is very effective, because the bias increase lowers the trapping energy barrier, and in a given ramp time, i.e., when the bias current approaches the critical value, the energy barrier vanishes and a switch event is indeed recorded, i.e., the escape is inevitable in a ramp. Moreover, the inspection methods through Josephson devices are demonstrated to be viable in both the cases of intrinsic [14] or extrinsic [15] unknown additional noise signals.

The fact that an additional Lévy noise source can affect the dynamics of a tunnel JJ was extensively studied theoretically in both the short [19, 20] and the long [21–26] junction case. The strategies used to study Lévy-induced "premature" switches from the superconducting metastable state depends on the junction type. In fact, in short JJs, the mean first-passage time as a function of the Lévy characteristics is studied, whereas in the long JJs the quantity of interest is the nonlinear relaxation time and the interplay between Lévy and thermal noise and the generation of solitons [27, 28] was also investigated [22, 23, 29].

The paper is organized as follows. In Sect. 2, the theoretical background used to describe the phase behavior of a short tunnel JJ is presented. Both the statistical properties of a Lévy and a Gaussian noise source and the power-law asymptotic behavior of the mean escape time, in the case of Lévy-distributed fluctuations, are briefly reviewed. In the same section, the theoretical results are shown and analyzed. Finally, we give the analytical estimate of the distributions of switching currents in the presence of a Lévy-noise source. In Sect. 3, conclusions are drawn.

2 Model and Results

As said in Introduction, the aim of this paper is to demonstrate how the combined action of Lévy-distributed and thermal current fluctuations can influence the distribution of the switching currents of a current-biased short Josephson tunnel junction [14, 15]. In the short junction regime the dimensions of the device are smaller than the characteristic length scale of the system, i.e., the Josephson penetration length [30], λ_J, and the response of a Josephson junction (JJ) can by studied through the resistively and capacitively shunted junction (RCSJ) model for the Josephson phase difference φ, which reads as follow [24, 25, 31]

$$\left(\frac{\Phi_0}{2\pi}\right)^2 C\frac{d^2\varphi}{dt^2} + \left(\frac{\Phi_0}{2\pi}\right)^2 \frac{1}{R}\frac{d\varphi}{dt} + \frac{d}{d\varphi}U = \left(\frac{\Phi_0}{2\pi}\right) I_N, \tag{1}$$

where C and R are the capacitance and the normal-state resistance of the device, respectively, and Φ_0 is the flux quantum. In the previous equation, I_N is the sum of the thermal noise, I_{th}, and the Lévy noise source, I_L. The term U is the so-called washboard potential

$$U(\varphi, i_b) = U_0\left[1 - \cos(\varphi) - i_b\varphi\right], \tag{2}$$

where i_b is the bias current normalized to the critical current I_c of the junction and $U_0 = (\Phi_0/2\pi)\, I_c$. This potential contains a sequence of minima and maxima, so that the phase is "confined" in a superconducting metastable state by the potential barrier

$$\Delta \mathcal{U}(i_b) = \frac{\Delta U(i_b)}{U_0} = 2\left[\sqrt{1-i_b^2} - i_b \arccos(i_b)\right]. \tag{3}$$

The current bias flowing through the junction represents the slope of the potential $U(\varphi, i_b)$. As the bias current is lower than the critical value, that is if $i_b < 1$, the phase-particle remains confined within a potential well, until a noise fluctuation pushes it out. When this occurs, the phase derivative increases, that is the device switches to the voltage state, since a voltage drop across the junction appears, according to the a.c. Josephson relation $V = (\Phi_0/2\pi)d\varphi/dt$ [17, 18]. If the bias current is increased slowly, the value at which the escape process occurs is called switching current, i_{SW}. Since this phenomenon is intrinsically stochastic, by repeating the measurement many times, we obtain a distribution of switching currents, which depends on the noise features affecting the switching dynamics of the phase. Indeed, we demonstrate in this work that the presence of a Lévy noise source significantly reshapes the switching current distribution (SCD) with respect to the pure Gaussian noise case.

We observe that the RCSJ model in (1) can be conveniently cast as

$$m\frac{d^2\varphi}{dt^2} + m\eta\frac{d\varphi}{dt} + U_0\frac{d}{d\varphi}u = U_0 i_N, \tag{4}$$

where the equivalent mass reads $m = (\Phi_0/2\pi)^2\, C$, the friction is $\eta = 1/(RC)$, and i_N is the current I_N normalized to the critical current I_C. In these units, the resonant frequency of the junction is $\omega_p = \sqrt{U_0/m}$ and the characteristic Josephson frequency [30] can be written as $\omega_c = \omega_p^2/\eta$. The normalized bias current is assumed to linearly increase, $i_b(t) = v_b t$, at a constant velocity $v_b = t_{\max}^{-1}$, with $t_{\max}$ being the maximum measurement time, so that when $t = t_{\max}$ the bias current reaches the critical value and the switching occurs also in the absence of noise. Specifically, in the following we assume $\eta = 0.1\omega_p$, $v_b = 10^{-7}\omega_p$, and the switching current distributions consist of a sequence of $N = 10^4$ bias current ramps.

In this work we consider the contemporaneous presence of a thermal noise source, with the usual Gaussian white-noise statistics, and a Lévy noise source $S_\alpha(\sigma, \beta, \lambda)$. The notation $S_\alpha(\sigma, \beta, \lambda)$ is usually adopted for the Lévy distributions [21–25], where $\alpha \in (0, 2]$ is called *stability index*, $\beta \in [-1, 1]$ is the *asymmetry parameter*, and $\sigma > 0$ and λ are a scale and a location parameter, respectively. The stability index characterizes the asymptotic long-tail power law for the distribution, which for $\alpha < 2$ is of the $|x|^{-(1+\alpha)}$ power type, while for $\alpha = 2$ coincides with the Gaussian distribution. In this work we consider only symmetric ($\beta = 0$), bell-shaped, standard ($\sigma = 1$ and $\lambda = 0$), stable distributions $S_\alpha(1, 0, 0)$, with $\alpha \in (0, 2]$. A Lévy noise source corresponds to stochastic processes that can exhibit very long *distance* in a single displacement, namely, a *flight*. In fact, the heavy tails that characterize a Lévy distribution cause the occurrence of events with large values of the variable of interest, whose probability is not negligible. These events correspond to the Lévy flights previously mentioned. Lévy noise, a generalization [32] of the Gaussian noise source [33–36], can be invoked to describe transport phenomena in different natural phenomena [37, 38], condensed matter systems [39, 40] and interdisciplinary applications [41, 42]. Results on the dynamics of systems driven by Lévy flights have been reviewed in [43, 44], and in [15] it is possible to find an extensive bibliography on examples and applications in which Lévy-distributed fluctuations are observed.

The statistical properties of the thermal current fluctuations, $i_{th} = I_{th}/I_c$, in normalized units, are given by

$$\left\langle i_{th}(\tilde{t})\right\rangle = 0, \qquad \text{and} \qquad \left\langle i_{th}(\tilde{t})i_{th}\left(\tilde{t}+\tilde{t}'\right)\right\rangle = 2\gamma_G(T)\delta\left(\tilde{t}'\right), \tag{5}$$

where the amplitude of the normalized correlator is connected to the physical temperature T through the relation

$$\gamma_G(T) = \frac{k_B T}{R}\frac{\omega_p}{I_c^2} = \frac{\eta}{\omega_p}\frac{k_B T}{U_0}. \tag{6}$$

It is worth stressing that, with the time normalization used in this work, the noise intensity γ_G is proportional to the ratio between the thermal energy and the Josephson coupling energy U_0.

Figure 1 shows the probability distribution functions (PDFs) and the cumulative distribution functions (CDFs) of the switching currents, computed in the presence of both Gaussian and Lévy noise contributions, at different values of the stability index $\alpha \in (0, 2]$. Here, the Gaussian and Lévy noise amplitudes are $\gamma_G = 10^{-3}$ and $\gamma_L = 5 \times 10^{-7}$, respectively.

First we look at the PDF for the pure Gaussian noise case, indicated by a red thick curve in Fig. 1a, which is characterized by a peak at high values of the bias current. In this case Lévy flights are missing and the Gaussian noise source induces switching events only at current values close to the critical current. Two key attributes of the SCDs due to thermally induced switching processes are the position and the width of the peak, both depending on the value of the noise intensity γ_G, that is, according to (6), on the value of the temperature T at which the junction resides. In fact, if thermal noise is taken into account, the phase particle can "hop out" of the washboard potential well and slip down the potential profile, with an escape rate Γ at temperature T, according to the Kramers theory [45], given by

$$\Gamma(i_b, T) = \frac{\omega_A}{2\pi}e^{-\frac{\Delta U(i_b)}{k_B T}}, \tag{7}$$

where ω_A is an attempt frequency. The dependence of ω_A on the dissipation is weak, and to a good approximation the rate can be simplified into

$$\Gamma(i_b, T) = \frac{1}{2\pi}\left(\omega_c \sqrt[4]{1 - i_b^2}\right) e^{-\frac{\Delta U(i_b)}{k_B T}}, \tag{8}$$

where the attempt frequency is reduced by the applied bias current. Thus, for any finite value of the bias current i_b an escape from the metastable superconducting state is inevitable, although it happens sooner for a shallow well.

The typical experimental procedure to measure activation events in JJ is to sweep the bias current upward from zero. Early in the sweep, the washboard potential well is deep; conversely, late in the sweep the well becomes shallow and the thermal activation is easier. In addition, as the bias current is swept, the plasma frequency decreases and so does the attempt rate. The interplay of the two factors determines the net escape rate as a function of the bias current.

The switching mechanism from a potential well depends on the temperature, so that at the so-called *crossover temperature* T_{cr} [46, 47] thermal activation becomes comparable to macroscopic quantum tunneling (MQT) rate. As MQT rate is temperature independent, the position and width of SCD peaks must both saturate at low temperatures, i.e., the SCD peaks at temperatures $T < T_{\mathrm{cr}}$ are practically superimposed and are placed at a bias value close to the critical current. Instead, at higher temperatures, $T > T_{\mathrm{cr}}$, thermally activated processes drive the switching dynamics, so that the SCD peaks shift to lower bias current values and become broader. For a thorough discussion about SCDs in both thermally activated and quantum tunneling regimes see [47]. In the following, we assume to work at a temperature high enough that thermal activation dominates and MQT processes can be altogether neglected.

When a Lévy noise component with $\alpha < 2$ is considered, the SCDs deviate significantly from the pure thermal noise case, see Fig. 1a. In fact, in this case Lévy flights can drive "premature" switches from the superconducting metastable state: while α decreases the low-current tail of the SCDs grows, i.e., the switching probability becomes sizable, at the expense of the SCD peak at high i_b values, which is indeed depleted. The position of this SCD peak is unaffected by the change of the Lévy parameter α, thus underlining its thermal origin.

The behavior of the CDFs, which represent the probability that i_{SW} takes a value less than or equal to i_b, highlights even more clearly the discrepancies with respect to the pure Gaussian case, see Fig. 1b. The CDFs for different α values are clearly arranged in well distinct curves: at a given bias i_b the CDFs increase while reducing α. In particular, while increasing i_b the CDF curves saturate at the value 1 as soon as the thermally activated switching events enter into play.

The numerical results presented so far can be supplemented by an analytical estimate for both the PDF and the CDF, in the case of a Lévy noise. To do this, first we recall that, for arbitrary spatial and energy scale, by rescaling time, energy, and space, the mean escape time τ for the Lévy statistics can be written in terms of the noise parameters as [15, 48]

$$\tau(\alpha, D) = \frac{\mathcal{C}_\alpha}{D^{\mu_\alpha}}, \tag{9}$$

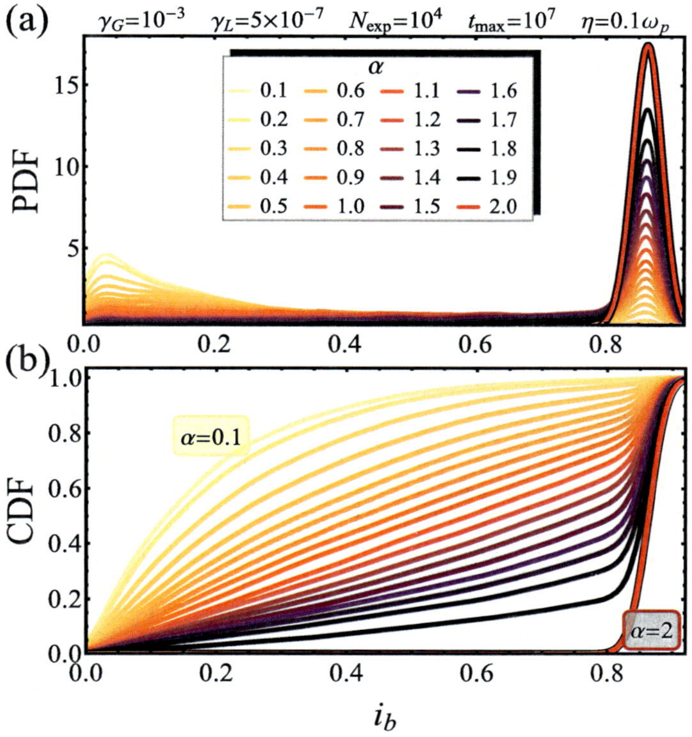

Fig. 1 **a** Probability distribution function (PDF) and **b** cumulative distribution function (CDF) for $\alpha \in [0.1, 2]$. The parameters of the simulations are: $\gamma_G = 10^{-3}$, $\gamma_L = 5 \times 10^{-7}$, $N_{\text{exp}} = 10^4$, $t_{\max} = 10^7$, and $\eta = 0.1\omega_p$. The legend in panel **a** refers to both panels

where both the power-law exponent μ_α and the coefficient $\mathcal{C}_\alpha$ depend on α and are supposed to have a universal behavior for overdamped systems [48–52].

By assuming $\mu_\alpha \simeq 1$ in the prefactor [48], the previous equation becomes

$$\tau(\alpha, D) = \left(\frac{\Delta x}{2}\right)^\alpha \frac{\mathcal{C}_\alpha}{D^{\mu_\alpha}}, \tag{10}$$

that is, the mean escape time only depends on the distance Δx between the minimum and maximum of the potential profile. This is remarkably different from the Gaussian noise case, where the probability to overcome the barrier depends exponentially on the barrier energy, see (8). In the specific case of a washboard potential, Δx depends on i_b according to the relation $\Delta x = \pi - 2 \arcsin i_b$. The CDF of i_{SW} as a function of i_b for a generic initial value of the bias ramp, i_0, can be therefore written as

$$\text{CDF}(i_b|i_0) = 1 - \text{Prob}\left[i_{SW} > i_b|i_0\right]. \tag{11}$$

Recalling also that the distribution of the escape times is exponential with rate $1/\tau(i_b)$ also for Lévy-flight noise [49], following the same logic of [53], the PDF associated to (11) as a function of the average escape time $\tau(i_b)$ reads

$$\mathcal{P}(i_b|i_0) = \mathcal{N}\,\frac{1}{v_b}\,\frac{1}{\tau(i_b)}\exp\left[-\frac{1}{v_b}\int_{i_0}^{i_b}\frac{1}{\tau(i)}di\right]. \tag{12}$$

Inserting in (12) the expression of $\tau(\alpha, D)$ given in (10), for the Lévy statistics (at the first order in i_b) we obtain the relation

$$\mathcal{P}(i_b|i_0) \propto \exp\left[-\left(\frac{2}{\pi}\right)^{\alpha}\frac{i_b D^{\mu_\alpha}}{\mathcal{C}_\alpha v_b}\right]. \tag{13}$$

The PDF of a current-biased JJ can be expressed in a closed compact form as

$$\mathcal{P}(i_b|i_0) = \frac{1}{\mathcal{N}}\frac{\mathrm{d}\mathcal{F}_\alpha}{\mathrm{d}i_b}\exp\left\{-\frac{D^{\mu_\alpha}}{\mathcal{C}_\alpha v_b}\left[\mathcal{F}_\alpha(i_b) - \mathcal{F}_\alpha(i_0)\right]\right\}. \tag{14}$$

Here, the nonlinear function $\mathcal{F}_\alpha$ reads

$$\begin{aligned}\mathcal{F}_\alpha(i_b) = \; & 2^\alpha\left\{\frac{\cosh^{-1}(i_b)}{2\left[\pi - 2\arcsin(i_b)\right]^\alpha}\left[E_\alpha\left(\cosh^{-1}(i_b)\right)\right.\right.\\ & \left.\left. - E_\alpha\left(-\cosh^{-1}(i_b)\right)\right] + i\frac{\pi^{1-\alpha}}{4}\left[E_\alpha\left(-\frac{i\pi}{2}\right) - E_\alpha\left(\frac{i\pi}{2}\right)\right]\right\},\end{aligned} \tag{15}$$

where $E_\alpha(\,)$ is the exponential integral with α argument, and $\mathcal{N}$ is a proper normalizing factor

$$\mathcal{N} = 1 - \exp\left[-\frac{D^{\mu_\alpha}}{\mathcal{C}_\alpha v_b}\left(\mathcal{F}_\alpha(1) - \mathcal{F}_\alpha(i_0)\right)\right]. \tag{16}$$

Thus, the corresponding CDF is

$$\mathrm{CDF}(i_b|i_0) = \frac{1}{\mathcal{N}}\left\{1 - \exp\left[-\frac{D^{\mu_\alpha}}{\mathcal{C}_\alpha v_b}\left(\mathcal{F}_\alpha(i_b) - \mathcal{F}_\alpha(i_0)\right)\right]\right\}. \tag{17}$$

This equation connects a property of the Lévy flights, i.e., the exponent α, with the accessible quantity of the switching-current distribution.

The validity of this analytical approach can be verified comparing the curves obtained via (17) and those numerically calculated in the case of a pure Lévy noise, that is imposing $\gamma_G = 0$. Thus, Fig. 2 shows a comparison between the Lévy induced "marginal" CDFs, i.e., restricted to the maximum bias value $i_b = 0.6$, obtained by solving (4) numerically (solid lines) and that analytically estimated from (17) (dashed line), for $\alpha \in [0.1\text{–}1.1]$ and $\gamma_L = 5\times10^{-7}$. In the considered limited range of i_b

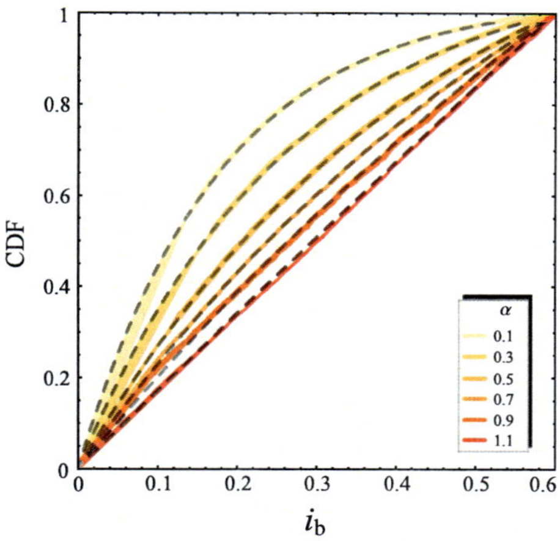

Fig. 2 Comparison between the curves obtained by fitting (17) (dashed gray lines) and the marginal CDFs, i.e., for $i_b \leq 0.6$, obtained by solving numerically (4) (solid lines). The values of the other parameters are: $\alpha \in [0.1, 1.1]$, $\gamma_L = 5 \times 10^{-7}$, $\gamma_G = 0$, $N_{\rm exp} = 10^4$, $t_{\rm max} = 10^7$, and $\eta = 0.1\omega_p$

values, the effects of the Gaussian noise contribution could be in any case safely ignored. This is why we set the Gaussian noise intensity to zero. Markedly, the agreement between computational and analytical results for $\alpha \leq 1$ is quite accurate.

3 Conclusions

In this work we investigate theoretically the switching current distributions in a short Josephson tunnel junction affected by both a Lévy and a Gaussian noise source, the latter depending on the temperature at which the system resides. If the bias current is linearly swept upward, the current value at which the system switches to the resistive state can be recorded. By repeating the measurement several times, a distribution of switching currents can be collected. We demonstrate that the analysis of both the PDFs and the CDFs of the switching current distributions allows to recognize the presence of a Lévy noise source, revealed by the appearance of anomalous premature switches in the low-current part of the distributions. Moreover, we show that it is possible to infer the specific characteristics of the Lévy noise source from the shape of the SCDs, also in the case of a thermal noisy background. Numerical simulations confirm the validity of the analytical expressions of both the PDF and the CDF, in the case of an escape process, from a metastable state of the washboard potential, driven by Lévy distributed fluctuations.

References

1. J.P. Pekola, Josephson junction as a detector of Poissonian charge injection. Phys. Rev. Lett. **93**, 206601 (2004)
2. J. Ankerhold, Detecting charge noise with a Josephson junction: a problem of thermal escape in presence of non-Gaussian fluctuations. Phys. Rev. Lett. **98**, 036601 (2007)
3. J.T. Peltonen, A.V. Timofeev, M. Meschke, T.T. Heikkilä, J.P. Pekola, Detecting non-Gaussian current fluctuations using a Josephson threshold detector. Phys. E **40**, 111 (2007)
4. B. Huard, H. Pothier, N.O. Birge, D. Esteve, X. Waintal, J. Ankerhold, Josephson junctions as detectors for non-Gaussian noise. Ann. Phys. **16**, 736 (2007)
5. H. Grabert, Theory of a Josephson junction detector of non-Gaussian noise. Phys. Rev. B **77**, 205315 (2008)
6. T. Novotnỳ, Josephson junctions as threshold detectors of full counting statistics: open issues. J. Stat. Mech. **P01050**(2009)
7. T. Peacock, P. Verhoeve, N. Rando, C. Erd, M. Bavdaz, B.G. Taylor, D. Perez, Recent developments in superconducting tunnel junctions for ultraviolet, optical & near infrared astronomy. Astron. Astrophys. Suppl. Ser. **127**, 497 (1998)
8. J.N. Ullom, D.A. Bennett, Review of superconducting transition-edge sensors for X-ray and gamma-ray spectroscopy. Supercond. Sci. Technol. **28**, 084003 (2015)
9. E.D. Walsh, D.K. Efetov, G.-H. Lee, M. Heuck, J. Crossno, T.A. Ohki, P. Kim, D. Englund, K.C. Fong, Graphene-based Josephson-junction single-photon detector. Phys. Rev. Appl. **8**, 024022 (2017)
10. L.S. Kuzmin, A.S. Sobolev, C. Gatti, D. Di Gioacchino, N. Crescini, A. Gordeeva, E. Il'ichev, Single photon counter based on a Josephson junction at 14 GHz for searching galactic axions. IEEE Trans. Appl. Supercond. **28**(2018)
11. C. Guarcello, A. Braggio, P. Solinas, F. Giazotto, Nonlinear critical-current thermal response of an asymmetric Josephson tunnel junction. Phys. Rev. Appl. **11**, 024002 (2019)
12. C. Guarcello, A. Braggio, P. Solinas, G.P. Pepe, F. Giazotto, Josephson-threshold calorimeter. Phys. Rev. Appl. **11**, 054074 (2019)
13. L.S. Revin, A.L. Pankratov, A.V. Gordeeva, A.A. Yablokov, I.V. Rakut, V.O. Zbrozhek, L.S. Kuzmin, Microwave photon detection by an Al Josephson junction. Beilstein J. Nanotechnol. **11**, 960–965 (2020)
14. C. Guarcello, D. Valenti, B. Spagnolo, V. Pierro, G. Filatrella, Anomalous transport effects on switching currents of graphene-based Josephson junctions. Nanotechnology **28**, 134001 (2017)
15. C. Guarcello, D. Valenti, B. Spagnolo, V. Pierro, G. Filatrella, Josephson-based threshold detector for Lévy-distributed current fluctuations. Phys. Rev. Appl. **11**, 044078 (2019)
16. C. Guarcello, G. Filatrella, B. Spagnolo, V. Pierro, D. Valenti, Voltage drop across Josephson junctions for Lévy noise detection. Phys. Rev. Res. **2**, 043332 (2020)
17. B.D. Josephson, Possible new effects in superconductive tunnelling. Phys. Lett. **1**, 251 (1962)
18. B.D. Josephson, The discovery of tunnelling supercurrents. Rev. Mod. Phys. **46**, 251 (1974)
19. G. Augello, D. Valenti, B. Spagnolo, Non-Gaussian noise effects in the dynamics of a short overdamped Josephson junction. Eur. Phys. J. B **78**, 225 (2010)
20. B. Spagnolo, P. Caldara, A. La Cognata, G. Augello, D. Valenti, A. Fiasconaro, A. Dubkov, G. Falci, Relaxation phenomena in classical and quantum systems. Acta Phys. Pol. B **43**, 1169 (2012)
21. C. Guarcello, D. Valenti, G. Augello, B. Spagnolo, The role of non-Gaussian sources in the transient dynamics of long Josephson junctions. Acta Phys. Pol. B **44**, 997 (2013)
22. D. Valenti, C. Guarcello, B. Spagnolo, Switching times in long-overlap Josephson junctions subject to thermal fluctuations and non-Gaussian noise sources. Phys. Rev. B **89**, 214510 (2014)
23. C. Guarcello, D. Valenti, A. Carollo, B. Spagnolo, Effects of Lévy noise on the dynamics of sine-Gordon solitons in long Josephson junctions. J. Stat. Mech.: Theory Exp. **054012**(2016)
24. B. Spagnolo, D. Valenti, C. Guarcello, A. Carollo, D. Persano Adorno, S. Spezia, N. Pizzolato, B. Di Paola, Noise-induced effects in nonlinear relaxation of condensed matter systems. Chaos Solitons Fractals 81(Part B), 412 (2015)

25. C. Guarcello, D. Valenti, B. Spagnolo, Phase dynamics in graphene-based Josephson junctions in the presence of thermal and correlated fluctuations. Phys. Rev. B **92**, 174519 (2015)
26. C. Guarcello, D. Valenti, A. Carollo, B. Spagnolo, Stabilization effects of dichotomous noise on the lifetime of the superconducting state in a long Josephson junction. Entropy **17**, 2862 (2015)
27. R. Parmentier, Fluxons in long Josephson junctions, in *Solitons in Action*. ed. by K. Lonngren, A.C. Scott (Academic Press, Cambridge, MA, USA, 1978), pp. 173–199
28. A. Ustinov, Solitons in Josephson junctions. Phys. D: Nonlinear Phenom. **123**, 315 (1998)
29. C. Guarcello, F. Giazotto, P. Solinas, Coherent diffraction of thermal currents in long Josephson tunnel junctions. Phys. Rev. B **94**, 054522 (2016)
30. A. Barone, G. Paternó, *Physics and Applications of the Josephson Effect* (Wiley, New York, 1982)
31. B. Spagnolo, C. Guarcello, L. Magazzú, A. Carollo, D. Persano Adorno, D. Valenti, Nonlinear relaxation phenomena in metastable condensed matter systems. Entropy 19 (2017)
32. A. Dubkov, B. Spagnolo, Fluct. Noise Lett. **5**, L267 (2005)
33. D. Valenti, L. Schimansky-Geier, X. Sailer, B. Spagnolo, Eur. Phys. J. B Condens. Matter Complex Syst. 50, 199 (2006)
34. D. Valenti, L. Tranchina, M. Brai, A. Caruso, C. Cosentino, B. Spagnolo, Ecol. Model. **213**, 449 (2008)
35. D. Valenti, L. Magazzú, P. Caldara, B. Spagnolo, Phys. Rev. B **91**, 235412 (2015)
36. G. Falci, A. La Cognata, M. Berritta, A. D'Arrigo, E. Paladino, B. Spagnolo, Phys. Rev. B **87**, 214515 (2013)
37. P. Kadanoff, Phys. Today **54**, 34 (2001)
38. N. Scafetta, B.J. West, Phys. Rev. Lett. **90**, 248701 (2003)
39. P. Barthelemy, J. Bertolotti, D.S. Wiersma, Nature **453**, 495 (2008)
40. S. Luryi, A. Subashiev, J. Geophys. Res. **21**, 1250001 (2012)
41. A. Dubkov, B. Spagnolo, Eur. Phys. J. B Condens. Matter Complex Syst. 65, 361 (2008)
42. A. La Cognata, D. Valenti, A.A. Dubkov, B. Spagnolo, Phys. Rev. E **82**, 011121 (2010)
43. A.A. Dubkov, B. Spagnolo, V.V. Uchaikin, Lévy flight superdiffusion: an introduction. Int. J. Bifurc. Chaos **18**, 2649 (2008)
44. V. Zaburdaev, S. Denisov, J. Klafter, Lévy walks. Rev. Mod. Phys. **87**, 483 (2015)
45. H.A. Kramers, Brownian motion in a field of force and the diffusion model of chemical reactions. Physica **7**, 284 (1940)
46. H. Grabert, U. Weiss, Crossover from thermal hopping to quantum tunneling. Phys. Rev. Lett. **53**, 1787 (1984)
47. J.A. Blackburn, M. Cirillo, N. Grønbech-Jensen, A survey of classical and quantum interpretations of experiments on Josephson junctions at very low temperatures. Phys. Rev. B **85**, 104501 (2012)
48. A.V. Chechkin, V.Y. Gonchar, J. Klafter, R. Metzler, Barrier crossing of a Lévy flight. Europhys. Lett. **72**, 348 (2005)
49. A.V. Chechkin, O.Y. Sliusarenko, R. Metzler, J. Klafter, Barrier crossing driven by Lévy noise: universality and the role of noise intensity. Phys. Rev. E **75**, 041101 (2007)
50. B. Dybiec, E. Gudowska-Nowak, P. Hänggi, Lévy-Brownian motion on finite intervals: mean first passage time analysis. Phys. Rev. E **73**, 046104 (2006)
51. B. Dybiec, E. Gudowska-Nowak, P. Hänggi, Escape driven by α-stable white noises. Phys. Rev. E **75**, 021109 (2007)
52. A.A. Dubkov, A. La Cognata, B. Spagnolo, The problem of analytical calculation of barrier crossing characteristics for Lévy flights. J. Stat. Mech.: Theory Exp. 2009, P01002 (2009)
53. T.A. Fulton, L.N. Dunkleberger, Lifetime of the zero-voltage state in Josephson tunnel junctions. Phys. Rev. B **9**, 4760 (1974)

A Fractional-Order Model of the Cardiac Function

Adrian-Josue Guel-Cortez and Eun-jin Kim

Abstract Improving the mathematical model of the cardiovascular system is an important aspect of the control and design of ventricular assist devices. In this work, through numerical simulations, we analyse the usage of fractional-order operators as a way to improve the circulation model. More specifically, we show that the use of fractional-order derivatives in the lumped circulation model can create different types of heart anomalous behaviours. This includes aortic regurgitation, mitral stenosis and ischaemic cardiomyopathy.

Keywords Cardiovascular system · Fractional derivatives · LVAD

1 Introduction

In 2015, 17.9 million people died around the world caused by cardiovascular disease and rose by 12.5% between 2005 and 2015 with a specific increase of ischaemic heart disease by 16.6% [1]. Even though heart transplantation is considered as the best therapy for patients with end-stage congestive heart failure [2], it is usually a delayed process that could last around 300 days or more on the average for potential recipients. For this reason, the medical community has increased emphasis on the use of ventricular assist devices (VADs) that can enhance the function of the natural heart while patients wait for heart transplantation. These ventricular assists devices are mechanical devices that help the heart with the pumping of the blood through the circulatory system (left ventricular assists device (LVAD)) or through the pulmonary

A.-J. Guel-Cortez (✉) · E. Kim
Centre for Fluid and Complex Systems, Coventry University, Priory Street,
Coventry CV1 5FB, UK
e-mail: guelcortea@uni.coventry.ac.uk

E. Kim
e-mail: ad3116@coventry.ac.uk

C. H. Skiadas et al. (eds.), *13th Chaotic Modeling and Simulation International Conference*, Springer Proceedings in Complexity,
https://doi.org/10.1007/978-3-030-70795-8_21

system (right ventricular assist device (RVAD)) [3–5]. There are currently different types of VADs [6, 7], but the usual consists of a rotary pump applied to the left ventricle [6].

To design proper VADs, different mathematical models for the cardiovascular system have been considered (for instance, see [8, 9]). These are categorised as zero-dimensional (0-D) or lumped parameter models and distributed parameter models or 1-D, 2-D and 3-D models [10]. The mathematical model must be able to properly characterise the cardiac cycle but also important aspects like heart failure (usually studied by pressure-volume (PV) loop analysis [11]), which is a relevant element in VADs' control design [12, 13]. Besides, even though higher dimension models may better describe the heart dynamics, we have to consider that it is a challenge to analyse control strategies for VADs when using such models [3]. Therefore, we require a mathematical model that could represent the heart complex dynamics while being as simple as possible.

An elementary way to describe the heart dynamics (including VADs) is by the use of Windkessel models which are represented through electrical or mechanical networks [3, 14]. Nonetheless, such simplicity has its drawbacks, for instance, the basic two-element Windkessel model explains aortic pressure decay in diastole, but it fails shortly in systole [10]. A way to overcome such drawback consists of adding more elements to the electrical representation [15–17]. Recently. in [18] a different approach has been proposed to solve this problem, it consists in the use of fractional-order operators (derivatives or integrals of non-integer order [19]) in the arterial Windkessel model. When talking about fractional-order operators, heterogeneous systems and phenomena exhibiting anomalous diffusion can be well fitted by using fractional-order derivatives [20, 21]. Examples of such systems are biological tissues [22] and large-scale complex networks [23, 24]. It is important to mention that fractional-order operators does not complicate control design, since there are numerous approaches for control of fractional-order systems that can be implemented (for instance, see [25]). This implies that fractional order operators, when added to the cardiovascular model are a feasible option that may better describe a wider range of real-case scenarios.

Considering the previous lines, in this work, we provide a qualitative analysis of the use of fractional-order operators in the lumped circulatory model by considering two different mathematical models from the literature. This analysis demonstrates that fractional-operators extend the capabilities of such models in the PV loop analysis. Specifically, we add fractional-order operators to modify the rate of change of the aortic, atrial and arterial pressures in the circulation model. These changes permit us to create different anomalous behaviours in the PV loops that resemble aortic regurgitation, mitral stenosis or ischaemic cardiomyopathy dynamics. The remainder of this paper is organised as follows: Sect. 2 presents the basic concepts and their connection with this work. In Sect. 3, our main results are provided. Finally, Sect. 4 contains our concluding remarks.

2 Preliminary Results

In the following subsections, we offer a description of the background knowledge that will be useful to the reminder of this work.

2.1 The Mathematical Model

We base our analysis on the mathematical models of the cardiovascular system with LVAD presented in [4, 5].

Firstly, in [4], a fifth-order lumped parameter electric circuit to reproduce the left ventricle hemodynamics of the heart is used. This model assumes that the right ventricle and pulmonary circulation are healthy and normal. Also, a simple first-order forced differential equation to represent the LVAD is considered (see Fig. 1). Furthermore, [4] uses a time-varying capacitance (compliance), which is derived by system identification procedures, to describe the contractual state of the left ventricle. On the other hand, [5, 26] presents a mathematical model of the cardiac function that integrates mechanical, electric and chemical activity on micro-scale sarcomere and macro-scale heart. More specifically, this model includes the fifth-order lumped parameter electric circuit as well as the electrical model for the LVAD from [4] but presents a different mathematical model of the mechanical, electric and chemical activity to describe the behaviour of the contractual state in the left ventricle.

Recalling from [5], the governing equations of the electric circuit shown in Fig. 1 are given by

$$\begin{aligned}
\frac{dV}{dt} &= \frac{1}{R_m}\left(P_R - P_V\right)\Theta\left(P_R - P_V\right) - \frac{1}{R_A}\left(P_V - m\right)\Theta\left(P_V - m\right) - \delta_p n,\\
\frac{dm}{dt} &= -\frac{1}{C_A}F_a + \frac{1}{C_A R_A}\left(P_V - m\right)\Theta\left(P_V - m\right) + \delta_p \frac{n}{C_A},\\
\frac{dF_a}{dt} &= \frac{m - P_S}{L_S} - \frac{R_C F_a}{L_S},\\
\frac{dP_R}{dt} &= \frac{-P_R + P_S}{C_R R_S} - \frac{1}{C_R R_M}\left(P_R - P_V\right)\Theta\left(P_R - P_V\right),\\
\frac{dP_S}{dt} &= \frac{P_R - P_S}{C_S R_S} + \frac{F_a}{C_S}.
\end{aligned} \tag{1}$$

where V stands for the left ventricular volume, m is the aortic pressure, n corresponds to the pump flow, F_a is the aortic total flow, P_R stands for the atrial pressure and P_S is the arterial pressure. Besides, the function $\Theta(u)$ is a Heaviside function which takes the non-zero value of 1 for $u > 0$ or 0 otherwise. Moreover, the equation governing the pump flow is

$$\frac{dn}{dt} = \frac{1}{L_*}\left[P_V - m - R_* n + \beta\omega^2\right]. \tag{2}$$

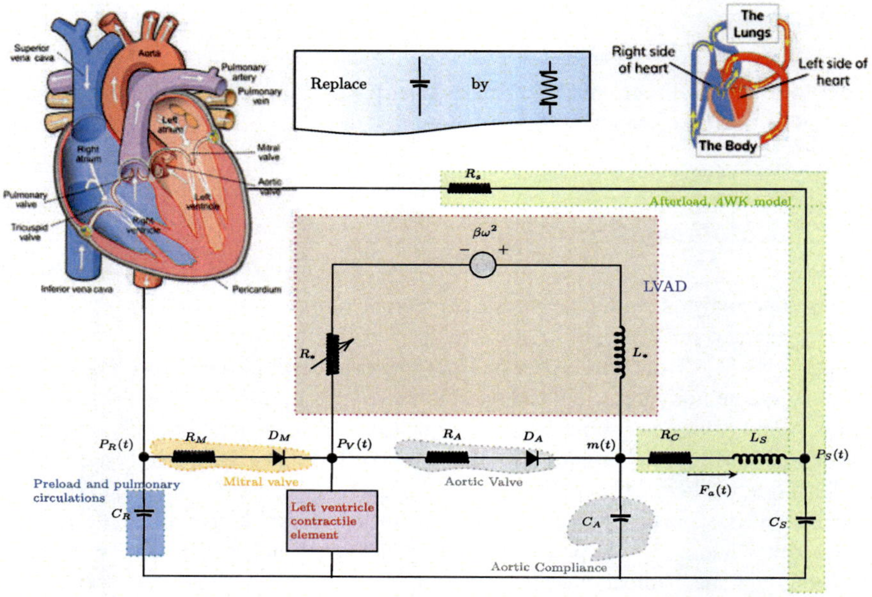

Fig. 1 Lumped parameter electric circuit used to reproduce the left ventricle hemodynamics of the heart and LVAD

In this work, we will consider no pump support. This can be set up by using $\delta_p = 0$. In addition, we use the same set of parameter values presented in [4] for its model and the same set for [5]'s model when considering a case that shows ectopic PV loop oscillations using $\mu_1 = 0.0024$ and $\mu_2 = 0.1584$ on its electric activity model.

2.1.1 Fractional-Order Windkessel Model

As we have mentioned in Sect. 1, the goal of this work is to study the addition of fractional order operators in the lumped parameter circulation model (1). Recently, [18, 27] proposes a fractional-order Windkessel model as an alternative to improve the systolic phase model description. Inspired on the same approach, in Fig. 2a, we introduce a two-element Windkessel model that uses a fractance element C_F^{α}. This fractance element is equivalent to an infinite tree of capacitors and resistors [28]. When this infinite tree is binary, it is proved to be of order $\alpha = \frac{1}{2}$ (for further details, see [23, 24, 29]). To change α's value, we can use the procedure described in [30] which consists of allowing more complicated fractal networks or recursive trees to be constructed (see Fig. 2b). Hence, using a fractional-order capacitor is equivalent to adding a fractal network in the circuit.

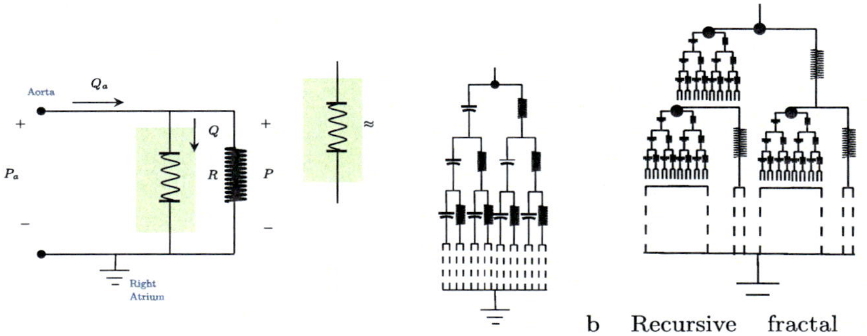

a Fractional-order Windkessel model.

b Recursive fractal tree model.

Fig. 2 Two-element fractional-order

For a better understanding of the mathematical description behind a fractional-order capacitor, consider the model for the fractional-order circuit shown in Fig. 2a. This model is given by

$$ {}_0\mathcal{D}_t^{\alpha} P_a = \frac{Q_a}{C_F^{\alpha}} - \frac{P_a}{RC_F^{\alpha}}. \tag{3}$$

Taking in to account expression (3). We have the following result

Proposition 1 *By using the Caputo definition of the fractional derivative operator* ${}_0\mathcal{D}_t^{\alpha}$ *of order* $0 < \alpha < 1$ *[31]. The time response of system* (3) *is given by*

$$\begin{aligned} P_a(t) = P_a(0)t^{\alpha-1}\mathbf{E}_{\alpha,\alpha}(-\frac{1}{RC_F^{\alpha}}t^{\alpha}) + \\ \frac{1}{C_F^{\alpha}}\int_0^t Q_a(t-\tau)\tau^{\alpha-1}\mathbf{E}_{\alpha,\alpha}(-\frac{1}{RC_F^{\alpha}}\tau^{\alpha})d\tau, \end{aligned} \tag{4}$$

where $\mathbf{E}_{\alpha,\alpha}(z)$ *is the Mittag-Leffler function of the complex value* z *[32].*

Proof The proof follows by applying the Laplace transform to (3) and obtaining the inverse Laplace transform $\mathscr{L}^{-1}[P_a(s)]$.

An important conclusion from Proposition 1 is that (4) describes the arterial pressure as a power low equation with a diffusive term. This kind of representation implies some challenges that have been recently discussed through various works (for instance, see [33, 34] and the references therein). Taking the previous lines into consideration, it is important to mention that instead of finding an analytical solution to our set of differential equations, in this work, we will offer a qualitative analysis by numerically solving our set of fractional order differential equations. The numerical solution will be computed by using the methods presented in [35], which consider that the fractional derivative of order $0 < \alpha \leq 1$ of a continuously differentiable real-valued function $x(t)$ is found by taking $x(t) = 0$ for all $t < 0$.

2.2 Pressure-Volume Loop Analysis in Cardiology

Since our analysis will focus on the behaviours obtained using fractional-order derivatives in left ventricular PV loops, it is important to define why PV loops are a critical feature to study. We can describe PV loop analysis as a reference method that offers unique insights into mechanical cardiac efficiency. This includes the understanding of the pathophysiology, diagnosis, and treatment of myocardial ischaemia, mitral and aortic regurgitation, mitral and aortic stenosis, and others (for further details, see [36, 37]).

In summary, as shown in Fig. 3, a PV loop plots the changes in ventricular pressure associated with the changes in volume occurring during a cardiac cycle. A full description of the different concepts related to volumes, pressures, and areas in a PV loop (shown in Fig. 3) can be found in [38]. Below, we describe some of the pressure-volume concepts that will be of main interest in this work

- The end-diastolic pressure volume relationship (EDPVR) (black dashed line). The slope of this line gives the elastance of the ventricle. An important value over this line corresponds to the End-diastolic volume (EDV) or preload, which helps to determine the initial value of the arterial elastance line (shown as a red dashed line in Fig. 3).

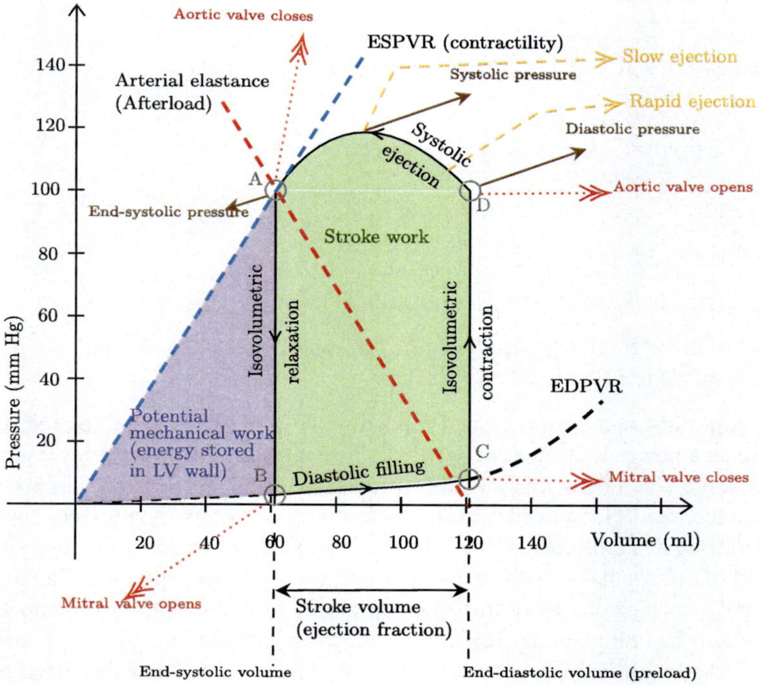

Fig. 3 PV loop diagram

- The stroke volume. This is the difference between the end-diastolic volume and the end-systolic volume. It is also known as the ejection fraction, i.e. the amount of blood to be ejected by the left ventricle to the circulatory system.
- Arterial elastance line (red dashed line). This line allow us to measure the afterload which is technically given by the pressure-volume relationship throughout the entire of ejection, but assumed to be the slope of such a line drawn from the x-axis value of EDV to the end systolic pressure value in point A.
- The slope of the End-systolic pressure volume-relationship (ESPVR) also known as contractility (blue dashed line). It also represents the elastance at end-systole E_{es}.
- The stroke work, which corresponds to the green shaded area of the PV loop.
- The four stages in the PV loop given by: **A**-Aortic valve closing, **B**-Mitral valve opening, **C**-Mitral valve closing and **D**-Aortic valve opening.

3 Main Results

In this section, we present the main results of our analysis. We start by describing how we modify (1) to include fractional-order dynamics. Then, we vary the fractional-order derivative orders and comment on their individual and grouped effects. Finally, we simulate the behaviour of the fractional-order circulation model when changing other parameters in the circuit.

3.1 The Fractional-Order Circulation Model

Consider the use of fractances in the left atrial, aortic and systemic compliances of the lumped parameter electric circuit shown in Fig. 1. Thus, equations for the left atrial pressure P_R, the aortic pressure m and the arterial pressure P_S in (1) are rewritten as

$$\begin{aligned}
\frac{d^{\alpha_m} m}{dt^{\alpha_m}} &= -\frac{1}{C_A} F_a + \frac{1}{C_A R_A} (P_V - m)\, \Theta\, (P_V - m) + \delta_p \frac{n}{C_A}, \\
\frac{d^{\alpha_R} P_R}{dt^{\alpha_R}} &= \frac{-P_R + P_S}{C_R R_S} - \frac{1}{C_R R_M} (P_R - P_V)\, \Theta\, (P_R - P_V), \\
\frac{d^{\alpha_S} P_S}{dt^{\alpha_S}} &= \frac{P_R - P_S}{C_S R_S} + \frac{F_a}{C_S}.
\end{aligned} \tag{5}$$

where $\alpha_m, \alpha_R, \alpha_S \in (0, 1]$ are the fractional orders of the time derivatives for m, P_R and P_S, respectively.

3.2 The Role of α_S

To understand the role of α_m, α_R and α_S in the circulation model, we linearly vary these parameters in [4]'s and [5]'s models. First, we analyse the effect of α_S. If we let $\alpha_m, \alpha_R = 1$ and only vary $\alpha_S \in (0, 1]$, in mathematical terms, we are adding an infinite memory and non-local operator to describe the rate of the arterial-pressure. Physically speaking, since the capacitor for the arterial compliance in the circuit tries to model the biggest part of the cardiovascular system, when adding a fractance (an infinite network of resistors and capacitors), we are trying to improve the description of the arterial system by implicitly adding an infinite number of elements.

Figure 4a, b show the process of decreasing α_S from 1 to 0.2. This process scarcely changes the stroke work but creates an attractor that shows a right-shift of the PV loop in both model's responses (with a greater factor in [5]'s model). Such a shift to the right is presented in ischaemic cardiomyopathy [37]. Besides, the slope of the ESPVR remains almost the same and the afterload slope shows small increments.

Moreover, a specific analysis in [5]'s model (Fig. 4b) shows that PV loops with ectopic behaviours using $\alpha_m = \alpha_R = \alpha_S = 1$ completely change when $\alpha_S < 0.6$. This change shows a little increase in afterload and EDV which is quite similar to the one presented in aortic regurgitation. Finally, when $\alpha < 0.3$, the PV loop shows a progressive right-shifting greater than in [4]'s model.

3.3 The Role of α_R

The value of α_R and α_m mathematically contain the same properties than α_S, but their role in the circulation model is different. If we consider α_R, this affects the rate of change in the atrial pressure. Here, Fig. 5 presents the simulations when

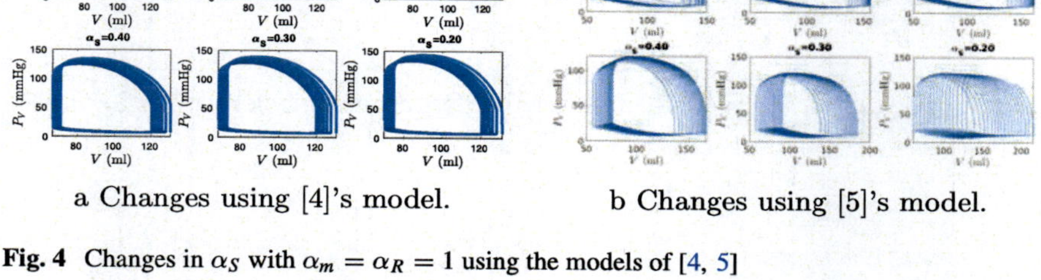

a Changes using [4]'s model. b Changes using [5]'s model.

Fig. 4 Changes in α_S with $\alpha_m = \alpha_R = 1$ using the models of [4, 5]

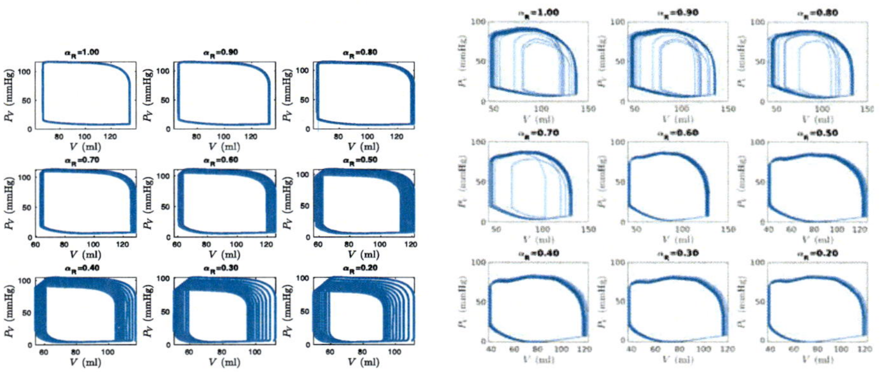

a Changes using [4]'s model. b Changes using [5]'s model.

Fig. 5 Changes in α_R with $\alpha_m = \alpha_S = 1$ using the models of [4, 5]

changing α_R in the models of [4, 5] while fixing $\alpha_m = \alpha_S = 1$. First, from Fig. 5a, we can see that for $\alpha_R < 1$ there is a gradual right-shift of the PV loop as in the case of α_S, but there is also a monotonically increasing preload that is a characteristic feature of aortic regurgitation. Analysing Fig. 5b, [5]'s model initially presented an anomalous behaviour using $\alpha_R = \alpha_m = \alpha_S = 1$ which is normalised when $\alpha_R < 0.7$ while showing a behaviour similar to the case of [4]'s model but in small proportions.

3.4 The Role of α_m

Since the aorta is the main artery that carries blood away from the heart to the rest of the body, aortic pressure plays an imperative role in the cardiovascular system. Also, the aortic pressure is affected by the ventricular pressure whose behaviour is modified by using different mechanic-electric models (to describe the ventricular contraction) or by the mitral valve functioning. In this context, the value of α_m plays a crucial role, because it directly affects the rate of change in the aortic pressure.

Here, we perform simulations of both [4, 5] models when changing α_m and fixing $\alpha_R = \alpha_S = 1$ (Fig. 6). For both models, we have an erratic PV loop behaviour that includes negative diastolic pressures when $\alpha_m < 0.6$. The presence of negative diastolic pressure values is a phenomenon that has been studied since long ago as an effect of mitral stenosis [39]. Nonetheless, since the simulation stops working when $\alpha_m < 0.47$, this erratic performance seems to be caused by numerical errors in both models. Such numerical inconsistencies may be fixed by using a smaller integration step but paying a high computational cost.

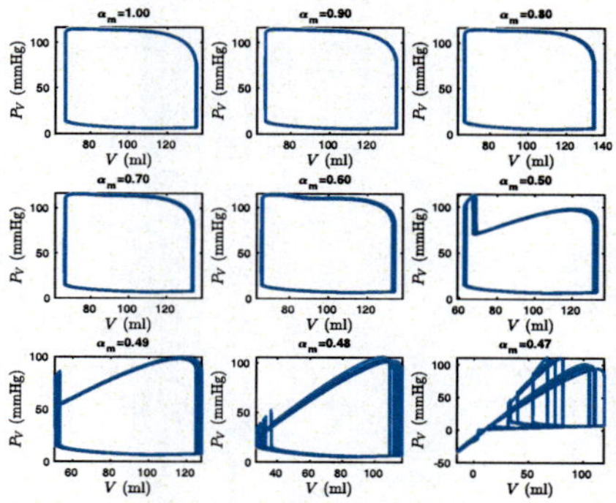

a Changes using [4]'s model.

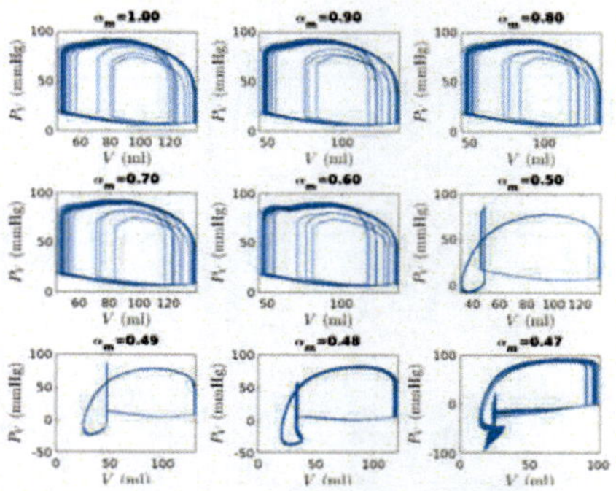

b Changes using [5]'s model.

Fig. 6 Changes in α_m with $\alpha_R = \alpha_S = 1$ using the models of [4, 5]

3.5 *Changing Fractional-Orders Equally and the Effect of Other Parameters*

As we have seen from previous sections, changing α_m, α_R and α_S separately while fixing the other parameters give an insight about their effect on the cardiovascular model. Here, in Fig. 7, we show the influence of adding equal and unequal fractional-order values. Such simulations show the presence of a slight right-shifting over time but with no highly anomalous behaviours. In this case, the use of fractional-order operators seems to transform the PV loop into a quasi-periodic attractor for [4]'s model and erase the ectopic oscillations while increasing the stroke volume for [5]'s model.

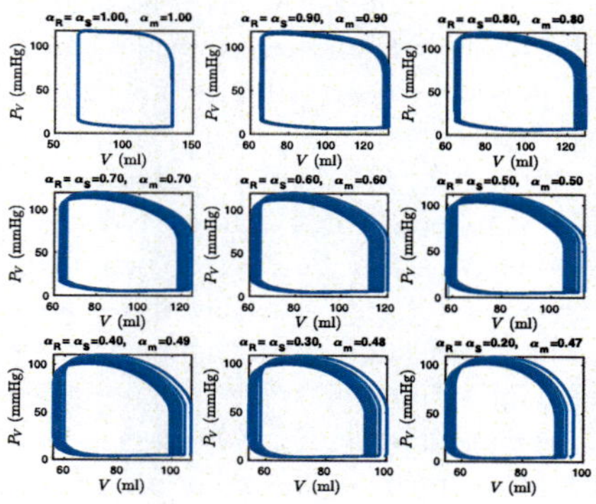

a Changes using [4]'s model.

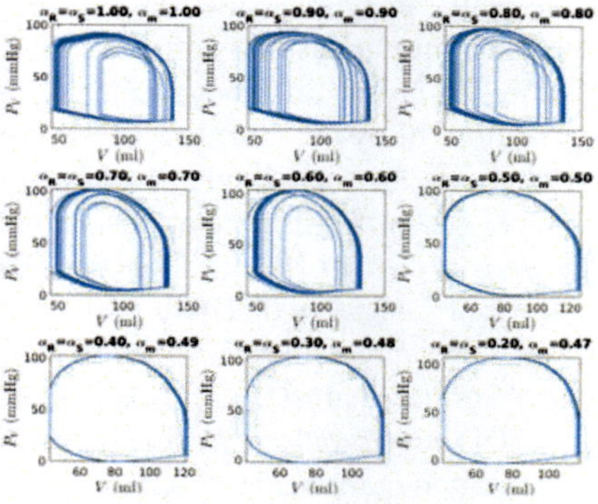

b Changes using [5]'s model.

Fig. 7 Changes in α_m, α_R, α_S using the models of [4, 5]

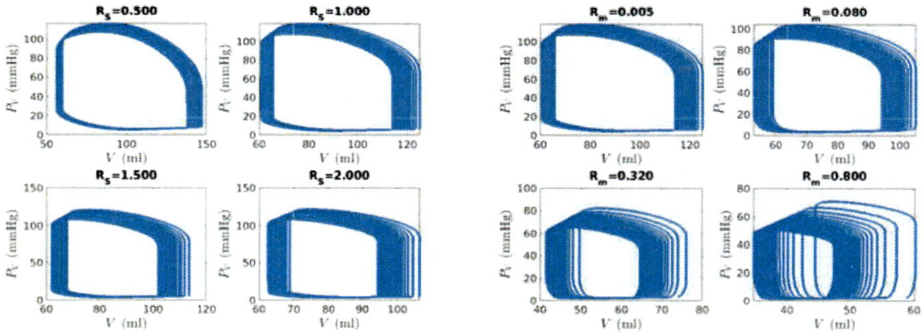

a Changes in R_S. b Changes in R_m.

Fig. 8 Changes in R_S and R_m with $\alpha_m = \alpha_R = \alpha_S = 0.7$ using [4]'s model

Finally, some of the parameters in the circulation model (1) are known to produce certain effects. For example, the value of R_m decrease/increase the stroke volume by changing the ESV and EDV positions while fixing ESPVR and changing R_S decrease/increase the stroke volume by fixing ESPVR but changing the afterload. Since this effect has been studied using [4]'s model, Fig. 8 only shows the effects of R_m and R_S on this model while using fixed $\alpha_m = \alpha_R = \alpha_S = 0.7$. These simulations show that the effects of R_m and R_S prevail while the fractional-order derivatives continue adding a right-shifting.

4 Concluding Remarks

In this work, we have introduced a qualitative analysis of the use of fractional-order operators in two cardiovascular models with distinct left ventricular contractile element models, by adding fractances instead of capacitors to represent aortic, atrial and arterial compliances. Moreover, through PV loop analysis, we have demonstrated that fractional-order derivatives can help to characterise heart anomalous behaviours. Specifically, it may help at describing aortic regurgitation, mitral stenosis and ischaemic cardiomyopathy. Future work includes the design of a parameter identification technique to fit the parameter's using real data, the inclusion of the LVAD model in the analysis and a controller's design methodology for the LVAD that considers the use of fractional-order derivatives in the model.

References

1. GBD 2015 Mortality and Causes of Death Collaborators, Global, regional, and national life expectancy, all-cause mortality, and cause-specific mortality for 249 causes of death, 1980-2015: a systematic analysis for the global burden of disease study 2015. Lancet (London, England) **388**(10053), 1459–1544 (2016)
2. M.R. Mehra et al., The 2016 international society for heart lung transplantation listing criteria for heart transplantation: a 10-year update. J. Heart Lung Transplant. **35**(1), 1–23 (2016)
3. M.A. Simaan, Rotary heart assist devices, in *Handbook of Automation* (2009)
4. M.A. Simaan, A. Ferreira, S. Chen, J.F. Antaki, D.G. Galati, A dynamical state space representation and performance analysis of a feedback-controlled rotary left ventricular assist device. IEEE Trans. Control Syst. Technol. **17**(1) (2009)
5. E. Kim, M. Capoccia, Mechano-electric effect and a heart assist device in the synergistic model of cardiac function. Math. Biosci. Eng. **17**(5), 5212–5233 (2020)
6. B. Long, J. Robertson, A. Koyfman, W. Brady, Left ventricular assist devices and their complications: a review for emergency clinicians. Am. J. Emerg. Med. **37**(8), 1562–1570 (2019)
7. D. Timms, A review of clinical ventricular assist devices. Med. Eng. Phys. **33**(9), 1041–1047 (2011)
8. L.W. Miller, J.G. Rogers, Evolution of left ventricular assist device therapy for advanced heart failure: a review. JAMA Cardiol. **3**(7), 650–658 (2018)
9. B. Gao, Y. Kang, Q. Zhang, Y. Chang, Biomechanical effects of the novel series LVAD on the aortic valve. Comput. Methods Programs Biomed. **197**, 105763 (2020)
10. M. Capoccia, Development and characterization of the arterial windkessel and its role during left ventricular assit device assitance. Artif. Organs **39**, E138–E153 (2015)
11. D.R. Warriner, A.G. Brown, S. Varma, P.J. Sheridan, P. Lawford, D.R. Hose, A. Al-Mohammad, Y. Shi, Closing the loop: modelling of heart failure progression from health to end-stage using a meta-analysis of left ventricular pressure-volume loops. PloS One **9**(12), e114153 (2014)
12. A.-H.H. AlOmari, A.V. Savkin, M. Stevens, D.G. Mason, D.L. Timms, R.F. Salamonsen, N.H. Lovell, Developments in control systems for rotary left ventricular assist devices for heart failure patients: a review. Physiol. Meas. **34**(1), R1 (2012)
13. C. Ozcan, A. Deshmukh, Atrial arrhythmias in patients with left ventricular assist devices. Curr. Opin. Cardiol. **35**(3), 276–281 (2020)
14. N. Westerhof, J.-W. Lankhaar, B.E. Westerhof, The arterial windkessel. Med. Biol. Eng. Comput. **47**(2), 131–141 (2009)
15. L. Marx, M.A.F. Gsell, A. Rund, F. Caforio, A.J. Prassl, G. Toth-Gayor, T. Kuehne, C.M. Augustin, G. Plank, Personalization of electro-mechanical models of the pressure-overloaded left ventricle: fitting of windkessel-type afterload models. Philos. Trans. R. Soc. A **378**(2173), 20190342 (2020)
16. U. Hackstein, S. Krickl, S. Bernhard, Estimation of ARMA-model parameters to describe pathological conditions in cardiovascular system models. Inform. Med. Unlocked 100310 (2020)
17. N.L. Kang. Relationship between blood pressure and flow rate in arteries using a modified windkessel model (2020). arXiv:2006.02590
18. M.A. Bahloul, T.M. Laleg-Kirati, Assessment of fractional-order arterial Windkessel as a model of aortic input impedance. IEEE Open J. Eng. Med. Biol. **1**, 123–132 (2020)
19. D. Valério, J. da Costa, *An Introduction to Fractional Control*, 01 2012
20. C. Ionescu, J.F. Kelly, Fractional calculus for respiratory mechanics: power law impedance, viscoelasticity, and tissue heterogeneity. Chaos Solitons Fractals **102**, 433–440 (2017)
21. M.A. Moreles, R. Lainez, Mathematical modelling of fractional order circuit elements and bioimpedance applications. Commun. Nonlinear Sci. Numer. Simul. **46**, 81–88 (2017)
22. R.L. Magin, *Fractional Calculus in Bioengineering*, vol. 2 (Begell House, Redding, 2006)
23. M. Sen, J.P. Hollkamp, F. Semperlotti, B. Goodwine, Implicit and fractional-derivative operators in infinite networks of integer-order components. Chaos Solitons Fractals **114**, 186–192 (2018)

24. A.-J. Guel-Cortez, M. Sen, B. Goodwine, Closed form time response of an infinite tree of mechanical components described by an irrational transfer function, in *2019 American Control Conference (ACC)*, pp. 5828–5833, July 2019
25. A.J. Guel-Cortez, C.-F. Méndez-Barrios, E.J. González-Galván, G. Mejía-Rodríguez, L. Félix, Geometrical design of fractional PDμ controllers for linear time-invariant fractional-order systems with time delay. Proc. Inst. Mech. Eng. Part I: J. Syst. Control Eng. **233**(7), 815–829 (2019)
26. E. Kim, M. Capoccia, Synergistic modelling of cardiac function with a heart assist device. Phys. Rev. E (2019)
27. M.A. Bahloul, T.-M. Laleg Kirati, Fractional-order modeling of the arterial compliance: an alternative surrogate measure of the arterial stiffness (2020)
28. M.S. Semary, M.E. Fouda, H.N. Hassan, A.G. Radwan, Realization of fractional-order capacitor based on passive symmetric network. J. Adv. Res. **18**, 147–159 (2019)
29. P. Yifei, Y. Xiao, L. Ke, Z. Jiliu, Z. Ni, Z. Yi, P. Xiaoxian, Structuring analog fractance circuit for 1/2 order fractional calculus, in *2005 6th International Conference on ASIC*, vol. 2, pp. 1136–1139, Oct 2005
30. J.F. Kelly, R.J. MacGough, Fractal ladder model and power law wave equations. J. Acoust. Soc. Am. **126**(4), 2072–2081 (2009)
31. A. Atangana, *Fractional Operators with Constant and Variable Order with Application to Geo-hydrology* (Academic Press, 2017)
32. A.K. Shukla, J.C. Prajapati, On a generalization of Mittag-Leffler function and its properties. J. Math. Anal. Appl. **336**(2), 797–811 (2007)
33. J. Sabatier, M. Merveillaut, R. Malti, A. Oustaloup, How to impose physically coherent initial conditions to a fractional system. Commun. Nonlinear Sci. Numer. Simul. **15**, 1318–1326 (2010)
34. J. Sabatier, Fractional-order derivatives defined by continuous kernels: are they really too restrictive? Fractal Fract. **4**(3) (2020)
35. R. Garrappa, Numerical solution of fractional differential equations: a survey and a software tutorial. Mathematics **6**(16) (2018)
36. F. Seemann, P. Arvidsson, D. Nordlund, S. Kopic, M. Carlsson, H. Arheden, E. Heiberg, Noninvasive quantification of pressure-volume loops from brachial pressure and cardiovascular magnetic resonance. Circ. Cardiovasc. Imaging **12**(1), e008493 (2019)
37. M.B. Bastos, D. Burkhoff, J. Maly, J. Daemen, C.A. den Uil, K. Ameloot, M. Lenzen, F. Mahfoud, F. Zijlstra, J.J. Schreuder, N.M. Van Mieghem, Invasive left ventricle pressure-volume analysis: overview and practical clinical implications. Eur. Heart J. **41**(12), 1286–1297 (2019)
38. D.E. Mohrman, L.J. Heller, *Cardiovascular Physiology*, 9th edn. (Lange Medical Books/McGraw-Hill, New York, USA, 2018)
39. H.N. Sabbah, D.T. Anbe, P.D. Stein, Negative intraventricular diastolic pressure in patients with mitral stenosis: evidence of left ventricular diastolic suction. Am. J. Cardiol. **45**(3), 562–566 (1980)

Vector Difference Equations, Substochastic Matrices, and Design of Multi-networks to Reduce the Spread of Epidemics

Harold M. Hastings and Tai Young-Taft

Abstract Cities have long served as nucleating centers for human development and advancement, c.f. (Smith in Cities: The First 6,000 Years, Penguin Books, London, 2020). Cities have facilitated the spread of both human creativity and human disease, and at the same time, efforts to minimize the spread of disease have influenced the design of cities, c.f. (E. Stinson, Health and Disease Have Always Shaped Our Cities. What Will Be the Impact of COVID-19? Architectural Digest, https://www.architecturaldigest.com/story/how-will-coronavirus-impact-cities, 2020. Accessed 2020-November 2.). The purpose of this paper is to explore the dynamics of epidemics on networks in order to help design a multi-network city of the future aimed at minimizing the spread of epidemics. In order to do this, we start with the SIR model (susceptible, infected, removed) on a network in which nodes represent cities or regions and edges are weighted by flows between regions. Since the goal is to stabilize the zero infections steady state, we linearize the discrete-time SIR model yielding difference equations for the dynamics of infections at each node and then include flows of infections from other nodes. This yields a vector difference equation $v_{new} = Mv$ for the spread of infections. We then generalize the concept of stochastic matrix in order to quantify the dynamics of this update equation. The entries of the update matrix M may vary in time, even discontinuously as flows between nodes are turned on and off. This may yield useful design constraints for a multi-network composed of weak and strong interactions between pairs of nodes representing interactions within and among cities.

Keywords Covid-19 · SIR model · Network · Matrix models · Cities

H. M. Hastings (✉) · T. Young-Taft
Bard College At Simon's Rock, 84 Alford Road, Great, Barrington, MA 01230, USA
e-mail: hhastings@simons-rock.edu

T. Young-Taft
e-mail: tyoungtaft@simons-rock.edu

C. H. Skiadas et al. (eds.), *13th Chaotic Modeling and Simulation International Conference*, Springer Proceedings in Complexity,
https://doi.org/10.1007/978-3-030-70795-8_22

1 Introduction

Cities have long served as nucleating centers for human development and advancement, c.f. [3–5, 21, 24, 28] and references therein. As Bettencourt et al. state: "Cities have long been known to be society's predominant engine of innovation and wealth creation, yet they are also its main source of crime, pollution, and disease. … Many diverse properties of cities from patent production and personal income to electrical cable length are shown to be power law functions of population size with scaling exponents, β, that fall into distinct universality classes. Quantities reflecting wealth creation and innovation have $\beta \approx 1.2 > 1$ (increasing returns), whereas those accounting for infrastructure display $\beta \approx 0.8 < 1$ (economies of scale)." [5]

Schläpfer et al. "show that both the total number of contacts and the total communication activity grow superlinearly with city population size, according to well-defined scaling relations and resulting from a multiplicative increase that affects most citizens. Perhaps surprisingly, however, the probability that an individual's contacts are also connected with each other remains largely unaffected. These empirical results predict a systematic and scale-invariant acceleration of interaction-based spreading phenomena as cities get bigger … a microscopic basis towards understanding the superlinear increase of different socioeconomic quantities with city size, that applies to almost all urban systems and includes, for instance, the creation of new inventions or the prevalence of certain contagious diseases." (emphasis added) [21].

Although Arcaute et al. [3] question these scaling rules, citing both the role of density and notable exceptions, e.g., the enormous intellectual productivity of Cambridge UK, they also support the central role of cities.

Finally, the there is a two-way relationship between cities and disease: as Stinson [26] observed, "Health and Disease Have Always Shaped Our Cities." For example, Olmstead and Vaud's development of Central Park (an 843 acre/ 341 hectare park) in New York City was in part driven by cholera epidemics in the early to mid-1800's.

Stinson then asked "What Will Be the Impact of COVID-19?", a question which motivated the thrust of the present paper. Perhaps the city of the future and the inter-urban network of the future will be in part physical, in part virtual, thus a milti-network.

We start with the SIR model (susceptible, infected, removed) (c.f. [23] for a review on a network. The SIR model has been widely used to stud the spread of Covid-19 (c.f. [9] and references therein). Biswas was et al. explored the early spread of Covid-19 in China with the SIR model on a network [6], obtaining a power law fit to the contacts. Here we develop a simple matrix/network extension of the SIR model near the desired zero infections steady state.

The SIR model considers three states: susceptible, infected, removed. We shall not distinguish here between "infected" and "infectious," since this distinction does not affect dynamics near the desired zero infections steady state. In addition, we simplify the model by assuming that infected individuals either gain long-term immunity or pass away. Under our simplifying assumptions we have a simple flow diagram (Fig. 1, below) and differential equation for the number infected (Eq. 1, below). (If immunity

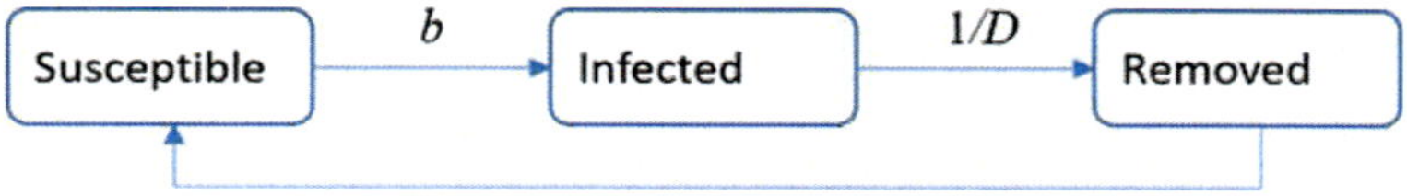

Fig. 1 Flow diagram for the SIR model. Arrows represent flows from susceptible infected (as susceptible individuals become infected), from infected to removed (as infected individuals become no longer infected, some of whom pass away and others recover), and from removed to susceptible (recovered individuals who lose or never had immunity). Since Covid-19 appears to show long-term immunity in most recovered individuals [11, 22], we may ignore flows from removed to susceptible. Flows into the infected compartment depend upon interactions between susceptible and infected individuals, parametrized by an effective transmission rate b; flows out of the infected department scale inversely with and the duration D of infection. See text, Eq. (1)

were only temporary, some of the immune population would become susceptible, and we would add a flow from the removed compartment to the susceptible compartment.)

Flows into and out of the infected state (compartment in engineering language) are given by the standard SIR equation

$$dI/dt = bIS - I/D \tag{1}$$

Here I denotes the number infected, S the number susceptible, b an effective transmission rate (described below) and D the duration of infection. Our goal here to make the steady state with zero infected ($I = 0$) Lyapunov stable equilibrium as in May and Anderson [20]. As they observed, "simple mathematical models of the transmission dynamics of HIV help to clarify some of the essential relations between epidemiological factors." We shall rewrite the flow from susceptible to infected in terms of probability p that a contact yields an infection, and the effective contact rate c between susceptible individuals and the probability s that a contact is susceptible, that is, $bS = pcs$. As illustrated in Fig. 2 below, the SIR model for infections then becomes

$$dI/dt = pcsI - I/D = (pcs - 1/D)I \tag{2}$$

We shall now explore the role of controls (masking, distancing, travel limits, shutdowns, etc.) aimed at controlling the cascading spread of Covid-19 through the use of a robust, simple model, namely a matrix/network extension of the above SIR model.

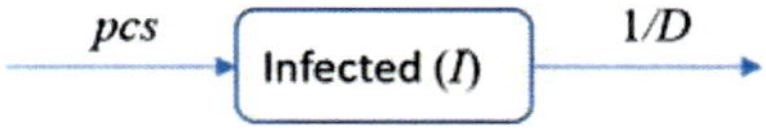

Fig. 2 Graphical representation of the SIR model (2). Flows are $pcsI$ (into the infected compartment) and $(1/D)I$ (out of the infected compartment) for a net flow of $(pcs - 1/D)I$

2 Linearizing the SIR Model

In order to develop the matrix model, we first linearize the SIR model (Eq. (2)) by assuming that the probability s that a contact is susceptible is constant. We then discretize the linearized SIR model by the Euler method, obtaining the difference equation

$$I(t + \mathrm{D}t) = I(t) + (pcs - 1/D)I(t)\Delta t = (1 + (pcs - 1/D)\Delta t)I(t) \quad (3)$$

Remarks.

(1) The parameter R_t (c.f. the basic reproductive rate R_0 in [20]) is readily obtained by setting the time step Δt in Eq. (3) equal to the duration of infection D. The following simple calculation then implies that

$$R_t = pcsD \quad (4)$$

is the growth rate of infections over the duration of infection D as the natural time scale:

$$I(t + D) = (1 + (pcsD - 1)I(t) = pcsD\, I(t) \quad (5)$$

(2) Eq. (6) yields a condition for herd immunity, namely, that the growth rate

$$R_t = pcsD < 1 \quad (6)$$

in which case the number of infected, I, decreases at least exponentially as

$$(\mathrm{pcsD})^{\mathrm{t}/D} \quad (7)$$

The required limit in growth rate R_t (Eq. (6)) can be can be achieved in several ways or by an appropriate combination thereof:

a. Reducing the probability p that a contact results in an infection through masking and/or physical (misnamed social) distancing. Of course, relaxing masking and/or physical distancing will increase p, potentially causing the growth rate R_t to increase above 1, ending (apparently temporary) herd immunity.
b. Reducing the effective contact rate c between susceptible and infected individuals, for example by making some contacts virtual and/or by otherwise reducing the incidence of "superspreading"/ "spreading" events in a multi-networked city of the future. May and Anderson [20] emphasized the role of the distribution of contact rates in determining the effective contact rate (for the spread of HIV in a male homosexual population):

$$c = \mu + \sigma^2/\mu \tag{8}$$

the mean + the ratio of the variance to the mean, which we rewrite perhaps more simply as

$$c = \mu(1 + (\sigma/\mu)^2) \tag{9}$$

The effective contact rate is thus increased by the square of the coefficient of variation, similarly emphasizing the role of variability in contact rates. Of course, eqns. (8) and (9) "blow up" if the contact rate has a fat tailed distribution with infinite variance.

Similarly, reducing contacts through travel restrictions (the effects of travel modeled by off-diagonal elements in a matrix update equation (eqns. (10ff), below).

c. Reducing the proportion of susceptibles s by vaccination.

3 Substochastic Matrices, Products with Non-negative Vectors and a Matrix/Network Model.

In this section we extend the linearized, discrete-time SIR model (Eq. (4)) to networks.

Notation. Since the symbol i is typically used for indexing vectors and matrices, we write $\boldsymbol{v}$ for the vector whose entries denote the number of infected individuals in nodes (cities, regions). We shall also follow standard notation I the identity matrix from here on, and M for the matrix of rate coefficients in our matrix/ network model. Thus diagonal entries are given by rate coefficients in the SIR model (Eq. (4)). Off-diagonal entries in M represent rates at which infections in one node/city/region generate infections in another node/city/region; see Fig. 3, below.

We obtain an update rule $\boldsymbol{v}(t + \Delta t) = \boldsymbol{v}(t) + M\boldsymbol{v}(t)$, or more simply

$$v_{\text{new}} = (I + M\Delta t)v \tag{10}$$

Here the diagonal elements represent local (within node/city/region) SIR interactions

$$M_{ii} = p_i c_i s_i - 1/D \tag{11}$$

and the off-diagonal elements M_{ij} represent the rate at which infections at node i generate additional infections at node j. Figure 4, below, is a graphical representation of the matrix/network SIR model.

Interpretation and consequences. Suppose $\boldsymbol{v}$ is the unit vector with a 1 in the jth row, representing one infection in node j of the network. Then $M\boldsymbol{v}$ is simply the jth

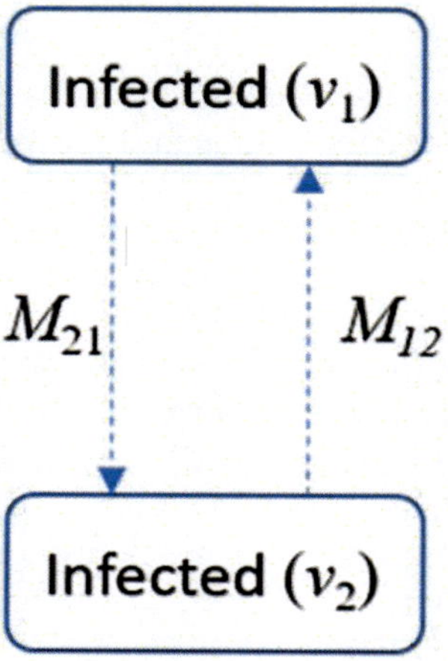

Fig. 3 Infections generated by connections in the matrix/network model. This small, "toy" model consists of two nodes, representing cities or regions. In this representation M_{21} represents the rate at which one infection in node 1 generates new infections at node 2, that is $M_{21}\Delta t$ new infections at node 2 after one time step Δt. Analogously M_{12} represents the rate at which each infection at node 2 generates new infections at node 1

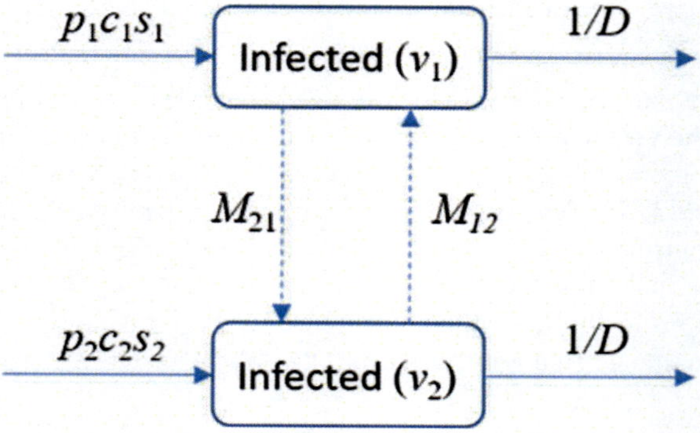

Fig. 4 Graphical representation of the matrix/network SIR model. This small, "toy" model consists of two nodes, representing cities or regions. As in Fig. 2, M_{21} represents the rate at which one infection in node 1 generates new infections at node 2, that is $M_{21}\Delta t$ new infections at node 2 after one time step Δt, analogously for M_{12}. In addition, the rate of change v_1, the number infected at node 1 is simply $(p_1c_1s_1 - 1/D)v_1 + M_{12}v_2$

column of M: $(Mv)_i = M_{ij}$. The jth column sum $\Sigma_i M_{ij}$ is the "rate constant" at which infections at node j drive a change in the number of infections throughout the entire network. As a consequence, if all column sums of the update matrix M are negative, then the total number of infections in the network will decrease after a time step Δt. This observation suggests that the column sums play a role in quantifying the growth or decay of infection in a matrix/network model analogous to the role of R_t in quantifying local dynamics: (column sums < 0) ~ ($R_t < 1$).

The analogy (column sums < 0) ~ ($R_t < 1$) can be simplified by rewriting the update equation $\boldsymbol{v}_{\text{new}} = (I + M\Delta t)\boldsymbol{v}$ (Eq. (10)) as

$$v_{\text{new}} = (I + M\Delta t)v = Av \tag{12}$$

where

$$A = I + M\Delta t \tag{13}$$

In the case of a trivial network (one node, no connections), and time step $\Delta t = D$, the duration of infection, A becomes a scalar and $A = R_t$ (Eq. (4), above). The stability criterion $R_t < 1$ then suggests the following definition (generalization).

Definition 1. A non-negative matrix is column-substochastic if its maximum column sum is less than 1.

Thus, whenever the update matrix M in the update rule is column-stochastic, one infection yields < 1 infections in total at the next time step, irrespective of location.

Note. The ℓ^1 norm of a non-negative vector is simply the sum of its entries. Here the ℓ^1 norm of $\boldsymbol{v}$ is simply the total number of infections.

We need a relatively straight-forward theorem on left multiplication by a column-substochastic matrix. We give a short proof (a straight-forward calculation) here since we did not find a proof in the literature.

Theorem 1. Left multiplication by a column-substochastic matrix with maximum column sum $\leq k$ reduces the ℓ^1 norm of a non-negative vector by a factor $\leq k$.

Proof. Let $\boldsymbol{v}$ be a non-negative vector and A a column-substochastic matrix. We calculate the ℓ^1 norm of $A\boldsymbol{v}$, namely $\Sigma_i\ (A\boldsymbol{v})_i$ as follows.

$$\Sigma_i(A\boldsymbol{v})_i = \Sigma_i(\Sigma_j A_{ij}\boldsymbol{v}_j) = \Sigma_{ij}A_{ij}\boldsymbol{v}_j = \Sigma_j(\Sigma_i A_{ij})\boldsymbol{v}_j \leq \Sigma_j k\boldsymbol{v}_j = k\Sigma_j\boldsymbol{v}_j \tag{14}$$

namely k times the ℓ^1 norm of the non-negative vector $\boldsymbol{v}$.**Done.**

Remarks. The Gerschgorin circle theorem yields an analogous bound for the eigenvalues of a non-negative matrix, based on row sums. However, in the spirit of Cohen and Newman's results for products of random matrices [7, 8], Theorem 1 has the advantage that it can be generalized to products of column-stochastic matrices, which themselves may vary in time.

Corollary 1. Repeated application of the update rule $v_{new} = Av$ to a non-negative vector v, represented by the product of a sequence of column-substochastic matrices $\{A(i)\}$, which may vary in time, and all with maximum row sum $\leq$ k < 1, namely.

$$v(t + n\Delta t) = A(n)A(n-1) \ldots A(2)A(1)v(t) \tag{15}$$

reduces the ℓ^1 norm of v by a factor of at most $k_.^{\ n}$.

Corollary 2. Thus, near the zero-infections steady state, absent introduction of infections from outside the network, the total number of infections decays at least as fast as $k_.^{\ n} = k_.^{\ t/\Delta t}$ (exponentially), that is, with time scale.

$$\tau \leq \Delta t \ln k \tag{16}$$

More generally we have the following.

Corollary 3. Repeated application of the update rule $v_{new} = Av$ $(A = I + M)$ to a non-negative vector v, represented by the product of a sequence of column-substochastic matrices which may vary in time, as in Eq. (15) above, reduces the ℓ^1 norm of v by a factor of at most $k_{g.m.}{}^n$ where $k_{g.m.}$ is the geometric mean of the corresponding maximum column sums.

Corollary 4. Thus, near the zero-infections steady state, provided that.

$$k_{g.m.} < 1 \tag{17}$$

and absent introduction of infections from outside the network, the total number of infections decays exponentially at least as fast as $k_{g.m.}{}^n = k_{g.m.}{}^{t/\Delta t}$, that is, with time scale

$$\tau \leq \Delta t \ln k_{g.m.} \tag{18}$$

Of course, Theorem 1 its corollaries approximate converses, based upon the following definition of column-superstochastic matrices. Since details are similar, here we simply state the converse and resulting time scales for growth without proof.

Definition 2. A non-negative matrix is column-superstochastic if its minimum column sum is greater than 1.

Theorem 2. Left multiplication by a column-superstochastic matrix with minimum column sum $\geq K$ increases the ^{1} norm of a non-negative vector by a factor $\geq K$.

Thus, repeated application of the update rule $\boldsymbol{v}_{\text{new}} = A\boldsymbol{v}$ to a non-negative vector $\boldsymbol{v}$, represented by the product of a sequence of column-substochastic matrices which may vary in time, and all with minimum row sum $\geq$ K > 1, as in Eq. (15) above, causes the number of infections to grow at least as fast as K^n (exponentially), that is with time scale

$$\tau \leq \Delta t \ln K \tag{19}$$

If we relax the criterion and consider only the geometric mean $K_{\text{g.m.}}$ of the corresponding maximum column sums, the number of infections grows at least as fast as $K_{\text{g.m.}}{}^n$ (exponentially), that is with time scale

$$\tau \leq \Delta t \ln K_{\text{g.m.}} \tag{20}$$

4 Discussion

In Sects. 2 and 3 we extended a linearized SIR model to a corresponding matrix network model, and extended the usual criterion $R_t < 1$ for infections to decline, ultimately stabilizing the zero infection steady state to a corresponding criterion for in the matrix/network model (Corollaries 1 and 2). Here we discuss that criterion and its consequences.

Recall the update rule (Eq. 12)

$$v_{\text{new}} = (I + M\Delta t)v = Av,$$

where $\boldsymbol{v}$ for the vector whose entries denote the number of infected individuals in nodes (cities, regions) in the network, the diagonal entries

$$M_{ii} = p_i c_i s_i - 1/D$$

represents SIR dynamics at node i, and the off-diagonal entries parametrize flows between regions. We developed a variety of stability criteria for driving the infections to 0, of which Corollary 1 is the simplest: the update matrices $A = I + M\Delta t$ are uniformly substochastic, with all column sums of all update matrices A uniformly bounded by a constant $k < 1$. This criterion combines the generation of new infections at each node with infection "exported" to other nodes: the following analog of R_t, namely 1 + the total number of infections generated throughout the network by a single infection at any node, less the recovery rate *1/D*, given by the corresponding column-sum of M, must be uniformly bounded by a constant $k < 1$. The bound on column sums can be considered as a granular version of a bound on R_t.

4.1 Consequences for Control and Network Design

We first discussed controlling the spread of coronavirus locally (at one node) by making $R_t = pcsD < 1$ at that node, through the combined effects of (a) Slowing the spread by reducing the probability that a contact results in an infection through masking and/or physical (misnamed social) distancing, (b) Reducing the effective contact rate between susceptible and infected individuals, for example by making some contacts virtual and/or by otherwise reducing the incidence of "superspreading"/ "spreading" events in a multi-networked city of the future, and (c) reducing the proportion of susceptibles by vaccination.

Our matrix/network models introduce more granularity: for stability an increasing rate of infections at a node, namely $p_j c_j s_j$, must be balanced by a similar reduction the rate of infections generated by infections at that node, the latter parametrized by the sum of off-diagonal elements in the corresponding column of M (or A since $A = M + I$), for example, through localizing or virtualizing interactions, restrictions on travel, or reduction of the size of events where interactions occur. Similar considerations apply in reducing the rate of growth of infections in cases where infections are growing (column sums of update matrices $A > 1$, analogs of $R_t > 1$) to reduce the strain on resources by "flattening the curve."

Moreover, the spread of Covid-19 seems to have the hallmarks of a cascading failure [10], here initially localized loci of infection which spread more rapidly than the system reacts to contain or mitigate the spread (described for example in [14]). Decentralized (modular) organization has been shown to promote survivability from "cascading failures in power grids." [16] discussed how decentralized (modular) organization and "reciprocal altruism" promote survivability from "cascading failures in power grids." The modular organization of the US (and perhaps any other) power grids, and lessons learned from cascading failure leading to the Northeast US blackout of 2003 [18] may provide ideas for a modular organization of networks in cities of the future, in which case signals provided by matrix models such as those discussed above (e.g., Eq. (9)) might provide fast enough warning signals to contain the spread of future epidemics from hot spots. For example, one might seek to avert the transition from substochastic dynamics (decay of infection) to superstochastic dynamics (growth of infection), both throughout the network, by introducing travel restrictions, reducing off-diagonal elements in the update matrix. See also the discussion of the "smart electric grid of the future" [2].

4.2 Limitations, Future Work and Conclusions

In the present paper have discussed only the simplest matrix/network extension of the classic SIR model, and looked only for qualitative correspondence to the data. In the future one should explore extensions to more general models and explorations of fitting data (and explanations for failure to fit the data, e.g., evolving social behavior, c.f. [9].

Moreover, one might explore what might be interesting non-linear extensions to the present discussion. The May-Wigner transition from instability to stability as a function of the strength of interactions in a network with linear interactions [1, 7, 8, 19] has also been observed in nonlinear dynamical systems [12, 17, 25]. Network structure also plays a role in stability, both small-world organization [27] and more complex network structures (c.f. [15]), and should be considered in understanding the role of social behavior and urban design in the spread of epidemics.

We also plan to extend our matrix/network model to a stochastic model to study the propagation and growth of random infections in a matrix/network model (c.f. [13]), and to develop a criterion for "circuit breakers" to limit cascading spread (see the discussion of the power grid, above). Both projects project may help in "flattening the curve" to help avoid overrunning the limits of medical resources.

5 Summary

In summary, we have developed a model for the spread of epidemics which over networks, which includes both local (within node/city/region) SIR dynamics and flows of infections between nodes. We have also provided a natural criterion for the zero infections steady state to be stable: expressed mathematically that all column sums are less than 1 (substochastic), and interpreted that one infection anywhere in the network yields less than one new infection (on average) over the entire network. This approach may be a small step towards models to help design "smart cities" of the future with an idea of preventing, containing, or at least mitigating future pandemics.

Acknowledgements We thank Jenny Magnes (Vassar College) for a helpful comments and edits on a prior version of this manuscript.

References

1. S. Allesina, S. Tang, The stability–complexity relationship at age 40: a random matrix perspective. Popul. Ecol. **57**(1), 63–75 (2015)
2. S.M. Amin, B.F. Wollenberg, Toward a smart grid: power delivery for the 21st century. IEEE Power Energ. Mag. **3**(5), 34–41 (2005)
3. E. Arcaute, E. Hatna, P. Ferguson, H. Youn, A. Johansson, M. Batty, Constructing cities, deconstructing scaling laws. J. Royal Soc. Interface **12**(102), 20140745 (2015)
4. L.M. Bettencourt, The origins of scaling in cities. Science **340**(6139), 1438–1441 (2013)
5. L.M. Bettencourt, J. Lobo, D. Helbing, C. Kühnert, G.B. West, Growth, innovation, scaling, and the pace of life in cities. Proc. Natl. Acad. Sci. **104**(17), 7301–7306 (2007)
6. K. Biswas, A. Khaleque, P. Sen, Covid-19 spread: Reproduction of data and prediction using a SIR model on Euclidean network (2020). arXiv preprint arXiv:2003.07063

7. J.E. Cohen, C.M. Newman, The stability of large random matrices and their products. Ann. Probabil. **12**(2), 283–310 (1984)
8. J.E. Cohen, C.M. Newman, When will a large complex system be stable? J. Theor. Biol. **113**, 153–156 (1985)
9. I. Cooper, A. Mondal, C.G. Antonopoulos, A SIR model assumption for the spread of COVID-19 in different communities. Chaos, Solitons Fractals **139**, 110057 (2020)
10. P. Crucitti, V. Latora, M. Marchiori, Model for cascading failures in complex networks. Phys. Rev. E **69**(4), 045104 (2004)
11. P. Figueiredo-Campos, B. Blankenhaus, C. Mota, A. Gomes, M. Serrano, S. Ariotti, ..., M. Veldhoen, Seroprevalence of anti-SARS-CoV-2 antibodies in COVID-19 patients and healthy volunteers up to six months post disease onset. Eur. J. Immunol. (2020). https://doi.org/https://doi.org/10.1002/eji.202048970
12. Y.V. Fyodorov, B.A. Khoruzhenko, Nonlinear analogue of the May−Wigner instability transition. Proc. Natl. Acad. Sci. **113**(25), 6827–6832 (2016)
13. H.M. Hastings, Stability of large systems. BioSystems **17**(2), 171–177 (1984)
14. H.M. Hastings, T. Young-Taft, C.-J. Tzen, Ecology, Economics, and Network Dynamics, Levy Economics Institute of Bard College, Annandale-on-Hudson NY, USA. Working paper 971, (2020)
15. S. Havlin, D.Y. Kenett, E. Ben-Jacob, A. Bunde, R. Cohen, H. Hermann, J.W. Kantelhardt, J. Kertész, S. Kirkpatrick, J. Kürths, J. Portugali, S. Solomon, Challenges in network science: Applications to infrastructures, climate, social systems and economics. Eur. Phys. J. Special Topics **214**(1), 273–293 (2012)
16. P. Hines, K. Balasubramaniam, E.C. Sanchez, Cascading failures in power grids. IEEE Potentials **28**(5), 24–30 (2009)
17. J.R. Ipsen, May-Wigner transition in large random dynamical systems. J. Stat. Mech: Theory Exp. **9**(093209), 2017 (2017)
18. E.J. Lerner, What's wrong with the electric grid? Physics Today. Available at: https://physicstoday.scitation.org/do/https://doi.org/10.1063/PT.5.5020/full/. 2013. Accessed 2020-November 4. (This article originally appeared in the October/November 2003 issue of The Industrial Physicist. The citations under "Further reading" have been updated.)
19. R.M. May, Will a large complex system be stable? Nature **238**(5364), 413–414 (1972)
20. R.M. May, R.M. Anderson, Transmission dynamics of HIV infection. Nature **326**(6109), 137–142 (1987)
21. M. Schläpfer, L.M. Bettencourt, S. Grauwin, M. Raschke, R. Claxton, Z. Smoreda, G.B. West, C. Ratti, The scaling of human interactions with city size. J. Royal Soc. Interface **11**(98), 20130789 (2014)
22. J. Seow, C. Graham, B. Merrick, S. Acors, S. Pickering, K.J. Steel,..., G. Betancor, Longitudinal observation and decline of neutralizing antibody responses in the three months following SARS-CoV-2 infection in humans. Nat. Microbiol. **2020** (2020). https://doi.org/10.1038/s41564-020-00813-8
23. D. Smith, L. Moore, The SIR model for spread of disease: the differential equation model. Loci. (originally Convergence.) (2004). https://www.maa.org/press/periodicals/loci/joma/the-sir-model-for-spread-of-disease-the-differential-equation-model. Accessed 2020-November 1
24. M.L. Smith, *Cities: The First 6,000 Years* (Penguin Books, London, 2020).
25. S. Sinha, S. Sinha, Evidence of universality for the May-Wigner stability theorem for random networks with local dynamics. Phys. Rev. E. **71**(2), 020902 (2005)
26. E. Stinson, Health and Disease Have Always Shaped Our Cities. What Will Be the Impact of COVID-19? Architectural Digest, https://www.architecturaldigest.com/story/how-will-coronavirus-impact-cities, 2020. Accessed 2020-November 2
27. D.J. Watts, S.H. Strogatz, Collective dynamics of 'small-world' networks. Nature **393**(6684), 440–442 (1998)
28. G.B. West, *Scale: The Universal Laws of Growth, Innovation, Sustainability, and the Pace of Life in Organisms, Cities, Economies, and Companies* (Penguin, New York, 2017).

Dynamics of the Charge Transfer Through a Memristor Between Two Initially Charged Cells

Aliyu Isah, A. S. Tchakoutio Nguetcho, Stéphane Binczak, and Jean-Marie Bilbault

Abstract In this paper, the behaviour of a memristor involved in CNN neighbourhood connection is investigated. The modified model of a charge-controlled memristor is considered, which gives us the desired continuity allowing to observe the behaviour of memristor for all initial conditions q_0. Detailed description of the system and analytical solution are provided.

Keywords Memristor · CNN · Neighborhood connection

1 Introduction

The discovery of the fourth basic circuit element—called *Memristor*—[1] unveiled numerous possibilities that are promising in overcoming the current challenges facing electronic designs and fabrication processes. The memory resistor or memristor in short, is postulated in 1971 but only realized in 2008 [2] as a two-terminal solid state device with its modeling description as:

$$\mathscr{V}(t) = M(x)\ \mathscr{I}(t), \tag{1a}$$

$$\frac{dx}{dt} = \kappa\ \mathscr{F}(x)\ \mathscr{I}(t), \tag{1b}$$

A. Isah (✉) · S. Binczak · J.-M. Bilbault
ImViA, Université de Bourgogne, BP 47870-21078, Dijon Cedex, France
e-mail: stbinc@u-bourgogne.fr

J.-M. Bilbault
e-mail: Jean-Marie.Bilbault@u-bourgogne.fr

A. Isah
Kano University of Science and Technology, BP 3244 Kano, Nigeria

A. S. Tchakoutio Nguetcho
LISSAS, Faculté des Sciences, Université de Maroua, BP 814 Maroua, Cameroon

C. H. Skiadas et al. (eds.), *13th Chaotic Modeling and Simulation International Conference*, Springer Proceedings in Complexity,
https://doi.org/10.1007/978-3-030-70795-8_23

$$M(x) = R_{off} - \delta R\, x\,; \quad x \in [0,\ 1]. \tag{1c}$$

Here: $\mathscr{V}(t)$ and $\mathscr{I}(t)$ are the voltage across and current through the device, respectively, $M(x)$ is the memristance as a function of the normalized state variable x, $\kappa = \frac{\mu_v R_{on}}{D^2}$ is a technological constant, $\mathscr{F}(x)$ is a window function for modeling the nonlinear dopant drift and $\delta R = R_{off} - R_{on}$.

The diversity of a memristor is a result of its nano-scaleability and memory capability. The most common reported and reliable application of a memristor is in the memory development [3, 4]. Moreover, other possible applications of a memristor include for example, but not only, cellular nonlinear networks CNNs [5, 6], neuromorphic computing [7], memristor as synapse using memristor bridge [8–10] and logic circuits [11, 12]. As memristor can be charge-controlled $\phi = \hat{\phi}(q)$ or flux-controlled $q = \hat{q}(\phi)$ depending on the nature of its input [13], a flux-controlled memristor is often used in the hitherto reported memristor based CNN applications. In this paper we consider rather a charge-controlled memristor in the CNN neighbourhood connection with the main focus on the memristor behaviour.

2 Memristor Models

In an ideal charge-controlled memristor, the state variable is a function of the charge flowing through it [3, 13, 14] and the instantaneous memristance $M(q)$ is determined by the charge as well. Hence, we have:

$$\text{Port equation: } \mathscr{V}(t) = M(q)\ \mathscr{I}(t) \text{ and state equation: } \frac{dq}{dt} = \mathscr{I}(t).$$

Setting $q_d = \dfrac{D^2}{\mu_v R_{on}}$ as the charge scaling factor [15, 16], then the normalized state variable can be expressed as $x = \dfrac{q}{q_d}$ and the equivalent expression of the memristance is defined from (1), as:

$$M(q) = \begin{cases} R_{off}, & \text{if } q \le 0 \\ R_{off} - \delta R\, \frac{q}{q_d}, & \text{if } 0 < q < q_d \\ R_{on}, & \text{if } q \ge q_d \end{cases} \tag{2}$$

Note that $q(t) = \displaystyle\int_{-\infty}^{t} i(t')\mathrm{d}t' = q_0 + \int_{0}^{t} i(t')\mathrm{d}t'$, where $q_0 = \displaystyle\int_{-\infty}^{0} i(t')\mathrm{d}t'$ is the initial charge reflecting the memory effect of the memristor [17]. The memristance function $M(q)$ of a charge–controlled memristor is analyzed with respect to the flowing charge through it. The memristance of a memristor varies between two extreme resistance limits—called the lower conducting region R_{off} and the higher conducting region R_{on}, thus the device exhibits bipolar resistance switching characteristics. Hence, the resistance does not change for $q \leq 0$ and $q \geq q_d$ and for a real memristor device this phenomenon is often modeled by using a window function imposing zero drift at the two extreme boundaries (i.e. $q = 0$ and $q = q_d$). In addition, it further affirms the passivity nature of memristance whereby:

$$P_M \geq 0 \;\;\Rightarrow\;\; M > 0,$$

where P_M is the power in the memristor whose memristance is M. However, (2) has discontinuities for the first derivative with respect to q at $q = 0$ and $q = q_d$ which cause angulation at these specific q values, see [18]. Furthermore, the study of memristance dynamics in the CNN neighborhood connection between pixels requires $M(q)$ and its derivative with respect to q to be continue at $q = 0$ or q_d in order to solve the system analytically. Hence, the idea is to modify the conventional TiO_2 memristor model given in (2) and to let it become suitable for all possible values of q.

As will be seen in Sect. 3, the memristance function must be continuous at $q = 0$ and $q = q_d$, this case is briefly described in [18]. However the detailed description is recalled in the following. The new model is deduced from a cubic polynomial function in q, having a continuous first derivative with respect to q, given by:

$$M(q) = a + b\,q + c\,q^2 + d\,q^3, \quad for\ q \in [0,\ q_d]. \tag{3}$$

Moreover, the derivative of $M(q)$ with respect to q i.e. $\frac{dM(q)}{dq}$ must be 0 for $q \leq 0$ and $q \geq q_d$. So, we get:

$$\frac{dM(q)}{dq} = b + 2c\,q + 3d\,q^2, \tag{4}$$

where a, b, c and d are constants determined by applying the conditions:

$$\begin{cases} for\ \ q < 0: & M(q) = a = R_{off}, \\ for\ \ q = 0: & \left.\frac{dM(q)}{dq}\right|_{q=0} = 0 \;\Rightarrow\; b = 0, \\ for\ \ q = q_d: & \left.\frac{dM(q)}{dq}\right|_{q=q_d} = 0 \text{ and } \left.M(q)\right|_{q=q_d} = R_{on}. \end{cases} \tag{5}$$

From (3)–(5), we get the constant coefficients as follows:

$$a = R_{off},\; b = 0,\; c = -\frac{3\,\delta R}{q_d^2} \text{ and } d = \frac{2\,\delta R}{q_d^3}$$

Hence, the model becomes:

$$M(q) = \begin{cases} R_{off}, & \text{if } q(t) \leq 0 \\ R_{off} - \frac{3\,\delta R}{q_d^2}\,q^2 + \frac{2\,\delta R}{q_d^3}\,q^3, & \text{if } 0 \leq q(t) \leq q_d \\ R_{on}, & \text{if } q(t) \geq q_d \end{cases} \tag{6}$$

with its charge derivative given by:

$$\frac{dM(q)}{dq} = \begin{cases} -\frac{6\ \delta R}{q_d^2} q + \frac{6\ \delta R}{q_d^3}\ q^2, & \text{if } 0 \leq q(t) \leq q_d \\ 0, & \text{if } q \leq 0 \ \text{ or } \ q \geq q_d \end{cases} \tag{7}$$

To determine the point of inflexion, the first and second derivatives of $M(q)$ with respect to q are obtained from (6), thus:

$$\frac{dM(q)}{dq} = 6\frac{\delta R}{q_d}\left(\frac{q^2}{q_d^2} - \frac{q}{q_d}\right) \text{ and } \frac{d^2M(q)}{dq^2} = 6\frac{\delta R}{q_d}\left(\frac{2q}{q_d^2} - \frac{1}{q_d}\right). \tag{8}$$

Therefore, one can see that $\dfrac{d^2M(q)}{dq^2} = 0$ if $q = \dfrac{q_d}{2}$, which is used in (8) to get:

$$\frac{dM(q)}{dq} = -\frac{3}{2}\frac{\delta R}{q_d},$$

and the corresponding memristance is obtained from (6), to be:

$$M(q)\Big|_{q=\frac{q_d}{2}} = \frac{1}{2}\big(R_{off} + R_{on}\big).$$

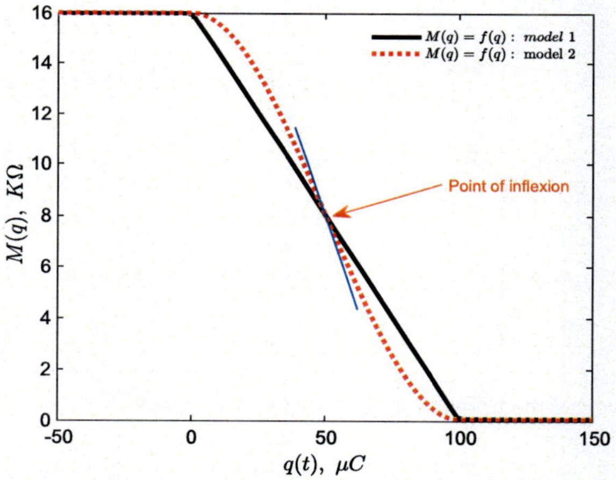

Fig. 1 Results comparison of memristances versus charge according to (2) and (6), as depicted by model 1 and 2 respectively. The parameter values used are [2]: $\mu_v = 10fm^2/V.s$, $D = 10nm$, $R_{off} = 16K\Omega$ and $R_{on} = 100\Omega$ which gives $q_d = 100\mu C$. The charge $q(t)$ may sometimes be negative or greater than q_d. Thus, $\frac{dM(q)}{dq}$ in (2) has discontinuity at $q = 0$ and $q = q_d$ as can be seen with angulation points in the black curve while $\frac{dM(q)}{dq}$ in (6) is continuous for all $q(t)$

The response of $M(q)$ according to (2) and (6) is given in Fig. 1. The red curve is for (6) which shows a rather better result due to its continuity for its first derivative at $q = 0$ and $q = q_d$. The system is normalized by calling $X = \dfrac{q(t)}{q_d}$ and $M = \delta R \mathscr{M}$. Therefore, (6) becomes:

$$\mathscr{M}(X) = \begin{cases} \frac{R_{off}}{\delta R}, & \text{if } X(t) \leq 0 \\ \frac{R_{off}}{\delta R} - 3\,X^2 + 2\,X^3, & \text{if } 0 \leq X(t) \leq 1 \\ \frac{R_{on}}{\delta R}, & \text{if } X(t) \geq 1 \end{cases} \tag{9}$$

3 Neighbourhood Connection with Memristor

The idea of considering a memristor in the CNN neighbourhood connections is reported [19, 20], where a flux-controlled model is used. Moreover, memristor-based CNNs have been proposed [5, 6], once again, using a flux-controlled memristor in the cell constitution. Cellular nonlinear network is implemented [21] where the cell composition is made of a linear capacitor and a nonlinear resistor R_{NL} (such as Fitzhugh-Nagumo), meanwhile the neighborhood connection is achieved using a linear resistor, see Fig. 2.

Memristor based CNNs have so many advantages which include fault tolerance [5], high pixel density and low power consumption due its nano nature in size. We consider a charge-controlled memristor in the neighbourhood connection (see Fig. 3),

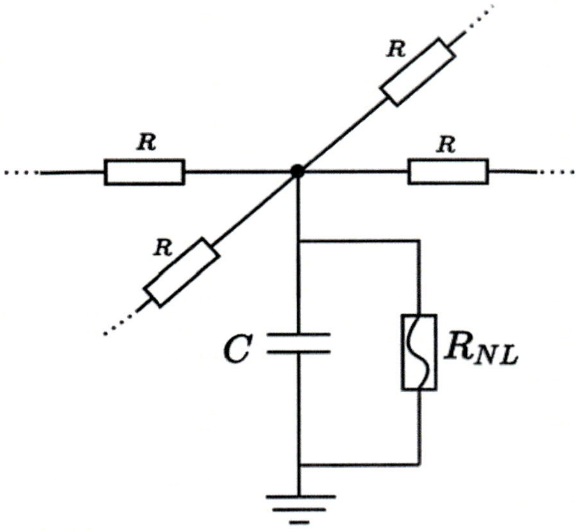

Fig. 2 2D CNN using resistor in the neighbourhood connection [21]

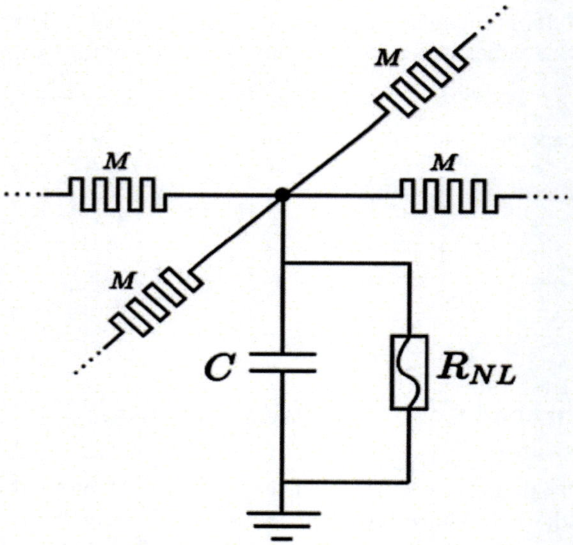

Fig. 3 2D CNN using memristor in the neighbourhood connection

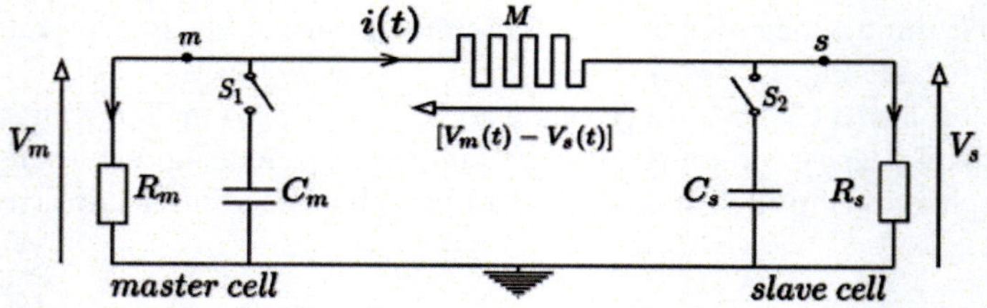

Fig. 4 Two cells system made up of master and slave cells coupled by a charge-controlled memristor. Initially, the cells are charged while the switches S_1 and S_2 are opened. However, by closing the switches, charges flow from one cell to the other according to the potential difference across them, until the system stabilizes when $V_m(t) = V_s(t) = \text{constant}$

simply as a replacement of the linear resistor in Fig. 2. Here we rather focused on the memristor behaviour. Thus, we consider two cells, namely: master and slave with setup as shown in Fig. 4. The method is accompanied by the study of memristor behaviour on the system response, such as the effect of initial charge q_0 on the system dynamics [17].

Closing the switches S_1 and S_2 simultaneously, we get the following Kirchhoff relationships:

$$i(t) = -C_m \frac{dV_m(t)}{dt} - \frac{V_m(t)}{R_m}, \tag{10}$$

$$i(t) = C_s \frac{dV_s(t)}{dt} + \frac{V_s(t)}{R_s}, \tag{11}$$

$$V_m(t) - V_s(t) = M(q)\,\dot{q}(t)\,, \tag{12}$$

while

$$i(t) = \dot{q}(t), \tag{13}$$

The time evolution of the voltage for each cell as well as the charge evolution can be obtained analytically from the system of equations (10)–(12). Therefore, the ***voltage evolution*** is obtained from (10)–(13), as follows:

$$\frac{\mathrm{d}}{\mathrm{d}t}(C_m V_m + C_s V_s) + \frac{C_m V_m}{R_m C_m} + \frac{C_s V_s}{R_s C_s} = 0. \tag{14}$$

Hence, we consider the case where the values of the parameters R_m, R_s, C_m and C_s would yield the time constant of the cells to be the same, so that $\tau_c = R_m C_m = R_s C_s$. Then (14) can be integrated and yield:

$$C_m V_m + C_s V_s = \lambda\, e^{-\frac{t}{\tau_c}}\,, \tag{15}$$

where $\lambda = C_m V_{m_0} + C_s V_{s_0}$ is a constant given by the initial conditions at time $t = 0$. Once again, (10) and (11) could be rewritten as:

$$R_m\, i(t) = -R_m C_m \frac{dV_m}{dt} - V_m, \tag{16}$$

$$R_s\, i(t) = R_s C_s \frac{dV_s}{dt} + V_s, \tag{17}$$

with $i(t) = \frac{dq}{dt} = \dot{q}(t)$, $V_m - V_s = M(q)\,\dot{q}$ given by (12) and letting $R_0 = R_m + R_s$, then from (16) and (17) we get:

$$\frac{dV_m}{dt} - \frac{dV_s}{dt} = -\frac{R_0}{\tau_c}\frac{dq}{dt} - \frac{1}{\tau_c}\left(M(q)\,\frac{dq}{dt}\right). \tag{18}$$

Equation (18) can be solved in 2 ways. Firstly, the equation can be solved directly, then taking into account (15), we find explicit expressions of $V_m(t)$ and $V_s(t)$ as function of the charge $q(t)$ flowing through the memristor. Secondly, (18) is solved using (12) for the study of charge transfer from one cell to the other through the memristor. With the same approach, and by taking into account the initial conditions V_{m_0}, V_{s_0} and q_0, and integrating (18), it gives:

$$V_m - V_s = (V_{m_0} - V_{s_0}) - \frac{R_0}{\tau_c}(q - q_0) - \frac{\eth}{\tau_c}, \tag{19}$$

where:

$$\eth = \int_{q_0}^{q} M(q')\,\mathrm{d}q'. \tag{20}$$

Solving (15) and (19) simultaneously, we get:

$$V_m = \frac{\lambda}{C_0}e^{-\frac{t}{\tau_c}} + \frac{C_s}{C_0}\left(V_{m_0} - V_{s_0}\right) - \frac{C_s R_0}{\tau_c C_0}(q - q_0) - \frac{C_s}{\tau_c C_0}\,\eth. \tag{21}$$

and

$$V_s = \frac{\lambda}{C_0}e^{-\frac{t}{\tau_c}} - \frac{C_m}{C_0}\left(V_{m_0} - V_{s_0}\right) + \frac{C_m R_0}{\tau_c C_0}(q - q_0) + \frac{C_m}{\tau_c C_0}\,\eth. \tag{22}$$

where $C_0 = C_m + C_s$.

To obtain the ***charge evolution***, (18) is rewritten as:

$$\frac{d}{dt}\left(V_m - V_s\right) = -\frac{R_0}{\tau_c}\frac{dq}{dt} - \frac{1}{\tau_c}\left(M(q)\,\frac{dq}{dt}\right),$$

and using (12), it becomes:

$$\left(R_0 + M(q)\right)\frac{dq}{dt} + \tau_c\,\frac{d}{dt}\left[M(q)\right]\frac{dq}{dt} + \tau_c\,M(q)\,\frac{d^2q}{dt^2} = 0.$$

With the identity: $\dfrac{dM(q)}{dt} = \dfrac{dM(q)}{dq} \times \dfrac{dq}{dt}$, then:

$$\left(R_0 + M(q)\right)\frac{dq}{dt} + \tau_c\,\frac{dM(q)}{dq}\left(\frac{dq}{dt}\right)^2 + \tau_c\,M(q)\,\frac{d^2q}{dt^2} = 0. \tag{23}$$

It is easier to handle normalized system. Therefore, (23) is normalized by considering $\tau = \dfrac{t}{\tau_c}$ while $Y = \dot{X} = \dfrac{dX}{d\tau}$ and $\dot{Y} = \ddot{X} = \dfrac{d^2X}{d\tau^2}$ are the first and second derivatives of X with respect to τ. Then (23) is to be studied in the phase plane (X,Y) using its normalized form:

$$\left(\frac{R_0}{\delta R} + \mathscr{M}\right)Y + \frac{d\mathscr{M}}{dX}Y^2 + \mathscr{M}\,\dot{Y} = 0, \tag{24}$$

where $\mathscr{M}$ is given in (9). Henceforth (9) and (24) are considered. For each case of $\mathscr{M}$ in (9), then (24) is solved to give the analytical expression required to obtain the phase portraits.

$If\;\; X \leq 0,\;\; \mathscr{M}(X) = \dfrac{R_{off}}{\delta R}$, then (24) becomes:

$$\begin{cases} \dot{Y} = -\dfrac{\gamma_1}{\gamma_2}Y, \\ \dot{X} = Y, \end{cases} \tag{25}$$

where $\gamma_1 = \dfrac{R_t}{2\delta R}$, $\gamma_2 = \dfrac{R_{off}}{2\delta R}$, $R_t = R_0 + R_{off}$ and $\dfrac{R_t}{R_{off}} = \dfrac{\gamma_1}{\gamma_2}$. Then (25) is solved to give:

$$H(X, Y) = \gamma_2 Y + \gamma_1 X = h, \tag{26}$$

where h is a constant corresponding to the conservative quantity $H(X, Y)$ determined by the system initial conditions: X_0, V_{m_0} and V_{s_0}, meanwhile Y_0 is obtained from (12), as:

$$Y_0 = \frac{\tau_c \left(V_{m_0} - V_{s_0}\right)}{q_d \cdot \delta R \cdot \mathscr{M}(X_0)}. \tag{27}$$

With $Y = \dfrac{dX}{d\tau}$, then (26) becomes:

$$X(\tau) = \left(X_0 - \frac{h}{\gamma_1}\right) e^{-\frac{\gamma_1}{\gamma_2}\tau} + \frac{h}{\gamma_1}. \tag{28}$$

where $X_0 = X(\tau = 0)$ and the normalized time for this case can be expressed from (28) as:

$$\tau = -\frac{\gamma_2}{\gamma_1} \ln \left(\frac{X(\tau) - \frac{h}{\gamma_1}}{X_{0_1} - \frac{h}{\gamma_1}} \right), \qquad \tau \in \big[-\infty,\ \tau_1 \big]. \tag{29}$$

where τ_1 is the time when $X(\tau) = 0$, and is given by:

$$\tau_1 = -\frac{\gamma_2}{\gamma_1} \ln \left(\frac{h}{h - \gamma_1 X_{0_1}} \right). \tag{30}$$

$If\ \ 0 \le X \le 1,\ \ \mathscr{M}(X) = \dfrac{R_{off}}{\delta R} - 3X^2 + 2X^3$, then (24) becomes:

$$\begin{cases} \dot{Y} = -\dfrac{\left(X^3 - \frac{3}{2}X^2 + \gamma_1\right) Y + 3(X^2 - X)Y^2}{X^3 - \frac{3}{2}X^2 + \gamma_2} \\ \dot{X} = Y \end{cases} \tag{31}$$

Thus, (31) becomes:

$$H(X, Y) = \left(X^3 - \frac{3}{2}X^2 + \gamma_2\right) Y + \frac{1}{4}X^4 - \frac{1}{2}X^3 + \gamma_1 X = h. \tag{32}$$

Similarly, with $Y = \dfrac{dX}{d\tau}$, then the analytical solution of (32) is given by:

$$\frac{P_3(X)}{P_4(X)} dX = -\frac{d\tau}{4} \quad \Rightarrow \tag{33}$$

$$\tau = \tau_1 - 4\Bigg[\ln\left[\left(\frac{X_1 - X}{X_1}\right)^{\alpha_1}\left(\frac{X_2 - X}{X_2}\right)^{\alpha_2}\right] + \ln\left[\frac{X^2 + \beta_1 X + \beta_2}{\beta_2}\right]^{\frac{\alpha_4}{2}}$$
$$+ \frac{2\alpha_3 - \alpha_4\beta_1}{\sqrt{4\beta_2 - \beta_1^2}}\left(\arctan\frac{2\left(X + \frac{\beta_1}{2}\right)}{\sqrt{4\beta_2 - \beta_1^2}} - \arctan\frac{\beta_1}{\sqrt{4\beta_2 - \beta_1^2}}\right)\Bigg], \quad (34)$$

where:

$$P_3(X) = X^3 - \frac{3}{2}X^2 + \gamma_2, \quad P_4(X) = X^4 - 2X^3 + 4\gamma_1 X - 4h,$$

$$\beta_1 = X_1 + X_2 - 2, \quad \beta_2 = \frac{-4h}{X_1 X_2},$$

$$\alpha_1 = \frac{X_1^3 - \frac{3}{2}X_1^2 + \gamma_2}{(X_1 - X_2)(X_1^2 + \beta_1 X_1 + \beta_2)},$$

$$\alpha_2 = \frac{X_2^3 - \frac{3}{2}X_2^2 + \gamma_2}{(X_2 - X_1)(X_2^2 + \beta_1 X_2 + \beta_2)},$$

$$\alpha_3 = \frac{\gamma_2 + \alpha_1\beta_2 X_2 + \alpha_2\beta_2 X_1}{X_1 X_2},$$

$$\alpha_4 = 1 - \alpha_1 - \alpha_2.$$

Meanwhile X_1 and X_2 are the two real roots of $P_4(X)$ given by

$$X_1 = \frac{1}{2}\left(1 + \sqrt{3 - U + 2\frac{(4\gamma_1 - 1)}{\sqrt{U}}} - \sqrt{U}\right),$$

$$X_2 = \frac{1}{2}\left(1 - \sqrt{3 - U + 2\frac{(4\gamma_1 - 1)}{\sqrt{U}}} - \sqrt{U}\right),$$

where $U = 1 + 2\sqrt[3]{\sqrt{\Delta} - Q} - 2\sqrt[3]{\sqrt{\Delta} + Q}$, $\Delta = Q^2 + \frac{8}{27}P^3$, $P = 2h - \gamma_1$ and $Q = h - \gamma_1^2$. The other two roots $P_4(X)$ are complex numbers. Recall that τ_1 is eventually the time where $X(\tau) = 0$, therefore at $X(\tau) = 1$, $\tau = \tau_2$ and is obtained from (34).

If $X \geq 1$, $\mathscr{M}(X) = \dfrac{R_{on}}{\delta R}$, then (24) becomes:

$$\begin{cases} \dot{Y} = -\dfrac{2\gamma_1 - 1}{2\gamma_2 - 1} Y \\ \dot{X} = Y \end{cases} \tag{35}$$

Then, (35) becomes:

$$H(X, Y) = \left(\gamma_2 - \frac{1}{2}\right) Y + \left(\gamma_1 - \frac{1}{2}\right) X + \frac{1}{4} = h. \tag{36}$$

With $Y = \dfrac{dX}{d\tau}$, it becomes:

$$X(\tau) = \frac{2h - \frac{1}{2}}{2\gamma_1 - 1} + \left(\frac{2\gamma_1 - 2h - \frac{1}{2}}{2\gamma_1 - 1}\right) e^{-\frac{2\gamma_1 - 1}{2\gamma_2 - 1}(\tau - \tau_2)}. \tag{37}$$

4 Discussion

The analytical solution gives the phase portraits characterizing the nature of the flowing charge through a memristor. An example of phase portraits for the described system is given in Fig. 5. The results are obtained for $R_{off} = 16K\Omega$, $R_{on} = 100\Omega$, $D = 10nm$ and $\mu_v = 10fm^2/V.s$. Then, it gives $q_d = 100\mu C$, $\gamma_1 = 6.792$ and $\gamma_2 = 0.503$. Each trajectory in Fig. 5 corresponds to a specific initial conditions set (X_0, V_{m_0}, V_{s_0}). The arrows point the direction of the system towards equilibrium and saturation. Therefore, for any given X_0, V_{m_0} and V_{s_0}, then Y_0 is to be obtained from (27). Hence, for $Y_0 > 0$ the system evolves from left to right and for $Y_0 < 0$ it goes from right to left as indicated by the arrow direction.

X_0 plays a significant role in the system dynamics because it determines the time evolution for any of the instances: $X \leq 0$, $0 \leq X \leq 1$ and $X \geq 1$. It is important to note that the state X_0 marks the history of the memristor. To capture all the possible occurrences of X_0, twelve possible scenarios are established [18] as enumerated underneath. Unfortunately, it is not possible for us to carry out the details analysis of each case due to length constraints. However, the description of each case is given in [18], as well as some comparison of system evolution with SPICE and numerical solution.

(1) $Y_0 > 0$:

A1. $X_0 \leq X < 0$.
A2. $X_0 < 0$ and $X(t \to \infty) < 1$.
A3. $X_0 < 0$ and $X(t \to \infty) > 1$.
A4. $0 < X_0 \leq X < 1$.
A5. $0 < X_0$ and $X(t \to \infty) > 1$.
A6. $1 < X_0 \leq X$.

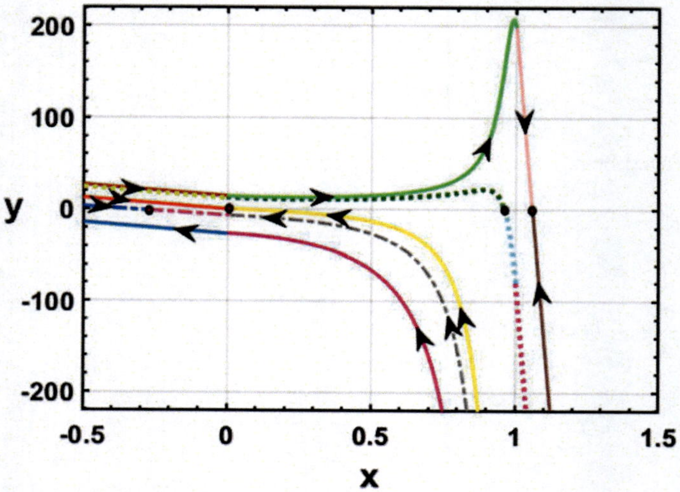

Fig. 5 Phase portraits showing the nature of charge transfer through a memristor for $Y_0 > 0$ (indicated by arrows from left to right) and $Y_0 < 0$ (indicated by arrows from right to left). Similar phase portraits are fully described in [18]

(2) $Y_0 < 0$:

B1. $1 < X \leq X_0$.
B2. $X_0 > 1$ and $0 < X(t \to \infty) < 1$.
B3. $X_0 > 1$ and $X(t \to \infty) < 0$.
B4. $0 < X \leq X_0 < 1$.
B5. $X_0 < 1$ and $X(t \to \infty) < 0$.
B6. $X \leq X_0 < 0$.

5 Conclusion

The nature of charge transfer through a charge-controlled memristor is investigated via a nonlinear equation characterizing the system dynamics. We provide a law for $M(q)$ which respects the continuity as required by the Kirchhoff's relationships. The new model avoids the use of a window function whose physical meaning is not intuitive. Additionally, we highlight the fundamental role of the memristor history, in showing that if all parameters are changed except q_0, the system behaviour is however quite different according to q_0.

Further study is ongoing to explore more contributions of memristor in cellular nonlinear networks reliable for image processing techniques.

References

1. L. Chua, Memristor-the missing circuit element. IEEE Trans. Circuit Theory **18**(5), 507–519 (1971)
2. D.B. Strukov, G.S. Snider, D.R. Stewart, R.S. Williams, The missing memristor found. Nature **453**(7191), 80–83 (2008)
3. L. Chua, Resistance switching memories are memristors. Appl. Phys. A **102**(4), 765–783 (2011)
4. M.J. Rozenberg, I.H. Inoue, M.J. Sanchez, Nonvolatile memory with multilevel switching: a basic model. Phys. Rev. Lett. **92**(17), 178302 (2004)
5. S. Duan, X. Hu, Z. Dong, L. Wang, P. Mazumder, Memristor-based cellular nonlinear/neural network: design, analysis, and applications. IEEE Trans. Neural Netw. Learn. Syst. **26**(6), 1202–1213 (2014)
6. M. Di Marco, M. Forti, L. Pancioni, Complete stability of feedback CNNs with dynamic memristors and second-order cells. Int. J. Circuit Theory Appl. **44**(11), 1959–1981 (2016)
7. B. Linares-Barranco, T. Serrano-Gotarredona, Memristance can explain spike-time-dependent-plasticity in neural synapses. Nat. Preced. 1–4 (2009)
8. S.H. Jo, T. Chang, I. Ebong, B.B. Bhadviya, P. Mazumder, W. Lu, Nanoscale memristor device as synapse in neuromorphic systems. Nano Lett. **10**(4), 1297–1301 (2010)
9. S.P. Adhikari, C. Yang, H. Kim, L.O. Chua, Memristor bridge synapse-based neural network and its learning. IEEE Trans. Neural Netw. Learn. Syst. **23**(9), 1426–1435 (2012)
10. H. Kim, M.P. Sah, C. Yang, T. Roska, L.O. Chua, Memristor bridge synapses. Proc. IEEE **100**(6), 2061–2070 (2011)
11. J. Borghetti, G.S. Snider, P.J. Kuekes, J.J. Yang, D.R. Stewart, R.S. Williams, 'Memristive' switches enable 'stateful' logic operations via material implication. Nature **464**(7290), 873–876 (2010)
12. P. Mazumder, S.M. Kang, R. Waser, Memristors: devices, models, and applications. Proc. IEEE **100**(6), 1911–1919 (2012)
13. L. Chua, Everything you wish to know about memristors but are afraid to ask. Radioengineering **24**(2), 319 (2015)
14. Y.V. Pershin, M. Di Ventra, A simple test for ideal memristors. J. Phys. D: Appl. Phys. **52**(1), 01LT01 (2018)
15. Y.N. Joglekar, S.J. Wolf, The elusive memristor: properties of basic electrical circuits. Eur. J. Phys. **30**(4), 661 (2009)
16. Z. Biolek, D. Biolek, V. Biolková, Computation of the area of memristor pinched hysteresis loop. IEEE Trans. Circuits Syst. II: Express Briefs **59**(9), 607–611 (2012)
17. A. Isah, A.T. Nguetcho, S. Binczak, J.M. Bilbault, Dynamics of a charge-controlled memristor in master-slave coupling. Electron. Lett. **4**(56), 168 (2020)
18. A. Isah, J.M. Bilbault, A.S.T. Nguetcho, S. Binczak, Memristor dynamics involved in cells communication for a 2D nonlinear network. IET Signal Process. (2020)
19. E. Lehtonen, M. Laiho, CNN using memristors for neighborhood connections, in *2010 12th International Workshop on Cellular Nanoscale Networks and their Applications (CNNA 2010)* (IEEE, 2010), pp. 1–4, Feb 2010

20. T. Prodromakis, C. Toumazou, A review on memristive devices and applications, in *2010 17th IEEE International Conference on Electronics, Circuits and Systems*, Athens, Greece, pp. 934–937, Dec 2010
21. J.C. Comte, P. Marquié, J.M. Bilbault, Contour detection based on nonlinear discrete diffusion in a cellular nonlinear network. Int. J. Bifurc. Chaos **11**(01), 179–183 (2001)

Effect of Synchronization on the Fractal Basin Boundary of the Duffing Oscillator

Aliasghar Janalizade and Mohammad R. Kolahchi

Abstract The forced damped Duffing equation is proven to have fractal basin boundaries given a particular set of parameters. We couple an oscillator in the regular region to one in the fractal region. This requires each equation to have its own parameters. We look for synchronization. For our particular coupling, the type of synchronization we find gives a particular relation in time between the positions, as well as the velocities. For the synchronized motion we find that the basin boundary could be smooth or fractal. As the coupling becomes stronger, the basin boundary becomes less fractal. Considering the coupling as part of the perturbation to the homoclinic orbits, we test the Melnikov criterion for having fractal basin boundaries, and find good agreement.

Keywords Synchronization · Fractal basin boundary · Homoclinic · Melnikov

1 Introduction

A chaotic attractor (cf. strange attractor) is formed when the dynamics is sensitive to the initial conditions (Grebogi et al. [1]). It was discovered that in driven dissipative systems, it is possible to have fractal basin boundaries so that sensitivity to the initial conditions remains, even when the dynamics is not chaotic (Moon and Li [2]). When the boundary is fractal, and not smooth, the sensitivity comes as the result of the uncertainty in the distance to the boundary (McDonald et al. [3]). The study of basins of attraction and the boundaries defining their territory have brought interesting results, the fractal character is one of them (Nusse and Yorke [4]).

A. Janalizade · M. R. Kolahchi (✉)
Department of Physics, Institute for Advanced Studies in Basic Sciences,
45137-66731 Zanjan, Iran
e-mail: kolahchi@iasbs.ac.ir

A. Janalizade
e-mail: ajanalizadeh@iasbs.ac.ir

C. H. Skiadas et al. (eds.), *13th Chaotic Modeling and Simulation International Conference*, Springer Proceedings in Complexity,
https://doi.org/10.1007/978-3-030-70795-8_24

The idea pursued by [2] was that the fractal basin boundaries develop, at least for the model of their study, en route to chaos. Their two-well potential was in fact what is present in the Duffing equation. They found in their numerical experiments that as the fractal basin boundary comes about, the stable manifold grows a finger that touches the unstable manifold. This is what the study of homoclinic orbits and homoclinic points entail, and has led to a theory for chaos (Smale [5]). The presence of transverse homoclinic points is needed for chaos, according to this theory. There is a method due to Melnikov that shows the existence of such points by defining a distance function between the stable manifold and the unstable function (Melnikov [6]). It is based on the perturbations of the homoclinic orbit (Brunsden and Holmes [7]). The result of Melnikov is used as a bound for the presence of fractal boundaries.

The notion of fractal boundaries is also present in conservative scattering systems, and undriven (i.e. autonomous) dissipative systems. In the latter case, the basin boundaries are as usual, and formed since the trajectory can converge to one of many fixed points. It is shown that in the necessarily transient dynamics, the basin boundaries can have several complex properties, among them being fractal (Motter et al. [8]). In the former Hamiltonian case with the total energy being conserved, attractors and their basin boundaries do not exist. For Hamiltonian systems instead, exit basins and their boundaries are defined. For two or more escapes fractal boundaries are possible (Bleher et al. [9]). For particular Hamiltonians, the boundary can have intricate properties, among them being fractal (Aguirre et al. [10]).

Here, we study the character of the fractal basin boundaries from a different standpoint–synchronization. Studying a phenomenon using synchronization as a novel tool is not new (Botha and Kolahchi [11]). In this study, the dynamics is governed by the Duffing equation. We know the parameters that result in fractal basin boundaries. We choose such parameters, and take an oscillator in the fractal basin boundary. This means that the initial conditions for the trajectory belong to the fractal space. For the set of parameters giving smooth boundaries, we take the same initial conditions. We study the possibility of the two trajectories becoming synchronized. For this we need to have a coupling between the two equations. We study the character of the synchronous basin boundary phase space as the strength of the coupling is varied. We also check the onset of the fractal basin boundaries with Melnikov's prediction, after calculating it for our particular coupling as part of the perturbation.

2 The Duffing Equation

The Duffing oscillator is given by (1),

$$\ddot{x} + \delta\dot{x} + \frac{1}{2}(-x + x^3) = g\ \cos(\omega t). \tag{1}$$

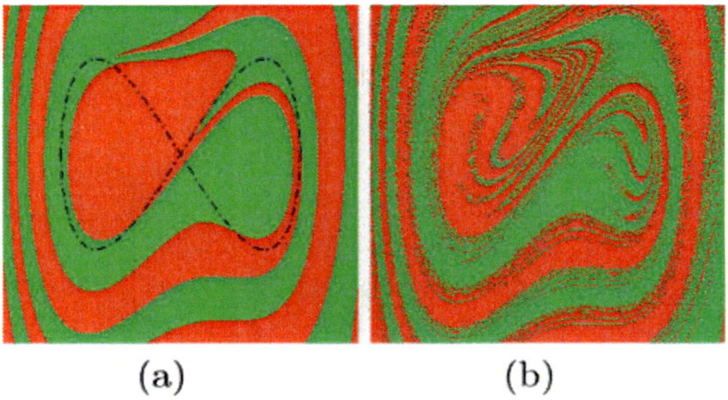

Fig. 1 The phase space according to (2) and (3) of the Duffing equation with **a** smooth and **b** fractal basin boundary with dimension 1.68

Here, we have a one dimensional oscillator with position x, under external force with amplitude g, and experiencing dissipation with strength δ [2]. The Hamiltonian is has a two-well potential, one with minimum at $x = 1$ (the right well), and the other with minimum at $x = -1$ (the left well). There is a peak separating them placed at $x = 0$. The separatrix for the two wells defines the homoclinic orbit too. There is a saddle point at the origin, and the homoclinic orbit connects the saddle to itself. The external force and dissipation disrupt the Hamiltonian. Equation (1) can also be written in this way,

$$\dot{x} = y, \tag{2}$$

$$\dot{y} + \delta y + \frac{1}{2}(-x + x^3) = g\ \cos(\omega t). \tag{3}$$

We can now think of (x, y) as the two dimensional phase space, and define the initial condition in terms of it.

In this velocity-position space, we find the locus of points that starting from them, are attracted to the right well, and so define the basin of attraction for the right potential well. Similarly, one can define the basin of attraction for the left potential well. We see this in Fig. 1a, green region attracted to the right well, and red to the left. The basin boundary is clearly seen as the boundary of the two regions. The unperturbed homoclinic orbit is the dotted line in Fig. 1a.

3 The Fractal Basin Boundary

For the dynamics given by (1), or equivalently by (2) and (3), there is a proof showing the presence of transverse homoclinic orbits, and hence chaos [7]. It is also possible to change the parameters of the dynamics so that the boundary is not smooth as in Fig. 1a, but is fractal as in Fig. 1b. A fractal is distinguished by its self-similarity under different scales. Now, to imagine a boundary with this property is a bit questioning.

Inside the fractal boundary which is not a simple smooth contour, and is spread on an area, we have a red region next to any green region, and vice versa. This already suggests sensitivity to the initial conditions. In a sense, the border separating the red basin from the green basin is not so clear. We can first determine the fractal dimension of the boundary, $1 < d < 2$. This may be given by the correlation exponent of Grassberger and Procaccia [12]. Then we ask how well we can know our distance to the boundary, ϵ. The question then becomes, what volume of space will this take up so that the number of initial conditions in this volume will result in uncertain solutions, and this is given by,

$$f \sim \epsilon^{D-d}. \tag{4}$$

For the smooth boundary on the plane, we have $d = 1$, and $D = 2$, so that the volume of initial conditions goes as ϵ, but for the fractal boundary with $D - d < 1$ this can become much larger–much higher sensitivity arises (Iansiti et al. [13]).

The fractal basin boundary brings sensitivity to the dynamics with no chaotic attractor present. We can ask about the synchronization properties of such basins of attraction, and study the phase space from this vantage point.

4 Synchronization

Systems with the notion of synchronization are either under the spell of each other, or an external agent. We study two Duffing oscillators which interact with each other as if a spring were connecting them. The important point is that one of the oscillators comes from the fractal basin boundary, and so has its parameters taken from the particular dynamics leading to it, as we have explained. The question we ask is whether the two oscillators can synchronize, and we take their motion in synch if they move keeping a constant relation with each other. The simplest such relation is if they move together. The constant relation means, in general, that if we define an order parameter for the interacting system, this parameter will have a given time dependence which repeats in time. Such cases are known, but for different systems, and under different conditions (Choi et al. [14]).

The two oscillators are given as,

$$\ddot{x}_1 + \delta_1 \dot{x}_1 + \frac{1}{2}(-x_1 + x_1^3) = g_1 \cos(\omega t) + k(x_2 - x_1), \tag{5}$$

$$\ddot{x}_2 + \delta_2 \dot{x}_2 + \frac{1}{2}(-x_2 + x_2^3) = g_2 \cos(\omega t) + k(x_1 - x_2). \tag{6}$$

The new parameter is k denoting the strength of coupling. The basins of attraction for the synchronized state is what we need to find, and this requires us to start both oscillators from the same initial conditions for position and velocity. That is, we need to know whether one point in the phase space leads to a synchronized state, and if so sweep the phase space in a similar fashion and find the basins of attraction for the two oscillators moving in synch. The basin boundary will eventually tell us whether sensitivity to initial state can exist when the two oscillators are synchronized. The coupling strength is the parameter to consider first, and indeed we find that both cases can happen, both smooth basin boundary as well as fractal basin boundary. The question of the condition to have synchronous behavior, and reasons for the details of the synchronous state, we shall take up elsewhere.

5 Results and Discussion

We have not been interested in ω, it seems. The Duffing equation is a crude approximation to the Josephson equation, which naturally brings the frequency of the driving force into play. We can then talk about the synchronizing effects between the external periodic force, and the internal periodic force; i.e. the $\sin(x)$ now approximated by its first two Fourier components. If we couple the junctions, as in an array, then the synchronizing effect can become much larger, resulting in constant voltage steps much larger than in case of a single Josephson junction. Such steps are called Shapiro steps (Benz et al. [15]).

Now, for our two coupled oscillators, the linear coupling is in fact adjusting the Duffing internal potential. For a single oscillator we know that increasing the amplitude of the force relative to friction can extend the fractal basin boundary, and bring its dimension d closer to an integer, here for the planar dimension we have $D = 2$. The uncertainty becomes much larger. We can compare Fig. 2a where the boundary is smooth, and Fig. 2b with larger amplitude and fractal basin boundary. It looks as if the fractal boundary had occupied the whole phase space. The amplitude of the applied force is much larger than the strength of dissipation, and only very close to the point attractors do we find areas where this fractal nature is not present. So the question is how the coupling can affect this response.

The oscillator from the phase space with no fractal boundary can synchronize with one from the phase space with fractal boundary. This may not be surprising. But can we move in the phase space and keep this character of synchrony with sensitivity to the initial condition; that is, keep the fractal boundary alive? The answer is positive, but it depends on the relative strength of the various interactions. In Fig. 2c, the coupling strength k is small enough to keep the fractal structure despite interaction with the oscillator from the smooth basin boundary phase space. If the coupling is stronger, the boundary becomes smooth; this we see in Fig. 2d, all other parameters are kept the same.

Before turning to the Melnikov method for a quantitative estimate, it helps to have an argument that emphasizes the role of synchronization. When synchronized, the two motions are correlated. To see this correlation in another way, we again look at Fig. 2. In Fig. 2a we have no coupling, and the strength of force is so that no fractal boundaries develop. When the coupling is strong enough, the correlation between the two oscillators in synch, Fig. 2d, has turned the basins with fractal boundaries, Fig.

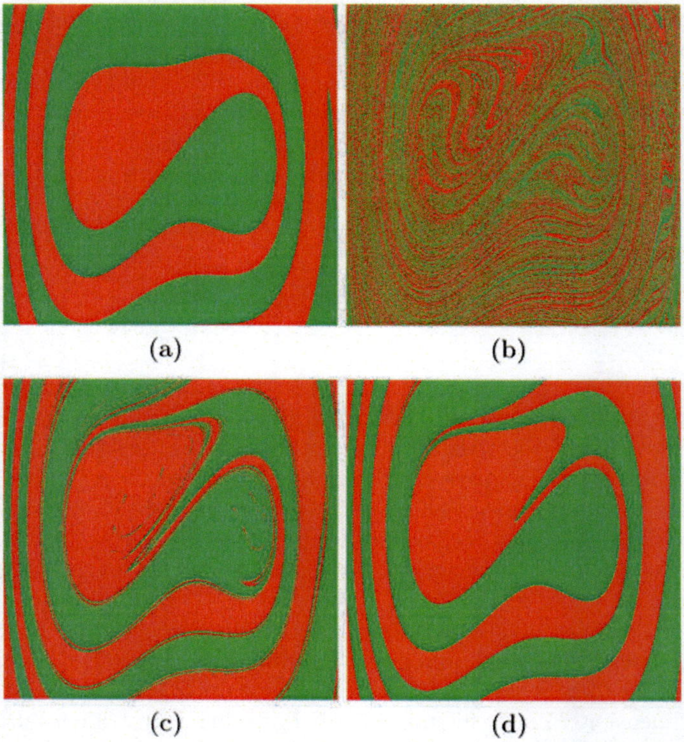

Fig. 2 In **a** with $g = 0.05$ and **b** with $g = 0.12$ and with dimension 1.75, there is no interaction; that is, the phase space shows the basin boundaries with $k = 0$. We then choose an oscillator from each with the same initial conditions, and after synchronizing them, we find in **c** with $k = 0.4$ fractal basin boundaries with dimension 1.65, and in **d** with $k = 0.9$ smooth boundaries. The phase space is defined as in Fig. 1

2b, into a map very similar to Fig. 2a. Now, if the coupling is made weaker, Fig. 2c, we see some of the fractal boundaries to have survived. This has happened very close to the single motion boundaries, where the uncertainty f is larger by comparison. So the correlation remains, but the influence of the oscillators on each other has switched roles. So overall, the uncertainty has lowered, or even disappeared, when a correlated motion has emerged.

Now, in the Melnikov method, the idea is to have an estimate for the distance between the stable manifold, and the unstable manifold. This is after the perturbation has disjointed the homoclinic orbit. So we need to add our interaction to the rest of the terms that have perturbed the Hamiltonian. Equation (7) gives the homoclinic orbits for the two oscillators, and the distance function comes as (8) (Moon [6]).

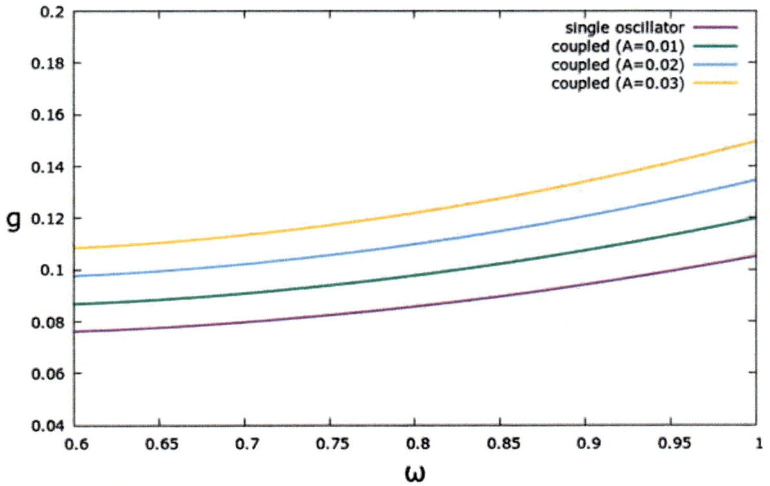

Fig. 3 Melnikov integral results considered with $\delta = 0.15$ for single oscillator, and coupled system with different values of A in (9)

$$\begin{cases} x = \pm\sqrt{2}\mathrm{sech}(t/2) \quad , \quad dx = \mp\frac{\sqrt{2}}{2}\mathrm{sech}(t/2)\mathrm{tanh}(t/2), \\ x' = \pm\sqrt{2}\mathrm{sech}(\frac{t-t'}{2}) \quad , \quad dx' = \mp\frac{\sqrt{2}}{2}\mathrm{sech}(\frac{t-t'}{2})\mathrm{tanh}(\frac{t-t'}{2}). \end{cases} \tag{7}$$

$$\begin{aligned} M(\lambda, t') = &-\frac{\sqrt{2}}{2}g\int_{-\infty}^{\infty}\mathrm{sech}(t/2)\mathrm{tanh}(t/2)\cos(\omega(\frac{t-\lambda}{2}))dt \\ &-\frac{1}{2}\delta\int_{-\infty}^{\infty}\mathrm{sech}^2(t/2)\mathrm{tanh}^2(t/2)dt \\ &-k\int_{-\infty}^{\infty}\mathrm{sech}(t/2)\mathrm{tanh}(t/2)(\mathrm{sech}(\frac{t-t'}{2})-\mathrm{sech}(t/2))dt. \end{aligned} \tag{8}$$

The last term in (8) gives the dependence on coupling k, and is denoted below as A. Setting the distance function equal to zero gives a lower bound as a necessary condition to have chaos, and it is also used as a bound for having fractal boundaries [2].

$$g > \frac{\sqrt{2}(\frac{1}{3}\delta + A)\cosh(\frac{\pi\omega}{\sqrt{2}})}{\pi\omega}. \tag{9}$$

We can plot (9) for different values of A; i.e., effectively for different couplings. (In finding the lower bound, λ and t' are determined.)

In Fig. 3 we have the Melnikov function for various couplings. For a given ω, it shows that as the coupling increases the threshold for fractal boundaries increases. This means that we need to start from a more fractal basin boundary if any of it is to survive after the coupling. As the amplitude g increases, the fractal dimension also increases, giving us a chance to have a higher coupling and still keep some of the fractal character for the basin boundaries of the synchronized system. In this sense, we find good agreement with the Melnikov analysis.

Finally, we give an enlarged view of Fig. 2c that shows the fractal basin boundaries for the coupled oscillators (Fig. 4).

6 Conclusions

For non-chaotic dynamics, usually when we look at the phase space for the basins of attraction, it clearly shows that an oscillator living in a given region is attracted to a given attractor, providing a way of labeling that oscillator. An oscillator living in

Fig. 4 Figure 2c, only enlarged

the fractal basin boundary of the phase space is not easy to label. This is because of the sensitivity to the placement of the oscillator, as it has become difficult to define the place of the border itself. So we have sensitivity to initial conditions even with no chaotic attractors. Here, we consider the coupling of two oscillators, one coming from the fractal basin boundary, the other coming from the motion with no fractal basin boundary. In other words, we have two Duffing equations, one with parameters resulting in fractal basin boundaries, the other not. The question is, what effect can synchronizing the two have on the basin boundaries? We find that the correlated motion brought about by synchronization can have a strong influence on the basin boundaries. When the coupling is weak enough, the fractal basin boundaries can still exist, but to a much less extent. This means the sensitivity has lowered to a great deal. It is also possible to have no fractal basin boundary, if the coupling is strong, so that the correlated motion is dictated by the oscillator from the dynamics with smooth boundaries.

Acknowledgements This paper was presented at the (virtual) Chaos2020 Conference in Florence, Italy. We thank Yury Shukrinov for helping us to participate in this virtual conference. We also thank André Botha for suggestions on the manuscript.

References

1. C. Grebogi, E. Ott, A. Yorke, Chaos, strange attractors, and fractal basin boundaries in nonlinear dynamics. Science **238**(4827), 632–638 (1987)
2. F.C. Moon, G.-X. Li, Fractal basin boundaries and homoclinic orbits for periodic motion in a two-well potential. Phys. Rev. Lett. **55**(14), 1439–1442 (1985)
3. S.W. McDonald, C. Grebogi, E. Ott, J.A. Yorke, Fractal basin boundaries. Physica **17D**, 125–153 (1985)
4. H.E. Nusse, J.A. Yorke, Basins of attraction. Science **271**, 1376–1380 (1996)
5. S. Smale, Bull. Am. Math. Soc. **73**(6), 747–817 (1967). S.H. Strogatz, *Nonlinear Dynamics and Chaos* (Addison-Wesley, 1995)
6. V.K. Melnikov, Trans. Mosc. Math. Soc. **12**, 1–57 (1963). F.C. Moon, *Chaotic and Fractal Dynamics* (Wiley-VCH, 2004)
7. V. Brunsden, P. Holmes, Power spectra of strange attractors near homoclinic orbits. Phys. Rev. Lett. **58**(17), 1699–1702 (1987)
8. A.E. Motter, M. Gruiz, G. Károlyi, T. Tél, Doubly transient chaos: generic form of chaos in autonomous dissipative systems. Phys. Rev. Lett. **111**, 194101 (2013)
9. S. Bleher, C. Grebogi, E. Ott, R. Brown, Fractal boundaries for exit in Hamiltonian dynamics. Phys. Rev. A **38**, 930 (1988)
10. J. Aguirre, J.C. Vallejo, Miguel A.F. Sanjuán, Wada basins and chaotic invariant sets in the Hénon-Heiles system. Phys. Rev. E **64**, 066208 (2001)
11. A.E. Botha, M.R. Kolahchi, Analysis of chimera states as drive-response systems. Sci. Rep. **8**, 1830 (2018)
12. P. Grassberger, I. Procaccia, Characterization of strange attractors. Phys. Rev. Lett. **50**, 346–349 (1983)
13. M. Iansiti, Qing Hu, R.M. Westervelt, M. Tinkham, Noise and chaos in a fractal basin boundary regime of a Josephson junction. Phys. Rev. Lett. **55**(7), 746–749 (1985)

14. M.Y. Choi, Y.W. Kim, D.C. Hong, Periodic synchronization in a driven system of coupled oscillators. Phys. Rev. E **49**(5), 3825–3832 (1994)
15. S.P. Benz, M.S. Rzchowski, M. Tinkham, C.J. Lobb, Fractional giant Shapiro steps and spatially correlated phase motion in 2D Josephson arrays. Phys. Rev. Lett. **64**(6), 693–696 (1990)

Anomalous Scaling in the Kinematic MHD Turbulence Under the Influence of Helicity

Eva Jurčišinová, Marian Jurčišin, and Richard Remecky

Abstract In the framework of the field theoretic renormalization group approach and the operator product expansion technique, the influence of the helicity (the spatial parity violation) on the anomalous dimensions of the leading composite operators, which drive the anomalous scaling of correlation functions of the magnetic field deep inside the inertial range, is investigated in the kinematic magnetohydrodynamic turbulence, i.e., in the model in which the Lorentz term is omitted in the stochastic Navier-Stokes equation and the magnetic field behaves as a passive vector quantity. It is shown that there is a quantitative difference between the role of the helicity in the problem of the passive magnetic (vector) field in the kinematic magnetohydrodynamic turbulence and the analogous problem of a passive vector advection in the Kazantsev-Kraichnan model with prescribed statistics of the velocity field. The analysis shows that under the influence of helicity the anomalous scaling is more pronounced, i.e. the anomalous dimensions are more negative, in the model of the kinematic magnetohydrodynamic turbulence driven by the stochastic Navier-Stokes equation.

Keywords Turbulence · Field theoretic renormalization group · Kinematic MHD · Helicity · Two-loop approximation

E. Jurčišinová · M. Jurčišin · R. Remecky (✉)
Institute of Experimental Physics, Slovak Academy of Sciences, Watsonova 47, 040 01 Košice, Slovakia
e-mail: remecky@saske.sk

E. Jurčišinová
e-mail: jurcisine@saske.sk

M. Jurčišin
e-mail: jurcisin@saske.sk

R. Remecky
Bogoliubov Laboratory of Theoretical Physics, Joint Institute for Nuclear Research, 141 980 Dubna, Moscow Region, Russian Federation

C. H. Skiadas et al. (eds.), *13th Chaotic Modeling and Simulation International Conference*, Springer Proceedings in Complexity,
https://doi.org/10.1007/978-3-030-70795-8_25

1 Introduction

One of the main and still open questions in the theory of fully developed turbulence is the problem of a fundamental understanding of the existence of deviations from the scaling predictions of the classical phenomenological Kolmogorov-Obukhov (KO) theory (see, e.g., Kolmogorov [1], Monin and Yaglom [2], McComb [3], Frisch [4], Sreenivasan and Antonia [5], Falkovich et al. [6], or Antonov [7]) in the framework of well-defined microscopic models. In this respect, it seems that very suitable models for such investigation are various microscopic models of the passive advection of some scalar or vector quantities, e.g. temperature, concentration of an impurity, magnetic field, etc., in turbulent environments which are modeled by random velocity fields defined in a proper mathematical way.

According to KO theory the statistical properties of random fields deep inside inertial interval are independent of the outer scale L (a typical scale at which the energy pumping into the dissipative system is continuously performed to maintain the steady state) as well as the inner scale l (a typical scale at which the strong energy dissipation starts). The assumption of the validity of these hypotheses (known as the first and the second Kolmogorov hypothesis) together with dimensional analysis leads to the scaling behavior of various correlation functions with well-defined scaling exponents. Note also that deep inside the inertial interval $l \ll r \ll L$ the system is fully random with very high values of the Reynolds number (in principle Re $\to \infty$) and some symmetries of the flow, lost during the transition from laminar to turbulent flow, are restored in the statistical way (Frisch [4]).

As an example, consider the following experimentally measured single-time two-point structure functions of the velocity field

$$S_N(r) = \langle [v_r(t,\mathbf{x}) - v_r(t,\mathbf{x}')]^N \rangle, \qquad r = |\mathbf{x} - \mathbf{x}'|, \tag{1}$$

where v_r denotes the component of the velocity field directed along the vector $\mathbf{r} = \mathbf{x} - \mathbf{x}'$. Following the hypotheses of the KO theory, the dimensional analysis leads to the scale invariant representation of these structure functions in the following simple form

$$S_N(r) = C \times (\bar{\epsilon} r)^{N/3}, \tag{2}$$

where $\bar{\epsilon}$ is the mean dissipation rate and C is a constant.

On the other hand, it is well known that both natural experiments as well as numerical simulations show the existence of some deviations from the simple scaling behavior presented in (2). The existence of deviations from the scaling behavior (2) is related to the nontrivial dependence of the structure functions (1) on the integral scale L, in contradiction with the first Kolmogorov hypothesis. As a result, the scaling behavior of the structure functions in the inertial range must be modified into the following form (see, e.g., Falkovich et al. [6] and Antonov [7] and references cited therein)

$$S_N(r) = (\bar{\epsilon} r)^{N/3} R_N(r/L), \tag{3}$$

with some unknown scaling functions $R_N(r/L)$. It is standardly assumed that the scaling functions $R_N(r/L)$ have power like asymptotic behavior in the region $r \ll L$ in the form

$$R_N(r/L) \sim (r/L)^{q_n}. \tag{4}$$

An effective technique to study such a problem is to use the so-called operator product expansion (OPE), also known as the short distance expansion, in the framework of the field theoretic renormalization group approach (see, e.g., Antonov [7], Zinn-Justin [8] and Vasil'ev [9], Adzhemyan et al. [10]). Application of the OPE technique leads to the following powerlike representation of the scaling functions (4)

$$R_N(r/L) = \sum_F C_F (r/L)^{\Delta_F}, \tag{5}$$

where the summation is implied over some class of composite operators F, Δ_F are their critical dimensions and C_F are some coefficients regular in L^{-1}. The singular dependence of the structure functions on L in the limit $L \to \infty$ together with nonlinearity of the exponents q_n as functions on N is called the anomalous scaling and it is evident that in the framework of the OPE technique the anomalous scaling behavior of the correlation functions must be related to the existence of the so-called "dangerous" composite operators in the OPE with negative critical dimensions since their presence leads to the singular behavior of the correlation functions in the limit $L \to \infty$. At the same time, the final asymptotic anomalous behavior is then determined by the most negative critical dimensions.

In this respect, the main aim of the present paper is to start the systematic investigation of the influence of the spatial parity violation (helicity) on the anomalous scaling behavior of the magnetic field correlation functions in the framework of the kinematic magnetohydrodymanic (MHD) turbulence, namely, to find the corresponding critical exponents that drive the scaling behavior of the single-time two-point correlation functions of a weak magnetic field (see, e.g., Antonov et al. [11], Antonov et al. [12], Hnatich et al. [13], Jurčišinová et al. [14], Antonov and Gulitskiy [15], Jurčišinová and Jurčišin [16], Jurčišinová and Jurčišin [17], Jurčišinová and Jurčišin [18], Jurčišinová et al. [19], or Jurčišinová et al. [20], as well as references cited therein).

2 The Kinematic MHD Turbulence

The advection of a passive (weak) magnetic field $\mathbf{b} \equiv \mathbf{b}(x)$, where $x \equiv (t, \mathbf{x})$, in the framework of the kinematic MHD turbulence is described by the following system of two stochastic equations for the fluctuating parts of the magnetic field and the velocity field $\mathbf{v} \equiv \mathbf{v}(x)$, respectively:

$$\partial_t \mathbf{b} + (\mathbf{v}\cdot\partial)\mathbf{b} = \nu_0 u_0 \Delta \mathbf{b} + (\mathbf{b}\cdot\partial)\mathbf{v} + \mathbf{f}^{\mathbf{b}}, \tag{6}$$

$$\partial_t \mathbf{v} + (\mathbf{v}\cdot\partial)\mathbf{v} = \nu_0 \Delta \mathbf{v} - \partial \mathcal{P} + \mathbf{f}^{\mathbf{v}}, \tag{7}$$

where $\partial_t \equiv \partial/\partial_t, \partial \equiv \partial/\partial x_i$, and $\Delta \equiv \partial^2$ is the Laplace operator. We follow the standard notation where the subscript 0 always denotes bare parameters of the unrenormalized theory with ν_0 as the bare viscosity coefficient, $\nu_0 u_0 = c^2/(4\pi\sigma)$ represents the magnetic diffusivity, c is the speed of light, σ is the conductivity, and $\mathcal{P}$ is the pressure. Due to the assumption of incompressibility both fields $\mathbf{b}$ and $\mathbf{v}$ are supposed to be divergence free vector fields, i.e., $\partial\cdot\mathbf{b} = 0$ and $\partial\cdot\mathbf{v} = 0$, respectively.

The last terms $\mathbf{f}^{\mathbf{b}}$ and $\mathbf{f}^{\mathbf{v}}$ in the system of equations (6)–(7) represent random forces of the stochastic model. The random force of the magnetic field $\mathbf{f}^{\mathbf{b}}$ represents the source of the magnetic field fluctuations to maintain the steady state of the system. On the other hand, the random force of the velocity field $\mathbf{f}^{\mathbf{v}}$ simulates the energy pumping into the system on large scales. We assume the Gaussian statistics with zero mean for both random forces with correlation functions in the following form

$$\mathcal{D}^b_{ij}(x;x') \equiv \langle f^b_i(x) f^b_j(x')\rangle = \delta(t-t') C_{ij}(|\mathbf{x}-\mathbf{x}'|/L), \tag{8}$$

$$\mathcal{D}^v_{ij}(x;x') \equiv \langle f^v_i(x) f^v_j(x')\rangle = \delta(t-t') \int \frac{\mathrm{d}^d\mathbf{k}}{(2\pi)^d} D_0 k^{4-d-2\varepsilon} R_{ij}(\mathbf{k}) e^{i\mathbf{k}\cdot(\mathbf{x}-\mathbf{x}')}, \tag{9}$$

where C_{ij} in (8) is a tensor function finite in the limit $L \to \infty$ and, in what follows, its detailed form is not essential while it satisfies the only condition that C_{ij} decreases rapidly for $|\mathbf{x}| \gg L$. On the other hand, the correlation function (9) of the random force $\mathbf{f}^{\mathbf{v}}$ is taken in the specific powerlike form suitable for the field theoretic renormalization group analysis. In (9), $\mathbf{k}$ is the wave vector, d is the spatial dimension, the positive amplitude D_0 is taken in the form $D_0 \equiv g_0 \nu_0^3 > 0$, where g_0 represents the bare coupling constant of the model, and the physical value of the formally small parameter $0 < \epsilon \leq 2$, which control the powerlike form of the energy pumping into the system, is $\epsilon = 2$.

The geometric properties of the energy pumping to the system is described by the form of the transverse (due to the assumption of incompressibility) tensor projector $R_{ij}(k)$ in the correlator (9) and, in our case with the assumption of the presence of the spatial parity violation, it contains two terms. The first term $P_{ij}(k) = \delta_{ij} - k_i k_j/k^2$ is the standard isotropic transverse projector and the second part, taken in the form $\imath\rho\epsilon_{ijl}\frac{k_l}{|k|}$, describes the presence of the helicity, where ϵ_{ijl} is the Levi-Civita symbol of rank 3 and parameter $0 \leq |\rho| \leq 1$ controls the amount of the spatial parity violation in the system. Thus, in what follows, the projector $R_{ij}(k)$ has the following explicit form

$$R_{ij}(k) = \delta_{ij} - k_i k_j/k^2 + \imath\rho\epsilon_{ijl}\frac{k_l}{|k|}. \tag{10}$$

3 Field Theoretic Formulation of the Model

Using the well-known theorem (Martin et al. [21]) the stochastic problem given by (6) and (7) with correlators of the random forces given in (8) and (9) can be transformed into the corresponding field theoretic model with double set of fields $\Phi = \{\mathbf{v}, \mathbf{b}, \mathbf{v}', \mathbf{b}'\}$ and with the action functional of the following form

$$\begin{aligned} \mathcal{S}(\Phi) =& \frac{1}{2}\int \mathrm{d}t_1 \mathrm{d}^d\mathbf{x_1}\mathrm{d}t_2\mathrm{d}^d\mathbf{x_2}\Big[v_i'(x_1)\mathcal{D}_{ij}^v(x_1,x_2)v_j'(x_2)+ \\ &+ b_i'(x_1)\mathcal{D}_{ij}^b(x_1,x_2)b_j'(x_2)\Big]+ \\ &+\int \mathrm{d}t\mathrm{d}^d\mathbf{x}\Big\{\mathbf{v}'[-\partial_t\mathbf{v}+\nu_0\Delta\mathbf{v}-(\mathbf{v}\cdot\nabla)\mathbf{v}]+ \\ &+\mathbf{b}'[-\partial_t\mathbf{b}+\nu_0 u_0\Delta\mathbf{b}+(\mathbf{b}\cdot\nabla)\mathbf{v}-(\mathbf{v}\cdot\nabla)\mathbf{b}]\Big\}. \end{aligned} \tag{11}$$

where $x_i = (t_i, \mathbf{x}_i)$ with $i = 1, 2$, $\mathbf{v}'$ and $\mathbf{b}'$ are auxiliary transverse fields and D_{ij}^b, D_{ij}^v are given in (8) and (9), respectively. The corresponding summations over dummy indices are performed.

The field theoretic model given by the action functional (11) corresponds to a standard Feynman diagrammatic technique with nonzero bare propagators for the magnetic field (in the momentum-frequency representation)

$$\Delta_{ij}^{b'b*} = \Delta_{ij}^{bb'} = \frac{P_{ij(\mathbf{k})}}{(-\imath\omega_k+\nu_0u_0k^2)}, \tag{12}$$

$$\Delta_{ij}^{bb} = \frac{C_{ij}(\mathbf{k})}{|-\imath\omega_k+\nu_0u_0k^2|^2} \tag{13}$$

and for the velocity field

$$\Delta_{ij}^{v'v*} = \Delta_{ij}^{vv'} = \frac{R_{ij}(\mathbf{k})}{(-\imath\omega_k+\nu_0k^2)}, \tag{14}$$

$$\Delta_{ij}^{vv} = \frac{g_0\nu_0^3k^{4-d-2\varepsilon}R_{ij}(\mathbf{k})}{|-\imath\omega_k+\nu_0k^2|^2}, \tag{15}$$

given by the quadratic part of the action functional (11), where $C_{ij}(\mathbf{k})$ is the Fourier transform of the function $C_{ij}(|\mathbf{x}-\mathbf{x}'|/L)$ in (8).

On the other hand, the model has two interaction vertices in the form

$$-\mathbf{b}'[(\mathbf{b}\cdot\partial)\mathbf{v}-(\mathbf{v}\cdot\partial)\mathbf{b}] = b_i'\mathcal{V}_{ijl}v_jb_l, \tag{16}$$

$$-\mathbf{v}'(\mathbf{v}\cdot\partial)\mathbf{v} = v_i'\mathcal{W}_{ijl}v_jv_l, \tag{17}$$

where tensor structures $\mathcal{W}_{ijl}$, $\mathcal{V}_{ijl}$ are given as follows (again in the momentum-frequency representation)

$$\mathcal{V}_{ijl} = \imath(\delta_{il} p_j - \delta_{ij} p_l), \tag{18}$$
$$\mathcal{W}_{ijl} = \imath(\delta_{ij} p_l + \delta_{il} p_j). \tag{19}$$

The graphical representation of all propagators and vertices of the model can be found, e.g., in [22].

The field theoretic renormalization group analysis of the model defined by the action functional (11) was performed in detail in [23]. There, the coordinates of the stable infrared (IR) fixed point of the model, which drives the scaling behavior of the model deep inside the inertial range, was found and discussed at the two-loop level of approximation. These results will be used in the next sections for the analysis of the influence of the helicity on the anomalous dimensions of the leading composite operators that drive the anomalous scaling of the magnetic field correlation functions in the inertial range.

4 Anomalous Dimensions of the Leading Composite Operators and the Anomalous Scaling

In the kinematic MHD turbulence phenomenologically interesting is the investigation of the inertial range behavior of the following single-time two-point correlation functions of the magnetic field

$$B_{N-m,m}(r) \equiv \left\langle b_r^{N-m}(t, \mathbf{x}) b_r^m(t, \mathbf{x}') \right\rangle, \quad r = |\mathbf{x} - \mathbf{x}'|, \tag{20}$$

built of two composite operators $b_r^{N-m}(t, \mathbf{x})$ and $b_r^m(t, \mathbf{x})$, where b_r denotes the component of the magnetic field directed along the vector $\mathbf{r} = \mathbf{x} - \mathbf{x}'$. After applying a general scaling representation for the two-point single-time quantities (see, e.g., Jurčišinová et al. [14] and references cited therein) together with the OPE technique one comes to the following final asymptotic inertial-range expression for the correlation functions (20):

$$B_{N-m,m}(r) \sim r^{\gamma^*_{N,0} - \gamma^*_{N-m,0} - \gamma^*_{m,0}}, \tag{21}$$

for even values of N and m,

$$B_{N-m,m}(r) \sim r^{\gamma^*_{N,0} - \gamma^*_{N-m,1} - \gamma^*_{m,1}}, \tag{22}$$

for even value of N and odd value of m, and

$$B_{N-m,m}(r) \sim r^{\gamma^*_{N,1} - \gamma^*_{N-m,0} - \gamma^*_{m,1}}, \tag{23}$$

for odd values of N and m, where $\gamma^*_{N,p}$ are the fixed point values of the anomalous dimensions of the corresponding composite operators

$$F_{N,p} = [\mathbf{n} \cdot \mathbf{b}]^p (\mathbf{b} \cdot \mathbf{b})^l, \quad N = 2l + p, \tag{24}$$

constructed solely from the magnetic field $\mathbf{b}(x)$, which give the most singular contributions in the OPE. Note also that the constant unit vector $\mathbf{n}$ defines the uniaxial anisotropy of the system (see, e.g., Antonov et al. [11] or Jurčišinová et al. [14] for details) represented, e.g., by a constant large-scale external magnetic field $\mathbf{B} = |\mathbf{B}|\mathbf{n}$ and introduced through the explicit form of the correlator (8) for the random force $\mathbf{f}^{\mathbf{b}}$.

Thus, to be able to study the influence of the helicity on the anomalous scaling of the correlation functions of the magnetic field, it is necessary to calculate the anomalous dimensions of the composite operators (24). Here, it is important to stress that, due to tensor structure of the corresponding single Feynman diagram (Jurčišinová et al. [14]), the anomalous dimensions $\gamma_{N,p}$ of the composite operators (24) are independent of the helicity at the one-loop level of approximation. It means that, to be able to study the influence of spatial parity violation on the scaling properties of the correlation functions (20), it is necessary to perform, at least, two-loop calculations. Namely, this is the aim of the present study.

In the two-loop level of approximation, the anomalous dimensions $\gamma_{N,p}$ can be written in the following general form

$$\gamma_{N,p} = \gamma_{N,p}^{(1)} \varepsilon + \gamma_{N,p}^{(2)} \varepsilon^2 + O(\varepsilon^3), \tag{25}$$

where $\gamma_{N,p}^{(1)}$ is the one-loop contribution to the anomalous dimension $\gamma_{N,p}$, which is independent of the helicity parameter ρ and $\gamma_{N,p}^{(2)} = \gamma_{N,p}^{(2)}(\rho)$ is the corresponding helicity dependent two-loop correction.

The simple one-loop contribution $\gamma_{N,p}^{(1)}$ was calculated, e.g., in Jurčišinová et al. [14] and has the following explicit form taken at the fixed point of the model

$$\gamma_{N,p}^{*(1)} = -\frac{(d+1)(N-p)(d+N+p-2) - 2N(N-1)}{3(d-1)(d+2)}. \tag{26}$$

It is important to stress here that this one-loop result is completely the same not only for the helical and non-helical cases of the model but is also the same as that obtained in the case of the analogous passive scalar problem [24], although the corresponding leading composite operators are different. It means that the problems of the anomalous scaling of passively advected scalar and vector (magnetic) field are completely equivalent at the one loop level of approximation even in the presence of the spatial parity violation of the turbulent Navier-Stokes velocity field. However, as our analysis shows, the situation is significantly different when the two-loop corrections are taken into account. As was shown in Gladyshev et al. [25], the anomalous dimensions of the leading composite operators, which drive the scaling behavior of the correlation functions of the scalar field passively advected by the helical stochastic Navier-Stokes equation, are independent of the helicity even at the two-loop level of approximation. On the other hand, as was already mentioned, the two-loop corrections to the anomalous dimensions of the leading composite operators $\gamma_{N,p}^{(2)}$ in the studied vector problem are helicity dependent. We shall not present their explicit form here since their general form is given by huge expression (even when taken at the fixed point) and therefore will be discussed elsewhere. Instead, the influence of the helicity on the fixed point values of the most important anomalous dimensions $\gamma^*_{N,0}$ for even values of N and $\gamma^*_{N,1}$ for odd values of N, which directly determine the asymptotic behavior of the correlation functions (20), is demonstrated in Figs. 1 and 2, where the explicit dependence of the total two-loop anomalous dimensions $\gamma^*_{2,0}$, $\gamma^*_{3,1}$, $\gamma^*_{4,0}$, and $\gamma^*_{5,1}$ on the helicity parameter ρ is shown for the spatial dimension $d = 3$ and for physically the most important value $\varepsilon = 2$. Of course, one can also expect serious influence of the helicity on the scaling behavior of the correlation functions (20). However, this problem will be analyzed elsewhere.

As follows from Figs. 1 and 2, the presence of the spatial parity violation in the turbulent environment significantly decreases the anomalous dimensions of the leading composite operators that drive the inertial range asymptotic behavior of the magnetic field correlation functions. This result is in qualitative accordance with that obtained in the framework of the helical Kazantsev-Kraichnan model of the kinematic MHD turbulence with the Gaussian statistics of the velocity field (see Jurčišinová and Jurčišin [18] and Jurčišinová et al. [19]), where it was shown that the presence of the helicity also leads to the more pronounced anomalous scaling of the passively advected magnetic field. However, the quantitative comparison of the two-loop results for the anomalous dimensions of the leading composite operators obtained in the present study to those obtained in the framework of the Kazantsev-

Fig. 1 The dependance of anomalous dimensions $\gamma^*_{2,0}$ and $\gamma^*_{3,1}$ on the parameter of helicity ρ for the spatial dimension $d = 3$ and $\varepsilon = 2$

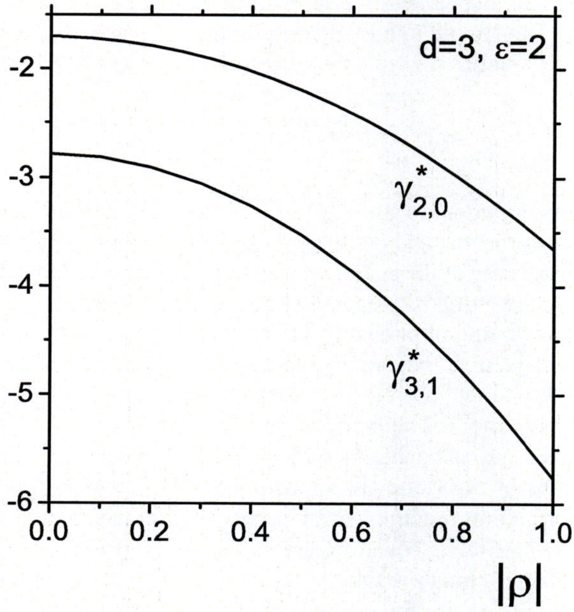

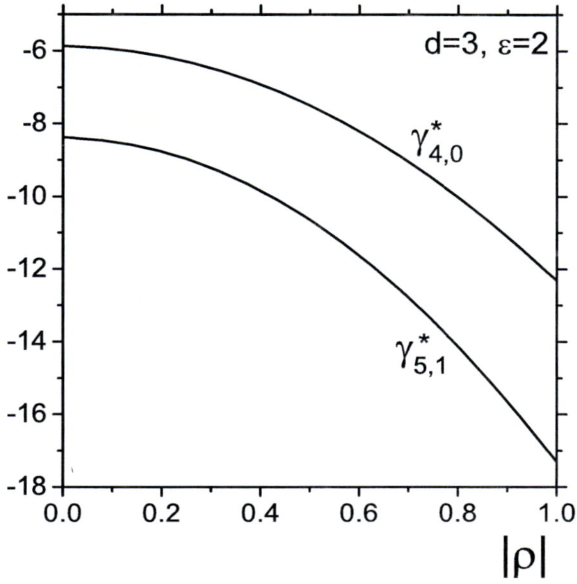

Fig. 2 The dependance of anomalous dimensions $\gamma^*_{4,0}$ and $\gamma^*_{5,1}$ on the parameter of helicity ρ for the spatial dimension $d = 3$ and $\varepsilon = 2$

Kraichnan model shows that stronger dependence on the helicity parameter ρ is observed in the genuine kinematic MHD turbulence investigated in the present paper. Again, a detailed full comparison of these two models will be given elsewhere.

5 Conclusion

Using the field theoretic renormalization group approach we have investigated the influence of the spatial parity violation (helicity) on the anomalous dimensions of the leading composite operators that drive the asymptotic inertial range scaling behavior of the correlation functions of the magnetic field in the model of fully developed kinematic MHD turbulence driven by the stochastic Navier-Stokes equation.

The influence of the helicity was investigated to the second order of the corresponding perturbation theory (the two-loop approximation) using the standard Feynman diagrammatic technique for the spatial dimension $d = 3$. The dependence of some anomalous dimensions on the parameter of helicity is demonstrated in Figs. 1 and 2, respectively.

The analysis shows stronger dependence of the total two-loop anomalous dimensions on the parameter of the helicity in the present kinematic MHD model in comparison to the analogous dependence obtained in the framework of the Kazantsev-Kraichnan model (see Jurčišinová and Jurčišin [18] and Jurčišinová et al. [19]), where the statistics of the velocity field is given directly by a specific Gaussian correlation function.

Acknowledgements The authors gratefully acknowledge the hospitality of the Bogoliubov Laboratory of Theoretical Physics of the Joint Institute for Nuclear Research, Dubna, Russian Federation. The work was supported by the VEGA grants No. 2/0065/17 and No. 2/0058/19 of Slovak Academy of Sciences and by the grant APVV-17-0020.

References

1. A.N. Kolmogorov, Dokl. Akad. Nauk SSSR, **30** 301 (1941). (AN Kolmogorov, Dokl. Akad. Nauk SSSR **31** 538 (1941), A.N. Kolmogorov, Dokl. Akad. Nauk SSSR **32** 16 (1941))
2. A.S. Monin, A.M. Yaglom, *Statistical Fluid Mechanics: Mechanics of Turbulence*, vol. 2 (MIT Press, Cambridge, 1975)
3. W.D. McComb, *The Physics of Fluid Turbulence* (Clarendon, Oxford, 1990)
4. U. Frisch, *Turbulence: The Legacy of A. N. Kolmogorov* (Cambridge University Press, Cambridge, 1995)
5. K.R. Sreenivasan, R.A. Antonia, Ann. Rev. Fluid Mech. **29**, 435 (1997)
6. G. Falkovich, K. Gawedzki, M. Vergassola, Rev. Mod. Phys. **73**, 913 (2001)
7. N.V. Antonov, J. Phys. A Math. Theor. **39**, 7825 (2006)
8. J. Zinn-Justin, *Quantum Field Theory and Critical Phenomena* (Clarendon, Oxford, 1989)
9. A.N. Vasil'ev, *Quantum-Field Renormalization Group in the Theory of Critical Phenomena and Stochastic Dynamics* (Chapman and Hall, Boca Raton, FL, 2004)
10. L.D. Adzhemyan, N.V. Antonov, A.N. Vasil'e, Phys. Rev. E **58**, 1823 (1998)
11. N.V. Antonov, A. Lanotte, A. Mazzino, Phys. Rev. E **61**, 6586 (2000)
12. N.V. Antonov, M. Hnatich, J. Honkonen, M. Jurčišin, Phys. Rev. E **68**, 046306 (2003)
13. M. Hnatich, J. Honkonen, M. Jurčišin, A. Mazzino, S. Šprinc, Phys. Rev. E **71**, 066312 (2005)
14. E. Jurčišinová, M. Jurčišin, R. Remecký, J. Phys. A Math. Theor. **42**, 275501 (2009)
15. N.V. Antonov, N.M. Gulitskiy, Phys. Rev. E **85**, 065301 (2012)
16. E. Jurčišinová, M. Jurčišin, J. Phys. A Math. Theor. **45**, 485501 (2012)
17. E. Jurčišinová, M. Jurčišin, Phys. Rev. E **88**, 011004 (2013)
18. E. Jurčišinová, M. Jurčišin, Phys. Rev. E **91**, 063009 (2015)
19. E. Jurčišinová, M. Jurčišin, M. Menkyna, Phys. Rev. E **95**, 053210 (2017)
20. E. Jurčišinová, M. Jurčišin, M. Menkyna, Eur. Phys. J. B **91**, 313 (2018)
21. P.C. Martin, E.D. Siggia, H.A. Rose, Phys. Rev. A **8**, 423 (1973)
22. E. Jurčišinová, M. Jurčišin, R. Remecký, Phys. Rev. E **84**, 046311 (2011)
23. E. Jurčišinová, M. Jurčišin, R. Remecký, P. Zalom, Phys. Rev. E **87**, 043010 (2013)
24. L.Ts. Adzhemyan, N.V. Antonov, J. Honkonen, T.L. Kim, Phys. Rev. E **71** 016303 (2005)
25. A.V. Gladyshev, E. Jurčišinová, M. Jurčišin, R. Remecký, Phys. Part. Nucl. **41**, 1023 (2010)

Pattern Formation of Limit Cycles for 2-D Generalized Logistic Maps

Shunji Kawamoto

Abstract The pattern formation of discrete limit cycles with chaotic dynamics is considered, for two-dimensional (2-D) generalized logistic maps. Firstly, the time-dependent chaos function with an amplitude function is proposed, and a 1-D generalized logistic map is presented as a population growth model. In particular, the number of newly infected people, due to the COVID-19, is modeled by the 1-D logistic map. Secondly, 2-D generalized logistic maps with a system parameter and amplitude functions are derived by extending the 1-D logistic map, and stable limit cycles with entrainment and synchronization are numerically calculated. Finally, the pattern formation of limit cycles is discussed for the 2-D generalized logistic map, which is restricted by the system parameter and amplitude functions, as one of non-equilibrium open systems.

Keywords Logistic map · 2-D logistic map · Time-dependent chaos function · Population growth · Limit cycle · Pattern formation · Non-equilibrium open system

1 Introduction

Over a long period of time, nonlinear dynamic phenomena, such as soliton, chaos and fractals, have been considered in the field of physics, chemistry, biology and engineering, and the papers and the books have appeared, in order to descrive the nonlinear science [1]. In the meantime, the nonlinear systems have been widely extended to medicine, optics, living systems, life science, neuro science and nonlinear demography [2–7].

In particular, one-dimensional (1-D) nonlinear difference equations have been shown to possess a rich spectrum of dynamical behaviors as chaos in many respects [8, 9], and strange attractor, limit cycle, entrainment and synchronization have been discussed, as chaotic dynamics [10–13]. At the same time, the limit cycle in 2-D space has been proposed for predator–prey populations in communities and in the field of

S. Kawamoto (✉)
Osaka Prefecture University, Sakai, Osaka, Japan
e-mail: skawamoto@sky.plala.or.jp

C. H. Skiadas et al. (eds.), *13th Chaotic Modeling and Simulation International Conference*, Springer Proceedings in Complexity,
https://doi.org/10.1007/978-3-030-70795-8_26

theoretical biology [14], and is reported to play a key role for chaotic dynamics of non-equilibrium open systems [15]. In addition, stable limit sycles have been observed in nonlinear oscillations [16], such as heart beating [17], daily rhythms in human body temperature [3], chemical reactions [18–20] and self-excited mechanical vibration [21].

Furthermore, the pattern formation in reaction–diffusion model has been studied [22], and the chaotic behaviors in predator–prey systems are considered with pattern formation conditions based on the bifurcation analysis [23].

During this decade, 1-D, 2-D and 3-D time-dependent solvable chaos maps and a nonlinear expansion method have been presented from the standpoint of chaos functions [24, 25]. After that, the 2-D maps corresponding to the FitzHugh-Nagumo (FHN) model, the Belousov-Zhabotinsky (BZ) reaction and reaction–diffusion systems are derived, and the bifurcation diagrams have been considered to descrive the chimera states and the generation of discrete limit cycles for population growth, neural cells and chemical cells [26–29]. Recently, a limit cycle analysis and the interaction of limit cycles for 2-D maps have been presented, as non-equilibrium open systems [30, 31].

The aim of this paper is to introduce amplitude functions and to consider the effect to the pattern formation of stable limit cycles with chaotic dynamics, for 2-D generalized logistic maps. In Sect. 2, the time-dependent chaos function with an amplitude function is proposed, and a 1-D generalized logistic map is presented as a population growth model. In particular, the number of newly infected people, due to the COVID-19, is modeled by the 1-D logistic map, and the MATLAB program is shown in Appendix. In Sect. 3, 2-D generalized logistic maps are derived from the 1-D generalized logistic map, and stable limit cycles with entrainment and synchronization are illustrated. Moreover in Sect. 3, the effect of the system parameter and the amplitude functions to the pattern formation of stable limit cycles is numerically considered, for the 2-D logistic map as one of non-equilibrium open systems. Conclusions are summarized in the last Section.

2 1-D Generalized Logistic Maps

From the following time-dependent chaos function;

$$x_n(t) = a(t)\sin^2(2^n t), \tag{1}$$

$$t \neq \pm m\pi/2^l \tag{2}$$

with an amplitude function *a(t)* and finite positive integers $\{l,\ m\}$, we find a one-dimensional (1-D) solvable chaos map;

$$x_{n+1}(t_{i+1}) = 4a(t_i)\sin^2(2^n t_i)\cos^2(2^n t_i)$$

$$= 4x_n(t_i)\left(1 - \frac{1}{a(t_i)}x_n(t_i)\right), \tag{3}$$

where the discrete time t_i satisfies the condition (2), and the time step Δt is given by $\Delta t \equiv t_{i+1} - t_i = \pi/431 \approx 0.0073$, in order to avoid the accumulation of round-off error and the periodicity caused by the iteration [26].

Then, the 1-D map (3) can be rewritten into a 1-D generalized time-dependent logistic map as

$$x_{n+1}(t_{i+1}) = \alpha x_n(t_i)\left(1 - \frac{1}{a(t_i)}x_n(t_i)\right) \tag{4}$$

with a system parameter α, $0 < \alpha \leq 4.0$ and the amplitude function $a(t_i) \neq 0$ of the chaos function (1).

As is known, the logistic map $x_{n+1} = \alpha x_n\ (1 - x_n)$ is a mapping with chaotic dynamics, and has been well discussed as a population growth model [32, 33], and in part as a discrete-time demographic model [34]. However, the logistic map has the pathological problem as a demographic model, and the Ricker model $x_{t+1} = x_t \exp[r(1 - x_t/k)]$ with two parameters $\{r, k\}$ is proposed for a number of fish and invertebrate populations, and has been considered for a population growth regulated by an epidemic disease [35, 36].

Recently, it is known that the COVID-19 caused by severe acute syndrome coronavirus has resulted in a pandemic, and the government of each country has adopted various measures to mitigate the outbreak [37]. In the following part of this section, we attempt to model the number of newly infected people, for example, in Japan [38], on the basis of the 1-D generalized time-dependent logistic map (4) with α and $a(t_i)$, which has the solution $x_n(t_i)$ consisting of chaotic time series.

In Fig. 1, a numerical result of the chaos function solution (1) and (2) to the 1-D

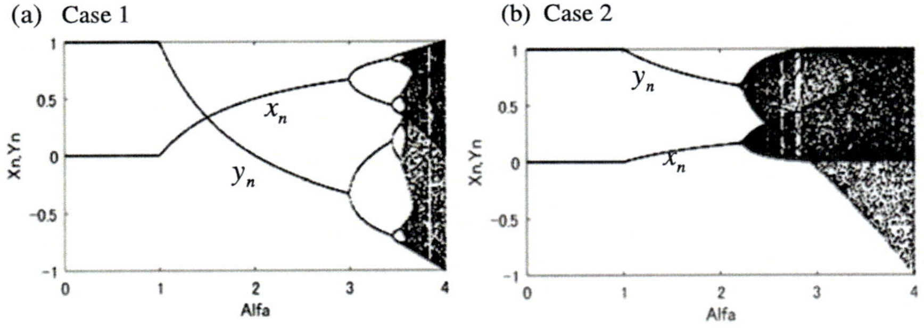

Fig. 1 The chaos function solution (1) with (2) to the 1-D map (3); Solution $x_n(t_i)$ with $a(t_i) = 1.0$, **b** Exponential amplitude function $a(t_i)$, **c** Solution $x_n(t_i)$ with $a(t_i)$, and **d** The number of newly infected people in Japan [38]

map (3) is compared with the number of newly infected people, due to the COVID-19; (a) Solution $x_n(t_i)$ with $a(t_i) = 1.0$, (b) Exponential amplitude function $a(t_i)$, (c) Solution $x_n(t_i)$ with $a(t_i)$, and (d) The number of newly infected people in Japan [38]. Here, it is important to note that the $a(t_i)$ of (b) is an exponential function, and has two peaks caused by external factors, such as the cruise ship called at the port on Feb. 3, 2020 and the termination of the state of emergency on May 25, 2020, respectively. Therefore, the two peaks arise after a few weeks, as shown in Fig. (d). The MATLAB program for Fig. 1a–c is presented in Appendix.

Furthermore, we show the numerical result obtained by iterating the 1-D generalized logistic map (4), at the system parameter $\alpha = 4.0, 3.9, 3.7, 3.6$, in Fig. 2. It is found that the stable time series in (a)-(d) are transformed from chaotic to non-chaotic, that is, chimera states, as α decreases [27].

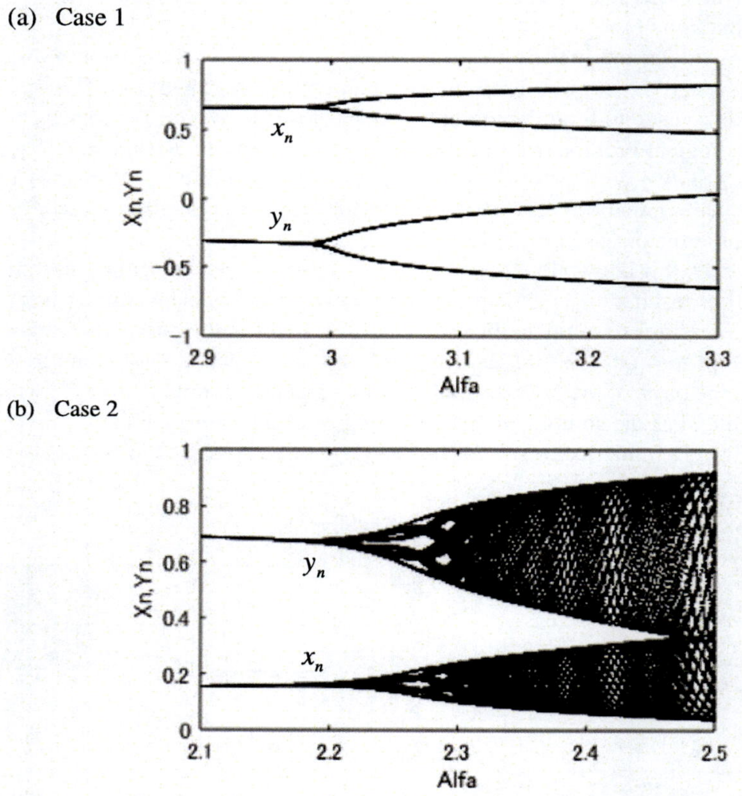

Fig. 2 Numerical solutions to the 1-D generalized time-dependent logistic map (4) with the amplitude function $a(t_i)$ and the system parameter α; **a** α = 4.0, **b** α = 3.9, **c** α = 3.7 and **d** α = 3.6

3 2-D Generalized Logistic Maps and Limit Cycles

In this section, we derive 2-D generalized time-dependent logistic maps by extending the 1-D generalized logistic map (4) with α and $a(t_i)$, and consider the effect of a system parameter and amplitude functions to the pattern formation of limit cycles, for the 2-D logistic maps.

According to the derivation of the 1-D solvable logistic map (3) with (1) and (2), we begin with the following chaos function solution;

$$x_n(t_i) \;=\; a_1(t_i)\sin^2(2^n t_i) \tag{5}$$

with an amplitude function $a_1(t_i) \neq 0$, where the discrete time t_i satisfies.

the condition (2). Then, we have

$$\begin{aligned} x_{n+1}(t_{i+1}) &= 4a_1(t_i)\sin^2(2^n t_i)\cos^2(2^n t_i) \\ &= 4x_n(t_i)\left(1 - \frac{1}{a_1(t_i)}x_n(t_i)\right), \end{aligned} \tag{6}$$

and by defining

$$y_n(t_i) \;\equiv\; a_2(t_i)\cos(2^n t_i) \tag{7}$$

with an amplitude function $a_2(t_i) \neq 0$, we find a condition from (5) and (7) as

$$\frac{1}{a_1(t_i)}x_n(t_i) \;+\; \frac{1}{a_2^2(t_i)}y_n^2(t_i) \;=\; 1. \tag{8}$$

By substituting (8) into the 1-D logistic map (6), and from the solutions (5) and (7), we find a 2-D generalized solvable logistic map;

$$x_{n+1}(t_{i+1}) \;=\; 4\left(\frac{1}{a_2^2(t_i)}\right)x_n(t_i)y_n^2(t_i), \tag{9}$$

$$y_{n+1}(t_{i+1}) \;=\; a_2(t_i)(1 - \left(\frac{2}{a_1(t_i)}\right)x_n(t_i)), \tag{10}$$

and by introducing a system parameter α, $0 < \alpha \leq 4.0$ in (9), we arrive at a 2-D generalized logistic map;

$$x_{n+1}(t_{i+1}) \;=\; \alpha\left(\frac{1}{a_2^2(t_i)}\right)x_n(t_i)y_n^2(t_i), \tag{11}$$

$$y_{n+1}(t_{i+1}) = a_2(t_i)\left(1 - \left(\frac{2}{a_1(t_i)}\right)x_n(t_i)\right) \qquad (12)$$

with amplitude functions $a_1(t_i)$ and $a_2(t_i)$. Therefore, the 2-D map (11) and (12) has chaos function solutions (5) and (7) at $\alpha = 4.0$. Here, it is important to note that the 2-D logistic map corresponds to the FHN model for neural cells, the BZ reaction and reaction–diffusion systems for chemical cells [26–29].

For limit cycles of the 2-D logistic map (11) and (12) with the system parameter α and amplitude functions $a_1(t_i)$ and $a_2(t_i)$, we show the following three Cases on $a_1(t_i)$ and $a_2(t_i)$ at $\alpha = 2.4$ in Fig. 3;

$$Case\ 1:\ a_1(t_i) = 1.0,\ \ a_2(t_i) = 1.0, \qquad (13)$$

$$Case\ 2:\ a_1(t_i) = 1.5,\ \ a_2(t_i) = 0.5, \qquad (14)$$

$$Case\ 3:\ a_1(t_i) = 2.0,\ \ a_2(t_i) = 0.2, \qquad (15)$$

where the amplitude functions are constants for simplicity. In Fig. 3a, discrete limit cycles of Case 1 (13) are illustrated as orbit solutions and sequential points, which converge to the limit cycle from an outside initial point for entrainment and an inside initial point for synchronization, respectively. Then, in Fig. 3b–c, limit cycles of Case 2 (14) and Case 3 (15) at $\alpha = 2.4$ are shown numerically. In order to compare the pattern of limit cycles for Cases 1, 2 and 3, and to understand about how the pattern formation depends on the system parameter and the amplitude functions, the limit cycles presented in Fig. 3 are shown on the same $x_n - y_n$ plane in Fig. 4. Similarly, Cases 1–3 at $\alpha = 2.6$ are illustrated in Figs. 5 and 6, for the pattern formation of stable limit cycles obtained by iterating the 2-D logistic map (11) and (12).

4 Conclusions

The 1-D generalized logistic map (4) is presented on the basis of the time-dependent chaos function solution with an amplitude function $a(t_i)$, and is modeled on the number of newly infected people, due to the COVID-19, as a population growth model. Moreover, the 2-D generalized logistic map (11) and (12) with a system parameter α and two amplitude functions $a_1(t_i)$ and $a_2(t_i)$ are derived by extending the 1-D logistic map (4). Then, we have found by iterating the 2-D map that the pattern of stable limit cycles depends on α, $a_1(t_i)$ and $a_2(t_i)$ essentially, as shown in Figs. 3–6. Particularly, as a restriction on stable limit cycles, the amplitude functions effect to the pattern formation of the limit cycles, and therefore the functions may

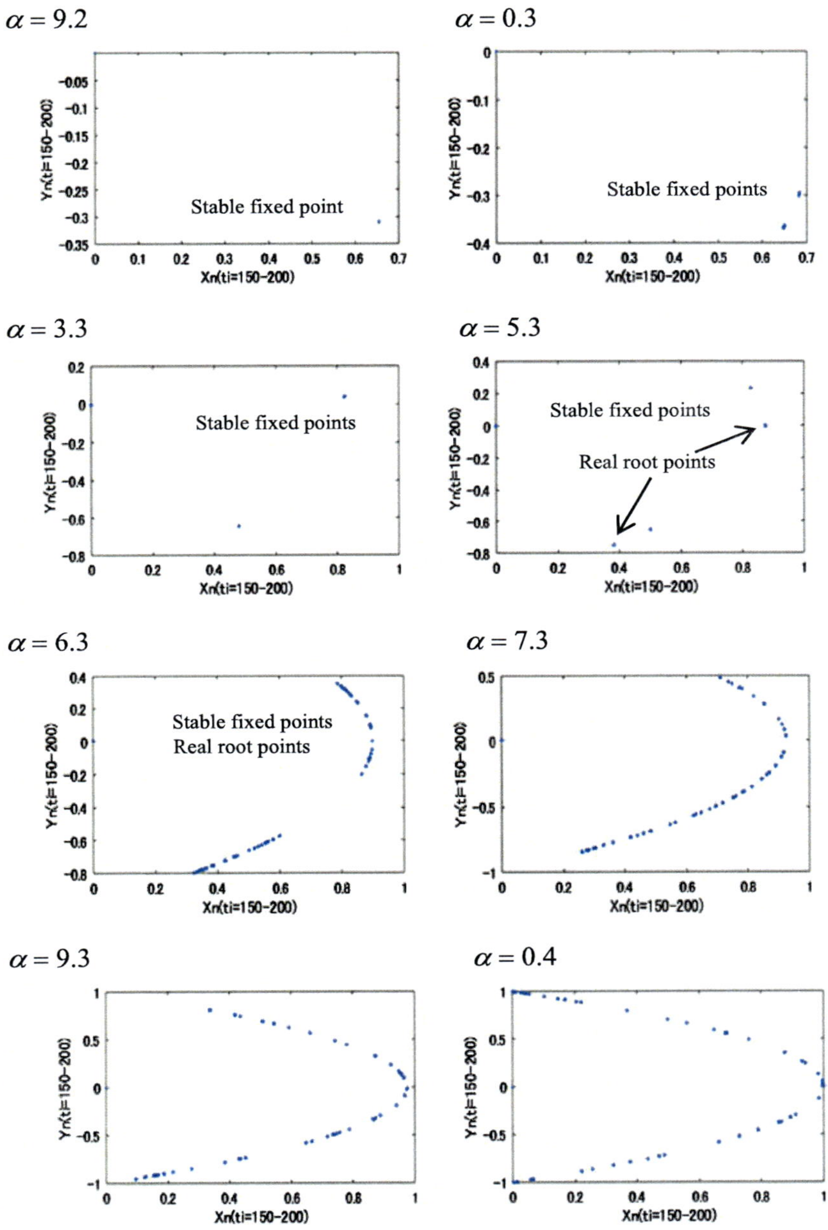

Fig. 3 Limit cycles of the 2-D logistic map (11) and (12) at $\alpha = 2.4$, and with amplitude functions: **a** Case 1 (13), **b** Case 2 (14) and **c** Case 3 (15)

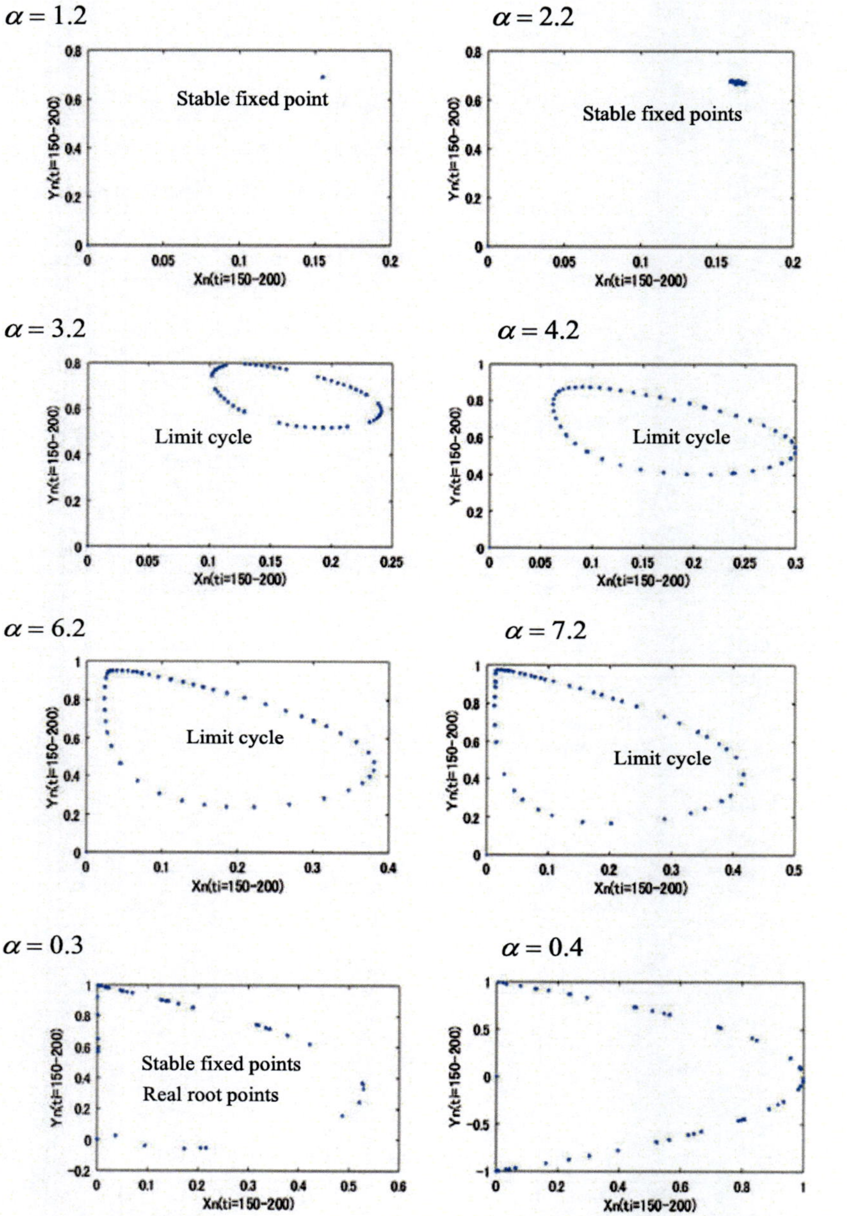

Fig. 4 The pattern formation of limit cycles shown in Fig. 3 for the 2-D logistic map (11) and (12), with amplitude functions of Cases 1–3 at $\alpha = 2.4$, and on the same x_n-y_n plane

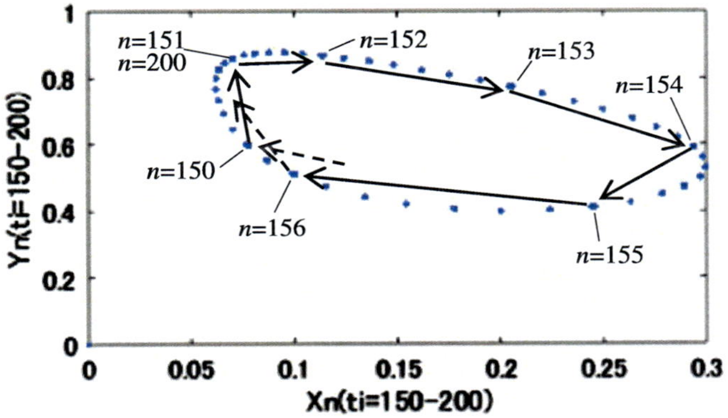

Fig. 5 Limit cycles of the 2-D logistic map (11) and (12) at $\alpha = 2.6$, and with amplitude functions: **a** Case 1 (13), **b** Case 2 (14) and **c** Case 3 (15)

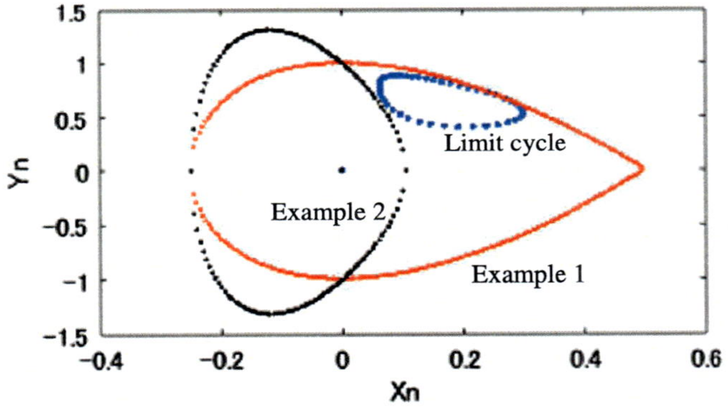

Fig. 6 The pattern formation of limit cycles shown in Fig. 5 for the 2-D logistic map (11) and (12), with amplitude functions of Cases 1–3 at $\alpha = 2.6$, and on the same x_n-y_n plane

correspond to the boundary conditions in physics. In addition, the stability analysis of the 2-D logistic map (11) and (12) may be a research subject to be born.

Appendix

```
% MATLAB program for 2-D bifurcation diagram Fig. 2 (b) by S. Kawamoto
% Initial conditions
ALFA=zeros(1, 400);
X=zeros(1, 200);
Y=zeros(1, 200);
XX=zeros(1, 200);
YY=zeros(1, 200);
X0=0.6;
Y0=0.1;
% System parameter ALFA and discrete limit cycles
figure('Position', [100 100 350 200])
for I=1:400
   ALFA(I)=2.1+0.001*I;
   for J=1
      X(I,J)=ALFA(I).*X0.*((Y0)^2);
      Y(I,J)=1-2*X0;
   end
   for J=2:200
      X(I,J)=ALFA(I).*X(I,J-1).*((Y(I, J-1))^2);
      Y(I,J)=1-2*X(I,J-1);
   end
  for J=150:200
     XX(J)=X(I,J);
     YY(J)=Y(I,J);
      plot(ALFA(I),XX(J),'k.','MarkerFaceColor','k','MarkerSize',4); hold on
      plot(ALFA(I),YY(J),'k.','MarkerFaceColor','k','MarkerSize',4); hold on
  end
end
xlabel('Alfa'); ylabel('xn, yn')
```

```
end
for I=2:200
    XX1(I)=X1(I-1, I-1);
    XX2(I)=X2(I-1, I-1);
end
% figures (a)-(c)
figure('Position', [100 100 350 100])
plot(TT, XX2, '-b.','MarkerFaceColor','b','MarkerSize', 7);
xlabel('ti'); ylabel('xn(ti)')
figure('Position', [100 100 350 100])
plot(TT, AA, '-b.','MarkerFaceColor','b','MarkerSize', 7);
xlabel('ti'); ylabel('a(ti)')
figure('Position', [100 100 350 200])
plot(TT, XX1, '-b.','MarkerFaceColor','b','MarkerSize', 7);
xlabel('ti'); ylabel('xn(ti)')
```

References

1. A. Scott, *Nonlinear Science* (Routledge, London, 2005).
2. E. Mosekilde, Y. Maistrenko, D. Postnov, *Chaotic Synchronization: Applications to Living Systems*. Nonlinear Science Vol. 42 (World Scientific, Singapore, 2002)
3. R.J. Bird, *Chaos and Life: Complexity and Order in Evolution and Thought* (Columbia University Press, New York, 2003).
4. E.M. Izhikevich, *Dynamical Systems in Neuroscience: The Geometry of Excitability and Bursting* (The MIT Press, London, 2007).
5. G.B. Ermentrout, D.H. Terman, *Mathematical Foundations of Neuroscience* (Springer, London, 2010).
6. C.H. Skiadas, C. Skiadas (eds.), *Handbook of Application of Chaos Theory* (Chapman and Hall/CRC Press, 2016)
7. C.H. Skiadas, C. Skiadas (eds.), *Demography of Population Health, Aging and Health Expenditures* (Springer Nature Switzerland AG, 2020)
8. F.C. Moon, *Chaotic and Fractal Dynamics* (Wiley, New York, 1992).
9. E. Ott, *Chaos in Dynamical Systems* (Cambridge University Press, Cambridge, 1993).
10. Y. Ueda, Random phenomena resulting from nonlinearity: In the system described by Duffing's equation. *Int. J. Non-linear Mechanics* **20**: 481–491, (1985): Translated from *Trans. IEEJ* 98-A: 167–173, (1978)
11. Y. Ueda, Randomly transitional phenomena in the system governed by Duffing's equation. J. Statistical Physics **20**, 181–196 (1979)
12. Y. Kuramoto, D. Battogtokh, Coexistence of coherence and incoherence in nonlocally coupled phase oscillators. Nonl. Phen. Compl. Syst. **5**, 380–385 (2002)
13. D.M. Abrams, S.H. Strogatz, Chimera states for coupled oscillators. Phys. Rev. Lett. **93**, 174102 (2004)
14. R.M. May, Limit cycles in predator-prey communities. Science **177**, 900–902 (1972)
15. G. Nicolis, I. Prigogine, *Self-Organization in Non-Equilibrium Systems* (John Wiley & Sons, New York, 1977).
16. S.H. Strogatz, *Nonlinear Dynamics and Chaos*, 2nd edn. (Westview Press, Boulder, 2015).
17. B. Van der Pol, J. Van der Mark, Frequency demultiplication. Nature **120**(3019), 363–364 (1927)
18. R.A. Fisher, The wave of advance of advantageous genes. Ann. Eugenics **7**, 355–369 (1937)
19. A.N. Zaikin, A.M. Zhabotinsky, Concentration wave propagation in two-dimensional liquid-phase self-oscillating system. Nature **225**, 535–537 (1970)
20. A.T. Winfree, Spiral waves of chemical activity. Science **175**, 634–636 (1972)
21. F.C. Moon, Chaotic vibrations of a magnet near a superconductor. Phys. Lett. **132A**(5), 249–252 (1988)
22. S. Kondo, T. Miura, Reaction-diffusion model as a framework for understanding biological pattern formation. Science **329**, 1616–1620 (2010)
23. T. Huang, H. Zhang, Bifurcation, chaos and pattern formation in a space- and time-discrete predator-prey system. Chaos, Solitons Fractals **91**, 92–107 (2016)
24. S. Kawamoto, 2-D and 3-D solvable chaos maps. Chaotic Model. Simul. (CMSIM) **1**, 107–118 (2017)
25. S. Kawamoto, Chaotic time series by time-discretization of periodic functions and its application to engineering. Chaotic Model. Simul. (CMSIM) **2**, 193–204 (2017)
26. S. Kawamoto, The FitzHugh-Nagumo model and 2-D solvable chaos maps. Chaotic Model. Simul. (CMSIM) **3**, 269–283 (2018)
27. S. Kawamoto, The FitzHugh-Nagumo model and spatiotemporal fractal sets based on time-dependent chaos functions, in *11th Chaotic Modeling and Simulation International Conference*, Springer Proceedings in Complexity, ed. by C.H. Skiadas, I. Lubashevsky (Springer Nature Switzerland AG, 2019), pp. 159–173

28. S. Kawamoto, Intermittency of chaos functions and the Belousov-Zhabotinsky reaction, in *12th Chaotic Modeling and Simulation International Conference*, Springer Proceedings in Complexity, ed. by C.H. Skiadas, Y. Dimotikalis (Springer Nature Switzerland AG, 2020), pp. 123–134
29. S. Kawamoto, Reaction-diffusion systems and propagation of limit cycles with chaotic dynamics, in *12th Chaotic Modeling and Simulation International Conference*, Springer Proceedings in Complexity, ed. by C.H. Skiadas, Y. Dimotikalis (Springer Nature Switzerland AG, 2020), pp. 135–149
30. S. Kawamoto, Limit cycle analysis for 2-D time-dependent logistic maps (submitted)
31. S. Kawamoto, Interaction of limit cycles for the FitzHugh-Nagumo model (submitted)
32. P.F. Verhulst, Mathematical researches into the law of population growth increase. Nouveaux Mémoires de l'Académie Royale des Sciences et Belles-Lettres de Bruxelles **18**, 1–42 (1845)
33. R.M. May, Biological populations with non-overlapping generations: Stable points, stable cycles, and chaos. Science **15**, 645–646 (1974)
34. D. Blanchet, Nonlinear demographic models and chaotic demo-dynamics. *Population*, 139–150 (1998)
35. W.E. Ricker, Stock and recruitment. J. Fish. Res. Bd. Can. **11**, 559–623 (1954)
36. R.M. May, Deterministic models with chaotic dynamics. Nature **256**, 165–166 (1975)
37. https://en.wikipedia.org/wiki/Coronavirus_disease_2019
38. https://en.wikipedia.org/wiki/COVID-19_pandemic_in_Japan

Interaction of Limit Cycles for the FitzHugh-Nagumo Model

Shunji Kawamoto

Abstract The interaction of discrete limit cycles related to the FitzHugh-Nagumo (FHN) model is discussed in this paper, from the standpoint of time-dependent chaos functions. Firstly, a two-dimensional (2-D) solvable chaos map corresponding to the FHN model is derived on the basis of chaos functions, and 2-D chaotic maps with one system parameter are presented for generating limit cycles. Secondly, two limit cycles of 2-D systems A and B are located on the x_n-y_n plane as nonlinear dynamics of neural cells, and the propagations are illustrated with a constant velocity in opposite direction. Finally, the interaction of two limit cycles is numerically considered by adding interaction terms in a 2-D system A&B composed of the 2-D systems. Then, the 2-D system A&B with two stable limit cycles is shown to have chaotic dynamics depending on the interaction terms, as a complex system of non-equilibrium open systems.

1 Introduction

For the study of nonlinear science [1], nonlinear dynamics have been considered widely in the field of physics, chemistry, biology, engineering and social sciences [2, 3]. In particular, one-dimensional (1-D) nonlinear difference equations are known to possess a rich spectrum of dynamical behavior as chaos in many respects, and the chaotic modeling and the chaos theory have been extended to medicine, optics, living systems, life science and neuro science [4–8].

In the meantime, the state of having many parts and being difficult to understand or to find an answer is called complexity, and how single parts organize spontaneously into complicated structures has been studied [9]. Later, complex system is stated to be a system composed of many components which may interact with each other [10]. As examples of complex systems, large scale natural systems such as atmosphere, climate, human brain and economic organizations have been presented, and

S. Kawamoto (✉)
Osaka Prefecture University, Sakai, Osaka, Japan
e-mail: skawamoto@sky.plala.or.jp

C. H. Skiadas et al. (eds.), *13th Chaotic Modeling and Simulation International Conference*, Springer Proceedings in Complexity,
https://doi.org/10.1007/978-3-030-70795-8_27

infrastructures such as power grids, transportation systems, information systems and communication networks have been considered [11–14]. The coexistence of coherence and incoherence in coupled phase oscillations has been proposed as chimera states [15, 16]. Moreover, the optimization of complex systems is presented, and the theory, algorithms and its applications are widely discussed.

In recent years, 2-D and 3-D chaos maps have been presented for the analysis of population growth, electrical oscillation, atmospheric convection and chemical reaction [17]. Then, the time-dependent chaos functions are introduced and applied to engineering with a nonlinear time series expansion [18]. After that, the FitzHugh-Nagumo (FHN) model [19, 20] has been considered from the standpoint of chaos functions for neural cells [21, 22], and the functions have been applied to bifurcation diagrams proposed in [23, 24] and the generation of limit cycles [25, 26] for reaction–diffusion systems [27, 28].

In this paper, Sect. 2 presents the derivation of 2-D chaotic maps with a system parameter corresponding to the FHN model, and Sect. 3 illustrates two limit cycles obtained numerically by iterating the 2-D maps on the x_n-y_n plane for chaotic dynamics, such as synchronization and entrainment, for neural cells. In Sect. 4, the propagation and the interaction of stable limit cycles with a constant velocity in opposite direction are considered by adding linear interaction terms with coupling constants in a combined 2-D system A&B, as a complex system based on the FHN model. Conclusions are summarized in the last section.

2 The FHN Model and 2-D Chaotic Maps

The forced Van der Pol oscillator is given by

$$\ddot{x} - \varepsilon(1 - x^2)\dot{x} + x = E_0 \sin(\omega t), \tag{1}$$

which represents a model for simple vacuum tube oscillator circuit with an external voltage source as a heart beating model, where $x = x(t)$ is the proposition coordinate function of time t, and $\{\varepsilon \neq 0, E_0, \omega\}$ are the system parameters [29]. By the Liénard transformation, we find a 2-D model as

$$\dot{x} = \varepsilon\left(x - \frac{1}{3}x^3 - y\right), \tag{2}$$

$$\dot{y} = \frac{1}{\varepsilon}x - \frac{1}{\varepsilon}E_0 \sin(\omega t), \tag{3}$$

which is known to have chaotic behaviors in the equivalent circuit with sinusoidal forcing [30]. Moreover, the FHN model [19, 20] is a 2-D simplification of the Hodgkin-Huxley model [31] of spike generation in squid giant axons, and is derived from (1) and expressed by

$$\dot{v} = v - \frac{1}{\varepsilon}v^3 - w + I(t), \tag{4}$$

$$\dot{w} = \frac{1}{\tau}(v + a - bw), \tag{5}$$

which has an electric circuit as one of a large class of nonlinear systems showing excitable and oscillatory behaviors, with the membrane potential v, the recovery variable w, the stimulus external current $I(t)$ and model parameters $\{a, b, \tau \neq 0\}$. Here, it should be noticed that the 2-D model (2) and (3) of the forced Van der Pol oscillator has the external voltage term in (3), and the FHN model (4) and (5) has the stimulus external current term in (4) [21].

On the other hand, from the following chaos solutions consisting of time-dependent chaos functions;

$$x_n(t) = a_1 \sin^2(2^n t) + b_1(t), \tag{6}$$

$$y_n(t) = a_2 \cos(2^n t) + b_2(t), \tag{7}$$

$$\frac{1}{a_1}(x_n(t) - b_1(t)) + \frac{1}{a_2^2}(y_n(t) - b_2(t))^2 = 1 \tag{8}$$

$$t \neq \pm\, m\pi/2^l \tag{9}$$

with nonzero coefficients $\{a_1, a_2\}$, functions $\{b_1(t), b_2(t)\}$ of time $t > 0$, then we have a 2-D solvable chaos map from (6–9) as

$$x_{n+1}(t_{i+1}) = \frac{4}{a_2^2}(x_n(t_i) - b_1(t_i))(y_n(t_i) - b_2(t_i))^2 + b_1(t_i), \tag{10}$$

$$y_{n+1}(t_{i+1}) = -2\left(\frac{a_2}{a_1}\right)x_n(t_i) + a_2 + b_2(t_i) + 2\left(\frac{a_2}{a_1}\right)b_1(t_i), \tag{11}$$

where the discrete time t_i satisfies the condition (9), and the passage from a point $(x_n(t_i), y_n(t_i))$ to the next one $(x_{n+1}(t_{i+1}), y_{n+1}(t_{i+1}))$ with the time step $\Delta t \equiv t_{i+1} - t_i$ can be considered as a 2-D mapping. It is interesting to note that the third-order nonlinear term of (10) is involved in (2) and (4), the terms b_1(t) in (6) and b_2(t) in (7) play a role of coodinate transformation for the solutions x_n(t) and y_n(t). Therefore, the term $b_1(t_i)$ in (10) is equivalent to the external current term $I(t)$ in (4), and the term $(b_2(t_i) + 2(a_2/a_1)b_1(t_i))$ in (11) is found to correspond to the external voltage source $E_0 \sin(\omega t)$ in (3) [21].

3 Stable Limit Cycles

In this Section, we rewrite the 2-D solvable chaos map (10) and (11), by introducing a system parameter ε, into the following 2-D chaotic systems A and B;

$$System\ A: \quad x_{1,n+1}(t_{i+1}) = \varepsilon_1\left(\frac{1}{a_2^2}\right)x_{1,n}(t_i)y_{1,n}^2(t_i), \tag{12}$$

$$y_{1,n+1}(t_{i+1}) = -2\left(\frac{a_2}{a_1}\right)x_{1,n}(t_i) + a_2, \tag{13}$$

$$System\ B: \quad x_{2,n+1}(t_{i+1})$$
$$= \varepsilon_2\left(\frac{1}{a_2^2}\right)(x_{2,n}(t_i) - b_1(t_i))(y_{2,n}(t_i) - b_2(t_i))^2 + b_1(t_i), \tag{14}$$

$$y_{2,n+1}(t_{i+1}) = -2\left(\frac{a_2}{a_1}\right)x_{2,n}(t_i) + a_2 + b_2(t_i) + 2\left(\frac{a_2}{a_1}\right)b_1(t_i) \tag{15}$$

with system parameters $\varepsilon_1, 0 < \varepsilon_1 \leq 4.0$ and $\varepsilon_2, 0 < \varepsilon_2 \leq 4.0$. Then, we obtain discrete limit cycles LC-A and LC-B for System A and System B, respectively.

Here, we set the LC-A and the LC-B on the $x_n - y_n$ plane as illustrated in Fig. 1a–b with initial points: (a) Inside initial points $(x_{1,0}, y_{1,0}) = (0.2, 0.3)$ and $(x_{2,0}, y_{2,0}) = (0.5, 0.3)$ for synchronization, and (b) Outside initial points $(x_{1,0}, y_{1,0}) = (0.2, 0.4)$ and $(x_{2,0}, y_{2,0}) = (0.5, 0.4)$ for entrainment, where the initial point of cell is indicated by a cross $\times$ (black) in Fig. 1. Then, we find that the distance $b_1(t_0)$ of initial points between LC-A and LC-B is given by $b_1(t_0) \equiv x_{2,0} - x_{1,0} = 0.3$ in Fig. 1a–b, and the limit cycles, which are shown by orbits and sequential points, possess chaotic dynamics of synchronization and entrainment, depending on the initial points. Here, we set $\varepsilon_1 = \varepsilon_2 = 2.3$, $a_1 = 1.0$, $a_2 = 0.5$ and the time step $\Delta t = \pi/431 \approx 0.0073$ for the numerical calculation [26].

Additionary, in the case of LC-B shifted on the Y_n axis from $y_n = 0.3$ to $y_n = 0.4$, the limit cycles LC-A and LC-B are located with the following initial points: (a) Inside initial points $(x_{1,0}, y_{1,0}) = (0.2, 0.3)$ and $(x_{2,0}, y_{2,0}) = (0.5, 0.4)$ for synchronization, and (b) Outside initial points $(x_{1,0}, y_{1,0}) = (0.2, 0.4)$ and $(x_{2,0}, y_{2,0}) = (0.5, 0.5)$ for entrainment as illustrated in Fig. 2a–b, where the initial point of cell is indicated by a cross $\times$ (black) in Fig. 2. Thus, it is found that we can locate limit cycles at any point on the $x_n - y_n$ plane, by setting initial points $\{(x_{1,0}, y_{1,0}), (x_{2,0}, y_{2,0})\}$ and external forces $\{b_1(t_0), b_2(t_0)\}$, in order to combine System A (12–13) and System B (14–15) in the next Section.

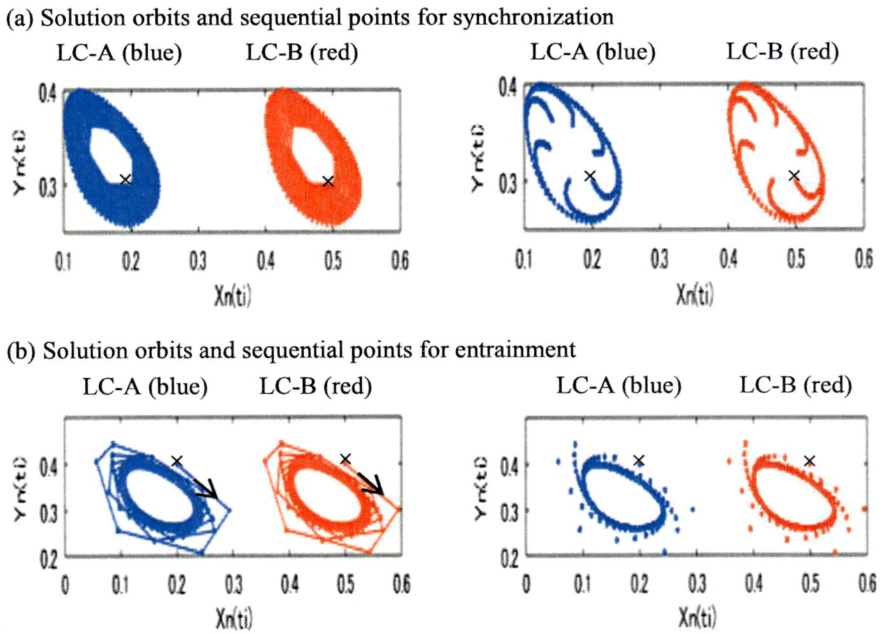

Fig. 1 Limit cycles LC-A (blue) and LC-B (red) located on the x_n-y_n plane: **a** Inside initial points $(x_{1,0}, y_{1,0}) = (0.2, 0.3)$ and $(x_{2,0}, y_{2,0}) = (0.5, 0.3)$, and **b** Outside initial points $(x_{1,0}, y_{1,0}) = (0.2, 0.4)$ and $(x_{2,0}, y_{2,0}) = (0.5, 0.4)$, and the initial point of cell indicated by a cross × (black) in (**a**) and (**b**)

4 Propagation and Interaction

We consider System A and System B with conditions $b_1(t_0) \equiv x_{2,0} - x_{1,0} = 0.3$ and $b_2(t_i) = 0$ for simplicity in this Section, and discuss the nonlinear dynamics of limit cycles LC-A and LC-B, which propagate and interact in opposite direction on the $x_n - y_n$ plane.

Under the conditions, we combine System A and System B into the following System A&B, in order to discuss the propagation and the interaction of limit cycles LC-A and LC-B by introducing a one-step propagation distance $v(\Delta t)$ in opposite direction with the constant velocity v and the time step $\Delta t = t_{i+1} - t_i$. Then, we find a combined system from (12–15) as.

System A&B:

$$x_{1,n+1}(t_{i+1}) = \varepsilon_1(\frac{1}{a_2^2})(x_{1,n}(t_i) - b_1(t_i))y_{1,n}^2(t_i) + b_1(t_i) + c_1(x_{2,n}(t_i) - x_{1,n}(t_i)), \tag{16}$$

$$y_{1,n+1}(t_{i+1}) = -2(\frac{a_2}{a_1})x_{1,n}(t_i) + a_2 + 2(\frac{a_2}{a_1})b_1(t_i), \tag{17}$$

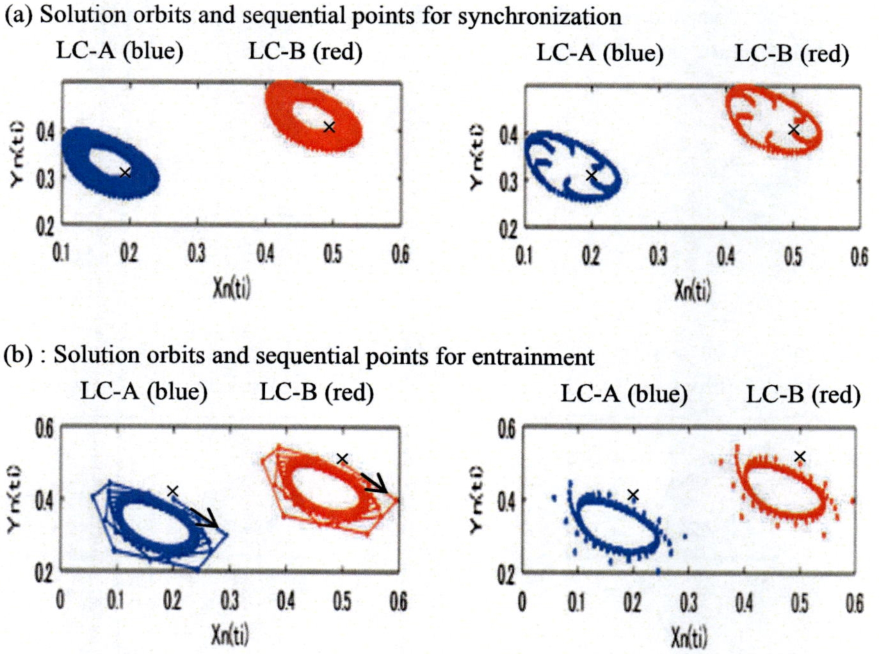

Fig. 2 Limit cycle LC-B (red) shifted on the y_n axis: **a** Inside initial points $(x_{1,0}, y_{1,0}) = (0.2, 0.3)$ and $(x_{2,0}, y_{2,0}) = (0.5, 0.4)$, and **b** Outside initial points $(x_{1,0}, y_{1,0}) = (0.2, 0.4)$ and $(x_{2,0}, y_{2,0}) = (0.5, 0.5)$, and the initial point of cell indicated by a cross x (black) in **(a)** and **(b)**

$$x_{2,n+1}(t_{i+1}) = \varepsilon_2\left(\frac{1}{a_2^2}\right)\left(x_{2,n}(t_i) - (b_0 - b_1(t_i))\right)y_{2,n}^2(t_i) + (b_0 - b_1(t_i)) + c_2\left(x_{1,n}(t_i) - x_{2,n}(t_i)\right), \tag{18}$$

$$y_{2,n+1}(t_{i+1}) = -2(\frac{a_2}{a_1})x_{2,n}(t_i) + a_2 + 2(\frac{a_2}{a_1})(b_0 - b_1(t_i)), \tag{19}$$

here we set $\varepsilon_1 = \varepsilon_2 = 2.3$, $a_1 = 1.0$, $a_2 = 0.5$, $b_0 \equiv b_1(t_0) = x_{2,0} - x_{1,0} = 0.3$, inside initial points $(x_{1,0}, y_{1,0}) = (0.2, 0.3)$ and $(x_{2,0}, y_{2,0}) = (0.5, 0.3)$, and outside initial points $(x_{1,0}, y_{1,0}) = (0.2, 0.4)$ and $(x_{2,0}, y_{2,0}) = (0.5, 0.4)$, for the numerical calculation. Therefore, it should be noticed that $b_1(t_0) = 0.3$ gives the travel distance for six steps of limit cycles LC-A and LC-B from the initial states {LC-A_1, LC-B_1} to the last ones {LC-A_7, LC-B_7} in opposite direction with one step distance $v(\Delta t) = 0.05$ and $\Delta t = \pi/431 \approx 0.0073$, and coefficients $\{c_1, c_2\}$ of linear interaction terms in (16) and (18) are coupling constants, respectively. Here, at each state of limit cycles; {LC-A_1,..., A_7} and {LC-B_1,..., B_7}, we carry out 200 iterations of System A&B, and the last points for the limit cycles at each state given by $(x_{1,200}, y_{1,200})$ and $(x_{2,200}, y_{2,200})$ are equal to the initial points for the next state of limit cycles, by adding one-step propagation distance $v(\Delta t) = \pm 0.05$ to the last points of limit cycles. The MATLAB program for System A&B (16–19) is presented in Appendix.

The propagation of limit cycles LC-A and LC-B with no-interaction $\{c_1 = 0, c_2 = 0\}$ is firstly shown in Fig. 3a–b: (a) Inside initial points for synchronization and (b) Outside initial points for entrainment are presented with six steps; $A_1 \to A_2 \to \cdots \to A_7$ (blue) and $B_1 \to B_2 \to \cdots \to B_7$ (red) for solution orbits and sequential points of limit cycles. It is found that the limit cycles propagate forward or backward in opposite direction, keeping the pattern and chaotic properties of limit cycle, respectively.

The interaction with coefficients $\{c_1 = 0.07, c_2 = 0.07\}$ is presented in Fig. 4a–b. As the System A&B has coefficients $\{c_1 = 0.07, c_2 = 0.07\}$ but different initial points $(x_{1,0}, y_{1,0}) = (0.2, 0.3)$ and $(x_{2,0}, y_{2,0}) = (0.5, 0.3)$, LC-A and LC-B depend on the difference of interaction terms $c_1(x_{2,n}(t_i) - x_{1,n}(t_i))$ and $c_2(x_{1,n}(t_i) - x_{2,n}(t_i))$, and the limit cycles are found to have a slight effect at each step.

Furthermore, for the case of coefficients $\{c_1 = 0.0, c_2 = 0.1\}$, LC-A (blue) has no interaction from LC-A_1 to LC-A_7 as shown in Fig. 5a–b. However, LC-B (red) has an effect at each step, and loose the pattern and chaotic properties of limit cycle from LC-B_1 to LC-B_7 gradually, in (a) Synchronization and (b) Entrainment.

5 Conclusions

In this paper, we derive a 2-D solvable chaos map based on time-dependent chaos functions, and present 2-D chaotic System A (12–13) and System B (14–15) with system parameters ε_1 and ε_2 for generating two limit cycles LC-A and LC-B, respectively. From the numerical result of the combined System A&B (16–19), it is found that we can locate LC-A and LC-B initially at any point on the x_n-y_n plane, and obtain the essential dynamics, that is, how LC-A and LC-B propagate in opposite direction and interact with each other in the complicated system, depending on the system parameters, the interaction coefficients and the velocity of limit cycles. In this regard, System A&B is a 2-D chaotic system with two limit cycles and being difficult to find an answer for the interactions, which may be a complex system on the basis of the FHN model for neural cells, as one of non-equilibrium open systems.

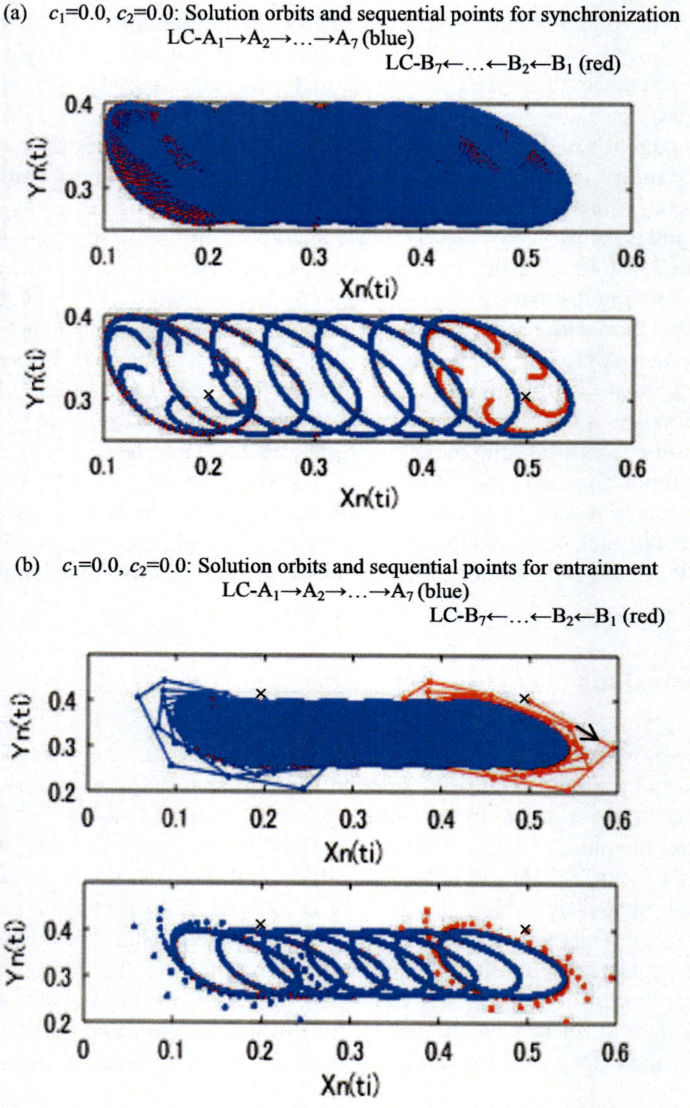

Fig. 3 Propagation of limit cycles LC-A (blue) and LC-B (red) with no-interaction $\{c_1 = 0.0, c_2 = 0.0\}$: **a** Inside initial points $(x_{1,0}, y_{1,0}) = (0.2, 0.3)$ and $(x_{2,0}, y_{2,0}) = (0.5, 0.3)$ for synchronization, **b** Outside initial points $(x_{1,0}, y_{1,0}) = (0.2, 0.4)$ and $(x_{2,0}, y_{2,0}) = (0.5, 0.4)$ for entrainment, and the initial point of cell indicated by a cross $\times$ (black) in **(a)** and **(b)**

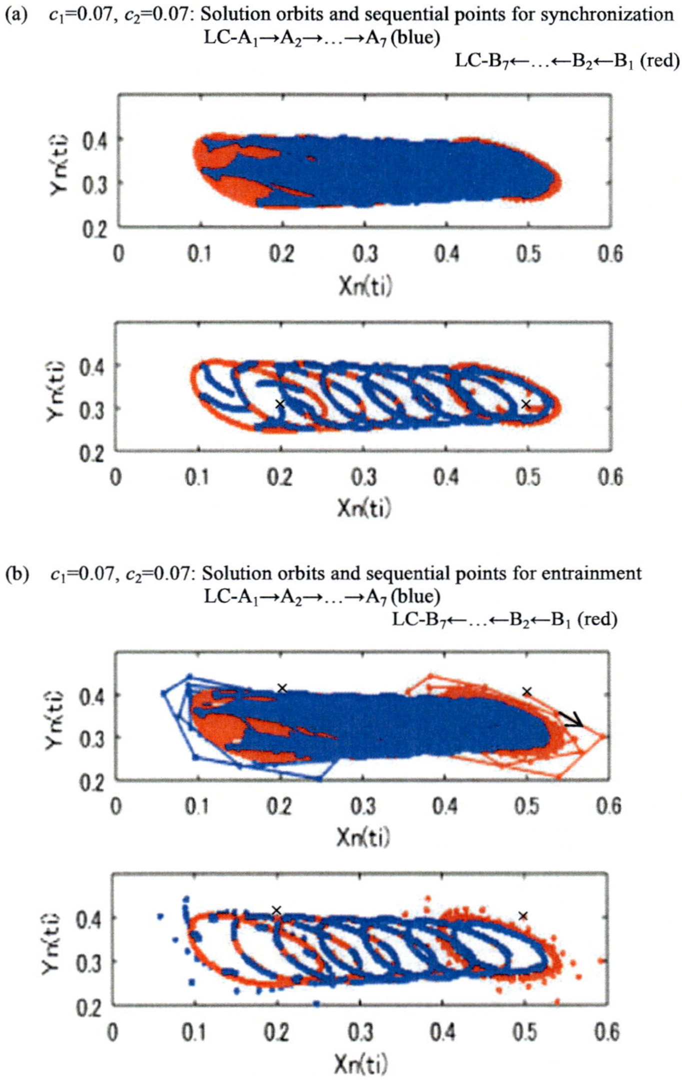

Fig. 4 Propagation of limit cycles LC-A (blue) and LC-B (red) with interaction $\{c_1 = 0.07, c_2 = 0.07\}$: **a** Inside initial points $(x_{1,0}, y_{1,0}) = (0.2, 0.3)$ and $(x_{2,0}, y_{2,0}) = (0.5, 0.3)$ for synchronization, **b** Outside initial points $(x_{1,0}, y_{1,0}) = (0.2, 0.4)$ and $(x_{2,0}, y_{2,0}) = (0.5, 0.4)$ for entrainment, and the initial point of cell indicated by a cross $\times$ (black) in (**a**) and (**b**)

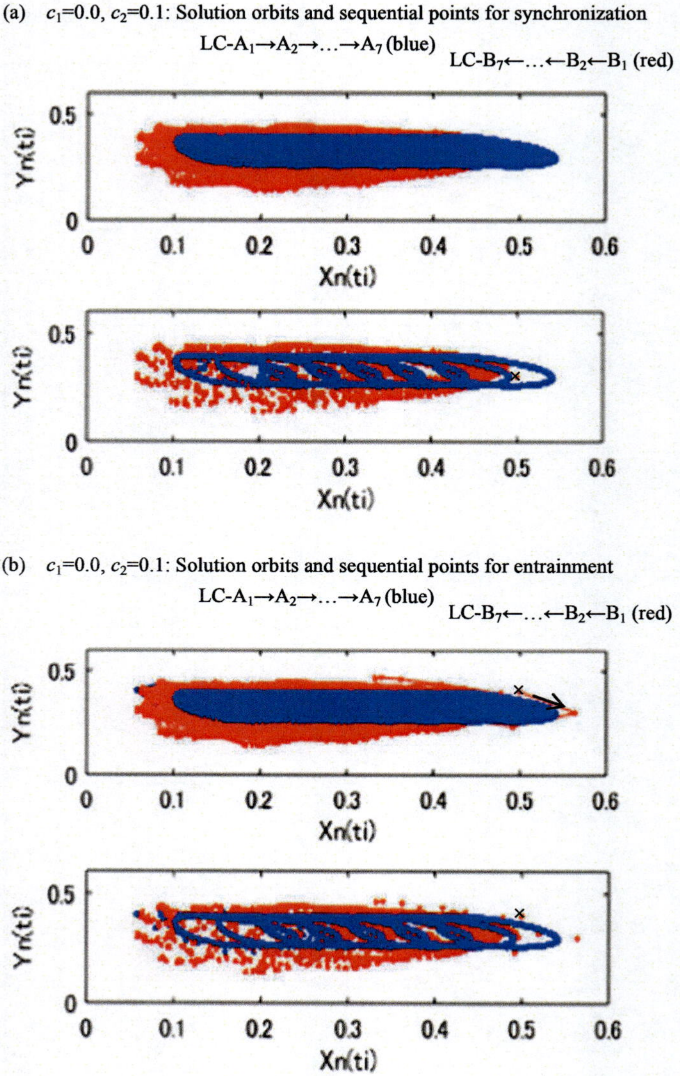

Fig. 5 Propagation of limit cycles LC-A (blue) and LC-B (red) with interaction $\{c_1 = 0.0, c_2 = 0.1\}$: **a** Inside initial points $(x_{1,0}, y_{1,0}) = (0.2, 0.3)$ and $(x_{2,0}, y_{2,0}) = (0.5, 0.3)$ for synchronization, **b** Outside initial points $(x_{1,0}, y_{1,0}) = (0.2, 0.4)$ and $(x_{2,0}, y_{2,0}) = (0.5, 0.4)$ for entrainment, and the initial point of cell indicated by a cross $\times$ (black) in (**a**) and (**b**)

Appendix

```
% MATLAB program for propagation and interaction of limit cycles by S. Kawamoto
% Initial conditions
X1=zeros(10, 199, 199);
Y1=zeros(10, 199, 199);
XX1=zeros(1, 200);
YY1=zeros(1, 200);
XXX1=zeros(10);
YYY1=zeros(10);
X0=zeros(10);
Y0=zeros(10);
X2=zeros(10, 199, 199);
Y2=zeros(10, 199, 199);
XX2=zeros(1, 200);
YY2=zeros(1, 200);
XXX2=zeros(10);
YYY2=zeros(10);
XX0=zeros(10);
YY0=zeros(10);
X10=0.2;
Y10=0.3;
X20=0.5;
Y20=0.3;
A1=1.0;
A2=0.5;
B=0.3;
C1=0.0;
C2=0.0;
EPSI1=2.3;
EPSI2=2.3;
VDELTAT=0.05;
% System parameters EPSI1 and EPSI2, and two limit cycles
figure('Position', [100 100 350 100])
for J=1
    X0(J)=X10;
    Y0(J)=Y10;
     XX0(J)=X20;
     YY0(J)=Y20;
     for I=1:199
         for N=1
             X1(J, I, N)=EPSI1*X0(J)*(Y0(J))^2/(A2)^2+C1*(XX0(J)-X0(J));
             Y1(J, I, N)=-2*(A2/A1)*X0(J)+A2;
             X2(J, I, N)=EPSI2*(XX0(J)-B)*(YY0(J))^2/(A2)^2+B+C2*(X0(J)-XX0(J));
            Y2(J, I, N)=-2(A2/A1)*XX0(J)+A2*(1+2*B/A1);
         end
```

```
        for N=2:I
            X1(J, I, N)=EPSI1*X1(J, I, N-I)*(Y1(J, I, N-1))^2/(A2)^2+C1*
                        (X2(J, I, N-1)-X1(J, I, N-1));
            Y1(J, I, N)=-2*(A2/A1)*X1(J, I, N-1)+A2;
            X2(J, I, N)=EPSI2*(X2(J, I, N-1)-B)*(Y2(J, I, N-1))^2/(A2)^2+B
                        +C2*(X1(J, I, N-1)-X2(J, I, N-1));
            Y2(J, I, N)=-2*(A2/A1)*X2(J, I, N-1)+A2*(1+2*B/A1);
        end
    end
    for I=1
        XX1(J, I)=X0(J);
        YY1(J, I)=Y0(J);
        XX2(J, I)=XX0(J);
        YY2(J, I)=YY0(J);
    end
    for I=2:200
        XX1(J, I)=X1(J, I-1, I-1);
        YY1(J, I)=Y1(J, I-1, I-1);
        XX2(J, I)=X2(J, I-1, I-1);
        YY2(J, I)=Y2(J, I-1, I-1);
    end
    for I=1:200
        XX1(I)=XX1(J, I);
        YY1(I)=YY1(J, I);
        XX2(I)=XX2(J, I);
        YY2(I)=YY2(J, I);
    end
    plot(XX2, YY2, '-r.','MarkerFaceColor','r','MarkerSize',7); hold on
    plot(XX1, YY1, '-b.','MarkerFaceColor','b','MarkerSize',7); hold on
    % The last points are the next initial points
    XXX1(J)=XX1(200)+VDELTAT;
    YYY1(J)=YY1(200);
    XXX2(J)=XX2(200)-VDELTAT;
    YYY2(J)=YY2(200);
end
% Limit cycles LC-A and LC-B
for J=2:7
    X0(J)=XXX1(J-1);
    Y0(J)=YYY1(J-1);
    XX0(J)=XXX2(J-1);
    YY0(J)=YYY2(J-1);
    for K=1:199
        for N=1
            X1(J, K, N)=EPSI1*(X0(J)-(J-1)*VDELTAT)*(Y0(J))^2/(A2)^2
                        +(J-1)*VDELTAT+C1*(XX0(J)-X0(J));
            Y1(J, K, N)=-2*(A2/A1)*X0(J)+A2*(1+2*(J-1)*VDELTAT/A1);
```

```
            X2(J, K, N)=EPSI2*(XX0(J)-(B-(J-1)*VDELTAT))*(YY0(J))^2/(A2)^2
                        +(B-(J-1)*VDELTAT)+C2*(X0(J)-XX0(J));
            Y2(J, K, N)=-2*(A2/A1)*XX0(J)+A2*(1+2*(B-(J-1)*VDELTAT)/A1);
          end
          for N=2:K
            X1(J, K, N)=EPSI1*(X1(J, K, N-1)-(J-1)*VDELTAT)
                        *(Y1(J, K, N-1))^2/(A2)^2+(J-1)*VDELTAT
                        +C1*(X2(J, K, N-1)-X1(J, K, N-1));
          Y1(J, K, N)=-2*(A2/A1)*X1(J, K, N-1)+A2*(1+2*(J-1)*VDELTAT/A1);
          X2(J, K, N)=EPSI2*(X2(J, K, N-1)-(B-(J-1)*VDELTAT))
                        *(Y2(J, K, N-1))^2/(A2)^2+(B-(J-1)*VDELTAT)
                        +C2*(X1(J, K, N-1)-X2(J, K, N-1));
          Y2(J, K, N)=-2*(A2/A1)*X2(J, K, N-1)+A2*(1+2*(B-(J-1)*VDELTAT)/A1);
        end
    end
    for K=1
        XX1(J, K)=X0(J);
        YY1(J, K)=Y0(J);
        XX2(J, K)=XX0(J);
        YY2(J, K)=YY0(J);
    end
    for K=2:200
        XX1(J, K)=X1(J, K-1, K-1);
        YY1(J, K)=Y1(J, K-1, K-1);
        XX2(J, K)=X2(J, K-1, K-1);
        YY2(J, K)=Y2(J, K-1, K-1);
    end
    for K=1:200
        XX1(K)=XX1(J, K);
        YY1(K)=YY1(J, K);
        XX2(K)=XX2(J, K);
        YY2(K)=YY2(J, K);
    end
    plot(XX2, YY2, '-r.','MarkerFaceColor','r','MarkerSize',7); hold on
     plot(XX1, YY1, '-b.','MarkerFaceColor','b','MarkerSize',7); hold on
     % The last points are the next initial points
     XXX1(J)=XX1(200)+VDELTAT;
     YYY1(J)=YY1(200);
     XXX2(J)=XX2(200)-VDELTAT;
     YYY2(J)=YY2(200);
end
xlabel('Xn(ti)'); ylabel('Yn(ti)')
```

References

1. A. Scott, *Nonlinear Science* (Routledge, London, 2005).
2. F.C. Moon, *Chaotic and Fractal Dynamics* (Wiley, New York, 1992).

3. H. Peitgen, H. Jurgens, D. Saupe, *Chaos and Fractals—New Frontiers of Science,* (Springer, New York, 1992)
4. E. Mosekilde, Y. Maistrenko, D. Postnov, *Chaotic Synchronization: Applications to Living Systems.* Nonlinear Science Vol. 42, (World Scientific, Singapore, 2002)
5. R.J. Bird, *Chaos and Life: Complexity and Order in Evolution and Thought* (Columbia University Press, New York, 2003).
6. E.M. Izhikevich, *Dynamical Systems in Neuroscience: The Geometry of Excitability and Bursting* (The MIT Press, London, 2007).
7. G.B. Ermentrout, D.H. Terman, *Mathematical Foundations of Neuroscience* (Springer, London, 2010).
8. C.H. Skiadas, C. Skiadas (ed.), *Handbook of Application of Chaos Theory* (Chapman and Hall/CRC Press, 2016)
9. M.M. Waldrop, *Complexity: The Emerging Science at the Edge of Order and Chaos* (SIMON & SCHUSTER PAPERBACKS, New York, 1993).
10. G. Nicolis, C. Rouvas-Nicolis, Complex systems. Scholarpedia **2**(11), 1473 (2007)
11. K. Aihara, T. Takabe, M. Toyoda, Chaotic neural networks. Phys. Lett. **A144**, 333–340 (1990)
12. D.J. Watts, S.H. Strogatz, Collective dynamics of 'small-world' networks. Nature **393**, 440–442 (1998)
13. K. Kaneko, I. Tsuda, *Complex Systems: Chaos and Beyond: A Constructive Approach with Applications in Life Sciences* (Springer, Berlin, 2000).
14. S.H. Strogatz, Exploring complex networks. Nature **410**, 268–276 (2001)
15. Y. Kuramoto, D. Battogtokh, Coexistence of coherence and incoherence in nonlocally coupled phase oscillators. Nonlin. Phenom. Comp. Syst. **5**, 380–385 (2002)
16. D.M. Abrams, S.H. Strogatz, Chimera states for coupled oscillators. Phys. Rev. Lett. **93**, 174102 (2004)
17. S. Kawamoto, 2-D and 3-D solvable chaos maps. Chaotic Model. Simul. (CMSIM) **1**, 107–118 (2017)
18. S. Kawamoto, Chaotic time series by time-discretization of periodic functions and its application to engineering. Chaotic Model Simul. (CMSIM) **2**, 193–204 (2017)
19. R. FitzHugh, Impulses and physiological states in theoretical models of nerve membrane. Biophys. J. **1**, 445–466 (1961)
20. J. Nagumo, S. Arimoto, S. Yoshizawa, An active pulse transmission line simulating nerve axon. Proc. IRE. **50**, 2061–2070 (1962)
21. S. Kawamoto, The FitzHugh-Nagumo model and 2-D solvable chaos maps. Chaotic Model. Simul. (CMSIM) **3**, 269–283 (2018)
22. S. Kawamoto. The FitzHugh-Nagumo model and spatiotemporal fractal sets based on time-dependent chaos functions, in *11th Chaotic Modeling and Simulation International Conference*, Springer Proceedings in Complexity, ed. by C.H. Skiadas, I. Lubashevsky (Springer Nature Switzerland AG, 2019), pp. 159–173
23. M.J. Feigenbaum, Quantitative universality for a class of nonlinear transformations. J. Stat. Phys. **19**(1), 25–52 (1978)
24. M.J. Feigenbaum, The universal metric properties of nonlinear transformations. J. Stat. Phys. **21**(6), 669–706 (1979)
25. R.M. May, Limit cycles in predator-prey communities. Science **177**, 900–902 (1972)
26. Y. Nishiuchi, T. Ueta, Bifurcation analysis of a simple 3D BVP oscillator and chaos synchronization of its coupled systems, in *Handbook of Application of Chaos Theory*, ed. by C.H. Skiadas, C. Skiadas (Chapman and Hall/CRC Press, 2016), pp. 145–154
27. S. Kawamoto, Intermittency of chaos functions and the Belousov-Zhabotinsky reaction, in *12th Chaotic Modeling and Simulation International Conference*, Springer Proceedings in Complexity, ed. by C.H. Skiadas, Y. Dimotikalis (Springer Nature Switzerland AG, 2020), pp. 123–134
28. S. Kawamoto, Reaction-diffusion systems and propagation of limit cycles with chaotic dynamics, in *12th Chaotic Modeling and Simulation International Conference*, Springer Proceedings in Complexity, ed. by C.H. Skiadas, Y. Dimotikalis (Springer Nature Switzerland AG, 2020), pp. 135–149

29. B. Van der Pol, J. Van der Mark, Frequency demultiplication. Nature **120**(3019), 363–364 (1927)
30. K. Tomita, Periodically forced nonlinear oscillators, in *Chaos*, ed. by A.V. Holden (Manchester University Press, Manchester, 1986), pp. 213–214
31. A.L. Hodgkin, A.F. Huxley, A quantitative description of membrane current and its application to conduction and excitation in nerve. J. Physiol. **117**, 500–544 (1952)

Stochastic Properties of an Inverted Pendulum on a Wheel on a Soft Surface

O. M. Kiselev

Abstract We study dynamics of the inverted pendulum on the wheel on a soft surface and under a proportional-integral-derivative controller. The behaviour of such pendulum is modelled by a system with a differential inclusion. If the system has a sensor for the rotational velocity of the pendulum, the tilt sensor and the encoder for the wheel then this system is observable. The using of the observed data for the controller brings stochastic perturbations into the system. The properties of the differential inclusion under stochastic control is studied for upper position of the pendulum. The formula for the time, which the pendulum spends near the upper position, is derived.

1 Introduction

The wheeled inverted pendulum (WIP) is a popular model for studies nor only dynamics and the system of the control for robotics equipments near instability positions. A list of contemporary works in this field is too large. Here we should mention studies for the derivation of the mathematical model for the WIP and the control synthesis, which one can see for example in the book [1], and the article [2]. The questions concerned an stability and control for WIP with two wheels were considered in [3] for the horizontal and in [4] for the inclined surface see Fig. 1.

We apply the proportional-integral-derivative (PID) controller to stabilize the WIP at the upper point. We should mention the PID controller is often used to the objects of different nature [5]. For example the PID controller can be used to stabilize the WIP on the hard horizontal, see [6, 7] and a review [8]. But the soft surface is more complicated to stabilize the WIP. The mathematical model for the WIP controlled

O. M. Kiselev (✉)
Institute of Mathematics, ul. Chernyshevskogo 112, Ufa 450008, Russia
e-mail: ok@ufanet.ru

Innopolis University, ul. Universitetskaya 1, Innopolis 420500, Russia

C. H. Skiadas et al. (eds.), *13th Chaotic Modeling and Simulation International Conference*, Springer Proceedings in Complexity,
https://doi.org/10.1007/978-3-030-70795-8_28

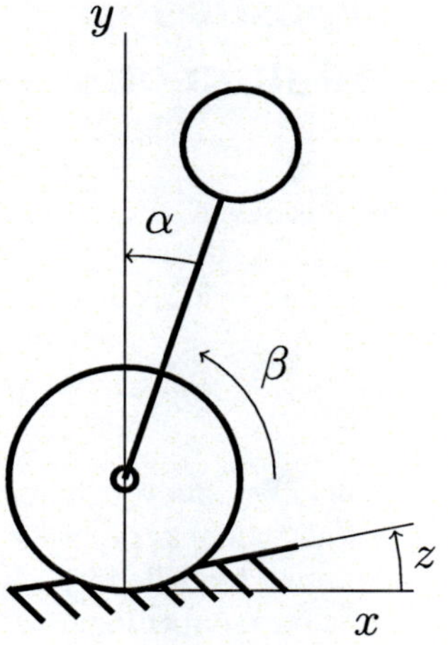

Fig. 1 The inverted pendulum on the wheel. Let us denote r is the radius of the wheel, l is the pendulum length, α is the angle of the pendulum turn, β is the angle of the wheel turn and z is the inclination angle of the surface

by the PID controller on the soft surface was offered and detailed studied in [9] (see also preprint [10]).

The major part of the PID controller is an observation value of the angle of the pendulum position. This angle can be found by gyroscope sensor. Usually the sensor obtains the value with a small stochastic error. The maximal amplitude and the dispersion for the error is standardized by specification of the sensor, see for example [11].

An additional sensor for the WIP is a tilt sensor. Such sensors are commonly used and have a detailed specification in which the interval of errors and the dispersion are pointed, see for example [12].

To obtain amount of the rotation of the wheel we use an encoder. If the wheel does not slip, the errors of the digital encoder appear due to round up only. Slipping brings additional errors to the value of the encoder. Therefore one of the problems for the controller is to detect the slipping.

In this work we show that the gyroscope and tilt for the pendulum and encoder for the wheel are enough to observe the state of the system in framework of the mathematical model.

However the noise of the sensors and the slipping bring stochastic errors into the value of the sensors. Therefore the controller which uses these values has stochastic perturbations. Hence the mathematical model with the digital PID controller is stochastic.

In Sect. 2 we describe the mathematical model of WIP on the soft surface with the PID controller. The dynamic model on the soft surface contains a rolling resistance. Pure mathematically this resistance is described by the differential inclusion, see [9].

In Sect. 3 one can find the formulas for the current state of the WIP. These formulas use the data obtained from the gyroscope, tilt sensor and encoder.

In Sect. 4 we discuss the stochastic properties of the data for the PID controller. The errors and dispersion of the data from the sensors are assumed as known form the technical notes.

In Sect. 5 we consider the properties of the WIP under the PID controller with stochastic perturbations. The perturbations appear in the control by using the data with the stochastic errors.

2 Dynamical System for the WIP

Here we will consider the moving of the WIP with the additional control torque on the wheel. The torque is denoted by u. Let us assume that the equipment moves on soft surface with inclination z. The value of z depends on the traversed path by the wheel and hence one can write $z = z(\beta)$. The mathematical model of such moving has the form (see [9], also preprint [10]):

$$\begin{aligned} \ddot{\alpha} &= \sin(\alpha) - (\cos(\alpha - z)\ddot{\beta} + \sin(\alpha - z)\dot{\beta}^2)\rho - 2\frac{\rho}{\zeta}u, \\ (\zeta + 2)\rho\,\ddot{\beta} &\in F(\alpha, \dot{\alpha}, \ddot{\alpha}, \dot{\beta}). \end{aligned} \tag{1}$$

Let us denote

$$f = -\sin(z) - \big(\ddot{\alpha}\cos(\alpha - z) - \dot{\alpha}^2\sin(\alpha - z)\big)\,\zeta + \frac{2}{\rho}u.$$

In formula (1) the map $F(\alpha, \dot{\alpha}, \ddot{\alpha}, \dot{\beta})$ has the form:

$$F(\alpha, \dot{\alpha}, \ddot{\alpha}, \dot{\beta}) = \begin{cases} f - \nu\,\mathrm{sgn}(\dot{\beta}), & \{\forall\dot{\beta} \neq 0\}; \\ (-\nu, \nu), & \{\dot{\beta} = 0\} \cup \{|f| \leq \nu\}; \\ f, & \{\dot{\beta} = 0\} \cup \{\{\alpha, \dot{\alpha}, \ddot{\alpha}\} \in \{|f| > \nu\}\}. \end{cases}$$

Here the parameters of the mathematical model are following: α is an angle of the pendulum turn, β is the angle of wheel turn, z is the current inclination of the soft surface, ν is the torque of the friction resistant, $\rho = r/l$ is the ratio of the wheel radius and the length of the pendulum, ζ is the ratio of pendulum mass and the rim mass.

The system (1) can be simplify for hard ($\nu = 0$) surface with the constant inclination ($z \equiv \epsilon$). As a result one gets the second order equation for α:

$$\begin{aligned} (\sin^2(\epsilon - \alpha)\zeta + 2)\ddot{\alpha} &= (2 + \zeta)\sin(\alpha) + \sin(\epsilon)\cos(\epsilon - \alpha) + \\ \frac{1}{2}\zeta\dot{\alpha}^2\sin(2(\epsilon - \alpha)) &- 2\left(\frac{1}{\rho}\cos(\epsilon - \alpha) + \left(1 + \frac{2}{\zeta}\right)\rho\right)u. \end{aligned} \tag{2}$$

The particular case for the WIP on the hard horizontal ($\epsilon = 0$) looks as follow:

$$(\sin^2(\alpha)\zeta + 2)\ddot{\alpha} = (2+\zeta)\sin(\alpha) - \tfrac{1}{2}\zeta\dot{\alpha}^2\sin(2\alpha) - 2\left(\tfrac{1}{\rho}\cos(\alpha) + \left(1+\tfrac{2}{\zeta}\right)\rho\right)u \tag{3}$$

The control torque with the PID controller has the following form:

$$u = k_1\alpha + k_2\dot{\alpha} + k_3 A, \quad \text{where,} \quad A \equiv \int^t \alpha(t)dt.$$

In this case system (1) has a particular solution:

$$\alpha \equiv 0, \qquad A = \frac{\operatorname{sgn}(\dot{\beta})\zeta\nu\rho}{(2k_3\zeta + 4k_3)\,\rho^2 + 2k_3\zeta},$$
$$\beta = \begin{cases} \beta_0 + \beta_1(t-t_0) - \dfrac{\zeta\nu\rho\operatorname{sgn}(\dot{\beta})}{(2k_3\zeta+4k_3)\,\rho^2+2k_3\zeta}\dfrac{(t-t_0)^2}{2}, & (t-t_0) < T; \\ \beta_0 + \beta_1 T - \dfrac{\zeta\nu\rho\operatorname{sgn}(\dot{\beta})}{(2k_3\zeta+4k_3)\,\rho^2+2k_3\zeta}\dfrac{T^2}{2}, & (t-t_0) \geq T, \end{cases} \tag{4}$$

where

$$T = \frac{1}{\beta_1}\frac{\zeta\nu\rho\operatorname{sgn}(\dot{\beta})}{(2k_3\zeta+4k_3)\,\rho^2+2k_3\zeta}, \quad \{t_0, \beta_0, \beta_1\} \in \mathbb{R}.$$

There exists the set of the parameters $\zeta, \rho, k_1, k_2, k_3$ when solution (4) is an attractor as $(t-t_0) < T$ [9] (see also preprint [10]).

In an ideal case the control should be defined by the current values of α, $\dot{\alpha}$, A, but for real equipment these parameters can be obtained using the sensors at the moment t_i, where $i \in \mathbb{N}$. As a result the control is a discrete function: $u(t) = u(t_i) = u_i$.

At the interval $t \in (t_i, t_{i+1})$ the control torque u_i is a constant. Such system has a first integral and can be integrate in quadratures.

For example the moving on the hard surface with the constant inclination (2) has a first integral at the interval $t \in (t_i, t_{i+1})$:

$$\mathcal{E}_i = \cos(\alpha)(\zeta+2) + \sin(\epsilon)\sin(\epsilon-\alpha) + \left(\tfrac{1}{2}\sin^2(\epsilon-\alpha)\zeta + 1\right)\dot{\alpha}^2 + \left(2\left(1+\tfrac{2}{\zeta}\right)\rho\alpha - \tfrac{2}{\rho}\sin(\epsilon-\alpha))\right)u_i. \tag{5}$$

This formula allows us to integrate $\dot{\alpha}$ at the interval $t \in (t_i, t_{i+1})$ and we can write the parameters of the system at $t = t_{i+1}$:

$$\alpha_{i+1} = F_1(\alpha_i, \dot{\alpha}_i, u_i, dt),$$
$$\dot{\alpha}_{i+1} = F_2(\alpha_i, \dot{\alpha}_i, u_i, dt),$$
$$A_{i+1} = F_3(\alpha_i, \dot{\alpha}_i, u_i, dt)$$

One can obtain the first integral for WIP on the hard horizontal surface (3) if one assumes $\epsilon = 0$.

The mathematical model for the WIP on the soft surface does not integrate obviously. Nevertheless this model can be written in the form:

$$\dot{A} = \alpha, \quad \dot{\alpha} = a,$$
$$\dot{a} = \sin(\alpha) - (\cos(\alpha - z)\dot{b} + \sin(\alpha - z)b^2)\rho - 2\frac{\rho}{\zeta}u,$$
$$\dot{\beta} = b, \quad \dot{b} \in \frac{1}{(2+\zeta)\rho}F(\alpha, a, \dot{a}, b).$$

One can obtain the numeric solution of this differential inclusion at the interval $t \in (t_i, t_{i+1})$. Let us define the map:

$$(A_n, \alpha_n, a_n, \beta_n, b_n) \to (A_{n+1}, \alpha_{n+1}, a_{n+1}, \beta_{n+1}, b_{n+1}).$$

Formally this map can be written like a discrete dynamical system:

$$\mathbf{X}_{n+1} = \mathbf{F}(\mathbf{X}_n), \quad \text{where} \quad \mathbf{X}_n = (A_n, \alpha_n, a_n, \beta_n, b_n, u_n).$$

3 Observability of the Mathematical Model for WIP

In this section we consider the set of the data necessary for the observability of the parameters of the mathematical model for WIP (1).

The angle of the tilt for the pendulum is defined by gyroscope. The gyroscope can be work in two different cases. The first one it defines the angle of the pendulum and the second one it define the angle velocity for the pendulum. In the second case one should integrates the angle velocity to obtain the pendulum angle. Below we will use the gyroscope in the mode of angle velocity. This means the value of the angle velocity $\dot{\alpha}$ is known at the moment of the measurement.

Besides the gyroscope we assume as existing the tilt sensor. This sensor define the linear acceleration of the pendulum in the plane of the moving of WIP.

Let us define the coordinates as (x, y), where x is the horizontal coordinate and y is the vertical one. The projections of the acceleration vector on the coordinate axes Ox and Oy one can write as follows:

$$\ddot{x} = \ddot{\beta} r \cos(z(\beta)) + \ddot{\alpha} l \sin(\alpha),$$
$$\ddot{y} = \ddot{\beta} r \sin(z(\beta)) + \ddot{\alpha} l \cos(\alpha) + g.$$

It is convenient to write these formulas in the form:

$$\ddot{\alpha} l \cos(\alpha + z) = -\ddot{x} \sin(z) + \ddot{y} \cos(z) - g \sin(z),$$
$$\ddot{\beta} r \cos(\alpha + z) = \ddot{x} \cos(\alpha) - \ddot{y} \sin(\alpha) + g \sin(\alpha).$$

One more sensor is the encoder. This sensor allows us to define the turn of the wheel. The data from the encoder allow to obtain the mean value of the angle velocity of the wheel as value of the difference between the current value of the turn angle of the wheel and another one value at previous measurement:

$$\dot{\beta} \sim \frac{\beta(t) - \beta(t - \Delta t)}{\Delta t}.$$

Let us consider the system for WIP on the horizontal surface (i.e. $z = 0$):

$$\ddot{\alpha} l \cos(\alpha) = \ddot{y}$$

$$\ddot{\beta} r \cos(\alpha) = \ddot{x} \cos(\alpha) - \ddot{y} \sin(\alpha) + g \sin(\alpha).$$

The value $a_1 = \dot{\alpha}$ is known from the sensor. Let us define by $a_2(t) = \ddot{y}/l$, $b_2 = \ddot{x}/r$, $\gamma = g/l$ and $b_1 = \dot{\beta}$. Then the dynamical system (1) can be written as the system of the trigonometric equation and the inclusion:

$$\frac{a_2}{\cos(\alpha)} = -\left(\cos(\alpha - z)\left(\frac{\sin(\alpha)\gamma}{\cos(\alpha)\rho} - \frac{a_2 \sin(\alpha)}{\cos(\alpha)\rho} + b_2\right) + b_1^2 \sin(\alpha - z)\right)\rho$$
$$- \frac{2u\rho}{\zeta} + \sin(\alpha), \quad (6)$$

$$(\zeta + 2)\left(\frac{\sin(\alpha)\gamma}{\cos(\alpha)\rho} - \frac{a_2 \sin(\alpha)}{\cos(\alpha)\rho} + b_2\right)\rho \in \begin{cases} f - \nu \operatorname{sgn}(b_1), & b_1 \neq 0; \\ (-\nu, \nu), & b_1 = 0 \cup |f| < \nu; \\ f, & b_1 = 0 \cup |f| \geq \nu; \end{cases} \quad (7)$$

where

$$f = \frac{2u}{\rho} - \left(\frac{a_2 \cos(\alpha - z)}{\cos(\alpha)} - a_1^2 \sin(\alpha - z)\right)\zeta - \sin(z).$$

The angle of the pendulum α and the control torque u are the unknown variables in the system (6) and (7).

If $b_1 \neq 0$ or $b_1 = 0 \cup |f| \geq \nu$ then the inclusion (7) turn to the following equation:

$$(\zeta + 2)\left(\frac{\sin(\alpha)\gamma}{\cos(\alpha)\rho} - \frac{a_2 \sin(\alpha)}{\cos(\alpha)\rho} + b_2\right)\rho =$$
$$\frac{2u}{\rho} - \left(\frac{a_2 \cos(\alpha - z)}{\cos(\alpha)} - a_1^2 \sin(\alpha - z)\right)\zeta - \sin(z) - \nu \operatorname{sgn}(b_1).$$

As a result one get the system of the equations for α, u. The control torque u can be easy found through the trigonometric functions of α and hence one get the trigonometric equation for α.

As $\{b_1 = 0\} \cup \{|f| < \nu\}$ the angle of the pendulum should be solution of the inequality:

$$\frac{-\nu}{\zeta + 2} - b_2\rho < (\gamma - a_2)\tan(\alpha) < \frac{\nu}{\zeta + 2} - b_2\rho.$$

Here one get the observed parameter u. To obtain the integral term A of the PID controller one should use the following formula:

$$A = u - \frac{k_1}{k_3}\alpha - \frac{k_2}{k_3}a_1.$$

Theorem 1 *Let one know the values of the acceleration* $(\ddot{x}, \ddot{y})$, *angle velocity of the pendulum* $\dot{\alpha}$ *and angle velocity of the wheel* $\dot{\beta}$, *then the observed dynamical system is solution of the trigonometric equation (6) and the inclusion (7).*

The equations for small values of $\nu, \alpha, \ddot{y}, \dot{\alpha}, \ddot{x}, \dot{\beta}$ and $(z \equiv 0)$ can be written in the following form:

$$a_2 \sim -\frac{2\rho u}{\zeta} + (1 - \gamma)\alpha - \rho b_2,$$

$$(\zeta + 2)(\rho b_2 + \gamma\alpha) \in \begin{cases} \sim \frac{2u}{\rho} - \zeta a_2 - \operatorname{sgn}(b_1)\nu, & b_1 \neq 0; \\ \sim (-\nu, \nu), & b_1 = 0 \cup |-\zeta a_2 + 2u/\rho| < \nu; \\ \sim \frac{2u}{\rho} - \zeta a_2, & b_1 = 0 \cup |-\zeta a_2 + 2u/\rho| \geq \nu. \end{cases} \tag{8}$$

Corollary 1 *The important case for the WIP on the hard horizontal is more simplest. In particular the angle* α *is the solution of the equation:*

$$\begin{aligned} \rho^2 \sin(\alpha)\,(2\gamma\zeta - 2a_2\zeta + 4\gamma - 4a_2) + \\ \zeta \sin(2\alpha)\left(-a_1^2\rho^2 + b_1{}^2\rho + \gamma - a_2 - 1\right) = \\ \rho^2 \cos(\alpha)\,(-2b_2\zeta\,\rho - 4b_2\,\rho - 2a_2\zeta) - \\ b_2 \cos(2\alpha)\zeta\rho - b_2\zeta\rho - 2a_2\,\zeta \end{aligned} \tag{9}$$

For small values of $\nu, \alpha, \ddot{y}, \dot{\alpha}, \ddot{x}, \dot{\beta}$ we get:

$$\alpha \sim -\frac{(\rho^2 + 1)\,\zeta\, a_2 + \left((\rho^3 + \rho)\,\zeta + 2\rho^3\right) b_2}{\left((\rho^2 + 1)\,\gamma - 1\right)\zeta + 2\rho^2\gamma},$$

$$u \sim -\frac{\left(\rho\,\zeta^2 + 2\rho\gamma\zeta\right) a_2 + \left(\rho^2\zeta^2 + 2\rho^2\zeta\right) b_2}{\left((2\rho^2 + 2)\,\gamma - 2\right)\zeta + 4\rho^2\gamma}.$$

4 The Observability and the Stochastic Properties

The value of the controlling torque at t_i is defined by the measurement of observed values of the parameters of the dynamic system. The obtained data from the sensors and the computed observed data at t_i will be denoted by variables with upper symbol $\breve{\bullet}$.

The absolute errors and the dispersion are known for typical sensors. Below we will assume that we know the standard deviation σ_j of the measured data at $t = t_j$.

Let the measured data be following:

$$\breve{a}_1 = \dot{\alpha} + \delta^{(1)},\ \breve{a}_2 = \ddot{y}/l + \delta^{(2)},\ \breve{b} = \beta + \delta^{(3)},\ \breve{b}_1 = \dot{\beta} + \delta^{(4)},\ \breve{b}_2 = \ddot{x}/r + \delta^{(5)}.$$

Here $\delta^{(i)}$ is stochastic error.

Let us consider the WIP on the hard horizontal ($\nu = 0$ and $z \equiv 0$). We will assume the errors are small and one can use a linear system for find observable values of α, u.

$$\frac{a_2}{\cos(\alpha)} = -\left(\cos(\alpha)\left(\frac{\sin(\alpha)\gamma}{\cos(\alpha)\rho} - \frac{a_2\sin(\alpha)}{\cos(\alpha)\rho} + b_2\right) + b_1^2\sin(\alpha)\right)\rho - \frac{2u\rho}{\zeta} + \sin(\alpha)$$

$$(\zeta + 2)\left(\frac{\sin(\alpha)\gamma}{\cos(\alpha)\rho} - \frac{a_2\sin(\alpha)}{\cos(\alpha)\rho} + b_2\right)\rho = \frac{2u}{\rho} - \left(a_2 - a_1^2\sin(\alpha)\right)\zeta$$

One can derive the equation for α:

$$\begin{aligned}&\{((-\gamma + a_1^2\cos(\alpha) + a_2)\zeta - 2\gamma + 2a_2)\rho^2 - \\ &b_1^2\cos(\alpha)\zeta\rho + a_2\cos(\alpha)\zeta\}\sin(\alpha) = \\ &(b_2\cos(\alpha)\zeta + 2b_2\cos(\alpha))\rho^3 + (\mathrm{sgn}(b_1)\cos(\alpha)\nu + a_2\cos(\alpha)\zeta)\rho^2 + \\ &b_2\cos(\alpha)^2\zeta\rho + a_2\zeta.\end{aligned} \tag{10}$$

For small $\ddot{y}, \alpha, \dot{\alpha}, \ddot{x}, \dot{\beta}, \nu$ we obtain the formulas for errors of the observed data α and A:

$$\breve{\alpha} \sim \alpha - \frac{((\zeta + 2)\rho^2 + \zeta)\rho}{(\zeta + 2)\gamma\rho^2 + (\gamma - 1)\zeta}\delta^{(5)} + \frac{(\rho^2 + 1)\zeta}{(\zeta + 2)\gamma\rho^2 + (\gamma - 1)\zeta}\delta^{(2)}$$

$$\begin{aligned}\breve{u} \sim u &+ \frac{(\gamma - 1)\zeta\nu\rho}{(2\gamma\zeta + 4\gamma)\,\rho^2 + (2\gamma - 2)\,\zeta}(\mathrm{sgn}(b_1 + \delta_4) - \mathrm{sgn}(b_1)) \\ &- \frac{(\zeta^2 + 2\gamma\zeta)\rho}{(2\gamma\zeta + 4\gamma)\,\rho^2 + (2\gamma - 2)\,\zeta}\delta^{(2)} - \frac{(\zeta^2 + 2\zeta)\rho^2}{(2\gamma\zeta + 4\gamma)\,\rho^2 + (2\gamma - 2)\,\zeta}\delta^{(5)},\end{aligned}$$

hence:

$$\breve{A} \sim \breve{u} - \frac{k_1}{k_3}\breve{\alpha} - \frac{k_2}{k_3}(\dot{\alpha} + \delta^{(1)}).$$

Remark While the WIP moves on the soft horizontal the stochastic layer appears near the hyperplane $\dot{\beta} = 0$. The width of this layer is $\min\{\delta^{(4)}\} \leq \dot{\beta} \leq \max\{\delta^{(4)}\}$. In this layer the stochastic error can be $\pm\nu$ when $|f| > \nu$. It is important the value of this errors defines by value of the rolling resistance for the wheel and does not depend on the error of the encoder.

The data with stochastic errors are used in the PID controller. As a result the stochastic perturbations appear in the mathematical model for the WIP (3) and in the systems (2) and (1). Therefore the mathematical model with the PID controller looks like the stochastic differential inclusion (1). In partial the results for the observed values of $(\beta, \dot{\beta})$ and α with stochastic errors are showed in the Figs. 2 and 3.

The remark about filtering data

The current values of the parameters of the dynamic system for WIP one can obtain by the different approaches.

The first one is the integration of the differential inclusion as the predetermined process. Such approach gives the errors at any step of the integration because of two causes. First of all this errors appear because of the errors in the initial data on the first step of integration. One more cause of the appearance of the errors is the inaccuracy of the mathematical model.

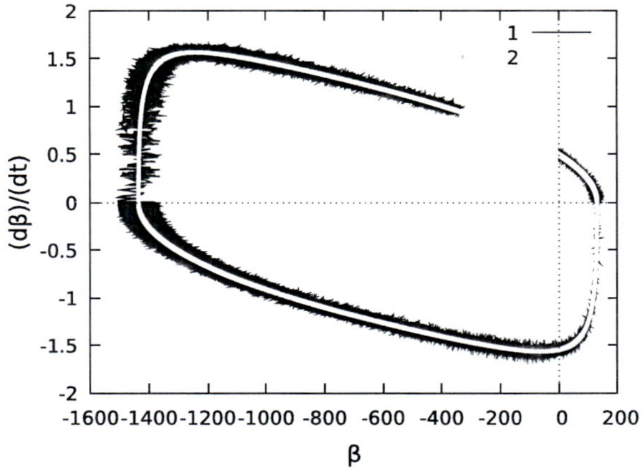

Fig. 2 The observed phase curve on the plane $(\beta, \dot{\beta})$ are black. This curve is obtained using the data from the sensors as $z \equiv 0$. The white line is the result of numeric solution for the system for WIP on soft surface. The feedback controller use the tilt sensor and the gyroscope (1). The parameters of the system are following: $\rho = 0.2$, $\zeta = 10$, $\nu = 0.05$, $\gamma = 1$, the PID coefficients are: $k_1 = 1.7$, $k_2 = 0.2$, $k_3 = 0.02$. The relative errors are uniform distributed data at the interval $(-0.02, 0.02)$. The dynamic system (1) solved at $A \sim 0.2385$, $\alpha = 0.02$, $\dot{\alpha} = 0$, $\beta = 0$, $\dot{\beta} = 0.5$ by Runge-Kutta method of fourth-order method with the step 0.1

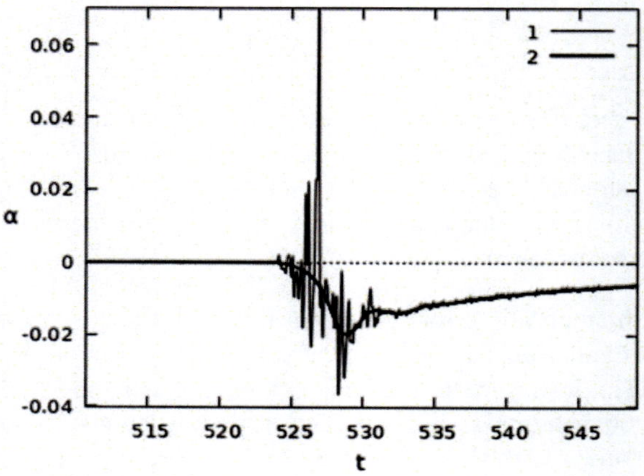

Fig. 3 On the picture one can see the result of numeric modelling for the angle of the pendulum of WIP on the soft horizontal. The horizontal axis defines values of the time and the vertical axis defines the value of the angle of the pendulum. The curve 2 is the numeric value and the curve 1 shows the model of the observed data. The observed data $\breve{\ddot{\alpha}}_i$ and $\breve{\dot{\alpha}}_i$ are modelled using current values of $\ddot{\alpha}_i$ and $\dot{\alpha}_i$ with the uniform distribution of the relative error at the interval $(-0.05, 0.05)$. The value of the angle $\breve{\alpha}_i$ is defined as the observed calculated using the angle acceleration and angle velocity from (10). The values $\breve{\ddot{\alpha}}$ and $\breve{\dot{\alpha}}$ were used to obtain $\breve{A}$ by integrating by the trapezoidal rule. The value of the control torque u_{i+1} was obtained at the interval $t \in (t_i, t_{i+1})$. The system of the equations was solved at the interval $t \in (t_i, t_{i+1})$ with the constant value of the control torque $u = u_{i+1}$ by Runge-Kutta fourth-order method with the step equals by 0.1

Another one approach is to use the observability of this system. This case does not needed to integrate the differential inclusion. But the errors appears in the current moment because of the errors of the measurement of the data using the sensors.

To minimize the quadratic deviation of the data one can combine the observed data and the forecast data using the deterministic mathematical model. Such algorithms are called as the filters. The filters for the linear system are well-knowing, see [13–15]. For non-linear smooth systems like the WIP on the hard surface is convenient the generalized Kalman's filter, see [16, 17]. But for the considered here case of WIP on the soft surface the generalized Kalman's filter is not appropriated because of non-linear the dynamical system in the neighbourhood of the hyper-surface $\dot{\beta} = 0$. One of the opportunity to use filtering in such case is the sigma-point filter, see [18, 19].

5 Stochastic Properties for WIP on Soft Horizontal

Let the interval dt between the moments of the measurements be small. Then one can see at the dynamic system as a determined dynamic system (1) with stochastic perturbation. The stochastic perturbation is contained in the control torque:

$$\tilde{u}_i = k_1\tilde{\alpha}_i + k_2\tilde{\dot{\alpha}} + k_0\tilde{A}_i.$$

In the work [9] (see also preprint [10]) it was shown that the unperturbed dynamical system with the PID controller has the attractor as $\text{sgn}(\dot{\beta}) = \pm 1$. This attractor is a line belonged the fifth-dimensional phase space: $(A, \alpha, \dot{\alpha}, \beta, \dot{\beta}) = (A_\pm, 0, 0, 0, \dot{\beta})$, where

$$\frac{2}{\rho}k_0 A_\pm = \pm\nu.$$

On this line the system for WIP is unstable and due to the perturbations crosses to the trajectory with changing of the sign of rotation of the wheel from $\text{sgn}(\dot{\beta}) = \pm 1$ to $\text{sgn}(\dot{\beta}) = \mp 1$. As a result the numeric modelling gives the trajectory like the hysteresis loop, see [9].

Let us consider here the impact of the stochastic perturbation on the stability for the hysteresis loop. The typical trajectory for the system with the stochastic perturbation is shown on the Fig. 4.

Theorem 2 *The line* $(A_\pm, 0, 0, 0, \dot{\beta})$ *as* $\text{sgn}(\dot{\beta}) = \pm 1$ *is the attractor for the stochastic system (1).*

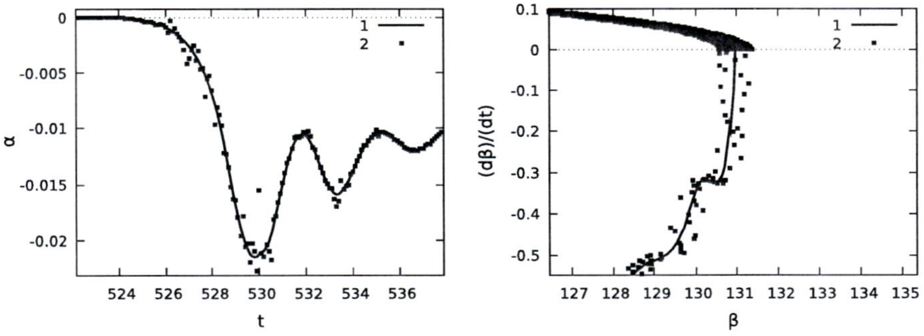

Fig. 4 In this picture one can see the result of the numeric modelling for the behaviour of the angle for the pendulum at $\xi = 10$, $\rho = 0.2$, $\nu = 0.05$, $\gamma = 1$, $k_1 = 1.7$, $k_2 = 0.2$, $k_3 = 0.02$. On the left picture the horizontal axis shows the time variable t and the vertical axis shows the angle of the pendulum α. On the right picture the horizontal axis shows β and the vertical axis shows $\dot{\beta}$. The step of the change of the control torque is 0.1. The line is the solution under the discrete control. The results of the measurements $\breve{\ddot{\alpha}}_i$, $\breve{\dot{\alpha}}_i$ and $\breve{\dot{\beta}}_i$ are modelled by the current values $\ddot{\alpha}_i$, $\dot{\alpha}_i$ and $\dot{\beta}_i$ with the uniform distribution of the relative errors at the interval $(-0.003, 0.003)$. The value of the angle $\breve{\alpha}_i$ is defined as the observed data through the angle acceleration and the angle velocity using (1). The value $\breve{A}$ is computed using $\breve{\alpha}$, $\breve{\dot{\alpha}}$ integrating by the trapezoidal method. It allows to obtain the control torque u_{i+1} at the interval $t \in (t_i, t_{i+1})$, where $t_{i+1} - t_i = dt$. At $t \in (t_i, t_{i+1})$ the system for the WIP on the soft horizontal is solved for the constant value the control torque $u = u_{i+1}$ by the Runge-Kutta method of the fourth order with the step equals 0.01

This theorem is corollary from the results of [9] concerning the stability of the line $(A_{\pm}, 0, 0, 0, \dot{\beta})$ as $\mathrm{sgn}(\dot{\beta}) = \pm 1$ for pure determined dynamical system for WIP under the PID controller and the theorem about stability under constantly perturbations [20].

The layer $|\dot{\beta}| \le \max\{\delta^{(3)}\}$ appears in the stochastic system near the hyper-plane $\dot{\beta} = 0$. In this layer the term $\nu \mathrm{sgn}(\dot{\beta})$ takes the random values $\pm\nu$ at $t \in (t_i, t_{i+1})$.

There exists the small neighbourhood $(\Delta_{\pm})$ near the unstable lines $(A_{\pm}, 0, 0, 0, \dot{\beta})$, where can be obtained four typical cases:

- Let $\dot{\beta} > 0$, $A < A_+$
  - and $\mathrm{sgn}\check{\dot{\beta}} = 1$, then the trajectory is kept in the neighbourhood of the line $(A_+, 0, 0, 0, \dot{\beta})$;
  - and $\mathrm{sgn}\tilde{\dot{\beta}} = -1$, then the trajectory is kept in the neighbourhood of the line $(A_+, 0, 0, 0, \dot{\beta})$.
- Let $\dot{\beta} < 0$, $A > A_-$
  - and $\mathrm{sgn}\check{\dot{\beta}} = -1$, then the trajectory is kept in the neighbourhood of the line $(A_-, 0, 0, 0, \dot{\beta})$;
  - and $\mathrm{sgn}\check{\dot{\beta}} = 1$, then the trajectory is kept in the neighbourhood of the line $(A_-, 0, 0, 0, \dot{\beta})$.

The sequence of the changes of the trajectories at the neighbourhoods of the lines $(A_{\pm}, 0, 0, 0)$ leads to the appearance of the hysteresis loop at the phase plane $(\beta, \dot{\beta})$, see Fig. 2.

Here it is important for applications the average time,which the WIP spends in the neighbourhood of the upper position.

The time between the sequence measurements is equal dt. Let the trajectory be in the neighbourhood $\Delta_{\pm}$ of the unstable line. The probability of $\mathrm{sgn}(\delta^{(3)}) = \pm 1$ in primary order as $\Delta_{\pm} \to 0$ equals $p_{\pm} \sim 1/2$. The average time for trajectory in this neighbourhood is following:

$$T_0 = dt \sum_{n=1}^{\infty} \frac{n}{2^n} = 2dt.$$

Theorem 3 *The average time spending at* Δ*-neighbourhood of the unstable lines* $(A_{\pm}, 0, 0, 0, \dot{\beta})$ *for the stochastic system (1) equals* $2dt$*, where* dt *is the time between the sequenced measurements of the state for the system.*

6 Conclusion

The system for the WIP with discrete control by the PID controller is stochastic due to the errors of the measurements. The stabilizing of the WIP on the soft surface leads to the appearance of the hysteresis loop in the plane of the phase variables β, $\dot{\beta}$. The average time spending near the upper position was calculated.

References

1. A.M. Formalskii, *Stabilisation and Motion Control of Unstable Objects*. De Gruyter Studies in Mathematical Physics, vol. 33 (2016)
2. Yu.G. Martynenko, A.M. Formal'skii, Controlled pendulum on a movable base. Mech. Solids **48**, 6–18 (2013)
3. K. Pathak, J. Franch, S.K. Agrawal, Velocity and position control of a wheeled inverted pendulum by partial feedback linearization. IEEE Trans. Robot. **21**, 505–513 (2005)
4. D.S. Nasrallah, H. Michalska, J. Angeles, Controllability and posture control of a wheeled pendulum moving on an inclined plane. IEEE Trans. Robot. **23**(3), 564–577 (2007)
5. K.J. Åström, T. Hågglund, *PID Controllers*, 2nd edn. (1994)
6. M.A. Ahmad, A.N.K. Nasir, R.M.T. Raja Ismail, Performance comparison between sliding mode control (SMC) and PD-PID controllers for a nonlinear inverted pendulum system. World Acad. Sci. Eng. Technol. **71**, 122–127 (2010)
7. A. Shimada, N. Hatakeyama, Movement control using zero dynamics of two-wheeled inverted pendulum robot, in *10th IEEE International Workshop on Advanced Motion Control* (2008), pp. 38–43
8. C.R. Halkyard, R.P.M. Chan, K.A. Stol, Review of modelling and control of two-wheeled robots. Annu. Rev. Control **37**, 89–103 (2013)
9. O.M. Kiselev, Stabilization of the wheeled inverted pendulum on a soft surface. Russ. J. Nonlinear Mech. **16**(3)
10. O.M. Kiselev, Stabilization of the wheeled inverted pendulum on a soft surface, arxiv:2006.05450
11. InventSense, *MPU-6000/MPU-6050 Product Specification*, 08 2013. Rev. 3.4
12. Freescale Semiconductor, $\pm 1.5\,g$, $\pm 6\,g$ *Three Axis Low-g Micromachined Accelerometer*, 04 2008. Rev. 0
13. K. Brammer, G. Siffling, *Kalman-Bucy-Filter, Deterministische Beobachtung und stochastische Filterung*. Methoden der Regelungstechnik
14. R.S. Bucy, R.E. Kalman, New results in linear filtering and prediction theory **83**, 95–108
15. R.E. Kalman, A new approach to linear filtering and prediction problems **82**(D), 35–45
16. J.W. Austin, C.T. Leondes, Statistically linearized estimation of reentry trajectories **17**, 54–61
17. A. Bertolini, M. Athans, R.P. Wishner, Suboptimal state estimation for continuous-time nonlinear systems from discrete noisy measurements **13**, 504–518
18. S.J. Julier, J.K. Uhlmann, A new extension of the Kalman filter to nonlinear systems, pp. 182–193
19. J.K. Uhlrnann, S.J. Julier, H.F. Durrant-Whyte, A new approach for filtering nonlinear systems, pp. 1628–1632
20. N.N. Krasovskii, *Nekotorye zadachi teorii ustoichivisti dvizheniya*. FizMatLit

Electrodynamics of Planar Reflecting Structures with Chiral Layers Based on Thin-Wire Helices and Fractal Elements

Dmitriy S. Klyuev, Anatoly M. Neshcheret, Oleg V. Osipov, and Alexander A. Potapov

Abstract The paper considers methods for calculating the electrodynamic characteristics of single and multi-layered planar chiral metamaterials based on thin-wire perfectly conducting helices and fractal elements placed in a dielectric container. The effective permittivity is determined by using the Maxwell Garnett model for a chiral metamaterial. In this paper the dispersion dependences of material parameters for the considered metamaterial are obtained. The problems of plane electromagnetic waves reflection from metastructures based on one and two planar layers of a chiral metamaterial are solved. The possibility of discrete-multi-frequency concentration of incident microwave energy at a number of resonant frequencies is proved. The matrix theory for the description of a multilayered chiral metamaterial is considered and the relations for the transmission matrices of the chiral layer based on helixes are obtained. It is proved that a two-layer chiral-dielectric metamaterial based on thin-wire conducting helices near a predetermined frequency allows performing a frequency-selective conversion of a normally incident electromagnetic energy flow into azimuthal scattering in the plane of the metastructure.

Abbreviations

CM	Chiral Metamaterial
EMF	Electromagnetic Field

D. S. Klyuev · A. M. Neshcheret · O. V. Osipov
Povolzhskiy State University of Telecommunications and Informatics, L. Tolstoy Str., 23, 443010 Samara, Russia
e-mail: klyuevd@yandex.ru

A. A. Potapov (✉)
V. A. Kotelnikov Institute of Radio Engineering and Electronics, Russian Academy of Sciences, Mokhovaya str., 11-7, 125009, RussiaMoscow
e-mail: potapov@cplire.ru

JNU-IREE RAS Joint Lab. of Information Technology and Fractal Processing of Signals, JiNan University, Guangzhou 510632, China

C. H. Skiadas et al. (eds.), *13th Chaotic Modeling and Simulation International Conference*, Springer Proceedings in Complexity,
https://doi.org/10.1007/978-3-030-70795-8_29

EMW Electromagnetic Wave
LCP Left-Circular Polarization
RCP Right-Circular Polarization
UHF Ultra-High Frequency

1 Introduction

Currently the research of so-called metamaterials, i.e. structures with electromagnetic properties atypical for natural media, is of great interest in the microwave and optical ranges. One of the most interesting from the point of view of using metamaterials is chiral media, which are composite structures consisting of a homogeneous container and conducting elements of a mirror-asymmetric shape placed in it. Since the 90s. XX century there is a huge number of scientific publications devoted to the studing of chiral media and structures based on them. The examples of chiral elements are flat helices, S-elements, gammadions, double open rings, Tellegen elements, etc. Significant contributions to the development of the theory of chiral metamaterials were made by S. A. Tretyakov, S. L. Prosvirnin, I. V. Semchenko, A. H. Sihvola, A.D. Shatrov, V. V. Shevchenko, I. V. Lindell, B. Z. Katzenelenbaum, A. J. Viitanen, A. Lakhtakia and many others [1–10]. The main applications of chiral media are frequency and polarization-selective microwave devices, polarization converters, low-reflection coatings, media with negative refraction, etc.

Interesting from the point of view of chiral elements using are thin-wire elements in the form of open rings with protruding ends (the Tellegen element), multi-turn helices, which represent a certain superposition of electric and magnetic dipoles. We know the results of research on the diffraction of plane electromagnetic waves on Tellegen elements (S. A. Tretyakov, F. Mariotte) [11], cylinders with helical conductivity (B. Z. Katzenelenbaum, A.D. Shatrov et al.) [7], where it is shown that polarizationally selective resonant phenomena are observed in structures in the form of long helices of a small radius compared to the wavelength and lattices based on them. Frequency and polarization selective properties of interaction with electromagnetic radiation are revealed in chiral media based on such elements. In most cases the theory of electric circuits is used to analyze the electromagnetic properties of helical elements and there are a small number of papers that offer strict electrodynamic methods. One of the problems is to build a dispersion model of a chiral element and a meta-media based on a set of such elements.

In the proposed work, a chiral metamaterial (CM) based on thin-wire multi-turn helices placed in a dielectric container is chosen as the research object. The interest is the construction of a mathematical model of a helical element based on its electrical characteristics (inductance and capacitance), as well as a model of the metamaterial itself based on a matrix of helical elements. As a result of constructing such a model, frequency dependences of the effective permittivity and the chirality parameter are obtained. Another interesting aspect is the studing of multilayered and periodically inhomogeneous chiral metastructures based on helix elements using a new dispersion model. From a practical point of view, we have obtained results concerning the possibility of using a chiral metamaterial for frequency selective redirection of the incident microwave energy flow by converting it into radiation in the metastructure plane at certain frequencies.

2 Metamaterial Dispersion Model

In [12], a mathematical model of a thin-wire ideally conducting helical element and a metamaterial is constructed on the basis of a set of such elements, taking into account the dispersion of material parameters and the interaction between neighboring elements. Based on the theory of circuits, the resonant frequencies of a thin-wire multi-turn chiral element were calculated through inductance and capacitance. In contrast to other works, when calculating the capacity, corrections related to the inter-turn capacity of the helix were calculated, and the interaction of the helix with neighboring elements was taken into account by calculating the inter-element capacity. Using such a low-frequency model of the element allowed us to take into account an arbitrary number of turns of the helical element and also its electromagnetic interaction with the surrounding helixes, which is extremely important due to the presence of spatial dispersion of the chiral media. Also, unlike most other models, the wire is not considered infinitely thin. The geometry of the helix element is shown in Fig. 1.

For Fig. 1 the following symbols are entered: N is number of turns; R is the radius of the helix; h is step of a helix; l is length of the helix in the expanded state and r is the radius of the thin wire.

After calculating the resonant frequency of the helical element, the metamaterial was described by the Maxwell Garnett formula taking into account the dispersion [10, 11]:

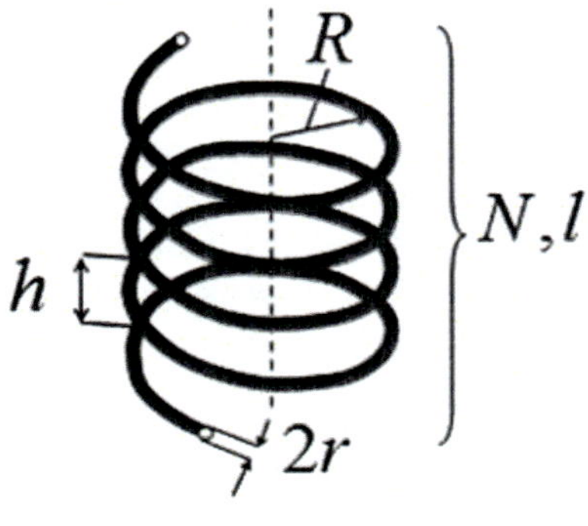

Fig. 1 Helix

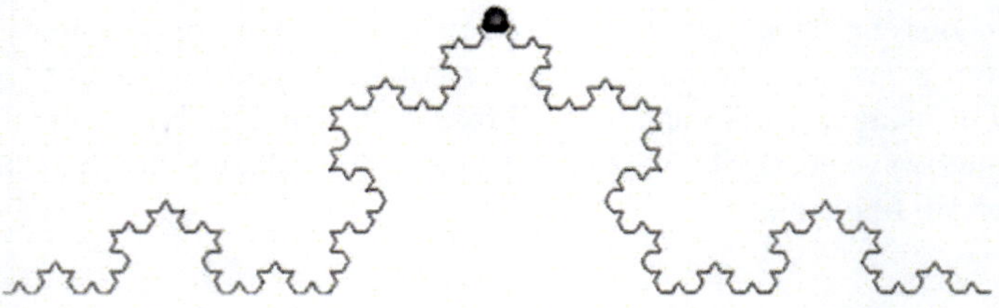

Fig. 2 Koch snowflakes

$$\frac{\varepsilon - \varepsilon_s}{\varepsilon + 2\varepsilon_s} = \alpha\frac{\varepsilon_r - \varepsilon_s}{\varepsilon_r + 2\varepsilon_s};\ \varepsilon_s(\omega) = \varepsilon_r + \frac{\beta_0^2}{\omega_0^2 - \omega^2};\ \chi(\omega) = A\frac{\beta_0^2\omega}{c(\omega_0^2 - \omega^2)}, \quad (1)$$

Where ε_r is the relative permittivity of the container; β_0 is a parameter that has a frequency dimension and is related to internal processes in the environment; c is speed of light; A is a parameter having dimension of length; α is volume concentration of micro-elements in the container.

By substituting an explicit expression for the resonant frequency ω_0 in (1), we obtain the values for the permittivity and the chiral parameter of the chiral metamaterial based on thin-wire helical elements with the dispersion in the framework of the proposed model. Further calculations showed that the using of the proposed low-frequency model is limited in frequency and the frequency limit of use is determined by the relationship between the linear dimensions of the helix elements and the wavelength of incident microwave radiation.

When creating a metamaterial, elements with fractal geometry can also be used instead of helices, such as Koch snowflakes (Fig. 2).

3 Solution of the Problem of Plane Electromagnetic Wave Reflection from a Chiral Metamaterial

The solution of the problem of plane electromagnetic wave (EMW) of linear polarization reflection from a planar layer of a chiral metamaterial based on a set of thin-wire helix elements in the framework of the proposed dispersion model was considered. The problem geometry is shown in Fig. 3. Region 2 in Fig. 3 is a planar layer of a chiral metamaterial with parameters $\varepsilon_2(\omega)$, μ_2 and $\chi_2(\omega)$; regions 1 and 3 are dielectric media. A plane EMW with perpendicular polarization at an angle of θ falls on the chiral layer. As a result of solving the electrodynamic problem, it is necessary to determine the coefficients of reflection (r_{ee}, r_{eh}) and transmission (t_{ee}, t_{eh}) of the main and cross-polarized components of the field. Here it is appropriate to note that when the electromagnetic field (EMF) interacts with chiral media, there is always cross-polarization, that is, radiation of the field with the components of vectors $\vec{\mathbf{E}}$, $\vec{\mathbf{H}}$, which are absent in the EMF structure of the incident wave.

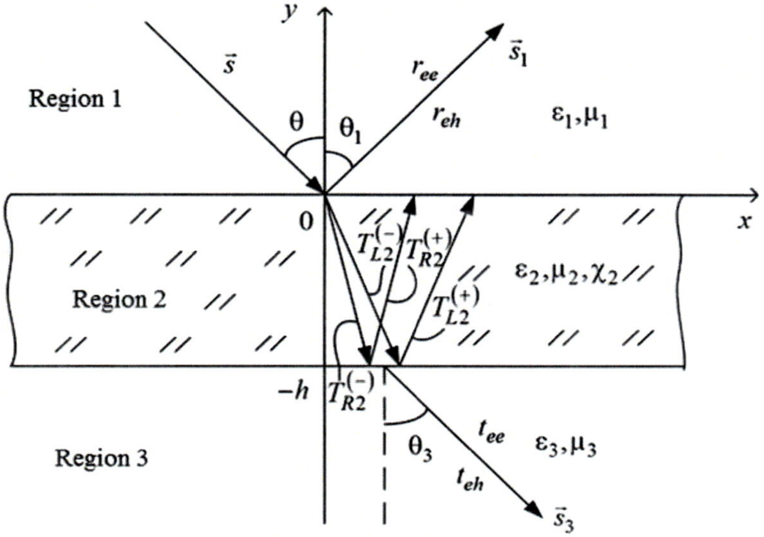

Fig. 3 Problem geometry

To describe the chiral metamaterial, material equations were used, according to the Lindell-Sivola formalism [1]:

$$\vec{\mathbf{D}} = \varepsilon(\omega)\,\vec{\mathbf{E}} - \mathrm{i}\,\chi(\omega)\,\vec{\mathbf{H}}, \quad \vec{\mathbf{B}} = \mu\,\vec{\mathbf{H}} + \mathrm{i}\,\chi(\omega)\,\vec{\mathbf{E}}, \tag{2}$$

where $\varepsilon(\omega)$, μ, $\chi(\omega)$ are the relative permittivity and permeability, and the chiral parameter of the metamaterial, taking into account the dispersion. The material (2) are written in the Gaussian system for the right forms of chiral elements for the harmonic dependence of the field vectors on time.

The problem was solved using the partial domain method. At the first stage, Maxwell's equations in differential form for EMF in chiral metamaterials were reduced by a well-known substitution method to the Helmholtz equations with respect to the so-called Beltrami fields:

$$\nabla^2\vec{\mathbf{E}}_{\mathrm{R}} + k_{\mathrm{R}}^2\vec{\mathbf{E}}_{\mathrm{R}} = 0; \quad \nabla^2\vec{\mathbf{E}}_{\mathrm{L}} + k_{\mathrm{L}}^2\vec{\mathbf{E}}_{\mathrm{L}} = 0, \tag{3}$$

where $\vec{\mathbf{E}}_{\mathrm{R,L}}$ are the electric field vectors of waves with right and left-circular polarizations in a chiral media (Beltrami fields); $k_{\mathrm{R,L}}$ are the wave numbers of the waves of right-circular polarization (RCP) and left-circular polarization (LCP), defined as

$$k_{\mathrm{R,L}} = k_0\left(\varepsilon^{1/2}(\omega)\mu^{1/2} \pm \chi(\omega)\right) \tag{4}$$

where k_0 is the wave number for a plane EMW in a vacuum.

The solution of (3) determines the EMF in the chiral metamaterial layer:

$$\vec{\mathbf{E}} = \vec{\mathbf{E}}_{\mathrm{R}} + \vec{\mathbf{E}}_{\mathrm{L}}, \quad \vec{\mathbf{H}} = \mathrm{i}\sqrt{\frac{\varepsilon}{\mu}}\left(\vec{\mathbf{E}}_{\mathrm{R}} - \vec{\mathbf{E}}_{\mathrm{L}}\right). \tag{5}$$

The electromagnetic field in the chiral layer 2 will be a superposition of 4 waves: 2 waves refracted into the layer from region 1 with RCP and LCP and 2 waves reflected from region 3 with RCP and LCP.

At the next stage of solving the problem from Maxwell's equations, EMF vectors in regions 1 and 2 are determined and then boundary conditions are used for tangential components of vectors $\vec{\mathbf{E}}$, $\vec{\mathbf{H}}$ at the two boundaries of the section "Region 1- Region 2" and "Region 2-Region 3". When substituting expressions for component vectors $\vec{\mathbf{E}}$, $\vec{\mathbf{H}}$ in regions 1, 2, 3 in boundary conditions:

$$\begin{gathered} \vec{\mathbf{E}}_{\tau}^{(1)}(y=0) = \vec{\mathbf{E}}_{\tau}^{(2)}(y=0); \quad \vec{\mathbf{H}}_{\tau}^{(1)}(y=0) = \vec{\mathbf{H}}_{\tau}^{(2)}(y=0); \\ \vec{\mathbf{E}}_{\tau}^{(2)}(y=-h) = \vec{\mathbf{E}}_{\tau}^{(3)}(y=-h); \quad \vec{\mathbf{H}}_{\tau}^{(2)}(y=-h) = \vec{\mathbf{H}}_{\tau}^{(3)}(y=-h), \end{gathered} \tag{6}$$

we obtain an inhomogeneous system of 8 linear algebraic equations with respect to unknown reflection and transmission coefficients:

$$\overleftrightarrow{\mathbf{A}}\vec{\mathbf{T}} = \vec{\mathbf{P}}, \tag{7}$$

where $\overleftrightarrow{\mathbf{A}}$ is a square matrix of size 8×8, the explicit type of elements of which is not given in the paper due to their bulkiness;

$$\begin{aligned} \vec{\mathbf{T}} &= \left[r_{\mathrm{ee}}\ r_{\mathrm{eh}}\ T_{\mathrm{R}}^{(+)}\ T_{\mathrm{R}}^{(-)}\ T_{\mathrm{L}}^{(+)}\ T_{\mathrm{L}}^{(-)}\ t_{\mathrm{ee}}\ t_{\mathrm{eh}}\right]; \\ \vec{\mathbf{P}} &= \left[0\ -1\ \sqrt{\varepsilon_1}\cos\theta/\sqrt{\mu_1}\ 0\ 0\ 0\ 0\ 0\right]. \end{aligned}$$

The coefficients of matrix $\overleftrightarrow{\mathbf{A}}$ are determined by the geometric parameters of the container and helix elements; by the material parameters of the chiral layer and regions 1 and 3, and also take into account the dispersion of $\varepsilon_2(\omega)$ and $\chi_2(\omega)$, and, as a result, the shape of inclusions.

4 Numerical Results

Figure 4 shows the frequency dependence of the reflected and transmitted power of the main and cross-polarized components of the field in the range from 1 to 10 GHz. The solid curves in Fig. 4 shows the dependencies of the transmitted main components power ($10\ \lg|t_{\mathrm{ee}}|^2$); dashed line are reflected main components power

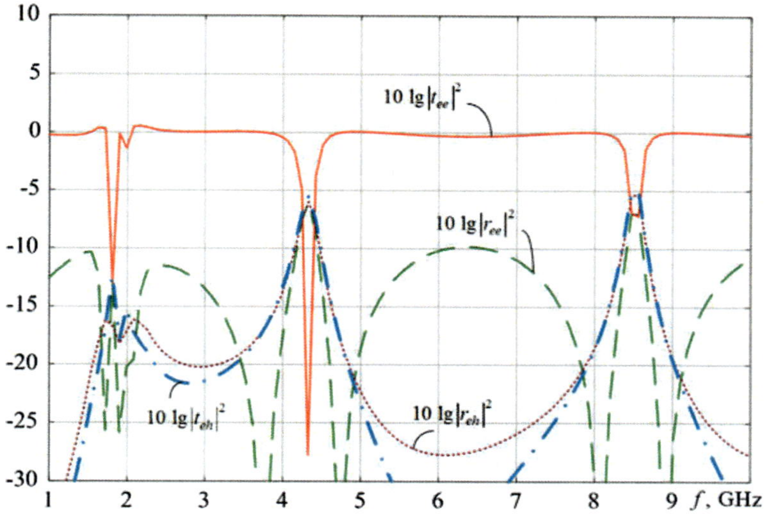

Fig. 4 Frequency dependence of capacity

(10 lg $|r_{ee}|^2$); dash-dotted lines are transmitted power of the cross-polarized components (10 lg $|t_{eh}|^2$); dotted lines are of the reflected power of cross-polarized components (10 lg $|r_{eh}|^2$). The incidence of the wave on the metastructure was considered normal. The calculation was performed for the following structure parameters:

$$R = 0.01\text{m},\ N = 3,\ r = 0.002\text{m},\ H = 0.05\text{m},\ d = 0.05\text{m}$$

At a frequency of 1.88 GHz is observed that the conditions for a better energy concentration of the incident radiation, as the levels of transmitted power main and cross-polarized components of the field are closely spaced in frequency local minimum (the attenuation levels of past capacity in the main and cross-polarized field more than 20 dB). Figure 4 shows that near the frequency of 1.88 GHz, resonant minima of the coefficients of transmission and reflection of the main component are observed on the characteristic. In this case the normally incident EMF energy flow is converted into an energy flow in the plane of the meta-material. At other frequencies the metastructure is completely transparent and the incident radiation passes through it almost without attenuation (near 0 dB). Thus, the structure can be interpreted as a frequency-selective microwave energy concentrator (hub) in the region near 1.88 GHz frequency. We also note that the studied metastructure is equivalent in properties to a natural crystal (or artificial Bragg lattice) in the optical range, namely, the frequencies of resonant minima of attenuation of the transmitted power of the main component of the field are calculated from the Wulf-Bragg condition. Thus, the possibility of discrete-multi-frequency concentration of incident microwave energy at a number of resonant frequencies is achieved.

5 Multilayered Metastructure

If the structure consists of several layers of metamaterial, each of which is described by the transfer matrix $\overleftrightarrow{\mathbf{Z}}_i(\omega)$, then the transfer matrix of the entire metastructure as a whole is represented as a product of the matrices of individual layers:

$$\overleftrightarrow{\mathbf{Z}}(\omega) = \prod_{i=1}^{N} \overleftrightarrow{\mathbf{Z}}_i(\omega), \tag{8}$$

where N is the number of layers of the metamaterial.

Expressions for the elements of the surface impedance tensor of a chiral metamaterial based on a set of thin-wire helical elements are obtained. The well-known Fresnel formulas for the chiral media without taking into account the dispersion are obtained as the use of the surface impedance tensor for CM and proof of their applicability.

The surface impedance tensor for CM has the following form:

$$\overleftrightarrow{\mathbf{Z}} = \begin{pmatrix} -\mathrm{i}\frac{\chi(\omega)}{\varepsilon(\omega)} & -\frac{k_z}{k_0\varepsilon(\omega)} \\ -\frac{k_0 n_c^2(\omega)}{k_z\varepsilon(\omega)} & -\mathrm{i}\frac{\chi(\omega)}{\varepsilon(\omega)} \end{pmatrix}, \tag{9}$$

where $n_c^2(\omega) = \varepsilon(\omega)\mu - \chi^2(\omega)$; k_z are the longitudinal component of the wave vector $\vec{\mathbf{k}}$ in the metamaterial.

Also using approximate boundary conditions for a thin chiral layer from [6, 14], we obtained a matrix of CM transmission that is infinitely extended along one of the coordinate axes. The matrix for a thin chiral layer has the following form:

$$\overleftrightarrow{Z}(\omega) = \begin{bmatrix} 1 & -k_0 h\chi(1-\beta^2) & 0 & -i\mu k_0(1-\beta^2) \\ -k_0 h\chi & 1 & i\mu k_0 h & 0 \\ 0 & -i\varepsilon k_0 h(1-\beta^2) & 1 & -k_0 h\chi(1-\beta^2) \\ -i\varepsilon k_0 h & 0 & -k_0 h\chi & 1 \end{bmatrix}, \tag{10}$$

where $\beta = k_x/(k_0 n_c)$; k_x are x is the component of the wave vector in CM.

As an example, the solution to the problem of reflection of plane electromagnetic waves linear polarization from a planar bilayer chiral-dielectric metamaterial based on helical elements within the dispersion model obtained in Chap. 1. In Fig. 5 the geometry of the problem is shown. Region 2 in Fig. 4 is a planar layer of a chiral metamaterial with parameters $\varepsilon_2(\omega)$, μ_2 and $\chi_2(\omega)$; region 3 is a homogeneous dielectric layer; regions 1 and 4 are dielectric media. A plane EMW with perpendicular polarization at an angle of θ falls on a two-layered structure. As a result of solving the electrodynamic problem, it is necessary to determine the coefficients of reflection (r_{ee}, r_{eh}) and transmission (t_{ee}, t_{eh}) of the field main and cross-polarized components.

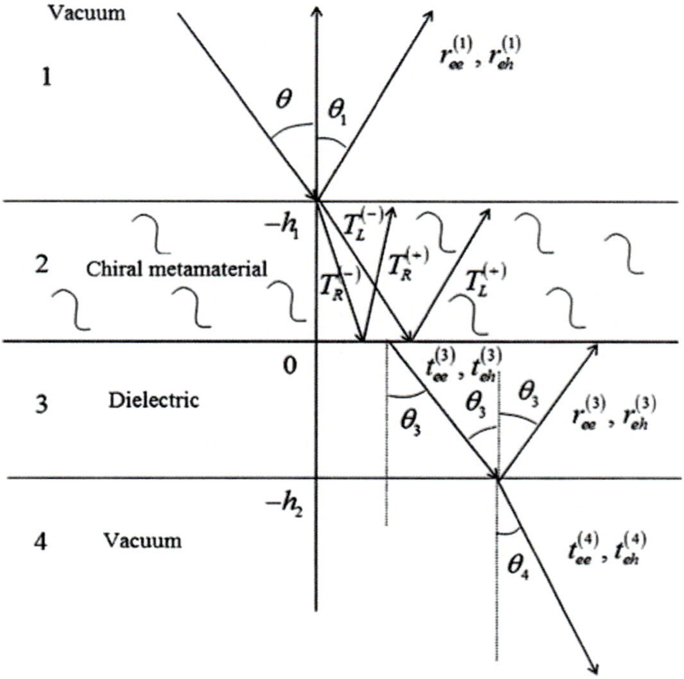

Fig. 5 Multilayered metastructure

The transmission matrix of the chiral layer 2 is determined by the ratio (10), while the transmission matrix of the dielectric layer 3 is obtained from (10) at $\chi = 0$. The transfer matrix of the entire metastructure is obtained by multiplying the matrices of the two layers (8).

At the last stage, the solution of the problem is reduced to a system of 12 linear algebraic equations with respect to unknown coefficients of reflection and transmission of waves in the outer regions and inner layers. All coefficients are shown in Fig. 5. The index "e" refers to the field of the main component, the index "h"—to the field of the cross-polarized component.

Figure 6 shows the frequency dependence of the reflected and transmitted power of the main component of the field in the range from 1 to 6 GHz. The dashed curves in Fig. 6 show the dependences of the transmittes power of the main component ($\alpha_{forw} = 10\ \lg \left|t_{\text{ee}}^{(4)}\right|^2$); solid lines show the reflected power of the main component ($\alpha_{rev} = 10\ \lg \left|r_{\text{ee}}^{(1)}\right|^2$). The insidence of the wave on the metastructure was considered normal in order for the degree of cross-polarization of the field to be insignificant. The dielectric layer has material parameters $\varepsilon = 1.8$, $\mu = 1$ and a thickness of 3 mm. The chiral layer is based on a 5 mm thick dielectric container: $\varepsilon = 1.8 - i \cdot 0.1$, $\mu = 1$. Chiral helical elements consist of 2 turns with a radius of 1 cm and are located at a distance of 5 cm from each other. The outer regions were considered

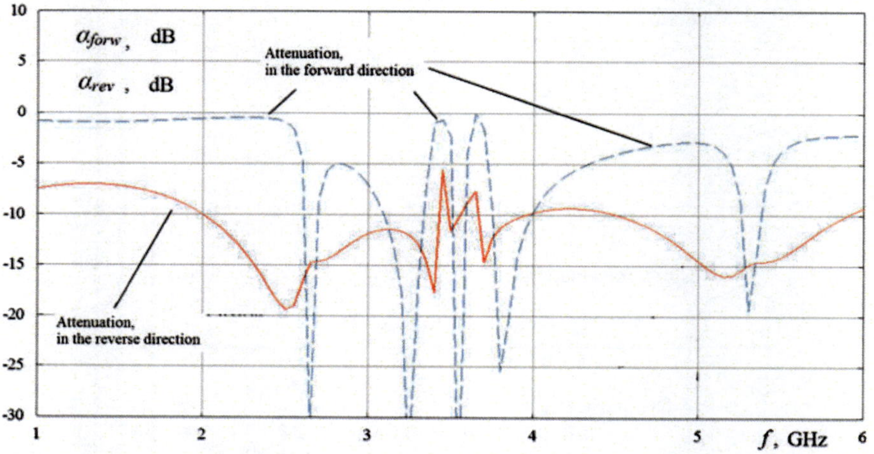

Fig. 6 Frequency dependence of the reflected and transmitted power

vacuum. As can be seen from Fig. 6, the dependencies show narrow resonant maxima of energy attenuation in the forward direction, at which it is possible to effectively convert normally incident microwave radiation into side scattering in the plane of the structure. From Fig. 6, it follows that at frequencies 2.7, 3.25, 3.6, 3.8 GHz, a situation is possible when the attenuation levels in the forward direction reach -30 dB.

In conclusion we will list the main conclusions from the results of research on the electromagnetic characteristics of planar chiral structures based on thin wire perfectly conducting elements of a helical shape:

1. A mathematical model based on the Maxwell–Garnett formula is proposed and the dispersion dependences of the permittivity and the chiral parameter for a chiral metamaterial based on a set of thin-wire ideally conducting elements of a helical shape are obtained.
2. It is proved that in a single-layer chiral metamaterial based on thin-wire multi-turn helical elements, it is possible to transform a normally incident microwave energy at certain frequencies into azimuthal scattering in the metamaterial plane.
3. The paper theoretically predicts the possibility of discrete-multi-frequency concentration of incident microwave energy using a single-layer chiral metastructure at a number of resonant frequencies.
4. It is proved that a two-layer chiral-dielectric metamaterial based on thin-wire conducting helices near a predetermined frequency allows for a frequency-selective conversion of a normally incident flow of electromagnetic energy into azimuthal scattering in the plane of the metastructure.

References

1. I.V. Lindell, A.H. Sihvola, S.A. Tretyakov, A.J. Viitanen, *Electromagneticwaves in Chiral and Bi-Isotropic Media* (Artech House, London, 1994), p. 291
2. A. Lakhtakia, V.K. Varadan, V.V. Varadan, Time-harmonic electromagnetic fields in chiral media, in *Lecture Notes in Physics* (Springer-Verlag, Berlin, Heidelberg and Boston, 1989), p. 121
3. A. Lakhtakia, V.V. Varadan, V.K. Varadan, Field equations, Huygens's principle, integral equations, and theorems for radiation and scattering of electromagnetic waves in isotropic chiral media. J. Optical Soc. Am. **V.5.**(2), 175–184 (1988)
4. J. Hanninen, M.I. Oksanen, S.A. Tretyakov, Vector circuit method for calculating reflect ion and transmission of electromagnetic waves in multilayer chiral structures. IEE Proceedings, Pari II **138**(6), 513–520 (1991)
5. A. Lakhtakia, V.V. Varadan, V.K. Varadan, Field equations, Huygens's principle, integral equations, and theorems for radiation and scattering of electromagnetic waves in isotropic chiral media. J. Optic. Soc. Am. **V.5.**(2), 175−184 (1988)
6. M.H. Umari, V.V. Varadan, V.K. Varadan, Rotation and dichroism associated with micro-wave propagation in chiral composite samples. Radio Sci. **V.26.**(5), 1327–1334 (1991)
7. F. Capolino, *Theory and Phenomena of Metamaterials* (Taylor & Francis, CRC Press, 2009), p. 992
8. B.Z. Katsenelenbaum, E.N. Korshunova, A.N. Sivov, A.D. Shatrov, Chiral electromagnetic objects. Phys. Usp. **40**(11), 1149–1160 (1997)
9. M.A. Buzova, D.S. Klyuev, M.A. Minkin et al., Solving the electrodynamic problem for a microstrip radiating structure with a chiral substrate. Tech. Phys. Lett. **6**, 491–493 (2018)
10. M.A. Buzova, A.N. Dementyev, D.S. Klyuev et al., Current distribution in a gyrotropic microstrip structure when excited by a plane electromagnetic wave. Dokl. Phys. **6**, 231–234 (2018)
11. I.V. Semchenko, S.A. Tretyakov, A.N. Serdyukov, Research on chiral and bianisotropic media in Byelorussia and Russia in the last ten years. Prog. Electromag. Res. 1996. V.12, 335 (1996)
12. S.A. Tretyakov, F. Mariotte, Maxwell Garnett modeling of uniaxial chiral composites with bianisotropic inclusions. J. Electromag. Waves Appl. **V.9.**(7/8), C.1011–1025 (1995)
13. D.S. Klyuev, A.M. Neshcheret, O.V. Osipov, A.A. Potapov, V. Sokolova Yu, Microstrip and fractal antennas based on chiral metamaterials in MIMO systems, in *Springer Nature Switzerland AG 2020: 12th Chaotic Modeling and Simulation International Conference, Springer Proceedings in Complexity* (2020), pp. 295–306
14. O.V. Osipov, T.A. Panferova, Approximate boundary conditions for thin chiral layers with curvilinear surfaces. J. Commun. Technol. Electron. **V. 55**(5), 532–534 (2010)

Investigation of Electromagnetic Characteristics of a Chiral Metamaterial Based on Mutually Orthogonal Helical and Fractal Elements

Dmitriy S. Klyuev, Anatoly M. Neshcheret, Oleg V. Osipov, and Alexander A. Potapov

Abstract The paper considers the solution of the problem of plane linearly polarized electromagnetic wave reflection from a planar layer of metamaterial, which is a matrix of thin-wire perfectly conducting elements in the form of two mutually orthogonal helix and fractal elements. A dispersion model is constructed for the considered metamaterial. The problem was solved using by the partial domain method and its solution was reduced to a system of linear algebraic equations with respect to unknown reflection and transmission coefficients. As a result of numerical analysis frequency-selective properties of the studied metamaterial were revealed. In particular, it is proved that the metastructure at certain frequencies may be used as a frequency selective screen for microwave radiation.

Abbreviations

CM	Chiral Metamaterial
EMF	Electromagnetic Field
EMW	Electromagnetic Wave
LCP	Left-Circular Polarization
RCP	Right-Circular Polarization
SLAE	System of Linear Algebraic Equations
UHF	Ultra-High Frequency

D. S. Klyuev · A. M. Neshcheret · O. V. Osipov
Povolzhskiy State University of Telecommunications and Informatics, L. Tolstoy Str., 23, 443010, RussiaSamara
e-mail: klyuevd@yandex.ru

A. A. Potapov (✉)
V. A. Kotelnikov Institute of Radio Engineering and Electronics, Russian Academy of Sciences, Mokhovaya str., 11-7, 125009 Moscow, Russia
e-mail: potapov@cplire.ru

JNU-IREE RAS Joint Lab. of Information Technology and Fractal Processing of Signals, JiNan University, Guangzhou 510632, China

C. H. Skiadas et al. (eds.), *13th Chaotic Modeling and Simulation International Conference*, Springer Proceedings in Complexity,
https://doi.org/10.1007/978-3-030-70795-8_30

1 Introduction

Currently, a significant interest in the microwave and optical ranges is the study of various types of metamaterials, that is, structures that have non-standard (for natural media) properties that occur when they interact with incident radiation [1]. In the microwave range metamaterials are composite structures consisting of a container and conductive elements. Of particular interest are the so-called chiral metamaterials based on a mirror-asymmetric shape elements [2]. Such elements can be three-dimensional or flat (helixes, S-formed elements, gammadions, double open rings, Tellegen elements, etc.). It is possible to create frequency and polarization-selective microwave devices, polarization converters, low-reflection coatings, media with negative refraction based on f chiral structures.

In 1948 Tellegen suggested that a new type of material could be created based on a combination of electric and magnetic dipoles [3]. The simplest element that has this property is an open ring with protruding ends (Tellegen element, canonical helix). We know the results of research on the diffraction of plane electromagnetic waves on Tellegen elements (S. A. Tretyakov, F. Mariotte) [4], cylinders with helical conductivity (B. Z. Katzenelenbaum, A.D. Shatrov et al.) [5], where it is shown that polarizationally selective resonant phenomena are observed in structures in the form of long helices of a small radius compared to the wavelength and lattices based on them. The possibility of using metamaterials based on thin-wire perfectly conducting helical elements for converting normally incident microwave radiation of a given frequency into scattering in the plane of the metamaterial was studied in detail.

In this paper we analyze the reflection of a plane electromagnetic wave of linear polarization from a planar layer of a metamaterial based on a double perpendicular thin-wire helix elements and also construct a dispersion model of such a metamaterial.

2 The Problem of Wave Reflection from a Planar Metamaterial

Consider the problem of determining the reflection and transmission coefficients when a plane electromagnetic wave incidents on a planar layer of a chiral metamaterial that is infinitely extended along the Oz axis. The problem geometry is shown in Fig. 1. Let a plane electromagnetic wave of linear perpendicular polarization incidents on a chiral metamaterial from a semi-infinite dielectric region 1 (ε_1 and μ_1 are relative dielectric and magnetic permittivity) at an angle θ.

Region 2 in Fig. 1 is a planar layer of a chiral metamaterial of thickness h (ε_2 and μ_2 are relative permittivity and permeability; χ_2 is the relative chiral parameter). The frequency dependencies of material parameters ε_2 and χ_2 are determined by the type of resonant elements and will be obtained in the next section.

A chiral metamaterial consists of double mutually orthogonal multi-turn thin-wire helices that are evenly placed in a planar container (Fig. 2). The set of two mutually

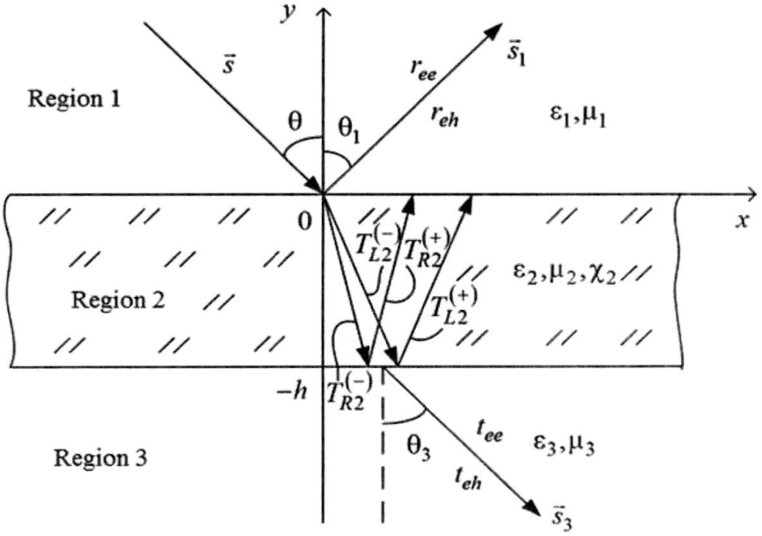

Fig. 1 Problem geometry

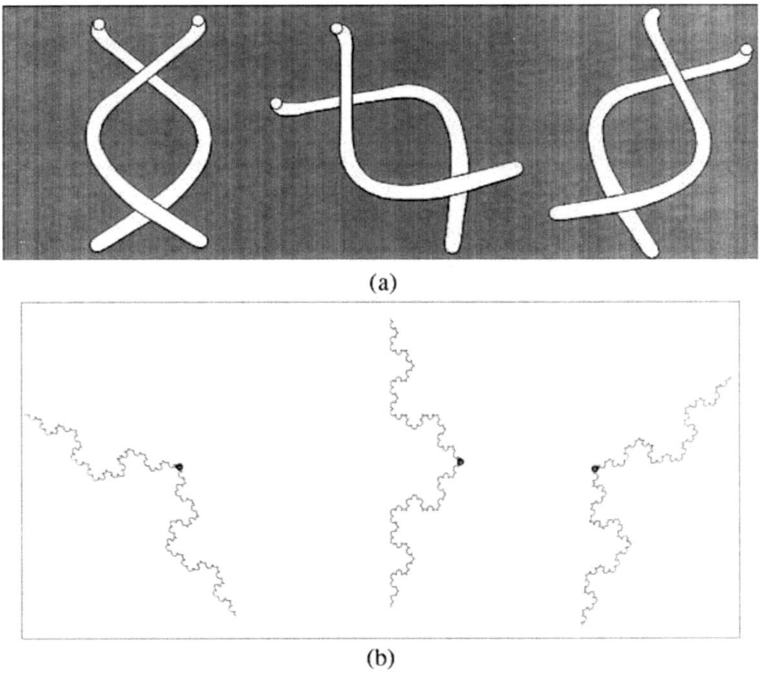

Fig. 2 Geometry of the metamaterial based on helices (**a**) and on fractal elements (Koch's Snowflake) (**b**)

orthogonal helices is called a combined helical element (Fig. 2a) and the set of two fractal elements is called a combined fractal element (Fig. 2b). It is assumed that two elements in a combined element may have different geometric parameters. In the case of a chiral media it is assumed that elements are oriented randomly in the container. The combined elements are evenly spaced in the container at a distance d from each other. For Fig. 2 shows examples of combined helical and fractal elements.

Region 3 in Fig. 1 is a dielectric (ε_3 and μ_3 are the relative permittivity and permeability).

The task is to derive the relations for calculating the reflection coefficients of the main (r_{ee}) and cross-polarized (r_{eh}) components of the field in region 1, as well as formulas for the transmission coefficients of the main (t_{ee}) and cross-polarized (t_{eh}) components of the field in region 3. It is appropriate to note that when the incident microwave radiation interacts with the chiral medium, the phenomenon of cross-polarization occurs [6], that is, in the structure of the reflected and transmitted waves, field components orthogonal to the components of the incident wave arise. In a chiral metamaterial, two waves with right (RCP) and left-circular (LCP) polarizations propagate [2, 6].

The chiral metamaterial 2 is described by material equations that simultaneously link the inductions and strengths of the electric and magnetic fields [2, 6]:

$$\begin{aligned} \vec{D}^{(2)} &= \varepsilon_2(\omega)\,\vec{E}^{(2)} \mp i\,\chi_2(\omega)\,\vec{H}^{(2)}\,, \\ \vec{B}^{(2)} &= \mu_2\,\vec{H}^{(2)} \pm i\,\chi_2(\omega)\,\vec{E}^{(2)}, \end{aligned} \tag{1}$$

where the upper and lower signs correspond to helix elements with right and left-hand twists, respectively. The material (1) are written in the Gaussian system.

The electromagnetic field in a chiral metamaterial is determined from a system of two related 2nd-order differential equations [1, 2]:

$$\begin{aligned} &\nabla^2\vec{E}^{(2)} + k_0^2\left(\varepsilon_2\mu_2 + \chi_2^2\right)\vec{E}^{(2)} - 2ik_0^2\mu_2\chi_2\vec{H}^{(2)} = 0\,; \\ &\nabla^2\vec{H}^{(2)} + k_0^2\left(\varepsilon_2\mu_2 + \chi_2^2\right)\vec{H}^{(2)} + 2ik_0^2\varepsilon_2\chi_2\vec{E}^{(2)} = 0\,, \end{aligned} \tag{2}$$

where $k_0 = \omega/c$ is the wave number for a plane wave in free space; ω is the circular frequency; c is the speed of light.

System of (2) using the standard representation in the form of Beltrami fields [2]

$$\vec{E}^{(2)} = \vec{E}_R + \vec{E}_L; \quad \vec{H}^{(2)} = i\sqrt{\frac{\varepsilon_2}{\mu_2}}\left(\vec{E}_R - \vec{E}_L\right), \tag{3}$$

is reduced to two homogeneous Helmholtz equations for waves RCP and LCP in a chiral metamaterial:

$$\nabla^2\vec{E}_{R,L} \pm k_{R,L}^2\vec{E}_{R,L} = 0, \tag{4}$$

where $\vec{E}_R$ is the electric field of the RCP wave; $\vec{E}_L$ is the electric field of the LCP wave; the wave numbers for RCP and LCP waves in a boundless chiral media are defined as follows:

$$k_{R,L}(\omega) = \frac{\omega}{c}\left[\sqrt{\varepsilon_2(\omega)\mu_2} \pm \chi_2(\omega)\right]. \quad (5)$$

The following expressions are obtained from solving (4) using representations (3) for the longitudinal components of the field vectors in a chiral metamaterial:

$$\begin{aligned} E_z^{(2)} &= T_R^{(-)} e^{-ik_R(\vec{s}_R,\vec{r})} + T_R^{(+)} e^{ik_R(\vec{s}_R,\vec{r})} + \\ &+ T_L^{(-)} e^{-ik_L(\vec{s}_L,\vec{r})} + T_L^{(+)} e^{ik_L(\vec{s}_L,\vec{r})}; \\ H_z^{(2)} &= \frac{i}{\eta_2}\begin{bmatrix} T_R^{(-)} e^{-ik_R(\vec{s}_R,\vec{r})} + T_R^{(+)} e^{ik_R(\vec{s}_R,\vec{r})} - \\ -T_L^{(-)} e^{-ik_L(\vec{s}_L,\vec{r})} - T_L^{(+)} e^{ik_L(\vec{s}_L,\vec{r})} \end{bmatrix}, \end{aligned} \quad (6)$$

where $\vec{s}_{R,L}^{(-)} = \{\sin\theta_{R,L}, -\cos\theta_{R,L}\}$ are the unit vectors along which the refracted waves propagated in the metamaterial 2; $\vec{s}_{R,L}^{(+)} = \{\sin\theta_{R,L}, \cos\theta_{R,L}\}$ are the unit vectors for reflected waves from region 3; $\theta_{R,L}$ are the angles of refraction of waves RCP and LCP, respectively; $\eta^{(2)} = \sqrt{\mu_2/\varepsilon_2}$ is the impedance of the chiral metamaterial; $T_R^{(-)}$ and $T_L^{(-)}$ are the transmission coefficients (on the field) waves of RCP and LCP in the metamaterial 2; $T_R^{(+)}$ and $T_L^{(+)}$ are reflection coefficients (on the field) waves of RCP and LCP from region 3 in the chiral layer.

For the case of an incident of a plane electromagnetic wave with perpendicular (s-) polarization, the following expressions are valid for the components of the field in the dielectric region 1, written taking into account cross-polarization:

$$\begin{aligned} E_z^{(1)} &= e^{-ik_1(\vec{s}_r,\vec{r})} + r_{ee} e^{-ik_1(\vec{s}_r,\vec{r})}; \\ H_x^{(1)} &= -\frac{\cos\theta}{\eta^{(1)}} e^{-ik_1(\vec{s}_r,\vec{r})} + r_{ee}\frac{\cos\theta}{\eta^{(1)}} e^{-ik_1(\vec{s}_r,\vec{r})}; \\ H_z^{(1)} &= r_{eh} e^{-ik_1(\vec{s}_r,\vec{r})}; \\ E_x^{(1)} &= -r_{eh}\eta^{(1)}\cos\theta\, e^{-ik_1(\vec{s}_r,\vec{r})}, \end{aligned} \quad (7)$$

where $\vec{s}_r = \{\sin\theta, \cos\theta\}$ is a unit vector that defines the direction of propagation of the reflected wave in region 1; r_{ee} is reflection coefficient of main components; r_{eh} is the reflection coefficient of cross-polarized components; $k_1 = k_0\sqrt{\varepsilon_1\mu_1}$ is the wave number for plane wave in dielectric region 1; $\eta^{(1)} = \sqrt{\mu_1/\varepsilon_1}$ is an impedance of the region 1.

When solving the problem it is assumed that a wave with a single amplitude of the electric field intensity incidents on the chiral layer.

For the components of the electromagnetic field vectors in region 3, taking into account cross-polarization, the following expressions can be written:

$$\begin{aligned} E_z^{(3)} &= t_{ee} e^{-ik_3(\vec{s}_3,\vec{r})}; \\ H_z^{(3)} &= t_{eh} e^{-ik_3(\vec{s}_3,\vec{r})}; \\ E_x^{(3)} &= t_{eh} \eta^{(3)} \cos\theta_3 \, e^{-ik_3(\vec{s}_3,\vec{r})}; \\ H_x^{(3)} &= -t_{ee} \frac{\cos\theta_3}{\eta^{(3)}} e^{-ik_3(\vec{s}_3,\vec{r})}; \end{aligned} \tag{8}$$

where $\vec{s}_3 = \{\sin\theta_3, -\cos\theta_3\}$ is a unit vector that defines the direction of propagation of transmitted wave in region 3; t_{ee} is the transmission coefficient of main components; t_{eh} is the transmission coefficient of cross-polarized components; $\eta^{(3)} = \sqrt{\mu_3/\varepsilon_3}$ is an impedance of region 3; $k_3 = k_0\sqrt{\varepsilon_3\mu_3}$ is wave number for plane wave in dielectric region 3.

At the last stage the boundary conditions are used for the tangential components of the electric and magnetic field vectors at the interface at $y = 0$ and $y = -h$:

$$\begin{aligned} &\vec{E}_\tau^{(1)}(y=0) = \vec{E}_\tau^{(2)}(y=0); \\ &\vec{H}_\tau^{(1)}(y=0) = \vec{H}_\tau^{(1)}(y=0); \\ &\vec{E}_\tau^{(2)}(y=-h) = \vec{E}_\tau^{(3)}(y=-h); \\ &\vec{H}_\tau^{(2)}(y=-h) = \vec{H}_\tau^{(3)}(y=-h). \end{aligned} \tag{9}$$

As a result of substituting expressions (6), (7) and (8) into boundary conditions (9) with respect to unknown reflection and transmission coefficients we obtain a non-uniform system of linear algebraic equations (SLAE):

$$\overleftrightarrow{A}\,\vec{T} = \vec{P}, \tag{10}$$

where $\overleftrightarrow{A}$ is a square matrix of size 8×8, the explicit form of elements of which is not given in the article due to their bulkiness;

$$\begin{aligned} \vec{T} &= \left[r_{ee}\ r_{eh}\ T_R^{(+)}\ T_R^{(-)}\ T_L^{(+)}\ T_L^{(-)}\ t_{ee}\ t_{eh} \right]; \\ \vec{P} &= \left[0\ -1\ \cos\theta_1/\eta^{(1)}\ 0\ 0\ 0\ 0\ 0 \right]. \end{aligned}$$

The coefficients of matrix $\overleftrightarrow{A}$ are determined by the geometric parameters of the container and the combined helical elements; by the material parameters of the chiral metamaterial and regions 1 and 3.

Similarly, the case of a wave incidence with parallel (p-) polarization is considered and the solution of the problem is reduced to a SLAE (10) with other coefficients of the matrix $\overset{\leftrightarrow}{A}$ and a vector column $\vec{P}$.

3 Dispersion Model of a Metamaterial Based on Combined Helical Elements

Let's consider a chiral metamaterial based on combined helical elements (Fig. 3). Each of the two subelements of the combined element is a thin-wire conducting helix consisting of N turns. Helix in combined dual element describes the following geometric parameters: N_i is number of turns; R_i is the radius of the helix loop; h_i is step of a helix; l_i is length of the helix in the expanded state; r_i is the radius of the wire; α_i is angle of winding of the helix; $i = 1, 2$ is the number of helices in the element.

To calculate the resonant frequency of the combined element, we use the Thomson formula:

$$\omega_0 = \frac{1}{\sqrt{LC}}, \tag{11}$$

where L is the inductance of the combined element; C is capacity of the combined element.

The inductance and capacitance of the combined element are determined by the corresponding characteristics of the single helices as follows:

$$C = C_1 + C_2; \;\; L = \frac{L_1 L_2}{L_1 + L_2}, \tag{12}$$

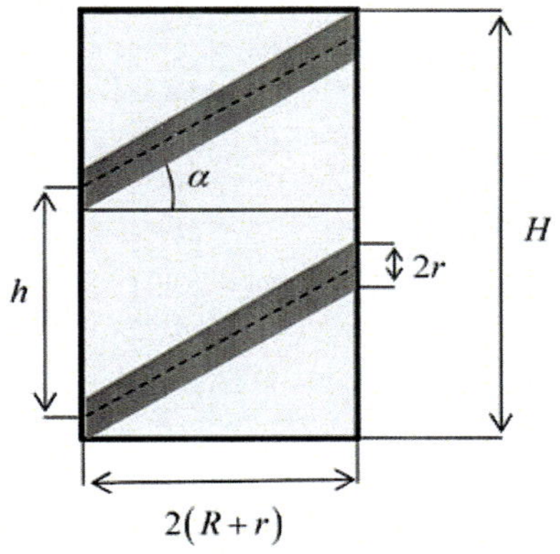

Fig. 3 Cross section of a single helix from a combined element

where L_i is the inductance of one of the helices; C_i is the capacitance of one of the helices.

The capacity of a single helix is defined as follows:

$$C_i = C_{wire\,i} + C_{inc\,i} + C_{ine\,i}, \tag{13}$$

that is, as the sum of the wire capacity, inter-coil capacity, and inter-element capacity, and as a result is expressed as follows:

$$C_i = \varepsilon_{c2}\left[\begin{array}{l}\dfrac{l_i}{18\ln\left(\frac{2l_i}{r_i}\right)-1}\cdot 10^{-11}+\\ +\dfrac{\pi\left[(R_i+2r_i)^2-R_i^2\right](N_i-1)}{h_i}+\dfrac{1}{d}\dfrac{N_i(R_i+r_i)}{\cos\alpha_i}r_i\end{array}\right]. \tag{14}$$

The inductance of a single helix is calculated using the formula:

$$L_i = \mu_{c2}\frac{\pi N_i^2 R_i^2}{l_i}. \tag{15}$$

In formulas (14) and (15) ε_{c2}, μ_{c2} are relative permittivity and permeability of the dielectric container.

Substituting the ratios (14), (15) in (12) and then in (11), we get the formula for the resonant frequency of the combined element.

The dispersion of the permittivity and the chirality parameter are determined by the following expressions [7]:

$$\varepsilon_2(\omega) = \varepsilon_{c2} + \frac{\beta_0^2}{\omega_0^2-\omega^2}; \quad \chi_2(\omega) = A_0\frac{\beta_0^2\omega}{c\left(\omega_0^2-\omega^2\right)}, \tag{16}$$

where A_0 and β_0 are parameters defined by the dimensions of the combined elements and the distance between them, respectively.

Substituting the expression for the resonant frequency of the combined element (11) in formulas (16), we obtain dispersion relations for the permittivity and relative chirality parameter for metamaterial based on thin-wire perfectly conducting elements in the form of mutually orthogonal helices.

4 Numerical Results

When analyzing the numerical characteristics, the main interest was the calculation of frequency dependencies of reflected ($10\ \lg|r_{ee}|^2$ and $10\ \lg|r_{eh}|^2$) and transmitted

($10\ \lg|t_{ee}|^2$ and $10\ \lg|t_{eh}|^2$) power (in dB). The container was modeled on styrofoam with a relative permittivity of $\varepsilon_{c2} = 1.5$.

Figure 4 shows the frequency dependence of the reflected and transmitted power of the main and cross-polarized components of the field in the range from 1 to 10 GHz.

The dashed curves in Fig. 4 show the dependences of the transmitted power of the main component ($10\ \lg|t_{ee}|^2$); solid lines show the reflected power of the main component ($10\ \lg|r_{ee}|^2$). The incidence of the plane wave on the metastructure was considered normal. The calculation was performed for the following values of the structure parameters:

$R_1 = 2R_2 = 0.01\,\text{m},\ \ N_{1,2} = 3,\ \ r_1 = 2r_2 = 0.002\,\text{m},$

$H_1 = 2H_2 = 0.05\,\text{m},\ \ d_1 = 2d_2 = 0.05\,\text{m}$

In this case both helices that are part of the structure of the combined element are identical.

The level of reflection and transmission of cross-polarized field components does not exceed −25 dB at normal incidence and is not shown on the graphs.

At a frequency of 1.18 GHz there is a sharp local decrease in the level of power transmitted through the metamaterial, that is, there is mainly a lateral scattering of normally incident electromagnetic energy. At this frequency the metastructure can act as a frequency-selective shield. At other frequencies the metastructure is completely transparent and the incident radiation penetrates through it almost without attenuation (near 0 dB).

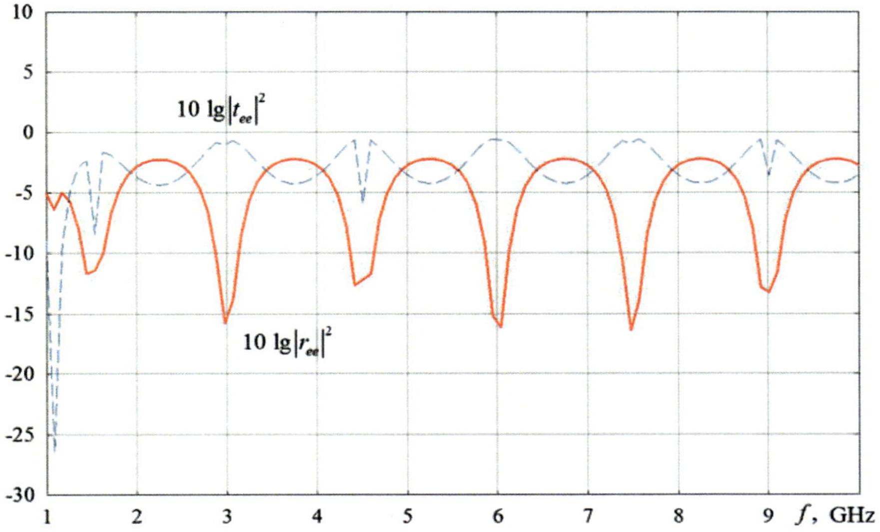

Fig. 4 Frequency dependences of the transmitted and reflected power for the case of identical helices in a combined element

It can be noted that the studied metastructure at a single frequency can serve as a frequency-selective protective screen that is not transparent to radiation near the main resonant frequency.

Let's now consider the case of different values of geometric parameters of helixes in the structure of the combined element.

Figure 5 shows the frequency dependence of the reflected and transmitted power of the main and cross-polarized components of the field in the range from 1 to 10 GHz.

The dashed curves in Fig. 5 show the dependences of the transmitted power of the main component ($10\ \lg|t_{ee}|^2$); solid lines show the reflected power of the main component ($10\ \lg|r_{ee}|^2$). The incidence of the wave on the metastructure was considered normal. The calculation was performed for the following structure parameters:

$$R_1 = 2R_2 = 0.01\,\text{m},\ \ N_{1,2} = 3,\ \ r_1 = 2r_2 = 0.002\,\text{m},$$

$$H_1 = 2H_2 = 0.05\,\text{m},\ \ d_1 = 2d_2 = 0.05\,\text{m}$$

As can be seen from Fig. 5, in the case of an asymmetric double helical element, there is a sharp frequency selectivity of the transmission of electromagnetic radiation through the metamaterial. In contrast to a symmetrical double helical element, a strong chirality occurs here and, as a result, a set of discrete frequencies occurs at which the electromagnetic wave does not penetrate through the metamaterial and is converted into lateral scattering. It can also be noted that in this case, the reflection and transmission of cross-polarized components of the field also increases to −15–18 dB. In addition, there are no frequencies starting from 2 GHz at which the level of attenuation of the main component of the field is close to 0 dB, which is associated with an increase in cross polarization. At resonant frequencies, the structure can play the role of a microwave shield.

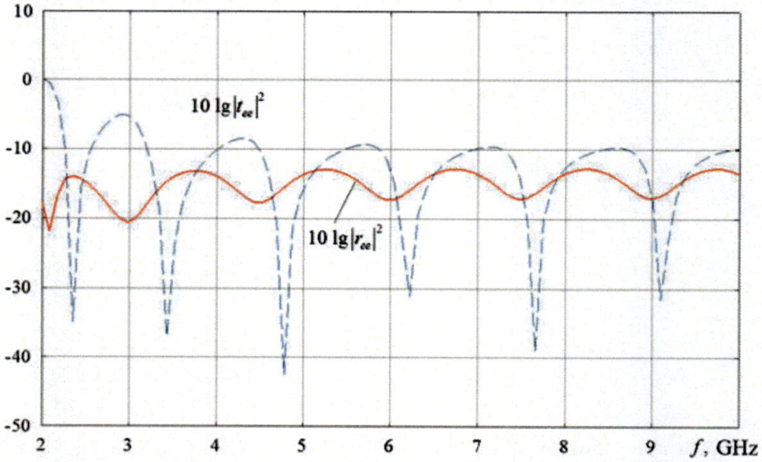

Fig. 5 Frequency dependences of the transmitted and reflected power for the case of not identical helices in a combined element

It was also proved that the studied metastructure is equivalent in properties to a natural crystal (or an artificial Bragg lattice) in the optical range, namely, the frequencies of resonant minima of attenuation of the passed power of the main component of the field are calculated from the Wulf-Bragg condition, taking into account the refraction of electronic waves in the crystal [8]:

$$2d\sqrt{\left(\varepsilon_2\mu_2 - \chi_2^2\right) - \cos^2\theta} = \nu\lambda, \tag{17}$$

where ν is the order of resonance; λ is wavelength; θ is angle of wave incidence.

5 Conclusion

Thus, based on the results of the work, the following conclusions can be drawn:

1. The presence of two identical thin-wire helices in the combined element reduces the effect of chirality and, as a result, the frequency selectivity of the entire metamaterial to almost zero.
2. The presence of two different geometric dimensions of helices in the combined element increases the degree of chirality of the entire metamaterial as a whole and there is a frequency selectivity of the transmission of a plane electromagnetic wave through the metamaterial and a sufficiently large cross-polarization of the field.
3. The studied metastructure in the case of various helices in the combined element is equivalent in properties to a natural crystal (or an artificial Bragg lattice) in the optical range.
4. The metastructure in the case of different helices in the combined element can serve as a frequency selective microwave shield.

Reference

1. F. Capolino, Theory and Phenomena of Metamaterials/F. Capolino (Taylor & Francis, CRC Press, 2009), p. 992
2. I.V. Lindell, Electromagnetic waves in chiral and bi-isotropic media, I.V. Lindell, A.H. Sihvola, S.A. Tretyakov, A.J. Viitanen (ed.) (Artech House, London, 1994), p. 291
3. B.D.H. Tellegen, The Gyrator, a New Electric Network Element / B.D.H. Tellegen. Philips Res. Reports **V. 3.**(81), 81–101 (1948)
4. S.A. Tretyakov, Maxwell Garnett modeling of uniaxial chiral composites with bianisotropic inclusions / S.A. Tretyakov, F. Mariotte. J. Electromag. Waves Appl. **V.9.**(7/8), C.1011–1025 (1995)
5. B.Z. Katsenelenbaum, E.N. Korshunova, A.N. Sivov, A.D. Shatrov, Chi-ral electromagnetic objects. Phys. Usp. **40**(11), 1149–1160 (1997)

6. A. Lakhtakia, Time-harmonic electromagnetic fields in chiral media. Lecture Notes in Physics / A. Lakhtakia, V.K. Varadan, V.V. Varadan (Springer-Verlag, Berlin, Heidelberg and Boston, 1989), p. 121
7. I.V. Semchenko, Research on chiral and bianisotropic media in Byelorussia and Russia in the last ten years / I.V. Semchenko, S.A. Tretyakov, A.N. Serdyukov. Progress Electromag. Res. **V.12.**, 335–370 (1996)
8. W.L. Bragg, The Diffraction of Short Electromagnetic Waves by a Crystal / W.L. Bragg. Proceedings of the Cambridge Philosophical Society **V. 17.**, 43 (1914)

On the Asymptotics of the Solution of a Klein–Gordon–Fock Equation with a Variable Coefficient of the Laplacian

Maria V. Korovina, Ilya N. Smirnov, and V. Yu. Smirnov

Abstract The problem of wave propagation in the medium whose velocity characteristics change under an external impact is considered. We investigate a three-dimensional case. The asymptotics of the solution for the Klein–Gordon–Fock operator with a variable coefficient of the Laplacian at infinity have been constructed. The case of holomorphic coefficients have been considered.

Keywords Asymptotic · Holomorphic coefficient · Laplace–Borel transform · Weighted spaces

1 Introduction

In this paper, the problem of wave propagation in dispersive three-dimensional media whose velocity characteristics change under an external impact is considered. The problem of this type and its physical applications were considered in [1, 17].

In this study, we consider the Klein–Gordon–Fock operator with a variable time-dependent coefficient in the term that contains the Laplacian:

$$\left(\frac{d}{dt}\right)^2 u(x,t) - a^0(t)\Delta u(x,t) + c^0(t)u(t,x) = 0. \tag{1}$$

M. V. Korovina · I. N. Smirnov (✉)
Lomonosov MSU, Moscow, Russia
e-mail: ismirnov@cs.msu.ru

M. V. Korovina
e-mail: betelgeuser@yandex.ru

V. Yu. Smirnov
Moscow Aviation Institute, Moscow, Russia
e-mail: vl-smirnov@mail.ru

C. H. Skiadas et al. (eds.), *13th Chaotic Modeling and Simulation International Conference*, Springer Proceedings in Complexity,
https://doi.org/10.1007/978-3-030-70795-8_31

Let us apply the variable separation method: $u(x,t) = Y(x)v(t)$. We get

$$\Delta Y(x) + \lambda Y(x) = 0, \tag{2}$$

$$\left(\left(\frac{d}{dt}\right)^2 + c^0(t)\right) v(t) - a^0(t)\lambda v(t) = 0. \tag{3}$$

Equation (2) is a classical Sturm–Liouville problem for the Laplace operator. According to [13], it can be reduced to solving a homogeneous Fredholm equation of the second kind with a symmetric weakly polar kernel. In the general case, the solution $Y(x)$ can be represented by using the Green's function

$$Y(x) = \lambda \int_D G(x, \hat{x}) Y(\hat{x}) dV_{\hat{x}},$$

where $G(x, \hat{x})$ is the Green's function corresponding to our problem. Such a representation allows us to consider that problem (2) is solved and proceed to solving equation (3), namely, to the problem of constructing the asymptotics of solutions to equation (3) in the neighborhood of a point infinitely distant in time.

Here, functions $a^0(t)$, $c^0(t)$ are regular at infinity, which means that there is the exterior of a circle $|t| > R$ such that the functions $a^0(t)$, $c^0(t)$ can be expanded in it into the divergent power series

$$a^0(t) = \sum_{j=0}^{\infty} \frac{a_j}{t^j}, \quad c^0(t) = \sum_{j=0}^{\infty} \frac{c_j}{t^j}.$$

The aim of our study is constructing the asymptotics of the solution to equation (1) at $t \to \infty$.

Let us make the substitution of variable $t = \frac{1}{r}$.

We get

$$\frac{d}{dt} v(t) = \frac{dv}{dr}\frac{dr}{dt} = -\frac{1}{t^2}\frac{dv}{dr} = -r^2 \frac{dv}{dr}.$$

Then, we rewrite equation (3) in the form:

$$\left(-r^2 \frac{d}{dr}\right)^2 v(r) + (c(r) - a(r)\lambda)\, v(r) = 0. \tag{4}$$

Generally speaking, infinity is an irregular singular point of equation (4). In the particular case where infinity is considered to be a regular singular point, the problem of constructing the asymptotics of solutions is solved. As is known, the asymptotics in the neighborhood of regular singular points are conormal (see, for example, [5]); namely, they have the form

$$\sum_j t^{s_j} \sum_{i=0}^{k} a_i^j \ln^i \frac{1}{t}, \qquad (5)$$

where a_i^j, s_j are some complex numbers.

The problem of constructing the asymptotics of solutions to ordinary differential equations in the neighborhood of an irregular singular point was ultimately formulated by Poincare [14] and remains unsolved in the general form until now.

The singular point for equation (4) is the point $r = 0$. It was shown in [3] that any homogeneous ordinary differential equation with holomorphic coefficients of order n can be represented in the form

$$\hat{H}u = H\left(r, -r^k \frac{d}{dr}\right)u = 0, \qquad (6)$$

where $\hat{H}$ is a differential operator with holomorphic coefficients with the symbol

$$H(r, p) = \sum_{i=0}^{n} a_i(r) p^i.$$

Here, $a_i(r)$ are holomorphic functions; $a_n(0) \neq 0$; a formula for calculating the minimum integer nonnegative value of k is obtained in [4]. Depending on this minimum value of k, the solutions of equation (6) have asymptotics of different types. If $k = 0$, equation (6) does not have singular points. In this case, the solutions are holomorphic functions. If $k = 1$, equation (6) has a singularity at the point $r = 0$; this singularity is regular. In other words, equation (6) in this case is the equation of Fuchsian type. At all other integer nonnegative k, the point $r = 0$ is an irregular singularity. Such singularities are called the cuspidal singularities.

Let us call the function $H_0(p) = H(0, p)$ the principal symbol of the differential operator $\hat{H}$. Nowadays, the most investigated equations are those whose principal symbol $H_0(p)$ has only simple roots. Let us call such equations the non-Fuchsian equations of the first type. All other equations are the non-Fuchsian equations of the second type.

The non-Fuchsian equations of the first type are considered, for instance, in [15] and in many other papers. It was shown in those studies that the asymptotics of solutions of non-Fuchsian equations of the first type can be represented in the form

$$\sum_{j=1}^{n} e^{\frac{\alpha_j}{r^k} + \sum_{i=1}^{k-1} \frac{a_{k-i}^j}{r^{k-i}}} r^{\sigma_j} \sum_{i=0}^{\infty} b_i^j r^i, \qquad (7)$$

where $\alpha_j, j = 1, \ldots, n$ are the roots of polynomial $H_0(p)$; a^j_{k-i} and σ_j are some complex numbers. The power series $\sum_{i=0}^{\infty} b^j_i r^i$ is asymptotic (generally speaking, divergent) power series.

In the case when the asymptotic expansion

$$u \approx u_1 + u_2 + \ldots + u_n = e^{\frac{\alpha_1}{r}} r^{\sigma_1} \sum_{i=0}^{\infty} b^i_1 r^i + e^{\frac{\alpha_2}{r}} r^{\sigma_2} \sum_{i=0}^{\infty} b^2_i r^i + \ldots + e^{\frac{\alpha_n}{r}} r^{\sigma_n} \sum_{i=0}^{\infty} b^n_i r^i \tag{8}$$

contains more than two terms, i.e., $n \geq 2$, then, the problem of interpretation of such an expression arises. There is the Stokes phenomenon whose gist is as follows: in any of the sectors of the complex plane, one of the terms of sum (7) is the dominant term of the asymptotics, while other terms are recessive; on another sector, the dominant term may become recessive and vice versa. At the boundary between the sectors, it is not clear which of the terms is dominant. Therefore, the recessive terms of asymptotics of the form (7) cannot be neglected. Hence, to construct uniform asymptotics of solutions, it is necessary to introduce a regular method for summation of divergent series. Summing up the asymptotic series obtained, we get an uniform asymptotics, namely, the asymptotic expansion that is valid in the entire vicinity of the irregular singular point. A more detailed consideration of uniform asymptotics and the summation method can be found in [16].

The method for summation of asymptotic series of the form (7) was proposed for the first time by Ecalle [2]. Summation is carried out with the help of the integral Laplace–Borel transform. Ecalle has invented resurgent analysis which provides a method for summation of formal power series. To clarify the concept of resurgent analysis, let us introduce the designation

The formal Laplace–Borel transform $\tilde{u}_1(p)$, $\tilde{u_2}(p) \ldots$ are the power series with respect to the dual variable p, which converge in the neighborhood of one of the roots $p = p_j$ of the principal symbol. At that, the inverse Borel transform provides a regular method for summing up these series. However, application of this transform in practice requires proving the fact that the functions $\tilde{u}_j(p)$ are infinitely extendable or, in other words, that the function $u_j(r)$ is resurgent (the definition of infinite extendability is presented below). To apply the methods of resurgent analysis, it is necessary to prove the resurgence of the solution. The infinite extendability of a solution is a differential equation with holomorphic coefficients under the condition of holomorphy of its right-hand side was proved in [6, 9]. This result makes it possible to use the resurgent analysis methods to construct uniform asymptotics of solutions to linear differential equations with holomorphic coefficients in the spaces of functions of exponential growth.

Using the result above, the authors of [6, 7, 9] have succeeded in constructing uniform asymptotics of solutions for equations such that the roots of the polynomial $H_0(p) = H(0, p)$ are simple. So, the problem of constructing uniform asymptotics of solution of the equations of this class for any order has been solved. Namely, the following theorem was proved.

Theorem 1 *Let function f be resurgent. Then all the solutions of the equation*

$$H\Big(r, -r^k \frac{d}{dr}\Big)u = f$$

are resurgent functions. If the polynomial $H_0(p)$ has simple roots at points $p_1, \ldots, p_m$, then the asymptotics of a solution of a homogeneous equation has the form

$$\sum_{j=1}^{n} e^{\frac{\alpha_j}{r^k} + \sum_{i=1}^{k-1} \frac{a^j_{k-i}}{r^{k-i}}} r^{\sigma_j} \sum_{i=0}^{\infty} b^j_i r^i,$$

where summation is carried out over all roots of polynomial $H_0(p)$.

The results of this theorem will be used in our study.

We also note papers [10–12], in which an asymptotic expansion of solutions of the basic boundary-value problems for the elasticity system and the biharmonic (polyharmonic) equation in the exterior of a compact set and in a half-space, including that in the form of a co-normal asymptotics, is obtained.

2 Definitions and Auxiliary Statements

In this section, we give definitions of some notion of resurgent analysis, which will be used below.

Let us designate by $S_{R,\varepsilon}$ the sector $S_{R,\varepsilon} = \{r| - \varepsilon < \arg r < \varepsilon, \quad |r| < R\}$.

Definition 1 Let us say that a function f is analytical on $S_{R,\varepsilon}$ and is of exponential growth with a rate of growth of at most k, if there are nonnegative constants C and α such that in the sector $S_{R,\varepsilon}$, the inequality

$$|f| < Ce^{a\frac{1}{|r|^k}}.$$

is fulfilled.

Let us designate by $E_k(S_{R,\varepsilon})$ the space of functions that are holomorphic in the domain $S_{R,\varepsilon}$ and of k-exponential growth at zero; by $E(\tilde{\Omega}_{R,\varepsilon})$ the space of holomorphic functions of exponential growth in the domain $\tilde{\Omega}_{R,\varepsilon}$; by $E(C)$, the space of entire functions of exponential growth.

The Laplace-Borel transform is the main technique used for constructing the theory of resurgent functions. Let us give the definition of the Laplace-Borel transform in such a form that is necessary for the purposes of this study.

Theorem 2 *Formulas*

$$\tilde{f} = B_k f = \int_0^{r_0} e^{-p/r^k} f(r) \frac{dr}{r^{k+1}}$$

and

$$B_k^{-1} \tilde{f} = \frac{k}{2\pi i} \int_{\tilde{\gamma}} e^{p/r^k} \tilde{f}(p) dp$$

define the mutually inverse mappings in the spaces

$$B_k : E_k(S_{R,\varepsilon}) \to E(\tilde{\Omega}_{R,\varepsilon})/E(C)$$

and

$$B_k^{-1} : E_k(\tilde{\Omega}_{R,\varepsilon})/E(C) \to E_k(S_{R,\varepsilon}).$$

Here, r_0 denotes an arbitrary point in the domain $S_{R,\varepsilon}$.

More details about the Laplace–Borel transform can be found in [16]. Now we can give a definition of a resurgent function. The contour $\tilde{\gamma}$ is shown in Fig. 1 in [6].

Note that for the k-Laplace–Borel transform, the following formulae are true:

$$B_k \circ \left(-\frac{1}{k} r^{k+1} \frac{\partial}{\partial r} \right) f(r) = p B_k f,$$

$$\frac{\partial}{\partial p} \circ B_k f = -B_k \left(\frac{1}{r^k} f(r) \right).$$

It was shown in [8] that if $k \in N$, then the equality

$$B_k r^k B_k^{-1} \left[\tilde{f} \right] = -\frac{1}{k} \int_{p_0}^{p} \tilde{f}(p') dp'$$

is satisfied. Now we can define a k-resurgent function.

Definition 2 The function $\tilde{f}$ is called infinitely extendable, if for any number R, there is a discrete set of points Z_R in C such that the function $\tilde{f}$ can be analytically extended from the initial domain of definition along any path with a length $< R$, which does not pass through Z_R.

Definition 3 The element f of the space $E_k(S_{R,\varepsilon})$ is called the k-resurgent function, if its k-Laplace–Borel transform $\tilde{f} = B_k f$ is infinitely extendable.

Theorem 3 *Let f be a resurgent function, then the solution of the equation*

$$H\left(-r^2\frac{d}{dr}, r\right)u = f$$

is a resurgent function. If the polynomial $H_0(p)$ has simple roots at the points $p_1, \ldots, p_m$, then the asymptotics of the solution of the homogeneous equation has the form

$$u(r) \approx \sum_{j=1}^{m} \exp\left(\frac{p_j}{r}\right) r^{\sigma_j} \sum_{i=0}^{\infty} b_i^j r^i, \tag{9}$$

where the sum is taken by the union of all the roots of the polynomial $H_0(p)$.

For equations with $k+1$-order degeneracy, where $k \in N$, namely, for equations of the form

$$H\left(r, -\frac{1}{k}r^{k+1}\frac{d}{dr}\right)u = 0$$

in the case when the roots of the principal symbol are simple, the asymptotics have the form

$$u(r) \approx \sum_{j=1}^{m} \exp\left(\frac{p_j}{r^k} + \sum_{i=1}^{k-1} \frac{\alpha_{k-i}^j}{r^{k-i}}\right) r^{\sigma_j} \sum_{i=0}^{\infty} b_i^j r^i. \tag{10}$$

In the case when $k+1 = \frac{m}{n}, m \in N, k \in N, m > k$, the asymptotics of the solution will be

$$u(r) \approx \sum_{j} \exp\left(\frac{p_j}{r^{\frac{m}{k}-1}} + \sum_{i=1}^{m-k-1} \frac{\alpha_{m-k-i}^j}{r^{\frac{m-i}{k}-1}}\right) t^{\sigma_j} \sum_{i=0}^{\infty} b_i^j t^i. \tag{11}$$

The proof of this theorem can be found in [6, 7, 9].

Property 1

$$B_k r^\sigma = \int_0^{r_0} e^{-p/r^k} r^\sigma \frac{dr}{r^{k+1}} =$$

$$= \int_0^{r_0} e^{-p/t} t^{\frac{\sigma}{k}} \frac{dr}{t^2} = \begin{cases} \dfrac{\pi}{\Gamma\left(\frac{\sigma}{k}\right)\sin \pi\frac{\sigma}{k}} p^{\frac{\sigma}{k}-1}, & \forall \dfrac{\sigma}{k} \notin Z, \\ \Gamma\left(1-\dfrac{\sigma}{k}\right) p^{\frac{\sigma}{k}-1}, & \dfrac{\sigma}{k} = 0, -1, \ldots, \\ (-1)^{\frac{\sigma}{k}} \dfrac{p^{\frac{\sigma}{k}-1}\ln p}{\left(\frac{\sigma}{k}-1\right)!}, & \dfrac{\sigma}{k} \in N \end{cases} \tag{12}$$

3 The Main Results

Let us introduce designations $b_j = c_j - \lambda a_j$, then $(c(r) - a(r)\lambda) = b_0 + rb_1 + r^2 \sum_{j=0}^{\infty} b_{j+2} r^j$ and equation (3) can be rewritten as follows:

$$\left(-r^2 \frac{d}{dr}\right)^2 v(r) + b_0 v(r) + r b_1 v(r) + r^2 g(r) v(r) = 0. \tag{13}$$

Here, $g(r) = \sum_{j=0}^{\infty} b_{j+2} r^j$. Let λ be such that $b_0 = c(0) - a(0)\lambda \neq 0$. Then, the principal symbol of the operator at the left-hand side of equation (3) is $p^2 + b_0$. Let us denote by p_i $i = 1, 2$ the roots of this polynomial. These roots are simple, therefore Theorem 3 is applicable; the asymptotics of the solution has the form (9). Let us construct it.

Let us find the asymptotics at the non-zero simple root p_1. Let us shift the root p_1 to zero by using the substitution of variables $v = e^{\frac{p_1}{r}} v_1$.

Since the equality

$$\left(-r^2 \frac{d}{dr}\right)^2 v + b_0 v = e^{\frac{p_1}{r}} \left(\left(r^2 \frac{d}{dr}\right)^2 - 2p_1 \left(r^2 \frac{d}{dr}\right) + p_1^2 + b_0 \right) v_1$$

holds, we can rewrite equation (13) in the form:

$$\left(-r^2 \frac{d}{dr}\right)^2 v_1 + 2p_1 \left(-r^2 \frac{d}{dr}\right) v_1 + r b_1 v_1(r) + r^2 g(r) v_1(r) = 0.$$

Let us perform one more substitution $v_1 = r^{\sigma_1} v_2$.

Since the equality

$$\left(r^2 \frac{d}{dr}\right)^2 v_1 = r^{\sigma_1} \left(\sigma_1(\sigma_1 + 1) r^2 + 2\sigma_1 r \left(r^2 \frac{d}{dr}\right) + \left(r^2 \frac{d}{dr}\right)^2 \right) v_2$$

is satisfied, we get the equation

$$\left(-r^2 \frac{d}{dr}\right)^2 v_2 + 2p_1 \left(-r^2 \frac{d}{dr}\right) v_2 + (2p_1\sigma_1 + b_1) r v_2(r) + r^2 g(r) +$$

$$+ \left(\sigma_1(\sigma_1 + 1) r^2 - 2\sigma_1 r \left(-r^2 \frac{d}{dr}\right) \right) v_2 = 0. \tag{14}$$

Let us set

$$\sigma_1 = -\frac{b_1}{2p_1} = -\frac{c_1 - \lambda a_1}{2p_1}, \tag{15}$$

Equation (13) can be rewritten as follows:

$$\left(-r^2\frac{d}{dr}\right)^2 v_2 + 2p_1\left(-r^2\frac{d}{dr}\right) v_2 + r^2 g(r) +$$

$$+\left(\sigma_1(\sigma_1+1)r^2 - 2\sigma_1 r\left(-r^2\frac{d}{dr}\right)\right) v_2 = 0.$$

Let us perform the Laplace–Borel transform. Using the designation $\hat{v} = Bv$, we get

$$p^2\hat{v}_2 + 2p_1 p\hat{v}_2 + 2\sigma_1 \int p\hat{v}_2 dp + \sigma_1(\sigma_1+1)) \int_1^p \int_1^{p_2} \hat{v}_2(p_1)dp_1 dp_2 +$$

$$+\int_1^p \int_1^{p_2} (B_1(g(r)v_2(r))(p_1)dp_1 dp_2 = f(p).$$

Here, f denotes an arbitrary holomorphic function. The principal symbol of the differential operator in equation (14) is $p(p+2p_1)$. The asymptotics of the solution of equation (15) in the neighborhood of the root $p = 0$ is constructed by the method of successive approximations (see [9]), which shows that the asymptotic behavior of the solution in the vicinity of the simple root $p = 0$ can be represented as follows:

$$\frac{A}{p} + \ln p \sum_{i=0}^{\infty} m_i p^i. \tag{16}$$

Here, A is some constant; the series $\sum\limits_{i=0}^{\infty} m_i p^i$ converges in some neighborhood of the point $p = 0$.

Let us find the inverse Laplace–Borel transform of function (16). We obtained from (12)

$$B_1^{-1}\left(\frac{A}{p} + \ln p \sum_{i=0}^{\infty} m_i p^i\right) = \sum_{i=0}^{\infty} A_i^1 r^i. \tag{17}$$

Here, $\sum\limits_{i=0}^{\infty} A_i^1 r^i$ stands for the corresponding asymptotic series. It follows from (17) that the asymptotic term for the function $v_1(r)$ corresponding to the root c_1 of the principal symbol in the neighborhood of $r = 0$ has the asymptotics $r^{\sigma_1}\sum_{i=0}^{\infty} A_i^1 r^i$; for the function $v(r)$, the corresponding asymptotic term has the form

$$e^{\frac{p_1}{r}} r^{\sigma_1} \sum_{i=0}^{\infty} A_i^1 r^i.$$

The asymptotic term corresponding to the root of p_2 is found in a similar way.

Thus, we have obtained that under the condition $b_0 = c(0) - a(0)\lambda \neq 0$, the asymptotics of the solution of equation (3) in a neighborhood of the point $r = 0$ has the form:

$$v \approx e^{\frac{p_1}{r}} r^{\sigma_1} \sum_{i=0}^{\infty} A_i^1 r^i + e^{\frac{p_2}{r}} r^{\sigma_2} \sum_{i=0}^{\infty} A_i^2 r^i.$$

Here, the numbers $p_i, i = 1, 2$ are the roots of the polynomial $p^2 + b_0$; the numbers σ_i, $i = 1, 2$ are determined by formula (13); $\sum_{i=0}^{\infty} A_i^j r^i$, $i = 1, 2$ are the corresponding asymptotic series.

Now, let $b_0 = c(0) - a(0)\lambda = 0$. Then, equation (3) can be rewritten as follows:

$$\left(-r^2 \frac{d}{dr}\right)^2 v + b_1 r v + r^2 g_2(r) v = 0. \tag{18}$$

Here $b_1 = c_1 - \lambda a_1$, $g_2(r) = \frac{\sum_{j=2}^{\infty}(c_j - \lambda a_j) r^j}{r^2} = \sum_{j=2}^{\infty}(c_j - \lambda a_j) r^{j-2}$.

Unlike the previous case, the principal symbol of (18) is equal to p^2 and has a multiple root at zero. Therefore, Theorem 3 is not applicable to this case. To use Theorem 3, it is necessary to transform equation (18).

Let $b_1 \neq 0$. We will construct the asymptotics for this case.

Since the equality

$$\left(-r^2 \frac{d}{dr}\right)^2 = r\left(r^{\frac{3}{2}} \frac{d}{dr}\right)^2 + \frac{1}{2} r^{1+\frac{1}{2}} \left(r^{\frac{3}{2}} \frac{d}{dr}\right)$$

is satisfied, equation (18) can be rewritten as follows:

$$\left(r^{-\frac{3}{2}} \frac{d}{dr}\right)^2 v - \frac{1}{2} r^{\frac{1}{2}} \left(r^{-\frac{3}{2}} \frac{d}{dr}\right) v + b_1 v + r g_2(r) v = 0. \tag{19}$$

Let us apply Theorem 3. The asymptotics of the solution has the form (11). In the same way as it was done above, we make a substitution of variables $v = r^{\sigma} v_1$.

It is easy to show that equation (19) can be rewritten in the form

$$\left(-2r^{\frac{3}{2}}\frac{d}{dr}\right)^2 v_1 + 2\left(\frac{1}{2}+2\sigma\right)r^{\frac{1}{2}}\left(2r^{\frac{3}{2}}\frac{d}{dr}\right)v_1 + 4b_1v_1 + 4r(g_2(r)+\sigma(\sigma+1))v_1 = 0.$$

Let us set $\sigma = -\frac{1}{4}$. Then the equation will take the form:

$$\left(-2r^{\frac{3}{2}}\frac{d}{dr}\right)^2 v_1 + 4b_1v_1 + 4r\left(g_2(r) - \frac{3}{16}\right)v_1 = 0. \tag{20}$$

The principal symbol of the differential operator in equation (20) is $p^2 + 4b_1$. We denote its roots by p_i, $i = 1, 2$. Let us apply Theorem 3. The asymptotics of the solution has the form (11). Since the roots of the principal symbol are simple, the asymptotics of the solutions can be constructed as in the previous case and have the form:

$$v \approx e^{\frac{p_1}{\sqrt{r}}} r^{-\frac{1}{4}} \sum_{i=0}^{\infty} A_i^1 r^{\frac{i}{2}} + e^{\frac{p_2}{\sqrt{r}}} r^{-\frac{1}{4}} \sum_{i=0}^{\infty} A_i^2 r^{\frac{i}{2}}.$$

Now let $b_1 = 0$, then equation (11) takes the form:

$$\left(-r^2\frac{d}{dr}\right)^2 v + r^2 g_2(r)v = 0. \tag{21}$$

In this case, the principal symbol is p^2 and has a multiple root of the second order at zero. Therefore, the method used previously is not applicable here. Since the equality

$$r^2\frac{d}{dr}r^2\frac{d}{dr} = r^2\left(r\frac{d}{dr}\right)\left(r\frac{d}{dr}\right) + r^3\frac{d}{dr},$$

holds, we can rewrite equation (21) in the form:

$$\left(r\frac{d}{dr}\right)^2 + r\frac{d}{dr} + g_2(r)v = 0. \tag{22}$$

This is a Fuchsian-type equation. The point $r = 0$ is a regular singular point of equation (22). The problem of constructing such solutions is well studied (see, for example, [5]). It is reasonable to search for the solution to this equation in the Sobolev weighted spaces $H^{k,\sigma}(0, \infty)$. The asymptotics of its solution is conormal. The reasoning above allows us to state the following theorem.

Theorem 4 *If $c_0 - a_0\lambda \neq 0$, then all the asymptotics of the solutions to equation (1) in the space of functions of exponential growth in the neighborhood of infinity with respect to variable t (at $t \to \infty$) have the form:*

$$u(x,t) \approx \left(\exp_{p_1 t} t^{\sigma_1} \sum_{i=0}^{\infty} A_i^1 t^{-i} + \exp_{p_2 t} t^{\sigma_2} \sum_{i=0}^{\infty} A_i^2 t^{-i} \right) Y(x),$$

where numbers p_i, $i = 1, 2$ *are the roots of the polynomial* $p^2 + c_0 - a_0\lambda$. *If* $c_0 - a_0\lambda = 0$, $c_1 - \lambda a_1 \neq 0$, *then the asymptotics of the solution has the form:*

$$u(x,t) \approx t^{\frac{1}{4}} \left(\exp_{p_1 \sqrt{t}} \sum_{i=0}^{\infty} A_i^1 t^{-\frac{i}{2}} + \exp_{p_2 \sqrt{t}} \sum_{i=0}^{\infty} A_i^2 t^{-\frac{i}{2}} \right) Y(x),$$

where the numbers p_i, $i = 1, 2$ *denote the roots of the polynomial* $p^2 + 4(c_1 - \lambda a_1)$; $\sum_{i=0}^{\infty} A_i^j r^i$, $j = 1, 2$ *are the corresponding asymptotic series. If* $c_0 - a_0\lambda = 0$, $c_1 - \lambda a_1 = 0$, *then the asymptotic behavior of the solution of equation (1) can be represented as the product of the function* $Y(X)$ *by the corresponding conormal asymptotics.*

Remark 1 If equation (2) has the form

$$\left(\frac{d}{dt}\right)^2 v + a^2 \lambda v = 0,$$

where $a \neq 0$ is a real constant, then all asymptotics of solutions to this equation in the space of the functions of exponential growth in the neighborhood of infinity with respect to variable t can be represented in the form

$$u(x,t) = (A_1 e^{c_1 t} + A_2 e^{c_2 t}) Y(x).$$

Here, $c_i, i = 1, 2$, are the roots of the polynomial $p^2 + a^2\lambda$; A_i are arbitrary constants.

The results of the article are presented at the 13*th* International Conference Chaotic Modeling, Simulation and Applications (CHAOS2020, June 9–12, 2020, Florence, Italy).

References

1. G.A. Askaryan, Effect of self-focusing. Usp. Fiz. Nauk **111**(10), 249–260 (1973). (in Russian)
2. J. Ecalle, Cinq applications des fonctions resurgentes. Prepub. Math. d'Orsayl **84**(62), 110 (1984)
3. D.S. Kats, Computation of the asymptotics of solutions for equations with polynomial degeneration of the coefficients. Differ. Equ. **51**(12), 1589–1594 (2015)
4. D.S. Kats, Coefficients of series in asymptotic expansions of solutions of equations with degenerations. Int. J. Open Inf. Technol. **4**(9), 1–7 (2016). (in Russian)
5. V.A. Kondrat'ev, Boundary value problems for elliptic equations in domains with conical or angular points. Trudy Moskov. Mat. Obshch. [Trans. Moscow Math. Soc.] **16**, 209–292 (1967)

6. M.V. Korovina, Existence of resurgent solutions for equations with higher-order degeneration. Differ. Equ. **47**(3), 346–54 (2011)
7. M.V. Korovina, Asymptotics solutions of equations with higher-order degeneracies. Doklady Math. **83**(2), 182–184 (2011)
8. M.V. Korovina, Asymptotics of solutions of equations with higher degenerations. Differ. Equ. **48**(5), 717–729 (2012)
9. M.V. Korovina, V.E. Shatalov, Differential equations with degeneration and resurgent analysis. Differ. Equ. **46**(9), 1267–1286 (2010)
10. O.A. Matevossian, Solutions of exterior boundary-value problems for the elasticity system in weighted spaces. Sb. Math. **192**(12), 1763–1798 (2001)
11. H.A. Matevossian, On solutions of mixed boundary-value problems for the elasticity system in unbounded domains. Izvestiya Math. **67**(5), 895–929 (2003)
12. H.A. Matevossian, On the mixed Neumann–Robin problem for the elasticity system in exterior domains. Russ. J. Math. Phys. **27**(2), 272–276 (2020)
13. I.G. Petrovskii, *Lectures on Partial Differential Equations* (GIFML, Moscow, 1961). (in Russian)
14. H. Poincare, Sur les integrales irregulieres des equations lineaires. Acta math. **8**, 295–344 (1886)
15. W. Sternberg, *Uber Die Asymptotische Integration von Differencialgleichungen* (Verlag von Julius Spriger, Berlin, 1920)
16. B. Sternin, V.E. Shatalov, *Borel-Laplace Transform and Asymptotic Theory. Introduction to Resurgent Analysis* (CRC Press, London, 1996)
17. V.I. Talanov, On self-focusing of wave beams in nonlinear media. Pis'ma Zh. Eksp. Teor. Fiz. **2**(5), 218–222, 1964. (in Russian)

A Model of Stabilization of Chaotic Wave Processes in Complex Dynamical Systems from the Point of View of the Matrix Decomposition Theory

Alexander M. Krot

Abstract A general model of the origin and evolution of chaotic wave processes in complex systems based on the proposed method of matrix decomposition of operators of nonlinear systems is developed in the article. The proposed model shows that the effect of self-organization in complex systems of different physical nature (for example, hydrodynamic, electronic and physiological ones) is based on the interaction of nonlinear processes of higher orders leading to stabilization (to the finite value) of the amplitude of chaotic wave process. Mathematically, this means the synchronous "counteraction" of nonlinear processes of even and odd orders in a general vector–matrix model of a complex system being in a chaotic mode. The implementation of the vector–matrix decomposition by means of computational experiments shows that the model of L. D. Landau describes the scenario of the occurrence of chaotic modes in complex systems quite well. It is noted that the regime of hard self-excitation of nonlinear oscillations in complex systems leads to the appearance of a chaotic attractor in the state-space. Moreover, the proposed vector–matrix model permits to find more general conditions for the origin and evolution of chaotic wave processes and, as a result, to explain the appearance of coherent nonlinear phenomena in complex systems.

Keywords Complex nonlinear dynamical system · State-space · Chaotic attractor · Matrix series in state-space · General vector–matrix model of chaotic wave processes · Mode of hard self-excitation of nonlinear oscillations · Stabilization of the amplitude of chaotic process

1 Introduction

The development of the theory of chaotic wave processes (in particular, the theory of turbulence in aerohydrodynamic flows) is important from the point of view of

A. M. Krot (✉)
Laboratory of Self-Organization System Modeling, United Institute of Informatics Problems, National Academy of Sciences of Belarus, Surganov Str. 6, 220012 Minsk, Belarus
e-mail: alxkrot@newman.bas-net.by

C. H. Skiadas et al. (eds.), *13th Chaotic Modeling and Simulation International Conference*, Springer Proceedings in Complexity,
https://doi.org/10.1007/978-3-030-70795-8_32

understanding the processes of self-organization in complex dynamic systems. L. D. Landau in his article "To the problem of turbulence" [1] developed the theory of initial turbulence, within the framework of which he showed that the initial instability of the laminar motion of a fluid is determined by a small perturbation of its velocity

$$\vec{v}_1 = A(t)\vec{f}(x, y, z), \tag{1}$$

where the time function $A(t)$ has the form:

$$A(t) = const \cdot e^{-i\Omega t}, \tag{2}$$

and Ω determines the spectrum of «eigen frequencies» (generally speaking, complex-valued ones, i.e.$\Omega = \omega + i\gamma$). Moreover, he emphasized [1]:

> Expressions (1), (2) for the corresponding function $\vec{v}_1(x, y, z, t)$ (when $\Omega = \Omega_1$) are suitable, however, only for a short period of time after the instant of failure of the stationary state, since the multiplier $e^{\gamma_1 t}$ grows rapidly with time. In reality, of course, the amplitude modulus $|A|$ of non-stationary motion does not grow unlimitedly, but tends to some finite limit.

As a continuation of this idea, a model of a discrete quasistationary linear dynamical system was proposed in [2] based on a generalized spectral representation in the basis of *eigenfunctions* corresponding to the eigenvalues of the operator of this system. As shown in [2], the class of such quasistationary dynamical systems and processes is characterized by operators invariant to the generalized shift represented by powers of the Frobenius matrix. Let us note that the term "invariance to the generalized shift" was first introduced by the founder of cybernetics Norbert Wiener to analyze non-stationary and non-linear dynamical systems and processes [3].

E. Lorenz [4], studying the behavior of a viscous fluid under convection (the Rayleigh–Benard flow), reduced the system of Navier–Stokes equations and thermal conductivity to a dynamical model using the Galerkin' method. As a result, the Lorenz model (described by three ordinary nonlinear differential equations) revealed a chaotic (strange) attractor in the state-space. Mathematically, the concept of a "chaotic attractor" has been formulated by D. Ruelle and F. Takens [5] as a key element in the interpretation of irregular behavior described by deterministic equations for understanding mainly turbulence. This led to a new area of researches called deterministic chaos. Subsequently, I. Prigogine [6], H. Haken [7] have been shown that a new order (self-organization) in a complex system of various physical nature arises through deterministic chaos (fluctuations), i.e. by a chaotic mode of functioning complex system. The progress in this area has contributed to intensive studies of various scenarios of the transition of the dynamics of complex systems from a periodic to a chaotic mode through quasiperiodicity, subharmonic cascade, intermittency, etc. (see, for example, [8]).

At the same time, despite the achieved results, problems remain concerning the *stabilization* of chaotic wave processes in complex systems during a long time under unchanged the control parameters (for example, characteristic Reynolds Re and

Rayleigh Ra numbers, etc. in the case of aerohydrodynamic systems). Indeed, as noted above, the amplitude modulus $|A|$ of unsteady motion does not grow unlimitedly, but tends to a certain finite limit [1, 9]. In this connection, the aim of this work is to study the mechanisms of amplitude stabilization and evolution of chaotic wave processes [10] in complex systems of various physical nature (mainly, in hydrodynamic, electronic and electrochemical systems).

2 Interrelation of Landau and Lorenz Models of Turbulence Description

So, according to the Landau model of initial turbulence "for the smallest times, when it is still applicable (2)", we have

$$d|A|^2/dt = 2\gamma_1|A|^2.$$

"This expression is essentially only the first term in the expansion in powers of A and A^*. As the modulus increases, the following terms of this expansion must be taken into account" [1]. This means that the amplitude equation of the form is valid:

$$d|A|^2/dt = 2\gamma_1\ |A|^2 - \alpha_L|A|^4, \tag{3}$$

where $|A|^2 = A \cdot A^*$, γ_1 is a damping constant, and α_L is the Landau constant [9]. In the case of $\alpha_L > 0$, the amplitude (3) describes such dynamical states of a moving fluid, when $\gamma_1 > 0$, i.e. if $\mathrm{Re} > \mathrm{Re}_c$, for the first time, an arbitrarily small disturbance becomes unstable against the background of the main motion (a system with *soft* self-excitation).

In the case of $\alpha_L < 0$, two terms is no longer enough in the expansion (3), so the following generalization [9] is true:

$$\frac{d|A(t)|^2}{dt} = 2\gamma_1|A|^2 - \alpha_L|A|^4 - \beta_L|A|^6, \tag{4}$$

where $\beta_L > 0$ is a constant. So, in the case of $\alpha_L < 0$, $\beta_L > 0$, the amplitude (4) characterizes such dynamical states that under $\mathrm{Re} > \mathrm{Re}_c$ a stationary motion cannot exist at all, since if $\mathrm{Re} = \mathrm{Re}_c$ then a perturbation jumps up to a finite amplitude. In the interval $Re'_c < \mathrm{Re} < \mathrm{Re}_c$, the main motion is *metastable*, i.e. it is stable with respect to infinitesimal perturbations, but unstable relative to perturbations of finite amplitude (a system with *hard* self-excitation [9]).

Following the Landau model (4) (after disruption of the stationary mode of fluid flow) having designated by $S = |A|^2$ the square of the amplitude of unsteady process, the equation is valid up to the third-order terms:

$$\dot{S} = 2\gamma_1 S - \alpha_L S^2 - \beta_L S^3 - ..., \tag{5}$$

where γ_1 is a damping coefficient, α_L is the Landau constant, and β_L is a positive or negative constant.

As already mentioned, the model of convection turbulence is described by the Lorenz system:

$$\begin{cases} \dot{u}_1 = au_2 - au_1; \\ \dot{u}_2 = -u_1 u_3 + cu_1 - u_2; \\ \dot{u}_3 = u_1 u_2 - bu_3, \end{cases} \tag{6}$$

where $a = \text{Pr}$, $b = 4\pi^2/(\pi^2 + q^2)$, $c = Raq^2/(\pi^2 + q^2)^3$ are the given parameters, Pr is the characteristic Prandtl number, Ra is the characteristic Rayleigh number, q is an integer [4, 8]. For example, in the case of $a = 10$, $b = 8/3$ under the value of the *control* parameter $c = 24.27$ [8], a chaotic attractor is observed in the state-space $(u_1, u_2, u_3) \in U$.

In the particular case, it is possible to pass from system (6) to one nonlinear equation with respect to the variable u_1, if the variable u_3 is considered by a constant, i.e. $\dot{u}_3 = 0$. Expressing from the third equation $u_3 = u_1 u_2/b$ and substituting this value in the second, and then u_2 in the first equation, we get:

$$\dot{u}_1 = -a\dot{u}_2 - au_1^2 u_2/b + acu_1 - au_1. \tag{7}$$

If we additionally assume a weak dependence of the variable u_2 on time, leading to the condition $\dot{u}_2 \approx 0$, then (7) takes the form:

$$\dot{u}_1 \approx a(c-1)u_1 - (a/b)u_2 u_1^2. \tag{8}$$

As follows from (8), the first equation from the Lorenz system (under the condition that the second $u_2\ (t)$ and third variables $u_3\ (t)$ are fixed in time) has the same form as the Landau (5) with $\beta_L = 0$. This means that the vector equation describing the Lorenz system generalizes the scalar Landau equation.

3 Development of a General Model of the Origin of Chaotic Wave Processes Using the Matrix Decomposition Method

The Landau theory, based on the intuitive logical conclusions and the results of Reynolds hydrodynamic experiments, is suitable for explaining the initial turbulence origin in a moving viscous fluid. However, generally speaking, chaotic wave processes occur in complex systems of various physical nature (for example, electrodynamic, chemical or physiological systems [6]). In this connection, we will try to build a general model of the origin and stabilization of chaotic wave processes using the matrix decomposition method in the state-space of a complex system. As known [6], the mechanisms of the occurrence of spatial and temporal structures in complex nonlinear dynamical systems (NDS) can be revealed from the analysis of a system of nonlinear partial differential equations of the form:

$$\frac{\partial u_i}{\partial t} = f_i\big(\{u_j\}, \{\nabla u_j\}, \{\nabla^2 u_j\}, ..., \{c_l\}\big), \quad i, j = 1, 2, ..., N, \tag{9}$$

where f_i is a nonlinear function, and $u_i = u_i(t, \vec{r})$ is a function of temporal and spatial variables, satisfying the initial and boundary conditions, ∇ is the Hamilton differential operator, and c_l are the system parameters.

Using the Galerkin method, the system of nonlinear partial differential equations can be reduced to a system of nonlinear ordinary differential equations [8], so that in the future we will not take into account the explicit dependence u_i on the spatial vector $\vec{\mathrm{r}}$, assuming that $u_i = u_i(t)$.

It is often necessary to study the behavior of the solution of (9) near a specific standard state $\{u_i^*\}$, moreover, $u_i^* = u_i^*(t)$ is considered as an unperturbed solution (9), constantly disturbed by external influences or internal fluctuations by the value $v_i = v_i(t)$ [6]. As a result, a new solution becomes instead u_i^*:

$$u_i = u_i^* + v_i. \tag{10}$$

From relations (9) and (10) we obtain a system of equations for $\{v_i\}$:

$$\frac{dv_i}{dt} = f_i\big(\{u_j^* + v_j\}\big) - f_i\big(\{u_j^*\}\big). \tag{11}$$

Near the reference point when $|v_i/u_i^*| << 1$, it is possible to linearize the function f_i in the vicinity of zero values $v_i = 0$, using the first terms of the multiple Taylor series [6]. Lyapunov's theorem establishes a connection between the stability of systems of (9) and (11): if the trivial solution ($v_i = 0$) of (11) is asymptotically stable, then u_i^* is an asymptotically stable solution to (9). However, the linear theory of stability does not allow us to predict the dynamics in the case of complex NDS.

In the vector–matrix form, the system of ordinary differential equations (obtained from (9)) can be considered as the Cauchy problem in the N-dimensional state-space U of complex NDS:

$$\dot{\vec{u}} = \vec{f}(\vec{u}(t), \ \vec{u}_0, \{c_l\}), \ \ \vec{u}(0) = \vec{u}_0, \ \ \vec{u}(t) \in U, \tag{12}$$

where $\vec{u}(t) = (u_1(t), \ldots, u_N(t))^T$, T is the transposition symbol, $\vec{u}_0$ is a vector of initial data, $\{c_l\}$ is a set of system parameters. The solution $\vec{u}(t)$ of (12) defines a certain curve in the state-space (phase space) $U = \Re^N$, called the *phase trajectory*. Similarly, we can write system (11) in the vector–matrix form:

$$\dot{\vec{v}} = \Delta \vec{f}\big(\vec{v}(t),\ \vec{u}^*, \{c_l\}\big), \tag{13}$$

where $\vec{v}(t) = (v_1(t), \ldots, v_N(t))^T$, $\Delta \vec{f}$ is an increment of the vector function, $\vec{u}^*$ is a vector of the unperturbed (standard) state, $\{c_l\}$ is a set of system parameters.

In [11–16], a matrix decomposition method for complex NDS operators based on a matrix series in the state-space has been developed, and numerous examples of the application of this method for the analysis of complex systems have been considered. Among them are attractors of complex NDS [13, 14], as well as the Hopfield artificial neural network and Chua circuit [15, 16]. According to this method, the increment of the vector function $\Delta \vec{f}$ of a complex NDS in the state-space is described by a matrix series of the form [11–13]:

$$\begin{aligned}\Delta \vec{f}(\vec{v}, \vec{u}^*) &= \vec{f}(\vec{u}^* + \vec{v}) - \vec{f}(\vec{u}^*)\\ &= L^{(1)}_{N\times N}\vec{v} + \frac{1}{2!}L^{(2)}_{N\times N^2}(\vec{v}\otimes\vec{v}) + \frac{1}{3!}L^{(3)}_{N\times N^3}(\vec{\mathrm{v}}\otimes\vec{\mathrm{v}}\otimes\vec{\mathrm{v}}) + \ldots\\ &= \sum_{k=1}^{\infty}\frac{1}{k!}L^{(k)}_{N\times N^k}\cdot\vec{v}^{\otimes k},\end{aligned} \tag{14}$$

where $L^{(k)}_{N\times N^k} = \underbrace{\left(\frac{\partial}{\partial \vec{v}^T}\otimes\left(\frac{\partial}{\partial \vec{v}^T}\otimes\ldots\otimes\left(\frac{\partial}{\partial \vec{v}^T}\right.\right.\right.}_{k}\otimes\vec{f}\big)\ldots\big)\big)_{\vec{u}^*}$ are the matrix kernels of the homogeneous nonlinear system operators, $\vec{\mathrm{v}}^{\otimes k} = \underbrace{(\vec{\mathrm{v}}\otimes\vec{\mathrm{v}}\otimes\ldots\otimes\vec{\mathrm{v}})}_{k}$ is the k-th Kronecker degree of the perturbation vector $\vec{\mathrm{v}}$, and $\vec{u}^*$ is considered as the standard state vector belonging to $\Re^N$.

Application of matrix expansion (14) to the right-hand side of (13) allows us to formulate the following theorem:

Theorem 1. The equation of dynamics of the perturbation vector $\vec{\mathrm{v}}$ of a smooth NDS near the standard state $\vec{u}^* \in \Re^N$ is described by the vector–matrix expansion:

$$\begin{aligned}\dot{\vec{v}} =& L^{(1)}_{N\times N}\vec{v} + \frac{1}{2!}L^{(2)}_{N\times N^2}(\vec{v}\otimes\vec{v}) + \frac{1}{3!}L^{(3)}_{N\times N^3}(\vec{v}\otimes\vec{v}\otimes\vec{v})\\ &+ \ldots + \frac{1}{k!}L^{(2)}_{N\times N^k}\cdot\vec{v}^{\otimes k} + \vec{R}_k(\vec{v}),\\ \dot{\vec{v}} =& L^{(1)}_{N\times N}\vec{v} + \frac{1}{2!}L^{(2)}_{N\times N^2}(\vec{v}\otimes\vec{v}) + \frac{1}{3!}L^{(3)}_{N\times N^3}(\vec{v}\otimes\vec{v}\otimes\vec{v})\\ &+ \ldots + \frac{1}{k!}L^{(k)}_{N\times N^k}\cdot\vec{v}^{\otimes k} + \vec{R}_k(\vec{v}),\end{aligned} \tag{15}$$

where $\vec{R}_k(\vec{v})$ is the remainder term of the Lagrange's form:

$$\vec{R}_k(\vec{v}) = \frac{1}{(k+1)!} L^{(k+1)}_{N\times N^{k+1}} \big|\vec{u}^* + \theta(\vec{v} - \vec{u}^*) \cdot \vec{v}^{\otimes k+1}, \quad 0 < \theta < 1.$$

Corollary (the synchronization condition in a complex system). If the mode of stochastic dynamics is theoretically reached in a complex system, then the following relation holds:

$$\sum_{m=1}^{\infty} \overline{\frac{1}{(2m)!} L^{(2m)}_{N\times N^{2m}} \cdot \vec{v}^{\otimes 2m}} = -\sum_{m=1}^{\infty} \overline{\frac{1}{(2m+1)!} L^{(2m+1)}_{N\times N^{2m+1}} \cdot \vec{v}^{\otimes 2m+1}},$$

where the upper line means the ensemble averaging.

Proof when a complex system demonstrates the chaotic dynamics like a stochastic system then following the stochastic processes theory can rewrite (15) in the form of Ito stochastic differential equation [7]:

$$d\vec{v} = L^{(1)}_{N\times N}\vec{v}\,dt + \frac{1}{2!}L^{(2)}_{N\times N^2}(\vec{v}\otimes\vec{v})dt + \frac{1}{3!}L^{(3)}_{N\times N^3}(\vec{v}\otimes\vec{v}\otimes\vec{v})dt + \ldots + \frac{1}{k!}L^{(k)}_{N\times N^k}\cdot\vec{v}^{\otimes k}dt + \ldots \tag{15a}$$

Then in Eq. (15a) $\vec{v}$ can be considered as a stochastic vector process $\vec{v}(\alpha, t)$ where α is a state parameter of stochastic system (elementary event) [3], i.e., $\alpha \in [0, 1]$ and t is time, $t \in [0, +\infty[$, and the mathematical expectation of increment of the stochastic process is equal to zero: $\mathrm{M}[d\vec{v}(\alpha, t)] = \overline{d\vec{v}(\alpha, t)} = 0$. In this case the complex system generates fluctuations, for which the average value of the state vector increment $\overline{d\vec{v}} = 0$. Taking into account the right-hand side of Eq. (15a) we obtain the corollary.

Comment. Under the incomplete compliance with the stochastic dynamics mode in a complex system, which is characteristic for a *deterministic* chaotic system, an approximate equality of the averaged signals from the outputs of kernels of even and odd orders occurs:

$$\sum_{m=1}^{\infty} \overline{\frac{1}{(2m)!} L^{(2m)}_{N\times N^{2m}} \cdot \vec{v}^{\otimes 2m}} \approx -\sum_{m=1}^{\infty} \overline{\frac{1}{(2m+1)!} L^{(2m+1)}_{N\times N^{2m+1}} \cdot \vec{v}^{\otimes 2m+1}}.$$

It is easy to see that the obtained (15) generalizes the Landau model (5) of initial turbulence; therefore, it can be considered as a general model for the emergence and evolution of chaotic wave processes in complex systems.

4 Implementation of a General Model for the Origin and Amplitude Stabilization of Chaotic Wave Processes

The performed computational experiments using the general model (15) of the appearance of chaotic wave processes for specific types of complex NDS indicate the fact that chaotic oscillations are observed only at certain relations between the linear $L^{(1)}_{N\times N}$, quadratic $L^{(2)}_{N\times N^2}$, and cubic $L^{(3)}_{N\times N^3}$, etc. kernels of the matrix series into the overall dynamics of a complex system.

Indeed, the model (15) of the appearance of chaotic wave processes in the above mentioned Lorenz NDS (6) has a rather simple form:

$$\dot{\vec{v}} = L^{(1)}_{3\times 3}\vec{v} + \frac{1}{2!}L^{(2)}_{3\times 9}(\vec{v}\otimes\vec{v}) = \dot{\vec{v}}^{(1)} + \dot{\vec{v}}^{(2)}. \tag{16}$$

In other words, the dynamics of the Lorenz system is described on the basis of only linear and quadratic kernels [12–14]:

$$L^{(1)}_{3x3}(\vec{u}^*) = \left[\begin{array}{c|c|c} -a & a & 0 \\ \hline -u_3^* + c & -1 & -u_1^* \\ \hline u_2^* & u_1^* & -b \end{array}\right] \tag{16a}$$

$$L^{(2)}_{3x9}(\vec{u}^*) = \left[\begin{array}{c|c|c|c|c|c|c|c|c} 0 & 0 & 0 & 0 & 0 & 0 & 0 & 0 & 0 \\ \hline 0 & 0 & -1 & 0 & 0 & 0 & -1 & 0 & 0 \\ \hline 0 & 1 & 0 & 1 & 0 & 0 & 0 & 0 & 0 \end{array}\right] \tag{16b}$$

Moreover, according to Corollary and Comment the computational modeling of signals from the outputs of linear and quadratic kernels in a chaotic state (when $a = 10$, $b = 8/3$, $c = 24.27$; $u_1^* = u_2^* = u_3^* = 0$) of Lorenz's NDS revealed a similarity of their evolution in time, but with the opposite sign and unequal amplitude (Fig. 1). In other words, when the Lorenz system operates in a chaotic mode, the output signals from the linear and quadratic kernels, being in antiphase, partially compensate each other, so that, as a whole, it leads to *stabilization of the amplitude* of the chaotic wave process to a finite value [10].

A similar study of the general model (15) of the origin of chaotic wave processes in Chua's electronic circuit [15, 16]:

$$\begin{cases} \dot{u}_1 = \alpha u_2 - A\alpha u_1^3 - C\alpha u_1; \\ \dot{u}_2 = u_1 - u_2 + u_3; \\ \dot{u}_3 = -\beta u_2, \end{cases} \tag{17}$$

when the system parameters are equal $\alpha = 15.6$, $\beta = 28$, $A = 0.002$, $C = -1.3$, $u_1^* = -1.5$, it has been shown [16] that the dynamics of the Chua system are precisely described on the basis of only linear, quadratic and cubic kernels:

$$L^{(1)}_{3\times 3}(\vec{u}^*) = \left[\begin{array}{c|c|c} -(3A\alpha u_1^{*2} + C\alpha) & \alpha & 0 \\ \hline 1 & -1 & 1 \\ \hline 0 & -\beta & 0 \end{array}\right]; \tag{17a}$$

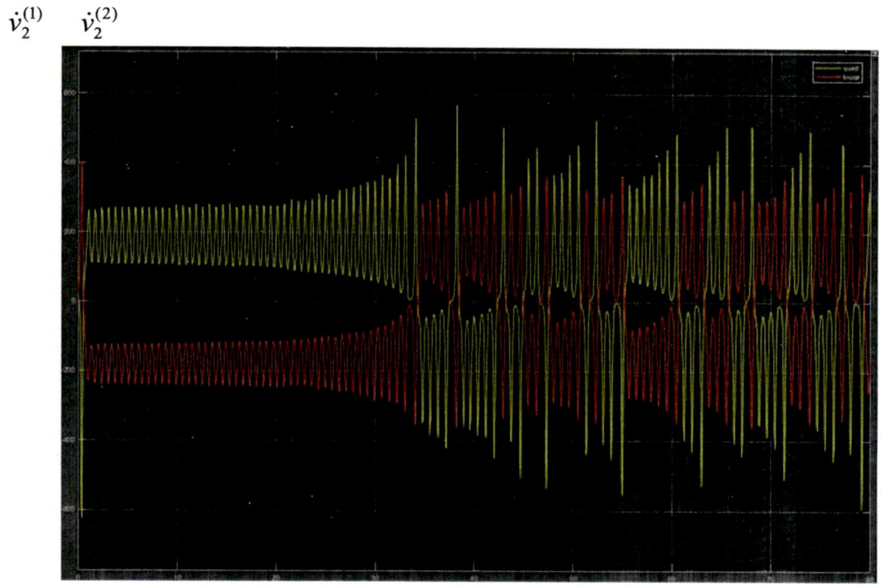

Fig. 1 Examples of signals generated by linear $L_{3x3}^{(1)}$ and quadratic $L_{3x9}^{(2)}$ kernels in accord with the model (16) of origin of chaotic wave processes in the complex NDS of Lorenz

$$L_{3\times 9}^{(2)}(\vec{u}^*) = \begin{bmatrix} -6A\alpha u_1^* & 0 & 0 & 0 & 0 & 0 & 0 & 0 & 0 \\ 0 & 0 & 0 & 0 & 0 & 0 & 0 & 0 & 0 \\ 0 & 0 & 0 & 0 & 0 & 0 & 0 & 0 & 0 \end{bmatrix}; \tag{17b}$$

$$L_{3\times 27}^{(3)}(\vec{u}^*) = \begin{bmatrix} -6A\alpha & 00000000000000000000000000 \\ 0 & 00000000000000000000000000 \\ 0 & 00000000000000000000000000 \end{bmatrix}, \tag{17c}$$

so, in this connection, the model (15) of chaotic wave processes in Chua system takes the form:

$$\dot{\vec{v}} = \dot{\vec{v}}^{(1)} + \dot{\vec{v}}^{(2)} + \dot{\vec{v}}^{(3)} = L_{3\times 3}^{(1)}\vec{v} + \frac{1}{2!}L_{3\times 9}^{(2)}(\vec{v} \otimes \vec{v}) + \frac{1}{3!}L_{3\times 27}^{(3)}(\vec{v} \otimes \vec{v} \otimes \vec{v}). \tag{18}$$

Like the Lorenz system, when the Chua circuit is functioning in a chaotic mode, the output signals from the cubic and quadratic kernels, being in antiphase, also partially compensate each other (Fig. 2). This manifests the *effect of self-organization* of processes in the Chua scheme [10]. Following to Corollary and Comment it consists in the interaction of nonlinearities of the 2nd and 3rd orders with their subsequent synchronization.

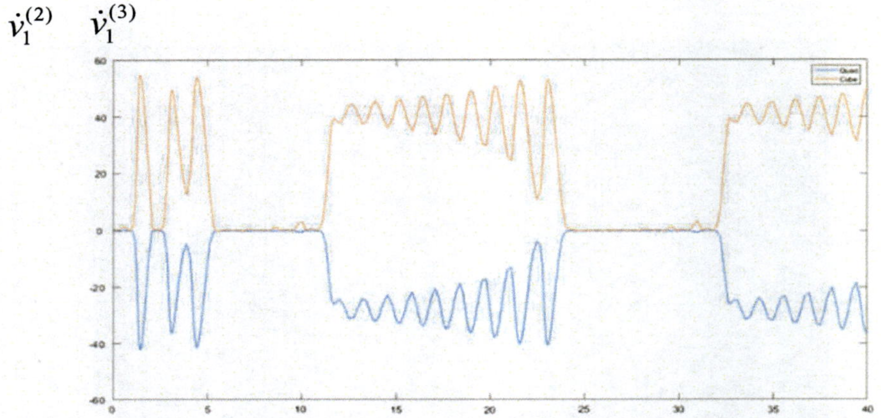

Fig. 2 Examples of signals generated by quadratic $L^{(2)}_{3\times 9}$ and cubic $L^{(3)}_{3\times 27}$ kernels in accord with the model (18) of origin of chaotic wave processes in complex NDS of Chua's type

Let us justify the synchronization effect of nonlinear oscillations in the Chua scheme within the framework of the general model (15). To this end, substituting (17a)–(17c) into (18), we write the equation for the evolution of the first component v_1 of a vector variable $\vec{v}$ from the state-space $\mathfrak{R}^3$ of the Chua scheme:

$$\dot{v}_1 = \alpha v_2 - \alpha(3Au_1^{*2} + C)v_1 - 3A\alpha u_1^* v_1^2 - A\alpha v_1^3. \tag{19}$$

Introducing the notation, as in (5) for the Landau model:$2\gamma_1 = -\alpha(3Au_1^{*2} + C)$, $\alpha_L = 3A\alpha u_1^*$, $\beta_L = A\alpha$, we can write (19) in the form:

$$\dot{v}_1 = \alpha v_2 + 2\gamma_1 v_1 - \alpha_L v_1^2 - \beta_L v_1^3. \tag{20}$$

Given the above modeling parameters, we obtain the inequalities: $2\gamma_1 = -\alpha(3Au_1^{*2} + C) > 0$, $\alpha_L = 3A\alpha u_1^* < 0$, $\beta_L = A\alpha > 0$, which fully correspond to the conditions for the hard self-excitation of the system within the framework of the Landau initial turbulence model [1, 9].

Under such conditions, a jump-like transition from a stationary regime of a complex NDS to an unsteady one is observed, accompanied by the appearance of two frequencies ω_1 and ω_2 defining two cycles in the state-space of the Chua system (Fig. 3). The obtained result also finds its explanation from the point of view of the Ruelle–Takens theory [5, 8]. Following this theory the power spectrum of signal from a complex NDS (as a function of the control parameter α in the case of Chua scheme) contains one frequency ω_1 after the first Hopf bifurcation, then after the second Hopf bifurcation it has two frequencies ω_1 and ω_2 but sometimes three frequencies (ω_1, ω_2 and ω_3), although the third frequency ω_3 may not detect before chaos identified [8]. When an additional frequency arises, insignificant perturbations can destroy regular

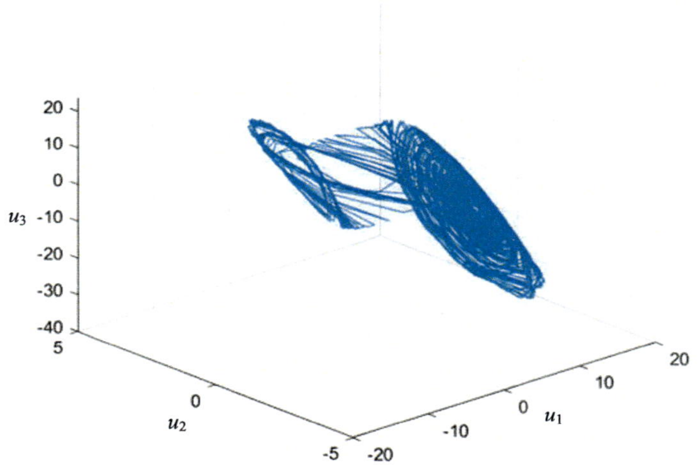

Fig. 3 Example of chaotic double-scroll attractor in the state-space of Chua scheme

cycles forming the torus T^3 and transform it into a chaotic attractor [5] (for example, the "double-scroll" attractor in the state-space of the Chua scheme (see Fig. 3)).

Let us note that, compared with the initial Landau turbulence model (5), the proposed model (15) applied to Chua circuit (18) gives a more general scalar (20) for the evolution of the first component v_1 of a vector variable $\vec{v}$, since it contains an extra term αv_2. This additional term characterizes the influence of the second component v_2 of the vector variable from the state-space $\Re^3$ which allows us to find more general conditions for the origin and evolution of chaotic wave processes in comparison with the Landau model even in the case of $2\gamma_1 < 0$ [10], for example, when $\alpha = 15.6$, $\beta = 28$, $A = 0.5131$, $C = -0.9255$, $u_1^* = -0.75$.

Thus, despite the difference in the physical phenomena described by the Lorenz and Chua systems, the dynamics of their behavior in the state-space has a similar scenario expressing in the synchronous "counteraction" of nonlinear processes of even and odd orders (see Figs. 1 and 2) generated by the corresponding kernels $L^{(2k)}_{N\times N^{2k}}$ and $L^{(2k+1)}_{N\times N^{2k+1}}$ in the general vector–matrix model (15) of complex NDS in a chaotic mode.

Now let us consider the general model (15) in the case of complex NDS of FitzHugh–Nagumo [17, 18]:

$$\begin{cases} \dot{u}_1 = cu_1 - \dfrac{c}{3}u_1^3 + c\,u_2 + cg; \\ \dot{u}_2 = \dfrac{a}{c} + \dfrac{b}{c}\,u_2 - \dfrac{1}{c}\,u_1, \end{cases} \tag{21}$$

where $a,\ b,\ c$ are the parameters of a complex NDS describing the active media of the nerve fiber (in particular, $a = 0.7$, $b = 0.8$, $c = 3.0$), g is a bifurcation characteristic, moreover, $-1.4023 < g < -0.3452$.Similarly to the Chua system, the dynamics of the complex FitzHugh–Nagumo system is also described by linear, quadratic and cubic kernels [12, 13]:

$$L^{(1)}_{2x2}(\vec{u}^*) = \left[\begin{array}{c|c} c - c\,u_1^{*2} & c \\ \hline -\frac{1}{c} & \frac{b}{c} \end{array}\right], \tag{21a}$$

$$L^{(2)}_{2x4}(\vec{u}^*) = \left[\begin{array}{c|c|c|c} -2cu_1^* & 0 & 0 & 0 \\ \hline 0 & 0 & 0 & 0 \end{array}\right], \tag{21b}$$

$$L^{(3)}_{2x8}(\vec{u}^*) = \left[\begin{array}{c|c|c|c|c|c|c|c} -2c & 0 & 0 & 0 & 0 & 0 & 0 & 0 \\ \hline 0 & 0 & 0 & 0 & 0 & 0 & 0 & 0 \end{array}\right], \tag{21c}$$

so that the model of the occurrence of chaotic wave processes in the FitzHugh–Nagumo system has the form:

$$\dot{\vec{v}} = \dot{\vec{v}}^{(1)} + \dot{\vec{v}}^{(2)} + \dot{\vec{v}}^{(3)} = L^{(1)}_{2\times2}\vec{v} + \frac{1}{2!}L^{(2)}_{2\times4}(\vec{v}\otimes\vec{v}) + \frac{1}{3!}L^{(3)}_{2\times8}(\vec{v}\otimes\vec{v}\otimes\vec{v}). \tag{22}$$

Taking into account (21a)–(21c), (22), the equation for evolution of the first component v_1 of the vector variable $\vec{v} \in \Re^3$ for FitzHugh–Nagumo NDS takes the form [10]:

$$\dot{v}_1 = cv_2 + c(1 - u_1^{*2})v_1 - cu_1^* v_1^2 - \tfrac{1}{3}cv_1^3. \tag{23}$$

Introducing notations similar to (5) of the Landau model: $2\gamma_1 = c(1 - u_1^{*2})$, $\alpha_L = cu_1^*$, $\beta_L = c/3$, for given parameters $a = 0.7$, $b = 0.8$, $c = 3.0$, and $u_1^* = -0.45$ we obtain the inequalities $2\gamma_1 > 0$, $\alpha_L < 0$, $\beta_L > 0$, corresponding to the condition of hard self-excitation of the system [9].

Figure 4 shows the attractors of FitzHugh–Nagumo NDS formed on the basis of the limit cycle for the above parameters and restored on the basis of directly numerical integration of system (21) and matrix decomposition (22).

Moreover, an *additional* control parameter u_1^* opens up new possibilities for studying the dynamics of FitzHugh–Nagumo NDS [10]. Indeed, at $u_1^* = 1.942908996$, a qualitatively different chaotic attractor is observed in the state-space of the FitzHugh–Nagumo system (see Fig. 5). Indeed, as seen from Fig. 5, with a set of parameters $b = 0.8$, $c = 3.0$, $u_1^* = 1.942908996$, an aperiodic mode arises similar to the aperiodic regime in the Belousov–Zhabotinsky chemical model [8].

Thus, computational experiments carried out with two parameters $u_1^* = -0.45$ and $u_1^* = 1.942908996$ show that a limit cycle in the first case and a chaotic attractor in the second case can be observed (see Figs. 4 and 5).

Following Corollary and Comment the signals from the outputs of kernels of the matrix decomposition of the FitzHugh–Nagumo system must demonstrate the

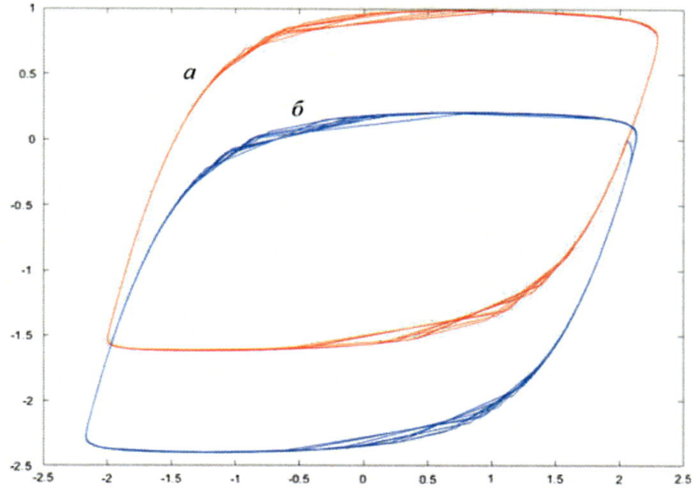

Fig. 4 Attractors of FitzHugh–Nagumo NDS built on the basis of matrix decomposition (**a**) and without its use (**b**)

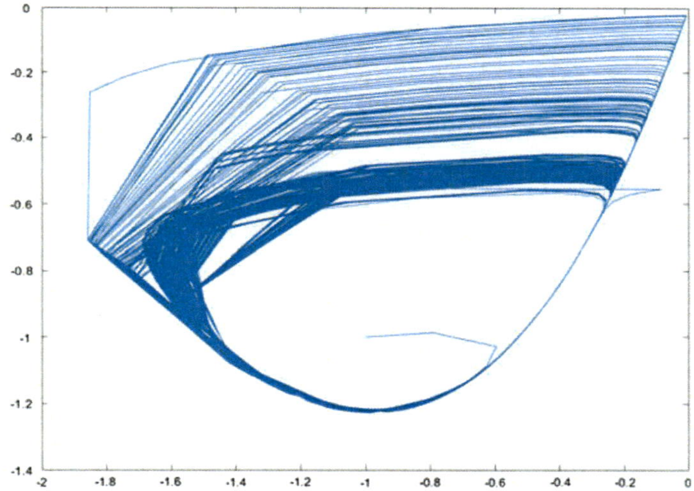

Fig. 5 The Fitz-Hugh chaotic attractor obtained under parameter values $b = 0.8$, $c = 3.0$, $u_1^* = 1.942908996$

general property of a complex NDS in the chaotic regime, i.e. the synchronous "counteraction" of nonlinear processes of even and odd orders (see Fig. 6).

Indeed, according to Fig. 6, the derivative of the first component $\dot{v}_1$ of the vector variable $\vec{v} \in \Re^3$ for the NDS of FitzHugh–Nagumo evolves in such a way that the output signals from the linear and cubic kernels are in-phase added to synchronously

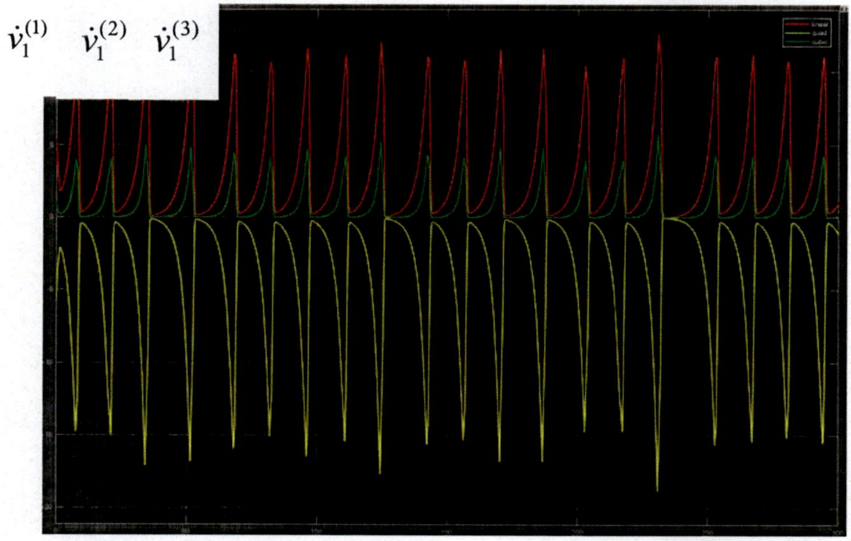

Fig. 6 Examples of signals generated by linear $L_{2\times2}^{(1)}$, quadratic $L_{2\times4}^{(2)}$ and cubic $L_{2\times8}^{(3)}$ kernels of the model (23) of origin of chaotic wave processes in the complex NDS of FitzHugh–Nagumo

"counteract" the output signal from the quadratic kernel. More precisely, during the period the signals from the kernels of odd orders $L_{2\times2}^{(1)}$, $L_{2\times8}^{(3)}$ add up and compensate for the signal from the even (second) kernel $L_{2\times4}^{(2)}$ which, as a whole, leads to stabilization of the amplitude of the chaotic wave process [10].

Thus, the application of the matrix decomposition method permits us to obtain new knowledge, for example, despite the difference in the physical phenomena described by the FitzHugh–Nagumo's, Chua's and Lorenz's systems, the dynamics of their behaviour in the state-space has a similar scenario. Moreover, using the matrix decomposition method for the NDS of FitzHugh–Nagumo a qualitatively different chaotic attractor has been revealed at $u_1^* = 1.942908996$ (Fig. 6) than the one presented in Fig. 4(a) when $u_1^* = -0.45$. To find out the reason for this, let us prove the following theorem:

Theorem 2 (a necessary condition for self-organization in 2D NDS). For a two-dimensional complex system, the following necessary conditions of instability (leading to self-organization in 2D NDS) are valid:

$$\mathrm{Sp}L_{2\times2}^{(1)} > 0, \quad if \quad \det L_{2\times2}^{(1)} > 0 \tag{24a}$$

$$\mathrm{Sp}L_{2\times2}^{(1)} < 0, \quad if \quad \det L_{2\times2}^{(1)} < 0, \tag{24b}$$

where $L^{(1)}_{2\times 2}$ is a matrix of linearized 2D NDS, Sp and det are the trace and the determinant of this matrix.

Proof we apply the bifurcation analysis technique to the kinetic model of FitzHugh–Nagumo, for which we rewrite (21) in general form:

$$\begin{cases} \dot{u}_1 = f_1(u_1, u_2); \\ \dot{u}_2 = f_2(u_1, u_2). \end{cases} \tag{25}$$

Let us consider the matrix of the linearized system $L^{(1)}_{2\times 2}$ corresponding to (21a):

$$L^{(1)}_{2\times 2} = \left[\begin{array}{c|c} l_{11} & l_{12} \\ \hline l_{21} & l_{22} \end{array}\right], \tag{26}$$

where $l^{(1)}_{ij} = \left[\frac{\partial f_i}{\partial u_j}\right]_{u_j^*}$, $i, j = 1, 2$, are elements of the matrix $L^{(1)}_{2\times 2}$. The trace and the determinant of linearized matrix (26) are found according to well-known formulas:

$$\mathrm{Sp}L^{(1)}_{2\times 2}(u_1^*) = l_{11} + l_{22}; \;\; \det L^{(1)}_{2\times 2}(u_1^*) = l_{11}l_{22} - l_{12}l_{21}. \tag{27}$$

It is known [6] that non-trivial solutions of a linearized system of equations exist when the condition is carried out:

$$\det\left[L^{(1)}_{2\times 2} - \omega E_{2\times 2}\right] = 0.$$

It defines the following characteristic equation:

$$\omega^2 - \mathrm{Sp}L^{(1)}_{2\times 2}\omega + \det L^{(1)}_{2\times 2} = 0. \tag{28}$$

To lose stability when the system leaves the stationary state, it is necessary that at least one root of (28) has a positive real part, i.e.

$$\mathrm{Re}\omega_i > 0, \tag{29}$$

where

$$\omega_i = \frac{1}{2}SpL^{(1)}_{2\times 2} \pm \sqrt{\left(\frac{1}{2}SpL^{(1)}_{2\times 2}\right)^2 - \det L^{(1)}_{2\times 2}} \;\; i = 1, 2. \tag{30}$$

As shown in [6], this occurs if the stationary state on the phase plane is described by an unstable node or focus, and this singular point is usually surrounded by a phase trajectory of the type of limit cycle. If inequality det $L^{(1)}_{2\times 2} > 0$ is fulfilled, then it follows from (28), (30) that in order to take place (29), the following condition must hold: $\mathrm{Sp}L_{2\times2}^{(1)} > 0$. This inequality determines the *necessary condition* for instability leading to the origin of the phenomenon of self-organization in the active medium [19]. As follows from (30), in the case of $\det L_{2\times2}^{(1)} < 0$ condition (29) is also satisfied if $\mathrm{Sp}L_{2\times2}^{(1)} < 0$, i.e. the necessary condition for self-organization takes the form (24b) which proves the theorem.

So, according to Theorem 2, the above mentioned reason is that, taking into account the diagonal elements $l_{11} = c\left(1 - u_1^{*2}\right)$ and $l_{22} = b/c$ of the linearized matrix and, NDS of FitzHugh—Nagumo for given parameters $a = 0.7$, $b = 0.8$, $c = 3.0$, and $u_1^* = 1.942908996$ satisfies the *second* necessary instability condition (24b) which leads to a change in inequalities in the Landau model: $\gamma_1 < 0$, $\alpha_L > 0$, $\beta_L > 0$. In contrast, with the initial choice of the parameters $a = 0.7$, $b = 0.8$, $c = 3.0$, and $u_1^* = -0.45$, FitzHugh–Nagumo NDS satisfies the first necessary instability condition (24a) corresponding to the condition of hard self-excitation of the system according to the Landau model: $\gamma_1 > 0$, $\alpha_L < 0$, $\beta_L > 0$ [9].

5 Conclusion

Despite the great interest in the processes of self-organization in complex systems, there is still no universal theoretical apparatus for their study which required the development of new mathematical models of NDS and fairly general methods of nonlinear analysis of their dynamic behavior.

In the present work, using the matrix decomposition of complex NDS operators [11–16], a general model (15) for origin and evolution of chaotic wave processes in complex systems is proposed. The proposed model shows that the effect of self-organization in a complex NDS of various physical nature (hydrodynamic, electronic, and electrochemical systems as examples) consists in the interaction of higher-order nonlinear processes leading to stabilization (to a finite value) of the amplitude of the chaotic wave process [10].

Mathematically, this is expressed in the synchronous "counteraction" of nonlinear processes of even and odd orders (see Figs. 1, 2, 6) generated by the corresponding kernels $L_{N\times N^{2k}}^{(2k)}$ and $L_{N\times N^{2k+1}}^{(2k+1)}$ in the general vector–matrix model (15) of complex NDS in a chaotic mode (see Theorem 1, Corollary and Comment). In addition, the proposed vector–matrix model permits us to find more general conditions of the origin and evolution of chaotic wave processes in comparison with the Landau initial turbulence model (for example, if $\gamma_1 < 0$). In other words, when the second necessary condition of instability (24b) is fulfilled in accordance with Theorem 2, an additional condition can be found for the appearance and evolution of chaotic wave processes in a complex NDS. Thus, upon attaining the complex NDS the chaotic mode, self-synchronization of oscillations is observed as a self-organization phenomenon.

References

1. L.D. Landau, To the problem of turbulence. Doklady Akademii nauk SSSR [Reports of the Academy of Sciences USSR] **44**(8), 339–342 (1944). ((in Russian))
2. A.M. Krot, On a class of discrete quasistationary linear dynamic systems. Sov. Phys. Dokl. **35**(8), 711–713 (1990)
3. N. Wiener, *Nonlinear Problems in Random Theory* (John Wiley & Sons, New York, 1958).
4. E.N. Lorenz, Deterministic nonperiodic flow. J. Atmosph. Sci. **20**(March), 130–141 (1963)
5. D. Ruelle, F. Takens, On the nature of turbulence. Commun. Math. Phys. **20**, 167–192 (1971)
6. G. Nicolis, I. Prigogine, *Self-Organization in Nonequilibrium Systems: From Dissipative Structures to Order through Fluctuation* (John Willey & Sons, New York, 1977).
7. H. Haken, *Advanced Synergetics: Instability Hierarchies of Self-Organizing Systems and Devices* (Springer-Verlag, Berlin Etc., 1983).
8. P. Bergé, Y. Pomeau, C. Vidal, *L'ordre Dans Le Chaos: Vers Une Approche Déterministe de La Turbulence* (Hermann, Paris, 1988).
9. L.D. Landau, E.M. Lifschitz, *Fluid Mechanics* (Pergamon, Oxford, 1959). XIII, p. 539
10. A.M. Krot, An evolutionary model of chaotic wave processes in complex dynamical systems based on the matrix decomposition theory. Dopov. Nac. akad. nauk Ukr. [*Reports of the National Academy of Sciences of Ukraine*] **9**, 12–19 (2019). https://doi.org/https://doi.org/10.15407/dopovidi2019.09.012
11. A.M. Krot, The decomposition of vector functions in vector-matrix series into state-space of nonlinear dynamic system, in *EUSIPCO-2000: Proc. X European Signal Processing Conference, Tampere, Finland, September 4–8, 2000. Vol. 3.* Tampere, 2000, pp. 2453–2456
12. A.M. Krot, Matrix decompositions of vector functions and shift operators on the trajectories of a nonlinear dynamical system. Nonl. Phenom. Compl. Syst. **4**(2), 106–115 (2001)
13. A.M. Krot, Analysis of attractors of complex nonlinear dynamical systems based on matrix series in the state-space. Informatica = Informatics**1**(1), 7–16 (2004) (in Russian)
14. A.M. Krot, The development of matrix decomposition theory for nonlinear analysis of chaotic attractors of complex systems and signals, in *DSP-2009: Proc. 16th IEEE International Conference on Digital Signal Processing, Thira, Santorini, Greece, July 5–7, 2009.* Santorini, 2009, pp. 1–5. https://doi.org/https://doi.org/10.1109/icdsp.2009.5201123
15. A.M. Krot, A. M, Bifurcation analysis of attractors of complex systems based on matrix decomposition theory, in *IEM 2011: Proc. of IEEE International Conference on Industrial Engineering and Management, Zhengzhou, China, August 12–14, 2011.* Wuhan, 2011, pp. 1–5. https://doi.org/https://doi.org/10.1109/icmss.2011.5999350
16. A.M. Krot, U.A. Sychou, The analysis of chaotic regimes in Chua's circuit with smooth nonlinearity based on the matrix decomposition method, in *Vestsi Natsyyanal'nai akademii navuk Belarusi. Seryya fizika-technichnych navuk = Proceedings of the National Academy of Sciences of Belarus. Physical-technical series*, 2018, vol. 63, no. 4, pp. 501–512 (in Russian). https://doi.org/https://doi.org/10.29235/1561-8358-2018-63-4-501-512
17. R. FitzHugh, Impulses and physiological states in theoretical models of nerve membrane. Biophys. J. **1**, 445–446 (1961)
18. J.S. Nagumo, S. Arimoto, S. Yoshisawa, An active pulse transmission line simulating nerve axon. Proc. IRE **50**, 2061–2070 (1962)
19. V.F. Dailudenko, A.M. Krot, Modeling of processes of self-organization in active media, in *Intelligent systems*, vol. 1. (IEC of National Academy of Sciences of Belarus, Minsk, 1998), pp. 32–45 (in Russian)

Application of Ultrawideband Chaotic Signals for Wireless Ranging

Lev Kuzmin and Elena Efremova

Abstract The results of an experimental estimation of the propagation time of ultra-wideband chaotic radio pulses occupying frequency band 2.5–3.5 GHz through a multipath wireless channel are described. Accuracy of the time of the arrival measurement of ultra-wideband chaotic radio pulses by their envelope is assessed in industrial and office environment, in order to solve the problem of wireless ranging and localization of objects in space.

Keywords Ultra-wideband signals · Chaotic signals · Chaotic radio pulses · Multipath propagation · Multipath fading · Localization · Time of arrival measurement

1 Introduction

Chaotic oscillations have a number of properties that permanently attract the attention of researchers. The use of chaotic signals as noise-like information carrier is interesting from the practical point of view in wireless communication systems Andreyev et al. [1], radar systems Liu et al. [2] and in the problem of wireless measurement of the distance between objects Beal et al. [3].

Measuring the time of radio signals propagation through a wireless channel is a vital scientific and practical task Zafari et al. [4]. One of the reasons is the development of robotic systems in which the objects must be localized using omni-directional radio systems in the unlicensed frequency range 2–10 GHz (centimeter wavelength range). Such systems are intended for use in industrial, residential and urban infrastructure, where the effects of multipath propagation are strong, which has significant effect on the accuracy. So, the problem of the accuracy of measuring radio signal propagation time in multipath channels is critical Hölzl et al. [5], Alarifi et al. [6]. In addition to

L. Kuzmin (✉) · E. Efremova
Kotelnikov Institute of Radio Engineering and Electronics, Mokhovaya st. 11/7, 125009 Moscow, Russia
e-mail: lvk@cplire.ru

C. H. Skiadas et al. (eds.), *13th Chaotic Modeling and Simulation International Conference*, Springer Proceedings in Complexity,
https://doi.org/10.1007/978-3-030-70795-8_33

traditional narrowband (NB) technologies Hölzl et al. [5] (WiFi, Bluetooth, ZigBee, where the distance is determined by the power of the received signal or the angle at the emitter), good candidates for solving this problem are noise-like and ultra-wideband (UWB) signals Alarifi et al. [6]. Their two useful basic features are the short time of coherence (narrow autocorrelation function of chaotic signal) and a large signal gain. Due to these features, one can to achieve potentially large processing gain in the receiver and high accuracy of estimating signal parameters (power or phase) after propagation through the channel. Moreover, the signal gain can very simply be varied technically during the chaotic signal formation, and it can be involved in the receiver.

Technical methods of evaluating signal parameters ultimately determine the capabilities of specific wireless technologies. In NB systems, the location accuracy does not exceed 1 m Hölzl et al. [5].

Theoretical estimates Skolnik [7], Cardinali et al. [8] show that for UWB signals the potentially achievable accuracy is rather high. Namely, with a 1.5 GHz bandwidth the ranging error varies (in order of magnitude) from 10^{-7} to 10^{-4} m for the distances up to 10 m Cardinali et al. [8]. From the physical viewpoint, the maximum distance is determined by the signal-to-noise ratio at the receiver input. For example, for the signal with 3 GHz central frequency (7.5 cm wavelength) and 1 GHz bandwidth, this corresponds to sub-millimeter measurement accuracy. Actually, this is the accuracy of optical directional measuring instruments.

At the same time, it is not always possible to implement theoretical limits in practice, both for technical reasons and for the reasons related with the signal propagation conditions in real wireless communication channels, such as the receiver noise and multipath propagation, which limit the ability of determining the real time of signal propagation.

An attempt to improve accuracy by means of the use of UWB signals encounters the effect of multipath propagation, which, in the case of ultra-short (US) pulses, manifests itself in distortion of the shape, amplitude and phase of the pulses. This restricts the potentially very high measurement accuracy of propagation time of US pulses ($\sim$100 ps). In UWB systems with US pulses, e.g., in UWB Decawave DWM-1000 or Ubisense modules, the indoor localization accuracy does not exceed 10 cm at distances up to 10 m Zafari et al. [4], Alarifi et al. [6], Poulose et al. [9], Wijaya et al. [10], Schroeer [11].

Therefore, it seems more appropriate to use a different approach based on measuring the signal propagation time by means of analyzing its envelope. Of course, in this case it is impossible to count on the sub-millimeter measurement accuracy, but technically it is much simpler. Moreover, in this case the restrictions imposed by the signal propagation conditions are also much weaker.

In the framework of this problem UWB chaotic radio pulses seem to be free from disadvantages of UWB US pulses in multipath channels. The envelope of UWB chaotic radio pulses changes only slightly when passing through a multipath channel, because of its noise-like nature Kuz'min et al. [12], Efremova et al. [13], and the pulse duration can be chosen much longer than the duration of the multipath channel response. This allows us to measure the signal propagation time by measuring the parameters of the envelope, not the carrier wave.

The goal of the work is to experimentally confirm the possibility of measuring time and distance using UWB chaotic microwave pulse envelope in real multipath channels and to propose a corresponding method.

2 Experimental Setup

The experiment was carried out in a laboratory room using UWB direct chaotic transceivers Dmitriev et al. [14] in accordance with the scheme in Fig. 1a. In the emitter T_X, a chaotic oscillation generator G_C formed Dmitriev et al. [15] a sequence $s(t)$ of chaotic radio pulses of duration $T_P = 166$ ns, power $P_P = 50$ mW, frequency band 2.5–3.5 GHz and repetition period $2T_P$, according to the modulation signal $m(t)$. After the channel C, the signal $s_C(t)$ was fed to the receiver E_D, which formed the envelope $e(t)$ of the signal $s_C(t)$. The distance d between the emitter and receiver was varied from 25 to 150 cm. In oscilloscope Osc the signal waveforms $m(t)$ and $e(t)$ were recorded from the moment of the leading-edge arrival of the $m(t)$ signal first pulse.

Following the procedure of Skolnik [7], let us estimate the distance measurement accuracy attainable at distance d. The shape of the chaotic radio pulse envelope is adequately approximated with a rectangular or trapezoidal pulse. Let us consider the rectangular pulse approximation.

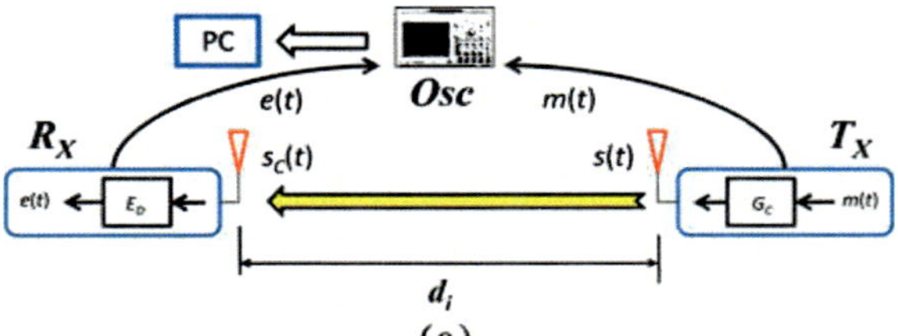

Fig. 1 Experimental setup (**a**) for measuring the time of chaotic radio pulse propagation from emitter Tx to receiver Rx at distance d_i (from $d_0 = 25$ cm to $d_8 = 150$ cm): $m(t)$ is the signal that modulates chaotic oscillator G_C, $e(t)$ is envelope of chaotic radio pulses at log-detector output E_D, C is wireless channel. Transceiver module (**b**)

The error δd in determining the distance is expressed by expression $\delta d = c\delta t$, where c is the light speed in vacuum and δt is the RMS measurement error of the pulse arrival time. According to Skolnik [7]

$$\delta t = \left(\frac{\tau}{2BE/N_0}\right)^{\frac{1}{2}} = \left(\frac{\tau N_0}{2BP_d\tau}\right)^{\frac{1}{2}} = \left(\frac{N_0}{2BP_d}\right)^{\frac{1}{2}}$$

where B is the bandwidth, E is the energy of one received pulse, N_0 is the noise spectral density, τ is the transmitted pulse duration, $P_d = E/\tau$ is the pulse power at distance d.

Let $Q_d \sim 1/d^n$ be the signal attenuation at distance d between the receiver and the emitter, and n be the attenuation index of the real wireless channel. Then Q_d (in decibel) can be calculated as $Q_d = Q_0 + 10n \lg d/d_0$, where Q_0 is the attenuation of the signal at a distance d_0.

Then the pulse power P_d [dB] at distance d from the emitter is determined by the following expression: P_d [dB] = P_{rad} [dB] $- Q_d$ [dB] = P_{rad} [dB] $- Q_0$ [dB]$-$ $10n \lg d/d_0$.

Due to antenna losses Q_{ant} [dB] and the use of low-noise-amplifier (LNA) with power gain Q_{amp} [dB] in the receiver, the expression should be supplemented: P_d [dB] = P_{rad} [dB] $-$ Q_0 [dB] $-$ Q_{ant} [dB] + Q_{amp} [dB] $-$ $10n \lg d/d_0$.

Hence,

$$\delta t(d) = \left(\frac{N_0}{2B} 10^{-(P_d\ [\mathrm{dB}]/10)}\right)^{1/2}$$

.

In the experiments, a transmitter Fig. 1b with 17-dBm (50 mW) emission power and 19-dB attenuator are used, so the emitted power $P_{rad} = -2$ dBm. The antennas losses are assumed to be $Q_{ant} = 6$ dB, the LNA power gain $Q_{amp} = 20$ dB. The distance d_0 was taken as 1 m. Attenuation in free space at a 1-m distance is $Q_0 = -41.3$ dB. Let us take the entire frequency band of the chaotic signal be $B = 1$ GHz, assuming there is no additional filtering in the receiver. Noise power spectral density at temperature 290 K is $N_0 = 4 \cdot 10^{-21}$ W/Hz. For example calculations give accuracy values of about 4 cm at 100 m. For a distance of 1.5 m, as in the experiment, the theoretical accuracy is about 0.6 mm.

The chaotic radio pulse propagation time is measured by means of comparing the signal waveforms $m(t)$ and $e(t)$. The time difference of the leading edges of the pulse envelopes of the transmitter signal $m(t)$ and the receiver signal $e(t)$ was evaluated for each pair of transmitted and received pulses. This method is the most attractive from the technical viewpoint and it is extensively discussed D'Andrea et al. [16], Dardari et al. [17], Liu et al. [18], Zwirello et al. [19].

Modulation signal $m(t)$ controls the base-emitter junction of the transistor generator G_C of UWB chaotic radio pulses. The transistor oscillator G_C gives chaotic radio pulses, the initial pieces of which are nearly identical Dmitriev et al. [15], so in the receiver, the leading edges of all pulse envelopes are similar for the same propagation conditions, to within the effect of noise.

The receiver is based on a UWB log-detector [20] and a low-noise amplifier with the signal gain 10^2. The receiver sensitivity is $3 \cdot 10^{-9}$ mW. Instantaneous value of envelope $e(d, t)$ at the receiver output is proportional to logarithm of instantaneous power of UWB chaotic signal at the receiver input. The envelope amplitude obeys $e(d, t) = 10\alpha lg P(d, t)/P_0$, where $P_0 = 1$ mW is reference power; α is log-detector slope; $P(d, t)$ is the signal power that came from the transmitter at distance d.

For the incoming power within the range $3 \cdot 10^{-9}$ to 10 mW, the detector gives the output signal amplitude $e(d, t)$ within the range 0.5 to 2 V, respectively. This is enough to have technical opportunity to receive UWB chaotic radio pulses with the power varying from 50 mW at the transmitter output ($d = 0$) to 0.05 mW ($d = 25$ cm) or to 0.0025 mW ($d = 150$ cm) at the receiver input. These power values correspond to the receiver SNR for which the negative effect of noise is negligibly small and only the interference due to multipath propagation must be taken into account.

For each transmitter pulse, the emission moment was determined by the leading edge as the moment τ_M the modulating signal $m(t)$ exceeded the threshold value V_M Fig. 2a. In the receiver, the pulse arrival time was recorded at the moment τ_E the signal $e(t)$ exceeded the value V_E, i.e., $m(\tau_M) = V_M$ and $e(\tau_E) = V_E$, respectively. The difference $\tau_E - \tau_M$ between these moments is equal to the pulse propagation time from the moment it was formed in the transmitter to the moment it appeared at the receiver output. This difference includes the processing time T_C for which the signal passes through the output circuits of the transmitter, the processing time of the receiver circuits and the propagation time through the space from antenna to antenna:

$$T_D = \tau_E - \tau_M = T_S + T_C$$

3 Results of the Experiment

Waveforms $m_i(t)$ and $e_i(t)$, $i = 0...8$, for distances $d_i = \{25, 30, 37.5, 50, 62.5, 75, 100, 125, 150\}$ cm between the emitter and receiver were recorded in the experiment. The waveforms $m_i(t)$ and $e_i(t)$ were sampled in the oscilloscope, sampling frequency $f_S = 2.5$ GHz. The time step 0.4 ns between the samples corresponds to 12 cm accuracy of distance.

To determine the pure pulse propagation time T_S through the wireless channel (from antenna to antenna) it is necessary to exclude the unknown time value T_C. To do this, the propagation times $T_S(\Delta d_i)$ through distance $\Delta d_i = d_i - d_0$ corresponding to the difference $T_S(\Delta d_i) = T_D(d_i) - T_D(d_0) = (\tau_E - \tau_M)_i - (\tau_E - \tau_M)_0 = T_S^{(i)} - T_S^{(0)}$ between the propagation time $T_D(d_0)$ for distance d_0 and $T_D(d_i)$ for d_i were evaluated.

Initial fragments of pulse waveforms $e_i(t)$ and $m_i(t)$ are depicted in Fig. 2b for distances $d_0 = 25$ and $d_8 = 150$ cm. The leading edge of the pulse rises for approx. 15 ns (Fig. 2b). The more the distance between the emitter and the receiver, the smaller the envelope amplitude. The trailing edge of the pulses is formed by the

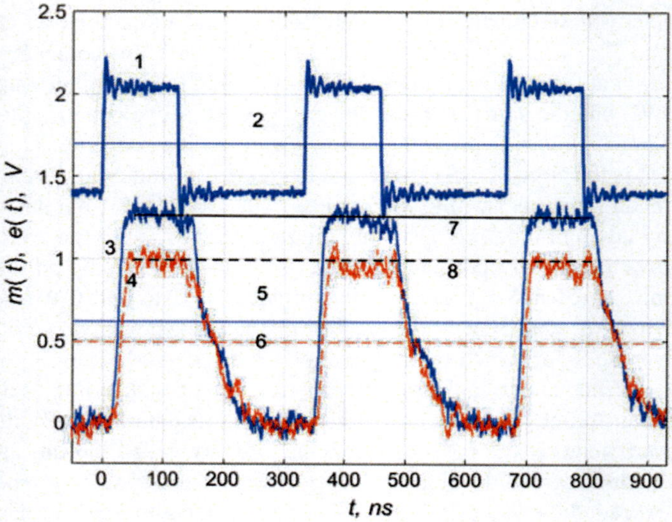

(a)

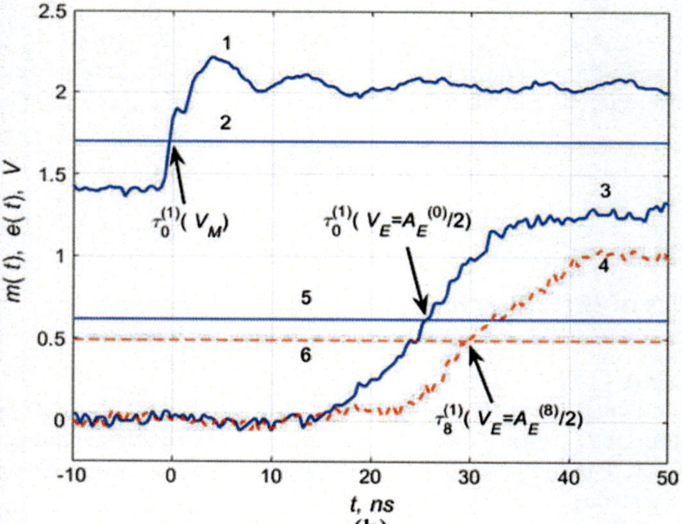

(b)

Fig. 2 Signal waveforms (**a**), (**b**) in the emitter and receiver: $m(t)$ is curve 1, $e(t)$ is curves 3 and 4 for $d_0 = 25$ cm and for $d_8 = 150$ cm, respectively; line 2 marks the threshold voltage to determine the emission moment; lines 5 and 6 are threshold voltage values for pulse envelopes in the receiver for $d_0 = 25$ cm and for $d_8 = 150$ cm, respectively; lines 7 and 8 mark the mean amplitude of pulse envelopes for distances $d_0 = 25$ cm and $d_8 = 150$ cm, respectively

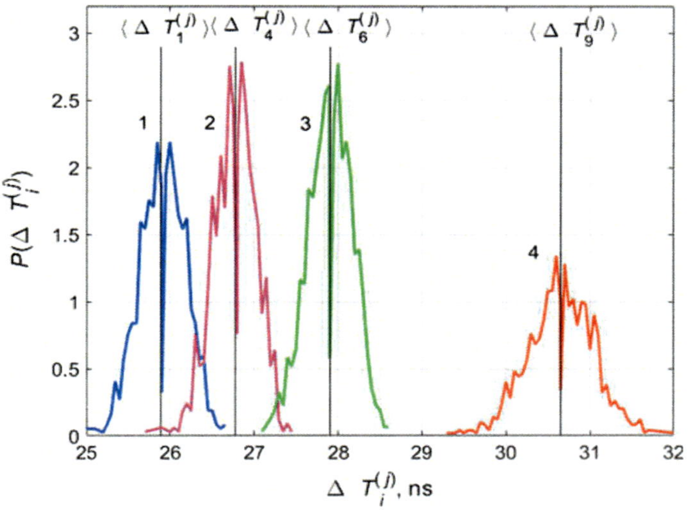

Fig. 3 Distribution of time of arrival values $P(\Delta T_i^{(j)})$ for distances $d_0 = 25$ cm (curve 1), $d_4 = 50$ cm (curve 2), $d_6 = 75$ cm (curve 3), $d_9 = 150$ cm (curve 4). Vertical lines mark average propagation time values $\Delta T_i = \langle \Delta T_i^{(j)} \rangle$

delayed paths. Values V_M and V_E were set at the half of the average amplitudes of pulse envelopes: $V_M = V_M^{(i)} = A_M^{(i)}/2$ and $V_E = V_E^{(i)} = A_E^{(i)}/2$. It allows us to compare the pulse propagation time for different distances by the time moments corresponding to the same pulse phase, namely, the moments when the pulse achieves half of its maximum amplitude. The envelope pulse amplitude is proportional to its power; the mean pulse amplitude corresponds to half the pulse power at a given distance from the source.

Moments $\tau_i^{(j)}(V_M)$ of the leading edge of each pulse of signal $m(t)$ and moments $\tau_i^{(j)}(V_E)$ for the signal $e(t)$ were determined in the experiment, where $j = 1 \ldots 1000$ corresponds to pulse numbers in signals $m_i(t)$ and $e_i(t)$. Distributions of time differences $\Delta T_i^{(j)}(d_i)$ estimated in the experiment are depicted in Fig. 3: $\Delta T_i^{(j)}(d_i) = \tau_i^{(j)}(V_E) - \tau_i^{(j)}(V_M)$. Vertical black lines denote the average values $\Delta T_i = \langle \Delta T_i^{(j)} \rangle$, where angle brackets mean averaging by j.

Distances $D_S(d_i, d_0)$ were estimated using average propagation time values ΔT_i (Fig. 4a): $D_S(d_i, d_0) = c(\Delta T_i - \Delta T_0)$, where c is light speed. Solid line corresponds to actual distance values. Errors of distance estimation $E_S(\Delta d_i) = D_S(d_i, d_0) - (d_i - d_0)$ are presented in Fig. 4b; errors vary from 1 to 15 cm.

The described method of measuring propagation time requires the knowledge of the threshold $V_E = A_E^{(i)}/2$, that depends on the amplitude of pulse envelope. Practically it is not convenient and it is much more interesting to measure the signal propagation time referring to a constant threshold value $V_E = T_E$ that does not depend on the pulse envelope amplitude $A_E^{(i)}$.

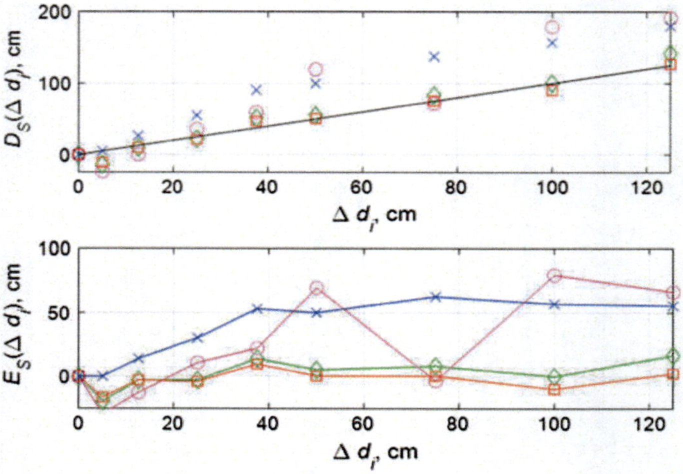

Fig. 4 Distance estimate $D_S(\Delta d_i)$ (**a**) and average error $E_S(\Delta d_i)$ (**b**) as functions of true distance Δd_i calculated from: average time delay between emitted and received pulses (diamonds) being in phase; average time delay between emitted and received pulses for constant threshold (crosses); minimum time propagation delay for the packet of pulses (squares); maximum cross-correlation function between emitted and received signals (circles)

Estimation of moments $\tau_i^{(j)}(T_E = 0.5\ V)$ and corresponding distances $D_S(d_i - d_0)$ for a constant threshold value $V_E = T_E = 0.5\ V$, that is not changed in the experiment for different sets of pulses, gives biased estimates (Fig. 4a, crosses). This result agrees with Zwirello et al. [19], where it was also shown that energy detection of UWB ultra short pulses with constant threshold gave biased propagation time.

Application of "threshold" method under condition of correct choice of signal reception phase solves the problem of estimating chaotic pulse propagation time by its envelope, but there arises a question: is it possible to measure the pulse propagation time for a series (a packet) of pulses, but not for each separate pulse? This approach seems to give more precise estimates. The following method is proposed.

Consider a value $\left(\tau_i^{(j)}(V_E) - \Delta\right) - \tau_i^{(j)}(V_M)$, where Δ is parameter. If Δ is exactly equal to the pulse propagation time, then above value is equal to zero: $\left(\tau_i^{(j)}(V_E) - \Delta\right) - \tau_i^{(j)}(V_M) = 0$. The sum of modules

$$T(\Delta) = \sum_{j=1}^{N} \mid \left(\tau_i^{(j)}(V_E) - \Delta\right) - \tau_i^{(j)}(V_M) \mid$$

is equal to zero $T(\Delta) = 0$, if the pulse propagation time for each pulse is equal to Δ; otherwise, in case of measurement errors, it is greater than zero. Let us find Δ_{opt}, that is the value minimizing $T(\Delta)$:

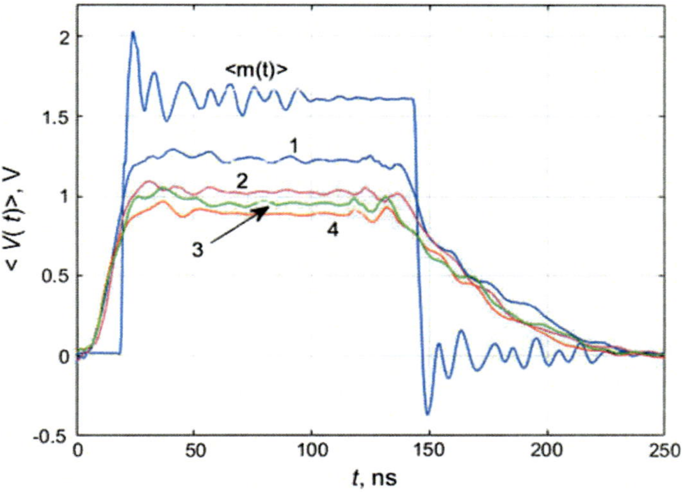

Fig. 5 Time-aligned envelope $e(t)$ and modulating $m(t)$ signal waveforms for the maximum of cross-correlation function for distances $d_0 = 25$ cm (curve 1), $d_4 = 50$ cm (curve 2), $d_6 = 75$ cm (curve 3) and $d_8 = 150$ cm (curve 4)

$$\Delta_{opt} = \arg\min_{\Delta} \sum_{j=1}^{N} \mid \left(\tau_i^{(j)}(V_E) - \Delta\right) - \tau_i^{(j)}(V_M) \mid$$

.

The found value Δ_{opt} is an estimate of pulse propagation time based on the series of N pulses. The results are depicted in Fig. 4a, b. The achieved accuracy is better than that based on the single pulse.

Finally, it is interesting to consider another limiting case. Let estimate the pulse propagation time using the cross-correlation technique widely accepted in physics. The propagation time T_D equals to the time shift between signals $m(t)$ and $e(t)$ for which the correlation function achieves the maximum:

$$T_D = \arg\max_{t} \int_{-\infty}^{+\infty} m(t-\tau)e(\tau)d\tau$$

.

Results of distance estimations based on the calculations of T_D are depicted in Fig. 4a, b. Here also, the bias of the pulse propagation time attracts attention. This is due to the amplitude properties of the correlated signals, i.e. the correlation function achieves the maximum if the correlated signals fit in shape, but the shape itself in the context of determining the propagation time has no physical meaning (Fig. 5).

4 Conclusion

The results confirm that application of UWB chaotic radio pulses in multipath channel allows us to achieve centimeter precision in wireless ranging. This precision is worse than the theoretical limit, but it is better than the oscilloscope precision. The described approaches to wireless ranging can be applied in real wireless transceiver based on UWB chaotic signals.

The obtained results show that the estimation error is essentially less than typical path delays of the multipath channel, that is about ≈ 1 m in terms of distance.

Authors thank Itskov V. V. and Petrosiyan M. M. for the help with the experiment.

This work is supported by the State Assignment of the Kotelnikov Institute of Radio Engineering and Electronics.

References

1. Yu.V. Andreyev, A.S. Dmitriev, E.V. Efremova, A.D. Khilinsky, L.V. Kuzmin, Qualitative theory of dynamical systems, chaos and contemporary communications. Int. J. Bifurc. Chaos **15**(11), 3639–3651 (2005)
2. L. Liu, R.X. Ma, H. Xu, W.K. Wang, B.J. Wang, J.X. Li, Experimental investigation of a UWB direct chaotic through-wall imaging radar using Colpitts oscillator, in *IET International Radar Conference*, Hangzhou (2015), pp. 1–6
3. A.N. Beal, S.D. Cohen, T.M. Syed, Generating and detecting solvable chaos at radio frequencies with consideration to multi-user ranging. Sensors **20**(3), 774–796 (2020)
4. F. Zafari, A. Gkelias, K.K. Leung, A survey of indoor localization systems and technologies. IEEE Commun. Surv. Tutor. **21**(3), 2568–2599 (2019)
5. M. Hölzl, R. Neumeier, G. Ostermayer, Localization in an industrial environment: a case study on the difficulties for positioning in a harsh environment. Int. J. Distrib. Sens. Netw. **11**(8), 1–11 (2015)
6. A. Alarifi, A. Al-Salman, M. Alsaleh, A. Alnafessah, S. Al-Hadhrami, M.A. Al-Ammar, H.S. Al-Khalifa, Ultra wideband indoor positioning technologies: analysis and recent advances. Sensors **16**(5), 707–742 (2016)
7. M.I. Skolnik, Theoretical accuracy of radar measurements. IRE Trans. Aeronaut. Navig. Electron. ANE-7 **4**, 123–129 (1960)
8. R. Cardinali, L. De Nardis, M.-G. Di Benedetto, P. Lombardo, UWB ranging accuracy in high- and low-data-rate application. IEEE Trans. Microw. Theory Tech. **54**(4), 1865–1875 (2006)
9. A. Poulose, O.S. Eyobu, M. Kim, D.S. Han, Localization error analysis of indoor positioning system based on UWB measurements, in *Eleventh International Conference on Ubiquitous and Future Networks (ICUFN)*, Zagreb, Croatia (2019), pp. 84–88
10. B. Wijaya, N. Deng, K. Jiang, R. Yan, D. Yang, Real-time adaptive UWB positioning system enhanced by sensor fusion for multiple targets detection, in *IEEE Intelligent Vehicles Symposium (IV)*, Paris, France (2019), pp. 1239–1246
11. G. Schroeer, A real-time UWB multi-channel indoor positioning system for industrial scenarios, in *International Conference on Indoor Positioning and Indoor Navigation (IPIN)*, Nantes (2018), pp. 1–5
12. L.V. Kuz'min, A.V. Grinevich, Method of blind detection of ultrawideband chaotic radio pulses on the background of interpulse interference. Tech. Phys. Lett. **45**, 831–834 (2019)
13. E.V. Efremova, A.S. Dmitriev, L.V. Kuzmin, Measuring the distance between an emitter and a receiver in the wireless communication channel by ultrawideband chaotic radio pulses. Tech. Phys. Lett. **45**, 853–857 (2019)

14. A.S. Dmitriev, E.V. Efremova, A.V. Kletsov, L.V. Kuz'min, A.M. Laktyushkin, V.Yu. Yurkin, Wireless ultrawideband communications and sensor networks. J. Commun. Technol. Electron. **53**, 1206–1216 (2008)
15. A. Dmitriev, E. Efremova, L. Kuzmin, N. Atanov, Forming pulses in non-autonomous chaotic oscillator. Int. J. Bifurc. Chaos **17**(10), 3443–3448 (2007)
16. A.N. D'Andrea, U. Mengali, R. Reggiannini, The modified Cramer-Rao bound and its application to synchronization problems. IEEE Trans. Commun. **42**(234), 1391–1399 (1994)
17. D. Dardari, C. Chong, M. Win, Threshold-based time-of-arrival estimators in UWB dense multipath channels. IEEE Trans. Commun. **56**(8), 1366–1378 (2008)
18. W. Liu, H. Ding, X. Huang, Z. Liu, TOA estimation in IR UWB ranging with energy detection receiver using received signal characteristics. IEEE Commun. Lett. **16**(5), 738–741 (2012)
19. L. Zwirello, T. Schipper, M. Jalilvand, T. Zwick, Realization limits of impulse-based localization system for large-scale indoor applications. IEEE Trans. Instrum. Meas. **64**(1), 39–51 (2015)
20. Analog Devices, *Data Sheet 1 MHz to 4 GHz, 80 dB, Logarithmic Detector/Controller*, http://www.analog.com/media/en/technical-documentation/data-sheets/ADL5513.pdf

The St Elmo's Fire: Its Formation and Measurement on Both Natural and Artificial Structures

V. J. Law and D. P. Dowling

Abstract Since the beginning of recorded history Saint Elmo's fire (SEF) has been observed and experienced by humans in and close to thunderstorms which are dissipating in strength, either at sea level or in mountainous regions. The systematic study of SEF in nature has proved difficult due to inadequate measurement equipment and the understanding of high-voltage circuit requirements for partial gas breakdown measurement. The understanding of how SEF is generated is not only important in developing protection systems for shipping and aircraft flying at high altitudes and landing, but also for bioelectronics. Arguably our understanding of SEF is largely based on historical events and the empirical mathematical construct of Peek's Law, which attempts to identify the visual inception voltage in terms of the minimum electrical field stress required for the generation of corona discharge at sharp protrusions. This paper examines how SEF is formed around water–ice particles (graupel), as well on the surface of dirigible airships and airplanes. The paper also compares these 'natural' mechanisms to those which are recreated under laboratory conditions, using high-voltage direct-current, (HVDC-), high voltage alternating-current HVAC (50 Hz), as well as using a Tesla coil (0.3 to 30 MHz). SEF has also been investigated for the stimulation of living insects; it can also be used for their eradication. The Tesla coil circuit has been demonstrated to be suitable for the creation of a minimum visual inception voltage on both these living insects (moths and beetles), as well as on larger artificial objects such as ship models.

Keywords St Elmo's fire · Plasmoid · Aircraft surface protrusion · Insects · Tesla coil

V. J. Law (✉) · D. P. Dowling
School of Mechanical and Materials Engineering, University College Dublin, Dublin 4, Belfield, Ireland

C. H. Skiadas et al. (eds.), *13th Chaotic Modeling and Simulation International Conference*, Springer Proceedings in Complexity,
https://doi.org/10.1007/978-3-030-70795-8_34

1 Introduction

The objective of this paper is to help explain the nature and impact of Saint Elmo's fire (SEF), which is a naturally occurring discharge observed in the Earth's lower atmosphere. A further objective is to compare the effect of this electrification process on both living and manmade objects. Human observations of SEF have been reported from ancient times to the present day. For example the ancient Greek's described eye-witness accounts of naturally occurring atmospheric luminous disturbances, within thunderstorms and volcanic eruptions. At the time these events were called Helena where one was seen, and Caster and Pollux (the twin brothers of Helen) when two appeared. In Roman times these 'brothers' were regarded as the patrons Saints of Mediterranean sailors, to whom they appeared as SEF on the masts and spars of sailing ships as the electrical storm disturbance, began to dissipate. These good omens being manifest as characteristic cracking or hissing sound with a blue/violet flame-like glow. Between the years 1610–1611, art emulates real life when William Strachey's account of the ill-fated *'Sea adventure'* voyage from the new world in 1610, is retold by William Shakespeare within the play *'The Tempest'* [1]. In this play, SEF takes on a more sinister role as the spirit *'Arial'* who manipulates the mariners off the ship. Further on in time, in the summer of 1786, Britain and mainland Europe experienced a major dust cloud, caused by a plume of ash that was carried by the prevailing Atlantic winds when the Icelandic volcanic Laki erupted. This led to an extronary number of violent lighting storms. Eye-witness accounts from the time, detailed how ball-lighting (BL) and fireball's (FB's) occurred. At least 17 events were recorded during the period from June 30th to August 31st in Britain and Ireland alone [2]. Modern measurements of the volcanic ash, from a Icelandic eruption in 2010, showed that the ash particles were typically SiO_2 with particle sizes in the nano and micron range [3, 4]. These particles spread over Scotland and mainland Europe, with larger particles falling closer to the source. In 1886, these atmospheric phenomenon started to be systematically complied and reported [5] as SEF [6], BL [7] and FB [8]. The latter two types proving to be more life threatening when compared to SEF. It is clear however that many of the eye-witness accounts from that time, were not scientifically formulated and possibly some were embellished. In addition it has become clear that BL has the ability to interfere with radio broadcasts and to transfer part of its information through a glass window pane with and without damage to the glass [4, and 8–11]. In the period 1899 to 1900, while working on wireless power transmission at Colorado Springs Nikola Telsa performed a series of electrical and photographic studies, aimed at artificially generated FB [12]. In his notes Tesla proposed two hypotheses, firstly, that a *'fireball is produced by sudden heating, to high incandescence, of a mass of air or other gas as the case may be, by the passage of a powerful discharge'*. And secondly, maintenance of the FB may come from an external source such as stray, or a second cooperating electric field. Today the term cooperating has become equivalent to a resonance condition. By 1928, the term for this electrical phenomenon began to be classified under the term 'plasma' (Greek: meaning mouldable substance), which considers an assembly of gas molecules that has some of its atoms or molecules temporally ionized or excited [13]. The optical emissions obtained from plasmas are related to the relaxation processes of the ionized and excited species. By 1952 the subclass 'plasmoid' was added by Winston H. Bostick, which defines a separate plasma-magnetic entity that may be ejected from the parent plasma [14].

More recent (1978), Challahan and Makin, in an attempt to partly explain nocturnal observations (1965–1968) of unidentified flying object (UFO) in the Utah Deseret-USA, performed a series of electrification experiments on moths and beetles that are naturally found in the Uintah basin [15]. From their experiments they proposed that a mass swarm of glowing (blue) insect could help explain some UFO sightings in the Utah basin.

Grigor'ev et al. [16] and Donoso et al. [17] have statically shown that buoyant and mobile BL and surface attached SEF originate from thunderstorm activity, but there is a need to differentiate between their electrification mechanisms. Today, there is a renewed need to look at SEF, in particular due to impact on aircraft static discharge, as well as the online misinterpretation of cockpit window Precipitation-static (P-Static) discharges as SEF. An experimentally generated example was formed through the use of single, dual and multiple array metal (1−2 mm diameter) aerial-antenna igniter obtained using a converted domestic microwave oven plasma reactors (DMOPR). This equipment was fabricated for the production of hydrogen [18], as well as functional carbon-based nanostructures [19]. Within these systems, dielectric heating of the liquid and microwave aerial-antenna igniter generate bubbles and as the bubbles grow (under enhanced electric field stress at the igniter tips) they become detached from the metal surface and become buoyant. Fast camera imaging of water bubble growth and calculation also suggest as the interface liquid film dries, electrification (gas breakdown) of the bubble gas occurs producing a plasmoid at approximately 0.1−0.3 kV cm^{-1} in the pressure range of 7−20 kPa. This electrification transition from bubble to plasmoid has many similarities to electrification of graupel in thunderclouds and insects flying through high electric fields near thunderclouds. Improving the understanding of these electrification processes is part of the motivation of this paper. A further objective is to examine the complexity of terminology of BL, plasmoid, and SEF which describe electrification around point-like objects that are either metal conductors or living biological entities, in this case insects. In addition to help to differentiate SEF, from the 2-dimenational surface Streamers that forms on insulating surfaces. This paper reviews atmospheric corona disturbance and nitrogen chemistry (Sect. 2), SEF formation via of graupel and electrification of thunderclouds (Sect. 3); Peek's Law that determines the visual inception electrical field strength for SEF production prior to gas breakdown (Sect. 4); storm activity leading to dirigible airship disasters (Sect. 5); aircraft encounter with atmospheric pressure electric fields (Sect. 6). Section 7 considers HVDC-, HVAC and Tesla coil stimulation of model sailing ships in the length range of 1 to 10 cm. Section 8 reports on how electric fields can be used to stimulation of in sects (Sect. 8.1) as well as to annihilate them (Sect. 8.2). Finally Sect. 9 provides discussion and summary of this work.

2 Atmospheric Corona Disturbance and Nitrogen Chemistry

Saint Elmo's fire (sometimes called brush discharge) is generally regarded to be produced at regions of high electrical stress around the tip of conducting tips and protrusions due to the geometric field enhancement where equal-potential lines become bunch up [19–21]. These partial discharges occur on elevated structures on the earth surface (Fig. 1) and beyond the cruising altitude of normal commercial aircraft (typically 10–12 km), close to the upper limit of the earth's troposphere. It is now accepted that the energy that initiates these discharges originate for the dc potential difference between thunderclouds and ground and, ac electrical fields generated by lighting within cloud-to-cloud and cloud-to-ground/sea [8–11]. Within this atmospheric altitude range, mathematical models that estimate the corona visible inception voltages are generally normalized to pressure. For example, fair weather air atmospheric pressure conditions (20 °C and 101.3 kPa) at sea level are normalized to 1 atm and decreases with altitude (10–12 km corresponding to 0.24 to 0.18 atm). An example of this normalization process can be found in Peek's Law [20], see Sect. 4.

Statistical analysis of over 200 SEF eyewitness observations under different geographical locations which indicated that approximately 62% of the observation had a blue/violet appearance, 20% appear to be white and 13% were red [15, 16]. Optical emission spectroscopy (OES) of SEF on aircraft reveals a similar blue/violet emission that can be attributed to the 2nd positive system of nitrogen (N_2) $(C^3\Pi_u^+ - B^3\Pi_g^+)$: < 18 eV [21–23]. In the laboratory at sea/ground level, the authors of this work has measured the OES of the 2nd positive N_2 system within a flowing air arc afterglow as a function of axial distance and determined the gas temperature varied from 500–300 K in the near-afterglow to >300 K in the far-afterglow [24].

3 SEF Formation via Graupel

In order to interpret the formation of SEF and to differentiate it from BL, their origins within the earth's weather system is considered as function optical emission of atmospheric air (mainly a mixture of N_2 and oxygen (O_2)) and graupel (a mixture of water and ice particles) surface chemistry.

First consider the convection of warmed air from the earth as it expands adiabatically as it rises through the troposphere until it reaches the stratosphere, where the sun's energy reheats the circulated air. This natural convection process allows the cloud to capture positive charged particles resulting in a initial electrification of the cloud. With increasing electrification a negative charge begins to be formed on the upper cloud boundary which then flows down outside to the base of the cloud. The accumulation of negative charge at the base of the cloud, now by convention called cumulonimbus, reinforces the cloud-ground/sea electric field. The electric field in this region is known to be of the order of 1–3 kV cm^{-1} that is not sufficient to overcome the dielectric strength of air. To achieve the required field strength an inductive charge process within clouds has been considered [25, 26] where water moisture (H_2O) is propelled to high altitudes by updroughts and cools to form graupel that undergoes a continuous dissociation–recombination process forming hydronium ions (H_3O^+) and hydroxyl ions (OH^-) intermediate products. This reversible reaction process is given in (1) where approximately 20% of the intermediate product ions are available for electrification.

$$2H_2O \overset{graupel}{\leftrightarrow} H_3O^+ + \mathrm{OH}^- \tag{1}$$

Under natural background acidic conditions, charge separation of the available ions then follows, where the H_3O^+ ions move into the vapor phase, and due to their buoyancy are lifted by updroughts to the top of the cloud leaving the larger and denser OH^- charged graupel to fall under gravity to bottom of the cloud. This dynamic process generates a potential difference between the top and bottom cloud boundaries. With increasing gravitational separation, the negative charged graupel forms a negative space-charge that enhances the pre-existing fairweather electric field between the cloud and ground/sea. When the charge attraction between the cloud bottom boundary and ground strengthens, electrons and negative charged ions shoot down from the cloud as stepped leaders to meet upward positive charged streamers to produce a lighting channel. As the enhanced electric field subsides, sufficient energy still remains to partially ionize N_2 molecules at the enhanced electrical fields at metal protrusions, at or, near ground level to produce the characteristic blue/violet appearance of SEF.

It has been proposed [26], that where a cluster (10 or more) of charged graupel particles fall to the ground in the absence of a metal protrusion SEF does not occur but a collective discharge action occurs breaking down the surrounding atmospheric air causing the production of buoyant BL. The characteristic optical emission of which ranges from yellow, through orange, to red $\lambda = 550-780$ nm) as indicated. The associated atomic and molecular ion spectra are: atomic-H-Balmer-α line ($\lambda =$ 656 nm), the 1st positive system nitrogen ($\lambda = 580$ and 654 nm), the O ($3p^5\mathrm{P} - 3s^5\mathrm{S}$) ($\lambda = 777$ nm) and the excited $NO_2{}^*$ molecule continuum ($\lambda = 450-800$ nm). Plus metastable neutral oxygen ($\lambda = 557.7$ nm). The emission lines and bands quenching as the graupel finally melts.

4 Peek's Law for a Single Metal Electrode (Protrusions and Tips)

Increasing the electrical stress around a single metal electrode tip (or protrusions) ultimately results in local air breakdown around the electrode. At this level of stress, the local air volume is weakly ionized followed by a rapid electron recombination back to the ground state discharge. On the milliseconds time scale the outer boundary of this volume, the ionization frequency (v_i) just balances the electron loss frequency (v_a) by attachment [27], see (2), and Fig. 3. Under these condition a static corona discharge, or SEF, appears attached to the electrode with the visual inception voltage being higher than the visual extinction voltage because, once started there are always electrons to ionize gas molecules [28].

$$V_i - V_a \approx 0 \tag{2}$$

Upon increasing the voltage stress level further (~5 kV cm^{-1}), the discharge extends outward to form multiple streamers flowing from the electrode, where breakdown is enhanced by the production of electrons at the head of streamer. If the voltage becomes large or a counter electrode is close by (1−10 cm) a conducting trail or channel may form producing a flashover discharge. If the applied voltage is maintained then sparks may be also formed. Increasing the voltage stress still further, creates bidirectional leaders are formed, which involve; space-charge and a gas heating (≥500 K) mechanisms, rather than corona onset alone.

Fig. 1 Engraving of SEF on the spars and masts of a sailing ship. Dr. G. Hartwig. (*The Aerial World*. (London 1886) [6]

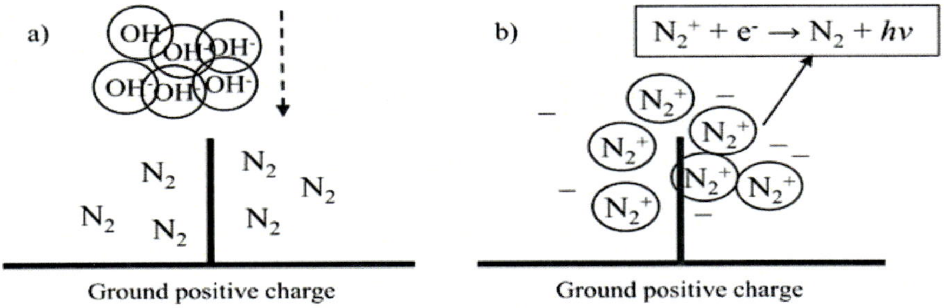

Fig. 2 A Diagrammatic representation of graupel falling, under gravity onto a conducting tip; **b** SEF formed around this conducting tip: The relaxation reaction (the blue/violet optical emission) is shown in the annotation box b. Vertical electric field conditions: cloud to ground/sea

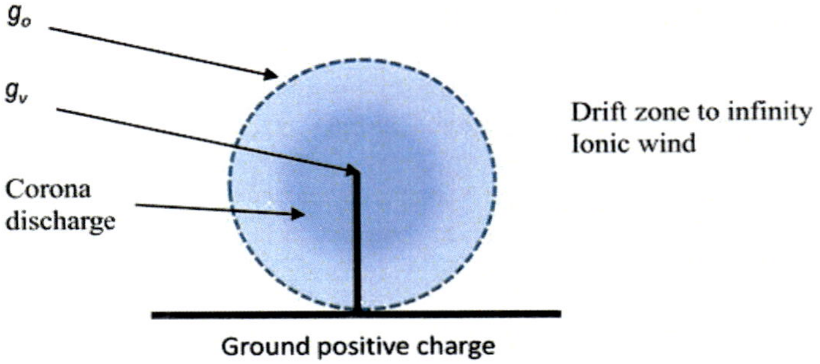

Fig. 3 Schematic of corona discharge boundary limits (g_v and g_o) for a single electrode and drift zone

From this sequence of increasing discharge energy states, it is reasonable to assume that SEF influences streamer and leader production. Table 1 (adopted from Gibson [30]) provides a guide to the inception voltage for the three different discharge types. The data shows that although the corona inception voltage for lighting rods has the lowest value for the three discharges (where the variation in the values is due to physical structure orientation of the rods [31]).

Peek's Law was originally proposed as an empirical formula for coaxial cylindrical configurations, parallel wires and spheres in the 1920s [19] and is still used today

Table 1 Corona inception electrical field values and characteristic temperature (K) for atmospheric discharge at ground/sea level

Parameter	Corona discharge 1–2 cm diameter grounded lighting rods	Streamer discharge	Leader discharge
Electric field (kV cm^{-1})	0.2–2.7	~5	~1–5
Gas temperature (K)	~300	~300	≥3000 K

[28–36]. Peek's empirical formula utilizes the local atmospheric condition and the surface condition of a conductor to estimate the corona visual inception voltage at local gas breakdown. For a manmade ac voltage source, see (3).

$$g_v = g_o \delta_m \left(1 + \frac{k}{\sqrt{\delta r_o}}\right) \quad \textit{Measured in units of kV cm}^{-1} \tag{3}$$

In (3), g_v is the voltage gradient (kV cm^{-1}) at the visual corona inception voltage; g_o is the disruptive electric gradient, for an ac voltage the value varies from: 27.2 k.cm^{-1} for a sphere, 30 kV.cm^{-1} for parallel wires, and 31 kV cm^{-1} for coaxial geometries. The parameter δ is the local relative air density (at sea level, δ = 1 under fair weather conditions and 0.9 to 0.8 for storm conditions),m is the surface roughness factor ($m = 1$ represents dry and smooth clean surface under laboratory conditions). For wet conditions, Peek found that the g_v fell sharply and considered this as a special case for m by substituting it with $g_o = 9$ kV cm^{-1}. The parameter k is an empirical dimension factor (0.301 to 0.308) and r_o is the tip geometry radius (cm). As energy is required to start a corona discharge the single electrode surface-to-space boundary limits requires that the surface electrical stress is raised to g_v so that at a finite distance away in space where $k\sqrt{r_o}$ is g_o air breakdown occurs. This conducting medium increases in volume beyond which ions and neutral drift to infinity causing an ionic wind. A simple schematic of this scenario for a single metal electrode is shown in Fig. 3.

Natural occurring disruptive electric gradients formed by near thunderstorms are generally accepted to be driven by direct current voltage source [20], therefore (3) may be rewritten as follows -

$$g_v = 21.9\delta m \left(1 + \frac{k}{\delta r_o}\right) \quad \textit{Measured in units of kV cm}^{-1} \tag{4}$$

where 21.9 is the route mean square (RMS) of the ac disruptive electric gradient for air (g_o). The parameters: g_v, k and r_O having the same meaning as in (3).

Given that Peek's Law, in its different forms, is an empirical mathematical construct for ac and dc applied voltages, parameters δ m k and r_o are varied to fit experimental observations. For example Mombello et al. [33] characterized contaminated conductors using the following parameters values: $g_o = 21.9$ and a range of m values: $m = 0.6$ (low), $m = 0.4$ (severe) and $m = 0.2$ (very severe) with a k value of 0.308. When studying lightning initiation, Bazelyan et al. [34, 35] used the following parameter values g_o, k and r_o value of 27.8, 0.54 and 1 cm, respectively. In 2018, Riba et al. [36] republished values of m based on the work of a Cigré report [37]. These values are given in Table 2, along with the Mombello et al. surface sand contamination index, where m for clean surface = 1, and decrease with the amount of surface sand contamination.

Again in 2018, Riba et al. [36] followed Peek's original work by reducing parameters $g_o\delta$ m to a single global parameter (b) for fitting purposes for metal spheres suspended above a ground plane. Such a data reduction technique limits the number

Table 2 Surface irregularity factor *m*

Surface irregularity factor	Surface classification	Surface contamination index
m = 1	Smooth and polished surface	Clean
m = 0.8–0.6	Dry weather	Low
m = 0.6–0.3	Extreme pollution, snowflakes, raindrops	Severe
m = 0.25	Heavy Rain	Very severe

of mechanism that can be modeled in this field of research. However the reduction technique does open up new fields of research where the radius of conducting objects, effects corona production. One such use of (3 and 4) is their application to the study of the dielectric extremities of month and beetle exoskeleton under which a conductive medium is held [15].

5 Storm and Thunderstorm Activity Leading to Airship Disasters

Another source of SEF has been extensively recorded in conjunction with airships filled with the lift gas hydrogen and helium. This section looks at some notable airships disasters between World War I (1914–1918) and World War (1939 to 1945). At the beginning of the inter war era, German's war reparations to the winning side supercharged the wide spread dirigible airship technology in both the militarily and commercial travel sectors even though airship susceptibility to ground-handling damage was well known. For example: hydrogen fires while the lift gasbags were being inflated; damage due to uncontrolled movement within the hangar; or lost from their mooring stations during stormy weather [38, 39]. Three airship disasters attributed to storm and thunderstorm activity are considered here; the French Navy Dixmude, originally built by Luftschiffbau Zeppelin as LZ-114 for the Imperial German Navy; the United States Navy Akron ZRS-4; and the Hindenburg LZ-129 that ended the dirigible airship adventure.

5.1 Dixmude Airship 1923

The French Navy Dixmude airship was the third and final 1918 'Height Climber' X-class Zeppelin design (working altitude 6000–6,400 m). To achieve the design altitude these airships required a significant weight lost by removal of part of the original R-class airframe, and removal of one of the original seven engines from the rear gondola. At the French naval air base Cuers-Pierrefeu near Toulon, the airship underwent a three-year reconstruction for low level flight, along with the replacement of its now perished H_2 lift goldbeater's skin gasbags.

After a number of trial flights in 1923, the Dixmude made its last flight in December, a return flight from Toulon-France to the Algerian oasis of Ain Salah. On her return, at 8.00 am on Thursday (some 50 h of flying time) the airship reported violent winds to the north and as the ship fought the winds further reports where sent stating that fuel was running low and two engines had broken down. The Dixmude had now become a 'free-balloon' and at the mercy of the winds. Her last radio message (02:08 Saturday morning 21st December) reported that they were following standard operating procedures to reel-in its radio antenna due to thunderstorm activity. Soon afterwards (02:30) railway workers and a hunter near Sciacca—Sicily reported a red flash in the Western night sky followed by burning objects falling in to the sea. In the morning, burnt airship wreckage was found and the charged corpse of the Dixmude Commandant and radio operator was found on the 26th December. With this news, many newspapers speculated that the Dixmude was struck by lightning causing the death of its 52 crew and passengers [40].

A year later, Hugo Eeckener (Manger of Luftschiffbau Zeppelin and later Commandant of the Graf Zeppelin) wrote in the '*Luftfahrt*' on the Dixmude disaster. In his article he states that the Dixmude was built to withstand routine lighting strikes, particularly at the nose and rear of the ship [41]. Furthermore, the burnt condition of the wreckage and corpse were consistent with a gasoline fire (*i.e. that the hydrogen in the gas cells was not set on fire by a flash of lightning*). He goes on to state that by the time of the disaster (some 60 h flying time) the Maybach engines would have been working well beyond their maintenance schedule (the military specification was 1–2 days (24–48 h)). Indeed Maybach refused to guarantee more than 48 h continuous use, especially for the crankshafts. In closing his report, Eeckener notes that the airship should not have be used for such a prolonged flight as both the airframe and the 6 engines were not originally intended for long duration flight, in particular flying in serve storms that are associated within the Mediterranean region in the winter months. It is most likely that the true cause of the airframe sudden and catastrophic failure be know, but high altitude automatic hydrogen valveing due to a violent up draft in connection with a lighting strike cannot be ruled out.

5.2 Akron Airship 1933

On the 4th April 1933, the worst airship disaster unfolded as the United States Navel helium filled dirigible flying aircraft carrier airship 'Akron' crashed at sea off the coast of New Jersey with the loss of 73 out of 76 crew. The great death toll being due to drowning or hypothermia as there was on life jacket on board [38]. The disaster happened whilst the airship was navigating at low altitude through a thunderstorm when tail section hit the water. Eeckener's comments on storm conditions comes to mind here, not only for this disaster, but also for its first (1932) crash [39].

5.3 *Hindenburg Airship1937*

At 3.27 EST on May 6th 1937 the 2nd worst dirigible airship disaster, with the lost 36 lives, unfolded at Lakehurst, New Jersey when the Hindenburg commenced its tethering procedure at the airship mooring mast. Herbert Morrison and Charles Nehlson's sound recording of the Hindenburg disaster, transmitted on the following day of the disaster, imprinted such public reflective memories (*Hindenburg syndrome* [43]) that would not allow hydrogen gas to be used in public transport for many decades. Both the US and German board of inquiries into the Hindenburg disaster picked out from the many ignition theories that a static discharge (spark) due the inclement weather at the time ignited the hydrogen gas. This partial conclusion may have been influenced by the knowledge that since the early 1900s static discharge was known to disrupt both aircraft radio and avionics. In addition an electric charge generated on the airframe can lead to exogenous charging. This electrical disturbance under all weather conditions has become known as P-Static. Alexander J. Dessler's critic of the incendiary-paint theory of the Hindenburg disaster [44] suggests that the water soaked outer surface skin material may have be factor in the spark ignition theory [48, 49] and by extension the Dixmude airship disaster: specifically the formation of SEF on the outer dielectric skin fabric which can allow an electrical capacitance leakage pathway(s) to the internal duralumin airframe, thereby providing the necessary conditions for P-static ignition. For the readers interest, Table 3 provides the maximum length and diameter, the calculated entire cylinder surface area and cylinder surface capacitance ($C = [(8 + 6.95(l/d)^{0.76}]\varepsilon_0 d/2$, where $\varepsilon_0 = 8.85 \times 10^{-12}$ F m^{-1}) for all three airships. Note however that the physical data does not reflect the electric field variation on individual panels or electrical stress points at the nose and the cruciform rear section of the airships.

A final note on note on Dixmude and Hindenburg disasters: Considering that, there was thunderstorm electrical activity present at the time it is surprising however, that neither Bain [40] nor Dressler et al. [44, 45] discussed or compared the two H_2

Table 3 Length (l), diameter (d) and entre surface area (A) and capacitance of the Dixmude, Akron and the Hindenburg

Airship	Max length (l = m)	Max diameter (d = m)	Entire cylinder surface area (A = km^2)	Right cylinder surface capacitance (F)
Dixmude	226.5	23.9	17.8	4.8×10^{-9}
Akron	239.3	40.5	33.4	6.2×10^{-9}
Hindenburg	245.3	41.2	36.2	6.3×10^{-9}

filled airships disasters. Moreover neither mentions Eeckener's *Luftfahrt* article [41]. Dressler only relies on physical chemistry evident and concludes with the possibility of corona discharge (brush discharge or SEF) was an energy ignition source. This line of thought is supported by Professor Mark Heald's eye witness accounting of seeing blue SEF-like flicking along the Hindenburg's top ridge minutes before the fire started.

6 Aircraft Encounter with Atmospheric Electric Fields

In order to quantify the level of electrical field the international Civil Aviation Organization has listed SEF within the aircraft encounter severity index at a value of 0, where the range of severity increases from 0 to 5. Index values of 4 to 5 are associated with encounters with volcanic ash, where an index value of 5 equates to engine failure or other damage leading to a crash [52]. Under these increasing severity conditions ash particles induce an electrostatic negative surface charge on the airframe surface, effect engine operation and interfere with radio communications. These in-flight charging conditions come under the generic term of P-static, see Fig. 4 and [53]. Over the years static discharger structures (or static wicks) installed at the wing trailing edge, and empennage, have been developed to alleviate charge build-up. This electrostatic surface phenomenon is mainly due to triboelectric charging as the aircraft moves through the ash cloud and is proportional to the drag presented by the aircraft frontal area and airspeed (typically ~10—400 μ A m^2 at 475–500 knots (880–926 km/h) for design purposes [25]), see Fig. 4. Ash cloud-aircraft interactions may occur near a volcanic eruption source, or some 100 s km away from the eruption [4].

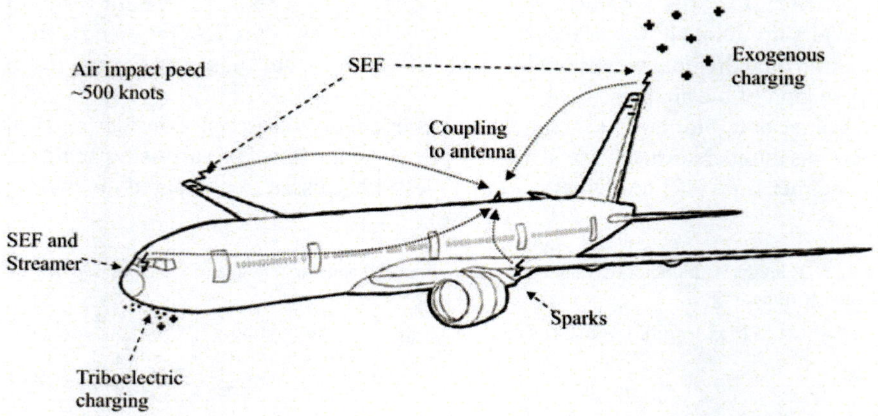

Fig. 4 Aircraft P-static locations and BL path (drawing modified from Nunes [53]). BL genially enters at the front aircraft and exits at the wings and the empennage

Table 4 Main parameter characteristics of SEF and surface streamer discharge

Parameter	SEF	Surface streamer discharge
Visual inception voltage field	0.2–2.7 kV cm^{-1}	3–5 kV cm^{-1}
Friction (induced by air flow)	No	Yes
Bonded metal protrusions	Yes	No
Insulator surface	No	Yes
Appearance and color	Blue glow	Blue and white spider-like appearance, mobile

On the Web, videos and commentary can be found that describes mysteriously rapidly moving spider-like discharges on aircraft cockpit windows. Generally these discharges are misinterpreted as SEF rather than surface streamer discharge, even though they have different visual properties. As previously stated in this work, SEF appears as a static glow attached to an electrical conducting tip or protrusion. The main characteristics that differentiate SEF from a surface streamer discharge on an insulator surface are listed in Table 4. The differentiation between these two types of discharge is that SEF occurs at aircraft extremities (wingtips, tail fins, and aerials), whereas surface streamer discharge occur on insulating surfaces (dielectric cockpit windows and non-electro-magnetic protected regions of the radome where each have high bulk resistivity > 10^{12} Ω–m). The application of metal diverter strips on radome protect the radar system by dissipating the received electrical charge onto the aircraft fuselage faraday cage, and then normally dissipated from the aircraft via the engine(s) and static wicks. However windows and radome may have low local surfaces resistivity due to contamination that enhance the local electrical field promoting streamers that can propagate or track along weaknesses to cause irreversible surface damage in accordance with an aircraft encounter severity index 2–3. At ground level and within research laboratories, similar surface damage caused by streamer and sparks are known to occur in atmospheric N_2 dielectric barrier discharge plasma, operated with a peak voltage of 2–8 kV in the 17–25 kHz range [49].

Since the 1940s, accounts of SEF horns on aircraft composite radome or as a streamer on cockpit windows have been documented, both of which may initiate BL formation within the Faraday cage of an aircraft and exiting via the wing or rear of the aircraft [10, 16, 55–57]. Reference [10] describes a thunderstorm accouter with a Lockheed C-130 Hercules transport plane, along with thin cloud encounter (~4500 m) with a Douglas C-133A Cargomaster plane, while [55] describes a thunderstorm encounter with an all metal airliner (Eastern Airlines EA539. In addition [56, 57] report on an in-flight (4500 m) encounter with a Boeing KC-97 Stratofreigther tanker. The wittiness's to all of these encounters describe seeing a luminous (gold or yellow-white) ball of approximately size 10–20 cm in diameter that slowly (5–10 s) passes horizontally from front to rear of the fuselage isle.

A commonly discussed cockpit BL window penetration mechanisms is observed for high flying aircraft is driven by and formed by atmospheric ions impinging and collecting on the dielectric window surface. In this a surface charge can produce an electric field on the other side of the window that is sufficient to sustain BL that is transient and a separate entity from the outside surface charge. Once established the BL is attracted (without coming into contact with any surface) to surface electric currents flowing within the electrically charged aircraft. The BL passes through the aircraft and may be dissipated at the static wicks located at the trailing edge of the wings and on the empennage. Under these formation conditions, BL can be considered to fulfill the criteria for being a plasma-magnetic entity or plasmoid.

By 2000, Varas and Rokne [58] uses the term plasmoid to describe BL when developing physics based software that color renders and sizes a plasmoid whilst it passes through a small opening. More recently Egrovo et al. [59] and Dubowsky et al. [60, 61] produced hydrated plasmoids with sub-second lifetimes using a high-voltage triggered impulse circuit. Their experiments shed light on the luminous colors of the plasmoids, but left open questions regarding the prolonged lifetimes observed in aircraft the fuselage.

7 Laboratory HVDC-, HVAC and Tesla Coil Stimulation of Model Sailing Ship Experiment

Section 5 considered large scale manmade object (100 s of meter in length). This section now considers the work by Tonmitr and Kaewrawang who used classical sailing ship models that where scaled down by 1:500 cm to give artificial cloud-to-ship length of 1–10 cm [21]. The aim of their research being the study early streamer air terminal devices to protect tall structures from lightning strike. For electrification purposes HVDC-, HVAC (50 Hz), and a Tesla coil (0.3 to 30 MHz) were used. Although little information on the high voltage circuit is given, the operational principles of the HVDC and Tesla circuit are similar to. The electrode is constructed as an upside-down cup with an open lattice to represents a thunder cloud and the gap distance is measured from the center of the electrode to the tip of the mast top of the ship model. The classical sailing ships are designed with three masts and are electrically boned to earth: however no detail of the ship's material is given.

Using Tonmitr and Kaewrawang's original inception voltage date and their respective computed voltage electric fields, the experiment outcome is analyzed using Peek's Law. Table 5 shows the original inception voltages with the new computed electric fields.

Figure 5 shows the inception voltages supplies as a function of electrode-mast gap distance for each circuit. Microsoft excel linear fitting for all three power supplies is also shown in Table 6. The trend line regression (goodness of fit) and backward extrapolation to the Y-axes origin are listed in Table 5. The linear fit assumes that the electric field is uniform throughout the measured gap distance. However, the real-world asymmetric geometry of the electrode-ship arrangement (voltage stress on the electrode with the ships mast at ground potential) the electric field becomes focused at short gap distance (0–5 cm) [20, 21]. Hence a Microsoft Excel linear fit algorithm

Table 5 HVDC-, HVAC and Tesla coil inception voltage as a function of electrode-mast gap distance, normalized electric field stress reference to kV cm^{-1}

Test subject	Electrode-to-ship gap (cm)	Corona inception voltage (kV) and electric field stress (kV cm^{-1})		
		DC-	AC (50 Hz)	Tesla coil
Model ship Scale = 1:500	5	19.5 (3.90)	24.4 (4.88)	10.10 (2.02)
	10	25.9 (2.59)	43.8 (4.38)	11.69 (1.16)
	15	32.8 (2.18)	57.4 (3.82)	19.94 (1.32)
	20	39.1 (1.95)	67.2 (3.36)	25.43 (1.27)
	25	45.7 (1.82)	73.0 (2.92)	23.93 (0.95)
	30	64.4 (2.14)	82.7 (2.75)	33.85 (1.12)

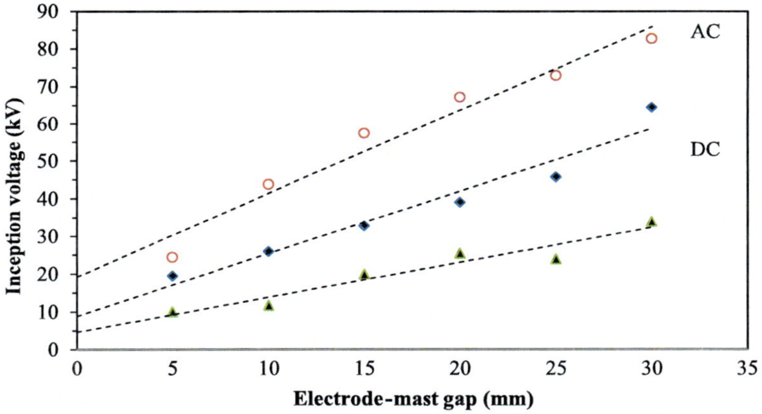

Fig. 5 Tonmitr and Kaewrawang's inception voltage date [21]. The trend lines are used as a guide and show the Y-axes intercept constant (Table 6)

Table 6 Figure 5 Excel line fittings parameters, goodness-of-fit (R^2) and line equation

Fitting parameter	Power supply	Excel regression (R^2)	Excel line equation
Linear fit	HVDC-	0.945	y = 1.6583x + 8.880
Linear fit	HVAC (50 Hz)	0.960	y = 2.2223x + 19.93
Linear fit	Tesla coil	0.925	y = 0.9198x + 4.727

is used to aid the reader's eye to show the theoretical intercept constant on the Y-axes. Given this, the HVAC supply provides the greatest inception voltage over the given gap distance (2.22 kV cm^{-1} between 5 to 30 mm), followed by the HVDC- circuit that provides electric field of 1.65 kV cm^{-1} over the same gap distances. Whereas the Tesla coil provides the least visual inception voltage with a radio frequency electric field of 0.91 kV cm^{-1}, again over the same gap distance.

The lower voltage disruptive gradient of the HVDC—supply may be partially understood by comparing the two Peek's Law (3 and 4) where a HVDC- (50 Hz) would be expected to produces a disruptive gradient a RMS (0.707) of the AC supply voltage gradient. As for the Tesla coil gradient, Tonmitr and Kaewrawang has suggested that this maybe due to a higher field utilization factor at radio frequency [21].

8 Laboratory Electric Field Stimulation and Annihilation of Insect Experiment

This section considers SEF formation on insects that have a characteristic length in the range of a few cm.

Specifically, the focus is on the analysis of live insects, which are subjected to electrification experiments, such as performed by Challahan and Mankin [15] as well as by Morar et al. [63].

8.1 Callahan and Mankin Experiments

Callahan and Mankin evaluated how an electric field stress when applied to living insects, would induce a visible glow on the dielectric exoskeleton which surrounds the conducting medium of the insect's body without causing death. The second aim was to determine whether a swarm of glowing insects could be seen in the night sky. Within the journal '*Applied Optics'* Tha Paw questions the validity of Callahan and Mankin's experiments and calculations [64], whom follow-up with rebuttals [65, 66]. In the discussion the focus was on whether a Tesla coil that operates in the medium to high frequency band (0.3 to 30 MHz) is suitable to mimic atmospheric electric fields induced by a thunderstorm, Callahan contest the electric fields produced by nearby thunderclouds is not pure dc in origin, but is also partly supported by energy absorption from radio or other electromagnetic energy form thundercloud region. In support of Callahan's use of the Tesla coil it is should be noted that in Peek's original atmospheric spark-over (flashover) experiments [20] used a similar spark gap transformer circuit that operates in the 100 kHz frequency range for the evaluation thunderstorms activity. Recently (2012) aircraft in-flight encounters with lighting and thunderstorms cause interference in the 10 MHz range [67].

The Paw also questioned the validity of the insect luminosity calculations. Mankin responses [66] by agreeing with Tha Paw that although the insects-swarm hypothesis is plausible, it is by no means completely validated, and may never be completely validated to everyone's satisfaction. The entomologist May Berebaum writing in *American Entomologist* some 34 years later, reopened Callahan and Mankin's insect swarm hypothesis by citing 2012 Television reports and numerous internet reports on possible insect swam seen at night [68]. For this reason this current work only examines the first aim of Callahan and Mankin's electrification experiments.

The first circuit used a Molectron HVDC supply provides a 0–20 kV impulse voltage across a capacitor consisting of 2 aluminum plates, each having an area of $A = 20\ \text{cm}^2$ and a separation distance $d = 1.9$ cm in which the insects were suspended between the plates using Duro rubber cement. However the circuit produced arcing (flashover) that resulted in the death of the insect. Therefore HVDC circuit was not deemed suitable for the purpose of SEF stimulation of living insects and no further experiments were performed. The second circuit used a Cenco Tesla coil: 0–10 kV operating at a resonant frequency with the medium to high frequency band where the resonant frequency is adjusted by altering the spark gap. The operating difference between circuit 1 and circuit 2 is that the HVDC produces high-voltage and high to medium current, while the Tesla coil produces a high-voltage and low-current that produces a lower energy transfer due to the loosely-coupled nature of the windings.

For the second circuit (Cenco Tesla coil) the insect was attached using approximately 1 cm of Duro rubber cement to the secondary winding tip of the Tesla coil (Fig. 6). The 2 MHz self-tuning bandwidth of the circuit was found sufficient to stimulate glow on living insects at electric field stress levels of approximately 2–3 kV cm^{-1} and continued as the voltage stress was reduced to 200–300 kV cm^{-1}.

The entomologists did not explain fully how electrical turning was performed. From an electrical engineering perspective, it is assumed that without an iron core the transformer primary and secondary windings are loosely coupled. The loose coupling efficiency means that 5–20% of the magnetic field of the primary winding passes through the secondary when it is open circuited. Thereby slowing the exchange of energy between the primary and secondary windings, which allows the oscillating

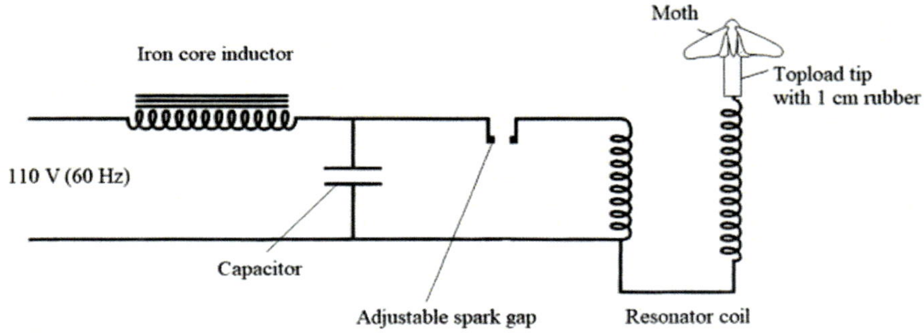

Fig. 6 A schematic of the Cenco Tesla coil, with a topload moth on the secondary winding tip

Table 7 Tesla coil experiment matrix

Insect species	description	Reason for SEF susceptibility	Visible inception electric field (kV cm^{-1})
Trichoplusia ni (Hübner); (Cabbage looper)	Moth (33–37 cm wingspan)	Insect ε′ ~2–3 [71], high static charge. 4 Mass night flights	2.1
Choristoneura fumiferana; (Eastern spruce budworm)	Moth (21–30 m wingspan)		2.1
Euthyrhynchus floridanus; (Stink bug)	Beetle (12–17 mm)	pointed projections on its elytra	2.1*
Tylocerina nodosus; (long-horned beetle)	Beetle (12–17 mm)	2 long antennae	2.1
Conotrachelus nenuphar; (Plum curculio)	Beetle (6–9 mm)	control	2.7

*The stink bug produced emission in the 350–450 nm (blue) spectral range at distance of 18 cm

energy to stay in the secondary longer before it returns to the primary and dissipate in the spark. At this point the topload capacitance (rubber and insect) frequency pulls the operating frequency downwards, but still remaining within the design frequency bandwidth of the circuit [69, 70].

Table 7 provides a matrix of the experiment variables (species name, main description of insect, reason for SEF susceptibility and SEF visible inception electric field strength. The tabulated data shows that the test species (except the Plum curculio) exhibited a visible inception electric field of 2.1 kV cm^{-1}. The luminesce originating at discrete external locations on insects body such as the distal tip of mandibles, ovipositors, antennae, and leg joints. Following on from Bostick's work these discrete luminesce regions may be classed as plasmoids as they originate from body of the host insect.

The Plum curculio however exhibited a higher visible inception electric field of 2.7 kV cm^{-1}. Although the authors give no reason for the higher visible inception electric field, they imply that insects of small body size and minimal protrusions are least susceptible to SEF; whereas insects with extensive protrusions (antennae, legs and wings etc.,) are more susceptible to SEF. However, in flight mode the authors suggested that a triboelectric (ancient Greek *tribo* meaning to 'rub') mechanism on the Lepidoptera (ancient Greek *lepís* 'scale' + *pterón* 'wing') would greatly enhance the production of SEF. In all cases insects survived the test and lived a nature full life span after the experiment. However when dead and dry insects underwent the same electrical stimulation no glow was observed, and it was only after these insect were immersed in water did a temporality glow occur.

8.2 *Morar Et Al Corona Discharge Insect Annihilation Experiments*

The aim of Morar et al. laboratory corona experiments was to identify the voltage levels that would annihilate pest insects using a HVAC (50 Hz) circuit [63]. Typically in this type of circuit an iron-core step-up transformer provides the high-voltage along with associated high-current. Therefore the necessary increase in voltage beyond the survivable voltage levels indentified by Callahan and Mankin needed to be surpassed implying additional current being supplied. In these annihilation experiments the HVAC was either directly applied to the insect that was positioned between the two electrodes separated a distance of 4 cm apart within a glass bell jar, or indirectly where the insects are exposed to ionized air within the same bell jar.

The insects studied where *Phorodum humuli* (Green malt bug, typically 2 to 2.6 mm in length) and where subject to various high voltage stress levels (10, 11.5, and 13 kV) and exposure times (10, 15 and 30 s). Table 8 provides a summarized matrix of the experimental variables and evaluation of the experimental outcome in terms of mortality as a function of time after the stimulation.

The results of this work showed that mortality rates for direct exposure to 2.5 to 3.25 kV cm^{-1} is observed, and for indirect exposure higher stress levels (4.7.5 to 5.25 kV cm^{-1}) is required. The authors also suggest ozone generated by the corona discharge within a sealed bell jar is reasonable for the annihilation of the Green malt bugs, although no ozone measurement levels was reported in this work.

Table 8 HVAC direct and indirect corona discharge experiments matrix [58]

Insect species	kV cm^{-1}	Exposure time (min)	Mortality (%) after 2 h	Mortality (%) after 24 h	Mortality (%) after 48 h
Direct exposure					
Phorodum humuli; (Green malt bug)	2.5	30	70	80	90
Phorodum humuli; (Green malt bug)	3.25	15	64	90	100
		30	75	100	100
Indirect exposure					
Phorodum humuli; (Green malt bug)	4.27	30	70	80	90
Phorodum humuli; (Green malt bug)	5.25	15	80	92	96
		30	96	100	100

9 Discussion

Since the beginning of recorded history SEF has been observed and experienced by humans that are close to thunderstorms, both at sea level and in mountain regions: many of which have been reenacted in the performing arts. Systematic study of naturally occurring SEF has proved difficult due to both the unpredictable timing of the event and the understanding of high voltage circuit requirements for partial air breakdown. The understanding of SEF generation is important for high power transmission line design, the developing protection system for marine shipping and airships and airplanes (flying at high altitude and when coming into land).

The three dirigible Airships discussed in this paper represent the worst airship disasters in term of life lost. With a high degree of certainty SEF, or another form of static discharge, did not have part in the helium filled USS Akron disaster. However, it cannot be said with certainty that SEF or another form of static discharge was the energy source for the ignition of the H_2 lift gas in the Dixmude and Hindenburg airships disasters.

Arguably our understanding of the complexity and interconnectivity of SEF and BL is largely based on observations of natural atmospheric disturbances and the interaction with living and non-living objects at, or near, ground level as well at high altitudes. The application OES, triboelectric charging, and Peek's empirical mathematical construct to these events provide an insight to the physical mechanism that generates SEF and BL. Bostick's generic term 'plasmoid' is found useful when describing the generation and motion of BL in aircraft fuselage and discrete SEF formations on insects.

This work highlights the use of different high-voltage sources for generating SEF. For example, when comparing the application of HVDC-, HVAC (50 Hz) circuits and the Tesla coil on test subject (model sailing ships and insect) the Tesla coil provides the lowest SEF visual inception voltage in terms of electric field stress (0.91 and 2.7 kV cm^{-1} depending on the test subject) before sufficient corona electrical current is drawn to cause flashover. The HVAC (50 Hz) circuit provides both the highest inception voltage and is shown to annihilate insect pest (Green malt bug). Paw's objection to the use of non dc high-voltages for SEF stimulation has been considered. However over the past 100 years it has been well known within the aircraft industry that the combination of BL and lightning introduce radio interference to dc electric field within cumulonimbus and the region between the cloud and ground level where, electrical stimulation of insect swarms have been observed. Thus, the use of HVAC and the Tesla coil is considered here. The use of the Tesla coil is of particular importance because HVDC and HVAC circuits normally employ a tightly-coupled soft-iron core in their step-up transformer which becomes fully magnetized when polarity reversal occurs at every half-cycle of the input waveform: resulting in large current flow at the switching point [72] that detrimental to the sensitivity of the inception voltage measure. The Tesla coil uses a loosely coupled primary and secondary winding thus avoiding magnetic saturation; therefore the coil can provide an extremely high voltage with low power output, which is a perquisite when indentifying the minimum SEF visible inception voltage. Furthermore the Tesla coil windings, with respect to tightly-coupled soft-iron core transformers, have an open construction that can be easily altered to the requirement of the application. A further feature of the Tesla coil is the medium to high frequency band of operation corresponds well with the electromagnetic interference emanating from the ionization and relaxation processes within the SEF and corona discharge. This would suggest that the Tesla coil is more suitable for inception voltage measurements when compared to HVDC or HVAC (50 Hz) circuits.

Acknowledgements The Authors declare that there is no conflict of interest regarding the publication of this paper.

References

1. W. Shakespeare, The Tempest. (Act 1, Scene 2). *First Folio*, ed. by E. Blount and I. Jaggard. (London. 1623)
2. M. Durand, J.G. Wilson, Ball lighting and fireballs during volcanic air pollution. Weather **61**(2), 40–43 (2006)
3. H.F. Dacre, A.L.M. Grant, R.J. Hogan, S.E. Belcher, D.J. Thomson, B. Devenish, F. Marenco, J.M. Haywood, A. Ansmann, I. Mattis and L. Clarisse. Evaluating the structure and magnitude of the ash plume during the initial phase of the 2010 Eyjafjallajökull eruption using lidar observations and NAME simulations. J. Geophys. Res. **116**, 15 (2011)
4. R.G. Harrison, K.A. Nicoll, Z. Ulanowski, T.A. Mather, Self-charging of the eyjafjallajökull volcanic ash plume. Environ. Res. Lett. **5**, 024004 (2010)
5. F. Arago, Sur le tonnerre. *Annuaire au Roi par le Bureau des Longitudes*. Notices Scientifiques. 221 (1838)
6. G. Hartwig, *The Aerial World* (London, 1886), p. 310
7. G. Hartwig, 19th century engraving depicting ball lightning. *The Aerial World* (London, 1886). https: //commons.wikimedia.org/wiki/File:Ball_lightning.png. (Accessed November 2019)
8. G. Hartwig, *The Aerial World* (London, 1886), p. 267
9. A.I. Grigor'ev, D. Grigor'eva, S.O. Shiryaeva, Ball lightning penetration into closed rooms: 43 eyewitness accounts. J. Sci. Expl. **6**(3), 261–279 (1992)
10. A. Meessen, Ball lightning: bubbles of electronic plasma oscillations. J. Unconven. Electromagn. Plasm. **4**, 163–179 (2012)
11. J.J. Lowke, D. Smith, K.E. Nelson, R.W. Crompton, A.B. Murph, Birth of ball lightning. J. Geophys. Res. **117**, D19107 (2012)
12. N. Tesla, *Colorado Springs Notes: 1899–1900*, ed. by A. Marincic (Nolit, Beograd, Yugoslavia, 1978)
13. I. Langmuir, Oscillations in ionized gases. Proc. Natl. Acad. Sci. U.S.A. **14**(8), 627–637 (1928)
14. W.H. Bostick, Experimental study of ionized matter projected across a magnetic field. Phys. Rev. **104**(2), 292–299 (1956)
15. P.S. Callahan, R.W. Mankin, Insects as unidentified flying objects. Appl. Opt. **17**(12), 3355–3360 (1978)
16. A.I. Grigor'ev, I.D. Grigor'eva, S.O. Shiryava, Ball lightning and St. elmo's fire as forms of thunderstorm activity. J. Sci. Exp. **5**(2), 163–190 (1991)
17. M. Donoso, J.L. Trueba, A.F. Rañada, The riddle of ball lightning: a review. Sci. World J. **6**, 254–278 (2006)
18. S. Nomura, H. Toyota, S. Mukasa, H. Yamashita, T. Maehara, A. Kawashima, Production of hydrogen in a conventional microwave oven. J. Appl. Phys. **106**(7), 073306 (2009)

19. Y. Hattori, S. Mukasa, H. Toyota, S. Normura, Electrical breakdown of microwave plasma in water. Curr. Appl. Phys. **13**, 1050–1054 (2013)
20. F.W. Peek, *Dielectric phenomena in high voltage engineering*. 2nd edition. (McGraw Book Company, Inc, New York, USA, 1920)
21. K. Tonmitr, A. Kaewrawang, Saint elmo's fire corona by using HVDC, HVAC and tesla coil. Marine Eng. Front. **1**(2), 19–23 (2013)
22. M. Bonsma, Weird and wonderful manifestations of electro-magnetism: past and present. Phys13news. Fall 2014, 3–5 (2014)
23. E.M. Wescott, D.D. Sentman, M.J. Heavner, T.J. Hallinan, D.L. Hampton and D.L. Osborn. The optical spectrum of aircraft St. elmo's fire. Geophy. Res. Lett. **23**(25), 3687–3690 (1996)
24. D.P. Dowling, F.T. O'Neill, V. Milosavljević, V.J. Law, DC pulsed atmospheric pressure plasma jet image information. IEEE Trans. Plasma Sci. **39**(11), 2326–2327 (2011)
25. C. Saunders, Charge separation mechanisms in clouds. J. Appl. Meteorol. 335–353 (1993)
26. T.V. Prevenslik, A unified theory for sprites, st. elmo's fire, and ball lighting. J. Metrol. **26**(260), 204–211 (2001)
27. S. Hughes, Green fireballs and ball lightning. Proc. R. Soc. A. **467**, 1427–1448 (2011)
28. M. Goldman, A. Goldman, R.S. Sigmond, The corona discharge, its properties and specific uses. Pure & Appi. Chem. **57**(9), 1353–1362 (1985)
29. J.-R. Riba, A. Morosini, F. Capelli, Comparative study of ac and positive and negative dc visual corona for sphere-plane gaps in atmospheric air. Energies. **11**(2671), 18 (2018)
30. A.S. Gibson, Jeremy A. Rioussety, V.P. Pasko, Minimum breakdown voltage for corona discharge in cylindrical and spherical geometries. Annual Research Journal, Electrical Engineering, Research Experience for Undergraduates. Penn State. **7**, 1–17 (2009)
31. N.L. Aleksandrov, E.M. Bazelyan, F. D'Alessandro, Y.P. Raizer, Numerical simulations of thunderstorm-induced corona processes near lightning rods installed on grounded structures. J. Electrostat. **64**(12), 802–816 (2006)
32. Gallimberti, G. Bacchiega, A. Bondiou-Clergerie, P. Lalande, Fundamental processes in long air gap discharges. C. R. Physique. **3**(10), 1335–1359 (2002)
33. E.E. Mombello, G. Rattá, H.D. Suárez, F.O. Torres, Corona loss characteristics of contaminated conductors in fair weather. Elect. Power Syst. Res. **59**, 21–29 (2001)
34. E.M. Bazelyan, N.L. Aleksandrov, Yu.P. Raizer, A.M. Konchakov, The effect of air density on atmospheric electric fields required for lightning initiation from a long airborne object. Atmos. Res. **86**(2), 126–138 (2007)
35. E.M. Bazelyan, Yu.P. Raizer, N.L. Aleksandrov, Corona initiated from grounded objects under thunderstorm conditions and its influence on lightning attachment. Plasma Sources Sci. Technol. **17**(024015), 17 (2008)
36. J.-R. Riba, W. Larelere, J. Rickmann, Voltage correction factors for air-insulated transmission lines operating in high-altitude regions to limit corona activity: a review. Energies. **11**(1908), 14 (2018)
37. Addendum to Doc.20: Interferences produced by corona effect of electric systems. https://e-cigre.org/publication/061-addendum-to-doc20--interferences-produced-by-corona effect-of-electric-systems (accessed January 2020)
38. #AIRBOYD #AvGeek. USS akron crash. https://www.youtube.com/watch?v=cLFLAj-9-vU (accessed June 2020)
39. British movietone. akron test ends in tragedy. https://www.youtube.com/watch?v=iJKF3RHKDtY (accessed June 2020)
40. Dixmude (airship) https://en.wikipedia.org/wiki/Dixmude_(airship) (accessed June 2020)
41. H. Eckener, Technical Memorandum No. 256. Lost of the Dixmude. Translated from the *Luftfahrt*. January 23, pp 1–3 (1924). https://digital.library.unt.edu/ark:/67531/metadc56906/m1/2/ (accessed March 2020)
42. Loss of the akron. Nature. 131, 499–500, (1933). https://www.nature.com/articles/131499c0.pdf (accessed June 2020)
43. C. O'Dell, Crash of the hindenburg (Herbert Morrison, reporting) (May 6, 1937). National Registry (2002)

44. A. Bain, Wm. D. Van Vorst, The hindenburg tragedy revisited: the fatal flaw found. Int. J. Hyd. Energy. **24**, 392–403 (1999)
45. R.H. Marriott, Radio range variations. Proc. IRE. **2**(3), 37–52 (1914)
46. R.A. Perala, A critical review of precipitation static research since the 1930's and comparison with aircraft charging by dust (2009). https://www.ema3d.com/wp-content/uploads/downloads/AEP_3.pdf (accessed June 2020)
47. Garcia-Hallo, D. Lemaire, N. Raveu, G. Peres, Method for p-static source location on aircraft using time domain measurements. Progress Electromagn. Res. C. **62**, 89–98 (2016)
48. A.J. Dressler, The hindenburg hydrogen fire: fatal flaws in the addison bain incendiary-paint theory, p. 21 (2004). https://spot.colorado.edu/~dziadeck/zf/LZ129fire.pdf (accessed June 2020)
49. A.J. Dressler, D.E. Overs, W.H. Appleby, The hindenburg fire: hydrogen or incendiary-paint. Buoyant Flight. **52**(2–3), 11 (2005)
50. D.H. Robinson, LZ-129 Hindenburg. 1st edition (Arco Publishing Co New York, 1964)
51. T. Terrel, LZ-129 hindenburg. Curious dragonfly LLC (2002). https://curiousdragonflycom.files.wordpress.com/2017/12/the-hindenburg.pdf (accessed June 2020)
52. A.Z. Shahneh, More frequent flight in the ash polluted sky using electrostatic filter. Int. J. Aeronaut. Aerospace Eng. **1**(1), 27–35 (2019)
53. R.R. Nunes, C. Christmann, H. Oertel, Deutscher Luft- und Raumfahrtkongress. Document ID: 370178 (2015)
54. V.J. Law, A. Ramamoorthy, D.P. Dowling, Real-time process monitoring during the plasma treatment of carbon weave composite materials. J. Mater. Sci. Eng. **B2**, 164–169 (2011)
55. R.C. Jennison, Ball lightning. Nature **224**, 89 (1969)
56. A.I. Grigor'ev, I.D. Grigor'eva, S.O. Shiryaeva, Ball lightning penetration into closed rooms: 43 eyewitness accounts. J. Sci. Expl. **6**(3), 261–279 (1992)
57. V.P. Torchigin, A.V. Torchigin, How the ball lightning enters the room through the window panes. Optik. **127**, 5876–5881 (2016)
58. P. Varsa, J. Rokne, Simulation of the ball lightning phenomenon. Comput. Graph. **30**, 485–493 (2006)
59. A.I. Egorov, S.I. Stepanov, G.B. Shabanov, Laboratory demonstrations of ball lighting. Phys. Usp. **47**(1), 107–109 (2007)
60. S.E. Dubowsky, D.M. Friday, K.C. Peters, Z. Zhao, R.H. Perry, B.J. McCall, Mass spectrometry of atmospheric-pressure ball plasmoids. Int. J. Mass Spectrom. **376**, 39–45 (2015)
61. S.E. Dubowsky, B. Deutsch, R. Bhargava, B.J. McCall, Infrared emission spectroscopy of atmospheric-pressure ball plasmoids. J. Mol. Spectrosc. **32**, 1–8 (2016)
62. L.A. Maglaras, A.L. Maglaras, The influence of the effect of grounding and corona current to the field strength the corona onset and the breakdown voltage of small air gaps. WSEAS Trans. Power Syst. **3**(3), 103–110 (2008)
63. R. Morar, I. Soarasan, S. Budu, I. Ghizdavu. M. Purca, L. Dascalescu, Corona discharge effects on some parasitical insects of cultured plants. J. Electr. 40–41, 669–673 (1997)
64. U.K. Tha Paw. Insects as unidentified flying objects: Comment. Appl. Optics **18**(16), 2723–2724 (1979)
65. P.S. Callahan, Insects as unidentified flying objects: authors' reply to comment. Appl. Optics **18**(16), 2724–2725 (1979)
66. R. Mankin, Insects as unidentified flying objects: authors' reply to comment. Appl. Opt. **18**(16), 2725–2726 (1979)
67. P. Laroche, P. Blanchet, A. Delannoy, F. Issac, Experimental studies of lightning strikes to aircraft. J. Aerospace Lab **5**(AL05–6), 1–13 (2012)
68. M. Berebaum, The truth (about insects) is out there. American Entomologist. 6–8, Spring (2013)
69. V.J. Law, W.G. Graham, D.P. Dowling, Plasma hysteresis and instability: a memory perspective, in *The Foundations of Chaos Revisited: From Poincaré to Recent Advancements, Understanding Complex Systems*, ed. by C. Skiadas. (Springer International Publishing Switzerland 2016). chap. 9, pp. 136–157. DOI https://doi.org/10.1007/978-3-319-29701-9

70. R.M. Craven, I.R. Smith, B.M. Novac, Magnetic coupling in tesla transformers. Appl. Phy. Res. **8**(6), 101–105 (2016)
71. P.S. Callahan, Insect bioelectronics: a theoretical and experimental study of insect spines as dielectric waveguides. IFAC Proce. **2**(4), 303–310 (1968)
72. V.J. Law, V. Milosavljević, N. O'Connor, F.L. Lalor S. Daniels, Handheld flyback driven coaxial dielectric barrier discharge: development and characterization. Rev. Sci. Inst. **79**(9), 094707 (2008)

Application of Microwave Oven Plasma Reactors for the Formation of Carbon-Based Nanomaterials

V. J. Law and D. P. Dowling

Abstract This paper reviews the literature on the formation of carbon-based nanomaterials using converted domestic microwave oven plasma reactors. The carbon-based compounds range from single and multi-walled carbon nanotubes, to onion-like nanostructures, fullerene, and graphene sheets. The microwave plasma process is performed using in-liquid containing plasma bubble (plasmoids) generated at an aerial-antenna igniter, susceptor surface ignition within gaseous plasma as well as the use of conventional gaseous plasmas. Based on the literature reports, the thermodynamic and kinetic plasma processing conditions are reviewed, along with process input criteria that include: applied microwave, hydrocarbon precursor, aerial-antenna igniter design, and susceptor material and sample collection. The use of microwave oven drilling (local thermal runaway) and reverse drilling that lead to the ejection of plasmoids which may give rise to the formation dusty plasma or fireball as a processing route for the formation of nanomaterials.

Keywords Aerial-antenna igniter · Plasmoids · Ball-lighting · Fireball · Microwave oven · Carbon-based nanoparticles

1 Introduction

Carbon-based particles may be manufactured in a variety of allotropes, but to be classed as nanoparticles (NP) the grain size <100 nm in at least one dimension. A brief list of the carbon-based allotropes includes fullerene and graphene, single and multi-walled carbon nanotubes (SWCNT and MWCNT), onion-like nanostructure's (OLNs) and carbon nanosheets [1, 2]. Their functionality and high-value is derived from their unique combination of chemical and physical properties (i.e., thermal and electrical conductivity, high mechanical strength, magnetic and optical properties) that have applications in structural engineering, lubrication, electronics, and the

V. J. Law (✉) · D. P. Dowling
School of Mechanical and Materials Engineering, University College Dublin, Belfield, Dublin 4, Ireland

C. H. Skiadas et al. (eds.), *13th Chaotic Modeling and Simulation International Conference*, Springer Proceedings in Complexity,
https://doi.org/10.1007/978-3-030-70795-8_35

biomedical sector. The aim of this paper is two-fold; firstly, to review how functional carbon-based nanomaterials (NMs) are selectively formed within a domestic microwave oven plasma reactor (DMOPR). Secondly to review the NMs formed within an ejected plasmoid that usually leads to the formation of a dusty plasma or 'fireball', which is generated by microwave 'drilling' using an open ended co-axial applicator, within a multi-mode cavity [3, 4].

To simplify the complex manufacturing processes of functional carbon-based allotropes the paper is divided into the following sections. Section 2 reviews 15 papers reporting the formation carbon-based NMs within the DMOPR. Sections 2.1 through to Sects. 2.5 considers the role of: microwave power, hydrocarbon precursors, aerial-antenna igniters, bubble formation and their transition to plasmoid at the electrode surfaces within microwave irradiated liquid, susceptors, chemical catalyst. Section 3 looks at NP formation at atmospheric pressure within microwave-excited plasmoid operating in a fireball mode. Section 4 provides a list of hydrogen and NM collection processes. Finally, Sect. 5 provides an overall summary of this paper.

2 Carbon-Based NMs Formed Within DMOPR

Batch syntheses of carbon-based NMs within converted DMOPRs have been widely reported [5–19]. Closely aligned to these plasma processes is hydrogen (H_2) generation within a DMOPR. The NMs and H_2 process may be formed in-liquid by the plasma [5, 6, and 15–19], on a solid-state susceptor surface (placed either within, or without a glass vial) exposed to microwave irradiation [7–11], or simply placed on a glass slide and then exposed to plasma [12–14]. For further reading the authors of this work have published three review papers on the following subjects: converting domestic microwave ovens into a plasma reactor [20], microwave plasma processing of organic compounds and biomaterials [21] and the use of the DMOPR as a rapid prototyping tool [22]. Before considering the above processes it is worth reviewing how the domestic microwave oven heats materials and how this view may be modified for the DMOPR.

For the domestic microwave oven it is recognized that the synthesis of organic materials is more rapid when compared to classic heating methods that rely on conduction and convection from the vessel wall, to center of the material that is being heated. In the microwave oven, the heating mechanism is through electromagnetic waves (microwave irradiation); causing molecular agitation and intermolecular friction generating heat within the material, so raising the temperature more rapidly compared conventional heating. This is practically true for polar liquids such as water (H_2O), methanol (CH_3OH) and ethanol (C_2H_5OH), whose chemical bonds are between atoms with very different electro negativities. In contrast microwave energy is absorbed weakly by hydrocarbons such as Toluene and Xylene, along with most organic polymers, all of whom exhibit low molecular polarity. These materials are said to be microwave transparent. To quantify molecule polarity of a material, in its ground state, the complex relative permittivity ($\varepsilon_r = \varepsilon_r' - j\varepsilon_r''$) is used. Where the real part (ε_r') is a dimensionless number and is a measure of a materials ability to couple with microwave energy. The relationship of ε' to the effective wavelength within a material at a given frequency is given in (1) [2, 20–23].

$$\lambda' = \frac{c}{f_o\sqrt{\varepsilon'}} \quad (1)$$

In this equation, c is the speed of light (3×10^8 m.s^{-1}) and f_o is the magnetron frequency (2.45×10^9 Hz). From this relationship it becomes apparent that non-uniform heating of materials due to their size and geometry is an issue. The imaginary part (ε_r") of the complex permittivity is also a dimensionless number that is used as a measure of the materials ability to be heated by absorb microwave energy (via direct current or Ohmic heating) and turned into heat. The ratio of ε_r"/ε_r'is called the dielectric loss tangent (*tan* δ) and describes the ability of the material to dissipate electromagnetic energy within a microwave cavity.

Since the early 2000s, it has been postulated that transition state (TS) of a chemical reaction that is subjected to microwave irradiation is effected by a combination of thermal effects (thermal runaway, hot spots and selective heating) and dielectric volume heating, plus species mobility and diffusion [2, 23]. One or more of these effects may enhance reactivity by reducing the activation energy (ΔG). In addition the ability to turn-on and -off dielectric volume heating enables product (P) selectivity to be controlled. See Fig. 1.

In this work the plasma domain is added to microwave domain effects detailed earlier. A generic process for the formation of NPs from a hydrocarbon precursor in microwave plasma is represented in (2). The transition state is represented by the subscript operator (arc or plasma depending upon the nature of the discharge) that contains thermal, microwave and plasma effects: any one, or more, may rate limit nano-compound selectivity and hence alter product yield. Under microwave plasma conditions, gas molecules undergo ionization and dissociation to form single charged species, the deposition of which becomes kinetically favorable. Microwave plasma experimental studies have also shown that the carbon particle size and relatively narrow particle size distribution is controlled (limited with respect to thermal

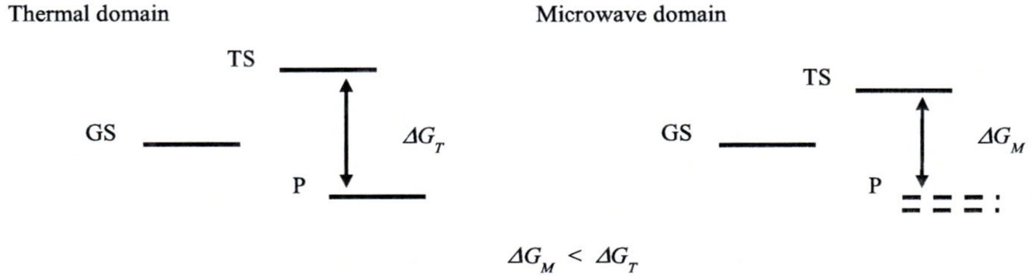

Fig. 1 A simple comparison the thermal and microwave reaction coordinates, where GS is the ground state, TS is the transition state and P is the product. The figure also infers that the enthalpy of the reactions have a negative change ($\Delta H < 0$)

chemical phase deposition), by the action of ion bombardment, which tends to inhibit particle coagulation and agglomeration [1].

$$Hyrodcarbon \underset{plasma}{\rightarrow} Carbon\ nanomaterial \quad (2)$$

Mukasa [24] and Hattori et al. [25, 26] have demonstrated that for an open-end transmission-line microwave reactor operating at 2.45, GHz, that 300 W and 40 W of input power is required to generate a plasma in *n*-dodecane and water, respectively. In Hattori's experiments the corresponding electric field breakdown was of the order of 10–30 kV.m^{-1}. Furthermore, the electrification process was found to occur in bubbles at a dielectric heated metal surface. As the bubbles are liberated from this surface, they float upwards due to their buoyancy. This process it repeated many times until the microwave power is turned-off.

Table 1 provides an overview of typical microwave power and chemical parameters reported for use in chemical synthesis from the literature. For ease of discussion the tabulated data is grouped into four broad phase states: classifications: In-liquid chemistry, ice/liquid chemistry, solid-state chemistry, and gaseous plasma chemistry. Common to all four classifications is that a chemical catalyst is sometime used; this is annotated with a superscript C (^{C}) alongside the reference number.

2.1 Microwave Power

Within most DMOPRs, a packaged cavity-magnetron (operating at a free-running frequency of $f_o = 2.45 \pm 0.1$ GHz ($\lambda_o \sim 12.2$ cm) is rectified at AC mains frequency (50/60 Hz) to produces a negative going pulsed square-wave DC voltage. Thus the duty-cycle $(D) = 50\%$, where the on-period (t_{on}), is equal to the off-period (t_{off}). To alter the time averaged power delivered to the oven's multi-mode cavity this rectified waveform is further pulse width modulated with T_{on} and T_{off} time periods of typically 30 and 30 to 60 s. Thus the applied power is at a maximum when using a continuously rectified waveform and a lower time averaged power level, when the pulse width modulation waveform envelope is used [20–22, and 27].

Table 1 (column 2) lists the reported powers levels which were reported in the reviewed papers. Power levels were reported in the range 300 to 1800 W, However as with all datasets collected from different research sources, experimental conditions are not reported in a consistence manner. In the body of work referenced here it would be expected that full power conditions would correspond to continuous mode, and power levels below this upper limit would be in the pulse width modulated mode. For example, references [6, 15, 17, 18 and 19] do differentiate between the oven's rated input power and the rated microwave maximum power output level, but none of the referenced papers report if continuous wave (CW) or pulse width modulated conditions is used. Moreover, reference [10] states a power of 1.8 kW is used, this power level would be expected to be the total rated AC mains input

Table 1 Typical examples from the literature microwave oven processing conditions used for chemical synthesis [5–19]

Reference	Stated power	Hydrocarbon precursor	Electrode /susceptor	Support substrate	Product	Rate or process time
In-liquid chemistry						
5	500 W	*n*-dodecane	Dual × 6	N/A	H2	~ 25 $ml.s^{-1}$
5	500 W	CH_3-OH C_2H_5-OH (95:5)	Dual × 7	Silicon	Diamond film	10 min plasma time
6^C	750 W	*n*-dodecane	Single × 7	Silica	1 L H2 + Carbides 22 vol%	~ 37.1 ml/s
6^C	750 W	Cyclohexane	Single × 7	Silica + Mo & Co	1 L H2 + Carbides 22 vol% + MWCNT	~ 37.1 ml/s
15	750 W	*n*-dodecane	Single × 6	N/A	1 L H2 + Carbides 22 vol% + CNT	~ 26 $ml.s^{-1}$
16	500 W	*n*-heptane Isooctane Decane Hexadecane	Single × 1	Bubble	varigrained NP	HC–CH Swan bands
17	750 W	*n*-dodecane	Single × 7*	Bubble	H2 + carbides and graphite	~ 10 $ml.s^{-1}$
Ice/liquid chemistry						
18	750 W	Cyclopentane hydrate	Single × 7	N/A	0.55 L H2 + carbides 9 vol%	~ 18 $ml.s^{-1}$
19	700 W	Methane hydrate	Single × 7	N/A	0.55 L H2 + carbides 55.5 vol%	~ 11 $ml.s^{-1}$
Solid-state chemistry						
7^C	700 W	Magnetite NP & Polystyrene	Aluminum or Graphite	N/A	Fe_3O_4 NP	~ 90 s plasma time
8	700 W	Polystyrene	Aluminum	Quartz tube	OLNs	~ 120 s plasma time
8	700 W	Polyethylene	Aluminum	Quartz tube	NS	~ 120 s plasma time
9^C	600 W	Polyethylene	Aluminum foil	Silicon	SWCNT & MWCNT	~ 1 h process time

(continued)

Table 1 (continued)

Reference	Stated power	Hydrocarbon precursor	Electrode /susceptor	Support substrate	Product	Rate or process time
10[C]	1.8 kW	Naphthalene	Graphite	Glass vial	OLNs	15 to 120 s plasma time
11[C]	600 W	Rice husks (powder)	Aluminum foil	Quartz tube	CNT	38 min plasma time
Gaseous plasma						
12[C]	900 W	Ethanol	Zinc salts	Glass slide	ZnO & Zn NP	60 to 120 s plasma time
13[C]	700 W	Ethanol	Zinc salts	Glass slide	ZnO NP	60 to 120 s plasma time
14	1 kW	Ethanol	Single × 1	Borosilicate glass	CNT	Not given
14	1 kW	Xylene Toluene	Single × 1	Borosilicate glass	OLNs	Not given

[c]= chemical catalyst used. * curved multiple aerial-antenna igniters used. N/A = Not applicable

power of a microwave oven that includes a grill. It should be also noted that many of the reported applied powers in Table 1 are considerably lower than those used in open-end transmission-line microwave reactors [24–26]. The authors therefore suggest that the stated power should be used as guide and further details should be sourced from the original reference.

2.2 *Hydrocarbon Precursors*

Liquid, clathrates, solid-state and gases hydrocarbon precursors listed in column 3 of Table 1 have all been used in DMOPR processing. This section provides an overview of carbon based nanomaterials synthesized from these precursors.

2.2.1 Liquid Hydrocarbon

In 2009 Normura et al. [6] reported DMOPR cracking of *n*-dodecane ($C_{12}H_{26}$; ε_r' ~ 2 at 2.45 GHz) for the simultaneous production of H_2, low-grade gaseous carbides (methane, ethylene and acetylene) plus the deposition of graphite. Using the percentage of each gas from chromatography product analyses allows a representative balance stoichiometric reaction equation to be written where a quantity of atomic hydrogen is trapped as interstitial impurities in solid graphite or between graphene sheets and where the cracking process proceeds by C–C bond β-scission (~408 kJ.mol^{-1}) rather than C-H bond cleavage (~418 kJ.mol^{-1}) [28, 29], see (3).

$$4C_{12}H_{26} \underset{In-liquid\ plasma}{\rightarrow} 37H_2 + CH_4 + C_2H_4 + 10C_2H_2 + 25C{\cdot}H_2\,graphite \quad (3)$$

In the same paper and [15] the authors used cyclohexane (C_6H_{12}; ε_r' ~ 2 at 2.45 GHz) resulting in SWCNT and MWCNT products. By 2013, Toyota et el [5] made further modifications to the method of generating H_2 from *n*-dodecane, in addition to forming diamond like films from a vapor mixture (95:5 ratio) of CH_3OH (tan $\delta = 0.659$ at 2.45 GHz) and C_2H_5OH (tan $\delta = 0.941$ at 2.45 GHz) [30]. The details of the metal aerial-antenna igniters used in [5, 6, 15 and 16] and the curved aerial-antenna igniter reported by Mochtar et al. [17], is described in Sect. 2.3.

Yu. A. Lebedvi et al. has used optical emission spectroscopy to study In-liquid plasma C_7-C_{16} alkanes (C_2Hn_{n2+2}) [16]. They found that just after solid NP formation (typically 100 nm in size) the In-liquid plasma emission contained Swan band emissions that are associated with ionized diatomic carbide fragments (C_2) that are potentially dissociated from the starting liquid alkane molecules. Using this information an estimated rotational and vibration temperature of 1700 ± 200 K and 7000 ± 2000 K were obtained. The addition of argon was also found to decrease the rotational gas temperature to 700 ± 100 K.

2.2.2 Hydrocarbon Clathrates

Hydrocarbon clathrates can form an important fuel resource for replacing petroleum and natural gas. Recently research into DMOPR batch processing cyclopetane hydrate [18] and methane hydrate [19] and have both been used to generate H_2 within DMOPR. These materials are composed of host ice/liquid water with an ice framework that has a stoichiometric number of *n* within which a guest hydrocarbon molecule is encapsulated, and where framework size varies to accommodate the guest molecule. For the methane (CH_4) guest molecule the water framework is a cubic structure with $n = 5.8$ to 6.1 H_2O molecules and is designated as a sI or CS-I structure [19]. For the larger cyclopentane (C_5H_{10}) guest molecule the host water framework is enlarged to $n = 17$ H_2O molecules and forms a sII or SC-II structure.

It can be surmised that the microwave plasma decomposition of the hydrates has two main reaction pathways. The first is by rapid plasma formation within bubbles that are directly for aerial-antenna igniter surfaces, and second by the slow microwave dielectric volume heating of water within the hydrate compound, which if unchecked will ultimately melt, to release hydrocarbons to the surface without going through the plasma process. This second decomposition pathway may explain the increase in carbides byproducts (9% by vol for C_5H_{10} and 55.5% by vol % for CH_4): within these carbide byproducts diatomic C_2H_4 molecule being the most abundant.

2.2.3 Solid-State Hydrocarbon

In this section, examples of where DMOPRs have been used for the fabrication of NMs from solid hydrocarbons precursors are described. For example, using aluminum foil (5 × 5 mm^2) as a plasma igniter with the polymer polystyrene $(C_8H_8)_n$: ε_r' = 2.5–2.6 at 2.45 GHz) has been used for synthesis of carbon NP, in addition to fibers with lengths of 100 s of nanometers, when magnetite is added to polystyrene [7]. When replacing polystyrene with polyethylene beads $(C_2H_4)_n$; ε_r' = 2.25 at 2.45 GHz) a 2-dimensional reduced graphene oxide (RGO) nanosheet is produced with a typical thickness of 3 nm [8]. This ability to select between 2-dimentional and 3-dimentional nanostructure has been attributed to the increase in the H/C ratio (1:1 to 2:1) per polymer unit, but these linear formula do not take into account the structural alteration of the polymer unit. In this case, an exchange of a phenyl group per polymer unit for a H atom. This structural change produces a reduction of the hydrocarbon precursor dielectric constant (2.5–2.6 to 2.25). Therefore it's reasonable to say that precursor structural effects need further investigation.

SWCNT and MWCNT have been synthesized using a DMOPR, from polyethylene resins (100 mg), placed on an aluminum foil (25 × 25 mm^2) in conjunction with a silicon substrate coated with the catalyst iron (III) nitrate nonahydrate $(Fe(NO_3)_3{\cdot}9H_2O)$. The resulting particles had diameters of 1.03–25.00 nm with a length of about 0.85 μm. [9]. The resin chemical structure was not reported and as resins are commonly used for coating extruded wire, it is reasonable to assume a copolymer was present. Under these complex conditions (resin and catalyst) it is again reasonable to assume the tan δ properties of the mixture are high to allow rapid volume heating and therefore decomposition of the resin and catalyst. It is also worth noting that the final product had iron catalyst particles embedded within the tubes walls.

Naphthalene $(C_{10}H_8)$ mixed with graphite has been reported to produce OLN that range from 10 s nm to a few microns in size within a DMOPR where the graphite is used as a microwave absorber and naphthalene as the hydrocarbon source [10]. To minimize arcing within the oven, 200 mg of the 1:2 graphite-naphthalene mixture is placed within a quartz vial and then placed in the oven cavity and irradiated for 1 min, where upon arcing occurs. Reaction (4) provides a representative reaction for this process. Outside the quartz vial they were able to deposit onion-like rings onto ceramic substrates.

$$C_{10}H_8 \underset{arc-plasma}{} OLN \tag{4}$$

It is been estimated that over 100 million tons per year of agriculture waste rice husk (RH) is generated from the production of rice, the majority of which is incinerated. This agriculture waste is rich in cellulose and lignin which may qualify it as an economic source of hydrocarbon precursor for the manufacture of carbon-based NMs. Initial prototype DMOPR studies have used RH (ground to a size 20 microns) mixed with a catalyst (FeO_4) placed on aluminum foil (25 × 25 mm^2) and then placed inside a DMOPR for processing to form and mixture of NP (typically 0.5 nm in size), SWCNT or possibly MWCNT, and tubular structures (fibers) with length of 100 s nm [11]. It is worth noting that raw rice husks have typical values of $\varepsilon_r' = 2.982$, $\varepsilon_r'' = 0.283$ and tan $\delta = 0.094$ at 12.4 GHz, and increase when mixed with CNT [31].

2.2.4 Gaseous Plasma

In this section the use of gaseous precursors are described for use in chemical synthesis in conjunction with DMOPR. In 2010, Irzh et al. reported DMOPR synthesis of Zinc oxide (ZnO) and its refinement to Zinc metal (Zn) NP with a size of 15 to 35 nm [12]. Their starting material is an ethanol solution of zinc nitrate ($Zn(NO_3)_2$) of 0.03, 0.5 and 1 M mixed with a carrier gas of argon (Ar) which is irradiated with microwaves that results in the production Ar^+ bombardment and volume dielectric heating the zinc compound. For example Zinc nitrate compound is reduced to ZnO (reaction 5a). Upon further microwave irradiation zinc metal NP are produced (reaction 5b). Product selectivity between ZnO and Zn appears to be by prolonged plasma irradiation rather than volume dielectric heating as the temperature at the oven walls reaches 230–250 °C, which below that of the thermal decomposition of ZnO to Zn [12].

$$Zn(NO_3)_2 \underset{Ar^+ plasma}{\rightarrow} ZnO\,NP + 2NO_2\,(gas) + O_2\,(gas) \tag{5a}$$

$$ZnO \underset{Ar^+ plasma}{\rightarrow} Zn\,metal\,NP + O\,(gas) \tag{5b}$$

In 2012, Raj et al. [13] reported on the remote plasma deposition of ZnO particles within a MOPR. Using oxygen (O_2) as the carrier gas, they found the morphology of the nano-compound can be altered by selecting the substrate material (glass–Si–Al_2O_3/Si), or by changing the Zinc salt concentration (0.03 to 0.16 M).

Singh et al. [14] demonstrated that using different liquid hydrocarbon solvents: ethanol (C_2H_5OH; H/C = 2.5:1), xylene ($C_6H_4(CH_3)_2$; H/C = 1.24:1) and toluene (C_7H_8; H/C = 1.14:1) mixed with H_2 at low pressure, can selectively produce either CNT or OLN. See reaction (6, 7 and 8).

$$C_2H_5OH\,(l) \underset{H_2\,plasma}{\rightarrow} CNT \tag{6}$$

$$C_6H_4(CH_3)_2\,(l) \underset{H_2\,plasma}{\rightarrow} OLN \tag{7}$$

$$C_7H_8\,l \underset{H_2\,plasma}{\rightarrow} OLN \tag{8}$$

2.3 *Passive Metal Plasma Igniters*

This section provides a general description as to the absorption of microwave energy within the oven's cavity when empty (unloaded) and when heating water in the cavity (loaded cavity). The terms 'unloaded' and 'loaded' refer to the *Q*-factor of the cavity [18–20]. Using this background information the role of passive metal plasma aerial-antenna igniters is expanded upon in Sects. 2.3.2 to 2.3.5.

2.3.1 Unloaded and Loaded Cavity Behavior

First consider what happens when irradiating thin metal objects, like (aluminum foil [7–11] and zinc pellets [12, 13], in an empty microwave oven. Under these near 'unloaded' conditions a number of reactions generally occur. Firstly, a portion of the wave energy penetrates the metal (typically 2 to 4 μm) and interacts with free elections with this surface region causing electrical currents to flow. For thin metals, the current that may produce sufficiently high voltage stress at sharp edges and surface irregularities of the metal to cause free electrons to be liberated and cause local gas breakdown, in the form of sparks and streamers, that produce high temperature hotspots. Secondly, the rest of the wave energy is reflected from the metal surface back into cavity. Thirdly, if the microwave irradiation continues, the reflected microwave energy can help to reinforce the microwave standing waveform within the cavity, to such a level that the reflected energy disrupts the operation of the magnetron, thus causing the oven to stop working, or in the extreme case setting fire to oven's internal circuits.

Now consider a polar organic solvent (in this case water) contained in a smooth-walled glass vessel that is placed in the oven's cavity and the microwave energy turned on. Under these 'loaded' cavity conditions, the water under goes rapid dielectric volume heating, that leads to superheating in the absence of any stirring of the organic liquid solvents and retardation of bubble nucleation sites at the vessel surface [23]. [*N. B. In mono-mode cavity reactors ceramic boiling chips that produce nucleation sites at their sharp edges and surfaces irregularities are used to prevent superheating effects* (*23*)]. Replacing the ceramic boiling chips with aluminum foil alters the surface-liquid reaction dynamics by limiting the superheating effect due to the addition of surface hotspots that initiate bubble nucleation, within which, and at sufficiently high voltage stress levels, gas breakdown occurs.

2.3.2 Microwave Aerial-Antenna Igniters

To extend the pressure ignition range within both gas and liquids, aluminum foil can be replaced with wires that have circular or rectangular cross-sections and have a microscopically rough surface that act as nucleation sites for bubble formation. The length of the wire is then (λ_m) matched to the oven's cavity magnetron output free

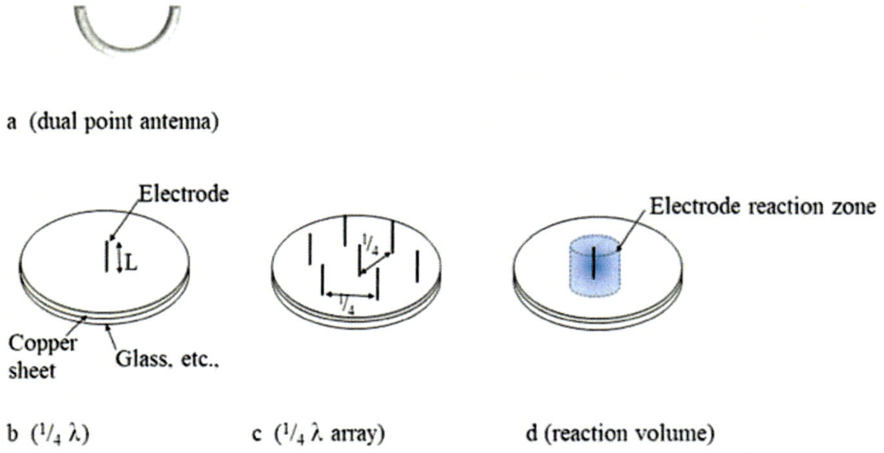

Fig. 2 **a–d** Examples of aerial-antenna plasma igniters

running frequency and wavelength (f_o = 12.4 ± 0.1 GHz; $\lambda_o \approx$ 12.2 cm) at the characteristic ε_r'of the medium that the microwave radiation is passing through. The physical length of the wire becomes matched to the electrical wavelength (λ_m) at f_o, see (9).

$$\lambda_m \approx \frac{C}{f_o\sqrt{\varepsilon_r'}} \tag{9}$$

The approximate expression in (5) is used as the cavity magnetron has a free running frequency bandwidth of 0.1 GHz, and C is the speed of light (2.99792 × 10^8 m.s^{-1}). In general three type of metal aerial-antenna igniter have been reported: U-shaped electrode with two exposed tips (Fig. 2a) [5], a wire with single exposed tip constructed perpendicular to a local ground plane (Fig. b) [14, 16], and an array of wires constructed perpendicular to a local ground plane (Fig. 2c) [6, 15 and 17–19]. See also Law and Dowling [20] and Satio et al. [32]. For single tip igniters placed within liquid n-dodecane (ε_r' = 1.78 to 2), λ_m approximates to 88.5 mm. However, it is found that a wire having a local ground plane the optimal wire length approximates to $\lambda_m/4$ (20 to 22 mm). The accepted assumption here is that a quarter wavelength structures generates maximum electrical stress as there is a 90 degree phase difference between the wire tip that forms and open circuit and the local ground. Lebedev, et al. [16] have estimated that these resonance ¼ wavelength structures can reach voltage stress levels of 2400 V. cm^{-1}.

Within references [5–7, 15, 17–22] it is generally considered that the electrodes have three well-defined roles: to confine the plasma to the immediate proximity of the electrode(s) tip, to function as a source for bubble and then plasmoid formation in which the heterogeneous reaction can take place, and in the case of manufacturing carbon nanomaterials, to provide a substrate on which the carbon material can grow on. Regarding the first role, Toyota et al. [5] has shown that the efficiency of plasma decomposition of *n*-dodecane increases with the number (*n*) of aerial-antenna igniters that have a separation of ¼ wavelength spacing up to $n = 6$ and falls at $n = 7$. A number of mechanisms have been proposed for this behavior: electromagnetic power loss by the resonant structure [5], and a rate-limiting process due to competing mass transport in and out of the aerial-antenna igniter reaction zone [20] (Fig. 2d). Both mechanisms have a degree globe volume, but do not include discrete bubble formation and their electrification. This aspect is explored further in Sect. 2.3.3.

2.3.3 Bubble Formation and Its Electrification at a Microwave Aerial-Antenna Igniter Surface

In 2007, Mukasa et al. reported upon fast camera (Photron Fastcam-512PCI: 400 fps) imaging of *n*-dodecane plasma bubbles at a pressure of to 100 hPa [24]. Later in 2010 [25] and 2013 [26] Hattori el al used a high-speed camera (GX-1 NAC: 2000 fps) to investigate bubble formation in water at 7 and 20 kPa. In each investigation an open-end coaxial microwave transmission-line was used to inject the microwave power. In reference [25–26] the open-end coaxial transition-line that had an inner metal conductor diameter of $a = 0.3$ to 1.1 cm and a dielectric of $b = 0.9$ cm to 1.5 cm diameter. Their water electric field stress calculations for 1 W and measured bubble radius are reconstructed in Fig. 3 using units of V.cm^{-1} and cm, respectively.

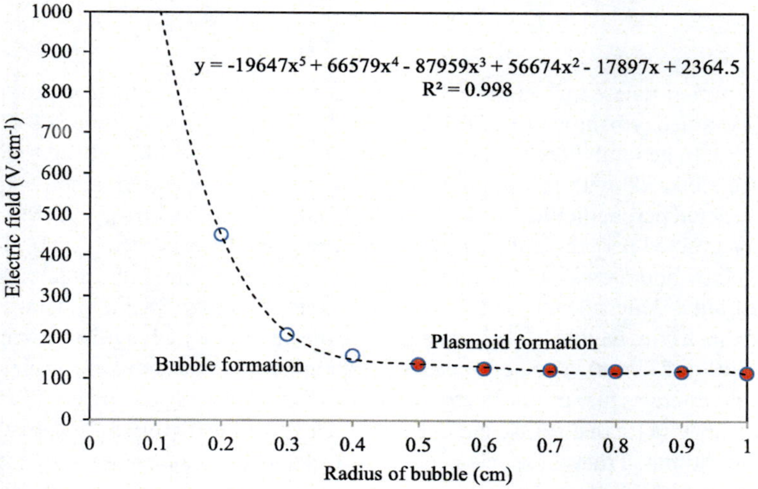

Fig. 3 Electric field stress at open-end transmission-line electrode as a dependency of bubble radius Hattori et al.[26]. Reported here as data points corresponding the following conditions: open circles = no plasmoid, and black circles = plasmoid. The plasmoid visual inception point is at 0.5 cm bubble radius. Annotated over the data points is an Excel polynomial 5th order curve fitting where $R^2 = 0.988$

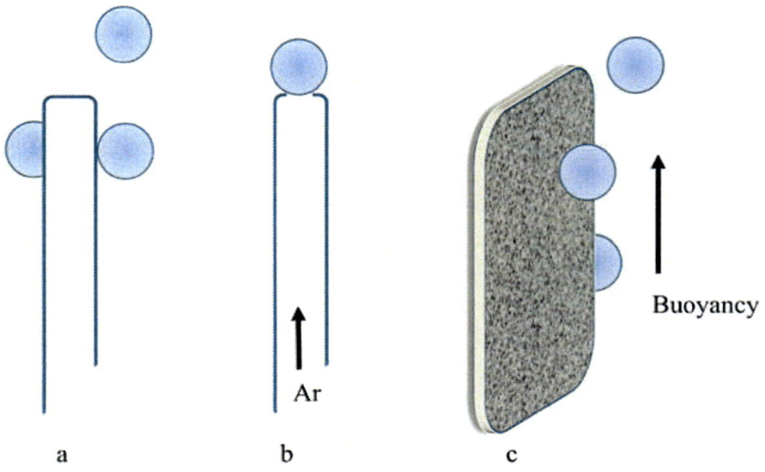

Fig. 4 **a–c** Bubble formation on aerial-antenna igniter (**a**), additive bubble formation on and through antenna (**b**), and additive bubble formation using the large area, porous quartz bubble control plate (**c**)

To represent the graph data points mathematically the Microsoft Excel curve fitting toolkit is used, and where it was found that a polynomial function with a order of 4–5 gave the best fit ($R^2 = 0.952$ and 0.998, respectively). Using this treatment it is seen that initial phase of water bubble growth the electric field rapidly falls is a point where the bubble becomes electrified to a produce a plasmoid. This visual plasmoid inception point approximates to 0.5 cm bubble diameter with a calculated electric field stress of the order of 150 $V.cm^{-1}$. Beyond this point the bubble growth rate slowly increases with reducing electric field. A number of possible mechanism leading to this electrification have been put forward [6, 15 and 24–26] which includes the following a number development stages. Firstly, in the initial stage of bubble growth where it adheres to the aerial-antenna igniter via a water film (bubble membrane) the bubble is heated by dielectric heating, leading a second stage that involves the internal vapor becoming supersaturated with respect to the surrounding liquid temperature. Thirdly, upon further dielectric heating the bubble membrane is vaporized (Fig. 4a) thereby allowing the supersaturated vapor to directly react with the rough metal surface, at which point the electric flied is of sufficient strength to breakdown the gas into electrons and ions that form the plasmoid.

2.3.4 Carbon-Based Nanoparticle Formation at a Microwave Aerial-Antenna Igniter Surface

When microwaves are used to irradiate a hydrocarbon liquid in the presence of an aerial-antenna igniter the production of H_2 and its low carbon carbides is obtained along with the condensation of carbon-based NMs on the metal igniter surface [5, 6, and 15–17]. The formation pathway of both gas and solid may be broadly understood by including a heterogeneous phase to the bubble-to-plasmoid transition process as outlined in Sect. 2.3.3. This multi-phase description is not intended to provide a conclusive picture, but rather a starting point for further experimentation. Given this caveat, it is generally acknowledged that within the vapor phase where rapidly changing vapor supersaturation conditions occur, particle nucleation is thermodynamically favorable for molecules to condense in the vapor and at solid surfaces, where on ongoing collisions lead to particle growth at these locations [33]. In the plasma state, the ionization and dissociation of the parent NM becomes kinetically favorable [23].

2.3.5 Additive or Supplementary Bubble Use

To improve power efficiency and stability of the targeted In-liquid reaction within a DMOPR, additive, or supplementary bubbles may be induced and where the buoyancy of the gas bubbles provides a means of gas–liquid mixing. For example, in 2014 Lebedev et al. [16] injected argon gas into the liquid hydrocarbon precursor through the aerial-antenna igniter to produces a stream of gas bubbles (Fig. 4b). Another approach is to re-circulate the hydrocarbon precursor liquid, with an initial charge of helium carrier gas, into a large area porous quartz bubble control plate [17] (Fig. 4c). This second (bubble control plate) approach provides a greater degree of bubble–liquid mixing and was originally investigated for a metal slot-antenna at the end of TE_{10}-mode waveguide for the production of bubbles within water [34].

2.4 *Microwave Susceptors*

The incorporation of susceptors into a microwave process is often termed 'Hybrid heating' [22]. This is because microwave susceptors have the ability to absorb electromagnetic radiation in the 2.45 GHz range and covert this energy into heat by re-emitting the energy in the form of conduction or infrared radiation. Table 1 demonstrates that the hybrid heating process is employed in the solid-state chemistry domain. Typically two microwave susceptor materials are used: aluminum block and foil ($\sigma = 3.69 \times 10^7$ Ω. m^{-1} at 2.45 GHz) and graphite ($\sigma \sim 3.69 \times 10^6$ Ω. m^{-1} at 2.45 GHz [35, 36]). Placing microwave transparent hydrocarbon precursor on aluminum block/foil or graphite provides a simple indirect thermal mechanism. However, the advantage of mixing the susceptor with the hydrocarbon precursor is that it enhances direct heating within the hydrocarbon precursor, but at the expense of chemical contamination, which may require post-process decontamination. To prevent susceptor contamination, separation between these materials by placing the precursor within a glass vial [10]. The use of a susceptor can lead to a thermal runaway due to the continuous increase of the microwave energy absorption with temperature rise, to avoid this scenario a system tailored to solid or powdered susceptor needs to be used.

2.5 *Chemical Catalyst*

Table 1 reveals that chemical catalysts are commonly used in the solid-state chemistry production of carbon based nanostructures. The catalyst material include: Magnetite (Fe_3O_4) [7], ($Fe(NO_3)_3{\cdot}9H_2O$) [9], and Ferrocene ($Fe(C_5H_5)$) [10, 11]. Magnetite also strongly interacts with the microwave electromagnetic field and therefore is also classed as a susceptor (tan $\delta = 0.02$ [32]. These materials not only provide nucleation sites for carbon growth but are also incorporated into the final nanostructure product which may [5], or not be desirable. In addition, porous silica, coated with transition metals Mo and Co have been used in cyclohexane In-liquid plasma production of H_2 and MWNCT [9].

3 Nanoparticles Formation Due to Drilling Within Single and Multi-Mode Cavity

In this section NP formation within a fireball generated in a purpose built multi-mode cavity plasma reactor [3, 37–41] and DMOPR [4] is considered as a separate microwave technology from that used for the synthesis of functional carbon-based nanostructures within the DMOPR. The difference in these two processing approaches is that a localized hot-spot (plasma) is produced by using an open-end coaxial transmission line (deployed as a monopole antenna), within a mono-mode microwave cavity to drill holes in a concrete, glass and metal surfaces [3, 4, 37 and 39] and subsequently within a DMOPR to drill into bone [4]. In references [3, 4, and 40–41] it has been proposed that the drilling action is caused by thermal runway process in the hot-spot; that is the energy released in the reaction has a positive feedback (Fig. 5a). It is also observed that when the drill bit is pulled out of the drilled hole [40, 41], a plasmoid is ejected (Fig. 5b) which subsequently forms a buoyant fireball as long the microwave irradiation is turned-on (Fig. 5c).

For a silicon substrate, fireballs are observed to have self-sustaining lifetimes of the order of 0.03 s that contains NPs with a mean size of 50 nm and mass density of 10^{-4} mg cm^{-3} as measured by synchrotron X-ray scattering. In addition these self-sustaining lifetimes are also consistent with dc generated water based plasmoids [42, 43], but are well beyond normal atmospheric plasma relaxation times (typically microseconds or less). However it is noted that that the self-sustaining lifetime are much shorter than self-sustaining lifetimes of naturally occurring ball-lighting, typically 1–2 s [44, 45].

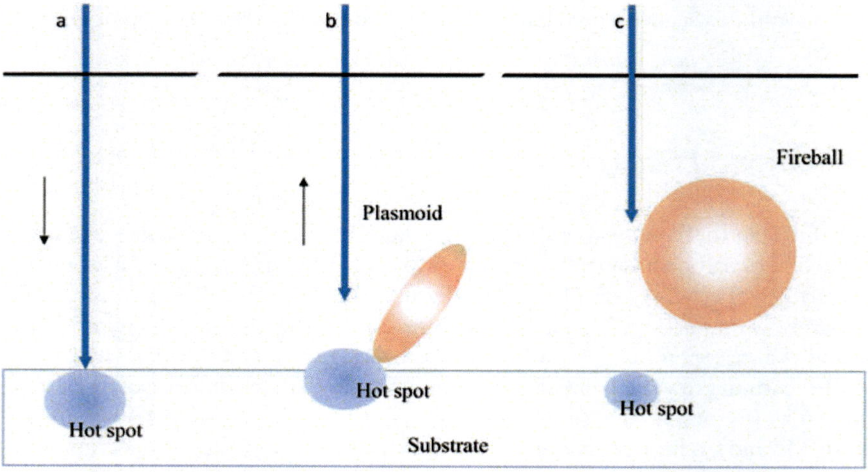

Fig. 5 **a–c** Microwave plasma drilling (**a**), microwave reverse drilling with the ejection of plasmoid (**b**), and the formation of buoyant self-sustaining fireball (**c**)

To account for the microwave generated fireball self-sustaining lifetime, Mitchell et al. [41] has invoked Abrahamson and Dinniss model of ball-lighting striking the ground in which silicon and carbon material are converted into an assortment of fused NP and dendrimer-like structure that contain silicon metal, silicon monoxide, and silicon carbide [46]. A fireball formed in this way may be considered to be classed as a dusty-plasma where the particles undergo an exothermic oxidation reaction ($\Delta H < 0$, see Fig. 1), the heat from which sustains the fireball. In the microwave fireball case however, the self-sustaining lifetime arises due to the oxidation reaction starts in the microwave illumination period and hence the fireballs lifetime is shortened accordingly.

4 Hydrogen and Nanomaterials Collection

In 2004, Sabot and Schlabach [1] considered that for advance commercial microwave plasma reactors, that the collection and purification stage after NM syntheses was poorly developed and was the main bottle neck in achieving high NM product yields. For the DMOPR processing papers [5–15 and 17–19] discussed here their primary aim was to present the proof-of-principle of H_2 and/or carbon-based NM synthesis. Thus emphasis on the collection and the purification stage was limited. For completeness this section provides brief overview of the collection processes reported in above papers.

In the case of DMOPR generation of H_2 and its low-carbon carbides, the gas was collected by displacement above the hydrocarbon precursor liquid [5, 6, and

15–19]. Chemical analyses of the gas was performed and reported upon, but drying and purification of the gas was not remarked upon.

For the extraction of Fe_3O_4 NP from the microwave plasma irradiated solid-phase magnetite -polystyrene mixture, the collection process is more complex. In this case, the raw product was ultrasonically washed in acetone for 5 min to allow the NP to disperse in the solvent and removed by filtering [7].

Post microwave irradiation extraction of nanosheets from solid polyethylene required an ethanol ultrasonic bath step for 10 min to allow the product to disperse in the solvent and be removed [6]. The NM was found to be unstable and require thermal treatment under an inert atmosphere to stabilize the crystallinity of the final product.

The DMOPR processed ZnO-Zn NP reported by Irzh et al. [12] where removed from their glass substrates washed in a 2-propanol ultrasonic bath for 30 min using under an argon atmosphere to disperse the final product in the solvent prior to analysis.

Extensive nanostructure the characterization of NMs microwave plasma treatment of naphthalene [10], RH [11], ZnO [13] and carbon-based NM [14], however little detail off the collection process was reported. It is also note that no NP collection process was reported in [36–39] as all particle size measurements were performed by Synchrotron X-Ray scattering.

5 Summary

This paper has reviewed the fundamental principles of rapid prototyping of functional carbon-based nanomaterials within converted DMOPRs. The nanomaterials have a grain size of < 100 nm, in at least one dimension, and range from single and multi-walled carbon nanotubes, onion-like nanostructures, fullerene, and graphene sheets. Based on reports from the technical literature; the review highlights the following complex heterogeneous reaction pathways:

- Liquid linear chain hydrocarbons (i.e. *n*-dodecane) preferentially undergo C–C β-scission to simultaneously produce H_2 and low-grade carbides and solid graphite at the metal aerial-antenna igniters. This observation differs from thermal (850–1100 K) cracking of *n*-dodcane where CO products are formed [28, 29]. Altering the hydrocarbon precursor from a linear structure to cyclo-, or phenyl group carbon-based structures (i.e. cyclohexane) produces CNT and MWCNT at the surface of metal aerial-antenna igniters. Electrification of the hydrocarbon precursors originate by vaporization of the liquid to produce gaseous bubbles at the surface at the metal aerial-antenna and then produces plasmoids cracked hydrocarbon reactive species that form gaseous and solid nanomaterials. Collection of the gaseous by-product is performed by displacement above the liquid hydrocarbon precursor.
- The production of nanomaterials by microwave irradiation of solid-state hydrocarbon mixed with catalyst and susceptors may be performed within a glass vial

where the plasma reaction takes place. The process is limited by the amount reactant within the sealed vial. In addition product collection and separation from the catalyst and susceptors requires further processing.

- Direct plasma interaction with solid-state hydrocarbon precursor that is mixed with catalyst and susceptors may be performed within plasma. As with the sealed vial process, product collection and separation requires further processing.
- Microwave plasma (DMOPR or single-mode waveguide) drilling and reverse drilling lead to a local thermal runaway and the ejection of plasmoid (fireball) with self-sustaining lifetimes of 0.03 s.

Acknowledgements This publication has emanated from research supported Science Foundation Ireland (SFI) I-Form Centre, under Grant. Number 16/RC/3872. The Authors declare that there is no conflict of interest regarding the publication of this paper.

References

1. D.V. Szabó, S. Schlabach, Microwave plasma synthesis of materials-from physics and chemistry to nanoparticles: a materials scientist's viewpoint. Inorganics **2**, 468–507 (2004)
2. A. Loupy, R.S. Varma, Microwave effects in organic synthesis: mechanistic and reaction medium considerations. Chem. Today. **24**(3), 36–39 (2006)
3. E. Jerby, V. Dikhtyar, O. Aktushev, U. Grosglick, Microwave drill. Science **298**, 57–589 (2002)
4. S. Das, A.K. Sharma, Microwave drilling of materials. Barc Newlett. **329**, 15–21 (2012)
5. H. Toyota, S. Nomura, S. Mukasa, A practical electrode for microwave plasma processes. Int. J. Mater. Sci. Appl. **2**(3), 83–88 (2013)
6. S. Nomura, H. Yamashita, H. Toyota1, S. Mukasa, Y. Okamura, Simultaneous production of hydrogen and carbon nanotubes in a conventional microwave oven. International Symposium on Plasma Chemistry (ISPC19). Bochum, Germany. 65, (2009)
7. P.H. Talemi, G.P. Simon, A simple microwave-based method for preparation of Fe_3O_4/carbon composite nanoparticles. Mater. Lett. **64**(15), 1684–1687 (2010)
8. P.H. Talemi, G.P. Simon, Preparation of graphene nanowalls by a simple microwave-based method. Carbon **48**(14), 3993–4000 (2010)
9. N. Kure, M.N. Hamidon, S. Azhari, N.S. Mamat, H.M. Yusoff, B.M. Isa, Z. Yunusa, Simple microwave-assisted synthesis of carbon nanotubes using polyethylene as carbon precursor. J. Nanomater. 2017, 4 (2017)
10. R. Bajpai, L. Rapoport, K. Amsalem, H.D. Wagner, Rapid growth of onion-like carbon nanospheres in a microwave oven. Crst EngComm. **18**(2), 230–239 (2016)
11. M. Asnawi, S. Azhari, M.N. Hamidon, I. Ismail, I. Helina, Synthesis of carbon nanomaterials from rice husk via microwave oven. J. Nanomater. 2018 (2018)
12. A. Irzh, I. Genish, L. Klein, L.A. Solovyov, A. Gedanken, Synthesis of ZnO and Zn nanoparticles in microwave plasma and their deposition on glass slides. Langmuir **26**(8), 5976–5984 (2010)
13. R.T. Raj, K. Rajeevkumar, Synthesis of ZnO nanostructures using domestic microwave oven based remote plasma deposition system. Nanosci. Nanotechnol. **2**(3), 66–70 (2012)
14. R. Singh, A.L.L. Jarvis, Microwave plasma enhanced chemical vapour deposition growth of carbon nanostructures. S. Afr. J. Sci. **106**(5–6), 4 (2010)
15. S. Nomura, H. Toyota, S. Mukasa, H. Yamashita, T. Maehara, A. Kawashima, Production of hydrogen in a conventional microwave oven. J. Appl. Phys. **106**(7), 073306 (2009)

16. Y.A. Lebedev, I.L. Epstein, V.A. Shakhatov, E.V. Yusupova, V.S. Konstantinov, Spectroscopy of microwave discharge in liquid C7–C16 hydrocarbons. High Temp. **52**(3), 319–327 (2014)
17. A.A. Mochtar, S. Nomura, S. Mukasal, H. Toyota, K. Kawamukai, K. Uegaito, F. Syahrial, A novel method for producing hydrogen from a hydrocarbon liquid using microwave in-liquid plasma. J. Energy Power Eng. **10**, 335–342 (2016)
18. S. Nomura, A.E.E. Putra, S. Mukasa, H. Yamashita, H. Toyota, Plasma decomposition of clathrate hydrates by 2.45 GHz microwave irradiation at atmospheric pressure. Appl. Phy. Exp. **4**(6), 099201 (2011)
19. I. Rahim, S. Nomura, S. Mukasa, H. Toyota, Decomposition of methane hydrate for hydrogen production using microwave and radio frequency in-liquid plasma methods. Appl. Therm. Eng. **90**, 120–126 (2015)
20. V.J. Law, D.P. Dowling, Converting a microwave oven into a plasma reactor: a review. IJCE. 2018, 12 (2018)
21. V.J. Law, D.P. Dowling, Domestic microwave oven and fix geometry waveguide applicator processing of organic compounds and biomaterials: a review. GJRE(A). **18**(2), Version 1.0. (2018)
22. V.J. Law, D.P. Dowling, The domestic microwave oven as a rapid prototyping tool. Analy. Chem. Indian J. **18**(2), 19 (2018)
23. L. Perreux, A. Loupy, A tentative rationalization of microwave effects in organic synthesis according to reaction medium and mechanism considerations. Tetrahedron **57**, 9199–9223 (2001)
24. S. Mukasa, S. Nomura, H. Toyota, Observation of microwave in-liquid plasma using high-speed camera. Jpn. J. Appl. Phys. **46**, 6015 (2007)
25. Y. Hattori, S. Mukasa, H. Toyota, S. Normura, Optimization and analysis of shape of coaxial electrode for microwave plasma in water. J. Appl. Phys. **107**, 063305 (2010)
26. Y. Hattori, S. Mukasa, H. Toyota, S. Normura, Electrical Breakdown of microwave plasma in water. Current Appl. Phys. **13**, 1050–1054 (2013)
27. V.J. Law, D.P. Dowling, Microwave oven plasma reactor moding and its detection. The 12th Chaotic modeling and simulation international conference. CHAOS 2019. Springer Proceedings in Complexity. Springer, Cham. chap 14. 157–179. ed C.H. Skiadas, Y. Dimotikalis (2020)
28. X. You, F.N. Egolfopoulos, H. Wang, Detailed and simplified kinetic models of *n*-dodecane oxidation: the role of fuel cracking in aliphatic hydrocarbon combustion. Proc. Combust. Inst. **32**, 493–410 (2009)
29. L. Zhao, T. Yang, R. Kasier, T.P. Troy, M. Ahmed, Combined experimental and computational study unimolecular decomposition of JP-8 fuel surrogates. J. Phys. Chem. A **121**, 1281–1291 (2017)
30. K.K. Rana, S. Rana, Microwave reactors: a brief review on its fundamental aspects and applications. Open Access Library J **1**, 20 (2014). http://dx.doi.org/https://doi.org/10.4236/oalib.1100686
31. Y.S. Lee, F. Malek, E.M. Cheng, Wei-Wen Liu, F. H. Wee, M. N. Iqbal, L. Zahid. M. S. Mezan, F. S. Abdullah, M. Othman, Dielectric properties of rice husk/carbon nanotubes composites in ku-band. 25–28, 75–78 (2014)
32. G. Saito, T. Akiyama, Nanomaterial. Synthesis using plasma generation in liquid. J. Nanomater. **2015**, 20 (2015)
33. M.T. Swihart, Vapor-phase synthesis of nanoparticles. Curr. Opin. Colloid Interface Sci. **8**, 127–133 (2003)
34. T. Ishijima, H. Sugiura, R. Satio, H. Toyada, H. Sugai, Efficient production of microwave bubble plasma in water for plasma processing in liquid. Plasma, Sources Sci. Technol. **19**(29), 015010 (2010)
35. M. Bhattacharya, T. Basak, A review on the susceptor assisted microwave processing of materials. Energy. **97**, 306–338 (2016)
36. M. Hotta, M. Hayashi, M.T. Lanagan, D.K. Agrawal, K. Nagata, Complex permittivity of graphite, carbon black and coal powders in the ranges of X-band frequencies (8.2 to 12.4 GHz) and between 1 and 10 GHz. ISIJ International. **51**(11), 1766–1772 (2011)

37. E. Jerby, Y. Meir, R. Jaffe, I. Jerby, Food cooking by microwave-excited plasmoid in air atmosphere. Proceedings of 14th International Conference on Microwave and High Frequency Heating. Nottingham, UK. 27–30 (2013)
38. V. Dikhtyar, E. Jerby, Fireball ejection from a molten hot spot to air by localized microwaves. Phys. Rev. Lett. **96**, 045002 (2006)
39. E. Jerby, Detachment of a microwave-excited plasmoid from molten glass. IEEE Trans. Plasma Sci. **39**(11), 2198–2199 (2011)
40. Y. Meir, E. Jerby, Z. Barkay, D. Ashkenazi, J.B. Mitchell, T. Narayanan, N. Eliaz, Jean-Luc LeGarrec, M. Sztucki, O. Meshcheryakov, Observations of ball-lightning-like plasmoids ejected from silicon by localized microwaves. Materials **6**, 4011–4030 (2013)
41. J.B.A. Mitchell, J.L. LeGarrec, M. Sztucki, T. Narayanan, V. Dikhtyar, E. Jerby, Evidence for nanoparticles in microwave-generated fireballs observed by synchrotron X-ray scattering. Phys. Rev. Lett. **100**. 065001 (2008)
42. S.E. Dubowsky, D.M. Friday, K.C. Peters, Z. Zhao, R.H. Perry, B.J. McCall, Mass spectrometry of atmospheric-pressure ball plasmoids. Int. J. Mass Spectrom. **376**, 39–45 (2015)
43. S.E. Dubowsky, B. Deutsch, R. Bhargava, B.J. McCall, Infrared emission spectroscopy of atmospheric-pressure ball plasmoids. J. Mol. Spectrosc. **32**, 1–8 (2016)
44. H. Ofuruton, Y.H. Ohtsuki, Experimental research on ball lightning. Il Nuovo Cimento C. **13**(4), 761–768 (1990)
45. A.L. Pirozerski, S.E. Emelin, Long-living plasmoids generation by high-voltage discharge through thin conducting layers. Proceedings of the 9th International Symposium on Ball Lightning, ed. G.C. Dijkhuis. Eindhoven, The Netherlands, pp. 180–190 (2006)
46. J. Abrahamson, J. Dinniss, Ball lightning caused by oxidation of nanoparticle networks from normal lightning strikes on soil. Nature **403**, 519–521 (2000)

The Relationship Between the Euler Characteristic and the Spectra of Graphs and Networks

Michał Ławniczak, Pavel Kurasov, Szymon Bauch, Małgorzata Białous, and Leszek Sirko

Abstract A relationship between the Euler characteristic of a quantum graph and its spectrum is a very new subject of the theoretical and experimental investigations. The Euler characteristic $\chi = |V| - |E|$, where $|V|$ and $|E|$ are the numbers of vertices and edges of a graph, determines the number β of independent cycles in it. The most important features of the graph spectrum, the number and density of the energy eigenvalues are determined by the graph total length $\mathcal{L}$ according to the Weyl's law. In the recent paper M. Ławniczak et al., Phys. Rev. E **101**, 052320 (2020), have shown theoretically and experimentally that to determine the Euler characteristic of a simple quantum graph without knowing its number of vertices and edges it is enough to measure a finite sequence of the lowest eigenenergies $\lambda_1, \ldots, \lambda_N$. In this article the above investigations are supported by the new theoretical calculations of the Euler characteristic.

1 Introduction

The very beginning of the graph theory is connected with the article [1] of Leonhard Euler, published in 1736, in which the problem of seven bridges in Konigsberg was considered. In 1936 Pauling [2] introduced the idea of graphs to physics to describe the motion of a quantum particle in a physical network. Now this approach is known as the quantum graph model. Quantum graphs are used to simulate e.g. mesoscopic quantum system [3, 4], quantum wires [5] and optical wave guides [6]. 12 years latter Feynman [7] introduced diagrams (graphs) as pictorial representation of the mathematical expressions describing the behavior and interaction of subatomic particles.

M. Ławniczak · S. Bauch · M. Białous · L. Sirko (✉)
Institute of Physics, Polish Academy of Sciences, Aleja Lotników 32/46, 02-668 Warsaw, Poland
e-mail: sirko@ifpan.edu.pl

P. Kurasov
Department of Mathematics, Stockholm University, 106 91 Stockholm, Sweden

C. H. Skiadas et al. (eds.), *13th Chaotic Modeling and Simulation International Conference*, Springer Proceedings in Complexity,
https://doi.org/10.1007/978-3-030-70795-8_36

The theory of quantum graphs is still being developed, see, e.g., [8–12]. The metric graph $\Gamma = (V, E)$ consist of edges $e \in E$ connected at the vertices $v \in V$. Each edge is an interval on the real line $\mathbb{R}$ having the length l_e and the vertices are defined as the unions of edge endpoints. The Laplace operator $L(\Gamma) = -\frac{d^2}{dx^2}$ acting in the Hilbert space of square integrable functions on the metric graph Γ is uniquely determined by this graph. If the graph has only Neumann (called also standard, natural, Kirchhoff) vertex boundary conditions: functions are continuous at vertices and the sums of their oriented derivatives at vertices are equal zero, then the Laplacian is self-adjoint and its spectrum is pure discrete [11]. The operator is also non-negative and has constant as the eigenvalue zero eigenfuction.

In this article we continue the presentation of the results on topology of quantum graphs and microwave networks, initiated in [13]. We will confirm that recovering of Euler characteristic of the graph without seeing it, i.e. knowing number of its edges and vertices, is possible on the basis of measuring a small number of the lowest eigenenergies of such a graph. It may be also possible to determine whether the graph is planar or fully connected. Our statements are supported by new theoretical calculations.

The measurements were done using microwave networks that simulate quantum graphs [14–19]. This is possible because the one-dimensional Schrödinger equation describing quantum graphs is formally equivalent to the telegrapher's equation for microwave networks [14, 17]. It should be noted that microwave networks are unique as they allow for the experimental simulation of quantum systems having, respectively, all three symmetry types considered in random-matrix theory (RMT): systems with preserved time reversal symmetry (TRS) represented by Gaussian orthogonal ensemble (GOE) [13–16, 18, 20]; systems with preserved TRS and half-spin represented by Gaussian symplectic ensemble (GSE) [21]; systems without TRS represented by Gaussian unitary ensemble (GUE) [14, 19, 22–26].

Recently, in experiments with the use of microwave networks, the usefulness of missing level statistics [25] and the existence of the graph that do not obey the standard Weyl's law, non-Weyl graphs, have been demonstrated [20].

2 The Spectrum Formula for the Euler Characteristic

The total length $\mathcal{L} = \sum_{e \in E} l_e$ and the Euler characteristic $\chi = |V| - |E|$ are the most important features of a metric graph $\Gamma = (V, E)$. The total length $\mathcal{L}$ determines asymptotically eigenvalues of a graph according to Weyl's law:

$$\lambda_n = \left(\frac{\pi}{\mathcal{L}}\right)^2 n^2 + \mathcal{O}(n), \tag{1}$$

where $\lambda_n = k_n^2$ (square of the wave vector) and $\mathcal{O}(n)$ is a function which in the limit $n \to \infty$ is bounded by a constant times n.

The Euler characteristic χ allows determining another important quantity characterizing the graph, which is the number of independent cycles β in it:

$$\beta = |E| - |V| + 1 \equiv 1 - \chi\,, \tag{2}$$

The number of independent cycles tells how much the graph differs from a tree and is equal to the number of the edges that must be removed to turn the graph into a tree.

It might seem that determining the total length and the Euler characteristic would require knowing of the whole spectrum, all eigenenergies of the graph. This natural mathematical approach leads to the following precise formulas [27, 28]:

$$\mathcal{L} = \pi \lim_{\mathrm{n}\to+\infty} \frac{n}{k_n} \tag{3}$$

$$\chi = X(t)|_{t\geq t_0} := 2 + 8\pi^2 \sum_{k_n\neq 0} \frac{\sin(k_n/t)}{(k_n/t)[(2\pi)^2 - (k_n/t)^2]}, \tag{4}$$

where $t_0 = 1/2l_{min}$, and l_{min} is the length of the shortest edge of a graph.

The knowledge of the whole spectrum allows reconstructing the metric graph if the lengths of its edges are rationally independent [29–31] and thus ensures a positive answer to the question posed by Kac [32] modified for quantum graphs as “Can one hear the shape of a graph?” [16, 29, 30].

In the real world of the physical measurements it is not possible to determine the entire spectrum of the tested system. In the case of microwave networks, the openness and internal absorption of the system limit the number of eigenfrequencies available in measurement to well below a thousand. There is also generally no guarantee that edge lengths are rationally independent, so it is very important to investigate whether the total length $\mathcal{L}$ and the Euler characteristic can be reconstructed directly from the limited experimental spectrum without precise information about the graph. This is much easier for the Euler characteristic because it is an integer (often negative) thus to determine it accurately it is enough to know the right-hand side of (4) with an accuracy better than 1/2. Therefore, in the rest of the article we will focus on the Euler characteristic.

The explicit formulas connecting the Euler characteristic with the spectrum of a quantum graph were derived in the articles [27, 28]. It turned out, however, that they converge very slowly requiring too much eigenenergies from the experimental point of view. The series in formula (4) converges much faster, what was shown together with the derivation of this formula in [13].

To estimate a minimum number of eigenenergies of the graph (resonances in the microwave networks) ensuring that the calculated value of χ will be within the limit $\pm 1/2$ of its true value we will analyze the following formula:

$$\epsilon = |X(t) - X_K(t)||_{t=t_0} = \left| 8\pi^2 \sum_{n=K+1}^{\infty} \frac{\sin(k_n/t_0)}{(k_n/t_0)[(2\pi)^2 - (k_n/t_0)^2]} \right|, \tag{5}$$

where

$$X_K(t) = 2 + 8\pi^2 \sum_{n=1}^{K} \frac{\sin(k_n/t)}{(k_n/t)[(2\pi)^2 - (k_n/t)^2]}. \tag{6}$$

To avoid any problem with the denominator in (5) we need

$$k_{K+1} > 2\pi t_0 \tag{7}$$

Taking into account the obvious inequality (see (1))

$$k_n^2 \geq \left(\frac{\pi}{\mathcal{L}}\right)^2 (n + 1 - |V|)^2, \tag{8}$$

we get that the condition (7) is fulfilled if

$$K > |V| - 1 + \frac{\mathcal{L}}{\ell_{\min}}. \tag{9}$$

Now ϵ can be estimated by

$$\begin{aligned}
\epsilon = |X(t_0) - X_K(t_0)| &\leq 8\pi^2 \sum_{n=K+1}^{\infty} \frac{|\sin(k_n/t_0)|}{(k_n/t_0)[(k_n/t_0)^2 - (2\pi)^2]} \\
&\leq 8\frac{(\mathcal{L}t_0)^3}{\pi} \sum_{n=K+1}^{\infty} \frac{1}{(n+1-|V|)[(n+1-|V|)^2 - 4\mathcal{L}^2 t_0^2]} \\
&\leq 8\frac{(\mathcal{L}t_0)^3}{\pi} \int_K^{\infty} \frac{dx}{(x+1-|V|)[(x+1-|V|)^2 - 4\mathcal{L}^2 t_0^2]} \\
&= \frac{\mathcal{L}t_0}{\pi} \log \frac{(K+1-|V|)^2}{(K+1-|V|)^2 - 4\mathcal{L}^2 t_0^2},
\end{aligned} \tag{10}$$

and finally we get:

$$K > |V| - 1 + 2\mathcal{L}t_0 \left[1 - \exp\left(\frac{-\epsilon\pi}{\mathcal{L}t_0}\right)\right]^{-1/2}. \tag{11}$$

When the total length $\mathcal{L}$ of the graph is much greater than its shortest edge the inequality (11) can be approximated by

$$K > |V| - 1 + \frac{2}{\sqrt{\epsilon\pi}} \left(\frac{\mathcal{L}}{2l_{min}}\right)^{3/2}. \tag{12}$$

3 Experimental Setup and Methodology of Measurements

To test the formula for the Euler characteristic (4) we carried out the one-port measurements of the scattering matrices $S(\nu)$ for planar and non-planar microwave networks simulating simple quantum graphs. A graph is simple if there is only one edge between any two vertices.

Figure 1 shows the pictures of the investigated microwave networks and the schemes of the quantum graphs that they simulate. The planar graph $\Gamma = (4, 6)$ with $|V| = 4$ and $|E| = 6$ (Fig. 1a) is simulated by the microwave network (Fig. 1b) of the total optical length $\mathcal{L} = 1.494 \pm 0.006$ m and $l_{min} = 0.155 \pm 0.001$ m. According to the definition its Euler characteristic $\chi = |V| - |E| = -2$, and $t_0 = \frac{1}{2l_{min}} \approx 3.23\ \mathrm{m}^{-1}$.

In the panels (c) and (d) of Fig. 1 we present a scheme of the quantum graph $\Gamma = (5, 10)$ and simulating it microwave network. The Euler characteristic $\chi = |V| - |E| = -5$. The network total optical length is $\mathcal{L} = 3.949 \pm 0.010$ m and the length of the shortest edge is $l_{min} = 0.202 \pm 0.001$ m, so $t_0 = 2.48\ \mathrm{m}^{-1}$.

In Fig. 2a the experimental setup is shown. This is a typical setup for such measurements and consists of the Agilent E8364B vector network analyzer (VNA) and

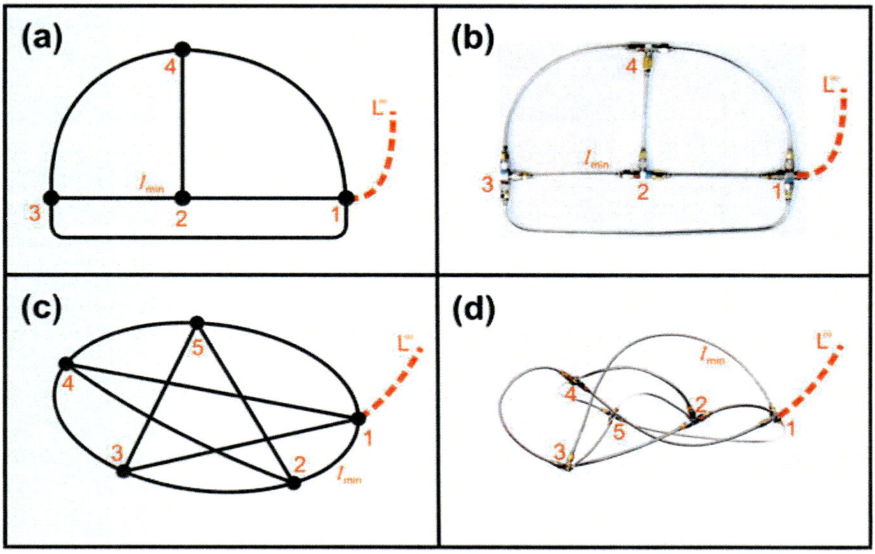

Fig. 1 Panels **a** and **b** show the schemes of a planar quantum graph $\Gamma = (4, 6)$ with $|V| = 4$ vertices and $|E| = 6$ edges and a microwave network with the same topology. Panels **c** and **d** show the schemes of a non-planar quantum graph $\Gamma = (5, 10)$ with $|V| = 5$ vertices and $|E| = 10$ edges and a microwave network with the same topology. The microwave networks were connected to the vector network analyzer with the flexible microwave cable, denoted in the figure by L^∞ which is equivalent to attaching an infinite lead to a quantum graph. The shortest edges of the graphs are also shown

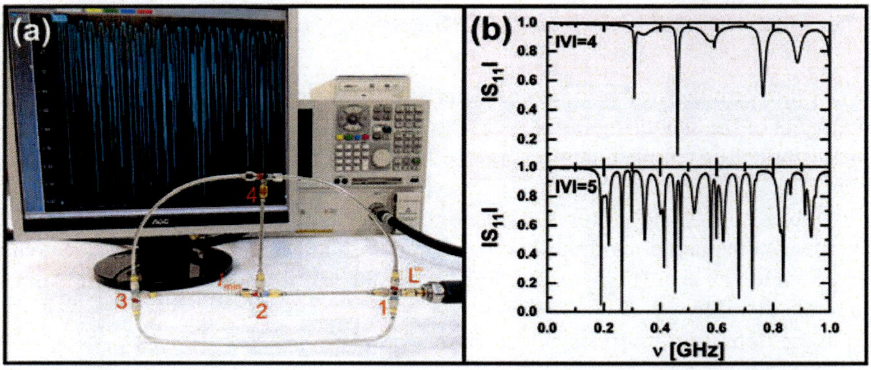

Fig. 2 **a** The experimental setup consisting of the microwave network simulating the quantum graph $\Gamma = (4, 6)$ connected by the high class flexible cable to the VNA. Fragments of the measured moduli of the scattering matrix $|S(\nu)|$ for the networks $\Gamma = (4, 6)$ with the total length $\mathcal{L} = 1.494$ m and $\Gamma = (5, 10)$ with $\mathcal{L} = 3.949$ m are shown in the upper and lower parts of the panel **b**, respectively. The increasing of the resonance density for the longer network is clearly seen

the HP 85133-616 high class flexible microwave cable that connects the VNA with measured network and is equivalent to attaching an infinite lead to the quantum graph [20]. Before each measurement, the VNA was calibrated with an Agilent 4691-60004 electronic calibration module to eliminate the cable influence on the measurement result.

Microwave networks are made of coaxial cables and junctions corresponding to edges and vertices of simulated quantum graphs. The coaxial cables consist of an inner conductor of a radius $r_1 = 0.05$ cm surrounded by dielectric material (Teflon) and an outer concentric conductor with an inner radius $r_2 = 0.15$ cm. The measured Teflon dielectric constant is $\varepsilon = 2.06$. Thus $\upsilon_{cut} = \frac{c}{\pi(r_1+r_2)\sqrt{\varepsilon}} = 33$ GHz is the cut-off frequency of the TE11 below which only the fundamental TEM can propagate in the cable [33, 34]. The physical length of the cables determine their optical length, which is also the edges length of the simulated quantum graphs, through relationship $l_{opt} = \sqrt{\varepsilon} l_{ph}$.

The measurements of the scattering matrix $S(\nu)$ were performed as function of the frequency (energy) in the range $0.2 - 11$ GHz. However, for the investigated networks, it was possible to identify the complete set of resonances in smaller, depending on their absorption, different ranges. To check the completeness of resonances sets the fluctuating part of the integrated spectral counting function $N_{fl}(\upsilon_i) = N(\upsilon_i) - N_{av}(\upsilon_i)$, that is the difference of the number of eigenfrequencies $N(\upsilon_i) = i$ for ordered frequencies $\upsilon_1 \leq \upsilon_2 \leq \cdots$ and the average number of eigenfrequencies $N_{av}(\upsilon_i)$ calculated for the analyzed frequency range. The resonances frequencies directly determined the real part of the wave vectors $\mathrm{Re}\, k_n = \frac{2\pi}{c}\nu_n$. The examples of the measured scattering matrix module $|S(\nu)|$ are shown in Fig. 2b.

4 Results

In Fig. 3 we present the calculation of the Euler characteristic performed for the planar graph $\Gamma = (4, 6)$ (Fig. 1a) which was simulated by the microwave networks (Fig. 1b) as a function of t for K equals 28 (solid red line), 106 (green broken line) and 171 (orange squares). In the calculation the formula (6) was applied. Since the identification of resonance positions in the experimental spectrum was possible up to the first 106 resonances (10.76 GHz) the results for $K = 171$ are obtained using numerically calculated spectrum of the graph $\Gamma = (4, 6)$. The black full line denotes the value of the Euler characteristic $\chi = -2$ and two black broken lines show the limits of the expected error of $\chi \pm \frac{1}{4}$. The brown vertical marker shows t_0 value. $K = 28$ was estimated from the inequality (11) as the minimum number of resonances needed to obtain the value of the Euler characteristic with the error within the limits $\epsilon = \pm 1/4$. It is easy to see that in fact 28 resonances are enough to get the correct value of the Euler characteristic. The red curve for $K = 28$ exhibits the clear plateau near the value -2 in the range $3\,\mathrm{m}^{-1} < t < 6\,\mathrm{m}^{-1}$. The Euler characteristics calculated for higher number of resonances exhibit such plateaus up to much bigger value of t, but it is irrelevant from the point of view of obtaining a correct value of it. This result is very important for experimental practice because it means that

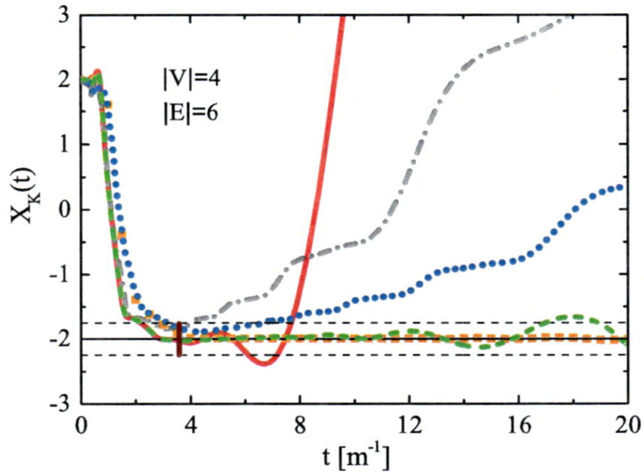

Fig. 3 The approximation function for the Euler characteristic $X_K(t)$ calculated for a planar microwave network with $|V| = 4$ vertices and $|E| = 6$ edges. The solid red line and broken green and red lines show the function $X_K(t)$ calculated from the formula (4) for the first $K = 28$ and $K = 106$ experimental, and $K = 171$ numerical (orange squares) resonances, respectively. The brown vertical marker shows the value of $t_0 = \frac{1}{2l_{min}}$ used for the evaluation of the required number of resonances $K = 28$ (see the formula (11)). For the comparison we also show the functions $X_K(t)$ (grey dot-dash and blue dot line) calculated from a much slowly converging formula (3) in [13] using the first $K = 28$ experimental and $K = 171$ numerical resonances. The black full line shows the expected value of the Euler characteristic $\chi = -2$. The black broken lines show the limits of the expected errors $\chi \pm 1/4$

it possible to gain significant information about quantum graphs topology from the measurements of a small number of theirs lowest eigenenergies. It should be also noticed that we don't need to know the total length $\mathcal{L}$ of the graph and the length of its shortest edge l_{min}, which are necessary to determine the convergence of (6), for the proper calculation of the Euler characteristic. In fact, until the plateau of $X_K(t)$ near certain integer is not observed we know that the number of resonances taken to calculation is insufficient and should be increased.

For comparison, in Fig. 3 we also show the Euler characteristic calculated using the formula (3) in [13], which is converging much more slowly. The grey dot-dash line denotes the $X_K(t)$ calculated using 28 experimental resonances and blue dot line is for 171 numerically calculated resonances. As expected these results approach the Euler characteristic value $\chi = -2$ of this graph definitely slower.

Another, crucial information about the quantum graph that can be obtained by analyzing the result of applying formula (4) concerns its planarity. According to Kuratowski's theorem [35] every non-planar graph contains as subgraphs K_5—the complete five-vertex graph or $K_{3,3}$—the complete bipartite graph with 3 plus 3 vertices. Such graphs have 6 or 4 cycles. The considered graph $\Gamma = (4, 6)$ has, formula (1), $\beta = 1 - \chi = 3$, so without seeing or having information about the number of its vertices and edges we know that the graph and the network simulating it are planar.

In Fig. 4 we demonstrate the calculation of the Euler characteristic performed for the non-planar graph $\Gamma = (5, 10)$ (Fig. 1c) which was simulated by the microwave network (Fig. 1d) as a function of t for K equals 74 (solid red line), 132 (green broken line) and 450 (orange squares). In this case we were able to determine the position of the first 132 resonances, up to about 5 GHz. A fragment of the network spectrum is shown in the lower part of Fig. 2b. Since the total length of this network is greater than the length of the planar network also the resonance density is greater in this case.

As in the case of Fig. 3, the black full line indicates the value of the Euler characteristic χ = -5, this line, and two black broken lines show the limits of the expected error of $\chi \pm\frac{1}{4}$. The brown vertical marker shows t_0 value. The resonances used in the calculations for K = 74 and 132 were taken from the measurements while for K = 450 they were obtained numerically. The value K = 74 was estimated from the inequality (11) as the minimum number of resonances necessary to obtain the value of the Euler characteristic with the error $\epsilon = \pm 1/4$. The examination of Fig. 4 shows that, similarly to the previous case, the minimum number of the resonances ($K = 74$) estimated from the inequality (11) is quite sufficient to determine the real value of the Euler characteristic.

Again, for comparison, we also show the Euler characteristic calculated using the formula (3) in [13]. The grey dot-dash line denotes the $X_K(t)$ calculated using 74 experimental resonances and blue dot line is for 450 numerically calculated resonances. It can be seen that in the case of the more complicated graph/network, it would be very difficult to obtain the real value of the Euler characteristic from this formula on the basis of experimental spectra. The calculation which used K = 450

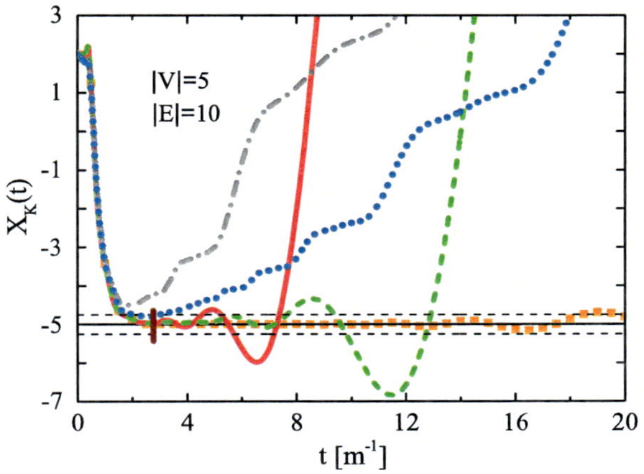

Fig. 4 The approximation function for the Euler characteristic $X_K(t)$ calculated for a non-planar microwave network with $|V| = 5$ vertices and $|E| = 10$ edges. The solid red line and broken green and red lines show the function $X_K(t)$ calculated from the formula (4) for the first $K = 74$ and $K = 132$ experimental, and $K = 450$ numerical (orange squares) resonances. The brown vertical marker shows the value of $t_0 = \frac{1}{2l_{min}}$ used for the evaluation of the required number of resonances $K = 74$ (see the formula (11)) For the comparison we again show the functions $X_K(t)$ (grey dot-dash and blue dot line) calculated from the much slowly converged formula (3) in [13] using the first $K = 28$ experimental and $K = 171$ numerical resonances. The black full line shows the expected value of the Euler characteristic $\chi = -2$. The black broken lines show the limits of the expected errors $\chi \pm 1/4$

eigenvalues, which barely allows this, would mean for this network that one has to identify all resonances in the frequency range 0–17 GHz, what taking into account the openness and absorption of microwave networks, is rather impossible. This indicates how crucial is to find a quickly converging formula.

Since the analyzed graph $\Gamma = (5, 10)$ is K_5 (five-vertex, fully connected) so according to Kuratowski's theorem it is non-planar. For both types, planar and non-planar, fully connected quantum graphs and microwave networks simulating them, it can be shown that:

$$|V| = \frac{3 + \sqrt{9 - 8\chi}}{2}. \quad (13)$$

Applying this formula to the investigated graphs we found that both of them are fully connected. In the case of the planar graph the Euler characteristic $\chi = -2$ and we get $|V| = 4$ and for K_5, $\chi = -5$, so $|V| = 5$. Thus knowing experimentally determined Euler characteristic we can find out whether the graph/network is fully connected without any additional information.

5 Conclusions

We extended our analysis of the Euler characteristic presented in [13] to higher frequency range using numerically calculated spectra of the analyzed graphs. On the basis of [13] and this work one can state that the Euler characteristic can be determined using a limited sequence of the lowest resonances of the microwave networks simulating quantum graphs. We show that from the experimental point of view the convergence of the formula for the Euler characteristic is a critical parameter. Moreover, we demonstrate that obtained in this way Euler characteristic may allow for identification whether a completely unknown graph/network is planar and also whether it is fully connected. In the case of the fully connected graphs and networks it is possible also to specify the number of their vertices and edges. Our results clearly show that the Euler characteristic is a new powerful tool for studying of simple quantum graphs and microwave networks.

Acknowledgements This work was supported in part by the National Science Centre, Poland, Grant No. 2016/23/B/ST2/03979, the Swedish Research Council (Grant D0497301) and the Center for Interdisciplinary Research (ZiF) in Bielefeld in the framework of the cooperation group on *Discrete and continuous models in the theory of networks*.

References

1. L. Euler, Comment. Acad. Sci. U. Petrop. **8**, 128 (1736)
2. L. Pauling, J. Chem. Phys. **4**, 673 (1936)
3. D. Kowal, U. Sivan, O. Entin-Wohlman, Y. Imry, Phys. Rev. B **42**, 9009 (1990)
4. Y. Imry, *Introduction to Mesoscopic Systems* (Oxford, NY, 1996)
5. J.A. Sanchez-Gil, V. Freilikher, I. Yurkevich, A.A. Maradudin, Phys. Rev. Lett. **80**, 948 (1998)
6. R. Mittra, S.W. Lee, *Analytical Techniques in the Theory of Guided Waves* (Macmillan, NY, 1971)
7. R.P. Feynman, Phys. Rev. **76**, 749 (1949)
8. P. Exner, P. Šeba, P. Šťovíček, J. Phys. A **21**, 4009 (1988)
9. T. Kottos, U. Smilansky, Phys. Rev. Lett. **79**, 4794 (1997)
10. R. Blümel, Yu. Dabaghian, R.V. Jensen, Phys. Rev. Lett. **88**, 044101 (2002)
11. G. Berkolaiko, P. Kuchment, *Introduction to Quantum Graphs*. Mathematical Surveys and Monographs, vol. 186 (2013), p. 270
12. Z. Pluhař, H.A. Weidenmüller, Phys. Rev. Lett. **112**, 144102 (2014)
13. M. Ławniczak, P. Kurasov, Sz. Bauch, M. Białous, V. Yunko, and L. Sirko, Phys. Rev. E **101**, 052320 (2020)
14. O. Hul, S. Bauch, P. Pakoński, N. Savytskyy, K. Życzkowski, L. Sirko, Phys. Rev. E **69**, 056205 (2004)
15. M. Ławniczak, O. Hul, S. Bauch, P. Šeba, L. Sirko, Phys. Rev. E **77**, 056210 (2008)
16. O. Hul, M. Ławniczak, S. Bauch, A. Sawicki, M. Kuś, L. Sirko, Phys. Rev. Lett. **109**, 040402 (2012)
17. M. Ławniczak, S. Bauch, L. Sirko, *Handbook of Applications of Chaos Theory*, ed. by C. Skiadas, C. Skiadas (CRC Press, Boca Raton, USA, 2016), p. 559
18. B. Dietz, V. Yunko, M. Białous, S. Bauch, M. Ławniczak, L. Sirko, Phys. Rev. E **95**, 052202 (2017)
19. M. Ławniczak, L. Sirko, Sci. Rep. **9**, 5630 (2019)

20. M. Ławniczak, J. Lipovský, L. Sirko, Phys. Rev. Lett. **122**, 140503 (2019)
21. A. Rehemanjiang, M. Allgaier, C.H. Joyner, S. Müller, M. Sieber, U. Kuhl, H.-J. Stöckmann, Phys. Rev. Lett. **117**, 064101 (2016)
22. V. Yunko, M. Białous, L. Sirko, Phys. Rev. E **102**, 012210 (2020)
23. M. Ławniczak, S. Bauch, O. Hul, L. Sirko, Phys. Rev. E **81**, 046204 (2010)
24. M. Allgaier, S. Gehler, S. Barkhofen, H.-J. Stöckmann, U. Kuhl, Phys. Rev. E **89**, 022925 (2014)
25. M. Białous, V. Yunko, S. Bauch, M. Ławniczak, B. Dietz, L. Sirko, Phys. Rev. Lett. **117**, 144101 (2016)
26. M. Ławniczak, M. Białous, V. Yunko, S. Bauch, B. Dietz, L. Sirko, Acta Phys. Pol. A **132**, 1672 (2017)
27. P. Kurasov, Arkiv för Matematik **46**, 95 (2008)
28. P. Kurasov, J. Funct. Anal. **254**, 934 (2008)
29. B. Gutkin, U. Smilansky, J. Phys. A **34**, 6061 (2001)
30. P. Kurasov, M. Nowaczyk, J. Phys. A **38**, 4901 (2005)
31. J. von Below, Can one hear the shape of a network?, in *Partial Differential Equations on Multistructures* (Luminy, 1999); Lect. Notes Pure and Appl. Math. **219**, (Dekker, 2001), p. 19
32. M. Kac, Am. Math. Mon. **73**, 1 (1966)
33. D.S. Jones, *Theory of the Electromagnetism* (Pergamon Press, Oxford, 1964), p. 254
34. N. Savytskyy, A. Kohler, Sz. Bauch, R. Blümel, L. Sirko Phys. Rev. E **64**, 036211 (2001)
35. K. Kuratowski, Fund. Math. (in French) **15**, 271 (1930)

Mathematical Definition of Vortex Boundary and Boundary Classification Based on Topological Type

Xiang Li, Qun Zheng, and Bin Jiang

Abstract Recently, new progress has been made in vortex recognition, and the definition of Liutex (Rortex) based on eigenvectors has established a relation between rotation axis and eigenvectors of velocity gradient tensor. Based on this relation, the mathematical condition of vortex boundary is given: the set of points with multiple roots in the characteristic equation of velocity gradient tensor in a flow field. In this way, the topological structure of critical point theory is applied to vortex boundary. According to whether the velocity gradient tensor can be diagonalized, there are shear boundary and non-shear boundary, while according to the positive, negative and zero of the double root, there would be stable boundary, unstable boundary and degenerate boundary. In order to define and classify the boundary, a mapping method is proposed, and we establish the relation between particle and fluid microelement. Under different decomposition modes, we can analyze the superposition of fluid deformation behavior more clearly by mapping image space, which can be used to compare Helmholtz decomposition with Liutex velocity gradient decomposition. Moreover, considering the simple shear, geometric meaning of Liutex can be explained through geometric relations, and it is believed that the dominant quantity rotation described by Liutex is circular symmetry. (National Science and Technology Major Project (2017-II-0006–0019, 2017-I-0009–0010))

X. Li (✉) · Q. Zheng · B. Jiang
College of Power and Energy Engineering, Harbin Engineering University, #145 Nantong, Harbin, Heilongjiang 150001, China
e-mail: hebutlx@163.com

Q. Zheng
e-mail: zhengqun@hrbeu.edu.cn

B. Jiang
e-mail: jiangbin_hrbeu@163.com

C. H. Skiadas et al. (eds.), *13th Chaotic Modeling and Simulation International Conference*, Springer Proceedings in Complexity,
https://doi.org/10.1007/978-3-030-70795-8_37

1 Introduction

The definition vortex of vortex involves the issue of a rigorous definition of vortex, which is not united yet. Wu proposed that the general goal of vortex definition should reflect its tubular motion pattern [1]. Among commonly used vortex criterion, including the Q criterion, the Delta criterion and the Lambda_ci criterion, only Liutex theory [2–4] satisfies the necessary conditions not to be broken by criterion [5]. Liutex, however, has not concentrated on the outer boundary of the vortex. There is always a simple shear component in the rotation represented by the vorticity [2, 6]. Liu points out that Lambda_ci is a measure of the absolute strength of the net rotation, in the plane perpendicular to eigenvector, and argues that the non-orthogonal coordinates do not describe the rotational momentum [12–15]. In this paper, the geometric and physical meanings of Liutex models are given, and it is believed that Liutex magnitude measures the symmetric rotation strength. We propose a mapping method, which can simulate the two-dimensional topology evolution to give the flow pattern evolution theory, and then give the definition and classification of vortex boundary.

2 Mapping Method

A mapping method is given in this paper to denote 2D fluid behaviors including rotation and deformations as points in a 2D space, which also holds in the plan perpendicular to eigenvector. Similarly, 3D behaviors as points in a 3D space. To be more specific, the method aims to map strain rate state onto particles behaviors. On the other hand, this method can explain directly by which Liutex magnitude is obtained through orthogonal rotation.

It is known that velocity tensor is decomposed into four behaviors of fluid mass, expansion, shear, rotation and angular deformation based on Stocks-Helmholtz decomposition (Fig. 1). However, this mapping is a built to reinterpret them, taking non-diagonal elements as the preimage, while the neighbor space of a point as the image. To achieve it, the expansion is out of consideration and we substitute the rest of three cases with the displacements of A and B (Fig. 1). Then supposing that

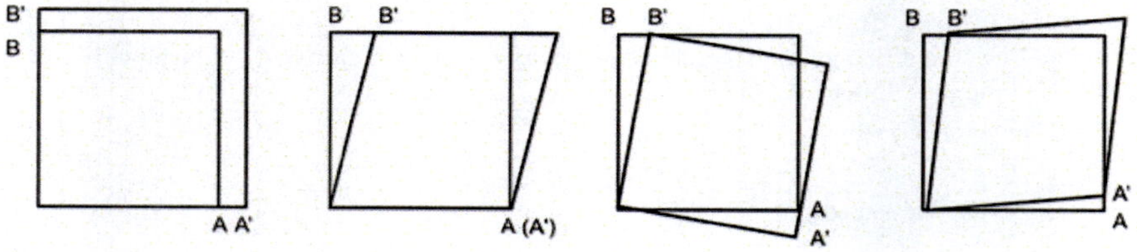

Fig. 1 Deformation of fluid mass

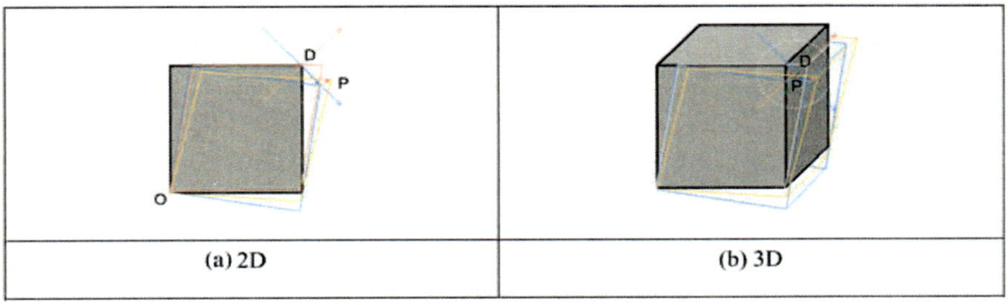

Fig. 2 Neighbor space of image D

time deviation is close to zero, then the strain rates can be expressed by displacements. Namely, combination of displacements of A and B is associated with velocity gradient tensor. Also, even though the expansion is out of consideration, it is enough to analyse Helmholtz decomposition.

Based on parallelogram law, image D is obtained (Fig. 2a). So all deformations now are combined and expressed. The image neighbor space of D is obtained, where the deformation is limited. It is a one-to-one mapping method in 2D from preimage onto image. For 3D, it's easy to develop as the law holds in every direction (Fig. 2b).

This method compares different decomposition and builds a connection between particles and a fluid mass. If points O, A, B is regarded as three typical particles of a fluid mass, it would explain how fluid particles represent a fluid mass, consequently to review the priority of continuum mechanics over particle mechanics.

The strain rates tensor and velocity tensor under Helmoholtz decomposition are mapped onto the neighber space of D, in Fig. 3a. The rotation axle and symmetry angular deformation axle divide the plane into four quadrants, where P get projections using Helmoholtz decomposition. However, when considering about other velocity decomposition, for example in some condition, deformations are contaminated by each other as mentioned before [2, 6]. In this condition, if it regards the simple shear as an independent basic deformation, namely the shear axis is considered, then there would be eight parts in the plane (Fig. 3b) where shaded region means rotation

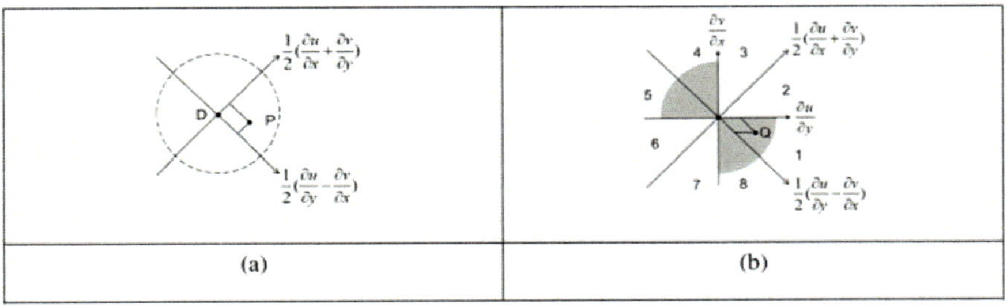

Fig. 3 Image space of the mapping with (a) and without (b) shear axles

component. Here, image points like Q are only projected on two closest axis. Thus, every point in the shaded area is consisted of rotation and simple shear.

This image space makes the components of behaviors more clearly under different decomposition manners. It makes which decomposition is more natural is an artificial selection.

3 Vortex Boundary Definition

In this paper, vortex boundary definition deduced from the geometry and physical meaning of Liutex magnitude given by mapping method. It begins with 2D plane perpendicular to eigenvector, a direction defined as the vortex axle in Liutex theory.

3.1 Physical Meaning of Liutex Magnitude

Liutex magnitude is a Galilain invariant obtained from two orthogonal rotation matrix, Q and P rotation acting on velocity tensor. Here P rotation is an orthogonal rotation on the plane perpendicular to left eigenvector, acting on velocity tensor as

$$\nabla \underset{\theta}{\vec{v}} = P \nabla \vec{v} P^{-1}$$

After the orthogonal transforming, the magnitude of Liutex is defined on the plane as the extreme value of non-diagonal element,

$$\begin{aligned}
\frac{\partial U}{\partial Y}|_{\theta} &= \alpha \sin(2\theta + \varphi) - \beta \\
\frac{\partial V}{\partial X}|_{\theta} &= \alpha \sin(2\theta + \varphi) + \beta \\
\frac{\partial U}{\partial X}|_{\theta} &= \alpha \cos(2\theta + \varphi) + \frac{1}{2}\left(\frac{\partial U}{\partial X} + \frac{\partial V}{\partial Y}\right) \\
\frac{\partial V}{\partial Y}|_{\theta} &= -\alpha \cos(2\theta + \varphi) + \frac{1}{2}\left(\frac{\partial U}{\partial X} + \frac{\partial V}{\partial Y}\right)
\end{aligned}$$

while the main diagonal elements are the equal, corresponding to rotation and isotropic expansion, respectively.

Both linear and shear strain rate are circular symmetry from the view of circular direction. As a result, Liutex extracts the same character in circular direction, called circular symmetry in 2D, and cylindrical in 3D, the physical meaning of Liutex. The circular symmetry is a kind of typical physical character of vortex, emphasizing circular direction rather than the position of each flow particle. So this paper claim that circular symmetry is the physical meaning of Liutex.

3.2 Geometry Meaning of Liutex Magnitude

From the mapping method above, it is easy to generate the physical meaning of α, deviation of circular symmetry. If one the axle (Fig. 3b) is changed to α, all cases of discriminant of Liutex magnitude can be obtained, as shown in Fig. 4a, where β axle is for rotation, shear axle for simple shear, and α axle for two cases: an isotropic expansion and asymmetry of angular deformation. While Liutex the magnitude of Rortex can be obtained by

$$R = 2(\beta - \alpha),$$

where

$$\alpha = \frac{1}{2}\sqrt{(\frac{\partial v}{\partial x} + \frac{\partial u}{\partial y})^2 + (\frac{\partial u}{\partial x} - \frac{\partial v}{\partial y})^2}$$

$$\beta = \frac{1}{2}(\frac{\partial u}{\partial y} - \frac{\partial v}{\partial x})$$

The geometric meaning of Liutex magnitude in shaded area is the length of projection on β axis., because of geometric relation shown in Fig. 4b, the projection of Q on β axle is β-α measuring the rotation. Thus, the mapping method considering about the simple shear help explain Liutex magnitude.

3.3 Vortex Boundary Definition

To consider about vortex boundary definition, the mathematical definition of vortex or rotation need to be given, which is not unified statement. Based on Liutex identification theory [12–23, 23–30], this part will give the mathematical discriminant of vortex boundary.

There are two characters of a gradient tensor, diagonalizable and undiagonalizable. For the first case, the gradient tensor is not diagonalizable to real diagonal matrix, when the velocity gradient tensor has one real eigenvalue and two complex conjugate eigenvalues. According to the eigenvector-based definition of Liutex, the 2×2 matrix on the plane perpendicular to the left eigenvector has the identical complex conjugate eigenvalues as the original velocity gradient tensor which can be expressed as

$$\lambda_1 = \lambda_{cr} + i\lambda_{ci}, \lambda_2 = \lambda_{cr} - i\lambda_{ci}$$

where [3],

$$\lambda_{ci}^2 = \beta^2 - \alpha^2$$

Second, the gradient tensor is diagonalizable to real diagonal matrix when there are three real distinct eigenvalues which means three linearly independent eigenvectors, main directions of compression and tension. Namely, there is an orthogonal rotation that transforms the deformation to simple expansion.

However, when $\beta^2 - \alpha^2 = 0$, there are two or three same eigenvalues of velocity gradient tensor, so there are also above two cases,. To be exact, the velocity is diagonalizable if geometric multiplicity is less than algebraic multiplicity, and if equal, it is not diagonalizable. The boundary of above two cases is the case of multiple root, including double or triple root of characteristic equation.

Correspondingly, in the image plane of the mapping method in Fig. 3b, the shaded area is for one real and two complex eigenvalues, the unshaded area for 3 real eigenvalues, and shear axle for same eigenvalues. Thus, the boundary of vortex is defined as the condition of multiple roots with $\beta^2 - \alpha^2 = 0$, as a result of admitting the Liutex theory. The boundary type can be classified more specifically using topology type of shear axle.

3.4 *Vortex Boundary Classification of Topology Type*

Establishing coordinates on moving points in the flow field, the origin of these coordinates are singular points on these local coordinates, for which the critical point theory is locally suitable. This part will classify every points topology type through a manner of critical points theory [8], so the classification of topology can be used on vortex boundary.

First, differential equations can be given as

$$\frac{dx}{dt} = X(x, y), \frac{dy}{dt} = Y(x, y)$$

where

$$X(x, y), Y(x, y) \in C^0(D), \mathrm{D} \subseteq R^2$$

Equations has different characters on singular points and ordinary points [9]. If $P_0(x_0,y_0)$is a singular point, the velocity is zero, otherwise, is not zero. However, every point can be transformed to a singular point through minus its velocity and establishing local coordinates on points. As a result of $d\vec{v} = \nabla\vec{v} \cdot dr$, the velocity tensor is substituted into

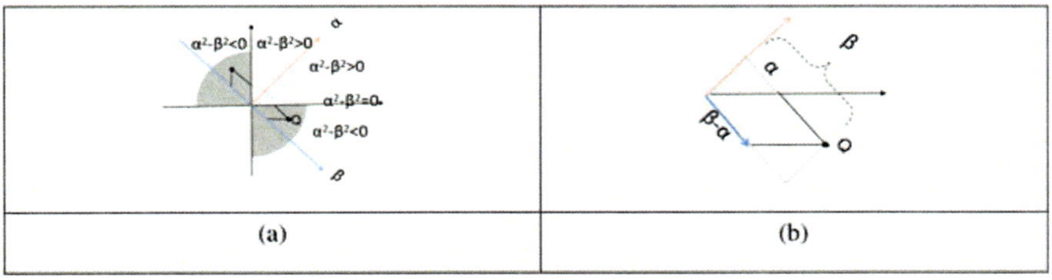

Fig. 4 Schematic diagram of Liutex (Rortex) magnitude

$$\frac{dx}{dt} = ax + by, \frac{dy}{dt} = cx + dy$$

where a, b, c and d are real numbers of velocity tensor. And the characteristic equation is

$$D(\lambda) = \lambda^2 + p\lambda + q = 0$$

where

$$p = -(a + d), q = ad - bc, \Delta = p^2 - 4q$$

The Fig. 5 is the topology of trace line of critical points as well as a kind of stream patterns of ordinary points with local coordinates. The topology doesn't change with the coordinate transformation, since they are determined by p, q and discriminant Δ while the discriminant of multiple roots in the above image space (Fig. 6) is $\Delta' = \alpha^2 - \beta^2$. So axles of the image makes the different cases of eigenvalues clearly.

From Fig. 6, on shear axle, there is $\Delta' = \Delta = 0$, where the vortex boundary is classified into two cases in terms of topology type. According to whether diagonalization can be, there are shear and non-shear boundary, for example, Fig. 6a, b. And according to the positive and negative of the double root, it can be classified into stable boundary and unstable boundary, as shown in Fig. 6c, d. In addition, if the multiple root is zero, the topology is degenerate; it is degenerate boundary, as shown in Fig. 6e, f.

4 Conclusion

Firstly, a mapping method, the foundation of the paper, is given to analyze fluid behaviors, and compare different velocity decomposition trying to explain how Liutex conflict with Helmholtz decomposition and showing how a fluid particle represent a

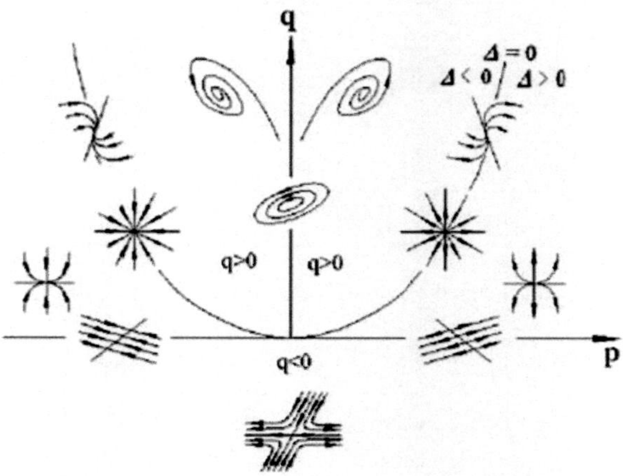

Fig. 5 Classification of critical points in two-dimensional linear systems [7]

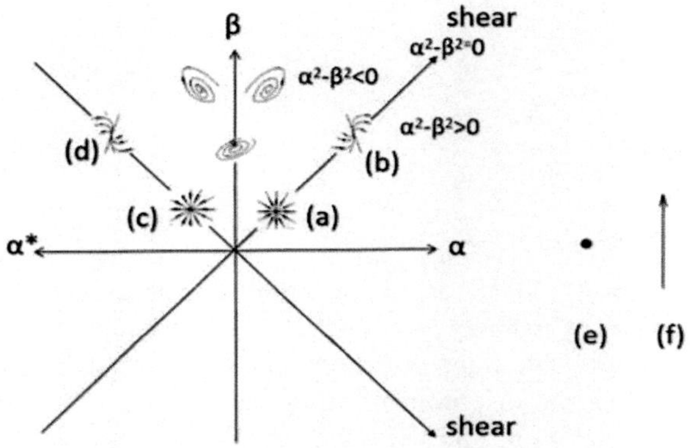

Fig. 6 Stream patterns of ordinary points and singular points in local coordinates

fluid mass. Secondly, circular symmetry in 2D, and cylindrical in 3D is the physical meaning of Liutex. It is a kind of typical physical character of vortex, emphasizing circular direction. And it explain the geometric meaning of Liutex magnitude through geometric relation by the mapping method. So, which velocity decomposition is more natural is an artificial selection.

We define the boundary of vortex as the condition of multiple roots of velocity gradient tensor with $\beta^2 - \alpha^2 = 0$, as a result of admitting the Liutex theory. The boundary type can be classified more specifically using topology type, by classifying every points topology type through a manner of critical points theory. According to whether diagonalization can be, there are shear and non-shear boundary. And according to the positive, negative and zero of the double root, there are stable boundary, unstable boundary and degenerate boundary.

References

1. Wu Jiezhi, Yang Yue, Thoughts on vortex dfinition [J] Acta aerodynamica sinica, 2020(1), 1–8 (2020)
2. C. Liu, Y. Gao, S. Tian et al., Rortex a new vortex vector definition and vorticity tensor and vector decompositions[J]. Phys. Fluids **30**(3) (2018)
3. W. Xu, Y. Gao, Y. Deng et al., An explicit expression for the calculation of the Rortex vector[J]. Phys. Fluids **31**(9), 095102 (2019)
4. Y. Wang, Y. Gao, C. Liu, Galilean invariance of Rortex. Phys. Fluids **30**, 111701 (2018). https://doi.org/https://doi.org/10.1063/1.5058939
5. V. Kolář, J. Šístek, Stretching response of Rortex and other vortex-identification schemes[J]. AIP Adv. **9**(10), 105025 (2019)
6. Li. Zhen, Z. Xi-Wen, He. Feng, Evaluation of vortex criteria by virtue of the quadruple decomposition of velocity gradient tensor[J]. Acta Phys. Sin. **05**, 249–255 (2014)
7. Wu Jiezhi, Introduction to vortex dynamics[M]. Higher Education Press (1993)
8. M.S. Chong, A.E. Perry, B.J. Cantwell, A general classification of three-dimensional flow fields[J]. Phys. Fluids A **2**(5), 765–777 (1990)
9. Zhang Zhifen, Qualitative theory of differential equations[M]. Science Press (1985)
10. J.-Z. Wu, H.-Y. Ma, M.-D. Zhou, *Vorticity and Vortices Dynamics* (Springer-Verlag, Berlin Heidelberg, 2006).
11. Y. Gao, C. Liu, An eigenvector-based definition of Rortex and comparison with eigenvalue-based vortex identification criteria (2018)
12. W. Xu, Y. Gao, Y. Deng, J. Liu, C. Liu*, An explicit expression for the calculation of the Rortex vector. Physics and Fluids, To appear (2019)
13. Y. Gao, Y. Yu, J. Liu, C. Liu, Explicit expressions for Rortex tensor and velocity gradient tensor decomposition. Physics of Fluids, to appear (2019)
14. J. Liu and C. Liu, Modified normalized Rortex/vortex identification method. Phys. Fluids **31**, 061704 (2019). https://doi.org/https://doi.org/10.1063/1.5109437
15. J. Liu, Y. Gao, C. Liu, An objective version of the Rortex vector for vortex identification. Phys. Fluids **31**, 065112 (2019). https://doi.org/https://doi.org/10.1063/1.5095624
16. Y. Gao, C. Liu, Letter: Rortex based velocity gradient tensor decomposition, Physics of Fluids. Phys. Fluids **31**, 011704 (2019). https://doi.org/https://doi.org/10.1063/1.5084739
17. X. Dong, Y. Gao, C. Liu, New normalized Rortex/vortex identification method. Phys. Fluids **31**, 011701 (2019). doi: https://doi.org/10.1063/1.5066016
18. X. Dong, Y. Gao, C. Liu, Study on vorticity structure in late flow transition. Phys. Fluids **30**, 104108 (2018)
19. Y. Gao, C. Liu, Rortex and comparison with eigenvalue-based vortex identification criteria. Phys. Fluids **30**, 085107 (2018). https://doi.org/https://doi.org/10.1063/1.5040112
20. S. Tian, Y. Gao, X. Dong, C. Liu, A definition of vortex vector and vortex. J. Fluid Mechan. **849**, 312–339 (2018). Cambridge University Press 2018. doi:https://doi.org/10.1017/jfm.2018.406
21. X. Dong, Y. Yan, Y. Yang, G. Dong, C. Liu, Spectrum study on unsteadiness of shock wave -vortex ring interaction. Phys. Fluids **30**, 056101 (2018). doi: https://doi.org/10.1063/1.5027299
22. X. Dong, S. Tian, C. Liu, Correlation analysis on volume vorticity and vortex in late boundary layer transition. Phys. Fluids **30**, 014105 (2018). JHD (Journal of Hydrodynamics) Papers
23. H. Xu, X. Cai, C. Liu, Liutex core definition and automatic identification for turbulence structures. Journal of Hydrodynamics, to appear (2019)
24. Y. Gao, J. Liu, Y. Yu, C. Liu, A Liutex based definition and identification of vortex core center lines. J. Hydrodyn **31**(2), 774–781 (2019)
25. J. Liu, Y. Deng, Y. Gao, S. Charkrit, C. Liu, Letter: Mathematical Foundation of Turbulence Generation-Symmetric to Asymmetric Liutex/Rortex. J Hydrodyn (2019)
26. J. Liu, Y. Gao, Y. Wang, C. Liu, Galilean invariance of Omega vortex identification method. J Hydrodyn (2019). https://doi.org/https://doi.org/10.1007/s42241-019-0024-2

27. J. Liu, Y. Gao, Y. Wang, C. Liu, Objective Omega vortex identification method. J Hydrodyn (2019). https://doi.org/https://doi.org/10.1007/s42241-019-0028-y
28. Y. Wang, Y. Gao, J. Liu, C. Liu, Explicit formula for the Liutex vector and physical meaning of vorticity based on the Liutex-Shear decomposition. J Hydrodyn (2019). https://doi.org/https://doi.org/10.1007/s42241-019-0032-2
29. C. Liu, Y. Gao, X. Dong, J. Liu, Y. Zhang, X. Cai, N. Gui, Third generation of vortex identification methods: Omega and Liutex/Rortex based systems. J. Hydrodyn. **31**(2), 1–19 (2019). https://doi.org/https://doi.org/10.1007/s42241-019-0022-4
30. Y. Zhang, X. Qiu, F. Chen, K. Liu, Yu-ning Zhang, Xiang-rui Dong, C. Liu, A selected review of vortex identification methods with applications. Int. J. Hydrodyn. **30**(5) (2018). Available online at https://link.springer.com, https://www.jhydrod.com/, https://doi.org/https://doi.org/10.1007/s42241-018-0112-8 with
31. Xiangrui Dong, Yiqian Wang, Xiaoping Chen, Yuning Zhang, C. Liu, Determination of epsilon for Omega vortex identification method. Int. J. Hydrodyn. **30**(4), 541–548 (2018, August)

On a Singular System of Coupled Nonlocal Fractional Boundary Value Problems

Rodica Luca

Abstract We study the existence and multiplicity of positive solutions for a system of Riemann-Liouville fractional differential equations, subject to coupled nonlocal boundary conditions which contain Riemann-Stieltjes integrals and various fractional derivatives, and the nonlinearities of the system are nonnegative functions and they may be singular at some points. We use the Guo-Krasnosel'skii fixed point theorem in the proofs of the main results.

Keywords Systems of Riemann-Liouville fractional differential equations · Nonlocal coupled boundary conditions · Singular nonlinearities · Existence of solutions · Multiplicity

1 Introduction

We consider the nonlinear system of fractional differential equations

$$\begin{cases} D_{0+}^{\alpha}u(t) + f(t, u(t), v(t)) = 0, & t \in (0, 1), \\ D_{0+}^{\beta}v(t) + g(t, u(t), v(t)) = 0, & t \in (0, 1), \end{cases} \tag{1}$$

supplemented with the coupled nonlocal boundary conditions

$$\begin{cases} u^{(j)}(0) = 0, \quad j = 0, \ldots, n-2, \quad D_{0+}^{\gamma_0}u(1) = \displaystyle\sum_{i=1}^{p} \int_0^1 D_{0+}^{\gamma_i}v(t)\, dH_i(t), \\ v^{(k)}(0) = 0, \quad k = 0, \ldots, m-2, \quad D_{0+}^{\delta_0}v(1) = \displaystyle\sum_{i=1}^{q} \int_0^1 D_{0+}^{\delta_i}u(t)\, dK_i(t), \end{cases} \tag{2}$$

R. Luca (✉)
Department of Mathematics, Gh. Asachi Technical University, 700506 Iaşi, Romania
e-mail: rluca@math.tuiasi.ro

C. H. Skiadas et al. (eds.), *13th Chaotic Modeling and Simulation International Conference*, Springer Proceedings in Complexity,
https://doi.org/10.1007/978-3-030-70795-8_38

where $\alpha,\ \beta \in \mathbb{R}$, $\alpha \in (n-1, n]$, $\beta \in (m-1, m]$, $n,\ m \in \mathbb{N}$, $n \geq 3$, $m \geq 3$, $p,\ q \in \mathbb{N}$, $\gamma_i \in \mathbb{R}$ for all $i = 0, \ldots, p$, $\delta_j \in \mathbb{R}$ for all $j = 0, \ldots, q$, $0 \leq \gamma_1 < \gamma_2 < \cdots < \gamma_p \leq \delta_0 < \beta - 1$, $\delta_0 \geq 1$, $0 \leq \delta_1 < \delta_2 < \cdots < \delta_q \leq \gamma_0 < \alpha - 1$, $\gamma_0 \geq 1$, D_{0+}^k denotes the Riemann-Liouville derivative of order k (for $k = \alpha,\ \beta,\ \gamma_i,\ i = 0, \ldots, p$, $\delta_j,\ j = 0, \ldots, q$), the nonnegative functions f and g may be singular at $t = 0$ and/or $t = 1$, and the integrals from the boundary conditions (2) are Riemann-Stieltjes integrals with H_i for $i = 1, \ldots, p$ and K_j for $j = 1, \ldots, q$ functions of bounded variation.

In this paper we present sufficient conditions for the nonlinearities f and g such that problem (1) and (2) has at least one or two positive solutions. In the proofs of the main results we use the Guo-Krasnosel'skii fixed point theorem of cone expansion and compression of norm type (see [13]). A positive solution of problem (1) and (2) is a pair of functions $(u, v) \in (C([0, 1], \mathbb{R}_+))^2$ satisfying (1) and (2), and $u(t) > 0$ for all $t \in (0, 1]$ or $v(t) > 0$ for all $t \in (0, 1]$, ($\mathbb{R}_+ = [0, \infty)$). The problem (1) and (2) is a generalization of the problem studied in [16]. Indeed, if $p = 1, q = 1$, $\gamma_0 = p_1, \gamma_1 = q_1, \delta_0 = p_2, \delta_1 = q_2$, H_1 is a step function given by $H_1(t) = \{0,\ t \in [0, \xi_1);\ a_1,\ t \in [\xi_1, \xi_2);\ a_1 + a_2,\ t \in [\xi_2, \xi_3); \ldots; \sum_{i=1}^N a_i,\ t \in [\xi_N, 1]\}$, and K_1 is a step function given by $K_1(t) = \{0,\ t \in [0, \eta_1);\ b_1,\ t \in [\eta_1, \eta_2);\ b_1 + b_2,\ t \in [\eta_2, \eta_3); \ldots; \sum_{i=1}^M b_i,\ t \in [\eta_M, 1]\}$, then the boundary conditions (2) become the multi-point boundary conditions (BC) from [16]. In recent years, fractional differential equations and systems of fractional differential equations, subject to various multi-point or Riemann-Stieltjes integral boundary conditions have been intensively studied (see for example, the papers [1–5, 14, 15, 17–19, 22, 23, 29, 31–33]. We also mention the books [7, 8, 20, 21, 26–28], and the papers [6, 9–12, 24, 25, 30], which present various applications of the fractional calculus in scientific and engineering fields.

2 Auxiliary Results

We consider the system of fractional differential equations

$$\begin{cases} D_{0+}^{\alpha} u(t) + \varphi(t) = 0, & t \in (0, 1), \\ D_{0+}^{\beta} v(t) + \psi(t) = 0, & t \in (0, 1), \end{cases} \tag{3}$$

with the boundary conditions (2), where $\varphi,\ \psi \in C(0, 1) \cap L^1(0, 1)$. We denote

$$\begin{aligned} \Delta = & \frac{\Gamma(\alpha)\Gamma(\beta)}{\Gamma(\alpha - \gamma_0)\Gamma(\beta - \delta_0)} - \left(\sum_{i=1}^{p} \frac{\Gamma(\beta)}{\Gamma(\beta - \gamma_i)} \int_0^1 s^{\beta - \gamma_i - 1}\, dH_i(s) \right) \\ & \times \left(\sum_{i=1}^{q} \frac{\Gamma(\alpha)}{\Gamma(\alpha - \delta_i)} \int_0^1 s^{\alpha - \delta_i - 1}\, dK_i(s) \right). \end{aligned} \tag{4}$$

By using similar arguments as those used in the proof of Lemma 2.2 from [31], we obtain the following lemma.

Lemma 1 *If $\Delta \neq 0$, then the unique solution $(u, v) \in C[0,1] \times C[0,1]$ of problem (3) and (2) is given by*

$$\begin{cases} u(t) = \int_0^1 \mathcal{G}_1(t,\vartheta)\varphi(\vartheta)\,d\vartheta + \int_0^1 \mathcal{G}_2(t,\vartheta)\psi(\vartheta)\,d\vartheta, & \forall t \in [0,1], \\ v(t) = \int_0^1 \mathcal{G}_3(t,\vartheta)\varphi(\vartheta)\,d\vartheta + \int_0^1 \mathcal{G}_4(t,\vartheta)\psi(\vartheta)\,d\vartheta, & \forall t \in [0,1], \end{cases} \tag{5}$$

where

$$\begin{aligned}
\mathcal{G}_1(t,\vartheta) &= g_1(t,\vartheta) + \frac{t^{\alpha-1}}{\Delta}\left(\sum_{i=1}^{p} \frac{\Gamma(\beta)}{\Gamma(\beta-\gamma_i)} \int_0^1 s^{\beta-\gamma_i-1}\,dH_i(s)\right) \\
&\quad \times \left(\sum_{i=1}^{q} \int_0^1 g_{1i}(\tau,\vartheta)\,dK_i(\tau)\right),
\end{aligned}$$

$$\begin{aligned}
\mathcal{G}_2(t,\vartheta) &= \frac{t^{\alpha-1}\Gamma(\beta)}{\Delta\Gamma(\beta-\delta_0)} \sum_{i=1}^{p} \int_0^1 g_{2i}(\tau,\vartheta)\,dH_i(\tau), \\
\mathcal{G}_3(t,\vartheta) &= \frac{t^{\beta-1}\Gamma(\alpha)}{\Delta\Gamma(\alpha-\gamma_0)} \sum_{i=1}^{q} \int_0^1 g_{1i}(\tau,\vartheta)\,dK_i(\tau), \\
\mathcal{G}_4(t,\vartheta) &= g_2(t,\vartheta) + \frac{t^{\beta-1}}{\Delta}\left(\sum_{i=1}^{q} \frac{\Gamma(\alpha)}{\Gamma(\alpha-\delta_i)} \int_0^1 s^{\alpha-\delta_i-1}\,dK_i(s)\right) \\
&\quad \times \left(\sum_{i=1}^{p} \int_0^1 g_{2i}(\tau,\vartheta)\,dH_i(\tau)\right),
\end{aligned} \tag{6}$$

for all $(t,\vartheta) \in [0,1] \times [0,1]$ and

$$\begin{aligned}
g_1(t,\vartheta) &= \frac{1}{\Gamma(\alpha)} \begin{cases} t^{\alpha-1}(1-\vartheta)^{\alpha-\gamma_0-1} - (t-\vartheta)^{\alpha-1}, & 0 \le \vartheta \le t \le 1, \\ t^{\alpha-1}(1-\vartheta)^{\alpha-\gamma_0-1}, & 0 \le t \le \vartheta \le 1, \end{cases} \\
g_{1i}(\tau,\vartheta) &= \frac{1}{\Gamma(\alpha-\delta_i)} \begin{cases} \tau^{\alpha-\delta_i-1}(1-\vartheta)^{\alpha-\gamma_0-1} - (\tau-\vartheta)^{\alpha-\delta_i-1}, \\ \quad 0 \le \vartheta \le \tau \le 1, \\ \tau^{\alpha-\delta_i-1}(1-\vartheta)^{\alpha-\gamma_0-1}, \quad 0 \le \tau \le \vartheta \le 1, \end{cases} \\
g_2(t,\vartheta) &= \frac{1}{\Gamma(\beta)} \begin{cases} t^{\beta-1}(1-\vartheta)^{\beta-\delta_0-1} - (t-\vartheta)^{\beta-1}, & 0 \le \vartheta \le t \le 1, \\ t^{\beta-1}(1-\vartheta)^{\beta-\delta_0-1}, & 0 \le t \le \vartheta \le 1, \end{cases} \\
g_{2j}(\tau,\vartheta) &= \frac{1}{\Gamma(\beta-\gamma_j)} \begin{cases} \tau^{\beta-\gamma_j-1}(1-\vartheta)^{\beta-\delta_0-1} - (\tau-\vartheta)^{\beta-\gamma_j-1}, \\ \quad 0 \le \vartheta \le \tau \le 1, \\ \tau^{\beta-\gamma_j-1}(1-\vartheta)^{\beta-\delta_0-1}, \quad 0 \le \tau \le \vartheta \le 1, \end{cases}
\end{aligned} \tag{7}$$

for all $i = 1, \ldots, q$ and $j = 1, \ldots, p$.

Based on the properties of the functions $g_1, g_2, g_{1i}, i = 1, \ldots, q, g_{2j}, j = 1, \ldots, p$ given by (7) (see Lemma 3 from [15], or Lemma 2.4 from [31]), we obtain below the following properties of the functions $\mathcal{G}_i, i = 1, \ldots, 4$.

Lemma 2 *Assume that $\Delta > 0$, $H_i, i = 1, \ldots, p$, $K_j, j = 1, \ldots, q$ are nondecreasing functions. Then the functions $\mathcal{G}_i, i = 1, \ldots, 4$ given by (6) have the properties:*

(a) $\mathcal{G}_i : [0, 1] \times [0, 1] \to [0, \infty), i = 1, \ldots, 4$ *are continuous functions;*
(b) $\mathcal{G}_1(t, \vartheta) \le \mathcal{J}_1(\vartheta)$ *for all* $(t, \vartheta) \in [0, 1] \times [0, 1]$*, where*

$$\mathcal{J}_1(\vartheta)=h_1(\vartheta) + \frac{1}{\Delta}\left(\sum_{i=1}^{p} \frac{\Gamma(\beta)}{\Gamma(\beta-\gamma_i)} \int_0^1 s^{\beta-\gamma_i-1} dH_i(s)\right)\left(\sum_{i=1}^{q} \int_0^1 g_{1i}(\tau, \vartheta) dK_i(\tau)\right), \forall \vartheta \in [0, 1],$$

$h_1(\vartheta) = \frac{1}{\Gamma(\alpha)}(1-\vartheta)^{\alpha-\gamma_0-1}(1-(1-\vartheta)^{\gamma_0}), \ \forall \vartheta \in [0, 1];$
(c) $\mathcal{G}_1(t, \vartheta) \ge t^{\alpha-1}\mathcal{J}_1(\vartheta)$ *for all* $(t, \vartheta) \in [0, 1] \times [0, 1]$*;*
(d) $\mathcal{G}_2(t, \vartheta) \le \mathcal{J}_2(\vartheta)$*, for all* $(t, \vartheta) \in [0, 1] \times [0, 1]$*, where*

$$\mathcal{J}_2(\vartheta) = \frac{\Gamma(\beta)}{\Delta\Gamma(\beta-\delta_0)} \sum_{i=1}^{p} \int_0^1 g_{2i}(\tau, \vartheta) dH_i(\tau), \quad \forall \vartheta \in [0, 1];$$

(e) $\mathcal{G}_2(t, \vartheta) = t^{\alpha-1}\mathcal{J}_2(\vartheta)$ *for all* $(t, \vartheta) \in [0, 1] \times [0, 1]$*;*
(f) $\mathcal{G}_3(t, \vartheta) \le \mathcal{J}_3(\vartheta)$ *for all* $(t, \vartheta) \in [0, 1] \times [0, 1]$*, where*

$$\mathcal{J}_3(\vartheta) = \frac{\Gamma(\alpha)}{\Delta\Gamma(\alpha-\gamma_0)} \sum_{i=1}^{q} \int_0^1 g_{1i}(\tau, \vartheta) dK_i(\tau), \quad \forall \vartheta \in [0, 1];$$

(g) $\mathcal{G}_3(t, \vartheta) = t^{\beta-1}\mathcal{J}_3(\vartheta)$ *for all* $(t, \vartheta) \in [0, 1] \times [0, 1]$*;*
(h) $\mathcal{G}_4(t, \vartheta) \le \mathcal{J}_4(\vartheta)$ *for all* $(t, \vartheta) \in [0, 1] \times [0, 1]$*, where*

$$\mathcal{J}_4(\vartheta)=h_2(\vartheta) + \frac{1}{\Delta}\left(\sum_{i=1}^{q} \frac{\Gamma(\alpha)}{\Gamma(\alpha-\delta_i)} \int_0^1 s^{\alpha-\delta_i-1} dK_i(s)\right)\left(\sum_{i=1}^{p} \int_0^1 g_{2i}(\tau, \vartheta) dH_i(\tau)\right), \ \forall \vartheta \in [0, 1],$$

$h_2(\vartheta) = \frac{1}{\Gamma(\beta)}(1-\vartheta)^{\beta-\delta_0-1}(1-(1-\vartheta)^{\delta_0}), \ \forall \vartheta \in [0, 1];$
(i) $\mathcal{G}_4(t, \vartheta) \ge t^{\beta-1}\mathcal{J}_4(\vartheta)$*, for all* $(t, \vartheta) \in [0, 1] \times [0, 1]$.

Lemma 3 *Assume that $\Delta > 0$, $H_i, i = 1, \ldots, p$, $K_j, j = 1, \ldots, q$ are nondecreasing functions, and $\varphi, \psi \in C(0, 1) \cap L^1(0, 1)$ with $\varphi(t) \ge 0$, $\psi(t) \ge 0$ for all $t \in (0, 1)$. Then the solution (u, v) of problem (3) and (2) given by (5) satisfies the inequalities $u(t) \ge 0$, $v(t) \ge 0$ for all $t \in [0, 1]$. Moreover, we have the inequalities $u(t) \ge t^{\alpha-1}u(\nu)$ and $v(t) \ge t^{\beta-1}v(\nu)$ for all $t, \nu \in [0, 1]$.*

Proof Under the assumptions of this lemma, by using relations (5) and Lemma 2, we deduce that $u(t) \ge 0$ and $v(t) \ge 0$ for all $t \in [0, 1]$. Besides, for all $t, \nu \in [0, 1]$, we obtain the following inequalities

$$
\begin{aligned}
u(t) &= \int_0^1 \mathcal{G}_1(t,\vartheta)\varphi(\vartheta)\,d\vartheta + \int_0^1 \mathcal{G}_2(t,\vartheta)\psi(\vartheta)\,d\vartheta \\
&\geq t^{\alpha-1}\left(\int_0^1 \mathcal{J}_1(\vartheta)\varphi(\vartheta)\,d\vartheta + \int_0^1 \mathcal{J}_2(\vartheta)\psi(\vartheta)\,d\vartheta\right) \\
&\geq t^{\alpha-1}\left(\int_0^1 \mathcal{G}_1(\nu,\vartheta)\varphi(\vartheta)\,d\vartheta + \int_0^1 \mathcal{G}_2(\nu,\vartheta)\psi(\vartheta)\,d\vartheta\right) = t^{\alpha-1}u(\nu), \\
v(t) &= \int_0^1 \mathcal{G}_3(t,\vartheta)\varphi(\vartheta)\,d\vartheta + \int_0^1 \mathcal{G}_4(t,\vartheta)\psi(\vartheta)\,d\vartheta \\
&\geq t^{\beta-1}\left(\int_0^1 \mathcal{J}_3(\vartheta)\varphi(\vartheta)\,d\vartheta + \int_0^1 \mathcal{J}_4(\vartheta)\psi(\vartheta)\,d\vartheta\right) \\
&\geq t^{\beta-1}\left(\int_0^1 \mathcal{G}_3(\nu,\vartheta)\varphi(\vartheta)\,d\vartheta + \int_0^1 \mathcal{G}_4(\nu,\vartheta)\psi(\vartheta)\,d\vartheta\right) = t^{\beta-1}v(\nu).
\end{aligned}
$$

□

3 Existence and Multiplicity of Positive Solutions

In this section we investigate the existence and multiplicity of positive solutions for problem (1) and (2) under various assumptions on the nonlinearities f and g which may be singular at $t = 0$ and/or $t = 1$. We present now the basic assumptions that will be used in our main results:

$(H1)$ $\alpha,\ \beta \in \mathbb{R}, \alpha \in (n-1, n], \beta \in (m-1, m], n,\ m \in \mathbb{N}, n \geq 3, m \geq 3, p,\ q \in \mathbb{N},\ \gamma_i \in \mathbb{R}$ for all $i = 0, \dots, p,\ \delta_j \in \mathbb{R}$ for all $j = 0, \dots, q,\ 0 \leq \gamma_1 < \gamma_2 < \dots < \gamma_p \leq \delta_0 < \beta - 1,\ \delta_0 \geq 1,\ 0 \leq \delta_1 < \delta_2 < \dots < \delta_q \leq \gamma_0 < \alpha - 1,\ \gamma_0 \geq 1$, $H_i : [0, 1] \to \mathbb{R},\ i = 1, \dots, p$ and $K_j : [0, 1] \to \mathbb{R},\ j = 1, \dots, q$ are nondecreasing functions, and $\Delta > 0$ (Δ is given by (4)).

$(H2)$ The functions $f,\ g \in C((0, 1) \times \mathbb{R}_+ \times \mathbb{R}_+, \mathbb{R}_+)$ and there exist the functions $\zeta_i \in C((0, 1), \mathbb{R}_+)$ and $\varphi_i \in C([0, 1] \times \mathbb{R}_+ \times \mathbb{R}_+, \mathbb{R}_+)$, $i = 1, 2$, with $0 < \int_0^1 (1-\tau)^{\alpha-\gamma_0-1}\zeta_1(\tau)\,d\tau < \infty$, $0 < \int_0^1 (1-\tau)^{\beta-\delta_0-1}\zeta_2(\tau)\,d\tau < \infty$ such that $f(t, x, y) \leq \zeta_1(t)\varphi_1(t, x, y),\ \ g(t, x, y) \leq \zeta_2(t)\varphi_2(t, x, y),\ \ \forall t \in (0, 1),\ \ x, y \in \mathbb{R}_+$.

By using Lemma 1, the pair (u, v) is a solution of problem (1) and (2) if and only if (u, v) is a solution for the nonlinear system of integral equations

$$
\begin{cases}
u(t) = \int_0^1 \mathcal{G}_1(t,\vartheta) f(\vartheta, u(\vartheta), v(\vartheta))\,d\vartheta + \int_0^1 \mathcal{G}_2(t,\vartheta) g(\vartheta, u(\vartheta), v(\vartheta))\,d\vartheta, \\
v(t) = \int_0^1 \mathcal{G}_3(t,\vartheta) f(\vartheta, u(\vartheta), v(\vartheta))\,d\vartheta + \int_0^1 \mathcal{G}_4(t,\vartheta) g(\vartheta, u(\vartheta), v(\vartheta))\,d\vartheta,
\end{cases}
$$

for all $t \in [0, 1]$.

We introduce the Banach space $\mathcal{Y} = C[0, 1]$ with supremum norm $\|x\| = \sup_{t\in[0,1]} |x(t)|$ and the Banach space $\mathcal{Z} = \mathcal{Y} \times \mathcal{Y}$ with the norm $\|(x, y)\|_{\mathcal{Z}} = \|x\| + \|y\|$. We define the cone $\mathcal{Q} \subset \mathcal{Z}$ by

$$\mathcal{Q} = \{(u, v) \in \mathcal{Z}, \ \ u(t) \geq 0, \ \ v(t) \geq 0, \ \ \forall t \in [0, 1]\}.$$

We consider now the operators $\mathcal{A}_1, \mathcal{A}_2 : \mathcal{Z} \to \mathcal{Y}$ and $\mathcal{A} : \mathcal{Z} \to \mathcal{Z}$ defined by

$$\mathcal{A}_1(u, v)(t) = \int_0^1 \mathcal{G}_1(t, \vartheta) f(\vartheta, u(\vartheta), v(\vartheta))\, d\vartheta + \int_0^1 \mathcal{G}_2(t, \vartheta) g(\vartheta, u(\vartheta), v(\vartheta))\, d\vartheta,$$
$$\mathcal{A}_2(u, v)(t) = \int_0^1 \mathcal{G}_3(t, \vartheta) f(\vartheta, u(\vartheta), v(\vartheta))\, d\vartheta + \int_0^1 \mathcal{G}_4(t, \vartheta) g(\vartheta, u(\vartheta), v(\vartheta))\, d\vartheta,$$

for all $t \in [0, 1]$ and $\mathcal{A}(u, v) = (\mathcal{A}_1(u, v), \mathcal{A}_2(u, v)), \ \ (u, v) \in \mathcal{Z}$.

The pair (u, v) is a solution of problem (1) and (2) if and only if (u, v) is a fixed point of operator $\mathcal{A}$.

Lemma 4 *Assume that* $(H1)$ *and* $(H2)$ *hold. Then* $\mathcal{A} : \mathcal{Q} \to \mathcal{Q}$ *is a completely continuous operator.*

Proof We denote by $M_1 = \int_0^1 \mathcal{J}_1(s)\zeta_1(s)ds$, $M_2 = \int_0^1 \mathcal{J}_2(s)\zeta_2(s)ds$, $M_3 = \int_0^1 \mathcal{J}_3(s)\zeta_1(s)ds$, $M_4 = \int_0^1 \mathcal{J}_4(s)\zeta_2(s)ds$. Using $(H2)$ and Lemma 2, we obtain $M_1 > 0$, $M_2 \geq 0$, $M_3 \geq 0$, $M_4 > 0$, and

$$\begin{aligned}
M_1 &= \int_0^1 \mathcal{J}_1(s)\zeta_1(s)\, ds = \int_0^1 \left[h_1(s) + \frac{1}{\Delta}\left(\sum_{i=1}^{p} \frac{\Gamma(\beta)}{\Gamma(\beta-\gamma_i)} \int_0^1 \tau^{\beta-\gamma_i-1}\, dH_i(\tau) \right) \right.\\
&\quad \left. \times \left(\sum_{i=1}^{q} \int_0^1 g_{1i}(\tau, s)\, dK_i(\tau) \right) \right] \zeta_1(s)\, ds \\
&\leq \int_0^1 \left[\frac{1}{\Gamma(\alpha)}(1-s)^{\alpha-\gamma_0-1} + \frac{1}{\Delta}\left(\sum_{i=1}^{p} \frac{\Gamma(\beta)}{\Gamma(\beta-\gamma_i)} \int_0^1 \tau^{\beta-\gamma_i-1}\, dH_i(\tau) \right) \right.\\
&\quad \left. \times \left(\sum_{i=1}^{q} \int_0^1 \frac{1}{\Gamma(\alpha-\delta_i)} \tau^{\alpha-\delta_i-1}(1-s)^{\alpha-\gamma_0-1}\, dK_i(\tau) \right) \right] \zeta_1(s)\, ds \\
&= \left[\frac{1}{\Gamma(\alpha)} + \frac{1}{\Delta}\left(\sum_{i=1}^{p} \frac{\Gamma(\beta)}{\Gamma(\beta-\gamma_i)} \int_0^1 \tau^{\beta-\gamma_i-1}\, dH_i(\tau) \right) \right.\\
&\quad \left. \times \left(\sum_{i=1}^{q} \int_0^1 \frac{1}{\Gamma(\alpha-\delta_i)} \tau^{\alpha-\delta_i-1}\, dK_i(\tau) \right) \right] \int_0^1 (1-s)^{\alpha-\gamma_0-1}\zeta_1(s)\, ds < \infty,
\end{aligned}$$

$$M_2 = \int_0^1 \mathcal{J}_2(s)\zeta_2(s)\,ds = \int_0^1 \frac{\Gamma(\beta)}{\Delta\Gamma(\beta-\delta_0)}\left(\sum_{i=1}^{p}\int_0^1 g_{2i}(\tau,s)\,dH_i(\tau)\right)\zeta_2(s)\,ds$$
$$\leq \int_0^1 \frac{\Gamma(\beta)}{\Delta\Gamma(\beta-\delta_0)}\left(\sum_{i=1}^{p}\int_0^1 \frac{1}{\Gamma(\beta-\gamma_i)}\tau^{\beta-\gamma_i-1}(1-s)^{\beta-\delta_0-1}\,dH_i(\tau)\right)\zeta_2(s)\,ds$$
$$= \frac{\Gamma(\beta)}{\Delta\Gamma(\beta-\delta_0)}\left(\sum_{i=1}^{p}\int_0^1 \frac{1}{\Gamma(\beta-\gamma_i)}\tau^{\beta-\gamma_i-1}\,dH_i(\tau)\right)\int_0^1 (1-s)^{\beta-\delta_0-1}\zeta_2(s)\,ds$$
$$< \infty,$$
$$M_3 = \int_0^1 \mathcal{J}_3(s)\zeta_1(s)\,ds = \int_0^1 \frac{\Gamma(\alpha)}{\Delta\Gamma(\alpha-\gamma_0)}\left(\sum_{i=1}^{q}\int_0^1 g_{1i}(\tau,s)\,dK_i(\tau)\right)\zeta_1(s)\,ds$$
$$\leq \int_0^1 \frac{\Gamma(\alpha)}{\Delta\Gamma(\alpha-\gamma_0)}\left(\sum_{i=1}^{q}\int_0^1 \frac{1}{\Gamma(\alpha-\delta_i)}\tau^{\alpha-\delta_i-1}(1-s)^{\alpha-\gamma_0-1}\,dK_i(\tau)\right)\zeta_1(s)\,ds$$
$$= \frac{\Gamma(\alpha)}{\Delta\Gamma(\alpha-\gamma_0)}\left(\sum_{i=1}^{q}\int_0^1 \frac{1}{\Gamma(\alpha-\delta_i)}\tau^{\alpha-\delta_i-1}\,dK_i(\tau)\right)\int_0^1 (1-s)^{\alpha-\gamma_0-1}\zeta_1(s)\,ds$$
$$< \infty,$$
$$M_4 = \int_0^1 \mathcal{J}_4(s)\zeta_2(s)\,ds = \int_0^1 \left[h_2(s) + \frac{1}{\Delta}\left(\sum_{i=1}^{q}\frac{\Gamma(\alpha)}{\Gamma(\alpha-\delta_i)}\int_0^1 \tau^{\alpha-\delta_i-1}\,dK_i(\tau)\right)\right.$$
$$\left.\times\left(\sum_{i=1}^{p}\int_0^1 g_{2i}(\tau,s)\,dH_i(\tau)\right)\right]\zeta_2(s)\,ds$$
$$\leq \int_0^1 \left[\frac{1}{\Gamma(\beta)}(1-s)^{\beta-\delta_0-1} + \frac{1}{\Delta}\left(\sum_{i=1}^{q}\frac{\Gamma(\alpha)}{\Gamma(\alpha-\delta_i)}\int_0^1 \tau^{\alpha-\delta_i-1}\,dK_i(\tau)\right)\right.$$
$$\left.\times\left(\sum_{i=1}^{p}\int_0^1 \frac{1}{\Gamma(\beta-\gamma_i)}\tau^{\beta-\gamma_i-1}(1-s)^{\beta-\delta_0-1}\,dH_i(\tau)\right)\right]\zeta_2(s)\,ds$$
$$= \left[\frac{1}{\Gamma(\beta)} + \frac{1}{\Delta}\left(\sum_{i=1}^{q}\frac{\Gamma(\alpha)}{\Gamma(\alpha-\delta_i)}\int_0^1 \tau^{\alpha-\delta_i-1}\,dK_i(\tau)\right)\right.$$
$$\left.\times\left(\sum_{i=1}^{p}\int_0^1 \frac{1}{\Gamma(\beta-\gamma_i)}\tau^{\beta-\gamma_i-1}\,dH_i(\tau)\right)\right]\int_0^1 (1-s)^{\beta-\delta_0-1}\zeta_2(s)\,ds < \infty.$$

By Lemma 2 we also deduce that $\mathcal{A}$ maps $\mathcal{Q}$ into $\mathcal{Q}$.

We prove now that $\mathcal{A}$ maps bounded sets into relatively compact sets. We assume $S \subset \mathcal{Q}$ is an arbitrary bounded set. Then there exists $L_1 > 0$ such that $\|(u,v)\|_{\mathcal{Z}} \leq L_1$ for all $(u,v) \in S$. By using the continuity of φ_1 and φ_2, we find that there exists $L_2 > 0$ such that $L_2 = \max\{\sup_{t\in[0,1],\,x,y\in[0,L_1]}\varphi_1(t,x,y),\ \sup_{t\in[0,1],\,x,y\in[0,L_1]}\varphi_2(t,x,y)\}$. By Lemma 2, for any $(u,v) \in S$ and $t \in [0,1]$, we obtain

$$\mathcal{A}_1(u,v)(t) \leq \int_0^1 \mathcal{J}_1(\vartheta)f(\vartheta,u(\vartheta),v(\vartheta))\,d\vartheta + \int_0^1 \mathcal{J}_2(\vartheta)g(\vartheta,u(\vartheta),v(\vartheta))\,d\vartheta$$

$$\leq \int_0^1 \mathcal{J}_1(\vartheta)\zeta_1(\vartheta)\varphi_1(\vartheta, u(\vartheta), v(\vartheta))\, d\vartheta + \int_0^1 \mathcal{J}_2(\vartheta)\zeta_2(\vartheta)\varphi_2(\vartheta, u(\vartheta), v(\vartheta))\, d\vartheta$$
$$\leq L_2 \int_0^1 \mathcal{J}_1(\vartheta)\zeta_1(\vartheta)\, d\vartheta + L_2 \int_0^1 \mathcal{J}_2(\vartheta)\zeta_2(\vartheta)\, d\vartheta = L_2(M_1 + M_2),$$
$$\mathcal{A}_2(u, v)(t) \leq \int_0^1 \mathcal{J}_3(\vartheta) f(\vartheta, u(\vartheta), v(\vartheta))\, d\vartheta + \int_0^1 \mathcal{J}_4(\vartheta) g(\vartheta, u(\vartheta), v(\vartheta))\, d\vartheta$$
$$\leq \int_0^1 \mathcal{J}_3(\vartheta)\zeta_1(\vartheta)\varphi_1(\vartheta, u(\vartheta), v(\vartheta))\, d\vartheta + \int_0^1 \mathcal{J}_4(\vartheta)\zeta_2(\vartheta)\varphi_2(\vartheta, u(\vartheta), v(\vartheta))\, d\vartheta$$
$$\leq L_2 \int_0^1 \mathcal{J}_3(\vartheta)\zeta_1(\vartheta)\, d\vartheta + L_2 \int_0^1 \mathcal{J}_4(\vartheta)\zeta_2(\vartheta)\, d\vartheta = L_2(M_3 + M_4).$$

Then $\|\mathcal{A}_1(u, v)\| \leq L_2(M_1 + M_2)$, $\|\mathcal{A}_2(u, v)\| \leq L_2(M_3 + M_4)$ for all $(u, v) \in S$, and hence $\mathcal{A}_1(S)$, $\mathcal{A}_2(S)$ and $\mathcal{A}(S)$ are bounded.

We will show next that $\mathcal{A}(S)$ is equicontinuous. By using Lemma 1, for $(u, v) \in S$ and $t \in [0, 1]$ we find

$$\mathcal{A}_1(u, v)(t) = \int_0^1 \left[g_1(t, \vartheta) + \frac{t^{\alpha-1}}{\Delta} \left(\sum_{i=1}^{p} \frac{\Gamma(\beta)}{\Gamma(\beta - \gamma_i)} \int_0^1 \tau^{\beta-\gamma_i-1}\, dH_i(\tau) \right) \right.$$
$$\left. \times \left(\sum_{i=1}^{q} \int_0^1 g_{1i}(\tau, \vartheta)\, dK_i(\tau) \right) \right] f(\vartheta, u(\vartheta), v(\vartheta))\, d\vartheta$$
$$+ \int_0^1 \frac{t^{\alpha-1}\Gamma(\beta)}{\Delta\Gamma(\beta - \delta_0)} \left(\sum_{i=1}^{p} \int_0^1 g_{2i}(\tau, \vartheta)\, dH_i(\tau) \right) g(\vartheta, u(\vartheta), v(\vartheta))\, d\vartheta$$
$$= \int_0^t \frac{1}{\Gamma(\alpha)} [t^{\alpha-1}(1 - \vartheta)^{\alpha-\gamma_0-1} - (t - \vartheta)^{\alpha-1}] f(\vartheta, u(\vartheta), v(\vartheta))\, d\vartheta$$
$$+ \int_t^1 \frac{1}{\Gamma(\alpha)} t^{\alpha-1}(1 - \vartheta)^{\alpha-\gamma_0-1} f(\vartheta, u(\vartheta), v(\vartheta))\, d\vartheta$$
$$+ \frac{t^{\alpha-1}}{\Delta} \left(\sum_{i=1}^{p} \frac{\Gamma(\beta)}{\Gamma(\beta - \gamma_i)} \int_0^1 \tau^{\beta-\gamma_i-1}\, dH_i(\tau) \right)$$
$$\times \int_0^1 \left(\sum_{i=1}^{q} \int_0^1 g_{1i}(\tau, \vartheta)\, dK_i(\tau) \right) f(\vartheta, u(\vartheta), v(\vartheta))\, d\vartheta$$
$$+ \frac{t^{\alpha-1}\Gamma(\beta)}{\Delta\Gamma(\beta - \delta_0)} \int_0^1 \left(\sum_{i=1}^{p} \int_0^1 g_{2i}(\tau, \vartheta)\, dH_i(\tau) \right) g(\vartheta, u(\vartheta), v(\vartheta))\, d\vartheta.$$

Then for any $t \in (0, 1)$ we deduce

$$(\mathcal{A}_1(u,v))'(t) = \int_0^t \frac{1}{\Gamma(\alpha)}[(\alpha-1)t^{\alpha-2}(1-\vartheta)^{\alpha-\gamma_0-1} - (\alpha-1)(t-\vartheta)^{\alpha-2}]$$
$$\times f(\vartheta, u(\vartheta), v(\vartheta))\, d\vartheta + \int_t^1 \frac{1}{\Gamma(\alpha)}(\alpha-1)t^{\alpha-2}(1-\vartheta)^{\alpha-\gamma_0-1} f(\vartheta, u(\vartheta), v(\vartheta))\, d\vartheta$$
$$+\frac{(\alpha-1)t^{\alpha-2}}{\Delta}\left(\sum_{i=1}^{p} \frac{\Gamma(\beta)}{\Gamma(\beta-\gamma_i)} \int_0^1 \tau^{\beta-\gamma_i-1}\, dH_i(\tau)\right)$$
$$\times \int_0^1 \left(\sum_{i=1}^{q} \int_0^1 g_{1i}(\tau,\vartheta)\, dK_i(\tau)\right) f(\vartheta, u(\vartheta), v(\vartheta))\, d\vartheta$$
$$+\frac{(\alpha-1)t^{\alpha-2}\Gamma(\beta)}{\Delta\Gamma(\beta-\delta_0)} \int_0^1 \left(\sum_{i=1}^{p} \int_0^1 g_{2i}(\tau,\vartheta)\, dH_i(\tau)\right) g(\vartheta, u(\vartheta), v(\vartheta))\, d\vartheta.$$

Hence for any $t \in (0,1)$ we obtain

$$|(\mathcal{A}_1(u,v))'(t)| \le \frac{1}{\Gamma(\alpha-1)} \int_0^t [t^{\alpha-2}(1-\vartheta)^{\alpha-\gamma_0-1} + (t-\vartheta)^{\alpha-2}]$$
$$\times \zeta_1(\vartheta)\varphi_1(\vartheta, u(\vartheta), v(\vartheta))\, d\vartheta + \frac{1}{\Gamma(\alpha-1)} \int_t^1 t^{\alpha-2}(1-\vartheta)^{\alpha-\gamma_0-1}$$
$$\times \zeta_1(\vartheta)\varphi_1(\vartheta, u(\vartheta), v(\vartheta))\, d\vartheta$$
$$+\frac{(\alpha-1)t^{\alpha-2}}{\Delta}\left(\sum_{i=1}^{p} \frac{\Gamma(\beta)}{\Gamma(\beta-\gamma_i)} \int_0^1 \tau^{\beta-\gamma_i-1}\, dH_i(\tau)\right)$$
$$\times \int_0^1 \left(\sum_{i=1}^{q} \int_0^1 g_{1i}(\tau,\vartheta)\, dK_i(\tau)\right) \zeta_1(\vartheta)\varphi_1(\vartheta, u(\vartheta), v(\vartheta))\, d\vartheta$$
$$+\frac{(\alpha-1)t^{\alpha-2}\Gamma(\beta)}{\Delta\Gamma(\beta-\delta_0)} \int_0^1 \left(\sum_{i=1}^{p} \int_0^1 g_{2i}(\tau,\vartheta)\, dH_i(\tau)\right) \zeta_2(\vartheta)\varphi_2(\vartheta, u(\vartheta), v(\vartheta))\, d\vartheta.$$

Therefore we find

$$|(\mathcal{A}_1(u,v))'(t)| \le L_2 \left[\frac{1}{\Gamma(\alpha-1)} \int_0^t [t^{\alpha-2}(1-\vartheta)^{\alpha-\gamma_0-1} + (t-\vartheta)^{\alpha-2}]\zeta_1(\vartheta)\, d\vartheta\right.$$
$$+\frac{1}{\Gamma(\alpha-1)} \int_t^1 t^{\alpha-2}(1-\vartheta)^{\alpha-\gamma_0-1}\zeta_1(\vartheta)\, d\vartheta$$
$$+\frac{(\alpha-1)t^{\alpha-2}\Gamma(\beta)}{\Delta}\left(\sum_{i=1}^{p} \frac{1}{\Gamma(\beta-\gamma_i)} \int_0^1 \tau^{\beta-\gamma_i-1}\, dH_i(\tau)\right)$$
$$\times \int_0^1 \left(\sum_{i=1}^{q} \int_0^1 g_{1i}(\tau,\vartheta)\, dK_i(\tau)\right) \zeta_1(\vartheta)\, d\vartheta$$
$$\left.+\frac{(\alpha-1)t^{\alpha-2}\Gamma(\beta)}{\Delta\Gamma(\beta-\delta_0)} \int_0^1 \left(\sum_{i=1}^{p} \int_0^1 g_{2i}(\tau,\vartheta)\, dH_i(\tau)\right) \zeta_2(\vartheta)\, d\vartheta\right]. \tag{8}$$

We denote

$$\begin{aligned}
\Xi_1(t) &= \frac{1}{\Gamma(\alpha-1)}\int_0^t [t^{\alpha-2}(1-\vartheta)^{\alpha-\gamma_0-1} + (t-\vartheta)^{\alpha-2}]\zeta_1(\vartheta)\,d\vartheta \\
&+\frac{1}{\Gamma(\alpha-1)}\int_t^1 t^{\alpha-2}(1-\vartheta)^{\alpha-\gamma_0-1}\zeta_1(\vartheta)\,d\vartheta, \\
\Xi_2(t) &= \Xi_1(t) + \frac{(\alpha-1)t^{\alpha-2}\Gamma(\beta)}{\Delta}\left(\sum_{i=1}^p \frac{1}{\Gamma(\beta-\gamma_i)}\int_0^1 \tau^{\beta-\gamma_i-1}\,dH_i(\tau)\right) \\
&\times\int_0^1\left(\sum_{i=1}^q\int_0^1 g_{1i}(\tau,\vartheta)\,dK_i(\tau)\right)\zeta_1(\vartheta)\,d\vartheta \\
&+\frac{(\alpha-1)t^{\alpha-2}\Gamma(\beta)}{\Delta\Gamma(\beta-\delta_0)}\int_0^1\left(\sum_{i=1}^p\int_0^1 g_{2i}(\tau,\vartheta)\,dH_i(\tau)\right)\zeta_2(\vartheta)\,d\vartheta.
\end{aligned}$$

We compute the integral of function Ξ_1, by exchanging the order of integration, and we deduce

$$\begin{aligned}
\int_0^1 \Xi_1(t)\,dt &= \frac{1}{\Gamma(\alpha)}\int_0^1 (1-\vartheta)^{\alpha-\gamma_0-1}(1+(1-\vartheta)^{\gamma_0})\zeta_1(\vartheta)\,d\vartheta \\
&\le \frac{2}{\Gamma(\alpha)}\int_0^1 (1-\vartheta)^{\alpha-\gamma_0-1}\zeta_1(\vartheta)\,d\vartheta < \infty.
\end{aligned}$$

For the integral of the function Ξ_2, we obtain

$$\begin{aligned}
\int_0^1 \Xi_2(t)\,dt &= \int_0^1 \Xi_1(t)\,dt + \frac{\alpha-1}{\Delta}\left(\int_0^1 t^{\alpha-2}\,dt\right) \\
&\times\left(\sum_{i=1}^p \frac{\Gamma(\beta)}{\Gamma(\beta-\gamma_i)}\int_0^1 \tau^{\beta-\gamma_i-1}\,dH_i(\tau)\right) \\
&\times\int_0^1\left(\sum_{i=1}^q\int_0^1 g_{1i}(\tau,\vartheta)\,dK_i(\tau)\right)\zeta_1(\vartheta)\,d\vartheta \\
&+\frac{\alpha-1}{\Delta}\left(\int_0^1 t^{\alpha-2}\,dt\right)\frac{\Gamma(\beta)}{\Gamma(\beta-\delta_0)}\int_0^1\left(\sum_{i=1}^p\int_0^1 g_{2i}(\tau,\vartheta)\,dH_i(\tau)\right)\zeta_2(\vartheta)\,d\vartheta \\
&\le \frac{2}{\Gamma(\alpha)}\int_0^1 (1-\vartheta)^{\alpha-\gamma_0-1}\zeta_1(\vartheta)\,d\vartheta + \frac{1}{\Delta}\left(\sum_{i=1}^p \frac{\Gamma(\beta)}{\Gamma(\beta-\gamma_i)}\int_0^1 \tau^{\beta-\gamma_i-1}\,dH_i(\tau)\right) \\
&\times\left(\sum_{i=1}^q \frac{1}{\Gamma(\alpha-\delta_i)}\int_0^1 \tau^{\alpha-\delta_i-1}\,dH_i(\tau)\right)\int_0^1 (1-\vartheta)^{\alpha-\gamma_0-1}\zeta_1(\vartheta)\,d\vartheta \\
&+\frac{\Gamma(\beta)}{\Delta\Gamma(\beta-\delta_0)}\left(\sum_{i=1}^p \frac{1}{\Gamma(\beta-\gamma_i)}\int_0^1 \tau^{\beta-\gamma_i-1}\,dH_i(\tau)\right) \\
&\times\int_0^1 (1-\vartheta)^{\beta-\delta_0-1}\zeta_2(\vartheta)\,d\vartheta.
\end{aligned}$$

Then we deduce

$$\int_0^1 \Xi_2(t)\,dt \le \left[\frac{2}{\Gamma(\alpha)} + \frac{\Gamma(\beta)}{\Delta}\left(\sum_{i=1}^{p}\frac{1}{\Gamma(\beta-\gamma_i)}\int_0^1 \tau^{\beta-\gamma_i-1}\,dH_i(\tau)\right)\right.$$
$$\left.\times\left(\sum_{i=1}^{q}\frac{1}{\Gamma(\alpha-\delta_i)}\int_0^1 \tau^{\alpha-\delta_i-1}\,dH_i(\tau)\right)\right]\int_0^1 (1-\vartheta)^{\alpha-\gamma_0-1}\zeta_1(\vartheta)\,d\vartheta$$
$$+\frac{\Gamma(\beta)}{\Delta\Gamma(\beta-\delta_0)}\left(\sum_{i=1}^{p}\frac{1}{\Gamma(\beta-\gamma_i)}\int_0^1 \tau^{\beta-\gamma_i-1}\,dH_i(\tau)\right)$$
$$\times\int_0^1 (1-\vartheta)^{\beta-\delta_0-1}\zeta_2(\vartheta)\,d\vartheta < \infty. \tag{9}$$

We obtain that $\Xi_2 \in L^1(0,1)$. Thus for any $\tau_1,\ \tau_2 \in [0,1]$ with $\tau_1 \le \tau_2$ and $(u,v) \in S$, by relations (8) and (9), we conclude

$$|\mathcal{A}_1(u,v)(\tau_1) - \mathcal{A}_1(u,v)(\tau_2)| = \left|\int_{\tau_1}^{\tau_2} (\mathcal{A}_1(u,v))'(t)\,dt\right| \le L_2\int_{\tau_1}^{\tau_2} \Xi_2(t)\,dt. \tag{10}$$

By (9) and (10) and the property of absolute continuity of the integral function, we deduce that $\mathcal{A}_1(S)$ is equicontinuous. By a similar technique, we obtain that $\mathcal{A}_2(S)$ is also equicontinuous, and then $\mathcal{A}(S)$ is equicontinuous. We apply now the Ascoli-Arzela theorem and we find that $\mathcal{A}_1(S)$ and $\mathcal{A}_2(S)$ are relatively compact, and so $\mathcal{A}(S)$ is also relatively compact. Hence $\mathcal{A}$ is a compact operator. In addition, we can prove that $\mathcal{A}_1$, $\mathcal{A}_2$ and $\mathcal{A}$ are continuous on $\mathcal{Q}$ (see Lemma 1.4.1 from [14]). Therefore $\mathcal{A}$ is a completely continuous operator on $\mathcal{Q}$. □

We define now the cone

$$\mathcal{Q}_0 = \{(u,v) \in \mathcal{Q},\ \ \min_{t\in[0,1]} u(t) \ge t^{\alpha-1}\|u\|,\ \ \min_{t\in[0,1]} v(t) \ge t^{\beta-1}\|v\|\}.$$

Under the assumptions $(H1)$ and $(H2)$, by using Lemma 3, we obtain that $\mathcal{A}(\mathcal{Q}) \subset \mathcal{Q}_0$, and then $\mathcal{A}|_{\mathcal{Q}_0} : \mathcal{Q}_0 \to \mathcal{Q}_0$ (denoted again by $\mathcal{A}$) is also a completely continuous operator.

Theorem 1 *Assume that $(H1)$ and $(H2)$ hold. If the functions f and g also satisfy the conditions*

$(H3)$ *There exist $\tau_1 \ge 1$ and $\tau_2 \ge 1$ such that*

$$\varphi_{10} = \lim_{\substack{x+y\to 0\\ x,y\ge 0}}\ \sup_{t\in[0,1]} \frac{\varphi_1(t,x,y)}{(x+y)^{\tau_1}} = 0 \ \ \textit{and}\ \ \varphi_{20} = \lim_{\substack{x+y\to 0\\ x,y\ge 0}}\ \sup_{t\in[0,1]} \frac{\varphi_2(t,x,y)}{(x+y)^{\tau_2}} = 0;$$

$(H4)$ *There exists $[\theta_1,\theta_2] \subset [0,1]$, $0 < \theta_1 < \theta_2 < 1$ such that*

$$\mathfrak{f}_{\infty}^{i} = \lim_{\substack{x+y\to\infty \\ x,y\geq 0}} \inf_{t\in[\theta_1,\theta_2]} \frac{f(t,x,y)}{x+y} = \infty \ \ or\ \ \mathfrak{g}_{\infty}^{i} = \lim_{\substack{x+y\to\infty \\ x,y\geq 0}} \inf_{t\in[\theta_1,\theta_2]} \frac{g(t,x,y)}{x+y} = \infty,$$

then problem (1) and (2) has at least one positive solution $(u(t), v(t))$, $t \in [0, 1]$.

Proof We consider the cone $\mathcal{Q}_0$. By $(H3)$, if $M_2 > 0$ and $M_3 > 0$, we obtain that for $\epsilon_1 = \min\left\{\frac{1}{4M_1}, \frac{1}{4M_3}\right\}$ and $\epsilon_2 = \min\left\{\frac{1}{4M_2}, \frac{1}{4M_4}\right\}$, there exists $R_1 \in (0, 1)$ such that

$$\varphi_i(t, x, y) \leq \epsilon_i (x + y)^{\tau_i}, \ \ \forall t \in [0, 1], \ \ x, y \geq 0, \ \ x + y \leq R_1, \ \ i = 1, 2. \quad (11)$$

For $r > 0$ we denote by $B_r = \{(u, v) \in \mathcal{Z}, \ \|(u, v)\|_{\mathcal{Z}} < r\}$, and by ∂B_r and $\overline{B}_r$ its boundary and its closure, respectively.

Then by (11) and Lemma 2, for any $(u, v) \in \partial B_{R_1} \cap \mathcal{Q}_0$ and $t \in [0, 1]$, we find

$$\begin{aligned}
&\mathcal{A}_1(u, v)(t) \leq \int_0^1 \mathcal{J}_1(\vartheta)\zeta_1(\vartheta)\varphi_1(\vartheta, u(\vartheta), v(\vartheta))\, d\vartheta \\
&\quad + \int_0^1 \mathcal{J}_2(\vartheta)\zeta_2(\vartheta)\varphi_2(\vartheta, u(\vartheta), v(\vartheta))\, d\vartheta \\
&\leq \epsilon_1 \int_0^1 \mathcal{J}_1(\vartheta)\zeta_1(\vartheta)(u(\vartheta) + v(\vartheta))^{\tau_1}\, d\vartheta + \epsilon_2 \int_0^1 \mathcal{J}_2(\vartheta)\zeta_2(\vartheta)(u(\vartheta) + v(\vartheta))^{\tau_2}\, d\vartheta \\
&\leq \epsilon_1 M_1 \|(u, v)\|_{\mathcal{Z}}^{\tau_1} + \epsilon_2 M_2 \|(u, v)\|_{\mathcal{Z}}^{\tau_2} \leq \epsilon_1 M_1 \|(u, v)\|_{\mathcal{Z}} + \epsilon_2 M_2 \|(u, v)\|_{\mathcal{Z}} \\
&\leq \tfrac{1}{4}\|(u, v)\|_{\mathcal{Z}} + \tfrac{1}{4}\|(u, v)\|_{\mathcal{Z}} = \tfrac{1}{2}\|(u, v)\|_{\mathcal{Z}},
\end{aligned}$$

$$\begin{aligned}
&\mathcal{A}_2(u, v)(t) \leq \int_0^1 \mathcal{J}_3(\vartheta)\zeta_1(\vartheta)\varphi_1(\vartheta, u(\vartheta), v(\vartheta))\, d\vartheta \\
&\quad + \int_0^1 \mathcal{J}_4(\vartheta)\varphi_2(\vartheta)\varphi_2(\vartheta, u(\vartheta), v(\vartheta))\, d\vartheta \\
&\leq \epsilon_1 \int_0^1 \mathcal{J}_3(\vartheta)\zeta_1(\vartheta)(u(\vartheta) + v(\vartheta))^{\tau_1}\, d\vartheta + \epsilon_2 \int_0^1 \mathcal{J}_4(\vartheta)\zeta_2(\vartheta)(u(\vartheta) + v(\vartheta))^{\tau_2}\, d\vartheta \\
&\leq \epsilon_1 M_3 \|(u, v)\|_{\mathcal{Z}}^{\tau_1} + \epsilon_2 M_4 \|(u, v)\|_{\mathcal{Z}}^{\tau_2} \leq \epsilon_1 M_3 \|(u, v)\|_{\mathcal{Z}} + \epsilon_2 M_4 \|(u, v)\|_{\mathcal{Z}} \\
&\leq \tfrac{1}{4}\|(u, v)\|_{\mathcal{Z}} + \tfrac{1}{4}\|(u, v)\|_{\mathcal{Z}} = \tfrac{1}{2}\|(u, v)\|_{\mathcal{Z}}.
\end{aligned}$$

So, we deduce $\|\mathcal{A}_1(u, v)\| \leq \frac{1}{2}\|(u, v)\|_{\mathcal{Z}}$, $\|\mathcal{A}_2(u, v)\| \leq \frac{1}{2}\|(u, v)\|_{\mathcal{Z}}$ for all $(u, v) \in \partial B_{R_1} \cap \mathcal{Q}_0$, and then

$$\|\mathcal{A}(u, v)\|_{\mathcal{Z}} \leq \|(u, v)\|_{\mathcal{Z}}, \ \ \forall\, (u, v) \in \partial B_{R_1} \cap \mathcal{Q}_0. \quad (12)$$

If $M_2 = 0$ and $M_3 \neq 0$, then we choose $\epsilon_1 = \min\{\frac{1}{4M_1}, \frac{1}{4M_3}\}$ and $\epsilon_2 = \frac{1}{2M_4}$; if $M_2 \neq 0$ and $M_3 = 0$, then we choose $\epsilon_1 = \frac{1}{2M_1}$ and $\epsilon_2 = \min\{\frac{1}{4M_2}, \frac{1}{4M_4}\}$; if $M_2 = M_3 = 0$, then we choose $\epsilon_1 = \frac{1}{2M_1}$ and $\epsilon_2 = \frac{1}{2M_4}$. In all these cases we obtain as above the inequality (12).

Next, in $(H4)$, we suppose that $\mathfrak{f}_{\infty}^{i} = \infty$. Then for $\epsilon_3 = 2(\theta_1^{\alpha-1} m_1 \min\{\theta_1^{\alpha-1}, \theta_1^{\beta-1}\})^{-1} > 0$, where $m_1 = \int_{\theta_1}^{\theta_2} \mathcal{J}_1(s)\, ds > 0$, there exists $C_1 > 0$ such that

$$f(t,x,y) \geq \epsilon_3(x+y) - C_1, \quad \forall t \in [\theta_1,\theta_2], \ x,y \geq 0. \tag{13}$$

Then by (13), for any $(u,v) \in \mathcal{Q}_0$ and $t \in [\theta_1,\theta_2]$, we find

$$\begin{aligned}
\mathcal{A}_1(u,v)(t) &\geq \int_{\theta_1}^{\theta_2} \mathcal{G}_1(t,\vartheta) f(\vartheta,u(\vartheta),v(\vartheta))\,d\vartheta + \int_{\theta_1}^{\theta_2} \mathcal{G}_2(t,\vartheta) g(\vartheta,u(\vartheta),v(\vartheta))\,d\vartheta \\
&\geq \int_{\theta_1}^{\theta_2} \mathcal{G}_1(t,\vartheta) f(\vartheta,u(\vartheta),v(\vartheta))\,d\vartheta \geq \theta_1^{\alpha-1}\int_{\theta_1}^{\theta_2} \mathcal{J}_1(\vartheta) f(\vartheta,u(\vartheta),v(\vartheta))\,d\vartheta \\
&\geq \theta_1^{\alpha-1}\int_{\theta_1}^{\theta_2} \mathcal{J}_1(\vartheta)[\epsilon_3(u(\vartheta)+v(\vartheta)) - C_1]\,d\vartheta \\
&\geq \theta_1^{\alpha-1}\epsilon_3 m_1 \min_{\vartheta\in[\theta_1,\theta_2]}(u(\vartheta)+v(\vartheta)) - \theta_1^{\alpha-1}m_1C_1 \\
&\geq \theta_1^{\alpha-1}\epsilon_3 m_1 (\min_{\vartheta\in[\theta_1,\theta_2]}u(\vartheta) + \min_{\vartheta\in[\theta_1,\theta_2]}v(\vartheta)) - \theta_1^{\alpha-1}m_1C_1 \\
&\geq \theta_1^{\alpha-1}\epsilon_3 m_1 (\theta_1^{\alpha-1}\|u\| + \theta_1^{\beta-1}\|v\|) - \theta_1^{\alpha-1}m_1C_1 \\
&\geq \theta_1^{\alpha-1}\epsilon_3 m_1 \min\{\theta_1^{\alpha-1},\theta_1^{\beta-1}\}\|(u,v)\|_{\mathcal{Z}} - C_2 \\
&= 2\|(u,v)\|_{\mathcal{Z}} - C_2, \quad C_2 = \theta_1^{\alpha-1}m_1C_1.
\end{aligned}$$

Then we deduce $\|\mathcal{A}_1(u,v)\| \geq 2\|(u,v)\|_{\mathcal{Z}} - C_2$ for all $(u,v) \in \mathcal{Q}_0$. We can choose $R_2 \geq \max\{C_2, 1\}$ and then we obtain

$$\|\mathcal{A}(u,v)\|_{\mathcal{Z}} \geq \|\mathcal{A}_1(u,v)\| \geq \|(u,v)\|_{\mathcal{Z}}, \quad \forall (u,v) \in \partial B_{R_2} \cap \mathcal{Q}_0. \tag{14}$$

We consider now in $(H4)$ that $\mathrm{g}_\infty^i = \infty$. Then for $\widetilde{\epsilon}_3 = 2(\theta_1^{\beta-1}m_4\min\{\theta_1^{\alpha-1}, \theta_1^{\beta-1}\})^{-1} > 0$, where $m_4 = \int_{\theta_1}^{\theta_2}\mathcal{J}_4(s)\,ds > 0$, there exists $\widetilde{C}_1 > 0$ such that

$$g(t,x,y) \geq \widetilde{\epsilon}_3(x+y) - \widetilde{C}_1, \quad \forall t \in [\theta_1,\theta_2], \ x,y \geq 0. \tag{15}$$

Then by (15), for any $(u,v) \in \mathcal{Q}_0$ and $t \in [\theta_1,\theta_2]$, we obtain

$$\begin{aligned}
\mathcal{A}_2(u,v)(t) &\geq \int_{\theta_1}^{\theta_2} \mathcal{G}_3(t,\vartheta) f(\vartheta,u(\vartheta),v(\vartheta))\,d\vartheta + \int_{\theta_1}^{\theta_2} \mathcal{G}_4(t,\vartheta) g(\vartheta,u(\vartheta),v(\vartheta))\,d\vartheta \\
&\geq \int_{\theta_1}^{\theta_2} \mathcal{G}_4(t,\vartheta) g(\vartheta,u(\vartheta),v(\vartheta))\,d\vartheta \geq \theta_1^{\beta-1}\int_{\theta_1}^{\theta_2} \mathcal{J}_4(\vartheta) g(\vartheta,u(\vartheta),v(\vartheta))\,d\vartheta \\
&\geq \theta_1^{\beta-1}\int_{\theta_1}^{\theta_2} \mathcal{J}_4(\vartheta)[\widetilde{\epsilon}_3(u(\vartheta)+v(\vartheta)) - \widetilde{C}_1]\,d\vartheta \\
&\geq \theta_1^{\beta-1}\widetilde{\epsilon}_3 m_4 \min_{\vartheta\in[\theta_1,\theta_2]}(u(\vartheta)+v(\vartheta)) - \theta_1^{\beta-1}m_4\widetilde{C}_1 \\
&\geq \theta_1^{\beta-1}\widetilde{\epsilon}_3 m_4 (\min_{\vartheta\in[\theta_1,\theta_2]}u(\vartheta) + \min_{\vartheta\in[\theta_1,\theta_2]}v(\vartheta)) - \theta_1^{\beta-1}m_4\widetilde{C}_1 \\
&\geq \theta_1^{\beta-1}\widetilde{\epsilon}_3 m_4 (\theta_1^{\alpha-1}\|u\| + \theta_1^{\beta-1}\|v\|) - \theta_1^{\beta-1}m_4\widetilde{C}_1 \\
&\geq \theta_1^{\beta-1}\widetilde{\epsilon}_3 m_4 \min\{\theta_1^{\alpha-1},\theta_1^{\beta-1}\}\|(u,v)\|_{\mathcal{Z}} - \widetilde{C}_2 \\
&= 2\|(u,v)\|_{\mathcal{Z}} - \widetilde{C}_2, \quad \widetilde{C}_2 = \theta_1^{\beta-1}m_4\widetilde{C}_1.
\end{aligned}$$

Then we conclude $\|\mathcal{A}_2(u,v)\| \geq 2\|(u,v)\|_{\mathcal{Z}} - \widetilde{C}_2$ for all $(u,v) \in \mathcal{Q}_0$. We can choose $R_2 \geq \max\{\widetilde{C}_2, 1\}$ and we obtain

$$\|\mathcal{A}(u,v)\|_{\mathcal{Z}} \geq \|\mathcal{A}_2(u,v)\| \geq \|(u,v)\|_{\mathcal{Z}}, \quad \forall (u,v) \in \partial B_{R_2} \cap \mathcal{Q}_0. \tag{16}$$

By Lemma 4, (12) and (14) (or (16)) and the Guo-Krasnosel'skii fixed point theorem, we deduce that $\mathcal{A}$ has a fixed point $(u, v) \in (\overline{B}_{R_2} \setminus B_{R_1}) \cap \mathcal{Q}_0$, that is $R_1 \leq \|(u, v)\|_{\mathcal{Z}} \leq R_2$ and $u(t) \geq t^{\alpha-1}\|u\|$ and $v(t) \geq t^{\beta-1}\|v\|$ for all $t \in [0, 1]$. Then $\|u\| > 0$ or $\|v\| > 0$, that is $u(t) > 0$ for all $t \in (0, 1]$ or $v(t) > 0$ for all $t \in (0, 1]$. Hence $(u(t), v(t))$, $t \in [0, 1]$ is a positive solution of problem (1) and (2). □

Theorem 2 *Assume that* $(H1)$ *and* $(H2)$ *hold. If the functions* f *and* g *also satisfy the conditions*

$(H5)$ $\varphi_{1\infty} = \lim_{\substack{x+y\to\infty\\ x,y\geq 0}} \sup_{t\in[0,1]} \dfrac{\varphi_1(t, x, y)}{x+y} = 0$ *and* $\varphi_{2\infty} = \lim_{\substack{x+y\to\infty\\ x,y\geq 0}} \sup_{t\in[0,1]} \dfrac{\varphi_2(t, x, y)}{x+y} = 0$;

$(H6)$ *There exist* $[\theta_1, \theta_2] \subset [0, 1]$, $0 < \theta_1 < \theta_2 < 1$, $\varsigma_1 \in (0, 1]$ *and* $\varsigma_2 \in (0, 1]$ *such that*

$$\mathfrak{f}_0^i = \lim_{\substack{x+y\to 0\\ x,y\geq 0}} \inf_{t\in[\theta_1,\theta_2]} \frac{f(t, x, y)}{(x+y)^{\varsigma_1}} = \infty \ \ or \ \ \mathfrak{g}_0^i = \lim_{\substack{x+y\to 0\\ x,y\geq 0}} \inf_{t\in[\theta_1,\theta_2]} \frac{g(t, x, y)}{(x+y)^{\varsigma_2}} = \infty,$$

then problem (1) and (2) has at least one positive solution $(u(t), v(t))$, $t \in [0, 1]$.

Proof We consider again the cone $\mathcal{Q}_0$. By $(H5)$ we deduce that for $\epsilon_4 \in \left(0, \frac{1}{2(M_1+M_3)}\right)$ and $\epsilon_5 \in \left(0, \frac{1}{2(M_2+M_4)}\right)$, there exist $C_3 > 0$ and $C_4 > 0$ such that

$$\varphi_1(t, x, y) \leq \epsilon_4(x+y) + C_3, \ \ \varphi_2(t, x, y) \leq \epsilon_5(x+y) + C_4, \ \ \forall t \in [0, 1], \ \ x, y \geq 0. \tag{17}$$

By using (17) and $(H2)$, for any $(u, v) \in \mathcal{Q}_0$, we find

$$\begin{aligned}
\mathcal{A}_1(u, v)(t) &\leq \int_0^1 \mathcal{J}_1(\vartheta)\zeta_1(\vartheta)\varphi_1(\vartheta, u(\vartheta), v(\vartheta))\, d\vartheta \\
&\quad + \int_0^1 \mathcal{J}_2(\vartheta)\zeta_2(\vartheta)\varphi_2(\vartheta, u(\vartheta), v(\vartheta))\, d\vartheta \\
&\leq \int_0^1 \mathcal{J}_1(\vartheta)\zeta_1(\vartheta)(\epsilon_4(u(\vartheta) + v(\vartheta)) + C_3)\, d\vartheta \\
&\quad + \int_0^1 \mathcal{J}_2(\vartheta)\zeta_2(\vartheta)(\epsilon_5(u(\vartheta) + v(\vartheta)) + C_4)\, d\vartheta \\
&\leq (\epsilon_4\|(u, v)\|_{\mathcal{Z}} + C_3)M_1 + (\epsilon_5\|(u, v)\|_{\mathcal{Z}} + C_4)M_2 \\
&= (\epsilon_4 M_1 + \epsilon_5 M_2)\|(u, v)\|_{\mathcal{Z}} + C_3 M_1 + C_4 M_2, \ \ \forall t \in [0, 1], \\
\mathcal{A}_2(u, v)(t) &\leq \int_0^1 \mathcal{J}_3(\vartheta)\zeta_1(\vartheta)\varphi_1(\vartheta, u(\vartheta), v(\vartheta))\, d\vartheta \\
&\quad + \int_0^1 \mathcal{J}_4(\vartheta)\zeta_2(\vartheta)\varphi_2(\vartheta, u(\vartheta), v(\vartheta))\, d\vartheta \\
&\leq \int_0^1 \mathcal{J}_3(\vartheta)\zeta_1(\vartheta)(\epsilon_4(u(\vartheta) + v(\vartheta)) + C_3)\, d\vartheta \\
&\quad + \int_0^1 \mathcal{J}_4(\vartheta)\zeta_2(\vartheta)(\epsilon_5(u(\vartheta) + v(\vartheta)) + C_4)\, d\vartheta \\
&\leq (\epsilon_4\|(u, v)\|_{\mathcal{Z}} + C_3)M_3 + (\epsilon_5\|(u, v)\|_{\mathcal{Z}} + C_4)M_4 \\
&= (\epsilon_4 M_3 + \epsilon_5 M_4)\|(u, v)\|_{\mathcal{Z}} + C_3 M_3 + C_4 M_4, \ \ \forall t \in [0, 1].
\end{aligned}$$

Then we deduce

$$\|\mathcal{A}_1(u,v)\| \le (\epsilon_4 M_1 + \epsilon_5 M_2)\|(u,v)\|_{\mathcal{Z}} + C_3 M_1 + C_4 M_2,$$
$$\|\mathcal{A}_2(u,v)\| \le (\epsilon_4 M_3 + \epsilon_5 M_4)\|(u,v)\|_{\mathcal{Z}} + C_3 M_3 + C_4 M_4,$$

and so

$$\begin{aligned}\|\mathcal{A}(u,v)\|_{\mathcal{Z}} &\le [\epsilon_4(M_1+M_3) + \epsilon_5(M_2+M_4)]\|(u,v)\|_{\mathcal{Z}} \\ &\quad + C_3(M_1+M_3) + C_4(M_2+M_4) \\ &< \|(u,v)\|_{\mathcal{Z}} + C_5, \quad C_5 = C_3(M_1+M_3) + C_4(M_2+M_4).\end{aligned}$$

We can choose $R_3 > 1$ such that

$$\|\mathcal{A}(u,v)\|_{\mathcal{Z}} \le \|(u,v)\|_{\mathcal{Z}}, \quad \forall (u,v) \in \partial B_{R_3} \cap \mathcal{Q}_0. \tag{18}$$

In $(H6)$, let us first consider $\mathfrak{f}_0^i = \infty$. We deduce that for $\epsilon_6 = (\theta_1^{\alpha-1} m_1 \times \min\{\theta_1^{\alpha-1}, \theta_1^{\beta-1})^{-1} > 0$, where $m_1 = \int_{\theta_1}^{\theta_2} \mathcal{J}_1(s)\,ds > 0$, there exists $R_4 \in (0,1]$ such that

$$f(t,x,y) \ge \epsilon_6 (x+y)^{\varsigma_1}, \quad \forall t \in [\theta_1,\theta_2], \;\; x,\, y \ge 0, \;\; x+y \le R_4. \tag{19}$$

Then by using (19), for any $(u,v) \in \partial B_{R_4} \cap \mathcal{Q}_0$ and $t \in [\theta_1,\theta_2]$ we obtain

$$\begin{aligned}\mathcal{A}_1(u,v)(t) &\ge \int_{\theta_1}^{\theta_2} \mathcal{G}_1(t,\vartheta) f(\vartheta,u(\vartheta),v(\vartheta))\,d\vartheta + \int_{\theta_1}^{\theta_2} \mathcal{G}_2(t,\vartheta) g(\vartheta,u(\vartheta),v(\vartheta))\,d\vartheta \\ &\ge t^{\alpha-1}\int_{\theta_1}^{\theta_2} \mathcal{J}_1(\vartheta)\epsilon_6 (u(\vartheta)+v(\vartheta))^{\varsigma_1}\,d\vartheta \ge \theta_1^{\alpha-1}\int_{\theta_1}^{\theta_2} \mathcal{J}_1(\vartheta)\epsilon_6(u(\vartheta)+v(\vartheta))\,d\vartheta \\ &\ge \theta_1^{\alpha-1}\epsilon_6 \int_{\theta_1}^{\theta_2} \mathcal{J}_1(\vartheta)\left(\min_{s\in[\theta_1,\theta_2]}(u(s)+v(s))\right)d\vartheta \\ &\ge \theta_1^{\alpha-1}\epsilon_6 \min_{s\in[\theta_1,\theta_2]}(u(s)+v(s)) \int_{\theta_1}^{\theta_2} \mathcal{J}_1(\vartheta)\,d\vartheta \\ &\ge \theta_1^{\alpha-1}\epsilon_6 m_1 \left(\min_{s\in[\theta_1,\theta_2]} u(s) + \min_{s\in[\theta_1,\theta_2]} v(s)\right) \\ &\ge \theta_1^{\alpha-1}\epsilon_6 m_1 (\theta_1^{\alpha-1}\|u\| + \theta_1^{\beta-1}\|v\|) \\ &\ge \theta_1^{\alpha-1}\epsilon_6 m_1 \min\{\theta_1^{\alpha-1},\theta_1^{\beta-1}\}(\|u\|+\|v\|) = \|(u,v)\|_{\mathcal{Z}}.\end{aligned}$$

Therefore $\|\mathcal{A}_1(u,v)\| \ge \|(u,v)\|_{\mathcal{Z}}$ for all $(u,v) \in \partial B_{R_4} \cap \mathcal{Q}_0$, and then

$$\|\mathcal{A}(u,v)\|_{\mathcal{Z}} \ge \|\mathcal{A}_1(u,v)\| \ge \|(u,v)\|_{\mathcal{Z}}, \quad \forall (u,v) \in \partial B_{R_4} \cap \mathcal{Q}_0. \tag{20}$$

In the case in which $\mathfrak{g}_0^i = \infty$ in $(H6)$, we deduce that for $\widetilde{\epsilon}_6 = (\theta_1^{\beta-1} m_4 \times \min\{\theta_1^{\alpha-1}, \theta_1^{\beta-1}\})^{-1} > 0$, where $m_4 = \int_{\theta_1}^{\theta_2} \mathcal{J}_4(s)\,ds > 0$, there exists $R_4 \in (0,1]$ such that

$$g(t,x,y) \ge \widetilde{\epsilon}_6 (x+y)^{\varsigma_2}, \quad \forall t \in [\theta_1,\theta_2], \;\; x, y \ge 0, \;\; x+y \le R_4. \tag{21}$$

Then by using (21), for any $(u,v) \in \partial B_{R_4} \cap \mathcal{Q}_0$ and $t \in [\theta_1,\theta_2]$ we obtain

$$
\begin{aligned}
\mathcal{A}_2(u,v)(t) &\geq \int_{\theta_1}^{\theta_2} \mathcal{G}_3(t,\vartheta) f(\vartheta, u(\vartheta), v(\vartheta))\, d\vartheta + \int_{\theta_1}^{\theta_2} \mathcal{G}_4(t,\vartheta) g(\vartheta, u(\vartheta), v(\vartheta))\, d\vartheta \\
&\geq t^{\beta-1} \int_{\theta_1}^{\theta_2} \mathcal{J}_4(\vartheta) \widetilde{\epsilon}_6 (u(\vartheta)+v(\vartheta))^{\varsigma_2}\, d\vartheta \geq \theta_1^{\beta-1} \int_{\theta_1}^{\theta_2} \mathcal{J}_4(\vartheta) \widetilde{\epsilon}_6 (u(\vartheta)+v(\vartheta))\, d\vartheta \\
&\geq \theta_1^{\beta-1} \widetilde{\epsilon}_6 \int_{\theta_1}^{\theta_2} \mathcal{J}_4(\vartheta) \left(\min_{s\in[\theta_1,\theta_2]} (u(s)+v(s)) \right) d\vartheta \\
&\geq \theta_1^{\beta-1} \widetilde{\epsilon}_6 \min_{s\in[\theta_1,\theta_2]} (u(s)+v(s)) \int_{\theta_1}^{\theta_2} \mathcal{J}_4(\vartheta)\, d\vartheta \\
&\geq \theta_1^{\beta-1} \widetilde{\epsilon}_6 m_4 \left(\min_{s\in[\theta_1,\theta_2]} u(s) + \min_{s\in[\theta_1,\theta_2]} v(s) \right) \\
&\geq \theta_1^{\beta-1} \widetilde{\epsilon}_6 m_4 (\theta_1^{\alpha-1} \|u\| + \theta_1^{\beta-1} \|v\|) \\
&\geq \theta_1^{\beta-1} \widetilde{\epsilon}_6 m_4 \min\{\theta_1^{\alpha-1}, \theta_1^{\beta-1}\} (\|u\|+\|v\|) = \|(u,v)\|_{\mathcal{Z}}.
\end{aligned}
$$

Hence $\|\mathcal{A}_2(u,v)\| \geq \|(u,v)\|_{\mathcal{Z}}$ for all $(u,v) \in \partial B_{R_4} \cap \mathcal{Q}_0$, and then

$$
\|\mathcal{A}(u,v)\|_{\mathcal{Z}} \geq \|\mathcal{A}_2(u,v)\| \geq \|(u,v)\|_{\mathcal{Z}}, \quad \forall (u,v) \in \partial B_{R_4} \cap \mathcal{Q}_0. \tag{22}
$$

By Lemma 4, (18) and (20) (or (22)) and the Guo-Krasnosel'skii fixed point theorem, we conclude that $\mathcal{A}$ has at least one fixed point $(u,v) \in (\overline{B}_{R_3} \setminus B_{R_4}) \cap \mathcal{Q}_0$, that is $R_4 \leq \|(u,v)\|_{\mathcal{Z}} \leq R_3$, which is a positive solution of problem (1) and (2). □

Theorem 3 *Assume that* $(H1)$, $(H2)$, $(H4)$ *and* $(H6)$ *hold. If the functions* f *and* g *also satisfy the condition*

$(H7)$ $D_0 \left(\max_{i=1,\ldots,4} M_i \right) < \frac{1}{4}$, *where* $M_1 = \int_0^1 \mathcal{J}_1(s)\zeta_1(s)\, ds$, $M_2 = \int_0^1 \mathcal{J}_2(s)\zeta_2(s)\, ds$, $M_3 = \int_0^1 \mathcal{J}_3(s)\zeta_1(s)\, ds$, $M_4 = \int_0^1 \mathcal{J}_4(s)\zeta_2(s)\, ds$, *and*

$$
D_0 = \max\left\{ \max_{t,x,y\in[0,1]} \varphi_1(t,x,y),\ \max_{t,x,y\in[0,1]} \varphi_2(t,x,y) \right\},
$$

then problem (1) and (2) has at least two positive solutions $(u_1(t), v_1(t))$, $(u_2(t), v_2(t))$, $t \in [0,1]$.

Proof If $(H1)$, $(H2)$ and $(H4)$ hold, then by the proof of Theorem 1, we deduce that there exists $R_2 > 1$ such that

$$
\|\mathcal{A}(u,v)\|_{\mathcal{Z}} \geq \|(u,v)\|_{\mathcal{Z}}, \quad \forall (u,v) \in \partial B_{R_2} \cap \mathcal{Q}_0. \tag{23}
$$

If $(H1)$, $(H2)$ and $(H6)$ hold, then by the proof of Theorem 2, we find that there exists $R_4 < 1$ (we can choose $R_4 < 1$) such that

$$
\|\mathcal{A}(u,v)\|_{\mathcal{Z}} \geq \|(u,v)\|_{\mathcal{Z}}, \quad \forall (u,v) \in \partial B_{R_4} \cap \mathcal{Q}_0. \tag{24}
$$

We consider now the set $B_1 = \{(u,v) \in \mathcal{Z},\ \|(u,v)\|_{\mathcal{Z}} < 1\}$. By $(H7)$, for any $(u,v) \in \partial B_1 \cap \mathcal{Q}_0$ and $t \in [0,1]$, we obtain

$$\begin{aligned}
\mathcal{A}_1(u,v)(t) &\le \int_0^1 \mathcal{J}_1(\vartheta)\zeta_1(\vartheta)\varphi_1(\vartheta,u(\vartheta),v(\vartheta))\,d\vartheta \\
&\quad+\int_0^1 \mathcal{J}_2(\vartheta)\zeta_2(\vartheta)\varphi_2(\vartheta,u(\vartheta),v(\vartheta))\,d\vartheta \\
&\le D_0\int_0^1 \mathcal{J}_1(\vartheta)\zeta_1(\vartheta)\,d\vartheta + D_0\int_0^1 \mathcal{J}_2(\vartheta)\zeta_2(\vartheta)\,d\vartheta = D_0M_1 + D_0M_2 < \frac{1}{2}, \\
\mathcal{A}_2(u,v)(t) &\le \int_0^1 \mathcal{J}_3(\vartheta)\zeta_1(\vartheta)\varphi_1(\vartheta,u(\vartheta),v(\vartheta))\,d\vartheta \\
&\quad+\int_0^1 \mathcal{J}_4(\vartheta)\zeta_2(\vartheta)\varphi_2(\vartheta,u(\vartheta),v(\vartheta))\,d\vartheta \\
&\le D_0\int_0^1 \mathcal{J}_3(\vartheta)\zeta_1(\vartheta)\,d\vartheta + D_0\int_0^1 \mathcal{J}_4(\vartheta)\zeta_2(\vartheta)\,d\vartheta = D_0M_3 + D_0M_4 < \frac{1}{2},
\end{aligned}$$

and so $\|\mathcal{A}_1(u,v)\| < \frac{1}{2}$ and $\|\mathcal{A}_2(u,v)\| < \frac{1}{2}$ for all $(u,v) \in \partial B_1 \cap \mathcal{Q}_0$. Then

$$\|\mathcal{A}(u,v)\|_{\mathcal{Z}} = \|\mathcal{A}_1(u,v)\| + \|\mathcal{A}_2(u,v)\| < 1 = \|(u,v)\|_{\mathcal{Z}}, \quad \forall\,(u,v) \in \partial B_1 \cap \mathcal{Q}_0. \tag{25}$$

Therefore, by (23) and (25) and the Guo-Krasnosel'skii fixed point theorem, we deduce that problem (1) and (2) has one positive solution $(u_1,v_1) \in \mathcal{Q}_0$ with $1 < \|(u_1,v_1)\|_{\mathcal{Z}} \le R_2$. By (24) and (25) and the Guo-Krasnosel'skii fixed point theorem, we conclude that problem (1) and (2) has another positive solution $(u_2,v_2) \in \mathcal{Q}_0$ with $R_4 \le \|(u_2,v_2)\|_{\mathcal{Z}} < 1$. So problem (1) and (2) has at least two positive solutions $(u_1(t),v_1(t)),\ (u_2(t),v_2(t)),\ t \in [0,1]$. □

4 Examples

Let $\alpha = \frac{10}{3}$ $(n=4)$, $\beta = \frac{9}{2}$ $(m=5)$, $p=1$, $q=2$, $\gamma_1 = \frac{1}{2}$, $\delta_0 = \frac{5}{4}$, $\delta_1 = \frac{1}{3}$, $\delta_2 = \frac{6}{5}$, $\gamma_0 = \frac{9}{4}$, $H_1(t) = \{1,\ t \in [0,1/3);\ 3,\ t \in [1/3,1]\}$, $K_1(t) = 4t$, $t \in [0,1]$, and $K_2(t) = \{1/5,\ t \in [0,1/2);\ 8/15,\ t \in [1/2,1]\}$.

We consider the system of fractional differential equations

$$\begin{cases} D_{0+}^{10/3}u(t) + f(t,u(t),v(t)) = 0, & t \in (0,1), \\ D_{0+}^{9/2}v(t) + g(t,u(t),v(t)) = 0, & t \in (0,1), \end{cases} \tag{26}$$

with the coupled nonlocal boundary conditions

$$\begin{cases} u(0) = u'(0) = u''(0) = 0, \quad D_{0+}^{9/4}u(1) = 2D_{0+}^{1/2}v\left(\frac{1}{3}\right), \\ v(0) = v'(0) = v''(0) = v'''(0) = 0, \\ \quad D_{0+}^{5/4}v(1) = 4\int_0^1 D_{0+}^{1/3}u(t)\,dt + \frac{1}{3}D_{0+}^{6/5}u\left(\frac{1}{2}\right). \end{cases} \tag{27}$$

We obtain here $\Delta \approx 12.905008 > 0$, so assumption $(H1)$ is satisfied. In addition, we find

$$g_1(t,s) = \frac{1}{\Gamma(10/3)} \begin{cases} t^{7/3}(1-s)^{1/12} - (t-s)^{7/3}, & 0 \le s \le t \le 1, \\ t^{7/3}(1-s)^{1/12}, & 0 \le t \le s \le 1, \end{cases}$$
$$g_{11}(\tau,s) = \frac{1}{2} \begin{cases} \tau^2(1-s)^{1/12} - (\tau-s)^2, & 0 \le s \le \tau \le 1, \\ \tau^2(1-s)^{1/12}, & 0 \le \tau \le s \le 1, \end{cases}$$
$$g_{12}(\tau,s) = \frac{1}{\Gamma(32/15)} \begin{cases} \tau^{17/15}(1-s)^{1/12} - (\tau-s)^{17/15}, & 0 \le s \le \tau \le 1, \\ \tau^{17/15}(1-s)^{1/12}, & 0 \le \tau \le s \le 1, \end{cases}$$
$$g_2(t,s) = \frac{1}{\Gamma(9/2)} \begin{cases} t^{7/2}(1-s)^{9/4} - (t-s)^{7/2}, & 0 \le s \le t \le 1, \\ t^{7/2}(1-s)^{9/4}, & 0 \le t \le s \le 1, \end{cases}$$
$$g_{21}(\tau,s) = \frac{1}{6} \begin{cases} \tau^3(1-s)^{9/4} - (\tau-s)^3, & 0 \le s \le \tau \le 1, \\ \tau^3(1-s)^{9/4}, & 0 \le \tau \le s \le 1, \end{cases}$$
$$h_1(\nu) = \tfrac{1}{\Gamma(10/3)}[(1-\nu)^{1/12} - (1-\nu)^{7/3}], \quad \nu \in [0,1],$$
$$h_2(\nu) = \tfrac{1}{\Gamma(9/2)}[(1-\nu)^{9/4} - (1-\nu)^{7/2}], \quad \nu \in [0,1],$$
$$\mathcal{G}_1(t,s) = g_1(t,s) + \tfrac{t^{7/3}\Gamma(9/2)}{81\varDelta}\left[4\int_0^1 g_{11}(\tau,s)\,d\tau + \tfrac{1}{3}g_{12}\left(\tfrac{1}{2},s\right)\right],$$
$$\mathcal{G}_2(t,s) = \tfrac{2t^{7/3}\Gamma(9/2)}{\varDelta\Gamma(13/4)}g_{21}\left(\tfrac{1}{3},s\right),$$
$$\mathcal{G}_3(t,s) = \tfrac{t^{7/2}\Gamma(10/3)}{\varDelta\Gamma(13/12)}\left[4\int_0^1 g_{11}(\tau,s)\,d\tau + \tfrac{1}{3}g_{12}\left(\tfrac{1}{2},s\right)\right],$$
$$\mathcal{G}_4(t,s) = g_2(t,s) + \tfrac{2t^{7/2}}{\varDelta}\left[\tfrac{2\Gamma(10/3)}{3} + \tfrac{\Gamma(10/3)}{3\Gamma(32/15)2^{17/15}}\right]g_{21}\left(\tfrac{1}{3},s\right),$$

for all $(t,s) \in [0,1] \times [0,1]$.

In addition we have

$$\mathcal{J}_1(\nu) = \begin{cases} h_1(\nu) + \frac{\Gamma(9/2)}{81\varDelta}\left\{\frac{2}{3}(1-\nu)^{1/12} - \frac{2}{3}(1-\nu)^3 + \frac{1}{3\Gamma(32/15)} \right. \\ \quad \left. \times \left[\left(\frac{1}{2}\right)^{17/15}(1-\nu)^{1/12} - \left(\frac{1}{2}-\nu\right)^{17/15}\right]\right\}, & 0 \le \nu < \frac{1}{2}, \\ h_1(\nu) + \frac{\Gamma(9/2)}{81\varDelta}\left[\frac{2}{3}(1-\nu)^{1/12} - \frac{2}{3}(1-\nu)^3 + \frac{1}{3\Gamma(32/15)} \right. \\ \quad \left. \times \left(\frac{1}{2}\right)^{17/15}(1-\nu)^{1/12}\right], & \frac{1}{2} \le \nu \le 1, \end{cases}$$
$$\mathcal{J}_2(\nu) = \frac{\Gamma(9/2)}{81\varDelta\Gamma(13/4)} \begin{cases} (1-\nu)^{9/4} - (1-3\nu)^3, & 0 \le \nu < \frac{1}{3}, \\ (1-\nu)^{9/4}, & \frac{1}{3} \le \nu \le 1, \end{cases}$$
$$\mathcal{J}_3(\nu) = \begin{cases} \frac{\Gamma(10/3)}{\varDelta\Gamma(13/12)}\left\{\frac{2}{3}(1-\nu)^{1/12} - \frac{2}{3}(1-\nu)^3 + \frac{1}{3\Gamma(32/15)} \right. \\ \quad \left. \times \left[\left(\frac{1}{2}\right)^{17/15}(1-\nu)^{1/12} - \left(\frac{1}{2}-\nu\right)^{17/15}\right]\right\}, & 0 \le \nu < \frac{1}{2}, \\ \frac{\Gamma(10/3)}{\varDelta\Gamma(13/12)}\left[\frac{2}{3}(1-\nu)^{1/12} - \frac{2}{3}(1-\nu)^3 + \frac{1}{3\Gamma(32/15)} \right. \\ \quad \left. \times \left(\frac{1}{2}\right)^{17/15}(1-\nu)^{1/12}\right], & \frac{1}{2} \le \nu \le 1, \end{cases}$$
$$\mathcal{J}_4(\nu) = \begin{cases} h_2(\nu) + \frac{1}{\varDelta}\left[\frac{2}{3}\Gamma(10/3) + \frac{\Gamma(10/3)}{3\Gamma(32/15)}\left(\frac{1}{2}\right)^{17/15}\right] \\ \quad \times \frac{1}{81}\left[(1-\nu)^{9/4} - (1-3\nu)^3\right], & 0 \le \nu < \frac{1}{3}, \\ h_2(\nu) + \frac{1}{\varDelta}\left[\frac{2}{3}\Gamma(10/3) + \frac{\Gamma(10/3)}{3\Gamma(32/15)}\left(\frac{1}{2}\right)^{17/15}\right] \\ \quad \times \frac{1}{81}(1-\nu)^{9/4}, & \frac{1}{3} \le \nu \le 1. \end{cases}$$

Example 1 We consider the functions

$$f(t,x,y)=\frac{(x+y)^c}{t^{\varrho_1}(1-t)^{\varrho_2}},\quad g(t,x,y)=\frac{(x+y)^d}{t^{\varrho_3}(1-t)^{\varrho_4}},\quad t\in(0,1),\quad x,y\geq 0, \tag{28}$$

with $c,\ d>1$ and $\varrho_i\in(0,1),\ i=1,\ldots,4$. We have $f(t,x,y)=\zeta_1(t)\varphi_1(t,x,y)$, $g(t,x,y)=\zeta_2(t)\varphi_2(t,x,y)$, where $\zeta_1(t)=\frac{1}{t^{\varrho_1}(1-t)^{\varrho_2}}$, $\zeta_2(t)=\frac{1}{t^{\varrho_3}(1-t)^{\varrho_4}}$ for all $t\in(0,1)$, and $\varphi_1(t,x,y)=(x+y)^c$, $\varphi_2(t,x,y)=(x+y)^d$ for all $t\in[0,1], x,\ y\geq 0$. Besides we have

$$\begin{aligned}
0&<\int_0^1(1-\tau)^{\alpha-\gamma_0-1}\zeta_1(\tau)\,d\tau=\int_0^1(1-\tau)^{1/12}\frac{1}{\tau^{\varrho_1}(1-\tau)^{\varrho_2}}\,d\tau\\
&=B\left(1-\varrho_1,\tfrac{13}{12}-\varrho_2\right)<\infty,\\
0&<\int_0^1(1-\tau)^{\beta-\delta_0-1}\zeta_2(\tau)\,d\tau=\int_0^1(1-\tau)^{9/4}\frac{1}{\tau^{\varrho_3}(1-\tau)^{\varrho_4}}\,d\tau\\
&=B\left(1-\varrho_3,\tfrac{13}{4}-\varrho_4\right)<\infty.
\end{aligned}$$

Therefore assumption $(H2)$ is satisfied. In addition, in $(H3)$, for $\tau_1=\tau_2=1$, we obtain $\varphi_{10}=\varphi_{20}=0$, and in $(H4)$ for $[\theta_1,\theta_2]\subset(0,1)$, we have $\mathfrak{f}^i_\infty=\infty$ (and $\mathfrak{g}^i_\infty=\infty$). Then by Theorem 1, we deduce that problem (26) and (27) with the singular nonlinearities (28) has at least one positive solution $(u(t),v(t))$, $t\in[0,1]$.

Example 2 We consider the functions

$$\begin{aligned}
f(t,x,y)&=\frac{c_0(t+2)}{(t^2+9)\sqrt{t}}\left[(x+y)^{\alpha_1}+(x+y)^{\alpha_2}\right],\quad t\in(0,1],\ x,y\geq 0,\\
g(t,x,y)&=\frac{d_0(3+\cos t)}{(t+2)^4\sqrt[3]{1-t}}\left(x^{\beta_1}+y^{\beta_2}\right),\quad t\in[0,1),\ x,\ y\geq 0,
\end{aligned} \tag{29}$$

where $c_0>0$, $d_0>0$, $\alpha_1>1$, $0<\alpha_2<1$, $\beta_1>0$, $\beta_2>0$. Here we have

$$\begin{aligned}
&\zeta_1(t)=\tfrac{1}{\sqrt{t}},\ t\in(0,1];\quad \varphi_1(t,x,y)=\tfrac{c_0(t+2)}{t^2+9}[(x+y)^{\alpha_1}+(x+y)^{\alpha_2}];\\
&\zeta_2(t)=\tfrac{1}{\sqrt[3]{1-t}},\ t\in[0,1);\quad \varphi_2(t,x,y)=\tfrac{d_0(3+\cos t)}{(t+2)^4}(x^{\beta_1}+y^{\beta_2}),
\end{aligned}$$

for all $t\in[0,1],\ x,y\geq 0$, and

$$\begin{aligned}
&\int_0^1(1-\tau)^{\alpha-\gamma_0-1}\zeta_1(\tau)\,d\tau=\int_0^1(1-\tau)^{1/12}\frac{1}{\tau^{1/2}}\,d\tau=B\left(\frac12,\frac{13}{12}\right)\approx 1.905<\infty,\\
&\int_0^1(1-\tau)^{\beta-\delta_0-1}\zeta_2(\tau)\,d\tau=\int_0^1(1-\tau)^{9/4}\frac{1}{(1-\tau)^{1/3}}\,d\tau=\frac{12}{35}<\infty.
\end{aligned}$$

Then assumption $(H2)$ is satisfied. For $\theta_1=\frac15$ and $\theta_2=\frac45$, we find $\mathfrak{f}^i_\infty=\infty$, and if we consider $\varsigma_1>\alpha_2$, $\varsigma_1\leq 1$, we have $\mathfrak{f}^i_0=\infty$, and then assumptions $(H4)$ and $(H6)$ are satisfied. By some computations we obtain $M_1\approx 0.33178183$, $M_2\approx 0.00112407$, $M_3\approx 0.18026609$, $M_4\approx 0.00939684$; in addition we have $D_0=$

$\max\left\{\frac{3c_0}{10}(2^{\alpha_1}+2^{\alpha_2}), \frac{d_0}{2}\right\}$. If $c_0 < \frac{5}{6M_i(2^{\alpha_1}+2^{\alpha_2})}$ for all $i = 1, \ldots, 4$, and $d_0 < \frac{1}{2M_i}$ for all $i = 1, \ldots, 4$, then the inequalities $D_0 M_i < 1$ for all $i = 1, \ldots, 4$ are satisfied (that is, assumption $(H7)$ is satisfied). For example, if $\alpha_1 = \frac{4}{3}$ and $\alpha_2 = \frac{3}{4}$, and $c_0 \leq 0.59$ and $d_0 \leq 1.5$, then the above inequalities are satisfied. By Theorem 3, we conclude that problem (26) and (27) with the singular nonlinearities (29) has at least two positive solutions $(u_1(t), v_1(t))$, $(u_2(t), v_2(t))$, $t \in [0, 1]$.

References

1. R.P. Agarwal, R. Luca, Positive solutions for a semipositone singular Riemann-Liouville fractional differential problem. Int. J. Nonlinear Sci. Numer. Simul. **20**(7–8), 823–832 (2019)
2. B. Ahmad, A. Alsaedi, S. Aljoudi, S.K. Ntouyas, A six-point nonlocal boundary value problem of nonlinear coupled sequential fractional integro-differential equations and coupled integral boundary conditions. J. Appl. Math. Comput. **56**(1–2), 367–389 (2018)
3. B. Ahmad, A. Alsaedi, S.K. Ntouyas, J. Tariboon, *Hadamard-Type Fractional Differential Equations, Inclusions and Inequalities* (Springer, Cham, Switzerland, 2017)
4. B. Ahmad, R. Luca, Existence of solutions for sequential fractional integro-differential equations and inclusions with nonlocal boundary conditions. Appl. Math. Comput. **339**, 516–534 (2018)
5. B. Ahmad, S.K. Ntouyas, A. Alsaedi, Sequential fractional differential equations and inclusions with semi-periodic and nonlocal integro-multipoint boundary conditions. J. King Saud Univ. Sci. **31**(2), 184–193 (2019)
6. A.A.M. Arafa, S.Z. Rida, M. Khalil, Fractional modeling dynamics of HIV and $CD4^+$ T-cells during primary infection. Nonlinear Biomed. Phys. **6**(1), 1–7 (2012)
7. D. Baleanu, K. Diethelm, E. Scalas, J.J. Trujillo, *Fractional Calculus Models and Numerical Methods*. Series on Complexity, Nonlinearity and Chaos (World Scientific, Boston, 2012)
8. S. Das, *Functional Fractional Calculus for System Identification and Controls* (Springer, New York, 2008)
9. Y. Ding, H. Ye, A fractional-order differential equation model of HIV infection of $CD4^+$ T-cells. Math. Comput. Model. **50**, 386–392 (2009)
10. V. Djordjevic, J. Jaric, B. Fabry, J. Fredberg, D. Stamenovic, Fractional derivatives embody essential features of cell rheological behavior. Ann. Biomed. Eng. **31**, 692–699 (2003)
11. J.R. Graef, L. Kong, Q. Kong, M. Wang, Uniqueness of positive solutions of fractional boundary value problems with non-homogeneous integral boundary conditions. Fract. Calc. Appl. Anal. **15**(3), 509–528 (2012)
12. Z.M. Ge, C.Y. Ou, Chaos synchronization of fractional order modified Duffing systems with parameters excited by a chaotic signal. Chaos Solitons Fractals **35**, 705–717 (2008)
13. D. Guo, V. Lakshmikantham, *Nonlinear Problems in Abstract Cones* (Academic Press, New York, 1988)
14. J. Henderson, R. Luca, *Boundary Value Problems for Systems of Differential, Difference and Fractional Equations. Positive Solutions* (Elsevier, Amsterdam, 2016)
15. J. Henderson, R. Luca, Existence of positive solutions for a singular fractional boundary value problem. Nonlinear Anal. Model. Control **22**(1), 99–114 (2017)
16. J. Henderson, R. Luca, Systems of Riemann-Liouville fractional equations with multi-point boundary conditions. Appl. Math. Comput. **309**, 303–323 (2017)
17. J. Henderson, R. Luca, A. Tudorache, Existence and nonexistence of positive solutions for coupled Riemann-Liouville fractional boundary value problems. Discrete Dyn. Nat. Soc. **2016**, Article ID 2823971, 1–12 (2016)

18. J. Henderson, R. Luca, A. Tudorache. Existence of positive solutions for a system of fractional boundary value problems, in *Differential and Difference Equations with Applications*, ed. by S. Pinelas, Z. Dosla, O. Dosly, P.E. Kloeden. ICDDEA, Amadora, Portugal, May 2015 (Springer, 2016), pp. 349–357
19. J. Jiang, L. Liu, Y. Wu, Positive solutions to singular fractional differential system with coupled boundary conditions. Commun. Nonlinear Sci. Numer. Simul. **18**(11), 3061–3074 (2013)
20. A.A. Kilbas, H.M. Srivastava, J.J. Trujillo, *Theory and Applications of Fractional Differential Equations*. North-Holland Mathematics Studies, vol. 204 (Elsevier Science B.V., Amsterdam, 2006)
21. J. Klafter, S.C. Lim, R. Metzler (eds.), *Fractional Dynamics in Physics* (World Scientific, Singapore, 2011)
22. R. Luca, Positive solutions for a system of fractional differential equations with p-Laplacian operator and multi-point boundary conditions. Nonlinear Anal. Model. Control **23**(5), 771–801 (2018)
23. R. Luca, A. Tudorache, Positive solutions to a system of semipositone fractional boundary value problems. Adv. Differ. Equ. **2014**(179), 1–11 (2014)
24. R. Metzler, J. Klafter, The random walks guide to anomalous diffusion: a fractional dynamics approach. Phys. Rep. **339**, 1–77 (2000)
25. M. Ostoja-Starzewski, Towards thermoelasticity of fractal media. J. Therm. Stress. **30**, 889–896 (2007)
26. I. Podlubny, *Fractional Differential Equations* (Academic Press, San Diego, 1999)
27. J. Sabatier, O.P. Agrawal, J.A.T. Machado (eds.), *Advances in Fractional Calculus: Theoretical Developments and Applications in Physics and Engineering* (Springer, Dordrecht, 2007)
28. S.G. Samko, A.A. Kilbas, O.I. Marichev, *Fractional Integrals and Derivatives. Theory and Applications* (Gordon and Breach, Yverdon, 1993)
29. C. Shen, H. Zhou, L. Yang, Positive solution of a system of integral equations with applications to boundary value problems of differential equations. Adv. Differ. Equ. **2016**(260), 1–26 (2016)
30. I.M. Sokolov, J. Klafter, A. Blumen, A fractional kinetics. Phys. Today **55**, 48–54 (2002)
31. A. Tudorache, R. Luca, Positive solutions for a system of Riemann-Liouville fractional boundary value problems with p-Laplacian operators. Adv. Differ. Equ. **2020**(292), 1–30 (2020)
32. Y. Wang, L. Liu, Y. Wu, Positive solutions for a class of higher-order singular semipositone fractional differential systems with coupled integral boundary conditions and parameters. Adv. Differ. Equ. **2014**(268), 1–24 (2014)
33. C. Yuan, Two positive solutions for $(n-1,1)$-type semipositone integral boundary value problems for coupled systems of nonlinear fractional differential equations. Commun. Nonlinear Sci. Numer. Simul. **17**(2), 930–942 (2012)

Energy Efficiency of Cortical Action Potential at Different Temperatures

Zheng Luo, Linyao Li, Dong Liang, Mengmeng Du, and Ying Wu

Abstract The brain is a complex system that generates human intelligence, whose information processing is remarkably energy consuming. The human brain accounts for merely 2% of body weight, yet consumes about 20% of metabolic energy. The issue of energy efficiency is therefore vital and worth studying, as abnormal energy metabolism of the brain will seriously impair physical health. Here we use the method of numerical simulation to quantitatively describe the energy consumption of the nervous system, especially that of cortical action potential generation. We add impacts of White Gaussian Noise and external stimulation current to the Hodgkin-Huxley-Style cortical neuronal model in order to simulate the noisy internal environment and synaptic transmission, derive formulas for measuring energy efficiency, and finally discuss three indicators including Information Rate (IR), Energy Consumption (EC) and Energy Efficiency (EE) affected by different factors. Our results indicate that the optimal temperatures at which cortical neurons' IR and EE reach a maximum both occur from 36.5 to 36.7 °C. The consistency between this temperature and the normal temperature of the human brain provides a strong theoretical support for the idea of optimization of the brain mechanism during human evolution.

Keywords Cortical neuronal model · Energy efficiency · Entropy of information · Energy consumption

---

Z. Luo · L. Li · D. Liang · M. Du · Y. Wu (✉)
State Key Laboratory for Strength and Vibration of Mechanical Structures, Xi'an Jiaotong University, Xi'an 710049, China
e-mail: wying36@xjtu.edu.cn

Z. Luo
e-mail: 3326653221@qq.com

M. Du
e-mail: dumm119@163.com

Z. Luo · D. Liang · M. Du · Y. Wu
School of Aerospace Engineering, Xi'an Jiaotong University, Xi'an 710049, China

L. Li
School of Mathematics and Statistics, Xi'an Jiaotong University, Xi'an 710049, China

C. H. Skiadas et al. (eds.), *13th Chaotic Modeling and Simulation International Conference*, Springer Proceedings in Complexity,
https://doi.org/10.1007/978-3-030-70795-8_39

1 Introduction

Mechanism of central nervous system diseases has always been a global hot issue ever since the beginning of the American BRAIN Project in 2013. These diseases are usually accompanied by abnormal energy expenditure and brain metabolic disorders, for example, epilepsy often happens with higher energy consumption [1], while lack of energy supply might cause the stroke [2]. Describing the energy consumption of neurons theoratically can not only reveal the mechanism of electrophysiological activities, but also provide a deeper understanding of the abnormal situations of the brain and nervous system.

Factors on the excitation of nervous system also attract lots of researchers. Noises from the internal environment as well as the internal environmental features such as temperature influences the ion channels directly [3, 4]. It is believed that quantitative measurements and numerical simulations of a single cortical neuron can lay the foundation for further researches.

2 Method

2.1 Neuronal Model

According to Yu's Hodgkin-Huxley-Style Cortical Neuronal model [5], the following equations describes the excitation of the cortical axon:

$$\begin{cases} C_m \dfrac{dV}{dt} = -g_L(V - V_L) - g_{Na} m h^3 (V - V_{Na}) - g_K n^4 (V - V_K) + I_e(t) \\ \tau_m \dfrac{dm}{dt} = -m + m_\infty,\ \tau_m = \dfrac{1}{\alpha_m + \beta_m},\ m_\infty = \dfrac{\alpha_m}{\alpha_m + \beta_m} \\ \tau_h \dfrac{dh}{dt} = -h + h_\infty,\ \tau_h = \dfrac{1}{\alpha_h + \beta_h},\ h_\infty = \dfrac{1}{1 + e^{\frac{V+60}{6.2}}} \\ \tau_n \dfrac{dn}{dt} = -n + n_\infty,\ \tau_n = \dfrac{1}{\alpha_n + \beta_n},\ n_\infty = \dfrac{\alpha_n}{\alpha_n + \beta_n} \end{cases} \quad (1)$$

where

V = the membrane voltage (mV);

I_e= the external membrane current density ($\mu A/cm^2$);

m,n,h = the sodium activation, the potassium activation, the sodium inactivation (dimensionless)[varying between 0 and 1];

C_m= 0.75 ($\mu F/cm^2$);

$g_{\mathrm{Na}} = 1500$, $g_{\mathrm{K}} = 400$, $g_{\mathrm{L}} = 3.3$ ($pS/\mu m^2$);

$V_{\mathrm{Na}} = 60$, $V_{\mathrm{K}} = -90$, $V_{\mathrm{L}} = -70$ (mV).

As the opening and closing of ion channel are affected by temperature, here is Φ to describe the effect:

$$\phi = Q_{10}^{(T-T_{base})/10}$$

$$\alpha_m(V) = \phi \cdot \frac{0.182(V+30)}{1-e^{-\frac{V+30}{8}}},\ \beta_m(V) = -\phi \cdot \frac{0.124(V+30)}{1-e^{\frac{V+30}{8}}},$$
$$\alpha_h(V) = \phi \cdot \frac{0.028(V+45)}{1-e^{-\frac{V+45}{6}}},\ \beta_h(V) = -\phi \cdot \frac{0.0091(V+70)}{1-e^{\frac{V+70}{6}}},$$
$$\alpha_n(V) = \phi \cdot \frac{0.01(V-30)}{1-e^{-\frac{V-30}{9}}},\ \beta_n(V) = -\phi \cdot \frac{0.002(V-30)}{1-e^{\frac{V-30}{9}}}. \quad (2)$$

where

$Q_{10} = 2.3$;

T_{base}= 23 °C.

In the actual living environment of neurons, internal noises from different sources are prevalent, including random opening and closing of ion channels, impacts of neighboring neurons and changes of internal environment etc. Therefore, we choose White Gaussian Noise (WGN) to simulate the internal noises.

Also, in order to simulate the synaptic input of the neuron, we consider a pulse-type stimulus current whose spiking follows a Poisson distribution, that is

$$I_{stim} = \sum_j I_{syn}^j(t) \quad (3)$$

inside which a single nerve impulse has the following form:

$$I_{syn}^j(t) = \begin{cases} I_0\left(t - t_s^j\right)e^{-\frac{t-t_s^j}{\tau}},\ t_s^j \le t \le t_c^j \\ 0,\ t_s^j \ge t \text{ or } t \ge t_c^j \end{cases} \quad (4)$$

where j represents the jth impulse, I_0 is the stimulus current intensity ($\mu A/(cm^2 \cdot ms)$), and represent the beginning and ending of the single impulse respectively, each impulse lasts for 8 *ms*, and $\tau = 2$ *ms*.

So the external current is the sum of WGN $\xi(t)$ and the pulse-type stimulus current.

$$I_e(t) = \xi(t) + I_{stim}(t) \quad (5)$$

When there is no synaptic input,

$$I_e(t) = \xi(t) \tag{6}$$

By integrating the model mentioned using Euler's Method, we simulate the excitation of cortical neurons successfully and obtain the spike trains(see in Sect. 3.1).

2.2 *Indicators*

Next, we design a series of indicators similar to Wang et al. [6] to measure the energy efficiency of cortical neurons.

(1) **Information Rate (IR)**

According to Strong [7], after getting a neural spike train, we can first discretize it into time bins in the size of $\Delta\tau(= 2\ ms)$. If the part of membrane potential is higher than the set threshold potential, we indicate that part of spike trains as a **letter 1**; if not, we name it **letter 0**. Here we get a binary letter train consisting of only 0 and 1.

Then we separate the letter train into words, for example, every 7 letters as a **word,** and calculate the probability of occurrence of each different word. We assume that the number of letters each word contains is k, so each word has a length of $T_s = k\Delta\tau$. For a spike train of $T\ ms$, there are $[T/T_S]$ complete words in total.

Finally we try to use this formula below to get the entropy of information H.

$$H_{total} = -\frac{1}{T_s}\sum_{i=1}^{n} P_i \log_2 P_i \tag{7}$$

It describes all the information encoded into the train, including that from noise and useful information from other neurons.

If the adjacent words are irrelevant, we can use the formula (7) directly. However, there still exists information among the relativity of adjacent words, thus making the calculated entropy lower than the actual entropy. In order to weaken the relativity, we should lengthen the words as much as possible. Yet as the overall experiment time is limited, too long word length will decrease the number of words sharply, thus making the statistics of the data meaningless. According to the knowledge of statistics, the entropy is proportional to $1/T_s$ in the case of a large T_s.Consequently, we choose to draw a H - $1/T_s$ diagram and use the method of linear extrapolation to obtain the actual entropy.

$$H_{total} = -\lim_{T_s\to\infty}\frac{1}{T_s}\sum_{i} P_i \log_2 P_i \tag{8}$$

To measure useful information amount, we repeat the experiment twice and get two neural spike trains. One with both noise and synaptic input, and one with only noise. The difference between the two is IR.

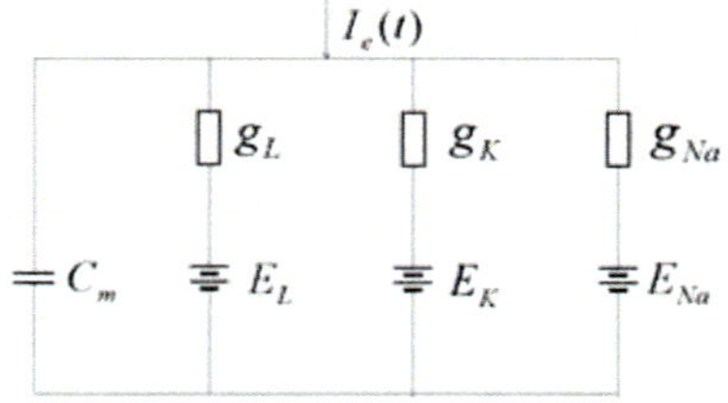

Fig. 1 The equivalent circuit of a Hodgkin-Huxley-Style neuron

$$I = H_{total} - H_{noise} \tag{9}$$

(2) **Energy Consumption (EC)**

It is now generally granted that a Hodgkin-Huxley-Style neuron has an equivalent circuit as shown in Fig. 1.

Moujahid et al. [8] put forward a method using this equivalent circuit to calculate the consumed energy. In the circuit, the total energy can be divided into two parts: energy stored in the battery and stored in the capacitor.

$$E(t) = \frac{1}{2}CV^2 + E_{Na} + E_K + E_L \tag{10}$$

Considering the change caused by ion currents and external stimulus current, we have

$$\frac{dE(t)}{dt} = VI_e(t) - g_{Na}m^3h(V - V_{Na})^2 - g_K n^4(V - V_K)^2 - g_L(V - V_L)^2 \tag{11}$$

where the minus signs represent the energy consumption, and I_e is assumed to provide energy for the neuron. After integrating the above formula over *t*, and then calculating its mean value, we can get the energy consumption rate E_{total}. When there is WGN only, we can get E_{noise} using the same method. Finally, the total energy consumption rate is

$$E = E_{total} - E_{noise} \tag{12}$$

(3) **Energy Efficiency (EE)**

Here we define EE as the information encoded by the consumption of unit energy per unit time during the excitation of a single cortical neuron, that is, the ratio of the information rate and the energy consumption. This indicator can effectively measure the neuron's working efficiency.

$$\varepsilon = \frac{I}{E} \tag{13}$$

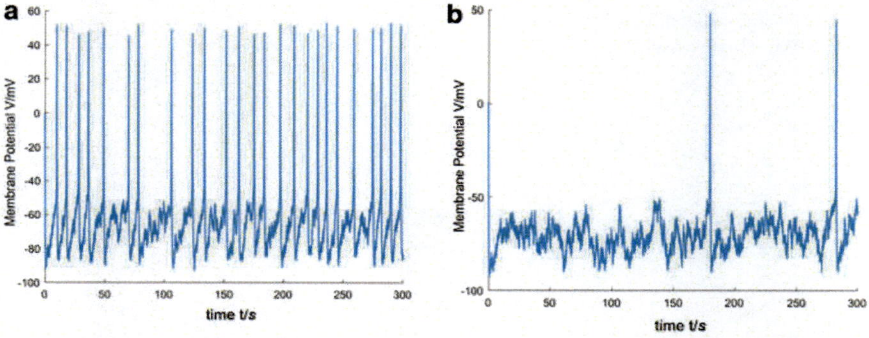

Fig. 2 **a** A spike train obtained with both the synaptic input and WGN **b** A spike train obtained with only WGN

3 Results and Discussion

3.1 *Spike Trains*

The spike trains of cortical neurons obtained by numerical simulation are shown as follows:

When regular external stimulus current and noise simultaneously occur, the cortical neuron acts in a orderly manner (Fig. 2a), but the frequency of excitation is different from the frequency of the regular synaptic input. This is because cortical neurons require a certain recovery time for each correspondence, which is called a refractory period, and as the synaptic input follows a Poisson distribution, the actual intervals between stimulus are not rigorously equal as well.

When there is noise only (Fig. 2b), the excitation of the neuron will become very irregular, and there might even be cases when no action potential is generated during the experiment at all. This is because WGN is irregular and is not strong enough to make the neuron generate action potentials frequently.

3.2 *IR, EC and EE*

After thousands of experiments, we summarize nine results as below, showing the effects of three factors on the indicators: IR, EC and EE (Fig. 3).

(1) **IR**

(a) **Effect of Noise Intensity**

Standard deviation of WGN can represent the magnitude of noise intensity. And WGN makes the results of each repeated experiment vary. Therefore, we choose two

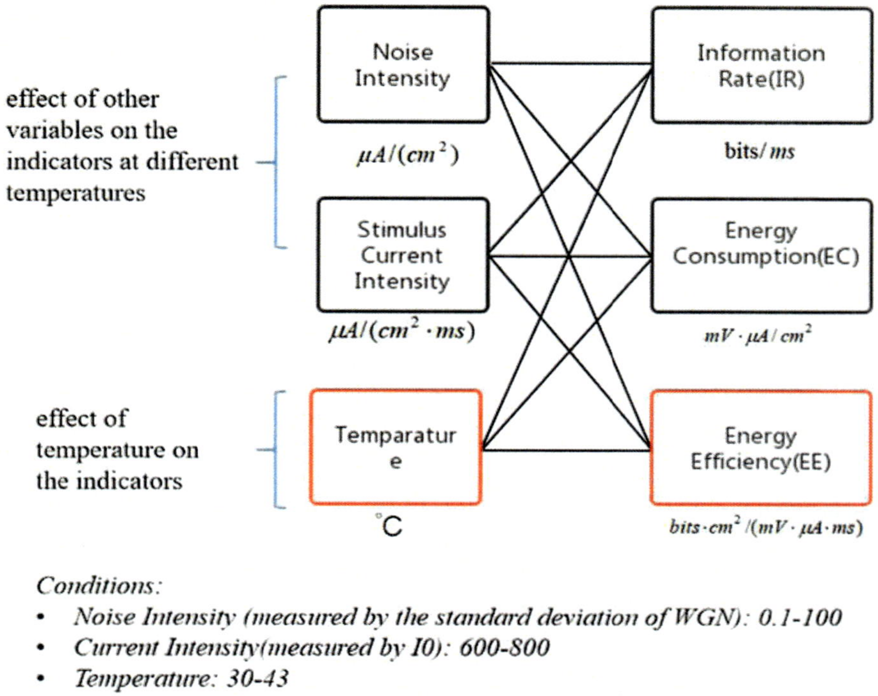

Fig. 3 The structure of results derived from experiments and the conditions of experiments

sets of digital features to reflect the dispersion of data, in which the standard deviation can be used to compare the degree of dispersion between the same indicators, while the coefficient of variation eliminates the influence of the measurement scale and dimension, and is suitable for the comparison of the degree of dispersion between different indicators. The coefficient of variation is defined as the ratio of the standard deviation of the original data to the average value of them, and is a normalized measure of the degree of dispersion.

From Table 1, it is easy to conclude that in general, the dispersion of IR increases as the noise intensity goes higher, yet it reaches the minimal value at the noise intensity of 10, which indicates the existence of a suitable noise intensity range makes the indicator most stable. The phenomenon coincides with the mainstream

Table 1 Effect of noise intensity on IR in ten repeated experiments

Noise intensity	0.1	1	10	100
Standard deviation of IR	18.97	26.77	**15.94**	26.95
Coefficient of variation of IR (%)	7.68	9.86	**7.58**	22.42

*Conditions: $I_0 = 600\,\mu A/(cm^2 \cdot ms)$, $T_0 = 37$ °C

stochastic resonance theory: noise with appropriate intensity has a positive effect on the information processing of neurons.

(b) **Effect of Stimulus Current Intensity**

The stimulus current intensity I_0 can represent the magnitude of the external stimulus current. As can be seen in Fig. 4, I_0 has a significant effect on the information rate. IR at different temperatures is significantly layered with the change in the external stimulus current. The stronger the external stimulus, the higher the information rate at different temperatures. As the intensity of external stimulation current increases by 200 μA/(cm$^2 \cdot$ ms), the overall information rate will increase by 30 bits/ms or so. This is because as the external stimulus current increases, the amount of information encoded also increases.

(c) **Effect of Temperature**

As is shown in Table 2, we sort the temperature by the order of information rate, and define the three temperature which is corresponding to the top three IR(the three highest IR) as the **Best Temperature**. In this table, the range of 36.5–36.7 °C is most frequently seen. Therefore by this measure, the optimal temperature in the normal range of the human body is 36.5–36.7 °C.

(1) **EC**
(a) **Effect of Noise Intensity**

From Table 3, it is easy to conclude that in general, the dispersion of EC increases as the noise intensity goes higher, yet it reaches the minimal value at the noise

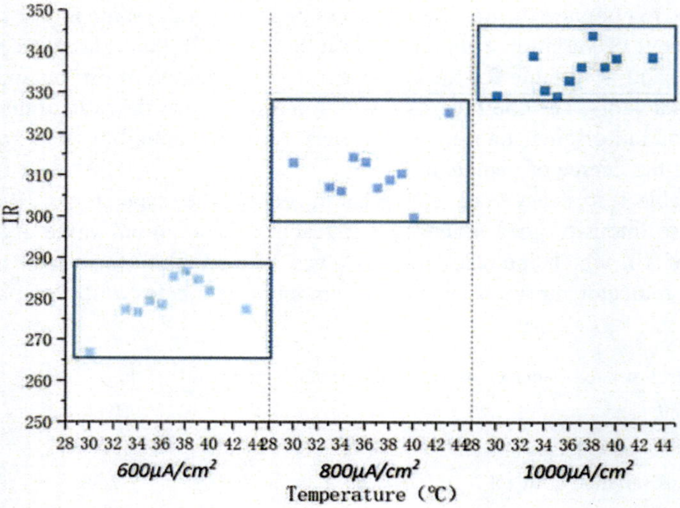

Fig. 4 Effect of stimulus current intensity on IR

Table 2 Best temperatures of IR in ten repeated experiments

IR	Current intensity	Best temperatures		
X = 10	600	36.5	36.7	36.6
	800	36.6	43.0	36.1
	1000	36.6	36.5	36.4
X = 100	600	36.5	36.6	36.0
	800	36.5	36.9	36.1
	1000	**36.5**	30.0	36.1

Table 3 Effect of noise intensity on EC in ten repeated experiments

Noise intensity	0.1	1	10	100
Standard deviation of EC	49.22	58.51	**45.75**	128.98
Coefficient of variation of EC (%)	1.31	1.56	**1.23**	3.50

*Conditions: $I_0 = 600\,\mu A/(cm^2 \cdot ms)$, $T_0 = 37$ °C

intensity of 10, which indicates the existence of a suitable noise intensity range makes the indicator most stable. The phenomenon coincides with the mainstream stochastic resonance theory: noise with appropriate intensity has a positive effect on the information processing of neurons.

(b) **Effect of Stimulus Current Intensity**

As can be seen in Fig. 5, I_0 has a significant effect on the energy consumption. EC at different temperatures is significantly layered with the change in the external stimulus current. The stronger the external stimulus, the higher the Energy Consumption at different temperatures. As the intensity of external stimulation current increases 200 μA/(cm^2 · ms), the overall information rate will increase by about 600 mV · $\mu A/cm^2$. The increased external stimulus current will consume more energy,which is unfavorable to the information processing.

(c) **Effect of Temperature**

As is shown in Table 4, we sort the temperature by the order of energy consumption, and define the three temperature which is corresponding to the top three EC(the three lowest EC) as the **Best Temperature.** In this table, the frequency of temperature does not have an obvious tendency, which can also be proved by Fig. 5. As the temperature changes, the energy consumption rate always fluctuates around a certain value.

(3) **EE**

(a) **Effect of Noise Intensity**

From Table 5, it is easy to conclude that in general, the dispersion of EE increases as the noise intensity goes higher, yet it reaches the minimal value at the noise

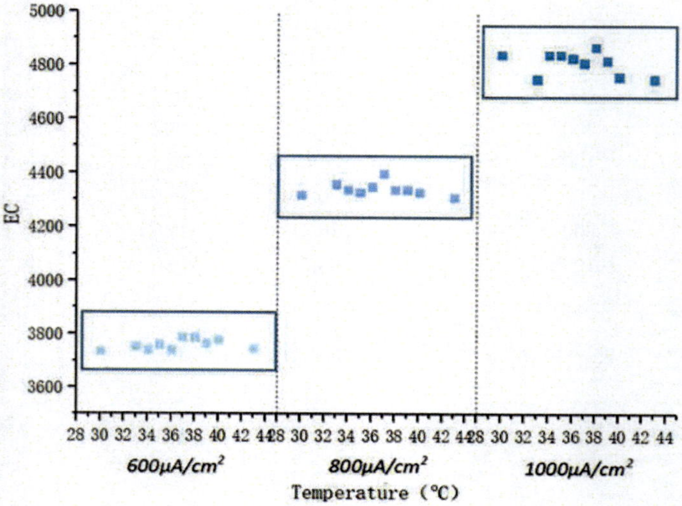

Fig. 5 Effect of stimulus current intensity on EC

Table 4 Best temperatures of EC in ten repeated experiments

EC	Current intensity	Best temperatures		
X = 10	600	30.0	34.0	36.0
	800	36.6	36.7	37.0
	1000	36.4	36.1	36.9
X = 100	600	36.8	36.3	40.0
	800	35.5	36.3	36.6
	1000	36.7	43.0	34.0

Table 5 Effect of noise intensity on EE in ten repeated experiments

Noise intensity	0.1	1	10	100
Standard deviation of EE($x10^{-2}$)	0.52	0.69	0.38	0.77
Coefficient of variation of EE (%)	7.28	9.46	5.03	23.47

*Conditions: $I_0 = 600\,\mu A/(cm^2 \cdot ms)$, $T_0 = 37$ °C

intensity of 10, which indicates the existence of a suitable noise intensity range makes the indicator most stable. The phenomenon coincides with the mainstream stochastic resonance theory: noise with appropriate intensity has a positive effect on the information processing of neurons.

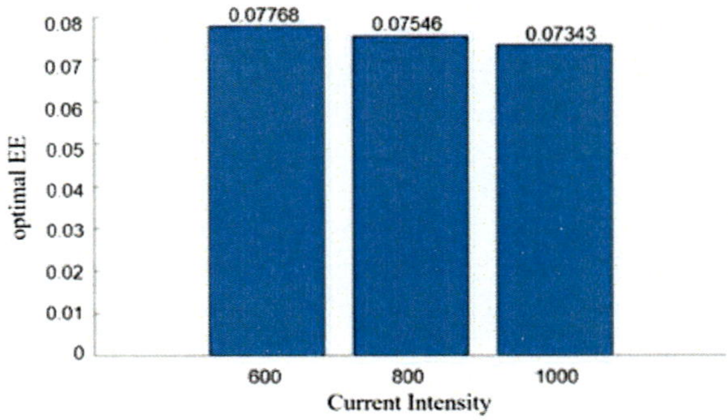

Fig. 6 Effect of stimulus current intensity on EE

(b) Effect of Stimulus Current Intensity

We define the highest stimulus current intensity at different temperatures as the optimal EE. As can be seen in Fig. 6, the stronger the external stimulus, the lower the optimal EE. As the intensity of external stimulation current increases 200 $\mu A/(cm^2{\cdot}ms)$, the overall energy efficiency will increase by about 0.002 $bits \cdot cm^2/(mV \cdot \mu A \cdot ms)$. The increased external stimulus current is unfavorable to the efficiency of the information processing.

$$*Conditions : \sigma = 10$$

(c) Effect of Temperature

As is shown in Table 6, we sort the temperature by the order of EE, and define the three temperature which is corresponding to the top three EE(the three highest EE) as the **Best Temperature**. In this table, the range of 36.5–36.7 °C is most frequently seen. Therefore by this measure, the optimal temperature in the normal range of the human body is 36.5–36.7 °C.

Table 6 Best temperature EE in ten repeated experiments

EE	Current intensity	Best temperatures		
X = 10	600	36.5	36.7	36.6
	800	36.6	36.5	43.0
	1000	36.6	36.5	36.3
X = 100	600	36.5	36.6	36.0
	800	36.5	36.1	36.7
	1000	36.7	36.5	30.0

4 Conclusions

Based on the Hodgkin-Huxley-Style Cortical Neuronal model with WGN and pulse-type stimulus current, we deduce the formula of measuring Energy Efficiency (EE), study the impacts of internal noises and the synaptic input of nervous system, and discuss the variation of different indicators. The results well reflects the actual situation of the electrophysiological activity of cortical neurons.

Via numerical simulation, we find that the best temperature range for Information Rate (IR) and Energy Efficiency (EE) is 36.5–36.7 °C,which is consistent with the normal temperature of the brain. The results gives a strong theoretical support for the optimization of the brain mechanism during human evolution. In general, the dispersion of data increases as the noise intensity goes higher, yet they all reach the minimal value at the noise intensity of 10, which indicates a result that coincides with stochastic resonance theory: the existence of a suitable noise intensity range which makes the indicators most stable. Meanwhile, the impact of stimulus current intensity is apparent: the larger stimulus current intensity is, the higher IR, the higher EC and the lower optimal-EE is.

Central nervous system diseases are usually accompanied by abnormal energy expenditure and brain metabolic disorders. We can effectively study a variety of dynamic processes of abnormal nervous system and try to reveal the mechanism of the onset or treatment of neurological diseases using the model given. For example, abnormal expression levels of Voltage-gated sodium channel (VGSC) is believed to be the cause of various kinds of primary epilepsy, and sodium ion channel inhibitors account for a large proportion of clinical antiepileptic drugs [9]. By adjusting the parameters including sodium ion channel conductance g_{Na} in the model, it is possible to realize phenomena such as excessive depolarization of the membrane potential and thus trying to explain the mechanism behind. Also, clinical experiments have shown that electrical stimulation has a significant effect on the rehabilitation of stroke patients. By changing the external stimulus current in this model and adjusting the neuronal membrane potential, it is also feasible to do further researches over the particular therapy [10].

References

1. H. Nersesyan, F. Hyder, D.L. Rothman et al., Dynamic fMRI and EEG recordings during spike-wave seizures and generalized tonic-clonic seizures in WAG/Rij rats. J. Cerebral Blood Flow Metabolism **24**(6), 589–599 (2004)
2. M. Chen, D. Guo, T. Wang, W. Jing et al., Bidirectional control of absence seizures by the basal ganglia: a computational evidence. PLoS Comput. Biol. **10**(3), (2014)
3. J.H. Willim, H.E. Donald, Effect of temperature on a voltage-sensitive electrical synapse in crayfish. J. Exp. Biol. **201**, 503–513 (1998)
4. A.L. Hodgkin, A.F. Huxley, A quantitative description of membrane current and its application to conduction and excitation in nerve. J. Physiol. **117**, 500–544 (1952)

5. Y. Yu, A.P. Hill, D.A. Mccormick et al., Warm body temperature facilitates energy efficient cortical action potentials. PLoS Comput. Biol. **8**(4), (2012)
6. L. Wang. Spike-Threshold and Energy Efficiency in Model Neurons(in Chinese) 2016
7. S. P. Strong, R. Koberle, R.R. de Ruyter van Steveninck,W. Bialek. Entropy and information in neural spike trains. Am. Phys. Soc. 1998,80(1)
8. A. Moujahid, A. d'Anjou, F.J. Torrealdea et al., Energy and information in Hodgkin-Huxley neurons. Phys. Rev. E **83**(3), (2011)
9. Z. Huimin, T. Huan, Z. Jingyu. Research progress of sodium channel in the pathogenesis of neurological diseases (in Chinese). Genomics Appl. Biol. 1–13[2020-05-28].http://kns.cnki.net/kcms/detail/45.1369.Q.20200519.1315.002.html
10. L. Xiao-jie, L. Xin-hui, Y. Li-he,W. Jin, L. Li-rong, C. Sheng-jun. Effect of Neurophysiological Stimulation on Spinal Motor Neurons in Stroke Patients (in Chinese). Prog. Mod. Biomed. 2020,20(08):1493–1496 + 1505

Traveling Waves and Spatio-Temporal Chaos in Nonlinear Partial Differential Equations

Nikolai Magnitskii

Abstract It is considered some nonlinear partial differential equations having physical, chemical or biological turbulence regimes, such as the Ginzburg–Landau, Kuramoto–Sivashinsky, Schrödinger, FitzHugh–Nagumo equations. It is shown, that all such systems of partial differential equations can have an infinite number of different stable wave solutions, travelling along the space axis with arbitrary speeds, and also an infinite number of different states of spatio-temporal chaos. These chaotic (turbulent) solutions are generated by cascades of bifurcations of cycles or tori and singular attractors according to the universal bifurcation Feigenbaum–Sharkovsky–Magnitskii (FShM) theory in the three-dimensional or four-dimensional systems of ordinary differential equations, to which the systems of partial differential equations can be reduced by self-similar change of variables. Examples of application of the theory to the description of processes of physical, chemical and biological turbulence are considered

Keywords Nonlinear PDE · Traveling waves · Spatio-temporal chaos · Turbulence · FShM-theory

1 Introduction

Wide class of physical, chemical, biological, ecological and economic processes and phenomena is described by reaction–diffusion systems of partial differential equations

$$u_t = D_1 u_{xx} + f(u, v, \mu), v_t = D_2 v_{xx} + g(u, v, \mu), 0 \le x \le l, \tag{1}$$

N. Magnitskii (✉)
Federal Research Center "Computer Science and Control" of RAS, Lomonosov Moscow State University, Moscow, Russia
e-mail: nikmagn@mail.ru

C. H. Skiadas et al. (eds.), *13th Chaotic Modeling and Simulation International Conference*, Springer Proceedings in Complexity,
https://doi.org/10.1007/978-3-030-70795-8_40

depending on scalar or vector parameter μ . Such system is very complex system. Behavior of its solutions depends on coefficients of diffusion and their ratio, length of space area and edge conditions. A special case of systems of the reaction–diffusion equations is the FitzHugh–Nagumo type system of equations

$$u_t = Du_{xx} + f(u, v, \mu), v_t = g(u, v, \mu), \tag{2}$$

describing the processes of chemical and biological turbulence. To the system of equations of the form (1) can also be reduced the well-known Kuramoto–Tsuzuki (or Time Dependent Ginzburg–Landau) equation

$$W_t = W + (1 + ic_1)W_{xx} - (1 + ic_2)|W|^2 W \tag{3}$$

with complex-valued function $W(x, t) = u(x, t) + iv(x, t)$.

It is well-known, that systems (1)–(3) can have periodic solutions, switching waves, traveling waves and traveling impulses, dissipative spatially nongomogeneous stationary structures, and also spatio-temporal chaos—irregular nonperiodic nonstationary nongomogeneous structures. The analysis of traveling waves and spatio-temporal chaos in systems (1)–(3) can be carried out by replacement $\xi = x - ct$ and transition to three-dimensional or four dimensional systems of ordinary differential equations with respect to the variable ξ. Thus the switching waves in the systems (1)–(3) are described by separatrixes of systems of ordinary differential equations going from their one singular point into another singular point, traveling waves and traveling impulses of the systems (1)–(3) are described by limit cycles and separatrix loops of singular points of systems of ordinary differential equations. And spatio-temporal chaos in the systems (1)–(3) is described by singular attractors of systems of ordinary differential equations in full accordance with the universal bifurcation Feigenbaum-Sharkovsky-Magnitskii (FShM) theory [1–5]. The greatest interest represents the fact that c is a bifurcation parameter, describing a speed of wave distribution along an axis x, which is not included obviously into initial system. This means, that system of a kind (1)–(3) with the fixed parameters can have infinitely number of various autowave solutions of any period traveling along a spatial axis with various speeds, and infinite number of modes of spatio-temporal chaos. Some of such systems including Kuramoto–Tsuzuki (3), a system describing chemical turbulence in autocatalytic chemical reaction

$$u_t = u_{xx} - \tfrac{1}{\varepsilon}u(1-u)\big(u - \tfrac{b+v}{a}\big), v_t = f(u)v,$$
$$f(u) = \begin{cases} 0, & 0 \le u < 1/3, \\ 1 - 6.75u(1-u)^2, & 1/3 \le u \le 1, \\ 1, & 1 < u, \end{cases}$$

and a system describing distribution of nervous impulses in a cardiac muscle

$$u_t = u_{xx} + \frac{1}{\varepsilon}u(1-u)\left(u - \frac{0.06+v}{0.75}\right), v_t = u^3 - v,$$

are studied in Magnitskii [2, 3] and Karamysheva and Magnitskii [6]. In this paper, we apply the described above approach to the analysis of traveling waves and space–time chaos in the Kuramoto–Sivashinsky equation and in the generalized nonlinear Schrödinger equation.

2 Traveling Waves and Spatio-Temporal Chaos in the Kuramoto–Sivashinsky Equation

The one-dimensional Kuramoto–Sivashinsky equation can be written in the differential form

$$u_t + uu_x + u_{xx} + u_{xxxx} = 0 \tag{4}$$

or in more general form

$$u_t + uu_x + \alpha u_{xx} + \beta u_{xxx} + u_{xxxx} = 0 \tag{5}$$

Equation (4) is widely used in many papers to describe wave processes in active and dissipative media when modeling the simplest turbulence processes, studying waves on the interface between two viscous liquids, describing wave phenomena in plasma in toroidal devices, analyzing the flame front behavior, etc [7].

2.1 Reduction to an ODE System by a Self-Similar Change of Variables

Consider (5) on the entire real line $u(x,t) : R \times R \to R$. We analyze its regular solutions by using the self-similar change of variables $\xi = x - ct$ reducing the (5) to the nonlinear ordinary differential equation

$$-cu' + \left(u^2\right)'/2 + \alpha u'' + \beta u''' + u'''' = 0, \tag{6}$$

where the derivatives are taken with respect to the variable ξ. Integrate (6) we obtain the equation

$$u''' + \beta u'' + \alpha u' - cu + \left(u^2\right)/2 + \delta = 0, \tag{7}$$

where δ is an arbitrary constant. We reduce (7) to the nonlinear system of three ordinary differential equations

$$u' = v, v' = -\alpha u - w, w' = (-c - \beta\alpha)u + u^2/2 - \beta w + \delta. \tag{8}$$

2.2 Dissipativity Domain and Singular Points

If F_i is the right-hand side of the i th equation of system (8), then

$$divF(u, v, w) = \partial F_1/\partial u + \partial F_2/\partial v + \partial F_3/\partial w = -\beta.$$

Consequently, system (8) is everywhere dissipative for $\beta > 0$. Equating the right-hand side of system (8) to zero, we find that for $\delta \geq 0$ and $c > \sqrt{2\delta}$, system (8) has two singular points

$$O_{\pm}(c \pm \sqrt{c^2 - 2\delta}, 0, -\alpha(c \pm \sqrt{c^2 - 2\delta}))$$

2.3 Analysis of Stability and of the Type of Singular Points

Matrix of the linearized right-hand side of the system (8) has the form

$$A = \begin{pmatrix} 0 & 1 & 0 \\ -\alpha & 0 & -1 \\ -c - \alpha\beta + u_* & 0 & -\beta \end{pmatrix},$$

where $u_* = c \pm \sqrt{c^2 - 2\delta}$, and its characteristic equation is.

$$det(A - \lambda I) = -\left(\lambda^3 + \beta\lambda^2 + \alpha\lambda \pm \sqrt{c^2 - 2\delta}\right) = 0.$$

For $c^2 - 2\delta = 0$ the characteristic equation becomes $\lambda\left(\lambda^2 + \beta\lambda + \alpha\right) = 0$ and has the roots $\lambda_1 = 0$, $\lambda_{23} = -\beta/2 \pm \sqrt{\beta^2/4 - \alpha}$. The latter two roots are either real negative (for $\beta^2/4 \geq \alpha$) or complex conjugate with negative real part (for $\beta^2/4 < \alpha$), and one has

$$\lambda_2\lambda_3 = (-\beta/2 - \sqrt{\beta^2/4 - \alpha})(-\beta/2 + \sqrt{\beta^2/4 - \alpha}) = \alpha > 0.$$

Since the product of roots of the characteristic equation is $\lambda_1\lambda_2\lambda_3 = -(\pm\sqrt{c^2 - 2\delta})$ by the Vieta theorem, we see that for small $c^2 - 2\delta > 0$ the root λ_1 at the singular point O_+, where the product of roots $\lambda_1\lambda_2\lambda_3 = -\sqrt{c^2 - 2\delta}$ is negative, is real negative as well, and hence O_+ is a stable node or a stable focus depending on which of the inequalities $\beta^2/4 \geq \alpha$ or $\beta^2/4 < \alpha$ holds. In a similar way, the root λ_1 at the singular point O_-, where the product of roots $\lambda_1\lambda_2\lambda_3 = \sqrt{c^2 - 2\delta}$ is positive, is real positive, and hence the point O_- is a saddle-node or a saddle-focus with a two-dimensional stable manifold and a one-dimensional unstable manifold.

It follows from this that it is most interesting to study possible cascades of bifurcations of the stable singular point O_+ for the case in which it is a stable focus and its first bifurcation is an Andronov–Hopf bifurcation of birth of a stable limit cycle. It is in this case that in system (8) there may exist all three cascades of bifurcations of stable limit cycles and infinitely many chaotic singular attractors in accordance with the Feigenbaum–Sharkovsky–Magnitskii universal bifurcation theory. And the Kuramoto–Sivashinsky (5) may have an infinite family of most complicated periodic and nonperiodic (chaotic) traveling waves or even traveling pulses.

Theorem 1 Let the parameters α and β be positive and satisfy $\beta^2/4 < \alpha$. Then the Andronov–Hopf bifurcation of a stable limit cycle from the stable focus O_+ occurs for $c > c_* = \sqrt{\beta^2\alpha^2 + 2\delta}$.

Proof A sufficient condition for the Andronov–Hopf bifurcation of a stable limit cycle from the stable singular focus O_+ is the passage of two complex conjugate roots of the characteristic equation at the singular point O_+ through the imaginary axis of the complex plane from left to right. At the transition point (bifurcation point), the real parts of the two complex conjugate roots of the characteristic equation are zero, and the third root remains real and negative. Therefore,$\lambda_1 < 0$, $\lambda_2 = i\omega$, $\lambda_3 = -i\omega$ at the bifurcation point. Then

$$\lambda_1(i\omega)(-i\omega) = -\sqrt{c^2 - 2\delta}, \lambda_1(i\omega) + \lambda_1(-i\omega) + \omega^2 = \alpha, \lambda_1 = -\beta$$

by the Vieta theorem, and we obtain $\omega = \sqrt{\alpha}$, $c_* = \sqrt{\beta^2\alpha^2 + 2\delta}$. The proof is complete.

It follows from Theorem 1 that the bifurcation parameter in system (8) is the parameter c which characterizes the perturbation propagation velocity along the x-axis in the Kuramoto–Sivashinsky equation and does not explicitly occur in the original (5). It is an extremely difficult problem to study the subsequent bifurcations occurring for large c by analytical methods, starting from the period doubling bifurcation of the limit cycle borned by the Andronov–Hopf bifurcation. To this end, one needs to determine the cycle multipliers analytically, which is possible in very rare cases, and determine the value of the parameter c at which all three multipliers are real numbers, two of them being $+1$ and -1 and the third one lying in the interval $(-1, 0)$. Therefore, the subsequent complication of the dynamics of solutions of system (5) and (8) will be investigated by numerical methods.

2.4 Scenario of Transition to Space–Time Chaos

We study system (8) numerically for the fixed parameter values $\alpha = 1.6, \beta = 0.57, \delta = 0$ and for increasing values of the bifurcation parameter $c > \sqrt{2\delta} = 0$. For $c \in (0, c_*)$, $c_* = \alpha\beta = 0.912$, the singular point O_+ of system (8) is a stable focus; at $c = 0.912$, a stable (asymptotically orbitally stable) limit cycle is created from this point as a result of an Andronov–Hopf bifurcation. This cycle exists until $c \approx 1.567$, at which a stable limit cycle of double period is created from it. As the parameter c increases, one observes a Feigenbaum cascade of cycle period doubling bifurcations in system (8). A cycle of period 4 is created at $c \approx 1.6786$, a cycle of period 8 is created at $c \approx 1.7025$, etc. For $c \approx 1.71$, system (8) has the simplest singular (chaotic) attractor, the Feigenbaum attractor, i.e., a nonperiodic trajectory that is the limit of cycles in the Feigenbaum cascade. As the parameter c increases further, one can see a sequence of stable cycles in accordance with the Sharkovskii order and then in accordance with the Magnitskii homoclinic order. For example, a cycle of period 6 of the subharmonic cascade is discovered for $c = 1.7235$; a cycle of period 7 for $c \approx 1.7405$; and a cycle of period 5 for $c = 1.75$ (Fig. 1a). The stable cycle C_4 of the homoclinic cascade, which makes four conditional revolutions about the saddle-focus, is discovered for $c = 1.7953$; the cycle C_5 for $c = 1.8189$ (Fig. 1b). It is well known (see Magnitskii [1, 2]) that the sequence of cycles of a complete homoclinic cascade must converge to the separatrix loop of the saddle-focus, but for this particular set of system parameter values there apparently exists no separatrix loop. To discover it in the space of four system parameters $(\alpha, \beta, \delta, c)$ is a complicated problem, because the manifold on which the loop exists has codimension two or even three.

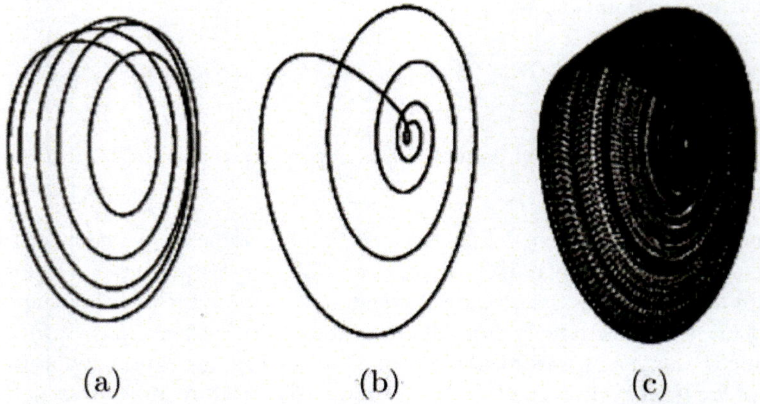

Fig. 1 Cycles of period 5 of the subharmonic (a) and homoclinic (b) cascades; singular attractors near the saddle-focus separatrix loop of ODE system (8) (c)

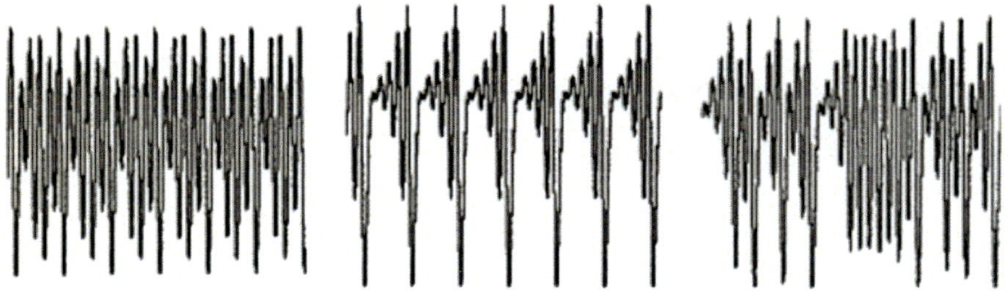

Fig. 2 Traveling periodic and chaotic waves in the Kuramoto–Sivashinsky (5) corresponding to the cycles and attractor of system (8) from Fig. 1

To the cycles thus discovered in system (8), there correspond traveling waves in the Kuramoto–Sivashinsky (5). Figure 2 illustrates such waves corresponding to the cycles and the singular attractor shown in Fig. 1.

3 Traveling Waves and Spatio-Temporal Chaos in Generalized Nonlinear Schrödinger, Equation

Consider the generalized nonlinear Schrödinger equation

$$i\frac{\partial\psi}{\partial t}+c_1\frac{\partial^2\psi}{\partial x^2}+c_2\psi+c_3|\psi|^2\psi=0 \tag{9}$$

with complex, in general case, parameters c_1, c_2, c_3. (9) describes, in particular, the propagation dynamics of surface plasmon polaritons on a metal- dielectric surface [8]. In this paper, we consider (9) in the case $c_3 = -c_2$. It is shown that, at certain values of parameters, (9) has an infinite number of different stable wave solutions running along the spatial axis with arbitrary velocities, and an infinite number of different modes of space–time chaos in full accordance with the universal Feigenbaum-Sharkovsky-Magnitskii theory. Moreover the bifurcation parameter is the value of the speed of propagation of traveling waves along the spatial axis, which is not explicitly included in the original equation.

3.1 Reduction to an ODE System Using Self-Similar Variable Substitution.

We present $\psi(x,t) = u(x,t) + iv(x,t)$, $c_1 = a + ib$, $c_2 = d + ie$, $c_3 = -c_2$ and consider (9) on the whole number axes: $u(x,t) : R \times R \to R, v(x,t) : R \times R \to R$.

The analysis of regular and chaotic solutions of (9) will be carried out using a self-similar change of independent variables $\xi = x - ct$, thereby reducing the original (9) with partial derivatives to a system of nonlinear ordinary differential equations

$$\begin{cases} cv' + au'' - bv'' + (du - ev)\big(1 - (u^2 + v^2)\big) = 0, \\ -cu' + av'' + bu'' + (eu + dv)\big(1 - (u^2 + v^2)\big) = 0, \end{cases} \tag{10}$$

where the derivative is taken over ξ. Solving system (10) with respect to the second derivatives u'', v'', we obtain the system of equations

$$\begin{cases} u'' = \big(bcu' - acv' + ((ad + be)u - (ae - bd)v)\big(u^2 + v^2 - 1\big)\big)/\big(a^2 + b^2\big), \\ v'' = \big(bcv' + acu' - ((bd - ae)u - (ad + be)v)\big(u^2 + v^2 - 1\big)\big)/\big(a^2 + b^2\big), \end{cases}$$

which we reduce to a nonlinear system of four ordinary differential equations

$$\begin{cases} u' = z, \\ z' = (bcz - acw + ((ad + be)u - (ae - bd)v)\big(u^2 + v^2 - 1\big))/(a^2 + b^2), \\ v' = w, \\ w' = (bcw + acz - ((bd - ae)u - (ad + be)v)\big(u^2 + v^2 - 1\big))/(a^2 + b^2). \end{cases} \tag{11}$$

3.2 Dissipativity Domain and Singular Points

If F_i is the right-hand side of the i th equation of system (11), then

$$divF(u, z, v, w) = \partial F_1/\partial u + \partial F_2/\partial z + \partial F_3/\partial v + \partial F_4/\partial w = 2bc/(a^2 + b^2).$$

Consequently, as $c > 0$, then system (11) is everywhere dissipative for $b < 0$. Equating the right-hand side of system (11) to zero, we find that for $bd \neq ae$ system (11) has a unique zero singular point $O(0, 0, 0, 0)$. To find the stability region of the singular point O, it is necessary to calculate the linearization matrix of system (11) at this point:

$$A = \begin{pmatrix} 0 & 1 & 0 & 0 \\ -\frac{(ad+be)}{a^2+b^2} & \frac{bc}{a^2+b^2} & -\frac{(bd-ae)}{a^2+b^2} & -\frac{ac}{a^2+b^2} \\ 0 & 0 & 0 & 1 \\ \frac{(bd-ae)}{a^2+b^2} & \frac{ac}{a^2+b^2} & -\frac{(ad+be)}{a^2+b^2} & \frac{bc}{a^2+b^2} \end{pmatrix}.$$

The characteristic equation corresponding to matrix A has the form:

$$det(A - \lambda I) = \lambda^4 - \frac{2bc}{a^2+b^2}\lambda^3 + \frac{2(ad+be)+c^2}{a^2+b^2}\lambda^2 - \frac{2ec}{a^2+b^2}\lambda + \frac{d^2+e^2}{a^2+b^2} = 0. \tag{12}$$

The necessity of the condition $b < 0, e < 0$ of stability of the singular point O follows from the necessary condition for the positivity of all the coefficients of the characteristic polynomial (12). If this condition is satisfied, then, by virtue of the Routh-Hurwitz criterion, a necessary and sufficient condition for the stability of the singular point O is the fulfillment of two inequalities:

$$\Delta_2 = -\frac{2bc}{a^2+b^2}\frac{2(ad+be)+c^2}{a^2+b^2} + \frac{2ec}{a^2+b^2} > 0,$$

$$\Delta_3 = -\frac{2ec}{a^2+b^2}\Delta_2 - \left(\frac{2bc}{a^2+b^2}\right)^2 \frac{d^2+e^2}{a^2+b^2} > 0.$$

It follows from the first inequality that $c^2 > ((b^2 - a^2)e + 2abd)/(-b)$, and from the second inequality that $c^2 > (bd - ae)^2/(be)$. And since $(bd - ae)^2 > -e\big((b^2 - a^2)e + 2abd\big)$, the necessary and sufficient condition for the stability of the singular point O is

$$b < 0, e < 0, c > \sqrt{(bd - ae)^2/(be)}. \tag{13}$$

Of greatest interest is the study of possible cascades of bifurcations of the singular point O in the case when the stability condition (13) is violated and the first bifurcation is the Andronov–Hopf bifurcation of the birth of a stable limit cycle. Exactly at this case, the second Andronov–Hopf bifurcation of birth of a stable two-dimensional torus, a cascade of two-dimensional torus bifurcations and the existence of an infinite number of chaotic singular toroidal attractors in accordance with the universal Feigenbaum–Sharkovsky–Magnitskii bifurcation theory are possible in system (11). In this case, the generalized nonlinear Schrödinger (9) can have an infinite family of the most complex periodic and non-periodic (chaotic) traveling waves up to traveling pulses.

3.3 Andronov–Hopf Bifurcation

Let us prove that if condition (13) is violated in system (11), the Andronov–Hopf bifurcation actually occurs, that is, the soft birth of a stable limit cycle from the singular point O.

Theorem 2 For $bd \neq ae$ and fixed negative values of the parameters $b < 0, e < 0$, the birth of a stable limit cycle from a stable focus O as a result of the Andronov–Hopf bifurcation occurs for $c < \sqrt{(bd - ae)^2/(be)}$.

Proof We will search a periodic solution to system (11) in the form

$$u = rcos(\omega t), v = rsin(\omega t), r = const. \tag{14}$$

Substituting the expected form of the solution in the system of (10), we obtain

$$\begin{cases} -r\omega^2 cos(\omega t) = (-bcr\omega sin(\omega t) - acr\omega cos(\omega t) + \\ ((ad + be)rcos(\omega t) + (bd - ae)rsin(\omega t))(r^2 - 1))/(a^2 + b^2), \\ -r\omega^2 sin(\omega t) = (-acr\omega sin(\omega t) + bcr\omega cos(\omega t) + \\ ((ad + be)rsin(\omega t) + (bd - ae)rcos(\omega t))(r^2 - 1))/(a^2 + b^2). \end{cases}$$

Equating the coefficients for $cos(\omega t)$ and $sin(\omega t)$, we obtain a system of equations for r and ω

$$-r\omega^2 = (-acr\omega + (ad + be)r(r^2 - 1))/(a^2 + b^2),$$

$$-bcr\omega + (bd - ae)r\left(r^2 - 1\right) = 0.$$

Expressing $r^2 - 1 = bc\omega/(bd - ae)$ from the second equation, we obtain from the first equation $\omega^2 - ce\omega/(bd - ae)$, which implies

$$\omega = -ce/(bd - ae), r^2 = 1 - bec^2/(bd - ae)^2.$$

Thus, for $c < \sqrt{(bd - ae)^2/(be)}$, system of (11) has as its solution a stable limit cycle (14), softly generated from the zero singular point, stable for $c > \sqrt{(bd - ae)^2/(be)}$. The theorem is proved.

It follows from Theorem 2 that the bifurcation parameter in the system of (11) is parameter c, which characterizes the magnitude of the velocity of propagation of perturbations along the x axis in the generalized nonlinear Schrödinger equation. The parameter c is not explicitly included in the original (9). The study of the following bifurcations in (11) with decreasing parameter values using analytical methods, starting with the bifurcation of the birth of a two-dimensional torus, doubling of its period and further bifurcations of the two-dimensional torus of a doubled period in accordance with the Feigenbaum-Sharkovsky-Magnitskii theory, is an extremely difficult task. To do this, it is necessary to find analytically the cycle multipliers, which is possible in very rare cases, and determine the value of the parameter c at which one of the multipliers is + 1, two complex conjugate multipliers lie on the unit circle, and the fourth real multiplier lies in the interval (–1, +

1). Therefore, a further study of the complication of the dynamics of solutions of system (9) and (11) will be carried out by numerical methods.

3.4 Scenario of Transition to Space–Time Chaos

We carry out a numerical study of system (11) for fixed values of the parameters $a = 2, b = -1, e = -1, d = 0$ and a decrease in the values of the bifurcation parameter c. For $c > \sqrt{(bd - ae)^2/(be)} = 2$, the singular point O of system (11) is a stable focus; for $c = 2$, a stable (asymptotically orbitally stable) limit cycle is generated as a result of Andronov–Hopf bifurcation, which exists up to the value $c \approx 1.367$, at which a stable two-dimensional torus is born from it as a result of the second Andronov–Hopf bifurcation. With a further decrease in the values of the parameter c in system (11), a Feigenbaum cascade of period doubling bifurcations of stable two-dimensional tori is observed. A stable two-dimensional torus of period two is born at $c \approx 1.243$, a two-dimensional torus of period 4 is born at $c \approx 1.223$, a two-dimensional torus of period 8 is born at $c \approx 1.218$ etc. At $c \approx 1.215$, in system (11) there is the first simplest singular (chaotic) attractor—the Feigenbaum toroidal attractor—a nonperiodic trajectory lying on the manifold that is the Cartesian product of the limit cycle and the Feigenbaum cyclic attractor. With a further decrease in the values of the parameter c, a sequence of stable two-dimensional tori is revealed in accordance with the Sharkovsky subharmonic order and then in accordance with Magnitskii homoclinic order. For example, a stable two-dimensional torus of period 6 of a subharmonic cascade is found at $c = 1.2141$, a two-dimensional torus of period 5—at $c = 1.2094$(Fig. 3a), a two-dimensional torus of period 3—at $c = 1.2046$ (Fig. 3b). A stable two-dimensional torus T_4 of the homoclinic cascade is detected

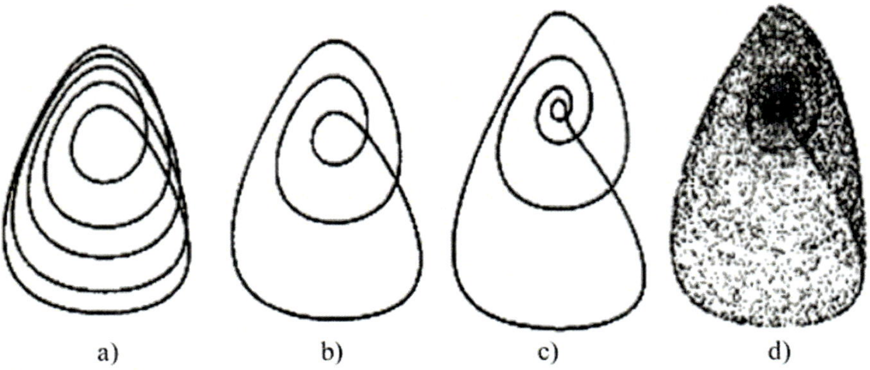

Fig. 3 Projections of Poincare sections: two-dimensional tori of periods five (**a**) and three (**b**) of a subharmonic cascade, a two-dimensional torus of period four of a homoclinic cascade (**c**) and a complex toroidal singular attractor (d) of the ODE system (11)

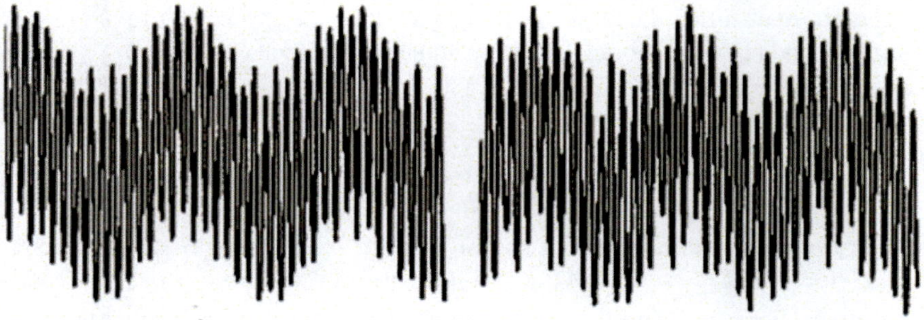

Fig. 4 Traveling periodic waves in the generalized nonlinear Schrödinger (9), corresponding to two-dimensional tori of periods two (a) and three (b) of a subharmonic cascade of bifurcations

at $c = 1.1883$ (Fig. 3c). As is known (see Magnitskii [1–3]), a sequence of two-dimensional tori of a complete homoclinic cascade of bifurcations must converge to a toroidal separatrix manifold, which is the Cartesian product of the limit cycle and the saddle-focus separatrix loop. However, for a given set of system parameter values, this separatrix manifold does not seem to exist. Its detection in the space of five system parameters (a, b, c, d, e) of the system is a separate difficult task, since this separatrix manifold has a codimension greater than unity. Figure 3 shows the projections of the Poincaré sections $v = 0$ onto the (z, u) plane: two-dimensional tori of periods five (a) and three (b) of a subharmonic bifurcation cascade, a two-dimensional torus of period four (c) of a homoclinic bifurcation cascade, and a complex toroidal singular attractor (d) of ODE system (11) lying in a neighborhood of a toroidal separatrix manifold.

Thus, it has been numerically established that, as the parameter c decreases after two Andronov–Hopf bifurcations in system (11), there realized a cascade of the Feigenbaum period doubling bifurcations of stable two-dimensional tori, a complete subharmonic cascade of bifurcations of stable two-dimensional tori in accordance with the Sharkovsky order and then incomplete homoclinic cascade of bifurcations of stable two-dimensional tori. The found two-dimensional tori of system (11) correspond to traveling waves in the generalized nonlinear Schrödinger (9). Figure 4 shows such waves corresponding to two-dimensional tori of period two at $c = 1.225$ and period three at $c = 1.2046$ of a subharmonic cascade of bifurcations.

4 Conclusions

An analytical and numerical analysis of the dynamics of wave solutions in the Kuramoto–Sivashinsky equation and the generalized nonlinear Schrödinger equation is performed in the paper. The birth conditions are found and solutions of the equations in the form of periodic and chaotic traveling waves are obtained. It is shown that the transition to space–time chaos in all cases is carried out in full accordance with the Feigenbaum-Sharkovsky-Magnitskii universal bifurcation theory through the subharmonic and homoclinic cascades of bifurcations of stable cycles and two-dimensional tori of systems of ordinary differential equations obtained from the original equations by self-similar change of variables. It is proved that the bifurcation parameter is the magnitude of the speed of propagation of traveling waves along the spatial axis, which is clearly not included in the original equations.

Paper is supported by Russian Foundation for Basic Research (grants 18-29-10008мк, 20-07-00066a).

References

1. N.A. Magnitskii, Theory of dynamical chaos. URRS, M. (2011) (in Russian)
2. N.A. Magnitskii, Universality of transition to chaos in all kinds of nonlinear differential equations, in *Nonlinearity, Bifurcation and Chaos - Theory and Application* (InTech, Rijeka, 2012), pp. 133–174
3. N.A. Magnitskii, Bifurcation theory of dynamical chaos, in *Chaos Theory* (InTech, Rijeka, 2018), pp. 197–215
4. N.M. Evstigneev, N.A. Magnitskii, FSM scenarios of laminar-turbulent transition in incompressible fluids, in *Nonlinearity, Bifurcation and Chaos, Theory and Applications* (InTech, 2012), pp. 251–280
5. N.M. Evstigneev, N.A. Magnitskii, Numerical analysis of laminar-turbulent bifurcation scenarios in Kelvin-Helmholtz and Rayleigh-Taylor instabilities for compressible flow, in *Turbulence* (InTech, Rijeka, 2017), pp. 29–59
6. T.V. Karamysheva, N.A. Magnitskii, Traveling waves, impulses and diffusion chaos in excitable media. Commun. Nonlinear Sci. Numer. Simul. **19**, 1742–1745 (2014)
7. N.A. Magnitskii, Traveling waves and space-time chaos in the Kuramoto-Sivashinsky equation. Differ. Equ. **54**(9), 1266–1270 (2018)
8. D.A. Burov, N.M. Evstigneev, N.A. Magnitskii, On the chaotic dynamics in two coupled partial differential equations for evolution of surface plasmon polaritons. Commun. Nonlinear Sci. Numer. Simul. **46**, 26–36 (2017). ELSEVIER

Study of Turbulent Transport in Magnetized Plasmas with Flow Using Symplectic Maps

Julio J. Martinell, Carolina A. Tafoya, and Jorge Torres

Abstract Test particle $E \times B$ transport due to a discrete spectrum of drift waves in two dimensions is studied using a Hamiltonian approach, which can be reduced to a 2D mapping. Finite Larmor radius (FLR) effects are included taking an average over the gyroperiod. The presence of poloidal flows is included in order to have the possibility of transport barrier formation. For large wave amplitudes regular particle orbits become chaotic which represents a type of Lagrangian turbulence leading to loss of particle confinement. Poloidal flows tend to decrease the chaos in some regions thus forming transport barriers. FLR effects also reduce chaos facilitating the formation of a barrier. This implies that fast particles are better confined. It is shown that when the particles have a thermal distribution, FLR effects lead to a non-Gaussian particle spatial distribution function, indicating an influence of non-locality. However, the transport preserves its diffusive scaling when there is no flow but including flow there can be a ballistic scaling of transport in poloidal direction if the flow is strong enough. When the background flow varies linearly with radius, the transport can be studied with a symplectic single-step map and the transport properties are given by the KAM theory. For non-monotonic flows a two-step map should be used and this leads to the appearance of transport barriers. The threshold of transport barrier formation is explored for different types of flow.

Keywords Symplectic maps · Chaotic transport · Transition to chaos · Fusion plasmas

J. J. Martinell (✉) · C. A. Tafoya · J. Torres
Instituto de Ciencias Nucleares, UNAM, A. Postal 70-543, México D.F., Mexico
e-mail: martinel@nucleares.unam.mx

C. A. Tafoya
e-mail: carolinaatt@ciencias.unam.mx

C. H. Skiadas et al. (eds.), *13th Chaotic Modeling and Simulation International Conference*, Springer Proceedings in Complexity,
https://doi.org/10.1007/978-3-030-70795-8_41

1 Introduction

Transport in magnetically confined plasmas is due mainly to turbulence produced by some type of drift waves. Unstable waves grow until saturation thus creating a distribution of frequencies characterized by a spectrum. Wave-particle interactions affect the dynamics of plasma particles producing cross-field transport. This turbulent or anomalous transport is the cause of the limited confinement observed in toroidal magnetic devices. The self-consistent study of transport and turbulence is a complicated process and it is usually done using gyro-kinetic equations. However the main properties of the transport can be deduced by assuming a given state of turbulence and following the particle dynamics in the turbulence fields. The wave-particle interaction for a test particle, which does not affect the background fields, can be used as indicator of what a plasma particle will experience in this turbulent field. For electrostatic turbulence, the most important effect on charged particles is the $E \times B$ drift and thus, the lowest order approximation is to follow the guiding center under the action of the E fields produced by the electrostatic turbulence. Given the spectrum of drift waves this study can be performed using a Hamiltonian approach [1, 2].

In addition to turbulence, nearly constant, larger scale electric fields may be present which usually are associated with macroscopic plasma flows. A radial electric field produces poloidal flows and if they have a radial shear they are known to have an effect on the level of turbulence. Then, under the $E \times B$ approach the combined effect of waves and flows can be studied. The type of wave spectrum is determinant for the resulting particle trajectories. A simple case that already retains the basic nonlinear interactions considers just two waves with different phase velocities. This has been studied previously in presence of a poloidal zonal flow, where the conditions for the establishment of global chaos, identified with turbulent transport, were analyzed [3].

The guiding center approximation can be improved by including finite Larmor radius effects (FLR) which is important to take into account the actual fields that the particles feel. This can be done by taking an average over the gyro-orbit and it has been shown that the principal effect is to produce chaos reduction [3]. This implies that the high energy particles, which have larger Larmor radius, have a slower transport than thermal particles.

While the turbulence problem has been studied widely using different approaches with various levels of complexity, the test particle model followed here is simple enough to understand the actual physics behind and captures the essential phenomena involved in the transport. This model has been used for a long time in plasmas and fluids [1, 4]. The main feature is that wave-particle interactions give rise to deterministic chaos in the particle orbits which can be interpreted as Lagrangian turbulence. This approach is suitable to study sheared flows which, in geostrophic fluids, has been used to study zonal flows [4]. These flows are concentrated in latitude and are known to reduce cross-flow transport leading to transport barrier formation. In the Hamiltonian description the transport barrier is identified with a surface in phase-space that is resilient to chaos and occurs around the maximum of the velocity profile where the shear vanishes [3]. This robust surface keeps its integrity until the turbulence level is large enough. In fusion plasmas, the concept of zonal flow is analogous in the sense that is refers to radially localized poloidal flows of zero frequency and independent of poloidal and toroidal angles. They usually allude to self-generated flows by the turbulence. However, in our not self- consistent study we will consider that a zonal flow is simply a poloidal velocity with a radial variation that is non-monotonic.

In this work we adopt the Hamiltonian test particle model, considering a special type of wave spectrum that has not been studied in this context. It contains an infinite number of waves with a uniform distribution of discrete frequencies. In addition it depends on the two dimensions across the magnetic field direction. This makes it more general than the two-wave spectrum studied previously [3]. This particular choice reduces the equations for the particle time evolution to a symplectic mapping when there are no macroscopic plasma flows [1]. In this conditions, it is quite convenient to study the statistical properties of transport by following an ensemble of particles for a period of time. Under certain conditions the transport can become non-diffusive as a result of FLR effects. When macroscopic flows are included the maps are modified in a way that in general they are no longer symplectic. Only for linear shear the map can be symplectic. For shear flows that have non-monotonic radial variation, we show here that the description has to be made in terms of two-step mappings. In the nomenclature of mapping theory, monotonic flows produce twist maps while non-monotonic flows give rise to non-twist maps. With this approach, the study of the appearance and destruction of transport barriers associated to sheared poloidal flows can be performed quite naturally, following the methods of nonlinear maps. For twist maps it is related with the stability of KAM tori in Hamiltonian systems. However, for non-twist flows KAM theory does not apply and they have to be studied with a different approach [5]. Since the presence of transport barriers is of uttermost importance in fusion these studies are most relevant.

The paper is organized as follows. In Sect. 2 the test particle model used is described in the $E \times B$ approximation and the corrections due to FLR are introduced. The representation for the waves is described leading to the iterative mapping for the particle evolution. The properties of the map without macroscopic flow are studied in Sect. 3 showing that the effect of FLR is to reduce the chaotic regions in the particle trajectories. Then, in Sect. 4 a macroscopic poloidal flow that is linearly increasing in the radial direction is introduced; it turns out that the transport parallel to the flow can become super-ballistic and it is shown that this is because the flow keeps increasing radially; for a radially limited flow the transport is only super-diffusive. Section 5 analyzes the effect of non-monotonic flows that can have a maximum at some radial position. It is shown that the transport can be strongly affected by the flow which can produce transport barriers that would lead to better plasma confinement. The properties of the barriers are studied in Sect. 6 both for monotonic and non-monotonic flows. Finally, the conclusions and a discussion of further work is presented in Sect. 7.

2 Test Particle Model

The guiding center velocity for a particle in a magnetized plasma in presence of an electric field is $\mathbf{v}_E = \mathbf{E} \times \mathbf{B}/B^2$ for which the equation of motion produces the following time evolution equation for the particle position

$$\frac{d\mathbf{r}}{dt} = \frac{\mathbf{E} \times \mathbf{B}}{B^2}$$

Taking a uniform magnetic field as an approximation, $\mathbf{B} = B_0\hat{\mathbf{z}}$, we can focus on the perpendicular motion since the velocity along the field is constant. Then, $\mathbf{r} = (x, y)$ is the test particle position in 2D. For an electrostatic field $\mathbf{E} = -\nabla\phi(x, y, t)$ the equation has the structure of a Hamiltonian dynamical system with $H = \phi$

$$\frac{dx}{dt} = -\frac{\partial\phi}{\partial y}, \qquad \frac{dy}{dt} = \frac{\partial\phi}{\partial x},$$

The approximation of zero Larmor radius is not good for energetic particles. FLR effects are included by taking the average over the gyro-orbit

$$\frac{dx}{dt} = -\left\langle\frac{\partial\phi}{\partial y}\right\rangle_\theta, \qquad \frac{dy}{dt} = \left\langle\frac{\partial\phi}{\partial x}\right\rangle_\theta$$

where the average, for the Larmor radius ρ, is defined as

$$\langle\Psi\rangle_\theta \equiv \frac{1}{2\pi}\int_0^{2\pi} \Psi\,(x + \rho\cos\theta, y + \rho\sin\theta)\, d\theta\,.$$

For the wave spectrum we assume a distribution with an infinite number of waves having the same wavenumber and amplitude A, given by

$$\phi = A \sum_{n=-\infty}^{\infty} \cos(x + \theta_n)\cos(y + \theta_n - nt)\,. \tag{1}$$

Using the identity for the Fourier representation of the delta function, $\sum_{n=-\infty}^{\infty}\cos(nt) = 2\pi\sum_{m=-\infty}^{\infty}\delta(t - 2\pi m)$, the equations of motion in terms of the new variables $x_\pm = x \pm y$ reduce to [1]

$$\frac{dx_+}{dt} = 2\pi A \sum_{m=-\infty}^{\infty} \sin(x_-)\delta(t - 2\pi m)\,.$$

$$\frac{dx_-}{dt} = -2\pi A \sum_{m=-\infty}^{\infty} \sin(x_+)\delta(t - (2m + 1)\pi).$$

This can be expressed as a two-step map

$$\begin{aligned} x_+(t_{2m}^+) &= x_+(t_{2m}^-) + 2\pi A \sin(x_-(t_{2m}^-)) \\ x_-(t_{2m}^+) &= x_-(t_{2m}^-) \\ x_+(t_{2m+1}^+) &= x_+(t_{2m+1}^-) \\ x_-(t_{2m+1}^+) &= x_-(t_{2m+1}^-) - 2\pi A \sin(x_-(t_{2m+1}^+)) \end{aligned} \quad (2)$$

This map first produces a displacement along the line $x - y = cons$ and in the second step it displaces along the line $x + y = cons$. Identifying t_{2m}^+ with t_{2m+1}^- it can be reduced to a one-step map

$$\begin{aligned} x_+^{n+1} &= x_+^n + 2\pi A \sin(x_-^n) \\ x_-^{n+1} &= x_-^n - 2\pi A \sin(x_+^{n+1}) \end{aligned}$$

Notice that this iterative map is exact and totally equivalent to solving the differential equations, since no discretization has been applied.

After gyro-averaging it takes the form

$$\begin{aligned} x_+^{n+1} &= x_+^n + 2\pi A J_0(\sqrt{2}\rho) \sin(x_-^n) \\ x_-^{n+1} &= x_-^n - 2\pi A J_0(\sqrt{2}\rho) \sin(x_+^{n+1}) \end{aligned} \quad (3)$$

where $J_0(x)$ is Bessel function of zero order.

3 Orbit Topology and Transport Properties Without Flow

The map (3) has been studied in [2] where it was shown that the phase space structure of the particle orbits is formed by periodic closed orbits as presented in Fig. 1 for a low amplitude $A = 0.01$ and zero Larmor radius zero. In this case most orbits are closed but actually there is always some level of chaos for any finite A near hyperbolic points, even though it may be unnoticed for very small wave amplitudes.

3.1 *Topology of Phase-Space*

The main results regarding the topological properties found in [2] are summarized here. When A increases, chaotic regions around X-points get larger. Eventually, the chaotic orbits dominate the whole phase space and only few closed orbits might remain around elliptic points. Figure 2(left) shows a case with $A = 0.55$, $\rho = 0$ where chaos permeates everywhere although some regions with closed orbits remain. This

Fig. 1 Phase space diagram for the particle orbits with $A = 0.01$, $\rho = 0$

corresponds to large transport losses since the central part of the confined plasma (left side) is chaotically connected to the edge (right side).

Now, the effect of FLR can be seen in Fig. 2(right) in which the Larmor radius ρ is increased ($A = 0.55$, $\rho = 0.9$). It is apparent that some closed orbits are reestablished. Then, FLR has the effect of reducing chaos and thus the transport, implying that particles with larger Larmor radius are better confined.

Another feature is that, as chaos increases there is O-point bifurcation which occurs for $A = A_c = \frac{A_{c0}}{J_0(\sqrt{2}\rho)} \approx A_{c0}(1 + \frac{\rho^2}{2})$, where $A_{c0} = 0.318$. It means that an orbit around the elliptic point becomes doubly periodic, jumping between two new O-points. This can be appreciated comparing Figs. 1 and 2.

3.2 *Distribution Functions (PDF)*

For $A \gg A_c$ there is global chaos in phase space and the particles experience random walk-like trajectories. In this conditions the collective transport can be described with statistical tools, following an ensemble of particles with random initial positions around a starting point. Recording the positions of all particles after a given time (i.e. number of iterations) one can construct histograms of the spatial particle distributions which represent the particle distribution function (PDF) in the 2D space. As for any random process the transport is diffusive, characterized by a Gaussian PDF along

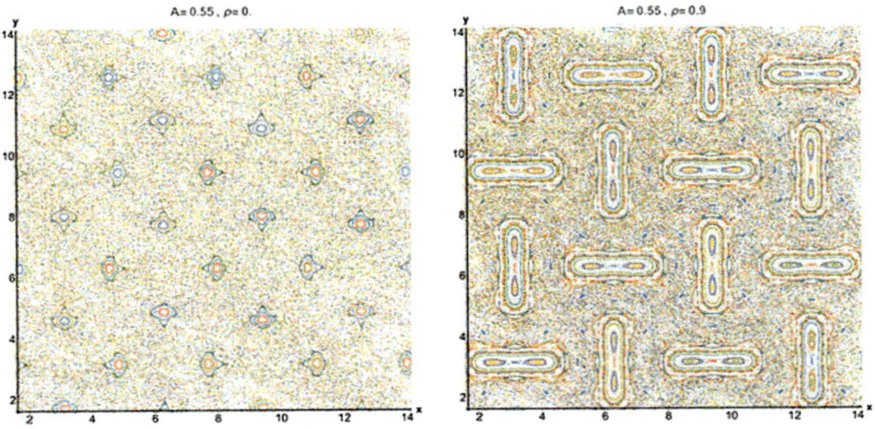

Fig. 2 Phase space diagrams for the particle orbits showing the effect of finite Larmor radius (left) with $A = 0.55$, $\rho = 0$ (right) with $A = 0.55$, $\rho = 0.9$

each coordinate: e^{-x^2/σ_x} for x coordinate, where the variance scales with time as $\sigma_x \sim t$. Similarly, the variance for the y coordinate $\sigma_y \sim t$. This result is obtained when all the particles have the same Larmor radius, which is shown in Fig. 3. The diffusion coefficient can be calculated from the map as

$$S \equiv \sum_{j=1}^{n} \langle (x^{j+1} - x^j)^2 \rangle_p = (A(\rho)^2/4)n.$$

with $A(\rho) = 2\pi A J_0(\sqrt{2}\rho)$. Therefore, taking the iteration number n as proxy for time, the diffusion coefficient for a given ρ is $D(\rho) = \frac{1}{2}(\pi A J_0(\sqrt{2}\rho))^2$, the quasi-linear value. To next order of approximation, $D(\rho)$ shows oscillations [6, 7] which in our case can be represented by $D = \pi^2 A(\rho)^2(1 + 2J_0(2\pi A(\rho)))$.

Now, in a plasma the particles have different velocities an therefore different Larmor radii. Thus, if the initial conditions for the Larmor radius are taken from a 2D Maxwellian distribution of the form

$$f(\rho) = (2\rho/\rho_{th}^2)\exp[-(\rho/\rho_{th})^2]. \quad (4)$$

the PDF is no longer Gaussian; it is found to be characterized by a kurtosis of about 5–7 (quite larger than 3 for a Gaussian). But the scaling $\sigma_y \sim t$ still holds. This means that the thermal distribution of FLR produces a non-local transport, but still with diffusive scaling. The underlying reason for the non-Gaussian PDF is that the superposition of two Gaussian functions of different width is no longer a Gaussian. Composing the ρ-dependent Gaussian with (4) one can find an analytical form for the non-Gaussian PDF. This can be compared with the numerically obtained PDF and the agreement is almost perfect as seen in Fig. 4a for $A = 1$ (see [2] for details).

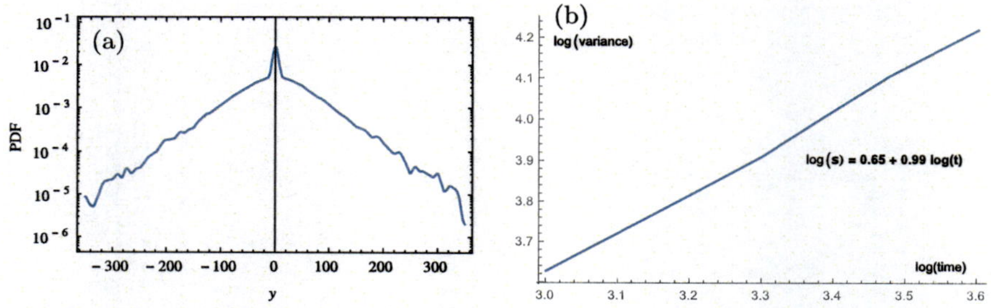

Fig. 3 **a** PDF of the poloidal (y) propagation in log scale showing the non-Gaussian distribution. **b** Time scaling of the variance displaying a diffusive scaling $\sigma \sim t$

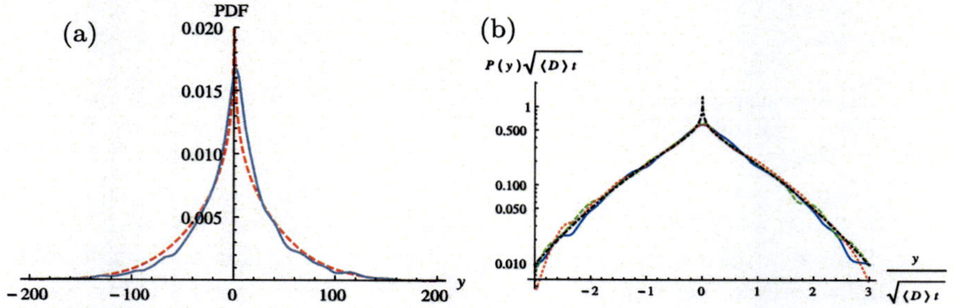

Fig. 4 **a** Numerical (continuous) and analytical (dotted) PDFs for $A = 1$ in linear scale. **b** Self-similar curves for y propagation PDF at various times for $\rho_{th} = 0.1$ in log scale

It is found that the PDF has the property of self-similarity;

$$P(y, t) = (\langle D \rangle t)^{-1/2} \mathcal{G}_{\rho_{th}}(y/(\langle D \rangle t)^{1/2}) \tag{5}$$

where the self-similar function $\mathcal{G}$ depends on the thermal radius ρ_{th}. The self-similar behavior can be seen in Fig. 4b for $\rho_{th} = 1$ which displays that the PDF at different times all have the same shape.

4 Presence of Poloidal Flows

A macroscopic poloidal flow along y is now included by adding a function of x to ϕ representing a radial electric field (i.e. replace $\phi(x, y) \rightarrow \phi_0(x) + \phi(x, y)$. It turns out that for the map to be symplectic it has to be kept as a two-step map as in (2). Adding the flow function $\phi_0(x)$ to ϕ the resulting two-step map is

$$
\begin{aligned}
x'_{+} &= x^{n}_{+} + \pi\Omega(x^{n}) \\
x'_{-} &= x^{n}_{-} + 2\pi A J_0(\sqrt{2}\rho)\sin(x'_{+}) - \pi\Omega(x^{n}) \\
x^{n+1}_{-} &= x'_{-} - \pi\Omega(x') \\
x^{n+1}_{+} &= x'_{+} - 2\pi A J_0(\sqrt{2}\rho)\sin(x^{n+1}_{-}) + \pi\Omega(x')
\end{aligned}
\tag{6}
$$

where $\Omega(x) = \phi_0'(x)$ is the flow velocity profile.

Depending on the form of function $\phi_0(x)$ the flow can be monotonic or non-monotonic. When $\Omega(x)$ is linear the map (6) has a non-zero Jacobian everywhere which implies that this is a twist map. If the variation of $\Omega(x)$ is non-monotonic, it has maxima or minima, where the derivative is zero. Thus the Jacobian is zero there and therefore the map is non-twist.

4.1 Linear Shear Flow

The simplest sheared flow is for a linear velocity profile, $\Omega(x) = Cx$ (obtained from $\phi_0(x) = Cx^2/2$). The phase-space structure includes invariant tori which have the effect of transport barriers when chaos starts developing. The tori depicted in Fig. 5 are actually the streamlines produced by the flow combined with the waves. The presence of flow destroys the symmetry in x and y; now the transport along the poloidal direction, y, is expected to be faster. The radial transport is hindered by the KAM tori. Only when they are destroyed, as global chaos is established, transport can be effective in taking particles to the plasma edge. This of course happens as A increases, as shown in Fig. 5.

The fixed points, $x^*_{\pm}$, can be obtained by solving the equations when $x^{n+1}_{\pm} = x^{n}_{\pm}$ which gives that they are located at $x^* = 0$, $y^* = k\pi$. It can also be seen that hyperbolic points become elliptic when $C^3/8 < A(\rho)[A(\rho) - C + C^2/4]$. Now, O-point bifurcation occurs at smaller values of A than before.

The streamlines that cross from top to bottom in Fig. 5 are the invariant KAM tori which are identified by a rotation number. According to KAM theory the tori with rational rotation numbers are destroyed first and the most robust are those with irrational rotation number. A transport barrier will subsist until the last KAM torus is destroyed. In the next section we address this problem.

The PDF of an ensemble of particles with an initial Maxwellian distribution of Larmor radii can be obtained as before. To proceed, one has to consider the regime where chaos has already destroyed the transport barriers, i.e. when global chaos is established. For the x propagation the results are the same as when there is no flow, for both the type of PDF and the transport scaling with time. Regarding the transport parallel to the flow, there are some differences. On the one hand, the thermal PDF is also non-Gaussian and has a self-similar property, according to the scaling [2]

$$
P(y,t) = (\langle D\rangle t)^{-3/2}\mathcal{G}_{\rho_{th}}(y/(\langle D\rangle t)^{3/2}). \tag{7}
$$

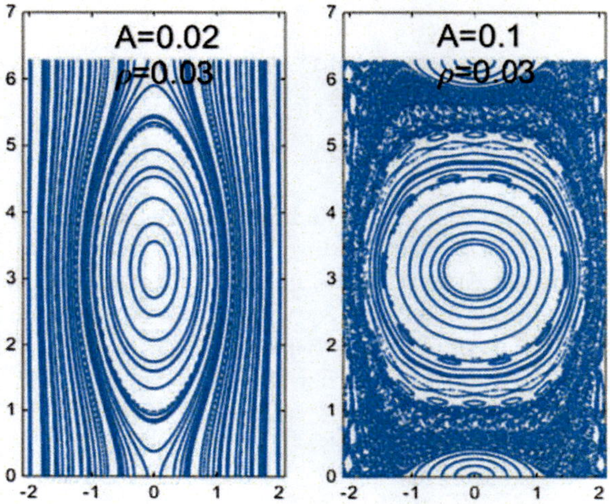

Fig. 5 Phase space of particle orbits with linear flow with $c = 0.5$ for $\rho = 0.03$ and $A = 0.02$ (left), $A = 0.1$ (right)

On the other hand, the scaling of the variance of the particle distribution is not diffusive. It scales as $\sigma \sim t^3$ in the direction of the flow. This may be surprising because it is super-ballistic which would imply some kind of acceleration. This cubic scaling is the reason for the different scaling in the self-similarity relation in (7) as compared with that in (5). The apparent acceleration is due to the fact that there is a cooperative effect between diffusive transport along x which takes particles to larger "radial" distances and y transport which gets faster at large x. Thus, the particles increase the velocity ("accelerate") as they move away from $x = 0$ and these dominate the statistics.

This is not physical since the velocity tends to infinity as $|x| \to \infty$. But this problem can be corrected when the shear is terminated at a given distance like in the flow depicted in Fig. 6. For this velocity profile the resulting scaling is intermediate between ballistic and diffusive (but still super-diffusive) $\sigma \sim t^\alpha$ with $1 < \alpha < 2$. The actual value of α depends on the width of the shear region. When the width gets narrower the transport is more diffusive-like ($\alpha = 1$) but wider velocity profiles produce ballistic-like transport.

5 Non-monotonic Shear Flow

In order to deal with non-monotonic flows one can either combine linear flows or use a single function with maxima or minima. In the following we consider both choices.

(a) **Piecewise linear map**. A non-monotonic flow can be constructed with piecewise linear functions.This is a direct extension of the linear flow although the problem is that at the border points there is a jump in the derivative. The type of profile that

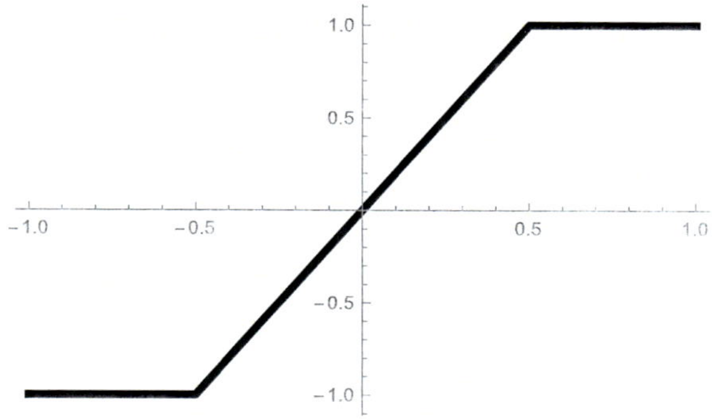

Fig. 6 Velocity profile for a truncated linear shear flow

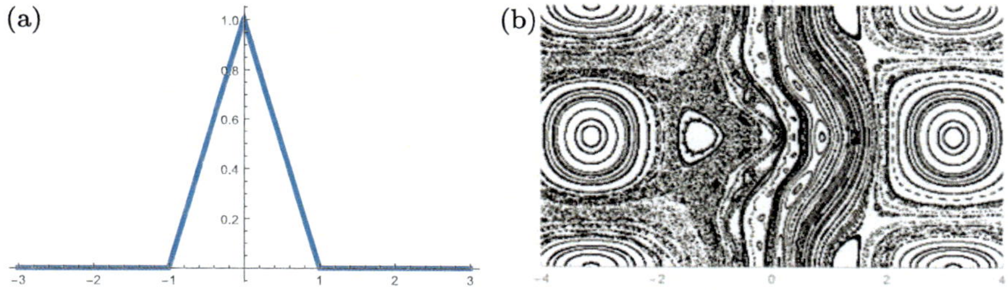

Fig. 7 **a** Triangular velocity profile. **b** Phase space diagram for this nonlinear shear flow

can model a zonal flow has a maximum at $x = 0$ and has a triangular shape. This is non-zero just in a region $-M < x < M$ and is defined by $\Omega(x) = C(M - |x^n|)$ when $x < M$ and zero otherwise. The velocity profile and the resulting phase space orbits are given in Fig. 7. It is seen that streamlines are now in the central region and these can act as transport barriers. They cannot be dealt with as KAM tori since now the map is non-twist and KAM theory does not apply. As usual, increasing A increments the chaotic region area and eventually the whole space is chaotic.

(b) **Gaussian profile**. A "zonal flow" can be represented by a Gaussian function $\Omega(x) = C\exp(-x^2)$ which is now continuous everywhere as well as its derivative. It is non-twist since the Jacobian $J = 0$ at $x = 0$. The velocity profile and the resulting phase space orbits are given in Fig. 8 which is similar to the triangular flow, as expected.

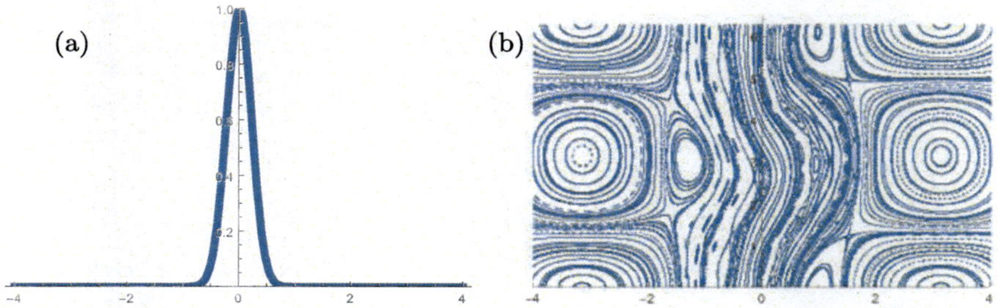

Fig. 8 **a** Gaussian velocity profile. **b** Phase space diagram for the two-step map with this shear flow

6 Transport Barriers

The presence of a sheared flow naturally introduces KAM surfaces which are the particle orbits that are carried by the flow and are seen as the lines in phase space diagrams crossing from the upper to the lower boundary. These act as transport barriers since no orbits can go radially (x direction) traveling across the KAM surfaces. In the chaotic regime some of the KAM surfaces remain stable usually separating chaotic regions completely. Only when an increase of A produces a large enough chaos level, the most robust surfaces can break down and the transport barriers cease to exist. In the case of the linear shear flows of Sect. 4 KAM theory can be used to determine the conditions for transport barrier break up. Criteria like Chirikov resonance overlap or Greene residue theorem allow to find the parameter threshold values for the establishment of global chaos and hence barrier destruction [7]. For the non-twist maps of non-monotonic flows, KAM theory does not apply; the relevant theory for this case tell us that the most robust surface is located by the velocity maximum, where the shear vanishes [5]. Then the destruction of this shearless surface is what determines establishment of global chaos, when transport barriers disappear. The level of chaos needed for its destruction can be determined using some established method such as the indicator points [8].

A more straightforward procedure (although less formal) to determine the existence and stability of a barrier is to evolve two initial points on both sides of the "barrier" (blue, red in Fig. 9). They are iterated for 10^5 steps. When the barrier is present and chaotic motion dominates, they fill the space on each side of barrier but they do not interpenetrate. As A is raised a value is reached where the two colors mix meaning that the barrier is broken. In this way one can determine the threshold values of the parameters A, ρ and C. The procedure can be used for twist ans non-twist maps.

For the case of twist map the destruction of KAM tori has to be considered. The most robust one is usually the one with a rotation number equal to the golden ratio. Figure 10 shows an example of central chaotic orbits, having the barriers on the sides. In the twist map there can be resurgences i.e. when the barrier disappears for

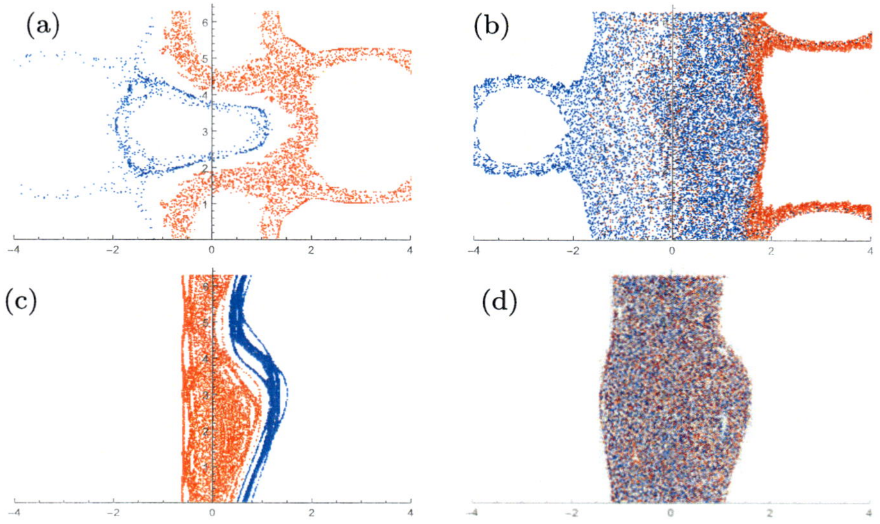

Fig. 9 Some cases for local chaos separated by transport barriers and their disappearance. **a** Presence of a central barrier, **b** no barrier present, **c** central and side barriers (banded chaos) and **d** only side barriers. This is for Gaussian flow

Fig. 10 Transport barriers for a monotonic flow are present on both sides of the central trapped orbits. These prevent particles from crossing radially to the plasma edge

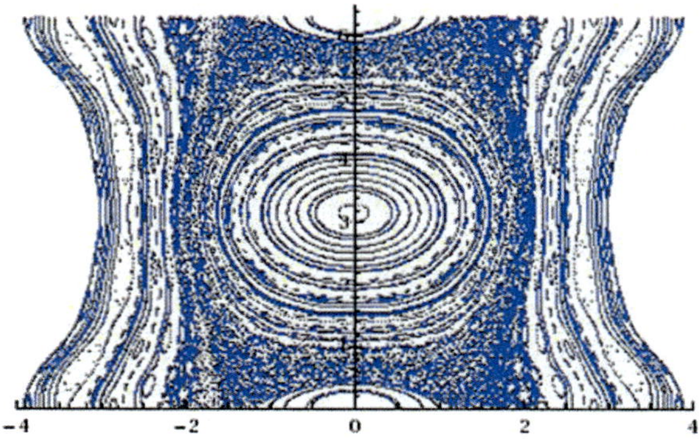

a critical value of $A = A_{th}$ it can appear again for some $A > A_{th}$. The method we used has some weaknesses since sometimes it seems that the barrier is present but when the number of iterations is increased there is an apparent barrier penetration. This is due to the Aubry-Mather theorem which says that KAM surfaces become Cantori allowing a very small number of points to cross. Therefore, once a possible threshold value has been found we have to increase the number of iterations to make sure it still holds.

For non-monotonic flows the criterion is related to the robustness of the shearless curve. These barriers are more robust than in twist maps. Figure 9 shows different situations that can arise in these maps. The first panel displays the case of a central

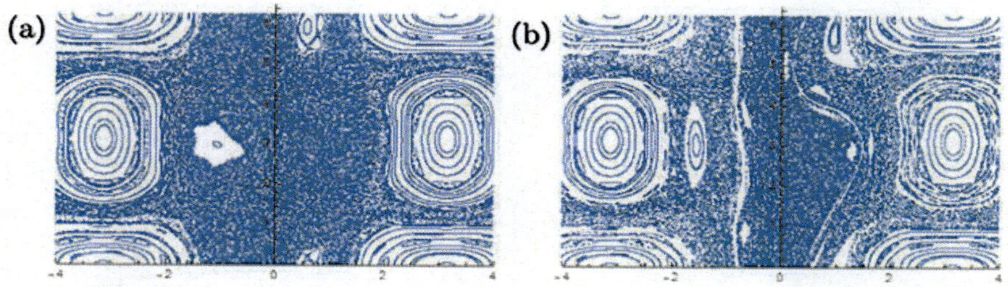

Fig. 11 Phase-space plots for a Gaussian velocity profile when $\rho = 0$, $A = 0.2$. The chaotic central region does not hold a barrier in panel **a** for $C = 0.3$, while in panel **b**, where $C = 0.8$, the barrier is present, identified by the white gap

barrier, while in the second panel the barrier has disappeared. There is also an additional pair of barriers that was not anticipated to occur; they appear as side barriers. This is seen in the lower panel of Fig. 9 where there is a coexistence of a central barrier and the two side barriers. This has been termed banded chaos in [5]. In the last panel the side barriers are preserved but not the central one.

Even though the physical behavior is expected to be the same for a Gaussian and a triangular velocity profile, it has been found that the behavior of the for the Gaussian flow is more regular because it does not have singular derivatives. The presence of a barrier can be more apparent in phase space plots for the Gaussian profile, like the one in Fig. 11 which shows the chaotic central region for two flow parameters (C). For $C = 0.3$, the flow is slow and the barrier is not present, while a barrier is clearly seen for a stronger flow ($C = 0.8$). A more formal analysis of the transport like the method of indicator points as applied to non-twist Hamiltonians [3, 8, 9] could be applied for the present case in order to obtain the exact threshold values of C for which the barrier is destroyed, for fixed A and ρ. It would be possible to build threshold diagrams plotting two map parameters, taken from (A, ρ, C) in this case, that display the regions with an without transport barriers. These diagrams usually have a fractal boundary between the two regions (bird wing diagrams) [8]. In order to do that one should apply the formal methods for a systematic study. We deter that study for a future work.

7 Conclusions

The study of anomalous transport in plasmas due to a discrete spectrum of drift waves in two dimensions has been performed including sheared flows by means of a test particle model. The description is reduced to a symplectic iterative map which can be efficiently used to simulate the evolution of an ensemble of particles in relatively short times. The wave spectrum chosen (1) has been used previously in [1] to predict the scaling $D \sim A^2$ of the diffusion coefficient with wave amplitude. Then, in [2] we studied the statistical properties of transport predicted by this spectrum including FLR effects, showing that these effects lead to non-local behavior indicated by non-Gaussian PDF. In the present study we extended the analysis for the cases where a shear flow is also present, analyzing the appearance of transport barriers.

Background flows modify the particle dynamics changing in turn the chaotic properties that produce the transport. A large poloidal shear flow drags the particles along the poloidal direction hindering the chaotic motion produced by the waves. The chaos reduction decreases the transport through a given surface and this is what is identified with a transport barrier. This is different from the usual concept of barrier formation based on shear flow turbulence reduction due to stretching and fragmentation of vortices. We have found that, although one would expect from the above argumentation to have stronger barriers for larger flows, in reality they are almost not affected by a velocity increment because the chaos level is also sensible to this parameter, tending to lessen the change. In addition, the type of transport can be modified by the flow; we show that the radial transport is still diffusive, alas non-local, as in the case without flow, but the poloidal transport becomes ballistic (for low chaos) or super-diffusive (for large chaos). However, the poloidal transport has no effect on the confinement and therefore it has no direct relevance on fusion experiments.

To study the transport barriers, which are extremely important for fusion plasmas, we considered two types of shear flows: (1) those for which the velocity increases linearly with radius (monotonic) and (2) those localized around some radial position where the velocity has a maximum (non-monotonic). We identify the later with zonal flows, although the full concept of a zonal flow is more extensive. In general, presence of flow requires the process to be described by a two-step mapping. For the twist maps (monotonic flow), feeble barriers can appear corresponding to the KAM surfaces that are more resilient. We showed qualitatively the presence of such barriers, showing themselves as boundaries separating chaotic regions, disappearing for large enough wave amplitude A. On the other hand, non-twist maps (non-monotonic flow) have a robust barrier near the shearless curve of the velocity. We have analyzed the behavior of this barrier as function of the relevant parameters (A, C, ρ) showing that it can appear when the velocity C is larger than a threshold value. We also found that there are two other barriers which were not expected a priori that can appear as side bands. This is related to the so-called banded chaos found in [4]. The next step is to study the threshold for transport barriers appearance in terms of the parameters of the flow and the waves using formal methods of nonlinear dynamics. For twist maps there are several criteria that can be used such as the Chirikov resonance overlap or the Greene's residue criterion [7]. For non-twist maps other methods like the indicator points which is based on symmetries of the map written as involutions [8] will be studied in a future work.

Acknowledgements This work was supported by project DGAPA-UNAM IN112118, Conacyt project A1-S-24157 and supercomputer project LANCAD-UNAM-DGTIC-104.

References

1. R. Kleva, J. Drake, Phys. Fluids **27**, 1686 (1984)
2. N. Kryukov, J.J. Martinell, D. del-Castillo-Negrete, J. Plasma Phys. **84**, 905840301 (2018)
3. J.J. Martinell, D. del-Castillo-Negrete, Phys. Plasmas **20**, 022303 (2013)
4. D. del-Castillo-Negrete, P.J. Morrison, Phys. Fluids A **5**, 948 (1993)
5. D. del-Castillo-Negrete, J.M. Greene, P.J. Morrison, Phys. D **91**, 1 (1996)
6. T.M. Antonsen, E. Ott, Phys. Fluids **24**, 1635–1640 (1981)
7. A.J. Lichtenberg, M.A. Lieberman, *Regular and Chaotic Dynamics*, 2nd edn. (Springer, 1992)
8. S. Shinohara, Y. Aizawa, Prog. Theor. Phys. **97**, 379 (1997)
9. D. del-Castillo-Negrete, J.J. Martinell, Commun. Nonlinear Sci. Numer. Simul. **17**, 2031 (2012)

Biharmonic Problems and Their Applications in Engineering and Technology

Hovik A. Matevossian, Giorgio Nordo, and Giovanni Migliaccio

Abstract In the present paper we study some properties of solutions of biharmonic problems. Namely, we study the Steklov, Steklov-type and Neumann boundary value problems for the biharmonic equation. For solving these biharmonic problems with application, in particular, to radar imaging, we need to solve the Dirichlet, Neumann and Cauchy boundary value problems for the Poisson equation using the scattering model. In order to select suitable solutions, we solve the Poisson equation with the corresponding boundary conditions, that is, some criterion function is minimized in the Sobolev norms. Under appropriate smoothness assumptions, these problems may be reformulated as boundary value problems for the biharmonic equation.

Keywords Biharmonic operator · Boundary value problems · Scattering model · Variational method

H. A. Matevossian (✉)
Federal Research Centre "Computer Science & Control" & Steklov Mathematical Institute, Russian Academy of Sciences, Vavilov str., 40-42, R 374, Moscow 119333, Russia

Moscow Avaiation Institute (National Research University), Volokolamskoe shosse, 4, Moscow 125993, Russia
e-mail: hmatevossian@graduate.org

G. Nordo
MIFT - Università di Messina, Viale Ferdinando Stagno d'Alcontres, 31, 98166 Messina, Italy
e-mail: giorgio.nordo@unime.it

G. Migliaccio
Università di Pisa, Largo Luccio Lazzarino 2, Pisa 56122, Italy

C. H. Skiadas et al. (eds.), *13th Chaotic Modeling and Simulation International Conference*, Springer Proceedings in Complexity,
https://doi.org/10.1007/978-3-030-70795-8_42

1 Introduction

Let $\Omega \subset \mathbb{R}^n$, $n \geq 2$, be a bounded Lipschitz domain with connected boundary $\partial\Omega$, and $\Omega \cup \partial\Omega = \overline{\Omega}$ is the closure of Ω.

In Ω we consider the following boundary value problems for the biharmonic equation:

$$\Delta^2 u = F, \qquad x \in \Omega \tag{1}$$

with the Steklov boundary conditions

$$\begin{cases} u = g_1 & \text{on } \partial\Omega, \\ \Delta u + \tau \frac{\partial u}{\partial \nu} = g_2 & \text{on } \partial\Omega, \end{cases} \tag{2}$$

or the Steklov–type boundary conditions

$$\begin{cases} \frac{\partial u}{\partial \nu} = h_1 & \text{on } \partial\Omega, \\ \frac{\partial \Delta u}{\partial \nu} + \tau u = h_2 & \text{on } \partial\Omega, \end{cases} \tag{3}$$

or the Neumann boundary conditions

$$\begin{cases} Mu \equiv \sigma \Delta u + (1-\sigma)\frac{\partial^2 u}{\partial \nu^2} = f_1 & \text{on } \partial\Omega, \\ Nu \equiv \frac{\partial \Delta u}{\partial \nu} + (1-\sigma)\frac{1}{2} \cdot \frac{\partial}{\partial t_{ij}}\left(\frac{\partial^2 u}{\partial \nu \partial t_{ij}}\right) = f_2 & \text{on } \partial\Omega, \end{cases} \tag{4}$$

where ν is the outer unit normal vector to the domain, and $T = \{t_{ij}\}$ various tangential directions to the Lipschitz boundary $\partial\Omega$, $\tau \geq 0$, $\tau \not\equiv 0$, and $\tau > 0$ on a set of positive $(n-1)$ – dimensional measure on $\partial\Omega$. The coefficient σ is a constant known as the Poisson ratio, $\frac{1}{n-1} < \sigma < 1$. A unique solution u (modulo linear functions) is obtained in the class of solutions with non-tangential maximal function of the second-order derivatives in $L^p(\partial\Omega)$. The biharmonic Neumann problem in Lipschitz domains was studied in detail in [36].

For $n = 2$, these problems and also the Neumann problem are related to the study of the transverse vibrations of a thin plate with a free edge and which occupies at rest a planar region of shape $\partial\Omega$. The coefficient σ represents the Poisson's ratio of the material that the plate is made of. For more details on the physical interpretation of the Neumann problem and on the Poisson's ratio σ, we refer, for example, to [4]. Note the paper [5], where the author studies the dependence of the vibrational modes of a plate subject to homogeneous boundary conditions upon the Poisson's ratio $0 < \sigma < \frac{1}{2}$, providing also a perturbation formula for the frequencies as functions of the Poisson's coefficient.

Elliptic problems with parameters in the boundary conditions are called Steklov problems from their first appearance in [37]. In the case of the biharmonic operator, these conditions were first considered in [3, 10, 33], who studied the isoperimetric properties of the first eigenvalue.

The standard elliptic regularity results are available in [7]. This monograph covers higher order linear and nonlinear elliptic boundary value problems, mainly with the biharmonic (polyharmonic) operator as leading principal part. Underlying models and, in particular, the role of different boundary conditions are explained in detail. As for linear problems, after a brief summary of the existence theory and L^p and Schauder estimates, the focus is on positivity. The required kernel estimates are also presented in detail.

In [6] and [7], the spectral and positivity preserving properties for the inverse of the biharmonic operator under Steklov and Steklov–type boundary conditions are studied. These are connected with the first Steklov eigenvalue. It is shown that the positivity preserving property is quite sensitive to the parameter involved in the boundary condition.

In [34], the dependence of the eigenvalues of the biharmonic operator subject to Neumann boundary conditions on the Poisson's ratio σ is studied. In particular, it is proved that the Neumann eigenvalues are Lipschitz continuous with respect to $\sigma \in [0, 1)$ and that all the Neumann eigenvalues tends to zero as $\sigma \to 1^-$. Moreover, is showed that the Neumann problem defined by setting $\sigma = 1$ admits a sequence of positive eigenvalues of finite multiplicity that are not limiting points for the Neumann eigenvalues with $\sigma \in [0, 1)$ as $\sigma \to 1^-$ and that coincide with the Dirichlet eigenvalues of the biharmonic operator.

Boundary value problems for a biharmonic (polyharmonic) equation and for the elasticity system in unbounded domains are studied in [12]–[28] in which the condition of the boundedness of the following weighted Dirichlet integral of solution is finite, namely

$$\int_\Omega |x|^a |\partial^\alpha u|^2 \, dx < \infty, \quad a \in \mathbb{R},$$

where $a \in \mathbb{R}$ is a fixed number and $|\partial^\alpha u|^2$ denotes the Frobenius norm of the Hessian matrix of u. In particular, in these papers has been studied the dimension of the space of the solutions to the boundary value problems for a biharmonic (polyharmonic) equation and for the elasticity system, providing explicit formulas which depends on n and a. This paper contains complete proofs of the results, partly presented in [29].

The behavior of solutions of the Dirichlet problem for the biharmonic equation as $|x| \to \infty$ was considered in [8, 9], where estimates for $|u(x)|$ and $|\nabla u(x)|$ as $|x| \to \infty$ were obtained under certain geometric conditions on the domain boundary.

Notation: $C_0^\infty(\Omega)$ is the space of infinitely differentiable functions in Ω with compact support in Ω; $H^m(\Omega)$ is the Sobolev space obtained by the completion of $C^\infty(\overline{\Omega})$ with respect to the norm

$$\|u; H^m(\Omega)\| = \left(\int_\Omega \sum_{|\alpha| \le m} |\partial^\alpha u|^2 dx \right)^{1/2}, \; m = 1, 2,$$

where $\partial^\alpha \equiv \partial^{|\alpha|}/\partial x_1^{\alpha_1} \ldots \partial x_n^{\alpha_n}$, $\alpha = (\alpha_1, \ldots, \alpha_n)$ is a multi-index, $\alpha_i \geq 0$ are integers, and $|\alpha| = \alpha_1 + \cdots + \alpha_n$; $\overset{\circ}{H}{}^m(\Omega)$ is the space obtained by the completion of $C_0^\infty(\Omega)$ with respect to the norm $\|u;\, H^m(\Omega)\|$. $\overset{\circ}{H}{}^m_{loc}(\Omega)$ is the space obtained by the completion of $C_0^\infty(\Omega)$ with respect to the family of semi-norms

$$\|u;\, H^m(\Omega \cap B_0(R))\| = \left(\int_{\Omega \cap B_0(R)} \sum_{|\alpha| \leq m} |\partial^\alpha u|^2 \, dx \right)^{1/2}$$

for all open balls $B_0(R) := \{x : |x| < R\}$ in $\mathbb{R}^n$ for which $\Omega \cap B_0(R) \neq \emptyset$. Finally $H^{1/2}(\partial\Omega)$ is the usual trace space on the boundary and $H^{-1/2}(\partial\Omega)$ is its dual (see, for ex., [1]).

2 Definitions and Auxiliary Statements

If we set $\sigma = 1$, the Neumann boundary conditions reads

$$\begin{cases} \Delta u = f_1 & \text{on} \quad \partial\Omega, \\ \frac{\partial \Delta u}{\partial \nu} = f_2 & \text{on} \quad \partial\Omega. \end{cases} \tag{5}$$

Note that the differential operator associated with problem (1), (5) is not a Fredholm operator. We also note that the boundary conditions (5) do not satisfy the so–called complementing conditions (see [2] and [7] for details), which are necessary conditions for the well–posedness of a differential problems.

Definition 1 A solution of the biharmonic equation (1) in Ω is a function $u \in H^2(\Omega)$ such that, for every function $\varphi \in C_0^\infty(\Omega)$, the following integral identity holds:

$$\int_\Omega \Delta u \, \Delta\varphi \, dx = \int_\Omega F \, \varphi \, dx, \quad F \in L^2(\Omega). \tag{6}$$

Definition 2 A function u is a solution of the Steklov problem (1), (2) with $g_1 = g_2 = 0$, if $u \in H^2(\Omega) \cap \overset{\circ}{H}{}^1(\Omega)$ such that for every function $\varphi \in H^2(\Omega) \cap \overset{\circ}{H}{}^1(\Omega)$, the following integral identity holds

$$\int_\Omega \Delta u \, \Delta\varphi \, dx + \int_{\partial\Omega} \tau \, \nabla u \, \nabla\varphi \, ds = 0. \tag{7}$$

Definition 3 A function u is a solution of the Steklov-type problem (1), (3) with $h_1 = h_2 = 0$, if $u \in H^2(\Omega)$, $\partial u/\partial\nu = 0$ on $\partial\Omega$, such that for every function $\varphi \in C_0^\infty(\mathbb{R}^n)$, $\partial\varphi/\partial\nu = 0$ on $\partial\Omega$, the following integral identity holds

$$\int_{\Omega} \Delta u \, \Delta \varphi \, dx - \int_{\partial \Omega} \tau \, u \, \varphi \, ds = 0. \tag{8}$$

Definition 4 A function u is a solution of the Neumann problem (1), (5) with $f_1 = f_2 = 0$, if $u \in H^2(\Omega)$ such that the integral identity (6) holds for every function $\varphi \in C_0^\infty(\Omega)$.

3 A Scattering Model

In the section we derive the mathematical model used for describing the radar process. In our parametrization the unknown is the height function H. As will be shown the height function is determined in two steps. In the first step $\mathfrak{L}(H)$, with $\mathfrak{L}$ a certain second-order differential operator, is determined. After retrieving H the equation $\mathfrak{L}(H) = f$ must be solved. To a good approximation the operator $\mathfrak{L}$ can be replaced by the Laplacian. So the second step simply consists of solving the Poisson equation over some smooth bounded domain, usually a rectangular region in the plane. The problem here is that no natural boundary conditions are available.

Here we will briefly discuss the mathematical inverse problem to be resolved in order to recover the ground topography height function from radar data. First cylindrical coordinates (r, φ, z) are introduced according to Fig. 1, where it is understood that the aircraft is flying at a constant speed along the z-axis. Further r denotes the distance from a point on the ground surface to the z-axis and φ is the angle between radius vector and a horizontal plane through the z-axis. Then the ground surface may be described by a function $H(r, z)$ through the equation

$$\frac{H(r, z)}{r} - \varphi = 0. \tag{9}$$

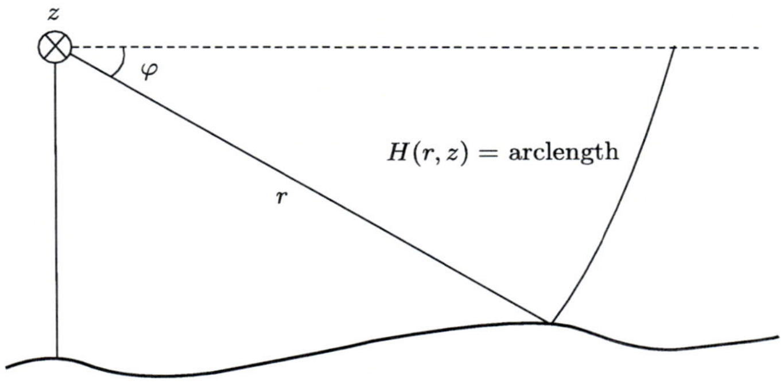

Fig. 1 The ground surface measured at a fixed aircraft position

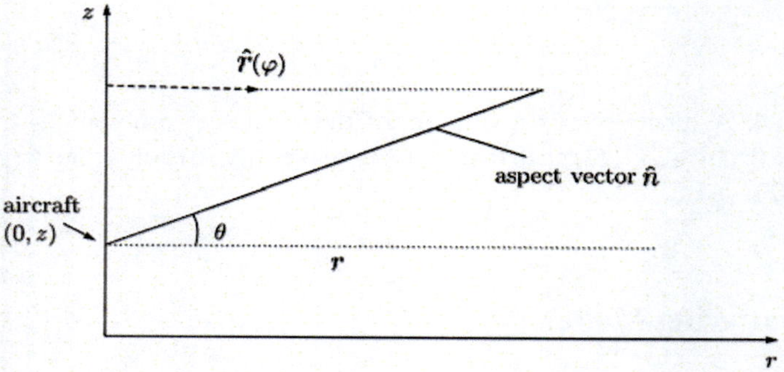

Fig. 2 The measuring geometry as seen from above

When r is large, $H(r, z)$ is approximately a Cartesian height function. Figure 2 shows a top view of the same scene. We have also indicated an aspect vector from the aircraft to some point on the ground, forming an angle θ with a vertical plane through the aircraft. Normalized to unit length, the aspect vector is denoted by $\hat{n}$. Accordingly

$$\hat{n} = \cos\theta\,\hat{r}(\varphi) + \sin\theta\,\hat{z}. \tag{10}$$

Here $\hat{r}(\varphi)$ denotes the cylindrical unit basis vector corresponding to the r-coordinate for the ground point as shown in the Fig. 2. For a point on the ground surface with coordinates (r, φ, z) we obtain, from (9), the following expression for the ground surface normal $\bar{m}$,

$$\bar{m} = grad\left(\frac{H(r,z)}{r} - \varphi\right) = \frac{\partial(H/r)}{\partial r}\hat{r} + \frac{1}{r}\frac{\partial H}{\partial z}\hat{z} - \frac{1}{r}\hat{\varphi}. \tag{11}$$

Let $\hat{m}$ denote the normalized normal. Then

$$\hat{m} \circ \hat{n} = \left(r\cos\theta\frac{\partial(H/r)}{\partial r} + \sin\theta\frac{\partial H}{\partial z}\right) \Big/ \sqrt{1 + r^2\left(\frac{\partial(H/r)}{\partial r}\right)^2 + \left(\frac{\partial H}{\partial z}\right)^2}. \tag{12}$$

Note that (r, φ, z) in (12) are related to the ground surface point and not to the position of the aircraft.

Let $(z_0, 0)$ be a position of the aircraft and R the distance to some point on the surface. According to Fig. 3 the coordinates (r, z) are then equal to $(R\cos\theta, z_0 + R\sin\theta)$. Next, to obtain a scattering model we will assume that the reflectivity from a ground surface element (see Fig. 4) is

$$\approx \frac{\hat{m} \circ \hat{n}}{R} dR\,d\theta. \tag{13}$$

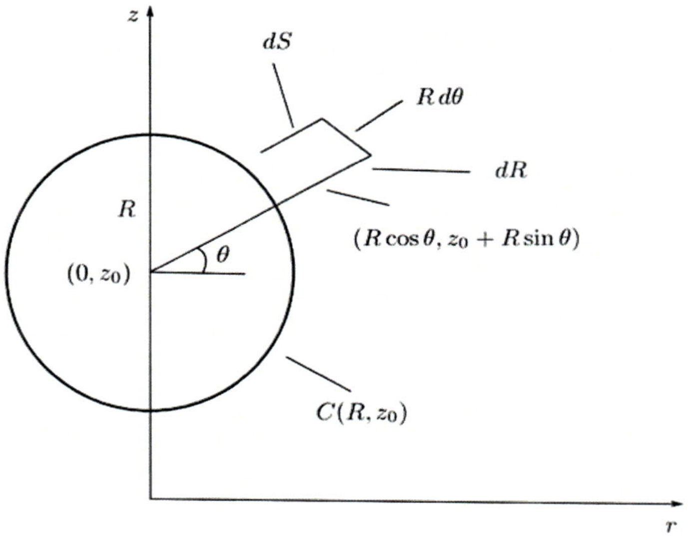

Fig. 3 The coordinate system used to describe an infinitesimal surface element, dS

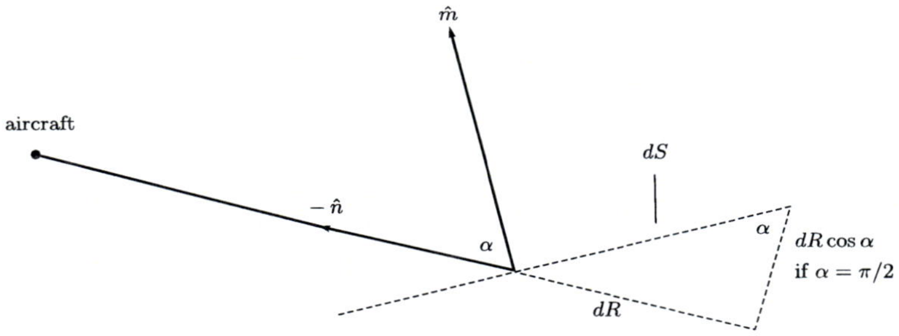

Fig. 4 The infinitesimal surface element, dS, as it is seen from the aircraft

From Fig. 4, where a vertical plane through $(z_0, 0)$ (the aircraft) and the ground point $(R\cos\theta, z_0 + R\sin\theta)$ is displayed, we conclude that the solid angle $d\Omega$ under which the surface element dS is seen from the antenna is approximately

$$\frac{dR\ \cos\alpha\ Rd\ d\theta}{R^2} = -\frac{\hat{m}\circ\hat{n}}{R}dR\,d\theta.$$

In expression (13) we are consequently assuming that the local reflectivity is proportional to the solid angle occupied by the infinitesimal surface element dS. The total reflected signal $G(R, z_0)$ from all points at a distance R from the antenna may now be obtained by integration over the circle $C(R, z_0) = \{(r, z) : r^2 + (z - z_0)^2 = R^2\}$ in Fig. 3.

$$G(R, z_0)dR = c \int_{-\pi}^{\pi} \frac{\hat{m} \circ \hat{n}(R \sin\theta, z_0 + R\cos\theta)}{R} d\theta dR$$

i.e.

$$RG(R, z_0) = c \int_{-\pi}^{\pi} \hat{m} \circ \hat{n}(R \sin\theta, z_0 + R\cos\theta) d\theta. \tag{14}$$

Assuming that $\hat{m} \circ \hat{n}$ is small (12) may be replaced by

$$\hat{m} \circ \hat{n} = r\cos\theta \frac{\partial(H/r)}{\partial r} + \sin\theta \frac{\partial H}{\partial z}.$$

By inserting this into (14) we get, after multiplying by R,

$$R^2 G(R, z_0) = c \int_{-\pi}^{\pi} \left(rR\cos\theta \frac{\partial(H/r)}{\partial r} + R\sin\theta \frac{\partial H}{\partial z} \right) d\theta.$$

Using the parametrization

$$r = Rcos\theta, \quad z = z_0 + Rsin\theta,$$

this may be rewritten as a curve integral over $C(R, z_0)$, with $dz = R\cos\theta d\theta$ and $dr = -R\sin\theta d\theta$,

$$R^2 G(R, z_0) = c \int_{C(R,z_0)} \left(r \frac{\partial(H/r)}{\partial r} dz - \frac{\partial H}{\partial z} dr \right). \tag{15}$$

By applying Green's formula we get

$$R^2 G(R, z_0) = c \iint_{D(R,z_0)} \mathfrak{L}(H)(r, z)\, dz\, dr, \tag{16}$$

where D is the disc,

$$D(R, z_0) = \{(r, z) : r^2 + (z - z_0)^2 \le R^2\}$$

and

$$\mathfrak{L}(H) = \frac{\partial}{\partial r}\left(r \frac{\partial(H/r)}{\partial r} \right) + \frac{\partial^2 H}{\partial z^2}. \tag{17}$$

The problem of finding the height function H from radar data $G(r, z)$ may now be divided into two parts:

(i) First solve the integral equation (16) for $\mathfrak{L}(H)(r, z) = f(r, z)$.

(ii) Next solve the partial differential equation

$$\mathfrak{L}(H) = f \tag{18}$$

for H. We note that if r is large and if $\hat{m} \circ \hat{n}$ is small it is reasonable to make the approximation

$$\mathfrak{L}(H) \approx \frac{\partial^2 H}{\partial r^2} + \frac{\partial^2 H}{\partial z^2} = \Delta H$$

so that (18) becomes Poisson's equation. To consider the first problem (i), both members in (16) are differentiated with respect to R. Then we get

$$\frac{1}{R}\frac{d}{dR}(R^2 G(R, z_0)) = c \int_{-\pi}^{\pi} \mathfrak{L}(H)(z_0 + R\cos\gamma, R\sin\gamma)\, d\gamma,$$

where the right-hand side is proportional to the average of $\mathfrak{L}(H)$ over the circle $C(R, z_0)$. Hence,

$$\mathfrak{L}(H)^{(F,F)}(\sigma, \omega) \sim |\omega| \left[\frac{1}{R}\frac{d}{dR}\{R^2 G(r, z)\}\right]^{(F,H_0)} (\sigma, \sqrt{\omega^2 + \sigma^2}). \qquad (19)$$

Here the notation (F, F) means that we have taken the Fourier transform with respect to both the variables and (F, H_0) means that we have taken Fourier transform with respect to the first variable and the Hankel-zero transform with respect to the second. After some calculations (19) may be rewritten

$$\mathfrak{L}(H)^{(F,F)}(\sigma, \omega) \sim |\omega|\sqrt{\omega^2 + \sigma^2}[RG(r, z)]^{(F,H_1)}(\sigma, \sqrt{\omega^2 + \sigma^2}). \qquad (20)$$

Formula (20) may now be used in order to recover the function $\mathfrak{L}(H)$ in spatial coordinates. Approximating $\mathfrak{L}(H)$ by ΔH we could rewrite (20) as

$$H^{(F,F)}(\sigma, \omega) \sim |\omega| \frac{1}{\sqrt{\omega^2 + \sigma^2}}[RG(r, z)]^{(F,H_1)}(\sigma, \sqrt{\omega^2 + \sigma^2}), \qquad (21)$$

where H_1 denotes that we have taken the Hankel-one transform with respect to the second variable. Then we could obtain H directly by a two timensional Fourier transform. However, our solution might be expected to have errors caused by, e.g. noisy radar data and errors caused by the particular numerical implementation of the inversion formula (19) (or (20)) and therefore we would rather prefer to divide the solution procedure into the two steps described above and to use the second step, the solution of Poisson's equation, so that we perform some kind of regularization of the final solution. Note also that by using (21) as our solution formula we have tacitly assumed periodic boundary conditions for the Poisson equation.

In the following we will treat part (ii) of the problem, where we wish to define a solution H to the equation

$$\Delta H = f.$$

4 Solution Concepts for the Poisson Equation

In this section we discuss different possibilities of defining a unique height function. Essentially our approach consists in minimizing some norm of the solution provided that it also satisfies the Poisson equation. In particular we consider the L^2- and H^1-norms. We also show how these two optimization problems may be reformulated as boundary value problems for the biharmonic equation. Note that the corresponding Poisson problem is well-posed unless $\sigma = 1$.

In the domain Ω for the Poisson equation we consider the following boundary value problems

$$\Delta u = f, \quad x \in \Omega \tag{22}$$

with the Dirichlet boundary condition

$$u = g \quad \text{on} \quad \partial\Omega, \tag{23}$$

or the Neumann boundary conditions

$$\nabla u \cdot \nu = h \quad \text{on} \quad \partial\Omega, \tag{24}$$

and the Cauchy boundary conditions

$$\begin{cases} u = g & \text{on} \quad \partial\Omega, \\ \nabla u \cdot \nu = h & \text{on} \quad \partial\Omega, \end{cases} \tag{25}$$

where ν is the outer unit normal vector to $\partial\Omega$.

The boundary operators are independent of any particular choice of orientation for the rectangular coordinate systems. Finally, for Ω a rectangular region in, e.g., the plane

$$\Omega = \{(x, y) : a < x < b, \, c < y < d\},$$

there may be the following boundary conditions

$$u(a, y) = u(b, y), \quad u(x, c) = u(x, d), \tag{26}$$

and with the periodic boundary conditions

$$u_x(a, y) = u_x(b, y), \quad u_y(x, c) = u_y(x, d), \tag{27}$$

$$\begin{cases} u(a, y) = u(b, y), \quad u(x, c) = u(x, d), \\ u_x(a, y) = u_x(b, y), \quad u_y(x, c) = u_y(x, d). \end{cases} \tag{28}$$

Provided g is smooth enough boundary conditions (23) define a unique solution of (22). For (24) and (25) the solution is determined up to a constant. It is also possible to use different mixtures of these three types of boundary conditions. Note that for cases (24) and (28) the following consistency conditions must hold, respectively:

$$\int_{\Omega} f\,dx = \int_{\partial\Omega} h\,ds \quad \text{for (24)}, \quad \int_{\Omega} f\,dx = 0 \quad \text{for (25)}.$$

We now consider a different way to select a solution to (22). Here we use a criterion function and optimize this criterion over the set of solutions to the Poisson equation. Scattering model of Sect. 3 shows the physical interpretation of function $u(x, y)$ is a surface function. We need to pick out the smoothest surface (in some sense) that fulfills (22), using the Sobolev space norms as criterion functions. Denote by $V_{f,i}$ the following set:

$$V_{f,i} = \{u \in H^i(\Omega) : \Delta u = f,\ f \in L^2(\Omega)\},\ i = 0, 1, 2, \tag{29}$$

where $H^0(\Omega) = L^2(\Omega)$.

The equality $\Delta u = f$ is to be interpreted in the sense of distributions. i.e.,

Definition 5 A solution of the Poisson equation (22) in Ω is a function $u \in H^1(\Omega)$ such that the following integral identity holds:

$$\int_{\Omega} u\,\Delta\varphi\,dx = \int_{\Omega} f\varphi\,dx, \quad \forall\varphi \in C_0^{\infty}(\Omega).$$

Lemma 1 *$V_{f,i}$ is a closed, convex and nonempty set of $H^i(\Omega)$.*

Proof. The convexity is due to the linearity of Δ. To verify that $V_{f,i}$, $i = 0, 1, 2$, is nonempty it suffices to verify that $V_{f,2}$ is nonempty.

We assume that $\Omega \subset (0, 2\pi)^n$. Extend f by taking $f = 0$ in $(0, 2\pi)^n \backslash \Omega$. Then $V_{f,2}$ contains the function

$$u = f_0|x|^2/(2n) - \sum_{m\neq 0} e^{imx}/|m|^2$$

assuming $f = \sum f_m e^{imx}$ and that m denotes a multi-index. To show that $V_{f,i}$ is closed we select a sequence $\{u_n\}_1^{\infty} \subset V_{f,i}$, such that $u_n \to u$ in $H^i(\Omega)$. Then $u_n \to u$ in L^2 and, by Cauchy's inequality

$$\left|\int_{\Omega} f\,\varphi\,dx - \int_{\Omega} u\,\Delta\varphi\,dx\right| = \left|\int (u_n - u)\,\Delta\varphi\,dx\right| \leq$$

$$\leq \int_{\Omega} |u_n - u|^2\,dx \int_{\Omega} |\Delta\varphi|^2\,dx \to 0, \quad \forall\varphi \in C_0^{\infty},$$

i.e. $\int_{\Omega} f\,\varphi\,dx = \int_{\Omega} u\,\Delta\varphi\,dx$ and $u \in V_{f,i}$ □

We recall some facts about Green's formula and of the normal derivatives of H^1 -functions [1]:

Preposition 1 *[1] If $v \in H^1(\Omega)$, $\Delta v \in L^2(\Omega)$ (in the sense of distribution), and $\varphi \in H^1(\Omega)$, then $(\nabla v \cdot \nu) \equiv \partial v/\partial \nu \in H^{-1/2}(\partial\Omega)$ is defined by*

$$\left(\frac{\partial v}{\partial \nu}, \varphi\right)_{H^{-1/2}(\partial\Omega), H^{1/2}(\partial\Omega)} = \int_\Omega \Delta v\, \varphi\, dx + \int_\Omega \nabla v\, \nabla\varphi\, dx.$$

This definition is justified by the fact the right hand site of the last above equality defines a bounded linear functional on $H^1(\Omega)$ and by the the following lemma:

Lemma 2 *If $v \in H^1(\Omega)$ and $\Delta v \in L^2(\Omega)$, then*

$$\int_\Omega \Delta v\, \varphi\, dx + \int_\Omega \nabla v\, \nabla\varphi\, dx = 0 \quad \text{for all} \quad \varphi \in \overset{\circ}{H}{}^1(\Omega).$$

Proof. Since $C_0^\infty(\Omega)$ is dense in $\overset{\circ}{H}{}^1(\Omega)$, it suffices to prove last equality for all $\varphi \in C_0^\infty(\Omega)$. Then by Green's formula and the definition of distributional derivatives, we have

$$\int_\Omega \Delta v\, \varphi\, dx + \int_\Omega \nabla v\, \nabla\varphi\, dx = \int_\Omega \Delta v\, \varphi\, dx - \int_\Omega v\, \Delta\varphi\, dx.$$

□

Note also the following well known lemmas for the Dirichlet and Neumann problems in Ω [11], i.e.

Lemma 3 *[11] Suppose $g \in H^{1/2}(\partial\Omega)$, $f \in L^2(\Omega)$. Then there exists a unique function $u \in H^1(\Omega)$ such that*

$$\begin{cases} \Delta u = f & \text{in } \Omega, \\ u = g & \text{on } \partial\Omega. \end{cases}$$

Lemma 4 *[11] Suppose that $g \in H^{-1/2}(\partial\Omega)$, $f \in L^2(\Omega)$ and that*

$$(g, 1)_{H^{-1/2}(\partial\Omega), H^{1/2}(\partial\Omega)} = \int_\Omega f\, dx.$$

Then there exists a unique function $u \in H^1(\Omega)$ such that

$$\begin{cases} \Delta u = f & \text{in } \Omega \text{ (in the sense of distributions)}, \\ u = g & \text{on } \partial\Omega \text{ (in the sense of Prepos. 1)}, \ \int_\Omega u\, dx = 0. \end{cases}$$

Let α be a multi-index and $\beta_1 > 0$ a given parameter. We consider the following optimization problems:

$$I_0(u) \equiv \min_{u \in V_{f,0}} \int_\Omega |u|^2 \, dx, \tag{30}$$

and

$$I_1(u) \equiv \min_{u \in V_{f,1}} \int_\Omega |u|^2 \, dx + \beta_1 \int_\Omega \sum_{|\alpha|=1} |\partial^\alpha u|^2 \, dx. \tag{31}$$

Theorem 1 *Problems* (30) *and* (31) *have unique solutions* u_0 *and* u_1*, respectively.*

Proof. The proof follows from Lemma 1 and the fact that we are minimizing Hilbert norms. □

For problems (30) and (31) we have the following results characterizing the solutions.

Theorem 2 *Let* $u_o = \Delta v$. *For the solution* u_0 *of the problems* (30), *where* $v \in H^2(\Omega, \partial\Omega) \cap \overset{\circ}{H}{}^1(\Omega)$ *is the unique solution of the Steklov biharmonic problem*

$$\begin{cases} \Delta^2 v = f & \text{in } \Omega, \\ v = \Delta v + \tau \frac{\partial v}{\partial \nu} = 0 & \text{on } \partial\Omega. \end{cases} \tag{32}$$

Proof. By a standard variational method, $u_0 \in L^2(\Omega)$ solves problem (30) if and only if $\Delta u_0 = f$ and

$$\int_\Omega u_0 \, \varphi \, dx = 0 \quad \text{for all} \quad \varphi \in L^2(\Omega), \quad \Delta\varphi = 0.$$

Assume first that u_0 solves problem (30). Let v be defined as the unique solution of the Dirichlet problem,

$$\begin{cases} \Delta v = u_0 & \text{in } \Omega, \\ v = 0 & \text{on } \partial\Omega. \end{cases}$$

If $\varphi, \ v \in H^1(\Omega)$ and $\Delta\varphi, \ \Delta v \in L^2(\Omega)$, we have the Green formula

$$\int_\Omega \Delta v \, \varphi \, dx - \int_\Omega v \, \Delta\varphi \, dx = \int_{\partial\Omega} (\nabla v \cdot \nu) \, \varphi \, ds - \int_{\partial\Omega} v \, (\nabla\varphi \cdot \nu) \, ds.$$

Now let $\varphi \in H^1(\Omega)$ be a harmonic function, $\Delta\varphi = 0$. Then we have

$$0=\int_{\Omega} u_0\,\varphi\,dx=\int_{\Omega}\Delta v\,\varphi\,dx=$$

$$\int_{\partial\Omega}(\nabla v\cdot\nu)\,\varphi\,ds-\int_{\partial\Omega} v\,(\nabla\varphi\cdot\nu)\,ds+\int_{\Omega} v\,\Delta\varphi\,dx,$$

that is

$$\int_{\partial\Omega}(\nabla v\cdot\nu)\,\varphi\,ds=0 \quad \text{for all such} \quad \varphi.$$

Since there exists a unique function $u\in H^1(\Omega)$ such that

$$\begin{cases}\Delta u=f, & f\in L^2(\Omega) \quad \text{in} \quad \Omega,\\ u=g, & g\in H^{1/2}(\partial\Omega) \quad \text{on} \quad \partial\Omega,\end{cases}$$

and $\varphi\big|_{\partial\Omega}$ may be chosen arbitrary in $H^{1/2}(\Omega)$, we conclude that $0=(\nabla v\cdot\nu)\in H^{-1/2}(\Omega)$. We have proved that $u_0=\Delta v\in L^2(\Omega)$, where v satisfies the Steklov biharmonic problem

$$\begin{cases}\Delta^2 v=f & \text{in} \quad \Omega,\\ v=\Delta v+\tau\,\frac{\partial v}{\partial\nu}=0 & \text{on} \quad \partial\Omega.\end{cases} \tag{33}$$

On the other hand we claim that (33) cannot have more than one solution $v\in H^1(\Omega)$ with $\Delta v\in L^2(\Omega)$. Indeed assume that (33) is satisfied and consider the function $\psi\in L^2(\mathbb{R}^n)$ defined by

$$\psi=\begin{cases}v(x) & \text{if} \quad x\in\Omega,\\ 0 & \text{if} \quad x\notin\Omega.\end{cases}$$

For arbitrary $\varphi\in C_0^\infty(\mathbb{R}^n)$ we have

$$\int_{\mathbb{R}^n}\psi\,\Delta\varphi\,dx=\int_{\Omega} v\,\Delta\varphi\,dx=$$

$$=\int_{\partial\Omega} v\,(\nabla\varphi\cdot\nu)\,ds-\int_{\partial\Omega}(\nabla v\cdot\nu)\,\varphi\,ds+\int_{\Omega}\varphi\,\Delta v\,dx,$$

i.e.

$$\int_{\mathbb{R}^n}\psi\,\Delta\varphi\,dx=\int_{\Omega}\varphi\,\Delta v\,dx$$

for all $\varphi\in C_0^\infty(\mathbb{R}^n)$.

Let now $h\in C_0^\infty(\mathbb{R}^n)$ be dined by

$$h(x)=\begin{cases}\Delta v & \text{if} \quad x\in\Omega,\\ 0 & \text{if} \quad x\notin\Omega.\end{cases}$$

We have proved that

$$\Delta\psi = h$$

in the sense of distributions. Using the Fourier transformation it follows that $\psi \in H^2(\mathbb{R}^n)$. Therefore $v \in H^2(\Omega)$, and v must be the unique solution in $H^2(\Omega)$ of (33), being the unique minimizer in $\overset{\circ}{H}{}^2(\Omega)$ of the coercive quadratic functional

$$J(v) \equiv \int_\Omega \left(\frac{1}{2}|\Delta v|^2 - fv\right) dx.$$

The proof is complete. □

Theorem 3 *Let* $u_1 = \Delta v$. *For the solution* u_1 *of the problems* (31), *where* $v \in H^2(\Omega)$, $(\nabla v \cdot \nu) = 0$, *is the unique solution of the Steklov-type biharmonic problem*

$$\begin{cases} \Delta^2 v = f & in \;\; \Omega, \\ \frac{\partial v}{\partial \nu} = \frac{\partial \Delta v}{\partial \nu} + \tau\, v = 0 & on \;\; \partial\Omega. \end{cases}$$

Proof. Assume that u_1 solves problem (31). Let v be defined as the unique solution in the class $\{\psi \in H^1(\Omega) : \Delta\psi \in H^1(\Omega)\}$ of the following biharmonic problem

$$\begin{cases} \Delta^2 v = f & \text{in} \;\; \Omega, \\ v = \beta_1 \Delta v, \quad \nabla v \cdot \nu = 0 & \text{on} \;\; \partial\Omega. \end{cases} \tag{34}$$

By standard variational method, $u_1 \in H^1(\Omega)$ solves problem (31), if and only if $\Delta u_1 = f$ and

$$\int_\Omega (u_1 \varphi + \beta_1 \nabla u_1 \cdot \nabla\varphi)\, dx = 0$$

for all $\varphi \in H^1(\Omega)$ and $\Delta\varphi = 0$ in Ω. Taking $\varphi = 1$, we observe that

$$\int_\Omega u_1\, dx = 0.$$

Let $v \in H^1(\Omega)$ be any solution of the Neumann problem

$$\begin{cases} \Delta v = u_1 & \text{in} \;\; \Omega, \\ \nabla v \cdot \nu = 0 & \text{on} \;\; \partial\Omega. \end{cases} \tag{35}$$

Applying Green's formula we have,

$$0 = \int_{\Omega} \varphi \, \Delta v \, dx + \beta_1 \int_{\Omega} \nabla(\Delta v) \nabla \varphi \, dx =$$

$$= \int_{\Omega} v \, \Delta \varphi \, dx + \int_{\partial\Omega} \varphi \, (\nabla v \cdot \nu) \, ds - \int_{\partial\Omega} v \, (\nabla \varphi \cdot \nu) \, ds +$$

$$+ \beta_1 \int_{\partial\Omega} \Delta v \, (\nabla \varphi \cdot \nu) \, ds - \beta_1 \int_{\Omega} \Delta v \, \Delta \varphi \, dx,$$

i.e.

$$\int_{\partial\Omega} (v - \beta_1 \Delta v) \, (\nabla \varphi \cdot \nu) \, ds = 0$$

for all $\varphi \in H^1(\Omega)$ and $\Delta \varphi = 0$ in Ω.

Since $(\nabla \varphi \cdot \nu) \in H^{-1/2}(\partial\Omega)$ may be chosen arbitrarily (Lemma 4) apart from the condition

$$\int_{\partial\Omega} (\nabla \varphi \cdot \nu) \, ds = 0,$$

it follows that, for some $C = \text{const}$, $v - \beta_1 \Delta v = C$ on $\partial\Omega$.

Now the solution v is uniquely defined up to an additive constant. This constant may be chosen so that $C = 0$.

We have proved that v defined by (35) satisfies (34), and that $v, \Delta v \in H^1(\Omega)$. It remains only to prove that the solution v of (34) is unique. To this end we introduce the function $\varphi = \beta_1 \Delta v - v$ and observe that v satisfies (34) if and only if $(v, \varphi) \in H^1(\Omega) \times \overset{\circ}{H}{}^1(\Omega)$ satisfies the system

$$\begin{cases} \beta_1 \Delta v - v = \varphi, \\ \beta_1 \Delta \varphi + \varphi = \beta_1^2 f - v \quad \text{with} \quad (\nabla v \cdot \nu) = 0 \quad \text{on} \quad \partial\Omega. \end{cases} \tag{36}$$

Next assume that v_1, v_2 satisfy (34), or $(v_1, \varphi_1), (v_2, \varphi_2)$ satisfy (36). From the previous argument it follows that $u_1 = \Delta v_1 = \Delta v_2$ is the unique solution of problem (31) so that $\Delta(v_1 - v_2) = 0$. From (36) we have

$$\begin{cases} \beta_1 \Delta(v_1 - v_2) - (v_1 - v_2) = \varphi_1 - \varphi_2, \\ \beta_1 \Delta(\varphi_1 - \varphi_2) + (\varphi_1 - \varphi_2) = v_1 - v_2. \end{cases}$$

This implies, $v_1 - v_2 = \varphi_1 - \varphi_2$ and $\Delta(\varphi_1 - \varphi_2) = 0$ with $\varphi_1 - \varphi_2 \in \overset{\circ}{H}{}^1(\Omega)$ whence we conclude that $\varphi_1 - \varphi_2 = 0$ and $v_1 - v_2 = 0$. The proof is complete. □

We conclude this section by a theorem relating the solution of problems (30) and (31). First we recall the following definition.

Definition 6 $\Omega \subset \mathbb{R}^n$ is called *star-shaped*, if there exists $x_0 \in \Omega$ such that for all $x \in \Omega$ the set $\{t \in \mathbb{R} : \ x_0 + t(x - x_0) \in \Omega\}$ is an interval.

Theorem 4 *Assume that $\Omega \subset \mathbb{R}^n$ is open, bounded and star-shaped. If $u_{1,\beta_1} \in H^1(\Omega)$ denotes the solution of problem (31) with the parameter $\beta_1 > 0$, and if $u_0 \in L^2(\Omega)$ denotes the solution of problem (30), then*

$$u_{1,\beta_1} \to u_0 \quad in \quad L^2(\Omega) \quad as \quad \beta_1 \to 0+ .$$

Proof. For $0 < \lambda < 1$ and x_0 chosen as in the previous definition, we take

$$\Omega_\lambda = \{x \in \mathbb{R}^n : \ x_0 + \lambda(x - x_0) \in \Omega\},$$

$$u_{0,\lambda}(x) = u_0(x_0 + \lambda(x - x_0)), \quad f_\lambda = f(x_0 + \lambda(x - x_0)).$$

Then [11],

$$\Delta u_{0,\lambda} = f_\lambda \quad \text{in} \quad \Omega_\lambda, \quad \Omega_\lambda \supset \overline{\Omega}, \quad u_{0,\lambda} \in H^2_{loc}(\Omega_\lambda).$$

Since $H^2_{loc}(\Omega_\lambda) \supset H^2(\Omega)$, it follows that $u_{0,\lambda} \in H^2(\Omega)$. Further it is rather easy to see that

$$\int_\Omega |u_{0,\lambda} - u_0|^2 \, dx \to 0,$$

and

$$\int_\Omega |f_\lambda - f|^2 \, dx \to 0 \quad \text{as} \quad \lambda \to 1.$$

Next define $v_\lambda \in \overset{\circ}{H}{}^1(\Omega)$ by

$$\Delta v_\lambda = f - f_\lambda \quad \text{in} \quad \Omega_\lambda.$$

Then

$$\int_\Omega |v_\lambda|^2 \, dx \le ||v_\lambda||_{H^1(\Omega)} \le C \int_\Omega |f - f_\lambda|^2 \, dx.$$

Consequently, taking $w_\lambda = u_{0,\lambda} + v_\lambda$, we have first,

$$w_\lambda \in H^1(\Omega), \quad \Delta w_\lambda = f \quad \text{in} \quad \Omega,$$

and hence,

$$\int_\Omega |w_\lambda - u_0|^2 \, dx \to 0 \quad \text{as} \quad \lambda \to 1.$$

Now, if $\varepsilon > 0$ is given, we may choose a λ close enough to 1, so that

$$\int_\Omega w_\lambda^2 \, dx < \int_\Omega u_0^2 \, dx + \varepsilon/2.$$

Further, by definition,

$$\int_{\Omega} u_{1,\beta_1}^2\, dx + \beta_1 \int_{\Omega} |\nabla u_{1,\beta_1}|^2\, dx \leq \int_{\Omega} w_{\lambda}^2\, dx + \beta_1 \int_{\Omega} |\nabla w_{\lambda}|^2\, dx.$$

Since

$$||w_{\lambda}||_{H^1(\Omega)} \leq C \int_{\Omega} |f|^2\, dx$$

we have, for sufficiently small β_1,

$$\int_{\Omega} u_{1,\beta_1}^2\, dx + \beta_1 \int_{\Omega} |\nabla u_{1,\beta_1}|^2\, dx \leq \int_{\Omega} u_0^2\, dx + \varepsilon.$$

It follows that,

$$\lim_{\beta_1 \to 0+} \sup \int_{\Omega} |u_{1,\beta_1}|^2\, dx \leq \int_{\Omega} |u_0|^2\, dx.$$

Further, for some sub-sequence of β_1, we have

$$u_{1,\beta_1} \to \widetilde{u} \quad \text{in} \quad L^2(\Omega) \quad \text{(weakly)},$$

$$\Delta\widetilde{u} = f \quad \text{in} \quad \Omega,$$

and

$$\int_{\Omega} |\widetilde{u}|^2\, dx \leq \lim_{\beta_1 \to 0+} \inf \int_{\Omega} |u_{1,\beta_1}|^2\, dx.$$

But then $\Delta\widetilde{u} = f$ and

$$\int_{\Omega} |\widetilde{u}|^2\, dx \leq \int_{\Omega} |u_0|^2\, dx$$

which, by definition of u_0, implies that $\widetilde{u} = u_0$. So,

$$u_{1,\beta_1} \to u_0 \quad \text{in} \quad L^2(\Omega) \quad \text{(weakly)}.$$

Next

$$\lim_{\beta_1 \to 0+} \sup \int_{\Omega} |u_{1,\beta_1} - u_0|^2\, dx =$$

$$= \lim_{\beta_1 \to 0+} \sup \int_{\Omega} |u_{1,\beta_1}|^2\, dx - 2 \lim_{\beta_1 \to 0+} \int_{\Omega} u_{1,\beta_1} u_0\, dx +$$

$$+ \int_{\Omega} |u_0|^2\, dx \leq \int_{\Omega} |u_0|^2\, dx - 2 \int_{\Omega} |u_0|^2\, dx + \int_{\Omega} |u_0|^2\, dx = 0.$$

Finally, since this strong limit u_0 is uniquely defined we may conclude, by a standard argument that

$$u_{1,\beta_1} \to u_0 \quad \text{in} \quad L^2(\Omega) \quad \text{as} \quad \beta_1 \to 0^+$$

without restriction to any subsequence. The proof is complete. □

Remark 1 All convex sets are star-shaped. Rectangles Ω appearing in our applications are thus star-shaped.

5 Appendix

Many complex engineering structures, such as the rotor blades of wind turbines and helicopters, are non-prismatic beamlike structures, which may be tapered, twisted and curved in their reference unstressed state and undergo large displacements of the reference centre-line's points, as well as in- and out-of-plane warping of the transverse cross-sections. Continuous efforts to better predict the mechanical behaviour of such structures, which are aimed at improving the performance in terms of structural efficiency and costs effectiveness, offer the opportunity to address some very interesting, challenging problems in the field of continuum and solid mechanics [30].

An important point in developing rigorous yet application-oriented mathematical models for such structures is an appropriate description of their motion. In general, a non-prismatic beamlike structure can be considered as a collection of deformable plane figures (referred to as the transverse cross-sections) along a suitable three-dimensional curve (called the reference centre-line). Each cross-sectional point in the reference state can moreover move to its position in the current state through a global rigid motion on which a local warping motion can be superimposed. The description of the motion of such structures can thus be performed by introducing two kinematic maps, herein called R_A and R_B, to identify the positions of the points of the mentioned structure in the reference and current states, as discussed in [30]–[32]. Specifically, the reference map R_B can be defined as follows

$$R_B(z_i) = R_{0B}(z_1) + x_\alpha(z_i)b_\alpha(z_1)$$

where R_{0B} denotes the position of the centre-line's points in the reference state, b_α are the vectors of the reference local frame in the plane of the reference cross-section, x_α identify the position of the points in the reference cross-section relative to the reference centre-line, and finally, z_i are three independent mathematical variables which do not depend on time. In particular, z_1 is equal to the reference arch-length s, and z_α belong to a bi-dimensional mathematical domain that is used to map the position of the points, x_α, of the cross-sections. Note that in this section α and β assume values 2 and 3, i and j take values 1, 2 and 3, while repeated indices are summed over their range.

In a similar manner, the current map R_A can be defined as follows

$$R_A(z_i, t) = R_{0A}(z_1, t) + x_\beta(z_i)a_\beta(z_1, t) + w_j(z_1, t)a_j(z_1, t)$$

where t is the considered evolution scalar real parameter (the time, for instance), R_{0A} denotes the position of the centre-line's points in the current state, while w_j are the components of the warping displacement fields with respect to the current local frames referred to as a_j.

These maps can be used to determine the gradient of transformation between the current and reference states and, successively, the Green–Lagrange strain tensor, as discussed in [30]. Given such strain tensor and a constitutive model, it is thus possible to determine the stress fields in the three-dimensional structure. The problem unknowns, such as the displacements of the centre-line's points and the warping fields, can then be determined as the solution of a set of balance equations deduced by a stationary condition of a suitable energy functional [30]. Specifically, the result of this procedure is a mathematical problem based on partial differential equations (PDEs) with Neumann-type boundary conditions the solution of which enables obtaining all unknowns of the problem, such as the warping fields w_k, the displacements of the centre-line's points, the Green–Lagrange strain fields and the corresponding stress fields as well. Further details can be found in [30] and [32].

The results of the article are presented at the 13th International Conference Chaotic Modeling, Simulation and Applications (CHAOS2020, June 9–12, 2020, Florence, Italy).

References

1. R.A. Adams, *Sobolev Spaces* (Academic Press, New York, 1975)
2. S. Agmon, A. Douglis, L. Nirenberg, Estimates near the boundary for solutions of elliptic partial differential equations satisfying general boundary conditions. I. Commun. Pure Appl. Math. **12**(4), 623–727 (1959)
3. F. Brock, An isoperimetric inequality for eigenvalues of the Stekloff problem. Z. Angew. Math. Mech. (ZAMM) **81**(1), 69–71 (2001)
4. R. Courant, H. Hilbert, *Methods of Mathematical Physics*, vol. I (Interscience Publishers Inc, New York, 1953)
5. R.J. Duffin, The influence of Poisson's ratio on the vibrational spectrum. SIAM J. Appl. Math. **17**, 179–191 (1969)
6. F. Gazzola, G. Sweers, On positivity for the biharmonic operator under Steklov boundary conditions. Arch. Rational Mech. Anal. **188**(3), 399–427 (2008)
7. F. Gazzola, H.-Ch. Grunau, G. Sweers, *Polyharmonic Boundary Value Problems: Positivity Preserving and Nonlinear Higher Order Elliptic Equations in Bounded Domains*. Lecture Notes Math. **1991** (Springer, 2010)
8. V.A. Kondratiev, I. Kopacek, O.A. Oleinik, On asymptotic properties of solutions of the biharmonic equation. Diff. Equ. **17**(10), 1886–1899 (1981)

9. V.A. Kondratiev, O.A. Oleinik, Estimates for solutions of the Dirichlet problem for biharmonic equation in a neighbourhood of an irregular boundary point and in a neighbourhood of infinity Saint-Venant's principle Proc. Royal Soc. Edinburgh **93A**(3–4), 327–343 (1983)
10. J.R. Kuttler, V.G. Sigillito, Inequalities for membrane and Stekloff eigenvalues. J. Math. Anal. Appl. **23**(1), 148–160 (1968)
11. J.-L. Lions, E. Magenes, *Non-homogeneous Boundary Value Problems and Applications*. I. (Springer, Berlin, 1972)
12. O.A. Matevosyan, The exterior Dirichlet problem for the biharmonic equation: Solutions with bounded Dirichlet integral. Math. Notes **70**(3), 363–377 (2001)
13. O.A. Matevossian, Solutions of exterior boundary value problems for the elasticity system in weighted spaces. Sbornik Math. **192**(12), 1763–1798 (2001)
14. H.A. Matevossian, On solutions of mixed boundary-value problems for the elasticity system in unbounded domains. Izvestiya Math. **67**(5), 895–929 (2003)
15. Matevosyan O. A. Solutions of the Robin problem for the system of elastic theory in external domains. J. Math. Sci. (N. Y.) **197**(3), 367–394 (2014)
16. H.A. Matevossian, On solutions of the Dirichlet problem for the polyharmonic equation in unbounded domains. P-Adic Numbers, Ultrametric Anal., Appl. **7**(1), 74–78 (2015)
17. O.A. Matevosyan, Solution of a mixed boundary value problem for the biharmonic equation with finite weighted Dirichlet integral. Diff. Equ. **51**(4), 487–501 (2015)
18. O.A. Matevossian, On solutions of the Neumann problem for the biharmonic equation in unbounded domains. Math. Notes **98**, 990–994 (2015)
19. O.A. Matevosyan, On solutions of the mixed Dirichlet-Navier problem for the polyharmonic equation in exterior domains. Russ. J. Math. Phys. **23**(1), 135–138 (2016)
20. O.A. Matevosyan, On solutions of one boundary value problem for the biharmonic equation. Diff. Equ. **52**(10), 1379–1383 (2016)
21. H.A. Matevossian, On the biharmonic Steklov problem in weighted spaces. Russ. J. Math. Phys. **24**(1), 134–138 (2017)
22. H.A. Matevossian, On solutions of the mixed Dirichlet-Steklov problem for the biharmonic equation in exterior domains. P-Adic Numbers, Ultrametric Anal., Appl. **9**(2), 151–157 (2017)
23. H.A. Matevossian, On the Steklov-type biharmonic problem in unbounded domains. Russ. J. Math. Phys. **25**(2), 271–276 (2018)
24. H.A. Matevossian, On the polyharmonic Neumann problem in weighted spaces. Complex Var. Elliptic Equ. **64**(1), 1–7 (2019)
25. Matevossian H. On the Mixed Dirichlet–Steklov-Type and Steklov-Type Biharmonic Problems in Weighted Spaces. Math. Comput. Appl. **24**(1), 25 (2019), 1–9
26. Matevossian H. A. Mixed Boundary Value Problems for the Elasticity System in Exterior Domains. Math. Comput. Appl. **24**(2), 58, 1–7 (2019)
27. H. Matevossian, On the Mixed Steklov–Neumann and Steklov-Type Biharmonic Problems in Unbounded Domains, in *IOP Conf. Ser.: Mater. Sci. Eng.* **683**(1), 012016 (2019), 10 pp
28. H.A. Matevossian, On the Mixed Neumann-Robin Problem for the Elasticity System in Exterior Domains. Russ. J. Math. Phys. **27**(2), 272–276 (2020)
29. H.A. Matevossian, G. Nordo, T. Sako, Biharmonic Problems and their Application in Engineering and Medicine, in *IOP Conf. Ser.: Mater. Sci. Eng.* **934** (1), 012065 (2020), 10 pp
30. G. Migliaccio, G. Ruta et al., Beamlike models for the analyses of curved, twisted and tapered horizontal-axis wind turbine (HAWT) blades undergoing large displacements. Wind Energ. Sci. **5**, 685–698 (2020). https://doi.org/10.5194/wes-5-685-2020
31. G. Migliaccio, G. Ruta, S. Bennati, R. Barsotti, Curved and Twisted Beam Models for Aeroelastic Analysis of Wind Turbine Blades in Large Displacement. In: Proc. XXIV AIMETA Conference (AIMETA 2019), *Lecture Notes in Mech. Engin.* (Springer, Cham, 2020) pp. 1785–1797. https://doi.org/10.1007/978-3-030-41057-5_144
32. G. Migliaccio, G. Ruta, Rotor blades as curved, twisted, and tapered beam-like structures subjected to large deflections. Engin. Struct. **222**, 111089 (2020). https://doi.org/10.1016/j.engstruct.2020.111089

33. Payne L. E. Some isoperimetric inequalities for harmonic functions. SIAM J. Math. Anal., **1**(3), 354–359 (1970)
34. L. Provenzano, A note of Neumann eigenvalues of the biharmonic operator. Math. Meth. Appl. Sci. **41**(3), 1005–1012 (2016). https://doi.org/10.1002/mma.4063
35. S.L. Sobolev, Some, Applications of Functional Analysis in Mathematical Physics, 3th ed., Nauka, Moscow, *Applications of Functional Analysis in Mathematical Physics* (Amer. Math. Soc, Providence, 1988), p. 1991
36. G. Verchota, The biharmonic Neumann problem in Lipschitz domains. Acta Math. Anal. **194**, 217–279 (1970)
37. Stekloff W. Sur les problemes fondamentaux de la physique mathematique. *Ann. Sci. de l'E.N.S.*, 3^{e} serie, **19** (1902), 191–259 et 455–490

On the Exterior Biharmonic Problem with the Steklov and Steklov-Type Boundary Conditions

Hovik A. Matevossian and Giovanni Migliaccio

Abstract We study the properties of generalized solutions in unbounded domains and the asymptotic behavior of solutions of elliptic boundary value problems at infinity. Moreover, we study the unique solvability of the mixed biharmonic problem with the Steklov and Steklov-type conditions on the boundary in the exterior of a compact set under the assumption that generalized solutions of this problem has a bounded Dirichlet integral with weight $|x|^a$. Depending on the value of the parameter a, we obtained uniqueness (non-uniqueness) theorems of this problem or present exact formulas for the dimension of the space of solutions.

Keywords Biharmonic operator · Steklov and Steklov-type boundary conditions · Dirichlet integral · Weighted spaces

1 Introduction

Let Ω be an unbounded domain in $\mathbb{R}^n$, $n \geq 2$, $\Omega = \mathbb{R}^n \setminus \overline{G}$ with the boundary $\partial\Omega \in C^2$, where G is a bounded simply connected domain (or a union of finitely many such domains) in $\mathbb{R}^n$, $0 \in G$, $\overline{\Omega} = \Omega \cup \partial\Omega$ is the closure of Ω, $x = (x_1, \ldots, x_n) \in \mathbb{R}^n$ and $|x| = \sqrt{x_1^2 + \cdots + x_n^2}$.

H. A. Matevossian (✉)
Federal Research Centre "Computer Science & Control" & Steklov Mathematical Institute, Russian Academy of Sciences, Vavilov str., 40-42, R 374, Moscow 119333, Russia
e-mail: hmatevossian@graduate.org

Moscow Avaiation Institute (National Research University), Volokolomskoe shosse, 4, Moscow 125993, Russia

G. Migliaccio
Università di Pisa, Largo Luccio Lazzarino 2, 56122 Pisa, Italy

C. H. Skiadas et al. (eds.), *13th Chaotic Modeling and Simulation International Conference*, Springer Proceedings in Complexity,
https://doi.org/10.1007/978-3-030-70795-8_43

In Ω we consider the following problems for the biharmonic equation

$$\Delta^2 u = 0 \tag{1}$$

with the Steklov boundary condition on Γ_1 and the Steklov-type boundary condition on Γ_2

$$u|_{\Gamma_1} = \left(\Delta u + \tau \frac{\partial u}{\partial \nu}\right)\bigg|_{\Gamma_1} = 0, \quad \frac{\partial u}{\partial \nu}\bigg|_{\Gamma_2} = \left(\frac{\partial \Delta u}{\partial \nu} + \tau u\right)\bigg|_{\Gamma_2} = 0, \tag{2}$$

where $\overline{\Gamma}_1 \cup \overline{\Gamma}_2 = \partial\Omega$, $\Gamma_1 \cap \Gamma_2 = \emptyset$, $\mathrm{mes}_{n-1}\,\Gamma_1 \neq 0$, $\nu = (\nu_1, \ldots, \nu_n)$ is the outer unit normal vector to $\partial\Omega$, $\tau \in C(\partial\Omega)$, $\tau \geq 0$, $\tau \not\equiv 0$, and $\tau > 0$ on a set of positive $(n-1)$—dimensional measure on $\partial\Omega$.

Elliptic problems with parameters in the boundary conditions have been called Steklov or Steklov-type problems since their first appearance in [30]. For the biharmonic operator, these conditions were first considered in [1, 10, 28], whose authors the isoperimetric properties of the first eigenvalue were studied.

Note that standard elliptic regularity results are available in [4]. The monograph covers higher order linear and nonlinear elliptic boundary value problems, mainly with the biharmonic or polyharmonic operator as leading principal part. The underlying models and, in particular, the role of different boundary conditions are explained in detail. As for linear problems, after a brief summary of the existence theory and L^p and Schauder estimates, the focus is on positivity. The required kernel estimates are also presented in detail.

In [3] and [4], the spectral and positivity preserving properties for the inverse of the biharmonic operator under Steklov and Navier boundary conditions are studied. These are connected with the first Steklov eigenvalue. It is shown that the positivity preserving property is quite sensitive to the parameter involved in the boundary condition. Moreover, positivity of the Steklov boundary value problem is linked with positivity under boundary conditions of Dirichlet and Navier type.

In [2], the boundary value problems for the biharmonic equation and the Stokes system are studied in a half space, and, using the Schwartz reflection principle in weighted L^q -space, the uniqueness of solutions of the Stokes system or the biharmonic equation is proved.

As is well known, if Ω is an unbounded domain, one should additionally characterize the behavior of the solution at infinity. As a rule, to this end, one usually poses either the condition that the Dirichlet (energy) integral is finite or a condition on the character of vanishing of the modulus of the solution as $|x| \to \infty$. Such conditions at infinity are natural and were studied by several authors (e.g., [8, 9]).

In the present note, this condition is the boundedness of the weighted Dirichlet integral:

$$D_a(u, \Omega) \equiv \int_\Omega |x|^a \sum_{|\alpha|=2} |\partial^\alpha u|^2 \, dx < \infty, \quad a \in \mathbb{R}.$$

In various classes of unbounded domains with finite weighted Dirichlet (energy) integral, one of the author [11]–[24] studied uniqueness (non-uniqueness) problem and found the dimensions of the spaces of solutions of boundary value problems for the elasticity system and the biharmonic (polyharmonic) equation.

By developing an approach based on the use of Hardy type inequalities [6, 8, 9], in the present note, we obtain a uniqueness (non-uniqueness) criterion for a solution of the mixed biharmonic problem with the Steklov and Steklov-type boundary conditions. To construct the solution, we use a variational method, that is, we minimize the corresponding functional in the class of admissible functions.

Notation: $C_0^\infty(\Omega)$ is the space of infinitely differentiable functions in Ω with compact support in Ω.

We denote by $H^m(\Omega, \Gamma)$, $\Gamma \subset \overline{\Omega}$, the Sobolev space of functions in Ω obtained by the completion of $C^\infty(\overline{\Omega})$ vanishing in a neighborhood of Γ with respect to the norm

$$||u; H^m(\Omega, \Gamma)|| = \left(\int_\Omega \sum_{|\alpha| \le m} |\partial^\alpha u|^2 dx \right)^{1/2}, \quad m = 1, 2,$$

where $\partial^\alpha \equiv \partial^{|\alpha|}/\partial x_1^{\alpha_1} \dots \partial x_n^{\alpha_n}$, $\alpha = (\alpha_1, \dots, \alpha_n)$ is a multi-index, $\alpha_i \ge 0$ are integers, and $|\alpha| = \alpha_1 + \dots + \alpha_n$; if $\Gamma = \emptyset$, we denote $H^m(\Omega, \Gamma)$ by $H^m(\Omega)$.

$\overset{\circ}{H}{}^m(\Omega)$ is the space obtained by the completion of $C_0^\infty(\Omega)$ with respect to the norm $||u(x); H^m(\Omega)||$;

$\overset{\circ}{H}{}^m_{loc}(\Omega)$ is the space obtained by the completion of $C_0^\infty(\Omega)$ with respect to the family of semi-norms

$$\|u; H^m(\Omega \cap B_0(R))\| = \left(\int_{\Omega \cap B_0(R)} \sum_{|\alpha| \le m} |\partial^\alpha u|^2 \, dx \right)^{1/2}$$

for all open balls $B_0(R) := \{x : |x| < R\}$ in $\mathbb{R}^n$ for which $\Omega \cap B_0(R) \neq \emptyset$.

Let $\binom{n}{k}$ be the (n, k) - binomial coefficient, $\binom{n}{k}$=0 for $k > n$.

2 Definitions and Auxiliary Statements

Definition 1 A solution of the homogenous biharmonic equation (1) in Ω is a function $u \in H^2_{loc}(\Omega)$ such that, for every function $\varphi \in C_0^\infty(\Omega)$, the following integral identity holds:

$$\int_\Omega \Delta u \, \Delta \varphi \, dx = 0.$$

Lemma 1 *Let u be a solution of equation (1) in Ω such that $D_a(u, \Omega) < \infty$. Then*

$$u(x) = P(x) + \sum_{\beta_0 < |\alpha| \le \beta} \partial^\alpha \Gamma(x) C_\alpha + u^\beta(x), \quad x \in \Omega, \tag{3}$$

where $P(x)$ is a polynomial, ord $P(x) < m_0 = \max\{2, 2 - n/2 - a/2\}$, $\beta_0 = 2 - n/2 + a/2$, *$\Gamma(x)$ is the fundamental solution of equation (1),* $C_\alpha =$ const, $\beta \ge 0$ *is an integer, and the function u^β satisfies the estimate:*

$$|\partial^\gamma u^\beta(x)| \le C_{\gamma\beta} |x|^{3-n-\beta-|\gamma|}, \ C_{\gamma\beta} = \text{const},$$

for every multi-index γ.

Remark 1 As is known [29], the fundamental solution $\Gamma(x)$ of the biharmonic equation has the form

$$\Gamma(x) = \begin{cases} C|x|^{4-n}, \ if\ 4 - n < 0 \text{ or } n \text{ is odd,} \\ C|x|^{4-n} \ln |x|, \ if\ 4 - n \ge 0 \text{ and } n \text{ is even.} \end{cases}$$

Proof of Lemma 1 Consider the function $v(x) = \theta_N(x) u(x)$, where $\theta_N(x) = \theta(|x|/N)$, $\theta \in C^\infty(\mathbb{R}^n)$, $0 \le \theta \le 1$, $\theta(s) = 0$ for $s \le 1$, $\theta(s) = 1$ for $s \ge 2$, while $N \gg 1$ and $G \subset \{x : |x| < N\}$. We extend v to $\mathbb{R}^n$ by setting $v = 0$ on $G = \mathbb{R}^n \setminus \overline{\Omega}$.
Then the function v belongs to $C^\infty(\mathbb{R}^n)$ and satisfies the equation

$$\Delta^2 v = f,$$

where $f \in C_0^\infty(\mathbb{R}^n)$ and supp $f \subset \{x : |x| < 2N\}$. It is easy to see that $D_a(v, \mathbb{R}^n) < \infty$.

We can now use Theorem 1 of [7] since it is based on Lemma 2 of [7], which imposes no constraint on the *sign* of σ. Hence, the expansion

$$v(x) = P(x) + \sum_{\beta_0 < |\alpha| \le \beta} \partial^\alpha \Gamma(x) C_\alpha + v^\beta(x),$$

holds for each a, where $P(x)$ is a polynomial of order ord $P(x) < m_0 = \max\{2, 2 - n/2 - a/2\}$, $\beta_0 = 2 - n/2 + a/2$, $C_\alpha =$ const and

$$|\partial^\gamma v^\beta(x)| \le C_{\gamma\beta} |x|^{3-n-\beta-|\gamma|}, \quad C_{\gamma\beta} = \text{const}.$$

Therefore, by the definition of v, we obtain (3). The proof of Lemma 1 is complete.

3 Main Results

Definition 2 By a solution of the mixed boundary value problem (1), (2) we mean a function $u \in H^2_{loc}(\Omega) \cap \overset{\circ}{H}{}^1_{loc}(\Omega, \Gamma_1)$, $\partial u/\partial \nu = 0$ on Γ_2, such that, for every function $\varphi \in C_0^\infty(\mathbb{R}^n) \cap \overset{\circ}{H}{}^1_{loc}(\Omega, \Gamma_1)$, $\partial \varphi/\partial \nu = 0$ on Γ_2, the following integral identity holds:

$$\int_\Omega \Delta u\, \Delta\varphi\, dx + \int_{\Gamma_1} \tau\, \nabla u\, \nabla\varphi\, ds - \int_{\Gamma_2} \tau\, u\, \varphi\, ds = 0. \tag{4}$$

Theorem 1 *The mixed problem (1), (2) with the condition $D(u, \Omega) < \infty$ has $n+1$ linearly independent solutions.*

Proof For any nonzero vector A in $\mathbb{R}^n$, we construct a generalized solution u_A of the biharmonic equation (1) with the boundary conditions

$$\begin{gathered} u_A(x)\big|_{\Gamma_1} = (Ax)\big|_{\Gamma_1}, \quad \left(\Delta u_A + \tau \frac{\partial u_A(x)}{\partial \nu}\right)\bigg|_{\Gamma_1} = \tau \frac{\partial (Ax)}{\partial \nu}\bigg|_{\Gamma_1}, \\ \frac{\partial u_A}{\partial \nu}\bigg|_{\Gamma_2} = \left(\frac{\partial \Delta u_A}{\partial \nu} + \tau\, u_A\right)\bigg|_{\Gamma_2} = 0, \end{gathered} \tag{5}$$

and the condition

$$\chi(u_A, \Omega) \equiv \begin{cases} \displaystyle\int_\Omega \left(\frac{|u_A|^2}{|x|^4} + \frac{|\nabla u_A|^2}{|x|^2} + |\nabla\nabla u_A|^2\right) dx < \infty \\ \qquad \text{for } n > 4, \\ \displaystyle\int_\Omega \left(\frac{|u_A|^2}{||x|^2 \ln |x||^2} + \frac{|\nabla u_A|^2}{||x| \ln |x||^2} + |\nabla\nabla u_A|^2\right) dx < \infty \\ \qquad \text{for } 2 \le n \le 4, \end{cases} \tag{6}$$

for $A, x \in \mathbb{R}^n$, where Ax denotes the standard scalar product of A and x.

Such a solution of problem (1), (5) can be constructed by the variational method [29], minimizing the functional

$$\Phi(v) = \frac{1}{2} \int_\Omega |\Delta v|^2\, dx$$

in the class of admissible functions $\left\{v\colon\ v \in H^2(\Omega),\ v(x)\big|_{\Gamma_1} = (Ax)\big|_{\Gamma_1},\ \left(\Delta v + \tau \frac{\partial v(x)}{\partial \nu}\right)\Big|_{\Gamma_1} = \tau \frac{\partial (Ax)}{\partial \nu}\Big|_{\Gamma_1},\ v \text{ is compactly supported in } \overline{\Omega}\right\}$.

The validity of condition (6) as a consequence of the Hardy inequality follows from the results in [8, 9].

Now, for any arbitrary number $e \neq 0$, we construct a generalized solution u_e of equation (1) with the boundary conditions

$$u_e\big|_{\Gamma_1} = e, \quad \left(\Delta u_e + \tau \frac{\partial u_e}{\partial \nu}\right)\Big|_{\Gamma_1} = 0, \quad \frac{\partial u_e}{\partial \nu}\Big|_{\Gamma_2} = \left(\frac{\partial \Delta u_e}{\partial \nu} + \tau\, u_e\right)\Big|_{\Gamma_2} = 0, \tag{7}$$

and the condition

$$\chi(u_e, \Omega) \equiv \begin{cases} \displaystyle\int_\Omega \left(\frac{|u_e|^2}{|x|^4} + \frac{|\nabla u_e|^2}{|x|^2} + |\nabla\nabla u_e|^2\right) dx < \infty \\ \qquad \text{for } n > 4, \\ \displaystyle\int_\Omega \left(\frac{|u_e|^2}{||x|^2 \ln |x||^2} + \frac{|\nabla u_e|^2}{||x| \ln |x||^2} + |\nabla\nabla u_e|^2\right) dx < \infty \\ \qquad \text{for } 2 \le n \le 4. \end{cases} \tag{8}$$

The solution of problem (1), (7) also is constructed by the variational method with the minimization of the corresponding functional in the class of admissible functions $\{v : v \in H^2(\Omega),\ v\big|_{\Gamma_1} = e,\ \left(\Delta v + \tau \frac{\partial v}{\partial \nu}\right)\big|_{\Gamma_1} = 0,\ \ v \text{ is compactly supported in } \overline{\Omega}\}$.

The condition (8) as a consequence of the Hardy inequality follows from the results in [8, 9].

Consider the function $v = (u_A - Ax) - (u_e - e)$.

Obviously, v is a solution of problem (1), (2):

$$\Delta^2 v = 0, \quad x \in \Omega,$$

$$v\big|_{\Gamma_1} = \left(\Delta v + \tau \frac{\partial v}{\partial \nu}\right)\Big|_{\Gamma_1} = 0, \quad \frac{\partial v}{\partial \nu}\Big|_{\Gamma_2} = \left(\frac{\partial \Delta v}{\partial \nu} + \tau\, v\right)\Big|_{\Gamma_2} = 0.$$

One can easily see that $v \not\equiv 0$ and $D(v, \Omega) < \infty$.

To each nonzero vector $\mathbf{A} = (A_0, A_1, \dots, A_n)$ in $\mathbb{R}^{n+1}$, there corresponds a nonzero solution $v_\mathbf{A} = (v_{A_0}, v_{A_1}, \dots, v_{A_n})$ of problem (1), (2) with the condition $D(v_\mathbf{A}, \Omega) < \infty$, and moreover,

$$v_\mathbf{A} = u_A - u_e - Ax + e.$$

Let $A_0, A_1, \dots, A_n$ be a basis in $\mathbb{R}^{n+1}$. Let us prove that the corresponding solutions $v_{A_0}, v_{A_1}, \dots, v_{A_n}$ are linearly independent. Let

$$\sum_{i=0}^{n} C_i v_{A_i} \equiv 0, \qquad C_i = \text{const}\,.$$

Set $W \equiv \sum_{i=1}^{n} C_i A_i x - C_0 e$. We have

$$W = \sum_{i=1}^{n} C_i u_{A_i} - C_0 u_e,$$

$$\int_{\Omega} |x|^{-2} |\nabla W|^2 \, dx < \infty, \quad n > 4,$$

$$\int_{\Omega} ||x| \ln |x||^{-2} |\nabla W|^2 \, dx < \infty, \quad 2 \le n \le 4.$$

Let us show that

$$W \equiv \sum_{i=1}^{n} C_i A_i x - C_0 e \equiv 0.$$

Let $T = \sum_{i=0}^{n} C_i A_i = (t_0, \dots, t_n)$, where $A_0 = -e$. Then

$$\int_{\Omega} |x|^{-2} |\nabla W|^2 \, dx = \int_{\Omega} |x|^{-2} (t_1^2 + \cdots + t_n^2) \, dx = \infty, \quad n > 4,$$

$$\int_{\Omega} ||x| \ln |x||^{-2} |\nabla W|^2 \, dx = \int_{\Omega} ||x| \ln |x||^{-2} (t_1^2 + \cdots + t_n^2) \, dx = \infty, \quad 2 \le n \le 4,$$

if $T \neq 0$.

Consequently, $T = \sum_{i=0}^{n} C_i A_i = 0$, and since the vectors $A_0, A_1, \dots, A_n$ are linearly independent, we obtain $C_i = 0,\ i = 0, 1, \dots, n$.

Thus, the mixed problem (1), (2) with the condition $D(u, \Omega) < \infty$ has at least $n + 1$ linearly independent solutions.

Let us prove that each solution u of problem (1), (2) with the condition $D(u, \Omega) < \infty$ can be represented as a linear combination of the functions $v_{A_0}, v_{A_1}, \dots, v_{A_n}$, i.e.

$$u = \sum_{i=0}^{n} C_i v_{A_i}, \qquad C_i = \text{const}.$$

Since $A_0, A_1, \dots, A_n$ is a basis in $\mathbb{R}^{n+1}$, it follows that there exists constants $C_0, C_1, \dots, C_n$ such that

$$A = \sum_{i=0}^{n} C_i A_i.$$

We set

$$u_0 \equiv u - \sum_{i=0}^{n} C_i v_{A_i}.$$

Obviously, the function u_0 is a solution of problem (1), (2), and $D(u_0, \Omega) < \infty$, $\chi(u_0, \Omega) < \infty$.

Let us show that $u_0 \equiv 0,\ x \in \Omega$. To this end, we substitute the function $\varphi(x) = u_0(x)\theta_N(x)$ into the integral identity (4) for the function u_0, where $\theta_N(x) = \theta(|x|/N)$, $\theta \in C^\infty(\mathbb{R})$, $0 \leq \theta \leq 1$, $\theta(s) = 0$ for $s \geq 2$ and $\theta(s) = 1$ for $s \leq 1$; then we obtain

$$\begin{aligned}\int_\Omega (\Delta u_0)^2 \theta_N(x)\,dx + \int_{\Gamma_1} \tau\,|\nabla u_0|^2 \theta_N(x)\,ds - \int_{\Gamma_2} \tau\,|u_0|^2 \theta_N(x)\,ds \\ = -J_1(u_0) - J_2(u_0) - J_3(u_0),\end{aligned} \tag{9}$$

where

$$J_1(u_0) = 2\int_\Omega \Delta u_0\,\nabla u_0\,\nabla\theta_N(x)\,dx, \quad J_2(u_0) = \int_\Omega u_0\,\Delta u_0\,\Delta\theta_N(x)\,dx,$$
$$J_3(u_0) = \int_{\Gamma_1} u_0\,\nabla u_0\,\nabla\theta_N(x)\,ds.$$

By applying the Cauchy–Schwarz inequality and by taking into account the conditions $D(u_0, \Omega) < \infty$ and $\chi(u_0, \Omega) < \infty$, one can easily show that $J_1(u_0) \to 0$, $J_2(u_0) \to 0$ and $J_3(u_0) \to 0$ as $N \to \infty$. Consequently, by passing to the limit as $N \to \infty$ in (9), we obtain

$$\int_\Omega (\Delta u_0)^2\,\theta_N(x)\,dx + \int_{\Gamma_1} \tau\,|\nabla u_0|^2 \theta_N(x)\,ds - \int_{\Gamma_2} \tau\,|u_0|^2 \theta_N(x)\,ds \to 0.$$

Using the integral identity

$$\int_\Omega (\Delta u_0)^2\,dx + \int_{\Gamma_1} \tau\,|\nabla u_0|^2\,ds - \int_{\Gamma_2} \tau\,|u_0|^2\,ds = 0,$$

we find that if u_0 is a solution of the homogeneous problem (1), (2), then $\Delta u_0 = 0$. Therefore, we have

$$\Delta u_0 = 0, \quad x \in \Omega,$$

$$u_0\big|_{\Gamma_1} = \left(\Delta u_0 + \tau\frac{\partial u_0}{\partial \nu}\right)\Big|_{\Gamma_1} = 0, \quad \frac{\partial u_0}{\partial \nu}\Big|_{\Gamma_2} = \left(\frac{\partial \Delta u_0}{\partial \nu} + \tau\,u_0\right)\Big|_{\Gamma_2} = 0.$$

Hence, it follows [5, Chap. 2] that $u_0 = 0$ in Ω. The relation

$$\int_{\partial\Omega} \tau\,(|\nabla u_0|^2 + |u_0|^2)\,ds = 0$$

implies that $u_0 \equiv 0$ on a set of a positive measure on $\partial\Omega$. The proof of the theorem is complete.

Theorem 2 *The mixed problem (1), (2) with the condition $D_a(u, \Omega) < \infty$ has:*
(i) the trivial solution for $n - 2 \leq a < \infty$, $n > 4$;
(ii) n linearly independent solutions for $n - 4 \leq a < n - 2$, $n > 4$;

(iii) $n+1$ *linearly independent solutions for* $-n \le a < n-4,\ n > 4;$

(iv) $k(r,n)$ *linearly independent solutions for* $-2r+2-n \le a < -2r+4-n,$ $r > 1,\ n > 4,$ *where*

$$k(r,n) = \binom{r+n}{n} - \binom{r+n-4}{n}.$$

The proof of Theorem 2 is based on Lemma 1 about the asymptotic expansion of the solution of the biharmonic equation and the Hardy type inequalities for unbounded domains [8, 9]. In case (iv), we need to determine the number of linearly independent solutions of the biharmonic equation (1), the degree of which not exceed the fixed number.

It is well know that the dimension of the space of all polynomials in $\mathbb{R}^n$ of degree $\le r$ is equal $\binom{r+n}{n}$ [27]. Then the dimension of the space of all biharmonic polynomials in $\mathbb{R}^n$ of degree $\le r$ is equal to

$$\binom{r+n}{n} - \binom{r+n-4}{n},$$

since the biharmonic equation is the vanishing of some polynomial of degree $r-4$ in $\mathbb{R}^n$. If we denote by $k(r,n)$ the number of linearly independent polynomial solutions of equation (1) whose degree do not exceed r and by $l(r,n)$ the number of linearly independent homogeneous polynomials of degree r, that are solutions of equation (1), then

$$k(r,n) = \sum_{s=0}^{r} l(s,n),$$

where

$$l(s,n) = \binom{s+n-1}{n-1} - \binom{s+n-5}{n-1}, \quad s > 0.$$

Further, we prove that the mixed problem (1), (2) with the condition $D_a(u,\Omega) < \infty$ for $-2r+2-n \le a < -2r+4-n$ has equally $k(r,n)$ of linearly independent solutions.

4 Appendix

Many complex engineering structures, such as the rotor blades of wind turbines and helicopters, are non-prismatic beamlike structures, which may be tapered, twisted and curved in their reference unstressed state and undergo large displacements of the reference centre-line's points, as well as in- and out-of-plane warping of the transverse cross-sections. Continuous efforts to better predict the mechanical behaviour of such structures, which are aimed at improving the performance in terms of structural efficiency and costs effectiveness, offer the opportunity to address some very interesting, challenging problems in the field of continuum and solid mechanics [25].

An important point in developing rigorous yet application-oriented mathematical models for such structures is an appropriate description of their motion. The description of the motion of such structures can thus be performed by introducing two kinematic maps, herein called R_A and R_B, to identify the positions of the points of the mentioned structure in the reference and current states, as discussed in [25, 26].

These maps can be used to determine the gradient of transformation between the current and reference states and, successively, the Green-Lagrange strain tensor, as discussed in [25]. Given such strain tensor and a constitutive model, it is thus possible to determine the stress fields in the three-dimensional structure. The problem unknowns, such as the displacements of the centre-line's points and the warping fields, can then be determined as the solution of a set of balance equations deduced by a stationary condition of a suitable energy functional [25]. As shown in [25, 26], the results of this procedure is a mathematical problem based on partial differential equations (PDEs) with Neumann-type boundary conditions the solution of which enable obtaining all unknowns of the problem, such as the warping fields w_k, the displacements of the centre-line's points, the Green-Lagrange strain fields and the corresponding stress fields.

The results of the article are presented at the 13th International Conference Chaotic Modeling, Simulation and Applications (CHAOS2020, June 9–12, 2020, Florence, Italy).

References

1. F. Brock, An isoperimetric inequality for eigenvalues of the Stekloff problem. Z. Angew. Math. Mech. (ZAMM) **81**(1), 69–71 (2001)
2. R. Farwig, A note on the reflection principle for the biharmonic equation and the Stokes system. Acta Appl. Math. **34**, 41–51 (1994)
3. F. Gazzola, G. Sweers, On positivity for the biharmonic operator under Steklov boundary conditions. Arch. Rational Mech. Anal. **188**(3), 399–427 (2008)
4. F. Gazzola, H.-Ch. Grunau, G. Sweers, *Polyharmonic Boundary Value Problems: Positivity Preserving and Nonlinear Higher Order Elliptic Equations in Bounded Domains*. Lecture Notes Math. vol. 1991 (Springer, 2010)
5. D. Gilbarg, N. Trudinger, *Elliptic Partial Differential Equations of Second Order* (Springer, Berlin, 1977)
6. YuV Egorov, V.A. Kondratiev, *On Spectral Theory of Elliptic Operators* (Birkhauser, Basel, 1996)
7. V.A. Kondratiev, O.A. Oleinik, On the behavior at infinity of solutions of elliptic systems with a finite energy integral. Arch. Rational Mech. Anal. **99**(1), 75–99 (1987)
8. V.A. Kondrat'ev, O.A. Oleinik, Boundary value problems for the system of elasticity theory in unbounded domains. Korn's inequalities. Russ. Math. Surveys **43**(5), 65–119 (1988)
9. V.A. Kondratiev, O.A. Oleinik, Hardy's and Korn's inequality and their application. Rend. Mat. Appl. Serie VII. **10**, 641–666 (1990)
10. J.R. Kuttler, V.G. Sigillito, Inequalities for membrane and Stekloff eigenvalues. J. Math. Anal. Appl. **23**(1), 148–160 (1968)

11. O.A. Matevosyan, The exterior Dirichlet problem for the biharmonic equation: Solutions with bounded Dirichlet integral. Math. Notes **70**(3), 363–377 (2001)
12. O.A. Matevossian, Solutions of exterior boundary value problems for the elasticity system in weighted spaces. Sbornik Math. **192**(12), 1763–1798 (2001)
13. H.A. Matevossian, On solutions of mixed boundary-value problems for the elasticity system in unbounded domains. Izvestiya Math. **67**(5), 895–929 (2003)
14. H.A. Matevossian, On solutions of the Dirichlet problem for the polyharmonic equation in unbounded domains. P-Adic Numbers, Ultrametric Anal. Appl. **7**(1), 74–78 (2015)
15. O.A. Matevosyan, Solution of a mixed boundary value problem for the biharmonic equation with finite weighted Dirichlet integral. Diff. Equations **51**(4), 487–501 (2015)
16. O.A. Matevossian, On solutions of the Neumann problem for the biharmonic equation in unbounded domains. Math. Notes **98**, 990–994 (2015)
17. O.A. Matevosyan, On solutions of the mixed Dirichlet-Navier problem for the polyharmonic equation in exterior domains. Russ. J. Math. Phys. **23**(1), 135–138 (2016)
18. O.A. Matevosyan, On solutions of one boundary value problem for the biharmonic equation. Diff. Equations **52**(10), 1379–1383 (2016)
19. H.A. Matevossian, On the biharmonic Steklov problem in weighted spaces. Russ. J. Math. Phys. **24**(1), 134–138 (2017)
20. H.A. Matevossian, On the Steklov-type biharmonic problem in unbounded domains. Russ. J. Math. Phys. **25**(2), 271–276 (2018)
21. H.A. Matevossian, On the polyharmonic Neumann problem in weighted spaces. Complex Var. Elliptic Equ. **64**(1), 1–7 (2019)
22. H.A. Matevossian, Mixed boundary value problems for the elasticity system in exterior domains. *Math. Comput. Appl.*, **24**:2 (2019), 58, 1–7
23. Matevossian H. On the Mixed Steklov–Neumann and Steklov-Type Biharmonic problems in unbounded domains. *IOP Conf. Ser.: Mater. Sci. Eng.*, **683** (2019), no. 1, 012016, 10 pp
24. H.A. Matevossian, On the mixed Neumann-Robin problem for the elasticity system in exterior domains. Russ. J. Math. Phys. **27**(2), 272–276 (2020)
25. G. Migliaccio, G. Ruta et al., Beamlike models for the analyses of curved, twisted and tapered horizontal-axis wind turbine (HAWT) blades undergoing large displacements. Wind Energ. Sci. **5**, 685–698 (2020). https://doi.org/10.5194/wes-5-685-2020
26. G. Migliaccio, G. Ruta, Rotor blades as curved, twisted, and tapered beam-like structures subjected to large deflections. Eng. Struct. **222**, 111089 (2020). https://doi.org/10.1016/j.engstruct.2020.111089
27. S.G. Mikhlin, *Linear Partial Differential Equations* (Vyssaya Shkola, Moscow, 1977). (in Russian)
28. L.E. Payne, Some isoperimetric inequalities for harmonic functions. SIAM J. Math. Anal. **1**(3), 354–359 (1970)
29. S.L. Sobolev, Some, Applications of Functional Analysis in Mathematical Physics, 3th ed., Nauka, Moscow, *Applications of Functional Analysis in Mathematical Physics* (Amer. Math. Soc, Providence, 1988), p. 1991
30. Stekloff W. Sur les problemes fondamentaux de la physique mathematique. *Ann. Sci. de l'E.N.S.*, 3^e serie, **19**, 191–259 et 455–490, 1902

Prediction of Qualitative Dynamics in Population Models Through Holling's Functional Responses

Ashutosh Maurya and Anupam Priyadarshi

Abstract Natural ecosystems are complex network of biotic and abiotic interactions of species and their biological, physical and chemical constituents. Over the century, mathematical models have played primary roles in understanding the mystery behind the ecosystems processes and interesting dynamics of natural ecosystems. The di-trophic predator-prey interactions are the basic building blocks for complex food web models (multiple trophic interactions). The pioneering work of Lotka-Volterra (Mem Accad Nazi LaIncci 2:31–113, 1926), explaining the abrupt deviations in species abundance and existence of oscillations in a simple predator-prey interaction. Studies in previous decades, it has been assumed that the functional responses are the main cause for chaotic and non-chaotic behavior. In this paper, we investigate how functional responses affect the system dynamics by using its different combinations in a simple two prey-one predator population model. Based on our present investigation, we concluded that the stabilizing properties of functional responses dominate oscillatory behavior.

Keywords Controlling limit cycles · Two parameter bifurcation comparison · Holling type functional responses · Stabilizing property · Dominating oscillatory property

1 Introduction

Food webs in ecosystems play a vital role to regulate species coexistence, species interactions and carbon, nitrogen cycles in natural ecosystems. They often assumed to be responsible for species control and ecosystem balance. The complex networks of biotic (predator-prey) and abiotic interactions in food webs are ubiquitously exist within trophic levels. Among the biotic interactions, predator-prey interactions exhibit very complex dynamics and hence, attracted the attention of theoretical,

A. Maurya (✉) · A. Priyadarshi
Department of Mathematics, Institute of Science, Banaras Hindu University, Varanasi, India
e-mail: ashutoshk.maurya@bhu.ac.in

C. H. Skiadas et al. (eds.), *13th Chaotic Modeling and Simulation International Conference*, Springer Proceedings in Complexity,
https://doi.org/10.1007/978-3-030-70795-8_44

experimental and field ecologists from the last century. Experiments and field observations are the basis for the development of theoretical concepts of ecological process through mathematical models. These mathematical models interpret the model dynamics using differential equations and predict the future happening based on their results [16]. The theoretical development of mathematical modeling assumed to be started from the pioneering work of Lotka and Volterra [17, 22], which explained the abrupt deviations in species abundance and predicted the existence of oscillations in a simple predator-prey model. The di-trophic predator-prey interactions are the basic building blocks for complex food web models of multiple trophic interactions [16]. The food-web systems exhibit more complex dynamics (quasi-periodicity, chaos) on increasing the number of species in its chain [1–3] which cannot be obtained by di-trophic food chain models [8].

Growth rates (prey increment with time [3, 4] and *functional responses* (the relationship between prey-capturing by per predator in unit time with prey density), frequently used in theoretical predator-prey interactions, are the two important factors to control the overall dynamics of food webs. The term functional response (FR) was first acknowledged by Soloman [12] and extensively deliberated by C.S. Holling using the *Disc equations* based on predator's handling (capturing, eating and digesting) ability [1–3]. Based on terrestrial experiments and artificial lab experiments [1–3], C.S. Holling proposed three types of functional responses namely Holling types I, II and III. Later on, Holling type IV FR is coined and identified by J.F. Andrews [7] in the culture of micro-organisms which is nonlinear and density dependent function. There are several other FRs such as Leslie-Gower, Ivlev etc., which are also frequently used in predator-prey interactions besides Holling type FR.

The Holling type I (H_I) FR is the simplest, in which predator's capture rate increases directly proportional to prey density till saturation. $H_I FR$ has been used in Lotka-Volterra's predator-prey model which produced neutral stable limit cycles [17, 22]. Experimentally, it is observed that the sea star's predation traits on juvenile scallops in aquatic region are H_I [10, 11]. It is also used in the filter-feeding zooplankton harvesting models. It's mathematical description is $k(x, y) = min(w_1 x, w_1)$. The Holling type II ($H_{II}$) is also similar in the sense that the rate of capture prey increases with increasing prey density but its saturation reached out slowly in comparison of H_I. It is frequently used in the population estimation of insects and parasitoids [6, 17]. It's mathematical description is $k(x, y) = \frac{w_1 x}{1+w_1 x+w_2 y}$. The Holling type III ($H_{III}$) exhibits S-shaped in which at low prey density, the capture rate exceeds and goes to saturation gradually like (H_{II}). It is widely used on population estimation of vertebrates [6, 17]. Its mathematical description is $k(x, y) = \frac{w_1 x^2}{1+w_1 x^2+w_2 y^2}$. The Holling type IV ($H_{IV}$) is dome-shaped and non-linear which is further proposed and modified by Sokol and Howell [23] and used for the population estimation on micro-organisms and mice estimations [6]. It's mathematical description is $k(x, y) = \frac{w_1 x}{1+w_1 x^2+w_2 y^2}$. The predator behaviour in the natural ecosystem is strongly associated with prey availability and associated functional response by which predator predates on prey [3, 5, 8, 14, 19, 20]. The functional response may induce oscillations in di-trophic predator-prey interaction (or chaos) in multi-trophic systems [6, 13, 21]. In food web systems, the predation rate associated with prey may vary due to change in physical habitat of species.

Studies of Holling's FR convey that the H_I gives the point stability and extinction while H_{II} and H_{IV} give periodic solution along with point stability and extinction with suitable parameter choices. H_{III} gives stability with persistence throughout parameter choice (see Fig. 1). We have extended their studies with the combined effects of different function responses in food web systems. We consider a simple food web consisting of two bottom prey and one predator on which combination of different functional responses are applied to infer the qualitative behaviour. We have studied the following questions: (1) Is the combination of functional responses gives oscillatory solutions with point stability and extinction? (2) What are the basic properties of this type of model? (3) What are the detailed of the dynamical behaviour of these systems? (3) What are the two-parameter bifurcation analysis of these system?

2 Background (Terminology)

Assuming x and y prey densities and z predator density, a simple food web model of one predator and two prey can be expressed with the following system of differential equations:

$$\begin{aligned} x' &= xg(x) - zk(x, y) \\ y' &= yi(y) - zj(x, y) \\ z' &= h(x, y, z) \end{aligned} \tag{1}$$

Here, prey x grows logistically in the absence of predator z as follows:

$$g(x) = (1 - x) \text{ such that } g(0) = 1 > 0,\ g_x = -1, \forall x \geq 0 \text{ and } g(1) = 0$$
$$i(y) = w_4(1 - y) \text{ such that } i(0) = w_4,\ i_y = -w_4, \forall y \geq 0 \text{ and } i(1) = 0$$

Where w_4 represent the maximum growth coefficient of prey y. The growth of predator z depends on several factors including prey availability, predator catching and handling ability, prey searching ability etc. Here, we investigate the model dynamics when predator grows according to Leslie-Gower type FR [8, 15] $h(x, y, z) = s_4 z(1 - \frac{s_3 z}{1+s_1 x+s_2 y})$. Where s_1, s_2 and s_3 are the coefficients of environmental carrying capacity which reduce predator population while is the intrinsic growth rate of the predator z. The identical participation of male and female in the growth of predator [9, 18, 19] gives $h(x, y, z) = s_4 z^2(1 - \frac{s_3}{1+s_1 x+s_2 y})$.

3 Model Formulations

The model formulations are described based on the following assumptions:

3.1 *Case 1: When both Preys Have the Same Functional Response*

Using identically H_I , H_{II} H_{III} and $H_{IV}FR$ in both bottom prey and keeping Leslie-Gower type predator, following food web models are formulated in (2–5) respectively. Many Authors studies propertise of these FR in their studies [6, 13, 21]. The detailed one parameter analysis of functional response related Models 2–5 is given in Fig. 1.

$$\begin{aligned} x' &= x(1-x) - w_1xz \\ y' &= y(1-y)w_4 - w_5yz \\ z' &= w_6z^2(1 - \frac{w_7}{1 + w_8x + w_9y}) \end{aligned} \tag{2}$$

$$\begin{aligned} x' &= x(1-x) - \frac{w_1xz}{1 + w_2x + w_3y} \\ y' &= y(1-y)w_4 - \frac{w_5yz}{1 + w_2x + w_3y} \\ z' &= w_6z^2(1 - \frac{w_7}{1 + w_8x + w_9y}) \end{aligned} \tag{3}$$

$$\begin{aligned} x' &= x(1-x) - \frac{w_1x^2z}{1 + w_2x^2 + w_3y^2} \\ y' &= y(1-y)w_4 - \frac{w_5y^2z}{1 + w_2x^2 + w_3y^2} \\ z' &= w_6z^2(1 - \frac{w_7}{1 + w_8x + w_9y}) \end{aligned} \tag{4}$$

$$\begin{aligned} x' &= x(1-x) - \frac{w_1xz}{1 + w_2x^2 + w_3y^2} \\ y' &= y(1-y)w_4 - \frac{w_5yz}{1 + w_2x^2 + w_3y^2} \\ z' &= w_6z^2(1 - \frac{w_7}{1 + w_8x + w_9y}) \end{aligned} \tag{5}$$

Where parameter w_1 ans w_5 are maximum attack rate at prey x and prey y respectively by predator z. Parameter w_2 and w_3 are half-saturation constant for prey x and prey y in absence of other prey respectively. Parameter w_4 is the distinguished logistic factor for prey y . Parameter w_6 is the growth rate for the predator z. Parameter w_7 is reduction in predator in the severe scarcity of prey x and prey y. Parameter w_8 and w_9 is prey preferences for predation of prey x and prey y respectively by the predator.

3.2 *Case 2: Different Functional Responses in Preys*

Due to different predation behaviour is happened for different prey species for common predator and also it is documented that if the physical habitat is changed then the same predator predation rate is altered on the same prey. Assuming H_I and H_{II} in the prey equation, the Model 1 becomes the Model 6. Similarly, the functional response H_I with combination H_{III} and $H_{IV}$is taken then Model 1 becomes Model 7–8 respectively. The FR H_{II} with combination H_{III} and H_{IV} are taken in the Model 1 becomes Model 9–10 and our last Model 11 has taken combination of the H_{II} and H_{IV} FR .

$$\begin{aligned} x' &= x(1-x) - w_1 xz \\ y' &= y(1-y)w_4 - \frac{w_5 yz}{1+w_3 y} \\ z' &= w_6 z^2 (1 - \frac{w_7}{1+w_8 x + w_9 y}) \end{aligned} \tag{6}$$

$$\begin{aligned} x' &= x(1-x) - w_1 xz \\ y' &= y(1-y)w_4 - \frac{w_5 y^2 z}{1+w_3 y^2} \\ z' &= w_6 z^2 (1 - \frac{w_7}{1+w_8 x + w_9 y}) \end{aligned} \tag{7}$$

$$\begin{aligned} x' &= x(1-x) - w_1 xz \\ y' &= y(1-y)w_4 - \frac{w_5 yz}{1+w_3 y^2} \\ z' &= w_6 z^2 (1 - \frac{w_7}{1+w_8 x + w_9 y}) \end{aligned} \tag{8}$$

$$\begin{aligned} x' &= x(1-x) - \frac{w_1 xz}{1+w_2 x + w_3 y} \\ y' &= y(1-y)w_4 - \frac{w_5 y^2 z}{1+w_2 x^2 + w_3 y^2} \\ z' &= w_6 z^2 (1 - \frac{w_7}{1+w_8 x + w_9 y}) \end{aligned} \tag{9}$$

$$
\begin{aligned}
x' &= x(1-x) - \frac{w_1 x z}{1 + w_2 x + w_3 y} \\
y' &= y(1-y)w_4 - \frac{w_5 y z}{1 + w_2 x^2 + w_3 y^2} \\
z' &= w_6 z^2 (1 - \frac{w_7}{1 + w_8 x + w_9 y})
\end{aligned} \tag{10}
$$

$$
\begin{aligned}
x' &= x(1-x) - \frac{w_1 x^2 z}{1 + w_2 x^2 + w_3 x^2} \\
y' &= y(1-y)w_4 - \frac{w_5 y z}{1 + w_2 x^2 + w_3 y^2} \\
z' &= w_6 z^2 (1 - \frac{w_7}{1 + w_8 x + w_9 y})
\end{aligned} \tag{11}
$$

4 Analysis

Theorem 1 *All the model formulated in (2–11) are bounded and dissipative. Further, all Model 2–11 are persistent.*

Proof The proof for individual models Models 2–11 can be expressed on similar lines. Taking $h(x, y, z)$ explicitly as Leslie Gower type FR, the following set of differential equations represent the Model 1 as:

$$
\begin{aligned}
x' &= x(1-x) - zf(x, y) \\
y' &= y(1-y)w_4 - zg(x, y) \\
z' &= w_6 z^2 (1 - \frac{w_7}{1 + w_8 x + w_9 y})
\end{aligned} \tag{12}
$$

From the above expression, we have $x' \leq x(1-x) \implies x(t) \leq \frac{1}{1+ke^{-t}} \forall t \geq 0$ Here $k = \frac{1}{x(0)} - 1$ is the constant of integration. Now, taking $t \to \infty$ yield $\lim_{x\to\infty} x(t) \leq 1 \forall t \geq 0$ similarly, we have $y'(t) \leq y(1-y)w_4 \implies y(t) \leq \frac{w_4}{w_4+k_1 exp(-w_4 t)} \forall t \geq 0$ with $k_1 = w_4(\frac{1}{y(0)} - 1)$ is the constant of integration. Now, taking $t \to \infty$, $\lim_{y\to\infty} y(t) \leq 1 \forall t \geq 0$ For predation functional response, Let $\phi(t) = x(t) + y(t) + \frac{z(t)}{\kappa_3}; \phi(0) \geq 0$, Then $\frac{d\phi}{dt} + k_4 \phi(t) \leq x(1 - x + \kappa_4) + y(1-y)w_4 + \zeta(x, y, z)$ where $\zeta(x, y, z) = w_6 z^2 (1 - \frac{w_7}{1+w_8 x + w_9 y}) + \frac{\kappa_4 z(t)}{\kappa_3}$. Using the maximum property of function, we get $\frac{d\phi}{dt} + k_4 \phi(t) \leq \frac{(1+w_4)}{4} + \kappa_4 + \zeta_0$ provided $w_6 a > w_7$ and $k_3 = a k_4^2 \forall t \geq 0$,

$$\phi(t) = \frac{(1 + w_4 + 4\kappa_4 + 4\zeta_0)}{4} - \frac{(1 + w_4 + 4k_4 - \phi_0)exp(-k_4 t)}{4}$$

$$\implies \lim_{t\to\infty} \phi(t) = \frac{(1 + w_4 + 4\kappa_4 + 4\zeta_0)}{4k_4}$$

This implies that solutions of system (12) are uniformly bounded for any initial value of ${\mathbb{R}_+}^3$. To show the system is dissipative, let there exist such that $(\gamma_1, \gamma_2, \gamma_3) > 0$ such that $\Omega(x_0, y_0, z_0) \subset {\mathbb{R}_+}^3 = \{(x, y, z) : 0 \le x \le \gamma_1, 0 \le y \le \gamma_2, 0 \le z \le \gamma_3\}$ for all $(x_0, y_0, z_0) \ge 0$ where $\Omega(x_0, y_0, z_0)$ is the omega-limit set of the orbit initiating at (x_0, y_0, z_0). Thus, the general form of the Model 1 is dissipative and bounded. Therefore, for any positive solution of the Model 1 can persist for longer time. Hence all **Models** 2–11 are dissipative and persist.

Equilibrium Points and Models Behavior

Theorem 2 *All Models 2–11 has the trivial equilibrium point* $E_0(0, 0, 0)$ *and the axial equilibrium points* $E_1(1, 0, 0)$, $E_2(0, 1, 0)$, *and the planar equilibrium point* $E_3(1, 1, 0)$. *The axial singularity* $E(0, 0, 1)$ *is biologically unfeasible and hence doesn't exist due to absence of prey.*

Theorem 3 *In the absence of one prey in Model 2, the following equilibrium points exist:* $E_{11}(\frac{w_7-1}{w_8}, 0, \frac{1-\bar{x}}{w_1})$ & $E_{12}(0, \frac{w_7-1}{w_9}, \frac{(1-\bar{y})w_4}{w_5})$ *and in the positive octant, the non-trivial equilibrium point* $E_{13}(\bar{x}, \bar{y}, \bar{z})$ *exists, where* $\bar{x} = \frac{w_7-1-w_9\bar{y}}{w_8}$, $\bar{y} = \frac{(1-\bar{y})w_4}{w_5}$ & $\bar{z} = \frac{1-\bar{x}}{w_1}$.

Theorem 4 *In the absence of one prey in the Model 3, the following equilibrium points exist:* $E_{21}(\frac{w_7-1}{w_8}, 0, \frac{1+w_2\bar{x}(1-\bar{x})}{w_1})$ & $E_{22}(0, \frac{w_7-1}{w_9}, \frac{(1-\bar{y})w_4(1+w_3\bar{y})}{w_5}$ *and in the positive octant, the non-trivial equilibrium point* $E_{23}(\bar{x}, \bar{y}, \bar{z})$ *exists, where* $\bar{x} = \frac{w_7-1-w_9\bar{y}}{w_8}$, $\bar{z} = \frac{(1-\bar{x})(1+w_2\bar{x}+w_3\bar{y})}{w_1}$ *and* $\bar{y}$ *can be calculated by a quadratic equation* $w_3\bar{y}^2 - w_3\bar{y} - (1 - \bar{y})(1 + w_2\bar{x}) + \frac{w_5\bar{z}}{w_4} = 0$.

Theorem 5 *In the absence of one prey in the Model 4, the following equilibrium points exist:* $E_{31}(\frac{w_7-1}{w_8}, 0, \frac{1+w_2\bar{x}^2(1-\bar{x})}{w_1\bar{x}})$ & $E_{32}(0, \frac{w_7-1}{w_9}, \frac{(1-\bar{y})w_4(1+w_3\bar{y}^2)}{w_5\bar{y}}$ *and in the positive octant, the non-trivial equilibrium point* $E_{33}(\bar{x}, \bar{y}, \bar{z})$ *exists, where* $\bar{x} = \frac{w_7-1-w_9\bar{y}}{w_8}$, $\bar{z} = \frac{(1-\bar{x})(1+w_2\bar{x}^2+w_3\bar{y}^2)}{w_1\bar{x}}$ *and* $\bar{y}$ *can be calculated by a quadratic equation* $w_3\bar{y}^3 - w_3\bar{y}^2 + \bar{y}(1 + w_2\bar{x}^2) + \frac{w_5\bar{z}}{w_4} - 1 - w_2\bar{x}^2 = 0$.

Theorem 6 *In the absence of one prey in the Model 5, the following equilibrium points exist:* $E_{41}(\frac{w_7-1}{w_8}, 0, \frac{1+w_2\bar{x}^2(1-\bar{x})}{w_1})$ & $E_{42}(0, \frac{w_7-1}{w_9}, \frac{(1-\bar{y})w_4(1+w_3\bar{y}^2)}{w_5}$ *and in the positive octant, the non-trivial equilibrium point* $E_{43}(\bar{x}, \bar{y}, \bar{z})$ *exists, where* $\bar{x} = \frac{w_7-1-w_9\bar{y}}{w_8}$, $\bar{z} = \frac{(1-\bar{x})(1+w_2\bar{x}^2+w_3\bar{y}^2)}{w_1}$ *and* $\bar{y}$ *can be calculated by a quadratic equation* $w_3\bar{y}^3 - w_3\bar{y}^2 - (1 - \bar{y})(1 + w_2\bar{x}^2) + \frac{w_5\bar{z}}{w_4} - 1 = 0$.

Theorem 7 *In the absence of one prey in the Mode 6, the following equilibrium points exist:* $E_{51}(\frac{w_7-1}{w_8}, 0, \frac{(1-\bar{x})}{w_1})$ & $E_{52}(0, \frac{w_7-1}{w_9}, \frac{(1-\bar{y})w_4(1+w_3\bar{y})}{w_5}$ *and in the positive octant, the non-trivial equilibrium point* $E_{53}(\bar{x}, \bar{y}, \bar{z})$ *exists, where* $\bar{x} = \frac{w_7-1-w_9\bar{y}}{w_8}$, $\bar{z} = \frac{(1-\bar{x})}{w_1}$ *and* $\bar{y}$ *can be calculated by a quadratic equation* $w_3\bar{y}^2 + (1 - w_3)\bar{y} + \frac{w_5\bar{z}}{w_4} - 1 = 0$.

Theorem 8 *In the absence of one prey in the Model 7, the following equilibrium points exist:* $E_{61}(\frac{w_7-1}{w_8}, 0, \frac{(1-\bar{x})}{w_1})$ & $E_{62}(0, \frac{w_7-1}{w_9}, \frac{(1-\bar{y})w_4(1+w_3\bar{y}^2)}{w_5\bar{y}}$ *and in the positive octant, the non-trivial equilibrium point* $E_{63}(\bar{x}, \bar{y}, \bar{z})$ *exists, where* $\bar{x} = \frac{w_7-1-w_9\bar{y}}{w_8}$, $\bar{z} = \frac{(1-\bar{x})}{w_1}$ *and* $\bar{y}$ *can be calculated by a quadratic equation* $w_3\bar{y}^3 - w_3\bar{y}^2 + \frac{w_5\bar{z}\bar{y}}{w_4} + \bar{y} - 1 = 0$.

Theorem 9 *In the absence of one prey in the Model 8, the following equilibrium points exist:* $E_{71}(\frac{w_7-1}{w_8}, 0, \frac{(1-\bar{x})}{w_1})$ & $E_{72}(0, \frac{w_7-1}{w_9}, \frac{(1-\bar{y})w_4(1+w_3\bar{y}^2)}{w_5}$ *and in the positive octant, the non-trivial equilibrium point* $E_{73}(\bar{x}, \bar{y}, \bar{z})$ *exists, where* $\bar{x} = \frac{w_7-1-w_9\bar{y}}{w_8}$, $\bar{z} = \frac{(1-\bar{x})}{w_1}$ *and* $\bar{y}$ *can be calculated by a quadratic equation* $w_3\bar{y}^3 - w_3\bar{y}^2 + \frac{w_5\bar{z}}{w_4} + \bar{y} - 1 = 0$.

Theorem 10 *In the absence of one prey in the Mode 9, the following equilibrium points exist:* $E_{81}(\frac{w_7-1}{w_8}, 0, \frac{(1-\bar{x})(1+w_2\bar{x})}{w_1})$ & $E_{82}(0, \frac{w_7-1}{w_9}, \frac{(1-\bar{y})w_4(1+w_3\bar{y}^2)}{w_5\bar{y}}$ *and in the positive octant, the non-trivial equilibrium point* $E_{83}(\bar{x}, \bar{y}, \bar{z})$ *exists, where* $\bar{x} = \frac{w_7-1-w_9\bar{y}}{w_8}$, $\bar{z} = \frac{(1-\bar{x})(1+w_2\bar{x})}{w_1}$ *and* $\bar{y}$ *can be calculated by a quadratic equation* $w_3\bar{y}^3 - w_3\bar{y}^2 + \frac{w_5\bar{z}\bar{y}}{w_4} + \bar{y} - 1 = 0$.

Theorem 11 *In the absence of one prey in the Mode 10, the following equilibrium points exist:* $E_{91}(\frac{w_7-1}{w_8}, 0, \frac{(1-\bar{x})(1+w_2\bar{x})}{w_1})$ & $E_{92}(0, \frac{w_7-1}{w_9}, \frac{(1-\bar{y})w_4(1+w_3\bar{y}^2)}{w_5\bar{y}}$ *and in the positive octant, the non-trivial equilibrium point* $E_{93}(\bar{x}, \bar{y}, \bar{z})$ *exists, where* $\bar{x} = \frac{w_7-1-w_9\bar{y}}{w_8}$, $\bar{z} = \frac{(1-\bar{x})(1+w_2\bar{x})}{w_1}$ *and* $\bar{y}$ *can be calculated by a quadratic equation* $w_3\bar{y}^3 - w_3\bar{y}^2 + \frac{w_5\bar{z}}{w_4} + \bar{y} - 1 = 0$.

Theorem 12 *In the absence of one prey in the Model 11, the following equilibrium points exist:* $E_{101}(\frac{w_7-1}{w_8}, 0, \frac{(1-\bar{x})(1+w_2\bar{x}^2)}{w_1\bar{x}})$ & $E_{102}(0, \frac{w_7-1}{w_9}, \frac{(1-\bar{y})w_4(1+w_3\bar{y}^2)}{w_5}$ *and in the positive octant, the non-trivial equilibrium point* $E_{103}(\bar{x}, \bar{y}, \bar{z})$ *exists, where* $\bar{x} = \frac{w_7-1-w_9\bar{y}}{w_8}$, $\bar{z} = \frac{(1-\bar{x})(1+w_2\bar{x}^2)}{w_1\bar{x}}$ *and* $\bar{y}$ *can be calculated by a quadratic equation* $w_3\bar{y}^3 - w_3\bar{y}^2 + \frac{w_5\bar{z}}{w_4} + \bar{y} - 1 = 0$.

The local behaviour of the above equilibrium points are stated in the following theorems:

Theorem 13 *The trivial equilibrium* $E_0(0, 0, 0)$ *in all Model 2–11 is always non-hyperbolic and unstable. There exist unstable subspace along* xy *plane and center subspace along z plane.*

Theorem 14 *The axial singularity $E_1(1, 0, 0)$ in all Model 2–11 is non-hyperbolic and saddle point. The y plane of the system is an unstable subspace, x plane of the system is stable subspace and z plane of the system is center subspace.*

Theorem 15 *The axial singularity $E_2(0, 1, 0)$ in all Model 2–11 is non-hyperbolic, saddle point. The x plane of the system is an unstable subspace, y plane of the system is stable subspace and z plane of the system is center subspace.*

Theorem 16 *The equilibrium $E_3(1, 1, 0)$ in all Model 2–11 is non-hyperbolic and stable. The system has stable subspace along $x - y$ plane and center subspace along z plane.*

Theorem 17 *The equilibrium points $E_4(\bar{x}, 0, \bar{z})$, $E_5(0, \bar{y}, \bar{z})$ and $E_6(\bar{x}, \bar{y}, \bar{z})$ are asymptotically stable in each Model 2–11 individually, if the Jacobian of each model at the equilibrium point satisfies the following conditions*

$$\begin{aligned} A_1 &= -(A_{11} + A_{22} + A_{33}) > 0 \\ A_2 &= A_{22}A_{33} + A_{11}A_{33} + A_{11}A_{22} - A_{12}A_{21} - A_{32}A_{33} - A_{31}A_{13} > 0 \\ A_3 &= A_{11}A_{23}A_{32} + A_{12}A_{21}A_{33} + A_{13}A_{31}A_{22} - A_{11}A_{22}A_{33} - A_{12}A_{31}A_{23} \\ &\quad - A_{32}A_{21}A_{13} > 0. \end{aligned} \tag{13}$$

The proofs of the above theorems are quite easy and hence left. Here, proof of Theorem 17 is stated:

Proof the general equilibrium point $E_6(\bar{x}, \bar{y}, \bar{z})$ put in RHS of model 1 and taking first order differentiation. We found the jacobian matrix $J_{x,y,z}$ for the model system (1) is

$$\begin{pmatrix} A_{11} & A_{12} & A_{13} \\ A_{21} & A_{22} & A_{23} \\ A_{31} & A_{32} & A_{33} \end{pmatrix}$$

where $A_{11} = 1 - 2x - w_1 z$, $A_{12} = 0$, $A_{13} = -w_1 x$, $A_{21} = 0$, $A_{22} = w_4 - 2w_4 y - w_5 z$, $A_{23} = -w_5 y$, $A_{31} = \frac{w_6 w_7 w_8 z^2}{1+w_8 x+w_9 z}^2$, $A_{32} = \frac{w_6 w_7 w_9 z^2}{1+w_8 x+w_9 z}^2$, $A_{33} = 2w_6 z(1 - \frac{w_7}{1+w_8 x+w_9 z})$ The characteristic polynomial for the Jacobean Matrix $\lambda^3 + A_1\lambda^2 + A_2\lambda + A_3 = 0$ where A_1, A_2, A_3 are mentenioned in the (13). The system is asymptotically stable if the eigenvalues are negative and $A_1 > 0$, $A_2 > 0$, $A_3 > 0$ and $A_1 A_2 - A_3 > 0$.

5 Numerical Simulations and Results

To observe the impact of different functional responses on one prey two predator interaction **Model** 2–11, the same parameters are used in each model simulations. Simulations results suggest that the qualitative behaviour of models are structurally robust (testing with other various parameter combinations), therefore following parameter combinations have been used to observe the global dynamics:

$$w_1 = 3.6, w_2 = 1.7, w_3 = 1.8, w_4 = 1.5, w_5 = 3.2, w_6 = 1, w_7 = 2.1, w_8 = 1.7, w_9 = 1.8 \tag{14}$$

6 One-Parameter Bifurcation Diagrams

We extensively carried out the numerical simulations and using the continuation algorithm, we drew one-parameter bifurcation diagrams with respect to the parameter w_4 for **Model** 2–11. The **Model** 2–4 are studied by many authors and our findings are in line with them [6, 13, 21] (see Fig. 1). We carried out the simulation of the **Model** 2–11 and found non- oscillation properties are found with persistency along with extinctions. Details bifurcation point and persistency range of all model are given in Table 1.

7 Two-Parameter Bifurcation Diagrams

The parameter region for the species coexistence, extinction of prey is observed in two-parameter bifurcation diagram in Figs. 2 and 3. Model 4 with H_{III} FR exhibit stabilizing effect on system behaviour and a possibility of avoiding the extinction of species. The extinction of prey or predator may not be possible in the Model 4 only which is a significant criteria of using these functional responses in predator-prey interactions. In the two-parameter bifurcation diagram for the Model 3, there are four regions, two regions are extinction region where preys are extinct while in between these two regions, all species survive and co-exists. In the middle of coexists region, a periodic region exists where all the populations are periodic. Similar behaviour is given by H-IV FR (Model 5) see in Fig. 2. While, the Model 2 shows three regions, the first region and third region gives extinction of one of the prey and center region gives coexistence. Similar behaviour is shown by Model 6, 8, 10. Model 7, 9, 11 give only two regions one is coexistence and other is extinction.

8 Discussion and Conclusions

For understanding the significant effect of different functional response in two prey and one predator, we applying a combination of functional response in the Model 1 and found there are only three significant qualitative behaviours which are well

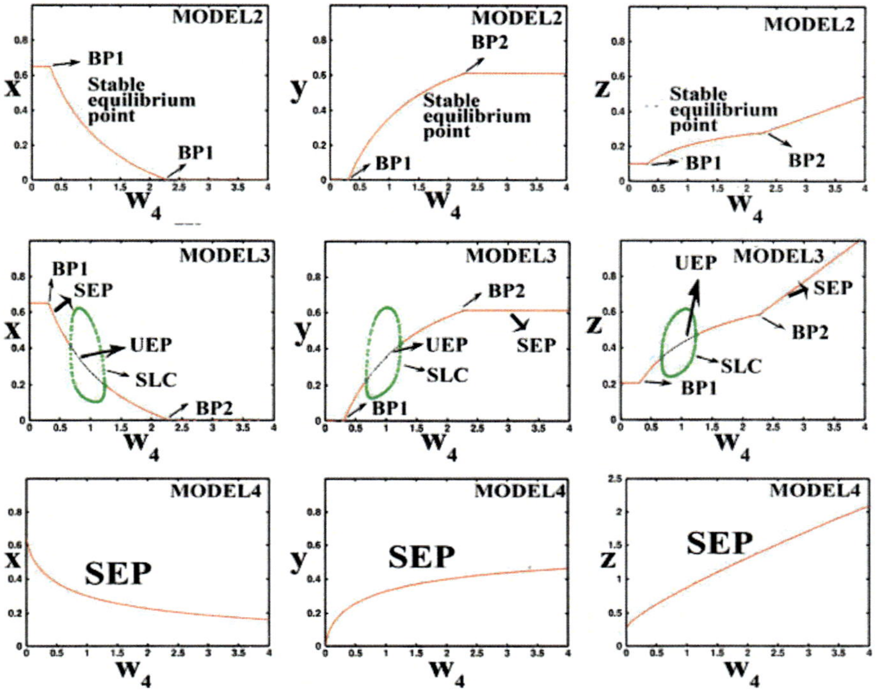

Fig. 1 One parameter bifurcation diagram for prey x, y and predator z with respect to the parameter w_4 for **Model** 2–4 are shown. Here red line is stable equilibrium point (SEP), green circle is stable limit cycle (SLC), Black colour is unstable equilibrium point(UEP). BP1= first branch point, BP2= second branch point, Model 2 is $H_I FR$ alone, Model 3 is $H_{II} FR$ alone, Model 4 is $H_{III} FR$ alone. Model 3 and Model 5 have similar qualitaive behaviour. Combination of FR (in Model 6–11) are similar qualitative dynamics like Model 2

Table 1 List of all branch point (First branch point (BP1) and Second branch point (BP2)), Hopf bifurcation point (First Hopf point (HB1) and Second Hopf point (HB2)), and persistence range found in the respective model

Model	BP1	BP2	HB1	HB2	Persistence range
2	0.3137	2.286	N.A.	N.A.	(0.3137–2.286)
3	0.3138	2.286	0.6805	1.235	(0.3138–2.286)
4	N.A.	N.A.	N.A.	N.A.	(0–4)
5	0.3138	2.286	0.6692	1.233	(0.3138–2.286)
6	0.3137	1.088	N.A.	N.A.	(0.3137–1.088)
7	0.8353	N.A.	N.A.	N.A.	(0–0.8353)
8	0.3137	1.367	N.A.	N.A.	(0.3137–1.367)
9	0.8354	N.A.	N.A.	N.A.	(0–0.8354)
10	0.6588	1.367	N.A.	N.A.	(0.6588–1.367)
11	0.8300	N.A.	N.A.	N.A.	(0.8300–4)

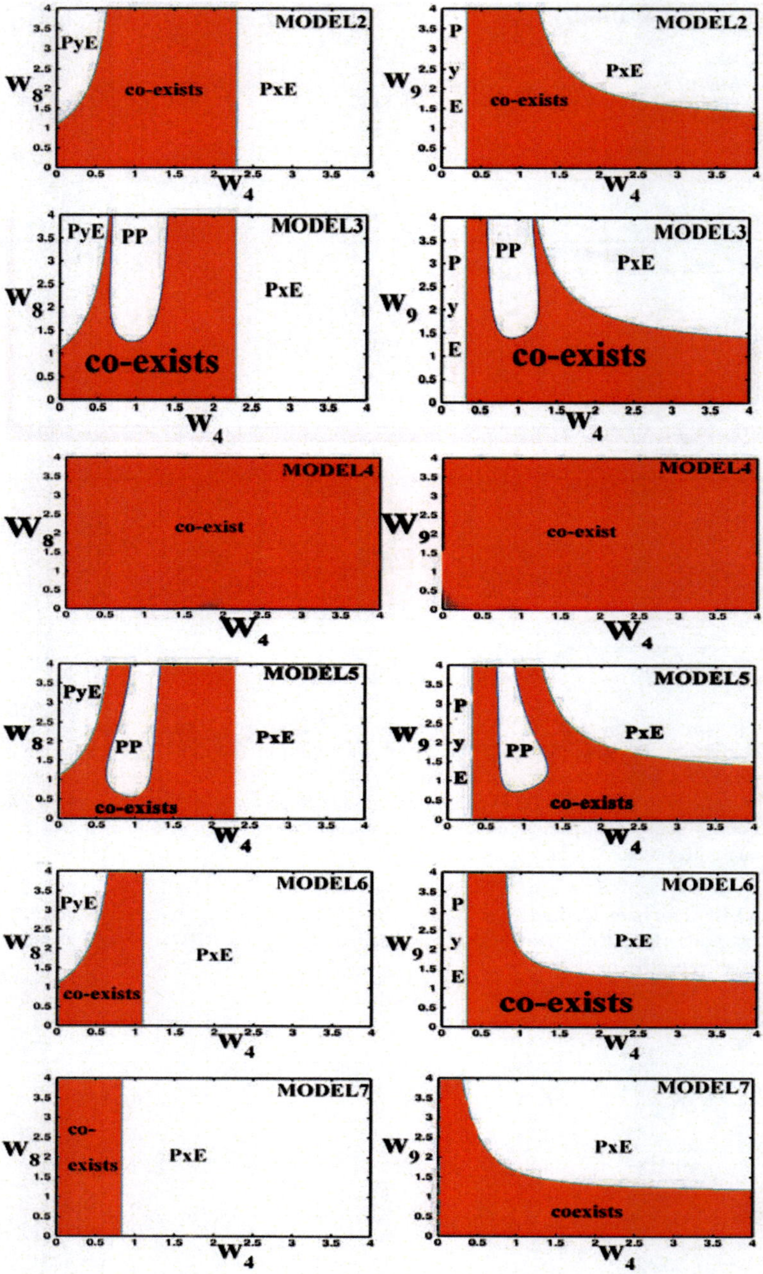

Fig. 2 Two-parameter bifurcation diagram with respect to w_8 & w_4 (left) and w_9 & w_4 right) for the Model 2–7 respectively. Here, PyE=prey y extinct, PP= periodic populations,PxE= prey x extinct

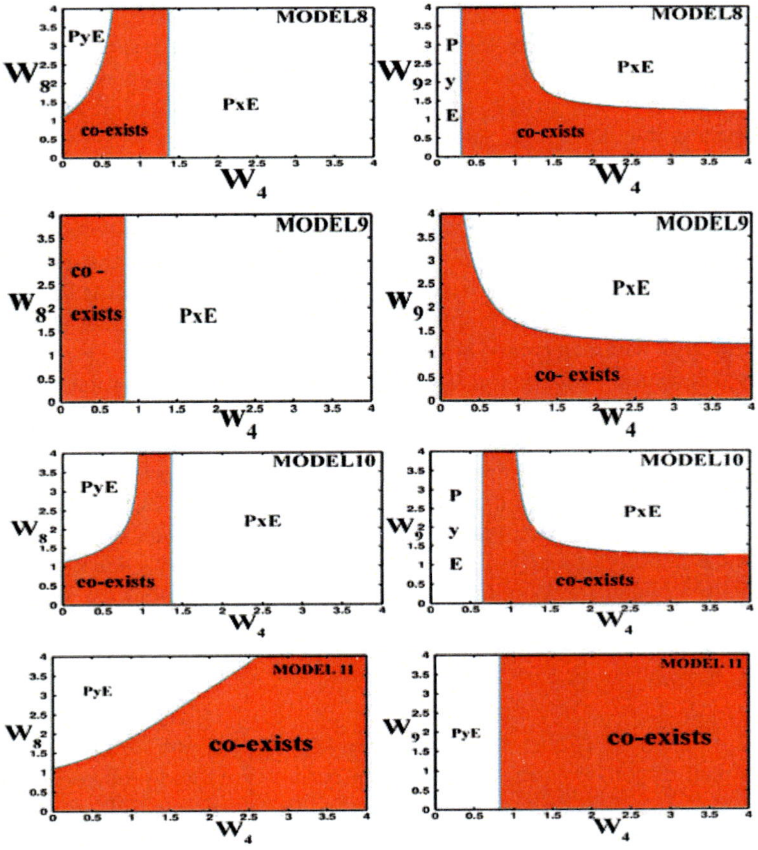

Fig. 3 Two-parameter bifurcation diagram with respect to w_8 & w_4 (left) and w_9 & w_4 right) for the Model 8–11 respectively. Here, PyE=prey y extinct, PxE= prey x extinct

explained in the result section. We have applied the combination of functional responses and found co-existence is possible for a wide range of parameter in every model but care should be taken as some parameter combination may lead to the extinction of some species except $H_{III}FR$ (Model 4). The oscillatory behaviour is observed in $H_{II}FR$ (Model 3) and $H_{IV}FR$ (Model 5) with system persistence, species co-existence in the form of limit cycles and extinction of one of the prey in these systems. We observed that $H_{II}FR$ (*Model* 3) and $H_{IV}FR$ (*Model* 5) most likely to produce periodic solution through supercritical or subcritical Hopf-bifurcation. Simple extinction and coexistence are observed in $H_I FR$ (Model 2).

The previous studies on FR in predator-prey models, provide an insight into system dynamics with the individual functional response and many experimental results require the use of multiple FR in their Mathematical Models. Therefore, it is essential to observe the impact of the combination of functional responses on simple predator-prey models and to compare the results. So, we have applied the combination of FRs in the same model and found that other FRs have stabilize the system by dampening the oscillations. Any FR is combined with $H_{II}FR$ (or $H_{IV}FR$), the oscillatory behaviour of the Model is disappeared and extinction of one of the prey occurred through the branch point. When H_IFR (or $H_{III}FR$) is taken with $H_{II}FR$ (or $H_{IV}FR$) then two threshold value is found with coalescing of both Hopf bifurcation point. Below the branch point, one prey y is extinct while above the branch point another prey x is extinct, in between these, species co-existence is found. Some distinguished differences like Hopf bifurcation points, branch points and specific population with parameter value are shown in Table 1. Here, we observe non- linear functional response $H_{II}FR$ (or $H_{IV}FR$) alone taken in system, model dynamics grants oscillations with extinction while the different combination of FR offering only stability and extinctions with the coalescence of Hopf bifurcation points. From this argument, we can say that non- linearity of functional responses are not always giving periodic solutions/ limit cycles. The FR H_I, H_{II} or H_{IV} alone produce two branch point while the combination with $H_{III}FR$ destroy one branch point and enhance the stability of the system. In overall systems, the prey's equations are varied through different functional response and predator's equations are same but we found the quantitative value of predator are varied rather than that of prey. i.e. shaping of system dynamics is main rooted by FR. FR is also the main cause of a different range of persistence value.

System equations deliberated above were kept very simple to make the effect of functional responses easier to analyze. In our two prey and predator model, we find H_I, H_{II}, H_{III} FR give distinguished qualitative behaviour and H_{II} and $H_{IV}FR$ dynamical behaviour are similar. The combination of Holling FR in model produce similar behaviour like Model 2. The combination of different FR in one model is not studied by others (in my knowledge) but found in experimental studies and we found non- linear FR have counter their oscillatry property and the model shows only co-existence and extinction.

Acknowledgements The first author also acknowledged to CSIR, New Delhi for providing Junior Research fellowship wide $fileno. - 9/013(0822)/2018 - EMR - 1$.

References

1. C.S. Holling, Some characteristics of simple types of predation and parasitism. Can. Entomol.**91**, 385–398 (1959)
2. C.S. Holling, The components of predation as revealed by a study of small mammal predation of the European pine sawfly. Can. Entomol. **91**, 293–320 (1959)
3. C.S. Holling, The functional response of invertebrate predators to prey density. Mem. Entomol. Soc. Can. **45**, 3–60 (1965)

4. G.F. Gause, *The Struggle for Existence Williams and Wilkins* (Baltimore, Maryland, USA, 1934)
5. J.A.J. Metz, M.W. Sabelis, J.H. Kuchlein, Sources of variation in predation rates at high prey densities: an analytic model and a mite example. Exp. Appl. Acarol. **5**, 187–205 (1988)
6. J.B. Collings, The effects of the functional response on the bifurcation behaviour of a mite predator-prey interaction model. J. Math. Biol. **36**, 149–168 (1997)
7. J.F. Andrews, A mathematical model for the continuous culture of microorganisms utilizing inhibitory substrates. Biotechnol. Bioeng. **10**, 707–723 (1968)
8. L.A. Real, The kinetics of functional response. Am. Nat. **111**, 289–300 (1977)
9. M.A. Aziz-Alaoui, Study of a Leslie-Gower-type tritrophic population model. Chaos Solitons Fractals **14**, 1275–93 (2002)
10. M.A. Barbeau, R.E. Scheibling, Behavioural mechanisms of prey size selection by sea stars and crabs preying on juvenile sea scallops. J. Exp. Mar. Bio.lEcol. **180**, 103–136 (1994)
11. M.A. Barbeau, R.E. Scheibling, B.G. Hatcher, Behavioural responses of predatory crabs and sea stars to varying density of juvenile sea scallops. Aquaculture **169**, 97–98 (1998)
12. M.E. Solomon, The natural control of animal populations. J. Anim. Ecol. **18**(1), 1–35 (1949)
13. N. Bairagi, P.K. Roy, J. Chattopadhyay, Role of infection on the stability of a predator-prey system with several response functions-A comparative study. J. Theor. Biol. **248**(1), 10–25 (2007)
14. P.A. Abrams, Functional responses of optimal foragers. Am. Nat. **120**, 382–390 (1982)
15. P.H. Leslie, J.C. Gower, The properties of a stochastic model for the predator—prey type of intersection between two species. Biometrica **47**, 219–234 (1960)
16. P. Turchin, *Complex population dynamics, A theoretical/empirical synthesis, Monograph in Population Biology*, vol. 35 (Princeton University Press, 2003)
17. R.J. Taylor, *Predation* (Chapman and Hall, New York, 1984)
18. R.K. Upadhayay, V. Rai, S.R.K. Iyengar, Species extinction problem: genetic vs ecological factors. Appl. Math. Model. **11**(13), 937–51 (2001)
19. S. Gakkhar, B. Singh, Complex dynamic behaviour in a food web consisting of two preys and a predator. Chaos, Solitons and Fractals **24**, 789–801 (2005)
20. T. Royama, A comparative study of models for predation and parasitism. Res. Popul. Ecol. **S1**, 1–90 (1971)
21. R.R. Vance, Predation and resource partitioning in one predator - two prey model Communities. Am. Nat. **112**, 987 (1978)
22. Volterra, Varizione e fluttuazionidelnumerod'individui in specie animaliconventi. Mem Accad. Nazi. LaIncci. **2**, 31–113 (1926)
23. W. Sokol, J.A. Howell, Kineties of phenol oxidation by washed cell. Biotechnol. Bioeng. **23**, 2039–2049 (1980)

Fractal Dimension of Braided Rivers from Detailed Two-Dimensional Hydrodynamic Simulations

Matteo Nicolini

Abstract As opposed to meandering or channelized fluvial beds, braided rivers are characterized by a morphological activity starting at very low flows, since fluxes are concentrated in a limited number of small channels. With increasing discharge, more channels are involved, up to the situation in which the complete alluvial plain is flooded. As a consequence, there is an intermediate range of flows for which pattern complexity is maximum and braided indices are highest, representing essential conditions for the coexistence of a large variety of habitats and for ecosystems prosperity. In this paper, a new methodology for a quantitative assessment of the complexity of braided rivers at a reach scale is introduced. It is based on the application of the box-counting algorithm to flooded areas identified through a two-dimensional (shallow water) hydrodynamic simulation model, in order to derive an estimate of the fractal dimension with varying flow rate. The identification of the range of discharges for which the fractal dimension is highest is of particular importance in river restoration projects. An application to the River Tagliamento (North-East Italy) is illustrated.

Keywords Morphodynamics · River restoration · Formative discharge · Box-counting algorithm · River Tagliamento

1 Introduction

Braided rivers are complex, non-linear systems characterized by chaotic dynamics. Their local properties, such as the solid transport and the number of channels in a given section, are spatially and temporally variable, their prediction being precluded in the mid-to-long term (Redolfi [16]).

Among others, the process of bifurcation plays a fundamental role in the formation and development of a network of braided channels. Gravel bed rivers with small width-to-depth ratio are characterized by stable equilibrium, while width-to-depth

M. Nicolini (✉)
Polytechnic Department of Engineering and Architecture, University of Udine, via del Cotonificio, 114 Udine, Italy
e-mail: matteo.nicolini@uniud.it

C. H. Skiadas et al. (eds.), *13th Chaotic Modeling and Simulation International Conference*, Springer Proceedings in Complexity,
https://doi.org/10.1007/978-3-030-70795-8_45

ratios above a critical threshold trigger an initial formation of alternate bars, which determine a sinuous pattern and then a first bifurcation. In these last decades, such processes have been investigated with analytical approaches, numerical simulations and laboratory experiments (Parker [15]; Ikeda [9]; Jaeggi [10]; Colombini et al. [4]; Tubino et al. [21]; Lanzoni [13]).

In particular, experimental studies have demonstrated the importance of the presence of alternate bars in the formation of a braiding pattern (Federici and Paola [7]; Bertoldi and Tubino [2]; Jang and Shimizu [11]).

The morphodynamics of such systems is generally described by a statistical approach, which identifies the dependence of some reach-averaged properties (like width, total and active channels, solid transport) on main controlling factors, like discharge, bed slope, sediment size and total river width (Ashmore [1]).

In the definition of reach-averaged properties, the most important issue is the identification of a proper scale length. In line with previous research, some authors have proposed at least ten times the river width (Egozi and Ashmore [6]), which is a value much higher than those related to meandering or channelized fluvial beds, typically of the order of the river width (Ikeda [9]; Seminara and Tubino [19]; Lanzoni [13]).

The identification of a characteristic length is difficult due to the fact that there are a lot of spatial and temporal scales coexisting together (Sapozhnikov and Foufoula-Georgiou [18]). Accordingly, some researchers (Sapozhnikov and Foufoula-Georgiou [17]; Walsh and Hicks [23]; Lane [12]) have proposed that fluvial patterns in braided rivers resemble those of self-similar fractals (or self-affine systems, if an isotropic character is not present).

One problem still open is the quantitative description of the pattern complexity: to this end, specific indices have been introduced, like the average number of channels in the characteristic length (Egozi and Ashmore [5]). Besides, not all channels are simultaneously morphologically active, since solid transport takes place only in a fraction of them. Bertoldi et al. [3] have shown that the total number of channels is well correlated with the dimensionless discharge, while the number of active channel is more dependent on the dimensionless stream power.

Another important controlling factor is the formation and growth of vegetation, whose influence on the planimetric configuration depends on the ratio between the growth time and the average interval of morphologically relevant inundations (Paola [14]). Gurnell et al. [8] have shown that such role is much more important in relatively low energy systems.

In the present paper we aim at a different approach for a quantitative description of such complex fluvial patterns, starting from detailed numerical simulations of the flow field and then analyzing the braiding patterns with conventional fractal analysis algorithms. An application to the River Tagliamento (North-East Italy) is presented. The results show that the values for which the fractal dimension is highest correspond to a narrow range of formative discharges.

2 Materials and Methods

2.1 Materials

The River Tagliamento is located in the North-East part of Italy, with a drainage basin of 2780 km^2 and a length of 178 km. Its upper part is characterized by the most extensive and connected length of dynamic and morphologically intact braided pattern within the Alps, leading to the most important braided reference system of the Alpine region (Tockner et al. [20]).

The area under investigation (Fig. 1) is between Venzone (where a historically relevant gage station is present) and Pinzano gorge (where the corridor width shrinks from 1 km down to 130 m), for a total length of 22.5 km and an area of nearly 30 km^2. The 100-year discharge is 4500 m^3/s.

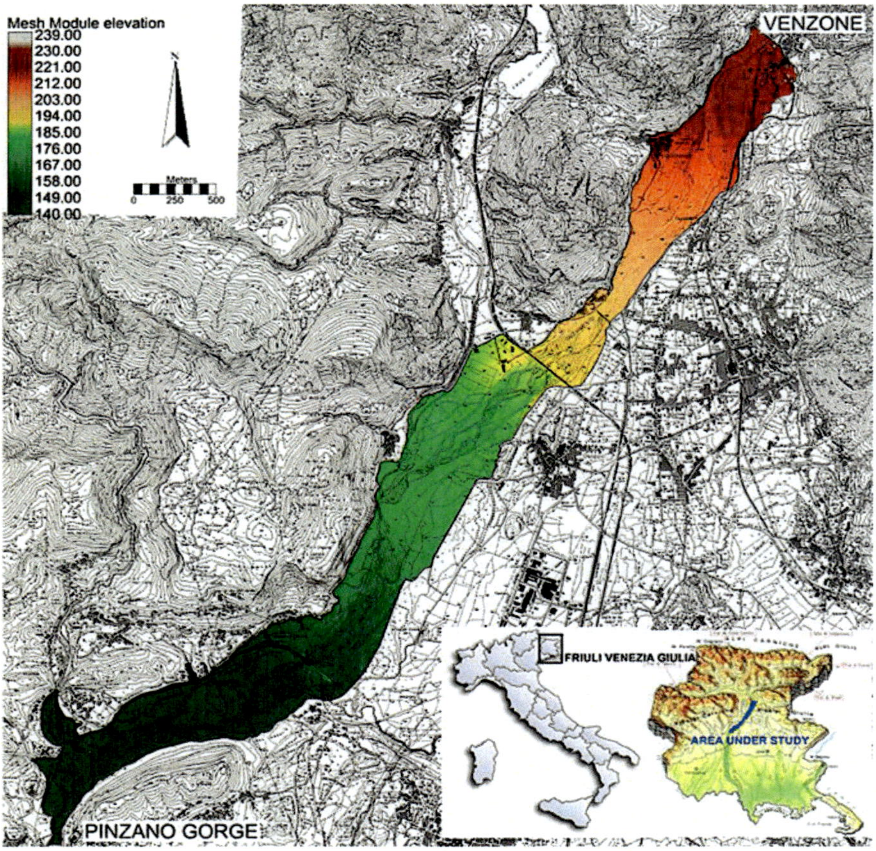

Fig. 1 General view of the area under analysis, with color map indicating terrain elevations

Lidar data available from Regione Autonoma Friuli Venezia Giulia with a resolution of 2 points/m^2 have been used for assigning elevations at mesh nodes, while aerial photographs (0.2 m resolution) have been analyzed in order to determine an estimate of water depths in those areas where water was present at the time of survey. It is important to put into evidence that lidar data do not actually discriminate between surface water elevation and underlying bottom elevation, thus not properly describing the geometry of flooded channels.

Several field campaigns and investigations allowed to determine local geometric characteristics of some defense structures (like groins, jetties and retaining walls) as well as bridge piers, which have been included in the computational mesh.

2.2 Methodology

The software SRH-2D has been adopted for the implementation of the hydrodynamic model. The software is free and can be downloaded from https://www.usbr.gov/. Last version is 3.2.4 (June 2019).

The software SMS (Surface Modeling System, distributed by Aquaveo™) has been used for the construction of the mesh, which is formed by 1,372,586 nodes and 1,557,718 cells, due to the fine discretization of the computational domain (cell sides range between 1 and 8 m). The inflow (upstream) boundary condition has been kept constant in each hydrodynamic simulation, and results have been saved after a sufficient transient time guaranteed steady-state conditions throughout the computational domain.

The freeware software Fractalyse has been chosen for the fractal analysis of simulated flooded areas (downloadable from https://www.fractalyse.org/). In particular, the procedure adopted is the well-known box-counting algorithm (Turcotte [22]), which allows to determine the fractal dimension of the braided patterns.

The overall methodology can be summarized in the following steps:

(1) Mesh generation and elevation assignment at nodes.
(2) Hydrodynamic simulations with very low discharges (typically 10–20 m^3/s), in order to identify the areas occupied by low flows.
(3) Subdivision of each aerial photograph image in RGB bands, and estimation of the water depth, H, obtained as: $H = a \bullet \ln(\lambda_R/\lambda_G)$, in which a is a calibration parameter and λ_R and λ_G are the red and green band intensities, respectively.
(4) Creation of a 'mask' covering only the areas occupied by low flows; in this way, mesh nodes affected by new elevation assignments are limited to those calculated in step 2 (this step is necessary because the algorithm in step 3 identifies fictitious water depths in some dry, vegetated areas).
(5) Assignment of new elevation at mesh nodes identified by the 'mask' of step 4; this is achieved by subtracting the water depths calculated in step 3 to the elevation originally present. In this way, only the wet areas as obtained by numerical simulations are altered by elevation changes.

(6) Hydrodynamic simulations with updated mesh node elevations and varying water discharges (from very low flows up to the condition in which the flood plain is flooded).
(7) Analysis of the planimetric flow patterns with box-counting algorithm and estimation of the fractal dimension of the flooded areas.

In particular, the calibration parameter a of step 3 has been determined starting from measured water depths just upstream a water intake for irrigation purposes (managed by Friulian Plain Bureau of Reclamation).

2.3 Box-counting Algorithm and Derivation of Fractal Dimension

In Euclidean geometry, a point is zero-dimensional, a line is one-dimensional, a plane is two-dimensional, and so on. The traditional meaning of the dimension of an object is that of giving the number of values needed to specify the position of a point on the object. Thus, one value needs to be given in order to specify the position of a point on a line, two values need to be specified for obtaining the position of a point on a plane, and so on.

Another meaning of 'dimension of an object' can be based on the idea of self-similarity. Of course, this must reproduce the traditional values when applied to classical Euclidean objects such as lines and planes. Consider how to give a self-similar description of a one-dimensional object, that is, a line segment: one way to do this is to say that a line segment of length, ℓ, consists of two copies of itself, each characterized by a length $\ell/2$. With the same reasoning, a filled-in square can be thought as four copies of itself, each having a side of length $\ell/2$.

In this way, two quantities characterize each of the self-similar shapes of the examples above:

- the number of self-similar copies, N;
- the edge length of the original relative to each copy, ϵ.

The following formula can be used to define the dimension D of an object:

$$D = \frac{\log N}{\log \epsilon} \tag{1}$$

Applying Eq. (1) to the examples introduced before, in the case of a line segment: $N = 2$, $\epsilon = 2$, $D = 1$; and for the filled square: $N = 4$, $\epsilon = 2$, $D = 2$.

Equation (1) gives a formula for calculating the dimension also of a fractal object: it is sufficient to know the number of self-similar copies, N, and the size of the original relative to each copy, ϵ. For complicated objects, like the braiding pattern of rivers, the box-counting algorithm can be summarized in the following steps:

(1) 'Cover' all the points in the object with boxes of edge-length ϵ_0, and count the number of these boxes, denoted as $N(\epsilon_0)$. In the case of braiding patterns, the boxes are squares.
(2) Repeat step (1) using boxes with edge-length $\epsilon_1 = \epsilon_0/2$. Then repeat again using $\epsilon_2 = \epsilon_1/2$, $\epsilon_3 = \epsilon_2/2$, and so on. Obviously, for each ϵ_i, there is also the corresponding $N(\epsilon_i)$.
(3) Theoretically, the dimension D is the number for which

$$\lim_{\epsilon \to 0} N(\epsilon) = A\epsilon^{-D} \tag{2}$$

where A is a constant. In practice, D may be estimated as

$$D = \frac{\log N(\epsilon_{i+1})/N(\epsilon_i)}{\log \epsilon_i/\epsilon_{i+1}} \tag{3}$$

The only difficulty is in selecting the value of i, and generally it is selected as large as possible in order to approximate the limit $\epsilon \to 0$. However, for real objects the boxes cannot be infinitely small, like in the present case, since it is inappropriate to make the covering boxes smaller than the size of a computational cell.

3 Results and Discussion

Several simulations have been run, with the discharge varying from 50 m^3/s up to 3000 m^3/s, this last value determining the inundation of all the floodplain and adjacent terraces.

For each discharge value, the simulation has been carried out for a sufficient time in order to establish steady state conditions. Figure 2 shows some results of the hydrodynamic model in terms of water depths obtained, respectively, for the discharge of 50, 300, 500 and 900 m^3/s.

Some level measurements allowed model calibration in the range of medium-to-high discharges, considering that a stage-discharge relationship is available at an intermediate section of the area under analysis. A previous physical model study identified such relationship for a control section where a bridge is present. In this way, Manning's roughness coefficients have been defined in order to reproduce the measured water levels within an acceptable tolerance.

From Fig. 2, it is evident how the number of flooded channels at low flows is quite limited, and the morphology of braiding is not very articulated. For higher discharges, the fluxes subdivide into more channels, and braiding complexity increases accordingly, until it reaches a maximum. The further increase in discharge rapidly determines a situation for which all the floodplain is interested by waters, and braiding complexity decreases.

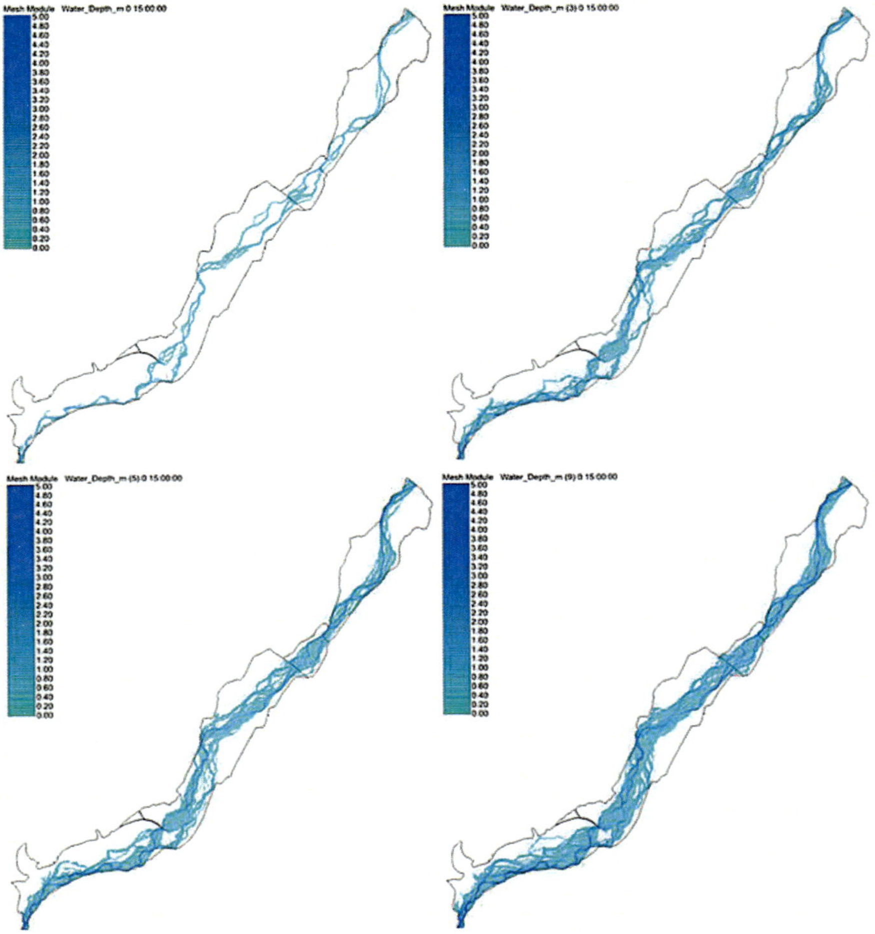

Fig. 2 Maps representing water depths for different discharge values, respectively 50 m^3/s (top left), 300 m^3/s (top right), 500 m^3/s (bottom left) and 900 m^3/s (bottom right)

Such behavior is well described by the fractal dimension depicted in Fig. 3, as a function of river discharge. In this case, the maximum value of nearly 1.29 is obtained for 500 m^3/s, corresponding to the return period of 4 months.

Figure 4 shows water depth maps for the southern zone of the area under study. The phenomenon described above is clearly captured by detailed two-dimensional hydrodynamic modelling.

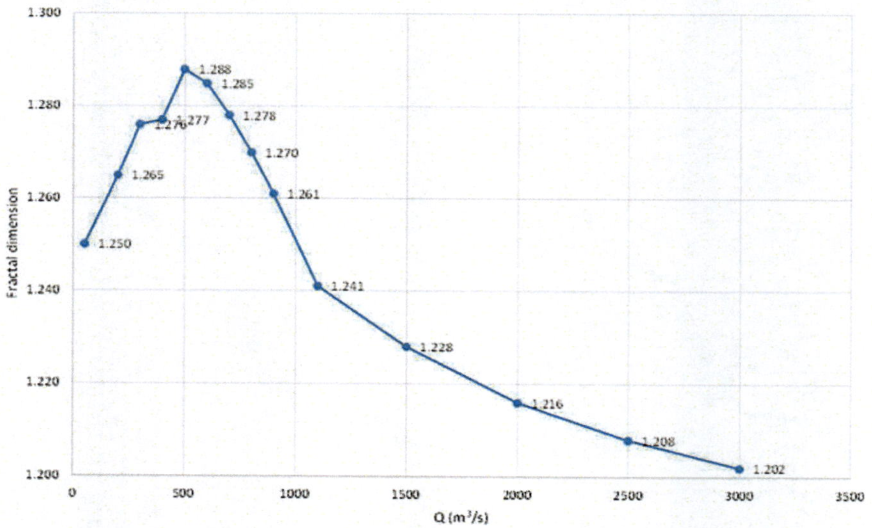

Fig. 3 Fractal dimension calculated through the application of the box-counting algorithm, as a function of varying river discharge

4 Conclusions

The paper has presented a methodology for quantifying the pattern complexity of steady-state planimetric configuration of braided rivers. Starting from the results obtained from detailed two-dimensional hydrodynamic simulations, the box-counting algorithm allows to determine the fractal dimension of the braided pattern.

The procedure is relevant for the a-priori evaluation of river management works, in particular those aimed at excavation, extraction and widening. The maintenance of a range of formative discharges at low-to-medium flows is very important for preserving ecological stability and the possibility of habitat evolution, which could be potentially endangered by failures in the restoration projects.

Fig. 4 Water depths for the southern zone of the computational area for varying discharges: from top to bottom, left to right: 50 m^3/s; 200 m^3/s; 300 m^3/s; 400 m^3/s; 500 m^3/s; 600 m^3/s; 700 m^3/s; 800 m^3/s

Acknowledgements The author wishes to thank Regione Autonoma Friuli Venezia Giulia (www.regione.fvg.it), the Friulian Plain Bureau of Reclamation (Consorzio di bonifica Pianura Friulana, www.bonificafriulana.it) and the International Centre of Mechanical Sciences (www.cism.it) for their support and the provision of materials and data related to River Tagliamento.

References

1. P. Ashmore, Morphology and dynamics of braided rivers, in *Treatise on Geomorphology, Volume 9: Fluvial Geomorphology*. ed. by J. Shroder, E. Wohl (Academic Press, San Diego, CA, 2013), pp. 289–312
2. W. Bertoldi, M. Tubino, Bed and bank evolution of bifurcating channels. Water Resour. Res. 41, 2005
3. W. Bertoldi, L. Zanoni, M. Tubino, Planform dynamics of braided streams. Earth Surf. Proc. Land. **34**, 547–557 (2009)
4. M. Colombini, G. Seminara, M. Tubino, Finite amplitude alternate bars. J. Fluid Mech. **181**, 213–232 (1987)
5. R. Egozi, P. Ashmore, Defining and measuring braiding intensity. Earth Surf. Proc. Land. **33**(14), 2121–2138 (2008)
6. R. Egozi, P. Ashmore, Experimental analysis of braided channel pattern response to increased discharge. J. Geophys. Res. 114, (2009)
7. B. Federici, C. Paola, Dynamics of channel bifurcations in noncohesive sediments. Water Resour. Res. **39**, 6 (2003)
8. A.M. Gurnell, W. Bertoldi, D. Corenblit, Changing river channels: the roles of hydrological processes, plants and pioneer fluvial landforms in humid temperate, mixed load, gravel bed rivers. Earth Sci. Rev. **111**, 129–141 (2012)
9. S. Ikeda, Prediction of alternate bar wavelength and height. J. Hydraul. Eng. **110**, 371–386 (1984)
10. M.N.R. Jaeggi, Formation and effects of alternate bars. J. Hydraul. Eng. **110**, 2 (1984)
11. C.L. Jang, Y. Shimizu, Bed and bank evolution of bifurcating channels. ASCE J. Hydraul. Eng. **131**, 7 (2005)
12. S. Lane. Approaching the system-scale understanding of braided river behavior. In: G.H. Sambrook Smith, J.L. Best, C.S. Bristow, G.E. Petts (eds.), Braided Rivers: Process, Deposits, Ecology and Management. Association of Sedimentologists Special Publication (vol. 36), Blackwell, Oxford, 339–359, (2006)
13. S. Lanzoni, Experiments on bar formation in a straight flume 1. Uniform sediment. Water Resour. Res. **36**(11), 3337–3349 (2000)
14. C. Paola, Modelling stream braiding over a range of scales, in *Gravel Bed Rivers V*. ed. by M.P. Mosley (New Zealand Hydrological Society, Wellington, 2001), pp. 11–46
15. G. Parker, On the cause and characteristic scales of meandering and braiding in rivers. J. Fluid Mech. **76**(3), 457–479 (1976)
16. M. Redolfi, Sediment transport and morphology of braided rivers: steady and unsteady regime. PhD Thesis, University of Trento and Queen Mary University of London, (2014)
17. V.B. Sapozhnikov, E. Foufoula-Georgiou, Self-affinity in braided rivers. Water Resour. Res. **32**(5), 1429–1439 (1996)
18. V.B. Sapozhnikov, E. Foufoula-Georgiou, Validation of braided-stream models: spatial state-space plots, self-affine scaling, and island shapes. Water Resour. Res. **34**(9), 2353–2364 (1998)
19. G. Seminara, M. Tubino, Weakly nonlinear theory for regular meanders. J. Fluid Mech. **244**, 257–288 (1992)
20. K. Tockner, J.V. Ward, D.B. Arscott, P.J. Edwards, J. Kollmann, A.M. Gurnell, G.E. Petts, B. Maiolini, The tagliamento river: a model ecosystem of European importance. Aquat. Sci. **65**, 239–253 (2003)
21. M. Tubino, R. Repetto, G. Zolezzi, Free bars in rivers. J. Hydraul. Res. **37**(6), 759–775 (1999)
22. D.L. Turcotte, Fractal and chaos in geology and geophysics. Cambridge University Press, (1992)
23. J. Walsh, D.M. Hicks, Braided channels: Self-similar or self-affine? Water Resour. Res. 38, (2002)

Simulations on the Peridynamic Equation in Continuum Mechanics

Sabrina Francesca Pellegrino

Abstract The peridynamic equation of motion consists in a second order in time partial integro-differential equation and is largely used in elastodynamics as it is able to model cracks avoiding the spatial partial derivatives. In this paper, we consider the linear model of peridynamics in a one-dimensional spatial domain. We review some numerical techniques to solve this equation and propose some new computational methods of higher order in space. Several numerical tests are given in order to validate our results.

Keywords Peridynamics · Non-local models · Quadrature formula · Trigonometric scheme

1 Introduction

Modeling fractures and damages is one of the major issue in the framework of continuum mechanics. The classical theory uses spatial derivatives to model the motion of a material subject to elastic stresses. So, it is not able to describe discontinuous phenomena like cracks and fractures, as partial derivatives are not defined on discontinuities, and, moreover, it cannot predict where the crack is located. Therefore, there is the need to develop a non-local theory able to use a unique equation both on or off a crack, see [3, 8, 9, 14, 23].

Recent studies show that differential operators of fractional orders may be introduced in order to depict the nature of such phenomena, see for instance [6, 15–20].

The peridynamic theory is a non-local generalization of the elasticity theory introduced by Silling in [34], and has attracted the attention of a growing number of researchers, as it addresses discontinuous problems. He proposed to model the motion of a material body using integro-differential partial equations, without involving spatial derivatives. The main assumption of the theory concerns the pres-

S. F. Pellegrino (✉)
Department of Mathematics, Università degli Studi di Bari Aldo Moro,
Via E. Orabona, 4, 70125 Bari, Italy
e-mail: sabrina.pellegrino@uniba.it

C. H. Skiadas et al. (eds.), *13th Chaotic Modeling and Simulation International Conference*, Springer Proceedings in Complexity,
https://doi.org/10.1007/978-3-030-70795-8_46

ence of an interacting force f between the particle x and the particle $\hat{x}$ belonging to V_x, which represents the peridynamic neighborhood of x. This basic assumption also suggests that peridynamics could be suitable for multiscale material modeling ([24, 32]).

We fix $[0, T]$, for some $T > 0$, as the time domain under consideration, and let $V \subset \mathbb{R}$ be the rest configuration of a material body having mass density $\rho : V \times [0, T] \to \mathbb{R}_+$. Then, the peridynamic equation is given by

$$\rho(x)u_{tt}(x,t) = \int_{V_x} f(\hat{x} - x, u(\hat{x}, t) - u(x,t))d\hat{x} + b(x,t), \quad x \in V, \quad T \in [0, T], \quad (1)$$

usually enriched by the initial conditions

$$u(x, 0) = u_0(x), u_t(x, 0) = v(x), \qquad x \in V, \qquad (2)$$

where u is the displacement field and b describes all the external forces acting on the material body. The integrand f represents the force density that the particle $\hat{x}$ exerts on the particle x and is called **pairwise force function**, see for instance [10, 34]. The interaction between x with all particle in the peridynamic neighborhood V_x is called **bond**.

We set

$$\xi = \hat{x} - x, \qquad \eta = u(\hat{x}, t) - u(x, t),$$

which denotes the relative position of two particles in the reference configuration and the relative displacement, respectively. Thus $\xi + \eta$ represents the current relative position vector, and we can observe that the pairwise force function f has to satisfy Newton's third law and the conservation of the angular momentum:

$$f(-\xi, -\eta) = -f(\xi, \eta), \qquad \eta \times f(\xi, \eta) = 0. \qquad (3)$$

It is reasonable to require the existence of a positive constant δ, called **horizon**, such that there are no interactions among particles having relative distance greater than δ, namely

$$f(\xi, \eta) = 0, \qquad \text{for } |\xi| > \delta \text{ and for every } \eta.$$

In what follows, we restrict our attention to the case of an homogeneous bar of infinite length, and in particular we focus on the **linear peridynamic** model

$$\rho(x)u_{tt}(x,t) = \int_{x-\delta}^{x+\delta} C(\hat{x} - x)\left(u(\hat{x}, t) - u(x,t)\right) d\hat{x} + b(x,t), \qquad x \in \mathbb{R}, \quad t \in [0, T], \qquad (4)$$

where the function C is a non-negative even function, i.e. $C(-\xi) = C(\xi)$, called **micromodulus function**.

In this paper, we survey some numerical techniques for the model and we propose an accurate spatial discretization accompanied to a trigonometric scheme for time integration. Additionally, we extend the methods to the non linear case.

The paper is organized as follows. Section 2 collects the main analytic results for the problem. In Sect. 3 we present some quadrature formula for the space discretization of the model. Section 4 is devoted to the time integration techniques. In Sect. 5 we extend the proposed methods to the non linear case. Section 6 shows the numerical tests, and, finally, Sect. 7 concludes the paper.

2 Analytical Results

In this section we recall the main theoretical results concerning the peridynamic equation. The well-posedness of the non linear model depends on the assumptions on the pairwise force function f, see [6, 10, 11].

Theorem 1 *((see [10])) Assume that* $u_0, v \in \mathcal{C}(\overline{V})$ *and* $b \in \mathcal{C}([0,T]; \mathcal{C}(\overline{V}))$. *If the pairwise force function* $f : \overline{B_\delta(0)} \times \mathbb{R}^d \to \mathbb{R}^d$ *is a continuous function, such that there exists a nonnegative function* $\ell \in L^1(B_\delta(0))$ *such that*

$$|f(\xi,\hat{\eta}) - f(\xi,\eta)| \le \ell(\xi)|\hat{\eta} - \eta|, \quad \textit{for all } \xi \in \mathbb{R}^d \textit{ and } \eta, \hat{\eta} \in \mathbb{R}^d,$$

then, the integral operator in (1) is well-defined and Lipschitz continuous, and the initial-value problem (1)-(2) is globally well-posed with solution $u \in C^2([0,T]; \mathcal{C}(\overline{V}))$.

For a **microelastic** material (see [34]), it is possible to derive the pairwise force function $f(\xi,\eta)$ from a scalar-valued function $w(\xi,\eta)$ called **pairwise potential function** (see [13]), such that

$$f(\xi,\eta) = \nabla_\eta w(\xi,\eta), \tag{5}$$

and the peridynamic equation (1) follows from the variational problem: find

$$u = \arg\min J(u)\,, \qquad J(u) = \int_0^T \int_V e(x,u(x,t),t)dxdt, \tag{6}$$

where $e = e_{kin} - e_{el} - e_{ext}$ is the Lagrangian density, and incorporates the kinetic energy density, the elastic energy density and the density due to the external force density, given respectively by

$$\begin{aligned} e_{kin} &= \frac{1}{2}\rho(x)\, u_t^2(x,t), \\ e_{el} &= \frac{1}{2}\int_V w(\hat{x}-x, u(\hat{x},t) - u(x,t))d\hat{x}, \\ e_{ext} &= -b(x,t)u(x,t). \end{aligned}$$

In particular, in the one-dimensional linear peridynamic model (4), the pairwise force function is given by

$$f(\xi, \eta) = C(\xi)\,\eta, \tag{7}$$

and the potential function is given by

$$w(\xi, \eta) = \frac{1}{2}C(\xi)\eta^2. \tag{8}$$

The following well-posedness result holds.

Theorem 2 *((see [13])) Let u_0, $v \in C^0(\mathbb{R})$ be the initial conditions. If the micro-modulus function C belongs to $C^2(\mathbb{R})$, then, for any $T > 0$, the initial value problem (4)-(2) is well-posed and the unique solution u belongs to $C^2([0, T]; C(\mathbb{R}))$.*

Moreover, if b is autonomous, namely $b(x, t) \equiv b(x)$, we can prove the conservation of the total energy of the system.

Theorem 3 *If the external force b does not depend on time, the total energy associated to (4), given by*

$$E(t) = \frac{1}{2}\int_V \rho\, |u_t(x, t)|^2\, dx + \frac{1}{2}\int_V\int_V w(\hat{x} - x, u(\hat{x}, t) - u(x, t))dx\, d\hat{x}, \tag{9}$$

is preserved, namely

$$\frac{d}{dt}E(t) = 0.$$

Proof. We set $\xi = \hat{x} - x$. Then, using (3) and (1) we have

$$\begin{aligned}
\frac{d}{dt}E(t) &= \int_V \rho\, u_t(x, t)u_{tt}(x, t)\, dx + \frac{1}{2}\int_V\int_V f(\xi, u(x, t) - u(x - \xi, t))u_t(x, t)\, dx\, d\xi \\
&- \frac{1}{2}\int_V\int_V f(\xi, u(x, t) - u(x - \xi, t))u_t(x - \xi, t)\, dx\, d\xi \\
&= \int_V \rho u_t(x, t)u_{tt}(x, t)\, dx + \frac{1}{2}\int_V\int_V f(\xi, u(x, t) - u(x + \xi, t))u_t(x + \xi, t)\, dx\, d\xi \\
&- \frac{1}{2}\int_V\int_V f(\xi, u(x, t) - u(x - \xi, t))u_t(x - \xi, t)\, dx\, d\xi \\
&= \int_V \rho u_t(x, t)u_{tt}(x, t)\, dx + \frac{1}{2}\int_V\int_V f(\xi, u(x, t) - u(x - \xi, t))u_t(x, t)\, dx\, d\xi \\
&+ \frac{1}{2}\int_V\int_V f(\xi, u(x, t) - u(x - \xi, t))u_t(x, t)\, dx\, dy \\
&= \int_V u_t(x, t)\left(\rho u_{tt}(x, t) + \int_V f(\xi, u(x, t) - u(x - \xi, t))\, d\xi\right) dx = 0.
\end{aligned}$$

The following result is related to the case of non autonomous external force.

Theorem 4 *((see [13])) If the external force is not autonomous, then, the Lagrangian density associated to the linear problem (4) satisfies the following inequality*

$$e_{kin}(t) + e_{el}(t) + \nu \int_0^t e^{\nu(t-s)} e_{ext}(s) ds$$
$$\leq e^{\nu t}(e_{kin}(0) + e_{el}(0)) + \frac{1}{2\nu} \int_0^t \int_{-\infty}^{\infty} \frac{e^{\nu(t-s)}}{\rho} |b(x,t)|^2 dx ds,$$

for all $\nu > 0$ *and* $t > 0$.

Additionally, in [6], the authors proved the well-posedness of the nonlinear peridynamic equation assuming very general constitutive assumptions in the framework of fractional Sobolev spaces.

Moreover, we can think to the linear one-dimensional peridynamic equation (4) as a non local version of the classical linear one-dimensional wave equation, (see for instance [4, 12]). Indeed, if we choose $u_0(x) = U \exp[(-x/L)^2]$, $v(x) = 0$, with U and L suitable constants, as initial conditions and the following micromodulus function

$$C(\hat{x} - x) = 4E \exp[-(\hat{x} - x)^2/l^2]/(l^3\sqrt{\pi}), \qquad \hat{x}, x \in \mathbb{R}, \tag{10}$$

where E denotes the Young modulus, and $l > 0$ a length-scale parameter, then for $l \to 0$, (4) becomes the wave equation of the classical elasticity theory:

$$\rho\, u_{tt}(x,t) = E u_{xx}(x,t) + b(x,t), \qquad x \in \mathbb{R},\ t \geq 0, \tag{11}$$

As a consequence, the parameter l can be seen as a degree of non locality.

3 Spatial Discretization of the Peridynamics

In this section we discretize in space the Eq. (4) by means of a quadrature formula.

Let $N > 0$ be an even integer and $\Delta x > 0$ be the spatial step size. We discretize the spatial domain $\mathbb{R}$ by a compact set $[-D, D]$, for some positive large constant D, and such interval by means of the points $x_j = -D + j\Delta x = -D + j\frac{2D}{N}$, for $j = 0, \dots, N$. We consider a quadrature formula of order s on these points:

$$\int_{-\infty}^{\infty} C(\hat{x} - x)(u(\hat{x},t) - u(x,t)) d\hat{x} \approx \Delta x \sum_{j=0}^{N} w_j\, C(x_j - x) \left(u(x_j,t) - u(x,t)\right), \tag{12}$$

where w_j are the weights of the formula.

Then, at the collocation points $x = x_i$ for $i = 0, \dots, N$, we approximate the Eq. (4) by

$$\rho u_{tt}(x_i,t) \approx \Delta x \sum_{j=0}^{N} w_j C(x_j - x_i)(u(x_j,t) - u(x_i,t)) + b(x_i,t), \qquad t \geq 0. \tag{13}$$

We define the **stiffness matrix** $K = (k_{ij})$, for $i, j = 0, \dots, N$ by

$$k_{ij} = \alpha_i \delta_{ij} - w_j C_{ij},$$

where $C_{ij} = C(x_j - x_i)$, $\alpha_i = \sum_{k=0}^{N} w_k C_{ik}$, and δ_{ij} is the Kronecker Delta.

The stiffness matrix K is not symmetric, in general, unless the weights are constant with respect to j, namely $w_j = w$ for every $j = 0, \dots, N$. In this case we obtain the **composite midpoint rule**: we approximate the spatial domain $(-\infty, \infty)$ by the interval $[-(N+1)\Delta x/2, (N+1)\Delta x/2]$ and the points of the discretization x_j^{MR} are taken as the midpoints of the sub-intervals $[-(N+1)\Delta x/2 + j\Delta x, -(N-1)\Delta x/2 + j\Delta x]$, for $j = 0, \dots, N$. For sufficiently smooth assumptions on C and u, this formula is of the second order of accuracy in space and the constant weights are equal to 1, (see for instance [13, 33]).

The **composite Gauss two points formula** is of the fourth order of accuracy and provides a symmetric stiffness matrix K. We fix $M > 0$ and consider a partition of the interval $[-D, D]$ given by the sequence $\tilde{x}_j = -D + j\Delta x$ for $j = 0, \dots, M$, where $\Delta x = 2\,D/M = (\tilde{x}_M - \tilde{x}_0)/M$. Then on each sub-interval $[\tilde{x}_{j-1}, \tilde{x}_j]$, the formula uses two points where the function $\psi(x)$ is evaluated:

$$\int_{\tilde{x}_0}^{\tilde{x}_M} \psi(x)dx \approx \frac{\Delta x}{2} \sum_{j=1}^{M} \left[\psi(m_j^-) + \psi(m_j^+)\right], \tag{14}$$

where

$$m_j = \frac{\tilde{x}_{j-1} + \tilde{x}_j}{2}, \qquad m_j^- = m_j - \frac{\Delta x}{2\sqrt{3}}, \qquad m_j^+ = m_j + \frac{\Delta x}{2\sqrt{3}}, \qquad j = 1, \dots, M.$$

Setting

$$x_j = \begin{cases} m^-_{\frac{j+1}{2}}, & \text{if } j \text{ is even}, \\ m^+_{\frac{j+1}{2}}, & \text{if } j \text{ is odd}, \end{cases} \qquad j = 0, \dots, N,$$

with $N = 2M - 1$, then we can rewrite the quadrature formula (14) in the following way

$$\int_{x_0}^{x_M} \psi(x)dx \approx \frac{\Delta x}{2} \sum_{j=1}^{M} \left[\psi(m_j^-) + \psi(m_j^+)\right] = \frac{\Delta x}{2} \sum_{j=0}^{N} \psi(x_j),$$

so, in this case the constant weights w_j are equal to $\frac{1}{2}$, for $j = 0, 1, \dots, N$.

Remark 1 Using the composite midpoint rule, or the composite Gauss two points formula, the stiffness matrix K is a positive and semi-definite with non-negative eigenvalues. In general K is not sparse because of the infinite horizon, however, in case of finite horizon $\delta > 0$ (see [5, 33]), that is $C(x - \hat{x}) = 0$, when $|\hat{x} - x| > \delta$, then K has a banded structure, and, in particular, the size of the band r depends on the horizon δ and on the space step Δx and is given by $r = \lfloor \delta / h \rfloor$.

3.1 *The Semidiscretized Problem*

Let x_j be the spatial nodes for $j = 0, \ldots, N$ and

$$U(t) = [U_0(t), U_1(t), \ldots, U_N(t)],$$

be an approximation of the solution, where $U_j(t) \approx u(x_j, t)$ for $j = 0, \ldots, N$.

We set

$$B(t) = \frac{1}{\rho}[b(x_0, t), \ldots, b(x_N, t)]^T.$$

Then, we can approximate the peridynamic equation (4) by the following second order differential system

$$U''(t) + \Omega^2\, U(t) = B(t), \tag{15}$$

with $\Omega^2 = \dfrac{\Delta x}{\rho} K$ (or $\Omega^2 = \dfrac{\Delta x\, w}{\rho} K'$, for K' depending only on the micromodulus function C), where K is a positive semi-definite matrix, and with the initial conditions

$$U_0 = [u_0(x_0), \ldots, u_0(x_N)]^T \quad \text{and} \quad V_0 = [v(x_0), \ldots, v(x_N)]^T.$$

The system (15) is equivalent to the following first order differential system

$$\begin{pmatrix} U' \\ V' \end{pmatrix} = \begin{pmatrix} 0 & I \\ -\Omega^2 & 0 \end{pmatrix} \begin{pmatrix} U \\ V \end{pmatrix} + \begin{pmatrix} 0 \\ B(t) \end{pmatrix}, \tag{16}$$

where $V = U'$, with the initial conditions U_0 and V_0. Therefore, we can write the exact solution of (16) as

$$\begin{pmatrix} U(t) \\ V(t) \end{pmatrix} = \exp(tA) \begin{pmatrix} U_0 \\ V_0 \end{pmatrix} + \int_0^t \exp[(t-s)A] \begin{pmatrix} 0 \\ B(s) \end{pmatrix} ds, \tag{17}$$

with $A = \begin{pmatrix} 0 & I \\ -\Omega^2 & 0 \end{pmatrix}$.

Remark 2 In order to avoid computational problems, particularly, when we will consider trigonometric schemes where the square root Ω of Ω^2 is required or the inverse of Ω is necessary, we can regularize the matrix Ω^2 by adding a diagonal matrix of the form $(\Delta x)^s\ I$, where s is the order of accuracy of the quadrature formula used. In this way, the matrix Ω^2 is symmetric and positive definite, so it admits a unique symmetric and definite-positive square root Ω.

4 Time Discretization of the Peridynamics

In this section we consider the time discretization of the semidiscretized system (16) obtained by applying a quadrature formula to the original problem. Let $\Delta t > 0$ be the time step and $t_n = n\Delta t$ be the partition of the time interval $[0, T]$, for $n = 0, \ldots, N_T$, where $N_T = \lfloor \frac{T}{\Delta t} \rfloor$, and let $U_n \approx U(t_n)$ and $V_n \approx U'(t_n)$.

We describe standard time discretization schemes, such as the Störmer-Verlet scheme, the implicit midpoint method, and a new procedure based on a **trigonometric** approach.

4.1 Störmer-Verlet Scheme

This is a symplectic, second order in time, explicit scheme:

$$\begin{cases} V_{n+\frac{1}{2}} = V_n + \frac{\Delta t}{2}[-\Omega^2 U_n + B(t_n)], \\ U_{n+1} = U_n + \Delta t\ V_{n+\frac{1}{2}}, \\ V_{n+1} = V_{n+\frac{1}{2}} + \frac{\Delta t}{2}[-\Omega^2 U_{n+1} + B(t_{n+1})]. \end{cases} \tag{18}$$

Because of its geometric properties, Störmer-Verlet method is widely used in the context of partial differential equations of wave propagation or peridynamic problems, (see [7, 25, 36]).

In [30], the authors perform the von Neumann analysis to study the stability of the Störmer-Verlet scheme.

Theorem 5 *((see [7])) Let $\Delta x, \Delta t > 0$ be the space and the time steps, respectively, and let $N > 0$, even, be the points' number used to discretize in space the linear model (4). If*

$$\Delta t < \sqrt{\frac{\rho}{\Delta x \sum_{j=-N/2}^{N/2} C_{ij}}}, \tag{19}$$

where $C_{ij} = C(x_i - x_j)$, then the Störmer-Verlet method (18) is numerically stable.

4.2 Implicit Midpoint Scheme

This is a symplectic implicit second order scheme:

$$\begin{cases} U_{n+1} = U_n + \frac{\Delta t}{2}(V_{n+1} + V_n), \\ V_{n+1} = V_n + \frac{\Delta t}{2}[-\Omega^2(U_{n+1} + U_n) + (B(t_n) + B(t_{n+1}))]. \end{cases} \tag{20}$$

Since this scheme is implicit, it allows us to consider larger time step values, and, as a consequence, it is linearly unconditionally stable.

4.3 Trigonometric Schemes

Thanks to the Variation of Constants formula, the solution in (17) can be rewritten as

$$\begin{cases} U(t) = \cos(t\Omega)U_0 + t\,\mathrm{sinc}(t\Omega)V_0 + \int_0^t (t-s)\mathrm{sinc}((t-s)\Omega)B(s)ds, \\ V(t) = -\Omega\sin(t\Omega)U_0 + \cos(t\Omega)V_0 + \int_0^t \cos((t-s)\Omega)B(s)ds, \end{cases} \tag{21}$$

where Ω is the unique positive definite square root of Ω^2, see Remark 2, and $\mathrm{sinc}(x) = \frac{\sin x}{x}$.

Therefore, applying a discretization of the Variation of Constants formula to the system (21), we find the following explicit numerical procedure

$$\begin{cases} U_{n+1} = \cos(\tau\Omega)U_n + \tau\,\mathrm{sinc}(\tau\Omega)V_n + \int_0^\tau (\tau-s)\,\mathrm{sinc}((\tau-s)\Omega)B(t_n+s)ds, \\ V_{n+1} = -\Omega\sin(\tau\Omega)U_n + \cos(\tau\Omega)V_n + \int_0^\tau \cos((\tau-s)\Omega)B(t_n+s)ds, \end{cases} \tag{22}$$

enriched by the initial conditions U_0 and V_0.

The scheme (22) requires the computation of the matrix functions $\cos(\tau\Omega)$ and $\mathrm{sinc}(\tau\Omega)$. The evaluation of $\cos(\tau\Omega)$ can be done by using a MATLAB routine, while, the computation of the $\mathrm{sinc}(\tau\Omega)$ matrix function is more delicate. A way to overcome this difficulty is to employ the series expression for $\mathrm{sinc}(\tau\Omega)$ but this could be too expensive or inaccurate [22]. So, one can try first to diagonalize the matrix function. Moreover, the computation of products of functions of matrices by vectors could be efficiently done by means of Krylov subspace methods, (see for instance [21, 26, 27]).

In [7], the authors show that the trigonometric method is unconditionally stable.

5 The One-Dimensional Non Linear Peridynamic Model

We propose a numerical approach to study the one-dimensional non linear model (1) for an homogeneous bar of infinite length. This approach allows us to extend to the non linear case the numerical methods proposed in the previous sections.

For an **isotropic** material, the general form of the pairwise force function is given by

$$f(\xi, \eta) = c\, s(|\xi|, |\eta|)\, \frac{\eta}{|\eta|}, \tag{23}$$

where c is a positive constant depending on the material and the horizon. The function

$$s(|\xi|, |\eta|) = \frac{|\eta| - |\xi|}{|\xi|},$$

describes the relative change of the Euclidean distance of the particles. Since f is discontinuous in the first argument, the order of accuracy of the implemented numerical schemes will reduce.

In order to apply the results of the previous sections, we assume that $|\eta| << 1$ and $f(\xi, \eta)$ is sufficiently smooth. We consider the integral form of the function $f(\xi, \cdot)$:

$$f(\xi, \eta) = f(\xi, 0) + \int_0^{\eta} \frac{\partial f(\xi, s)}{\partial \eta}(\eta - s)ds,$$

and then we apply an accurate quadrature formula

$$f(\xi, \eta) \approx f(\xi, 0) + \sum_{r=1}^{m} w_r \frac{\partial f(\xi, s_r)}{\partial \eta}(\eta - s_r),$$

where w_r are the weights while s_r are the nodes of this formula. In general this approach leads to implicit methods, in fact, if we use the trapezoidal formula

$$f(\xi, \eta) \approx f(\xi, 0) + \frac{\eta}{2}\left[\frac{\partial f(\xi, 0)}{\partial \eta} + \frac{\partial f(\xi, \eta)}{\partial \eta}\right], \tag{24}$$

we derive a second order implicit method. Instead, if $f(\xi, \eta)$ is sufficiently smooth, we can derive an explicit scheme by using a Taylor expansion

$$f(\xi, \eta) \approx f(\xi, 0) + C_1(\xi)\eta + \ldots + C_s(\xi)\eta^s, \tag{25}$$

where

$$C_i(\xi) = \frac{\partial^i f(\xi, 0)}{\partial \eta^i}, \qquad i = 1, \ldots, s.$$

6 Simulations

In this section we present some numerical tests to confirm our results. We start with the linear model (4), assuming that the material body is not subject to external forces, namely $b(x, t) = 0$. For simplicity, we consider the case of a constant density $\rho(x) = 1$. We take (10) as micromodulus function and we choose $u_0(x) = e^{-(x/L)^2}$ $x \in \mathbb{R}$ and $v = 0$ as initial conditions.

The choice of this micromodulus function is justified by the fact that its decay at infinity makes possible to consider a bounded computation domain. Moreover, in this setting the exact solution for (4) is given by

$$u^*(x,t) = \frac{2}{\sqrt{\pi}} \int_0^\infty \exp(-s^2) \cos(2sx) \cos\left(2t\sqrt{1-\exp(-s^2)}\right) ds, \qquad (26)$$

see for instance [35].

We denote by $\mathbf{u}^*(t) = (u^*(x_0,t), ..., u^*(x_N,t))^T$ the exact solution vector at time t and at the points of the spatial discretized domain.

In order to perform an error study and to show the orders of accuracy of the decribed methods, we define $\mathbf{e}_k$ as

$$\mathbf{e}_k = \|\mathbf{u}(t_k) - \mathbf{u}^*(t_k)\|_\infty := \max\left\{|u(x_i,t_k) - u^*(x_i,t_k)| : i = 0,\dots,N,\right\},$$

and then, for each method, we take the maximum error in the time interval $[0, T]$, namely

$$||\mathbf{e}||_\infty := \max\{\mathbf{e}_k : k = 1,\dots,N_T\}.$$

We denote by MT, MSV, MMI and GT the methods consisting of the Midpoint+Trigonometric method, the Midpoint+Störmer-Verlet method, the Midpoint+Implicit Midpoint method and the Gauss two points+Trigonometric method, respectively.

6.1 Test 1: Comparison Between MT, MSV, MMI and GT Methods

In this section we study the performance of the MT, MSV, MMI and GT methods, by varying the time and space steps, by computing the error between the exact and the numerical solution and studying the rate of convergence.

Figure 1 shows the numerical solution computed by MSV method, while Table 1 summarizes the errors of the different methods by varying the spatial and time discretization steps. The term R_n denotes the ratio between the errors corresponding to Δx and $\Delta x/2$, thus, $\log_2(R_n)$ represents the convergence order of the methods.

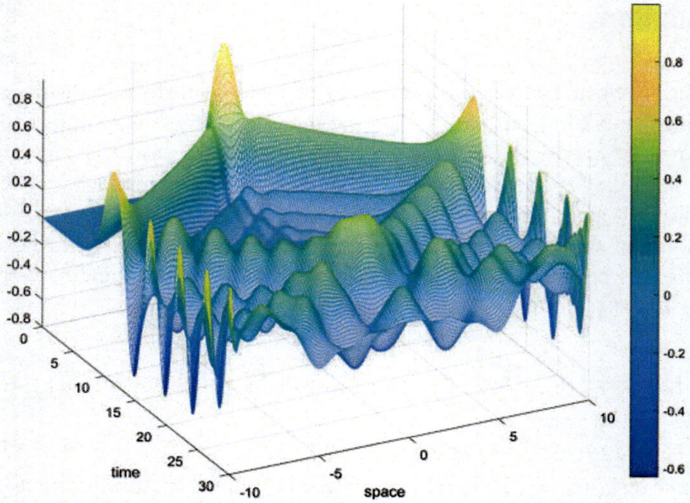

Fig. 1 With reference to Test 1: the numerical solution obtained by the MSV method. The parameters for the simulations are $\Delta x = \Delta t = 0.1$, $N = 200$, $N_T = 300$

Table 1 With reference to Test 1: the comparison among MSV, MT, MMI and GT methods by varying Δx, Δt, N and N_T

Methods	$\Delta x = \Delta t$	N	N_T	$\|\|\mathbf{e}\|\|_\infty$	$\log_2 (R_n)$
MSV	0.100	200	30	1.2911×10^{-3}	–
	0.050	400	60	3.2340×10^{-4}	1.9971
	0.025	800	120	8.0821×10^{-5}	2.0004
MT	0.100	200	30	5.9276×10^{-3}	–
	0.050	400	60	1.1126×10^{-3}	2.3959
	0.025	800	120	2.1350×10^{-4}	2.3992
MMI	0.100	200	30	2.5754×10^{-3}	–
	0.050	400	60	6.4621×10^{-4}	1.9946
	0.025	800	120	1.6106×10^{-4}	2.0043
GT	0.100	400	30	1.4940×10^{-4}	–
	0.050	800	60	9.3380×10^{-6}	3.9998
	0.025	1600	120	5.8300×10^{-7}	4.0015

The last column of Table 1 confirms that the methods MSV, MT, MMI are of the second order of accuracy while GT is of the fourth order. The method MSV is computationally less expensive than the others, but it has a bounded stability region, see Table 2.

Table 2 With reference to Test 1: the maximum error for the methods MSV, MT and MMI for different choices of Δx, Δt, N and N_T

Methods	Δx	Δt	N	N_T	$\|\|\mathbf{e}\|\|_\infty$
MSV	0.100	0.100	200	300	1.0543
	0.050	0.200	400	150	2.6300×10^{168}
	0.025	0.400	800	75	4.3600×10^{131}
MT	0.100	0.100	200	300	1.0941
	0.050	0.200	400	150	1.1081
	0.025	0.400	800	75	1.2987
MMI	0.100	0.100	200	300	1.0923
	0.050	0.200	400	150	1.0925
	0.025	0.400	800	75	8.2060×10^{-1}

6.2 *Test 2: Comparison Between MSV and MMI in the Nonlinear Case*

Now we focus on the non linear case. In particular, we assume that f has the following form

$$f(\xi, \eta) = \begin{cases} c\frac{|\xi+\eta|-|\xi|}{|\xi|}\frac{\xi+\eta}{|\xi+\eta|}, & \text{if } 0 < |\xi| \leq \delta, \\ 0, & \text{if } |\xi| > \delta, \end{cases} \qquad c > 0,$$

which has a singularity in $\xi = 0$. One can find an exact solution for this problem in [29].

Table 3 depicts the maximum errors by varying the spatial and time discretization steps. We can see how all methods reduce their order of accuracy to 1. The reason of such reduction relies on the singularity of the pairwise force function f.

Table 3 With reference to Test 2: the comparison among the performance of MSV and MMI methods in the nonlinear case by varying Δx, Δt, N and N_T

Methods	Δx	Δt	N	N_T	$\|\|\mathbf{e}\|\|_\infty$	$\log_2(R_n)$
MSV	0.1000	0.0100	10	1000	5.4590×10^{-2}	–
	0.0500	0.0050	20	2000	2.7285×10^{-2}	1.0007
	0.0250	0.0025	40	4000	1.3605×10^{-2}	1.0007
MMI	0.1000	0.0100	10	1000	5.3895×10^{-2}	–
	0.0500	0.0050	20	2000	2.7281×10^{-2}	0.9819
	0.0250	0.0025	40	4000	1.3603×10^{-2}	1.0036

7 Conclusions and Perspectives

In this paper, we have reviewed numerical spatial discretization of higher order together with time integration techniques applied to a linear peridynamic model. Moreover, we have extended such techniques to the nonlinear model.

In future we would apply spectral techniques to both the linear and the nonlinear model following the results obtained in [7, 25] and we will extend the results to space domains of dimension greater than 1, using finite element or volume methods or mimetic finite difference methods, see [1, 2, 28, 31].

Acknowledgements This paper has been partially supported by PRIN2017. The author is member of the Gruppo Nazionale per l'Analisi Matematica, la Probabilità e le loro Applicazioni (GNAMPA) of the Istituto Nazionale di Alta Matematica (INdAM).

References

1. L. Beirao Da Veiga, L. Lopez, V. Vacca, Mimetic finite difference methods for Hamiltonian wave equations in 2D. Comput. Math. Appl. **74**(5), 1123–1141 (2017)
2. B. Berardi, F. Difonzo, L. Lopez, A mixed MoL-TMoL for the numerical solution of the 2D Richards' equation in layered soils. Comput. Math. Appl. **79**(7), 1990–2001 (2020)
3. M. Berardi, F. Difonzo, M. Vurro, L. Lopez, The 1D Richards' equation in two layered soils: a Filippov approach to treat discontinuities. Adv. Water Resour. **115**, 264–272 (2018)
4. H.R. Beyer, B. Aksoylu, F. Celiker, On a class of nonlocal wave equations from applications. J. Math. Phys. **57**, 062902 (2016)
5. F. Bobaru, M. Yang, S. Alves, F. Silling, E. Askari, J. Xu, Convergence, adaptive refinement, and slaning in 1D peridynamics. Int. J. Numer. Mech. Eng. **77**:852–877 (2009)
6. G.M. Coclite, S. Dipierro, F. Maddalena, E. Valdinoci, Wellposedness of a nonlinear peridynamic model. Nonlinearity (to appear)
7. G.M. Coclite, A. Fanizzi, L. Lopez, F. Maddalena, S.F. Pellegrino, Numerical methods for the nonlocal wave equation of the peridynamics. Appl. Numer. Math. **155**, 119–139 (2020)
8. G.M. Coclite, F. Paparella, S.F. Pellegrino, On a salt fingers model. Nonlinear Anal. **176**, 100–116 (2018)
9. E. Dal Santo, C. Donadello, S. F. Pellegrino, M.D. Rosini, Representation of capacity drop at a road merge via point constraints in a first order traffic model. ESAIM: M2AN **53**(1):1–34 (2019)
10. E. Emmrich, D. Puhst, Well-posedness of the peridynamic model with Lipschitz continuous pairwise force function. Commun. Math. Sci. **11**(4), 1039–1049 (2013)
11. E. Emmrich, D. Puhst, Survey of existence results in nonlinear peridynamics in comparison with local elastodynamics. Comput. Methods Appl. Math. **15**(4), 483–496 (2015)
12. E. Emmrich, O. Weckner, Numerical simulation of the dynamics of a nonlocal, inhomogeneous, infinite bar. J. Comput. Appl. Mech. **6**(2), 311–319 (2005)
13. E. Emmrich, O. Weckner, The peridynamic equations and its spatial discretization. Math. Model. Anal. **12**(1), 17–27 (2007)
14. A.C. Eringen, D.G.B. Edelen, On nonlocal elasticity. Int. J. Eng. Sci. **10**(3), 233–248 (1972)
15. R. Garrappa, Exponential integrators for time-fractional partial differential equations. Eur. Phys. J. Special Topics **222**, 1915–1927 (2013)
16. R. Garrappa, On some generalizations of the implicit Euler method for discontinuous fractional differential equations. Math. Comput. Simul. **95**, 213–228 (2014)

17. R. Garrappa, Numerical evaluation of two and three parameter Mittag-Leffler ffunctions. SIAM J. Numer. Anal. **53**(3), 1350–1369 (2015)
18. R. Garrappa, Trapezoidal methods for fractional differential equations: theoretical and computational aspects. Math. Comput. Simul. **110**, 96–112 (2015)
19. R. Garrappa, I. Moret, M. Popolizio. Solving the time-fractional Schrödinger equation by Krylov projection methods. J. Comput. Phys. **293**:115–134 (2015)
20. R. Garrappa, M. Popolizio, Computing the matrix Mittag-Leffler function with applications to fractional calculus. J. Sci. Comput. **77**, 129–153 (2018)
21. V. Grimm, M. Hochbruck, Rational approximation to trigonometric operators. BIT **48**(2), 215–229 (2008)
22. Nicholas J. Higham, Matthew I. Smith, Computing the matrix cosine. Numer. Algorithms **34**(1), 13–26 (2003)
23. E. Kröner, Elasticity theory of materials with long range cohesive forces. Int. J. Solids Struct. **3**, 731–742 (1967)
24. R. Lipton, Dynamic brittle fracture as a small horizon limit of peridynamics. J. Elast. **117**, 21–50 (2014)
25. L. Lopez, S.F. Pellegrino, A spectral method with volume penlization for a nonlinear peridynamic model. Int. J. Numer Methods Eng. **122**(3), 707–725. https://doi.org/10.1002/nme.6555
26. L. Lopez, V. Simoncini, Analysis of projection methods for rational function approximation to the matrix exponential. SIAM J. Numer. Anal. **44**(2), 613–635 (2006)
27. L. Lopez, V. Simoncini, Preserving geometric properties of the exponential matrix by block Krylov subspace methods. BIT Numer. Math. **46**(4), 813–830 (2006). Dec
28. L. Lopez, V. Vacca, Spectral properties and conservation laws in mimetic finite difference methods for PDEs. J. Comput. Appl. Math. **292**(15), 760–784 (2016)
29. E. Madenci, E. Oterkus, *Peridynamic Theory and Its Applications* (Springer, New York, 2013)
30. K.W. Morton, D.F. Mayers, *Numerical Solution of Partial Differential Equations* (Cambridge University Press, Cambridge, 1994)
31. S.F. Pellegrino, On the implementation of a finite volumes scheme with monotone transmission conditions for scalar conservation laws on a star-shaped network. Appl. Numer. Math. **155**, 181–191 (2020)
32. D. Qiang, T. Yunzhe, T. Xiaochuan, A peridynamic model of fracture mechanics with bond-breaking. J. Elasticity (2017)
33. S. Silling, E. Askari, A meshfree based on the peridynamic model of solid mechanics. Comput. Struct. **83**(17–18), 1526–1535 (2005)
34. S.A. Silling, Reformulation of elasticity theory for discontinuities and long-range forces. J. Mech. Phys. Solids **48**(17–18), 175–209 (2000)
35. O. Weckner, R. Abeyaratne, The effect of long-range forces on the dynamics of a bar. J. Mech. Phys. Solids **53**(3), 705–728 (2005)
36. M. Zaccariotto, T. Mudric, D. Tomasi, A. Shojaei, U. Galvanetto, Coupling of FEM meshes with Peridynamic grids. Comput. Methods Appl. Mech. Eng. **330**, 471–497 (2018)

Mechanism of Formation for Fluctuation Phenomena

E. Prozorova

Abstract The paper proposes a deterministic mechanism for the formation of Brownian motion. The movement of molecules, even under equilibrium conditions, leads to a constant displacement of the center of inertia, which creates a moment and leads to the emergence of additional force. The result is a possible accumulation of molecules in separate areas. Having arisen as a result of collisions of slower molecules than average ones, they are forced to move together for some time, moellely no changing their position. The action of the moment manifests itself in all known processes. In the kinetic theory based on the Lagrange and Liouville equations, the motion of the axis of inertia in the process of rearrangement and motion of particles is neglected. The concept of a derivative in terms of finite values of such quantities as the mean free path, time between collisions, etc. has features. At small mean free paths, the motion of the center of inertia contributes to the equation of state. A new algorithm is proposed for calculating the force entering the Langevin equation and the equation of state for a liquid. The listed issues are.

Keywords Conference · CHAOS · Chaotic modeling · CMSIM style

1 Introduction

Fluctuations are called random deviations of physical systems from their equilibrium state (or physical processes—from their steady flow). Fluctuations exist both in no equilibrium states and in unsteady processes; in their absence, relaxation would be a "smooth" process and they could be described by single-valued functions of time. The presence of thermal fluctuations causes random deviations of real processes from such a "smooth" flow [1–7]. Diffusion and Brownian motion occur due to the chaotic thermal motion of molecules, and as a result are described by similar mathematical rules.

E. Prozorova (✉)
Mathematical-Mechanical Department St. Petersburg State University, Av. 28, Peterhof 198504, Russia
e-mail: e.prozorova@spbu.ru

C. H. Skiadas et al. (eds.), *13th Chaotic Modeling and Simulation International Conference*, Springer Proceedings in Complexity,
https://doi.org/10.1007/978-3-030-70795-8_47

The difference between them is that during diffusion, a molecule always moves in a straight line until it collides with another molecule, after which it changes its trajectory. A Brownian particle does not “fly free”, but undergoes very small and frequent, as it were, “tremors”, as a result of which it randomly moves here and there. Fluctuation effect was first explained by A. Einstein and then by M. Smolukhovsky. The theoretical studies of M. Smoluchowski, which differed from the works of A. Einstein only by a slightly less rigor, but greater clarity. It consisted in the fact that the diffusion force should be equal to the viscous value Stokes drag force.

$$\frac{n_p F_s}{c_s} = D_E \frac{\partial n_p}{\partial x}$$

Here F_s force, n_p—particle concentration,c_s—mass concentration, $D_e = \frac{kT}{(6\pi\eta_2 R_{0)}}$, R_0—- particle radius, η_2—viscosity. Later the theory was developed on the basis of the Langevin and Fokker–Planck equations. The evolution of a Brownian particle (fluctuation) is determined by its interaction with the environment, which is always collective. In the kinetic representation, the evolution of a system of Brownian particles is described by a nonlocal equation for the n-particle distribution function. Now the Langevin and Fokker–Planck equations are obtained from the Liouville equation for specially selected models of integral kernels using of phenomenological conservation laws [6, 7]. The Langevin and Fokker – Planck equations refer to the stochastic approximation of particle motion. Since the equation contains a resistance force arising during the translational individual movement of particles, then the speed of movement of the particle in time τ will change. The visible part of the particle motion will depend on the characteristics of the instruments. In addition, collective effects will play a major role. Like the Liouville equation, the Langevin equation does not take into account particle collisions. The possibilities of calculating the motion of particles by these equations are limited.

In the general case, the connection between macroscopically observable quantities and fluctuations of the corresponding dynamical variables is established by solving the dynamic Liouville equation. However, the equation does not take into account possible dissipative processes associated with particle collisions. At the macrolevel, equations are more consistently obtained for the distribution function (the Fokker–Planck and Boltzmann equations).

In the kinetic theory based on the Lagrange and Liouville equations, the motion of the axis of inertia in the process of rearrangement and motion of particles is neglected. This means that the contribution of the angular momentum (force), which leads to collective processes, is not taken into account [8–12] The influence of the moment is confirmed by the work [13], which is devoted to the calculation of the conductivity of a nonideal fully ionized plasma under the assumption that there are no straight sections of the electron trajectory. The experimental data turned out to be possiblesatisfy for the velocity correlator only with the involvement of the angular momentum.

The exact molecular theory, which gives results that are in satisfactory agreement with experiment, can be applied only in the special case when the potential corresponding to the force arising from the interaction of neighboring molecules depends only on the distance between them. But this assumption is valid only for liquids consisting of monoatomic molecules.

In recent years, the molecular modeling method has been widely used [7, 14, 15]. The main version of modern theory is random disturbances. The theory of Brownian motion is approximate. And although in most practically important noted that cases the existing theory gives satisfactory results, in some cases it may require refinement. It should be the classical Boltzmann equation does not comply with the law of conservation of angular momentum as and another equations of kinetic theory and statistic mechanics. This is clearly seen if we multiply equation of speed on radius-vector of the particle to get angular momentum. Even with central interaction we get different values for the non-equilibrium conditions. In numerical calculations by the difference scheme using the grid pitch that is smaller the mean free path and with the ideology of a closed volume not obtain the influence of the angular momentum due to the absence of collisions. A mechanism for the occurrence of fluctuations has not been proposed. The work is limited to the case of "simple" media, by which we mean gases and liquids, consisting of point molecules that do not contain internal degrees of freedom. Now we suggest the reason of fluctuation effect.

2 Lagrangian Function for the Collective Interaction

In classical mechanics, kinetic theory and statistical mechanics, the role of the angular momentum and, therefore, the moment of force is underestimated. Let us consider three interacted among themselves (Fig. 1) particles

$$\boldsymbol{r}_c = \frac{\boldsymbol{m}_1\boldsymbol{r}_1 + \boldsymbol{m}_2\boldsymbol{r}_2 + \boldsymbol{m}_3\boldsymbol{r}_3}{\boldsymbol{m}_1 + \boldsymbol{m}_2 + \boldsymbol{m}_3}$$

$$r_{c+\Delta c} = \frac{m_1(r_1 + \dot{r}_1\Delta t) + m_2(r_2 + \dot{r}_2\Delta t) + m_{3(}r_3 + \dot{r}_3\Delta t)}{m_1 + m_2 + m_3}$$

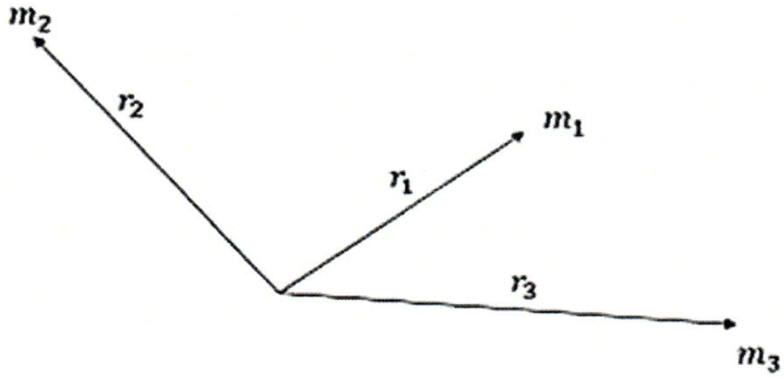

Fig. 1 Interaction of three parts

At the next moment in time, the position will change under the action of the force arising in connection with the new position of the center of inertia. Thus, the new position of the molecules will create a new force. The same result can be obtained by counting the angular momentum, from which to determine the effective force and speed of the center of inertia.

At equilibrium, or at small strains, but under no equilibrium thermodynamic effects and perturbations lead to an uneven distribution of the physical parameters and the role of collective effects, that determined by the growing influence of the angular momentum. In addition, when these strains change position of the center of mass of elementary volume, that is sign for changing Lagrangian function.

$\frac{dL}{dt} = \sum_i \left[\frac{\partial L}{\partial q_i}\dot{q}_i + \frac{\partial L}{\partial \dot{q}_i}\ddot{q}_i\right] + \sum_i \left[\frac{\partial L}{\partial (q_i - a)}(\dot{q}_i - \dot{a}) + \frac{\partial L}{\partial (\dot{q}_i - \dot{a})}(\ddot{q}_i - \ddot{a})\right]$,

$\boldsymbol{a} = \sum_i \frac{m_i \boldsymbol{r}_i}{m_i}$, for electrical interaction $\boldsymbol{a} = \sum_i \frac{e_i \boldsymbol{r}_i}{e_i}$. q_i—generalized coordinate, $\dot{q}_i$ –generalized speed. The classical Liouville equation.

$\frac{d}{dt}\frac{\partial L}{\partial \dot{q}_i} - \frac{\partial L}{\partial q_i} = 0 (i = 1, 2 \ldots)$, $L = L(q_i, \dot{q}_i, t)$ – Lagrange function. The whole theory is developed for a force of the form $\mathbf{F} = -\frac{\partial \mathrm{U}}{\partial \mathbf{R}}$, $U = U(\boldsymbol{R})$.$\boldsymbol{R}$- radius vector.

In view of the time we are invited to consider force formula.

$\boldsymbol{F} = \boldsymbol{F}_0 + \nabla\left((\boldsymbol{R} - \boldsymbol{a}) \times \frac{\partial U}{\partial \boldsymbol{R}}\right)$, $\boldsymbol{R}$ – the current radius. This formula is transformed with the permutability derivatives and directions of forces in the formula

$$\boldsymbol{F} = \boldsymbol{F}_0 + \nabla\left((\boldsymbol{R} - \boldsymbol{a}) \cdot \frac{\partial \mathrm{U}}{\partial \boldsymbol{R}}\right).$$

For the N partial distribution function, it was [7]

$\frac{\partial F_N}{\partial t} + \sum_{i=1}^{N} \dot{\xi}_i \cdot \frac{\partial F_N}{\partial x_i} + \frac{1}{m}\sum_{i=1}^{N} \frac{\partial}{\partial \xi_i} \cdot (X_i F_N) = 0$. Now.

$\frac{\partial F_N}{\partial t} + \frac{\partial L}{\partial a}\dot{\boldsymbol{a}} + \sum_{i=1}^{N} (\dot{\boldsymbol{\xi}}_i - \dot{\boldsymbol{a}}) \cdot \frac{\partial F_N}{\partial (x_i - \boldsymbol{a})} + \frac{1}{m}\sum_{i=1}^{N} \frac{\partial}{\partial (\xi_i - \dot{\boldsymbol{a}})} \cdot (X_i F_N) = 0.$

Usually, however, such as the Hamiltonian system of two interacting molecules after separation of the center of mass is represented as the sum of the Hamiltonians of isolated molecules $H_0 = H_A + H_B$ operator and their electrostatic interaction [16]

$$H = H_0 + H_B$$

$$H_B = -\sum_{a=1}^{n_A}\sum_{j=1}^{N_B}\frac{Z_a}{r_{aj}} - \sum_{b=1}^{n_B}\sum_{j=1}^{N_B}\frac{Z_b}{r_{bj}} + \sum_{i=1}^{N_A}\sum_{j=1}^{N_B}\frac{1}{r_{ij}} + \sum_{a=1}^{n_A}\sum_{b=1}^{n_B}\frac{Z_a Z_b}{R_{ab}},$$

where the indices A, B numbered core indices i, j—the electrons of molecules A, B, respectively, the atomic units. In the rarefied gas unusual situation arises when for describing the derivatives we use the limit of the ratio of the increment function to the increment argument. It turns out that for recording the time derivative of the final terms. We have mean free path (rarefied gas) taking into account only the high-speed components, as slow collisions do not have time to occur.

It is interesting to compare the derivatives for discrete and continuous descriptions. First, consider the relaxation process (in time).

Here and then $\boldsymbol{\xi}_i$ is the velocity of the molecule, u is the velocity of the elementary volume $\boldsymbol{p}_i = \boldsymbol{\xi}_i - \boldsymbol{u}$, is the intrinsic velocity of the molecule, r is the coordinate. The velocity $\boldsymbol{\xi}_i$ of molecules is included in the definition of the Boltzmann equation and in the calculations for the models used as an independent variable.

$f = f(t, \boldsymbol{r}(t), \boldsymbol{\xi}(t))$. We represent the distribution function as.

$f = \frac{\sum_{i=1}^{n} \delta(\boldsymbol{r}_i - \boldsymbol{r})}{\sum_{i=1}^{N} \delta(\boldsymbol{r}_i - \boldsymbol{r})}$, that is.

$f = \frac{n}{N}$, where n is the number of molecules in an elementary volume, N is the number of molecules in a perturbed volume. We consider$n \le N$. Then

$$\frac{\partial f}{\partial t}|_{r=const} = \frac{\partial}{\partial t}\frac{\sum_{i=1}^{n} \delta(\boldsymbol{r}_i - \boldsymbol{r})}{\sum_{i=1}^{N} \delta(\boldsymbol{r}_i - \boldsymbol{r})}.$$

Let us consider the dependence $\delta(\boldsymbol{r}_i - \boldsymbol{r})$- on t only as $(\boldsymbol{r}_i - \boldsymbol{r})$.

Consideration gets more complicated when there are cross-border flows.

If there are no streams

$$\frac{\boldsymbol{F}_1}{\boldsymbol{F}_3} - \frac{\boldsymbol{F}_2}{\boldsymbol{F}_4} = \frac{\sum_{i=1}^{n} \delta(\boldsymbol{r}_i - \boldsymbol{r}) + \sum_i^n \Delta t \frac{\partial \delta(\boldsymbol{r}_i - \boldsymbol{r})}{\partial t} + \ldots}{\sum_{i=1}^{N} \delta(\boldsymbol{r}_i - \boldsymbol{r}) + \sum_i^N \Delta t \frac{\partial \delta(\boldsymbol{r}_i - \boldsymbol{r})}{\partial t} + \ldots} - \frac{\sum_{i=1}^{n} \delta(\boldsymbol{r}_i - \boldsymbol{r})}{\sum_{i=1}^{N} \delta(\boldsymbol{r}_i - \boldsymbol{r})} \approx$$

$$\approx \left(\frac{\sum_{i=1}^{n} \delta(\boldsymbol{r}_i - \boldsymbol{r}) + \sum_i^n \Delta t \frac{\partial \delta(\boldsymbol{r}_i - \boldsymbol{r})}{\partial t} + \ldots}{\sum_{i=1}^{N} \delta(r_i - r)} \left(1 - \frac{\sum_i^N \Delta t \frac{\partial \delta(\boldsymbol{r}_i - \boldsymbol{r})}{\partial t} + \ldots}{\sum_{i=1}^{N} \delta(\boldsymbol{r}_i - \boldsymbol{r})}\right)\right.$$

$$\left. - \frac{\sum_{i=1}^{n} \delta(\boldsymbol{r}_i - \boldsymbol{r})}{\sum_{i=1}^{N} \delta(r_i - r)}\right) \approx \frac{\sum_i^n \Delta t \frac{\partial \delta(\boldsymbol{r}_i - \boldsymbol{r})}{\partial t} + O((\Delta t)^2}{\sum_{i=1}^{N} \delta(\boldsymbol{r}_i - \boldsymbol{r})}.$$

Thus, when solving the Boltzmann equation, we obtain the dependence of the time derivative only through the derivatives of the macroparameters. This hypothesis is used in the theory of rarefied gas when constructing a solution to the Boltzmann equation by the Chapman-Enskiy method.

2. Taking into account flows across the border. We have

$$\frac{F_1}{F_3} - \frac{F_2}{F_4} = \frac{\sum_{i=1}^{n} \delta(\boldsymbol{r}_i - \boldsymbol{r}) + \sum_i^n \Delta t \frac{\partial \delta(\boldsymbol{r}_i - \boldsymbol{r})}{\partial t} + \sum_j \frac{\boldsymbol{p}_j}{\boldsymbol{m}} \delta(\boldsymbol{r}_j - \boldsymbol{r}) + \sum_j \frac{\boldsymbol{p}_j}{\boldsymbol{m}} \Delta t \frac{\partial \delta(\boldsymbol{r}_j - \boldsymbol{r})}{\partial t} + \ldots}{\sum_{i=1}^{N} \delta(\boldsymbol{r}_i - \boldsymbol{r}) + \sum_i^N \Delta t \frac{\partial \delta(\boldsymbol{r}_i - \boldsymbol{r})}{\partial t} + \sum_j \frac{\boldsymbol{p}_j}{\boldsymbol{m}} \delta(\boldsymbol{r}_j - \boldsymbol{r}) + \sum_j \frac{\boldsymbol{p}_j}{\boldsymbol{m}} \Delta t \frac{\partial \delta(\boldsymbol{r}_j - \boldsymbol{r})}{\partial t} + \ldots} - \frac{\sum_{i=1}^{n} \delta(\boldsymbol{r}_i - \boldsymbol{r})}{\sum_{i=1}^{N} \delta(\boldsymbol{r}_i - \boldsymbol{r})}.$$

$1/\Delta$t, the time derivative begins to prevail. Thus, the role of borders is increasing. The distribution function does not provide correct parameter values. Here, only the molecular dynamics method with a very small time step is correct.

2. Calculate the derivatives with respect to space in a rarefied gas

$$(F(t+\Delta t) - F(t))/\Delta t$$
$$\approx \frac{-div\left(\sum_{j=1}^{n_2} \frac{p_j}{m}\delta(r_i-r) + \sum_{i=1}^{n} \frac{p_j}{m}\delta(r_i-r) + \ldots\right)}{(\sum_{i=1}^{N}\delta(r_i-r) + \sum_{j=1}^{N_1}\frac{p_j}{m}\Delta_k\delta\big(r_j(t)-r\big) + \ldots)}.$$
$$\sum_{i=N}^{n+\Delta n}\delta(r_j-r)) = -div\sum_{j=1}^{n_2}\frac{p_j}{m}\delta\big(r_j-r\big).$$

3 Effect of Angular Momentum for a Discrete Environment

Knowledge of virial coefficients is necessary in various practical problems [25]. In previous works, the effect of angular momentum on physical parameters in the kinetic theory and continuum mechanics was discussed. It can be assumed that density fluctuations are associated with inhomogeneity of the velocity distribution at temperatures greater than zero Kelvin degrees, which in turn is associated with the movement of the inertia axis of elementary volumes. Here we consider the algorithm for calculating the additional force associated with the action of the angular momentumin a discrete medium. The formula for determining the center of gravity of the system of material points.

$$r_c = \frac{\sum_{i=1}^{k} m_i r_i}{\sum_{i=1}^{k} m_i}.$$

Center of inertia at different times for identical molecules in an elementary volume

$$\frac{\sum_{i=1}^{n} m\delta(r_i-r_c)r_i}{Nm} = r_c, \quad \frac{\sum_{i=1}^{n} m\delta(r_i+\Delta r_i-r_c+\Delta r_c)(r_i+\Delta r_i)}{(N+\Delta N)m} \approx \widetilde{r_c}.$$

Angular momentum is

$$\widetilde{p_i}\times(r_i+\Delta r_i-\widetilde{r_c}) - p_i\times r_i = L_i,$$
$$\frac{\big(p_i+\Delta p_i\big)\times(r_i+\Delta r_i-r_c-\Delta r_c) - p_i\times(r_i-r_c)}{\Delta t} = \frac{dL_i}{dt}.$$

$$\frac{dL}{dt} = M,\ F = \frac{dM}{dr}.$$

$$\frac{d\boldsymbol{M}_i}{d\boldsymbol{r}} = \frac{d}{d\boldsymbol{r}}\frac{\boldsymbol{p}_i(\Delta\boldsymbol{r}_i - \Delta\boldsymbol{r}_c)}{\Delta t} = \frac{d}{d\boldsymbol{r}}\boldsymbol{p}_i\big(\Delta\boldsymbol{p}_i - \Delta\boldsymbol{p}_c\big).$$

Then we get an additional force

$$\boldsymbol{F}_i^{ad} \approx \frac{\sum_{i=1}^{n}((\Delta\boldsymbol{r}_i - \Delta\boldsymbol{r}_c)\frac{\partial\delta}{\partial r}\boldsymbol{r}_i + \Delta\boldsymbol{r}_i\delta(\boldsymbol{r}_i - \boldsymbol{r}_c)\boldsymbol{r}_i)}{\boldsymbol{N}}\Delta\boldsymbol{p}_i^k = \boldsymbol{F}_i^{ad}\Delta\boldsymbol{t}.$$

$\boldsymbol{p}_i^{k2}\Delta t^2 + \boldsymbol{p}_j^{k2}\Delta t^2 \le (\boldsymbol{r}+\boldsymbol{\sigma})_i^{k2} - (\boldsymbol{r}+\boldsymbol{\sigma})_j{}^{k2}$. The number of molecules that form dimers.

$\boldsymbol{p}_i^{k2}\Delta t^2 + \boldsymbol{p}_j^{k2}\Delta t^2 \ge \boldsymbol{E}$. The number of dimers formed.

Here $\boldsymbol{p}_j^k$ is the contribution of the component k. Here we determine the number of dimers that form equilibrium conditions. Pressure.

$$\sum_{i=1}^{n}\frac{(\boldsymbol{p}_i + \Delta\boldsymbol{p}_i)^2}{\boldsymbol{m}} + \sum_{N}^{n+\Delta n}\frac{\boldsymbol{p}_i{}^2}{\boldsymbol{m}} = \mathrm{p}.$$

The first term is responsible for the usual thermodynamic pressure, taking into account the change in speed from the action of the moment, the second term is responsible for the pressure caused by dimers. From this it can be seen that the pressure change is non-monotonic. Langevin equation taking into account the influence of the angular momentum is

$\frac{dV}{dt} = -\frac{\zeta_{\frac{\prime\prime}{v}}}{m}\mathrm{v} + \frac{1}{m}\frac{dM}{dr}$, where $\zeta_{\frac{\prime\prime}{v}}$—coefficient of friction of the selected particle, m is the mass of the particle, M is the moment of force acting on the particle.

The classical Langevin equation for one particle.

$\frac{dV}{dt} = -\frac{\zeta_{\frac{\prime\prime}{v}}}{m} + \frac{1}{m}F(t)$, where $F(t)$ is a random force

A Markov Gaussian process is considered with the condition that the average for an ensemble of particles ⟨F = 0. In our version, this condition is fulfilled by virtue of the fulfillment of the theorem on the conservation of the moment in a closed volume. For equallibrium condition, this is true.

In conclusion, we note that in the construction of statistical theories of an equilibrium liquid in the Clausius theory, for pressure,

$$p - nkT = -\frac{1}{6}\rho^2\int v(\boldsymbol{r})g(\boldsymbol{r})d\boldsymbol{r},$$

where $v(r)$ is the power of the intermolecular interaction force $(d\varphi(r))\ dln(r)$, $\varphi(r)$– potential,

$g(r) = d\varphi(r)/(dln(r))$, and $g(r)$ is the pair distribution function; i.e. the value of the same structure (dimensional) as the term in the equations with allowance for the angular momentum, that is $M = (r - r_0)\times\frac{d\varphi(r)}{dr}$, r_0. center of mass position. So

$$p - nkT = -\frac{1}{6}\rho^2\int(v(\boldsymbol{r}) + r\frac{dM}{dr})g(\boldsymbol{r})d\boldsymbol{r}.$$

The theoretical calculation of the binary correlative function for liquids by the methods of statistical mechanics is associated with great difficulties that have not yet been overcome. Calculation based on experimental studies of X-ray scattering is also possible only for a small circle of simple liquids, the molecular structure of which has been well studied. Nevertheless, using the existing relations of the statistical theory of fine-structure fluctuations and the results of X-ray studies of simple liquids, one can get an idea of a number of features that distinguish fine-structure concentration fluctuations from thermodynamic ones.

Additional force should contribute at high temperatures. For inert gases at medium and low temperatures, the contribution is negligible. For rarefied gas the angular momentum is value of first order. For the water molecules we have the potential for interaction of dipole and so it have some maximum and minimum, but angular momentum is main correlation effect for point molecules and another interaction is values of smaller. Then it seems probable that for pressure is to be sufficient if liquefied gases are simple, for example Ar, of two virial coefficients and binary interaction of particles. The effect of a small distortion of circular orbits will be small due to the incommensurability the time of the rotation and displacement times of the molecules. For water, additional components will arise in connection with the asymmetry of the molecules and the perturbation of the basic potential by the interaction of the hydrogen parts with each other and with the nucleus of the second molecule.

4 Boltzmann Kinetic Equation

For nonequilibrium states, the Boltzmann kinetic equation is widely used. The equation is derived in two ways. One—based on conservation laws, for the second, the starting point is a chain of coupled equations proposed by Bogolyubov [17]. As a first approximation, the Boltzmann equation is obtained. In this case considering fluctuations, it is theoretically possible to take into account both collisions between particles and collisions between particles and gas molecules, considering a mixture of two components: particles and gas. Solving the equation even for a gas presents significant difficulties. In addition, the equilibrium distributin function does not satisfy the equation [18] and it does not preserve the angular momentum [9]. The latter is verified by vector multiplication of the moment for the velocity equation by the radius vector.

The classical derivation of the Boltzmann equation consists in recording the balance of particles through the relation for the single-particle distribution function.

$$f\left(t+dt,\boldsymbol{r}+\xi_j dt,\xi_j+\boldsymbol{F}_j dt\right)\mathrm{drd}\xi_\mathrm{j}=f\left(\boldsymbol{r},\xi_j,\mathrm{t}\right)d\boldsymbol{r}d\xi_j+\left(\frac{\partial f}{\partial t}\right)_{coll}dt.$$

Often the latter is written in the form

$$f\big(t+dt, \boldsymbol{r}+\boldsymbol{\xi}_j dt, \boldsymbol{\xi}_j+\boldsymbol{F}_j dt\big)drd\xi_j = \mathrm{f}\big(\boldsymbol{r}, \boldsymbol{\xi}_j, \mathrm{t}\big) + (\widetilde{\frac{\partial \mathrm{f}}{\partial \mathrm{t}}})_{\mathrm{coll}}\ \mathrm{dt}.$$

That is r, the radius vector; x- coordinate of the point; ξ- the velocity of the point,m—the molecular weight, and, according to the definition of the distribution function f_N, the probability of finding the system at the points (x, ξ) in the intervals $dx_i d\xi_i$ is

$$f_N(t, x_1, \ldots, x_N, \xi_1, , \xi_N)dx_1 \ldots dx_N d\xi_1 \ldots d\xi_N.$$

The new Boltzmann equation can be written as follows with momentum

$$\frac{df}{dt} = \frac{\partial f}{\partial t} + \boldsymbol{\xi}_i \cdot \left[\frac{\partial f}{\partial \boldsymbol{r}_i}\right] + \boldsymbol{\xi}_i \cdot \frac{\partial}{\partial \boldsymbol{r}_i}\left[\boldsymbol{r}_j \frac{\partial f}{\partial \boldsymbol{r}_j}\right] - \frac{F}{m}\frac{\partial f}{\partial \boldsymbol{\xi}_i} = I.$$

Where $(\frac{\partial f}{\partial t})_{coll}$, $(\widetilde{\frac{\partial f}{\partial t}})_{coll}$ – are the collision integrals that record in different phase spaces. Externally, these equalities are identical, but the second relation is satisfied on the interaction times of the molecules and all interactions are correlated. For gas dynamic problems, the characteristic length of an elementary volume for which equality is written is equal to 10^{-8} cm, and the requirement of a large number of particles in an elementary volume is not satisfied for altitudes of 120–300 km in the terrestrial atmosphere. Indeed, the required minimum size is 10^{-3} cm. Since,$N = \pi R^2 \cdot \xi \cdot \tau \cdot n$ where R is the radius of the cylinder of elementary volume; τ is the mean time of free movement, then for statistical independence the number N of particles must be at least 100.cm. Then $\pi R^2 \cdot 10^4 \cdot 10^{12} \cdot 10^{-5} = 10^2$, that is $R = 10^{-3}$ cm.

Functionally, the Boltzmann equation is invariant with respect to the choice of macro parameters of the distribution function. It is necessary to compare the equilibrium distribution function with macroparameters taken from the Euler and Navier–Stokes equations. The difference will give us a small increment functions. We find that for the Euler equations (zero approximation of the Chapman-Enskog) the difference is zero. There are differences to the first approximation. The first approximation is responsible for the tangential component (p_{ij} tensor of viscous stresses). Euler equations are obtained with the use of locally-equilibrium distribution function. Consequently, they are responsible for the normal component of the velocity values regardless of macroparameters. Upon receipt of the first order correction of the terms included in the final decision of the Chapman-Enskog leave only after integration over the phase velocity ξ. The integrals are taken from f function, i.e. for (ρu). Consider.

$$\frac{Df_0}{dt} = \frac{1}{n} f_0 \frac{\partial n}{\partial t} + \frac{3}{2}\frac{1}{T} f_0 \frac{\partial T}{\partial t} + \frac{mc^2}{2kT^2} f_0 \frac{\partial T}{\partial t} + f_0\left(\frac{m}{kT}(\xi - \boldsymbol{u})\frac{\partial \boldsymbol{u}}{\partial t}\right) + \xi$$
$$\cdot \left\{\frac{1}{n} f_0 \frac{\partial n}{\partial x} + \left(-\frac{3}{2}\right)\frac{1}{T} f_0 \frac{\partial T}{\partial x} + \frac{mc^2}{2kT^2} f_0 \frac{\partial T}{\partial x} + f_0\left(\frac{m}{kT}(\xi - \boldsymbol{u})\frac{\partial \boldsymbol{u}}{\partial x}\right)\right\}$$

$$= 2J\left(f_0, f_0\varphi^k\right) = \int f_0 f_1^0\left(\varphi_1^{(k)'} + \varphi^{(k)'} - \varphi_1^{(k)} - \varphi^{(k)}\right) gbdbd\varepsilon d\xi_1\xi = 0.$$

In classic case

$$\frac{\partial f_0}{\partial t}|_{t=0} = f_0\left\{\frac{m}{kT}\left(c_i c_j - \frac{1}{3}c^2\delta_{ij}\right)\frac{\partial u_i}{\partial t} + \frac{1}{2T}\frac{\partial T}{\partial t}c_i\left[\left(\frac{m}{kT}\right)c^2 - 5\right]\right\}.$$

The Boltzmann equation was wrote for full function and have the local equilibrium function and addition item. The tangent velocity component is obtained because off **ξ** have arbitrary direction of velocity relative position of coordinate axes.

$$\int \boldsymbol{n}\cdot(\boldsymbol{\tau}\cdot f\boldsymbol{\xi})\boldsymbol{dsd\xi} = \int div(\boldsymbol{\tau}\cdot f\boldsymbol{\xi})\boldsymbol{dxd\xi}$$

τf give us addition item. Besides local equilibrium function f_0 we have addition item $\frac{p_{ij}}{2p}\left(\frac{m}{2T}\right)c_i c_j - \frac{q_i}{p}\left(\frac{m}{kT}\right)\left(1 - \frac{c^2}{5}\frac{m}{kT}\right)c_i]$. Main account gives derivatives of local equilibrium function. These items definite the self-diffusion and thermo-diffusion which were foretold by S. V. Vallander [19]. The second derivative is result item $c_i \cdot \frac{\partial f}{\partial r_i}$.

Examples from kinetic theory a) the problem of kinetic theory. Gas in stationary force field with potential φ (analogue of the problem [20]):

$$\xi_i\frac{\partial f}{\partial x_i} + \xi_i\frac{\partial}{\partial x_i}x_i\frac{\partial f}{\partial x_i} - \frac{1}{m}\frac{\partial\varphi}{\partial x_i}\frac{\partial f}{\partial \xi_i} = J(f, f).$$

ξ_i the phase velocity of the coordinates x , y.z; f—distribution function, J(f, f)—the collision integral. Classic distribution is$f = A(x)e^{-B(x)\xi^2}$. In this case, we have the old results, B = Const. For A (x) we have the equation.

$$\frac{dA}{dx_i} + \frac{d}{dx_i}x_i\frac{dA}{dx_i} + 2\frac{A\cdot B}{m}\frac{\partial\varphi}{\partial x_i} = 0.$$

Then we have $f = n_0\left(\frac{m}{2\pi kT}\right)^{3/2}e^{-\frac{\varphi}{kT}}e^{-\frac{m}{2kT}\xi^2}$.

General Maxwell distribution has the form. $f = n\left(\frac{m}{2\pi kT}\right)^{\frac{3}{2}}\exp\left\{-\frac{m}{2kT}c^2\right\}, c = \xi - u$.

The modified Boltzmann equation

$$\xi_i\frac{\partial f}{\partial x_i} + \xi_i\frac{\partial}{\partial x_i}x_i\frac{\partial f}{\partial x_i} - g_i\frac{\partial f}{\partial \xi_i} = J(f, f).$$

$$g = \frac{X}{m} - \text{accelerationofmolecules}$$

Apply to the solution of the old algorithm$\ln f = \gamma_0 + \gamma_i\xi_i + \gamma_4\xi^2$..
Then we get the equation of the old and the new equation.

$$\frac{\partial\gamma_0}{\partial t}+g_i\gamma_i=0,$$

$$\frac{\partial\gamma_i}{\partial t}+2g_i\gamma_4+\frac{\partial\gamma_0}{\partial x_i}+\frac{\partial\gamma_0}{\partial x_i}+\frac{1}{2}x_i\frac{\partial\gamma_0}{\partial x_i^2}^2+\frac{\partial}{\partial x_i}x_i\frac{\partial\gamma_0}{\partial x_i}=0,$$

$$\frac{\partial\gamma_4}{\partial t}\delta_{ij}+\frac{1}{2}\left(\frac{\partial\gamma_i}{\partial x_j}+\frac{\partial\gamma_j}{\partial x_i}\right)+\frac{1}{2}\left(\frac{\partial\gamma_i}{\partial x_j}+\frac{\partial\gamma_j}{\partial\xi_i}\right)$$
$$+\frac{1}{2}*\frac{1}{2}(x_i+x_j)\left(\frac{\partial\gamma_i}{\partial x_j}+\frac{\partial\gamma_j}{\partial x_i}\right)\frac{\partial\gamma_0}{\partial x_i}$$
$$+\frac{1}{2}(x_i+x_j)\frac{1}{2}\left(\frac{\partial}{\partial x_j}\left(\frac{\partial\gamma_i}{\partial x_j}+\frac{\partial\gamma_j}{\partial x_i}\right)+\frac{\partial}{\partial x_i}\left(\frac{\partial\gamma_i}{\partial x_j}+\frac{\partial\gamma_j}{\partial x_i}\right)\right)=0,$$

as before $\frac{\partial\gamma_4}{\partial x_i}=0$, $T=\text{const}$.
Thus, an exact solution of the modified Boltzmann equation was received.

5 Conclusion

Brownian motion (fluctuations) is a consequence and evidence of the existence of thermal motion and is involved in many physical processes. The mathematical theory of describing such a motion is currently probabilistic in nature and says nothing about the causes and mechanism of the phenomenon. Now we account for reason this effect. On the example of the interaction of three particles, a new position of the center of inertia is established. It is proposed to calculate the corresponding driving force using the moment, which makes it possible to apply the procedure for calculating the force in the interaction of many particles. A model is proposed for including this force to calculate the virial coefficient and to calculate the force in the Langevin equation.

References

1. R. Balescu, Equilibrium and nonequilibrium statistic mechanics. A Wiley-Intersciences Publication John Willey and Sons. New-yourk-London (1975)
2. P. Resibois, M. De Lener, Classical kinetic theory of fluids. John Wiley and Sons. New-York, London (1977)
3. N.G. Van Kampen. Stochastic processes in physics and chemistry. North-Holland (1984)
4. Physics of Simple Liquids. Edited by H.N.V. Temperley, I.S. Rowlinson, G.S. Rushbrooke. Amsterdam, 1068. 308.
5. L5. Boltzmann, Selected Works. Moscow: Nauka (1984)
6. D.N. Zubarev, Nonequilibrium statistical thermodynamics. Nauka, Moscow (1971)
7. V.Y. Rudyak, Statistical aerohydromechanics of homogeneous and heterogeneous media, vol. 2 (NGASU, Novosibirsk) (2005)

8. E.V. Prozorova, Features of the rarefied gas description in terms of a distribution function. APhM2018.IOP Conf. Series: Journal of Physics: Conf. Series 1250 (2019) 012023.IOP Publishing. https://doi.org/10.1088/1742-6596/1250/1/012023
9. E. Prozorova, Influence the form of writing conservation laws in computation. JP J. Heat Mass Transf. (2019). Accepted 23 Nov 2019
10. E. Prozorova, The effect of angular momentum and Ostrogradsky-Gauss theorem in the equations of mechanics WSEAS Transactions on Fluid Mechanics. https://doi.org/10.37394/232013.2020.15.2
11. E. Prozorova, The influence of the no symmetric stress tensor on the flow separation WSEAS Transactions on Applied and Theoretical Mechanics. https://doi.org/10.37394/232011.2020.15.9
12. E. Prozorova, Consequences of the Ostrogradsky-Gauss theorem for numerical simulation in aeromechanics. Elsever Granthaalayah 8(6) (2020). https://doi.org/10.29121/granthaalayah.v8.i6.2020.549
13. A.L. Khomkin, A.S. Shumikhin, Plasma frequency, parabolic paths and conductivity of non-ideal fully ionized plasma. High Temp. Therm. Phys. RAS 58(3) (2020)
14. A.N. Morozov, A.V. Skripkin, Application of integral transforms to a description of the Brownian motion by a non-Markovian random process. Russ. Phys. J. 52(2) (2009)
15. A.N. Morozov, A.V. Skripkin, Spherical particle Brownian motion in viscous medium as non-Markovian random process. Phys. Lett. A. 375 (2011)
16. I.G. Kaplan, Introduction to the theory of intermolecular interactions. Science (1982)
17. N.N. Bogolyubov. Problems of dynamic theory in statistical physics. Kiev.T.2 (1970)
18. V.V. Vedgnyapin, S.Z. Adjiev. Entropy according to Boltzmann and Poincaré. Advances in Mathematical Sciences. 69(6) (2014)
19. S.V. Vallander. The equations for movement viscosity gas. DAN SSSR V. LXX III, N1 (1951)
20. M.N. Kogan, The dynamics of the rarefied gases. M.: Nauka (1967)

Analysis of the Logistic and Skew Tent Map for Smart Coupling over a Finite Field

Zongchao Qiao, Ina Taralova, Mazen Saad, and Safwan El Assad

Abstract Chaotic maps have been proved to be efficient in the design of pseudo-random number generator (PRNG). However, the great majority of chaotic maps uses real numbers. Due to the high sensitivity of chaos and the finite feature of hardware, when a PRNG based on real numbers is numerically implemented, quantization and round-off errors may occur and lead to security breach. Besides, initial conditions and parameters of chaotic maps constitute the seed of a PRNG and pseudo-chaotic behavior has to be guaranteed for all initial conditions. Logistic map and skew tent map are supposed to exhibit good chaos in well defined parameters, but in some particular initial conditions, their trajectories will be trapped into the fixed points and lose the chaos quality. For this, we analyze inverse maps to find all these unexpected seeds (fixed points and their preimages) that should be avoided carefully. And to overcome the drawbacks caused by real numbers, a robust PRNG scheme based on a smart coupling of integer chaotic maps over a finite field is proposed in this paper. Simulation results indicate the PRNG can produce pseudo-random numbers that can be used in encryption systems or other engineering applications.

Keywords Chaotic map coupling · Logistic map · Skew tent map · Finite field · Pseudo-random number · Encryption

---

Z. Qiao (✉) · I. Taralova
Ecole Centrale de Nantes, LS2N UMR CNRS, 6004 Nantes, France
e-mail: zongchao.qiao@ec-nantes.fr

I. Taralova
e-mail: ina.taralova@ec-nantes.fr

M. Saad
Ecole Centrale de Nantes, Laboratoire de Mathématiques Jean Leray, UMR CNRS, 6629 Nantes, France
e-mail: mazen.saad@ec-nantes.fr

S. El Assad
Université de Nantes/Polytech Nantes, IETR UMR CNRS, 6164 Nantes, France
e-mail: safwan.elassad@univ-nantes.fr

C. H. Skiadas et al. (eds.), *13th Chaotic Modeling and Simulation International Conference*, Springer Proceedings in Complexity,
https://doi.org/10.1007/978-3-030-70795-8_48

1 Introduction

Pseudo-random number generators (PRNGs) are vital components for a plethora of applications, from noise simulation in statics and control, to secure information transmission and cryptography [1].

The pseudo-random feature implies that randomness is combined with perfect reproducibility. The seed of a PRNG determines uniquely its output sequence, and guarantees its reproducibility: the same seed will generate the same output sequence, and a different seed is supposed to generate another uncorrelated sequence with features close to true random numbers [2].

This property is crucial for most applications and it makes the PRNG appear on the list of the most used tools in engineering, economics and physics, etc. As an example, considering the application in control domain, the PRNG can be used to simulate erroneous sensor measurements (external noise), un-modeled dynamics (internal noise). The robustness of different control laws should be compared for identical "noise" provided by the PRNG, otherwise the obtained results may be biased and eventually unreliable. Equivalently, in cryptography, decryption must use the identical key stream (pseudo-random numbers) as the encryption in order to recover the same original message.

Numerical methods used in PRNG design are considered to be insecure and have heavy calculation burden [3]. At the same time, the deterministic feature, random-like behavior combined with a great sensitivity to the initial conditions render chaotic maps perfect candidates for PRNG design [4]. The initial conditions and parameters constitute the seed of a PRNG.

Designing a chaotic PRNG requires a careful and wise choice of the individual maps to compose the generator. One-dimensional (1-D) chaotic maps have advantages of simple structure and lower computational cost [5]. But they cannot be used alone as PRNG owing to their not long enough periods, attainable map functions and uneven distributions, etc. Hence, an efficient algorithm to make the most use of the chaotic maps to design PRNGs is another important issue. In the open literature, based on multiple chaotic maps, some effective methods have been proposed to enhance the chaos property, such as coupling different chaotic maps [6, 7], integrating chaotic maps [4, 8], multiplexing mechanism [9, 10], permutation approaches [1, 11], linear feedback shift register operations [12].

However, most of the proposed PRNGs are defined by floating-point notation expression, which are slow data transfer and inefficient resource utilization from a hardware perspective [3]. Furthermore, because of the extremely sensitivity of the chaotic systems to the initial conditions and parameters, the chaos properties are also strongly affected by the data type of the chaotic systems when applied to hardware implementation. Therefore, due to the finite nature of the machine number set, these systems may lose the chaotic characteristics because quantization, truncation or round-offs are required when they are numerically realized under a finite precision [13]. Thus, they are not secure enough to be applied into practical situations.

To overcome these drawbacks, Elmanfaloty and Abou-Bakr [3] brought out a solution of using fixed-point precision expression and proposed a binary chaotic PRNG with 32-bits fraction length. Considering the integer arithmetic is more hardware friendly with reduced resources utilization than the decimals arithmetic, we aim to design PRNG using integer finite precision numbers.

Pseudo-random numbers play an important role in cryptosytems which demand not only the randomness, but also a high sensitivity to its seed (secret key for a cryptosystem), especially in the stream cipher whose security depends mainly on its key stream (pseudo-random numbers provided by a PRNG) [14]. Thus, we can propose a new PRNG from the perspective of cryptosystem, but note that the PRNG is not limited to this kind of application.

In our previous works, we have proposed reliable PRNGs based on the 1-D integer chaotic maps for stream cipher and block cipher cryptosystems [15, 16]. They all have avoided the degradation security problem caused by the finite field implementation of the real field chaotic systems and have achieved high security and reliability. They also can be used in other engineering applications.

In this paper, to pursue a new efficient and general coupling method to improve the randomness of the 1-D logistic map and skew tent map, we introduce a smart coupling algorithm based on integer chaotic maps and design a robust PRNG scheme over an N-bits (N = 32) integer finite field. It is known that a chaotic attractor possesses an infinity of unstable dense periodic orbits [17, 18]. In particular, the fixed points may represent a problem if the "randomly chosen" initial conditions coincide with a periodic point, even though the latter is unstable. This is clearly the case, if starting from any point of the unstable orbit, the trajectory will remain locked at the same periodic point. Both logistic and skew tent maps are non invertible maps. But to overcome this unexpected case and ensure pseudo-chaotic behavior for all initial conditions in PRNG design, we can analyze their inverse maps just to reveal all the possible initial conditions and their preimages that lead to the fixed points. Then, based on this analysis, the logistic map and the skew tent map over an N-bits integer finite-state space are reformulated. Meanwhile, in this process, the fixed point problem is solved. After that, a coupling matrix, the kernel of the PRNG design, is applied to break the original orbits of the 1-D chaotic maps for avoiding undesirable dynamic behavior and enhancing the scheme complexity. Finally, statistical and security tests are applied to evaluate the cryptographic properties of the PRNG.

The paper is organized as follows. Section 2 analyzes the logistic map and skew tent map from the inverse function point of view, reformulates their expressions over an N-bits integer field and briefly investigates their qualities for PRNG design. The proposed PRNG and the coupling performance are discussed in Sect. 3. Section 4 analyzes the cryptographic properties of the proposed PRNG. Section 5 gives the conclusion.

2 1-D Chaotic Map over Integer Finite Field

2.1 *Logistic Map*

Logistic map is a well-known classical chaotic map defined over a real number domain ranging from 0 to 1, which is given as below:

$$x(n+1) = \mu x(n)\,(1 - x(n)) \tag{1}$$

where $\{x(n), n = 1, 2, 3...\}$ represents the iteration state and $x(0) \in (0, 1)$ is the initial condition; the parameter $\mu \in (0, 4]$ controls the chaotic behavior. The Lyapunov exponent is the largest when $\mu = 4$, indicating the logistic map reaches the complete chaos.

The delayed phase space of the logistic map when $\mu = 4$ is shown in Fig. 1 (in blue), where the solid red line means $x(n+1) = x(n)$. The intersections are two unstable fixed points: 0, $\frac{\mu-1}{\mu}$, which are 0 and $\frac{3}{4}$ when $\mu = 4$. If the initial value is $\frac{3}{4}$, even though the parameter μ equals 4, all the following iterations will be trapped into the fixed point $\frac{3}{4}$. This is an undesirable case when one intends to use the chaos features to design PRNG or encryption purposes. In addition to this, preimages (backward iterates) of the fixed points can cause the fixed point problem as well, which can be seen from Fig. 1. The preimages of 0 are 0, $\frac{1}{2}$ and 1, while the preimages of $\frac{3}{4}$ are $\frac{1}{4}$ and $\frac{3}{4}$. These values all lead to the fixed points. It also can be observed from Fig. 2, which plots 800 iterations $x(n)$ ($n = 1, 2, ...800$) versus the different initial conditions $x(0)$: for specific $x(0)(0, 0.25, 0.5, 0.75, 1)$, the corresponding iterations are locked into the fixed points, thus there exists no chaos even for $\mu = 4$. To unearth all preimages of the fixed points, we analyze the inverse map.

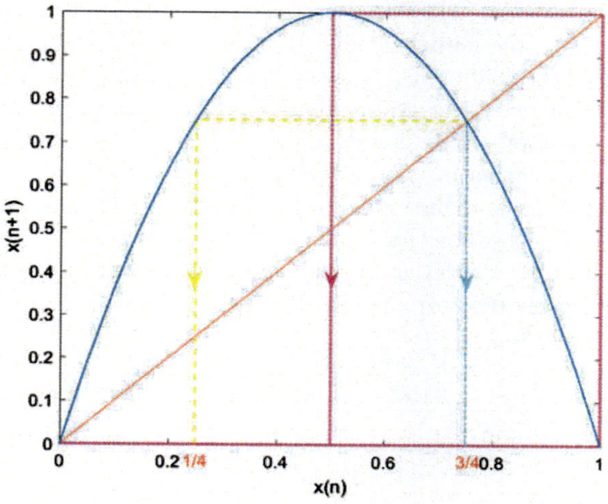

Fig. 1 Delayed phase space and preimages of logistic map over a real domain ($\mu = 4$)

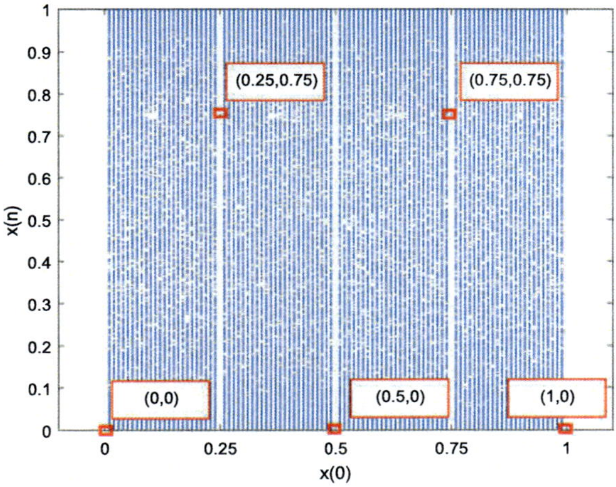

Fig. 2 Iterations $x(n)$ of logistic map versus its initial conditions $x(0)$ over a real domain ($\mu = 4$)

In the literature, plenty of papers have investigated the logistic map, but very few of them analyze the inverse map. However, it is very important to ascertain chaotic behavior so that the trajectories don't get locked into the fixed point. Because the unstable fixed points for the iterated map behave as stable fixed points by the inverse map. In other words, the preimages of the unstable fixed points converge towards the fixed points under forward iterations. Therefore, not only the fixed points, but also their preimages have to be avoided as a seed in order to guarantee chaotic behavior for $\mu = 4$.

The preimages of logistic map can be obtained by (2)

$$x(n-1) = \frac{\mu \pm \sqrt{\mu^2 - 4\mu x(n)}}{2\mu} \tag{2}$$

The preimages $(x(n-1), x(n-2), \ldots)$ of the fixed points $(x(n) = 0, \frac{3}{4})$ are summarized in Table 1. We can find that, if ignoring the irrational numbers in the range of (0, 1), the initial conditions to be avoided are $\frac{1}{4}, \frac{1}{2}$ and $\frac{3}{4}$.

Logistic map redefined over the N-bits integer finite filed when $\mu = 4$ is given by (3).

$$X(n+1) = \begin{cases} 2^N - 1, & X(n) \in [2^{N-1} - 16, 2^{N-1} + 16]\, or\ X(n) = \frac{3}{4} \times 2^N \\ \left\lfloor \frac{X(n) \times \left(2^N - X(n)\right)}{2^{N-2}} \right\rfloor, & otherwise \end{cases} \tag{3}$$

where symbol $\lfloor \cdot \rfloor$ means that each element in it rounds to the nearest integer less than or equal to the element; $\{X(n), n = 1, 2, 3...\}$ is the produced chaotic sequence by iterations and all values are integers ranging in $\left[1, 2^N - 1\right]$.

The delayed phase space of (3) is displayed in Fig. 3, where there obviously exist two unstable fixed points: $X(n) = 0, \frac{3}{4} \times 2^N$. Similar to the analysis of Fig. 1, the

Table 1 Preimages of the fixed points of logistic map

$x(n)$	$x(n-1)$	$x(n-2)$	$x(n-3)$	...
0	0	0	0	...
			1	...
		1	$\frac{1}{2}$	...
			1	...
	1	$\frac{1}{2}$	$\frac{2\pm\sqrt{2}}{4}$ (irrational value)	–
		1	$\frac{1}{2}$	...
			1	...
$\frac{3}{4}$	$\frac{1}{4}$	$\frac{2\pm\sqrt{3}}{4}$ (irrational value)	–	–
	$\frac{3}{4}$	$\frac{1}{4}$	$\frac{2\pm\sqrt{3}}{4}$ (irrational value)	–
		$\frac{3}{4}$	$\frac{1}{4}$	...
			$\frac{3}{4}$	...

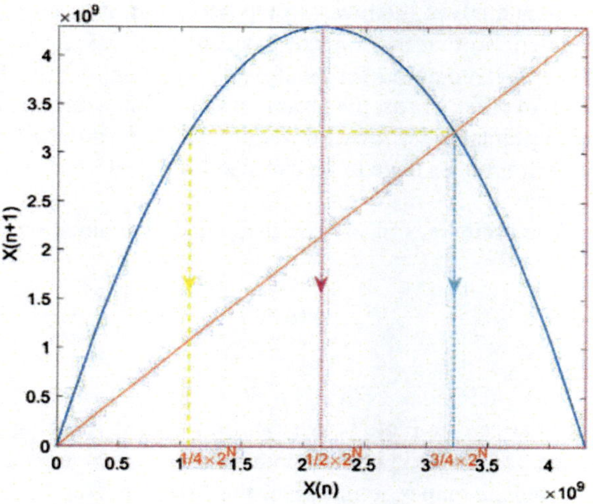

Fig. 3 Delayed phase space and preimages of logistic map over the 32-bits integer field ($\mu = 4$)

preimages of the fixed points in the range of $[1, 2^N - 1]$ are $\frac{1}{4} \times 2^N$, $\frac{1}{2} \times 2^N$ and $\frac{3}{4} \times 2^N$ that should be avoided carefully. Considering $\frac{1}{4} \times 2^N$ produces $\frac{3}{4} \times 2^N$ after one iteration, (3) just needs to deal with the value $\frac{3}{4} \times 2^N$. Due to the finite precision definition, neighboring points of $\frac{1}{2} \times 2^N$ in the range of $[\frac{1}{2} \times 2^N - 16, \frac{1}{2} \times 2^N + 16]$ also can produce the value 2^N which is not in the range of $[1, 2^N - 1]$, thus, these neighboring points also need to be avoided. This explains the reason for the expression of (3).

2.2 Skew Tent Map

Skew tent map is derived from the classical tent map but it achieves better statistical performances. Skew tent map defined in real domain (0, 1) is given by (4).

$$x(n+1) = \begin{cases} \frac{x(n)}{p}, & 0 < x(n) < p \\ \frac{1-x(n)}{1-p}, & p \leqslant x(n) < 1 \end{cases} \tag{4}$$

where $\{x(n), n = 1, 2, 3...\}$ represents the iteration state and $p \in (0, 1)$ is the control parameter.

The delayed phase space diagram of the skew tent map is shown in Fig. 4 where the solid red line reveals the unstable fixed points: 0 and $\frac{1}{2-p}$. From the inverse function point of view, the preimages have the following iteration relation:

$$x(n-1) = \{p \times x(n), 1 - (1-p)\,x(n)\} \tag{5}$$

The skew tent map over the N-bits integer field is given by (6):

$$X(n+1) = \begin{cases} \left\lfloor 2^N \times \frac{X(n)}{P} \right\rfloor, & 0 < X(n) < P \\ \left\lfloor 2^N \times \frac{2^N - X(n)}{2^N - P} \right\rfloor, & P < X(n) < 2^N \\ 2^N - 1, & otherwise \end{cases} \tag{6}$$

where the iterated state is $X(n) \in \mathbf{N}_+$ and $X(n) \in [1, 2^N - 1]$; $P \in \mathbf{N}_+$ is the control parameter and $P \in [1, 2^N - 1]$.

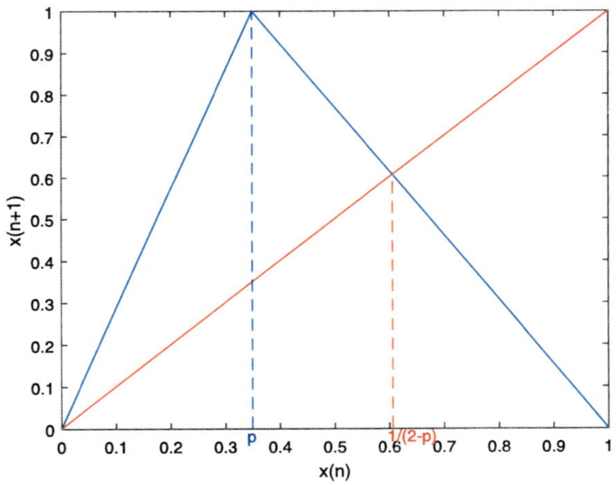

Fig. 4 Delayed phase space of skew tent map over a real domain

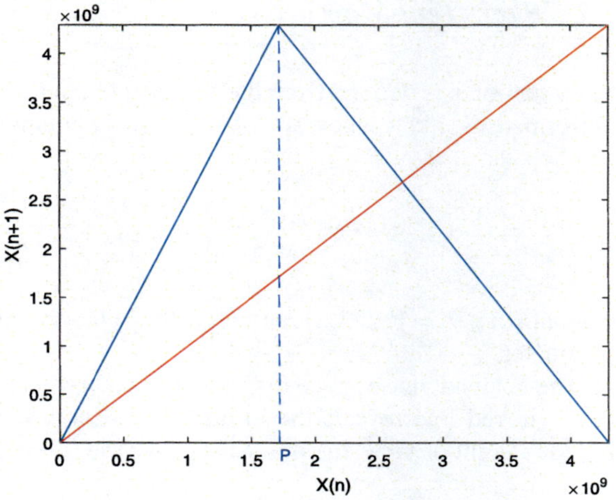

Fig. 5 Delayed phase space of skew tent map over the 32-bits integer field

The unstable fixed point 0 does not belong to the region of definition, while the other fixed point in the mapping shown in Fig. 5 must satisfy one condition: the equation of $X_{fixedpoint} = 2^N \times \frac{2^N - X_{fixedpoint}}{2^N - P}$ should have integer solutions $X_{fixedpoint}$. It is a bit complicated to analyze the preimages of the fixed point without restricting the parameter P. We analyze the parameter P in the range of $[1, 2^{32} - 1]$ in turn and find when using specific P, there exist preimages that can lead to the fixed point, for instance, if $P = 262140$, $X_{fixedpoint} = 2147549185$. In order to prevent the trajectory from being trapped into the fixed point, although this rarely happens, we add the following statement in the algorithm:

$$X(n+1) = X(n+1) - 1, if \ X(n+1) = X(n) \tag{7}$$

2.3 *Analysis for Encryption Purposes*

2.3.1 Lyapunov Exponent

The Lyapunov exponent characterizes the stability of a chaotic motion by measuring the average exponential divergence between two nearby trajectories. If the Lyapunov exponents have a positive value, the chaotic map shows chaotic behavior and the larger this value is, the better the chaotic performances are [19].

The Lyapunov exponent of the logistic map ($\mu = 4$) is 0.6931. Figure 6 demonstrates that the estimated Lyapunov exponents of the skew tent map are always positive for the range of interest of parameter P, exhibiting the maximum 0.6939 for $P = \frac{1}{2} \times 2^N$. At this parameter value, the chaoticity of skew tent map is equivalent to that of the logistic map.

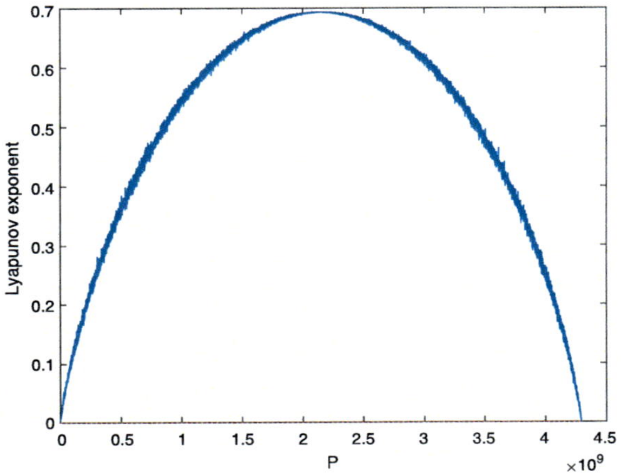

Fig. 6 Estimated Lyapunov exponents of skew tent map

2.3.2 Histogram

Histograms of the chaotic maps (3), (6), and (7) defined over 32-bits integer field are plotted in Figs. 7 and 8 in 1000 classes, where 2×10^6 values are generated for each map but the first 10^6 are considered transient and removed; the red lines mean the average values in every 10 classes.

We can observe that the piece-wise linear map (skew tent map) shows more uniform distribution features than the logistic map, and therefore the skew tent map appears more suitable for encryption purposes.

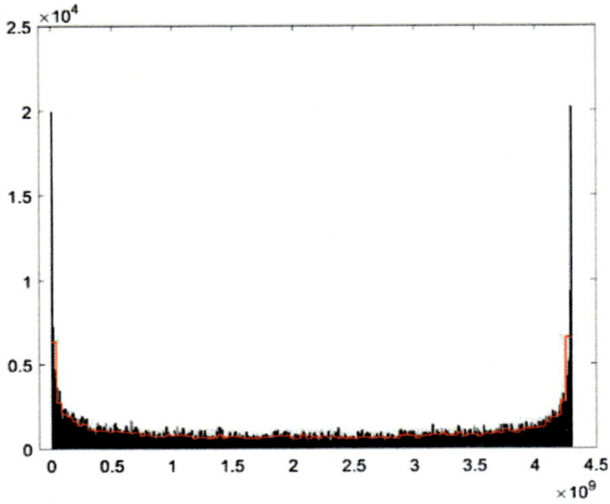

Fig. 7 Histogram of logistic map

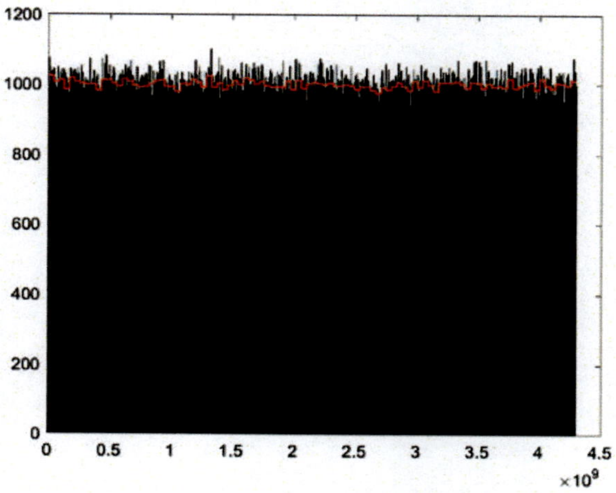

Fig. 8 Histogram of skew tent map

2.3.3 Key Space Contribution

In encryption applications, a large secret key space of PRNG is necessary to resist the brute-force attack and it is considered to be secure if the key space is equal or greater to 2^{128} [20]. Unlike the statistical and security performances, which only can be tested after completing the design of PRNG, key space needs to be taken into account when we are conceiving a new PRNG scheme for cryptosystems for secure transmission or storage of information (medical data, etc.).

Initial conditions and parameters form the key space. For logistic map, only N-bits initial value can be taken into key space, while for the skew tent map, besides the N-bits initial value, N-bits parameter P can be counted into key space as well. From this point of view, skew tent is able to provides N more bits of key space than the logistic map.

3 Proposed PRNG

In this section, we first give the proposed PRNG scheme whose core is a smart coupling structure. This coupling is inspired by the idea of weak coupling over a real number domain in our previous work [21, 22]. Then, we analyze the coupled effect using logistic map and skew tent map. Considering the key space issue in encryption system, two skew tent maps are chosen to construct the PRNG and the performance analyses are given in Sect. 4.

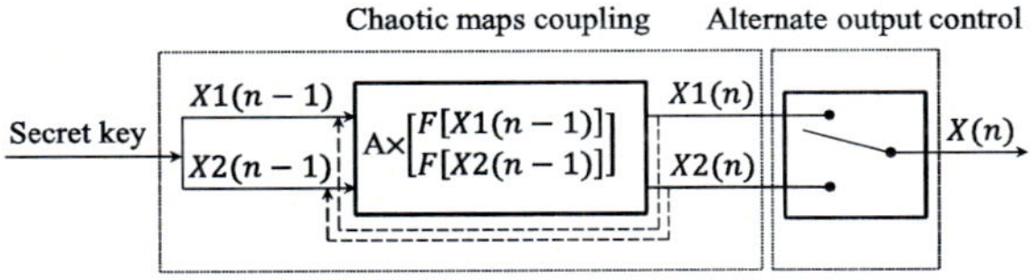

Fig. 9 The proposed PRNG scheme

3.1 Proposed PRNG Scheme

The proposed PRNG scheme is shown in Fig. 9.

It mainly contains two operations: first, a new proposed smart chaotic maps coupling is used to break the original chaotic orbits, enhance the complexity and improve the chaotic property; the alternate output control aims to mix the coupled numbers and increase the unpredictability.

The coupling matrix A is defined as follows:

$$A = \begin{bmatrix} 17 - e1 & e1 \\ 2e1 & 31 - 2e1 \end{bmatrix} \tag{8}$$

$e1 \in [1, 2^4 - 1]$ is a coupling control parameter.

The coupling process is described as below:

$$\begin{bmatrix} X1\,(n) \\ X2\,(n) \end{bmatrix} = A \times \begin{bmatrix} F\,[X1\,(n-1)] \\ F\,[X2\,(n-1)] \end{bmatrix} \tag{9}$$

where F represents the chaotic functions and it can be a similar type of chaotic map or two different kinds of maps; $X1(n-1)$ and $X2(n-1)$ are the previous states of the current states $X1(n)$ and $X2(n)$.

The final output chaotic sequence X is controlled by selecting the intermediate outputs $X1$ and $X2$ alternately (another switching law is also possible):

$$X\,(n) = \begin{cases} X1\,(n)\,,\ when\ mod\,(n, 2) = 1 \\ X2\,(n)\,,\ when\ mod\,(n, 2) = 0 \end{cases} \tag{10}$$

3.2 Coupling Performance

The behavior of the final output sequence depends highly on the coupling performance. Here, we first couple two different maps: logistic map (3) and skew tent map (6 and 7), and analyze the coupling performance in terms of the statistical histogram distribution and delayed phase space behavior.

From the histograms of intermediate outputs $X1$ and $X2$ with length of 3125000 (distributed in 1000 classes) shown in Figs. 10 and 11, we can observe that the

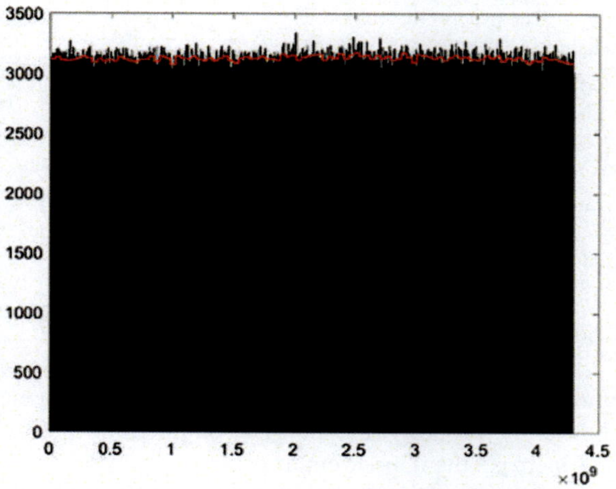

Fig. 10 Histogram of $X1$

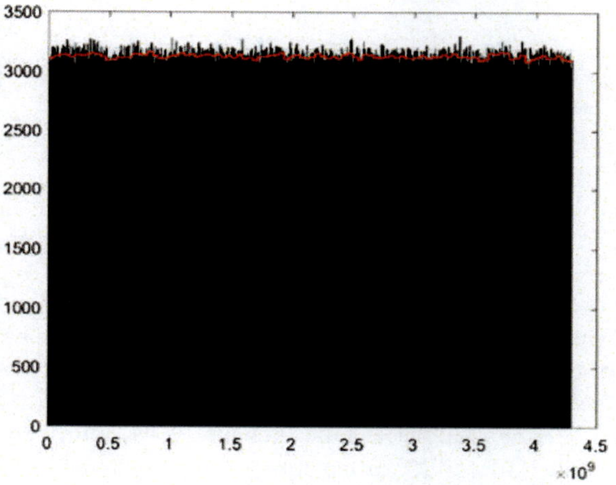

Fig. 11 Histogram of $X2$

sequence after the coupling algorithm is able to achieve a uniform distribution. According to the delayed phase space diagrams shown in Figs. 12 and 13, the coupling matrix can hide the generating function effectively, which is required for most applications in security.

If logistic map and skew tent map are used to design the PRNG, the key space contains the initial values of these two maps (each is in 32 bits), a parameter P (32 bits) for the skew tent map and a control parameter $e1$ (4 bits) for the coupling matrix A. Thus, the key space is 2^{100} in total, which is not large enough for encryption purposes. Hence, this coupling combination can be used to design PRNG, but needs to parallel the coupling scheme to expand the key space for encryption applications.

Skew tent map contributes more key space than logistic map. Apart from this advantage, the skew tent map has an approximately uniform distribution, which

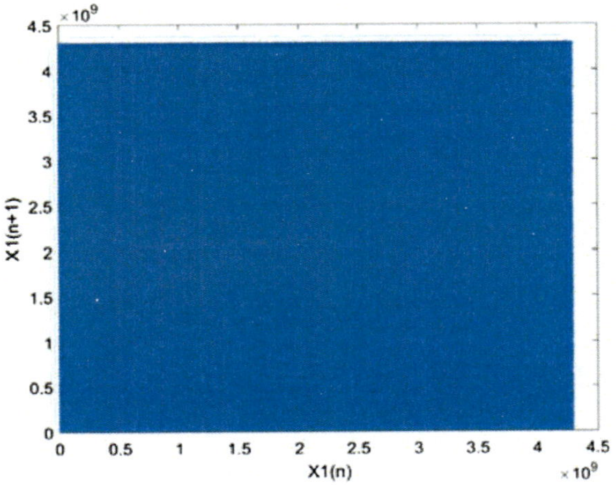

Fig. 12 Delayed phase space of $X1$

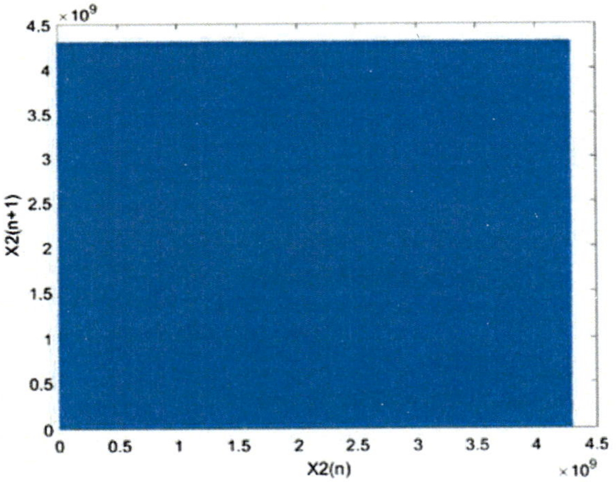

Fig. 13 Delayed phase space of $X2$

outperforms most of the well-known chaotic maps. Thus, we use two skew tent maps to design the PRNG for cryptosystem.

The secret key of this PRNG contains the initial conditions $(Xs1(0), Xs2(0))$, the parameters $(P1, P2)$ for skew tent maps, and the coupling control parameters $e1$. Thus, the key size is :

$$|K| = |Xs1\,(0)| + |P1| + |Xs2\,(0)| + |P2| + |e1| = 132\; bits \tag{11}$$

where $|Xs1\,(0)| = |P1| = |Xs2\,(0)| = |P2| = 32\; bits$ and $|e1| = 4\; bits$.

Therefore, the key space of this proposed PRNG is 2^{132}, which is large enough to make the brute-force attack infeasible if the PRNG is used for cryptography.

The coupling performances of $X1$,$X2$ when using skew tent maps are as good as when coupling the logistic map and skew tent map. The final output is the one which will be exploited, so in the following analyses, we just give the performance test results of the final output chaotic sequence X.

4 Performance Analysis

The seed and the output of PRNG are called also the secret key and key stream respectively in encryption applications. To guarantee high security, the key stream must be random enough to ensure no statistical information is exposed to hackers so that unauthorized ones cannot deduce the inner states and even recover the secret key. Thus, the PRNG should have random statistical performance and good security property. This section analyzes these performances by delayed phase space graph, histogram and χ^2 test, key sensitivity and NIST test. In these tests, each sequence has 3125000 values (3125000×32 bits $= 100 \times 10^6$ bits). All simulations are conducted in MATLAB (R2017b) and each secret key is randomly created.

4.1 *Delayed Phase Space*

The delayed phase space of the final output sequence X is drawn in Fig. 14, where 3125000 values are generated and the last 10^6 values are plotted. Contrary to the easily identified mapping function of the original chaotic map shown in Fig. 5, the final output sequence is distributed randomly in the delayed phase space and shows more complex dynamical behavior thanks to the coupling and alternate output control operations. Thus, it is impossible for potential attackers to analyze the iteration trajectory and they cannot find a hint of which chaotic map we use in the PRNG.

4.2 *Histogram*

A basic requirement for a robust PRNG is that the generated chaotic sequence have a uniform distribution. The histogram of the chaotic sequence X with length of 3125000 is drawn in Fig. 15 in 1000 classes, which shows visually that the generated sequence is uniformly distributed in the whole definition field.

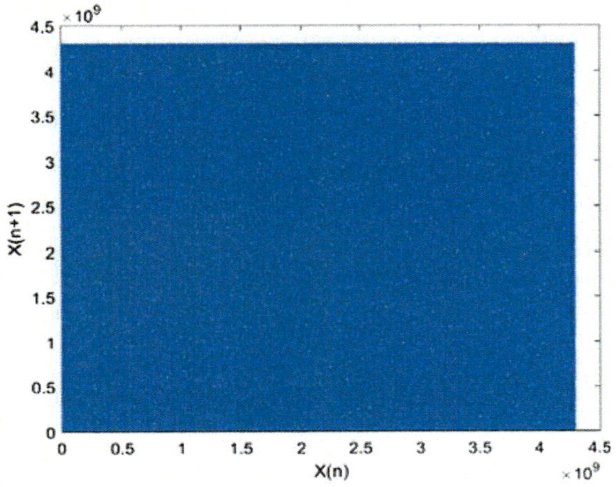

Fig. 14 Delayed phase space of X

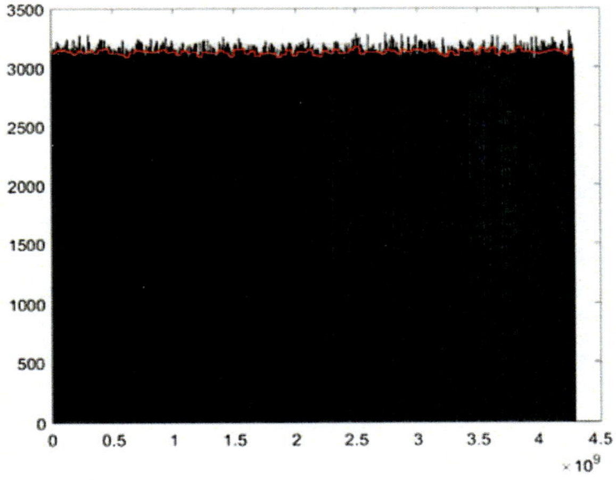

Fig. 15 Histogram of X

4.3 χ^2 Test

To analyze the uniformity property more precisely, the χ^2 test is applied. The experimental value χ^2_{exp} is calculated by (12):

$$\chi^2_{\text{exp}} = \sum_{i=0}^{K-1} \frac{(O_i - E_i)^2}{E_i} \tag{12}$$

where $K = 1000$ is the number of classes, O_i is the number of observed values in the i-th class and E_i is the expected number in a uniform distribution. The theoretical value $\chi^2_{theo}(K, \alpha)$ equals to 1073.64 which is obtained for a threshold $\alpha = 0.05$. If $\chi^2_{exp} < \chi^2_{theo}(K, \alpha)$, the test sequence can be considered to have a uniform distribution.

Here, we use 100 different secret keys to produce 100 chaotic sequences. Each contains 3125000 values, hence, $E_i = 3125000/1000$. χ^2_{exp} is calculated for each sequence. The average $\chi^2_{exp} = 1004.49$ that is smaller than $\chi^2_{theo}(K, \alpha)$. Thus, χ^2_{exp} test has confirmed the uniformity of the output chaotic sequence.

4.4 Key Sensitivity

The generated chaotic sequence should show high sensitivity to the secret key (seed). This property is necessary and important for resisting differential attack. The key sensitivity can be measured by Hamming Distance (HD) given as follows:

$$HD(X, Y) = \frac{1}{Nb} \times \sum_{k=1}^{Nb} (X(k) \oplus Y(k)) \tag{13}$$

where X and Y are two output chaotic sequences from the proposed PRNG whose secret keys are just one bit (randomly chosen) different; Nb is the bit length in a sequence and $\oplus$ represents the XOR operator.

Here, we use 100 different secret keys to produce 100 pairs of X and Y. Then, 100 HDs are computed by (13). The average HD is 49.9989 which is very close to the optimal HD value 50% (bit change probability). This result means the high secret key sensitivity is achieved.

4.5 NIST Test

NIST (National Institute of Standard and Technology) test is a suite of tests which is widely used to measure sequences for randomness. We apply the NIST test on the produced sequence ($3125000 * 32\ bits = 100 \times 10^6\ bits$). The results in Table 2 demonstrate the generated chaotic sequence has passed the NIST test successfully, which verifies the output sequence of the PRNG is pseudo-random.

Table 2 Results of NIST test

Test	P-value	Proportion	Results
Frequency test	0.798	99.000	Passed
Block-frequency test	0.290	97.000	Passed
Cumulative-sums test	0.765	98.500	Passed
Runs test	0.679	99.000	Passed
Longest-run test	0.494	99.000	Passed
Rank test	0.475	100.000	Passed
FFT test	0.658	100.000	Passed
Nonperiodic-templates	0.502	98.973	Passed
Overlapping-templates	0.924	97.000	Passed
Universal	0.658	99.000	Passed
Approximty entropie	0.964	100.000	Passed
Random-excursions	0.441	98.182	Passed
Random-excursions-variant	0.328	98.788	Passed
Serial test	0.906	99.500	Passed
Linear-complexity	0.154	99.000	Passed

5 Conclusion

In this paper, a smart coupling of chaotic maps has been proposed to design a robust PRNG. The specific coupling over a finite field has been inspired from the idea of weak coupling in the case of infinite field maps. Combined with the output control operation, the proposed PRNG structure can break the orbits of original chaotic maps, enhance the nonlinear dynamics and increase the complexity effectively. In addition, the used chaotic maps have been redefined over the N-bits finite field and the problem of locking into a fixed point (or its preimages) has been solved thanks to the analysis of the inverse maps, which overcome the degradation security problem caused by applying the real domain defined chaotic maps into finite hardware implementations, and thus, ensure a high reliability to use them in PRNG design. Conducted simulations results have demonstrated that the proposed PRNG is able to produce pseudo-random numbers with good randomness and cryptographic properties. Therefore, this PRNG can be used in the design of cryptosystems or any other pseudo-random generator required applications.

References

1. D. Lambić, M. Nikolić, Pseudo-random number generator based on discrete-space chaotic map. Nonlinear Dyn. **90**(1), 223–232 (2017). (Oct 1)
2. O. Jallouli, S. El Assad, M.A. Taha, M. Chetto, R. Lozi, D. Caragata, An efficient pseudo chaotic number generator based on coupling and multiplexing techniques, in *International Conference on Emerging Security Information, Systems and Technologies (SECURWARE 2016)* 2016 Jul 24, pp. 35–40
3. R.A. Elmanfaloty, E. Abou-Bakr, Random property enhancement of a 1D chaotic PRNG with finite precision implementation. Chaos, Solitons and Fractals **1**(118), 134–144 (2019)
4. X.Y. Wang, Z.M. Li, A color image encryption algorithm based on Hopfield chaotic neural network. Opt. Lasers Eng. **1**(115), 107–118 (2019)
5. M.A. Murillo-Escobar, C. Cruz-Hernández, L. Cardoza-Avendaño, R. Méndez-Ramírez, A novel pseudo random number generator based on pseudo randomly enhanced logistic map. Nonlinear Dyn. **87**(1), 407–425 (2017)
6. Z. Hua, F. Jin, B. Xu, H. Huang, 2D logistic-Sine-coupling map for image encryption. Signal Process. **1**(149), 148–161 (2018)
7. M.L. Sahari, I. Boukemara, A pseudo-random numbers generator based on a novel 3D chaotic map with an application to color image encryption. Nonlinear Dyn. **94**(1), 723–744 (2018)
8. M. Wang, X. Wang, Y. Zhang, S. Zhou, T. Zhao, N. Yao, A novel chaotic system and its application in a color image cryptosystem. Opt. Lasers Eng. **1**(121), 479–494 (2019)
9. Z. Qiao, I. Taralova, S. El Assad, A robust pseudo-chaotic number generator for cryptosystem based on chaotic maps and multiplexing mechanism, in *International Conference for Internet Technology and Secured Transactions (ICITST'2019)*, Dec 2019, London, United Kingdom. [Qiao et al] 4 p. hal-02430572
10. O. Jallouli, S. El Assad, M. Chetto, R. Lozi, Design and analysis of two stream ciphers based on chaotic coupling and multiplexing techniques. Multimed. Tools Appl. **77**(11), 13391–13417 (2018)
11. M. François, T. Grosges, D. Barchiesi, R. Erra, Pseudo-random number generator based on mixing of three chaotic maps. Commun. Nonlinear Sci. Numer. Simul. **19**(4), 887–895 (2014)
12. H. Xu, X. Tong, X. Meng, An efficient chaos pseudo-random number generator applied to video encryption. Optik **127**(20), 9305–9319 (2016)
13. B. Yang, X. Liao, Period analysis of the logistic map for the finite field. Sci. China Inf. Sci. **60**(2), 022302 (2017)
14. D. Lambić, Security analysis and improvement of the pseudo-random number generator based on quantum chaotic map. Nonlinear Dyn. **94**(2), 1117–1126 (2018)
15. Z. Qiao, I. Taralova, S. El Assad, Efficient pseudo-chaotic number generator for cryptographic applications. Int. J. Intell. Comput. Res. (IJICR). **11**(1), 1041–1048 (2020). https://doi.org/10.20533/ijicr.2042.4655.2020.0126
16. Z. Qiao, S. El Assad, I. Taralova, Design of secure cryptosystem based on chaotic components and AES S-Box. AEU-Int. J. Electron. Commun. **5**, 153205 (2020)
17. K. Burns, B. Hasselblatt, The Sharkovsky theorem: a natural direct proof. Am. Math. Mon. **118**(3), 229–244 (2011)
18. T.Y. Li TY, J.A. Yorke, Period three implies chaos, in *The Theory of Chaotic Attractors* (Springer, New York, NY, 2004), pp. 77–84
19. W.K. Lee, R.C. Phan, W.S. Yap, B.M. Goi, SPRING: a novel parallel chaos-based image encryption scheme. Nonlinear Dyn. **92**(2), 575–593 (2018)
20. F. *Ö*zkaynak, Brief review on application of nonlinear dynamics in image encryption. Nonlinear Dyn. **1**;92(2), 305–313 (2018)
21. O. Garasym, R. Lozi, I. Taralova, Robust PRNG based on homogeneously distributed chaotic dynamics. J. Phys.: Conf. Ser. **692**, 012011 (2016)
22. Lozi R, Taralova I, From chaos to randomness via geometric under sampling. ESAIM: Proc. Surv. **46**, 177–195 (2014)

Information Measures and Synchronization in Regular Ring Lattices with Discontinuous Local Dynamics

J. Leonel Rocha and S. Carvalho

Abstract We study information measures and synchronization in complete dynamical networks of maps, with local identical chaotic dynamical systems. The network topologies are regular ring lattices which are characterized by circulant matrices and the conditional Lyapunov exponents are explicitly determined. For discontinuous local dynamics, some properties of the mutual information rate and the Kolmogorov-Sinai entropy are established, depending on the topological entropy of the individual chaotic nodes and on the synchronization interval. It is proved that as large as the network topology is, measured by its network topological entropy and directly related with the network order, the information measures studied increase or decrease, according to the network order in relation to the synchronization interval. Some numerical studies are included.

Keywords Mutual information rate · Kolmogorov-Sinai entropy · Synchronization · Complete networks · Discontinuous dynamics · Lyapunov exponents · Topological order · Circulant matrix

1 Introduction

In the last decades, several authors have dedicated their investigation to the study of the information theory and its applications. The amount of information produced by a network may be measured by the mutual information rate. This measure together with the Kolmogorov-Sinai entropy are expressed in terms of the conditional

J. Leonel Rocha
CEAUL. ADM, ISEL-Engineering Superior Institute of Lisbon, Polytechnic Institute of Lisbon, Rua Conselheiro Emídio Navarro 1, 1959-007 Lisboa, Portugal
e-mail: jrocha@adm.isel.pt

S. Carvalho (✉)
CEAFEL. ADM, ISEL-Engineering Superior Institute of Lisbon, Polytechnic Institute of Lisbon, Rua Conselheiro Emídio Navarro 1, 1959-007 Lisboa, Portugal
e-mail: sonia.carvalho@isel.pt

C. H. Skiadas et al. (eds.), *13th Chaotic Modeling and Simulation International Conference*, Springer Proceedings in Complexity,
https://doi.org/10.1007/978-3-030-70795-8_49

Lyapunov exponents. On the other hand, it is well known that chaotic systems can be synchronized. The recognized potential for communications systems has driven this phenomenon to become a distinct subfield of nonlinear dynamics.

Information theory and synchronization are directly related in a network. Motivated by the theoretical and practical connection between the information measures and the phenomenon of synchronization, our purpose in this paper is to analyze the relations between the mutual information rate, the Kolmogorov-Sinai entropy and the synchronization in a space of complete dynamical network of maps of order $N \in \mathbb{N} \setminus \{1\}$. The discontinuous local dynamics considered at each node establish the topological, metrical and chaotic complexity of the network that is being studied. Discontinuous dynamical systems are recurrently found in physical systems, which are also used in various applications in engineering, economic, biological and ecological models, among others, see, for example, [1, 9, 16]. The study of discontinuous dynamics in synchronization phenomena has also attracted the attention of several researchers, see [10] and the other works of this same volume and issue.

The paper is organized as follows: In Sect. 2 are presented preliminar definitions and results. We start Sect. 3 with the analysis of the case where the local dynamics are given by discontinuous piecewise linear maps with slope $s > 1$. We obtain explicit expressions for the synchronization interval and for the parallel and transversal Lyapunov exponents. For this case it is proved that to stabilize the synchronized states, it suffices to require that the transversal Lyapunov exponent is negative. Some properties of the mutual information rate and the Kolmogorov-Sinai entropy, depending on the slope s and the synchronization interval are established. We also study the approach to a topological invariant associated with the dynamics between the nodes of the complete network: the network topological entropy. In this context is established a topological order: it is proved that as large as the network topology, measured by its network topological entropy and directly related with the network order, the information measures studied increases or decreases, according to the network order in relation to the synchronization interval. Numerical simulations are performed to obtain more information and complement the theoretical results presented. Finally, in Sect. 4, we discuss our work and provide some conclusions.

2 Preliminars

An active channel is usually described by an active network constructed using $N \in \mathbb{N} \setminus \{1\}$ elements that have some intrinsic dynamics and can be characterized by classical dynamical systems, such as chaotic oscillators, neurons, phase oscillators, and so on. Throughout this work we will consider a family of complex networks of chaotic dynamical systems defined by complete networks of order N with $\frac{N(N-1)}{2}$ edges and discontinuous local dynamics. These networks of N identical chaotic dynamical oscillators or units, are described by a connected and unoriented graph $G = (V, E)$, where V represents the vertices (nodes), and E the edges of G, with no loops and no multiple edges, where every vertex of G has degree $N-1$. The space of complete dynamical networks with N nodes will be denoted by K_N.

Consider A the adjacency matrix of K_N and $D = diag(N-1, \dots, N-1)$, then $L = [l_{ij}] = A - D$ represents the laplacian matrix of the complete graph and is written in the following form,

$$L = \begin{bmatrix} -(N-1) & 1 & 1 \dots & 1 \\ 1 & -(N-1) & 1 \dots & 1 \\ \dots & \dots & \dots\dots & \dots \\ 1 & 1 & \dots\ 1 & -(N-1) \end{bmatrix}.$$

The dynamics of these N coupled oscillators can be expressed by the following system of differential equations,

$$\dot{x}_i = f(x_i) + \sigma \sum_{j=1}^{N} l_{ij} x_j, \tag{1}$$

where f is a vector-valued map describing the dynamics of the nodes, $\sigma > 0$ is the coupling strength or parameter and $i = 1, 2, ..., N$.

However, the state equations of the complex network given by (1), can be rewritten in the discretized form as,

$$x_i(k+1) = f(x_i(k)) + \sigma \sum_{j=1}^{N} l_{ij} f(x_j(k)), \tag{2}$$

which is also known as a complex dynamical network of maps, see, for example, [8, 12]. Let f' be the derivative of f, then the jacobian matrix of this dynamical network K_N is written as follows,

$$J = \begin{bmatrix} f' - (N-1)\sigma f' & \sigma f' & \dots & \sigma f' \\ \sigma f' & f' - (N-1)\sigma f' & \dots & \sigma f' \\ \dots & \dots & \dots & \dots \\ \sigma f' & \sigma f' & \dots & f' - (N-1)\sigma f' \end{bmatrix}.$$

Every matrix associated with a complete network K_N has a certain regularity, so we are able to determine its spectra and the associated eigenspaces. Let $\mu_1 < \mu_2 \le \dots \le \mu_N$ and $\lambda_1 < \lambda_2 \le \dots \le \lambda_N$ be the eigenvalues of the laplacian and the jacobian matrices of K_N, respectively. Notices that the matrices A, L and J are irreducible matrices.

1. Both matrices L and J are circulant matrices, so they are diagonalizable and have the same eigenspaces. Let $x^{(N)} = (1, 1, \dots, 1)$, this is an eigenvector of every circulant matrix, and it is associated with the eigenvalue $\mu_1 = 0$ and $\lambda_1 = f'$, respectively, that is equal to the row sum of each matrix.

2. *The other eigenvectors of a circulant matrix.*
Let $\omega_N = \exp^{\frac{2\pi i}{N}}$ be one of the N-th complex roots of 1. It is known that, for $1 \leq k \leq N$,

$$x^{(k)} = \left(\omega_n^{0k}, \omega_n^{1k}, \ldots, \omega_N^{(N-1)k}\right)$$

is an eigenvector of every circulant matrix C where every row has the elements $\{c_1, \ldots, c_N\}$. In particular, if $k = N$ we obtain $x^{(N)} = (1, 1, \ldots, 1)$. Moreover, the eigenvalue associated with $x^{(k)}$, considering the regularity of C and ω_N^k, is equal to $\sum_{j=1}^{N} c_j \omega_N^{jk}$.

Let us recall an important property of the sum of the N complex roots of the unit, i.e.,

$$\sum_{j=1}^{N} \omega_N^{jk} = \begin{cases} N, & \text{if } k \equiv 0\,(\text{mod } N) \\ 0, & \text{otherwise} \end{cases}.$$

We can also state that, the laplacian matrix L has exactly two eigenvalues $\mu_1 = 0$, a simple root, and $\mu_2 = -N$, with multiplicity $N - 1$, and the jacobian matrix J has also two eigenvalues $\lambda_1 = f'$, also a simple root, and $\lambda_2 = f'(1 - N\sigma)$, with multiplicity $N - 1$. Notice that, in the context of the study of information measures, the eigenvalue λ_1 measures the exponential divergence of nearby trajectories in the direction of the synchronization manifold and the eigenvalue λ_2 measures the exponential divergence of nearby trajectories in the direction transversal to the synchronization manifold, see [2, 3].

In an active network, every pair of elements form a communication channel and the rate with which information is exchanged between these elements, a transmitter S_i and a receiver S_j, is given by the mutual information rate, represented by $I_C(S_i, S_j) = \lambda_{\parallel}^{+} - \lambda_{\perp}^{+}$, where $\lambda_{\parallel}^{+}$ denotes the positive Lyapunov exponents, associated to the synchronization manifold, and $\lambda_{\perp}^{+}$ denotes the positive Lyapunov exponents, associated to the transversal manifold, see [2]. The Kolmogorov-Sinai entropy, denoted by H_{KS}, gives a suitable way of obtaining the entropy production of a dynamical system. It also provides a global measure of the amount of information that can be simultaneously transmitted among the network. For systems with a measurable (the trajectory is bounded to a finite domain) and ergodic (average quantities can be calculated in space and time) invariant (with respect to time translations of the system and to smooth transformations) natural measure, that is smooth along the unstable manifold, the Kolmogorov-Sinai entropy is obtained by the sum of the positive Lyapunov exponents, see [2, 3, 6, 13]. Regarding the case of complete network K_N, where every node is connected with all the others, each node is only one connection apart from any other and there is just one single transversal Lyapunov exponent. Thus, according to the dynamical network given by (2) and [3], we have the following definitions for the information measures analyzed in this paper,

$$I_C = \begin{cases} \lambda_{\parallel} - \lambda_{\perp}, & \text{if } \lambda_{\perp} > 0 \\ \lambda_{\parallel}, & \text{if } \lambda_{\perp} \leq 0 \end{cases} \tag{3}$$

and

$$H_{KS} = \begin{cases} \lambda_{\parallel} + \lambda_{\perp}, & \text{if } \lambda_{\perp} > 0 \\ \lambda_{\parallel}, & \text{if } \lambda_{\perp} \leq 0 \end{cases}. \tag{4}$$

Other central point of our investigation is related with the synchronization in the space of complete networks K_N and its relations with the information measures I_C and H_{KS}, just mentioned in (3) and (4), respectively. Following the results presented in [8], a dynamical network given by (2), having in each node identical chaotic nodes ($\chi(f) > 0$), synchronizes in the following interval,

$$\sigma_1 = \frac{1 - e^{-\chi(f)}}{|\mu_2|} < \sigma < \frac{1 + e^{-\chi(f)}}{|\mu_N|} = \sigma_2, \tag{5}$$

where $0 = \mu_1 < |\mu_2| \leq \ldots \leq |\mu_N|$ are the eigenvalues of the laplacian matrix L and $\chi(f)$ is the Lyapunov exponent of each individual n-dimensional node, see also [5, 6]. Notice that, if each local dynamical node is chaotic, then the Lyapunov exponent $\chi(f)$ is positive. Throughout this work, the synchronization interval of K_N will be denoted by $I_\sigma =]\sigma_1, \sigma_2[$.

3 Local Dynamics: Discontinuous Piecewise Linear Maps with Slope $s > 1$

In this section we consider the space of all the complete dynamical network of maps K_N, given by (2), where the local dynamics in each node is defined by a discontinuous piecewise linear one-dimensional (1D) map, $f : I = [b_1, b_2] \subset \mathbb{R} \to I$, with $|I| = 1$ represents the amplitude of the interval I, such that there exist points $b_1 = d_0 < d_1 < \ldots < d_p < d_{p+1} = b_2$, where f has constant slope $s > 1$ everywhere in each subinterval $I_i =]d_i, d_{i+1}[$, $i = 0, \ldots, p$. Generally, the discontinuous piecewise linear map is defined by,

$$f(x) = s\,x + a_i \pmod 1, \ \forall x \in I_i \text{ and } a_i \in \mathbb{R}. \tag{6}$$

In this context the map f has $p - 1$ discontinuity points: $d_1, d_2, \ldots, d_p$, see Fig. 1. Thus, throughout this section we consider the following parameters space,

$$\Sigma^+ = \left\{(N, s, \sigma) \in \mathbb{R}^3 : N \in \mathbb{N} \setminus \{1\}, s > 1, \sigma > 0\right\}. \tag{7}$$

Since each complete dynamical network K_N has identical chaotic nodes and $|\mu_2| = |\mu_N| = N$, then the synchronization interval is nonempty, for all $s > 1$. Moreover, from (5), the synchronization interval may be expressed in terms of the topological entropy of f, i.e., the chaoticity of the dynamics of the node map f is measured by $h_{top}(f) = \chi(f) = \log|s|$, see [5, 11].

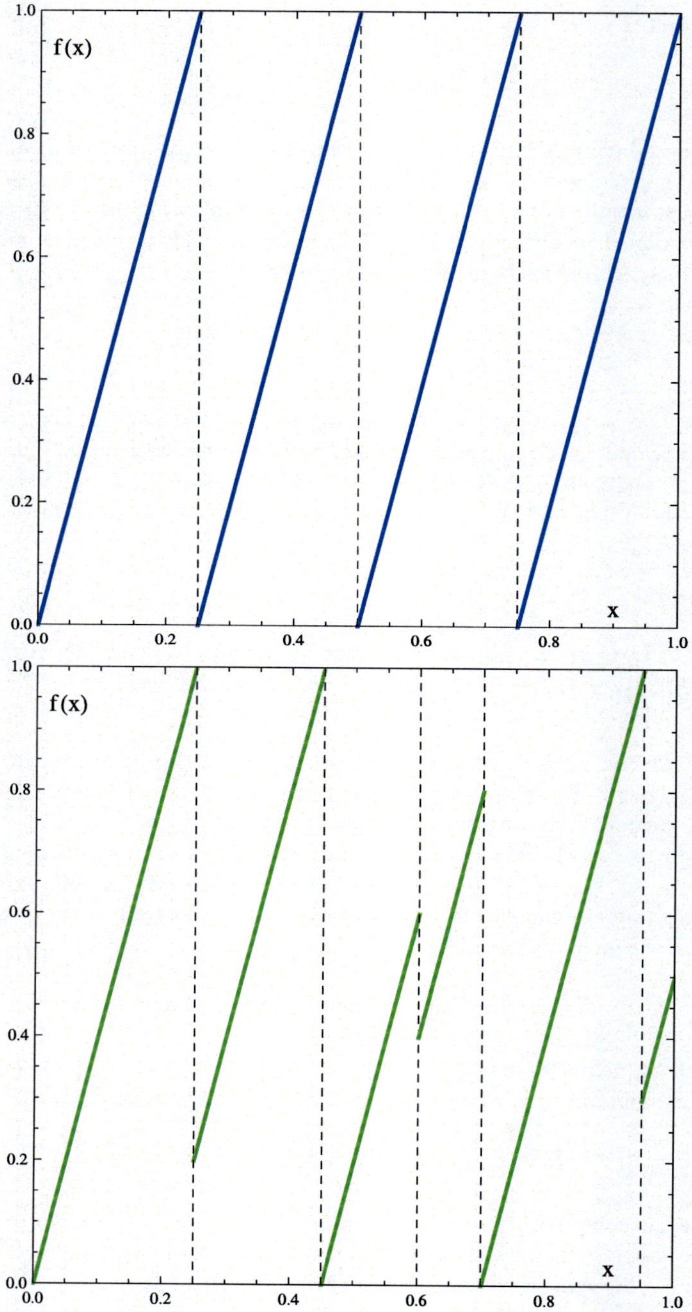

Fig. 1 Graphics of discontinuous piecewise linear 1D map $f(x) = 4x + a_i$, $x \in [0, 1]$

Property 1 *Consider the* (K_N, Σ^+) *space of complete dynamical networks, given by (2). Let* f *be the discontinuous piecewise linear map with slope* $s > 1$ *everywhere, given by (6). The synchronization interval of* K_N*, defined by (5), is given by,*

$$\sigma_1 = \frac{s-1}{Ns} < \sigma < \frac{s+1}{Ns} = \sigma_2, \ \forall s > 1. \tag{8}$$

Consequently, if the dynamics of the individual nodes, defined by f, of the complete dynamical network K_N are fixed, then the amplitude of the synchronization interval $|I_\sigma|$ decrease, as larger is the order N of the complete dynamical network.

3.1 Synchronization and Information Measures

Establishing that the local dynamics f is a discontinuous piecewise linear map with slope $s > 1$ everywhere, given by (6), we have noticed that the jacobian matrix J has only two distinct eigenvalues, $\lambda_1 = s$ and $\lambda_2 = s(1 - N\sigma)$, with multiplicity $N - 1$. So, the parallel Lyapunov exponent is given by,

$$\lambda_{\parallel} = \int_I \ln|\lambda_1| \, d\bar{\mu} = \ln(s), \tag{9}$$

where $|I| = 1$ represents the amplitude of the interval I and $\bar{\mu}$ is a measurable and ergodic invariant natural measure. The transversal Lyapunov exponent is given by,

$$\lambda_{\perp} = \int_I \ln|\lambda_2| \, d\bar{\mu} = \ln|s(1 - N\sigma)|. \tag{10}$$

Notice that for each complete dynamical network K_N, there is a single transversal Lyapunov exponent.

The following proposition stablishes that to stabilize the synchronized states, it suffices to require that the transversal Lyapunov exponent is negative, with piecewise linear maps f with slope $s > 1$ as local chaotic dynamics.

Proposition 1 *Consider the* (K_N, Σ^+) *space of complete dynamical networks, given by (2). Let* f *be the discontinuous piecewise linear map with slope* $s > 1$ *everywhere, given by (6),* I_σ *be the synchronization interval, given by (8), and* $I_{\lambda_\perp^-}$ *be the interval where* $\lambda_\perp < 0$*, with* $\lambda_\perp$ *given by (10).* $\forall s > 1$*, it is verified that:*

(i) $I_\sigma \equiv I_{\lambda_\perp^-} \neq \emptyset$;
(ii) there exists $\sigma > 0$ *such that the synchronized states of (2) stabilize exponentially, i.e.,* $x_1(k) = x_2(k) = \ldots = x_N(k) \to s(k)$*, as* $k \to \infty$.

Proof Consider $(N, s, \sigma) \in \Sigma^+$ and $\lambda_\perp \in I_{\lambda_\perp^-}$, according to (10) we have that,

$$\ln|s(1-N\sigma)| < 0 \Leftrightarrow |s(1-N\sigma)| < 1 \Leftrightarrow$$
$$-1 \leq s(1-N\sigma) < 1 \Leftrightarrow \frac{s-1}{Ns} < \sigma < \frac{s+1}{Ns},\ \forall s > 1.$$

Thus, from Property 1 the result of item (i) is proved, see Fig. 2.

Given that the chaoticity of the node map f is measured by $\chi(f) = \log(s) > 0$, $\forall s > 1$, from condition given by (5), it follows that the inequality

$$\frac{1-e^{-\chi(f)}}{|\mu_2|} < \frac{1+e^{-\chi(f)}}{|\mu_N|} \Leftrightarrow \frac{s-1}{Ns} < \frac{s+1}{Ns}$$

should be satisfied for the existence of a coupling strenght $\sigma > 0$. In this context can be defined a ratio $\frac{1}{R} := \frac{\mu_2-\mu_N}{\mu_1-\mu_2}$, where $0 = \mu_1 < |\mu_2| \leq \ldots \leq |\mu_N|$ are the eigenvalues of the laplacian matrix L, which measures the distance from the first eigenvalue to the main part of the spectral density $\rho(\mu)$ normalized by the extension of the main part, see [8]. From straightforward and simple calculations it is proved that,

$$\frac{2e^{-\chi(f)}}{1-e^{-\chi(f)}} > \frac{1}{R} \Leftrightarrow \frac{2}{s-1} > 0,\ \forall s > 1.$$

Thus, considering the previous conditions, we can state that for all $s > 1$ there exists a coupling strenght $\sigma > 0$ such that the synchronized states of (2) stabilize exponentially. Proposition 1 is thus proved. ■

The results in Proposition 1 bring up to the discussion the complete synchronization versus the negativity of the conditional or transversal Lyapunov exponents. For more details on this classic discussion, see, for example, [6–8, 12, 15] and references therein. The negativity of the conditional Lyapunov exponents is a necessary condition for the stability of the synchronized state, see also [4]. To illustrate these results see the numerical cases shown in Fig. 2.

Taking into account the expressions of the parallel Lyapunov exponent and the transversal Lyapunov exponent, given by (9) and (10), respectively, the information measures defined by (3) and (4) are explicitly written by the following expressions:

$$I_C = \begin{cases} \ln\left(\frac{1}{|1-N\sigma|}\right), & \text{if } \lambda_\perp > 0 \\ \ln(s), & \text{if } \lambda_\perp \leq 0 \end{cases} \tag{11}$$

and

$$H_{KS} = \begin{cases} \ln\left(s^2|1-N\sigma|\right), & \text{if } \lambda_\perp > 0 \\ \ln(s), & \text{if } \lambda_\perp \leq 0 \end{cases}. \tag{12}$$

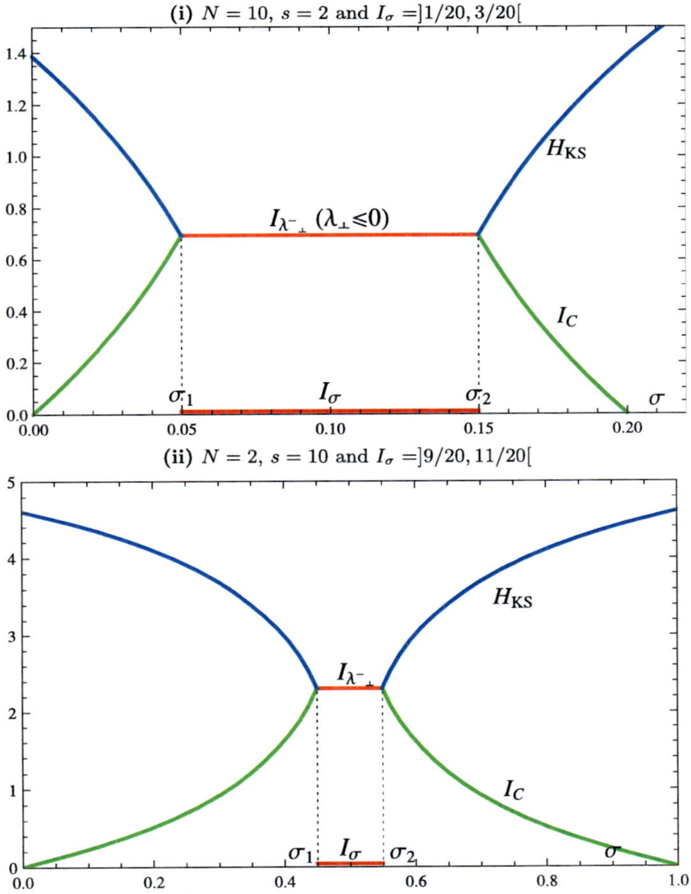

Fig. 2 Numerical simulation for Propositions 1 and 2, where to stabilize the synchronized states of the dynamical network, given by (2), it suffices to require that all transversal Lyapunov exponents be negative

The next proposition establishes some properties of the mutual information rate and the Kolmogorov-Sinai entropy, depending on the synchronization interval I_σ.

Proposition 2 *Consider the* (K_N, Σ^+) *space of complete dynamical networks, given by (2). Let* f *be the discontinuous piecewise linear map with slope* $s > 1$ *everywhere, given by (6),* I_σ *be the synchronization interval, given by (8), and* $I_{\lambda_\perp^-}$ *be the interval where* $\lambda_\perp < 0$*, with* $\lambda_\perp$ *given by (10).* $\forall s > 1$*, it is verified that:*

(i) if $\sigma \in I_\sigma$*, then* $I_C = H_{KS}$*;*
(ii) if $\sigma \notin I_\sigma$ *and* $\sigma < \sigma_1$*, then* I_C *increases and* H_{KS} *decreases;*
(iii) if $\sigma \notin I_\sigma$ *and* $\sigma > \sigma_2$*, then* I_C *decreases and* H_{KS} *increases.*

Proof Considering the definitions of I_C and H_{KS}, given by (11) and (12), respectively, it is verified that $I_C = H_{KS}$ if and only if $\lambda_\perp \leq 0$. On one hand, we have proved, in Proposition 1 (i), that $I_\sigma \equiv I_{\lambda_\perp^-} \neq \emptyset$, $\forall s > 1$. Thus, item (i) is proved.

The result of item (ii) is a consequence of the previous argumentation, i.e., if $\sigma \notin I_\sigma$ and $\sigma < \sigma_1$, then follows that $I_C \neq H_{KS}$, $\forall s > 1$. In this region, considering the monotony of logarithmic function, we also have that $I_C = \ln\left(\frac{1}{|1-N\sigma|}\right)$ increases, for all coupling strenght $\sigma > 0$ under the conditions required by hypothesis, and $H_{KS} = \ln\left(s^2|1 - N\sigma|\right)$ decreases, for all σ under the same conditions. This proves the claim (ii).

The proof of item (iii) is similar to the proof of item (ii), the monotony of I_C and H_{KS} follows as discussed above, $\forall\sigma \notin I_\sigma$ and $\sigma > \sigma_2$. See also Fig. 2. This completes the proof of Proposition 2. ■

Under the conditions of Proposition 2, we can establish that $I_C \leq H_{KS}$, $\forall\sigma > 0$, see also [2].

3.2 *Topological Order and Information Measures*

To end this section we will approach a topological invariant associated with the dynamics between the nodes of the complete dynamical network K_N: the network topological entropy. In this work we use the network topological entropy concept used in [13, 14]. Let G be the connected and unoriented graph associated to the complete dynamical network K_N of order $N \in \mathbb{N} \setminus \{1\}$. Considering that the adjacency matrix A of the complete dynamical network K_N is irreducible, then the Perron-Frobenius Theorem states that the network topological entropy of K_N is given by,

$$h_{top}(K_N) = h_{top}(G) = \ln(\lambda_A) = \ln(N - 1), \tag{13}$$

where λ_A is the Perron eigenvalue of A. Clearly, the topological entropy of a complete dynamical network K_N is characterized by its order N.

Proposition 3 *Let (K_N, Σ^+) be the space of complete dynamical networks, given by (2), with a fixed local dynamics given by a discontinuous piecewise linear map f with slope $s > 1$ everywhere, given by (6). It is verified that if the network topological entropy $h_{top}(K_N)$ increases, such that $\sigma \notin I_\sigma$, then:*

(i) the mutual information rate I_C increases;
(ii) the Kolmogorov-Sinai entropy H_{KS} decreases.

Proof Consider that the local dynamics of two complete dynamical networks K_N and K_{N+1} are fixed, i.e., $\chi(f) = \log(s) > 0$ is constant, $\forall s > 1$. Considering the definition of the mutual information rate I_C, given by (11), with $\sigma \notin I_\sigma$, let $u_N = \ln\left(\frac{1}{|1-N\sigma|}\right)$, with $N \in \mathbb{N} \setminus \{1\}$. The following statements holds,

$$u_{N+1} - u_N = \ln\left(\frac{1}{|1-(N+1)\sigma|}\right) - \ln\left(\frac{1}{|1-N\sigma|}\right) = \ln\left|\frac{1-N\sigma}{1-(N+1)\sigma}\right| > 0.$$

This means that the mutual information rate I_C increases, if the network topological entropy $h_{top}(K_N)$ increases. Now under the same assumptions, if we consider the definition of the Kolmogorov-Sinai entropy H_{KS}, given by (12) with $\sigma \notin I_\sigma$ and let $v_N = \ln\left(s^2|1-N\sigma|\right)$, with $N \in \mathbb{N} \setminus \{1\}$. It can be easily verified that,

$$v_{N+1} - v_N = \ln\left(s^2|1-(N+1)\sigma|\right) - \ln\left(s^2|1-N\sigma|\right) = \ln\left|\frac{1-(N+1)\sigma}{1-N\sigma}\right| < 0.$$

This proves that the Kolmogorov-Sinai entropy H_{KS} decreases, when the network topological entropy $h_{top}(K_N)$ increases. Thus, the desired results are proved. ■

It is interesting to note that this result is related with the structural complexity of the dynamical network analyzed, this means that as larger as the network topology is, measure by its network topological entropy $h_{top}(K_N)$ and directly related with the network order $N \in \mathbb{N} \setminus \{1\}$, the mutual information rate I_C and the Kolmogorov-Sinai entropy H_{KS} increases or decreases, according to order N in relation to $\sigma \notin I_\sigma$, see Fig. 3.

Remark 1 Under the conditions of Proposition 3 and according to Property 1, we can establish that the amplitude of the synchronization interval $|I_\sigma|$ decrease, as larger is the network topological entropy $h_{top}(K_N)$ of the complete dynamical network.

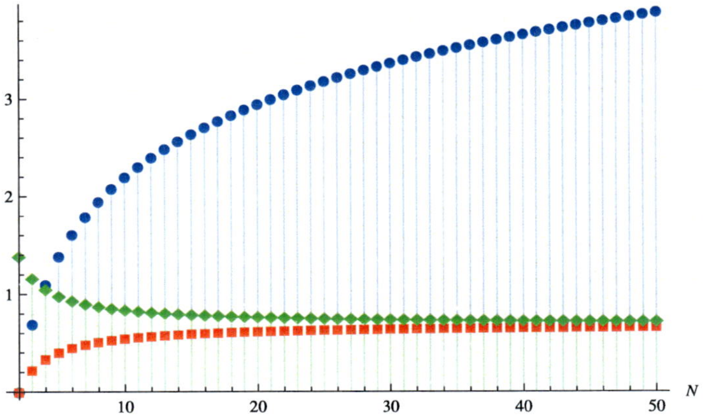

Fig. 3 Numerical simulation for Proposition 3: blue points are $h_{top}(K_N)$, red squares are I_C, green lozenges are H_{KS}, where it is considered $s = 2$ and $N \in \{2, 3, \ldots, 50\}$

4 Conclusion and Discussion

In this paper we have considered the space of complete dynamical networks K_N, an extreme case of a ring lattice with maximal degree, of identical chaotic dynamical oscillators or nodes, in which each oscillator is coupled linearly and symmetrically with their neighbours, see (2). The topology of the networks K_N is characterized by a circulant matrix of order $N \in \mathbb{N} \setminus \{1\}$, which provides a certain regularity in the network. Furthermore, we consider discontinuous local dynamics: piecewise linear maps f with positive slope $s > 1$. The chaoticity of the local dynamics is measured by the topological entropy of f, i.e., $h_{top}(f) = \log |s|$. Several measures have been considered in the field of information theory. In our paper we have considered the mutual information rate and the Kolmogorov-Sinai entropy. However, the synchronization is vital for modern methods of digital communication that rely on the synchronous operation of many subsystems. So, our main concern was to determine explicit expressions for these measures, properties between them and relations with the synchronization interval I_σ. The results presented in Propositions 1 and 2 fully serve this purpose. Finally, in Proposition 3 it is addressed the structural complexity of the complete dynamical networks K_N, using the network topological entropy $h_{top}(K_N)$. We proved that as large as the network topological entropy $h_{top}(K_N)$ is, the mutual information rate I_C and the Kolmogorov-Sinai entropy H_{KS} increase or decrease, according to the network order N in relation to $\sigma \notin I_\sigma$. Therefore, a topological order was established regarding the studied information measures.

Clearly, with the fixed coupling topology in the networks K_N and variation of the discontinuous local dynamics f in the nodes, the complexity of our analysis increases due to measure theory issues. For future work, on one hand we would like to generalize our study for the case where $|s| > 1$ dependening on the amplitudes of the subintervals with slope $s > 1$ and slope $s < -1$. On the other hand, we would like to analyze this problem in other networks of regular lattices rings, starting with the minimal degree, the cycles of order N. In this context, we finish this work with some open questions: There are sufficient conditions to guarantee the negativity of the conditional Lyapunov exponents, for different slopes of f? Under what conditions the chaotic signals transmitted through filters produce an output with higher dimension, due to the appearance of a fractal set?

Acknowledgements Research funded by the project IPL—MISRedes, IDI&CA 2019, FCT—Fundação para a Ciência e a Tecnologia, Portugal, through the projects UIDB/00006/2020 (CEAUL), UIDB/04721/2020 (CEAFEL) and ISEL.

Conflict of Interest The authors declare that they have no conflict of interest.

References

1. V. Avrutin, L. Gardini, I. Sushko, F. Tramontana, *Continuous and Discontinuous Piecewise-Smooth One-Dimensional Maps: Invariant Sets and Bifurcation Structures* (World Scientific, Singapore, 2019)
2. M.S. Baptista, J. Kurths, Chaotic channel. Phys. Rev. E **72**, 045202R (2005)
3. M.S. Baptista, R.M. Szmoski, R.F. Pereira, S.E. Souza Pinto, Chaotic, informational and synchronous behaviour of multiplex network. Sci. Rep. **6 (22617)**, 1–9 (2016)
4. S. Boccaletti, J. Kurths, G. Osipov, D.L. Valladares, C.S. Zhou, The synchronization of chaotic systems. Phys. Rep. **366**, 1–101 (2002)
5. A. Caneco, J.L. Rocha, C. Grácio, Topological entropy in the synchronization of piecewise linear and monotone maps. Coupled Duffing oscillators. Int. J. Bifurc. Chaos **19 (11)**, 3855–3868 (2009)
6. A. Caneco, J.L. Rocha, Synchronization and information transmission in networks, in *Proceedings of ECIT 2012, EDP Sciences, ESAIM: Proceedings and Surveys*, vol. 46, eds. by W. Jarczyk et al. (2014), pp. 111–124
7. J. Cao, J. Lu, Adaptive synchronization of neural networks with or without time-varying delays. Chaos **16**, 013133 (2006)
8. X. Li, G. Chen, Synchronization and desynchronization of complex dynamical network: an engineering viewpoint. IEEE Trans. Circuits Syst. I **50**(11), 1381–1390 (2003)
9. A.C.J. Luo, *Discontinuous Dynamical Systems* (Springer, New York, 2012)
10. A.C.J. Luo, Discontinuous dynamical systems and synchronization. Eur. Phys. J. Spec. Top. **228**(6), 1383–1384 (2019)
11. J. Milnor, W. Thurston, On iterated maps of the interval. *Lect. Notes Math.* (Springer) **1342**, 465–563 (1988)
12. L.M. Pecora, T.L. Carroll, Driving systems with chaotic signals. Phys. Rev. A **44**, 2374–2383 (1991)
13. J.L. Rocha, A. Caneco, Mutual information rate and topological order in networks, Chaotic Modeling and Simulation. Int. J. Nonlinear Sci. **4**, 553–562 (2013)
14. J.L. Rocha, C. Grácio, S. Fernandes, A. Caneco, Spectral and dynamical invariants in a complete clustered network. Appl. Math. Inf. Sci. **9**(5), 2367–2376 (2015)
15. J.W. Shuai, K.W. Wong, L.M. Cheng, Synchronization of spatiotemporal chaos with positive conditional Lyapunov exponents. Phys. Rev. E **56**, 2272 (1997)
16. F. Tramontana, F. Westerhoff, Piecewise-linear maps and their application to financial markets. Front. Appl. Math. Stat. **2**(10), 1–10 (2016)

How Chaotic Dynamics Drive a Vintage Grill-Room Spit

Julio Rodriguez and Max-Olivier Hongler

Abstract In 1943 Yves Rocard published an amazing book entitled "Dynamique Générale des Vibrations". Among a collection of mechanical devices that are studied with great care in the book, the so-called Bouasse-Sarda's *"tournebroche"* (rotary grill spit) is a fascinating two degrees of freedom device. What makes it interesting is that it stylises a wide class of forced parametric oscillators exhibiting a rich range of complex dynamical behaviours including fully deterministic chaos, a dynamic concept not popular in the forties and hence not yet discussed in Y. Rocard's book. Today's numerical tools offer a new possibility to revisit this amazingly simple but rich dynamical system.

Highlights: In the present paper, we propose an extensive, didactical and updated investigation of the nonlinear dynamical system studied by Y. Rocard more than 75 years ago.

1 Preliminary

Our study starts in the Spring of 2006 on a sunny Saturday, when one of the authors (MOH) rambled through Geneva's flea market. Among a chaotic pile of secondhand books, suddenly a well-stocked pile of scientific and engineering publications stuck out. The bookseller was so happy to get rid of such "boring material" that the whole stock was purchased for virtually nothing. Among them, "Optique Géométrique Supérieure" by H. Bouasse (1917), a version of the "Vibration Problems in Engineer-

J. Rodriguez (✉)
Swissgrid AG, Bleichemattstrasse 31, 5001 Aarau, Switzerland
e-mail: Julio.Rodriguez@swissgrid.ch

M.-O. Hongler
Ecole polytechnique fédérale de Lausanne, STI-DO, BM 5132, Station 17, 1015 Lausanne, Switzerland
e-mail: max.hongler@epfl.ch

C. H. Skiadas et al. (eds.), *13th Chaotic Modeling and Simulation International Conference*, Springer Proceedings in Complexity,
https://doi.org/10.1007/978-3-030-70795-8_50

ing" by S. Timoschenko (1937) and a remarkable contribution entitled "Dynamique Générale des Vibrations" by Y. Rocard published in Laval (Mayenne, France) 1943, during the French occupation. In Rocard's opus, we discovered a truly intriguing two degrees of freedom device called the "*tournebroche*" (i.e. rotating spit) de Bouasse-Sarda. More than 75 years after Rocard's publication, we revisit this pretty fascinating "*tournebroche*" dynamics and show how it naturally offers the possibility to exhibit chaotic evolution.

2 Introduction

The dynamic response of mechanical devices driven by external energy sources is often discussed under the implicit assumtion that the energy source itself is fully insensitive to behaviour of the driven device. It is assumed that the energy source is large enough to ensure that all feedback response of the driven system to the source are negligible. While such approximation is fully legitimate when a clear energy-scale distinction between the system itself and its driver can be made,[1] in many other circumstances the system-driver's retro-actions may generate complex and unsuspected responses that cannot be ignored. The suggestive Bouasse-Sarda (B-S) "*tournebroche*" device sketched in Fig. 1 which is also called the B-S regulator offers a rich and intuitive illustration of the underlying complexity of the clear and consistent distinction between a system and its driving unit. From Fig. 1, we may view the spring as the driven system and the mass m together with the pulley as playing the role of an energy source driving the oscillatory mechanism. Depending on the size of the crankshaft a in the B-S device, there is a simple way to continuously increase the spring-pulley retro-action and hence infer the resulting evolution. As we shall discuss, this a-priori deceptively simple system uncovers a very rich variety of dynamical behaviours which include limit cycles, bistability, the now famous Feigenbaum cascade with period-doubling bifurcations ultimately leading to a fully developed deterministic chaos and antimonotonicity.

3 Equations of the Motion

The dynamical system sketched in Fig. 1 consists of a shaft or a rotating spit with, on one side, a crank of radius a and, on the other side of the shaft, a drum of radius ρ with inertial moment I. A cable is fastened around the drum on which suspends a mass m. A mass M is attached to a spring with stiffness K. The spring is then fixed to

[1] As a paradigmatic illustration, you may invoke the thermodynamic model of someone diving into a pool: the diver's body temperature is lowered by the pool's environment, yet one usually neglects the temperature increase of the pool's content. From a mechanical perspective, this idea of the interaction between a system and the force acting on it, is studied in Roseau [17], and as an illustration to the theoretical framework, the B-S regulator is used.

Fig. 1 The Bouasse-Sarda's rotary spit drawn by Y. Rocard in his 1943 book entitled: *"Dynamique Générale des Vibrations"*

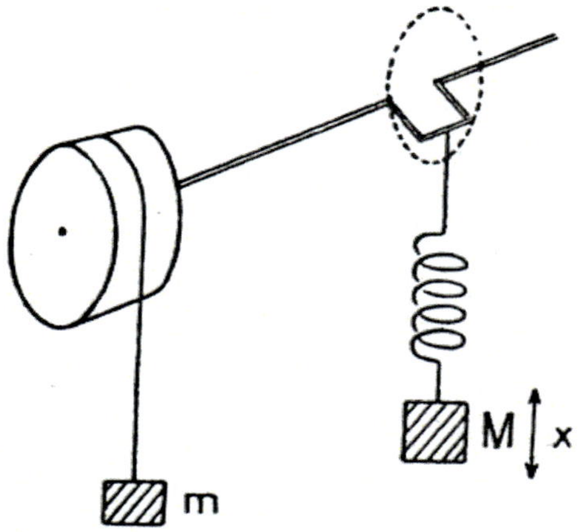

the crank. For this two degrees of freedom device $(\theta(t), x(t)) \in \mathbb{R}^2$, the Lagrangian $\mathcal{L} = T(\dot{\theta}, \dot{x}) - V(\theta, x)$ is given by:

$$\begin{aligned} T(\dot{\theta}, \dot{x}) &= (\mathsf{I} + m\rho^2)\frac{\dot{\theta}^2}{2} + \mathsf{M}\frac{\dot{x}^2}{2}, \\ V(\theta, x) &= -(mg\rho\theta + \mathsf{M}gx) + \frac{\mathsf{K}(x - a\sin(\theta))^2}{2}, \end{aligned} \tag{1}$$

leading to the time evolution[2]:

$$\begin{cases} (\mathsf{I} + m\rho^2)\ddot{\theta} + h\dot{\theta} = mg\rho + \mathsf{K}(x - a\sin(\theta))a\cos(\theta), \\ \mathsf{M}\ddot{x} + f\dot{x} = \mathsf{M}g - \mathsf{K}(x - a\sin(\theta)). \end{cases} \tag{2}$$

Equation (2) involves a set of nine positive control parameters $(\mathsf{I}, m, \mathsf{M}, \rho, \mathsf{K}, f, h, g, a)$, with gravitational acceleration g. Energy dissipation is implemented via a couple of viscous damping mechanisms adjusted by the parameters $f \geq 0$ and $h \geq 0$. We emphasise that in Y. Rocard's book Rocard [15], the dynamical system involves only friction on the spring (i.e. $h = 0$). The second equation in (2) is simply a $\theta(t)$ forced harmonic oscillator for which the response is explicitly calculable provided that signal $\theta(t)$ itself is given. In particular, for small damping parameter f, one expects a resonance peak at frequency $\omega_0 := \sqrt{\mathsf{K}/\mathsf{M}}$, which arises whenever ω_0 belongs to the Fourier spectrum of the signal $\theta(t)$.

To proceed, we rewrite (2) in terms of the (dimensionless) coordinate $x(t) = au(t) + (\mathsf{M}g/\mathsf{K})$ and obtain:

[2] For detailed calculations, see Appendices 7.1 and 7.2.

$$\begin{cases} \mathsf{J}\ddot{\theta} + h\dot{\theta} = \underbrace{g(\mathsf{M}a\cos(\theta) + m\rho)}_{=:H_a^{(1)}(\theta)} + \underbrace{\mathsf{K}a^2(u - \sin(\theta))\cos(\theta)}_{=:H_a^{(2)}(\theta,u)}, \\ \mathsf{M}\ddot{u} + f\dot{u} = -\mathsf{K}(u - \sin(\theta)), \end{cases} \tag{3}$$

with $\mathsf{J} := \mathsf{I} + m\rho^2$.

Note that for vanishing crankshaft (i.e. $a = 0$), the system (3) degenerates into a couple of independent one degree of freedom dynamics. For a small crankshaft $a \gtrsim 0$ (i.e. up to first order in a and thus omitting the $H_a^{(2)}(\theta, u)$ term), the (θ, u) variables are partially coupled. Indeed, we have a special two degrees of freedom system for which the $\theta(t)$ evolution can be calculated independently of $u(t)$. Observe that up to first order in a, the $\theta(t)$ evolution describes either the dynamics of a point particle evolving inside a tilted washboard potential or, alternatively, an anharmonic pendulum subject to a constant external torque. Finally, for general a, the action of the feedback type term $H_a^{(2)}(\theta, u)$ drastically modifies the evolution. The u variable now affects the driving mechanism $\theta(t)$ itself in (3).

4 Study of the Dynamics

4.1 Static Equilibrium and Associated Stability Issues

The fixed points $\mathcal{P}^* := (\theta^*, u^*)$ of the dynamics (3) read:

$$\begin{aligned}(\theta^*, u^*) &= \left(\cos^{-1}(-\tfrac{m\rho}{\mathsf{M}a}), \sin(\cos^{-1}(-\tfrac{m\rho}{\mathsf{M}a}))\right), \\ &= \left(\cos^{-1}(-\tfrac{m\rho}{\mathsf{M}a}), \sqrt{1 - (\tfrac{m\rho}{\mathsf{M}a})^2}\right),\end{aligned} \tag{4}$$

and hence $\mathsf{M}a \geqslant m\rho$ is required for the existence of $\mathcal{P}^*$. Observe that $\mathcal{P}^*$ does not depend on g.

Stability of $\mathcal{P}^*$: The stability of $\mathcal{P}^*$ is studied via the linearisation of the dynamics in the $\mathcal{P}^*$-neighbourhood. By adding a couple of conjugate variables $(\eta, v) := (\dot{\theta}, \dot{u})$, we rewrite the coupled second-order differential system (3) as an equivalent set of four first-order differential equations, and hence the first variational equation is:

$$\begin{pmatrix} \dot{\epsilon}_\theta \\ \dot{\epsilon}_\eta \\ \dot{\epsilon}_u \\ \dot{\epsilon}_v \end{pmatrix} = \underbrace{\begin{pmatrix} 0 & 1 & 0 & 0 \\ -\mathsf{J}^{-1}(\mathsf{M}gas^* + \mathsf{K}(ac^*)^2) & -\mathsf{J}^{-1}h & \mathsf{J}^{-1}\mathsf{K}a^2c^* & 0 \\ 0 & 0 & 0 & 1 \\ \mathsf{M}^{-1}\mathsf{K}c^* & 0 & -\mathsf{M}^{-1}\mathsf{K} & -\mathsf{M}^{-1}f \end{pmatrix}}_{:=DF(\theta^*,u^*)} \begin{pmatrix} \epsilon_\theta \\ \epsilon_\eta \\ \epsilon_u \\ \epsilon_v \end{pmatrix} \tag{5}$$

with $c^* := \cos(\theta^*)$ and $s^* := \sin(\theta^*)$ and perturbations $\epsilon_\theta, \epsilon_\eta, \epsilon_u, \epsilon_v$ around $(\theta^*, 0, u^*, 0)$. The characteristic polynomial $P(\lambda)$ relevant to infer the stability property of $\mathcal{P}^*$ is given by:

$$P(\lambda) := \text{Det}\left[DF(\theta^*, u^*) - \lambda \mathbb{I}_4\right] = 0, \tag{6}$$

with $\mathbb{I}_4$ standing for the (4×4) identity matrix. A direct calculation yields:

$$\begin{aligned} P(\lambda) = \lambda^4 + (\mathsf{J}^{-1}h + \mathsf{M}^{-1}f)\lambda^3 + \\ (\mathsf{J}^{-1}\mathsf{M}^{-1}hf + \mathsf{M}^{-1}\mathsf{K} - DF_{2,1}(\theta^*, u^*))\lambda^2 + \\ (\mathsf{J}^{-1}\mathsf{M}^{-1}h\mathsf{K} - \mathsf{M}^{-1}f DF_{2,1}(\theta^*, u^*))\lambda + \\ \mathsf{J}^{-1}\mathsf{K}ga\sin(\theta^*). \end{aligned}$$

Note that all coefficients of $P(\lambda)$ are positive. Provided we have:

$$(\mathsf{J}^{-1}h + \mathsf{M}^{-1}f) > 0,$$

direct application of the Routh-Hurwitz criterion shows that all real parts of the eigenvalues of the linearised dynamics around $\mathcal{P}^*$ are negative and hence $\mathcal{P}^*$ is stable.

In summary, when $\mathsf{M}a \geqslant m\rho$, a fixed point $\mathcal{P}^*$ given by (4) exists and its stability (i.e. stable focus) is ensured by the viscous dampers (and this even if only a single damper is active).

4.2 Dissipation Free Dynamics–Harmonic Undamped Oscillations

In the absence of dissipation (i.e. $f = h = 0$), the characteristic polynomial $P(\lambda)$ is bi-quadratic, yielding a couple of eigenfrequencies (ν_-, ν_+):

$$\begin{cases} \nu_\pm = \frac{1}{2}\left[r_1 \pm \sqrt{(r_1^2 - 4r_2)}\right], \\ r_1 = \mathsf{M}^{-1}\mathsf{K} - DF_{2,1}(\theta^*, u^*), \\ r_2 = \mathsf{J}^{-1}\mathsf{K}ga\sin(\theta^*). \end{cases} \tag{7}$$

and the fixed point $\mathcal{P}^*$ is a center. For large inertial moment J (i.e. J >> M and J >> K), one verifies that $r_2 \to 0$, and we obtain in this limiting case an oscillation with frequency $\nu_+ \cong \sqrt{\mathsf{K}/\mathsf{M}}$ of M, together with a very low frequency $\nu_- \cong 0$ oscillation of m around the stationary point $\mathcal{P}^*$.[3]

[3] Remember that the linearising assumption is valid only for motions close to $\mathcal{P}^* = (\theta^*, u^*)$. For arbitrary J, K, M, m, the resulting 2-D harmonic system may enter into resonance, thus precluding the possibility to use linear approximations.

4.3 "Washboard"-Like Dynamics

For a small crankshaft parameter a, we may approximately neglect $H_a^{(2)}(\theta, u)$ in (3), which is of order a^2. In this subsection, $h > 0$. The resulting dynamics for the variable $\theta(t)$ describes a classical particle evolving with friction inside a tilted washboard potential. Alternatively, it can also be viewed as an anharmonic pendulum subject to constant external torque Coullet et al. [4]:

$$\ddot{\theta}(t) + \frac{h}{\mathsf{J}}\dot{\theta}(t) = \frac{\mathsf{M}ag}{\mathsf{J}}\cos(\theta(t)) + \frac{mg\rho}{\mathsf{J}}. \tag{8}$$

The time rescaling $\tau(t) := \sqrt{\frac{\mathsf{M}ag}{\mathsf{J}}}\,t$ together with the phase shift $\phi(\tau(t)) := \theta(t) - \frac{\pi}{2}$ enable us to write:

$$\phi'' + \beta\phi' + \sin(\phi) = \gamma, \tag{9}$$

where $'$ now stands for the τ-derivative and $\beta := \frac{h}{\sqrt{\mathsf{JM}ag}}$ and $\gamma := \frac{m\rho}{\mathsf{M}a}$. The dynamical system given by (9) is currently discussed in the context of Josephson's junctions dynamics.[4]

In particular, the trade-off between the damping factor β and the external torque γ is summarised in Fig. 2.[5] Note that in (9), one has explicitly $\gamma = \frac{m\rho}{\mathsf{M}a}$ and this is coherent with the condition for the existence of $\mathcal{P}^*$) (i.e. $1 > \frac{m\rho}{\mathsf{M}a}$). Hence, when $\gamma > 1$, only periodic solutions exists, consistent with the fact that are no fixed points.

We emphasise that since (9) describes a single degree of freedom system, then up to first order in a (i.e. small crankshafts), no chaotic behaviour can possibly be observed for the $\theta(t)$-evolution. As a consequence, the $u(t)$-motion itself, which describes the evolution of a forced harmonic oscillator with damping, does not exhibit a chaotic behaviour either. This follows from the fact that the $\theta(t)$ motion results from simple quadratures.[6] Summarising, for small crankshafts a, the two degrees of freedom (θ, u) system does not exhibit chaotic behaviour. As a consequence, **up to first order in a, the largest Lyapunov exponent always remains negative.**

4.4 The Rocard's Quasi-uniform Regime

We now focus on the general situation (i.e. also considering the influence of the retro-action $H_a^{(2)}(\theta, u)$). We follow the lines originally drawn by Y. Rocard himself and note that this quasi-uniform regime has been later re-discussed in Panovko and Gubanova [14] for $h = 0$ and in Blekhman and Dzhanelidze [2], Colombo [3], Roseau [17], Fridman [6], Fridman [7] for $h > 0$. Rocard assumed that $\theta(t) \cong \omega t$, and so the

---

[4] See, for example, the recent review Blackburn et al. [1].

[5] Figure 2 is directly reproduced from Fig. 6 in Coullet et al. [4].

[6] See, for example, Chap. 2 in Lakshmanan and Rajasekar [11].

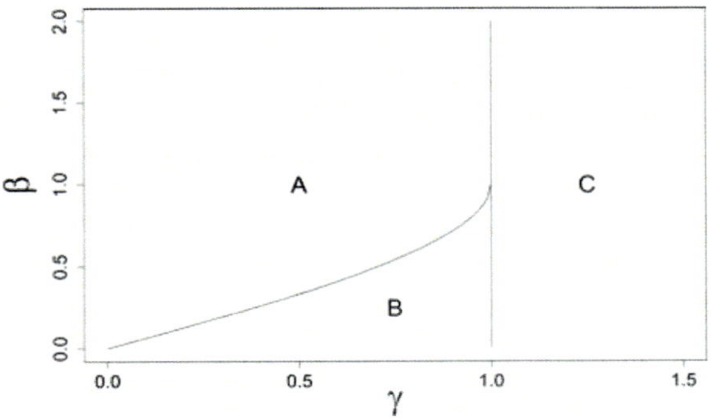

Fig. 2 Evolution regimes as a function of the damping β and the external torque γ: A = only the stationary solution is stable, B = both the stationary and the periodic solution are stable, C = only the periodic solution is stable. This diagram is reproduced from Coullet et al. [4]

$u(t)$-motion follows the permanent regime of an harmonically driven linear oscillator, namely:

$$u(t) = \frac{\frac{\mathsf{K}}{\mathsf{M}}\sin(\omega t+\varphi)}{\sqrt{(\frac{\mathsf{K}}{\mathsf{M}}-\omega^2)^2+(\frac{f}{\mathsf{M}})^2\omega^2}} =: \mathsf{A}\sin(\omega t+\varphi), \tag{10}$$

with $\tan(\varphi) = \frac{-\frac{f}{\mathsf{M}}\omega}{\frac{\mathsf{K}}{\mathsf{M}}-\omega^2}$. The pulsation[7] ω (i.e. the dominant frequency at which both drum and spring oscillate) is determined by solving the polynomial equation explicitly given by:

$$\frac{2\mathsf{M}^2(mg\rho - h\omega)}{f(\mathsf{K}a)^2} = \frac{\omega}{(\frac{\mathsf{K}}{\mathsf{M}}-\omega^2)^2+(\frac{f}{\mathsf{M}})^2\omega^2}. \tag{11}$$

For the sake of completeness, (11) is re-derived in Appendix 7.3. The solutions to (11) are the roots of a polynomial of degree 5 (for $h > 0$) or degree 4 (for $h = 0$), namely

$$\begin{aligned} &\text{for } h > 0 : 0 = h\omega^5 - mg\rho\omega^4 + h(p^2-2q)\omega^3 - (p^2-2q)mg\rho\omega^2 \\ &\qquad\qquad +(q^2(h+\tfrac{a^2 f}{2}))\omega - q^2 mg\rho, \\ &\text{and} \\ &\text{for } h = 0 : 0 = \omega^4 + (p^2-2q)\omega^2 - \tfrac{q^2a^2f}{2mg\rho}\omega + q^2, \end{aligned} \tag{12}$$

with $p = \frac{f}{\mathsf{M}}$ and $q = \frac{\mathsf{K}}{\mathsf{M}}$. Depending on the value of $a \geqslant 0$, (11) may have up to three real solutions for $h > 0$, or two for $h = 0$. To see this, start with the RHS of (11) and define the function

$$\alpha(\omega) := \frac{f(\mathsf{K}a)^2}{2\mathsf{M}^2}\left(\frac{\omega}{(\frac{\mathsf{K}}{\mathsf{M}}-\omega^2)^2+(\frac{f}{\mathsf{M}})^2\omega^2}\right) \tag{13}$$

for positive ω. As such, the function α is positive and exhibits a resonance-like curve shape: it takes on the value zero at $\omega = 0$, increases to its maximum value at

[7] Adopting Rocard's terminology (the French-English translation is the authors').

$\omega = \omega_0$ (in the neighbourhood of $\sqrt{\frac{\mathsf{K}}{\mathsf{M}}}$), and then, as ω increases further, converges asymptotically to zero. The solutions of (11) are the intersections between the curve given by α and the line $mg\rho - h\omega$ (i.e. a decreasing line for $h > 0$ or a straight line, parallel to the ω-axis, for $h = 0$). See, for example, Fig. 3a (for $h > 0$) and Fig. 3b (for $h = 0$). One still needs to prove that there is only one point of inflection after α attained its maximum. This is done in Appendix 7.4.

For small values of a, there is only one real root for $h > 0$ and only complex roots for $h = 0$. As a increases, there exists a particular value of a for which the line $mg\rho - h\omega$ intersects α not far from its maximum value. In the case $h = 0$, the intersection is at the maximum value of α itself. This a_0 and the corresponding ω_0 are explicitly given by

$$a_0 = \sqrt{\frac{2mg\rho}{fq^2}\left(4\omega_0^3 + 2(p^2 - 2q)\omega_0\right)},$$

with $\omega_0 = \left(\frac{-(p^2-2q)+\sqrt{(p^2-2q)^2+12q^2}}{6}\right)^{\frac{1}{2}}$. See Appendix 7.5 for details. Table 1 (for $h > 0$) and Table 2 (for $h = 0$) present a summary of the different possible real roots for the polynomial in (12) according to the value of a.

Having determined how many real solutions (11) has with respect to a, the next question is: under Rocard's assumption, towards which solution of (11) does the B-S system converge?

Table 1 Real roots of the polynomial in (12) for $h > 0$ with respect to $a \geqslant 0$ and for positive parameters m, M, ρ, K, f and g

$a = 0$	$\omega_3 = \frac{mg\rho}{h} \in \mathbb{R}$
$]0, a_1[$	$\omega_3 \in \mathbb{R}$
$a = a_1$	$\omega_1 = \omega_2$ $\omega_3 \neq \omega_1, \quad \omega_j \in \mathbb{R}, j = 1, 2, 3$
$]a_1, a_2[$	$\omega_1 \neq \omega_2, \quad \omega_2 \neq \omega_3,$ $\omega_3 \neq \omega_1, \quad \omega_j \in \mathbb{R}, j = 1, 2, 3$
$a = a_2$	$\omega_1 \neq \omega_2$ $\omega_2 = \omega_3, \quad \omega_j \in \mathbb{R}, j = 1, 2, 3$
$]a_2, +\infty[$	$\omega_1 \in \mathbb{R}$

Table 2 Real roots of the polynomial in (12) for $h = 0$ with respect to $a \geqslant 0$ and for positive parameters m, M, ρ, K, f and g

$[0, a_0[$	No real roots
$a = a_0$	$\omega_0 = \omega_1 = \omega_2 \in \mathbb{R}$
$]a_0, +\infty[$	$\omega_1 \neq \omega_2, \quad \omega_j \in \mathbb{R}, j = 1, 2$

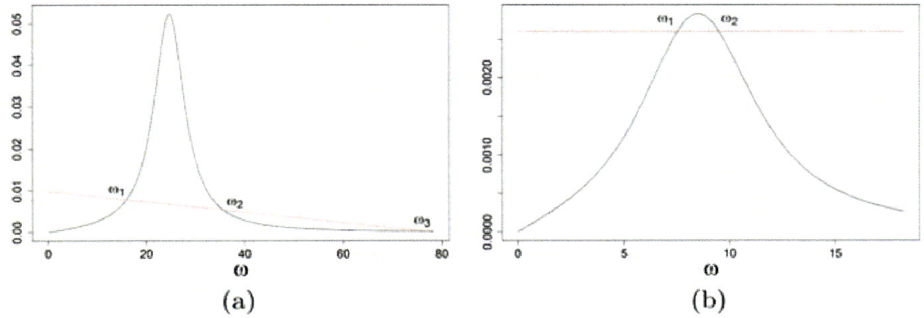

Fig. 3 In black, function α as defined in (13) with respect to ω. The red dashed line is $mg\rho - h\omega$. In **a** $a = 0.05$ and in **b** $a = 0.04$, while the values of the other parameters are those of the first line in Table 4 for (**a**) and the third line in Table 4 for (**b**)

To answer this question, let us focus on the case $h = 0$ and follow Rocard's reasoning (here $a > a_0$). For this, let $\omega_1 < \sqrt{\frac{\mathsf{K}}{\mathsf{M}}}$ and $\omega_2 > \sqrt{\frac{\mathsf{K}}{\mathsf{M}}}$ be the two real solutions of (11). Since ω_1 solves (11), we have $\alpha(\omega_1) = mg\rho$. On the other hand, from (3), the quasi-uniform assumption $\theta(t) \cong \omega t$ implies:

$$\mathsf{J}\ddot{\theta} \cong 0 = mg\rho + g\mathsf{M}a\cos(\theta(t)) + H_a^{(2)}(\theta(t), u(t)),$$

with $u(t)$ given by (10) and $\theta(t) \cong \omega t$, so that:

$$-\alpha(\omega_1) = -mg\rho = g\mathsf{M}a\cos(\theta(t)) + H_a^{(2)}(\theta(t), u(t)).$$

Hence,

$$\mathsf{J}\ddot{\theta} \cong 0 = mg\rho - \alpha(\omega_1),$$

that is, the drum's angular acceleration is approximately $mg\rho - \alpha(\omega_1)$. Suppose now that the drum is slightly accelerated (i.e. $\omega_1 \to \omega_1 + \delta$, with $\delta > 0$). By the shape of the "resonance" curve α, we have $\alpha(\omega_1) < \alpha(\omega_1 + \delta)$, implying that $mg\rho - \alpha(\omega_1 + \delta) < 0$, and therefore $\mathsf{J}\ddot{\theta} \cong mg\rho - \alpha(\omega_1 + \delta) < 0$. This implies that the angular velocity decreases and so the nominal pulsation ω_1 tends to be recovered. The same reasoning holds when the drum is slightly decelerated. A similar analysis for the solution ω_2 can be performed. Table 3 summarises the results for the four possibilities.

Table 3 Summary of the perturbation analysis on the theoretical pulsations ω_1 and ω_2

$\omega_1 + \delta$	$\alpha(\omega_1 + \delta) \uparrow$	$\mathsf{J}\ddot{\theta} \cong mg\rho - \alpha(\omega_1 + \delta) < 0$	ω_1 recovered
$\omega_1 - \delta$	$\alpha(\omega_1 - \delta) \downarrow$	$\mathsf{J}\ddot{\theta} \cong mg\rho - \alpha(\omega_1 - \delta) > 0$	ω_1 recovered
$\omega_2 + \delta$	$\alpha(\omega_2 + \delta) \downarrow$	$\mathsf{J}\ddot{\theta} \cong mg\rho - \alpha(\omega_2 + \delta) < 0$	ω_2 not recovered
$\omega_2 - \delta$	$\alpha(\omega_2 - \delta) \uparrow$	$\mathsf{J}\ddot{\theta} \cong mg\rho - \alpha(\omega_2 - \delta) > 0$	ω_2 not recovered

Therefore, the pulsation ω_1 persists under slight perturbations, which is not the case for ω_2. A similar analysis is also presented in Panovko and Gubanova [14]. When $h > 0$ and depending on the initial conditions, the dynamical system may converge towards two different pulsations ω, thus allowing for bistability. See Roseau [17] for further details.

4.5 Route to Chaotic Evolution–Feigenbaum Cascade of Period-Doubling Bifurcations

In the general case where not only the $H_a^{(2)}(\theta, u)$ retro-action is active but also $\theta(t) \neq \omega t$, the previous Rocard's quasi-uniform regime becomes unstable itself. Therefore, further discussions are to be based on numerical solutions, i.e. an impossible approach at Y. Rocard's time.[8] As will be shown in the next section, for a whole range of control parameters, the B-S dynamics may exhibit a deterministic chaotic evolution, a behaviour obviously not presented in Rocard's 1943 opus. Specifically, we shall numerically unveil that for a sequence of critical crankshaft values a_k for $k = 1, 2, 3, \ldots$, a nowadays classic Feignebaum's cascade of period-doubling bifurcations emerge from the dynamics. The sequence of $\{a_k\}$ converges to an accumulation point a_∞ with the Feigenbaum's universal constant $\mathcal{F}$:

$$\mathcal{F} := \lim_{k\to\infty} \frac{a_{k-1} - a_{k-2}}{a_k - a_{k-1}} \cong 4.6692\ldots. \tag{14}$$

5 Numerical Simulations

All simulations numerically integrate (2) with initial conditions:

$$\begin{array}{ll} x(0) = \mathsf{A}\sin(\varphi) + \frac{\mathsf{M}g}{\mathsf{K}} & \dot{x}(0) = \omega \mathsf{A}\cos(\varphi) \\ \theta(0) = 0 & \dot{\theta}(0) = \omega \end{array} \tag{15}$$

with $\mathsf{A} = \frac{\frac{\mathsf{K}a}{\mathsf{M}}}{\sqrt{(\frac{\mathsf{K}}{\mathsf{M}}-\omega^2)^2+(\frac{f}{\mathsf{M}})^2\omega^2}}$, $\tan(\varphi) = \frac{-\frac{f}{\mathsf{M}}\omega}{\frac{\mathsf{K}}{\mathsf{M}}-\omega^2}$, and ω the smallest real solution of (11), if not otherwise stated. These initial conditions correspond to Rocard's approximated solution in (10), expressed here in the coordinates for (2). The values for the parameters are in Table 4.

Parameter a [m] is given according to the numerical experiment. The value of $g = 1.625$ corresponds to the gravity acceleration on the moon.

---

[8] For an overview of historical events concerning chaotic dynamical systems see Chap. 4 in Lorenz [12], Sect. 2.5 in Skiadas [18] or in Sect. 10.4 in Ginoux [8].

Table 4 Numerical values of the parameters for different numerical experiments

I [kg][m]^2	m [kg]	M [kg]	ρ [m]	K [N]/[m]	f	h	g [m]/[s]^2
0.03	0.1	0.02	0.01	12.5	0.15	0.000125	9.8
0.0001	0.5	0.5	0.1	50	0.01	0.03	9.8
0.0001	0.1	0.03	0.016	2.5	0.2	0	1.625

5.1 Numerical Analysis of the Quasi-uniform Regime

We investigate how $\dot{\theta}$ behaves with respect to a once the B-S system converges to its equilibrium state. For this, the value of the parameters are as in the first line of Table 4 and for each $a \in \{\frac{6j}{1000} | j = 0, \ldots, 50\}$ (i.e. 51 equidistant point in the interval [0, 0.30]), we run the numerical simulation for a time length of $T = 500$. The max, min and mean value of $\dot{\theta}$ for the last 50 time unites are calculated. The results are presented in Fig. 4: the 51 equidistant values of a (on the x-axis) are plotted with their corresponding mean value of $\dot{\theta}$ (dots in red), and max and min values (dots in gray). The underlying black line is the predicted pulsation, calculated by Rocard's method (i.e. solving (11)). In Fig. 4a, the smallest real solution of (11) is taken for the initial conditions in (15). On the other hand, in Fig. 4b, the largest real solution of (11) is taken for the initial conditions, as long as (11) has three real solutions, and then, as a increases, the smallest real solution is used to determine the initial conditions.

One can clearly appreciate the precision of Rocard's method to predict the pulsation ω, and this even for the case $h > 0$. The prediction is accurate before "l'accrochage" (i.e. for small a) and still valid for a large range of frequencies (form approximately 20 to 5). As a increases, $\dot{\theta}$ oscillates with increasing amplitude around the mean value, hence the increasing difference between max and min values of $\dot{\theta}$. For a larger than approximately $a = 0.26$, the B-S stops oscillating. Hence the red dots with value zero for both plots in Fig. 4. Here, the predicated pulsation is no longer valid.

5.2 Period-Doubling Bifurcations with h > 0

We chose here the crank's length a as control parameter and observe the period-doubling bifurcations on variable x. This case $h > 0$ has the advantage that one can clearly visualize period-doubling bifurcations with variables x and $\dot{x}$. With parameter values as in the second line of Table 4, Fig. 5 shows, in its first column (a, c, e and g), the last 7 time units of the numerical integration on [0, 40] for the variables x and $\dot{x}$. The second column in Fig. 5b, d, f and h displays (also for the last 7 time units) variables x in black and $\sin(\theta)$ in dashed blue (with an appropriate amplitude and additional constant for comparison's sake). For both columns, each plot shows

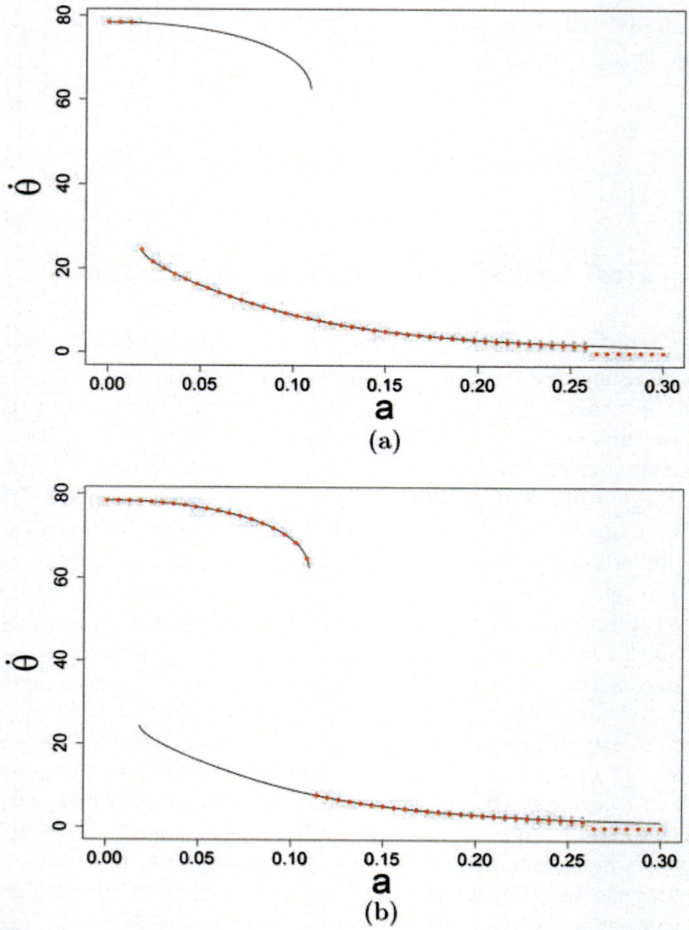

Fig. 4 Mean (red dots) and max & min (gray dots) of $\dot{\theta}$ against parameter a and predicted pulsation (black line) with parameters as in the first line of Table 4. The black line is the pulsation determined by solving (11)

a specific value of a. Clearly, as a increases, the x variable bifurcates: for $a = 0.06$, x is qualitatively similar to a sin curve and the pair $(x, \dot{x})$ converges towards a limit cycle (i.e. a closed curve without any intersection on the $x - y$ plan). For $a = 0.07$, the variable x bifurcates to a period-two oscillating regime: the pair $(x, \dot{x})$ converges towards a closed curve in $\mathbb{R}^2$ with one intersection. This continues for $a = 0.08$ with a period-four oscillating regime, and for $a = 0.081$ with a period-eight oscillating regime, and so on. Note that in Fig. 5a, the variable θ is already in a period-two oscillating regime.

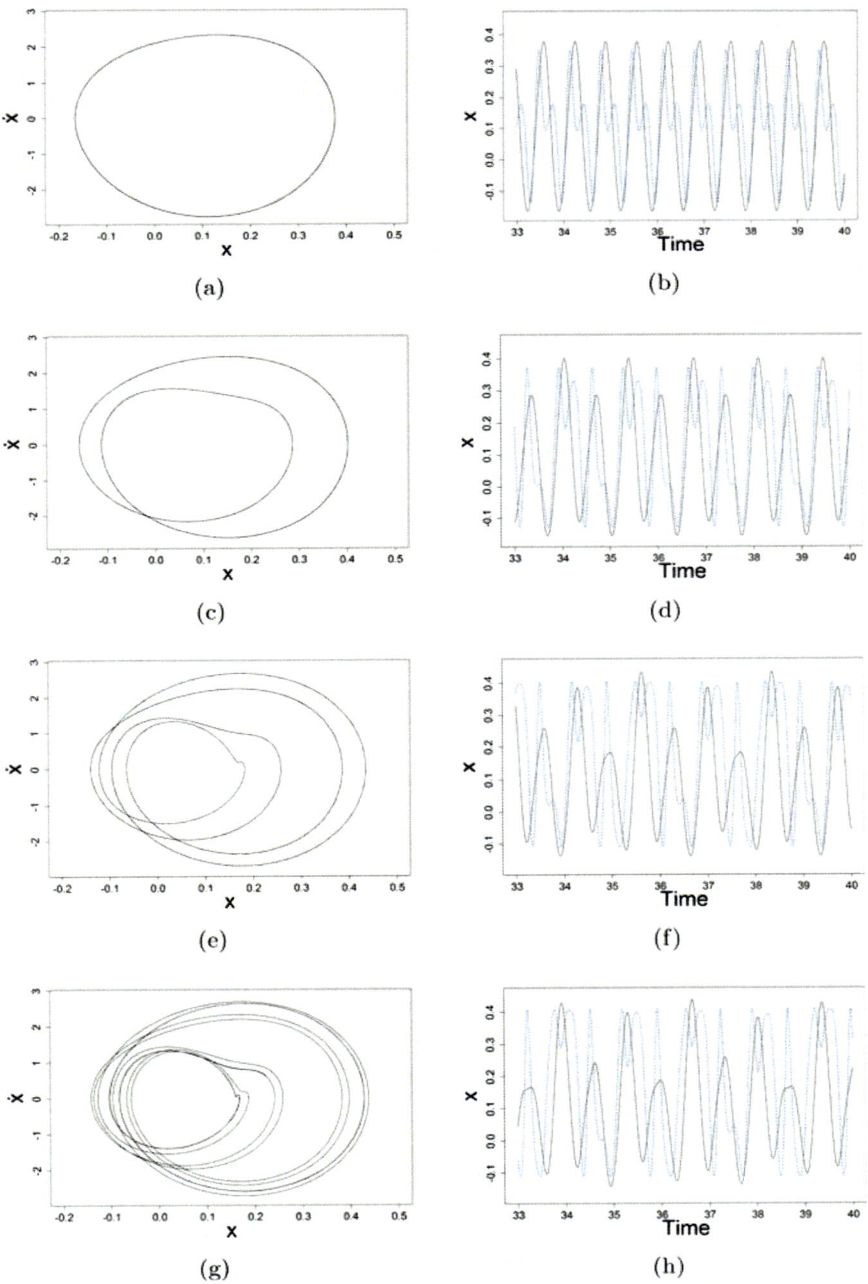

Fig. 5 Time evolution of the $x - y$ representation and of variables x (black) and $\sin(\theta)$ (dashed blue), with parameter value $a = 0.06$ (**a** and **b**), $a = 0.07$ (**c** and **d**), $a = 0.08$ (**e** and **f**) and $a = 0.081$ (**g** and **h**). The numerical values of the parameters are those of the second line in Table 4

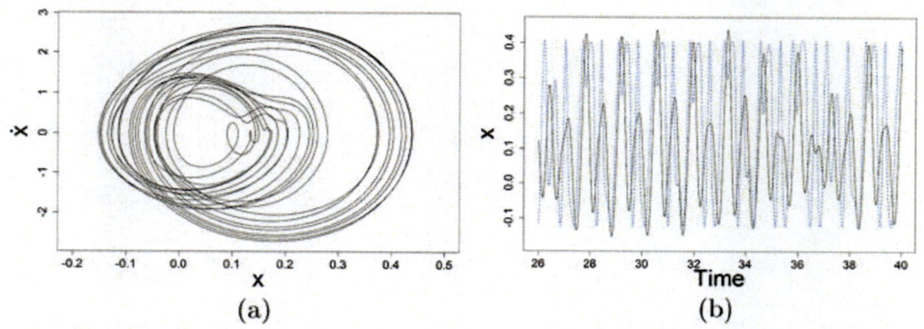

Fig. 6 Time evolution of the $x - y$ representation and of variables x (black) and $\sin(\theta)$ (dashed blue), with parameter value $a = 0.083$ (Figures (a) & (b)). The numerical integration is also on time $t \in [0, 40]$ but, in order to appreciate the non-periodicity of this regime better, the last 14 time units are shown. The numerical values of the parameters are those of the second line in Table 4

As a increases further, the system becomes chaotic. This is numerically shown in Fig. 6 when $a = 0.083$. The numerical integration is also on $[0, 40]$ but, in order to appreciate the non-periodicity of this regime better, the last 14 time units are shown.

In Fig. 7a, the Feigenbaum's cascade of period-doubling bifurcation is displayed for the B-S system with $h > 0$. The different values of the crank a are shown on the x-axis, while the y-axis shows the x values (when $\dot{x} = 0$) for which the B-S converges to (for the corresponding crank value a). Figure 7b shows a zoom out from Fig. 7a, so one can observe how the x variable (basically the amplitude) depends on a.

5.3 *Period-Doubling Bifurcations with* $h = 0$

As in Sect. 5.2, period-doubling bifurcations may also appear for B-S with $h = 0$. We again chose here the crank's length a as control parameter and observe the period-doubling bifurcations on variable x. Figure 8a shows yet another Feigenbaum's cascade of period-doubling bifurcation, and here for a B-S system with $h = 0$ as given in Y. Rocard's book Rocard [15]. The numerical values of the parameters are those of the third line in Table 4. As in Fig. 7, the different values of the crank a are shown on the x-axis, while the y-axis shows the x values (when $\dot{x} = 0$) for which the B-S converges to (for the corresponding crank value a). To the best of the authors' knowledge, this is the first observation of deterministic chaos in a B-S system for $h = 0$. Period-doubling bifurcations in the case for $h > 0$ (as discussed in Sect. 5.2) have already been observed in Rodriguez [16].

Figure 8b zooms out from Fig. 8a and one can see how the variable x depends on parameter a, suggesting an antimonotonicity behaviour: the creation and annihilation of periodic orbits. This feature is further developed in the next section.

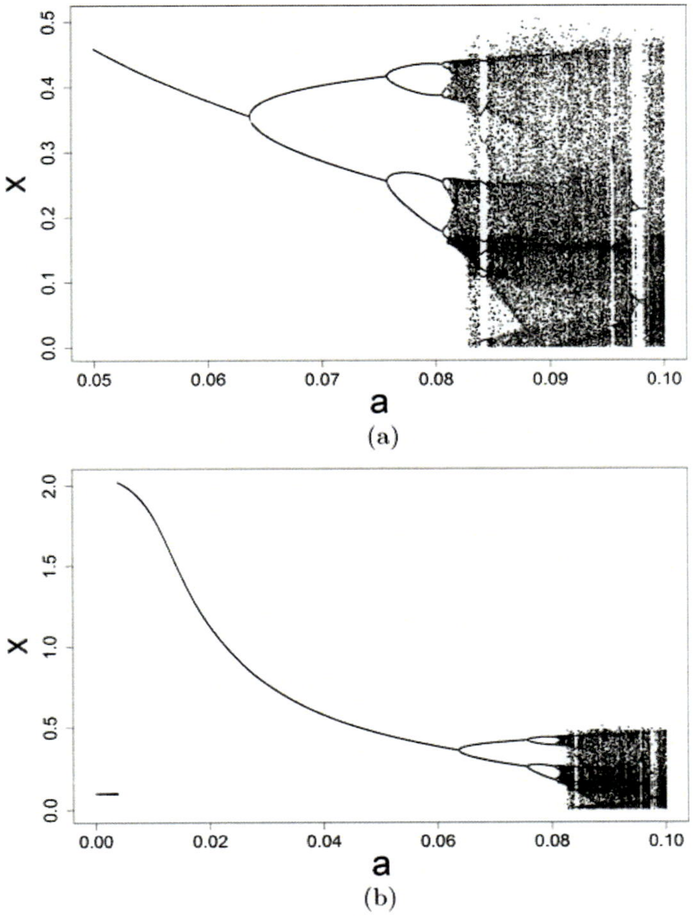

Fig. 7 In **a**, x (when $\dot{x} = 0$) is plotted against parameter a. **b** is a zoom out to show the dynamical behaviour before the the period-doubling cascade (i.e. for smaller values of a). The numerical values of the parameters are those of the second line in Table 4

5.4 Antimonotonicity with h = 0

The phenomenon of antimonotonicity, as reported in the seminal works of Kan and Yorke [9], Dawson et al. [5] and Kan et al. [10]), is observed for a B-S with parameter values as in the third line of Table 4, and for which the inertia I is taken as the second control parameter. In Fig. 9, similar bifurcation diagrams as in Fig. 8 are presented, while here each plot has a different value for I. Figure 9a has the largest I value and one can see a period-two bifurcation being created and then annihilated as parameter a gets larger. Further decreasing the value of I shows, in Fig. 9b, yet another period-

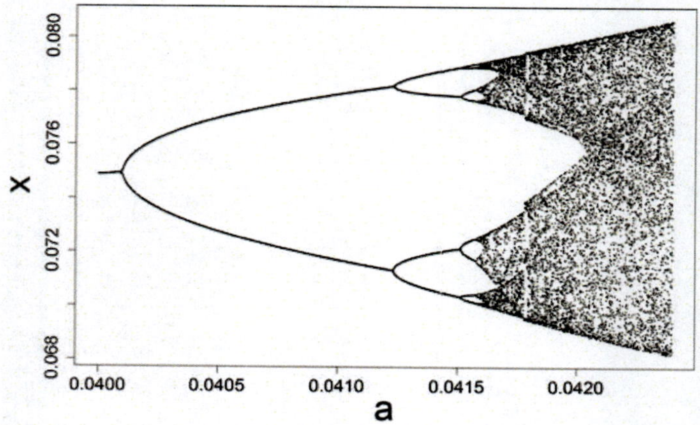

(a)

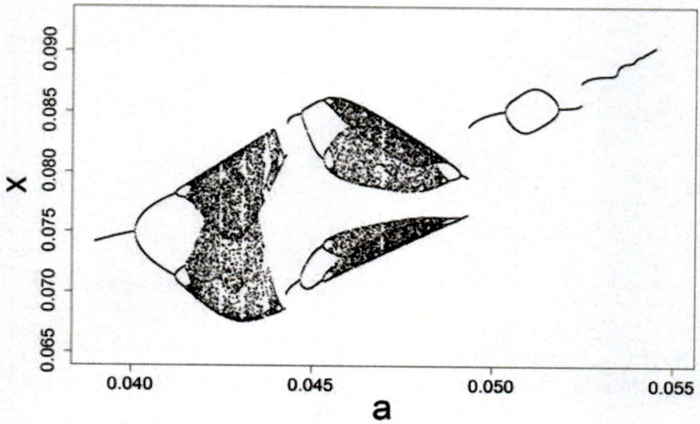

(b)

Fig. 8 In **a**, x (when $\dot{x} = 0$) is plotted against parameter a. **b** is a zoom out to show the dynamical behaviour beyond the period-doubling cascade (i.e. for larger values of a). The numerical values of the parameters are those of the third line in Table 4

two bifurcation: that is, the x variable goes from a period-one, to a period-two, to a period-four and then back to a period-two and finally to a period-one, as a increases in value. This phenomenon continues on as I decreases as shown in Fig. 9b, c.

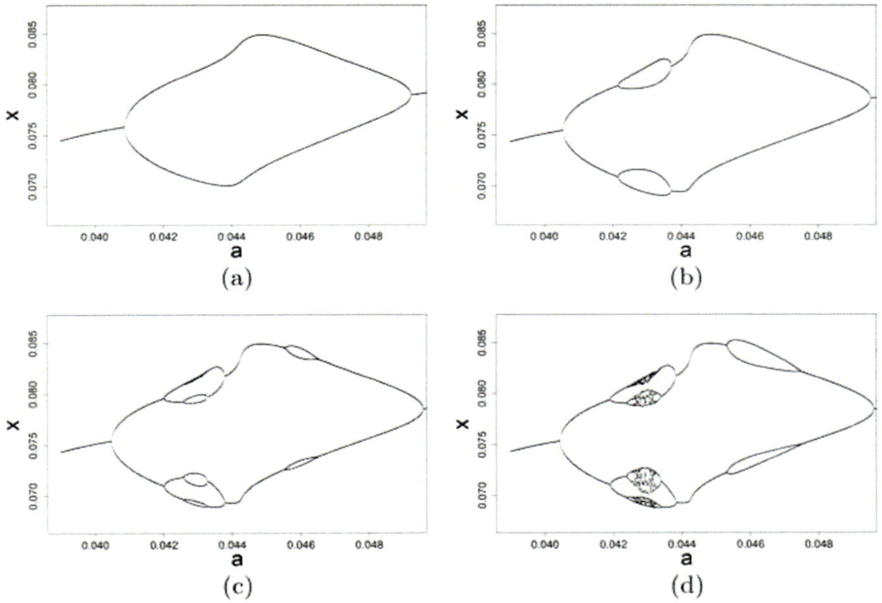

Fig. 9 In all four Figures, x (when $\dot{x} = 0$) is plotted against parameter a. In **a** $I = 0.000115$, in **b** $I = 0.000109$, in **c** $I = 0.0001078$ and in **d** $I = 0.000107$. The other numerical values for the parameters are those of the third line in Table 4

5.5 Numerical Estimation of Feigenbaum's Universal Constant

In our numerical experiments, we numerically determine the locations of the first six bifurcations thresholds. These are presented in Table 5 (when $h > 0$ and for parameter values as in the second line of Table 4) and in Table 6 (when $h = 0$ and for parameter values as in the third line of Table 4).

Table 5 Numerically determined bifurcation points for parameter a for Fig. 7a (when $h > 0$ and for parameter values as in the second line of Table 4)

n	Period	Bifurcation parameter (a_n)	Ratio $\frac{a_{k-1}-a_{k-2}}{a_k-a_{k-1}}$
1	2	0.06367045	–
2	4	0.07569875	–
3	8	0.08059755	2.455356
4	16	0.08106015	10.589710
5	32	0.08115485	4.884900
6	64	0.08117475	4.758794
7	128	0.08117895	4.738095

Table 6 Numerically determined bifurcation points for parameter a for Fig. 8a (when $h = 0$ and for parameter values as in the third line of Table 4)

n	Period	Bifurcation parameter (a_n)	Ratio $\frac{a_{k-1}-a_{k-2}}{a_k-a_{k-1}}$
1	2	0.04009785	–
2	4	0.04123185	–
3	8	0.04151235	4.042781
4	16	0.04157945	4.180328
5	32	0.04159425	4.533784
6	64	0.04159745	4.625000
7	128	0.04159815	4.571429

6 Conclusions

Yves Rocard used to say: “La physique c’est toujours un petit peu faux”.[9] Indeed, Rocard’s idea to determine the pulsation being an approximation is “a bit wrong”—however, one must recognise that Rocard’s method is remarkably accurate over a wide range of parameters! Let us emphasise that Rocard had the right intuition despite the fact that numerical simulations were not available in his time. Today’s tools not only enable us to confirm Rocard’s result, but allow us to study the dynamical system when one relaxes the assumption $\theta(t) \cong \omega t$.

In this contribution, we have numerically shown that the B-S system exhibits period-doubling bifurcations when increasing the values of parameter a and this, even when $h = 0$. Furthermore, we have shown that for small a, the B-S follows “washboard”-like dynamics, implying that it will converge only towards a periodic regime. Only once a is large enough may the B-S have chaotic behaviour.

There are several questions that could be treated as follow up: (1) for what value of a do both systems (drum and spring) converge towards a common pulsation ω for arbitrary initial conditions, and, (2), how does the accuracy of the predicated pulsation ω depend on the inertia I, in particular, in what way does friction h influence the quality of Rocard’s method to predict ω.

Acknowledgements The authors would like to thank Dr. M. Anderegg for his constructive remarks that were very helpful to this contribution.

[9] See, for example, Lurçat [13]. In English: “Physics is always a bit wrong”.

7 Appendix

7.1 Euler-Lagrange Equations

Applying the Euler-Lagrange equations to $L(\theta, x, \dot{\theta}, \dot{x}) = T(\dot{\theta}, \dot{x}) - V(\theta, x)$

$$\frac{\mathrm{d}}{\mathrm{d}t}(\frac{\partial L}{\partial \dot{\theta}}) = \frac{\partial L}{\partial \theta}, \qquad \frac{\mathrm{d}}{\mathrm{d}t}(\frac{\partial L}{\partial \dot{x}}) = \frac{\partial L}{\partial x},$$

we have

$$\begin{aligned}
\tfrac{\partial L}{\partial \dot{\theta}}(\theta, x, \dot{\theta}, \dot{x}) &= \tfrac{\partial T}{\partial \dot{\theta}}(\dot{\theta}, \dot{x}) &&= (\mathsf{I} + m\rho^2)\dot{\theta}, \\
\tfrac{\partial L}{\partial \dot{x}}(\theta, x, \dot{\theta}, \dot{x}) &= \tfrac{\partial T}{\partial \dot{\theta}}(\dot{\theta}, \dot{x}) &&= \mathsf{M}\dot{x}, \\
\tfrac{\partial L}{\partial \theta}(\theta, x, \dot{\theta}, \dot{x}) &= -\tfrac{\partial V}{\partial \theta}(\theta, x) &&= -\big(- mg\rho + \mathsf{K}(x - a\sin(\theta))(-a\cos(\theta))\big), \\
\tfrac{\partial L}{\partial x}(\theta, x, \dot{\theta}, \dot{x}) &= -\tfrac{\partial V}{\partial x}(\theta, x) &&= -\big(\mathsf{K}(x - a\sin(\theta)) - \mathsf{M}g\big),
\end{aligned}$$

and so the dynamical system is given by

$$\begin{aligned}
(\mathsf{I} + m\rho^2)\ddot{\theta} + h\dot{\theta} &= mg\rho + \mathsf{K}(x - a\sin(\theta))a\cos(\theta), \\
\mathsf{M}\ddot{x} + f\dot{x} &= -\mathsf{K}(x - a\sin(\theta)) + \mathsf{M}g.
\end{aligned}$$

7.2 Deriving the Equations of Motion

We assume the reader is familiar with the modelling of a spring of stiffness K on which a mass M is attached to it:

$$\mathsf{M}\ddot{x} + f\dot{x} = -\mathsf{K}x + \mathsf{M}g,$$

and the modelling of a drum on which a constant torque $mg\rho$ is applied on it:

$$(\mathsf{I} + m\rho^2)\ddot{\theta} + h\dot{\theta} = mg\rho.$$

To derive the equations of motion of the B-S regulator, we must adapt the force of the spring and add an additional torque to the drum, both terms accounting for the coupling. Figure 10 shows the radius of the crank a on to which the spring is attached. The referential for the spring's position is given for $\theta = 0$ (i.e. the doted line for which, by definition, $x = 0$). By definition, upwards from the referential line is considered as a negative displacement and downwards from the referential line is a positive displacement. For θ, clockwise direction is defined as positive. As one can see, for different values of θ, one mus readjust x's reference position. Accordingly, the force of the spring on the mass M is given by $-\mathsf{K}(x - a\sin(\theta))$.

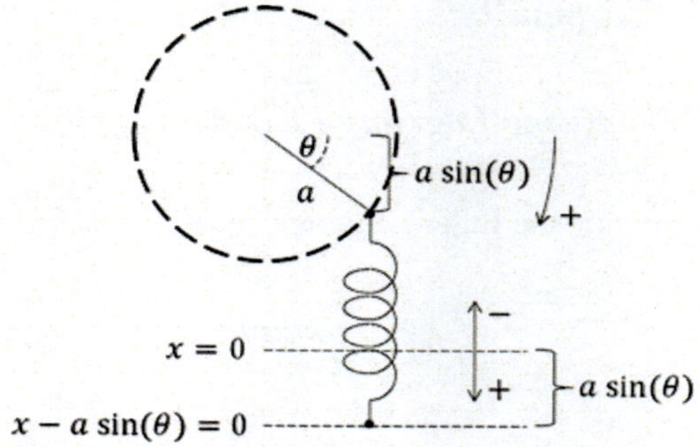

Fig. 10 Determining the force of the spring on mass M according to the value of θ

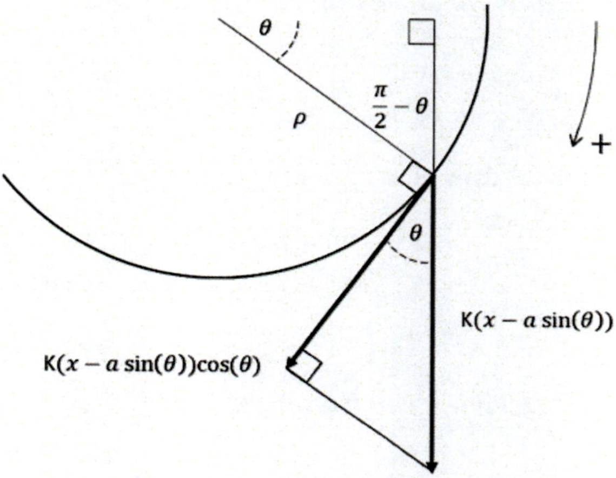

Fig. 11 The additional torque acting on the drum

The drum (viewing the flat surface of the cylinder) is sketched in Fig. 11. The force of the spring on mass M acts on the drum. The component of this force that acts on the drum as a torque is determined by elementary geometry, as shown in Fig. 11. In Panovko and Gubanova [14], a similar derivation is presented.

7.3 *Deriving (11)*

We here derive (11). By hypothesis, in this regime $\theta(t) \cong \omega t$, then $\ddot{\theta}(t) \cong 0$, enabling us write:

$$0 = H_a^{(1)}(\omega t) + H_a^{(2)}(\omega t, u(t)) - h\omega.$$

Since we are in a stationary regime, the dissipation due to damping has to be counterbalanced by the potential energy delivery over one cycle. Accordingly, over a full cycle, we have:

$$0 = \int_0^{\frac{2\pi}{\omega}} H_a^{(1)}(\omega t) + H_a^{(2)}(\omega t, u(t)) - h\omega dt,$$
$$= \int_0^{\frac{2\pi}{\omega}} H_a^{(1)}(\omega t) + H_a^{(2)}(\omega t, u(t))dt - (h\omega)\frac{2\pi}{\omega}.$$

For the two other integrals, we have

$$\int_0^{\frac{2\pi}{\omega}} H_a^{(1)}(\omega t)dt = \int_0^{\frac{2\pi}{\omega}} g(\mathsf{M}a\cos(\omega t) + m\rho)dt,$$
$$= g\big(\mathsf{M}a \underbrace{[\frac{\sin(\omega t)}{\omega}]_0^{\frac{2\pi}{\omega}}}_{=0} + m\rho\frac{2\pi}{\omega}\big),$$
$$= mg\rho\frac{2\pi}{\omega}.$$

$$\int_0^{\frac{2\pi}{\omega}} H_a^{(2)}(\omega t, u(t))dt = \int_0^{\frac{2\pi}{\omega}} \mathsf{K}a^2\big(\mathsf{A}\sin(\omega t + \varphi) - \sin(\omega t)\big)\cos(\omega t)dt,$$
$$= \mathsf{K}a^2\big(\int_0^{\frac{2\pi}{\omega}} \mathsf{A}\sin(\omega t + \varphi)\cos(\omega t)dt - \underbrace{\int_0^{\frac{2\pi}{\omega}} \sin(\omega t)\cos(\omega t)dt}_{[\frac{\sin(\omega t)^2}{2\omega}]_0^{\frac{2\pi}{\omega}} = 0}\big),$$
$$= \mathsf{AK}a^2 \int_0^{\frac{2\pi}{\omega}} \sin(\omega t + \varphi)\cos(\omega t)dt,$$
$$= \mathsf{AK}a^2 \int_0^{\frac{2\pi}{\omega}} \sin(\omega t)\cos(\varphi)\cos(\omega t) + \cos(\omega t)^2\sin(\varphi)dt,$$

and so we have:

$$\mathsf{AK}a^2\cos(\varphi)\int_0^{\frac{2\pi}{\omega}} \sin(\omega t)\cos(\omega t)dt = \mathsf{AK}a^2\cos(\varphi)[\frac{\sin(\omega t)^2}{2\omega}]_0^{\frac{2\pi}{\omega}} = 0,$$

and

$$\mathsf{AK}a^2\sin(\varphi)\int_0^{\frac{2\pi}{\omega}} \cos(\omega t)^2 dt = \mathsf{AK}a^2\sin(\varphi)[\frac{t}{2} + \frac{\sin(2\omega t)}{4\omega}]_0^{\frac{2\pi}{\omega}},$$
$$= \mathsf{AK}a^2\sin(\varphi)\frac{\frac{2\pi}{\omega}}{2}.$$

Therefore, we have

$$(h\omega - mg\rho)\frac{2\pi}{\omega} = \mathsf{A}\mathsf{K}a^2 \sin(\varphi)\frac{\frac{2\pi}{\omega}}{2},$$

$$(h\omega - mg\rho) = \frac{1}{2}\mathsf{A}\mathsf{K}a^2 \sin(\varphi) = \frac{\frac{(\mathsf{K}a)^2}{\mathsf{M}} \sin(\varphi)}{2\sqrt{(\frac{\mathsf{K}}{\mathsf{M}} - \omega^2)^2 + (\frac{f}{\mathsf{M}})^2\omega^2}},$$

$$\frac{2\mathsf{M}(h\omega - mg\rho)}{(\mathsf{K}a)^2} = \frac{\sin(\varphi)}{\sqrt{(\frac{\mathsf{K}}{\mathsf{M}} - \omega^2)^2 + (\frac{f}{\mathsf{M}})^2\omega^2}}.$$

At this stage, let us note that for the above equation to be consistent (i.e. so that $\sin(\varphi)$ is defined), one must satisfy (for given m, M, ρ, K, f, h, g, a and ω)

$$\left|\frac{2\mathsf{M}(h\omega - mg\rho)}{(\mathsf{K}a)^2}\sqrt{(\frac{\mathsf{K}}{\mathsf{M}} - \omega^2)^2 + (\frac{f}{\mathsf{M}})^2\omega^2}\right| \leqslant 1.$$

Now, since

$$\sin(\varphi) = \frac{-\frac{f}{\mathsf{M}}\omega}{\sqrt{(\frac{\mathsf{K}}{\mathsf{M}} - \omega^2)^2 + (\frac{f}{\mathsf{M}})^2\omega^2}},$$

we finally end with

$$\frac{2\mathsf{M}^2(mg\rho - h\omega)}{f(\mathsf{K}a)^2} = \frac{\omega}{(\frac{\mathsf{K}}{\mathsf{M}} - \omega^2)^2 + (\frac{f}{\mathsf{M}})^2\omega^2}.$$

7.4 *Point of Inflections Analysis*

For $\omega \geqslant 0$, the function to analyse is (see (13))

$$\alpha(\omega) = \frac{f(\mathsf{K}a)^2}{2\mathsf{M}^2}\Big(\frac{\omega}{(q - \omega^2)^2 + p^2\omega^2}\Big)$$

with $p = \frac{\mathsf{f}}{\mathsf{M}} > 0$, $q = \frac{\mathsf{K}}{\mathsf{M}} > 0$ and $a > 0$. Proving that there is only one point of inflection after α attained its maximum is equivalent to showing that the function α'' changes sign only once in $]\omega_0, +\infty[$, where ω_0 is the unique positive number such that $\alpha(\omega_0)$ is the maximum value of α in $\mathbb{R}_{\geqslant 0}$.

The first derivative of α with respect to ω is

$$\alpha'(\omega) = \frac{f(\mathsf{K}a)^2}{2\mathsf{M}^2}\Big(\frac{-3\omega^4 - (p^2 - 2q)\omega^2 + q^2}{\big((q - \omega^2)^2 + p^2\omega^2\big)^2}\Big)$$

and with the substitution $\omega(s) := \sqrt{sq}$ (i.e. $sq = \omega^2$) and $r = \frac{p^2}{q}$, we have

$$\alpha'(\omega(s)) = \frac{f(\mathsf{K}a)^2}{2\mathsf{M}^2q^2}\Big(\frac{-3s^2 - (r-2)s + 1}{(s^2 + (r-2)s + 1)^2}\Big).$$

Define the function β as $\beta(s) := \alpha'(\omega(s))$ as well as $\beta_n(s) := -3s^2 - (r-2)s + 1$ and $\beta_d(s) := s^2 + (r-2)s + 1$ so that

$$\alpha'(\omega(s)) = \beta(s) = \frac{f(\mathsf{K}a)^2}{2\mathsf{M}^2q^2}\frac{\beta_n(s)}{\beta_d(s)^2}.$$

Differentiating the above expression with respect to s leads to

$$\frac{d\big(\alpha'(\omega(s))\big)}{ds} = \alpha''(\omega(s))\omega'(s) = \beta'(s)$$

where α'' is the second derivative with respect to ω, and ω' and β' are, respectfully, the first derivative with respect to s. Since $\omega'(s) = \frac{\sqrt{q}}{2\sqrt{s}}$ and

$$\begin{aligned}\beta'(s) &= \frac{f(\mathsf{K}a)^2}{2\mathsf{M}^2q^2}\frac{\big(\beta_d(s)\beta_n'(s) - 2\beta_n(s)\beta_d'(s)\big)}{\beta_d(s)^3}\\ &= \frac{f(\mathsf{K}a)^2}{2\mathsf{M}^2q^2}\Big(\frac{6s^3 + 3(r-2)s^2 + ((r-2)^2 - 10)s - 3(r-2)}{(s^2 + (r-2)s + 1)^3}\Big)\end{aligned}$$

then the second derivative of α with respect to ω is

$$\alpha''(\omega(s)) = \Big(\frac{2\sqrt{s}}{\sqrt{q}}\Big)\frac{f(\mathsf{K}a)^2}{2\mathsf{M}^2q^2}\Big(\frac{6s^3 + 3(r-2)s^2 + ((r-2)^2 - 10)s - 3(r-2)}{(s^2 + (r-2)s + 1)^3}\Big).$$

With theses calculations, we show that α has a unique maximum for $\omega \geqslant 0$. Setting the first derivate of α to equal zero (i.e. $\alpha'(\omega) = 0$) and with the substitution $x := \omega^2$, we have

$$0 = 3x^2 + (p^2 - 2q)x - q^2 \tag{16}$$

giving the two roots

$$x_{1,2} = \frac{-(p^2 - 2q) \pm \sqrt{(p^2 - 2q)^2 + 12q^2}}{6}.$$

Since $\omega \geqslant 0$, we are only interested in the positive root, namely:

$$\omega_0 := \sqrt{x_1} = \Big(\frac{-(p^2 - 2q) + \sqrt{(p^2 - 2q)^2 + 12q^2}}{6}\Big)^{\frac{1}{2}}$$

We still need to show that $\alpha''(\omega_0) < 0$. For this, let s_0 such that $\omega(s_0) = \omega_0$. Then $\beta(s_0) = 0$ and by (16), $\beta_n(s_0) = 0$. This leads to $s_0 = \frac{-(r-2)+\sqrt{(r-2)^2+12}}{6}$. Therefore,

$$\beta'(s_0) = \frac{f(\mathsf{K}a)^2}{2\mathsf{M}^2 q^2} \frac{\beta_n'(s_0)}{\beta_d(s_0)^2}$$

and since $\beta_n'(s) = -6s - (r-2)$, then $\beta_n'(s_0) = -\sqrt{(r-2)^2 + 12} < 0$. Hence, the function α has one maximum for $\omega \geqslant 0$.

We now analyse the inflections points. For this we need to analyse the signs of $\alpha''(\omega)$ for $\omega \geqslant 0$. This means, we need to study the signs of $\beta'(s)$. There are two cases for $r - 2$ to investigate. For both cases $s \geqslant 0$.

Case :$r \geqslant 2$. Here the denominator of β' is strictly positive because $(r - 2)$ is positive (i.e. $s^2 + (r - 2)s + 1 > 0$). According to Descartes' rule of signs, the numerator of β' has only one positive root. Therefore, β' changes sign only once in $\mathbb{R}_{\geqslant 0}$. Since the function α has a its maximum attained by $\omega_0 > 0$ and has 0 as asymptote when ω tends towards $+\infty$, then this unique change of sign for β' corresponds to the one point of inflection of α taking place after ω_0.

Case :$0 < r < 2$. Here the denominator of β' is strictly positive because it has no real roots: the discriminant of $s^2 + (r - 2)s + 1$ is $(r - 2)^2 - 4 < 0$. According to Descartes' rule of signs, the numerator of β' may either have

(a) two positive roots or
(b) no roots at all.

Case (b) is not possible, since the function α possesses a maximum and has 0 as asymptote (i.e. the function α must have at least one inflection point since α' is zero at ω_0, decreases in value and then converges to zero when ω tends towards $+\infty$). In case a), since there are two positive roots, these must be on either side of ω_0 (the positive number that maximises the value of α), since at ω_0, the function α'' is strictly negative (i.e. $\alpha''(\omega_0) < 0$) and the sign of the dominant term of the polynomial in the numerator is positive. Hence, one point of inflection after ω_0 (i.e. after attaining the maximum of α).

7.5 *Determining the Particular a_0 with $h = 0$*

From Appendix 7.4, we know that the function α (see (13)) attains its maximum value for

$$\omega_0 = \sqrt{\frac{-(p^2 - 2q) + \sqrt{(p^2 - 2q)^2 + 12q^2}}{6}}.$$

Substituting the values a_0 and ω_0 in (11), leads to:

$$\frac{2\,mg\rho\mathsf{K}^2}{f(\mathsf{K}a_0)^2} = \frac{\omega_0}{\omega_0^4 + (p^2 - 2q)\omega_0^2 + q^2},$$

$$2\,mg\rho\mathsf{K}^2\big(\omega_0^4 + (p^2 - 2q)\omega_0^2 + q^2\big) = \omega_0(f\mathsf{K}^2)a_0^2,$$

$$a_0^2 = \frac{2\,mg\rho}{fq^2}\Big(\frac{\omega_0^4 + (p^2 - 2q)\omega_0^2 + q^2}{\omega_0}\Big).$$

Since $3\omega_0^4 + (p^2 - 2q)\omega_0^2 - q^2 = 0$ (see (16)), then

$$\begin{aligned} a_0^2 &= \frac{2\,mg\rho}{fq^2}\Big(\frac{\omega_0^4 + (p^2 - 2q)\omega_0^2 + \big(3\omega_0^4 + (p^2 - 2q)\omega_0^2\big)}{\omega_0}\Big), \\ &= \frac{2\,mg\rho}{fq^2}\Big(\frac{4\omega_0^4 + 2(p^2 - 2q)\omega_0^2}{\omega_0}\Big), \\ &= \frac{2\,mg\rho}{fq^2}\big(4\omega_0^3 + 2(p^2 - 2q)\omega_0\big). \end{aligned}$$

Therefore, the critical crankshaft is

$$a_0 = \sqrt{\frac{2\,mg\rho}{fq^2}\big(4\omega_0^3 + 2(p^2 - 2q)\omega_0\big)},$$

with $\omega_0 = \Big(\frac{-(p^2-2q)+\sqrt{(p^2-2q)^2+12q^2}}{6}\Big)^{\frac{1}{2}}$.

References

1. J.A. Blackburn, M. Cirillo, N. Grønbech-Jensen, A survey of classical and quantum interpretations of experiments on Josephson junctions at very low temperatures. Phys. Rep. **611**, 1–33 (2016)
2. I.I. Blekhman, G.Y. Dzhanelidze, The Bouasse Sarda dynamic regulator. Bull. Acad. Sci. USSR, Div. Tech. Sci. **10** (1955)
3. G. Colombo, Teoria del regolatore di Bouasse e Sarda. Rendiconti del Seminario Matematico della Università di Padova **28**, 338–347 (1958)
4. P. Coullet, J.M. Gilli, M. Monticellli, N. Vandenberghe, A damped pendulum forced with a constant torque. Am. J. Phys. **73**, 1122–1128 (2005)
5. S.P. Dawson, C. Grebogi, J.A. Yorke, I. Kan, H. Koçak, Antimonotonicity: inevitable reversals of period-doubling cascades. Phys. Lett. A **162**(3), 249–254 (1992)
6. M. Fridman, The transient and steady states of the Bouasse-Sarda regulator. J. Sound Vibr. **123**, 341–356 (1988)
7. M. Fridman, The method of multiple scales applied to the Bouasse-Sarda regulator. J. Sound Vibr. **138**, 135–154 (1990)
8. J.-M. Ginoux, *History of Nonlinear Oscillations Theory in France (1880–1940)* (Springer, 2017)
9. I. Kan, J.A. Yorke, Antimonotonicity: concurrent creation and annihilation of periodic orbits. Bull. Am. Math. Soc. **23**(2), 469–476 (1990)

10. I. Kan, H. Koçak, J.A. Yorke, Antimonotonicity: concurrent creation and annihilation of periodic orbits. Ann. Math. **136**(2), 219–252 (1992)
11. M. Lakshmanan, S. Rajasekar, *Nonlinear Dynamics, Integrability* (Springer, Chaos and Patterns, 2003)
12. E.N. Lorenz, *The Essence of Chaos* (University of Washington Press, 1995)
13. F. Lurçat, *Le Chaos*. Que sais-je? (2002)
14. Y.G. Panovko, I.I. Gubanova, *Stability and Oscillation of Elastic Systems: Modern Concepts, Paradoxes and Errors*. (NASA Technical Translation, 1973)
15. Y. Rocard, *Dynamique Générale des Vibrations*. (Masson et Cie, 1943)
16. J. Rodriguez, *Networks of Self-Adaptive Dynamical Systems*. PhD thesis, Ecole Polythechnique Fédérale de Lausanne, 2011
17. M. Roseau, On the coupling between a vibrating mechanical system and the external forces acting on it. Int. J. Non-Linear Mech. **17**, 211–216 (1982)
18. C.H. Skiadas, *The Foundations of Chaos Revisited: From Poincaré to Recent Advancements* (Springer, 2016)

Localized Waves in Silicates. What Do We Know from Experiments?

F. Michael Russell, Juan F. R. Archilla, and Santiago Medina-Carrasco

Abstract Since the latest review about solitary localized waves in muscovite, called quodons, (F.M. Russell in Quodons in Mica. Springer, Cham, pp. 475–559, 2015a [1], F.M. Russell in Quodons in Mica. Springer, Cham, pp. 3–33, 2015b [2]) there have been many developments, specially from the point of view of experiments, published in several journals. The breakthrough hypothesis that was advanced in that review that dark tracks were produced by positive electrical charge moving in a localized wave, either transported by swift particles or by nonlinear localized waves, has been confirmed by experiments in muscovite and other silicates. In this paper we review the experimental results, some already published and some new, specially the phenomenon of charge transport without an electric field, called hyperconductivity. We also consider alternative explanations as phase transitions for other tracks. We also attempt to describe numerical simulations that have confirmed the order of magnitude of quodons energy and calculations underway to determine more properties of electron and hole transport by quodons.

1 Introduction

The existence of localized waves in silicates layers were first proposed in 1994 [3]. This was an important step in a long story of research about the nature of tracks in muscovite mica since 1967 [4, 5]. A scientific review [2] and a longer historical

F. M. Russell
School of Computing and Engineering, University of Huddersfield, HD1 3DH Huddersfield, UK

J. F. R. Archilla (✉)
Group of Nonlinear Physics, Universidad de Sevilla, ETSII, Avda. Reina Mercedes s/n, 41012 Sevilla, Spain
e-mail: archilla@us.es

S. Medina-Carrasco
X-Ray Laboratory, CITIUS, Universidad de Sevilla, Avda Reina Mercedes 4B, 41012 Sevilla, Spain
e-mail: sanmedi@us.es

C. H. Skiadas et al. (eds.), *13th Chaotic Modeling and Simulation International Conference*, Springer Proceedings in Complexity,
https://doi.org/10.1007/978-3-030-70795-8_51

review [1] were published in 2015. The hypothesis that *quodons*, i.e., quasi on-dimensional lattice excitations, transport electric charge was proposed in those reviews but not developed.

This hypothesis was a fundamental change that led to new theory, new interpretation of previous results about tracks in muscovite, and specially to experiments that confirmed and modified the theory. Therefore, we have thought that it was time for a new review that provided a comprehensive and brief summary of the state of knowledge and the challenges in front of the research.

The research can be divided in three stages that are interconnected.

1. **Tracks**: Tracks by swift particles.
2. **Quodons**: Tracks by lattice excitations or *quodons*.
3. **Hyperconductivity**: Quodons with electric charge and *hyperconductivity*.

Here we present the beginning and end of the three stages and some of the highlights. Later, we will explain in detail some key aspects.

1.1 Tracks *by Swift Particles*

This stage starts in 1967 with the observation in mica of dark tracks of charged particles from neutrino interactions [4] and finishes in 1993 with an explanation of track formation by release of lattice energy [6] and the description of semi-transparent tracks in mica related with positron dark tracks [7]. Dark tracks are made out of magnetite and some shorter semi-transparent tracks are made out of the mineral epidote.

Note that tracks are also the result of experiments similar to particle tracks in a bubble chamber. They are experiments that nature has made and have been conserved as a fossil in muscovite crystals. They have been done at temperatures, pressure and specially time scales outside of the possibilities of physicists.

1.2 *Tracks by Lattice Excitations or* Quodons

From the very beginning [4] it has been observed that only 0.1% of the dark tracks in muscovite were produced by swift particles, while the rest lie along the close-packed direction within the cation layers and therefore are related with the crystal structure. This second stage starts with the calculation of nonlinear forces between potassium ions and using them to obtain an approximate KdV equation for lattice displacements. The KdV equation supports soliton solutions [8], therefore the majority dark lines in muscovite could be produced by lattice-solitons. These results were presented at a conference in 1994 [3] and extended the following year [9, 10].

Interestingly, in the same year 1994, it was attempted to observe lattice-solitons by bombarding silicon with 0.8 MeV Ar^+ and detecting the ejection of an atom [11].

The experiment failed, perhaps among other reasons because it used silicon which is not layered and have a complicated structure for soliton propagation as the nearest neighbours do not form straight lines.

These lattice excitations were named *qodons* in 1995 [10] and later *quodons* in 1998 [12]. This was an acronym for quasi one-dimensional excitations, a descriptive term which also recognized that the actual type of excitation was not well known. It is worth noting that the term lattice-soliton was changed to *breather*. Breathers differ from solitons in having an internal vibration and smaller energy and were starting to be thoroughly studied [13, 14].

The highlight of this stage is probably the success of another experiment in 2007 [15] similar in design to the previous one [11]. In this case a mica monocrystal was bombarded with alpha particles and it was possible to detect the ejection of atoms at the opposite side of the sample along the direction of close-packed lines within the cation layers.

This stage finishes in 2015 with two comprehensive reviews, a shorter and scientifically oriented one [2] and a longer historical review oriented to the non-specialist [1]. But in these two reviews the next stage is also hinted.

1.3 Quodons with Electric Charge and Hyperconductivity

It was well known that most tracks in muscovite were produced by the recoil of potassium atoms after beta decay [16]. In 2015, a thorough analysis of the decay modes of ^{40}K [17–19] showed that 90% of decays left a charge behind, and this charge was positive except in 0.001% of positron decays, when it was negative. Then, it was realized that dark tracks by swift particles were produced only by positive particles and that the thickness of, for example, positron tracks, at sonic speed, when they were about to stop, were similar to quodon tracks. These two observations led to the deduction that quodons have electric charge, and dark tracks, positive charge [20]. This hypothesis was already introduced at the previous reviews and it was later extended in [21].

This profound change in the quodon concept provided something to measure easily, electric current, when quodons were excited by particle bombardment without an electric field, a phenomenon called *hyperconductivity*. Experiments were successful and also were able to explain new properties of quodons [22, 23] in muscovite and other layered silicates.

2 Important Points

In this section we concentrate in some important points which illustrate either fossil tracks or experimental results or experiment setup.

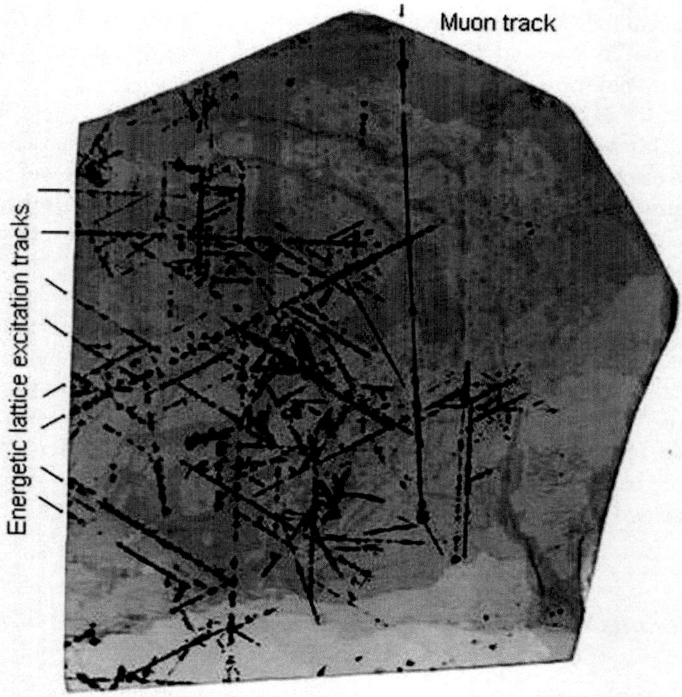

Fig. 1 A sheet of mica muscovite showing many majority tracks due to lattice excitations within the hexagonal structure of the cation layer and a muon track in an unrelated direction

2.1 First Encounter with Dark Tracks in Muscovite

It is important to emphasize that the main author of this research F.M. Russell has been all his career dedicated to high energy physics, first at Harwell Laboratory,[1] then at Oak Ridge National Laboratory (ORNL) in the U.S.A, and thereafter at the Rutherford Appleton Laboratory (RAL) in the U.K. In this way, when in 1963 at a museum in North Caroline,[2] he found himself in front of a specimen of muscovite with abundant dark tracks, he recognized the striking similarity with the tracks of swift particles in bubble chambers. A similar sheet is presented in Fig. 1.

---

[1] Atomic Energy Research Establishment near Harwell, Oxfordshire, U.K.

[2] Museum of North Carolina Minerals, Spruce Pine, North Caroline, U.S.A.

2.2 How Were the Swift Particles Identified?

There were different methods, but perhaps the clearest is the kinkiness of those dark tracks. Charged swift particles when entering in matter experience scattering with the matter ions. The probability of scattering at a given angle can be calculated by Rutherford law and the angles can be can be seen and measured with a microscope and the results compared with given particles. An example can be seen in Fig. 2, comparing the second difference, basically the scattering angle, of some track in mica with positrons in photographic film [25], taking into account the difference in mass and density of the scattering ions.

2.3 Which Particle Tracks Were Identified?

The particles that produce dark tracks in muscovite and could be identified were positive muons, i.e., antimuons, which are the particles that can be produced deep underground after neutrino interaction [4, 5, 16, 26], positrons from ^{40}K decay and antimuon decay [7, 16, 27–29]. Protons can be recognized by the short length of

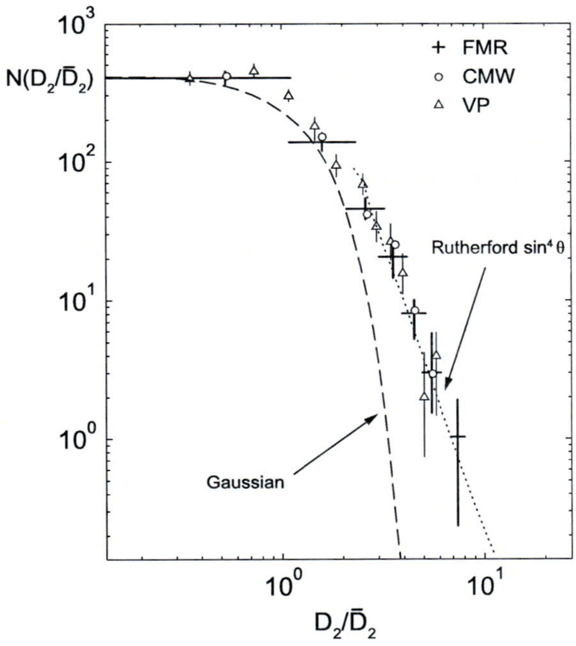

Fig. 2 Probability of scattering at given angles for tracks corresponding to positrons in muscovite [+], compared with positrons in photographic film [Δ] and by Wolfendale's group [o]. The results fit closely to the Rutherford Law, thus strongly supporting the hypothesis that the lines are tracks of charged particles. Data from [16, 24, 25] for VP, CMW and FMR, respectively. Reproduced with permission from: Russell [16] Copyright © 1988, Elsevier

the tracks corresponding to non-relativistic speed [2, 30]. Also, alpha particles can be discriminated from the multiple scattering events, proof of their large energy and mass [2, 26].

The remarkable fact that all the particles that produce dark tracks were positive was used in 2015 to recognize that the large majority of quodons that produce dark tracks have also positive charge [1, 2, 20].

2.4 *How Were the Tracks Produced?*

There is not enough energy to produce the dark tracks, this means that the source of energy is already in the lattice, in the form of a metastable state [10, 11, 28, 29].

Natural crystals of muscovite mica contain various impurities, especially iron, incorporated during their growth. It has been found that this can lead to a unique situation, as a crystal cools following growth, during which minute perturbations of the crystal can be recorded and stored indefinitely. Although muscovite is a common mineral in rocks, large crystals grow only in pegmatites associated with magmas at temperatures of about 500°C and under high pressure at about 5 km underground [31]. Inevitably, large single crystals of good quality are rare but they are of special interest because of the information they have been found to contain. A common feature of micas is their ease of cleavage, in the (001)-plane. The black material forming the patterns is the iron oxide mineral magnetite, so named because it is ferro-magnetic.

As a crystal cools slowly at high temperature it tries to reach a lower energy state by expelling the magnetite at the weakest part of the lattice, the cleavage plane. The magnetite grows epitaxially, centred in the potassium sheet and grows in the directions of structural weakness. These are the principal crystallographic directions, which are easily determined by percussion figures [1]. This has been confirmed by both optical and electron microscopy. In fact, the distortion of the lattice is readily seen by observing the local region of the intrusive magnetite by reflected light or by surface interferometry. Contrary to the basic assumption in [32] of global bi-stability of structure there is no evidence for this in the observed patterns involving magnetite.

2.5 *Two Different Recording Processes*

It has been found that there are two different recording processes leading to the observed patterns, involving different impurities. The dominant process leading to magnetite is triggered by passage through the crystal of a positive charge in the vicinity of the potassium sheets. This can result from a positively charged, high-energy, muon created in a neutrino interaction within the Earth or by direct penetration of a cosmic ray. Another source is from electron-positron showers arising from a high-energy gamma interaction. The flight-paths of these particles are influenced by channelling and diffraction scattering due to the pronounced layered structure of muscovite [16]. The most informative source, however, is from the rare decay channel of ^{40}K creating positrons [7]. Study of the fossil tracks of these positrons has shown that the origin of the nucleation sites for triggering magnetite growth does not involve ionization of the lattice. For relativistic positrons from this source a fossil track results even when the rate of energy loss is less than 1 eV per 10,000 atoms along the flight path. The rate of energy loss increases as a positron slows down, leading to an increase in the amount of magnetite formed. Due to anisotropy of the mechanical properties of the layered structure this increase shows as a widening of the magnetite ribbon delineating the flight-path. This suggests that the recording process is of a chemical nature, with the probability for an impurity ion migrating to the flight path increasing as the positron's speed decreases.The dominant source of the long ribbons of magnetite arising from moving positive charges is the dominant decay channel of ^{40}K, in which an electron is emitted. These energetic electrons do not initiate fossil magnetite tracks. However, they leave a positive charge at the decay site that can be trapped and carried by a mobile lattice excitation arising from the recoil motion of the decayed nucleus. These mobile, non-dissipative, highly localized excitations move at slightly sub-sonic speed, leading to magnetite ribbons of width of similar width to those due to nearly stopped positrons [2]. The last known source of swift positively-charged ions is from atomic cascades arising from nuclear scattering of relativistic particles.

The second and much rarer recording process involves formation of the mineral epidote, which requires an excess of calcium during crystal growth. These fossil tracks arise from the emission of a positron, leaving a negative charge at the decay site, which is trapped and transported by the mobile recoil excitation. This leads to a ribbon of transparent epidote that is not intrusive in the potassium sheets [7]. The formation process of the epidote is poorly understood and might involve a bi-stable crystal state [32]. It is hoped that this explanation of the origin of the fossil magnetite-ribbon tracks might encourage study of the formative process for the fossil epidote tracks, as this has the potential for ballistic, low-loss, transport of electrons in layered insulators [7, 22, 23, 33].

3 How Was the Experiment in Lattice-Excitations or Quodons Done?

The highlight of the research on lattice excitations, sometimes called lattice-solitons, breathers or quodons in this context was the experiment in 2007 [15]. Alpha particles were sent at an angle with the muscovite sheet and therefore with the potassium layer to prevent the possibility of transmission and it was detected at the other side of the monocrystal corresponding to low Miller indexes, the ejection of an atom from the surface. The atom was detected because it was ionized by an electric field and the charge detected. Ejection of atoms from a silicate surface needs energies of 7–8 eV, however, it is not necessary that a quodon had that energy as the passage of a

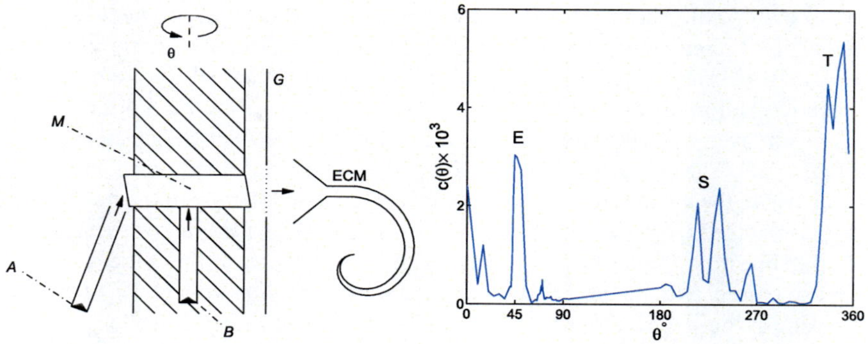

Fig. 3 **Left**: Setup of the quodon experiment. A: alpha source, M: mica monocrystal, ECM: electron channel multiplier, G: grid, B: alternative position for the alpha source. **Right**: Outcome of the quodon experiment: Plot of the angular dependence of the ECM count rate. T: test, S: sputtering from the front face, E: peak from ejected atoms at the rear face in the [0 1 0] direction. Reproduced with permission from: Russell and Eilbeck [15]. Copyright © 2007, EPLA

vibrational energy in the vicinity of the surface is enough to increase the probability of ejection [34]. Both the setup and the outcome can be seen in Fig. 3.

4 How it Was Demonstrated Hyperconductivity, i.e., that Quodons Carry Charge?

Hyperconductivity is defined as the transport of charge in absence of an electric field. The charge is transported by nonlinear excitations which have their own energy and momentum from the cause that created them. Due to the combination of nonlinearity and discreteness they travel long distances in atomic terms with little attenuation.

An experiment was set up quite similarly to the previous ones. The way to excite lattice excitations or quodons was also by sending alpha particles, due to its simplicity. The hypothesis was that alpha particles would produce many quodons and some proportion of them would propagate to the other side of the sample and in this way a current could be measured. Muscovite is a very good insulator but there was the possibility that the surface and certainly the ionized air would transport charge. To discard this effect the two sides of the sample were connected and therefore the potential difference among both contacts would be zero and also the electric field would be zero. Lattice excitation or quodons would travel due to their initial energy and momentum.

The experiment was a success but with some unexpected results. Instead of a steady current after the alpha gate was open, the current showed a peak, but then it would diminish to a small limiting value. The phenomenon was soon explained: there are not free carriers in muscovite band structure, the available charge is the

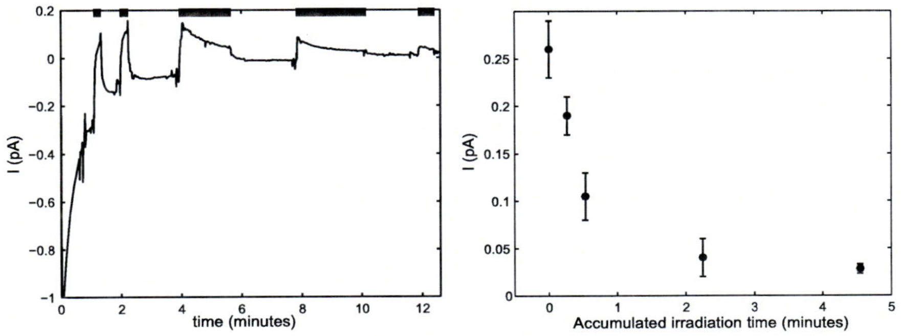

Fig. 4 **Left**: Plot of the hyperconductivity current, the time intervals marked in black at the top correspond to the opening of the alpha gate. **Right**: Hypercurrent as a function of the accumulated time of alpha exposure, showing an exponential decrease corresponding to the depletion of the charge reservoir from ^{40}K beta decay. Reproduced with permission from: Russell et al. [22]. Copyright © 2017, EPLA

one obtained after beta decay of ^{40}K, mainly positive after β^-, i.e., the emission of an electron is the dominant branch, but also some negative charge after β^+ positron emission. This reservoir is depleted in some minutes, and the remaining current is exactly the flux of electric charge brought by the alpha flux [22]. The current peaks and their decrease can be seen in Fig. 4.

5 What Properties of Hyperconductivity and Quodons Were Deduced from Experiments?

More experiments in hyperconductivity [23] were able to deduce a number of facts:

- Other layered silicates as lepidolite, phlogopite, chrysolite and both natural and synthetic fluorphlogopite supported hyperconductivity and thus the propagation of quodons. However, a layered silicate as biotite with similar structure does not support it. It was not found in unrelated materials that could be used in quodon technology as PTFE, quartz, borosilicate glass and epoxy resin.
- Hyperconductivity is not sensitive to minor crystal defects and can even anneal some of them. It can also pass through some interfaces.
- Hyperconductivity is not affected by magnetic fields up to 1.1 T.
- Quodons have very long flight paths, this can be deduced by comparing the drop in hypercurrent when the alpha bombardment is stopped. In a good crystal the hypercurrent continues to flow some seconds, while in a crystal with many defects, the hypercurrent stops almost immediately as can be seen in Fig. 5

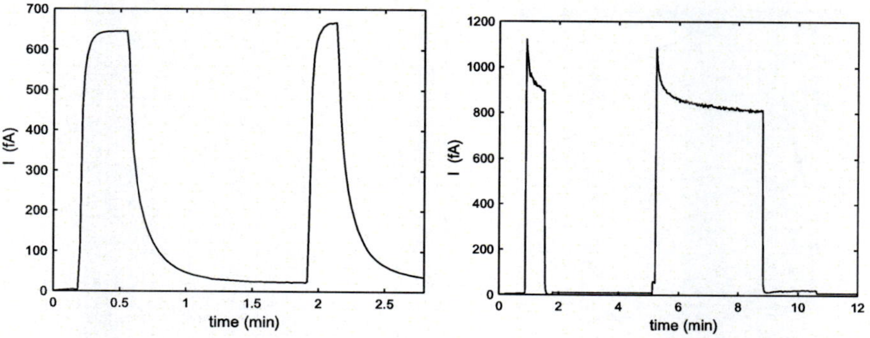

Fig. 5 Plot of the hypercurrent corresponding to two intervals of alpha exposure in a previously depleted crystal. **Left**: Crystal of lepidolite of good quality. Note the soft decay of the hypercurrent after the alpha flow is stopped. **Right**: Crystal of phlogopite of bad quality. Note the abrupt decrease of the hypercurrent after alpha irradiation is stopped. Reproduced with permission from: Russell, F.M. et al. [23]. Copyright © 2019, EPLA

6 What Types of Quodons Are There?

There is no clear information from the experiments, however from the fossil tracks as seen in Fig. 6, it can be deduced:

- There are positive quodons, negative quodons and probably neutral quodons. Negative quodons can be seen as an epidote track in exactly the opposite direction from a positron track and therefore corresponding to the recoil of the nucleus of ^{40}K after β^+ decay. Neutral quodons can be deduced from intermittent dark tracks, which seem like quodons loosing an regaining positive charge.

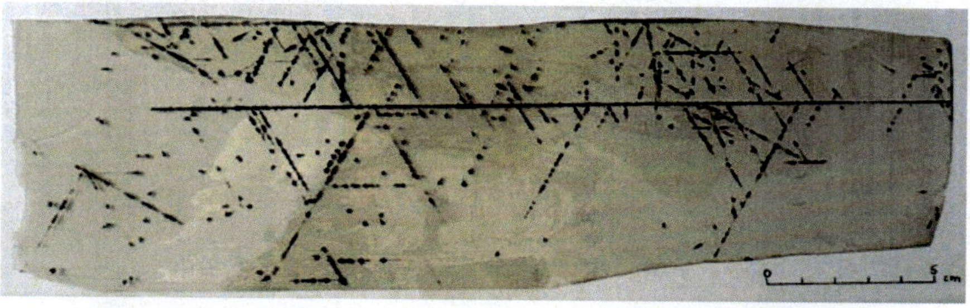

Fig. 6 A sheet of mica muscovite showing a quodon primary track and many secondary tracks scattered from it. Also it is possible to see the intermittency in the secondary tracks along the close packed direction of the cation layer. This is interpreted as a quodon loosing and regaining a positive charge. Reproduced with permission from: Russell and Eilbeck [30]. Copyright © 2011, AIMS

- There are some more energetic quodons that produce straighter and thicker dark tracks and some less energetic quodons because they appear often as weaker dark tracks scattered from a primary track. As both types are dark, it is deduced that both have positive charge. They might have different nature, maybe primary tracks could be crowdions or kinks as they transport charge in an ionic crystal and have large energies of 20–30 eV [17, 35–38]. Secondary tracks, could be interpreted as breathers, because they have good mobility in mica models with little or no radiation, with energies of 0.2–0.3 eV [12, 39, 40] and recently they have been shown to scatter in different close-packed directions [41]. However, breathers do not transport charge and if they couple to a charge their properties and physical description change completely. Certainly, breathers could correspond to neutral quodons. A model for lattice excitations coupled to a hole or electron has been constructed for muscovite, but the properties of localized excitations using it are still under study [42, 43].

7 Alternative Explanations of Tracks

There have not been many alternative explanations of tracks in muscovite. It was suggested that the majority of dark lines corresponds to dislocations because they lie in the close-packed directions, but without further proof [24]. Arguments against dislocations are that they should appear along crystal fractures, which does not occur [16] and that dark tracks do not continue to the edge of the crystal specimen as it should occur with dislocations [2]. Recently an interesting explanation based on phase transition in a bistable lattice has been proposed [32]. The research was based on the observation[3] that the pitch of the on-site potential and the equilibrium distance of the interatomic potential should be different in a real material. This bring about the existence of different stable configurations, and the authors found a switching wave between configurations that propagates longitudinally along the direction of atomic chains. They used a Frenkel-Kontorova 2D system with morse interaction potential. There was no attempt to relate their findings with physical magnitudes and to explain the coloration of lines or the kinkiness of the swift particle tracks. Also, the hyperconductivity experiments were not explained and the charge of the ions in the cation layer were not taken into account as explained in the article. Nevertheless, it opens a new path to understand some of the phenomena observed in muscovite and other layered silicates, particulary epidote tracks, which are not produced by swift particles.

[3] J.F.R. Archilla, private communication (2019).

8 Summary

In this article we have tried to present an updated review of the research in nonlinear waves in layered silicates, particularly, but not only, in mica muscovite. We have attempted to make clear for the non specialist which are the main experimental facts and their interpretation, leaving many details to the references. The main results are that some dark tracks in muscovite can be related to swift positive particles, that many other tracks along atom chain direction of the cation layers can be interpreted as lattice excitations, called quodons. Most quodons carry positive charge although some may have negative charge or none. This was demonstrated by hyperconductivity experiments, that is, the transport of charge in the absence of an electric field. Variants of hyperconductivity experiments allowed for the deduction of many properties of quodons. Other interpretations of dark tracks may be complementary and be useful to understand some of the tracks.

Acknowledgements JFRA thanks a travel grant from VIPPITUS 2020 and projects PAIDI 2019/FQM-280 and MICINN PID2019-109175GB-C22. SM-C acknowledges grant from PPITUS-2018.

References

1. F.M. Russell, I saw a crystal, in *Quodons in Mica*, ed. by J.F.R. Archilla, et al. Springer Ser. Mat. Sci., vol. 221 (Springer, Cham, 2015a), pp. 475–559
2. F.M. Russell, Tracks in mica, 50 years later, in *Quodons in Mica*, ed. by J.F.R. Archilla, et al. Springer Ser. Mat. Sci., vol. 221 (Springer, Cham, 2015b), pp. 3–33
3. D.R. Collins, F.M. Russell, Computer modelling studies of solitons in layered crystals, in *Proceedings of the 6th Joint EPS-APS International Conference on Physics Computing*, ed. by R. Gruber, F. Tomasini (Lugano, Manno, Switzerland, 22–26 August 1994), pp. 243–246
4. F.M. Russell, The observation in mica of tracks of charged particles from neutrino interactions. Phys. Lett. **25B**, 298–300 (1967)
5. F.M. Russell, Tracks in mica caused by electron showers. Nature **216**, 907–909 (1967)
6. F.M. Russell, J.W. Steeds, Track formation by release of lattice energy. Nucl. Tracks. Rad. Meas. **22**(1–4), 65–70 (1993)
7. J.W. Steeds, F.M. Russell, W.J. Vine, Formation of epidote fossil positron tracks in mica. Optik **92**, 149–154 (1993)
8. D.J. Korteweg, G. de Vries, On the change of form of long waves advancing in a rectangular canal, and on a new type of long stationary waves. Phil. Mag. **5**(39), 422–443 (1895)
9. F.M. Russell, D.R. Collins, Lattice-solitons and non-linear phenomena in track formation. Rad. Meas. **25**, 67–70 (1995)
10. F.M. Russell, D.R. Collins, Lattice-solitons in radiation damage. Nucl. Instrum. Meth. B **105**, 30–34 (1995)
11. D. Schlößer, K. Kroneberger, M. Schosnig, F.M. Russell, K.O. Groeneveld, Search for solitons in solids. Rad. Meas. **23**(1), 209–213 (1994)
12. J.L. Marín, J.C. Eilbeck, F.M. Russell, Localized moving breathers in a 2D hexagonal lattice. Phys. Lett. A **248**(2–4), 225–229 (1998)
13. S. Flach, Obtaining breathers in nonlinear Hamiltonian lattices. Phys. Rev. E **51**(4), 3579–3587 (1995)

14. R.S. MacKay, S. Aubry, Proof of existence of breathers for time-reversible or Hamiltonian networks of weakly coupled oscillators. Nonlinearity **7**, 1623 (1994)
15. F.M. Russell, J.C. Eilbeck, Evidence for moving breathers in a layered crystal insulator at 300 K. EPL **78**, 10004 (2007)
16. F.M. Russell, Identification and selection criteria for charged lepton tracks in mica. Nucl. Tracks. Rad. Meas. **15**, 41–44 (1988)
17. J.F.R. Archilla, Y.A. Kosevich, N. Jiménez, V.J. Sánchez-Morcillo, L.M. García-Raffi, A supersonic crowdion in mica, in *Quodons in mica*, ed. by J.F.R. Archilla, et al. Springer Ser. Mat. Sci., vol. 221 (Springer, Cham, 2015), pp. 69–96
18. J.A. Cameron, B. Singh, Nuclear data sheets for A=40. Nucl. Data Sheets **102**(2), 293–513 (2004)
19. X. Mougeot, R.G. Helmer, LNE-LNHB/CEA—Table de Radionucléides, K-40 tables (2012). http://www.nucleide.org
20. F.M. Russell, Charge coupling to anharmonic lattice excitations in a layered crystal at 800 K (2015). arXiv:1505.03185 [cond-mat.supr-con]
21. J.F.R. Archilla, F.M. Russell, On the charge of quodons. Lett. Mater. **6**, 3–8 (2016)
22. F.M. Russell, J.F.R. Archilla, F. Frutos, S. Medina-Carrasco, Infinite charge mobility in muscovite at 300 K. EPL **120**, 46001 (2017)
23. F.M. Russell, A.W. Russell, J.F.R. Archilla, Hyperconductivity in fluorphlogopite at 300 K and 1.1 T. EPL **127**(1), 16001 (2019)
24. R. Craig, E. Mamidzhanian, A.W. Wolfendale, Ancient cosmic ray tracks in mica? Phys. Lett. B **26**, 468–470 (1968)
25. L. Voyvodic, E. Pickup, Multiple scattering of fast particles in photographic emulsions. Phys. Rev. **85**, 91–100 (1952)
26. F.M. Russell, Duration of sensitive period for track recording in mica. Nature **217**, 51–52 (1968)
27. F.M. Russell, Positive charge transport in layered crystalline solids. Phys. Lett. A **130**, 489–491 (1988)
28. F.M. Russell, Channelled lepton tracks and practical perspectives. Nucl. Tracks. Rad. Meas. **19**(1–4), 955–957 (1991)
29. F.M. Russell, Decorated track recording mechanisms in muscovite mica. Nucl. Tracks. Rad. Meas. **19**, 109–113 (1991)
30. F.M. Russell, J.C. Eilbeck, Persistent mobile lattice excitations in a crystalline insulator. Discret. Contin. Dyn. Syst. S **4**(5), 1267–1285 (2011)
31. W.A. Deer, R.A. Howie, J. Zussman, *An Introduction to the Rock-Forming Minerals*, 3rd edn. (The Mineralogical Society, London, 2013)
32. K.A. Krylova, E.A. Korznikova, A.S. Semenov, D.V. Bachurin, S.V. Dmitriev, Linking tracks in mica crystals with phase transitions in a bistable lattice. Eur. Phys. J. B **93**(23), 16001 (2020)
33. F.M. Russell, Transport properties of quodons in muscovite and prediction of hyperconductivity, in *Nonlinear Systems*, vol. 2, ed. by J.F.R. Archilla et al. (Springer, Cham, 2018), pp. 241–260
34. V.I. Dubinko, P.A. Selyshchev, J.F.R. Archilla, Reaction-rate theory with account of the crystal anharmonicity. Phys. Rev. E **83**, 041124 (2011)
35. J.F.R. Archilla, Yu.A. Kosevich, N. Jiménez, V.J. Sánchez-Morcillo, L.M. García-Raffi, Ultra-discrete kinks with supersonic speed in a layered crystal with realistic potentials. Phys. Rev. E **91**, 022912 (2015)
36. J.F.R. Archilla, Y. Zolotaryuk, Y. A. Kosevich, Y. Doi, Nonlinear waves in a model for silicate layers. Chaos **28**(8), 083119 (2018)
37. J. Bajars, J.C. Eilbeck, B. Leimkuhler, Numerical simulations of nonlinear modes in mica: Past, present and future, in *Quodons in Mica*, ed. by J.F.R. Archilla, et al. Springer Ser. Mater. Sci., vol. 221, (Springer, Cham, 2015), pp. 35–67
38. Q. Dou, J. Cuevas, J.C. Eilbeck, F.M. Russell, Breathers and kinks in a simulated crystal experiment. Discret. Contin. Dyn. Syst. S **4**(5), 1107–1118 (2011)

39. J.F.R. Archilla, Y. Doi, M. Kimura, Pterobreathers in a model for a layered crystal with realistic potentials: Exact moving breathers in a moving frame. Phys. Rev. E **100**(2) (2019)
40. J. Bajars, J.C. Eilbeck, B. Leimkuhler, Nonlinear propagating localized modes in a 2D hexagonal crystal lattice. Physica D **301–302**, 8–20 (2015)
41. Bajars, J., Eilbeck, J.C., Leimkuhler, B.: 2D mobile breather scattering in a hexagonal crystal lattice (2020). arXiv:2007.12079 [nlin.PS]
42. J.F.R. Archilla, Y. Doi, M. Kimura, Energy and charge transport in a silicate layer, in *Proceedings of NOLTA 2019: the 2019 International Symposium in Nonlinear Theory and its Applications*, (Kuala Lumpur, Malaysia, December 2–6, 2019), p. 453, IEICE, Malaysia
43. J.F.R. Archilla, Y. Doi, M. Kimura, A tight-binding model for charge transport in silicate layers, in *Proceedings of NOLTA 2020: the 2020 International Symposium in Nonlinear Theory and its Applications*, (Virtual, November 16–19, 2020), p. 185, IEICE, Japan

Software Realization, Analysis and Experimental Investigation of Equivalent Inductance

Volodymyr Rusyn, Christos H. Skiadas, and Aceng Sambas

Abstract Circuit realization of the inductance equivalent that contains two operational amplifiers, one capacitor, and four resistors is presented. The mathematical equation that allow convert inductor value to resistance of potentiometer is shown. Computer modeling results of the algorithm for calculate inductance was realized in the modern software LabView. Experimental results of realization of the equivalent of inductance are presented. The designed layout was applied for chaotic Chua's generator.

1 Introduction

Chaotic theory used in many areas, such as biology [1–3], ecology [4, 5], economy [6–8], optics [9], mathematics [10, 11], memristor [12, 13], security communication systems [14, 15], etc. Many different electronics circuits generated chaotic oscillations [16–28]. One of the circuit element must be used inductor. However, there are many problems with product or buying inductor with non-standard nominal values.

In Fig. 1 shows electrical scheme that allows change inductor to operational amplifier realization.

The equation that allow convert equivalent inductance can be computed as

V. Rusyn (✉)
Department of Radio Engineering and Information Security, Yuriy Fedkovych Chernivtsi National University, Chernivtsi, Ukraine
e-mail: rusyn_v@ukr.net

C. H. Skiadas
ManLab, Technical University of Crete, Chania, Crete, Greece
e-mail: skiadas@cmsim.net

A. Sambas
Department of Mechanical Engineering, Universitas Muhammadiyah Tasikmalaya, Tasikmalaya, Indonesia

C. H. Skiadas et al. (eds.), *13th Chaotic Modeling and Simulation International Conference*, Springer Proceedings in Complexity,
https://doi.org/10.1007/978-3-030-70795-8_52

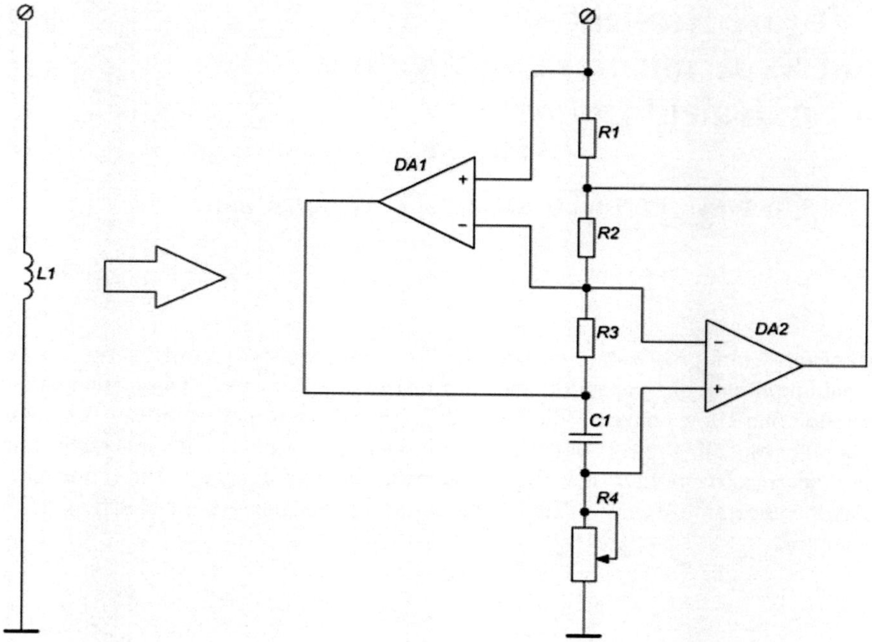

Fig. 1 Operational amplifier realization of simulated inductor

$$L_{eq} = \frac{R1R3R4C1}{R2}, \tag{1}$$

where R1-R4—values of resistances, C1—value of capacitance.

In this work, by using the Laboratory Virtual Instruments Engineering Workbench (LabVIEW), we developed an algorithm that automatically convert value of inductance to value of resistance of potentiometer R4.

The paper is organized as follows. In Sect. 2, algorithm that realize inductance equivalent, LabView software interface and practical realization are presented. In the following section, the result of experimental investigation is presented. Inductor-free simplified Chua's chaotic circuit is easily extended, and similar dynamical behaviors are exhibited through the corresponding numerical simulations and hardware experiments. The conclusionsare summarized in the last section.

2 Software and Practical Realization of the Equivalent of Inductance

Algorithm of equivalent of inductance was realized in LabView. This is a graphical programming platform that helps engineers implement all stages of development of large and small projects: from prototype creation to final testing. In this development environment, the best integration of software and hardware components with the latest computer technologies is combined today.

LabView contains all the tools for solving currentand upcoming challenges with enormous potential forinnovation, future success and effectiveness.

LabView includes powerful multi-function tools forconducting any types of measurements and development of any applications. With these tools, engineersand scientists can work in the widest range of applications and spend much less time developing. Thanks to this, LabView is a development environment for solving a wide range of research, performance enhancementsand innovations.

Figure 2 demonstrate algorithm of equivalent of inductance that was realized in LabView.

Figure 3 shows program interface for calculate of resistance of resistor R4.

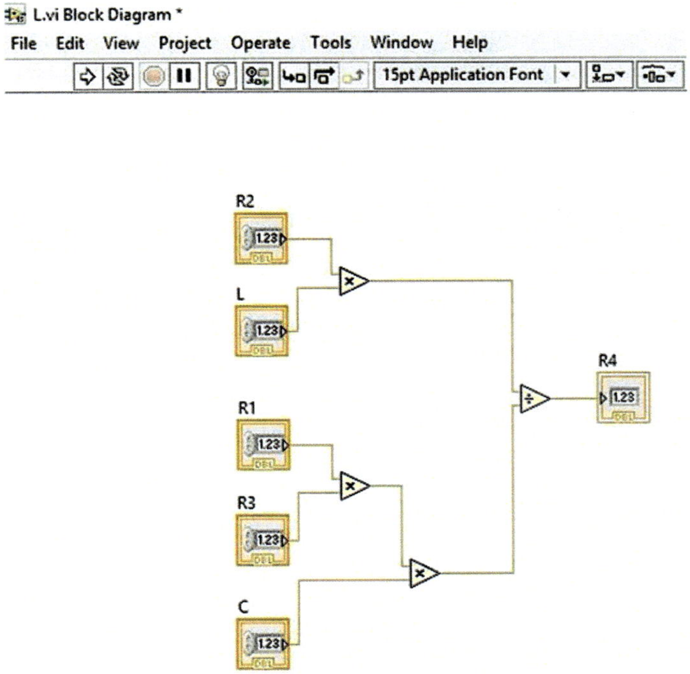

Fig. 2 Software realization of algorithm

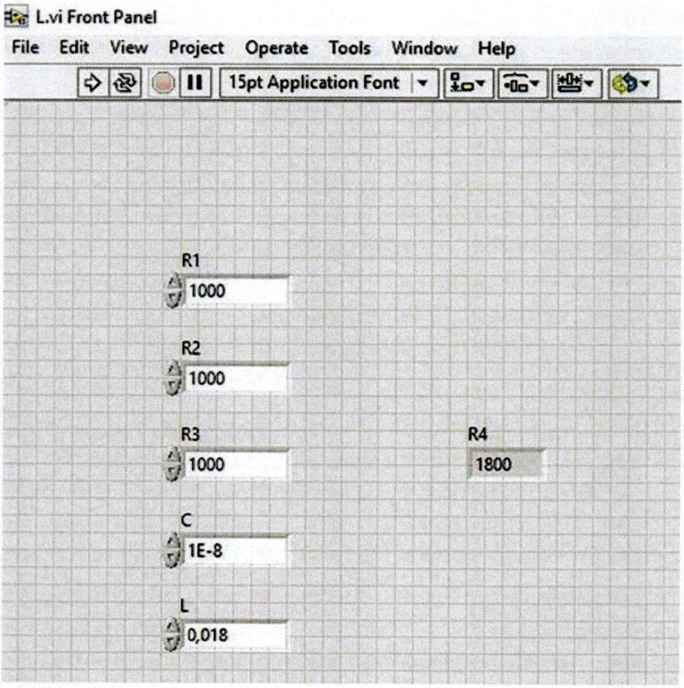

Fig. 3 Software interface for calculate of resistance of resistor R4

For example, we have next values of components:R1 = R2 = R3 = 1000 Ω, capacitor C = 10 nF, L = 18 mH. These values we inserted in special windows of L-EQ calculator and get value for resistor R4 = 1800 Ω. Practical realization and results are shown in Fig. 4, i.e. inductance Leq = 17.9 mH and R0 = 0.23 Ω. Voltage source −12 V. For power supply of the circuit was used laboratory DC power supply Hantek HT3003PB. For measurementof inductance was used LCR meter UNI-T.

If we changed resistance of resistor R4, we get inductance equivalent equal 20 mH. This experimental result shows in Fig. 5.

3 Experimental Investigation of the Equivalent of Inductance

After the advent of Chua's chaotic circuit, numerous works have been reported on different realization schemes of this circuit. We consider a realization of the double scroll chaotic Chua's attractor given by the following set of (rescaled) three coupled ODEs:

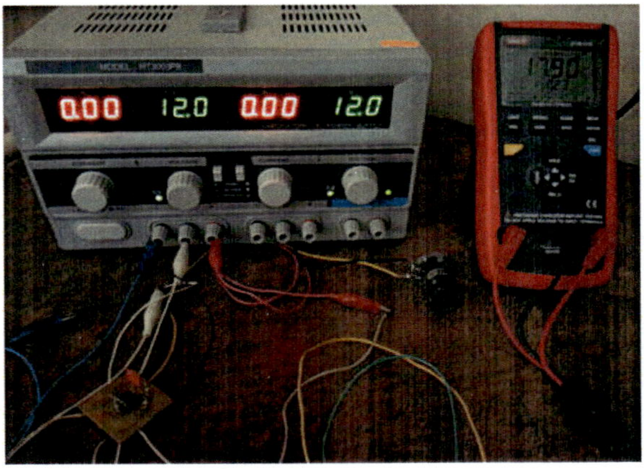

Fig. 4 Practical realization of the equivalent of inductance (L = 17.9 mH)

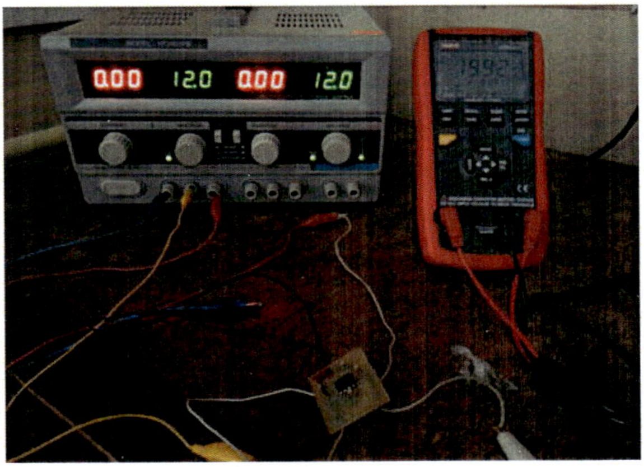

Fig. 5 Practical realization of the equivalent of inductance (L = 19.92 mH)

$$\begin{cases} \frac{dx}{dt} = \alpha(y - x - g(x)), \\ \frac{dy}{dt} = x - y + z, \\ \frac{di_L}{dt} = -\beta y, \end{cases} \tag{2}$$

where α = 10, β = 14.87, g(x)—piecewise linear function.

The circuit realization of the above is displayed in Fig. 6, with component values: capacitors C1 = 100 nF, C2 = 10 nF, DA1—operational amplifier TL082, powered by a 12 V, GB –voltage source, inductor L1 = 18 mH, resistors R1 = 1.71 kΩ, R2 = 47 kΩ, R3 = R4 = 3.3 kΩ, R5 = 47 kΩ, R6 = R7 = 290, R8 = 1.2 kΩ, diodes VD1, VD2–1N4148.

The experimental results are captured by TektronixTDS 1002 digital oscilloscope.

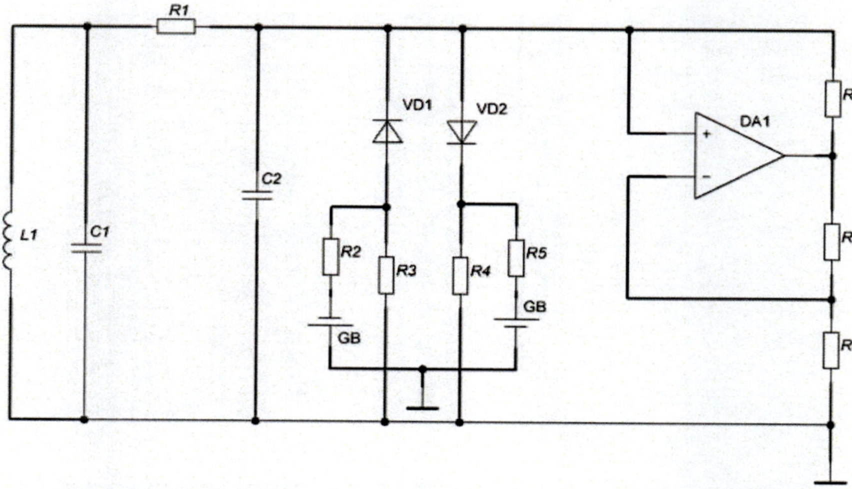

Fig. 6 The classical chaotic Chua's generator

Fig. 7 Chaotic attractor

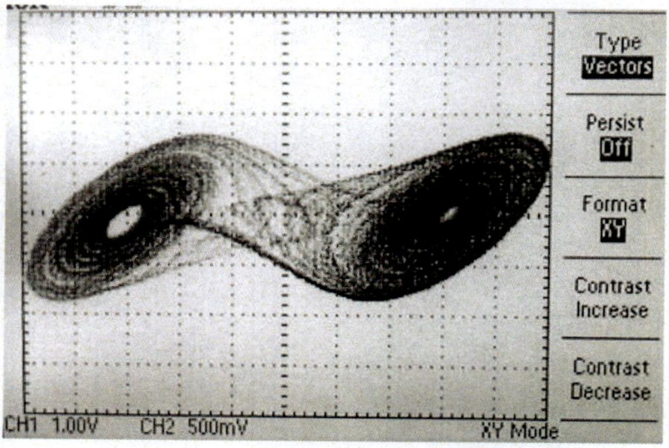

Figures 7 and 8 shows chaotic attractor and timeseries that was realized practically. Figures 9 and 10 shows spectra of the classical chaotic Chua's generator.

4 Conclusions

Designed L-EQ calculator allows convert of inductance values to resistance using some algorithm. Circuit realization of the inductance equivalent that contains two operational amplifiers, one capacitor, and four resistors is presented. The mathematical equation that allow convert inductor value to resistance of potentiometer is

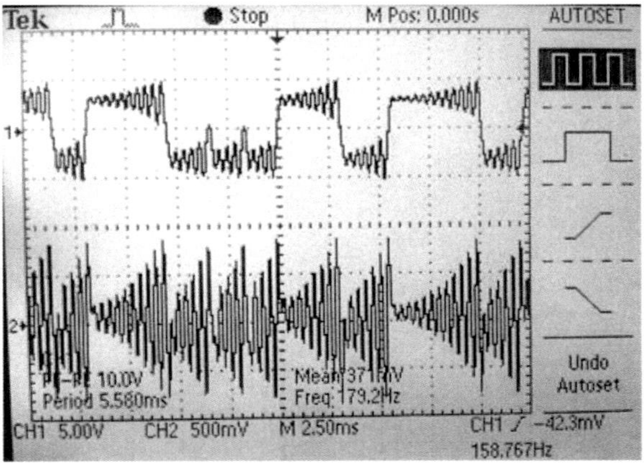

Fig. 8 The x-signal (upper) and the y-signal (lower) timeseries realized practically. Their non-periodic nature is evident

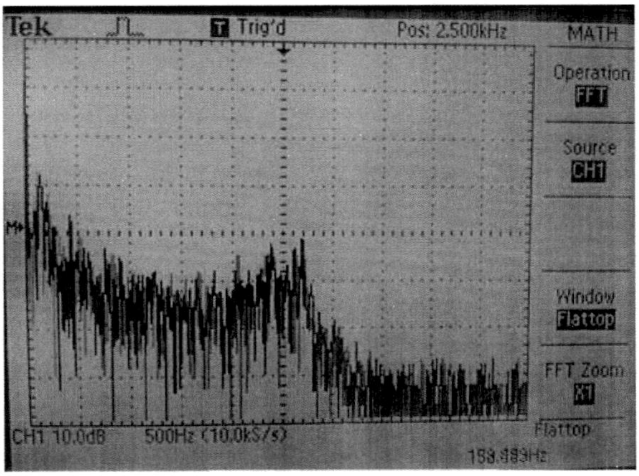

Fig. 9 The spectral distribution of V_{C1}

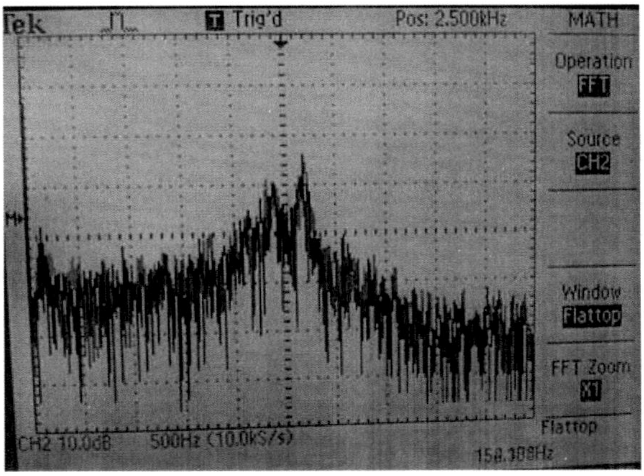

Fig. 10 The spectral distribution of V_{C2}

shown. Computer modeling results of the algorithm for calculate inductance was realized in the modern software LabView. Experimental results of realization of the equivalent of inductance are presented. The designed layout was applied for chaotic Chua's generator. Chaotic attractor, timeseries and s spectral distributions are also presented.

References

1. A. Rhif, S. Vaidyanathan, A. Sambas, A fish biology chaotic system and its circuit design. IOP Conf. Ser. J. Phys. **1179** 012011 (2019)
2. V. Hajnova, L. Pribylova, Two-parameter bifurcations in LPA model. J. Math. Biol. **75**(5), 1235–1251 (2017)
3. S. Vaidyanathan, M. Feki, A. Sambas, C.H. Lien, A new biological snap oscillator: its modelling, analysis, simulations and circuit design. Int. J. Simul. Process Model. **13**(5), 419–432 (2018)
4. L. Pribylova, Regime shifts caused by adaptive dynamics in prey-predator models and their relationship with intraspecific. Ecol. Complex. **36**, 48–56 (2018)
5. L. Pribylova, A. Peniaskova, Foraging facilitation among predators and its impact on the stability of predator-prey dynamics. Ecol. Complex. **29**, 30–39 (2017)
6. V. Rusyn, O. Savko, Modeling of chaotic behavior in the economic model. in *CHAOS 2015–8th Chaotic Modeling and Simulation International Conference, Proceedings 2015*, 705–712 (2015)
7. S. Vaidyanathan, A. Sambas, S. Kacar, U. Cavusoglu, A new finance chaotic system, its electronic circuit realization, passivity based synchronization and an application to voice encryption. Nonlinear Eng. **8**(1), 193–205 (2019)
8. L. Pribylova, Bifurcation routes to chaos in an extended Van der Pol's equation applied to economic models. Electron. J. Differ. Equ. **53**, 1–21 (2009)
9. V. Rusyn, A. Samila, Ch. Skiadas, Computer modeling and practical realization of chaotic circuit with a light-emitting diode. in *Fourteenth International Conference on Correlation Optics*, Chernivtsi, pp. 113690D (2020)
10. C.H. Skiadas, Exact solutions of stochastic differential equations: Gompertz, generalized logistic and revised exponential. Method. Comput. Appl. Probab. **12**(2), 261–270 (2010)
11. C.H. Skiadas, C. Skiadas, Chaotic modelling and simulation: Analysis of chaotic models, attractors and forms. In: Taylor & Francis Group, pp. 1–345, LLC (2008)
12. L. Chua, Memristor—the missing circuit element. IEEE Trans. Circuit Theory **18**(5), 507–519 (1971)
13. V. Rusyn, S. Khrapko, Memristor: Modeling and research of information properties. Springer Proc. Complex. 229–238 (2019)
14. A. Sambas, W.S.M. Sanjaya, M. Mamat, P.R. Putra, A.T. Azar, Mathematical modelling of chaotic jerk circuit and its application in secure communication system. Stud. Fuzziness Soft. Comput. **337**, 133–153 (2016)
15. A. Sambas, M.W.S. Sanjaya, M. Mamat, H. Diyah, Design and analysis bidirectional chaotic synchronization of rossler circuit and its application for secure communication. Appl. Math. Sci. **7**(1), 11–21 (2013)
16. C.-H. Lien, S. Vaidyanathan, A. Sambas, M. Mamat, W.S.M. Sanjaya, A new two-scroll chaotic attractor with three quadratic nonlinearities, its adaptive control and circuit design. IOP Conf. Ser. Mater. Sci. Eng. **332**, 1, 012010 (2018)
17. V.B. Rusyn, Modeling and research of chaotic rossler system with LabView and multisim software environments. Bulletin of National Technical University of Ukraine Kyiv Polytechnic Institute. Series Radiotechnique Radioapparatus Building, **59**, 21–28 (2014)

18. V. Rusyn, A. Stancu, L. Stoleriu, Modeling and control of chaotic multi-scroll jerk system in LabView. Bulletin of National Technical University of Ukraine Kyiv Polytechnic Institute. Series Radiotechnique Radioapparatus Building, **62**, 94–99 (2015)
19. A. Sambas, S. Vaidyanathan, M. Mamat, W.S. Mada Sanjaya, A six-term novel chaotic system with hidden attractor and its circuit design. Stud. Syst. Decis. Control **133**, 365–373 (2018)
20. V.B. Rusyn, I. Pavlyukevich, L. Pribylova, H.C. Skiadas, Design, modeling and research of the new non-autonomous chaotic generator. Visnyk NTUU KPI Seriia - Radiotekhnika Radioaparatobuduvannia **77**, 13–16 (2019)
21. V. Rusyn, M. Kushnir, O. Galameiko, Hyperchaotic control by thresholding method. in *Modern Problems of Radio Engineering, Telecommunications and Computer Science - Proceedings of the 11th International Conference*, TCSET'2012, art. No. 6192785, p. 67 (2012)
22. V. Rusyn, S. Subbotin, A. Sambas, Analysis and experimental realization of the logistic map using Arduino Pro Mini. CEUR Workshop Proc. **2608**, 300–310 (2020)
23. V. Rusyn, Ch. Skiadas, Threshold method for control of chaotic oscillations. Springer Proceedings in Complexity. Springer, pp. 217–229 (2020)
24. V. Rusyn, Modeling and research information properties of rucklidge chaotic system using LabView. in *CHAOS 2017—Proceedings: 10th Chaotic Modeling and Simulation International Conference*, 739–744 (2017)
25. V.B. Rusyn, L. Pribylova, D.-G. Dimitriu, Control of the modified chaotic Chua's circuit using threshold method. Visn. NTUU KPI. Radioteh. radioaparatobuduv. **75**, 61–65 (2018)
26. V. Rusyn, M.A. Mohamad, D. Purwandari, M. Mamat, J. Titaley, B. Pinontoan, Chaotic and controlling regimes of a new modified Chua's generator. J. Adv. Res. Dyn. Control Syst. **12**(02), 556–561 (2020)
27. V. Rusyn, M. Mohamad, J. Titaley, N. Nainggolan, M. Mamat, Design, computer modelling, analysis and control of the new chaotic generator. J. Adv. Res. Dyn. Control Syst. **12**(02), 2306–2311 (2020)
28. V. Rusyn, M. Sadli, M. Mamat, M.S. Mujiarto, W.S. , Computer modelling of a new simple chaotic generator. J. Phys. Conf. Ser. **1477**, 022010 (2020)

Dynamics of a Bertrand Duopoly Game with Differentiated Goods, Heterogeneous Expectations and Relative Profit Maximization

Georges Sarafopoulos and Kosmas Papadopoulos

Abstract In this article the authors investigate the dynamics of an oligopoly game in which, they consider a nonlinear Bertrand-type duopoly game with differentiated goods and heterogeneous expectations. In this study the case, where managers have a variety of attitudes toward relative performance that are indexed by their type is investigated. In this game they suppose a linear demand and cost functions. The game is modeled with a system of two difference equations. Existence and stability of equilibria of the system are studied. It is revealed that the models gives more complex, chaotic and unpredictable trajectories, as a consequence of change in the parameter k of speed of the player's adjustment, the parameter d of the horizontal product differentiation and the relative profit parameter μ. The chaotic features are justified numerically via computing Lyapunov numbers and sensitive dependence on initial conditions.

Keywords Bertrand duopoly game · Relative profit maximization · Discrete dynamical system · Nash equilibrium · Stability · Bifurcation diagrams · Lyapunov numbers · Strange attractors · Chaotic Behavior

1 Introduction

An Oligopoly is a market structure between monopoly and perfect competition, where there are only a few number of firms in the market producing homogeneous products. The dynamic of an oligopoly game is more complex because firms must consider not only the behaviors of the consumers, but also the reactions of the competitors i.e. they form expectations concerning how their rivals will act. Cournot, in 1838

G. Sarafopoulos (✉) · K. Papadopoulos
Department of Economics, Democritus University of Thrace, University Campus, 69100 Komotini, Greece
e-mail: gsarafop@econ.duth.gr

K. Papadopoulos
e-mail: kpapa@econ.duth.gr

C. H. Skiadas et al. (eds.), *13th Chaotic Modeling and Simulation International Conference*, Springer Proceedings in Complexity,
https://doi.org/10.1007/978-3-030-70795-8_53

has introduced the first formal theory of oligopoly. In 1883 another French mathematician Joseph Louis Francois Bertrand modified Cournot game suggesting that firms actually choose prices rather than quantities. Originally Cournot and Bertrand models were based on the premise that all players follow naive expectations, so that in every step, each player (firm) assumes the last values that were taken by the competitors without estimation of their future reactions. However, in real market conditions such an assumption is very unlikely since not all players share naive beliefs. Therefore, different approaches to firm behavior were proposed. Some authors considered duopolies with homogeneous expectations and found a variety of complex dynamics in their games, such as appearance of strange attractors [2, 6, 10, 27, 36, 39, 52]. Also models with heterogeneous agents were studied [3, 5, 6, 20, 48, 51].

In the real market producers do not know the entire demand function, though it is possible that they have a perfect knowledge of technology, represented by the cost function. Hence, it is more likely that firms employ some local estimate of the demand. This issue has been previously analyzed by Baumol and Quandt [9], Puu [35], Naimzada and Ricchiuti [32], Askar [7, 8]. Bounded rational players (firms) update their strategies based on discrete time periods and by using a local estimate of the marginal profit. With such local adjustment mechanism, the players are not requested to have a complete knowledge of the demand and the cost functions [4, 33, 51, 8], Sarafopoulos [39, 40], Sarafopoulos et al. [41–44].

In this paper we study the dynamics of a Bertrand- type duopoly with differentiated goods where each firm behaves with heterogeneous expectations strategies. We show that the model gives more complex chaotic and unpredictable trajectories as a consequence of change in three parameters, the speed of players' adjustment, the parameter of horizontal product differentiation and the relative profit parameter. The paper is organized as follows: In Sect. 2, the dynamics of the duopoly game with heterogeneous expectations, linear demand and cost functions for two players are analyzed. We set first player as bounded rational and the second as a naïve player. The existence and local stability of the equilibrium points are also analyzed. In Sect. 3 numerical simulations are used to verify the algebraic results of Sect. 2 plotting the bifurcation diagrams of the game's system and to show the complex dynamics via computing Lyapunov numbers, and sensitive dependence on initial conditions.

2 The Game

2.1 *The Construction of the Game*

In this study we assume that in the two companies there is a separation between ownership and management, so there is a possibility that the managers who make decisions for the company to decide at the expense of their company trying to increase the profits of the competitor. Also, we consider heterogeneous players and more specifically, we consider that the Firm 1 chooses the price of its product in a rational way, following an adjustment mechanism (bounded rational player), while the Firm 2 decides with naïve way by selecting a price that maximizes its output (naïve player). We consider a simple Bertrand-type duopoly market where firms (players) produce differentiated goods and offer them at discrete-time periods on a common market. Price decisions are taken at discrete time periods t = 0, 1, 2,... At each period t, every firm must form an expectation of the rival's strategy in the next time period in order to determine the corresponding profit-maximizing prices for period t + 1. We suppose that q_1, q_2 are the production quantities of each firm. Also, we consider that the preferences of consumers represented by the equation:

$$U(q_1, q_2) = \alpha(q_1 + q_2) - \frac{1}{2}\left(q_1^2 + q_2^2 + 2dq_1q_2\right) \tag{1}$$

where α is a positive parameter (α > 0), which expresses the market size and $d \in (-1, 1)$ is the parameter that reveals the differentiation degree of products. For example, if $d = 0$ then both products are independently and each firm participates in a monopoly. But, if $d = 1$ then one product is a substitute for the other, since the products are homogeneous. It is understood that for positive values of the parameter d the larger the value, the less diversification we have in both products. On the other hand negative values of the parameter d are described that the two products are complementary and when $d = -1$ then we have the phenomenon of full competition between the two companies. The inverse demand functions (as functions of quantities) coming from the maximizing of (1) are given by the following equations:

$$p_1(q_1, q_2) = \alpha - q_1 - dq_2 \; and \; p_2(q_1, q_2) = \alpha - q_2 - dq_1 \tag{2}$$

The direct demand functions (as functions of prices):

$$q_1(p_1, p_2) = \frac{\alpha(1-d) - p_1 + dp_2}{1-d^2} \; and \; q_2(p_1, p_2) = \frac{\alpha(1-d) - p_2 + dp_1}{1-d^2} \tag{3}$$

In this work we suppose that both players follow the same linear cost function, which is described by the following equation:

$$C_i(q_i) = c \cdot q_i \tag{4}$$

and c > 0 is the same marginal cost for two firms.

With these assumptions the profits of the firms are given by:

$$\Pi_1(p_1, p_2) = p_1q_1 - C_1(q_1) = (p_1 - c) \cdot \frac{\alpha(1-d) - p_1 + dp_2}{1-d^2} \tag{5}$$

and

$$\Pi_2(p_1, p_2) = p_1 q_1 - C_2(q_2) = (p_2 - c) \cdot \frac{\alpha(1-d) - p_2 + dp_1}{1-d^2} \tag{6}$$

Then the marginal profits at the point of the strategy space are given by:

$$\frac{\partial \Pi_1}{\partial p_1} = \frac{\alpha(1-d) + c - 2p_1 + dp_2}{1-d^2}, \ \frac{\partial \Pi_1}{\partial p_2} = \frac{d(p_1 - c)}{1-d^2} \tag{7}$$

and

$$\frac{\partial \Pi_1}{\partial p_2} = \frac{\alpha(1-d) + c - 2p_2 + dp_1}{1-d^2}, \ \frac{\partial \Pi_2}{\partial p_1} = \frac{d(p_2 - c)}{1-d^2} \tag{8}$$

As it is noticed both managers care about the maximization of a utility function that contains a percentage of opponent company's profits (generalized relative profit function), which is given by:

$$U_i = (1 - \mu_i) \cdot \Pi_i + \mu_i \cdot (\Pi_i - \Pi_j) = \Pi_i - \mu_i \cdot \Pi_j \tag{9}$$

where $\mu \in [0, 1]$ is the percentage that the player i takes into account the opponent company's prifots. So, the marginal utility of the player i is given by the following equation:

$$\frac{\partial U_i}{\partial p_i} = \frac{\partial \Pi_i}{\partial p_i} - \mu_i \cdot \frac{\partial \Pi_j}{\partial p_i} \tag{10}$$

and the marginal utilities for each player are:

$$\frac{\partial U_1}{\partial p_1} = \frac{\alpha(1-d) + c(1 + \mu d) - 2p_1 + d(1-\mu)p_2}{1-d^2} \tag{11}$$

and

$$\frac{\partial U_2}{\partial p_2} = \frac{\alpha(1-d) + c(1 + \mu d) - 2p_2 + d(1-\mu)p_1}{1-d^2} \tag{12}$$

The first player is characterized as bounded rational player. According to the existing literature it means that he decides his price following a mechanism that is described by the equation:

$$\frac{p_1(t+1) - p_1(t)}{p_1(t)} = k \cdot \frac{\partial U_1}{\partial p_1}, \ k > 0 \tag{13}$$

Through this mechanism the player increases his level of adaptation when his marginal utility is positive or decreases his level when his marginal utility is negative, where k is the speed of adjustment of player, it is a positive parameter (k > 0), which gives the extend variation of price of the company 1, following a given utility signal.

The second player chooses this price that maximizes his utility function (naïve player). So, his strategy is given by the equation:

$$p_2(t+1) = \arg\max_{y} U_2(p_1(t), p_2(t)) \tag{14}$$

The dynamical system of the players is described by:

$$\begin{cases} p_1(t+1) = p_1(t) + k \cdot p_1(t) \cdot \dfrac{\partial U_1}{\partial p_1} \\ p_2(t+1) = \dfrac{\alpha(1-d) + c(1+\mu d) + d(1-\mu) \cdot p_1(t)}{2} \end{cases} \tag{15}$$

We will focus on the dynamics of this system to the parameter k, d and μ.

2.2 *Dynamical Analysis*

The dynamical analysis of the discrete dynamical system involves finding equilibrium positions and studying them for stability. The ultimate goal of this algebraic study is to formulate a proposition that will be the stability condition of the Nash Equilibrium position. Finally, these algebraic results are verified and visualized doing some numerical simulations using the program of Mathematica.

2.2.1 The Equilibrium Positions

The equilibriums of the dynamical system (15) are obtained as the nonnegative solutions of the algebraic system:

$$\begin{cases} p_1^* \cdot \dfrac{\partial U_1}{\partial p_1} = 0 \\ p_2^* = \dfrac{\alpha(1-d) + c(1+\mu d) + d(1-\mu) \cdot p_1^*}{2} \end{cases} \tag{16}$$

which is obtained by setting: $p_1(t+1) = p_1(t) = p_1^*$ and $p_2(t+1) = p_2(t) = p_2^*$.

- If $p_1^* = 0$ and $\frac{\partial U_2}{\partial p_2} = 0$ then: $p_2^* = \frac{\alpha(1-d)+c(1+\mu d)}{2}$ and the equilibrium position is the point:

$$E_1 = \left(0, \frac{\alpha(1-d)+c(1+\mu d)}{2}\right) \tag{17}$$

- If $\frac{\partial U_1}{\partial p_1} = \frac{\partial U_2}{\partial p_2} = 0$ then the following system is obtained:

$$\begin{cases} \alpha(1-d)+c(1+\mu d)-2p_1^*+d(1-\mu)p_2^*=0 \\ \alpha(1-d)+c(1+\mu d)-2p_2^*+d(1-\mu)p_1^*=0 \end{cases} \tag{18}$$

and the nonnegative solution of this algebraic system will give the Nash Equilibrium position $E_* = (p_1^*, p_2^*)$ where:

$$p_1^* = p_2^* = \frac{(2+d-\mu d)\cdot[\alpha(1-d)+c(1+\mu d)]}{4-d^2(1-\mu)^2} \tag{19}$$

2.2.2 Stability of Equilibrium Points

To study the stability of the equilibrium positions we need the Jacobian matrix of the dynamical system (15) which is the matrix:

$$J(p_1^*, p_2^*) = \begin{bmatrix} f_{p_1} & f_{p_2} \\ g_{p_1} & g_{p_2} \end{bmatrix} \tag{20}$$

where:

$$\begin{aligned} f(p_1, p_2) &= p_1 + k\cdot p_1 \cdot \frac{\partial U_1}{\partial p_1} \\ g(p_1, p_2) &= \frac{\alpha(1-d)+c(1+\mu d)+d(1-\mu)\cdot p_1}{2} \end{aligned} \tag{21}$$

and as a result the Jacobian matrix of game's discrete dynamical system (15) is the following matrix:

$$J(p_1^*, p_2^*) = \begin{bmatrix} 1+k\cdot\left(\frac{\partial U_1}{\partial p_1}+p_1^*\cdot\frac{\partial^2 U_1}{\partial p_1^2}\right) & k\cdot p_1^*\cdot\frac{\partial^2 U_1}{\partial p_1 \partial p_2} \\ \frac{d(1-\mu)}{2} & 0 \end{bmatrix} \tag{22}$$

For the E_1 the Jacobian matrix becomes as:

$$J(E_1) = \begin{bmatrix} 1+k\cdot\frac{\partial U_1}{\partial p_1} & 0 \\ \frac{d\cdot(1-\mu)}{2} & 0 \end{bmatrix} \overset{A=1+k\cdot\frac{\partial U_1}{\partial p_1}}{\underset{B=\frac{d\cdot(1-\mu)}{2}}{=}} \begin{bmatrix} A & 0 \\ B & 0 \end{bmatrix} \tag{23}$$

with $Tr = \text{A}$ and $Det = 0$.

From the characteristic equation of $J(E_1)$, we find the nonnegative eigenvalue:

$$r_1 = Tr = 1 + k \cdot \frac{\partial U_1}{\partial p_1} \tag{24}$$

it's clearly seems that $|r_1| > 1$ and the E_1 equilibrium is unstable.

For the E_* the Jacobian matrix becomes as:

$$J(E_*) = \begin{bmatrix} 1 - 2k \cdot (1 + c) \cdot q_1^* & -k \cdot (1 - \mu) \cdot q_1^* \\ \frac{\mu - 1}{2(1+c)} & 0 \end{bmatrix} \tag{25}$$

with

$$Tr = 1 - 2k \cdot (1 + c) \cdot q_1^* \; and \; Det = -k \cdot \frac{(1-\mu)^2}{2(1+c)} \cdot q_1^* \tag{26}$$

To study the stability of Nash equilibrium we use three conditions that the equilibrium position is locally asymptotically stable when they are satisfied simultaneously:

$$\begin{array}{ll} (i) & 1 - Det > 0 \\ (ii) & 1 - Tr + Det > 0 \\ (iii) & 1 + Tr + Det > 0 \end{array} \tag{27}$$

It's easy to find that the first condition (i) is always satisfied:

$$1 - Det > 0 \Leftrightarrow 1 + k \cdot p_1^* \cdot \frac{d^2 \cdot (1-\mu)^2}{2\left(1 - d^2\right)} > 0 \tag{28}$$

Also, the condition (ii) gives:

$$1 - Tr + Det > 0 \Leftrightarrow k \cdot p_1^* \cdot \frac{\left[4 - d^2 \cdot (1-\mu)^2\right]}{2\left(1 - d^2\right)} > 0 \tag{29}$$

and it's always satisfied because $\frac{\left[4 - d^2 \cdot (1-\mu)^2\right]}{2\left(1-d^2\right)} > 0$.

Finally, the condition (iii) becomes as:

$$1 + Tr + Det > 0 \Leftrightarrow k \cdot p_1^* \cdot \frac{\left[4 + d^2 \cdot (1-\mu)^2\right]}{2\left(1 - d^2\right)} - 2 < 0 \tag{30}$$

Proposition *The Nash equilibrium of the discrete dynamical system (15)is locally asymptotically stable if:*

$$k \cdot p_1^* \cdot \frac{\left[4 + d^2 \cdot (1 - \mu)^2\right]}{2\left(1 - d^2\right)} - 2 < 0$$

where

$$p_1^* = \frac{(2 + d - \mu d) \cdot [\alpha(1 - d) + c(1 + \mu d)]}{4 - d^2 \cdot (1 - \mu)^2}$$

3 Numerical Simulations

3.1 Stability Spaces

At first the 3D stability space Fig. 1 is made including the main three parameters we will focus on, the parameters k (speed of adjustment), the parameter d (product's differentiation degree) and μ (relative profit parameter). This three-dimensional space is obtained by the stability condition that is described above in Proposition, setting specific values for the other parameters $\alpha = 5$ and $c = 1$. Also, the two-dimensional stability region for a couple of these three parameters are presented that

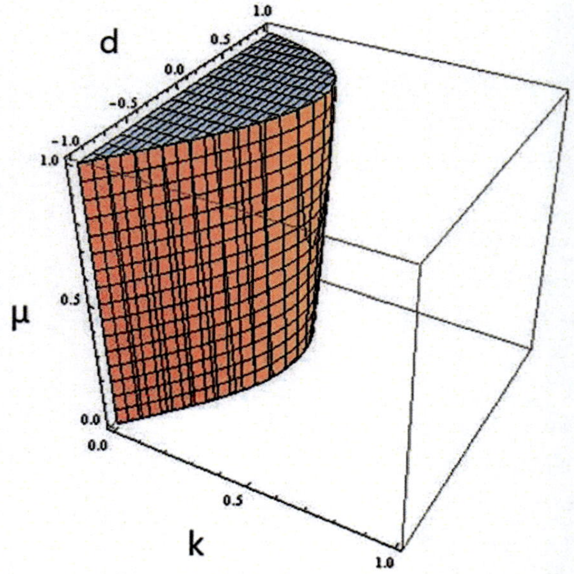

Fig. 1 Three-dimensional stability space between the parameters k, d and μ for α = 5 and c = 1

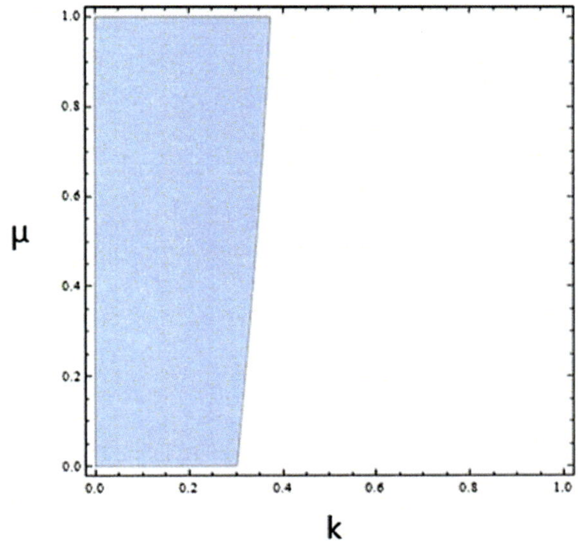

Fig. 2 Two-dimensional stability region between the parameters k (horizontal axis) and μ (vertical axis) for $\alpha = 5$, $c = 1$ and $d = 0.50$

is resulted setting specific values of for one of the three main parameters. Specifically, the stability regions between the parameters k and μ Fig. 2 also, between the parameters d and μ Fig. 3 and between the parameters k and d Fig. 4 are presented. A useful result using the Fig. 2 is that for small values of the parameter k (speed of adjustment) there is a locally asymptotically stable Nash Equilibrium for every value of the parameter μ (relative profit parameter) into the interval [0,1]. It means that for these values of the other parameters α, c, d and k the parameter μ cannot destabilize the economy.

3.2 *Focusing on the Parameter k*

In this section some numerical simulation including bifurcation diagrams, strange attractors, Lyapunov numbers graph and Sensitive dependence on initial conditions are presented focusing on the parameter k when the other parameters are fixed taking the values: $\alpha = 5$, $c = 1$, $\mu = 0.30$ and $d = 0.50$. At first, the Nash Equilibrium for the values of these parameters becomes as:

$$p_1^* = p_2^* \simeq 2.21 \Rightarrow E_*(p_1^*, p_2^*) \equiv E_*(2.21, 2.21) \tag{31}$$

and for the stability condition it means that the parameter k must take values into the interval:

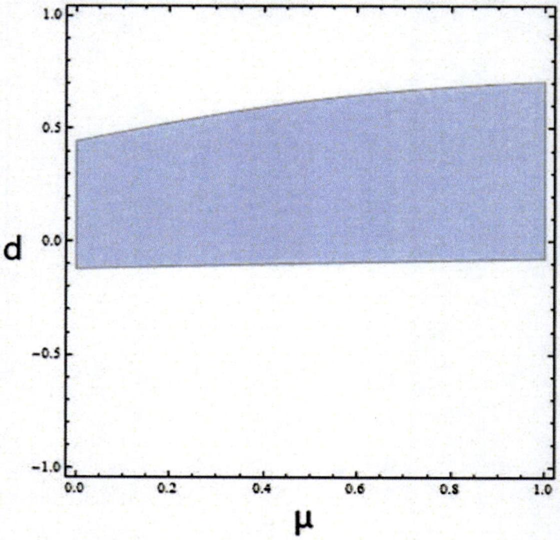

Fig. 3 Two-dimensional stability region between the parameters μ (horizontal axis) and d (vertical axis) for $\alpha = 5$, $c = 1$ and $k = 0.315$

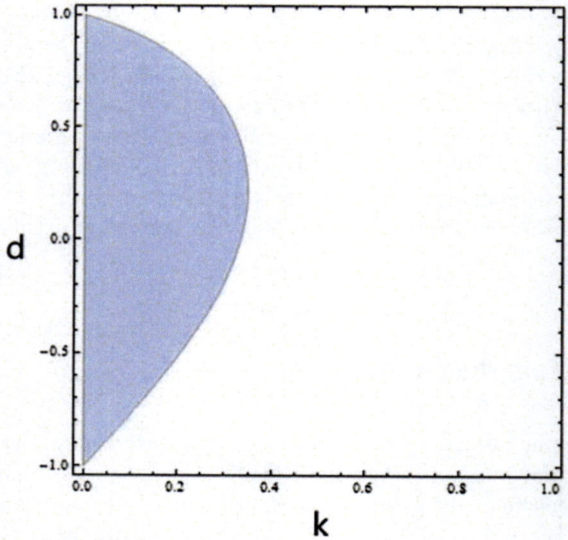

Fig. 4 Two-dimensional stability region between the parameters k (horizontal axis) and d (vertical axis) for $\alpha = 5$, $c = 1$ and $\mu = 0.20$

$$k \in (0, 0.32) \tag{32}$$

This algebraic result is verified by the bifurcation diagrams of p_1^* Fig. 5 and p_2^* Fig. 6 with respect to the parameter k. As it seems there is a locally asymptotically stable orbit until the value of 0.32 for the parameter k and after this value doubling period bifurcations are appeared and finally, for higher values of the parameter k the system's behavior becomes chaotic and unpredictable.

.

This chaotic trajectory can create strange attractors Fig. 8 for a higher value of the parameter k like 0.47, outside the stability space. Also, computing the Lyapunov numbers Fig. 9 for this value of the parameter k and setting the same fixed values for the other parameters α, c, μ and d it seems that they are getting over the value of 1 as an evidence for the chaotic trajectory.

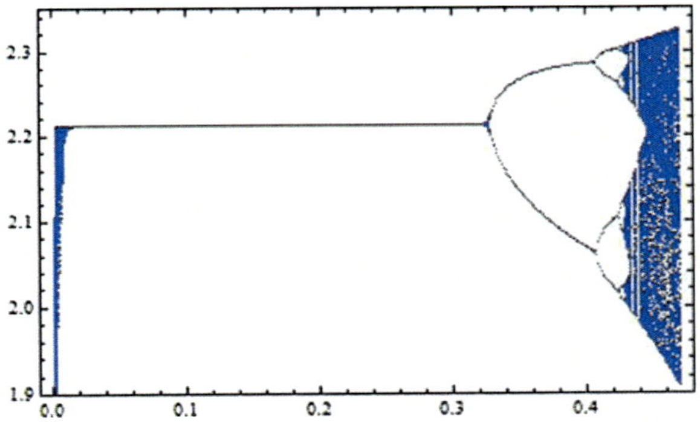

Fig. 5 Bifurcation diagram with respect to the parameter k against the variable p_1^* with 400 iterations of the map (15) for $\alpha = 5$, $c = 1$, $d = -0.50$ and $\mu = 0.30$

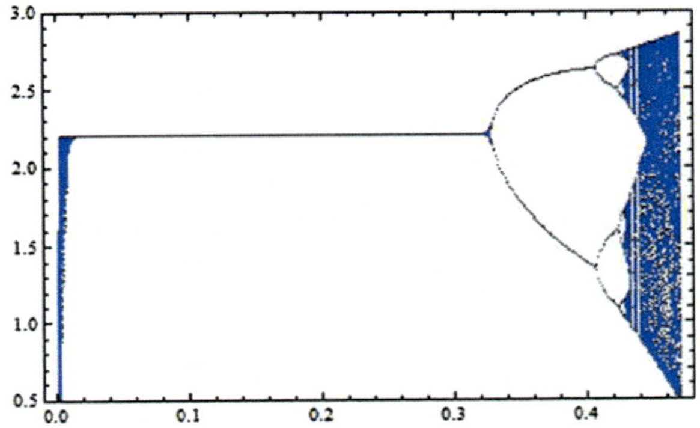

Fig. 6 Bifurcation diagram with respect to the parameter k against the variable p_2^* with 400 iterations of the map (15) for $\alpha = 5$, $c = 1$, $d = -0.50$ and $\mu = 0.30$

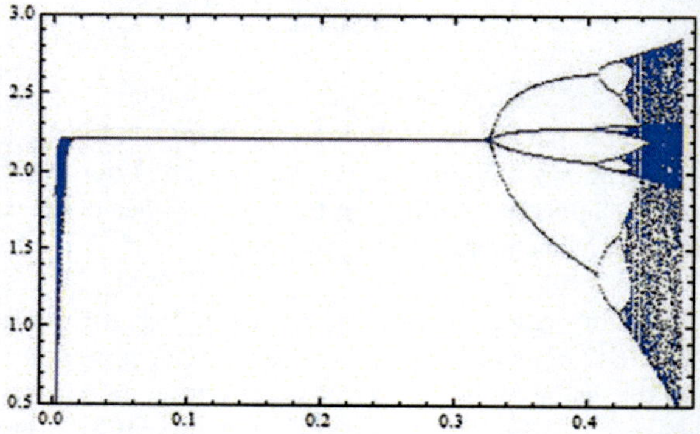

Fig. 7 The two previous bifurcation diagrams of Figs. 5 and 6 in one

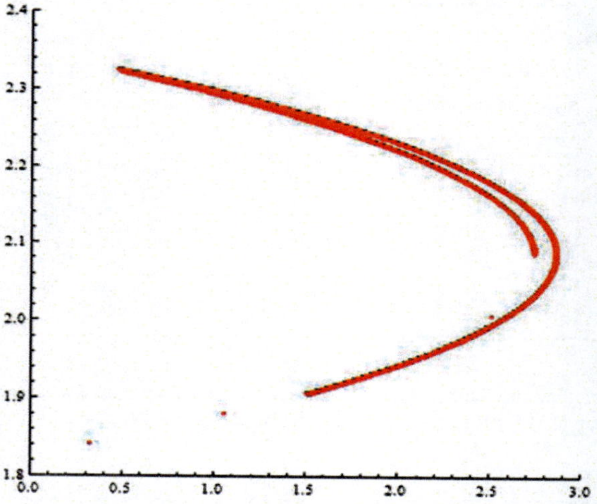

Fig. 8 Phase portrait (strange attractor) of the orbit of (0.1, 0.1) with 8000 iterations of the map (15) for $\alpha = 5$, $c = 1$, $d = -0.50$, $\mu = 0.30$ and $k = 0.47$

This chaotic trajectory makes the system sensitive on initial conditions, which means that only a small change on a coordinate may change completely the system's behavior. For example, choosing two different initial conditions (0.1, 0.1) Fig. 10 and (0.101,0.1) Fig. 11 with a small change at the p_1^*-coordinate and plotting the time series of system it seems that at the beginning the time series are indistinguishable, but after a number of iterations, the difference between them builds up rapidly.

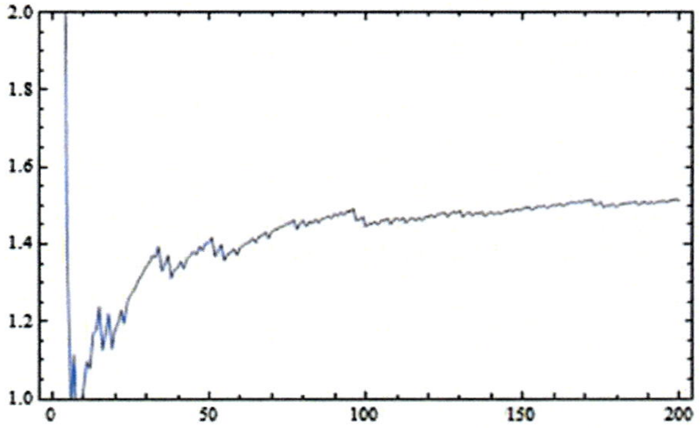

Fig. 9 Lyapunov numbers of the orbit of (0.1, 0.1) with 8000 iterations of the map (15) for $\alpha = 5$, $c = 1$, $d = -0.50$, $\mu = 0.30$ and $k = 0.47$

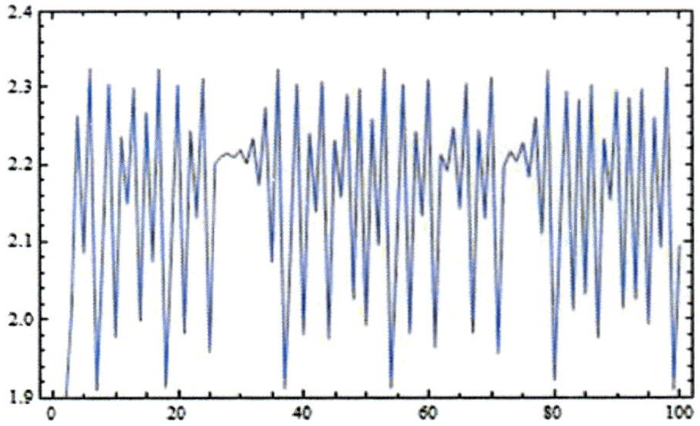

Fig. 10 Sensitive dependence on initial conditions for p_1^*-coordinate plotted against the time: the orbit of (0.1, 0.1) of the system (15) for $\alpha = 5$, $c = 1$, $d = -0.50$, $\mu = 0.30$ and $k = 0.47$

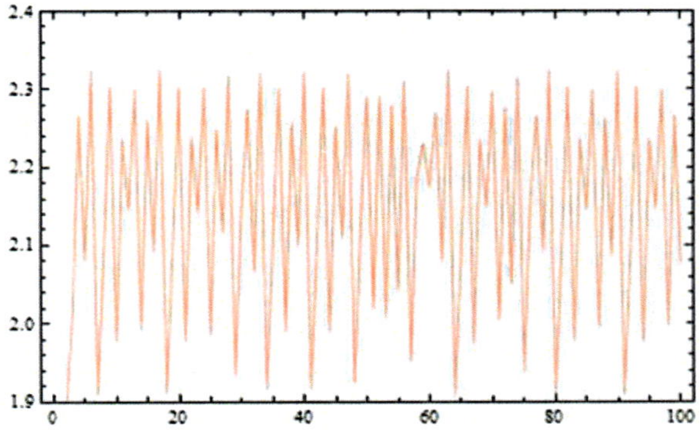

Fig. 11 Sensitive dependence on initial conditions for p_1^*-coordinate plotted against the time: the orbit of (0.101, 0.1) of the system (15) for $\alpha = 5$, $c = 1$, $d = -0.50$, $\mu = 0.30$ and $k = 0.47$

3.3 Focusing to the Parameter d

Using the stability region of Fig. between the parameters μ (horizontal axis) and d (vertical axis) for α = 5, c = 1 and k = 0.315, it seems that when the parameter d takes values into a close interval there is a stable Nash equilibrium for every value of the parameter μ. For example, setting the value of 0.20 to the parameter μ, a stable Nash Equilibrium is appeared into the interval (−0.20,0.55) for the parameter d. This indication is verified by the following bifurcation diagrams of p_1^* Fig. 12 and p_2^* Fig. 13 with respect to the parameter d.

Setting large and small values to the parameter d, strange attractors Figs. 15 and 19 and Lyapunov numbers Figs. 16 and 20 higher than the number of 1 are appeared showing the chaotic trajectories and unpredictable behavior of the system of (15) for these values of the parameter d = 0.79 and d = −0.50 outside the stability space.

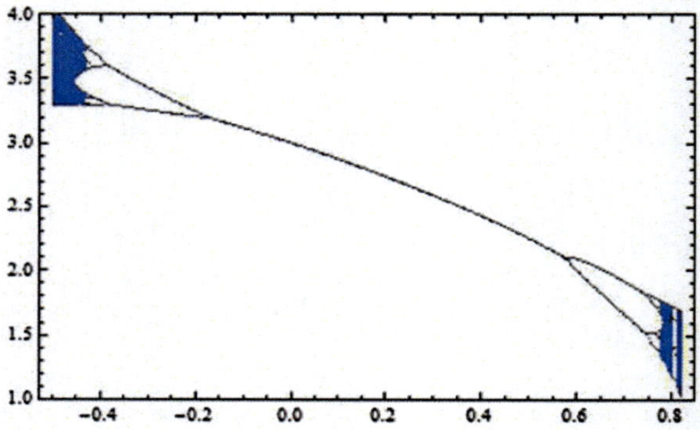

Fig. 12 Bifurcation diagram with respect to the parameter d against the variable p_1^* with 400 iterations of the map (15) for α = 5, c = 1, k = 0.30 and μ = 0.20

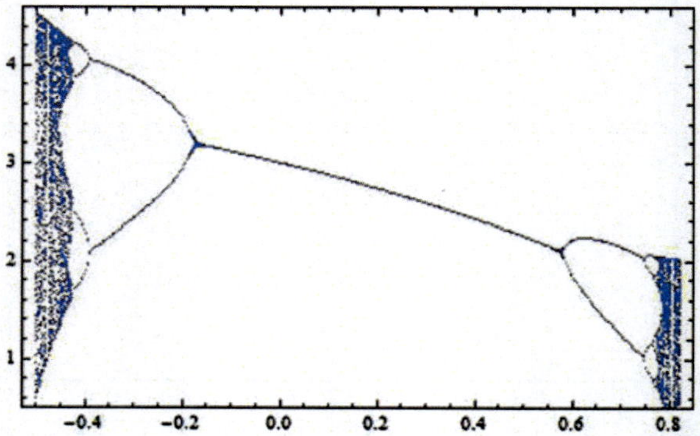

Fig. 13 Bifurcation diagram with respect to the parameter d against the variable p_2^* with 400 iterations of the map (15) for α = 5, c = 1, k = 0.30 and μ = 0.20

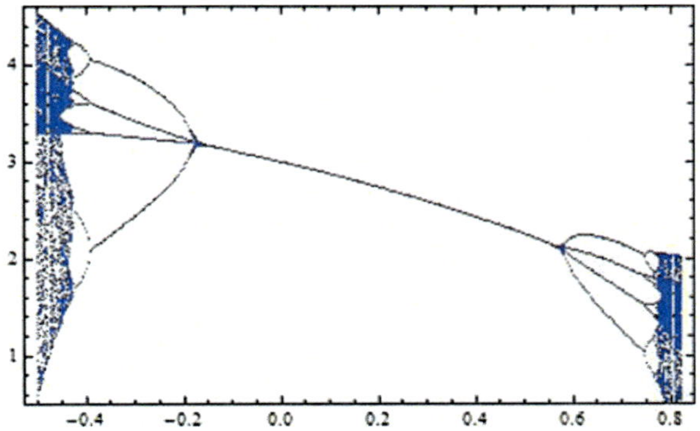

Fig. 14 The two previous bifurcation diagrams of Figs. 12 and 13 in one

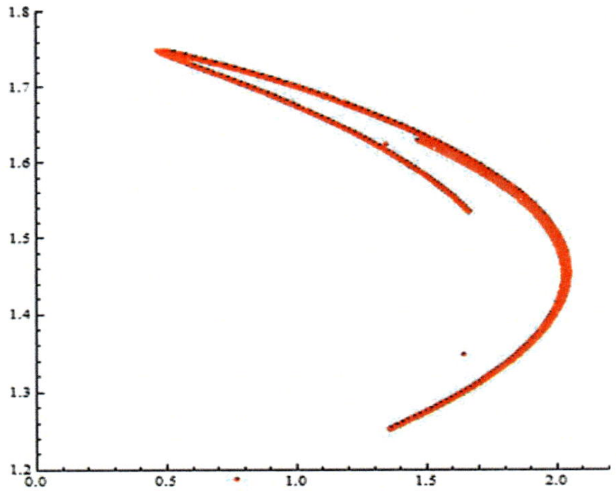

Fig. 15 Phase portrait (strange attractor) of the orbit of (0.1, 0.1) with 8000 iterations of the map (15) for $\alpha = 5$, $c = 1$, $k = 0.30$, $\mu = 0.20$ and $d = 0.79$

Finally, for these values of the parameter d the system becomes sensitive on initial conditions a result that is revealed by the sensitive dependence on initial conditions with a small change at the first coordinate and plotting the time series of system of (15) for $d = 0.79$ Figs. 17 and 18 and for $d = -0.50$. Figures 21 and 22.

3.4 *Focusing to the Parameter* μ

Using the same methods of numerical simulations focusing to the parameter μ (relative profit parameter) it is shown that small values of this parameter can destabilize

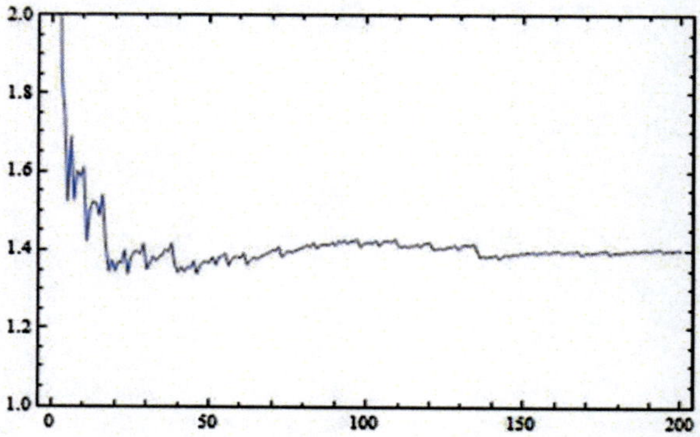

Fig. 16 Lyapunov numbers of the orbit of (0.1, 0.1) with 8000 iterations of the map (15) for α = 5, c = 1, k = 0.30, μ = 0.20 and d = 0.79

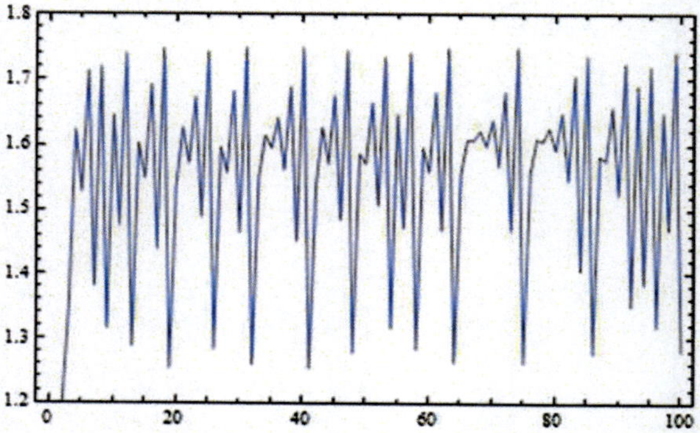

Fig. 17 Sensitive dependence on initial conditions for p_1^*-coordinate plotted against the time: the orbit of (0.1, 0.1) of the system (15) for α = 5, c = 1, k = 0.30, μ = 0.20 and d = 0.79

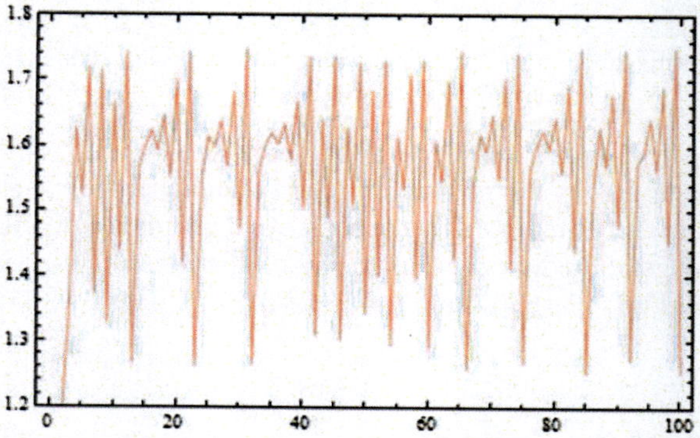

Fig. 18 Sensitive dependence on initial conditions for p_1^*-coordinate plotted against the time: the orbit of (0.101, 0.1) of the system (15) for α = 5, c = 1, k = 0.30, μ = 0.20 and d = 0.79

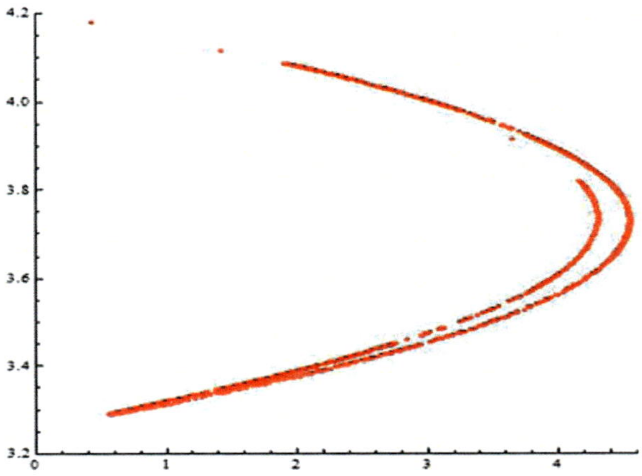

Fig. 19 Phase portrait (strange attractor) of the orbit of (0.1, 0.1) with 8000 iterations of the map (15) for $\alpha = 5$, $c = 1$, $k = 0.30$, $\mu = 0.20$ and $d = -0.50$

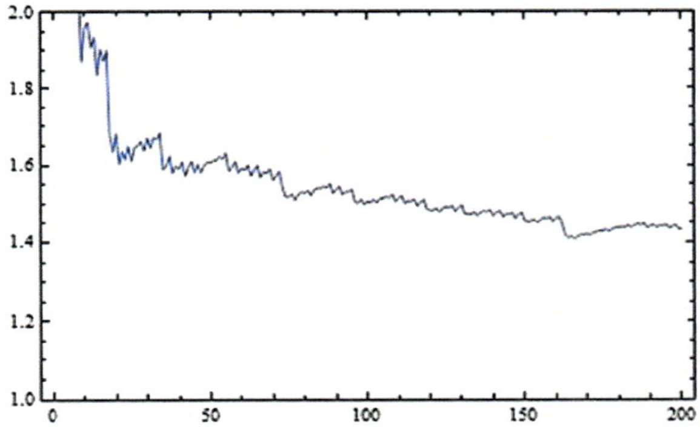

Fig. 20 Lyapunov numbers of the orbit of (0.1, 0.1) with 8000 iterations of the map (15) for $\alpha = 5$, $c = 1$, $k = 0.30$, $\mu = 0.20$ and $d = -0.50$

the economy through doubling bifurcation diagrams Figs. 23, 24 and 25 and strange attractors Fig. 26 and Lyapunov numbers Fig. 27 higher than the number of 1 are appeared. Also, the system of (15) becomes sensitive on small changes of initial conditions for small values of the parameter μ ($\mu = 0.05$) and it is revealed by the Figs. 28 and 29 of time series at first setting to the system the initial conditions of (0.1,0.1) and secondly of (0.101,0.1). As it seems, at the beginning the time series are indistinguishable, but after a number of iterations, the difference between them builds up rapidly.

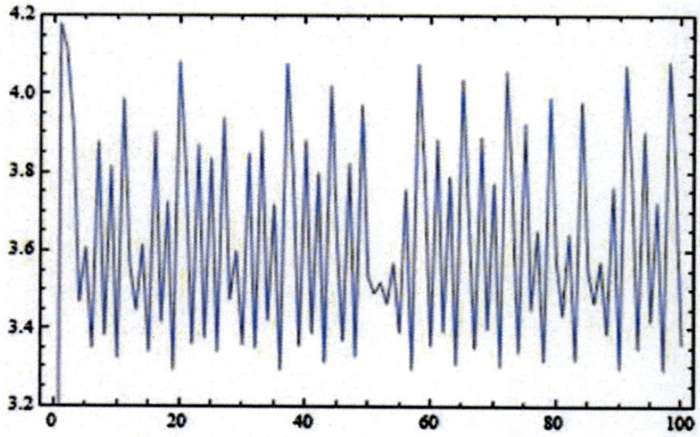

Fig. 21 Sensitive dependence on initial conditions for p_1^*-coordinate plotted against the time: the orbit of (0.1, 0.1) of the system (15) for α = 5, c = 1, k = 0.30, μ = 0.20 and d = −0.50

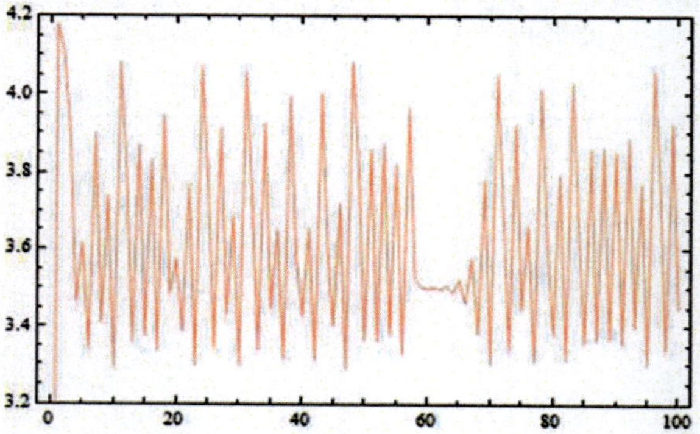

Fig. 22 Sensitive dependence on initial conditions for p_1^*-coordinate plotted against the time: the orbit of (0.101, 0.1) of the system (15) for α = 5, c = 1, k = 0.30, μ = 0.20 and d = −0.50

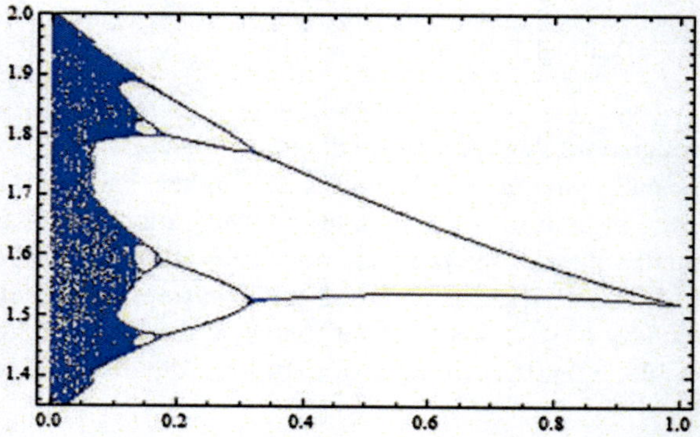

Fig. 23 Bifurcation diagram with respect to the parameter d against the variable p_1^* with 400 iterations of the map (15) for α = 5, c = 1, k = 0.315 and d = 0.74

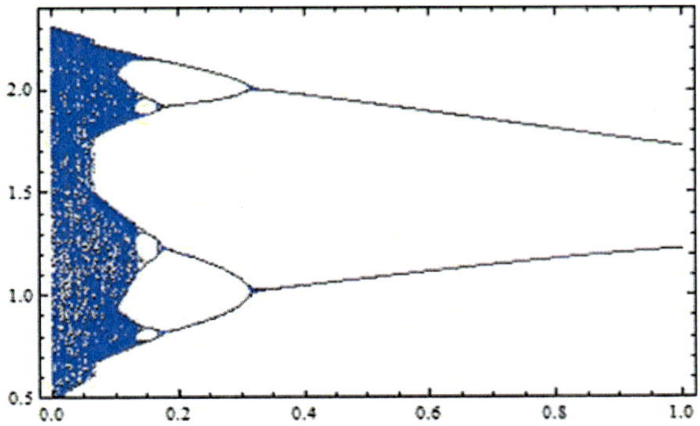

Fig. 24 Bifurcation diagram with respect to the parameter d against the variable p_2^* with 400 iterations of the map (15) for $\alpha = 5$, $c = 1$, $k = 0.315$ and $d = 0.74$

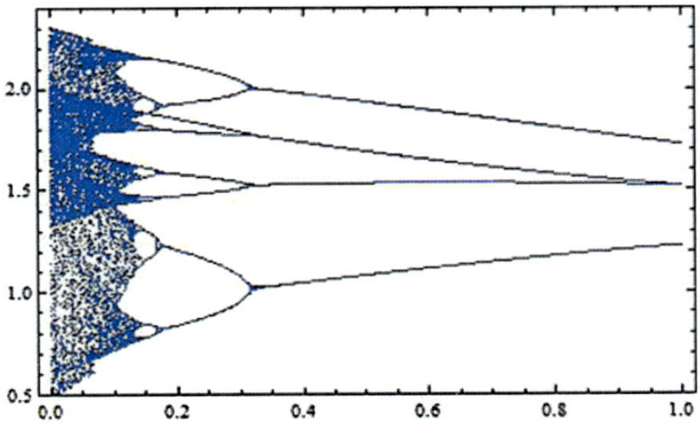

Fig. 25 The two previous bifurcation diagrams of Figs. 23 and 24 in one

4 Conclusions

In this paper we analyzed the dynamics of a differentiated Bertrand duopoly with heterogeneous expectations, linear demand and cost functions. By assuming that at each time period each firm maximizes its expected relative profit under different expectations, a discrete dynamical system was obtained. Existence and stability of equilibrium of this system are studied. We showed numerically that the model gives chaotic and unpredictable trajectories. The main result is that a lower and higher degree of product differentiation relative may destabilize the Bertrand–Nash equilibrium. Also, this instability can be appeared for higher values of the speed of adjustment and lower values of the relative profit parameter. Finally, we showed also that for lower values of the speed of adjustment the equilibrium is stable for each value of the relative profit parameter.

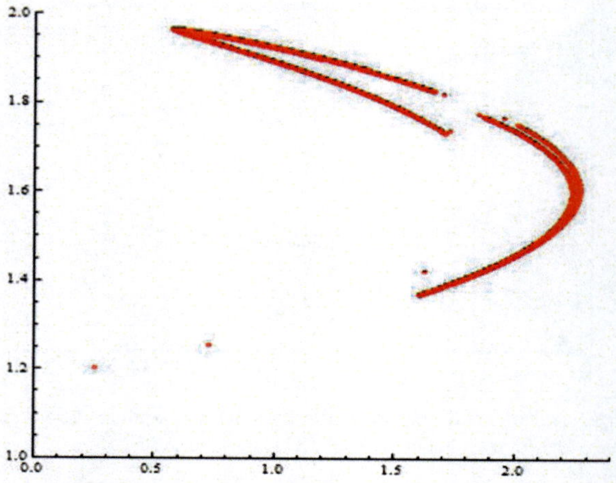

Fig. 26 Phase portrait (strange attractor) of the orbit of (0.1, 0.1) with 8000 iterations of the map (15) for $\alpha = 5$, $c = 1$, $k = 0.315$, $d = 0.74$ and $\mu = 0.05$

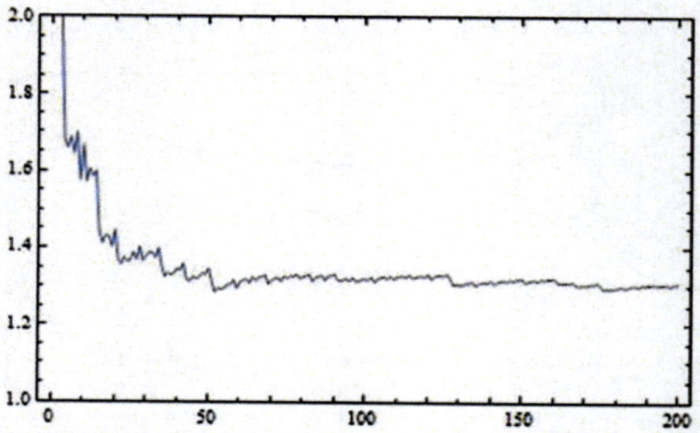

Fig. 27 Lyapunov numbers of the orbit of (0.1, 0.1) with 8000 iterations of the map (15) for $\alpha = 5$, $c = 1$, $k = 0.315$, $d = 0.74$ and $\mu = 0.05$

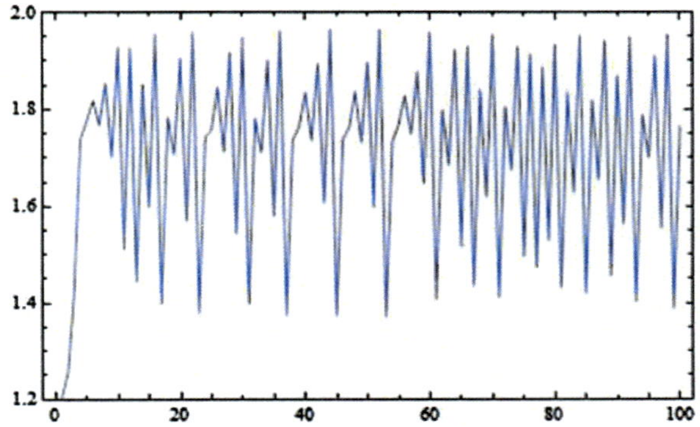

Fig. 28 Sensitive dependence on initial conditions for p_1^*-coordinate plotted against the time: the orbit of (0.1, 0.1) of the system (15) for $\alpha = 5$, $c = 1$, $k = 0.315$, $d = 0.74$ and $\mu = 0.05$

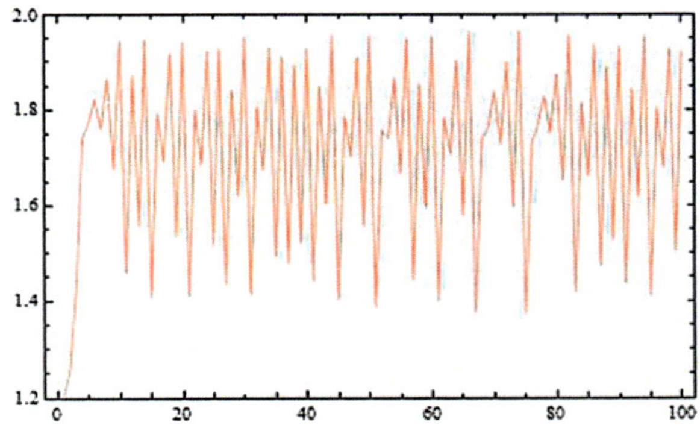

Fig. 29 Sensitive dependence on initial conditions for p_1^*-coordinate plotted against the time: the orbit of (0.101, 0.1) of the system (15) for $\alpha = 5$, $c = 1$, $k = 0.315$, $d = 0.74$ and $\mu = 0.05$

References

1. H.N. Agiza, Explicit stability zones for Cournot games with 3 and 4 competitors. Chaos Solitons Fract. **9**, 1955–1966 (1998)
2. H.N. Agiza, On the analysis of stability, bifurcation, chaos and chaos control of Kopel map. Chaos, Solitons Fractals **10**, 1909–1916 (1999)
3. H.N. Agiza, On the stability, bifurcations, chaos and chaos control of Kopel map. Chaos Solitons Fract. **11**, 1909–1916 (2004)
4. H.N. Agiza, A.A. Elsadany, Chaotic dynamics in nonlinear duopoly game with heterogeneous players. Appl. Math. Comput. **149**, 843–60.
5. H.N. Agiza, A.A. Elsadany, Nonlinear dynamics in the Cournot duopoly game with heterogeneous players. Phys. A **320**, 512–524 (2003)
6. H.N. Agiza, A.S. Hegazi, A.A. Elsadany, Complex dynamics and synchronization of duopoly game with bounded rationality. Math. Comput. Simulat. **58**, 133–146 (2002)
7. S.S. Askar, On complex dynamics of monopoly market. Econ. Model. **31**, 586–589 (2013)

8. S.S. Askar, Complex dynamic properties of Cournot duopoly games with convex and log-concave demand function. Oper. Res. Lett. **42**, 85–90 (2014)
9. W.J. Baumol, R.E. Quandt, Rules of thumb and optimally imperfect decisions. Am. Econ. Rev. **54**(2), 23–46 (1964)
10. G.I. Bischi, M. Kopel, Equilibrium selection in a nonlinear duopoly game with adaptive expectations. J. Econ. Behav. Org. **46**, 73–100 (2001)
11. G.I. Bischi, F. Lamantia, L. Sbragia, Competition and cooperation in natural resources exploitation: an evolutionary game approach, in *Game practice and the environment*. ed. by C. Cararro, V. Fragnelli (Edward Elgar, Cheltenham, 2004), pp. 187–211
12. G.I. Bischi, A. Naimzada, Global analysis of a dynamic duopoly game with bounded rationality, in *Advances in dynamic games and applications*, ed. by J.A. Filar, V. Gaitsgory, K. Mizukami (Birkhauser, Basel, 2000), vol. 5, pp. 361–385
13. G.I. Bischi, A.K. Naimzada, L. Sbragia, Oligopoly games with local monopolistic approximation. J. Econ. Behav. Organ. **62**(3), 371–388 (2007)
14. G.I. Bischi, M. Kopel, Equilibrium selection in a nonlinear duopoly game with adaptive expectations. J. Econ. Behav. Organ. **46**, 73–100 (2001)
15. G.I. Bischi, A. Naimzada, Global analysis of a dynamic duopoly game with bounded rationality, in *Advances in dynamic games and applications*, eds. by J.A. Fila, V. Gaitsgory, K. Mizukami (Birkhauser, Basel, 1999), vol. 5, pp. 361–385
16. A. Cournot, *Researches into the Mathematical Principles of the Theory of Wealth* (Irwin, Homewood (IL), 1963).
17. R. Day, *Complex Economic Dynamics* (MIT Press, Cambridge, 1994)
18. A.K. Dixit, Comparative statics for oligopoly. Internat. Econ. Rev. **27**, 107–122 (1986)
19. A.K. Dixit, A model of duopoly suggesting a theory of entry barriers. Bell J Econ **10**, 20–32 (1979)
20. W.J. Den Haan, The importance of the number of different agents in a heterogeneous asset-pricing model. J. Econ. Dyn. Control **25**, 721–746 (2001)
21. A.A. Elsadany, Dynamics of a Cournot duopoly game with bounded rationality based on relative profit maximization. Appl. Math. Comput. **294**, 253–263 (2017)
22. S. Elaydi, *An Introduction to Difference Equations*, 3rd edn. (Springer, New York, 2005)
23. J. Ma, Z. Guo, The influence of information on the stability of a dynamic Bertrand game. Commun. Nonlinear Sci. Numer. Simul. **30**, 32–44 (2016)
24. C.H. Hommes, Heterogeneous agent models in economics and finance, in *Handbook of Computational Economics, Agent-Based Computational Economics*,eds. by L. Tesfatsion, K.L. Judd (Elsevier Science B.V, 2006), vol. 2, pp. 1109–1186
25. G. Gandolfo, *Economic Dynamics* (Springer, Berlin, 1997).
26. Y. Gao, Complex dynamics in a two dimensional noninvertible map. Chaos Solitons Fract. **39**, 1798–1810 (2009)
27. M. Kopel, Simple and complex adjustment dynamics in Cournot duopoly models. Chaos Solitons Fract. **12**, 2031–2048 (1996)
28. M. Kulenonic M, O. Merino O, *Discrete Dynamical Systems and Difference Equations with Mathematica* (Chapman and Hall/Crc, 2002)
29. A. Medio, G. Gallo, *Chaotic Dynamics: Theory and Applications to Economics* (Cambridge University Press, Cambridge (MA), 1995)
30. A. Medio, M. Lines, Introductory notes on the dynamics of linear and linearized systems, in *Nonlinear Dynamical Systems in Economics*, ed. by M. Lines (CISM, Springer, Wien, NewYork, 2005), pp 1–26
31. A. Medio, M. Lines, *Nonlinear Dynamics. A Primer* (Cambridge University Press, Cambridge (MA), 2001)
32. A.K. Naimzada, G. Ricchiuti, Complex dynamics in a monopoly with a rule of thumb. Appl. Math. Comput. **203**, 921–925 (2008)
33. A. Naimzada, L. Sbragia, Oligopoly games with nonlinear demand and cost functions: two boundedly rational adjustment processes. Chaos Solitons Fractals **29**, 707–722 (2006)

34. Y. Nakamura, Price versus quantity in a mixed duopoly: The case of relative profit maximization. Econ. Model. **44**, 37–43 (2015)
35. T. Puu, The chaotic monopolist. Chaos, Solitons Fractals **5**(1), 35–44 (1995)
36. T. Puu, The chaotic duopolists revisited. J Econ. Behav. Org. **37**, 385–394 (1998)
37. T. Puu, Chaos in duopoly pricing. Chaos Solitons Fract. **1**, 573–581 (1991)
38. T. Puu, Complex oligopoly dynamics, in *Nonlinear Dynamical Systems in Economics.*, ed. by M. Lines (Springer Wien NewYork: CISM, 2005), pp 165–86
39. G. Sarafopoulos, On the dynamics of a duopoly game with differentiated goods. Procedia Econ. Finance **19**, 146–153 (2015)
40. G. Sarafopoulos, Complexity in a duopoly game with homogeneous players, convex, log linear demand and quadratic cost functions. Procedia Econ. Finance **33**, 358–366 (2015)
41. G. Sarafopoulos, K. Papadopoulos, On a Cournot duopoly game with differentiated goods, heterogeneous expectations and a cost function including emission costs. Sci. Bull. Econ. Sci. **1**(1), 11–22 (2017)
42. G. Sarafopoulos, K. Papadopoulos, On the stability of a Cournot dynamic game under the influence of information. Knowledge E Publishing, ISSN: 2518–668X, 131–144 (2018)
43. G. Sarafopoulos, K. Papadopoulos, Chaos in oligopoly models. Int. J. Product. Manag. Assess. Technol. **7**(1), 50–76 (2019)
44. G. Sarafopoulos, K. Papadopoulos, Complexity in Bertrand duopoly game with heterogeneous players and differentiated goods, in *Economic and Financial Challenges for Eastern Europe*, Article 2 (Springer, Berlin, 2019), pp. 15–26
45. A. Satoh, Y. Tanaka, Relative Profit Maximization and Equivalence of Cournot and Bertrand Equilibria in Asymmetric Duopoly, MPRA Paper no. 56031.
46. H. Sedaghat, *Nonlinear Difference Equations: Theory with Applications to Social Science Models* (Kluwer Academic Publishers (now Springer), 2003)
47. N. Singh, X. Vives, Price and quantity competition in a differentiated duopoly. Rand J. Econ. **15**, 546–554 (1984)
48. F. Tramontana, Heterogeneous duopoly with isoelastic demand function. Econ. Model. **27**, 350–357 (2010)
49. F. Westerhoff, Nonlinear expectation formation, endogenous business cycles and stylized facts. Stud. Nonlinear Dyn. Econ. **10**(4) (Article 4) (2016)
50. W. Wu, Z. Chen, W.H. Ip, Complex nonlinear dynamics and controlling chaos in a Cournot duopoly economic model. Nonlinear Anal. Real World Appl. **11**, 4363–4377 (2010)
51. J. Zhang, Q. Da, Y. Wang, Analysis of nonlinear duopoly game with heterogeneous players. Econ. Model. **24**, 138–148 (2007)
52. J. Zhang, Q. Da, Y. Wang, The dynamics of Bertrand model with bounded rationality. Chaos, Solitons Fractals **39**, 2048–2055 (2009)

On a Cournot Dynamic Game with Cost Uncertainty and Relative Profit Maximization

Georges Sarafopoulos and Kosmas Papadopoulos

Abstract In this paper, a Cournot duopoly model with homogeneous goods is examined with uncertain cost function. A random linear cost function is introduced in this model for the first player. The case of homogeneous expectations is studied. The existence and uniqueness of the equilibrium are obtained. The asymptotic behavior of the equilibrium point is also investigated. Complete stability and bifurcation analysis are provided. The obtained theoretical results are verified by numerical simulations.

Keywords Cournot duopoly game · Cost uncertainty · Relative profit maximization · Discrete dynamical system · Nash equilibrium · Stability · Bifurcation diagrams · Lyapunov numbers · Strange attractors · Chaotic behavior

1 Introduction

An Oligopoly is a market structure between monopoly and perfect competition, where there are only a few number of firms in the market producing homogeneous products. The dynamic of an oligopoly game is more complex because firms must consider not only the behaviors of the consumers, but also the reactions of the competitors i.e. they form expectations concerning how their rivals will act. Cournot, in 1838 has introduced the first formal theory of oligopoly. In 1883 another French mathematician Joseph Louis Francois Bertrand modified Cournot game suggesting that firms actually choose prices rather than quantities. Originally Cournot and Bertrand models were based on the premise that all players follow naive expectations, so that in every step, each player (firm) assumes the last values that were taken by the competitors without estimation of their future reactions. However, in real market conditions

G. Sarafopoulos (✉) · K. Papadopoulos
Department of Economics, Democritus University of Thrace, University Campus, 69100 Komotini, Greece
e-mail: gsarafop@econ.duth.gr

K. Papadopoulos
e-mail: kpapa@econ.duth.gr

C. H. Skiadas et al. (eds.), *13th Chaotic Modeling and Simulation International Conference*, Springer Proceedings in Complexity,
https://doi.org/10.1007/978-3-030-70795-8_54

such an assumption is very unlikely since not all players share naive beliefs. Therefore, different approaches to firm behavior were proposed. Some authors considered duopolies with homogeneous expectations and found a variety of complex dynamics in their games, such as appearance of strange attractors (Agiza [1]; Agiza et al. [4]; Agliari et al. [5, 6]; Bischi and Kopel [11]; Kopel [18]; Puu [23]; Sarafopoulos [24, 25]; Sarafopoulos et al. [28]; Zhang et al. [32]). Also models with heterogeneous agents were studied (Agiza and Elsadany [2, 3]; Den Haan [12]; Fanti and Gori [15]; Hommes [17]; Sarafopoulos et al. [26, 27, 29]; Tramontana [30]; Zhang et al. [31]).

In the real market producers do not know the entire demand function, though it is possible that they have a perfect knowledge of technology, represented by the cost function. Hence, it is more likely that firms employ some local estimate of the demand. This issue has been previously analyzed by Baumol and Quandt [9]; Puu [22]; Naimzada and Ricchiuti [20]; Askar [7]; Askar [8]. Bounded rational players (firms) update their strategies based on discrete time periods and by using a local estimate of the marginal profit. With such local adjustment mechanism, the players are not requested to have a complete knowledge of the demand and the cost functions (Agiza and Elsadany [2]; Naimzada and Sbragia [21]; Zhang et al. [32]; Askar [8]; Bischi et al. [10, 11]).

In this paper we study the dynamics of a Cournot-type duopoly with homogeneous goods where each firm behaves with homogeneous expectations. We show that the model gives more complex chaotic and unpredictable trajectories as a consequence of change in the speed of players' adjustment. The paper is organized as follows: In Sect. 2, the dynamics of the duopoly game with homogeneous expectations, linear demand and cost functions and relative profit functions for two players are analyzed. A cost uncertainty is introduced into first player's utility function. We set both players as bounded rational players. The existence and local stability of the equilibrium points are also analyzed. In Sect. 3 numerical simulations are used to verify the algebraic results of Sect. 2 plotting the bifurcation diagrams of the game's system and to show the complex dynamics via computing Lyapunov numbers, and sensitive dependence on initial conditions.

2 The Game

2.1 *The Construction of the Game*

In this study we assume that in the two companies there is a separation between ownership and management, so there is a possibility that the managers who make decisions for the company to decide at the expense of their company trying to increase the profits of the competitor. Also, we consider homogeneous players and more specifically, we consider that both firms choose the quantity of their productions in a rational way, following an adjustment mechanism (bounded rational players). We consider a simple Cournot-type duopoly market where firms (players) produce the same good and offer it at discrete-time periods on a common market. Production decisions are taken at discrete time periods t = 0, 1, 2,... At each period t, every firm must form an expectation of the rival's strategy in the next time period in order to determine the corresponding profit-maximizing prices for period t + 1. We suppose that q_1, q_2 are the production quantities of each firm. Also, we consider that the preferences of consumers represented by the equation:

$$U(q_1, q_2) = \alpha(q_1 + q_2) - \frac{1}{2}\left(q_1^2 + q_2^2 + 2dq_1q_2\right) \tag{1}$$

where α is a positive parameter (α > 0), which expresses the market size and $d \in [-1, 1]$ is the parameter that reveals the differentiation degree of products [13]. For example, if $d = 0$ then both products are independently and each firm participates in a monopoly. But, if $d = 1$ then one product is a substitute for the other, since the products are homogeneous. It is understood that for positive values of the parameter d the larger the value, the less diversification we have in both products. On the other hand negative values of the parameter d are described that the two products are complementary and when $d = -1$ then we have the phenomenon of full competition between the two companies. The inverse demand functions (as functions of quantities) coming from the maximizing of (1) are given by the following equations (assuming d = 1):

$$p_1(q_1, q_2) = \alpha - q_1 - q_2 \text{ and } p_2(q_1, q_2) = \alpha - q_2 - q_1, (d = 1) \tag{2}$$

In this work we suppose that the first player's cost function contains an uncertainty by which the marginal cost (linear cost function) is equal to the combination between the parameters:$c_1, c_2 > 0$, which is described by the following equation:

$$C_1(q_1) = [p \cdot c_1 + (1 - p) \cdot c_2] \cdot q_1 \tag{3}$$

where $p \in [0, 1]$, is the positive uncertainty cost parameter.

On the other hand the second player uses a simple linear cost function that its marginal cost is equal to $c_1 > 0$ and it is described by the equation:

$$C_2(q_2) = c_1 \cdot q_2 \tag{4}$$

With these assumptions the profits of the firms are given by:

$$\Pi_1(q_1, q_2) = p_1 \cdot q_1 - C_1(q_1) = [\alpha - q_1 - q_2 - p \cdot c_1 - (1 - p) \cdot c_2] \cdot q_1 \tag{5}$$

and

$$\Pi_2(q_1, q_2) = p_2 \cdot q_2 - C_2(q_2) = [\alpha - c_1 - q_1 - q_2]q_2 \tag{6}$$

Then the marginal profits at the point of the strategy space are given by:

$$\frac{\partial \Pi_1}{\partial q_1} = \alpha - p \cdot c_1 - (1-p) \cdot c_2 - 2q_1 - q_2, \frac{\partial \Pi_1}{\partial q_2} = -q_1 \tag{7}$$

and

$$\frac{\partial \Pi_2}{\partial q_2} = \alpha - c_1 - q_1 - 2q_2, \frac{\partial \Pi_2}{\partial q_1} = -q_2 \tag{8}$$

As it is noticed both managers care about the maximization of a utility function that contains a percentage of opponent company's profits (generalized relative profit function), which is given by:

$$U_i = (1-\mu_i) \cdot \Pi_i + \mu_i \cdot (\Pi_i - \Pi_j) = \Pi_i - \mu_i \cdot \Pi_j \tag{9}$$

where $\mu \in [0, 1]$ is the percentage that the player i takes into account the opponent company's prifots. So, the marginal utility of the player i is given by the following equation:

$$\frac{\partial U_i}{\partial q_i} = \frac{\partial \Pi_i}{\partial q_i} - \mu_i \cdot \frac{\partial \Pi_j}{\partial q_i} \tag{10}$$

and the marginal utilities for each player are:

$$\frac{\partial U_1}{\partial q_1} = \alpha - p \cdot c_1 - (1-p) \cdot c_2 - 2q_1 - (1-\mu)q_2 \tag{11}$$

and

$$\frac{\partial U_2}{\partial q_2} = \alpha - c_1 - (1-\mu)q_1 - 2q_2 \tag{12}$$

Both players are characterized as bounded rational players. According to the existing literature it means that they decide their productions following a mechanism that is described by the equation:

$$\frac{q_i(t+1) - q_i(t)}{q_i(t)} = k \cdot \frac{\partial U_i}{\partial q_i}, k \leq 0 \tag{13}$$

Through this mechanism the player increases his level of adaptation when his marginal utility is positive or decreases his level when his marginal utility is negative, where k is the speed of adjustment of player, it is a positive parameter (k > 0), which gives the extend variation of production quantity of the each company, following a given utility signal.

The dynamical system of the players is described by:

$$\begin{cases} q_1(t+1) = q_1(t) + k \cdot q_1(t) \cdot \dfrac{\partial U_1}{\partial q_1} \\ q_2(t+1) = q_2(t) + k \cdot q_2(t) \cdot \dfrac{\partial U_2}{\partial q_2} \end{cases} \quad (14)$$

We will focus on the dynamics of this system to the parameter k.

2.2 *Dynamical Analysis*

The dynamical analysis of the discrete dynamical system involves finding equilibrium positions and studying them for stability. The ultimate goal of this algebraic study is to formulate a proposition that will be the stability condition of the Nash Equilibrium position. Finally, these algebraic results are verified and visualized doing some numerical simulations using the program of Mathematica.

2.2.1 The Equilibrium Positions

The equilibriums of the dynamical system (14) are obtained as the nonnegative solutions of the algebraic system:

$$\begin{cases} q_1^* \cdot \dfrac{\partial U_1}{\partial q_1} = 0 \\ q_2^* \cdot \dfrac{\partial U_2}{\partial q_2} = 0 \end{cases} \quad (15)$$

which is obtained by setting: $q_1(t+1) = q_1(t) = q_1^*$ and $q_2(t+1) = q_2(t) = q_2^*$.

- If $q_1^* = q_2^* = 0$ then the boundary equilibrium position is the point:

$$E_0 = (0, 0) \quad (16)$$

- If $q_1^* = 0$ and $\frac{\partial U_2}{\partial q_2} = 0$ then: $q_2^* = \frac{\alpha - c_1}{2}$ and the equilibrium position is the point:

$$E_1 = \left(0, \frac{\alpha - c_1}{2}\right) \quad (17)$$

- If $q_2^* = 0$ and $\frac{\partial U_1}{\partial q_1} = 0$ then: $q_1^* = \frac{\alpha - p \cdot c_1 - (1-p) \cdot c_2}{2}$ and the equilibrium position is the point:

$$E_2 = \left(\frac{\alpha - p \cdot c_1 - (1-p) \cdot c_2}{2}, 0 \right) \tag{18}$$

- If $\frac{\partial U_1}{\partial q_1} = \frac{\partial U_2}{\partial q_2} = 0$ then the following system is obtained:

$$\begin{cases} \alpha - p \cdot c_1 - (1-p) \cdot c_2 - 2q_1^* - (1-\mu) \cdot q_2^* = 0 \\ \alpha - c_1 - (1-\mu) \cdot q_1^* - 2q_2^* = 0 \end{cases} \tag{19}$$

and the nonnegative solution of this algebraic system will give the Nash Equilibrium position $E_* = \left(q_1^*, q_2^*\right)$ where:

$$q_1^* = \frac{\alpha(1+\mu) + (1-\mu-2p) \cdot c_1 - 2(1-p) \cdot c_2}{4-(1-\mu)^2} \tag{20}$$

and

$$q_2^* = \frac{\alpha(1+\mu) - (2-p+p \cdot \mu) \cdot c_1 + (1-p) \cdot (1-\mu) \cdot c_2}{4-(1-\mu)^2} \tag{21}$$

This means that:

$$\alpha(1+\mu) + (1-\mu-2p) \cdot c_1 - 2(1-p) \cdot c_2 > 0 \tag{22}$$

and

$$\alpha(1+\mu) - (2-p+p \cdot \mu) \cdot c_1 + (1-p) \cdot (1-\mu) \cdot c_2 > 0. \tag{23}$$

2.2.2 Stability of Equilibrium Points

To study the stability of the equilibrium positions we need the Jacobian matrix of the dynamical system Eq. (15) which is the matrix:

$$J\left(q_1^*, q_2^*\right) = \begin{bmatrix} f_{q_1} & f_{q_2} \\ g_{q_1} & g_{q_2} \end{bmatrix} \tag{24}$$

where:

$$\begin{aligned} f(q_1, q_2) &= q_1 + k \cdot q_1 \cdot \frac{\partial U_1}{\partial q_1} \\ g(q_1, q_2) &= q_2 + k \cdot q_2 \cdot \frac{\partial U_2}{\partial q_2} \end{aligned} \tag{25}$$

and as a result the Jacobian matrix of game's discrete dynamical system Eq. (14) is the following matrix:

$$J\left(q_1^*, q_2^*\right) = \begin{bmatrix} 1 + k \cdot \left(\frac{\partial U_1}{\partial q_1} + q_1^* \cdot \frac{\partial^2 U_1}{\partial q_1^2}\right) & k \cdot q_1^* \cdot \frac{\partial^2 U_1}{\partial q_1 \partial q_2} \\ k \cdot q_2^* \cdot \frac{\partial^2 U_2}{\partial q_2 \partial q_1} & 1 + k \cdot \left(\frac{\partial U_2}{\partial q_2} + q_2^* \cdot \frac{\partial^2 U_2}{\partial q_2^2}\right) \end{bmatrix} \tag{26}$$

For the E_0 the Jacobian matrix becomes as:

$$J(E_0) = \begin{bmatrix} 1 + k \cdot \frac{\partial U_1}{\partial q_1} & 0 \\ 0 & 1 + k \cdot \frac{\partial U_2}{\partial q_2} \end{bmatrix} \overset{A=1+k\cdot\frac{\partial U_1}{\partial q_1}}{\underset{B=1+k\cdot\frac{\partial U_2}{\partial q_2}}{=}} \begin{bmatrix} \mathrm{A} & 0 \\ 0 & B \end{bmatrix} \tag{27}$$

with $Tr = \mathrm{A} + B$ and $Det = A \cdot B$.
From the characteristic equation of $J(E_0)$, we find the nonnegative eigenvalues:

$$r_1 = A = 1 + k \cdot \frac{\partial U_1}{\partial q_1} \text{ and } r_2 = B = 1 + k \cdot \frac{\partial U_2}{\partial q_2} \tag{28}$$

it's clearly seems that $|r_1|, |r_2| > 1$ and the E_0 equilibrium is unstable.
For the E_1 the Jacobian matrix becomes as:

$$J(E_1) = \begin{bmatrix} 1 + k \cdot \frac{\partial U_1}{\partial q_1} & 0 \\ -k \cdot (1 - \mu) q_2^* & 1 - 2k \cdot q_2^* \end{bmatrix} \overset{C=1+k\cdot\frac{\partial U_1}{\partial q_1}}{\underset{E=1-2k\cdot q_2^*}{=}} \begin{bmatrix} C & 0 \\ D & E \end{bmatrix} \tag{29}$$

with $Tr = C + E$ and $Det = C \cdot E$.
From the characteristic equation of $J(E_1)$, we find the nonnegative eigenvalue:

$$r_1 = C = 1 + k \cdot \frac{\alpha(1 + \mu) + (1 - \mu - 2p) \cdot c_1 - (1 - p) \cdot c_2}{2} \tag{30}$$

it's clearly seems that $|r_1| > 1$, because:
$\alpha(1 + \mu) + (1 - \mu - 2p) \cdot c_1 - 2(1 - p) \cdot c_2 > 0$ Eq. (22) and the E_1 equilibrium is unstable.
For the E_2 the Jacobian matrix becomes as:

$$J(E_2) = \begin{bmatrix} 1 - 2k \cdot q_1^* & -k \cdot (1 - \mu) q_1^* \\ 0 & 1 + k \cdot \frac{\partial U_2}{\partial q_2} \end{bmatrix} \overset{F=1-2k\cdot q_1^*}{\underset{H=1+k\cdot\frac{\partial U_2}{\partial q_2}}{=}} \begin{bmatrix} F & G \\ 0 & H \end{bmatrix} \tag{31}$$

with $Tr = F + H$ and $Det = F \cdot H$.
From the characteristic equation of $J(E_2)$, we find the nonnegative eigenvalue:

$$r_2 = H = 1 + k \cdot \frac{\alpha(1 + \mu) - (2 - p + p \cdot \mu) \cdot c_1 + (1 - p) \cdot (1 - \mu) \cdot c_2}{2} \tag{32}$$

it's clearly seems that $|r_2| > 1$, because:
$\alpha(1+\mu) - (2 - p + p \cdot \mu) \cdot c_1 + (1-p) \cdot (1-\mu) \cdot c_2 > 0$ Eq. (23) and the E_2 equilibrium is unstable.

For the E_* the Jacobian matrix becomes as:

$$J(E_*) = \begin{bmatrix} 1 + k \cdot q_1^* \cdot \frac{\partial^2 U_1}{\partial q_1^2} & k \cdot q_1^* \cdot \frac{\partial^2 U_1}{\partial q_1 \partial q_2} \\ k \cdot q_2^* \cdot \frac{\partial^2 U_2}{\partial q_2 \partial q_1} & 1 + k \cdot q_2^* \cdot \frac{\partial^2 U_2}{\partial q_2^2} \end{bmatrix} \tag{33}$$

with

$$Tr = 2 - 2k \cdot q_1^* - 2k \cdot q_2^* \tag{34}$$

and

$$Det = 1 - 2k \cdot q_1^* - 2k \cdot q_2^* + \left[4 - (1-\mu)^2\right] \cdot k^2 \cdot q_1^* \cdot q_2^* \tag{35}$$

To study the stability of Nash equilibrium we use three conditions that the equilibrium position is locally asymptotically stable when they are satisfied simultaneously [14, 16, 19]:

$$\begin{array}{ll} (i) & 1 - Det > 0 \\ (ii) & 1 - Tr + Det > 0 \\ (iii) & 1 + Tr + Det > 0 \end{array} \tag{36}$$

The condition (i) gives:

$$1 - Det > 0 \Leftrightarrow 2k\left(q_1^* + q_2^*\right) - \left[4 - (1-\mu)^2\right] \cdot k^2 \cdot q_1^* \cdot q_2^* > 0 \tag{37}$$

It's easy to find that the first condition (i) is always satisfied:

$$1 - Tr + Det > 0 \Leftrightarrow \left[4 - (1-\mu)^2\right] \cdot k^2 \cdot q_1^* \cdot q_2^* > 0 > 0 \tag{38}$$

because: $\left[4 - (1-\mu)^2\right] > 0$.
Finally, the condition (iii) becomes as:

$$1 + Tr + Det > 0 \Leftrightarrow \left[4 - (1-\mu)^2\right] \cdot q_1^* \cdot q_2^* \cdot k^2 - 4\left(q_1^* + q_2^*\right) \cdot k + 4 > 0 \tag{39}$$

Proposition: The Nash equilibrium of the discrete dynamical system Eq. (15) is locally asymptotically stable if:

$$2k\left(q_1^* + q_2^*\right) - \left[4 - (1-\mu)^2\right] \cdot k^2 \cdot q_1^* \cdot q_2^* > 0$$

and

$$\left[4-(1-\mu)^2\right]\cdot q_1^*\cdot q_2^*\cdot k^2-4\left(q_1^*+q_2^*\right)\cdot k+4>0.$$

3 Numerical Simulations Focusing on the Parameter k

From the condition (i) focusing on the parameter k we take the following inequality:

$$0<k<\frac{2\left(q_1^*+q_2^*\right)}{\left[4-(1-\mu)^2\right]\cdot q_1^*\cdot q_2^*} \tag{40}$$

The condition (iii) is the following:

$$\left[4-(1-\mu)^2\right]\cdot q_1^*\cdot q_2^*\cdot k^2-4\left(q_1^*+q_2^*\right)\cdot k+4>0$$

And its discriminant is positive:

$$\Delta=16\left[\left(q_1^*-q_2^*\right)^2+(1-\mu)^2\right]>0 \tag{41}$$

so the condition (iii) is satisfied if:

$$k\in(0,k_1)\cup(k_2,+\infty) \tag{42}$$

where:

$$k_{1,2}=\frac{4\left(q_1^*+q_2^*\right)\pm\sqrt{\Delta}}{2\left[4-(1-\mu)^2\right]\cdot q_1^*\cdot q_2^*} \tag{43}$$

are its two positive roots.

To provide some numerical evidence for the chaotic behavior of the system Eq. (14), as a consequence of change in the parameter k (the speed of adjustment), we present various numerical results here to show the chaoticity, including its bifurcations diagrams, strange attractors, Lyapunov numbers and sensitive dependence on initial conditions.

In order to study the local stability properties of the equilibrium points, it is convenient to take specific values for the other parameters: $\alpha=5$, c1 = 1, c2 = 0.5 and p = μ = 0.5. So, as a result we find that $q_1^*\simeq 1.73$ and $q_2^*\simeq 1.57$ and the stability condition becomes as:

$$0<k<0.48 \tag{44}$$

This algebraic result is verified by the bifurcation diagrams of q_1^* (Fig. 1) and q_2^* (Fig. 2) with respect to the parameter k. As it seems there is a locally asymptotically stable orbit until the value of 0.48 for the parameter k and after this value doubling period bifurcations are appeared and finally, for higher values of the parameter k the system's behavior becomes chaotic and unpredictable (Fig. 3).

Fig. 1 Bifurcation diagram with respect to the parameter d against the variable q_1^* with 400 iterations of the map Eq. (15) for $\alpha = 5$, $c_1 = 1$, $c_2 = 0.50$, $p = 0.50$ and $\mu = 0.50$

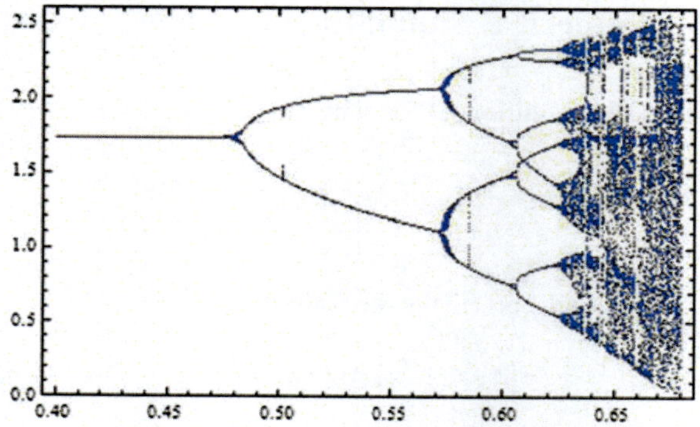

Fig. 2 Bifurcation diagram with respect to the parameter d against the variable q_2^* with 400 iterations of the map Eq. (15) for $\alpha = 5$, $c_1 = 1$, $c_2 = 0.50$, $p = 0.50$ and $\mu = 0.50$

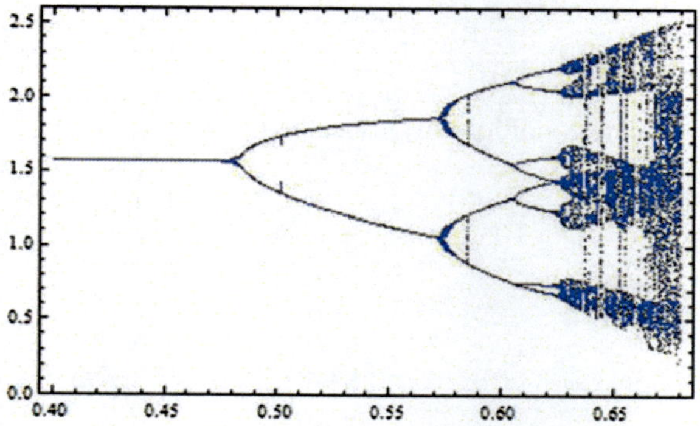

Fig.3 The two previous bifurcation diagrams of Figs. 1 and 2 in one

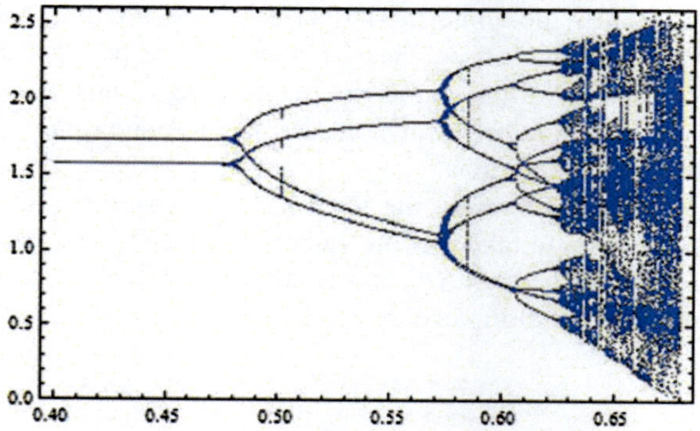

This chaotic trajectory can create strange attractors (Fig. 4) for a higher value of the parameter k like 0.675, outside the stability space. Also, computing the Lyapunov numbers (Fig. 5) for this value of the parameter k and setting the same fixed values for the other parameters α, c_1, c_2, p and μ it seems that they are getting over the value of 1 as an evidence for the chaotic trajectory.

This chaotic trajectory makes the system sensitive on initial conditions, which means that only a small change on a coordinate may change completely the system's behavior. For example, choosing two different initial conditions (0.1,0.1) (Fig. 6) and (0.101,0.1) (Fig. 7) with a small change at the q_1^*-coordinate and plotting the time series of system it seems that at the beginning the time series are indistinguishable, but after a number of iterations, the difference between them builds up rapidly.

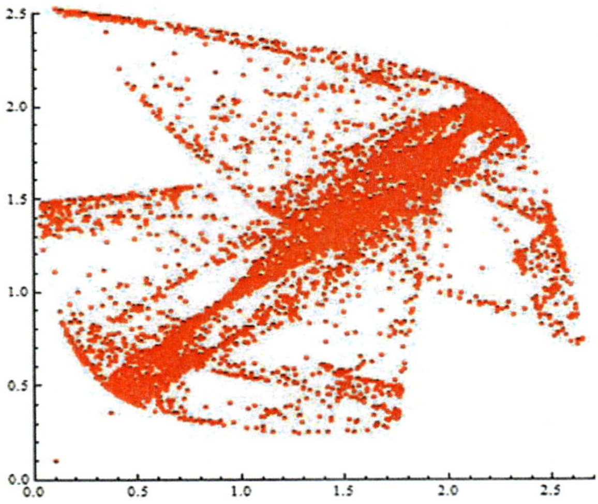

Fig. 4 Phase portrait (strange attractor) of the orbit of (0.1,0.1) with 8000 iterations of the map Eq. (15) for $\alpha = 5$, $c_1 = 1$, $c_2 = 0.50$, $p = 0.50$, $\mu = 0.50$ and $k = 0.675$

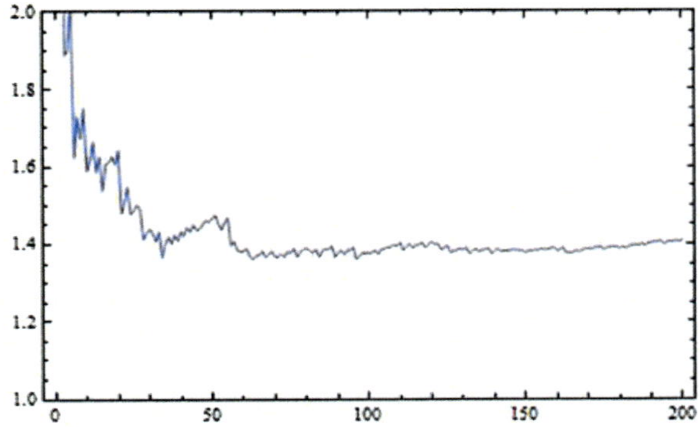

Fig. 5 Lyapunov numbers of the orbit of (0.1,0.1) with 8000 iterations of the map Eq. (15) for $\alpha = 5$, $c_1 = 1$, $c_2 = 0.50$, $p = 0.50$, $\mu = 0.50$ and $k = 0.675$

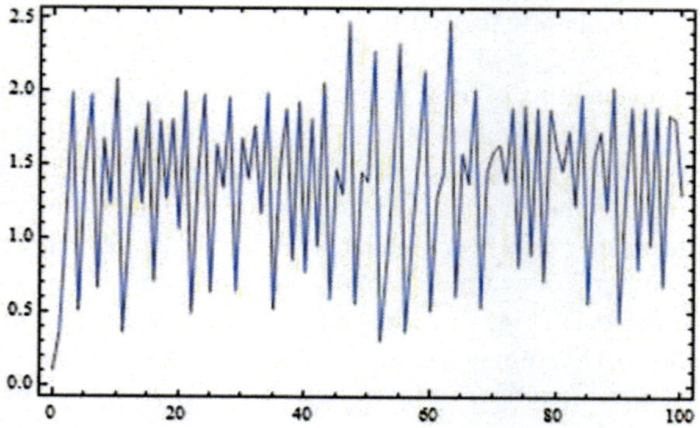

Fig. 6 Sensitive dependence on initial conditions for q_1^*-coordinate plotted against the time: the orbit of (0.1,0.1) of the system Eq. (15) for $\alpha = 5$, $c_1 = 1$, $c_2 = 0.50$, $p = 0.50$, $\mu = 0.50$ and $k = 0.675$

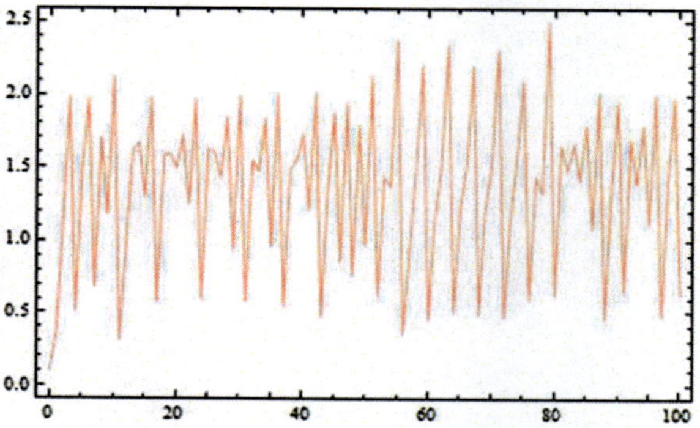

Fig. 7 Sensitive dependence on initial conditions for q_1^*-coordinate plotted against the time: the orbit of (0.101,0.1) of the system Eq. (15) for $\alpha = 5$, $c_1 = 1$, $c_2 = 0.50$, $p = 0.50$, $\mu = 0.50$ and $k = 0.675$

4 Conclusions

In this paper we analyzed the dynamics of a differentiated Cournot duopoly with homogeneous expectations, linear demand and cost functions. An uncertainty of the first firm's cost function is introduced. By assuming that at each time period each firm maximizes its expected relative profit under the same expectations, a discrete dynamical system was obtained. Existence and stability of equilibrium of this system are studied. We showed numerically that the model gives chaotic and unpredictable trajectories. The main result is that higher values of the speed of adjustment may destabilize the Cournot–Nash equilibrium.

References

1. H.N. Agiza, On the analysis of stability, bifurcation, chaos and chaos control of Kopel map. Chaos Solitons Fract. **10**, 1909–1916 (1999)
2. H.N. Agiza, A.A. Elsadany, Chaotic dynamics in nonlinear duopoly game with heterogeneous players. Appl. Math. Comput. 149 843–60 (2004)

3. H.N. Agiza, A.A. Elsadany, Nonlinear dynamics in the Cournot duopoly game with heterogeneous players. Phys. A **320**, 512–524 (2003)
4. H.N. Agiza, A.S. Hegazi, A.A. Elsadany, Complex dynamics and synchronization of duopoly game with bounded rationality. Math. Comput. Simulat. **58**, 133–146 (2002)
5. A. Agliari, L. Gardini, T. Puu, Some global bifurcations related to the appearance of closed invariant curves. Math. Comput. Simulat. **68**(3), 201–219 (2005)
6. A. Agliari, L. Gardini, T. Puu, Global bifurcations in duopoly when the cournot point is destabilized via a subcritical neimark bifurcation. Int. Game Theory Rev. **08**(01), 1–20 (2011)
7. S.S. Askar, On complex dynamics of monopoly market. Econ. Model. **31**, 586–589 (2013)
8. S.S. Askar, Complex dynamic properties of Cournot duopoly games with convex and log-concave demand function. Oper. Res. Lett. **42**, 85–90 (2014)
9. W.J. Baumol, R.E. Quandt, Rules of thumb and optimally imperfect decisions. Am. Econ. Rev. **54**(2), 23–46 (1964)
10. G.I. Bischi, M. Kopel, Equilibrium selection in a nonlinear duopoly game with adaptive expectations. J. Econ. Behav. Org. **46**, 73–100 (2001)
11. G.I. Bischi, A. Naimzada, Global analysis of a dynamic duopoly game with bounded rationality, in *Advances in dynamic games and applications*, vol. 5, ed. by J.A. Filar, V. Gaitsgory, K. Mizukami (Birkhauser, Basel, 2000), pp. 361–385
12. W.J. Den Haan, The importance of the number of different agents in a heterogeneous asset-pricing model. J. Econom. Dynam. Control **25**, 721–746 (2001)
13. A.A. Elsadany, Dynamics of a Cournot duopoly game with bounded rationality based on relative profit maximization. Appl. Math. Comput. **294**, 253–263 (2017)
14. S. Elaydi, *An Introduction to Difference Equations*, 3rd edn. (Springer-Verlag, New York, 2005)
15. L. Fanti, L. Gori, The dynamics of a differentiated duopoly with quantity competition. Econ. Model. **29**(2):421–427 (2012)
16. G. Gandolfo, *Economic Dynamics*, (Springer, Berlin, 1997)
17. C.H. Hommes, *Heterogeneous agent models in economics and finance*, in: L. Tesfatsion, K.L. Judd (Eds.), Handbook of Computational Economics, Agent-Based Computational Economics, vol. 2, Elsevier Science B.V: 1109–1186, (2006)
18. M. Kopel, Simple and complex adjustment dynamics in Cournot duopoly models. Chaos Solitons Fract. **12**, 2031–2048 (1996)
19. M. Kulenonic, O. Merino, *Discrete Dynamical Systems and Difference Equations with Mathematica,* Chapman & Hall/Crc., (2002)
20. A.K. Naimzada, G. Ricchiuti, Complex dynamics in a monopoly with a rule of thumb. Appl. Math. Comput. **203**, 921–925 (2008)
21. A. Naimzada, L. Sbragia, Oligopoly games with nonlinear demand and cost functions: Two boundedly rational adjustment processes. Chaos Solitons Fractals **29**, 707–722 (2006)
22. T. Puu, The chaotic monopolist. Chaos, Solitons Fractals **5**(1), 35–44 (1995)
23. T. Puu, The chaotic duopolists revisited. J Econom. Behav. Org. **37**, 385–394 (1998)
24. G. Sarafopoulos, On the dynamics of a duopoly game with differentiated goods. Procedia Econ. Finan. **19**, 146–153 (2015)
25. G. Sarafopoulos, Complexity in a duopoly game with homogeneous players, convex, log linear demand and quadratic cost functions. Procedia Econ. Finan. **33**, 358–366 (2015)
26. G. Sarafopoulos, K. Papadopoulos, On a Cournot duopoly game with differentiated goods, heterogeneous expectations and a cost function including emission costs. Sci. Bull. Econ. Sci. **1**(1), 11–22 (2017)
27. G. Sarafopoulos, K. Papadopoulos, On the stability of a Cournot dynamic game under the influence of information, *Knowledge E Publishing*, ISSN: 2518–668X, 131–144, (2018)
28. G. Sarafopoulos, K. Papadopoulos, Chaos in oligopoly models. Int. J. Prod. Manage. Assess. Technol. **7**(1), 50–76 (2019)
29. G. Sarafopoulos, K. Papadopoulos, Complexity in Bertrand duopoly game with heterogeneous players and differentiated goods, *Economic and Financial Challenges for Eastern Europe*, Article 2 (p. 15–26), Springer, (2019)

30. F. Tramontana, Heterogeneous duopoly with isoelastic demand function. Econ. Model. **27**, 350–357 (2010)
31. J. Zhang, Q. Da, Y. Wang, Analysis of nonlinear duopoly game with heterogeneous players. Econ. Model. **24**, 138–148 (2007)
32. J. Zhang, Q. Da, Y. Wang, The dynamics of Bertrand model with bounded rationality. Chaos, Solitons Fractals **39**, 2048–2055 (2009)

Driven Intrinsic Localized Modes in Soft Nonlinear Microscopic and Macroscopic Lattices

Masayuki Sato, Masato Sakai, and A. J. Sievers

Abstract The possibility that large amplitude, localized vibrational excitations can exist in periodic physical lattices with nonlinear intersite forces was discovered over thirty years ago. The energy profiles of these intrinsic localized modes (ILMs) resemble those of localized vibrational modes at defects in a harmonic lattice. Described here are a variety of experiments on driver locked ILMs for two soft nonlinear lattices: an atomic spin array and an electrical nonlinear transmission line. CW locked ILMs in the quasi-1D antiferromagnet $(C_2H_5NH_3)_2CuCl_4$ have been found at frequencies below the antiferromagnetic resonance by employing four-wave mixing emission. A discrete step structure is observed in the emission signal as well as repeatable nonlinear ILM switching and hysteresis. These findings are compared with locked ILMs and large amplitude lattice spatial modes (LSMs) that have been measured for a driven 1-D nonlinear cyclic electric transmission line, where the nonlinear element is a saturable capacitor. Interestingly, by tuning the driver frequency away from the modal spectrum an LSM can be continuously converted into ILMs and vice versa. As a consequence, the resultant electrical energy distribution for the experimental soft nonlinear cyclic array can be either balanced or unbalanced.

Keywords Intrinsic Localized Mode · Lattice Spatial Mode · Spin Wave · Electric Transmission Line · Soft Nonlinearity

M. Sato (✉) · M. Sakai
Graduate School of Natural Science and Technology, Kanazawa University, Kakumamachi, Kanazawa 920-1192, Japan
e-mail: msato@kenroku.kanazawa-u.ac.jp

A. J. Sievers
Laboratory of Atomic and Solid State Physics, Cornell University, Ithaca, NY 14853-2501, USA
e-mail: ajs19@cornell.edu

C. H. Skiadas et al. (eds.), *13th Chaotic Modeling and Simulation International Conference*, Springer Proceedings in Complexity,
https://doi.org/10.1007/978-3-030-70795-8_55

1 Introduction

Although vibrational impurity modes in lattices have a long history [1], a major advance in the subject of excitations in discrete nonlinear lattices was the discovery that, even in the absence of impurities, some localized vibrations can be stabilized by lattice discreteness [2]. These excitations which, like impurity modes, extend only over the lattice constant scale are generically called “intrinsic localized modes” (ILMs) since they involve no disorder but come in many different shapes [3]. They are also called “lattice solitons” [4] or “discrete breathers” [5] because of similarities to some exact soliton solutions in nonlinear continuum theories. These excitations have been formalized in terms of a number of useful existence and stability criteria [6]. Both theory and a variety of experiments have been reviewed in the literature [7, 8] with experiments on microscopic antiferromagnetic lattices [9–14] and macroscopic electrical nonlinear transmission lines (ENTL) [15–21] playing important roles. At first glance experiments on these two kinds of nonlinear systems would appear to be unrelated since the antiferromagnet is a quantum system while the ENTL is a classical one, but in certain cases both can be treated with classical equations of motion. The resulting dynamics is that both systems transmit energy over specific frequency bands. For the antiferromagnet it is spin wave energy that is transmitted, while for the ENTL it is electromagnetic. With regard to nonlinear properties of the two dynamical systems they can be designed to match as well. For the antiferromagnetic lattice both the spin exchange energy and the magnetic anisotropy energy have soft intersite and onsite nonlinearities so that the uniform antiferromagnetic resonance mode frequency decreases with increasing spin amplitude. By introducing a soft nonlinear capacitor in each cell of the ENTL, such as MOS-FETs, the low frequency uniform electromagnetic mode of this band pass filter will also decrease in frequency with increasing driving amplitude.

The purpose of this review is to examine the observed ILM findings for two driven, soft nonlinear lattices with damping, one a magnetic crystal with a microscopic lattice constant and the other an ENTL with a macroscopic one. From these very different experimental techniques on different systems come ILM findings that are remarkably similar.

2 Observation of Driven Antiferromagnetic ILMs

2.1 *Spin Wave Dynamics of $(C_2H_5NH_3)_2CuCl_4$*

How can an antiferromagnetic spin 1/2 system be compatible with a classical dynamical description? Below a Neél temperature of $T_N = 10.2$ K the spin 1/2 Cu^{2+} ions of $(C_2H_5NH_3)_2CuCl_4$ are oriented along the a-crystal axis, in alternate sheets of strong ferromagnetically coupled spins with weak antiferromagnetic coupling between adjacent sheets as illustrated in Fig. 1. [12] At 1.4 K the ratio of the interlayer spin

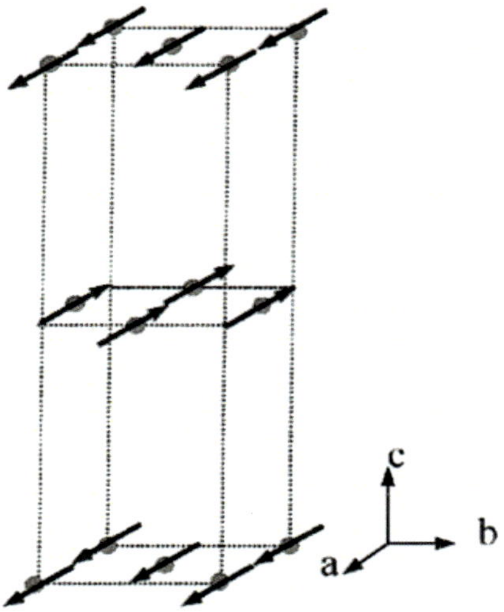

Fig. 1 Lattice and spin structure of $(C_2H_5NH_3)_2CuCl_4$. Circles denote Cu^{2+} ions and arrows indicate the spin configuration within the antiferromagnetic state. Only Cu^{2+} ions are shown for this layered, face centered, orthorhombic compound. The easy, second easy, and hard spin axes are labeled the a, b, and c crystal directions, respectively [12]

exchange field to the intralayer one is 1.5×10^{-3} so the spins in a given layer are strongly aligned in the same direction, thus the low frequency spin dynamics can be approximated by a 1D two sublattice antiferromagnet with each layer represented by a macroscopic classical spin. Due to the easy axis anisotropy and the weak antiferromagnetic interaction between these classical spins the lower and upper frequency antiferromagnetic resonance (AFMR) modes are polarized along the c- and b- crystal axes, respectively.

The classical equation of motion for the normalized spin $\vec{S}_n$ at the nth sheet site, including damping and driving terms, is [9]

$$\frac{d}{dt}\vec{S}_n = -\gamma S_n \times H_n - \gamma\lambda\vec{S}_n \times \left(\vec{S}_n \times \vec{H}_n\right) \tag{1}$$

where γ is the gyromagnetic ratio and λ the Landau damping parameter. The magnetic field acting on the nth macroscopic spin is

$$\vec{H}_n = -2J\left(\vec{S}_{n-1} + \vec{S}_{n+1}\right) - 2\overset{\leftrightarrow}{A}\cdot\vec{S}_n + h_{c0}\hat{e}_c \cos\omega t \tag{2}$$

with J the nearest neighbor antiferromagnetic exchange constant, $\overset{\leftrightarrow}{A}$ the anisotropy field tensor, and h_{c0} the amplitude of the ac driving field along the c axis crystal direction. For this easy plane anisotropy case the two small amplitude dispersion

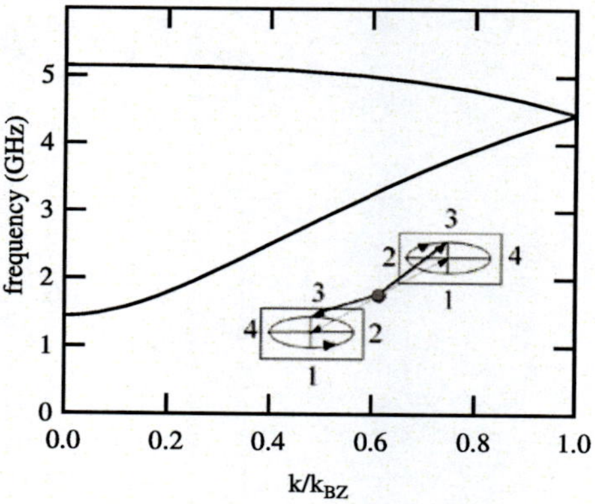

Fig. 2 Spin-wave dispersion curve for the antiferromagnet $(C_2H_5NH_3)_2CuCl_4$. Upper and lower branches along the c axis. The inset shows the uniform mode spin motion for the lower AFMR mode, which has a net ac magnetization only along the c axis. ILMs may be expected to occur below the minimum frequency of the dispersion curve [12]

curves are shown in Fig. 2 with the resulting uniform mode frequencies in the GHz range.

The insert shows the polarization of the lowest frequency AFMR mode with a linearly polarized transverse ac moment generated in the 1–3 direction but not in the 2–4 direction. It is the rod-shaped sample that has been studied in pulse and CW experiments.

2.2 *Experimental Procedure*

In magnetic solids with the AFMR involving spins on the order of 10^{22} per cc the question is how to distinguish ILM dynamics from this background? The answer is not to rely on a linear experimental technique but, since ILMs are nonlinear, to make use of a detection technique that involves a nonlinearity. Both pulse and CW four-wave techniques have been used to observe magnetic ILMs in this antiferromagnet [11–14]. The CW technique is outlined here. To illustrate the experimental approach, consider the AFMR absorption spectra shown in Fig. 3.

A weak probe with frequency f_p is swept across the AFMR absorption line, which occurs at about 1.38 GHz at $t = 0$. Next the powerful driver with frequency f_D nearby is switched on as shown in Fig. 3a. This driver pulls the absorption line to lower frequencies. The closer the driver frequency is to the AFMR the larger the effect as shown in Fig. 3b. Figure 3c illustrates the point where the AFMR becomes unstable. So far all we see is the linear absorption spectrum. To access the nonlinearity associated with the ILM a four-wave mixing technique was employed. The resulting power emitted by the ILM alone, $P_{ILM}^{(3)}$, is detected at the spectrum analyzer frequency $f_{sp} = 2f_D - f_p$. Further analysis shows that among other elements the four wave mixing signal has the form [12]

$$\sqrt{P_{ILM}^{(3)}} \propto n_{ILM} P_D, \tag{3}$$

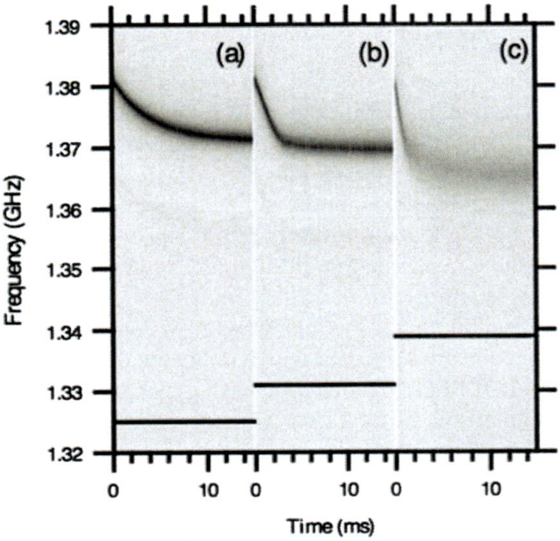

Fig. 3 Pulling the AFMR absorption frequency and capturing ILMs with a cw driver. **a** Initially the AFMR absorption line is at 1.38 GHz and the driver, 1.325 GHz, is switched on at t = 0. The frequency gap, Δf, between them continuously changes over a 15-ms time interval. **b** In this case Δf is now small enough so that the change in the AFMR frequency now occurs over 3 ms. **c** Here Δf is sufficiently small that the AFMR mode becomes unstable and ILMs are transferred to the driver frequency [22]

where n_{ILM} is the number of ILMs emitting and P_D is the driving power. Because n_{ILM} has integer values this relation will identify steps in the square root of the emitted power as the ILMs appear and disappear. Figure 4 illustrates the appearance of such a step.

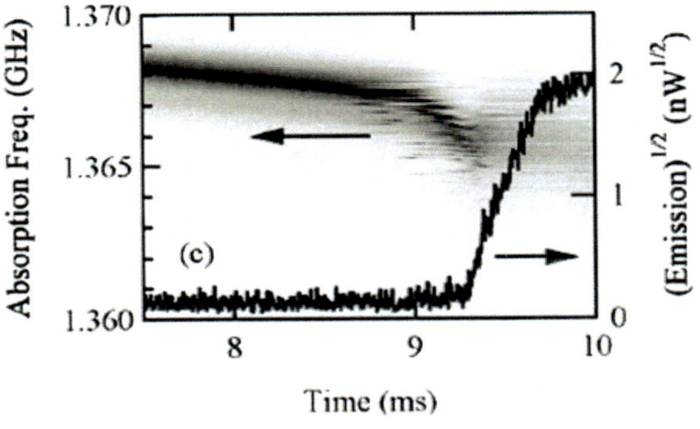

Fig. 4 Illustrates the turn on of the emission when the AFMR becomes instable. Both the absorption spectrum and the emission spectrum are superimposed on the same figure. Here the breakup of the absorption pattern occurs at around 9 ms while simultaneously the emission grows rapidly in strength, as ILMs become locked at the driver frequency [14]

2.3 Observing Steady State Magnetic ILMs

A time sweep of the emission spectrum does not have sufficient resolution to pick out individual ILMs so a more refined technique is required. With ILMs locked to the driver it is also possible to change the frequency gap, Δf, between the AFMR and the driver, by changing the AFMR frequency. Described here is the technique of tuning by changing the sample temperature. Since the anisotropy energy depends on the sublattice magnetization it depends on the temperature so both positive and negative manipulation of the frequency gap Δf is possible by simply sweeping the sample temperature.

By monitoring the four-wave emission as a function of sample temperature variation one finds the data shown in Fig. 5a. It maps out two hysteresis loops, each for a single ILM [14]. Slowly increasing the sample temperature decreases the sublattice magnetization, decreases Δf and generates a locked ILM as shown by the dotted curve (arrow pointing to the right). Increasing the sample temperature produces the solid curve, arrow to the left, ending with the destruction of the ILM. In Fig. 5a the

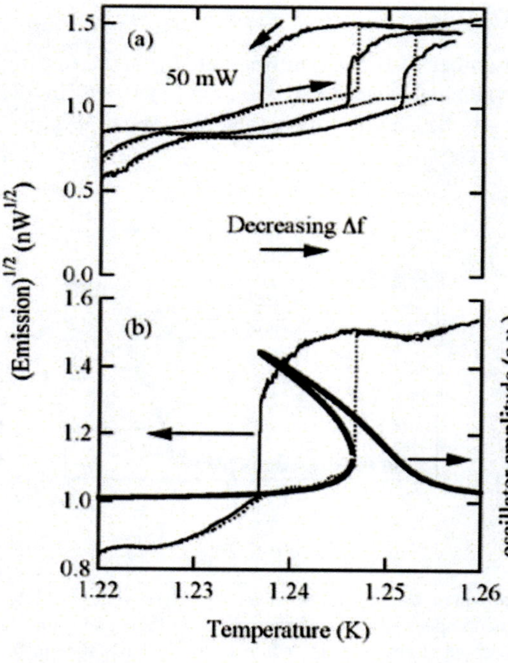

Fig. 5 Observation of locked ILM switching by tuning sample temperature. (a) Increasing the temperature decreases the frequency gap, Δf (dotted lines for three different driving powers). Solid lines are for decreasing temperatures. (b) Comparison of the 50 mW data with those expected for a driven nonlinear oscillator (thick solid line) [14]

results for three different driver power levels are displayed with the lowest temperature emission steps up and down for a 50 mW driver. Figure 5b presents an expanded view of the lowest temperature sweep described in (a). Superimposed on that trace is the amplitude pattern for the switching of a nonlinear oscillator model to qualify the hysteretic switching that has been observed. Interestingly it is found that the step heights do not change significantly as the power of the locking driver is changed.

3 Electric Lattice with Soft, Saturable, Nonlinearity

3.1 Experimental Setup

Figure 6a shows the components within a unit cell of a 32 element electric nonlinear transmission line (ENTL). The nonlinear capacitor consists of two anti-paralleled N-channel MOS-FETs (IRFU-120). When the gate is negatively biased, holes in a P-semiconductor (for the N-channel FET) are accumulated below the gate between an oxide and semiconductor, and the boundary works as a conductive sheet. The capacitance between the gate and source electrodes is large with its value limited by the thickness of the oxide. When the gate is positively biased, the semiconductor is inverted to N-type and the boundary layer forms a conductive sheet. The capacitance between the drain and the gate is again large with its value limited by the thickness of the oxide. The two anti-paralleled capacitances together, identified as $C(V)$, are plotted as $1/C$ versus the applied DC bias in Fig. 6b. Since the capacitance increases with absolute voltage, the resonance frequency decreases with increasing amplitude (soft nonlinearity); however, the decrement saturates abruptly when the voltage becomes large [23, 24]. One might consider a drawback of the MOS-FET to be its saturation property; however, we suggest that such behavior gives rise to an electrical

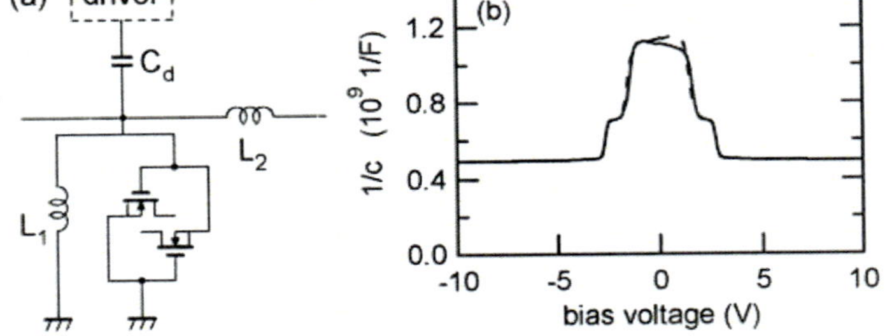

Fig. 6 Design of the nonlinear element for ENTL. Two MOS-FETs (IRFU-120) equal a nonlinear capacitor together with coil L_1(200 μH) to form a resonant circuit. L_2(200 μH) provides coupling to the next element. The driver for each element is via coupling capacitor C_d(70pF). **b** Inverse versus capacitance of the element illustrating the nonlinearity. Small AC voltage applied together with the DC bias voltage

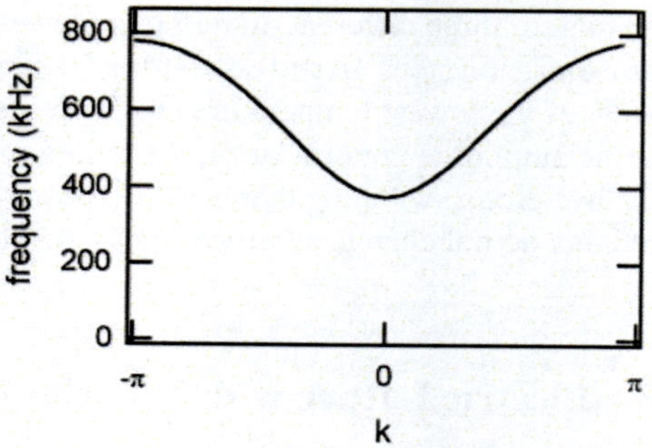

Fig. 7 Linear dispersion curve for the ENTL lattice. The circuit contains the element in Fig. 6a. ILMs are expected below the minimum frequency of the dispersion curve

analogue of a well studied mechanical nonlinear system containing piecewise linear springs in which the restoring force has a change in slope.

For this cyclic ENTL with an onsite capacitive nonlinearity $C(V_n)$ the dynamical equation of motion for the lattice voltage V_n as a function of time becomes

$$(C(V_n) + C_d)\frac{d^2V_n}{dt^2} + \left.\frac{dC(V)}{dV}\right|_{V_n}\left(\frac{dV_n}{dt}\right)^2 + \frac{V_n}{L_1} + \frac{L_1\omega_0}{Q}\frac{dV_n}{dt}$$
$$+ \frac{1}{L_2}(2V_n - V_{n+1} - V_{n-1}) = C_d\frac{d^2}{dt^2}V_d \quad (4)$$

where Q is the quality factor and $C(V_n)$ is the differential capacitance $C(V) = dq/dV$ and q is the stored charge. The other parameters are defined in Fig. 6a.

Figure 7 shows the linear band frequencies as a function of wave number. Because of the soft nonlinearity, an ILM can be generated below the bottom of the extended wave band. This shape is similar to that of the lower spin wave dispersion curve shown in Fig. 2.

3.2 Experimental Results

Figure 8 shows the driver frequency dependence of two kinds of nonlinear excitations: large amplitude lattice spatial modes [25] (LSMs) and ILMs for the soft nonlinear lattice containing the nonlinear element of Fig. 6a. At each driver frequency, a snap shot of the spatial voltage pattern at the maximum voltage versus time was captured. Those snapshots were ordered as a function of frequency resulting in Fig. 8a–d. Arrows indicate sweep direction of the driver frequency. Panel (e) demonstrates how the amplitude increases with decreasing driving frequency. At a frequency larger than 350 kHz in panels (a) and (b), a noisy pattern is observed. As the frequency is lowered, a 4 peak-LSM appears, followed by a noisy pattern, then a 3 peak-LSM, a

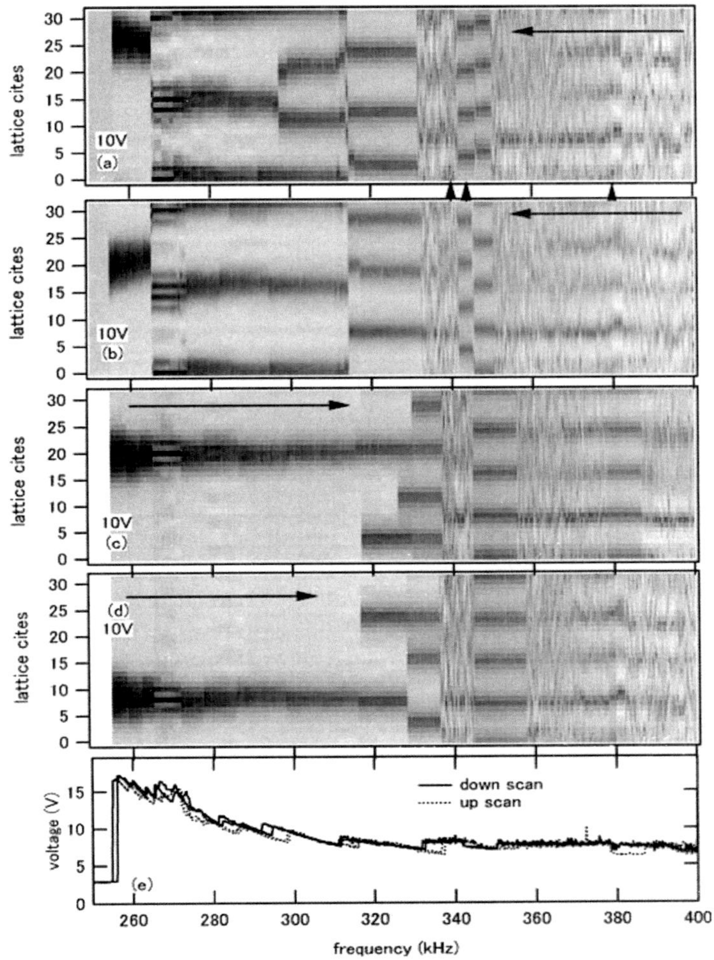

Fig. 8 Measured frequency dependence of driver locked LSMs and ILMs for the soft saturable 32 element nonlinear lattice. Four panel sweeps illustrate conversions. Driver frequency decreases in panels (**a**) and (**b**), and increases in panels (**c**) and (**d**). Snap shots at a moment when voltage is the maximum are mapped as a function of the driver frequency. Panel (**e**) displays maximum voltage as a function of frequency. Single excitation profiles such as in panel (**a**) 265–255 kHz, and (**c**) 255–317 kHz are due to ILMs. Other strip patterns at higher frequencies are due to LSMs

2 peak-LSM, and finally a single ILM is observed. LSM patterns are made from a few cyclic component waves, so that their peaks are mostly equidistance.

In addition, the LSM widths are observed to increase with decreasing frequency. This width dependence is causes by the saturable nonlinearity. Because of the lattice discreteness, the widening accompanies symmetry change of the peak from cite-center to bond-center or vise-versa. For example, the transition from cite-centered to bond-centered shapes is observed at 297.5 kHz. Panel 8(b) shows a sequential measurement. Similar patterns to those in 8(a) are observed.

Panels 8(c) and 8(d) display scan-up measurement results starting from the single ILM state. Hysteresis is observed for ILM and LSM generation. Panels (c)–(d) indicate that a single ILM is stable at frequencies around 255–317 kHz, while an LSM is stable at frequencies larger than 317 kHz. At a frequency below 317 kHz, there may be a stable two ILM state. Two peaks in panels (a)–(b) at these frequencies may be an LSM seeded ILM array.

The noisy pattern and 4-peaked LSM are very similar between Fig. 8a–d. This signature indicates that the wave position is strongly influenced by lattice irregularity, because the wave samples the entire lattice. On the other hand, the single ILM position shifts, because the ILM feel only nearby impurities.

3.3 Analysis

To illustrate how the LSM is generated the multi-channel oscilloscope voltage data has been transformed in (time, space) domain into the (frequency, wavenumber) domain. Figure 9a, b, c show the results for driver frequencies indicated by the horizontal arrows in the panels and also identified by the three vertical arrow heads at bottom of Fig. 8a. The 2D-FT log amplitude is displayed (gray-scale), where darker indicates larger amplitude. The solid curves (red) in frame (a)–(b) represent the superimposed linear dispersion curve. Excitation point is at $k = 0$ (uniform mode) in this wavenumber-frequency space. Because of the low Q of the system it is not necessary that this frequency and the driver frequency be coincident. In addition, the nonlinearity and uniformly excited lattice ensures that the rest of the dispersion curve is shifted down as well. All other displayed intensities come from this one driving source. In panels (a) and (c), a faint gray intensity curve is apparent across the (k, ω) range (dashed red curve in (a) is a guide to eye). This intensity display is from the shifted nonlinear dispersion curves, excited by the noisy pattern displayed in Fig. 8a. The two dark spots that appear on the dashed curve at the driver frequency in Fig. 9a are from secondary excitation waves generated by a four-wave mixing process $2(k = 0, \Omega) \Rightarrow (k_{+}, \Omega) + (k_{-}, \Omega)$. This determines the number of LSM peaks in real space. For example, the spot locations are at $k = \pm 4 \times (2\pi/32)$ in panel (a), producing a 4 peak-LSM by mixing with the $k = 0$ uniform excitation, where $(2\pi/32)$ is the k-space unit for the $N = 32$ lattice. All other spots are excited by successive nonlinear wave generation processes.

In panel 9(b), the intensity spots are only at the driver frequency. In real space, a clean 4 peak-LSM is found. Here it is hard to identify a nonlinearly shifted dispersion curve in the absence of the noisy pattern. The secondary spot positions are the same as those in panel (a). In panel (c), the secondary spots are shifted to $k = \pm 3 \times (2\pi/32)$. These are for 3-peaked LSM. A clean 3-peaked LSM is observed at the slightly lower frequency 330 kHz.

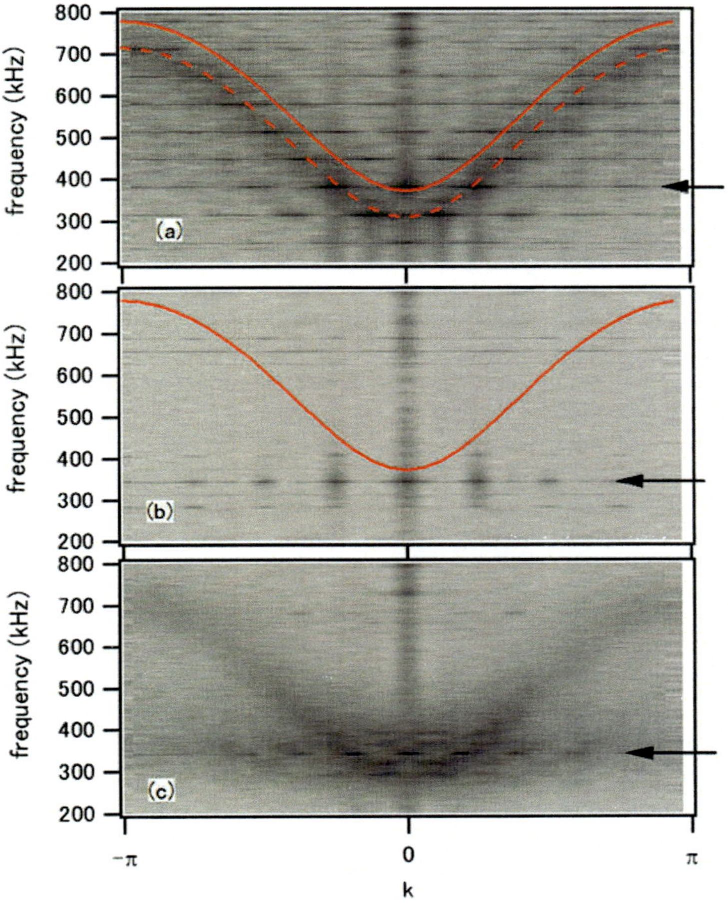

Fig. 9 2D-Fourier transform maps of voltage records for the ENTL system. 32 channel oscilloscope data from Fig. 8a are transformed into the frequency and wavenumber domain by time- and spatial-Fourier transform. Three panels are identified by driver frequencies: **a** 380 kHz, **b** 343.8 kHz and **c** 340 kHz. (See horizontal arrows.) Same driver frequencies are also identified at the bottom of Fig. 8a. Log amplitude is displayed to emphasis small signals. Solid curves (red) in panels (**a**)–(**b**) are the superimposed linear dispersion curve. In panel (**a**) and (**c**), intensity from shifted dispersion curve is faintly observed. (Dashed red in (**a**) is a guide to the eye.) Because of random excitation in Fig. 8a at those frequencies, normal modes on the cyclic dispersion curve are excited and observed in this 2D-FT figure. In panel (**b**), the shifted nonlinear dispersion curve is not seen, since there are no such random noise vibrations. The wave number spots are produced by LSM formation at the driver frequency. Such wavenumber spots are seen also in panels (**a**) and (**c**)

The number of LSM peaks decreases with decreasing driver frequency, because the shift of the nonlinear dispersion curve relative to the driver frequency becomes smaller, and two secondary spot spacing in k-space becomes smaller. With a large number of lattice elements the transition from an LSM to an ILM array could happen at a still lower frequency, as the interaction decreases between neighboring peaks. In this case, many ILMs may remain. In Fig. 8, because of the saturation of the nonlinearity the peak widths become wider as the frequency decreases. To maintain the driver excitation, the frequency difference between the LSM (or ILMs) and the driver must be compensated by the nonlinearity.

The number of LSM peaks is roughly lattice size divided by wavelength determined by the secondary excitation wave. The maximum number of LSM peaks depends on the largest wavenumber of the four-wave mixing process, that is, the larger the wavenumber, the more peaks. The rule of thumb is the larger the spectral band width, the smaller the secondary wave number, if the frequency shift remains the same.

4 Discussion and Conclusions

Both soft nonlinear lattices described here, the antiferromagetic and the electric transmission line, are essentially simple 1D systems, one with a microscopic lattice constant and the other with a macroscopic one. Although in each case the experimental techniques used to observe and explore the properties of ILMs have been quite different fundamentally similar results have been observed. In both systems the experiments consisted of frequency locking an ILM to a driver. For the magnetic system, because these excitations are strongly nonlinear, four-wave mixing emission spectroscopy is an ideal way to enhance the ILM signal over that produced by the more numerous plane wave spin excitations. The result is countable ILMs for an atomic spaced system. Experimentally, it is easiest to keep track of the difference between the AFMR and driver frequency as the important ILM parameter, since one had the flexibility to vary either the AFMR frequency, via the sample temperature, or to vary directly the driver frequency. Precision measurements allow the properties of single ILMs to be studied and their hysteresis curves to be measured. Surprisingly the step heights do not change significantly with locking driver power.

More extensive information about the general dynamic properties of a soft nonlinear 1D lattice has been obtained by studying a cyclic ENTL where each lattice element is monitored. Such a driven nonlinear transmission line, with periodic boundary conditions and elements that contain a saturable nonlinear capacitor, has been used to generate ILMs below the modal spectrum. An ILM switching hysteresis signature, similar to that observed for the antiferromagnet, is a natural feature. An additional finding is the observation of LSMs within the spectrum. The most dramatic feature is that by simply changing the driver frequency the spectrum can evolve continuously from an LSM pattern distributed around the ring, with a successive decrease in the number of LSM-peaks, to multiple ILMs localized on a few sites, and finally to a single ILM. A four-wave mixing process plays a key role in determining the resulting LSM signature. As a consequence, the resultant AC energy distribution for the experimental soft nonlinear cyclic array can either be balanced or unbalanced.

Acknowledgements MS is supported by JSPS Kakenhi Grant No. JP16K13716. AJS is supported by a Podell Endowment Award for Research and Scholarship.

References

1. I.M. Lifshitz, Some problems of the dynamic theory of non-ideal crystal lattices. Nuovo Cim. Suppl. **3**, 716 (1956)
2. A.S. Dolgov, The localization of vibrations in a nonlinear crystal structure. Sov. Phys. Solid State **28**, 907 (1986)
3. A.J. Sievers, S. Takeno, Intrinsic localized modes in anharmonic crystals. Phys. Rev. Lett. **61**, 970 (1988)
4. D.N. Christodoulides, R.I. Joseph, Discrete self-focusing in nonlinear arrays of coupled waveguides. Opt. Lett. **13**, 794 (1988)
5. S. Flach, C.R. Willis, Discrete breathers. Phys. Repts. **295**, 182 (1998)
6. S. Aubry, Breathers in Nonlinear Lattices: Existence, Linear Stability and Quantization. Physica D **103**, 201 (1997)
7. D.K. Campbell, S. Flach, Y.S. Kivshar, Localizing energy through nonlinearity and discreteness. Phys. Today **57**, 43 (2004)
8. S. Flach, A. Gorbach, Discrete breathers-Advances in theory and applications. Phys. Repts. **467**, 1 (2008)
9. R. Lai, A.J. Sievers, Nonlinear nanoscale localization of magnetic excitations in atomic lattices. Phys. Repts. **314**, 147 (1999)
10. U.T. Schwarz, L.Q. English, A.J. Sievers, Experimental generation and observation of intrinsic localized spin wave modes in an antiferromagnet. Phys. Rev. Lett. **83**, 223 (1999)
11. M. Sato, A.J. Sievers, Direct observation of the discrete character of intrinsic localized modes in an antiferromagnet. Nature **432**, 486 (2004)
12. M. Sato, A.J. Sievers, Counting discrete emission steps from intrinsic localized modes in a quasi-1D antiferromagnetic lattice. Phys. Rev. B **71**, 214306 (2005)
13. L.Q. English, M. Sato, A.J. Sievers, Nanoscale intrinsic localized modes in an antiferromagnet. J. Appl. Phys. **89**, 6707 (2001)
14. J.P. Wrubel, M. Sato, A.J. Sievers, Controlled switching of intrinsic localized modes in a one-dimensional lattice. Phys. Rev. Lett. **95**, 264101 (2005)
15. R. Hirota, K. Suzuki, Theoretical and experimental studies of lattice solitons in nonlinear lumped networks. Proc. IEEE **61**, 1483 (1973)
16. K. Fukushima, M. Wadati, Y. Narahara, Envelope solitons in a new nonlinear transmission line. J. Phys. Soc. Jpn. **49**, 1953 (1980)
17. P. Marquie, J.M. Bilbault, M. Remoissenet, Observation of nonlinear localized modes in an electrical lattice. Phys. Rev. E **51**, 6127 (1995)
18. L.Q. English, F. Palmero, A.J. Sievers, P.G. Kevrekidis, D.H. Barnak, Traveling and stationary intrinsic localized modes and their spatial control in electrical lattices. Phys. Rev. E **81**, 046605 (2010)
19. L.Q. English et al., Nonlinear localized modes in two-dimensional electrical lattices. Phys. Rev. E **88**, 022912 (2013)

20. F. Palmero, L. Q. English, X.-L. Chen, et al., Experimental and numerical observation of dark and bright breathers in the band gap of a diatomic electrical lattice. Phys. Rev. E **99**, 032206 (2019)
21. M. Remoissenet, *Waves Called Solitons*, 3rd edn. (Springer, Berlin, 1999), p. 76
22. M. Sato, A.J. Sievers, Experimental and numerical exploration of intrinsic localized modes in an atomic lattice. J. Biol. Phys. **35**, 57 (2009)
23. W. Shi, S. Shige, Y. Soga, M. Sato, A.J. Sievers, Intrinsic localized modes in a nonlinear electrical lattice with saturable nonlinearity. EPL **103**, 30006 (2013)
24. S. Shige, K. Miyasaka, W. Shi, Y. Soga, M. Sato A.J. Sievers, Experimentally observed evolution between dynamic patterns and intrinsic localized modes in a driven nonlinear electrical cyclic lattice, EPL **121**, 30003(2018); Erratum: EPL **126**, 29901 (2019)
25. V.M. Burlakov, Interference of mode instabilities and pattern formation in anharmonic lattices. Phys. Rev. Lett. **80**, 3988 (1996)

Tilted Nadaraya-Watson Regression Estimator

Mohammad T. Shakeri, Farzaneh Boroumand, Hassan Doosti, Nino Kordzakhia, and Mahdi Salehi

Abstract In nonparametric statistics the tilting techniques are sustainably used for adjusting an empirical distribution by replacing uniform distribution of weights by general multinomial distribution. In this paper a tilting approach has been used for minimizing "the distance" to an infinite order (IO) regression estimator, a comparator regression function estimator. We also provide the simulation study results illustrating the tilted version of the Nadaraya-Watson (N-W) estimator performs better than its classical analog (the N-W estimator) in terms of Median Integrated Squared Error (MISE). In addition, the performance of the tilted N-W regression function estimator has been examined using the Italy's COVID-19 deaths data.

Keywords Nadaraya-Watson estimator · Tilted Nadaraya-Watson estimator · Infinite order estimator · Kernel estimator · Trapezoidal kernel · Cross-validation function · MISE · ISE

M. T. Shakeri · F. Boroumand (✉)
Department of Biostatistics, Health School, Mashhad University of Medical Sciences, Mashhad, Iran
e-mail: farzaneh.boroumand@mq.edu.au

M. T. Shakeri
e-mail: shakerimt@mums.ac.ir

F. Boroumand · H. Doosti · N. Kordzakhia
Department of Mathematics and Statistics, Faculty of Science and Engineering, Macquarie University, NSW, Australia
e-mail: Hassan.Doosti@mq.edu.au

N. Kordzakhia
e-mail: Nino.Kordzakhia@mq.edu.au

M. Salehi
Department of Mathematics and Statistics, University of Neyshabur, Neyshabur, Iran
e-mail: salehi.sms@neyshabur.ac.ir

C. H. Skiadas et al. (eds.), *13th Chaotic Modeling and Simulation International Conference*, Springer Proceedings in Complexity,
https://doi.org/10.1007/978-3-030-70795-8_56

1 Introduction

Let the regression model be

$$Y_i = r(X_i) + \epsilon_i, \quad 1 \le i \le n,$$

where $(Y_1, X_1), (Y_2, X_2), ..., (Y_n, X_n)$, are the data pairs, the design variable $X \sim f_X$, X and ϵ are independent, ϵ_i's are independent and identically distributed (iid) errors with zero mean $E(\epsilon) = 0$ and variance $E(\epsilon^2) = \sigma^2$. The regression function $r(x)$ is unknown. The kernel, local polynomial regression and orthogonal series methods are commonly used for estimating an unknown regression function. The kernel and local polynomial regression function estimators are also known as smoothers, see Wasserman [1], Fan and Gijbels [2], Hall and Racine [3] and the references within.

In this paper, we propose a tilting method which represents an optimised modification of Nadaraya-Watson estimator. An estimator $\hat{r}_n$ of r is a linear smoother if

$$l(x) = (l_1(x), ..., l_n(x))^T$$

and

$$\hat{r}_n = l(x)^T Y = \sum_{i=1}^{n} l_i(x_i) Y_i,$$

where l_i are weights and for all x, $\sum_{i=1}^{n} l_i(x) = 1$. The $l_i(x)$ for Nadaraya-Watson estimator is defined as

$$l_i(x) = \frac{K(\frac{X_i - x}{h})}{\sum_{j=1}^{n} K(\frac{X_j - x}{h})}, h > 0. \tag{1}$$

In (1), K is the weighting function that assigns the values to the design points X_i according to proximity to x. The Nadaraya-Watson (N-W) estimator, which is the kernel estimator, depends on the bandwidth parameter $h > 0$, so-called smoothing parameter. As the bandwidth increases the kernel estimator tends to a flat function, due to this property the N-W estimator is often referred to as the locally constant estimator.

In the tilting approach, the empirical distribution is being adjusted by replacing the equal weights $1/n$ by p_i, where $p_i \ge 0$ and $\sum_i^n p_i = 1$, [4]. The tilting method is sustainably used for an unknown density function estimation. Grenander [5] proposed tilting-based method by imposing some restrictions on density estimators. The empirical likelihood-based methods and distance measure approach are used for estimation of tilting parameters within regression function estimators. The empirical likelihood-based method is a semi-parametric method that allows to find a parameter through estimating equations. Owen [6] was the first who proposed the empirical likelihood method as an alternative to likelihood ratio tests. Chen [7], Zhang [8], Müller et al. [9] used the empirical likelihood-based methods for estimating tilting parameters. In the distance measure approach, the tilted estimator is obtained by minimizing some distance, subject to constraints. Hall and Presnell [10], Carroll et al. [11], Doosti and Hall [12], Doosti et al. [13] used various distance measures for estimating density functions. Namely, in Doosti et al. [13] have introduced a cross-validation function tailored to this estimation problem. They had shown that the proposed density function estimator performs better than the conventional kernel-based estimators.

The aim of this study is to introduce the tilted nonparametric N-W regression function estimator which is obtained by minimizing the distance to a comparator estimator. In this paper, an infinite order flat-top kernel estimator is selected as the comparator estimator. The infinite order flat-top kernel estimator, also known as the trapezoidal kernel, can be defined through Fourier transform which is flat near the origin and is infinitely differentiable.

2 Notation and Preliminary Results

Definition 1 Let λ be the Fourier transform of kernel K, and we select g (g is not unique) to make $\lambda(s)$, $\lambda^2(s)$, ans $s\lambda(s)$ integrable. For a fixed constant $c > 0$

$$\lambda(s) = \begin{cases} 1, & \mid s \mid \leq c \\ g(\mid s \mid), & \mid s \mid > c \end{cases}.$$

The flat-top kernel is

$$K(x) = \frac{1}{2\pi} \int_{-\infty}^{\infty} \lambda(s) e^{-isx} ds. \tag{2}$$

The infinite order regression estimator was introduced by McMurry and Politis [14] in the form of linear smoother

$$\check{r}_n = \sum\nolimits_{i=1}^{n} \check{l}_i(x_i) Y_i,$$

$$\check{l}_i(x) = \frac{K_{IO}(\frac{X_i - x}{h})}{\sum_{j=1}^{n} K_{IO}(\frac{X_j - x}{h})}.$$

K_{IO} refers to any kernel which fulfils the definition 1. The idea behind using an infinite order kernel for estimating r_n is that these type of kernels reduce the bias asymptotically at the rate $O(h^k)$, [14]. The trapezoidal kernel satisfies Definition 1

$$K_T(x) = \frac{2(\cos(x/2) - \cos(x))}{\pi x^2}.$$

The Fourier transform of the trapezoidal kernel K_T is

$$\lambda(s) = \begin{cases} 1 & \mid s \mid \leq 1/2, \\ 2(1- \mid s \mid) & 1/2 < \mid s \mid \leq 1, \\ 0 & \mid s \mid > 1. \end{cases}$$

We denote $\hat{r}_n(.|\theta)$, the tilted estimator, where $\theta = (h, p)$ is the vector of unknown parameters. Our objective is to estimate θ by minimizing the distance measure between $\hat{r}_n(.|\theta)$ and $\check{r}$. For preserving the convergence rate of $\check{r}_n$, in this paper, we use the L_2-metric as the distance between these two estimators, [12].

3 Numerical Study

This section contains the results of the numerical study carried out for analysing the performance of the tilted N-W estimators. The exponential regression function, $r(x) = x + 4exp(-2x^2)/\sqrt{2\pi}$, has been used with the uniform design density and normally distributed error terms for generation of 500 data sets. The relative performance of the tilted N-W estimator has been assessed with respect to varying sample sizes and standard deviation levels. The assessment has been made by comparing the Median Integrated Squared Error (MISE) and ISE, the latter belonging to [−2,2]. The cross-validation function method has been employed for an optimal bandwidth selection for Nadaraya-Watson estimator, [1]. For an infinite order estimator, the bandwidths were selected using the rule of thumb introduced by McMurry and Politis in [14] which is available form R-package 'iosmooth'. In fact, the bandwidth for tilted N-W estimator has been estimated within the proposed numerical procedure.

In Table 1, we provides the MISEs for the simulated data. Evidently, for fixed sample size, as the variance increases the tilted N-W estimators perform better than others. For larger sample sizes, in contrast, the N-W estimator, outperforms the tilted N-W estimator. However, for smaller sample sizes and the moderate standard deviation levels, the tilted N-W estimator remains, at some extent, superior to the conventional estimators.

4 Real Data

In this section, the tilted N-W estimator along with two other kernel-based estimators are being used for a curve fitting to real data. We shall apply the tilted N-W estimator approach to Italy's COVID-19 daily deaths data from 23 February 2020 to 6 May 2020 downloaded from https://www.ecdc.europa.eu.

Table 1 MISE for the Infinite Order (IO) estimator with the trapezoidal kernel, N-W estimator and tilted N-W estimator with 4 (p4) and 10 (p10) weighting nodes

n	σ	IO	NW	NW p4	NW p10
60	0.3	0.1559	**0.0663**	0.1308	0.1529
	0.5	0.1980	**0.1398**	0.1724	0.1953
	0.7	0.2515	**0.2152**	0.2316	0.2492
	1	0.3588	0.3697	**0.3418**	0.3650
	1.5	0.6530	0.6281	**0.6197**	0.6520
	2	1.0524	0.9892	**0.9871**	1.0597
100	0.3	0.1195	**0.0442**	0.1034	0.1191
	0.5	0.1432	**0.0914**	0.1253	0.1426
	0.7	0.1781	**0.1443**	0.1607	0.1766
	1	0.2490	0.2324	**0.2305**	0.2469
	1.5	0.4165	0.4366	**0.4041**	0.4144
	2	0.6487	0.6780	**0.6107**	0.6371
200	0.3	0.0991	**0.0232**	0.0891	0.0997
	0.5	0.1089	**0.0470**	0.0993	0.1086
	0.7	0.1256	**0.0822**	0.1172	0.1253
	1	0.1577	**0.1299**	0.1533	0.1589
	1.5	0.2401	0.2542	**0.2386**	0.2416
	2	0.3568	0.3878	**0.3464**	0.3534
1000	0.3	0.0801	**0.0058**	0.0776	0.0800
	0.5	0.0823	**0.0125**	0.0790	0.0825
	0.7	0.0853	**0.0207**	0.0801	0.0845
	1	0.0922	**0.0356**	0.0830	0.0917
	1.5	0.1080	**0.0716**	0.0972	0.1074
	2	0.1294	0.1286	**0.1209**	0.1308

Table 2 MSEs for Nadaraya-Watson, infinite order, and tilted Nadaraya-Watson estimators

	Infinite order estimator	Nadaraya-Watson estimator	Tilted Nadaraya-Watson estimator
MSE	3819.803	3778.425	3667.398

The interest in COVID-19 death rate modelling among scientists is growing rapidly since the outbreak of the pandemic began [15–17]. Along with the tilted N-W estimator, we applied the N-W, and IO estimators. The tilted N-W estimator performed the best in terms of the Mean Square Errors (MSE). Table 2 provides the MSE for each estimator. In terms of minimising the MSE, the tilted N-W estimator ranked first, followed by N-W and IO estimators resulting in the relative improvement of 3% and 4%, respectively. Slightly, improved performance of the tilted N-W

Fig. 1 Fitting N-W, IO, and tilted N-W regression curves to the logarithm of the COVID-19 deaths data

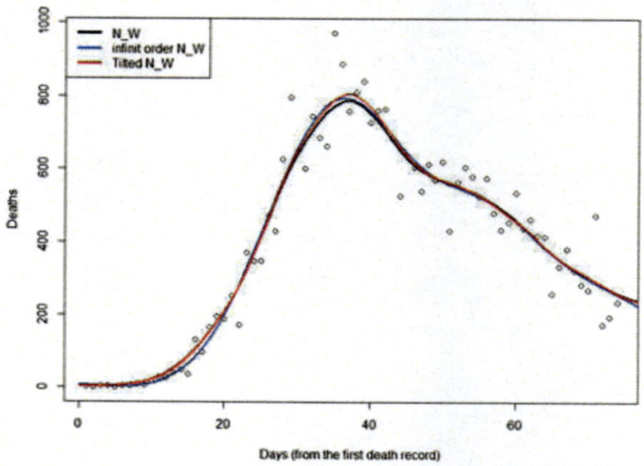

estimator is attributed to the lower MSE components at the edges versus other kernel-based regression function estimators which are generally known for so-called "edge effect", [19] (Fig. 1).

Acknowledgements This research was undertaken with the assistance of resources and services from the National Computational Infrastructure (NCI), which is supported by the Australian Government. This research forms part of the corresponding author's PhD thesis approved by the ethics committee of Mashhad University of Medical Sciences with the project code 971017.

References

1. L. Wasserman, All of Nonparametric Statistics (Springer Science and Business Media, 2006)
2. J. Fan, I. Gijbels, *Local Polynomial Modelling and its Applications: Monographson, Statistics and Applied Probability*, vol. 66 (CRC Press, 1996)
3. P.G. Hall, J.S. Racine, Infinite order cross-validated local polynomial regression. J. Econometrics **185**(2) (2015)
4. P.G. Hall, Q. Yao, Data tilting for time series. J. Royal Stati. Soc.: Ser. B (Statistical Methodology) **65**(2) (2003)
5. U. Grenander, On the theory of mortality measurement: part ii. Scand. Actuarial J. **1956**(2), 125–153 (1956)
6. A.B. Owen, Empirical likelihood ratio confidence intervals for a single functional. Biometrika **75**(2), 237–249 (1988)
7. S.X. Chen, Empirical likelihood-based kernel density estimation. Aust. J. Stat. **39**(1), 47–56 (1997)
8. B. Zhang, A note on kernel density estimation with auxiliary information. Commun. Stat.-Theory Methods **27**(1), 1–11 (1998)
9. U.U. Müller, A. Schick, W. Wefelmeyer, Weighted residual-based densityestimators for nonlinear autoregressive models. Statistica Sinica 177–195 (2005)
10. P.G. Hall, B. Presnell, Intentionally biased bootstrap methods. J. Royal Stat. Soc.: Ser. B (Statistical Methodology) **61**(1), 143–158 (1999)

11. R.J. Carroll, A. Delaigle, P. Hall, Testing and estimating shape-constrainednonparametric density and regression in the presence of measurement error. J. Am. Stat. Assoc. **106**(493), 191–202 (2011)
12. H. Doosti, P.G. Hall, Making a non-parametric density estimator more attractive,and more accurate, by data perturbation. J. Royal Stat. Soc.: Ser. B (Statistical Methodology)
13. H. Doosti, P.G. Hall, J. Mateu, Nonparametric tilted density function esti-mation: a cross-validation criterion. J. Stat. Plann. Infer. **197**, 51–68 (2018)
14. T.L. McMurry, D.N. Politis, Nonparametric regression with infinite orderflat-top kernels. J. Nonparametric Stat. **16**(3–4), 549–562 (2004)
15. R. Verity, Estimates of the severity of coronavirus disease 2019: a model-basedanalysis. The Lancet infectious diseases (2020)
16. C. Anastassopoulou, Data-based analysis, modelling and forecasting of the COVID-19 outbreak. PloS One **15**(3), e0230405 (2020)
17. J. M. Nichols, Modelling daily death patterns in the COVID - 19 Pandemic using Linear Regression, Fast Fourier Transforms, ODE Solver and Generalized Poissonian Distributions in FORTRAN, *Working paper Texas AM University (TAMU)* (15 July 2020)
18. P.G. Hall, J.S. Racine, Infinite order cross-validated local polynomial regression. J. Econometrics **185**(2), 510–525 (2015)
19. P.G. Hall, T.E. Wehrly, A geometrical method for removing edge effects from kernel-type nonparametric regression estimators. J. Am. Stat. Assoc. **86**(415), 665–672 (1991)

Is Weather Chaotic? Coexisting Chaotic and Non-chaotic Attractors Within Lorenz Models

Bo-Wen Shen, R. A. Pielke Sr., X. Zeng, J.-J. Baik, S. Faghih-Naini, J. Cui, R. Atlas, and T. A. L. Reyes

Abstract The pioneering study of Lorenz in 1963 and a follow-up presentation in 1972 changed our view on the predictability of weather by revealing the so-called butterfly effect, also known as chaos. Over 50 years since (Lorenz in J. Atmos. Sci. 20:130–141, [1]) study, the statement of "weather is chaotic" has been well accepted. Such a view turns our attention from regularity associated with Laplace's view of determinism to irregularity associated with chaos. Here, a refined statement is suggested based on recent advances in high-dimensional Lorenz models and real-world global models. In this study, we provide a report to: (1) Illustrate two kinds of attractor coexistence within Lorenz models (i.e., with the same model parameters but with different initial conditions). Each kind contains two of three attractors including point, chaotic, and periodic attractors corresponding to steady-state, chaotic, and limit cycle solutions, respectively. (2) Suggest that the entirety of weather possesses the dual nature of chaos and order associated with chaotic and non-chaotic processes, respectively. Specific weather systems may appear chaotic

B.-W. Shen (✉) · S. Faghih-Naini · J. Cui · T. A. L. Reyes
San Diego State University, San Diego, USA
e-mail: bshen@sdsu.edu

R. A. Pielke Sr.
CIRES and ATOC, University of Colorado At Boulder, Boulder, USA

X. Zeng
The University of Arizona, Tucson, USA

J.-J. Baik
Seoul National University, Seoul, South Korea

S. Faghih-Naini
University of Bayreuth, Bayreuth, Germany

Friedrich-Alexander University Erlangen-Nuremberg, Erlangen, Germany

J. Cui
North Carolina State University, Raleigh, USA

R. Atlas
University of Maryland, Baltimore County, Baltimore, USA

C. H. Skiadas et al. (eds.), *13th Chaotic Modeling and Simulation International Conference*, Springer Proceedings in Complexity,
https://doi.org/10.1007/978-3-030-70795-8_57

or non-chaotic within their finite lifetime. While chaotic systems contain a finite predictability, non-chaotic systems (e.g., dissipative processes) could have better predictability (e.g., up to their lifetime). The refined view on the dual nature of weather is neither too optimistic nor pessimistic as compared to the Laplacian view of deterministic unlimited predictability and the Lorenz view of deterministic chaos with finite predictability.

1 Introduction

Is weather chaotic? A view that weather is chaotic was proposed and is recognized based on the pioneering work of Lorenz [1] who first introduced the concept of deterministic chaos. Defined as aperiodic solutions that display sensitive dependence on initial conditions (ICs), chaos is also known as the butterfly effect. In a follow-up conference presentation in 1972 [2], the concept of sensitivity to ICs was further discussed by addressing whether a butterfly's flap may lead to a chain of responses that remotely generates a tornado. Since then, the butterfly effect has come to be known as a metaphor for indicating the huge impact of a tiny perturbation on the formation of a tornado. The original [1, 3] study and the 1972 presentation, as well as the 1969 study [4], laid the foundation for chaos theory that is viewed as the third scientific achievement in the twentieth century, after relativity and quantum mechanics, inspiring numerous studies in multiple fields, including earth science, mathematics, philosophy, physics, etc. [5].

While periodic solutions were a main focus until the [1] study, non-periodic solutions have increasingly received attention over the past 50 years. Lorenz's discovery has led to the statement of "weather is chaotic" and to a paradigm shift in the view of finite predictability from the Laplacian view of unlimited deterministic predictability. The idea of finite predictability for chaotic weather has prompted a search for the upper limit of predictability that was determined as 2 weeks based on analyses of unstable solutions from simplified models and data [4]. With the above being said, our current view on the chaotic nature of weather and a predictability limit of 2 weeks are based on the understanding of chaotic (as well as unstable) solutions obtained from elegant but simple models. To facilitate discussions, we define two kinds of predictability, including (1) intrinsic predictability that is dependent only on flow itself and (2) practical predictability that is limited by the imperfect initial conditions and/or (mathematical) formulas [6, 7].

Chaotic solutions are just one type of solution that occurs over finite intervals of time-independent parameters within the Lorenz model. To reveal the true nature of weather, we should take into consideration other types of solutions within original Lorenz models and newly developed generalized Lorenz models [7–24]. For example, in addition to chaotic solutions, two types of non-chaotic solutions indeed appear over different intervals of parameters within the Lorenz model [21]. Furthermore, recent studies using a generalized high-dimensional Lorenz model [16, 25–27] showed that chaotic and non-chaotic solutions may coexist within the same model parameters but for different ICs [28, 29]. Thus, it is important to understand whether or not and how other types of solutions and their coexistence may help illustrate a more comprehensive view on the nature of weather, and to improve our understanding of predictability associated with different types of solutions. Specifically, we may ask whether the statement of "weather is chaotic" that exclusively considers chaotic solutions is scientifically precise.

To address the above, here, we first provide a review of major solutions using the Lorenz model (LM), including three types of solutions or three attractors in Sect. 2. In this study, a specific type of solution is referred to as an "attractor", defined as the smallest attracting point set that cannot be decomposed into two or more subsets with distinct regions of attraction [22]. We then summarize our recent findings for two kinds of attractor coexistence (i.e., with the same model parameters but with different initial conditions) using a newly developed, generalized, high-dimensional LM (GLM) [26] in Sect. 3. Section 4 is presented in order to support the findings for two kinds of attractor coexistence using the original LM with different parameters. Based on an analysis of the LM and the GLM and a brief review of previous studies (e.g., regarding 40-day intra-seasonal oscillations, coexisting solutions at two time scales, etc.), we suggest a refined view on the dual nature of weather in Sect. 5. Additional support for this view is also presented by the review of prior studies. Concluding remarks are provided in Sect. 6.

2 The Lorenz Model [1]

In his 1963 study, Prof. Lorenz presented an elegant system of three ordinary differential equations (ODEs) derived from the governing equations for the Rayleigh–Benard convection [1, 30]. The system describes the time evolution of three variables, X, Y, and Z, as follows:

$$\frac{dX}{d\tau} = \sigma Y - \sigma X, \tag{1}$$

$$\frac{dY}{d\tau} = -XZ + rX - Y, \tag{2}$$

$$\frac{dZ}{d\tau} = XY - bZ. \tag{3}$$

Here, τ is the dimensionless time. Three time-independent parameters include the Prandtl number (σ), the normalized Rayleigh number (r), also called the heating parameter, and a function of the ratio between the vertical and horizontal scales of the convection (b). (X, Y, Z) represent the amplitudes of the three Fourier modes for dynamic and thermodynamic variables. The system contains three types of physical processes, including buoyancy/heating terms (represented by σY and rX), dissipative terms (represented by $-\sigma X$, $-Y$, and $-bZ$), and nonlinear processes (indicated by $-XZ$ and XY). With the exception of the heating parameter (r), the following parameters are kept constant: $\sigma = 10$ and $b = 8/3$. Control and parallel runs are performed in order to reveal the difference (or divergence) of two solutions. The only difference between control and parallel runs is that a parallel run includes tiny perturbations ($\in = 10^{-10}$) or finite perturbations ($\in = -0.9$) in initial conditions.

Using the state variables X, Y, and Z as coordinates, a phase space can be constructed for an analysis of solutions. An orbit or a trajectory is defined as the time varying components of solutions within the phase space. The dimension[1] of the phase space is equal to the number of time-dependent variables or to the number of ODEs. Thus, Eqs. (1), (2) and (3) with three variables are referred to as a three-dimensional Lorenz model (3DLM). High-dimensional LMs contain more than three variables.

2.1 Lorenz's Chaotic and Non-chaotic Attractors

Depending on the competitive or collective impact of nonlinear processes and linear buoyancy/heating and dissipative processes, various types of solutions (i.e., different attractors) appear within the Lorenz model. Historically, the dependence of their appearance on the strength of heating measured by the parameter (r) has been a focus. Steady-state, chaotic, and nonlinear oscillatory solutions have been shown to occur under conditions of weak, moderate, and strong heating, respectively [21, 33].[2] In Fig. 1, the three different types of solutions are shown using $r = 20$, 28, and 350, respectively. The top panels display solutions for control runs within the *X*-*Y* space, while bottom panels display the time evolution of the Y components for both control and parallel runs. For a steady-state solution, its orbit eventually approaches a single point, that is, a non-trivial equilibrium point within the *X*-*Y* space (Fig. 1a), appearing as a point attractor; and its amplitude remains constant over time after arriving at the equilibrium point. Mathematically, equilibrium points, also called critical points, are defined as solutions of the time-independent nonlinear system (e.g., no time derivatives in Eqs. (1), (2) and (3) [35].[3] When the heating parameter exceeds the critical value of $r_c = 24.74$, the 3DLM with $r = 28$ displays the so-called chaotic solution or a chaotic attractor with irregular oscillations. The solution's boundary within the *X*-*Y* space appears as a tilted "8" pattern. Interestingly, when heating becomes larger (e.g., $r = 350$), the system produces a nonlinear periodic solution known as a limit cycle solution or a periodic attractor, as shown in Fig. 1c and f.

[1] The term "dimension" is conventionally used for a system of ODEs [31, 32]. In this study, the 5DLM, 7DLM, and 9DLM are referred to as high-dimensional or high-order Lorenz models [12].

[2] Similar findings for the dependence of various solutions (i.e., chaotic and limit cycle solutions) on the strength of heating were also reported using a two-layer, quasi-geostrophic model that describes the finite-amplitude evolution of a single baroclinic wave by Pedlosky and Frenzen [34].

[3] In our 5D-, 7D-, and 9DLMs, we can obtain closed form solutions of trivial and non-trivial equilibrium points and use them to verify the numerical solutions of equilibrium points.

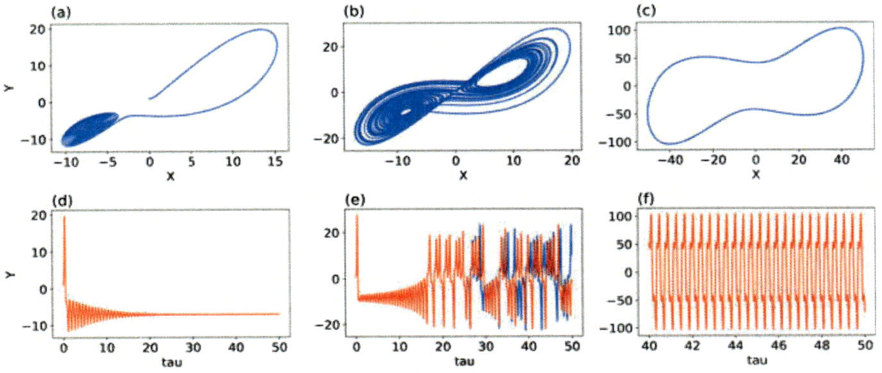

Fig. 1 Three types of solutions within the 3DLM. Left, middle, and right panels displays steady-state, chaotic, and limit cycle solutions at small, moderate, and large heating parameters (i.e., $r = 20$, 28, and 350), respectively. The solutions are categorized into a point attractor, a chaotic attractor, and a periodic attractor, respectively. Top panels show orbits within the X-Y space and bottom panels depict the time evolution of Y. Blue lines provide solutions from control runs. To display results from parallel runs, red lines are added in the bottom panels. Sensitive dependence on initial conditions is shown in panel (**e**) with two visible lines. Two panels, **b** and **e**, are reproduced from Shen [38]

Additional details on the characteristics of nonlinear oscillatory solutions may be found in earlier studies [21, 36, 37] and/or recent studies [16, 26, 38]. Below, the impact of a tiny initial perturbation on three attractors, including a point attractor, a chaotic attractor, and a periodic attractor, is further discussed.

Parallel runs with a tiny initial perturbation ($\in= 10^{-10}$) are compared to control runs in order to reveal the difference of initial, nearby trajectories within the phase space. For steady-state and nonlinear oscillatory solutions, control and parallel runs produce almost identical results, only appearing in red, for example, in Fig. 1d and f. The runs indicate insignificant impacts by a tiny initial perturbation. In other words, steady-state and nonlinear oscillatory solutions are insensitive to a tiny change in ICs. In comparison, within the chaotic regime, two solution orbits whose starting points are very close to each other display very different time evolutions, as clearly shown in blue and red in Fig. 1e. The phenomenon is called the sensitive dependence of solutions on ICs and only appears within a chaotic solution.

2.2 *Boundedness and Divergence of Chaotic Trajectories*

Within the chaotic regime, a sensitive dependence of solutions on ICs is referred to as the butterfly effect (BE, e.g., [39, 40]). As shown in Fig. 2a, the term "butterfly" was partly used due to its geometric pattern within the *Y-Z* space [39]. A butterfly pattern with a finite size and varying curvatures within the phase space also qualitatively suggests an important feature of solution boundedness. Therefore, BE means that a

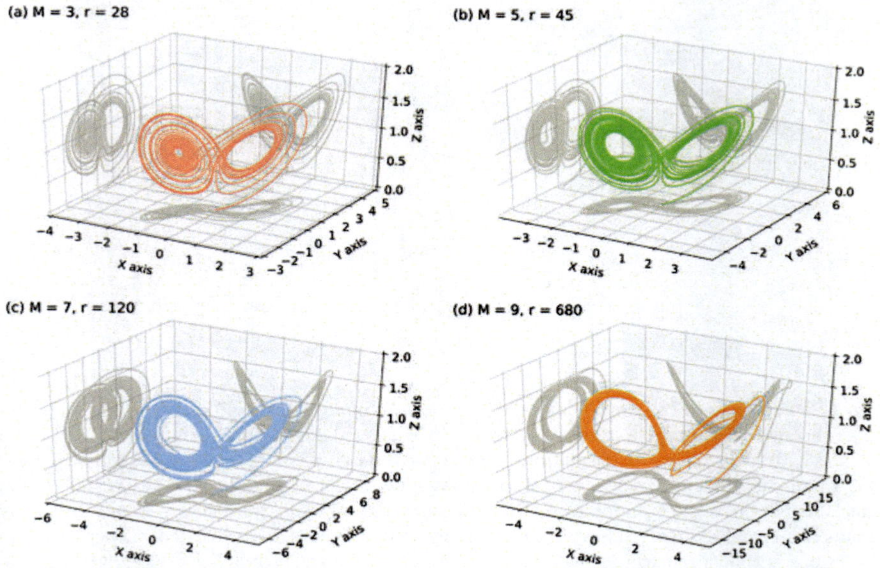

Fig. 2 Chaotic solutions in the X-Y-Z phase space within the 3D, 5D, 7D, and 9D Lorenz models (LMs). Panels **a–c** use the same initial conditions with $Y = 1$ and the remaining as zero, while panel **d** uses an IC with 100 for all variables. Variables (X, Y, Z) are normalized by $2\sqrt{r-1}$, $2\sqrt{r-1}$, and $(r-1)$, respectively. A larger heating parameter is required for the onset of chaos in a higher-dimensional LM. Also see the detailed analysis of solutions in Shen [38, 41]

tiny change in an IC can produce a very different time evolution of a solution for three variables (X, Y, Z). However, the separation (or divergence) of two orbits should be bounded by the size of a butterfly pattern.

The average separation rate (i.e., an average rate of divergence) of nearby trajectories has been quantitatively measured using the Lyapunov exponent (LE, [42–44]). A positive LE suggests an exponential rate in the averaged separation of two infinitesimally nearby trajectories over an infinite period of time (e.g., Eqs. (25) and (26) of [7]). Chaotic solutions within the 3DLM, as well as high-dimensional LMs, have a positive LE. Since the LE is defined as a long-term averaged separation, researchers often misinterpret the divergence of two nearby, but finitely separated, chaotic trajectories within the 3DLM as continuing over time and lasting forever. The misunderstanding also makes people believe that an unconstrained solution is due to the divergent nature of chaos [45]. In fact, in addition to a positive LE, solution boundedness is another major feature of a chaotic system. Due to solution boundedness, a trajectory should recurve within the phase space [45] Therefore, time-varying (local) growth rates along a chaotic orbit are observed [44] and may become negative, as indicated by a negative finite time LE [46–49]. In other words, the infinite-time limit in the definition of an LE does not imply a monotonically increasing separation between two nearby trajectories over a long period of time. Two initial nearby trajectories can quickly separate and reach the bound of their separation.

3 The Generalized Lorenz Model

The 3DLM produces three different attractors and each attractor exclusively appears within the phase space, depending on the interval of system parameters. The 3DLM with a single-type solution suggests that either chaos or order exclusively exists. By comparison, two different solutions may coexist and dominate system dynamics in a separate region (i.e., a different subspace) within the phase space within the same model, and with the same parameters, but with different ICs. Attractor coexistence has mainly been studied using conservative Hamiltonian systems [45], but can also be found in the forced dissipative 3DLM [33, 50, 51]. Below, we first discuss two kinds of attractor coexistence using the GLM, and then apply the GLM to understand whether the 3DLM can also possess two kinds of attractor coexistence.

Based on our recent studies [7, 26], we successfully developed a GLM that: (1) is derived based on partial differential equations for the Rayleigh–Benard convection;[4] (2) allows a large number of modes, say M modes, where M is an odd number greater than three; and (3) produces aggregated negative feedback[5] that is accumulated from the feedback of various smaller-scale processes, yielding a larger effective dissipation in higher dimensional LMs [26, 38]. *As a result of aggregated negative feedback, a higher-dimensional LM requires a larger critical value for the Rayleigh parameter* (r_c) *for the onset of chaos.* For example, the r_c for the 5DLM, 7DLM, and 9DLM are 42.9, 116.9, and 679.8, respectively, as compared to a r_c of 24.74 for the 3DLM [26]. Figure 2 displays chaotic solutions obtained from the 3D, 5D, 7D, and 9D LMs with different heating parameters. Therefore, an initial tiny perturbation with the same strength may play a different role within the GLM with a different value of M. Such a feature shows a dependence on the number of selected modes. Namely, it depends on the degree of spatial complexity associated with a various number of modes of the GLM.

3.1 Two Kinds of Attractor Coexistence

The GLM with M = 5 or M = 7 (i.e., 5DLM or 7DLM) also produces three different types of solutions, including a steady-state, chaotic, and limit cycle/torus.[6] More importantly, the GLM with M = 9 (i.e., 9DLM) displays two kinds of attractor coexistence, each with two different attractors. For the first kind of coexistence, both chaotic and steady-state solutions occur concurrently using the same model

[4] By comparison, chaotic models in Lorenz [52–54] were not derived from physics-based partial differential equations.

[5] Negative feedback can be found within the so-called Lorenz-Stenflo system that extends the 3DLM with one additional ODE containing one additional mode that takes rotation into consideration [55–57].

[6] A torus is defined as a composite motion with two (or more) oscillatory frequencies whose ratio is irrational [8].

and the same parameters. The only difference is their ICs. Such a coexistence shares properties similar to that of the 3DLM but appears over a wider range of the Rayleigh parameter (e.g., $679.8 < r < 1{,}058$), as compared to the small interval (e.g., $24.06 < r < 24.74$) for the 3DLM.

In addition to the first kind of attractor coexistence, the 9DLM is able to produce the second kind of attractor coexistence, consisting of nonlinear, periodic (i.e., limit cycle) orbits, and steady-state solutions at large Rayleigh parameters (e.g., $r = 1{,}600$). The new kind of coexistence was recently documented in Reyes and Shen [16], Shen [26] and Shen et al. [27]. Additionally, coexisting two periodic solutions were documented using the 9DLM with $r = 1{,}120$ [26]. As a result, when system parameters change at a large time scale (e.g., at climate time scales), different kinds of attractor coexistence may alternatively or concurrently appear, leading to complexities that better resemble real weather and climate.

3.2 *Two Kinds of IC Dependence and Final State Sensitivity*

Depending on system parameters, ICs and the dimension of the model (say the value of M within the GLM), a modeling system may contain one or more attractors[7] within the phase space. Since different attractors coexist, we expect different kinds of solution dependence on ICs, as illustrated using the 9DLM with $r = 680$ that produces the coexistence of chaotic and non-chaotic orbits. Control runs apply three sets of ICs at different locations within the phase space: close to the non-trivial equilibrium point, near the origin (i.e., a saddle point), and at point (100, 100, 100, 100, 100, 100, 100, 100, 100). For parallel runs, a finite-amplitude perturbation ($\epsilon = -0.9$) is added into the ICs. In Fig. 3, solutions of the control runs are shown in blue, while results of parallel runs are displayed in green, red, or orange. Top panels display the time evolution of the Y components, while bottom panels present solutions within the *X*-*Y* space. The model with $r = 680$ produces the coexistence of steady-state and chaotic orbits, displaying a dependence on ICs. For the first case (Fig. 3a and d) with the IC that is close to the non-trivial equilibrium point, the orbit moves toward the equilibrium point, producing steady-state solutions. Since the orbit spirals into the non-trivial equilibrium point within the *X*-*Y* space, it is also called a spiral sink solution. For the second case (Fig. 3b and e) where an IC is close to a saddle point at the origin but away from the non-trivial equilibrium point, solutions still approach the same non-trivial equilibrium point as a steady-state solution, while initially displaying a different time evolution as compared to the first case. On the other hand, for the third case (Fig. 3c and f), the model produces a chaotic solution, different from the steady-state solution. A comparison between control and parallel runs suggests that an initial perturbation only has a short-term impact on the initial

[7] The coexistence of chaotic and quasi-periodic orbits has been recently documented in a modified Lorenz system by Saiki et al. [58].

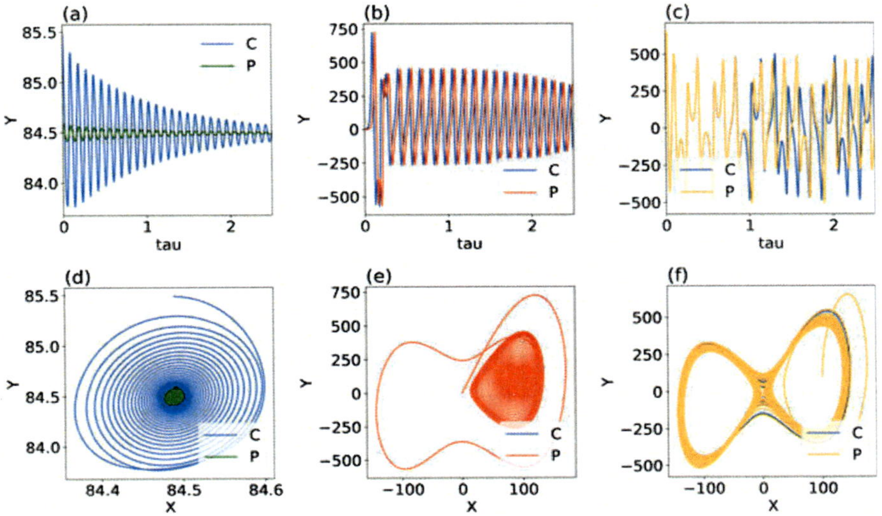

Fig. 3 Solutions of the GLM with M = 9 and r = 680. Initial conditions for the three cases are placed near the non-trivial critical point (**a**, **d**), the origin (i.e., trivial critical point) (**b**, **e**), and at (100, 100, 100, 100,100, 100, 100, 100,100) (**c**, **f**). Top panels show the time evolution of Y for $t \in [0, 2.5]$, while bottom panels display the corresponding solutions $t \in [0, 10]$ within the X-Y space. Control and parallel runs are denoted by 'C' and 'P', respectively. A finite-amplitude perturbation ($\in = -0.9$) is added into the parallel runs. Two panels, (**c**) and (**f**), are reproduced from Shen [38]

transient evolution of steady-state solutions but can lead to a very different evolution for chaotic solutions.[8]

When coexisting chaotic and regular attractors from 256 different initial conditions are plotted within the X-Y phase space, Fig. 4 clearly shows that chaotic and non-chaotic orbits occupy two different regions (or two different subspaces). Additional details on the spatial distribution of 256 initial conditions may be found in Shen et al. [27]. As a result of the different regions of attraction for coexisting attractors, final state sensitivity may appear [59] when ICs begin near the boundary of two different attractors. Such a sensitivity creates a different challenge for prediction.

3.3 *Finite and Deterministic Predictability*

The rate of a growing initial error with time has been used to determine predictability, suggesting a finite predictability in chaotic (or unstable) systems. Such a growth

[8] Such a dependence on initial conditions, close to (or away from) the non-trivial equilibrium point, can be shown by the following YouTube video for a double pendulum (between 1:00 and 1:20). https://www.youtube.com/watch?v=LfgA2Auyo1A.

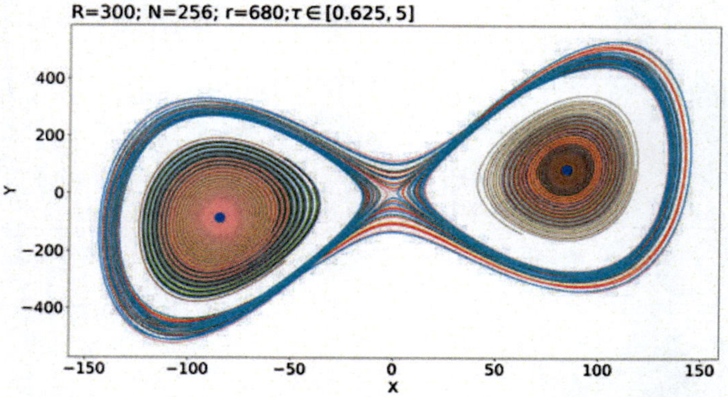

Fig. 4 Coexistence of chaotic and non-chaotic orbits starting with 256 different initial conditions (ICs) for $T \in [0.625, 5]$. Chaotic orbits recurrently return back to the saddle point at the origin. Non-chaotic orbits eventually approach one of two stable critical points as shown in large blue dots. Chaotic and non-chaotic orbits occupy different regions of attraction within the phase space

rate is proportional to the divergence of two nearby trajectories measured using a Lyapunov exponent. Within chaotic regimes of the 3DLM, as well as within the GLM that contains one positive LE and solution boundedness, time-varying divergence and convergence of nearby trajectories yields time-varying growth rates and, thus, time-varying predictability. Estimated predictability over a short period should display a dependence on various initial states.[9] By comparison, when non-chaotic (i.e., steady-state or nonlinear periodic) solutions appear as a single type of solution or coexist with another type of solution, their predictability should be deterministic (unlimited). Stated conservatively, the non-chaotic solution should remain predictable until it is changed by time varying parameters that represent heating or dissipations. As a result, very different intrinsic predictability may appear and depend on ICs within a system that possesses the coexistence of chaotic and non-chaotic attractors.

4 Attractor Coexistence Within the 3DLM

Within chaotic solutions of the 3DLM that has no stable equilibrium points, a tiny perturbation can always lead to a very different time evolution. Stated alternatively, within the chaotic regime, the system, in the absence of energy sinks for steady-state solutions, does not have a mechanism for completely removing the impact of a tiny perturbation on state variables. By comparison, within the GLM with M = 9, or higher, that possesses coexisting chaotic and steady-state solutions, a tiny initial perturbation may play a very different role. A tiny perturbation may have no

[9] As a result, we agree with Prof. Arakawa that the predictability limit is not necessarily a fixed value [46].

long-term impact when it appears to be associated with a steady-state solution that approaches one of the stable equilibrium points, suggesting that the perturbation eventually dissipates. On the other hand, a tiny perturbation may lead to a large impact on the time evolution of the chaotic solution. As a result, the 9DLM with a dual role for a tiny initial perturbation over a wide range of the heating parameter, as well as other features such as hierarchical scale dependence, is more realistic than the classical 3DLM with typical parameters. On the other hand, we may ask whether the 3DLM with different parameters may also produce two kinds of attractor coexistence, providing additional support to the findings of the GLM.

Next, we first discuss the coexistence of the 3DLM with typical parameters that include $\sigma = 10$. We then address the question of whether $\sigma = 10$ is a magic choice. As simply shown in the animation, https://goo.gl/scqRBo, the 3DLM with the same parameters, including r = 24.4, $\sigma = 10$, and b = 8/3, but with different ICs, produces two types of solutions that include chaotic or steady-state solutions, yielding the first kind of attractor coexistence. However, such a coexistence only appears over a very small range of r, giving the length of an interval less than 0.7 (i.e., $24.06 < r < r_c = 24.74$), and, thus, its characteristics and potential role in revealing the nature of weather has not been well appreciated.

For the past 50 years, although various types of solutions for Lorenz [6] have been documented, chaotic solutions have been the main focus. As discussed in the main text, since chaotic solutions appear over a finite range of parameters, their applicability in revealing the nature of weather depends on the realism of not only the models employed but also model parameter values. In his book in 1993, Lorenz humbly expressed that it may not have been possible for him to discover the butterfly-pattern solution if a realistic value of $\sigma = 1$ was used, as shown below:

> *I was lucky in more ways than one. An essential constant of the model is the Prandtl number – the ratio of the viscosity of the fluid to the thermal conductivity. Barry had chosen the value 10.0 as having the order of magnitude of the Prandtl number of water. As a meteorologist, he might well have chosen to model convection in air instead of water, in which case he would probably have used the value 1.0. With this value the solutions of the three equations would have been periodic, and I probably would never have seen any reason for extracting them from the original seven.*

Therefore, one may wonder how fortunate Prof. Lorenz was and whether a realistic value of $\sigma = 1$ may have influenced our view on the nature of weather. We make an attempt of addressing the question by analyzing a GLM with M = 9 and examining a 3DLM with $\sigma = 1$. As discussed in Shen [26], the GLM with M = 9 has stable, non-trivial equilibrium points for all $r > 1$ when $\sigma = 10$ and b = 8/3. To have stable, non-trivial equilibrium points for $\sigma = 1$ within the 3DLM, we chose b = 2/5. Such a choice leads to two kinds of attractor coexistence, a unique feature first identified within the 9DLM [26]. With $\sigma = 1$ in the 3DLM, the first kind of coexistence includes chaotic and steady-state solutions at a moderate heating parameter (e.g., r = 170, as shown in Fig. 5). Table 1 lists initial conditions for the results provided in Fig. 5. Thus, chaotic solutions may still appear within the 3DLM for a realistic value of $\sigma = 1$, but they coexist with steady-state solutions. The appearance of chaotic solutions depends not only on the range of the heating parameter but also on the

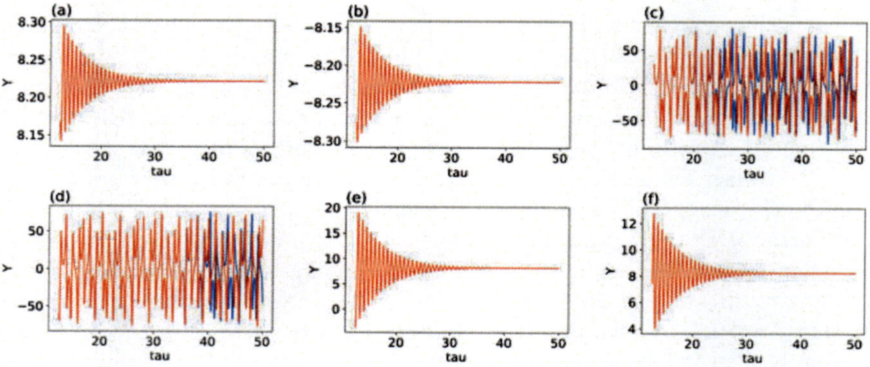

Fig. 5 A co-existence of chaotic (**c**, **d**) and non-chaotic (**a**, **b**, **e**, **f**) solutions using the same parameters for $\sigma = 1$, b = 0.4, and $r = 170$ within the 3DLM. Blue and red lines display solutions from the control and parallel runs, respectively. Initial conditions for the results in six panels are listed in Table 1

Table 1 Initial conditions (ICs) for revealing the coexistence of two attractors for $\sigma = 1$, $b = 0.4$, and $r = 170$ within the 3DLM. $X_c = Y_c = \sqrt{b(r-1)}$ and $Z_c = (r-1)$. The six rows provide the ICs for Fig. A1

X	Y	Z
X_c	$Y_c + 1$	Z_c
$-X_c$	$-Y_c + 1$	Z_c
0	1	0
−76.72346293	37.62433028	−146.96230812
−27.75526885	167.67883615	3.66782724
136.44623635	99.45689394	−19.76741851

ICs. Additionally, the second kind of coexistence that consists of a limit cycle and a steady-state solution appears at a large heating parameter (e.g., r = 250, not shown).

Both traditional and new model configurations with $(\sigma, b) = (10, 8/3)$ and $(1, 2/5)$, respectively, can produce chaotic solutions. For the traditional configuration that has been well applied in numerous studies since Lorenz [1], all three equilibrium points are unstable when $r > 24.74$. The stability of the three equilibrium points for $\sigma = 10$, as well as for $\sigma = 1$, is illustrated in Fig. 6. The non-existence of stable equilibrium points within the chaotic regime makes it easier to obtain chaotic solutions. However, no tiny, initial perturbation can completely lose its impact within the chaotic regime. We may interpret this as a finding that a tiny, initial perturbation cannot completely dissipate (before leading to a large impact). By comparison, for the new configuration of $\sigma = 1$, while the origin is still a saddle point, the two, non-trivial equilibrium points are stable (Fig. 6b). The existence of stable equilibrium points enables the coexistence of chaotic and steady-state solutions, the latter of which has no long-term memory regarding a tiny, initial perturbation.

As a result of coexistence for $\sigma = 1$ within the 3DLM, a proper choice of initial conditions is required in order to simulate a chaotic solution. Without knowing this,

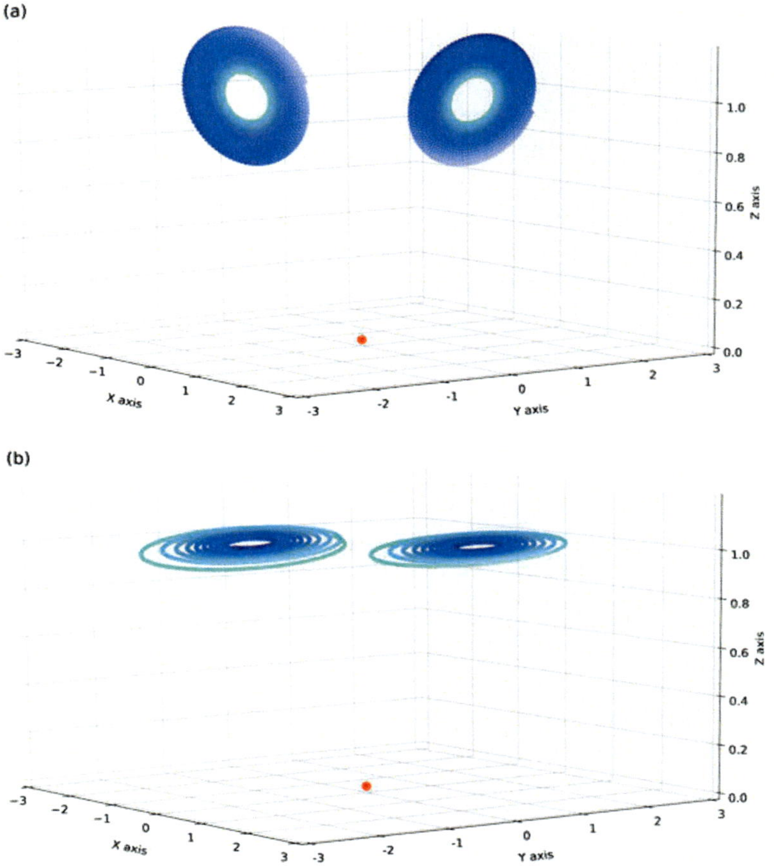

Fig. 6 Local behavior near the two non-trivial critical points for the 3DLM with $\sigma = 10$ (**a**) and $\sigma = 1$ (**b**). Lighter blue dots indicate the locations of orbits at earlier times. A red dot indicates the origin, which is a saddle point. Orbits in panel (**a**) spiral away from the non-trivial critical points while orbits in panel (**b**) spiral toward the non-trivial critical points

Prof. Lorenz thought it may have been impossible to obtain a "strange" solution if $\sigma = 1$ was first used in the Saltzman [30] model, giving no motivation for him to work on the 3DLM. In other words, the value of $\sigma = 10$ used in the original study [30] was indeed a "fortunate" choice so that an unexpected irregularly oscillatory solution could be revealed, inspiring Prof. Lorenz to develop the 3DLM in order to discover interesting chaotic features. However, on the other hand, we now understand that such a configuration can only depict a partial picture for the nature of weather. Based on our results and analysis, a realistic system should include physical processes for (some of) the tiny disturbances in order to completely dissipate. Since it produces the coexistence of chaotic and steady-state solutions and since the steady-state solution has no long-term memory of tiny perturbations, the 3DLM with the new configuration of $\sigma = 1$ satisfies the objective. Such a system, which is similar to the 9DLM that produces two kinds of coexisting attractors, provides a more realistic view on the true nature of weather than the original 3DLM with a typical configuration. The above results support the idea that two kinds of attractor coexistence should be taken into consideration to reveal the nature of weather.

5 A Refined View on the Nature of Weather

Within the forced dissipative 3DLM, chaotic solutions appear within a finite range of parameters (e.g., heating parameter), bounded on one side by stable, steady-state solutions and on the other side by nonlinear periodic solutions. Since climate and weather involve open systems [60], an assumption of constant parameters within numerical simulations using the 3DLM, as well as high-dimensional LMs, is not realistic [61]. Time varying parameters that lead to different attractors should be used in models for realistic climate or weather [19]. For example, when a moderate heating becomes weaker (or stronger), a steady-state solution (or a limit cycle) may appear. Since regular and chaotic solutions may alternatively appear, chaotic solutions alone may not be able to represent the entirety of weather.

Additionally, our results show that chaotic and non-chaotic solutions may coexist and two kinds of attractor coexistence may alternatively appear within the 9DLM using time varying parameters. The analysis suggests a need to refine our view of weather by taking the dual nature associated with attractor coexistence into consideration. To this end, we suggest, contrary to the traditional view that weather is chaotic, that weather is, in fact, a superset that consists of both chaotic and non-chaotic processes, including both order and chaos.

5.1 Vacillation, Coexisting Two LCs, and Coexisting Two Time-Scale Orbits

The (potential) occurrence of a regular nonlinear periodic solution (i.e., limit cycle) in the atmosphere was first illustrated by laboratory experiments using dishpans. Based on experiments by Lorenz [39], Fultz [62] and Hide [63] suggested three types of solutions, including: (1) steady state solutions, (2) irregular chaotic solutions, and (3) vacillation. "Amplitude vacillation" is defined as a solution whose amplitude grows and periodically decays in a regular cycle [3, 64, 65]. Studies by Pedlosky [66], Smith [67], and Smith and Reilly [68] found that *amplitude vacillation can be viewed as a limit cycle solution.* By conducting a study for observational characteristics of low-frequency variability, Ghil and Robertson [69] suggested that *40-day, intra-seasonal oscillations* may arise from a bifurcation off the blocking flow and *may be represented by a limit cycle with a period of 40 days.*

As discussed earlier, we showed that the 3DLM with a realistic value of $\sigma = 1$ also generates two kinds of attractor coexistence. Additionally, the coexistence of two stable limit cycle solutions was documented using the Lorenz 1984 model [15, 70–75] that also contains three types of solutions, including steady state, periodic solutions, and chaotic solutions. Using a seasonally varying forcing term with a time scale of 12 months, Lorenz [71] showed that chaos appears during winter (within a specific range of parameters) and two coexisting LCs during summer (within a different range of parameters). Such numerical results also support the view of the dual nature of chaos and order that alternatively appear. The above results suggest that once summer begins and has been observed, a better predictability for a limit cycle solution may be expected during each cycle of the solution in summer, as compared to that in winter. More recently, Lucarini and Bodai [76] applied a multistable system with coexisting attractors to reveal the bistability of the climate system with both positive and negative feedback [76, 77].

Coexisting solutions at two time scales, that are not the same as the coexisting attractors discussed above, have also been documented within the scientific literature. Related studies additionally support the refined view on the nature of weather. For example, co-existence of fast and slow manifolds has been discussed by Curry et al. [78], Lorenz [79, 80] and Lorenz and Krishnamurthy [81]. Both types of solutions in Lorenz [79] are non-chaotic. By comparison, fast and slow "variables" that are chaotic may also coexist within coupled systems [82, 83] In fact, an analysis using a singular perturbation method [84] indicates that the GLM also possesses the coexistence of slow and fast variables that correspond to large and very small spatial modes (e.g., Eqs. (2) and (4) of [27] in a high-dimension phase space). A current trend is to include time-varying parameters to increase the complexities of low order systems [85]. It can be shown that a higher dimensional Lorenz model (e.g., 7DLM) can be viewed as a lower-dimensional Lorenz model (e.g., 5DLM) with a periodic forcing, suggesting that the complexities of spatial mode–mode interaction may lead to the temporal complexities.

5.2 Error Saturations and Computational Chaos

In real-world weather models, the appearance of (fully) chaotic solutions may be indicated by error saturations, defined as follows. A logistic equation has been used to describe the evolution of root mean square (rms) average forecast error for ensemble runs [52, 86–89]. Given an initial condition with a small value, the solution of the logistic equation has time varying, non-negative growth rates (e.g., growing at an initial larger growth rate, then at a nonlinear smaller growth rate, and eventually approaching a constant defined as a saturated error that has a zero growth rate). The occurrence of error saturation at a fully nonlinear stage indicates a comparable number of members with positive and negative error growth rates at a given time. Such a result is consistent with the features of a positive LE and solution boundedness associated with a specific chaotic solution.

The error growth model with non-negative growth rates may describe the statistical behavior of the system within which the majority of small errors tends to grow. *By comparison, the error growth model cannot accurately represent the initial, transient evolution of the rms averaged forecast error associated with large ensemble members with periodic or decaying components whose growth rates are small.* For periodic solutions such as vacillation [87], an ensemble averaged error may grow (or decay) with time when a large (or small) ensemble number of growing errors and a small (or large) ensemble number of decaying errors are averaged. As a result, when oscillatory waves were simulated, their rms errors may oscillate with time rather than become saturated. For example, oscillatory rms errors appeared after 40-day simulations in Fig. 5 of Liu et al. [90] who performed global simulations using the Community Atmosphere Model [91]. An additional example can be found in 30-day simulations of multiple African Easterly Waves (AEWs) using a global mesoscale model that produced oscillatory correlation coefficients [38].

On the other hand, it should be noted that error saturations may appear in association with computational chaos that is a numerical artifact. For example, Lorenz [92] presented several cases in order to show that while differential equations of a model may possess nonlinear limit cycle solutions, the corresponding discrete version of the model with large time steps produces a sensitive dependence of solutions on the initial condition, referred to as computational chaos. As a result, the appearance of error saturations (as well as positive LE) that appear within numerical models does not necessarily represent the chaotic nature of weather. *Due to the appearance of computational chaos, an estimate of a practical predictability limit using saturation errors should be interpreted with caution, as it does not necessarily represent an intrinsic predictability limit for real weather.*

6 Concluding Remarks

The statement of "weather is chaotic" has been introduced to indicate the chaotic nature of weather with a finite intrinsic predictability. The statement has also been cited to embrace a practical predictability limit of 2 weeks [93]. The finite intrinsic and practical predictability are indeed largely derived from the chaotic and unstable solutions of Lorenz models [1, 4]. In other words, the current view of "weather is chaotic" does not take into consideration other types of solutions within original Lorenz models and new types of solutions within newly developed generalized Lorenz models.

In this study, we first applied the aforementioned models in order to reveal three types of solutions and two kinds of attractor coexistence, indicating different intrinsic predictability for different solutions. We then suggested a refined view on the dual nature of chaos and order in weather. In contrast to the current view that focuses on chaotic solutions with a predictability limit (of 2 weeks), our refined view suggests that coexisting chaotic and non-chaotic systems can have different intrinsic predictability. *The refined view may unify the theoretical understanding of different predictability within Lorenz models with recent numerical simulations of advanced global models that can simulate large-scale tropical waves beyond two weeks* [38, 94].

The refined view with a duality of chaos and order is fundamentally different from the Laplacian view of deterministic predictability and the Lorenz view of deterministic chaos. The refined view that is not too optimistic nor too pessimistic suggests both potential and challenges. For non-chaotic processes with steady-state or nonlinear periodic solutions [95, 96], their intrinsic predictability is deterministic (e.g., up to the lifetime of a dissipative solution or the time scale of the forcing) and their practical predictability can be continuously increased by improving the accuracy of the model and the initial conditions. For limit cycle solutions that may be associated with computational chaos, accurate simulations with better predictability, as compared to chaotic solutions, can be obtained by increasing temporal resolutions and/or removing redundant dissipations. To reveal longer predictability or better estimates on predictability in model and observation data, we will focus on developing schemes for the detection of chaotic and non-chaotic solutions [16, 28, 97] and examining the roles of butterfly effects in multiscale simulations using high-resolution global models.

In addition to the chaotic nature of weather with a finite predictability, another major influential impact of the 3DLM is that the sensitive dependence on initial condition, referred to as the butterfly effect of the first kind, has been inaccurately metaphorized to indicate the ability of a butterfly flap in creating a tornado [98], referred to as the butterfly effect of the second kind [7]. To understand their roles in reality and numerical models, the two different kinds of butterfly effects are being analyzed based on a comprehensive review of historical literature and recent understanding of chaos dynamics.

Acknowledgements We thank reviewers of the manuscript, the editor, and Drs. M. Alexander, R. Anthes, B. Bailey, J. Buchmann, D. Durran, M. Ghil, F. Judt, B. Mapes, Z. Musielak, T. Krishnamurti (Deceased), C.-D. Lin, T. Palmer, J. Pedlosky, J. Rosenfeld, R. Rotunno, I. A. Santos, C.-L. Shie, S. Vannitsem, and F. Zhang (Deceased) for valuable comments and discussions. We appreciate the eigenvalue analysis provided by Mr. N. Ferrante. We are grateful for support from the College of Science at San Diego State.

References

1. E.N. Lorenz, Deterministic nonperiodic flow. J. Atmos. Sci. **20**, 130–141 (1963)
2. E.N. Lorenz, Predictability: Does the flap of a butterfly's wings in Brazil set off a tornado in Texas? in *Proceedings of 139th Meeting of AAAS Section on Environmental Sciences, New Approaches to Global Weather: GARP* (Cambridge, MA, AAAS), 5 pp (1972). https://eaps4.mit.edu/research/Lorenz/Butterfly_1972.pdf

3. E.N. Lorenz, The mechanics of vacillation. J. Atmos. Sci. **20**, 448–464 (1963)
4. E.N. Lorenz, The predictability of a flow which possesses many scales of motion. Tellus **21**, 289–307 (1969)
5. J. Gleick, *Chaos: Making a New Science* (Penguin, New York, 1987), p. 360
6. E.N. Lorenz, The predictability of hydrodynamic flow. Trans. N.Y. Acad. Sci., Ser. II, **25**(4), 409–432 (1963)
7. B.-W. Shen, Nonlinear feedback in a five-dimensional Lorenz model. J. Atmos. Sci. **71**, 1701–1723 (2014). https://doi.org/10.1175/JAS-D-13-0223.1
8. S. Faghih-Naini, B.-W. Shen, Quasi-periodic in the five-dimensional non-dissipative Lorenz model: the role of the extended nonlinear feedback loop. Int. J. Bifurc. Chaos **28**(6) 1850072 (20 pages) (2018). https://doi.org/10.1142/S0218127418500724
9. C.C. Felicio, P.C. Rech, On the dynamics of five- and six-dimensional Lorenz models. J. Phys. Commun. **2**, 025028 (2018)
10. J. Guckenheimer, R.F. Williams, Structural stability of Lorenz attractors. Publ. Math. IHES. **50**, 59 (1979)
11. W.M. Macek, Nonlinear dynamics and complexity in the generalized Lorenz system. Nonlinear Dyn. **94**, 2957–2968 (2018). https://doi.org/10.1007/s11071-018-4536-z
12. S. Moon, B.-S. Han, J. Park, J.M. Seo, J.-J. Baik, Periodicity and chaos of high-order Lorenz systems. Int. J. Bifurc. Chaos **27**(11) 1750176 (11 pages) (2017). https://doi.org/10.1142/S0218127417501760.
13. S. Moon, J.M. Seo1, B.-S. Ha, J. Park, J.-J. Baik, A physically extended Lorenz system. Chaos **29**, 063129 (2019). https://doi.org/10.1063/1.5095466
14. Z.E. Musielak, D.E. Musielak, K.S. Kennamer, The onset of chaos in nonlinear dynamical systems determined with a new fractal technique. Fractals **13**, 19–31 (2005)
15. R.A. Pielke, X. Zeng, Long-term variability of climate. J. Atmos. Sci. **51**, 155–159 (1994)
16. T. Reyes, B.-W. Shen, A recurrence analysis of chaotic and non-chaotic solutions within a generalized nine-dimensional lorenz model. Chaos, Soliton. Fract. **125**(2019), 1–12 (2019). https://doi.org/10.1016/j.chaos.2019.05.003
17. D. Roy, Z.E. Musielak, Generalized Lorenz models and their routes to chaos. I. energy-conserving vertical mode truncations. Chaos Soliton. Fract. **32**, 1038–1052 (2007)
18. B.-W. Shen, Solitary waves, homoclinic orbits, and nonlinear oscillations within the non-dissipative Lorenz model, the inviscid pedlosky model, and the kdv equation, in *The 13th Chaos International Conference (CHAOS2020)*, 9–12 June 2020. https://doi.org/10.1007/978-3-030-70795-8_58 (in press)
19. B.-W. Shen, R.A. Pielke Sr., X. Zeng, J.-J. Baik, S. Faghih-Naini, J. Cui, R. Atlas, Is weather chaotic? Coexistence of chaos and order within a generalized Lorenz model. Bull. Am. Meteorol. Soc. 2021;102(1):E148–58. https://doi.org/10.1175/BAMS-D-19-0165.1
20. S. Smale, Mathematical problems for the next century. Math. Intell. **20**(2), 7–15 (1998)
21. C. Sparrow, *The Lorenz Equations: Bifurcations, Chaos, and Strange Attractor*. (Springer, New York, 1982). 269 pp. Appl. Math. Sci.
22. J.C. Sprott, X. Wang, G. Chen, Coexistence of point, periodic and strange attractors. Int. J. Bifurc. Chaos **23**(5) (2013). https://doi.org/10.1142/S0218127413500934
23. W. Tucker, A rigorous ODE Solver and Smale's 14th problem. Found. Comput. Math. **2**, 53–117 (2002)
24. Q. Yang, G. Chen, A chaotic system with one saddle and two stable node-foci. Int. J. Bifurc. Chaos **18**(5), 1393–1414 (2008)
25. T. Reyes, B.-W. Shen, A recurrence analysis of multiple African easterly waves during summer 2006, in *Current Topics in Tropical Cyclone Research, Anthony Lupo, IntechOpen*, 2020. https://doi.org/10.5772/intechopen.86859
26. B.-W. Shen, Aggregated negative feedback in a generalized Lorenz model. Int. J. Bifurc. Chaos **29**(3), 1950037 (2019). https://doi.org/10.1142/S0218127419500378
27. B.-W. Shen, T.A.L Reyes, S. Faghih-Naini, Coexistence of chaotic and non-chaotic orbits in a new nine-dimensional Lorenz model, in: C. Skiadas, I. Lubashevsky (Eds.) *11th Chaotic Modeling and Simulation International Conference. CHAOS 2018. Springer Proceedings in Complexity* (Springer, Cham). https://doi.org/10.1007/978-3-030-15297-0_22

28. J.C. Sprott, A. Xiong, Classifying and quantifying basins of attraction. Chaos **25**, 083101 (2015). https://doi.org/10.1063/1.4927643
29. J.C. Sprott, J.A. Vano, J.C. Wildenberg, M.B. Anderson, J.K. Noel, Coexistence and chaos in complex ecologies. Phys. Lett. A **335**(2005), 207–212 (2005)
30. B. Saltzman, Finite amplitude free convection as an initial value problem. J. Atmos. Sci. **19**, 329–341 (1962)
31. M. Hirsch, S. Smale, R.L. Devaney, *Differential Equations, Dynamical Systems, and an Introduction to Chaos*, 3rd edn. (Academic Press, 2013), 432 pp
32. J.M.T. Thompson, H.B. Stewart, *Nonlinear Dynamics and Chaos*, 2nd edn. (John Wiley & Sons, LTD, 2002), p. 437
33. P.G. Drazin, *Nonlinear Systems* (Cambridge), 333pp (1992)
34. J. Pedlosky, C. Frenzen, Chaotic and periodic behavior of finite-amplitude baroclinic waves. J. Atmos. Sci. **37**, 1177–1196 (1980)
35. J. Guckenheimer, P. Holmes, *Nonlinear Oscillations, Dynamical Systems, and Bifurcations of Vector Fields* (Springer, New York, 1983), p. 459
36. T. Shimizu, Analytical form of the simplest limit cycle in the Lorenz model. Physica **97A**, 383–398 (1979)
37. S.H. Strogatz, *Nonlinear Dynamics and Chaos. With Applications to Physics, Biology, Chemistry, and Engineering* (Westpress view, Boulder, CO, 2015), p. 513
38. B.-W. Shen, On the predictability of 30-day global mesoscale simulations of African easterly waves during summer 2006: a view with the generalized Lorenz model. Geosciences **9**, 281 (2019). https://doi.org/10.3390/geosciences9070281
39. E.N. Lorenz, *The Essence of Chaos* (University of Washington Press, Seattle, 1993), p. 227
40. E.N. Lorenz, *2008: The Butterfly Effect* (University of Rome, Rome, Premio Felice Pietro Chisesi e Caterina Tomassoni Award Lecture, April 2008).
41. B.-W. Shen, Hierarchical scale dependence associated with the extension of the nonlinear feedback loop in a seven-dimensional Lorenz model. Nonlin. Proces. Geophys. **23**, 189–203 (2016). https://doi.org/10.5194/npg-23-189-2016
42. A. Wolf, J.B. Swift, H.L. Swinney, J.A. Vastano, Determining Lyapunov exponents from a time series. Physica **16D**, 285–317 (1985)
43. X. Zeng, R. Eykholt, R.A. Pielke, Estimating the Lyapunov-exponent spectrum from short time series of low precision. Phys. Rev. Lett. **66**, 3229–3232 (1991)
44. X. Zeng, R.A. Pielke Sr., R. Eykholt, Chaos theory and its applications to the atmosphere. Bull. Atmos. Meteorol. Soc. **74**(4), 631–644 (1993)
45. R.C. Hilborn, *Chaos and Nonlinear Dynamics. An Introduction for Scientists and Engineers*, 2nd ed. (Oxford University Press, New York, 2000), pp. 650
46. B. Bailey, Quantifying the predictability of noisy space-time dynamical processes. Stat. Interf. **4**, 535–549 (2011)
47. R.O. Ding, J.P. Li, Nonlinear finite-time Lyapunov exponent and predictability. Phys. Lett. **354A**, 396–400 (2007)
48. B. Eckhardt, D. Yao, Local Lyapunov exponents in chaotic systems. Physica D **65**, 100–108 (1993)
49. J.M. Nese, Quantifying local predictability in phase space. Physica D. **35**, 237–250 (1989)
50. E. Ott, *Chaos in Dynamical Systems*, 2nd edn. (Cambridge University Press, 2002), 478pp
51. J. Yorke, E. Yorke, Metastable chaos: the transition to sustained chaotic behavior in the Lorenz model. J. Stat. Phys. **21**, 263–277 (1979)
52. E.N. Lorenz, Predictability—A problem partly solved. Seminar on Predictability, vol. I, ECMWF (1996) (also, published as Lorenz (2006))
53. E.N. Lorenz, Designing chaotic models. J. Atmos. Sci. **62**, 1574–1587 (2005)
54. E.N. Lorenz, Predictability – A problem partly solved, in T. Palmer, R. Hagedorn (eds.), *Predictability of Weather and Climate*). (Cambridge University Press, Cambridge, 2006), pp. 40–58. https://doi.org/10.1017/CBO9780511617652.004
55. J. Park, H. Lee, Y.-L. Jeon, J.-J. Baik, Periodicity of the Lorenz-Stenflo equations. Phys. Scr. **90**, 065201 (2015)

56. J. Park, B.-S. Han, H. Lee, Y.-L. Jeon, J.-J. Baik, Stability and periodicity of high-order Lorenz-Stenflo equations. Phys. Scr. **91**, 065202 (2016)
57. J.C. Xavier, P.C. Rech, Regular and chaotic dynamics of the Lorenz-Stenflo system. Int. J. Bifurc. Chaos **20**, 145–152 (2010)
58. Y. Saiki, E. Sander, J.A. Yorke, Generalized Lorenz equations on a three-sphere. Eur. Phys. J. Spec. Top. **226**, 1751–1764 (2017)
59. C. Grebogi, S.W. McDonald, E. Ott, J.A. Yorke, Final state sensitivity: an obstruction to predictability. Phys. Lett. A **99**(9), 415–418 (1983)
60. K. McGuffie, A. Henderson-Sellers, *The Climate Modeling Primer*, 4th edn. (John Wiley & Sons, Ltd, 2014), p. 439
61. J.D. Daron, D.A. Stainforth, 2015: On quantifying the climate of the nonautonomous Lorenz-63 model. Chaos **25**, 043103 (2015). https://doi.org/10.1063/1.4916789
62. D. Fultz, R.R. Long, G.V. Owens, W. Bohan, R. Kaylor, J. Weil, Studies of thermal convection in a rotating cylinder with some implications for large-scale atmospheric motion. Meteorol. Monographs **21**(4) (1959). (American Meteorological Society)
63. R. Hide, Some experiments on thermal convection in a rotating liquid. Quart. f. Roy. Meteorol. Soc. **79**, 161 (1953)
64. M. Ghil, P. Read, L. Smith, 2010: Geophysical flows as dynamical systems: the influence of Hide's experiments. Astron. Geophys. **51**(4), 428–435 (Aug. 2010). https://doi.org/10.1111/j.1468-4004.2010.51428.x
65. M. Ghil, S. Childress, *Topics in Geophysical Fluid Dynamics: Atmospheric Dynamics, Dynamo Theory, and Climate Dynamics* (Springer, New York, 1987), p. 485
66. J. Pedlosky, Limit cycles and unstable baroclinic waves. J. Atmos. Sci. **29**, 53–63 (1972)
67. R.K. Smith, A note on a theory of vacillating baroclinic waves. J. Atmos. Sci. **32**, 2027 (1975)
68. R.K. Smith, J.M. Reilly, On a theory of amplitude vacillation in baroclinic waves: some numerical solutions. J. Atmos. Sci. **34**, 1256–1260 (1977)
69. M. Ghil, A.W. Robertson, "Waves" vs. "particles" in the atmosphere's phase space: a pathway to long-range forecasting? PNAS **99**(suppl 1) 2493–2500 (2002). https://doi.org/10.1073/pnas.012580899
70. E.N. Lorenz, Irregularity: a fundamental property of the atmosphere, in *Crafoord Prize Lecture, presented at the Royal Swedish Academy of Sciences, Stockholm, September 28, 1983.* Tellus, 36A, 98–110 (1984) (LM84)
71. E.N. Lorenz, Can chaos and intransitivity lead to interannual variability? Tellus **42A**, 378–389 (1990)
72. C. Masoller, A.C. Sicardi Schifino, L. Romanelli, Regular and chaotic behavior in the new Lorenz system. Phys. Lett. A **167**(1992), 185–190 (1992)
73. L. Veen van, Time scale interaction in low-order climate models. Utrecht University Repository. (PhD Dissertation) (2002)
74. L. Veen, Baroclinic flow and the Lorenz-84 mode. Int. J. Bifurc. Chaos **13**, 2117 (2003). https://doi.org/10.1142/S0218127403007904
75. H. Wang, Y. Yu, G. Wen, Dynamical analysis of the Lorenz-84 atmospheric circulation model. J. Appl. Math. **2014**, 296279, 15 pp (2014). https://doi.org/10.1155/2014/296279
76. V. Lucarini, T. Bodai, 2019: Transitions across melancholia states in a climate model: reconciling the deterministic and stochastic points of view. Phys. Rev. Lett. **122**, 158701 (2019)
77. I.R. Garashchuk, D.I. Sinelshchikov, A.O. Kazakov, N.A. Kudryashov, Hyperchaos and multistability in the model of two interacting microbubble contrast agents editors-pick. Chaos **29**, 063131 (2019). https://doi.org/10.1063/1.5098329
78. J.H. Curry, S.E. Haupt, M.N. Limber, Truncated systems, initialization, and the slow manifold. Tellus **47A**, 145161 (1995)
79. E.N. Lorenz, On the existence of a slow manifold. J. Atmos. Sci. **43**, 154471557 (1986)
80. E.N. Lorenz, The slow manifold. What is it? J. Atmos. Sci. **49**, 24492451 (1992)
81. E.N. Lorenz, V. Krishnamurthy, On the nonexistence of a slow manifold. J. Atmos. Sci. **44**, 29402950 (1987)

82. L. Mitchell, G.A. Gottwald, Data assimilation in slow-fast systems using homogenized climate models. J. Atmos. Sci. **69**, 1359–1377 (2012). https://doi.org/10.1175/JAS-D-11-0145.1
83. M. Peña, E. Kalnay, 2004: Separating fast and slow modes in coupled chaotic systems. Nonlin. Proces. Geophys. **11**, 319–327 (2004). https://doi.org/10.5194/npg-11-319-2004
84. C.M. Bender, S.A. Orszag, *Advanced Mathematical Methods for Scientists and Engineers* (McGraw-Hill, New York, 1978), p. 593
85. V. Lucarini, Stochastic resonance for nonequilibrium systems. Phys. Rev. E **100**, 062124 (2019). https://doi.org/10.1103/PhysRevE.100.062124
86. E. Kalnay, *Atmospheric Modeling, Data Assimilation and Predictability* (Cambridge, New York, 2003), 369 pp
87. E.N. Lorenz, Atmospheric predictability as revealed by naturally occurring analogues. J. Atmos. Sci. **26**, 636–646 (1969)
88. C. Nicolis, 1992: Probabilistic aspects of error growth in atmospheric dynamics. Q. J. R. Meteorol. Soc. **118**, 553–568 (1992)
89. F. Zhang, Y.Q. Sun, L. Magnusson, R. Buizza, S.-J. Lin, J.-H. Chen, K. Emanuel, What Is the predictability limit of midlatitude weather? J. of Atmos. **76**, 1077–1091 (2019)
90. H.-L. Liu, F. Sassi, R.R. Garcia, Error growth in a whole atmosphere climate model. J. Atmos. Sci. **66**, 173–186 (2009)
91. W.D. Collin et al., Description of the NCAR Community Atmosphere Model (CAM3). NCAR Tech. Note NCAR/TN-464+STR, 226pp (2004)
92. E.N. Lorenz, Computational chaos: a prelude to computational instability. Physica **35D**, 299–317 (1989)
93. Lewis, Roots of ensemble forecasting. Mon. Weather Rev. **133**(7), 1865–1885 (2005)
94. F. Judt, Atmospheric predictability of the tropics, middle latitudes, and polar regions explored through global storm-resolving simulations. J. Atmos. Sci. **77**, 257–276 (2020). https://doi.org/10.1175/JAS-D-19-0116.1
95. B.-W. Shen, Homoclinic orbits and solitary waves within the non-dissipative Lorenz model and KdV equation. Int. J. Bifurc. Chaos. **30**, 2050257–1–2050315 (2020). https://doi.org/10.1142/S0218127420502570
96. B.-W. Shen, On periodic solutions in the non-dissipative Lorenz model: the role of the nonlinear feedback loop. Tellus A **70**, 1471912 (2018). https://doi.org/10.1080/16000870.2018.1471912
97. J. Cui, B.-W. Shen, A Kernel Principal Component Analysis of Coexisting Attractors within a Generalized Lorenz Model. Chaos, Solitons & Fractals, 146. https://doi.org/10.1016/j.chaos.2021.110865
98. R. Pielke, The real butterfly effect (2008). https://pielkeclimatesci.wordpress.com/2008/04/29/the-realbutterfly-effect/

Solitary Waves, Homoclinic Orbits, and Nonlinear Oscillations Within the Non-dissipative Lorenz Model, the Inviscid Pedlosky Model, and the KdV Equation

Bo-Wen Shen

Abstract In contrast to the conventional view that weather is chaotic, a revised view on the dual nature of chaos and order in weather and climate has recently been proposed. The revised view is based on the findings of attractor coexistence using the classical and generalized Lorenz models, as well as promising 30 day simulations using a high-resolution global model. To provide additional support, this study further illustrates mathematical universalities between the Lorenz and Pedlosky models whose solutions represent very different physical processes, including small-scale convection and large-scale quasi-geostrophic baroclinic waves. A comparison amongst the non-dissipative Lorenz model, the inviscid Pedlosky model, the KdV equation, and other systems is additionally provided in order to reveal the same form of these solutions for solitary waves and homoclinic orbits, and to propose a generic conservative system with two families of oscillatory solutions. The generic system is then applied to illustrate the role of nonlinearity in producing stable critical points for stabilizing the system. An analysis of the generic system (i.e., the non-dissipative Lorenz model) and the Lorenz model that retains one or three dissipative terms reiterates the collective impact of multiple dissipative terms with nonlinearity (as well as the heating term) on the appearance of unstable critical points and irregular and chaotic responses. The results suggest the need for a systematic approach for examining the impact of new (stable) components on the local and global stability of the new coupled system.

Keywords Baroclinic wave · Homoclinic orbits · Solitary waves · Nonlinear oscillations · Lorenz model · KdV equation · Nonlinear pendulum equation · Pedlosky model · Quasi-geostrophic system

B.-W. Shen (✉)
Department of Mathematics and Statistics, San Diego State University, 5500 Campanile Drive, San Diego, CA 92182, USA
e-mail: bshen@sdsu.edu
URL: http://bwshen.sdsu.edu

C. H. Skiadas et al. (eds.), *13th Chaotic Modeling and Simulation International Conference*, Springer Proceedings in Complexity,
https://doi.org/10.1007/978-3-030-70795-8_58

1 Introduction

Over a span of 50 years, pioneering studies by Prof. Lorenz of MIT [20, 22, 23] have changed our view on the predictability of weather by revealing the so-called butterfly effect, also known as chaos [1, 10, 24]. Studies by Prof. Lorenz and other chaos researchers have yielded a paradigm shift in fundamental research activity: from regularity associated with Laplace's view of determinism to irregularity associated with Lorenz's view of chaos, indicating finite predictability and a potential role for free will. The statement of "weather is chaotic" has been well accepted within the scientific community. However, in contrast to the above, a revised view that focuses on the dual nature of both order and chaos in weather and climate has been proposed [46, 48, 52, 53]. Such a view is based on publications using a real-world, high-resolution global model over the past 10 years (e.g., [47, 54–56]), and recent findings using the theoretical Lorenz model [20] and various generalized Lorenz models (e.g., [5, 7, 13, 16, 27, 28, 35, 37, 41–46, 48]), as well as real-world data analysis (e.g., [47, 49, 50, 64]). Major findings include: (1) realistic short-term climate (30-day) simulations that are beyond two weeks and (2) the coexistence of chaotic and non-chaotic processes.

The revised view suggests the potential for extending predictability beyond the Lorenz theoretical limit of two weeks and calls for a need to revisit predictability problems by analyzing both chaotic and regular solutions (i.e., weather systems). Previous studies have demonstrated that theoretical mathematical models are powerful tools for revealing the nature of real weather and for understanding the impact of new model components on system stability. By extending earlier studies, my main goal here is to provide additional support to the revised view on the duality of weather by comparing the Lorenz model [20] and the Pedlosky model [31–34], which are derived from very different physical models. I, also, present mathematical universalities amongst the non-dissipative versions of the two models and four additional physical systems, including the Korteweg-de Vries (KdV) equation (e.g., [2–4, 18, 19, 26, 63]) and the Nonlinear Schrodinger (NLS) equation (e.g., [12]) in order to propose a generic system for predictability studies. I then illustrate how the inclusion of multiple dissipative terms into the generic system may lead to chaotic responses.

The concept of duality in weather is supported by attractor coexistence within the Lorenz model [52, 53, 65] and the generalized Lorenz model [46, 47]. However, mathematical simplicity and the specific physical nature of the convection problem within Lorenz-type models still makes people wonder about the reality and applicability of the revised view. To provide additional support for the revised view, a comparison of differences and similarities in the Lorenz and Pedlosky models is provided. As discussed in Sect. 2 and Appendix A, the Pedlosky model that consists of a 2nd-order ordinary differential equation (ODE) and a first-order ODE was derived from a set of partial differential equations (PDEs) for a two-layer quasi-geostrophic (QG) flow. Here, quasi-geostrophic dynamics indicate a significant impact due to Earth's rotation measured by the Coriolis parameter. As compared to convection that appears at small temporal and spatial scales, quasi-geostrophic flows possess very different physical processes at large temporal and spatial scales. Despite different physical processes at small and large scales, here, I show that the two systems of ODEs for the Lorenz and Pedlosky models are mathematically identical when system parameters are properly selected. Such mathematical universalities indicate the applicability of findings using Lorenz models within the Pedlosky model, a system governed by QG dynamics.

While chaotic solutions have been a focus over the past several decades, the revised view reiterates the importance of regular solutions. As discussed in earlier studies (e.g., [46–48] and references therein), regular steady-state and limit cycle solutions may appear within the three-dimensional Lorenz model (3DLM, [20]) and two types of oscillatory solutions within the non-dissipative Lorenz model (3D-NLM, e.g., [45]). The latter (i.e., 3D-NLM) may qualitatively represent characteristics of the former (i.e., 3DLM) at sufficiently large heating parameters (e.g., [45, 59]). Additionally, homoclinic orbits that connect stable and unstable manifolds at a saddle point can be found in both the 3DLM and 3D-NLM. Within the 3D-NLM, analytical solutions of homoclinic orbits that represent a separatrix that separates small- and large-cycle oscillations were obtained. The results suggest that the existence of homoclinic orbits may be a good indicator for the existence of two families of regular oscillations within non-dissipative systems. On the other hand, while homoclinic orbits may coexist with oscillatory orbits, depending on the initial conditions, their presence tends to introduce irregular responses and chaos within non-dissipative and/or dissipative systems (e.g., [11, 48]).

Earlier studies (e.g., [9]) suggested that improving our understanding of homoclinic orbits and nonlinear oscillatory solutions has the potential to address the transition between two types of symmetric and unsymmetric vacillations. Vacillations were discovered using dishpan experiments in the laboratory during the 1950s [8, 14] and may be viewed as a limit cycle solution (e.g., [32, 57, 58]). In comparison, for regular weather systems, roll clouds may appear in the form of solitary-type solutions. The solitary-wave solution appears as a result of the balance between nonlinearity and dispersion (e.g., [36, 38, 66, 67]). A recent study (e.g., [48]) demonstrated that solitary-type solutions in a traveling-wave coordinate, appearing in the form of hyperbolic secant ($sech$) and hyperbolic secant squared ($sech^2$) functions, can be found in the two different components of homoclinic orbits within the 3D-NLM. Additionally, mathematical universalities amongst the 3D-NLM, the KdV, and the NLS equations were presented. Here, all of the features (i.e., solutions and ODEs) are reviewed and compared to those of the Pedlosky model, as well as a simplified nonlinear pendulum equation, in order to propose a generic system for predictability studies.

The paper is organized as follows. In Sect. 2, a review of the Lorenz and Pedlosky models is first provided. Mathematical similarities for equations and solutions amongst the non-dissipative Lorenz model (3D-NLM), the inviscid Pedlosky model, the NLS, the Duffing, the KdV, and simplified pendulum equations are then presented in order to propose a generic system, which is a conservative, 2nd-order ODE. In Sect. 3, a stability analysis of the generic system with zero or multiple dissipative terms is performed to reveal the impact of additional terms (e.g., nonlinear and/or dissipative terms) on the system's stability. Concluding remarks are provided at the end. Appendix A presents a two-layer, quasi-geostrophic system that provides a foundation for the Pedlosky model. A brief introduction to the nonlinear pendulum equation is provided in Appendix B.

2 The Lorenz and Pedlosky Models and Other Systems

In this section, I first show that the full version of the Lorenz and Pedlosky models produces the same set of ODEs when their parameters are properly selected. I then present mathematical universalities amongst the 3D-NLM, the inviscid Pedlosky model, the NLS, the Duffing, the KdV, and the simplified pendulum equations (e.g., [15, 60–62]).

2.1 *The Lorenz Model and Its Non-dissipative Version*

The Lorenz model consists of the following three first-order ODEs [20]:

$$\frac{dX}{d\tau} = \sigma Y - \sigma X, \tag{1}$$

$$\frac{dY}{d\tau} = -XZ + rX - \epsilon Y, \tag{2}$$

$$\frac{dZ}{d\tau} = XY - bZ. \tag{3}$$

Here, X, Y, and Z are the state variables. The above system contains four time-independent parameters: ϵ, σ, r, and b. Compared to the original Lorenz model with three parameters, an additional parameter ϵ is introduced to illustrate the role of positive or negative dissipation in changing the stability of the system (to be discussed in Sect. 3.3). When $\epsilon = 1$, the above system becomes the original 3DLM. The second and third parameters are called the Prandtl number and the normalized Rayleigh number [40], also referred to as a heating parameter. The fourth parameter "b" represents the ratio of the horizontal and vertical scales of the convection cell.

By retaining one dissipation term of $(-\epsilon Y)$ and neglecting the other two dissipative terms (i.e., $-\sigma X$ and $-bZ$), Eqs. (1)–(3) can be transformed into the following single ODE for X:

$$\frac{d^2X}{d\tau^2} + \epsilon\frac{dX}{d\tau} + \left(\frac{X^2}{2} - (\sigma r + C_e)\right) X = 0, \tag{4}$$

and the ODE for a conservative quantity of C_e:

$$\frac{dC_e}{d\tau} = 0. \tag{5}$$

Here, $C_e = C(X^2/2 - \sigma Z) = constant$ even when $\epsilon \neq 0$. Such a constant can be determined by its initial value, yielding $C_e = C_{ic}$. Below, Eq. (4) with $\epsilon = 0$ is first analyzed in order to examine Eq. (4) with $\epsilon \neq 0$ in Sect. 3.3. With $\epsilon = 0$, Eq. (4) that represents the 3D-NLM (e.g., [29, 30, 48]) becomes:

$$\frac{\mathrm{d}^2 X}{\mathrm{d}\tau^2} + \left(\frac{X^2}{2} - (\sigma r + C_{ic})\right) X = 0. \tag{6}$$

Additional details can be found in [45, 48].

A general form of the 2nd order ODE that represents Eq. (6) is written as follows:

$$\frac{d^2 U}{d\tau^2} + G(U) = 0, \tag{7}$$

here, $G(U)$ represents a function of U.
Multiplying both sides by U' and integrating with respect to τ, we can obtain:

$$\frac{1}{2}\left(\frac{dU}{d\tau}\right)^2 + \int G(U)dU = E_0; \tag{8}$$

here, E_0 is an integration constant. The first and second terms on the left side may be viewed as mathematical kinetic energy and potential energy, respectively [17]. As a result, Eq. (8) represents a conservative quantity for total energy. To facilitate discussions below, Eq. (8) is referred to as the energy form of Eq. (7). Applying the same idea, we can derive the following equation from Eq. (6):

$$\left(\frac{\mathrm{d}X}{\mathrm{d}\tau}\right)^2 - (\sigma r + C_{ic})X^2 + \frac{X^4}{4} = E_1, \tag{9}$$

where E_1 is constant. Note that based on classifications of domain average potential energy and available potential energy (APE) within the 3DLM and 3D-NLM (e.g., [42, 48], the mathematical energy in Eq. (9) may represent the sum of the rescaled domain average kinetic energy and APE, the latter of which is not the same as the domain average potential energy.

As a result of $Z = C_{ic} - X^2/(2\sigma)$ in Eq. (5), the following equation can be obtained from Eq. (6) of the 3D-NLM with $C_{ic} = 0$:

$$\frac{d^2 Z}{d\tau^2} + 3\sigma Z^2 - 4\sigma r Z = 0. \tag{10}$$

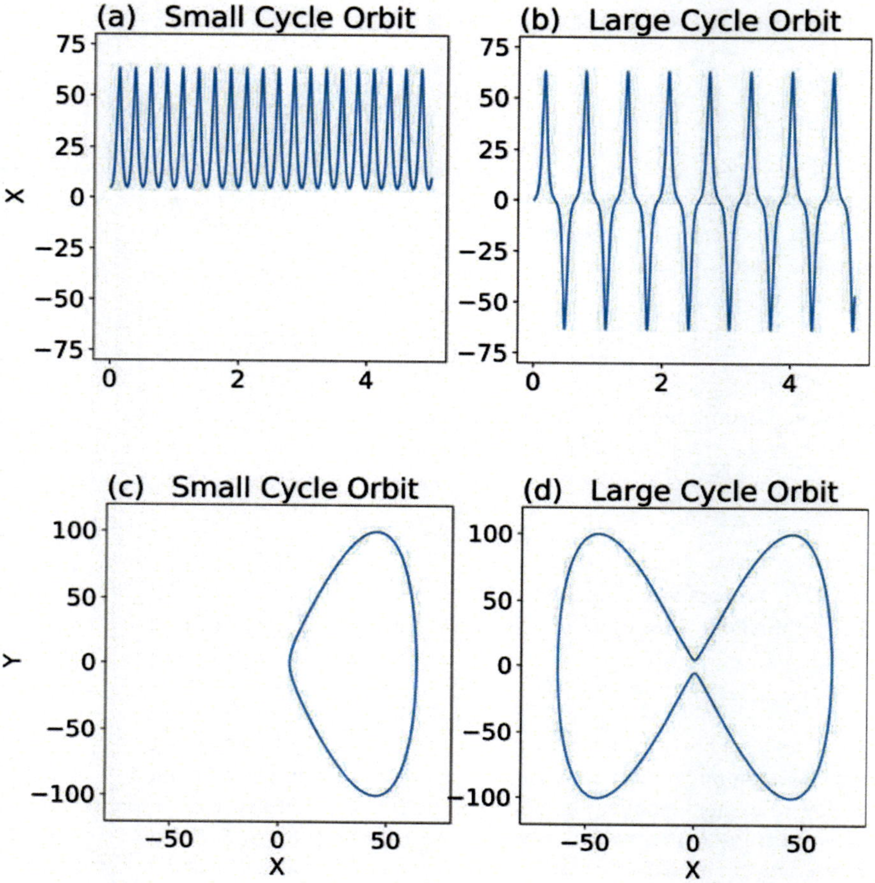

Fig. 1 Two types of oscillatory solutions within the 3D-NLM. Panels **(a)**–**(c)** illustrate small-cycle solutions with an initial condition of $(X, Y, Z) = (5, 0, 0)$. Panels **(b)**–**(d)** show large-cycle solutions with an initial condition of $(X, Y, Z) = (0, 5, 0)$

Note that such a C_{ic} is selected for a comparison with the KdV equation below.

As discussed in a previous study [45], the 3D-NLM produces three different types of solutions, including two kinds of oscillatory solutions and homoclinic orbits. Numerical solutions for oscillations with small and large cycles are displayed in Fig. 1, indicating that the X component of the large-scale oscillation varies between positive and negative values, and that the small-scale cycle is of one sign.

In comparison, an analytical solution of the homoclinic orbit that connects unstable and stable manifolds at the saddle point is provided in Fig. 2. Here, the homoclinic orbit is decomposed into two components (i.e., the "contracting component" and "expanding component") in order to obtain their analytical solutions. The "contracting component" of the homoclinic orbital solution is defined for $\tau \in [0, \infty)$, with an

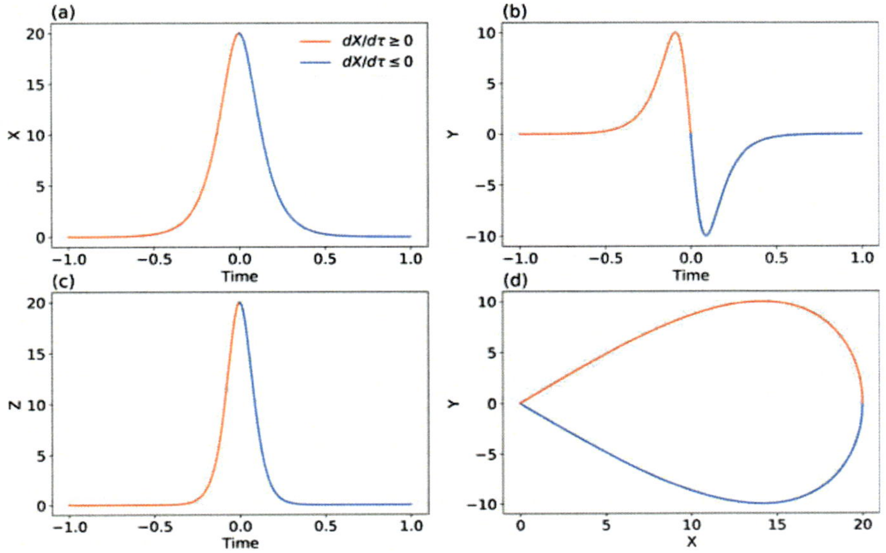

Fig. 2 Solutions of the homoclinic orbit for X (**a**), Y (**b**), and Z (**c**) within the 3D-NLM for $\sigma = r = 10$. The homoclinic orbit within the $X - Y$ space is shown in panel (**d**). Blue and red lines represent the monotonically decreasing and increasing components of X (as well as Z) for the homoclinic orbit, respectively

initial condition of $(X, Y, Z) = (2\sqrt{\sigma r}, 0, 2r)$. The solution is written as follows:

$$X(\tau) = \frac{4\sqrt{\sigma r}}{e^{\sqrt{\sigma r}\tau} + e^{-\sqrt{\sigma r}\tau}} = 2\sqrt{\sigma r} sech(\sqrt{\sigma r}\tau), \tag{11a}$$

$$Y(\tau) = -4r\frac{e^{\sqrt{\sigma r}\tau} - e^{-\sqrt{\sigma r}\tau}}{\left(e^{\sqrt{\sigma r}\tau} + e^{-\sqrt{\sigma r}\tau}\right)^2} = -(2r) tanh(\sqrt{\sigma r}\tau) sech(\sqrt{\sigma r}\tau), \tag{11b}$$

$$Z(\tau) = \frac{X^2(\tau)}{2\sigma} = (2r) sech^2(\sqrt{\sigma r}\tau). \tag{11c}$$

From the above equations, the X and Z components are expressed in terms of hyperbolic secant and secant squared functions, respectively.

Since the first derivative of the X (as well as Z) solution is non-positive (i.e., $dX/d\tau \leq 0$) for $\tau \in [0, \infty)$, the solution in Eqs. (11a) (as well as (11c)) indeed represents the monotonically decreasing component of the homoclinic orbit. As a result, the term "contracting component" is used to describe the solution in the $X - Y$ phase space, since Y is not a monotonic function for $\tau \in [0, \infty)$. The contracting component begins at $(X, Y, Z) = (2\sqrt{\sigma r}, 0, 2r)$ for $\tau = 0$ and approaches the origin for $\tau \to \infty$.

The solution for the expanding component of the homoclinic is obtained using the invariant property, as follows. Since the 3D-NLM is invariant under $\tau \to -\tau$ and

$Y \to -Y$ (e.g., [61]), the above solution $(X(\tau), -Y(\tau), Z(\tau))$ in backward time $\tau \in (-\infty, 0]$ represents the expanding component of the homoclinic solution. As shown in red in Fig. 2, such a component begins at the origin for $\tau \to -\infty$ and then moves to the point $(X, Y, Z) = (2\sqrt{\sigma r}, 0, 2r)$ for $\tau = 0$. The corresponding X and Z components are monotonically increasing functions of time.

2.2 *The Pedlosky Model*

In this subsection, I mainly compare the inviscid Pedlosky model (e.g., Eqs. 2.12a, b of [32]) and the non-dissipative Lorenz model, while Appendix A presents the viscid version of the Pedlosky model and compares it with the dissipative Lorenz model. As discussed in Appendix A, the Pedlosky model was derived from partial differential equations (PDEs) for a two-layer quasi-geostrophic system that possesses baroclinic instability [30–34, 39]. Such an instability appears in the presence of large-scale horizontal temperature (density) gradients associated with the vertical shear of horizontal winds. Although the Lorenz and Pedlosky models were derived from very different PDEs with different physical processes, Appendix A shows that two systems of ODEs can be mathematically identical when time-independent parameters are properly selected, as listed in Table 1.

Below, major features of the inviscid Pedlosky model and the 3D-NLM are presented and compared. The inviscid version with $\eta = 0$ in Eqs. (30)–(31) of Appendix A is written as:

Table 1 Mathematical similarities of the Lorenz and Pedlosky models that refer to Eqs. (25)–(27) of [20] and Eq. (6.9) of [31], respectively. The two models are documented in Eqs. (1)–(3) and Eqs. (30)–(31), respectively, in this study. [30–32], and [34] are denoted as P70, P71, P72, and PF80, respectively. The Pedlosky model was proposed for studying nonlinear baroclinic waves with finite amplitudes

Model	The Lorenz model	The Pedlosky model	References
Dynamics of the PDEs	Rayleigh-Benard convection	Nonlinear Baroclinic wave	
dissipative or viscid	$\frac{dX}{d\tau} = -\sigma X + \sigma Y$	$\frac{d^2R}{d\tau^2} + \alpha\eta\frac{dR}{d\tau} - R + R(R^2 - D) = 0$	Eq. 6.9 of P71;
	$\frac{dY}{d\tau} = -XZ + rX - Y$ $\frac{dZ}{d\tau} = XY - bZ$	$\frac{dD}{d\tau} + \eta D + \beta\eta R^2 = 0$	Eq. 2.11a, b of PF80
	$r = \frac{1+\sigma}{\sigma}$	$R^2 = \frac{X^2}{2}; D = \frac{1}{2}X^2 - \sigma Z$ $\alpha = \frac{1+\sigma}{b}; \beta = \frac{2\sigma - b}{b}; \eta = b$	
non-dissipative or inviscid	$\frac{d^2X}{d\tau^2} - (\sigma r + \frac{C_1}{C_o})X + \frac{X^3}{2} = 0$	$\frac{d^2R}{d\tau^2} - (1 + D_o)R + R^3 = 0$	Eq. 6.8 of P70
		$\frac{dD}{d\tau} = 0$	
	$\left(\frac{dX}{d\tau}\right)^2 - (\sigma r + \frac{C_1}{C_o})X^2 + \frac{X^4}{4} = 0$	$(\frac{dR}{d\tau})^2 - (1 + D_o)R^2 + \frac{R^4}{2} = E_2$	Eq. 3.7 of P72
	C_1 and C_2 are constants	D_o and E_2 constants	

$$\frac{d^2R}{d\tau^2} - (D_0 + 1)R + R^3 = 0, \tag{12}$$

$$\frac{dD}{d\tau} = 0. \tag{13}$$

Here, D_0, that denotes an initial value of D, is a constant. The two ODEs are now decoupled and the coefficient of the linear (forcing) term becomes time independent. Interestingly, Eq. (12) is mathematically identical to Eq. (6) of the 3D-NLM that produces three types of solutions. Below, without loss of generality, we may assume a non-negative C_{ic} in Eq. (6) and D_0 in Eq. (12) (e.g., any initial value for X and a zero value for Z). As compared to the viscid version, the linear term that represents the forcing term has a negative constant coefficient, yielding a negative stiffness under the inviscid condition, indicating instability. Note that "stiffness" is defined as the "coefficient" of the linear term (i.e., Λ in Eq. (32)).

Following discussions that derive Eq. (8) from Eq. (7), the energy form of Eq. (12) is written as follows:

$$(\frac{dR}{d\tau})^2 - (1 + D_o)R^2 + \frac{R^4}{2} = E_2; \tag{14}$$

here, E_2 is an integration constant. Equation (14) for the Pedlosky model is mathematically equivalent to Eq. (9) of the 3D-NLM. The above equation was documented as Eq. (3.7) in [32], who obtained solutions as the elliptic functions (dn) and (cn) for $E_2 < 0$ and $E_2 \geq 0$, respectively. The two elliptic functions are displayed in Fig. 3. Since dn is a one sign function, a 2D $R - R'$ phase portrait (i.e., $R = dn$ versus $R' = d(dn)/d\tau$) only appears in the first and fourth quadrant, referred to as solutions with small cycle oscillations. In comparison, for the solution of the cn function, a 2D $R - R'$ phase portrait displays orbits appearing in all of the four quadrants, classified as large cycle oscillations. As a result, the solutions provided in Fig. 3 are consistent with the solutions in Fig. 1. Since the limit for both the cn and dn functions is a hyperbolic secant function $(sech)$, a homoclinic orbit in the $R - R'$ space can be obtained within the Pedlosky model. Such an orbit is a separatrix that separates the small-cycle (e.g., dn) and large-cycle (e.g., cn) solutions.

2.3 The NLS, Duffing, and KdV Equations

The above discussions presented mathematical similarities between the 3D-NLM and the inviscid Pedlosky model, as well as similarities between the full 3DLM and Pedlosky models. A recent study [48] that demonstrated mathematical universalities amongst the 3D-NLM, the Duffing, the KdV, and the NLS is summarized below for a comparison.

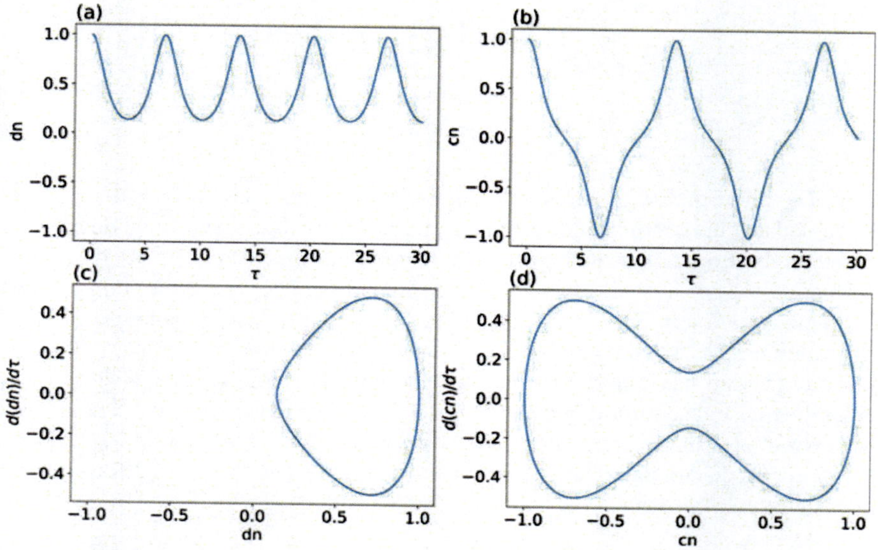

Fig. 3 Jacobi elliptic functions dn (**a, c**) and *cn* (**b, d**) representing solutions of the inviscid Pedlosky model for "negative" and positive energy, respectively. Top panels show time evolutions of *dn* and *cn*, respectively. Bottom panels show the phase portrait in the two-dimensional $V - \dot{V}$ phase space, here, V is either *dn* or *cn* and $\dot{V}$ represents the time derivative of V

Table 2 A comparison of the 3D-NLM, the invisid Pedlosky model, the Duffing, the NLS, and the KdV equations. *cn* and *sech* represent the Jacobi elliptic and hyperbolic secant functions, respectively. Please see details in the main text

3D-NLM	Other models	Solutions
The Equation for X'' $\frac{d^2X}{d\tau^2} - (\sigma r + \frac{C_1}{C_o})X + \frac{X^3}{2} = 0$	The invisid Pedlosky Model $\frac{d^2R}{d\tau^2} - (1 + D_0)R + R^3 = 0$	*cn* and *dn*
The Equation for X'' the same as the above	The Duffing Equation $\frac{d^2g}{d\tau^2} + \delta\frac{dg}{d\tau} + \alpha g + \beta g^3 = \gamma cos(\omega\tau)$ $\delta = 0, \gamma = 0, \alpha = -(\sigma r + \frac{C_1}{C_o}), \beta = 1/2$	*cn*
The Equation for $(X')^2$ $\left(\frac{dX}{d\tau}\right)^2 - \sigma r X^2 + \frac{X^4}{4} = 0$	The Nonlinear Schrodinger Equation $\left(\frac{dh}{dx}\right)^2 + \delta h^2 + \frac{\gamma}{2}h^4 = 0$ $\delta < 0, \gamma > 0$	*sech*
The Equation for Z'' $\frac{d^2Z}{d\zeta^2} + 3Z^2 - 4rZ = 0$ $\zeta = \sqrt{\sigma}\tau$	The Korteweg-de Vries Equation $\frac{d^2f}{d\xi^2} + 3f^2 - cf = 0$ $c = 4r$	$sech^2$

As listed in Table 2, the 1st order ODE (Eq. 9) and the 2nd order ODE (Eq. 6) for the X component share similarities with the following two mathematical systems, respectively:

(i) the Nonlinear Schrodinger equation:

$$\left(h'\right)^2 + \delta h^2 + \frac{\gamma}{2} h^4 = 0, \quad \delta < 0, \gamma > 0; \tag{15}$$

(ii) the Duffing equation (with no dissipation nor external forcing):

$$g'' + \alpha g + \beta g^3 = 0, \quad \alpha < 0, \beta > 0. \tag{16}$$

For the Z component of the 3D-NLM, its 2nd order ODE (e.g., Eq. 10 of the 3D-NLM with $C_{ic} = 0$) is mathematically identical to the following KdV equation in a traveling wave coordinate (e.g., [12, 60]):

(iii) the KdV equation:

$$f'' + 3f^2 - cf = 0, \quad c \in R. \tag{17}$$

In addition, [48] previously illustrated how analytical solutions of the homoclinic orbit for the Z component of the 3D-NLM (Eq. 11c) can aid in obtaining solutions of the KdV equation in the form of a hyperbolic sech squared function.

Based on the above discussions of various physical systems, which are summarized in Tables 1 and 2, Eq. (6) of the 3D-NLM may be viewed as a generic system that represents the above systems (i.e., the inviscid Pedlosky, the Duffing, the NLS, as well as the KdV equations). Below, I further illustrate similarities between Eq. (6) and a nonlinear pendulum equation near its unstable equilibrium point.

2.4 A Simplified Nonlinear Pendulum Equation

In this subsection, I present a simplified version of the nonlinear pendulum equation and compare it with the aforementioned systems in order to propose a generic system that retains common mathematical features. Then, the dynamics of the pendulum system can be applied in order to understand the major dynamics of the aforementioned systems.

The mathematical equation of a nonlinear pendulum and a stability analysis of a stable equilibrium point are provided in Appendix B. Here, I further analyze the stability of the unstable equilibrium point within the pendulum system. By introducing a new variable (Θ) with a phase lag of 180°, $\Theta = \theta - \pi$, I obtain the following equation from Eq. (36) with no damping term:

$$\frac{d^2\Theta}{d\tau^2} - sin(\Theta) = 0. \tag{18}$$

When $sin(\Theta)$ is approximated by a Taylor series expansion of $(\Theta - \Theta^3/6)$, the above equation becomes:

$$\frac{d^2\Theta}{d\tau^2} - \Theta + \frac{\Theta^3}{6} = 0. \tag{19}$$

Equation (19) shares the same mathematical form as that in Eq. (6) of the 3D-NLM, as well as in Eq. (12) of the inviscid Pedlosky model.

Following discussions that lead to Eq. (8) from Eq. (7), we have the following potential energy (PE) functions for the full system in Eq. (18) and the simplified nonlinear system with a cubic term in Eq. (19), respectively:

$$PE_{full} = \int G(\Theta)d\Theta = \int -sin(\Theta)d\Theta = cos(\Theta) + K_1, \tag{20}$$

$$PE_{cubic} = \int \left(-\Theta + \frac{\Theta^3}{6}\right) d\Theta = -\frac{1}{2}\Theta^2 + \frac{1}{24}\Theta^4 + K_2. \tag{21}$$

Similarly, the potential energy for the linear version of the equation (i.e., no cubic term) is written as follows:

$$PE_{linear} = -\frac{1}{2}\Theta^2 + K_3. \tag{22}$$

Here, K_1, K_2, and K_3 are integration constants and set to zero in order to simplify discussions.

Here, the potential energy functions are used to qualitatively reveal the relationship between the extrema of potential energy and the stability of equilibrium points. As shown in Fig. 4, all of the three PE functions display one local maximum that corresponds to an unstable equilibrium point. Within the full and nonlinear systems, their PE functions additionally include two local minima, yielding two stable equilibrium points. The appearance of stable and unstable equilibrium points is consistent with those in Figs. 1 and 3.

The simplified nonlinear system (i.e., Eq. 19) shares dynamics that are fundamentally similar to that within the full system (i.e., Eq. 18). However, as shown in Fig. 4, differences in the exact location of the stable equilibrium points yield different periods within the two systems. Despite minor differences, it can be said that the linear system (with no cubic term) produces locally unstable solutions that lead to global instability and that the (simplified) nonlinear system enables the coexistence of local instability and global stability.

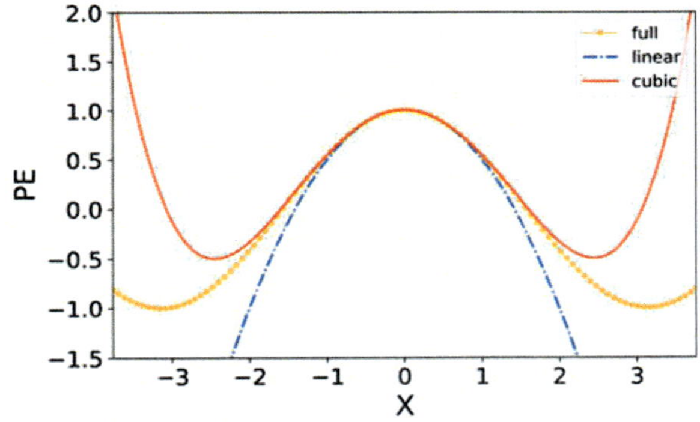

Fig. 4 Potential energy functions for the full and simplified pendulum equations near the unstable equilibrium point in Eqs. (20–22)

3 Discussions and Analysis

Based on the above discussions, a generic system is proposed representing all of the aforementioned systems, including the 3D-NLM, the inviscid Pedlosky model, the Duffing, the NLS, the KdV, and the simplified pendulum equations. The generic system is then used to reveal the role of nonlinearity and its collective role with dissipations, as well as the forcing term, in changing system stability.

3.1 A Generic System and Its Stability

Mathematical universalities amongst the aforementioned systems indicate that the Lorenz model (in particular the non-dissipative version) is representative for different physical systems. Here, by considering conservative systems, I propose the following nonlinear 2nd order ODE as a generic system:

$$\frac{d^2 V}{d\tau^2} - pV + qV^3 = 0, \tag{23}$$

here, p and q are assumed to be non-negative constants. In general, p may also depend on a systems' initial state (e.g., C_{ic} in Eq. (6) for the 3D-NLM and D_0 in Eq. (12) for the inviscid Pedlosky model). With a comparison to dissipative systems, Appendix A presents the condition under which parameter p may vary with time.

The "alternative form" of Eq. (23) in terms of total energy is written as follows:

$$\left(\frac{dV}{d\tau}\right)^2 - pV^2 + q\frac{V^4}{2} = E_V; \tag{24}$$

here, E_V is an integration constant. Equations (23) and (24) are equivalent to Eqs. (6) and (9) of the 3D-NLM, respectively. Equation (24) represents the relationship

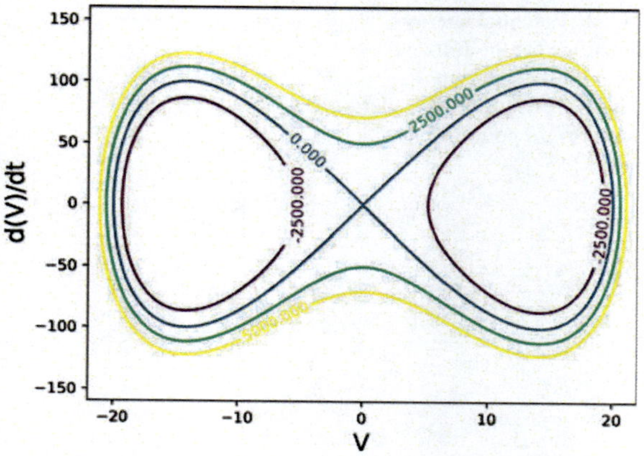

Fig. 5 Contour lines of total energy defined on the left hand side of Eq. 24. Contour lines with positive and negative values correspond to large- and small-cycle oscillations

between the state variable V and its derivative V'. Since it is common to view both of these variables as state variables in order to construct a 2D $V - V'$ phase space, a contour analysis of Eq. (24) may qualitatively illustrate major features of the solutions. As shown in Fig. 5, the left hand side of Eq. (24) produces 2D contour lines with positive and negative values that are associated with large-cycle and small-cycle oscillations, respectively. Zero contour lines (that begin at the origin and return back to the origin) are associated with homoclinic orbits that act as boundaries between two types of oscillatory solutions.

By tracing the energy source in the aforementioned systems, we can show that the linear term (pV) comes from one or more of the following:

(1) the heating term within the 3D-NLM;
(2) baroclinic instability within the inviscid Pedlosky model;
(3) initial conditions within the KdV equation; and
(4) (gravitational) potential energy (with a local maximum) within the pendulum equation.

The above are consistent with the fact that a generic system containing only the linear term produces unstable and stable solutions with exponential growth and decay, respectively (i.e., $exp\left(\sqrt{p}\tau\right)$ and $exp\left(-\sqrt{p}\tau\right)$). Below, a summary on a linear stability analysis using eigenvalues of the system's Jacobian matrix is provided in order to reveal the impact of nonlinearity and dissipation on the local stability of critical points.

3.2 Destabilization by Nonlinearity Within the Generic System

One common approach for analyzing the stability of Eq. (23) is discussed below. First, we convert the equation into a system of the first-order ODEs as follows:

$$\frac{dV}{d\tau} = W \tag{25}$$

$$\frac{dW}{d\tau} = -pV + qV^3 \tag{26}$$

Defined as steady state solutions to the system, the three critical points are $(V, W) = (0, 0)$ and $\left(\pm\sqrt{p/q}, 0\right)$. When the nonlinear cubic term (qV^3) is neglected, the reduced system is linear and only contains the trivial critical point. Secondly, we compute the so-called Jacobian matrix and determine its eigenvalues at each of the critical points. Finally, by analyzing the eigenvalues, we classify all of the critical points as (spiral) sources, (spiral) sinks, or saddle points. Such a procedure is referred to as a linear stability analysis. For the system in Eqs. (25)–(26), the trivial critical point can be shown to be a saddle with positive and negative eigenvalues and the two non-trivial critical points can be shown to be centers with pure imaginary eigenvalues.

Although an unstable solution may begin near the saddle point as a result of the instability associated with the linear term, the growth of the unstable solution may be constrained by stable centers introduced by inclusion of the nonlinear term. In other words, the nonlinear term produces new stable critical points in order to limit the growth of solutions. As a result, an orbit of Eq. (23) may be locally unstable (near the saddle point) but still remains globally stable. Features of local instability and global stability are supported by the appearance of three types of solutions in Figs. 1 and 2, and the potential energy extrema in Fig. 4.

As a brief summary, the above analysis suggests that the linear term acts as a forcing term to produce (local) instability, and the nonlinear term introduces new stable critical points to limit the growth of the solution, leading to global stability.

3.3 The Role of Dissipations: A Comparison of the Generic System with the Lorenz and Pedlosky Models

Here, applying the generic system as a baseline, I discuss how to examine the impact of additional terms and/or components on the stability of the new system. The generic system in Eq. (23) represents the same dynamics in the 3D-NLM, the inviscid Pedlosky model, the KdV equation, etc. While two regular, oscillatory solutions with small and large cycles dominate the system, homoclinic orbits that appear as a separatrix for the two types of oscillatory solutions possess an interesting feature of sensitive dependence on initial conditions (SDIC), as shown in Fig. 4 of [48]. However, such a feature may be viewed as limited chaos in contrast to the full chaos within dissipative systems, as discussed by [24]. Below, the impact of dissipations on chaotic responses (i.e., full chaos) are examined by comparing the 3D-NLM (i.e., the generic system) and the 3DLM.

The 3DLM in Eqs. (1–3) includes three different dissipative terms, (i.e., $-\sigma X$, $-\epsilon Y$, and $-bZ$). Interestingly, as discussed in Appendix B, the viscid Pedlosky model in Eqs. (A3)-(A4) also contains "three" dissipative terms that are linear functions of the scaled dissipative coefficient η. Below, by comparing the control case with no dissipative term, I discuss the following three parallel cases with $\epsilon = 1$:

(1) only one dissipative term $-\epsilon Y$ (i.e., $-Y$) retained;
(2) three dissipative terms retained with $b = 2\sigma$; and
(3) three dissipative terms that satisfy the inequality $\sigma > b + 1$.

The first case with a single positive dissipative term produces three critical points that remain in the same locations as those within the non-dissipative system (i.e., 3D-NLM). Under this condition, the quantity $\left(\frac{1}{2}X^2 - \sigma Z\right)$ is still conservative (i.e., Eq. (5)). (Note that this quantity indeed represents the sum of the domain averaged kinetic and potential energy (which is different from available potential energy) [41, 42].) While the trivial critical point is still a saddle point, two non-trivial critical points remain stable but become spiral sinks. As a result, additional dissipation further stabilizes the system. Specifically, (locally) unstable solutions may begin near the saddle point but eventually approach the spiral sinks, becoming steady-state point attractors. Since the appearance of non-trivial critical points depends on the inclusion of nonlinearity (as well as the linear forcing term), the stabilization associated with spiral sinks is indeed caused by the collective impact of the dissipative and nonlinear terms (as well as the linear forcing term).

For the second case that includes three dissipative terms with a special choice of $b = 2\sigma$, we obtain the following equation by simplifying $X\times$ Eq. (1) $- \sigma\times$ Eq. (3):

$$\frac{d}{dt}\left(\frac{1}{2}X^2 - \sigma Z\right) = -2\sigma\left(\frac{1}{2}X^2 - \sigma Z\right). \tag{27}$$

The above suggests that the quantity $\left(\frac{1}{2}X^2 - \sigma Z\right)$ exponentially decays with time. As a result, the 3DLM with $b = 2\sigma$ produces stable solutions, consistent with the analysis in Appendix A.

Finally, for the third case with $\sigma > b + 1$ that presents the condition for the appearance of the Hopf bifurcation, the system produces a pair of complex eigenvalues with a positive real part, leading to unstable spiral points when $r > 24.74$ with the choice of $\sigma = 10$ and $b = 8/3$. Under such conditions, all of the critical points within the 3DLM are indeed unstable, as listed for a comparison with the control case and the first case in Table 3. As a result, chaotic orbits appear in the presence of the instability associated with unstable critical points. Interestingly, given the fact that individual orbits may grow with time, boundedness of the solutions, as well as the structural stability of the system (i.e., a global property), still exists at the chaotic

Table 3 The stability of critical points within the 3D-NLM (i.e., the generic system) and 3DLM. Locations of the non-trivial critical points (CP) are the same for the control and first parallel cases (i.e., in the second and third columns), but are different for the third parallel case (i.e., in the fourth column)

Critical points	3D-NLM	3D-NLM & $\epsilon \neq 0$	3DLM
trivial CP	saddle	saddle	saddle
non-trivial CPs	center	(spiral) sinks	unstable spiral points

regime. Note that the location of non-trivial critical points is changed when all three of the dissipative terms are added into the 3D-NLM. The Y component at each of the non-trivial critical points is no longer zero within the 3DLM, while it is zero within the 3D-NLM.

The above discussions lead to the following key point: the inclusion of multiple dissipative terms can destabilize a system. As a result, I hypothesize that after multiple dissipative terms or a new stable modeling component is added into a system, the new system is not necessarily more stable than the original system.

4 Concluding Remarks

Based on recent studies that documented promising 30-day simulations using a real-world global model and two kinds of attractor coexistence using classical and generalized Lorenz models, a revised view on the dual nature of chaos and order in weather and climate has been proposed [46, 52, 53]. To provide additional support to the revised view, I presented mathematical universalities amongst the dissipative and non-dissipative Lorenz, the Pedlosky, and other models. Various types of non-chaotic solutions were reviewed and analyzed using different methods in order to reveal duality within the theoretical models. Additionally, based on common dynamics amongst the selected systems, a generic conservative system with a nonlinear 2nd-order ODE was proposed in order to reveal the role of nonlinearity and additional dissipative terms in stabilizing or destabilizing systems. Major findings are summarized below.

I first presented mathematical universalities between the Lorenz and Pedlosky models that were derived from different PDEs for small-scale convection and large-scale, quasi-geostrophic baroclinic waves, respectively. As a result, findings discovered in one model (e.g., the Lorenz model with attractor coexistence) are applicable to the other model (e.g., the Pedlosky model).

The revised view challenges the validity of the statement "weather is chaotic" and suggests a new research direction for detecting both regular and chaotic solutions in model simulations and real-world observations. This study reviewed non-chaotic solutions of the non-dissipative Lorenz model and the inviscid Pedlosky model and compared them to the solutions of the Duffing, the NLS, and the KdV equations. The non-dissipative or inviscid model represents the full version of the model with sufficiently strong forcing (i.e., a large heating parameter for the Lorenz model, e.g., [59]). As summarized in Tables 1 and 2, mathematical universalities include the same ODE amongst different systems, the same solution between the X component of the 3D-NLM and the scaled amplitude of the inviscid Pedlosky model, and the same solution between the Z component of the 3D-NLM and the KdV equation.

Non-chaotic solutions and their major features were also analyzed and represented using various methods, including numerical integrations of the model, a calculation of the Jacobi elliptic functions, an analysis of the potential energy functions, a contour line analysis of total energy, and a linear stability analysis near each of the critical points. Derived from the above analysis, major features of solutions within the selected systems include:

(i) three types of solutions, including two families of oscillatory solutions with small and large cycles and homoclinic orbits (e.g., in Figs. 1 and 2);
(ii) large-cycle oscillations associated with elliptic *cn* functions and small-cycle oscillations associated with elliptic *dn* functions that are of one sign (e.g., in Fig. 3);
(iii) an unstable saddle point appearing in association with the maximum of the potential energy and two stable points associated with the minimum of potential energy (e.g., in Fig. 4);
(iv) homoclinic orbits obtained as the limit of the elliptic *cn* and *dn* functions, appearing as a separatrix that separates small- and large-cycle oscillations;
(v) the X component of homoclinic solutions within the 3D-NLM and the amplitude of the solitary wave envelope within the NLS that share the same mathematical form of the hyperbolic secant ($sech$) function; and the Z component of homoclinic solutions within the 3D-NLM and the solitary solution of the KdV that have the same mathematical form of the hyperbolic secant squared ($sech^2$) function.

The aforementioned common features amongst the selected systems lead to a generic conservative system that possesses two important physical processes: a linear forcing and nonlinearity (i.e., Eq. 23). The generic system was compared to a simplified nonlinear pendulum equation in order to reveal the role of nonlinearity in stabilizing the system and was then compared to the dissipative Lorenz model (and the viscid Pedlosky model as well) in order to examine the impact of one or more dissipative processes on system stability. A linear stability analysis of Eq. (23) within the generic system indicated that the linear term ($-pV$) is associated with the presence of a saddle point at the origin and that the inclusion of the nonlinear cubic term (qV^3) introduces two, non-trivial, stable critical points that limit the growth of unstable solutions and, thus, leads to global stability. However, as illustrated using a "different" system in Appendix B, inclusion of the nonlinear term may produce global unphysical instability. In comparison, the impact of dissipation was examined using the generic system (i.e., 3D-NLM) and the 3DLM that only contains one or three dissipative terms. For the control case (with no dissipative terms) and the three parallel cases presented in Sect. 3.3, the first two parallel cases revealed a stabilization role by additional dissipative terms, while the third parallel case with three dissipative terms that satisfy the inequality $\sigma > b + 1$ displayed a destabilization role in producing non-trivial unstable points and, thus, chaotic responses.

In summary, (i) the inclusion of nonlinearity may stabilize or destabilize the system by introducing additional stable or unstable critical points, respectively; and (ii) when multiple dissipative terms are added into the system, their collective impact may stabilize or destabilize the new system. An implication of the findings is that when additional nonlinearity and/or dissipative components are added into a real world modeling system, the new system may produce regular simulations at short time scales but yield irregular or even chaotic simulations associated with the presence of new unstable critical points at long time scales. As a result, developing a systematic approach for examining the collective impact of multiple dissipative terms and nonlinearity on both local and global stability is desired, and will be the subject of a future study.

Acknowledgements I thank anonymous reviewers, M. Dunster, M. Ghil, B. Hunt, and J. Pedlosky for valuable comments and discussions, and Ms. S. Faghih-Naini for her help in verifying derivations.

Appendix A: The Pedlosky Model and a Two Layer Quasi-geostrophic System

Here, I provide a brief introduction to the Pedlosky model that is based on the following two-layer quasi-geostrophic (QG) system for studying weakly nonlinear baroclinic wave-mean interactions [31–33]:

$$\left[\frac{\partial}{\partial t} + \frac{\partial \Psi_1}{\partial x}\frac{\partial}{\partial y} - \frac{\partial \Psi_1}{\partial y}\frac{\partial}{\partial x}\right]\left(\nabla^2 \Psi_1 - F(\Psi_1 - \Psi_2)\right) = \gamma \nabla^2 \Psi_1 \tag{28}$$

$$\left[\frac{\partial}{\partial t} + \frac{\partial \Psi_2}{\partial x}\frac{\partial}{\partial y} - \frac{\partial \Psi_2}{\partial y}\frac{\partial}{\partial x}\right]\left(\nabla^2 \Psi_2 - F(\Psi_2 - \Psi_1)\right) = \gamma \nabla^2 \Psi_2 \tag{29}$$

The above system is applied to a three-dimensional domain that consists of a channel in the two-dimensional $x - y$ plane and two layers with an initially equal depth in the vertical direction. The 2D channel is periodic in the x direction and has a finite width of L in the y direction. Two vertical layers are denoted by different subscripts (i.e., 1 and 2). A constant but different density is applied in each of the layers. The difference of densities (i.e., the vertical gradient of density) leads to a baroclinicity that may act as an energy source for unstable solutions. Ψ_1 and Ψ_2 represent streamfunctions that define velocities in the x and y directions for layers 1 and 2, respectively. F is a rotational Froude number that is a function of the Coriolis parameter, (reduced) gravitational acceleration, the length of the domain, and the depth of the layer, all of which are constants. In this study, an assumption of a constant Coriolis parameter is made in order to simplify discussions. The two terms that involves γ and appear on the right hand side represent dissipative terms.

In a series of studies by Prof. Pedlosky, the following Pedlosky model was derived (e.g., Eqs. 2.12a, b of [32]):

$$\frac{d^2R}{d\tau^2} + \alpha\eta\frac{dR}{d\tau} - (D+1)R + R^3 = 0, \tag{30}$$

$$\frac{dD}{d\tau} + \eta D + \beta\eta R^2 = 0. \tag{31}$$

The above system includes one second-order ODE and one first-order ODE for two state variables are R and D. R represents the scaled amplitude of the streamfunction (or wave solution). While the mathematical definition of the state variable D was given in Eq. 2.13 of [32], its physical impact on the system's solution is provided below in order to facilitate discussions. Three time-independent parameters include α and β, which are functions of horizontal wavenumbers of the solutions, and η that is a rescaled dissipative coefficient. As a result, three terms that involve $\alpha\eta$, η, and $\beta\eta$ represent the dependence of dissipation on scales.

An Alternative Form of the Lorenz Model by Marzec and Spiegel

Marzec and Spiegel [25] transformed the Lorenz model into the following system with 2nd-order and 1st-order ODEs [6]:

$$\frac{d^2X}{d\tau^2} + (\sigma+1)\frac{dX}{d\tau} + \Lambda X + \frac{1}{2}X^3 = 0, \tag{32}$$

$$\frac{d\Lambda}{d\tau} = -b\Lambda - b\sigma(r-1) + (\sigma - \frac{1}{2}b)X^2, \tag{33}$$

where $\Lambda = \sigma(Z - r + 1) - \frac{1}{2}X^2$. By comparing this system to Eqs. (30)–(33), we define $D = \frac{1}{2}X^2 - \sigma Z$ and $R = X/\sqrt{2}$, leading to $\Lambda = -(D + \sigma r - \sigma)$. Furthermore, Eqs. (32) and (33) become:

$$\frac{d^2R}{d\tau^2} + (\sigma+1)\frac{dR}{d\tau} - (D + \sigma r - \sigma)R + R^3 = 0, \tag{34}$$

$$\frac{dD}{d\tau} + bD + (2\sigma - b)R^2 = 0. \tag{35}$$

Under a proper choice of parameters (i.e., $b = \eta$, $(2\sigma - b) = \beta\eta$, $(\sigma + 1) = \alpha\eta$, and $(\sigma r - \sigma) = 1$), also listed in Table 1, the Pedlosky model with Eqs. (30) and (31) is mathematically identical to the Lorenz model with Eqs. (34) and (35). On the other hand, the model requires different intervals of parameters for the onset of chaos (e.g., a smaller σ). Additionally, a special case with $b = 2\sigma$ decouples Eq. (35) from Eq. (34) and yields an exponential decaying solution of $D = D_0 exp(-b\tau)$, here D_0 is an initial condition. This special case is further discussed in Sect. 3.3.

Following the interpretation of [25], the solution of Eq. (30) may be viewed as oscillations of a damped spring whose time varying stiffness $-(D + 1)$ is governed by Eq. (31). Here, the time evolving stiffness, appearing as the coefficient of the linear term in Eq. (30), may be positive or negative, indicating the competing impact of forcing, dissipation, and nonlinearity, as discussed in the main text.

Appendix B: A Nonlinear Pendulum Equation

Here, I provide a brief introduction to the nonlinear pendulum equation and a short note on the stability of a stable equilibrium point in the system. Figure 6 displays a pendulum consisting of a weightless rod of length L with an attached bob with a mass of m. The other end of the rod is supported at a point on a wall. The point of support is denoted as the origin. As a result, the mass of the pendulum is free to oscillate or rotate. The time varying position of the mass is determined by the angle, θ, between the rod and the downward vertical direction. As shown in Fig. 6, such an angle is measured in the counterclockwise direction. As a result, the natural position is $\theta = 0$ and the inverted position where the pendulum bob is vertical, with the weight in the up position, is $\theta = \pi$. The two positions are referred to as the lower and upper equilibrium point, respectively, and classified as either stable or unstable equilibrium points, respectively.

The time varying angle (θ) of the nonlinear pendulum with a dissipative term is governed by the following equation:

$$\frac{d^2\theta}{d\tau^2} + k\frac{d\theta}{d\tau} + sin(\theta) = 0; \tag{36}$$

here, k represents the dissipative coefficient. For the non-dissipative system (i.e., $k = 0$), the above system has a stable and unstable point at $\theta = 2n\pi$ and $\theta = (2n + 1)\pi$, respectively. Here n is an integer.

While the stability of the unstable critical point at $\theta = \pi$ is analyzed in the main text, the stability of the stable critical point at $\theta = 0$ is discussed below. Using a Taylor series expansion with $sin(\theta) = \theta$ or $sin(\theta) = \theta - \theta^3/6$, Eq. (36) can be approximated by:

$$\frac{d^2\theta}{d\tau^2} + \theta = 0, \tag{37}$$

or

$$\frac{d^2\theta}{d\tau^2} + \theta - \frac{\theta^3}{6} = 0. \tag{38}$$

Within Eq. (37) that contains a linear term, only one critical point at $(\theta, d\theta/d\tau) = (0, 0)$, which is a center, is found. The critical point is stable and is associated with a simple harmonic oscillation. In comparison, within Eq. (38) that includes a cubic nonlinear term, two additional critical points appear at $(\theta, d\theta/d\tau) = (\pm\sqrt{6}, 0)$. These non-trivial critical points are saddle points that are unstable. As a result, the simplified nonlinear system that has a limited degree of nonlinearity may be locally stable but globally unstable when initial kinetic energy is sufficiently strong. The features are illustrated in Fig. 7. Global instability in the simplified nonlinear system (Eq. (38)) is not consistent with that in the full pendulum system in Eq. (36), since

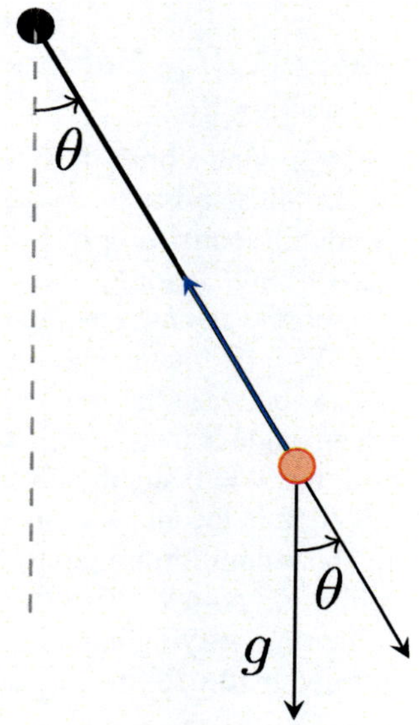

Fig. 6 A pendulum consisting of a weightless rod of length L and a bob with a mass of m. The bob and the point of support are marked with a red and black dot, respectively. The parameter "g" denotes the gradational force. The angle θ is measured in the counterclockwise direction. Stable and unstable equilibrium points appear at $\theta = 0$ and $\theta = \pi$, respectively

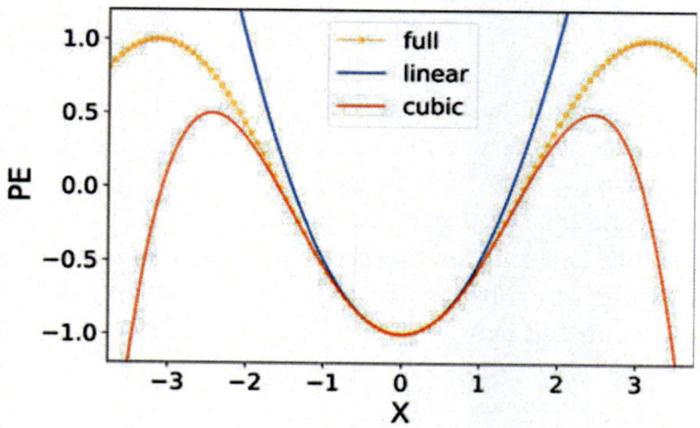

Fig. 7 Potential energy functions for the full and simplified pendulum equations near the stable equilibrium point in Eqs. (37)–(38)

the successive appearance of stable and unstable critical points within the full system (i.e., at $\theta = 2n\pi$ and $\theta = (2n + 1)\theta$) can still limit growth of the unstable solution, yielding global stability. However, such an issue with global instability may appear within a complicated system (e.g., climate models).

References

1. R. Anthes, Turning the tables on chaos: is the atmosphere more predictable than we assume? UCAR Mag. Spring/Summer (2011). https://news.ucar.edu/4505/turning-tables-chaos-atmosphere-more-predictable-we-assume. Accessed 23 Nov 2020
2. N.J. Balmforth, Solitary waves and homoclinic orbits. Annu. Rev. Fluid Mech. **27**(1), 335–373 (1995)
3. J. Boussinesq, Essai sur la theorie des eaux courantes (http://gallica.bnf.fr/ark:/12148/bpt6k56673076). Memoires presentes par divers savants. Acad. des. Sci. Inst. Nat. France **XXIII**, 1–680 (1877)
4. J.P. Boyd, Dynamical methodology: solitary waves. Encycl. Atmos. Sci. (Second Ed.) **2015**, 417–422 (2015)
5. J.H. Curry, J.R. Herring, J. Loncaric, S.A. Orszag, Order and disorder in two- and three-dimensional Benard convection. J. Fluid Mech. **147**, 1–38 (1984)
6. P.G. Drazin, *Nonlinear Systems* (Cambridge, 1992), 333 pp.
7. S. Faghih-Naini, B.-W. Shen, Quasi-periodic in the five-dimensional non-dissipative Lorenz model: the role of the extended nonlinear feedback loop. Int. J. Bifurc. Chaos **28**(6), 1850072 (20 pages) (2018). https://doi.org/10.1142/S0218127418500724
8. D. Fultz, R.R. Long, G.V. Owens, W. Bohan, R. Kaylor, J. Weil, Studies of thermal convection in a rotating cylinder with some implications for large-scale atmospheric motion, in *Studies of Thermal Convection in a Rotating Cylinder with Some Implications for Large-Scale Atmospheric Motions*. Meteorological Monographs, vol. 4 (American Meteorological Society, Boston, MA, 1959). https://doi.org/10.1007/978-1-940033-37-2_1
9. M. Ghil, S. Childress, *Topics in Geophysical Fluid Dynamics: Atmospheric Dynamics, Dynamo Theory, and Climate Dynamics* (Springer, New York, 1987), p. 485
10. J. Gleick, *Chaos: Making a New Science* (Penguin, New York, 1987), p. 360
11. J. Guckenheimer, P. Holmes, *Nonlinear Oscillations, Dynamical Systems, and Bifurcations of Vector Fields* (Springer, New York, 1983), p. 459
12. R. Haberman, *Applied Partial Differential Equations, with Fourier Series and Boundary Valule Problems*, 5th edn. (Pearson Education, Inc. 2013), p. 756
13. K.B. Hermiz, P.N. Guzdar, J.M. Finn, Improved low-order model for shear flow driven by Rayleigh-Benard convection. Phys. Rev. E **51**, 325–331 (1995)
14. R. Hide, Some experiments on thermal convection in a rotating liquid. Quart. J. Roy. Meteorol. Soc. **79**, 161 (1953)
15. M. Hirsch, S. Smale, R.L. Devaney, *Differential Equations, Dynamical Systems, and an Introduction to Chaos*, 3rd edn. (Academic Press, 2013), p. 432
16. L.N. Howard, R.K. Krishnamurti, Large-scale flow in turbulent convection: a mathematical model. J. Fluid Mech. **170**, 385–410 (1986)
17. D.W. Jordan, S. Smith, *Nonlinear Ordinary Differential Equations: An Introduction for Scientists and Engineers*, 4th edn. (Oxford University Press, New York, 2007), p. 560
18. D.J. Korteweg, G. de Vries, On the change of long waves advancing in a rectangular canal, and on a new type of long stationary waves. Phil. Mag. (Series 5) **39**, 422–433 (1895)
19. J. Lighthill, *Waves in Fluids*. (Cambridge University Press, 1978), p. 504
20. E.N. Lorenz, Deterministic nonperiodic flow. J. Atmos. Sci. **20**, 130–141 (1963a)
21. E.N. Lorenz, The mechanics of vacillation. J. Atmos. Sci. **20**, 448–464 (1963b)

22. E.N. Lorenz, The predictability of a flow which possesses many scales of motion. Tellus **21**(289), 307 (1969)
23. E.N. Lorenz, Predictability: does the flap of a butterflys wings in Brazil set off a tornado in Texas?, in *Proceedings of the 139th Meeting of AAAS Section on Environmental Sciences, New Approaches to Global Weather: GARP*, (Cambridge, MA, AAAS, 1972), p. 5
24. E.N. Lorenz, *The Essence of Chaos* (University of Washington Press, Seattle, 1993), p. 227
25. C.J. Marzec, E.A. Spiegel, Ordinary Differential Equations with Strange Attractors. SIAM J. Appl. Math. **38**, 387 (1980). https://doi.org/10.1137/0138034
26. J.W. Miles, The Korteweg-de Vries equation: a historical essay. J. Fluid Mech. **106**, 131–147 (1981). https://doi.org/10.1017/S0022112081001559
27. S. Moon, B.-S. Han, J. Park, J.M. Seo, J.-J. Baik, Periodicity and chaos of high-order Lorenz systems. Int. J. Bifurc. Chaos **27**(11), 1750176 (11 pages) (2017). https://doi.org/10.1142/S0218127417501760
28. Z.E. Musielak, D.E. Musielak, K.S. Kennamer, The onset of chaos in nonlinear dynamical systems determined with a new fractal technique. Fractals **13**, 19–31 (2005)
29. Y. Nambu, Generalized Hamiltonian dynamics. Phys. Rev. D **7**, 2403 (1973). https://doi.org/10.1103/PhysRevD.7.2405
30. J. Pedlosky, Finite amplitude baroclinic waves. J. Atmos. Sci. **27**, 15–30 (1970)
31. J. Pedlosky, Finite-amplitude baroclinic waves with small dissipation. J. Atmos. Sci. **28**, 587–597 (1971)
32. J. Pedlosky, Limit cycles and unstable baroclinic waves. J. Atmos. Sci. **29**, 53–63 (1972)
33. J. Pedlosky, *Geophysical Fluid Dynamics*, 2nd edn. (Springer, NY, 1987), p. 710
34. J. Pedlosky, C. Frenzen, Chaotic and periodic behavior of finite-amplitude baroclinic waves. J. Atmos. Sci. **37**, 1177–1196 (1980)
35. T. Reyes, B.-W. Shen, A recurrence analysis of chaotic and non-chaotic solutions within a generalized nine-dimensional Lorenz model. Chaos, Soliton. Fractals. **125**(2019), 1–12 (2019). https://doi.org/10.1016/j.chaos.2019.05.003
36. J.W. Rottman, R. Grimshaw, Atmospheric internal solitary waves, in *Environmental Stratified Flows*, ed. by R. Grimshaw. Environmental Fluid Mechanics, chap. 3, (Springer, New York, 2002), pp. 6188
37. D. Roy, Z.E. Musielak, Generalized Lorenz models and their routes to chaos. I. Energy-conserving vertical mode truncations. Chaos Soliton. Fract. **32**, 1038–1052 (2007)
38. J.S. Russell report on waves, in *Report of the 14th Meeting of the British Association for the Advancement of Science* (1844), p. 311390
39. R.M. Samelson, Periodic orbits and disturbance growth for Baroclinic waves. J. Atmos. Sci. **58**, 436–450 (2001)
40. B. Saltzman, Finite amplitude free convection as an initial value problem. J. Atmos. Sci. **19**, 329–341 (1962)
41. B.-W. Shen, Nonlinear feedback in a five-dimensional Lorenz model. J. Atmos. Sci. **71**, 1701–1723 (2014). https://doi.org/10.1175/JAS-D-13-0223.1
42. B.-W. Shen, Nonlinear feedback in a six-dimensional Lorenz model: impact of an additional heating term. Nonlin. Process. Geophys. **22**, 749–764 (2015). https://doi.org/10.5194/npg-22-749-2015
43. B.-W. Shen, Hierarchical scale dependence associated with the extension of the nonlinear feedback loop in a seven-dimensional Lorenz model. Nonlin. Process. Geophys. **23**, 189–203 (2016). https://doi.org/10.5194/npg-23-189-2016
44. B.-W. Shen, On an extension of the nonlinear feedback loop in a nine-dimensional Lorenz model. Chaotic Model. Simul. (CMSIM) **2**(147157), 2017 (2017)
45. B.-W. Shen, On periodic solutions in the non-dissipative Lorenz model: the role of the nonlinear feedback loop. Tellus A **70**, 1471912 (2018). https://doi.org/10.1080/16000870.2018.1471912
46. B.-W. Shen, Aggregated negative feedback in a generalized Lorenz model. Int. J. Bifurc. Chaos **29**(3), 1950037 (20 pages) (2019). https://doi.org/10.1142/S0218127419500378
47. B.-W. Shen, On the predictability of 30-day global mesoscale simulations of multiple African easterly waves during summer 2006: a view with a generalized Lorenz model. Geosciences **9**(7), 281 (2019). https://doi.org/10.3390/geosciences9070281

48. B.-W.* Shen, Homoclinic orbits and solitary waves within the non-dissipative Lorenz model and KdV equation. Int. J. Bifurc. Chaos (2020). https://doi.org/10.1142/S0218127420502570
49. B.-W.* Shen, S. Cheung, J.-L.F. Li, Y.-L. Wu, S.S. Shen, Multiscale processes of Hurricane Sandy (2012) as revealed by the parallel ensemble empirical mode decomposition and advanced visualization technology. Adv. Data Sci. Adapt. Anal. **08**, 1650005 (22 pages) (2016). https://doi.org/10.1142/S2424922X16500054
50. B.-W. Shen, S. Cheung, Y. Wu, F. Li, D. Kao, Parallel Implementation of the Ensemble Empirical Mode Decomposition (PEEMD) and its application for earth science data analysis. Comput. Sci. Eng. **19**(5), 49–57 (2017). https://doi.org/10.1109/MCSE.2017.3421555
51. B.-W. Shen, T. Reyes, S. Faghih-Naini, Coexistence of chaotic and non-chaotic orbits in a new nine-dimensional Lorenz model, in *11th Chaotic Modeling and Simulation International Conference, CHAOS 2018*, (Springer Proceedings in Complexity, Springer, Cham) ed. by C. Skiadas, I. Lubashevsky (2019). https://doi.org/10.1007/978-3-030-15297-0_22
52. B.-W. Shen, R.A. Pielke Sr., X. Zeng, J.-J. Baik, S. Faghih-Naini, J. Cui, R. Atlas, Is weather chaotic? coexistence of chaos and order within a generalized Lorenz model. Bull. Am. Meteorol. Soc. **102**(1), E148–58 (2021). https://doi.org/10.1175/BAMS-D-19-0165.1
53. B.-W. Shen, R.A. Pielke Sr., X. Zeng, J.-J. Baik, S. Faghih-Naini, J. Cui, R. Atlas, T.A. Reyes, Is weather chaotic? coexisting chaotic and non-chaotic attractors within Lorenz models, in *The 13th Chaos International Conference (CHAOS2020)* (2020). https://doi.org/10.1007/978-3-030-70795-8_57 (in press, this volume)
54. B.-W. Shen, R. Atlas, O. Reale, S.-J. Lin, J.-D. Chern, J. Chang, C. Henze, J.-L. Li, Hurricane forecasts with a global mesoscale-resolving model: preliminary results with hurricane Katrina (2005). Geophys. Res. Lett. **33**, (2006). https://doi.org/10.1029/2006GL026143
55. B.-W. Shen, W.-K. Tao, W. Lau, R. Atlas, Predicting tropical cyclogenesis with a global mesoscale model: hierarchical multiscale interactions during the formation of tropical cyclone Nargis (2008). J. Geophys. Res. **115**, D14102 (2010). https://doi.org/10.1029/2009JD013140
56. B.-W. Shen, W.-K. Tao, M.-L. Wu, African easterly waves in 30-day high-resolution global simulations: a case study during the 2006 NAMMA period. Geophys. Res. Lett. **37**, L18803 (2010). https://doi.org/10.1029/2010GL044355
57. R.K. Smith, A note on a theory of vacillating Baroclinic waves. J. Atmos. Sci. **32**, 2027 (1975)
58. R.K. Smith, J.M. Reilly, On a theory of amplitude vacillation in baroclinic waves: some numerical solutions. J. Atmos. Sci. **34**, 1256–1260 (1977)
59. C. Sparrow, The Lorenz equations: bifurcations, chaos, and strange attractors, in *Applied Mathematical Sciences* (Springer, New York, 1982), p. 269
60. J.C. Sprott, *Elegant Chaos* (World Scientific, Singapore, 2010), p. 285
61. S.H. Strogatz, *Nonlinear Dynamics and Chaos: With Applications to Physics, Biology, Chemistry, and Engineering* (Westpress view, Boulder, CO, 2015), p. 513
62. J.M.T. Thompson, H.B. Stewart, *Nonlinear Dynamics and Chaos*, 2nd ed. (John Wiley and Sons, 2002), 437 pp.
63. G.B. Whitham, *Linear and Nonlinear Waves* (John Wiley & Sons Inc., New York, 1974), p. 636
64. Y.-L. Wu, B.-W. Shen, An evaluation of the parallel ensemble empirical mode decomposition method in revealing the role of downscaling processes associated with African easterly waves in tropical cyclone genesis. J. Atmos. Oceanic Technol. **33**, 1611–1628 (2016). https://doi.org/10.1175/JTECH-D-15-0257.1
65. J. Yorke, E. Yorke, Metastable chaos: the transition to sustained chaotic behavior in the Lorenz model. J. Stat. Phys. **21**, (1979)
66. N.J. Zabusky, M.A. Porter, Soliton. Scholarpedia **5**(8), 2068 (2010). https://doi.org/10.4249/scholarpedia.2068
67. N.J. Zabusky, M.D. Kruskal, Interaction of solitons in a collisionless plasma and the recurrence of initial states. Phys. Rev. Let. **15**, 240–243 (1965)

Overview of Scenarios of Transition to Chaos in Nonideal Dynamic Systems

Aleksandr Shvets

Abstract A number of deterministic dynamic systems that are nonideal according to the Sommerfeld-Kononenko classification are considered. In particular, pendulum, hydrodynamic, and electroelastic systems with limited excitation are considered. The scenarios of transitions to chaos that are possible in the above systems are analyzed. We study both the transitions "regular attractor - chaotic attractor" and the transitions "chaotic attractor of one type - chaotic attractor of another type". In particular, the "chaos - hyperhaos" and "hyperhaos - hyperhaos" transitions are studied. Ten scenarios of transition to chaos are analyzed in detail. Some of the scenarios were widely known, while others are very unusual and are revealed only in nonideal dynamic systems.

Keywords Nonideal dynamic system · Scenario of transition to chaos · Chaotic attractor

1 Introduction

A prominent place among dynamical systems is occupied by so-called nonideal systems or systems with limited excitation. For the first time such systems originated in the experiments of A. Sommerfeld in the early twentieth century [1, 2]. But as a established scientific direction, the theory of systems with limited excitation was formed after the publication Kononenko [3] in which he introduced a clear axiomatics and constructed mathematical models for a wide range of problems. The theory of systems with limited excitation explores the interaction of vibrational systems with excitation sources of their oscillations. Within the framework of this theory, it is assumed that oscillation excitation sources have a power comparable to the power consumed by the oscillatory load. In this case, the operation of the energy

A. Shvets (✉)
National Technical University of Ukraine "Igor Sikorsky Kyiv Polytechnic Institute", Kyiv, Ukraine
e-mail: alex.shvets@bigmir.net

C. H. Skiadas et al. (eds.), *13th Chaotic Modeling and Simulation International Conference*, Springer Proceedings in Complexity,
https://doi.org/10.1007/978-3-030-70795-8_59

source depends on the regime of oscillation load and influence of the source cannot be expressed as a predetermined explicit time function. Whereas in the traditional mathematical modeling of the oscillatory system, idealized sources of excitation of unlimited power are considered. In many cases, the "ideal" approach is fundamentally wrong, which in practice leads to gross errors in describing the dynamics of both the oscillatory system and the source of excitation [4–7]. The use of limited excitation models becomes even more relevant in our time, when humanity faces the problems of global energy conservation, which requires the maximum minimization of the power of the applied excitation sources.

The discovery of deterministic chaos stimulated the emergence of a new direction in theory of systems with limited excitation associated with the search for chaotic modes of interaction of oscillatory systems with sources of excitation. Of particular interest are those chaotic regimes whose appearance is associated with a nonlinear interaction between the oscillatory system and the excitation source, and not with their autonomous properties.

In the papers [8–11] describe the occurrence of chaotic attractors in a number of deterministic nonideal dynamical systems whose chaotization is fundamentally impossible when considering cases of ideal (unlimited) excitation.

2 Mathematical Models of Considered Nonideal Systems

In studying the occurrence of deterministic chaos in dynamical systems, it is of great interest to identify and describe scenarios of transition to chaos. Moreover, both scenarios of transitions from regular attractors to chaotic, and scenarios of transitions between chaotic attractors of different types. Some of these scenarios are widespread and implemented in many dynamic systems. Such scenarios include, for example, the Feigenbaum's scenario and the Manneville-Pomeau scenario. Other scenarios were described relatively recently and the question of their prevalence requires further study.

In this paper, we analyze the scenarios of transition to chaos identified and described in a number of nonideal dynamical systems. The implementation of transitions to chaos considered in such systems: a pendulum—an excitation source,

$$\begin{aligned}
\frac{dy_1}{d\tau} &= Cy_1 - y_2y_3 - \frac{1}{8}(y_1^2y_2 + y_2^3),\\
\frac{dy_2}{d\tau} &= Cy_2 + y_1y_3 + \frac{1}{8}(y_1^3 + y_1y_2^2) + 1,\\
\frac{dy_3}{d\tau} &= Dy_2 + Ey_3 + F,
\end{aligned} \tag{1}$$

a spherical pendulum—an excitation source,

$$
\begin{aligned}
\frac{dy_1}{d\tau} &= Cy_1 - [y_3 + \frac{1}{8}(y_1^2 + y_2^2 + y_4^2 + y_5^2)]y_2 - \frac{3}{4}(y_1y_5 - y_2y_4)y_4 + 2y_2,\\
\frac{dy_2}{d\tau} &= Cy_2 + [y_3 + \frac{1}{8}(y_1^2 + y_2^2 + y_4^2 + y_5^2)]y_1 - \frac{3}{4}(y_1y_5 - y_2y_4)y_5 + 2y_1,\\
\frac{dy_3}{d\tau} &= D(y_1y_2 + y_4y_5) + Ey_3 + F,\\
\frac{dy_4}{d\tau} &= Cy_4 - [y_3 + \frac{1}{8}(y_1^2 + y_2^2 + y_4^2 + y_5^2)]y_5 + \frac{3}{4}(y_1y_5 - y_2y_4)y_1 + 2y_5,\\
\frac{dy_5}{d\tau} &= Cy_5 + [y_3 + \frac{1}{8}(y_1^2 + y_2^2 + y_4^2 + y_5^2)]y_4 + \frac{3}{4}(y_1y_5 - y_2y_4)y_2 + 2y_4,
\end{aligned} \tag{2}
$$

an analog generator—a piezoceramic transducer,

$$
\begin{aligned}
\frac{d\xi}{d\tau} &= \zeta,\\
\frac{d\zeta}{d\tau} &= -\xi + \alpha_1\zeta + \alpha_2\zeta^2 - \alpha_3\zeta^3 + \alpha_4\beta,\\
\frac{d\beta}{d\tau} &= \gamma,\\
\frac{d\gamma}{d\tau} &= \alpha_5\xi + \alpha_6\zeta - \alpha_0\beta - \alpha_7\gamma,
\end{aligned} \tag{3}
$$

a tank with a liquid—an excitation source,

$$
\begin{aligned}
\frac{dp_1}{d\tau} &= \alpha p_1 - \left[\beta + \frac{A}{2}(p_1^2 + q_1^2 + p_2^2 + q_2^2)\right]q_1 + B(p_1q_2 - p_2q_1)p_2,\\
\frac{dq_1}{d\tau} &= \alpha q_1 + \left[\beta + \frac{A}{2}(p_1^2 + q_1^2 + p_2^2 + q_2^2)\right]p_1 + B(p_1q_2 - p_2q_1)q_2 + 1,\\
\frac{d\beta}{d\tau} &= N_3 + N_1\beta - \mu_1 q_1,\\
\frac{dp_2}{d\tau} &= \alpha p_2 - \left[\beta + \frac{A}{2}(p_1^2 + q_1^2 + p_2^2 + q_2^2)\right])q_2 - B(p_1q_2 - p_2q_1)p_1,\\
\frac{dq_2}{d\tau} &= \alpha q_2 + \left[\beta + \frac{A}{2}(p_1^2 + q_1^2 + p_2^2 + q_2^2)\right]p_2 - B(p_1q_2 - p_2q_1)q_1.
\end{aligned} \tag{4}
$$

The derivation of systems of (1)–(4) is given in monograph [12], in which phase variables and parameters of these systems are described in detail. Note that the techniques for detecting, classifying, and investigating the properties of attractors of systems (1)–(4) are described in [13, 14].

3 Scenarios of Transition to Chaos

Further we enumerate and describe the scenarios for the transition to chaos in systems (1)–(4).

3.1 *Feigenbaum's Scenario*

The most widespread scenario of transition to chaos through an infinite cascade of bifurcations of doubling the period of limit cycles [15–17]. The transition to chaos in the Feigenbaum's scenario is observed in almost all dynamic systems. In particular, the appearance of chaotic attractors according to the Feigenbaum's scenario takes place in all systems (1)–(4).

3.2 *Intermittency by Manneville–Pomeau*

Another widespread scenario of transition to chaos was first described in papers [18–20]. The transition from the limit cycle to the chaotic attractor occurs in one bifurcation. As a result of this bifurcation, the limit cycle disappears and a chaotic attractor arises in the system. The motion of trajectories along the attractor consists of two phases—laminar and turbulent. In the laminar phase, the trajectory makes quasiperiodic movements in a small neighborhood of the disappeared limit cycle, and in the turbulent phase, it moves away to distant (relatively the disappeared cycle) regions of the phase space. Note that the transition from the laminar phase to the turbulent one and vice versa is unpredictable. The described transition to chaos through intermittency is also observed in all systems (1)–(4).

Next, we consider scenarios that are generalizations and combinations of a cascade of bifurcations of period doubling and intermittency.

3.3 *Generalized Intermittency "Chaos–Chaos"*

The Manneville–Pomeau's scenario describes the transition "limit cycle – chaotic attractor". A natural complication of this scenario is the scenario of transition "a chaotic attractor of one type – a chaotic attractor of another type". This scenario was first described in [13] and later called "generalized intermittency". In further publications [14, 21], this scenario was analyzed and substantiated in more details.

Briefly, this scenario can be described as follows. A some chaotic attractor exists in the system, which disappears when the bifurcation parameter reaches a certain value and a chaotic attractor of another type of is born in the system. The motion

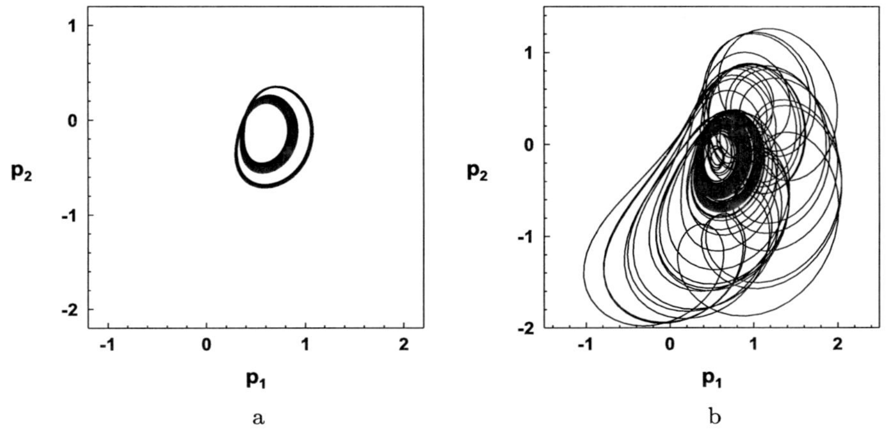

Fig. 1 Generalized intermittency

of the trajectories along the new attractor consists of two alternating phases. At one of these phases, the trajectory makes chaotic walks in a small neighborhood of the trajectories of the disappeared attractor. Then, at an unpredictable point in time, the trajectory leaves the region of localization of the disappeared attractor and leaves in the remote regions of the phase space. After some time, the movement of the trajectory again begins to resemble movement near of trajectories the disappeared chaotic attractor. The first of these phases was called coarse-grained (rough) laminar phase. The second is as before called turbulent phase. Note the duration of both the coarse-grained (rough) laminar and turbulent phases is unpredictable as are the moments of times of transition from one phase to another. An illustration of such a scenario is Fig. 1. Here in Fig. 1a shows the projection of the phase portrait before the bifurcation point, and in Fig. 1b after the bifurcation point. The densely black part of the projection in the central region of Fig. 1b is the laminar phase of intermittency, and the more "sparse" part of this figure is the turbulent phase.

In fact, in this scenario a disappearing chaotic attractor plays the role of a disappearing limit cycle from the classical intermittency scenario.

3.4 Generalized Intermittency "Hyperchaos–Hyperchaos"

Recall that an attractor is called hyperchaotic if it has at least two positive Lyapunov's characteristic exponents. That is, in the phase space there are at least two directions along which the trajectories belonging to the hyperchaotic attractor diverge. Hyperchaotic attractors can exist only in dynamical systems whose phase space dimension is at least four.

This scenario was discovered and described in detail in the papers [22–24]. This scenario is similar of the scenario described in item 3.3. The only difference is that the transitions from a hyperchaotic attractor of one type to a hyperchaotic attractor of another type are considered. Also, in such scenarios, transitions of the "chaos–hyperhaos" type are possible.

3.5 Symmetry and Intermittency in Feigenbaum's Scenario

This scenario is an unusual combination of Feigenbaum's scenario and intermittency. We briefly describe this scenario based on the results of the papers [21, 22].

First, there are two stable limit cycles in the system. These cycles are symmetric with respect to one or another coordinate axis. Each of the limit cycles has its own basin of attraction. Then, with a change of the bifurcation parameter, infinite cascades of bifurcations of doubling the periods of these cycles simultaneously begin. Moreover, all bifurcations of doubling of each of the cycles occur at the same value of the bifurcation parameter. This cascade of doubling bifurcations ends with the simultaneous appearance of two symmetric chaotic attractors. Each of these attractors has its own basin of attraction.

With a further change of the parameter of bifurcation, phase portraits are "glued together" and only one chaotic attractor remains in the system. The arising chaotic attractor has a symmetric structure of the phase portrait. The motion of a typical trajectory of a chaotic attractor can be conditionally divided into two phases. In the first of these phases, the trajectory is located in the localization region of one of the disappeared chaotic attractors from time to time approaching the boundary of such a region. That is, in this phase, the trajectory is in one of the symmetric parts of the arising chaotic attractor. Then, at an unpredictable moment of time, the trajectory passes into the localization region of the second of the disappeared chaotic attractors, that is, into another symmetric region of the arising chaotic attractor. This is the second phase of the trajectory. At an unpredictable moment of time, the trajectory again returns to the first symmetric region of the chaotic attractor. This process of transition from one symmetric region of the attractor to another is repeated an infinite number of times. Note that the duration of stay the trajectory in one of the symmetric regions of the attractor is unpredictable.

Thus, an unusual combination of the Feigenbaum's scenario (an infinite cascade of bifurcations of doubling of limit cycles) and intermittency (unpredictable intermittency between symmetric parts of the phase portrait of the arising chaotic attractor) takes place.

3.6 Intermittency with Two Laminar Phases

We will describe such a scenario schematically. A necessary condition for the implementation of this scenario of transition to chaos is the simultaneous existence in the system of two symmetric stable limit cycles. Further changes in any parameter of bifurcation lead to the disappearance of both limit cycles and the birth of a chaotic attractor. In this case, the contours of the arising chaotic attractor in their form are two "united" symmetric limit cycles. The onset of chaos has many features typical of intermittency. However, in this case, moving the trajectory in the attractor includes three phases, two laminar and one turbulent. In the first laminar phase, the trajectory makes quasiperiodic movements in a small neighborhood of one of the "stuck together" limit cycles. At an unpredictable moment of time, a turbulent surge occurs and the trajectory leaves for a region of the phase space that is distant from the neighborhood of the disappeared cycle. Moreover, after the completion of the turbulent phase, the trajectory can either return to the first laminar phase of motion or go to the second laminar phase, which corresponds to quasiperiodic movements in a small neighborhood of the second of the disappeared limit cycles. Such a process of motion of a trajectory along an attractor of the form "one of the laminar phases – the turbulent phase – one of the laminar phases" is repeated an infinite number of times. Moreover, both the time moments of the transition of the trajectory into the turbulent phase and the "switching" between two laminar phases are unpredictable. Thus, the transition to chaos resembles the classic scenario of Manneville-Pomeau. However, unlike the classical scenario, we have not one, but two laminar phases of trajectory moving.

A graphic illustration of this scenario is shown in Fig. 2. In Fig. 2a, b show projections of symmetric limit cycles. In Fig. 2c shows the distribution of the invariant measure in the projection of the phase portrait of a chaotic attractor. Bold sections Fig. 2c correspond to two laminar phases of the trajectory motion. The paler areas in Fig. 2c correspond to the turbulent phase.

Note that for the first time such a scenario was described in paper [22].

3.7 Generalized Intermittency with Two Coarse-Grained (Rough) Laminar Phases

This scenario is in many ways similar to the scenario described in item "3.5. Symmetry and intermittency in Feigenbaum's scenario". For describing the beginning of this scenario, we can simply repeat the description of the scenario adduced in item 3.5. The differences begin from the moment of arising two symmetric chaotic attractors, which arise at the same value of the bifurcation parameter and have different attraction basins. With a further change in the bifurcation parameter, two symmetric chaotic attractors are combined into one chaotic attractor. The most significant difference from the scenario from item 3.5. is that the motion of the trajectory along the attractor consists of three phases. At two of these phases, the trajectory makes

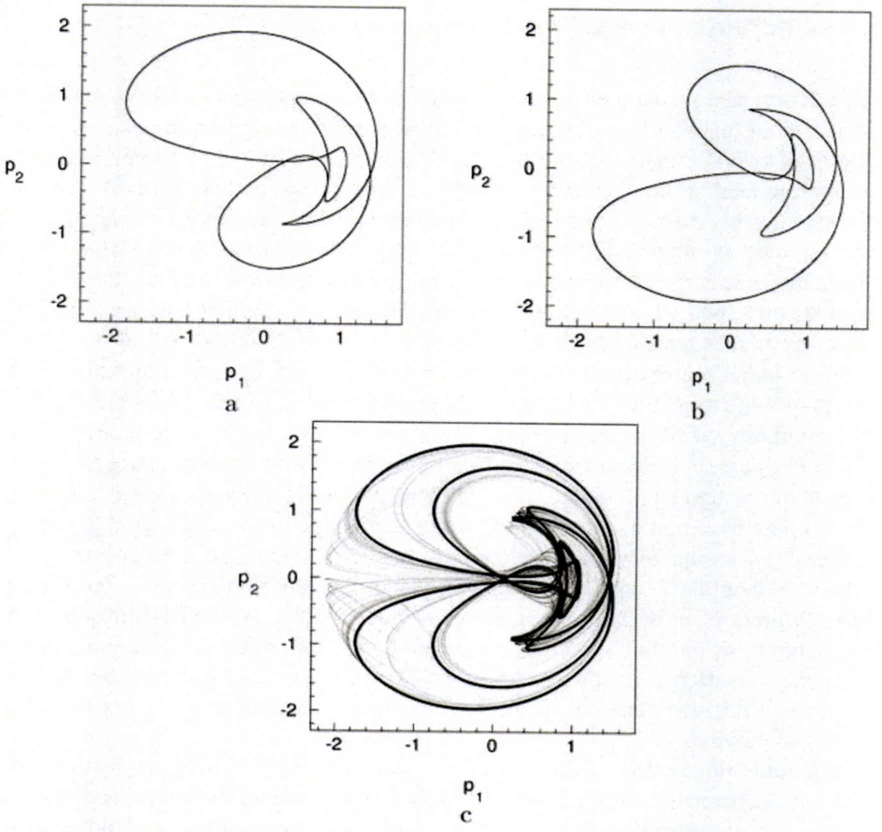

Fig. 2 Intermittency with two laminar phases

chaotic walks in a small neighborhood of the trajectories of the disappeared symmetric chaotic attractors. Such phases of movement are called coarse-grained (rough) laminar. The third phase of the movement is the departure of the trajectory into remote areas—this is the turbulent phase. The description of the sequence of transitions from one phase to another almost literally repeats such a description given in item "3.6. Intermittency with two laminar phases" . Only everywhere should the word "laminar" be replaced by "coarse-grained (rough) laminar".

A graphic illustration of ending this scenario is shown in Fig. 3a–c. In Fig. 3a, b are shown projections of phase portraits of symmetric chaotic attractors. Figure 3c is shown a chaotic attractor, which occurs after the disappearance of a pair of symmetric chaotic attractors. Three phases of the trajectory of the attractor are clearly visible. These are two coarse-grained (rough) laminar phases (two dark regions in the upper right and lower parts of the figure) and a turbulent phase (the lighter region in the left part of the figure).

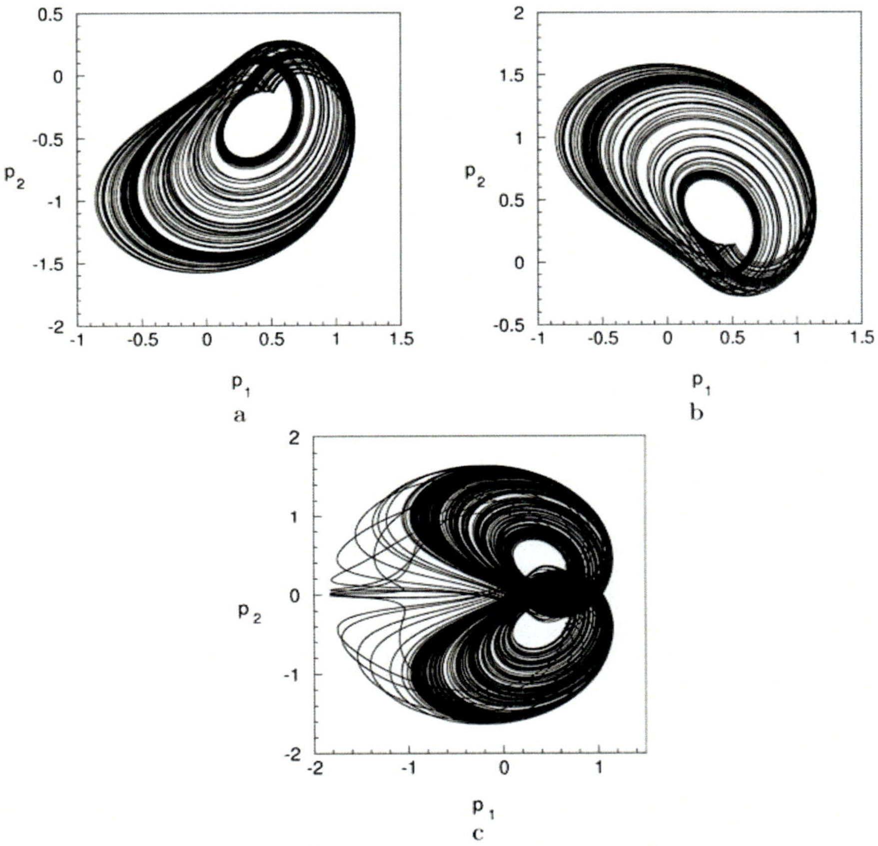

Fig. 3 Intermittency with two rough laminar phases

3.8 Generalized Intermittency "Hyperchaos–Hyperchaos" with Two Coarse-Grained (Rough) Laminar Phases

This scenario has been discovered and described relatively recently and published in [23, 24]. Qualitatively, this scenario is similar to the scenario given in the previous item "3.7. Generalized intermittency with two coarse-grained (rough) laminar phases". It begins with the appearance of two symmetric stable limit cycles. As a result of further changes in the bifurcation parameter, two hyperchaotic symmetric attractors arise, which then disappear and a combined hyperchaotic attractor is born in the system. The disappearing symmetric hyperchaotic attractors form two coarse-grained (rough) laminar phases of the final hyperchaotic attractor of this scenario.

Probably two more varieties of such a scenario are possible. The arising symmetric attractors will be chaotic, and the final attractor will be hyperchaotic. Conversely, symmetric attractors will be hyperchaotic, the final attractor will be chaotic.

We emphasize that the implementation of the scenarios described in items 3.3–3.8 was found only for systems with a five-dimensional phase space.

3.9 Atypical Change of Sequences Scenarios

The Feigenbaum's scenario and intermittency by Manneville-Pomeau are the main scenarios of the transition to chaos in dynamical systems. Moreover, in majority of dynamic systems, both of these scenarios are observed. As a rule, with an increase (decrease) in the value of the bifurcation parameter, the following sequence of transitions to chaos takes place. Cascade of bifurcations of period doubling—chaos—window of periodicity—cascade of bifurcations of period doubling—chaos—window of periodicity and so on. On the other hand, with a decrease (increase) in the value of the bifurcation parameter, the following sequence takes place: window of periodicity—chaos through intermittency—window of periodicity—chaos through intermittency—window of periodicity and so on [25, 26]. In systems (1), (3), (4) the situation was observed when, with an increase (decrease) in the value of the bifurcation parameter, transition to chaos is possible both according to the Feigenbaum's scenario and through intermittency.

We illustrate this with help of the phase-parametric characteristic of system (3) is shown in Fig. 4. In system (3) there are numerous transitions from limit cycles to chaotic attractors, as well as the destruction of chaotic attractors and the occurrence of limit cycles. All such transitions are clearly visible on the built bifurcation tree. Separate "branches" of this tree correspond to limit cycles, and densely dark areas correspond to chaotic attractors. The splitting points of the branches of the bifurcation tree are clearly visible in Fig. 4. In these points the bifurcations of the period doubling of the limit cycle occur. The threshold points are also clearly visible, during the passage of which an endless cascade of period doubling bifurcations ends with the appearance of a chaotic attractor, that is, a transition to chaos occurs according to the Feigenbaum's scenario. In turn, here is also possible a hard transition to chaos, in only one bifurcation, through Manneville-Pomeau intermittency. As can be seen from Fig. 4, the transition to chaos according to the Feigenbaum's scenario occurs both with increasing and decreasing values of the bifurcation parameter. A similar situation occurs for the transition to chaos through intermittency. Thus, there is some symmetry in the alternation of scenarios of transition to chaos. Such symmetry is atypical for dynamical systems.

Note that a similar symmetry effect of the scenarios of transition to chaos was established for nonideal systems (1), (4) in papers [27, 28].

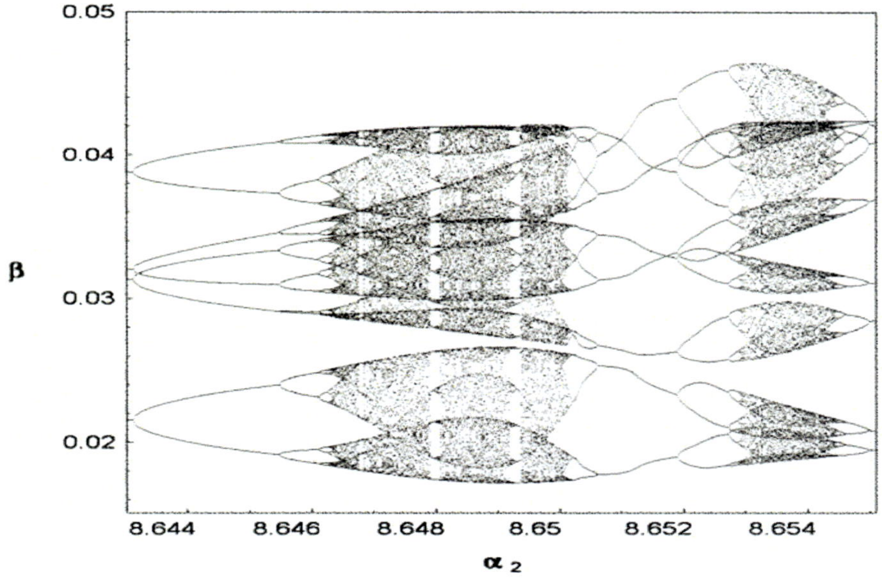

Fig. 4 Symmetry in transition to chaos

4 Conclusion

The considered nonideal dynamic systems (1)–(4) are characterized by extremely diverse dynamic behavior. In these systems, there are possible all types of regular attractors: equilibrium positions, limit cycles, invariant tori. Also, various types of chaotic attractors were found in these systems, and various types of hyperchaotic attractors were revealed in systems (3), (4). In addition to all the scenarios of transition to chaos inherent in nonlinear dynamics as a whole, a number of unusual scenarios of transition to chaos were discovered and described in these systems. In further studies, the construction and study of the attraction basins of attractors of such systems can be of great interest. No less interesting will be an attempt to discover the described unusual scenarios of the transition to chaos in other dynamical systems.

References

1. A. Sommerfeld, Beitrage zum dynamischen Ausbau der Festigkeitslehre. Physikalische Zeitschrift **3**, 266–271 (1902)
2. A. Sommerfeld, Beitrage zum dynamischen ausbau der festigkeislehre. Zeitschrift des Vereins Deutscher Ingenieure **46**, 391–394 (1902)
3. V.O. Kononenko, *Vibrating System with a Limited Power-Supply* (Iliffe, London, 1969)

4. T.S. Krasnopol'skaya, Self-excitation of mechanical oscillations by an electrodynamic vibrator. Sov. Appl. Mech. **13**, 187–191 (1977)
5. K.V. Frolov, T.S. Krasnopol'skaya, Sommerfeld effect in systems without internal damping. Sov. Appl. Mech. **23**, 1122–1126 (1987)
6. T.S. Krasnopolskaya, Acoustic chaos caused by the Sommerfeld effect. J. Fluids Struct. **8**(7), 803–815 (1994)
7. T.S. Krasnopolskaya, Chaos in acoustic subspace raised by the Sommerfeld-Kononenko effect. Meccanica **41**(3), 299–310 (2006)
8. T.S. Krasnopol'skaya, AYu. Shvets, Prorerties of chaotic oscillations of the liquid in cylindrical tanks. Prikladnaya Mekhanika **28**(6), 52–61 (1992)
9. T.S. Krasnopol'skaya, A.Y. Shvets, Chaotic oscillations of a spherical pendulum as an example of interaction with energy source. Int. Appl. Mech. **28**, 669–674 (1992)
10. A.Y. Shvets, Deterministic chaos of a spherical pendulum under limited excitation. Ukr. Math. J. **59**, 602–614 (2007)
11. J.M. Balthazar, J.L. Palacios Felix, et al., Nonlinear interactions in a piezoceramic bar transducer powered by a vacuum tube generated by a nonideal source. J. Comput. Nonlinear Dyn. **4**(1), 1–7, 011013 (2009)
12. T.S. Krasnopolskaya, A.Y. Shvets, *Regular and Chaotical Dynamics of Systems with Limited Excitation* (R&C Dynamics, Moscow, 2008)
13. T.S. Krasnopolskaya, A.Y. Shvets, Chaotic surface waves in limited power-supply cylindrical tank vibrations. J. Fluids Struct. **8**(1), 1–18 (1994)
14. T.S. Krasnopolskaya, A.Y. Shvets, Dynamical chaos for a limited power supply for fluid oscillations in cylindrical tanks. J. Sound Vibr. **322**(3), 532–553 (2009)
15. M.J. Feigenbaum, Quantative universality for a class of nonlinear transformations. J. Stat. Phys. **19**(1), 25–52 (1978)
16. M.J. Feigenbaum, The universal metric properties of nonlinear transformations. J. Stat. Phys. **21**(6), 669–706 (1979)
17. M.J. Feigenbaum, The transition to aperiodic behavior in turbulent systems. Comm. Math. Phys. **77**(1), 65–86 (1980)
18. P. Manneville, Y. Pomeau, Different ways to turbulence in dissipative dynamical systems. Physica D. Nonlinear Phenom **1**(2), 219–226 (1980)
19. Y. Pomeau, P. Manneville, Intermittent transition to turbulence in dissipative dynamical systems. Comm. Math. Phys. 74(2), 189–197 (1980)
20. P. Berge, Y. Pomeau, C.H. Vidal, *Order Within Chaos* (Wiley, New York, 1984)
21. A.Y. Shvets, V.O. Sirenko, Peculiarities of Transition to chaos in nonideal hydrodynamics systems. Chaot. Model. Simulat. (CMSIM) J. **2**, 303–310 (2012)
22. A. Shvets, V. Sirenko, Complicated scenarios of transitions to deterministic chaos in non-ideal dynamic systems, in *Nonlinear Dynamics-2016 (ND-KhPI2016): Proceedings of 5th International Conference, dedicated to the 90th anniversary of Academician V. L. Rvachev* (2016), pp. 222–229
23. A. Shvets, V. Sirenko, Hyperchaos in oscillating systems with limited excitation, in *11th Chaotic Modeling and Simulation International Conference. CHAOS 2018. Springer Proceedings in Complexity*, ed. by C. Skiadas, I. Lubashevsky (Springer, Cham, 2019), pp. 265–273
24. A.Y. Shvets, V.A. Sirenko, Scenarios of transitions to hyperchaos in nonideal oscillating systems. J. Math. Sci. **243**(2), 338–346 (2019)
25. V. Afraimovich, S.B. Hsu, *Lectures on Chaotic Dynamical Sestems* (Sommerville, International Press, 2003)
26. S.P. Kouznetsov, *Dynamic Chaos* (Physmatlit, Moscow, 2006)
27. A.Y. Shvets, A. Makaseyev, Delay factors and chaotization of non-ideal pendulum systems, *CHAOS 2012—5th Chaotic Modeling and Simulation International Conference, Proceedings* (2012), pp. 565–574
28. A. Shvets, Donetskyi. Transition to deterministic chaos in some electroelastic systems, in *11th Chaotic Modeling and Simulation International Conference. CHAOS 2018. Springer Proceedings in Complexity*, ed. by C. Skiadas, I. Lubashevsky (Springer, Cham, 2019), pp. 257–264

Identification of Hidden and Rare Attractors in Some Electroelastic Systems with Limited Excitation

Aleksandr Shvets and Serhii Donetskyi

Abstract Mathematical models of a deterministic dynamic system of the type analog generator–piezoelectric transducer are considered taking into account the influence of delay. A technique for searching for hidden and rare attractors of such system is proposed. Two approaches to the study of systems with delay are analyzed. The transformations of hidden attractors into self-excited ones, and rare attractors into non rare ones and vice versa are studied. The pairs of regular attractor—chaotic attractor are studied in point of view of their qualifications in terms of "hidden" and "rare". Symmetry was revealed in the scenarios of the transition from regular attractors to chaotic attractors. The effect of delay on the regular and chaotic dynamics of the system is investigated.

Keywords Hidden attractor · Rare attractor · Scenario of transition to chaos · Delay

1 Introduction

A dynamic system consisting of a piezoceramic transducer excited by an vacuum tube generator of limited power is considered. Such systems are widely used in various technical devices. Many aspects of the dynamic behavior of such systems were studied in papers [1–4, 6–9]. The existence of various types of steady-state modes of interaction between the generator and the converter was revealed. In particular, chaotic and hyperchaotic regimes of interaction were discovered and the features of transitions from regular regimes to chaotic one were described. It was shown that dynamic chaos in that systems arise solely due to the interaction between the generator and the converter, and their individual characteristics.

A. Shvets (✉) · S. Donetskyi
National Technical University of Ukraine "Igor Sikorsky Kyiv Polytechnic Institue", Kyiv, Ukraine
e-mail: alex.shvets@bigmir.net

C. H. Skiadas et al. (eds.), *13th Chaotic Modeling and Simulation International Conference*, Springer Proceedings in Complexity,
https://doi.org/10.1007/978-3-030-70795-8_60

However, earlier studies almost did not take into account such an important factor as the delay in the impact of the generator on transducer and delay of the inverse effect of the transducer on generator. Note that delay may be present in real systems due to limited speed signals: waves of compression, tension, bending, current and electrical voltage, as well as many other factors. In some cases influence of delay does not lead to significant changes in the dynamic behavior of researched systems. In other cases delay leads not only to significant quantitative changes of characteristics steady-state movement, but to completely qualitatively changes in the type of steady-state regimes.

2 The Mathematical Model

Consider a system consisting of a piezoceramic transducer, the source of excitation of which is an analog vacuum tube generator. Assume that the piezoceramic transducer has the shape of a circular cylinder and placed in an acoustic environment. Transducer and generator form an electrical circuit through a transmitting transformer. Let e_g and E_g be the variable and constant components of the grid generator lamp voltage, accordingly. Denote by $V(t)$ the electric voltage applied to the electrodes of the transducer. Introduce variable $\psi(t)$ by the formula

$$\psi(t) = \int_0^t (e_g + E_g)dt$$

Then, the equation describing the electrical oscillations of the generator has the form [6–8]:

$$\ddot{\psi}(t) + \omega_0^2\psi(t) = a_1\dot{\psi}(t) + a_2\dot{\psi}^2(t) - a_3\dot{\psi}^3(t) - a_4V(t-\rho), \tag{1}$$

here

$$\begin{aligned} \omega_0^2 &= \frac{R_a + R_c}{R_cL_cC_c};\ a_0 = a_1 - \frac{M^2R_c}{L_cC_cLR_a^2};\ a_2 = \frac{3M_cI_3E_g}{L_cC_c};\ a_3 = \frac{M_cI_3}{L_cC_c}; \\ a_4 &= \frac{2MM_c}{LR_aL_cC_c};\ a_1 = \frac{M_c}{L_cC_c}\left(I_1 - \frac{R_aR_cC_c - L_c}{R_a(M_c - DL_c)} + \frac{R_cL_1}{R_a^2M_c} - 3I_3E_g^2\right). \end{aligned} \tag{2}$$

A detailed description of all the electromagnetic parameters of the generator, contained in the equation (2) is given in [7, 8, 10]. Constant non-negative parameter ρ introduced for accounting delay of influence the impact of the transducer on the generator.

In turn, the equation describing the voltage oscillations $V(t)$, taking into account the delay of the signal of the generator on transducer, can be written as [6–8]:

$$\ddot{V}(t) + \omega_1^2V(t) = a_5\psi(t-\rho) + a_6\dot{\psi}(t) - a_7\dot{V}(t), \tag{3}$$

here,

$$\omega_1^2 = \frac{2h}{LS\epsilon_{33}(1-k^2)}; \; a_5 = -\frac{M\omega_1^2 R_c(R_a+R_c)}{2M_c R_a L_c}; \; a_6 = -\frac{M\omega_1^2 R_c}{2M_c R_a}; \\ a_7 = \frac{k^2}{\eta_0 hS(1-k^2)}; \; k = \frac{d_{33}}{\sqrt{\epsilon_{33}s_{33}}}, . \tag{4}$$

The parameters d_{33}, s_{33} and ϵ_{33} are constant coefficients of the theory of longitudinal deformations of a piezoelectric element described in [10]. Note that the presence of delay in real devices "generator–piezoceramic transducer" may be associated with territorial remoteness, sometimes quite significant, subsystems of the specified device. This leads to a delay feedback from one subsystem to another for the reasons stated in the Introduction.

Thus, the system of equations with retarded argument (1, 3) describes the interaction process of a piezoceramic transducer with a source of its excitation, an analog generator. We pass to the new system with the dimensionless variables according to the following formulas:

$$\xi(\tau) = \frac{\psi(\tau)\omega_0}{E_g}, \; \frac{d\xi(\tau)}{d\tau} = \zeta(\tau), \; \beta(\tau) = \frac{V(\tau)}{E_g}; \\ \frac{d\beta(\tau)}{d\tau} = \gamma(\tau), \; \tau = \omega_0 t. \tag{5}$$

Then we get the following system of equations

$$\begin{aligned} \frac{d\xi(\tau)}{d\tau} &= \zeta(\tau); \\ \frac{d\zeta(\tau)}{d\tau} &= -\xi(\tau) + \alpha_1\zeta(\tau) + \alpha_2\zeta^2(\tau) - \alpha_3\zeta^3(\tau) - \alpha_4\beta(\tau-\delta); \\ \frac{d\beta(\tau)}{d\tau} &= \gamma(\tau); \\ \frac{d\gamma(\tau)}{d\tau} &= -\alpha_0\beta(\tau) + \alpha_5\xi(\tau-\delta) + \alpha_6\zeta(\tau) - \alpha_7\gamma(\tau). \end{aligned} \tag{6}$$

here

$$\alpha_0 = \frac{\omega_1^2}{\omega_0^2}, \; \alpha_1 = \frac{a_0}{\omega_0}, \; \alpha_2 = \frac{a_2 E_g}{\omega_0}, \; \alpha_3 = \frac{a_3 E_g^2}{\omega_0}, \\ \alpha_4 = \frac{a_4}{\omega_0}, \; \alpha_5 = \frac{a_5}{\omega_0^3}, \; \alpha_6 = \frac{a_6}{\omega_0^2}, \; \alpha_7 = \frac{a_7}{\omega_0}, \; \delta = \omega_0\rho. \tag{7}$$

The function $\beta(\tau)$ corresponds to the signal propagated transducer into the acoustic medium, and the function $\xi(\tau)$ describes internal processes in the generator. Delays that are present in the system of equations (6) may lead to qualitative changes of the steady-state regimes of interaction. Particularly, delay can lead, as to emergence of new regular and chaotic attractors, as to disappearance of such attractors, existing in the system in the absence of delay. Consider these processes in some concrete examples.

3 Simulation at Absence of Delay

Firstly, consider the case of absence of delay in system (6), i.e. $\delta = 0$. In such case the system (6) is a nonlinear system of differential equations of fourth order, so all researches were carried out by means of various numerical methods. The technique for such calculations was developed and described in detail in [5, 6, 11–13].

Recently, new classification of attractors of dynamic systems has been proposed, which give the definition of self-excited, hidden and rare attractors [14–16]. We briefly recall the definitions of such types of attractors. An attractor is called self-exciting if there is an equilibrium position such that any neighborhood of it intersects with basin of attraction of the attractor. An attractor is called hidden if it is not self-excited.

For most dynamic systems, the main focus is the study of self-excited attractors. This is due to the fact that currently detection of the existence of hidden attractors is, generally speaking, extremely challenging. However, attractors of dynamical systems are not limited by self-excited attractors. Occurrence in specific dynamic system (device, structural element) of a hidden attractor can completely change the expected behavior of the system and make it impossible to perform its intended operational functions. Moreover, developers of a particular system may not guess that they are functioning of the system occurs in the regime of a hidden attractor.

The attractor is called rare if it is located in the phase space nearby to other attractor herewith phase volume (measure) of its basin of attraction significantly less compared to the basin of attraction of attractor near which it is located. Or simply rare attractor is attractor that has an extremely small basin of attraction. It is clear that the probability of a trajectory entering this attractor is small.

The main goal of this work is to detect attractors of the system (6) and to identify their types, taking into account the classification proposed in [14–16]. Note that the study of rare attractors of system (6) was started in paper [17].

For revealing hidden attractors of the system (6), following algorithm can be proposed. Firstly, the trajectory of system is calculated using the Runge-Kutta method with correction of a variable step of numerical integration according to Dormand-Prince [18], at that is selected relatively large local error $O(10^{-4}) - O(10^{-5})$ for Dormand-Prince procedure and any initial conditions for the system (6) are specified. Next, for chosen trajectory, after ending of transition process, Lyapunov's characteristic exponents (LCEs) are calculated [19]. On the basis of LCEs identification of type of attractor are made. However, a paradoxical situation may arise. In our case such situation is absence of zero exponent in LCE spectrum for an attractor, which is not an equilibrium position. This situation may happen due to presence in the system hidden (rare) attractor. After decreasing the local error up to $O(10^{-7}) - O(10^{-8})$

the problem with absence of non zero LCE goes away and trajectories of system over time might belong to hidden (rare) attractors.

Let parameters of the system be equal $\alpha_0 = 0.995, \alpha_1 = 0.0535, \alpha_3 = 9.95, \alpha_4 = 0.103, \alpha_5 = -0.0604, \alpha_6 = -0.12, \alpha_7 = 0.01$ and leave parameter α_2 as bifurcation one. For so chosen values the parameters, system (6) has single unstable equilibrium position

$$\xi = 0, \zeta = 0, \beta = 0, \gamma = 0,$$

so-called zero equilibrium position [6, 8, 17].

At $\alpha_2 = 8.925$ chaotic attractor is the only attractor of the system in the neighborhood of the zero equilibrium position. This attractor is self-excited attractor. At increasing value of bifurcation parameter up to $\alpha_2 = 8.94$ another attractor arises, namely, the limit cycle. In system (6), in the neighborhood of the zero equilibrium position, two attractors begin to exist simultaneously. One of them is a chaotic attractor, and the second is a limit cycle. Phase portrait projections of coexisting attractor are presented in Fig. 1a. Moreover, the chaotic attractor (black attractor) is still a self-excited attractor. In turn, the limit cycle (red attractor) is both a hidden attractor and a rare attractor. The limit cycle is a hidden attractor because it is not a self-excited attractor. And it is rare attractor, because it has very small basin of attraction.

In the future, one of the attractors of a pair of simultaneously existing attractors will be conventionally called black attractor, and the other—red. A part of the phase-parametric characteristics of the coexisting pair of attractors (black and red) is shown in Fig. 2. For both attractors, as for black attractor as for red attractor, separate "branches" of this trees correspond to limit cycles. The densely black regions correspond to chaos of the black attractor, and the densely red regions correspond to the quasiperiodic regimes of the red attractor. In Fig. 3 an enlarged fragment of the middle part of the phase-parametric characteristics is shown. Figures 2 and 3 give a clear view of bifurcations in system (6) in a selected region of the space of its parameters. So there are numerous bifurcations of the black attractor of the type

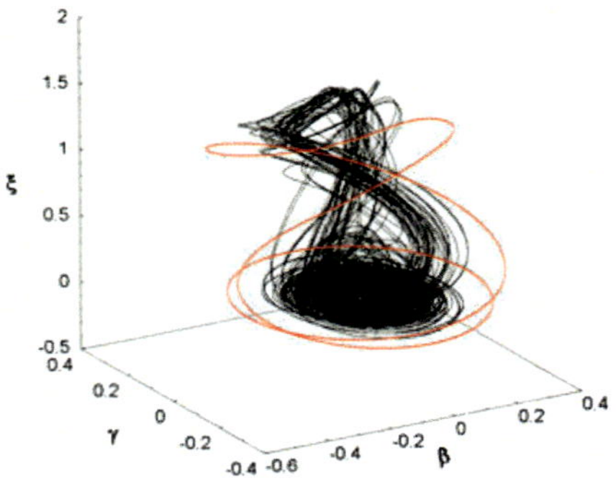

(a)

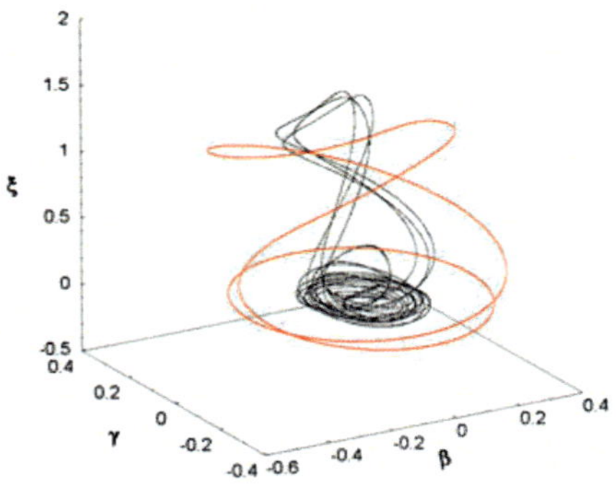

(b)

Fig. 1 Phase portrait projections at: $\alpha_2 = 8.958$ (**a**); $\alpha_2 = 8.959$ (**b**)

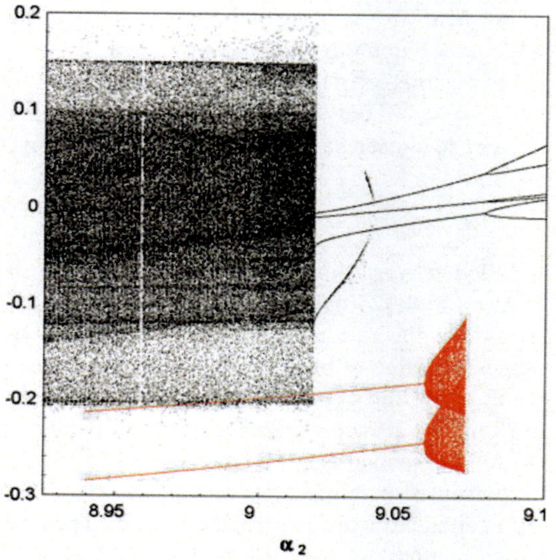

Fig. 2 Phase-parametric characteristic for $\alpha_2 \in [8.925; 9.1]$

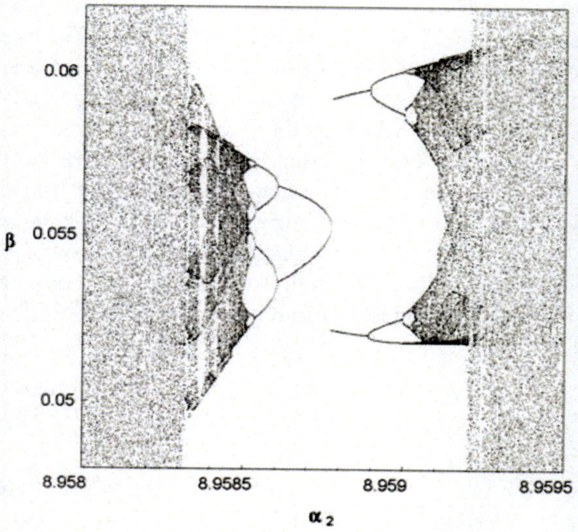

Fig. 3 Phase-parametric characteristic in window of periodicity at $\alpha_2 \in [8.958; 8.9595]$

"cycle–chaos–cycle–chaos", etc. However, in such transitions, the black attractor will always be a self-excited attractor. In turn, for the red attractor, in fact, there is only one rigid bifurcation "cycle–invariant torus". Moreover, the red attractor, both being a limit cycle and an invariant torus, also constantly remains a hidden attractor and a rare attractor.

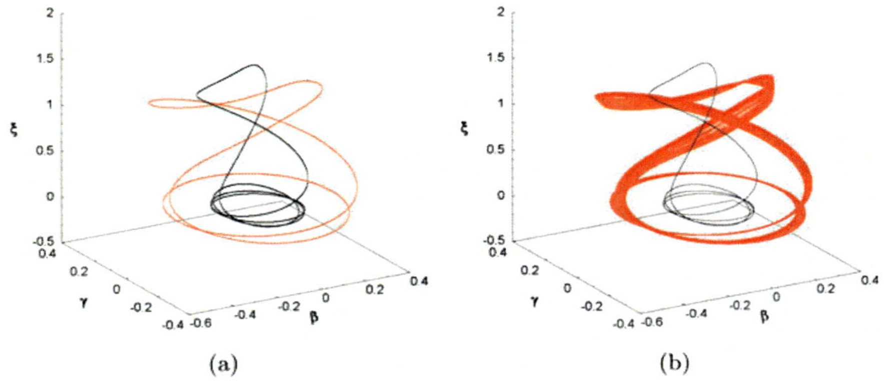

Fig. 4 Phase portrait projections at: $\alpha_2 = 9.04$ (**a**); $\alpha_2 = 9.07$ (**b**)

Another interesting feature of the alternation of scenarios of transition to the chaos of the black attractor is visible in Fig. 3. The transition to chaos according to the Feigenbaum's scenario through the endless cascade of periods doubling of limit cycles, occurs at increasing of bifurcation parameter α_2 and at decreasing of bifurcation parameter α_2. The same feature is inherent in the transition to chaos through intermittency. That is, there is some symmetry in the alternation of scenarios of transition to chaos. This situation, in general, is atypical for dynamical systems [19], however it seems to be natural for the system (6), since similar symmetry were already found before in paper [17].

We also illustrate these bifurcations using phase portraits of attractors. At $\alpha_2 = 9.02$ the black chaotic attractor turns into limit cycle through one rigid bifurcation. In turn red attractor remains periodic. This situation holds until $\alpha_2 = 9.06$ when new bifurcation occurs with red limit cycle which turns it into invariant torus and remains up so to its disappearance at $\alpha_2 = 9.072$. As we have already noted, a black attractor will always be a self-excited attractor. The red attractor will always be a hidden attractor and at the same time will be a rare attractor. Projections of phase portraits of pairs "self-exited limit cycle—hidden and rare limit cycle" and "self-exited limit cycle—hidden and rare invariant torus" are pictured in Fig. 4a, b correspondingly.

4 Methods of Transformation Systems with Delay

We consider methods for transforming systems with a delayed argument into systems of ordinary differential equations. First one is based on assumption that delay factor δ is sufficiently small, so that we can write

$$\beta(\tau-\delta)\approx\beta(\tau)-\delta\cdot\frac{d\beta(\tau)}{d\tau}=\beta(\tau)-\delta\cdot\gamma(\tau);$$
$$\xi(\tau-\delta)\approx\xi(\tau)-\delta\cdot\frac{d\xi(\tau)}{d\tau}=\xi(\tau)-\delta\cdot\zeta(\tau).$$

Substituting the obtained expressions into the system of equations (6), we obtain

$$\begin{aligned}
\frac{d\xi(\tau)}{d\tau}&=\zeta(\tau);\\
\frac{d\zeta(\tau)}{d\tau}&=-\xi(\tau)+\alpha_1\zeta(\tau)+\alpha_2\zeta^2(\tau)-\alpha_3\zeta^3(\tau)-\alpha_4\beta(\tau)+\alpha_4\delta\cdot\gamma(\tau);\\
\frac{d\beta(\tau)}{d\tau}&=\gamma(\tau);\\
\frac{d\gamma(\tau)}{d\tau}&=-\alpha_0\beta(\tau)+\alpha_5\xi(\tau)-\alpha_5\delta\cdot\zeta(\tau)+\alpha_6\zeta(\tau)-\alpha_7\gamma(\tau).
\end{aligned}\tag{8}$$

The system of equation (8) is transformed in a system of ordinary differential equations. Delay δ is in the system (8) as additional parameter. This approach is applicable for systems with constant delay and with variable delay.

A more accurate approximation method is applicable only to systems with a constant delay of δ. Let us divide segment $[-\delta; 0]$ into m equal parts and introduce such new functions.

$$\beta\left(\tau-\frac{i\delta}{m}\right)=\beta_i(\tau),\qquad \xi\left(\tau-\frac{i\delta}{m}\right)=\xi_i(\tau),\qquad i\in\{0,\ldots,m\}.$$

Then, using difference approximation of derivative, we turn system (6) into system

$$\begin{aligned}
\frac{d\xi_0(\tau)}{d\tau}&=\zeta_0(\tau);\\
\frac{d\zeta_0(\tau)}{d\tau}&=-\xi_0(\tau)+\alpha_1\zeta_0(\tau)+\alpha_2\zeta_0^2(\tau)-\alpha_3\zeta_0^3(\tau)-\alpha_4\beta_m(\tau);\\
\frac{d\beta_0(\tau)}{d\tau}&=\gamma_0(\tau);\\
\frac{d\gamma_0(\tau)}{d\tau}&=-\alpha_0\beta_0(\tau)+\alpha_5\xi_m(\tau)+\alpha_6\zeta_0(\tau)-\alpha_7\gamma_0(\tau);\\
\frac{d\xi_i(\tau)}{d\tau}&=\frac{m}{\delta}\cdot\Big(\xi_{i-1}(\tau)-\xi_i(\tau)\Big),\qquad i\in\{1,\ldots,m\};\\
\frac{d\beta_i(\tau)}{d\tau}&=\frac{m}{\delta}\cdot\Big(\beta_{i-1}(\tau)-\beta_i(\tau)\Big),\qquad i\in\{1,\ldots,m\}.
\end{aligned}\tag{9}$$

System (9) is a system of ordinary differential equations of $(2m+4)$-th order. The delay δ is introduced as additional parameter of this system.

It should be noted that solutions $\xi, \zeta, \beta, \gamma$ of the system (6) are approximated by solutions $\xi_0, \zeta_0, \beta_0, \gamma_0$ of the system (9) respectively. And $\xi_0 \to \xi,\ \zeta_0 \to \zeta,\ \beta_0 \to \beta,\ \gamma_0 \to \gamma$ as $m \to \infty$.

Thus, we can study the influence of delay on the dynamic behavior of the generator-transducer system using either a system of equations (8) or a system of equations (9). Such studies are carried out using a number of numerical methods according to the technique described in [6, 11, 13]. The application of the second approach to reduce the system of differential equations with delay to the system of differential equations without delay allows, generally speaking, to obtain more accurate results at studying the dynamics of the "generator-transducer" system. However, this significantly increases the duration of computer calculations and complicates the procedure for creating the appropriate computer programs. So there must be a balance between computational speed and accuracy.

We will find such balance for a number of concrete cases. Assume that $\alpha_0 = 0.995$, $\alpha_1 = 0.0535$, $\alpha_3 = 9.95$, $\alpha_4 = 0.103$, $\alpha_5 = -0.0604$, $\alpha_6 = -0.12$, $\alpha_7 = 0.01$. As bifurcation parameters we use the parameters α_2, δ.

Comparison results for the two used methods are shown in Fig. 5.

In Fig. 5a the projection of the phase portrait of the limit cycle of system (8), constructed at $\alpha_2 = 9.075$, $\delta = 0.01$ is shown. Accordingly, in Fig. 5c, e the projection of the phase portrait of the limit cycle of system (9), constructed at the same values of α_2, δ are shown. Figure 5c is constructed at $m = 3$ and the Fig. 5e is constructed at $m = 30$. In Fig. 5b, d, f the projections of the phase portrait of the chaotic attractor constructed at $\alpha_2 = 9.075$, $\delta = 0.04$ are shown. In Fig. 5b the chaotic attractor of system (8) is shown. Accordingly in Fig. 5d the chaotic attractor of system (9) ($m = 3$) is shown and in Fig. 5f the chaotic attractor of system (9) ($m = 30$) is shown.

Note that the identification of the type of attractor (limit cycle or chaotic attractor) was carried out on the basis of calculation and analysis of the LCE spectrum. As can be seen in Fig. 5, all constructed phase portraits almost coincide. However, the duration of computer calculations increases significantly when applying the second method of transformation a system with delay (6) to a system without delay (9). At m = 30, the duration of computer calculations by the second method is more than 5000 times the duration of computer calculations by the first method. Moreover, the constructed phase portraits and Lyapunov's characteristic exponents practically coincide. Therefore all further analysis of influence of the delay was carried out by the first method.

5 Influence of the Delay on the Type of Attractor

We investigate the influence of delay on the appearance and disappearance of various attractors of the "generator-transducer" system. As in the previous section, we assume that $\alpha_0 = 0.995$, $\alpha_1 = 0.0535$, $\alpha_2 = 9.075$, $\alpha_3 = 9.95$, $\alpha_4 = 0.103$, $\alpha_5 = -0.0604$, $\alpha_6 = -0.12$, $\alpha_7 = 0.01$. As bifurcation parameters we use the delay δ.

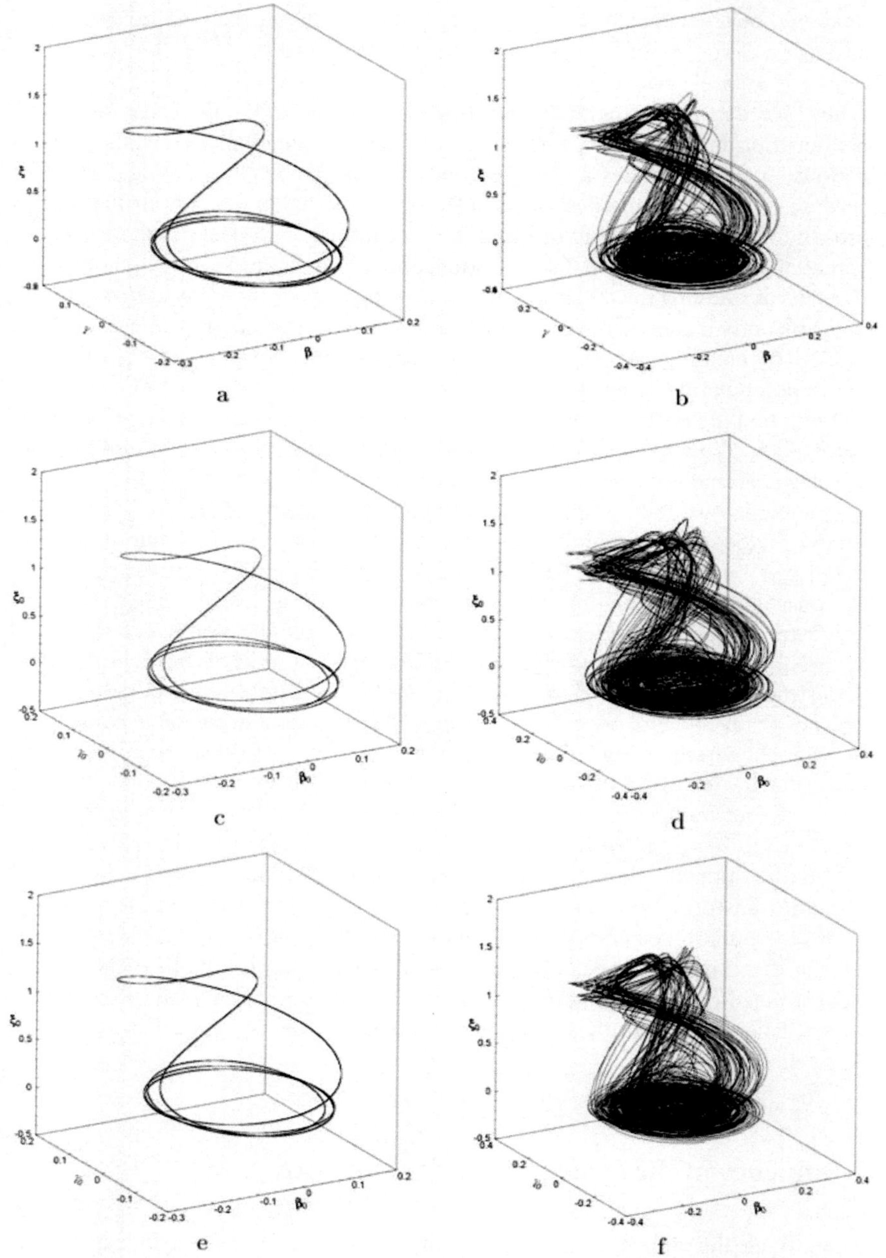

Fig. 5 Phase portrait projections for system (8) at: delta = 0.01 (**a**), delta = 0.04 (**b**); for system (9) at m =3, delta = 0.01 (**c**), delta = 0.04 (**d**); for system (9) at m = 30, delta = 0.01 (**e**), delta = 0.04 (**f**)

In Fig. 6, phase parametric characteristics of a pair of attractors coexisting in the system (6) are constructed. As previously conditionally we will call them black and red attractor.

In the absence of delay in the system, there is only one black attractor. This attractor is a stable limit cycle. In addition, this attractor will be the self-excited attractor. However, even at a negligible value of the delay $\delta = 0.0015$, another attractor is born in the system, namely, the invariant torus (red region in Fig. 6). This invariant torus is both a hidden attractor and a rare attractor. At increasing delay, at $\delta = 0.005$ for red attractor the bifurcation "torus–cycle" is taken place. The invariant torus is destroyed and a resonant limit cycle is born in the system. With a further increase of the delay, in the selected interval of the change of the delay, no red attractor bifurcations occur anymore. The new resonant limit cycle will continue to be both a hidden attractor and a rare attractor.

Next, we consider the bifurcations of the black attractor. As can be seen from Fig. 6 at $\delta = 0.032$ through one rigid bifurcation the limit cycle disappears and a chaotic attractor arises in the system. Such a chaotic attractor exists at the vast majority of $\delta > 0.032$. To this chaotic attractor corresponds the densely black region in Fig. 6 and this attractor is self-excited attractor in this region. In addition, small "slots" are visible in this thickly black area. As a rule, such "slots" correspond to periodicity windows in chaos. However, here the situation is much more interesting. At the values of delay corresponding to such "slots", the chaotic attractor does not disappear and does not turn into a limit cycle. It will still be chaotic attractor, but not self-excited attractor. This attractor turns into hidden attractor, since trajectory that starts in the neighborhood of equilibrium position skips black attractor and approach to red limit cycle. Moreover, the phase portrait of such hidden attractor

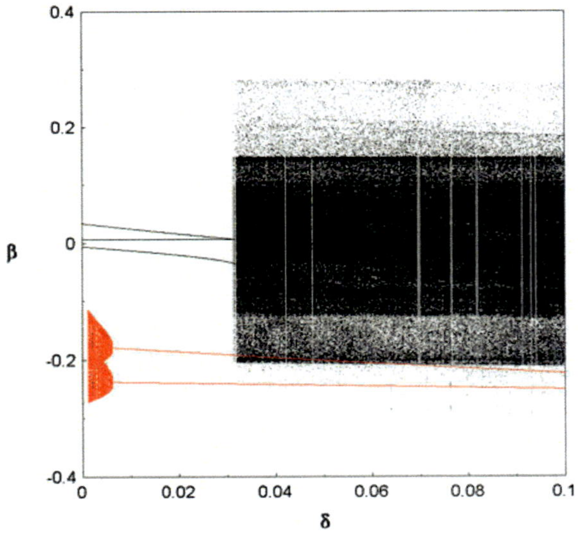

Fig. 6 Phase-parametric characteristic at $\alpha_2 = 9.075$ for $\delta \in [0; 0.1]$

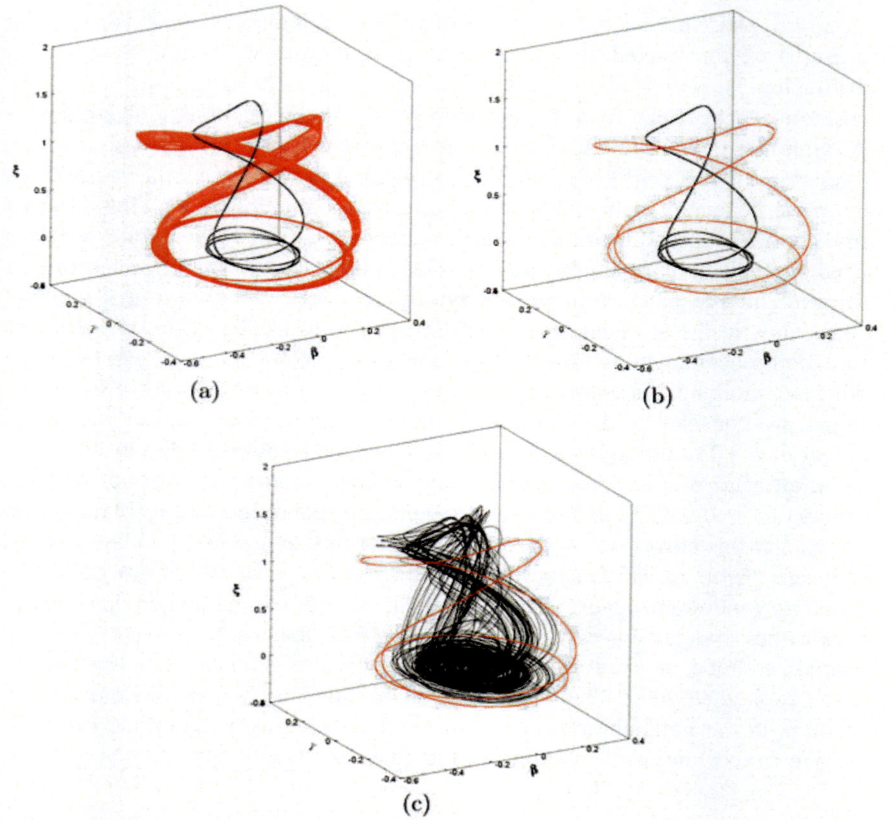

Fig. 7 Phase portrait projections at: $\delta = 0.0015$ **(a)**; $\delta = 0.01$ **(b)**; $\delta = 0.035$ **(c)**

is not practically distinguishable from the phase portraits of self-excited chaotic attractor. Thus, at the values of delay corresponding to the "slots" in Fig. 6, a rare and self-excited limit cycle and a hidden chaotic attractor simultaneously exist in the system.

Projections of phase portrait of mentioned pairs of attractor ("limit cycle–torus", "limit cycle–limit cycle", "limit cycle–chaos") are shown in Fig. 7.

6 Conclusion

In the space of parameters of the "generator-piezoceramic transducer" system, regions were discovered in which two attractors coexist simultaneously. Moreover, these attractors can be both regular and chaotic. These attractors are located in close proximity to one another. At absence of delay one of the coexisting attractors is always a self-excited attractor, and the second attractor is always hidden and rare. It is shown that the presence of delay in the system can contribute to the appearance and disappearance of attractors of various types. The possibility of simultaneous coexistence in the system of a hidden chaotic attractor and a rare and self-excited limit cycle is established.

References

1. T.S. Krasnopol'skaya, Self-excitation of mechanical oscillations by an electrodynamic vibrator. Sov. Appl. Mech. **13**, 187–191 (1977)
2. K.V. Frolov, T.S. Krasnopol'skaya, Sommerfeld effect in systems without internal damping. Sov. Appl. Mech. **23**, 1122–1126 (1987)
3. T.S. Krasnopolskaya, Acoustic chaos caused by the Sommerfeld effect. J. Fluids Struct. **8**(7), 803–815 (1994)
4. T.S. Krasnopolskaya, Chaos in acoustic subspace raised by the Sommerfeld-Kononenko effect. Meccanica **41**(3), 299–310 (2006)
5. T.S. Krasnopol'skaya, A.Y. Shvets, Chaotic oscillations of a spherical pendulum as an example of interaction with energy source. Int. Appl. Mech. **28**, 669–674 (1992)
6. T.S. Krasnopolskaya, A.Y. Shvets, Deterministic chaos in a system generator—piezoceramic transducer. Nonlinear Dyn. Syst. Theor. **6**(4), 367–387 (2006)
7. T.S. Krasnopolskaya, A.Y. Shvets, *Regular and chaotical dynamics of systems with limited excitation* (R&C Dynamics, Moscow, 2008)
8. A.Y. Shvets, T.S. Krasnopolskaya, *Hyperchaos in piezoceramic systems with limited power supply*, vol. 6, in *IUTAM Symposium on Hamiltonian Dynamics, Vortex Structures, Turbulence. IUTAM Bookseries*, ed. by A.V. Borisov, V.V. Kozlov, I.S. Mamaev, M.A. Sokolovskiy (Springer, Dordrecht, 2008), pp. 313–322
9. J.M. Balthazar, J.L. Palacios Felix et al. *Nonlinear interactions in a piezoceramic bar transducer powered by a vacuum tube generated by a nonideal source*. J. Comput. Nonlinear Dyn., **4**(1), 1–7, 011013 (2009)
10. B.A. Auld, *Acoustic Fields and Waves in Solids* (Wiley, New York, 1973)
11. A. Yu. Shvets, Deterministic chaos of a spherical pendulum under limited excitation. Ukr. Math. J. **59**, 602–614 (2007)
12. T.S. Krasnopolskaya, AYu. Shvets, Dynamical chaos for a limited power supply for fluid oscillations in cylindrical tanks. J. Sound Vibr. **322**(3), 532–553 (2009)
13. A. Yu. Shvets, V.A. Sirenko, Scenarios of transitions to hyperchaos in nonideal oscillating systems. J. Math. Sci. **243**(2), 338–346 (2019)
14. G. Leonov, N. Kuznetsov et al., Localization of hidden Chua's attractors. Phys. Lett. A **375**(23), 2230–2233 (2011)
15. G. Leonov, N. Kuznetsov, Hidden attractors in dynamical systems. Int. J. Bifurcat. Chaos **23**(1), 1330002 (2013)
16. A. Chudzik, P. Perlikowski et al., Multistability and rare attractors in van der Pol-Duffing oscillator. Int. J. Bifurcat. Chaos **21**(7), 1907–1912 (2011)
17. A. Shvets, S. Donetskyi. *Transition to deterministic chaos in some electroelastic systems*, in *11th Chaotic Modeling and Simulation International Conference. CHAOS 2018. Springer Proceedings in Complexity*, ed. by C. Skiadas, I. Lubashevsky (Springer, Cham, 2019), pp. 257–264

18. E. Hairer, S.P. Norsett, G. Wanner, *Solving Ordinary Differential Equations. Nonstiff Problems* (Springer, Berlin, 1987)
19. S.P. Kouznetsov, *Dynamic Chaos* (Physmatlit, Moscow, 2006)
20. N.A. Magnizkiy, S.V. Sidorov, *New Methods of Chaotic Dynamics* (Editorial URSS, 2004)
21. A.A. Samarskii, A.V. Gulin, *Computational Methods*, vol. 430 (Moscow, Nauka, 1989)

Rules and Regulations of Potential Impact of Acoustic Factors from High-Speed Railway Lines on Environment and Human Body During Construction of New Facilities

Vladimir Yu. Smirnov, Oksana I. Kos, and Elena A. Eseva

Abstract High speeds on railway lines have a negative impact on the environment and the human body, primarily due to the vibration effect. For a comfortable stay of a person, it is necessary that the vibration effect does not exceed the maximal admissible effect. It was carried out analysis of standard technical documents, establishing standards in the area of vibration impact on the environment. The main direction of improvement of the regulatory documentation is the harmonization of sanitary-hygienic and technical complexes.

Keywords High-speed railway lines · Vibration impact · Sanitary norms · Sanitary-epidemiological rules · Engineering and environmental surveys

1 Introduction

Railway transport is traditionally considered one of the most environmentally friendly types of transport. In recent years, Russia has been developing projects for the construction of high-speed railway lines (HSRL), including in the direction of the Ural region, to Kazan and to Saint Petersburg. The speed of trains on the HSRL will be from 250 to 400 km/h, while keeping a low cost of transportation. Currently, the President of the Russian Federation has approved the project of high-speed highway Moscow—Saint Petersburg.

V. Yu. Smirnov (✉) · O. I. Kos (✉)
Moscow Aviation Institute (National Research University), Moscow, Russia
e-mail: vl-smirnov@mail.ru

E. A. Eseva
Russian University of Transport (MIIT), Moscow, Russia
e-mail: EsevaEA@rambler.ru

C. H. Skiadas et al. (eds.), *13th Chaotic Modeling and Simulation International Conference*, Springer Proceedings in Complexity,
https://doi.org/10.1007/978-3-030-70795-8_61

2 Documents Establishing the Requirements for the Regulation of Vibration

High speeds on railway lines have a negative impact on the environment and the human body, primarily due to the vibration effect. In this regard, more attention is paid to the evaluation and regulation of vibration impact.

In accordance with Article 26 of the federal law of December 30, 2009, N384-FZ "in the design documentation of a civil building and structure, measures must be provided, so that vibration in the building and structure does not harm people's health" [1].

GOST R 2041-2012 establishes terms and definitions, used in the field of vibration evaluation [2].

Vibration sources in accordance with GOST R ISO 14837-1-2007 are "vehicle, wheels, rails, track, support structure" rail vehicles, wheels, track, support structure GOST R ISO 14837-1-2007 [3], artificial structures during the passage of a rail vehicle GOST 31185-2002 [4].

The propagation paths of vibration in accordance with GOST R ISO 14837-1-2007 are the condition of the soil, the distance to the object of impact.

The object of impact in accordance with GOST R ISO 14837-1-2007 is "the foundation, type of construction".

One of the propagation paths of vibration is that the soil can be represented by an infinitely rigid elastic medium.

In infinite solid elastic media, vibration is propagated through the following mechanisms:

- compression waves (longitudinal waves) with the movement of particles in the form of oscillations in the direction of motion;
- shear waves (transversal waves) with the movement of particles in the form of oscillations in the plane perpendicular to the direction of wave propagation;
- Rayleigh waves, which are surface waves with elliptic motion of particles in a vertical plane extending in the direction of wave propagation;
- Love waves (surface waves), propagating in the direction of motion.

As the train speed increases, the vibration impact on the environment increases [5].

The main document establishing the requirements for the regulation of vibration are the federal sanitary norms CH 2.2.4/2.1.8.562-96 [6]. Since January 1, 2017 sanitary-epidemiological rules and standards SanPiN 2.2.4.3359-16 are put into effect [7], improved vibration requirements in workplaces, including public buildings.

It distinguishes sanitary and technical standardization.

In the first case, the vibration parameters of the workplaces and the surface of contact with the limbs of workers are limited, based on physiological requirements, which redxuce the possibility of vibration disease.

In the second case, the vibration parameters are limited, taking into account not only the specified requirements, but technically achievable level of vibration for this type of machines to date.

The main direction of improving the regulatory documentation is to establish compliance with sanitary and technical complexes. Rationing vibration from subway trains is carried out in accordance with the code of rules SP 120.13330.2012 [8]. The methodology for vibrations evaluation from the subway (metro trains) is given in the code of rules SP 23-105-2004 [6]. Measurement of vibration generated internally in railway tunnels by the passage of trains GOST 31185-2002 [4].

The main calculated normalized vibration parameters (root-mean-square values of vibration accelerations) determined during testing of high-speed trains are performed taking into account the temporary nature of the vibrations acting.

3 The Vibration Parameters

To assess the value of vibration acceleration and vibration velocity, the rms characteristic was taken, since other power-law averages do not reflect the real effect of vibration acceleration and vibration velocity on the environment and the human body, for example, the arithmetic mean is close to zero. Root mean square value vibration acceleration takes into account both the peak value and the vibration acceleration value fluctuating around zero.

1. For acceptance tests

In the case of integral assessment of frequency, the normalized parameter is the corrected value of vibration velocity L_ϑ or their logarithmic levels (LU), measured using correction filters or calculated by the formulas:

$$\boldsymbol{L}_\vartheta = \sqrt{\sum_{j=1}^{n} \frac{\boldsymbol{L}_{\vartheta j}^2 t_j}{\boldsymbol{T}}} \tag{1}$$

where

$L_{\vartheta j}$—root-mean-square value of vibration velocity or general vibration at the j-th measurement;

j = 1, 2,..., n, where n—number of measurements;

t_j—the duration of the j-th dimension;

T—total measurement time.

Causes of errors in measuring the rms values of vibration velocity and vibration acceleration: cable connections, the influence of electromagnetic fields, triboelectric phenomena, zero level shift.

Let us determine the average error of the root-mean-square value of the vibration velocity:

$$m = \frac{\delta}{\sqrt{n}} \tag{2}$$

$$\delta^2 = \overline{L_{\vartheta j}^2} - \left(\overline{\mathrm{L}_{\vartheta \mathrm{j}}}\right)^2 \tag{3}$$

$$\delta^2 = \frac{\sum L_{\vartheta j}^2 * t_j}{T} - \left(\sqrt{\frac{\sum L_{\vartheta j}^2 * t_j}{T}}\right)^2 \tag{4}$$

$$m = \frac{\sqrt{\frac{\sum L_{\vartheta j}^2 * t_j}{T} - \left(\sqrt{\frac{\sum L_{\vartheta j}^2 t_j}{T}}\right)^2}}{\sqrt{T}} \tag{5}$$

where

L_ϑ—root-mean-square value of vibration velocity or general vibration at the j-th measurement;

j = 1, 2,…, n, where n—number of measurements;

t_j—the duration of the jth dimension;

T—total measurement time.

Let's perform interval estimation of the general arithmetic mean. Interval boundaries $\overline{L_{\vartheta j_{\text{геН}}}}$ relatively $\overline{L_{\vartheta j}}$ we will calculate by the formula:

$$\overline{L_{\vartheta j}} - t_\alpha * m < \overline{L_{\vartheta j_{\text{геН}}}} < \overline{L_{\vartheta j}} + t_\alpha * m, \tag{6}$$

where

j = 1, 2,…, n, where n—number of measurements;

t_α—the tabular value of the Student's criterion,

m—root-mean-square error.

Since the values of the local vibration velocity L_υ, m s vibration differ greatly in intensity, it is much more convenient to consider the equivalent logarithmic levels of the root mean square values of vibration acceleration, $L_{\upsilon w\ (w)}$ dB is determined by the formula:

$$L_\vartheta = 10lg \sum_{j=1}^{n} 10^{0,1L_u} j \tag{7}$$

where

$L_{\vartheta j}$—weighting factors for the i-th frequency band, respectively, for absolute values or their logarithmic levels, determined for local vibrations,

n—is the number of frequency bands (1/3 or 1/1 octaves) in the normalized frequency range.

Root-meansquare of local vibration acceleration a_{hw} and whole-body vibration a_w, m/s^2 is determined by the formula:

$$a_{hw(w)} = \sqrt{\frac{1}{T}\sum_{j=1}^{n} a_{hw(w)j}^{2} * t_j} \quad (8)$$

where $a_{hw(w)j}$—root-meansquare of local vibration acceleration or whole-body vibration at the j-th measurement;

i = 1, 2,…, n, where n—number of measurements;

t_j—the duration of the j-th dimension;

T—total measurement time.

Let's determine the mean error of the root-mean-square value:

$$m = \frac{\delta}{\sqrt{n}} \quad (9)$$

$$\delta^2 = \overline{a_{hw(w)j}^{2}} - \left(\overline{a_{hw(w)j}^{2}}\right)^2 \quad (10)$$

$$\delta^2 = \frac{\sum a_{hw(w)j}^{2} * t_j}{T} - \left(\sqrt{\frac{\sum a_{hw(w)j}^{2} * t_j}{T}}\right)^2 \quad (11)$$

$$m = \frac{\sqrt{\frac{\sum a_{hw(w)j}^{2} * t_j}{T} - \left(\sqrt{\frac{\sum a_{hw(w)j}^{2} t_j}{T}}\right)^2}}{\sqrt{T}}, \quad (12)$$

Let's perform interval estimation of the general arithmetic mean. Interval boundaries $\overline{a_{hw(w)\tau\text{eH}}}$ relatively $\overline{a_{hw(w)}}$ we will calculate by the formula:

$$\overline{a_{hw(w)}} - t_\alpha * m < \overline{a_{hw(w)\tau\text{eH}}} < \overline{a_{hw(w)}} + t_\alpha * m \quad (13)$$

where

j = 1, 2,…, n, where n—number of measurements;

t_α—the tabular value of the Student's criterion,

m—root-mean-square error.

Since the values of the vibration acceleration of the local a_{hw}, ms^2 vibrations differ greatly in intensity, it is much more convenient to consider the equivalent logarithmic levels of the root-mean-square values of vibration accelerations, $L_{ahw\,(w)}$ dB is determined by the formula:

$$La_{hw(w)} = 10 * lg\left(\frac{a_{hw(w)}}{a_0}\right)^2 \quad (14)$$

where $a_{hw(w)}$—root-mean-square value of vibration acceleration of local or general vibration, m/s^2;

a_0—reference level, a_0 = 10-6 m/s^2.

2. For periodic tests

Equivalent corrected value of vibration acceleration of local a_{hw} and general vibration a_w, m/s^2, for an 8-h working day

$$A_l = k_l \left[\frac{1}{T_0} \int_{T_0} a^2_{hwj}(t)dt \right]^{1/2} \quad (15)$$

where a_{hwj}—root-mean square value of the vibration acceleration of the total vibration during the i-th working operation);

i = 1, 2,…, n, where n is the number of working operations;

j—index indicating the direction of measurement (evaluation) of vibration (x, y or z);

$k_j = 1, 4$ in x- and y- directions and $k_j = 1$ for z-direction;

T_i—time of i-th working operation;

T_0—basic working day duration equal to 8 h.

Let's determine the mean error of the root-mean-square value:

$$m = \frac{\delta}{\sqrt{n}} \quad (16)$$

$$\delta^2 = \overline{a^2_{hw,8h(w)j}} - \left(\bar{a}_{hw,8h(w)j}\right)^2 \quad (17)$$

$$\delta^2 = \frac{\sum a^2_{hw,8h(w)j} * t_j}{T} - \left(\sqrt{\frac{\sum a^2_{hw,8h(w)j} * t_j}{T}}\right)^2 \quad (18)$$

$$m = \frac{\sqrt{\frac{\sum a^2_{hw,8h(w)j} * t_j}{T} - \left(\sqrt{\frac{\sum a^2_{hw,8h(w)j} t_j}{T}}\right)^2}}{\sqrt{T}}, \quad (19)$$

Let's perform interval estimation of the general arithmetic mean. Interval boundaries $\overline{a_{hw(w)\text{теН}}}$ relatively $\overline{a_{hw(w)}}$ we will calculate by the formula:

$$\overline{a_{hw,8h(w)}} - t_\alpha * m < \overline{a_{hw,8h(w)\text{теН}}} < \overline{a_{hw,8h(w)}} + t_\alpha * m \quad (20)$$

where

j = 1, 2,…, n, where n—number of measurements;

t_α—the tabular value of the Student's criterion,

m—root-mean-square error.

The equivalent corrected level of vibration acceleration $a_{hw,8h}$ dB is determined by the formula:

$$a_{hw,8h} = k_j * \sqrt{\frac{1}{T_0}\sum_{j=1}^{n} a_{wj}^2 * T_j} \quad (21)$$

where a_{hwj}—root-mean-square of vibration acceleration of local vibration during the ith working operation, calculated by the formula (21);

$j = 1, 2, \ldots, n$, where n is the number of working operations;

$k_j = 1.4$ for the x and y directions and $k_j = 1$ for the z direction;

T_j—time of the i-th work operation;

T_0—basic working day duration equal to 8 h.

Equivalent corrected level of vibration acceleration $La_{hw(w)}$,8 h, dB is determined by the formula:

$$La_{hw(w),8h} = 10 * lg\left(\frac{a_{hw(w)}}{a_0}\right)^2 \quad (22)$$

where $a_{hw(w)}$—root-mean-square value of vibration acceleration of local or general vibration, m/s^2;

a_0—reference level, a_0 = 10-6 m/s^2.

4 Indicators of Normative Values

The calculated root-mean square of the vibration acceleration, equivalent corrected vibration acceleration values, Corrected values of vibration accelerations in octave or third-octave band or their logarithmic levels are rated as unsatisfactory if one of the indicators exceeds the standard value. Threshold limit values of equivalent corrected vibration accelerations in accordance with Sanitary Rules and Norms 2.2.4.3359-16 [7] are presented in Table 1 and Table 2, maximum permissible values of corrected vibration accelerations for local vibration in accordance with SN 2.2.4/2.1.8.566-96 [10] are presented in Table 3, maximum permissible values for the whole-body vibration in accordance with SP 2.5.1336-03 [11] for transport operating conditions are presented in Table 4, for transport and technological operating conditions are in Table 5.

Vibration can cause discomfort to people, depending on its frequency: through the vibration of its body and its parts in the range from 1 to 80 Hz. A person, in addition to feeling the vibrations of his body, can perceive vibrations also tactilely in a wider range of frequencies.

Sanitary regulations SP 2.5.1336-03 [11] set vibration levels on the floor and on the surface of the seats in passenger and service compartments of coach-sleeping car, on the floor and the surface of the seats in passenger compartment with seats should not exceed the parameters in Table 6.

Measurement of total vibration and analyze of its effects on humans. Part 2. Vibration in buildings 31191.2-2004 [12].

Table 1 .

Average geometric frequencies of octave bands (Hz)	Weighting coefficient values			
	For vibration acceleration		For vibration velocity	
	Ki	L_{k_r}	Ki	L_{k_r}
8	1.0	0	0.5	−6
16	1.0	0	1.0	0
31.5	0.5	−6	1.0	0
63	0.25	−12	1.0	0
125	0.125	−18	1.0	0
250	0.063	−24	1.0	0
500	0.0315	−30	1.0	0
1000	0.016	−36	1.0	0

Table 2 .

Type of vibration	Category of vibration	Direction	Correction	Regulatory equivalent corrected values and levels of vibration acceleration	
				m/s^2	dB
Local		X L, Y L, Z L	Wh	2.0	126
General	Transport	Zwb	Wk	0.56	115
		Xwb, Ywb	Wd	0.4	112
	Transport and technological	Zwb	Wk	0.28	109
		Xwb, Ywb	Wd	0.2	106

Table 3 .

Centre frequencies of octave bands (Hz)	Maximum permissible vibration acceleration of local vibration in XL, YL, ZL -directions	
	Root-mean square of vibration acceleration (m/s^2)	Logarithmic levels (dB)
8.0	1.4	123
16.0	1.4	123
31.5	2.8	129
63.0	5.6	135
125.0	11.0	141
250.0	22.0	147
500.0	45.0	153
1000.0	89.0	159

Table 4 .

Centre frequencies of octave bands (Hz)	Maximum permissible vibration levels for transport operating conditions			
	Vertically in Z0—directions		Horizontally in X0, Y0—directions	
	Root-mean square of vibration acceleration (m/s^2)	Logarithmic levels, dB	Root-mean square of vibration acceleration (m/s^2)	Logarithmic levels, dB
1.0	0.63	116	0.23	107
1.25	0.56	115	0.23	107
1.6	0.5	114	0.23	107
2.0	0.45	113	0.23	107
2.5	0.40	112	0.28	109
3.15	0.36	111	0.36	111
4.0	0.32	110	0.45	113
5.0	0.32	110	0.56	115
6.3	0.32	110	0.71	117
8.0	0.32	110	0.90	119
10.0	0.36	111	0.70	117
12.5	0.40	112	0.50	114
16.0	0.45	113	0.40	112
20.0	0.50	114	0.36	111
25.0	0.56	115	0.40	112
31.5	0.63	116	0.45	113
40.0	0.71	117	0.50	114

It is convenient to use integral weighted It is convenient to use integral weighted characteristics to assess the vibration inside buildings in terms of living comfort and the likelihood of complaints from their inhabitants. The obtained vibration parameter allows us to characterize a specific accommodation inside the building from the standpoint of its suitability for living.

5 Conclusions

The study of the physical factor (vibration impact) is an integral part of complex engineering and environmental studies, which are performed to assess the state of the environment, the forecast of possible changes in the environment under the influence of anthropogenic impact, the environmental study for construction of high-speed rail lines and to ensure favorable conditions for people.

Table 5 .

Centre frequencies of 1/3 octave bands (Hz)	Maximum permissible levels of vibration for transport and technological operating conditions in X0, Y0, Z0—directions	
	Root-mean square of vibration acceleration (m/s^2)	Logarithmic levels (dB)
1.6	0.25	108
2.0	0.224	107
2.5	0.20	106
3.15	0.178	105
4.0	0.158	104
5.0	0.158	104
6.3	0.158	104
8.0	0.158	104
10.0	0.20	106
12.5	0.25	108
16.0	0.315	110
20.0	0.40	112
25.0	0.50	114
31.5	0.63	116
40.0	0.80	118
50.0	1.00	120
63.0	1.25	122
80.0	1.60	124

Rules and regulations, governing the assessment of the impact of the acoustic factor on the environment and the reaction of people, need to be improved.

The main direction of improvement is the harmonization of sanitary-hygienic and technical complexes of regulatory and legal documentation [9].

Table 6 Maximum permissible vibration levels in passenger cars (on the floor and on the surface of the seat)

Centre frequencies of 1/3 octave bands (Hz)	Root-mean square of vibration acceleration (m/s^2)	
	Vertically in Z—directions	Horizontally in X, Y—directions
1.0	0.22	0.10
1.25	0.20	0.10
1.6	0.18	0.10
2.0	0.16	0.10
2.5	0.14	0.12
3.15	0.12	0.16
4.0	0.11	0.20
5.0	0.11	0.25
6.3	0.11	0.31
8.0	0.11	0.40
10.0	0.14	0.50
12.5	0.18	0.63
16.0	0.22	0.80
20.0	0.28	1.00
25.0	0.35	1.25
31.5	0.45	1.60
40.0	0.56	2.00
50.0	0.71	2.50
63.0	0.90	3.15
80.0	1.12	4.00

References

1. Federal Law, Technical Regulations on the Safety of Buildings and Structures. Dated December 30, 2009 No. 384-FZ (as amended on July 2, 2013) (2009)
2. State Standard P 2041-2012, Mechanical vibration, shock and condition monitoring —Vocabulary (IDT) (Standartinform Publication, Moscow 2014) (In Russian)
3. State Standard ISO 14837-1:2005, Mechanical vibration—ground-borne noise and vibration arising from rail systems—Part 1: general guidance (Standartinform Publication, Moscow 2008) (In Russian)
4. State Standard 31185-2002, Mechanical vibration—measurement of vibration generated internally in railway tunnels by the passage of trains (Standartinform Publication, Moscow, 2009) (In Russian)
5. V.V. Krylov, Focusing of ground vibrations generated by high-speed trains, in *Proceedings of ISMA–2014 Including USD*. Ground vibration (2014), 2007–2015
6. SP 23-105-2004, Vibration assessment for design, construction and operation of metro units (Moscow, 2004) (In Russian)
7. Sanitary Rules and Norms SanPin 2.2.4.3359-16 Sanitary and epidemiological requirements for physical factors in the workplace (Moscow, 2016) (In Russian)

8. Code of practice SP 120.13330.2012 Subways. With Amendment No. 1 and Amendment No. 2. Moscow, 2012 (In Russian)
9. SN 2.2.4/2.1.8.562-96, Noise at Workplaces, in Residential and Public Spaces, and in Areas of Residential Development (Moscow, 1997) (In Russian)
10. SN 2.2.4/2.1.8.566-96, The sanitary norms of industrial vibration, vibration of residential and public buildings (Moscow, 1997) (In Russian)
11. SP 2.5.1336-03, Sanitary rules for the design, manufacture and reconstruction of locomotives and special rolling stock of railway transport (Moscow, 2003) (In Russian)
12. State Standard 31191.2-2004, Vibration and shock. Measurement and evaluation of human exposure to whole-body vibration. Part 2. Vibration in buildings (Standartinform Publication, Moscow 2008) (In Russian)

Connecting Bernoulli and Schrödinger Equations and Its Impact on Quantum-Mechanic Wave Function and Entanglement Problems

Siavash H. Sohrab

Abstract An invariant model of Boltzmann statistical mechanics is applied to derive invariant Schrödinger equation of quantum mechanics from invariant Bernoulli equation of hydrodynamics. The results suggest new perspectives regarding quantum mechanics wave function and its collapse, stationary versus propagating wave functions, and wave-particle duality. The invariant hydrodynamic model also leads to the definition of generalized shock waves in "supersonic" flows at molecular-, electro-, and chromo-dynamic scales with (*Mach, Lorentz, and Michelson*) numbers exceeding unity. The invariant internal hydro-thermo-diffusive structure of such generalized "shock" waves are described.

1 Introduction

It is well known that our universe involves statistical fields at five major scales that are approximately separated by a factor of 10^{-17}, beginning at exceedingly small Planck scale of 10^{-35} m, electrodynamics 10^{-17} m, molecular-dynamics 10^{0} m, astrophysics 10^{17} m, and finally cosmology 10^{35} m, with each statistical field having a characteristic "atomic" particle graviton, electron, molecule, star, and galaxy. Schematic diagram in Fig. 1 shows hierarchies of such statistical fields from photonic to cosmic scales. Under the assumption of weak interactive forces known as ideal gas, Boltzmann statistical mechanics governs the generalized thermodynamics associated with such statistical fields of diverse scales. Studies on generalized Boltzmann statistical mechanics and turbulent phenomena that are common universal features shared by stochastic quantum fields [1–17] and classical hydrodynamic fields [18–30] resulted in recent introduction of a scale-invariant model of statistical mechanics and its applications to thermodynamics [31, 32], fluid mechanics [33, 34], and quantum mechanics [35–37] at intermediate, large, and small scales.

S. H. Sohrab (✉)
Department of Mechanical Engineering, Robert R. McCormick School of Engineering and Applied Science, Northwestern University, 2145 Sheridan Road, Evanston, IL 60208-3111, USA
e-mail: s-sohrab@northwestern.edu

C. H. Skiadas et al. (eds.), *13th Chaotic Modeling and Simulation International Conference*, Springer Proceedings in Complexity,
https://doi.org/10.1007/978-3-030-70795-8_62

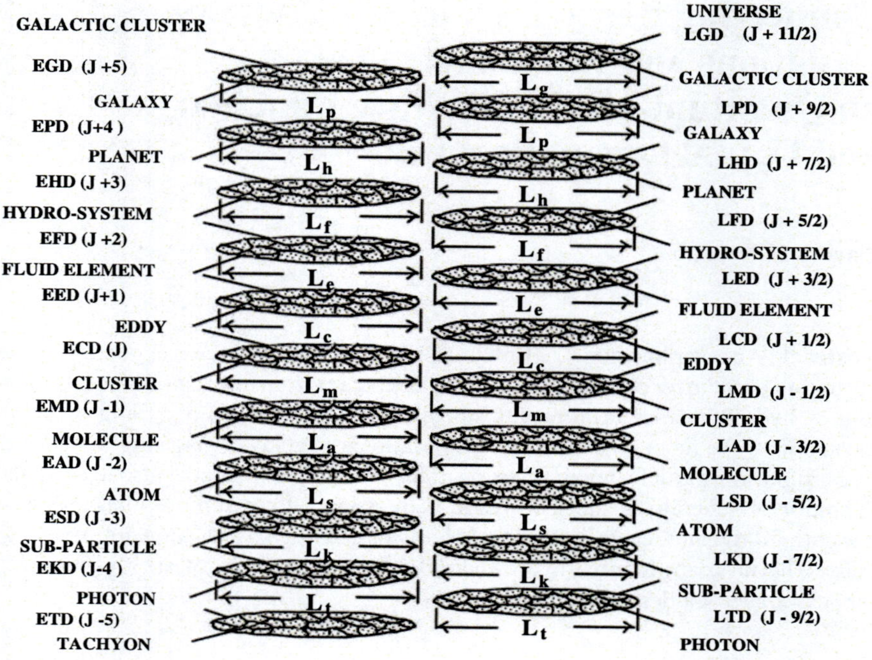

Fig. 1 A scale-invariant model of statistical mechanics. Equilibrium-β-Dynamics on the left-hand-side and non-equilibrium Laminar-β-Dynamics on the right-hand-side for scales β = g, p, h, f, e, c, m, a, s, k, and t as defined in [35]. Characteristic lengths of (system, element, "atom") are $(L_\beta, \lambda_\beta, \ell_\beta)$ and λ_β is the mean-free-path

In the present study, after a brief description of a scale-invariant model of statistical mechanics, the invariant forms of conservation equations are presented. Next, derivation of invariant Schrödinger equation from invariant Bernoulli equation for potential incompressible flow is discussed. The nature of quantum mechanics wave functions for both time-independent and time-dependent Schrödinger equations respectively associated with time-periodic stationary versus propagating states are identified. Also, the objective (real) versus subjective (imaginary) aspects of wave function [3] in connection to particle localization and Born probabilistic interpretation are studied.

The invariant forms of conservation equations result in introduction of an invariant definition of Mach number leading to a hierarchy of generalized normal "shock" waves from photonic to cosmic scales. The internal hydro-thermo-diffusive structure of such shock waves is examined and some of its implications to dissipation in quantum gravity and black holes are discussed.

2 Scale–Invariant Model of Boltzmann Statistical Mechanics

The scale-invariant model of statistical mechanics for equilibrium galactic-, planetary-, hydro-system-, fluid-element-, eddy-, cluster-, molecular-, atomic-, subatomic-, kromo-, and tachyon-dynamics corresponding to the scale β = g, p, h, f, e, c, m, a, s, k, and t is schematically shown on the left hand side of Fig. 1.

For each statistical field, one defines particles that form the background fluid and are viewed as point-mass or "*atom*" of the field. Next, the *elements* of the field are defined as finite-sized composite entities composed of an ensemble of "*atoms*". Finally, ensemble of a large number of "*elements*" is defined as the statistical "*system*" at that particular scale. The most-probable element of scale β is identified as the "atom" (system) of the next higher $\beta + 1$ (lower $\beta - 1$) scale.

Following the classical methods [19, 38–42], the invariant definitions of the density ρ_β, and the velocity of *atom* $\mathbf{u}_\beta$, *element* $\mathbf{v}_\beta$, and *system* $\mathbf{w}_\beta$ at the scale β are given as [36]

$$\rho_\beta = n_\beta m_\beta = m_\beta \int f_\beta du_\beta, \quad \mathbf{u}_\beta = \mathbf{v}_{w\beta-1} \tag{1}$$

$$\mathbf{v}_\beta = \rho_\beta^{-1} m_\beta \int \mathbf{u}_\beta f_\beta du_\beta, \quad \mathbf{w}_\beta = \mathbf{v}_{w\beta+1} \tag{2}$$

Similarly, the invariant definitions of the peculiar and diffusion velocities are introduced as

$$\mathbf{V}'_\beta = \mathbf{u}_\beta - \mathbf{v}_\beta \quad , \quad \mathbf{V}_\beta = \mathbf{v}_\beta - \mathbf{w}_\beta \quad , \quad \mathbf{V}_\beta = \mathbf{V}'_{\beta+1} \tag{3}$$

Following the classical methods [19, 38–40], the scale-invariant forms of mass, thermal energy, linear and angular momentum conservation equations at scale β are given as [33, 34]

$$\frac{\partial \rho_{i\beta}}{\partial t_\beta} + \nabla.(\rho_{i\beta}\mathbf{v}_\beta) = \Re_{i\beta} \tag{4}$$

$$\frac{\partial \varepsilon_{i\beta}}{\partial t_\beta} + \nabla.(\varepsilon_{i\beta}\mathbf{v}_\beta) = 0 \tag{5}$$

$$\frac{\partial \mathbf{p}_{i\beta}}{\partial t_\beta} + \nabla.(\mathbf{p}_{i\beta}\mathbf{v}_\beta) = -\nabla.\mathbf{P}_{ij\beta} \tag{6}$$

$$\frac{\partial \boldsymbol{\pi}_{i\beta}}{\partial t_\beta} + \nabla.(\boldsymbol{\pi}_{i\beta}\mathbf{v}_\beta) = \rho_{i\beta}\boldsymbol{\omega}_\beta.\nabla\mathbf{v}_\beta \tag{7}$$

involving the *volumetric density* of thermal energy $\varepsilon_{i\beta} = \rho_{i\beta}\,\tilde{h}_{i\beta}$, linear momentum $\mathbf{p}_{i\beta} = \rho_{i\beta}\,\mathbf{v}_{i\beta}$, and angular momentum $\boldsymbol{\pi}_{i\beta} = \rho_{i\beta}\,\boldsymbol{\omega}_{i\beta}$. Also, $\Re_{i\beta}$ is the chemical reaction rate and $\tilde{h}_{\beta}$ is the absolute enthalpy [32].

It is noted that the time coordinates in (4–7) also have a scale subscript β. In a recent study [43], the nature of physical space and time was investigated and the concepts of *internal spacetime* versus *external space and time* were introduced. Assuming that a statistical field at scale β is in thermodynamic equilibrium with the physical space at scale (β − 1) within which it resides, both fields will have a homogenous constant temperature $T_{\beta} = T_{\beta-1}$ defined in terms of Wien wavelength of particle thermal oscillations as [32]

$$m_{\beta}u_{\beta}^{2} = m_{\beta-1}v_{w\beta-1}^{2} = kT_{\beta-1} = k\lambda_{w\beta-1} \tag{8a}$$

Hence, constant internal measures of (extension $\lambda_{w\beta}$, duration $\tau_{w\beta}$) will be associated with every "point" of space at temperature $T_{\beta} = T_{\beta-1}$. For example, at cosmic scale β = g one employs internal (ruler, clock) of the lower scale of astrophysics β = s to define *external space and time* coordinates defined as [43]

$$(x_{\beta}, y_{\beta}, z_{\beta}) = (N_{x\beta}, N_{y\beta}, N_{z\beta})\lambda_{wx\beta-1}, \quad t_{\beta} = N_{t\beta}\tau_{w\beta-1} \tag{8b}$$

with the four numbers $(N_{x\beta}, N_{y\beta}, N_{z\beta}, N_{t\beta})$ being independent numbers.

3 Derivation of Invariant Schrödinger Equation from Invariant Bernoulli Equation

The connection between energy spectrum of photon gas given by *Planck* [44] distribution and both energy and dissipation spectrum of isotropic stationary turbulence has been recognized [35, 36]. In a recent study [35], the gap between problems of quantum mechanics and turbulence was investigated through connections between Cauchy, Euler, Bernoulli equations of hydrodynamics, Hamilton–Jacobi equation of classical mechanics, and Schrödinger equation of quantum mechanics. In a more recent investigation on foundation of classical thermodynamics [32] it was shown that stochastic definitions of Planck $h = h_k \equiv m_k\lambda_{rk}c$ and Boltzmann $k = k_k \equiv m_k\nu_{rk}c$ universal constants involve the speed of light identified as *root-mean-square speed of photons* $c = v_{rk}$ in Casimir [45] vacuum. The new insights into the statistical nature of both Planck and Boltzmann universal constants as well as the definition of absolute temperature [32] suggest a slightly modified derivation of Schrödinger equation [35, 46] discussed in the following.

For potential flow $\nabla \times \mathbf{v}_{\beta} = 0$ with velocity $\mathbf{v}_{\beta} = -\nabla\Phi_{\beta}$ [35], (6) leads to invariant Bernoulli equation

$$-\frac{\partial m\Phi_\beta}{\partial t'} + \frac{(\nabla m\Phi_\beta)^2}{2m} + \tilde{V}_\beta = 0 \tag{9}$$

where the atomic potential energy is $\tilde{V}_\beta = (\mathrm{p}_\beta/\rho_\beta)m_\beta = \mathrm{p}_\beta/\mathrm{n}_\beta = p_\beta\hat{\mathrm{v}}$. By (3), local velocity in an arbitrary direction is expressed in terms of the most probable or Wien velocity of the lower scale and peculiar velocity as perturbation

$$\mathbf{v}_{\mathrm{x}\beta} = \mathbf{u}_{\mathrm{x}\beta} - \varepsilon\mathbf{V}'_{\mathrm{x}\beta} = \mathbf{v}_{\mathrm{wx}\,\beta-1} - \varepsilon\mathbf{V}'_{\mathrm{x}\beta}, \quad \varepsilon \ll 1 \tag{10}$$

In absence of vorticity $\nabla \times \mathbf{v}_\beta = \nabla \times \mathbf{v}_{\mathrm{w}\beta-1} = \nabla \times \mathbf{V}'_\beta = 0$ equation (10) gives

$$m_\beta\Phi_\beta = m_\beta\Phi_{\mathrm{w}\beta-1} - \varepsilon m_\beta\Phi'_\beta \tag{11}$$

Comparison of (9) with Hamilton–Jacobi equation of classical mechanics [2] leads to the definition of invariant action [36]

$$\mathrm{S}_\beta(x'_\beta, t'_\beta) = -m_\beta\Phi_\beta = -m_\beta\Phi_{\mathrm{w}\,\beta-1} + \varepsilon m_\beta\Phi'_\beta = \mathrm{S}_{\mathrm{w}\,\beta-1} - \varepsilon\mathrm{S}'_\beta \tag{12}$$

and quantum mechanics *wave function* Ψ_β defined as

$$\Psi_\beta(x'_\beta, t'_\beta) = -\mathrm{S}'(x'_\beta, t'_\beta)/m_\beta = \Phi'_\beta \tag{13}$$

Substitution from (12)–(13) into Bernoulli equation (9) and separation of zeroth and first power of ε leads to [35]

$$\frac{\partial S_{\mathrm{w}\,\beta}}{\partial t'_\beta} + \frac{(\nabla S_{\mathrm{w}\,\beta})^2}{2m} + \tilde{V}_\beta = 0 \tag{14}$$

$$\frac{\partial\Psi_\beta}{\partial t'_\beta} + (\mathrm{v}_{\mathrm{wx}'})_{\beta-1}\nabla\Psi_\beta + \frac{\varepsilon}{2}(\nabla\Psi_\beta)^2 = 0 \tag{15}$$

To reveal the "*stationary states*" of the system one moves to coordinate system moving at the most-probable speed

$$z'_\beta = x'_\beta - (v_{\mathrm{wx}'})_{\beta-1}t'_\beta \tag{16}$$

The solution of (14) results in conservation of energy due to internal and peculiar translational motions

$$\tilde{E}_{\mathrm{t}\beta} = \tilde{T}_{\mathrm{t}\beta} + \tilde{V}_\beta \tag{17}$$

where

$$\tilde{E}_{t\beta} \equiv m_\beta v_{wx\,\beta-1}^2, \quad \tilde{T}_{t\beta} \equiv m_\beta v_{wx\,\beta-1}^2/2, \quad \tilde{V}_\beta \equiv p\hat{v} = m_\beta V_{x\beta}'^2/2 \tag{18}$$

The equality of translational kinetic and potential energies $\tilde{T}_{t\beta} = \tilde{V}_{t\beta}$ is in accordance with Boltzmann equipartition principle. In (18) the velocities refer to periodic motions in direction of a single translational coordinate say (x+, x−) and (17) corresponds to *atomic translational enthalpy* [32]

$$\hat{h}_{t\beta} = \hat{u}_{t\beta} + p_\beta \hat{v}_\beta = 2kT_\beta \tag{19}$$

where $\hat{u}_{t\beta} = mv_{wx}^2 = mv_{wx+}^2/2 + mv_{wx-}^2/2 = mv_{wx+}^2$ is *atomic internal translational kinetic energy* [32].

As described in [32], the conventional assumption of particle undergoing translational motion along three degrees of freedom (x, y, z) is not appropriate since particle cannot *simultaneously* move in three independent coordinate directions. Also, according to Clausius [47], the kinetic energy due to random rotational and vibrational motions of particles cannot be properly neglected. Therefore, as discussed in [32], the conventionally assumed random translational kinetic energy in (y+, y−) and (z+, z−) directions are instead respectively attributed to particle rotational (θ+, θ−) and vibrational (r+, r−) kinetic energies [32]

$$\hat{u}_{ty\beta} \equiv m_\beta v_{wy\beta}^2 \Rightarrow \hat{u}_{r\beta} \equiv I_\beta \varpi_{\theta\beta}^2 = kT_\beta \tag{20a}$$

$$\hat{u}_{tz\beta} \equiv m_\beta v_{wz\beta}^2 \Rightarrow \hat{u}_{v\beta} \equiv k_\beta r_\beta^2 = kT_\beta \tag{20b}$$

According to (19–20), particles have four *simultaneously independent* degrees of freedom namely, translational, rotational, vibrational, and potential. Boltzmann principle of equipartition of energy requires that all four degrees of freedom have the same energy resulting in atomic internal energy and atomic enthalpy of ideal gas respectively defined as [32]

$$\tilde{E}_\beta = \hat{u}_\beta = \hat{u}_{t\beta} + \hat{u}_{r\beta} + \hat{u}_{v\beta} = 3kT_\beta \tag{21}$$

$$\tilde{H}_\beta = \hat{h}_\beta = \hat{u}_\beta + p_\beta \hat{v}_\beta = 4kT_\beta \tag{22}$$

Total internal energy, potential energy, and enthalpy are respectively $N_\beta \tilde{E}_\beta = N_\beta \hat{u}_\beta = U_\beta$, $N_\beta \tilde{V}_\beta = p_\beta V_\beta$, and $N_\beta \tilde{H}_\beta = N_\beta \hat{h}_\beta = H_\beta$ such that [32]

$$H_\beta = U_\beta + p_\beta V_\beta \tag{23}$$

In summary, by the above procedure Bernoulli equation in "three dimension" accounts for three types of *internal kinetic energies* $\tilde{E}_\beta$ as well as potential energy $\tilde{V}_\beta = p_\beta \hat{v}_\beta$ that is also a *kinetic energy* due to random *external peculiar motion* of particles. Another perspective concerning the results in (19) and (20) is to view the particle as a small cylindrical object with periodic oscillations in axial (z+, z−), angular (θ+, θ−), and radial (r+, r−) directions in a cylindrical coordinate (z, θ, r) system.

Next, Bernoulli equation in the first order of ε is considered. In the limit $\varepsilon \to 0$, taking time derivative of (15) and substituting for $\partial\Psi_\beta/\partial t'_\beta$ in the resulting equation from (15) itself leads to the wave equation [35]

$$\frac{\partial^2\Psi_\beta}{\partial t_\beta'^2} = \mathrm{v}_{\mathrm{w}\,\beta-1}^2 \nabla_{z'}^2 \Psi_\beta \tag{24}$$

Since wave function in (24) guides the motion of "particle" that is a singularity on the wave, one moves to the adjacent lower scale (β − 1) and introduces *space and time coordinates* [35, 43]

$$\zeta = (z'_\beta - z'_o)/\varepsilon, \quad \tau = (t'_\beta - t'_o)/\varepsilon \tag{25}$$

It is important to emphasize that the time and space coordinates (25) are based on internal *spacetime* governed by thermodynamic temperature $T_{\beta-1}$ as discussed in [43]. Internal wavelength and frequency are not independent and wave number $k_\beta \equiv 2\pi/\lambda_\beta$ and angular frequency $\omega_\beta \equiv 2\pi\nu_\beta$ must follow the relations

$$k_{\mathrm{w}\beta} = 3k_{\mathrm{r}\beta}, \quad \omega_{\mathrm{w}\beta} = \sqrt{3}\omega_{\mathrm{r}\beta} \tag{26}$$

in order to satisfy the relationship between root-mean–square and most-probable speeds $\mathrm{v}_{\mathrm{r}+\beta} = \sqrt{3}\mathrm{v}_{\mathrm{w}\beta}$ [43]. Therefore, one introduces the scaled space and time coordinates

$$\xi_\beta = 3\zeta_\beta, \quad t_\beta = \sqrt{3}\tau_\beta \tag{27}$$

Substitution from (27) into (24) leads to the wave equation

$$\frac{\partial^2\Psi_\beta}{\partial t_\beta^2} = \mathrm{c}_{\beta-1}^2 \nabla_\xi^2 \Psi_\beta \tag{28}$$

with root-mean-square speed or speed of "sound" defined as $\mathrm{c}_\beta^2 = \mathrm{v}_{\mathrm{r}+\beta}^2 = 3\mathrm{v}_{\mathrm{w}\beta}^2$.

The separated product solution of (28) is the complex wave function

$$\Psi_\beta = \Phi'_\beta(\zeta, \tau) = \mathrm{e}^{i3k_r\zeta}\mathrm{e}^{-\mathrm{i}\sqrt{3}\omega_r\tau} = \Phi'_\beta(\xi, t) = \mathrm{e}^{ik_r\xi}\mathrm{e}^{-i\omega_r t} \tag{29}$$

with frequency $\omega_{\mathrm{r}\beta-1}$ associated with *stationary state* thus the vanishing of time dependence. In other words, because at stationary state the mean velocity must be zero, the frequency relation $\hbar^2\omega_r^2 = 2\hbar^2\omega_{r+}^2$ must hold in order to maintain the stationary internal kinetic energy given by the relation from (21).

The wave (28) and the solution in (29) after substitution from (23) for internal energy $\tilde{U}_\beta$ result in the *invariant time-independent Schrödinger equation* [35, 37, 48]

$$\frac{\hbar^2}{2m_\beta}\nabla_\xi^2\psi_\beta + (\tilde{H}_\beta - \tilde{V}_\beta)\psi_\beta = 0 \tag{30}$$

By equations (13, 27–30) ψ_β represent stationary state of spatial geometry of velocity potential $\Phi'_\beta(\xi)$ governing the peculiar motion of particles. Clearly, any interference with the stationary field by a measuring device will disturb the velocity potential and hence lead to "*collapse of the wave function*" ψ_β.

The time-independent Schrödinger equation is next employed to define a new time-dependent wave function

$$\Psi_\beta(\xi, t) = \Phi'_\beta(\xi, t) = \psi_\beta(\xi)e^{-2i\omega_r t'_\beta} = \psi_\beta(\xi)e^{-(4/3)i\omega_{r+}t_\beta} \tag{31}$$

involving a new time coordinate $t'_\beta = (2/3)^{1/2}\tau_\beta = (\sqrt{2}/3)t_\beta$. The multiplicative factor of two in frequency is because the period of traveling wave is half of that of stationary wave. The wave function in Eq. (31) with the factor of (4/3) multiplying the frequency $\omega_{r+\beta}$ results in total atomic thermal energy or atomic enthalpy $(4/3)\hbar\omega_{r+\beta} = (4/3)mv_{r+\beta}^2 = 4mv_{w\beta}^2 = 4kT_\beta = \tilde{H}_\beta$. In view of the definition of invariant Planck constant $h_\beta = m_\beta\lambda_{r\beta}v_{r\beta} = h$ [32], Eq. (31) gives energy and momentum *operators* [49]

$$i\hbar_\beta\frac{\partial\Psi_\beta}{\partial t_\beta} = \tilde{H}_\beta\Psi_\beta \tag{32a}$$

$$-i\hbar_\beta\frac{\partial\Psi_\beta}{\partial\xi_\beta} = \overline{p}_{r\beta}\Psi_\beta \tag{32b}$$

where $\overline{p}_{r\beta} = m_\beta v_{r\beta}$ is the root-mean-square momentum. Multiplying (30) by the new time-dependent part $e^{-(4/3)i\omega_{r+}t_\beta} = e^{-i\tilde{H}t_\beta/\hbar}$ from (31) and substitution from (32a) leads to the *invariant time-dependent Schrödinger equation* [35, 48]

$$i\hbar_\beta\frac{\partial\Psi_\beta}{\partial t_\beta} + \frac{\hbar_\beta^2}{2m_\beta}\nabla_\xi^2\Psi_\beta - \tilde{V}_\beta\Psi_\beta = 0 \tag{33}$$

Therefore, the energy in (32a) corresponding to the classical Hamiltonian is the atomic enthalpy $\tilde{H}_\beta = (4/3)mc_\beta^2 = (4/3)(3mv_{w\beta}^2) = 4kT_\beta$ that is the sum of the atomic *internal energy* $\tilde{E}_\beta = mv_{r+\beta}^2 = mc_\beta^2 = 3mv_{w\beta}^2 = 3kT_\beta$ and *atomic potential energy* $\tilde{V}_\beta = mV_{x\beta}'^2 = p_\beta\hat{v}_\beta = \tilde{U}_\beta/3 = kT_\beta$. Hence, enthalpy as the sum

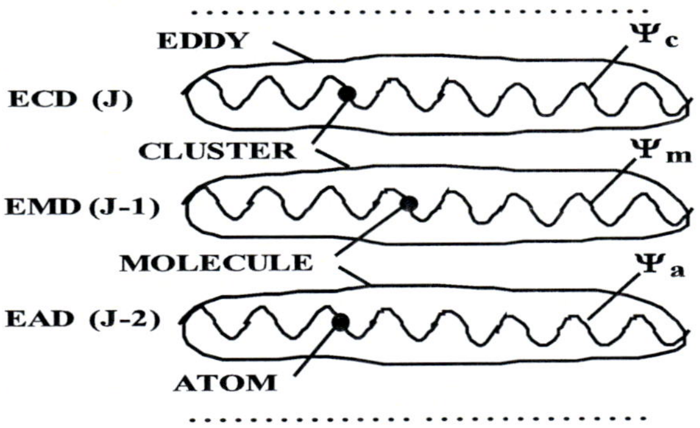

Fig. 2 Macroscopic wave functions Ψ_β or de Broglie guidance waves at (ECD), (EMD), and (EAD) scales that guide particles identified as wave-packets or de Broglie matter-waves [46]

of kinetic or "*electromagnetic*" energy $\tilde{U}_\beta$ and potential or "*gravitational*" energy $\tilde{V}_\beta$ of equilibrium radiation in enclosures is in exact agreement with the pioneering prediction of Hasenöhrl [50, 51]

$$\tilde{H}_\beta = \tilde{E}_\beta + \tilde{V}_\beta = mc_\beta^2 + (1/3)mc_\beta^2 = (4/3)mc_\beta^2 \tag{34}$$

However, as opposed to $\tilde{E}_\beta = (p_x^2 + p_y^2 + p_z^2)/2m$ in classical model [49], the atomic internal energy is now defined as the kinetic energy associated with internal translational, rotational, and vibrational degrees of freedom [32] in accordance with (21).

Since the most probable element at scale β is the entire system of statistical field at lower scale $\beta - 1$ (see Fig. 1), once again one finds a velocity potential $\Phi'_{\beta-1}$ hence a new wave function $\Psi_{\beta-1} = \Phi'_{\beta-1}$ Therefore, in harmony with de Broglie [2, 3] picture of quantum mechanics, motion of "*particle*" or "*wave-packet*" is guided by an external wave function as shown in Fig. 2.

As an example, let us consider stationary fluid corresponding to the field of equilibrium cluster-dynamics $\beta = c$ where the "atom" is a cluster $\mathbf{u}_c = \mathbf{v}_{wm}$ that by (2) is the most-probable element of the adjacent lower scale of equilibrium molecular-dynamics $\beta - 1 = m$. The random motion of clusters accounts for the phenomena of Brownian motions as discussed in [35]. The molecules as sub-particles of $\beta = c$ field are confined within the most-probable molecular cluster that is stabilized by an external force induced by *Poincaré stress* [35] and follow the wave (28) hence Schrödinger (30). It is important to emphasize that the wave (28) for quantum mechanics wave function Ψ_c by definition (13) is the velocity potential of the peculiar particle velocity in ECD field. Therefore, in harmony with de Broglie picture of quantum mechanics [2, 3], the "outer" scale $\beta = c$ wave function $\Psi_\beta = \Psi_c$ guides the motion of particle or molecule identified as wave-packet as shown in Fig. 2.

In order to connect the quantum mechanics wave function to *particle localization*, one moves to the stationary coordinates in (16) and obtains for the first perturbation of density $\rho'_\beta = \rho_{o\beta} + \varepsilon\rho_\beta$ from continuity (4) in the absence of chemical reactions

$$\frac{\partial \rho_\beta}{\partial \tau_\beta} + \mathrm{v_w}.\nabla_\zeta \rho_\beta = 0 \tag{35}$$

Taking time derivative of (35) and substituting for $\partial \rho_\beta / \partial \tau_\beta$ in the resulting equation from (35) itself one obtains the wave equation propagating at $\mathrm{v}_{\mathrm{w}\beta-1}$ similar to (24) that after the introduction of scaled coordinates in (27) leads to the density wave equation [52]

$$\frac{\partial^2 \rho_\beta}{\partial t_\beta^2} = c_{\beta-1}^2 \nabla_\xi^2 \rho_\beta \tag{36}$$

Hence, under *stationary states*, density hence particle localization correlate with quantum mechanics wave function Ψ_β. Indeed, by (35) and (36) it can be shown that $\rho_\beta^{1/2}$ also satisfies a wave equation similar to (36) that when combined with (28) through cross-multiplication result in a new solution hence a modified quantum mechanics wave function $\Psi_\beta = \rho_\beta^{1/2}\Psi_\beta$ such that $\boldsymbol{\Psi}_\beta \boldsymbol{\Psi}_\beta^* = \rho_\beta$ in harmony with the classical result [49]. Therefore, both objective and subjective aspect of quantum mechanics wave function discussed by de Broglie [3] are clarified. This is because density is the real hence *objective* part of $\boldsymbol{\Psi}_\beta$ that accounts for particle localization. On the other hand, the complex i.e. imaginary part of $\boldsymbol{\Psi}_\beta$ is its *subjective* part that accounts for normalization hence success of Born [53] probabilistic interpretation of $\boldsymbol{\Psi}_\beta$.

The results shown in Fig. 2 and the *objective* versus *subjective* aspects of $\boldsymbol{\Psi}_\beta$ discussed above also resolve the *wave-particle duality* problem in quantum mechanics. This is because particle that appears as a local singularity is actually a de Broglie *wave packet* at scale $\beta - 1$ that is embedded within and is guided by an "external" complex hence virtual wave function $\boldsymbol{\Psi}_\beta$ associated with velocity potential of peculiar atomic motions in the background *space* composed of atoms of the same scale. The adjective "*external*" is because the velocity potential $\Psi_\beta = \Phi'_\beta$ refers to peculiar motion at outer scale β.

According to the new paradigm of physical foundation of quantum mechanics, each equilibrium statistical field is composed of a spectrum of cluster or wave-packet sizes containing "atoms" with velocity, speed, and energy respectively following Gauss, Maxwell–Boltzmann, and Planck distribution functions. For example, the statistical field of equilibrium-electro-dynamics ESD (Fig. 1) takes place within and is in thermodynamic equilibrium with the background physical space that is the field of EKD or Casimir vacuum. In view of Maxwell–Boltzmann distribution function, the spectrum of "atomic" clusters must remain *stochastically stationary* by the principle of detailed balance. Transition of an electron from a small fast-oscillating "atom" (high energy-level-j) to a large slowly-oscillating "atom" (low energy-level-i) will results in emission of a "sub-particle" that is a photon γ_{ji} to carry away the excess energy given by Bohr [54] frequency formula $\Delta\varepsilon_{ji\beta} = h(\nu_{j\beta} - \nu_{i\beta})$as schematically shown in Fig. 3a,

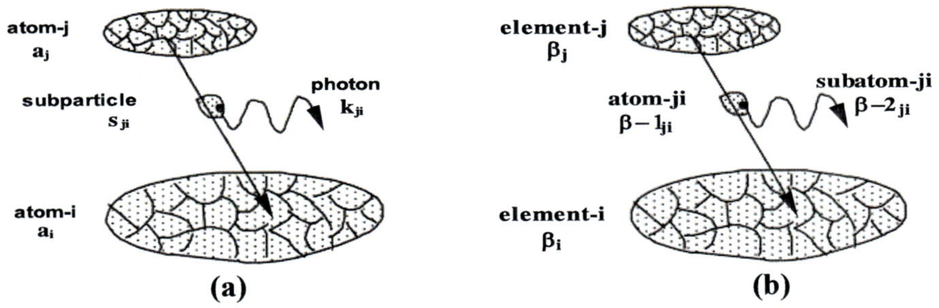

Fig. 3 Transition of electron s_{ij} from atom-j to atom-i leading to emission of photon k_{ij} (b) Generalized transitions [36]

Therefore, *stochastically stationary* sizes of particle clusters (energy levels) are identified as Bohr *stationary-states* of quantum mechanics [54] and must satisfy the *stationarity criteria* imposed by Maxwell–Boltzmann distribution. A generalized scale-invariant concept of "atomic" transitions is shown in Fig. 3b. For example, at cosmic scales $\beta = g$ (Fig. 1), transition of an "atom" i.e. galaxy from a small rapidly-oscillating galactic cluster (high-energy-level j) to a large slowly-oscillating cluster (low-energy-level i) results in emission of a star s_{ij} that constitutes a "subatomic" particle of cosmic field [43]. Such quantum transitions between spectrums of particle clusters (Fig. 3) are in harmony with quantum transitions between different "cells" in recent *cellular automaton* model of quantum mechanics [55].

Finally, we examine the influence of the nonlinear term in (15) by taking the time derivative of (15) and substituting for $\partial\Psi_\beta/\partial\tau_\beta$ in the result from (15) itself, and introducing internal coordinates from (25) to obtain the non-linear equation

$$\frac{\partial^2\Psi}{\partial\tau^2} = v_{w\beta-1}^2\nabla_\zeta^2\Psi + 2v_{w\beta-1}\nabla_\zeta\Psi\nabla_\zeta^2\Psi + (\nabla_\zeta\Psi)^2\nabla_\zeta^2\Psi \tag{37}$$

Comparisons of numerical analysis of (37) for double-slit problem with results obtained from non-linear Schrödinger equations involving what de Broglie called *quantum potential* in de Broglie-Bohm [56] model of quantum mechanics will be most interesting.

4 Scale-Invariant Model of Normal Shock and Its Hydro-Thermo-Diffusive Structure

The wave (28) and (36) correspond to "stochastically stationary" equilibrium state with coordinate system $z'_\beta = x'_\beta - v_{w\beta}t'_\beta$ involving the most probable speed v_{w_β}. Parallel to density wave in (36), one obtains a momentum wave equation from momentum conservation (6) that after introduction of coordinates in (27) results in

$$\frac{\partial^2 \mathbf{v}_\beta}{\partial \mathbf{t}_\beta^2} = c_{\beta-1}^2 \nabla_\xi^2 \mathbf{v}_\beta \tag{38}$$

The momentum waves in (38) propagate at the speed of "sound" or root-mean-square speed $c_{\beta-1} = v_{r^+\beta-1} = \sqrt{3} v_{w\beta-1}$ of molecules [32]. Similar procedures applied to conservation (4)–(5) lead to density, temperature (thermal), and pressure waves [52]

$$\frac{\partial^2 f_\beta}{\partial t_\beta^2} = c_{\beta-1}^2 \nabla_\xi^2 f_\beta, \quad f_\beta = \rho_\beta, T_\beta, p_\beta \tag{39}$$

The derivation of (39) involves the assumption of ideal gas $p_\beta = \rho_\beta R_\beta T_\beta$ with gas constant as ratio of universal gas constant and molecular weight $R_\beta = R^o/\tilde{w}_\beta$ and the absence of mass and heat diffusivities.

Since each statistical field in the hierarchy (Fig. 1) has a root-mean-square speed $v_{r\beta}$ and usually a much faster "atomic" speed u_β, in view of (39) one may associate a "wave" and a "particle" speed with each statistical field [43, 52]

$$v_{r^+\beta} = c_\beta \quad \textbf{Wave speed} \tag{40a}$$

$$u_\beta = v_{w\beta-1} \quad \textbf{Particle emission speed} \tag{40b}$$

For example, in statistical field of ECD at scale $\beta + 1 = c$ the sound waves occur at the adjacent lower LMD scale $\beta = m$ and follow (39) and hence

$$v_{r^+m} = c_m \quad \textbf{Acoustic speed} \tag{41a}$$

$$u_m = v_{wa} \quad \textbf{Particle emission speed} \tag{41b}$$

where the speed of "*sound*" waves in standard atmospher is about $v_{r^+\beta} = v_{r^+m} = c_m \approx 340\,\text{m/s}$[57]. The velocity of particle emission in (41b) on the other hand is the speed of single molecule $u_m = v_{wa} \approx 1200\,\text{m/s}$ that is the speed of typical detonation wave [58].

At the scale LKD (Fig. 1) physical space is identified as Casimir vacuum [45] and is considered to be a compressible fluid, Planck compressible ether [59] as discussed in [60]. Lorentz perceptions about the medium of space as Aristotle or Huygens ether [60] is further described in the following quotation by Verhulst [61] from Lorentz 1915 lecture at the Royal Academy of Sciences in Amsterdam:

> Why can we not speak of the ether instead of vacuum? Space and time are not symmetric; a material point can at different times be at different spots, but not in different places at the same time

Therefore, it is reasonable to anticipate that density waves in (35) will be connected to waves of *space curvature* thus corresponding to the recently observed *gravitational waves* first discussed by Poincaré [62]. It is ironic that in their 1935 paper submitted to *Physical Review*, Einstein and Rosen denied the existence of gravitational waves but later changed their opinion due to discovery of the error by Robertson [63]. Although gravitational waves travel at the speed of light $v_{gw} = v_{r+k} = c_k = c = 2.9978 \times 10^8$ m/s, *gravitational emission (gravitational radiation)* [52] propagates at exceedingly larger speed of tachyon waves estimated as $c_t = 7.7 \times 10^{13}$ c making the entire universe *causally connected* [52, 60]. This is in harmony with the perceptions of Laplace who, as was noted by Poincaré [64], believed that the speed of gravitational signal is a million times faster than that of light. Such superluminal signals could resolve the *entanglement* problem by providing for *ontological* description of long-distance correlations between entangled particles such as photons [55].

The scale-invariant definition of the speed of "sound" waves described above leads to invariant dimensionless number v_β/c_β called (*Mach, Lorentz, Michelson*) numbers ($Ma = v/c_m$, $Lo = v/c_e$, $Mi = v/c_k$) associated with (supersonic, super-electronic, and superluminal) flows [43]. Thus, supersonic flow of gas ($Ma > 1$), super-electronic flow of plasma ($Lo > 1$), and superluminal flow of gravitons ($Mi > 1$) lead to the formation of (*Mach, Lorentz,* and *Poincaré-Minkowski*) cones as illustrated in Fig. 4 [43, 65].

As a result, statistical field of scale β will be separated from the statistical field at adjacent lower scale of $\beta - 1$ by a surface of discontinuity called *shock wave* [65] as schematically shown in Fig. 5.

In Fig. 5, a stationary body at B is being approached by supersonic flow from the left. Hence, the fluid with properties ($c_{\beta-1}$, $\rho_{\beta-1}$, $T_{\beta-1}$, $p_{\beta-1}$) to the left of the shock is "supersonic" and that to the right of the shock with properties (c_β, ρ_β, T_β, p_β) is subsonic. For example, gaseous supersonic flow at LAD scale $\beta = a$ with signal speed $\mathbf{v}_{r+a} = c_a > c_m$ arrives at point A of the shock and a subsonic flow $\mathbf{v} \le c_m$ at LMD scale $\beta = m$ leaves the shock wave at point C. Similarly, but at much higher speeds encountered in cosmology, a superluminal flow at LTD scale $\beta = t$ with wave speed

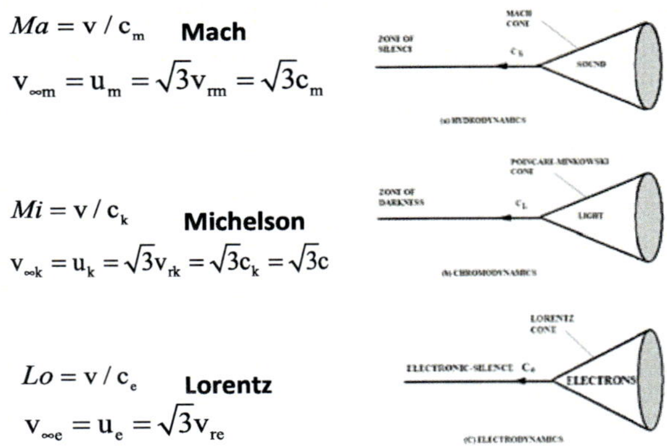

Fig. 4 "Supersonic" flows at (molecular-, chromo-, and electro-dynamics) scales leading to the formation of (*Mach, Poincaré-Minkowski, Lorentz*) cones [43]

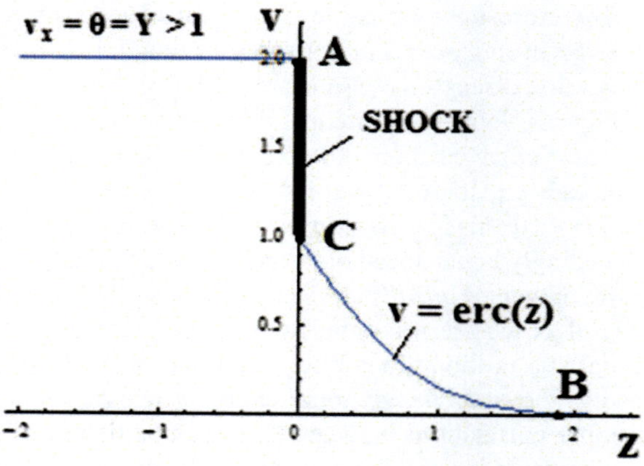

Fig. 5 Shock in polyatomic gas with thin and thick layers corresponding to type C shock structure of Taniguchi et al. [66]

$v_{r+t} = c_t > c_k = c$ arrives at point A of the *event-horizon* or "shock" surrounding a black hole at point B and a luminal flow with wave speed $v_{r+k} = c_k$ at LKD scale $\beta = k$ leaves the shock at point C. This invariant model of generalized shock waves is in harmony with Unruh [67] "*dumb-hole*" model of black hole.

The nature of jump-like transition across diverse types of generalized "shock waves" is interesting. In classical gas dynamics, the hydro-thermo-diffusive structure of normal shock will be governed by (39), with the reduced temperature Θ profile predicted as [65]

$$\Theta = -\mathrm{erf}(\xi) = \mathrm{erf}(0.2 - y) \tag{42}$$

in close agreement with the experimental data [68] shown in Fig. 6.

It is reasonable to anticipate that similar error-function type profiles (Fig. 6) would govern the dimensionless velocity, density, pressure, and temperature inside generalized "shock" waves in the hierarchy of statistical fields shown in Fig. 1. We next

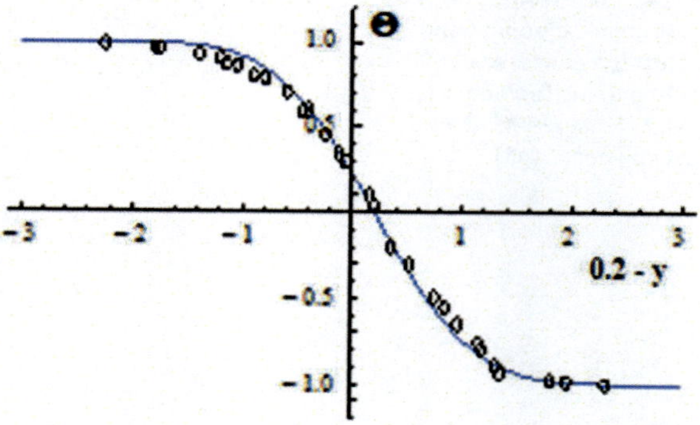

Fig. 6 Comparison between measured normalized wire temperature Θ versus position (0.2–y) in normal shock [68] and theory [65]

examine the nature of a superluminal shock wave at stochastic chromodynamics scale.

As discussed in Sect. 2, absolute thermodynamic temperature identified as Wien wavelength $T_\beta = \lambda_{w\beta}$ of thermal oscillations [32] also defines the internal measures of extension and duration $(\lambda_{w\beta}, \tau_{w\beta})$ called *spacetime* [43]. For example, crossing a shock wave from supersonic laminar-atomic-dynamic (LAD) side with $T_a = \lambda_{wa}$ to the subsonic laminar-molecular-dynamic (LMD) side $T_m = \lambda_{wm}$ corresponds to transition from one Euclidean space with larger λ_{wa} (high temperature) to another Euclidean space with smaller λ_{wm}(low temperature). However, the description of shock internal structure necessarily involves varying temperatures (Fig. 5) hence *variable-measure* or *non-Euclidean* geometry.

Following Poincaré [69] description of hyperbolic space, one expresses temperature as $T_\beta = (R^2 - r^2)^{1/2}$ with square root added to account for the dimension of absolute temperature (meter) as Wien wavelength of thermal oscillations [32]. If one postulates that space "curvature" be inversely related to absolute temperature $\kappa_\beta \propto 1/T_\beta$ it will lead to $(\kappa \to 0, \kappa \to \infty)$ at $T_\beta = (\infty, 0)$ resulting in formation of (white hole, black hole) at (center, circumference) of *Poincaré disk* [43, 69]. Also, as discussed in Sect. 2, thermodynamic temperature defines internal spacetime leading to Poincaré [62] and Minkowski [70] 4-dimesional spacetime $(x_\beta, y_\beta, z_\beta, t_\beta)$. However, 6 additional *compactified* dimensions could be associated with 3-rotational and 3-vibrational *internal degrees of freedom* in (20a, 20b). Therefore, the total number of dimensions required for the description of each statistical field (Fig. 1) including the physical space or Casimir [45] vacuum will be $4 + 6 = 10$ in harmony with models of superstring theories [71].

By the equation of state $p_\beta = \rho_\beta R_\beta T_\beta$ at constant pressure, density is also inversely related to absolute temperature. Because pressure can be viewed as *volumetric potential energy density* of the field [32], negative values of pressure, often assumed in inflationary models of cosmology, are expected to be nonphysical. In view of finite value of Casimir [45] zero-point energy, it is reasonable to anticipate a finite positive pressure of Casimir vacuum in accordance with modified van der Waals equation of state [72]

$$p_{rg} = p_{ra} - p_{rv} = \frac{1}{Z_c}\left[\frac{T_r}{\tilde{v}_r - 1/3} - \frac{9}{8\tilde{v}_r^2} + Z_c - \frac{3}{8}\right] \tag{43}$$

Since critical compressibility factor of all substances are smaller than that of van der Waals fluid $Z_{c,vdw} = 3/8$, in the limit $\tilde{v}_r \to \infty$ (43) leads to positive reduced vacuum pressure [72]

$$p_{rv} = (3/8)/Z_c - 1 \geq 0 \tag{44}$$

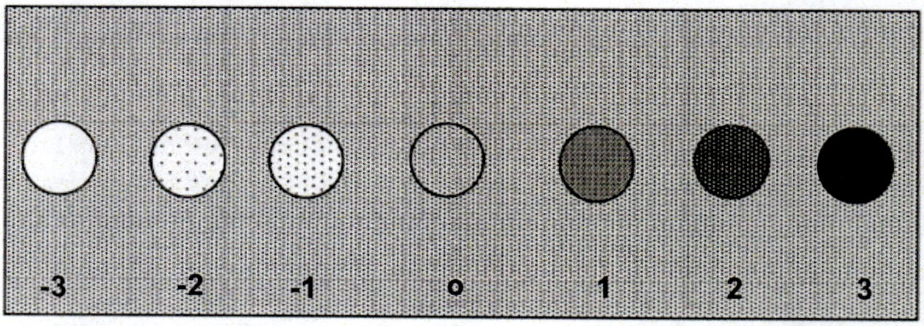

Fig. 7 Different degrees of rarefaction and compression of Casimir vacuum identified as a compressible fluid. **(−3)** White hole $\rho_{WH} = 0$**(−2, −1)** Anti-matter $\rho_{AM} < \rho_v$ **(0)** Casimir vacuum $\rho = \rho_v$ **(1, 2)** Matter $\rho_{MA} > \rho_v$ **(3)** Black hole $\rho_{BH} = \infty$ [43, 60]

A schematic diagram of physical space as states of a compressible fluid from infinite rarefaction (*white hole* WH) to infinite compression (*black hole* BH) corresponding to pressure range $0 = p_{WH} < p_{am} < p_v < p_m < p_{BH} = \infty$ is shown in Fig. 7.

Therefore, if one introduces the concept of space "scalar curvature" as deviation of space density from the density of Casimir [45] vacuum $\kappa \equiv \rho - \rho_v$, one finds that $(\kappa > 0, \kappa = 0, \kappa < 0)$ could be respectively associated with (*Riemannian, Euclidian, Lobachevskian*) space. The hydrodynamic model of chromodynamics shock waves in compressible space (Fig. 7) is in harmony with the perceptions of 't Hooft [73] concerning quantum gravity as a dissipative deterministic dynamic system. The application of the model to derive Schrödinger (33) at photonic scale of Casimir vacuum $\beta = k$ suggests that density waves in (39) correspond to *gravitational waves* that propagate at the speed of light [32, 43, 52, 60].

5 Concluding Remarks

A scale-invariant model of Boltzmann statistical mechanics was applied to derive invariant Schrödinger equation from invariant Bernoulli equation by way of Hamilton–Jacobi equation of classical mechanics. The nature of time-dependent and time-independent Schrödinger equations and the corresponding quantum mechanics wave functions in connection to *stationary states* of the system were described.

The invariant hydrodynamic model resulted in generalized shock waves and introduction of a scale-invariant definition of Mach number v_β/c_β. Hence, "supersonic" flows at molecular-, electro-, and chromo-dynamics scales result in formation of (…, *Mach, Lorentz, Poincaré–Minkowski, …*) cones separating "super-sonic" from "subsonic" sides. The internal hydro-thermo-diffusive structure of generalized shock waves were discussed. Finally, a model of space curvature as deviation of density from Casimir vacuum density $\kappa = \rho_\beta - \rho_v$ was introduced leading to $(\kappa > 0, \kappa = 0, \kappa < 0)$ corresponding to (*Riemannian, Euclidean, Lobachevskian*) space. The results are in harmony with quantum gravity considered as dissipative deterministic dynamic system [73].

Acknowledgements This research was in part supported by NASA grant No. NAG3-1863.

References

1. L. de Broglie, interference and corpuscular light. Nature **118**, 441–442 (1926); Sur la Possibilité de Relier les Phénomènes d'Interférence et de Diffraction à la Théorie des Quanta de Lumière, C. R. Acad. Sci. Paris, 183, 447-448, 1927; La Structure Atomique de la Matière et du Rayonnement et la Mécanique Ondulatoire **184**, 273–274 (1927); Sur le Rôle des Ondes Continues en Mécanique Ondulatoire **185**, 380–382 (1927)
2. L. de Broglie, *Non-Linear Wave Mechanics: A Causal Interpretation* (Elsevier, New York, 1960)
3. L. de Broglie, The reinterpretation of wave mechanics. Found. Phys. **1**(5), 5–15 (1970)
4. E. Madelung, Quantentheorie in hydrodynamischer form. Z. Physik. **40**, 332–326 (1926)
5. E. Schrödinger, Über die Umkehrung der Naturgesetze. Sitzber Preuss Akad Wiss Phys-Math Kl **193**, 144–153 (1931)
6. R. Fürth, Über Einige Beziehungen zwischen klassischer Staristik und Quantenmechanik. Z. Phys. **81**, 143–162 (1933)
7. D. Bohm, A suggested interpretation of the quantum theory in terms of "Hidden Variables" I. Phys. Rev. **85**(2), 166–179 (1952)
8. T. Takabayasi, On the foundation of quantum mechanics associated with classical pictures. Prog. Theor. Phys. **8**(2), 143–182 (1952)
9. D. Bohm, J.P. Vigier, Model of the causal interpretation of quantum theory in terms of a fluid with irregular fluctuations. Phys. Rev. **96**(1), 208–217 (1954)
10. E. Nelson, Derivation of the Schrödinger Equation from Newtonian Mechanics. Phys. Rev. **150**(4), 1079–1085 (1966)
11. E. Nelson, *Quantum Fluctuations* (Princeton University Press, Princeton, New Jersey, 1985)
12. L. de la Peña, New foundation of stochastic theory of quantum mechanics. J. Math. Phys. **10**(9), 1620–1630 (1969)
13. L. de la Peña, A.M. Cetto, Does quantum mechanics accept a stochastic support? Found. Phys. **12**(10), 1017–1037 (1982)
14. A.O. Barut, Schrödinger's Interpretation of ψ as a Continuous Charge Distribution. Ann. der Phys. **7**(4–5), 31–36 (1988)
15. A.O. Barut, A.J. Bracken, Zitterbewegung and the internal geometry of the electron. Phys. Rev. D **23**(10), 2454–2463 (1981)
16. J.P. Vigier, De Broglie, Waves on Dirac aether: a testable experimental assumption. Lett. Nuvo Cim. **29**, 14, 467–475 (1980); C. Cufaro Petroni, J.P. Vigier, Dirac's aether in relativistic quantum mechanics. Found. Phys. **13**, 2, 253–286 (1983); J.P. Vigier, Derivation of inertia forces from the Einstein-de Broglie-Bohm (E.d.B.B) Causal stochastic interpretation of quantum mechanics. Found. Phys. **25**, 10, 1461–1494 (1995)
17. F.T. Arecchi, R.G. Harrison, *Instabilities and Chaos in Quantum Optics* (Springer, Berlin, 1987)
18. O. Reynolds, On the dynamical theory of incompressible viscous fluid and the determination of the criterion. Phil. Trans. Roy. Soc. A **186**(1), 123–164 (1895)
19. D. Enskog, Kinetische Theorie der Vorgange in Massig Verdunnten Gasen, Almqvist and Wiksells Boktryckeri-A.B., Uppsala, 1917. English translation: G. S. Brush, *Kinetic Theory* (Pergamon Press, New York, 1965), pp. 125–225

20. G.I. Taylor, Statistical theory of turbulence-parts I-IV. Proc. Roy. Soc. A **151**(873), 421–478 (1935)
21. T. Kármán, L. Howarth, On the statistical theory of isotropic turbulence. Proc. Roy. Soc. A **164**(917), 192–215 (1938)
22. H.P. Robertson, The invariant theory of isotropic turbulence. Proc. Camb. Phil. Soc. **36**, 209–223 (1940)
23. A.N. Kolmogoroff, Local structure on turbulence in incompressible fluid. C. R. Acad. Sci. U. R. S. S. **30**, 301–305 (1941); A refinement of previous hypothesis concerning the local structure of turbulence in a viscous incompressible fluid at high reynolds number. J. Fluid Mech. **13**, 82–85 (1962)
24. A.M. Obukhov, On the distribution of energy in the spectrum of turbulent flow. C. R. Acad. Sci. U. R. S. S. **32**, 19–22 (1941); Some specific features of atmospheric turbulence. J. Fluid Mech. **13**, 77–81 (1962)
25. S. Chandrasekhar, Stochastic problems in physics and astronomy. Rev. Mod. Phys. **15**(1), 1–89 (1943)
26. S. Chandrasekhar, *Stochastic, Statistical, and Hydrodynamic Problems in Physics and Astronomy*. Selected Papers, vol. 3 (University of Chicago Press, Chicago, 1989), pp. 199–206
27. W. Heisenberg, On the theory of statistical and isotropic turbulence. Proc. Roy. Soc. A **195**, 402–406 (1948); Zur Statistischen Theorie der Turbulenz. Z. Phys. **124**, 7–12, 628–657 (1948)
28. G.K. Batchelor, *The Theory of Homogeneous Turbulence* (Cambridge University Press, Cambridge, 1953)
29. L.D. Landau, E.M. Lifshitz, *Fluid Dynamics* (Pergamon Press, New York, 1959)
30. H. Tennekes, J.L. Lumley, *A First Course in Turbulence* (MIT Press, 1972)
31. S.H. Sohrab, Some implications of a scale invariant model of statistical mechanics to classical and relativistic thermodynamics. Int. J. Thermo. **17**(4), 233–248 (2014)
32. S.H. Sohrab, On a scale-invariant model of statistical mechanics and the laws of thermodynamics. J. Energy Resources and Technol. **138**(3), 1–12 (2016)
33. S.H. Sohrab, Invariant forms of conservation equations and some examples of their exact solutions. J. Energy Resour. Technol. **136**, 1–9 (2014)
34. S.H. Sohrab, Solutions of modified equation of motion for laminar flow across (within) rigid (liquid) sphere and cylinder and resolution of stokes paradox. AIP Conf. Proc. **1896**, (2017)
35. S.H. Sohrab, Quantum theory of fields from planck to cosmic scales. WSEAS Trans. Math. **9**, 734–756 (2010)
36. S.H. Sohrab, On a scale invariant model of statistical mechanics, kinetic theory of ideal gas, and riemann hypothesis. Int. J. Mod. Commun. Tech. Res. **3**(6), 7–37 (2015)
37. S.H. Sohrab, Scale invariant model of statistical mechanics and quantum nature of space, time, and dimension. Chaot. Model. Simul. (CMSIM) **3**, 231–245 (2016)
38. R.S. de Groot, P. Mazur, *Nonequilibrium Thermodynamics* (North-Holland, 1962)
39. H. Schlichting, *Boundary-Layer Theory* (McGraw Hill, New York, 1968)
40. F.A. Williams, *Combustion Theory*, 2nd edn. (Addison Wesley, New York, 1985)
41. J.O. Hirschfelder, C.F. Curtiss, R.B. Bird, *Molecular Theory of Gases and Liquids* (Wiley, New York, 1954)
42. S. Chapman, T.G. Cowling, *The Mathematical Theory of Non-Uniform Gases* (Cambridge University Press, Cambridge, 1953)
43. S.H. Sohrab, An invariant model of Boltzmann statistical mechanics and some of its implications to thermodynamics and quantum nature of space and time. WSEAS Tran. Appl. Theo. Mech. **13**, 199–212 (2018)
44. M. Planck, On the Law of the Energy Distribution in the Normal Spectrum. Ann. der Phys. **4**, 553–558 (1901)
45. H.B.G. Casimir, On the attraction between two perfectly conducting plates. Proc. K. Ned. Akad. Wet. **51**, 793–795 (1948)
46. S.H. Sohrab, Universality and statistical nature of turbulence, quantum mechanics, and chaos, in *9th CHAOS Conference Proceedings*, 23–26 May 2016, Senate House, University of London, UK (2016)

47. R. Clausius, Über die Art der Bewegung, welche wir Wärme nennen. Annalen der Phyzik **176**(3), 353–380 (1957); Philos. Mag. **14**,108-127 (1957)
48. E. Schrödinger, Quantization as a problem of proper values, Part I. Ann. der Phys. **79**, 4 (1926)
49. W. Schommers, Evolution of Quantum mechanics, in *Quantum Theory and Pictures of Reality*, ed. by W. Schommers (Springer, Berlin, 1989), pp. 1–48
50. F. Hasenöhrl, Zur Theorie der Strahlung in bewegten Körpern. Ann. der Phys. **16**, 589–592 (1905)
51. S.H. Sohrab, Invariant laws of thermodynamics and validity of Hasenöhrl mass-energy equivalence formula m = (4/3) E/c^2 at photonic, electrodynamic, and cosmic scales. Bull. Amer. Phys. Soc. **62**, 1, 124 (2017)
52. S.H. Sohrab, Implications of a scale invariant model of statistical mechanics to nonstandard analysis and the wave equation. WSEAS Trans. Math. **3**(5), 93 (2008)
53. M. Born, Zur Quantenmechanik der Stoßvorgänge. Z. Physik **37**, 863 (1926)
54. B.L. van der Waerden, Towards quantum mechanics, in *Sources of Quantum Mechanics*, ed. by B.L. van der Waerden (Dover, New York, 1967), pp. 1–59
55. G. 't Hooft, *The Cellular Automaton Interpretation of Quantum Mechanics* (Springer, 2016)
56. J.T. Cushing, *Quantum Mechanics* (The University of Chicago Press, Chicago, 1994)
57. K.A. Krout, S.H. Sohrab, On the speed of sound. Int. J. Thermodyn. **19**(1), 29–34 (2016)
58. B. Lewis, A chain reaction theory of the rate of explosion in detonating gas mixtures. J. Amer. Chem. Soc. **52**(8), 3120 (1930)
59. H.A. Lorentz, *Selected Works of L.H. Lorentz*, vol. 5, ed. by N.J. Nersessian, H.F. Cohen (Palm Publications, Nieuwerkerk, 1987), p. 7
60. S.H. Sohrab, Invariant model of statistical mechanics, quantum mechanics, and physical nature of space and time, in *8th CHAOS Conference Proceedings, 26–29 May 2015, Henri Poincaré Institute,* Paris, France (2015)
61. F. Verhulst, *Henri Poincaré, Impatient Genius* (Springer, 2012), p. 63
62. H. Poincaré, Sur la Dynamique de l'Electron. C. R. Acad. Sci. Paris **140**, 1504–1508 (1905); Rend. Circ. Mat. Palermo **21**, 9–175 (1906)
63. D. Kenneflick, Einstein Versus the Physical Review, Physics Today, September 43–48 (2005)
64. H. Poincaré, The Present and the Future of Mathematical Physics, Address delivered before the Section of Applied Mathematics of the *International Congress of Arts and Science*, St. Louis, September 24, 1904; Bull. Amer. Math. Soc. **37**(1) 25–38 (2000)
65. S.H. Sohrab, On a scale invariant model of statistical mechanics and derivation of invariant forms of conservation equations. WSEAS Trans. Heat Mass Trans. **9**(15), 169–194 (2014)
66. S. Taniguchi, T. Arima, T. Ruggeri, M. Sugiyama, Thermodynamic theory of the shock wave structure in a rarified polytrophic gas: beyond the Bethe-Teller theory. Phys. Rev. E **89**, (2014)
67. W. Unruh, Sonic analogue of black holes and the effects of high frequencies on black hole evaporation. Phys. Rev. D. **51**, 2827 (1995)
68. F.S. Sherman, A low-density wind tunnel study of shock wave structure and relaxation phenomena in gases. NACA Tech. Notes **3298** (1955)
69. H. Poincaré, *Science and Hypothesis* (Dover, New York, 1952), p. 65
70. H. Minkowski, Space and Time, in *Theory of Relativity* (Dover, New York, 1952), p. 75
71. G. 't Hooft, *In Search of the Ultimate Building Blocks* (Cambridge University Press, 1998), p. 161
72. S.H. Sohrab, Modified van der Waals equation of state. WSEAS Trans. Biol. Biomed. **1**(4), 422–424 (2004)
73. G. 't Hooft, Quantum gravity as a dissipative deterministic system. Class. Quant. Grav. **16**, 3263–3279 (1999)

The Physics of Evolution and Breaking Symmetry

Vyacheslav Somsikov

Abstract Paper is devoted to explaining the nature of symmetry breaking of dynamical classical and quantum systems in the framework of evolutionary physics. A brief explanation of the deterministic mechanism of irreversibility is presented. The nature of the non-potential forces, which leads to symmetry breaking, is analyzed. The concept of evolutionary nonlinearity and the deterministic symmetry breaking based on the motion equation for the structural particle and modified Schrödinger equation is discussed. The nature of the potential, which follows from evolutionary nonlinearity and leads to violation of symmetry in classical and quantum systems, is considered.

Keywords Symmetry · Irreversibility · Evolution · Principle of least action · Entropy · Quantum systems · Mechanics · Dynamics

1 Introduction

The world is a collection of open nonequilibrium dynamic systems. Therefore, this world is characterized by the evolutionary processes of the emergence of new structures of matter, phase transitions, the appearance and disappearance of objects of the Universe, catastrophic phenomena on Earth, the climate change, and so on. To understand these processes, it is necessary to have a theoretical foundation to answer questions such as how, why and in what direction nature is developing. For this, first of all, it is necessary to understand the essence of the physical nature of evolution [1–5]. However, there is a big obstacle to this understanding [5]. It is a fact that the basic laws of physics, its formalisms are reversible. Therefore, modern physics describes the stationary world rather well, but does not always adequately describe the irreversible processes of evolution. This is also the reason for the lack of unity in various fields of physics, for example, mechanics, thermodynamics, statistical physics, quantum mechanics.

V. Somsikov (✉)
Al-Farabi Kazakh National University, Almaty 050040, Kazakhstan
e-mail: vmsoms@rambler.ru

C. H. Skiadas et al. (eds.), *13th Chaotic Modeling and Simulation International Conference*, Springer Proceedings in Complexity,
https://doi.org/10.1007/978-3-030-70795-8_63

The search for an explanation of irreversibility first led to its probabilistic mechanism. It is based on the property of exponential instability of phase trajectories of Hamiltonian systems and the hypothesis of the existence of random external fluctuations [4]. The probabilistic mechanism of irreversibility significantly helped in the development the kinetical and statistical methods of irreversible processes, in understanding the nature of chaos, in explaining spontaneous symmetry breaking in quantum systems [6].

The need to describe irreversible processes occurring in different forms of matter with symmetry breaking forces us to develop empirical methods for their description. Today these methods are based primarily on the Hamilton formalism. Although these methods are extremely complex and diverse, their essence reduces to perturbation theory by adding the corresponding empirical terms to the Hamiltonian [6, 7]. In addition, In addition, empirical equations of physical kinetics and statistical physics are actively used today to describe nonequilibrium evolutionary systems.

Although good agreement with experiments was obtained based on these methods, they do not reveal the nature of symmetry breaking and the mechanism of irreversibility. Moreover, these methods encounter problems in determining the scope of their use, in explaining the correctness of applying the Hamilton reversible formalism to solve irreversible problems in explaining cause-effect relationships in the corresponding processes [6, 8]. It does not answer key questions about the evolution of the picture of the world. For example, the following questions remain open: how does order emerge from chaos, how does nature choose the path "from simple to complex," how symmetry breaking leads to evolution, where evolution is directed. This is due, in particular, to the fact that probabilistic laws and principles are incompatible with the fundamental laws of physics. Indeed, the solution of problems concerning with the construction of an evolutionary picture of the world cannot but rely on the ideas of determinism, the cognoscibility of the world, its uniqueness, and also the closeness of the laws of physics. However, these ideas exclude probability in their starting positions. Therefore, the search for the mechanism of irreversibility within the framework of the determined laws of physics did not stop even after the discovery of the probabilistic mechanism of irreversibility [4, 5].

Only taking into account the role of the body structure in its dynamics, the deterministic mechanism of irreversibility (DMI) was discovered [9]. DMI has opened up the possibility of building a "physics of evolution". The task of the « physics of evolution" is to describe the key processes of evolution: the emergence, development and destruction of systems, the definition of the principles for constructing an evolutionary picture of the world "from simple to complex" based on the fundamental laws of the physicist.

A method for constructing the physics of evolution was found as follows. First of all, the model of a structured particle (SP) replaced the unstructured material point model in classical mechanics, where the SP is an equilibrium system of potentially interacting material points. Then, according to the principle of dualism of symmetry, the energy of the system was presented as the sum of the energy of motion and internal energy. After that, from this form of energy, the equation of motion of the system was obtained. This equation is irreversible. Irreversibility is determined by the nonlinear terms of this equation, which are proportional to the gradients of external forces. These terms provide the linking of the elements of the symmetry groups of micro and macro variables, which determine the change in the internal energy and the system's motion energy respectively. Thanks to these conditions, the motion energy is converted into internal energy when the total energy is conserved. Thus, due to the interaction of dynamic symmetry groups arising in the presence of gradients of external forces, irreversibility appears. In turn, DMI has opened the possibility of building a "physics of evolution".

In this paper, based on the deterministic mechanism of irreversibility in classical mechanics, an alternative explanation of the nature of symmetry breaking for quantum mechanics is proposed. For this, the properties of the equation of motion for a system of material points are first explained. Then the DMI mechanism and the concept of evolutionary nonlinearity is analyzed. It is shown how the free energy function follows from the evolutionary nonlinearity and how symmetry breaking in classical and quantum systems follows from this function. Then, the universal nature of symmetry breaking for nonequilibrium dynamical systems of classical and quantum mechanics is discussed.

2 System's Motion Equation

The key idea that led to the emergence of DMI was the idea of the need to take into account the influence of body structure on its dynamics. This idea arose as a result of studying the interaction of the simplest systems of elastically colliding disks [9, 10]. As a result of studying disks systems, the question arose of how to take into account the role of the body structure in its dynamics. As it turned out, this should be done based on the principle of dualism of symmetry. According to this principle, the motion of the structured bodies is determined not only by the symmetry of space, as in the case of an unstructured material point, but also by the internal symmetry of the body. The second question was how to take into account that these types of symmetry determine the motion of the system. As it turns out, this question is solved by representing the total energy of the system as the sum of the motion energy and internal energy. The body's motion equation was found from this form of energy when its model was represented in the form of systems of potentially interacting material points [9, 10]. The generality of this representation of the body model follows from the fact that, as a rule, all bodies with a good degree of approximation can be represented by a set of potentially interacting material points. As a result, the system's motion equation was obtained in the framework of the laws of classical mechanics without using Hamilton's reversible formalism, since this formalism is not applicable to the description of dissipative systems [11, 12].

Thus, the equation of motion of the system was obtained based on fundamental laws and principles. These laws and principles include the laws of conservation of energy, momentum, the Galileo principle, and the principle of dualism of symmetry.

Using the principle of dualism of symmetry solved two problems. Firstly, this principle took into account the role of system symmetry in its dynamics. Secondly, its use made it possible to understand that symmetry breaking within the framework of formalisms of classical mechanics is impossible. This approach to the derivation of the system's motion equation completely has justified himself.

The system's motion equation and the DMI mechanism resulting from it were explained in sufficient detail in previous works [9, 10]. Therefore, the key stages of its construction were emphasized below, and then its main properties are described.

In accordance with the principle of dualism of symmetry, there are exists two groups of variables that determine the total energy of the system [10]. Variables that define internal energy are micro variables. The variables that determine the energy of motion SP are macro variables. These micro and macro variables belonging to two different symmetry groups are independent [12]. As it turned out, the total energy in these variables breaks down into the internal energy of the system and its energy of motion.

In accordance with the law of conservation of energy, the sum of the energy of the system's motion and internal energy is invariant along its trajectory, but each of these types of energy is not an invariant of the system's motion. A violation of the symmetry of time is associated with a violation of the invariance of the system's motion energy.

The system's motion equation is obtained by differentiating the energy with respect to time, and then by summing the scalar values of the changes in the energies of each material point. This equation has the form [12, 13]:

$$M_N \dot{V}_N = -F_N^0 - \mu V_N, \tag{1}$$

where $F_N^0 = -\sum_{i=1}^N F_i^0$; F_i^0—is external force acting on the i-th material point; $\mu = \dot{E}_N^{\text{int}}/(V_N^{max})^2$; F_{ij}—is the strength of interaction i and j material points; $F_{ij}^0 = F_i^0 - F_j^0$; $\dot{E}_N^{\text{int}} = \sum_{i=1}^{N-1} \sum_{j=i+1}^N \tilde{v}_{ij}(m\dot{\tilde{v}}_{ij} + F_{ij}^0 + NF_{ij})$; $V_N^{max} + \dot{E}_N^{\text{int}}/F_N^0 = 0$.

The first term on the right side of (1) determines the potential external forces that act on the center of mass. These forces change the system's velocity.

The second term is nonlinear and bisymmetric, since it depends on micro and macro variables simultaneously. This term determines non-potential forces that change of the system's internal energy. The work of these forces is non-equal to zero only when the field of external forces is non-homogeneous. This term is called evolutionary nonlinearity [14]. It is responsible for the origin of the DMI. In the general case, this term links two symmetry groups of micro- and macro variables and determines the mutual transformation of the system's motion energy and internal energy.

Equation (1) is true for systems consisting from any number of material points. In the approximation of a solid body, (1) becomes the reversible Newton's motion equation.

The canonical Lagrange and Hamilton equations do not take into account evolutionary nonlinearity and therefore these equations are reversible and cannot be used to analyze the dissipative dynamics of a body [11, 15]. This is due to the use of golonomic restrictions in obtaining them. However, extended Hamiltonian, which was constructed based on (1), can be used to analyze the SP's dissipative dynamics. The phase trajectory can be determined in the dual phase space [10, 16].

In the general case, from (1) we will have that *the body is at rest or motion at a constant velocity if the nonlinear forces are equal to external forces.*

In according with the principle Galilee, in evenly moving system it is impossible to determine the fact of its motion, because dynamic equations do not depend from the choice of an inertial coordinate system. In according with (1) a similar situation arises at the system's motion in the homogeneous field of forces, when we have: $F_{ij}^0 = 0$. For clarification this statement, let us take as example a star system with planets moving with uniform acceleration in an external gravitational field of forces. If someone moves in the gravitational field of one of the planets, then by the nature of these motions it is impossible to establish inside the system that system as a whole moves with acceleration. This conclusion is consistent with the D'Alembert equation and Einstein's ideas [11].

Analysis of system dynamics using (1) performed in micro and macro variables. As a result, the role of the body structure in its dynamics was taken into account. Therefore, we will say that (1) gives a "***complete description***" of the dynamics of bodies in an inhomogeneous force field.

Of course, solving (1) for a system with a large number of elements is a difficult task. However, studying the properties of this equation makes it possible to determine the general properties of the behavior of dissipative systems and establish how statistical and thermodynamic laws follow from the basic laws of classical mechanics [9, 12].

3 Irreversibility of the Nonequilibrium Systems

Below we consider the processes of equilibrating nonequilibrium systems using (1). In the approximation of local thermodynamic equilibrium with a sufficient degree of generality, the nonequilibrium system can be specified by a set of SP's moving in relative to each other [8, 17]. In this case, SP is already playing the role of system elements. The motion of each SP in an inhomogeneous field of forces, created by all SP is determined by (1) [11]. In this case, the proof of equilibration is reduced to the proof that the energy of the relative motions of the SP is irreversibly converted into their internal energy. Thus, in order to prove irreversibility, it is necessary to show that such transformations take place. To do this, we estimate the energy flows for SP [8, 14].

Obviously, for the nonequilibrium system consisting from a set of SP, the mechanism of the formation of direct and reverse energy flows is associated with the mutual transformation of the energies of the relative motions of the SP and their internal energies. Hence, the proof of the irreversibility of the dynamics of the nonequilibrium system is reduced to the proof that the influx of internal energy of the SP is greater than the outflow.

Let us ΔE^{tr} is the energy of the relative motion of the SP, which is transformed into its internal energy. According to (1), ΔE^{tr} is determined by a bilinear term whose value is equal to the second order of smallness. Therefore we can write: $\Delta E^{tr} \sim \chi^2$, where χ is any small parameter. If it so, then $\Delta E^{tr}/E^{\text{int}} \ll 1$ and the equilibrium violation of the SP can be neglected. Let us notice that the value ΔE^{tr} has a second order of smallness also in according with the estimations of an increment of entropy [18].

Thus, as it follows from (1), the SP dynamics in a weak inhomogeneous field of external forces is irreversible. Indeed, in such a field of force the changes of SP's internal energy is a second order smallness, and the violation of the SP equilibrium can be neglected. However, according Galileo principle, the system's motion energy cannot increase due to of internal energy of the system, which is in equilibrium. Therefore, we have the decreases of the SP's motion energy along its trajectory. In this case, irreversibility takes place. The internal motions of the material points that determine the internal energy are also irreversible due to the dependence of internal energy from the time.

If the SP's interaction forces or their gradients are the great enough, the equilibrium of the SP can be disturbed. Then each SP, like a nonequilibrium system, can be represented as a set of equilibrium subsystems moving relative to each other. In this case, for increment of the SP's internal energy one can write: $\Delta E^{tr} = \Delta E^{tr}_{ins} + \Delta E^{h}$, where ΔE^{tr}_{ins} is the increment of the energy of the relative motions and ΔE^{h} is the increment of the subsystem's internal energies. That is, $\Delta E^{tr}_{ins} < \Delta E^{tr}$.

The energy of the equilibrium subsystems cannot be transformed into their motion energy. Therefore, we will consider that only energy of relative motions of subsystems can be transformed back into the motion energy of the SP. Let us denote this reverse flow of the SP's internal energy, as ΔE^{tr}_{ret}. According to (1), the value ΔE^{tr}_{ret} is determined by the bilinear function of the subsystems variables, which determined its motion energies and the internal energies. These are terms of the second order of smallness of their micro and macro variables. But because: $\Delta E^{tr} \sim \chi^2$, we will have that $\Delta E^{tr}_{ret} \sim \chi^4$. Thus, the reverse flow of the internal energy of the SP, ΔE^{tr}_{ret},

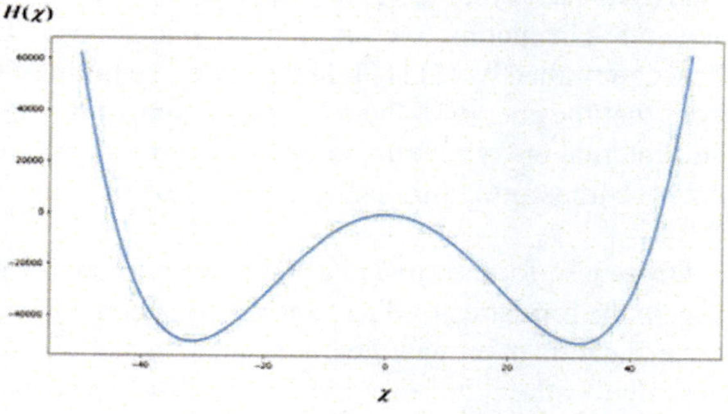

Fig. 1 The graph of the formula 2

into of its motion energy during their strong interactions cannot to be more, than the fourth order of smallness. In this case, a potential, which corresponds to the change of the SP motion energy, can be defined by the next equation:

$$H = \alpha\chi^2 - \beta\chi^4 \tag{2}$$

Here the α, β constants can be determined using of (1).

Figure 1 shows a graph of (2). For values:$|\chi| \prec \chi_0$, where $\pm\chi_0$ are the roots of (2), the irreversibility or violation of symmetry of time takes place. For stationarity of the non-equilibrium system it is necessary to fulfill the equality: $H = 0$.

If to take any complex system with a large number of hierarchical levels, then for its stationarity this equality should hold for each hierarchical level [13]. The study of these conditions goes beyond the laws of classical mechanics, because for this study it is necessary to take into account Planck radiation and different quantum effects [8].

To prove and understand the nature of system's equilibration, we studied the change in the internal energy of the system during its motion in a stationary external inhomogeneous field of forces depending on the number of particles in the system. This was done numerically based on (1) [16, 19].

It turned out that for some initial conditions and for sufficiently small systems, the internal energy could not only increase, but also decrease. The calculations showed that for an oscillator with N = 2 which moved in an inhomogeneous field of external forces, depending on the initial phase of its oscillations, the internal energy can be converted into energy of motion [13]. With an increase in the number of particles in the system, the part of internal energy, which can be converted into the energy of movement of the system decreases, and when N > 100, internal energy can only increase.

When $N > 10^3$ there is no increase in the increase in internal energy [19]. Thus, $N \sim 10^3$ determines the range of applicability of the thermodynamic description for the system. This is in consistent with [15] where stated, that the irreversibility is qualitative: the more particles in the system, the more irreversibly it behaves.

The calculations showed that the magnitude of fluctuations of the internal energy of the system due to a changes in the initial conditions for given energy values and for a given number of elements in the system obeys the law: $\delta E^{tr} \sim 1/\sqrt{N}$ [19]. This corresponds to the statistical law of fluctuations of quadratic functions [18]. Because the above statistical laws for dynamical systems are derived basing on the deterministic (1), it can be argued that they follow from the deterministic laws of physics. A similar conclusion was made in [20]. It follows that the fundamental laws of physics determine the scope of statistical laws. And if it so, then probabilistic laws can be considered as possible simplifications for analysis of the complex systems. This view coincides with the well-known position of Leibniz and Einstein [21].

Thus, in accordance with (1) from a mathematical point of view, the DMI mechanism can be explained by the fact that there is a connection for vectors from different symmetry groups. For SP, these are the groups of symmetry SP and space symmetry. This linking is determined by bilinear terms since these terms depend on variables belonging to different symmetry groups. These bilinear terms determine the conversion of the body's motion energy into its internal energy and lead to a violation of the conservation law of the motion energy, when the sum of the internal energy and the motion energy is preserved. Bilinear terms arise when the body moves in an external inhomogeneous field of forces. The forces between SPs play the role of these forces in nonequilibrium systems. Thus, the energy of the relative motion SP is transformed into the internal energy of the chaotic motion of the elements SP. This is the essence of the second law of thermodynamics. Equation (2) characterizes the efficiency of converting the energy of movement of the system into internal energy.

The existence of dissipation is a necessary condition for the formation of attractors [22]. However, dissipation is possibly only for structured bodies due to the transformation of the motion energy into their internal energy. Hence the conclusion about the infinite divisibility of matter [10, 23]. This means that according to the laws of classical mechanics, matter should be an infinite hierarchy of systems. That is, any selected part of the matter is a system.

Let us accept the condition of the matter's infinite divisibility, demand the unity of the picture of the world and the evolutionary origin of matter. In this case, the basic element of matter should be an open nonequilibrium dynamic system [12]. If this is so, then the following explanation of the Heisenberg uncertainty principle can be proposed [24, 25]. In according with (1), the trajectory of an element depends on its structure. This means that its trajectory, determined based on the canonical formalism of classical mechanics, has uncertainty. This uncertainty is determined by the value: $\Delta E_N^{tr} \Delta t > 0$. In any interaction of particles, or in any measurement process, this uncertainty will occur. The attractiveness of this explanation is that it reinforces the position of the principle of know ability of the world [26].

Below we consider how to explain the spontaneous symmetry breaking for quantum systems.

4 DMI and Spontaneous Symmetry Breaking

The modern theory of quantum systems is based on the canonical Hamiltonian formalism. The scope of this formalism is limited to holonomic systems that are invariant with respect to the direction of time. Therefore, within the framework of the Hamiltonian formalism, only non-dissipative reversible systems can be studied [4, 15, 27, 28]. Indeed, how it is follows from (1), the dissipation is due to the nonlinear conversion of its motion energy into internal energy. However, the use of the condition of the holonomicity of connections in deriving the Lagrange and Hamilton equations precludes the possibility of describing such a transformation. Therefore, the search for an explanation of the mechanisms of the spontaneous symmetry breaking for phase transitions led to a phenomenological model that describes the thermodynamics and kinetics of superfluidity [29].

The explanation of the mechanism of spontaneous symmetry breaking was based on the so-called order parameter. Using this parameter, the empirical type of the free energy function in the phase transition region was determined. Thus, the restriction of the canonical Hamiltonian formalism related to its reversibility is circumvented empirically, for example, taking into account the additional terms of the expansion of the potential function in the Hamiltonian. The final solution to the problem is based on the idea of the existence of an external infinitely small influence on the system [6, 29, 30]. Therefore, in this case we also have a probabilistic explanation of spontaneous symmetry breaking. Let us briefly recall the essence of this explanation of the nature of symmetry breaking in the theory of phase transitions.

Landau and Ginsburg offered the mathematical description of spontaneous violation of symmetry in 1937 [29]. They explained this effect to infinitesimal fluctuations in the values of operating parameters near an unstable line of phase transitions. They called these parameters scalar order parameters. They themselves determined symmetry breaking due to the expansion of free energy, representing it in the phase transition region as follows [6, 29]:

$$F(\varphi) = F_0 + V(a_2\varphi^2/2 + a_4\varphi^4/4 - h\varphi), \tag{3}$$

where $F(\varphi)$ is the thermodynamic potential (Gibbs energy); φ is the scalar order parameter; a_2, a_4, h are the empirical coefficients.

Equation (3) is similar as (2). This expression is sometimes called the "Mexican hat", based on its graphic form (see Fig. 1). To explain the spontaneous symmetry breaking in superfluidity and superconductivity, Ginzburg and Landau used the effective electron wave function. It performed the role of a two-component order parameter: $\psi(r) = |\psi(r)| \exp[i\phi(r)]$ [6]. In accordance with this, properties of the superconductor were specified by the wave function, which depends on the magnetic field, $B(r)$. It was determined by the functional of the free energy [6]:

$$\begin{aligned} F(\psi) = & F_{n0} + \int dr\Big\{|B|^2/(8\pi) + a\big|\psi^2\big| + b\big|\psi^4\big| \\ & + \sum_{\alpha} 1/(2m)\big|(-i\hbar\nabla_\alpha - (q/c)A_\alpha)\psi(r)^2\big|\Big\} \end{aligned} \tag{4}$$

Here F_{n0} is the free energy in the normal state, $B(r) = rot(A)$, q, m is the effective charge and mass of superconducting electrons.

Bogolyubov proposed an explanation of the physical nature of the two-component wave function, as well as the theory of superconductivity and superfluidity. He showed the unity of the phenomena of superfluidity and superconductivity, which was subsequently confirmed experimentally [6]. As it turned out later, a similar scalar potential as formula (2) is acceptable not only for spontaneous symmetry breaking during phase transitions. A similar type of expansion of thermodynamic potentials or scalar functions is also used to describe the violation of superfluidity, superconductivity, particle formation processes, the appearance of mass, etc. [2, 6, 27, 31].

Let us consider how a deterministic explanation of symmetry breaking in quantum systems can be introduced. The dynamics of quantum systems is determined by the canonical Schrödinger equation. This equation, like Newton's equation of motion, is reversible. To describe the irreversible dynamics of quantum systems, the Schrödinger equation was modified [25]:

$$\{i\hbar\frac{\partial}{\partial t}+\frac{\hbar^2}{2M}\nabla_R^2-U(\tilde{r}_1,\tilde{r}_2,\ldots\tilde{r}_N,R,t)$$
$$+\sum\nolimits_{i=1}^{N}[\frac{\hbar^2}{2m}\nabla_{\tilde{r}_i}^2-W_{\text{int}}(\tilde{r}_i)]\}\psi(\tilde{r}_1,\tilde{r}_2,\ldots\tilde{r}_N,R,t)=0 \quad (5)$$

Here R—coordinates of the center of mass of system; $\tilde{r}_i$—these are coordinates i—particles concerning the center of mass of system; $i = 1, 2, 3 \ldots N$; N—is a number of particles; M—is a system's mass; m—is a mass of i-th particle; U, W_{int} external and internal potential energy consequently; wave function ψ represents work of wave functions for each particle and for all system when the external field is homogeneous.

Equation (5) was obtained in the same way as in classical mechanics, using the principle of dualism of symmetry. In connection with this principle, the energy of a quantum system should be represented by the sum of the motion energy and internal energy. For this, the Hamiltonian in (5), in accordance with the *complete description* method, was written in micro and macro variables, as the sum of the Hamiltonians for the internal dynamics of the elements of the system and for the dynamics of the system as a whole. In this case, irreversibility in quantum systems will be explained by evolutionary nonlinearity.

It is important that in the *complete description* of the extended Schrödinger equation there is no singular bifurcation point for (2), since this point is a region of the space of micro variables. In this point, the motions of the body's elements affect on the system's motion. From this, it is clear that with a macroscopic description of the system's motion in the bifurcation point, which was used in the general Schrödinger equation, we lose the determinism of the description [9, 13]. Thus, spontaneous symmetry breaking can be explained due to the linking of micro- and macro variables through the terms of evolutionary nonlinearity and due to instability of the system's motion at the bifurcation point, where the motions of the body's elements determine the system's motion. This allows us to conclude that, despite the difference in the processes of symmetry breaking in classical and quantum mechanics, its nature is universal, and the mechanism of spontaneous symmetry breaking in quantum mechanics can be explained in the same way as in classical mechanics. This conclusion is based on the condition of infinite divisibility of matter [26, 27]. Therefore, symmetry-breaking processes can be studied using the *complete description.* This description can be implemented based on the motion (1) for classical systems or based on (5) for quantum systems [25].

In accordance with (1), micro and macro variables are linked due to the inhomogeneous field of external forces. This will be determined by the terms of evolutionary nonlinearity, should depend simultaneously on micro and macro variables. These terms leads to a violation of symmetry of time, since the motion energy of the system due to these terms is converted into internal energy of the elements.

The conclusion that the nature of irreversibility in quantum mechanics should be similar to the nature of irreversibility in classical mechanics is confirmed by the similarity of the form of potential (2) for classical mechanics and the potential in (3), used to explain spontaneous symmetry breaking in quantum mechanics.

Thus, the differences between explanation of the symmetry breaking in quantum systems and DMI for systems of classical mechanics are as follows. To explain the spontaneous symmetry breaking in quantum systems for superconductivity, superfluidity, etc., the method of second quantization of systems was used. It was assumed that the state of the system is determined by the available statistical states of its elements. Therefore, due to the empirical shape of the transition matrix, used to explain spontaneous symmetry breaking, it was necessary to use the hypothesis of the existence of external arbitrarily small oscillations.

To explain the symmetry breaking in classical mechanics the method of *complete description* was used. In this explanation of symmetry breaking, the deterministic motions of system elements play the role of random external influence. Due to the instability in the bifurcation point, the micro processes leads to macro processes.

The advantage of a *complete description* of quantum systems is that it basically takes into account the role of the internal dynamics of quantum systems in their dynamics and can help to understand and find analytically the creation and destruction operators of the corresponding particles, for example, in the case of supersymmetres [30].

5 Conclusions

The canonical Hamiltonian formalism is built based on Newton's reversible mechanics, when the conditions of holonomic restrictions and the potentiality of all collective forces acting on the systems are met [32, 33]. Therefore, this formalism is reversible [34]. In this case, the irreversibility problem is usually solved by adding a small perturbing term to the Hamiltonian [17, 18]. Such an empirical solution of the irreversibility problem does not follow from closed equations of dynamics and is equivalent to a probabilistic explanation of symmetry breaking processes. However, in the framework of mechanics of structured bodies, a deterministic way of describing these processes becomes possible.

The motion equation of a structured body is obtained by representing the body as a system of potentially interacting material points. In according with the principle of dualism of symmetry, the dynamics of this system is determined not only by the symmetry of space, but also by symmetry of the system. Therefore, to determine its dynamics, the system's energy should be presented in the form of sums of the system's motion energy and its internal energy. Such representation is realized in the spaces of two independent groups of micro and macro variable. Micro variables determine the system's internal energy, and macro variables determine the system's motion energy in space. The violation of symmetry for the system is caused by the conversion of the system's motion energy into its internal energy, when the system moves in an inhomogeneous field of forces. Such a transformation is defined by bilinear terms of evolutionary nonlinearity, which connects the dynamics of the elements of the system with its dynamics, as a whole. In according with the extended Schrödinger equation, violation of symmetry in quantum systems will also be determined by bilinear terms of evolutionary nonlinearity.

It turned out that evolutionary nonlinearity for systems of classical mechanics is determined by the potential, which coincides with the potential that determines the spontaneous symmetry breaking in quantum systems. This allows us to conclude that the symmetry breaking mechanism is universal for classical and quantum systems. If so then the nature of symmetry breaking in quantum and classical systems can be studied as a result of a *complete description* of the dynamics of systems using micro- and macro-descriptions of dynamics based on the principle of dualism of symmetry.

In mathematical language, this means that for deterministic irreversibility or, in other words, for breaking deterministic symmetry, at least two symmetry subgroups are needed. These two subgroups of symmetry form a complete symmetry group, which determines the dynamics of the body. In this case, the invariant of motion is the total energy, which is equal to the sum of the energies corresponding to these two subgroups of symmetry. When a body moves in an inhomogeneous field of forces between these subgroups of symmetry, an interaction occurs, that is, the engagement of the elements of the symmetry from different subgroups. As a result, the motion energy is converted into internal energy. This means a violation of symmetry or irreversibility.

References

1. E. Wigner, Symmetry and conservation laws. UFN **84**(4) (1964); Events, laws of nature and principles of invariance. UFN **85**(4) (1965); Violation of symmetry in physics. UFN **89**(3) (1966)
2. F. Wilchek, The origin of mass. Mod. Phys. Lett. A **21**(9) (2006)
3. N. Bor, The problem of causality in atomic physics. UFN **147**(2) (1985)
4. G.M. Zaslavsky, *Stochasticity of Dynamical Systems* (Nauka, Moscow, 1984)
5. I. Prigogine, *From the Existing to the Arising* (Nauka, Moscow, 1980)
6. D.V. Shirkov, 60 years of broken symmetries in physics (From Bogolyubov's theory of superfluidity to the Standard Model). UFN **179**(6) (2009)
7. J. Mabillard, P. Gaspard, Microscopic approach to the macrodynamics of matter with broken symmetries. arXiv:2005.14012v1[cond-mat.stat-mech] 28 May 2020
8. Y.B. Rumer, M.S. Rivkin, *Thermodynamics* (Stat. Phys. Kinet., Nauka, Moscow, 1977)
9. V. Somsikov, Deterministic mechanism of irreversibility. JAP, **14**(3) (2018)
10. V.M. Somsikov, *To the Basics of Physics Evolution* (Nauka, Almaty, 2016)
11. G. Goldstein, *Classical Mechanics* (Nauka, Moscow, 1975)
12. V.M. Somsikov, Transition from the mechanics of material points to the mechanics of structured particles. Mod. Phys. Lett. B **4** (2016)
13. V.M. Somsikov, Deterministic irreversibility and the matter structure. JAP **16** (2019)

14. V.M. Somsikov, Non-linearity of dynamics of the non-equilibrium systems. World J. Mech. **2**(7) (2017)
15. O. Penrose, Reversibility and irreversibility. ("PDE and Materials", report no.44/2006 of the Mathematisches Forschungsintitut Oberwolfach (ed. J.M. Ball, R.D. James, S. Muller)) (2006)
16. V.M. Somsikov, Deterministic irreversibility mechanism and basic element of matter, in *Proceedings of 12th Chaotic Modeling and Simulation. International Conference* (Springer, 2020)
17. L.D. Landau, E.M. Lifshits, *Physical Kinetics* (Nauka, Moscow, 1979)
18. L.D. Landau, E.M. Lifshits, *Statistical Physics* (Nauka, Moscow, 1976)
19. V.M. Somsikov, A.B. Andreev, On criteria of transition to a thermodynamic description of system dynamics. Russian Phys. J. **58**(11) (2016)
20. F. Baldovin, L.G. Moyano, C. Tsallis, Boltzmann-Gibbs thermal equilibrium distribution descends from Newton laws: a computational evidence. arXiv:cond-mat/0402635 v1 25 Feb 2004
21. L. Peliti, R. Rechtman, Einstein's approach to statistical mechanics: the 1902–04 papers. J. Stat. Phys. **167** (2017)
22. Yu.A. Loskutov, Charm of chaos. UFN **150**(12) (2010)
23. V.M. Somsikov, The method of the description of dynamics nonequilibrium systems within the frames of classical mechanics. arX: physics/ 0703242 v1 29 September 2007
24. S. Boughn, Wherefore Quantum Mechanics? arXiv:1910.08069[physics.hist-ph]
25. V.M. Somsikov, Limitation of classical mechanics and ways it's expansion. PoS (Baldin ISHEPP XXII-047), in *XXII International Baldin Seminar on High Energy Physics Problems*, 15–20 JINR, Dubna, (2014)
26. V. Famourzadeh, M. Sefidkhosh, Straddling between determinism and randomness: Chaos theory vis-à-vis Leibniz. arXiv:1909.13635v1[physics.hist-ph] 30 Aug 2019
27. J. Bernstein, A question of mass. Am. J. Phys. **79**(1) (2011)
28. V.G. Zelevinsky, Lectures on quantum mechanics (Nsk., Sib. Univer. Publ. house, 2002)
29. L.D. Landau, To the theory of phase transitions. I. JETP, **7**, (1937); Landau L.D. To the theory of phase transitions. II JETP, **7**, (1937)
30. L.E. Gendenstein, I.V. Krive, Supersimmetriya in quantum mechanics. UFN **146**(4) (1985)
31. P.W. Higgs, How was it possible to circumvent the Goldstone theorem. UFN. **85**(10) (2015)
32. C. Lanczos, *Variational Principles of Mechanics* (Mir, Moscow, 1962)
33. N.E. Martínez-Pérez, C. Ramírez, Symmetry breaking in non conservative systems. arXiv: 1602.05255v1 [physics.class-ph] 17 Feb 2016
34. H.G. Callaway, Fundamental physics, partial models and time's arrow, in *Proceedings of the 2015 Conference on Model-based Reasoning* (Springer, 2016)

Problems of Creating an Evolutionary Picture of the World

Vyacheslav Somsikov and Svetlana Azarenka

Abstract The paper is devoted to study, how the physics of evolution allows developing an evolutionary picture of the world. Here we briefly examine the basic concepts of the world's picture and how these concepts can find development based on the physics of evolution. For this purpose, the next questions will be analysed: how physics of evolution leads to the conclusion about the infinite divisibility of matter; how nature solves the problem of the static state of matter, when motion is a way of existence of matter; how symmetry and its violation determine the evolution of matter; what are the principles of building "from simple to complex", etc. As a result, we show how taking into account the structure of matter in its dynamics leads to the possibility of describing evolutionary processes in nature. This means the possibility of constructing a deterministic evolutionary picture of the world within the framework of the laws of physics.

Keywords Irreversibility · Entropy · Chaos · Phase space · Physics of evolution

1 Introduction

Creating a picture of the world is the main task of science. However, many problems arise on the way to its creation. The most important of them is the problem of knowability of nature. This problem is most clearly manifested in the clash of ideas of reductionism and holism.

Proponents of reductionism believe that all phenomena in nature are knowable in principle, and there is a finite set of fundamental laws, the knowledge of which allows you to create a picture of the world, moving from "simple to complex" [1, 2]. Proponents of holism, on the contrary, believe that the properties of the "whole" do

V. Somsikov (✉)
Al-Farabi Kazakh National University, Almaty 050040, Kazakhstan
e-mail: vmsoms@rambler.ru

S. Azarenka
Almaty Academy of the Ministry of Internal Affairs of the Republic of Kazakhstan, Nur-Sultan, Kazakhstan

C. H. Skiadas et al. (eds.), *13th Chaotic Modeling and Simulation International Conference*, Springer Proceedings in Complexity,
https://doi.org/10.1007/978-3-030-70795-8_64

not follow from the properties of its elements. In addition, there was even an opinion that the "golden age" of science was over, and its further development is possible only on the path of expanding knowledge without intensification. As evidence, they use examples such as a misunderstanding of the relationship between the laws of inert matter and living matter. In their opinion, these relations cannot be established within the framework of the laws of physics. All this causes heated debate between supporters of reductionism and holism [3–5].

The problem of the cognizability of nature is directly related to the problem of the principle of causality in physics. Here the development of a picture of the world is faced with a problem related to the fact that the causality principle is still not even among the fundamental principles of physics. This occurs mainly because the fundamental laws of physics are reversible, and natural processes are usually irreversible [6]. Consequently, not only the physical picture of the world does not correspond to the principles of causality and its unity, but also physics itself represents a multitude of separate areas of knowledge that are weakly interconnected. These are, for example, classical mechanics, thermodynamics, and quantum mechanics. As a result, physics explains how the world works, but does not answer questions about how the world develops, in what direction the processes of its evolution are going, and what determines these directions [6–8].

The problem of irreversibility has arisen since the creation of Newtonian mechanics. In the process of solving this problem, Boltzmann et al., Scientists discovered its probabilistic mechanism, according to which evolutionary processes are random [9, 10]. If this is so, then it is not clear how to build a physical picture of the world based on the laws of physics, how to understand the emergence of organized structures of matter from chaos. It also means that on the way to the development of knowledge, insurmountable epistemological problems arise [11–13]. Therefore, the search for a solution to the problem of irreversibility within the framework of the laws of physics was continued.

Finally, a **deterministic irreversibility mechanism (DMI)** has been proposed. DMI has opened up the possibility of building physics of evolution. The task of "physics of evolution" is to study the processes of evolution of matter and to develop methods for constructing its evolutionary models within the framework of the fundamental laws of physics [14].

In a previous paper, we examined the question of how DMI strengthens the positions of determinism in physics and the cognoscibility of a picture of the world [11]. The purpose of this paper is to show how the physics of evolution opens up opportunities for constructing a deterministic evolutionary picture of the world, how it affects the development of philosophical concepts that underlie the modern picture of the world.

2 Basic Concepts of a Picture of the World and Physics of Evolution

Here we briefly examine the fundamental concepts of the picture of the world that are related to the physics of evolution, and how these concepts can find their development on the basis of physics of evolution.

The key concepts on which the picture of the world is based arose in antiquity. One of the first fundamental concepts was related to the question of what everything consists of. Many ancient philosophers, as a result of observations of nature, came to the conclusion that matter consists of elementary particles or indivisible bricks. The father of this idea can be considered Democritus [13, 15]. He claimed that matter is composed of atoms. Subsequently, modern ideas appeared for discretizing the structure of matter, its fractality and self-similarity [16]. Today we are witnessing the discovery of an increasing number of components of matter, and so far this limit has not been discovered. In this regard, **we will show here how the physics of evolution leads to the conclusion about the infinite divisibility of matter**.

On the one hand, the Universe, the Milky Way, stars, as we see, are static. However, on the other hand, according to Heraclitus, we have, that "everything flows, everything changes" [15]. That is, the world is evolving. In connection with it there was an opinion that **motion is a way of existence of matter**. It was strengthened in the works of Galilee, Newton, and Leibniz. They found the fundamental laws of motion. They introduced modern concepts of energy, acceleration, angular momentum, etc., characterizing matter in its dynamics [17]. Here **we show how nature solves the problem of the static nature of matter, when motion is a way of existence of matter**.

The rate of change in the position of matter in space is determined by the concept of time. Therefore, it is impossible to describe matter without using space-time concepts. The relationship between the concepts of matter, dynamics, space and time is established using the concept of symmetry. The concept of symmetry appeared in connection with Plato [18]. According to Plato, symmetry is the cornerstone of the picture of the world. Thanks to the dynamics, matter takes such diverse forms that are determined by the interaction of the elements of matter in accordance with the symmetry of space and time. **The problem of conformity of form and content, provided that both form and content are in constant evolution, stands in the way of the further development of the picture of the world**.

However, the violation of symmetry is also characteristic of nature, as well as its conservation. Apparently, the second law of thermodynamics is the first law in the history of physics, which is associated with the violation of symmetry. In recent decades, it has been discovered that in quantum physics spontaneous symmetry breaking also occurs. Until recently, these symmetry violations, one way or another, were explained in a probabilistic manner. But this contradicts the principle of causality. Moreover, in quantum mechanics there is the Heisenberg uncertainty principle, which actually means the existence of a limit to the cogniscability of a picture of the world [19]. These problems lead to questions: **how symmetry, its violation, determine the evolution of matter, what is the nature of symmetry breaking**.

We find the first fundamental laws of logical thinking, fundamental concepts about the world around us, such as matter, force, motion, space and time, at Aristotle [15]. He believed that the world is one, and the laws of its development are universal. But how to connect the unity of the world and the universality of the laws of its development with the existence of many weakly coordinated sections of physics? Here we discuss, **how this problem is related to the physics of evolution by the example of the relationship of the laws of classical mechanics, thermodynamics and quantum mechanics**.

A millennium after Aristotle, the laws of classical mechanics were discovered. These laws determine the motion of material objects. They also define concepts such as acceleration and energy [15]. According to Galileo, Descartes, Newton, not the velocity of the body is proportional to the force, as Aristotle claimed, but to its acceleration [17, 20]. However, experience shows that the acceleration of the body becomes equal to zero when the force acting on it, becomes equal to the force of friction. This corresponds to the irreversible mechanics of Aristotle. Newton, on the contrary, sought to eliminate friction in order to reveal the essence of the law of motion, regardless of the various properties of bodies and the environment. Thanks to a model of a body in the form of an unstructured particle, he discovered that acceleration, not speed, is proportional to force. A natural question arises, **how to combine the mechanics of Aristotle and Newton?**

An important problem in the development of knowledge concerns the principles of constructing a scientific picture of the world and the limitations of cognizability of the world. Two points of view can be distinguished here. From the point of view of those who adhere to the positions of cognizability of the picture of the world, there must be principles that allow movement in the direction of construction of knowledge from "simple to complex". In this case, the whole picture of the world should be based on fundamental laws that give rise to the whole variety of known empirical laws. Reductionists support this view [1, 2]. In particular, Weinberg believes that the "theory of everything" should be based on laws that make it possible to understand the whole picture of the world. Holists hold the opposite point of view [3, 4]. They argue that the whole contains "new", not arising from the laws of its elements. They say: "The whole is not the sum of its parts" [20]. This raises important questions: **is it possible to build a holistic picture of the world within the framework of the laws of physics, are there physical principles for building a picture of the world "from simple to complex"**.

That is, we approach the problem of the existence of principles for constructing systems based on their elements in the framework of the fundamental laws of physics.

Below we will try to show how the problems that were shown in this chapter are related to the physics of evolution.

3 DMI and Physics of Evolution

It is hardly possible to understand how the physics of evolution expands the possibilities of developing a picture of the world, if we do not briefly explain the nature of DMI. It is most convenient to begin an explanation of the nature of DMI based on the concepts of symmetry. Indeed, without breaking the symmetry of time that leads to DMI, evolution itself is impossible [18, 21].

1. Principle of dualism of symmetry. As already noted, symmetry is a key concept for a picture of the world [18]. Indeed, the principle of least action, which determines the harmony of the world, follows from the types of symmetries of bodies. This is because the different types of symmetry of time and space correspond to the invariants of dynamics, in particular energy, momentum. These invariants determine the properties and laws of the dynamics of bodies. For example, the dynamics of a structureless **material point (MP)** as the simplest model of a body that Newton used to reveal the essence of the laws of dynamics is determined only by the symmetries of space and time. MP energy is determined in accordance with their symmetries. The motion equation of MP follows from its energy [14].

Boltzmann showed that the body model, within the framework of the laws of classical mechanics, is well approximated by a system of potentially interacting MPs. And knowing the energy of the MP system, in principle, you can find the system's motion equation.

Usually, the system's motion equation is determined using the canonical formalisms of classical mechanics. Canonical formalisms were built on the assumption that all collective forces are potential, since the forces between elements of the system are potential. This assumption is confirmed by the fact that the total forces of moving systems are potential. But it turned out that the dynamics of systems determined by the canonical formalisms of classical mechanics is reversible, as is the motion of a single MP [22]. Any search for a solution to the problem of irreversibility within the framework of canonical formalisms led to a probabilistic mechanism of irreversibility. However, this mechanism is excluded the existence of causal relationships. Therefore, the search for a deterministic solution of the problem of irreversibility was continued.

A study of the simplest systems of elastically colliding disks led to the conclusion that irreversibility is possible only for systems interacting with each other and that the fundamental laws of physics did not prohibit irreversibility. This led to the assumption that the reversibility of Hamiltonian systems, which follows from the formalisms of classical mechanics, is associated with the restrictions used in their construction. This assumption was confirmed. It became clear why all attempts to find DMN in the framework of formalisms of classical mechanics were unsuccessful [21, 23].

Thus, the first key idea providing a breakthrough in solving the problem of irreversibility was the idea of the need to take into account the role of the body structure in its dynamics. The essence of this idea is easy to understand by the example of body motion on a surface with friction. Aristotle was guided by this example. Boltzmann also tried to find the mechanism of irreversibility. He relied on the statistical methods he developed. However, these methods are incompatible with the deterministic laws of classical mechanics.

Systems, unlike MP, also have symmetry. Then the equation of motion of the system should also depend on their symmetry. This means the need to take into account the symmetry of bodies in their dynamics. The fact that **the dynamics of bodies is determined by the symmetries of the body and the symmetries of the surrounding space was the second idea necessary to solve the problem of irreversibility. It was called the principle of dualism of symmetry (PDS)**. The essence of PDS is that the state of the body, the nature of its interaction with external objects, its dynamics and evolution are determined by both the symmetries of the external world and the symmetries of the body. Based on the PDS and the dual representation of energy, as a sum of internal energy and of the motion energy, the body's motion equation was obtained. This equation is irreversible, and DMI follows from this equation. Thus, it turns out that the property of the irreversibility of their system arose from the reversible properties of the dynamics of elements in a deterministic way. The question arises: **how can the irreversibility of the MP's system motion arise when the motion of each MP is reversible?**

It was shown that irreversibility is connected with the fact that the motion energy, which determines the trajectory of the system in space, is transformed into the internal energy [21]. The system's motion energy we will call the energy of "order". We will call the internal energy of an equilibrium body the energy of "chaos". That is, the system's motion energy turns into "chaos" energy in an inhomogeneous space, but the energy of "chaos" cannot be converted into energy of the "order". This is a process of breaking the symmetry of time, since the invariance of the body's motion energy is violated. **But why this process is irreversible? The answer on this question is hidden in the nature of the forces, which transform the system's motion energy into internal energy**.

The dissipative nature of the forces that transform the energy of motion into internal energy follow from the MP's system motion equation. For each MP, forces are determined through the efficiency of the transformation of potential energy into its kinetic energy. The fact that these forces are potential ensures the reversibility of this energy conversion. For a structured body, forces are determined through the efficiency of the transformation of external energy. In this case, the external energy is transformed into both the motion energy and internal energy. Consequently, in the case of a structural body, forces are divided into two classes: forces that determine the change in the energy of motion of the body, and forces that determine the change in internal energy. The forces performing the work on moving the system are equal to the sum of the external forces acting on the elements of the system. They are potential. This corresponds to classical mechanics [22]. It should be emphasized that here the internal energy is associated with dynamic parameters, and not with the thermodynamic, as temperature and pressure of the body at rest

The forces that change the internal energy of a system are made up of two parts. One part is the potential forces of interaction of the elements of the system. Another part of the forces performing work on changing internal energy is proportional to the gradients of external forces. **Non-potential friction forces arise as a gradient of**

potential external forces! This is the mechanism of the emergence of non-potential friction forces from potential forces [23]. The difference in the collective forces that determine the system's motion and the change in internal energy is related to the difference between internal energy and the motion energy. Internal energy also exists if the system is at rest, due to the continuous motions of the elements relative to its center of mass. The system's motion energy exists only when it moves in space. This energy does not depend on the internal motion of the system's elements. The internal energy of an equilibrium system is the energy of "pure chaos". The word "pure" means that when a system is divided into subsystems, these subsystems are in equilibrium and do not have the energy of relative motions.

Without taking into account the role of body structures in their dynamics, irreversibility cannot be explained. But without dissipation, attractors and structures do not arise [16]. **And if the world evolved "from simple to complex", then this means that the primary element of matter had to have a structure**. The assumption of the need to take into account the structural nature of bodies already at the first stages of describing open systems also follows from statistical considerations [24].

The development of nonlinear dynamics has shown that the appearance of various forms of matter or attractors is possible only in the presence of dissipation. The dissipation occurs only in the interaction of systems. Therefore, to describe the evolutionary processes, it is necessary to take into account the openness of bodies [12, 16]. If we also take into account the infinity of divisibility of matter and accept its evolutionary origin, we conclude that the structureless elements of matter cannot arise and exist. This inevitably leads to the conclusion that **the main element of matter should be an open nonequilibrium dynamic system (ONDS), and the matter itself is a hierarchy of ONDS** [23].

The dynamics of the ONDS is described using the extended formalism of classical mechanics, which takes into account the role of the structure of systems in their dynamics. Such a formalism is obtained in the same way as canonical formalisms from the D'Alembert equation, but instead of the Newton's motion equation, equations of motion for systems of MP is used [14, 25].

The most important concept that derives directly from PDS is D-entropy. D-entropy determines how internal energy changes due to change of the motion energy. Unlike the thermodynamic concept of entropy, D-entropy is applicable for both large and small systems. D-entropy can be used to analyze the processes of occurrence, evolution, destruction of new systems. D-entropy reveals the epistemological significance of existing types of entropy, since it directly connects the dynamics and states of the system, implements the relationship of "order" and "chaos".

For a deeper understanding of the role of the body's structure in the mechanism of violation of time symmetry, the oscillator motion in an inhomogeneous force field has been numerically studied [26]. As a result, a previously unknown effect of passage of an oscillator through a potential barrier was discovered. The effect occurs when its motion energy is less than the energy of the barrier, but when the sum of internal energy and the motion energy is greater than the energy of the barrier. It turned out that such a passage of an oscillator through a potential barrier is determined by its phase. The nature of this effect cannot be established without taking into account the PDS [14, 27].

2. Physics of evolution in the world picture. Below we discuss the contribution that the "physics of evolution" can make to the evolutionary picture of the world.

The simplest ONDS can be represented by a set of equilibrium subsystems, which we called SP, and SP, in turn, can be represented by a set of potentially interacting MPs. Thus, ONDS is the third step in bringing the body model closer to reality. SP mechanics arise from MP mechanics, and mechanics of ONDS arise from SP mechanics. **This means that there are principles for constructing a model of matter "from simple to complex."** Here are some of these principles [23].

1. *The principle of the relationship of the laws of systems and their elements.* ONDS mechanics are built based on fundamental laws that apply to their elements. These are the laws of conservation of energy, momentum. The energy of structureless particles has only the motion energy. However, the ONDS energy consists from the motion energy and internal energy. Changes in these energies occur so that their sum is conserved. The nature of the change in the internal energy of the ONDS obeys the second law of thermodynamics. That is, **from the fundamental laws of element dynamics follows the empirical law of the dynamics of ONDS**.

The presence of a nonlinear interrelation between the laws of adjacent hierarchical levels for MP, SP, ONDS, determined by the "evolutionary nonlinearity" of the SP motion equation [23], suggests that such a relationship should exist for all hierarchical levels of matter. It means opportunity of constructing the entire hierarchical picture of matter for all its hierarchical levels, if the laws of behavior of one of the hierarchical levels of matter are known. Thus, the reductionism is valid for any hierarchical level of matter.

2. *The parameters of the upper hierarchical levels of ONDS are determined based on the parameters of the lower levels.* The parameters of the upper hierarchical level are constructed based on the parameters of the lower hierarchical level. For example, parameters characterizing the dynamics of MP systems are based on parameters that determine the dynamics of MP. MP parameters are coordinates, velocity, mass. For ONDS, there are also these parameters, but for ONDS, the mass is the sum of the MP masses. The ONDS coordinates are determined by its center of mass. The coordinates and speeds of the ONDS center of mass are determined through the coordinates and velocities of the MPs included in them.

New concepts also appear for ONDS. For example, dissipative forces are determined by gradients of external potential forces. They lead to the concept of D-entropy, which characterizes the change in the internal energy of the ONDS. D-entropy leads to the concept of entropy in thermodynamics. The concepts of the thermodynamics and statistical physics are appeared from here: temperature, pressure, density, distribution function. The emergence of new concepts for ONDS leads to a modification of the methods and techniques for their description. For example, instead of the phase space, it is more convenient to use the SD-space for ONDS analysis. This modification of the phase space is dictated by PDS.

3. *The evolution of ONDS is the result of double symmetry breaking.* At each hierarchical level of ONDS, evolution is determined by breaking the symmetries of

this level and the symmetry of the lower hierarchical level. That is, the violation of symmetry is always associated with the interaction of the adjacent hierarchical level of matter.

The process of symmetry breaking in quantum systems is characterized by bifurcation [16, 23]. The essence of bifurcation is that a change in the topology of the system occurs at special points in the phase space. Probabilistic laws were used to solve bifurcation problems. However, according to the mechanics of the SP, an analytical method for solving them follows from the condition of infinite divisibility of matter [23]. Indeed, if we take into account the infinite divisibility of matter, then the bifurcation point will become a region of micro variables. **The description of the system at the micro level eliminates the peculiarity of the macro-description of the dynamics of the body at the bifurcation point!** The possibility of a deterministic micro-description of the dynamics of a system at the bifurcation point means that the use of probabilistic models can be considered as coarsening of models of bodies and theories. This coarsening allows us to describe processes in the absence of knowledge about the initial data's. The physical laws themselves determine the region of permissible coarsening. This mechanism of symmetry breaking at the bifurcation point indicates its universality, both for classical and for quantum mechanics, since it is always associated with the interaction of the upper level with the lower level [23].

3. The conditions of the ONDS existence. Although "motion is the way of existence of matter", in practice we often deal with stationary objects. Let us assume that **the ONDS is stationary if the characteristic time of its existence is much longer than the characteristic times of internal processes that ensure this existence**. From the point of view of physics, stationarity means that at all physical points of the ONDS, the values of its parameters do not change during the characteristic times of internal processes.

The way in which stationary ONDS exists is the balance of incoming and outgoing flows of matter, energy, entropy at all hierarchical levels [21, 28]. Bernard's convection cell is one of the simplest stationary objects. It exists due to the flow of matter, which transfers energy from the heated region to the cold region. This flow is created by the temperature difference at the boundaries of the gas or liquid [7]. The larger the temperature gradient, the smaller the structure of convective cells. With a sufficient value of the gradient, turbulence appears.

The stationarity of the complex ONDS requires stationarity at all its hierarchical levels. If for the existence of a Bernard's cell it is sufficient to maintain a temperature gradient, then for the existence of more complex ONDS, for example, a living cell, a balance of flows of various types of matter and energy is necessary. In this case, the flow of matter entering the ONDS is a similar of combination of ONDS lower hierarchical levels. Therefore, all levels of ONDS can exist only through interactions between themselves and with the outside world. It follows from PDS that these interactions are determined by the symmetries of both body and space.

In fact, the stationarity of the ONDS cannot be absolutely. During a time long enough, its structure will change. This **time, which determines the ONDS lifetime, can be called evolutionary**. For complex systems, the lifetime will be determined by the existence of various channels of energy conversion between different hierarchical levels of ONDS. The connection between the steps of this hierarchical ladder of the ONDS is determined by *evolutionary nonlinearity*, which also responsible for symmetry breaking.

Then, the greater the gradient of external forces, the deeper the hierarchy of systems is violated [21]. This corresponds to the established by Einstein and other laws, according to which, the deeper the energy levels of an atom, the more short-wave photon it can be excited.

The existence and evolution of two adjacent hierarchical levels of matter, MP and SP, can be described in the framework of the laws of classical mechanics. However, the processes connecting the more distant hierarchical levels of matter are much more complicated. Since the number of hierarchical levels of matter is infinite, a complete description of evolution is a task of enormous complexity. Nevertheless, **the existence of deterministic interrelations of laws for the two adjacent hierarchical levels allows us to state that such relationships exists for all hierarchical levels of matter**.

One of important example for used of the physics of evolution is the problem of the Universe. Indeed, the physics of evolution can be directly used to solve some problems of astrophysics, since it allows one to calculate the energy fluxes in the Universe during the motion of galaxies, stars, planets, in inhomogeneous fields of gravity forces, particles flows. Today, astrophysics is based on the Newtonian motion equation and on the Einstein equation, which is a relativistic analogue of the Newton motion equation. In many ways, the contradictions between the results of observations and these equations compel us to introduce hypotheses about hidden matter, about dark energy, etc. This may be because Newton's motion equation does not take into account the role of matter structures in their dynamics. This disadvantage can be eliminated by using the equations of physics of evolution [27]. For example, taking into account changes in the internal energy of a star when this star moves in inhomogeneous gravitational fields will give corrections to the energy balance of stars due to gravitational friction [23, 29]. According to the physics of evolution, for the existence of a nonequilibrium Universe, it must expand. This will provide it with the negentropy necessary to maintain nonequilibrium processes, for organize new structures and compensate for the growth of D-entropy in the Universe.

Thus, the physics of evolution includes the mechanics of systems, the extended formalisms of classical mechanics, the principles of the relationship of the hierarchical steps of the ONDS, obtained from the analysis of the properties of the SP's motion equations [14, 23].

4 Physics of Evolution and Philosophical Principles

The essence of the physics of evolution has been explained in previous sections. In this section we will consider the philosophical problems associated with the physics of evolution.

1. Unity and the struggle of opposites. A search for DMI led to PDS. From PDS we came to the dualism of energy, according to which the invariant of the body's motion is the sum of the body's motion energy and the internal energy. Thus, each of these energies can change, but its sum is invariant. The concept of "symmetry of ideal chaos" is associated with the internal energy of an equilibrium system. This symmetry corresponds to the absence of relative motion for all subsystems, the combination of which can represent this system. It follows that the equilibrium of the system means that the sum of the moments of the subsystems relative to the center of mass of this system is always zero. Thus, we have "chaos", which we associate with the internal energy of the equilibrium system. In addition, we have an "order", which we associate with the body's motion due to the body's motion energy.

That is, the nature of DMI is due to the transition of the body's motion energy into the internal energy of the chaotic motion of their elements. Hence, the evolution of matter, the formation of its structures are due to the struggle of two opposites—"chaos" and "order". Chaos plays the role of a "black hole", providing the absorption of the body's motion energy. This is the destructive role of chaos. However, on the other hand, the existence of "chaos" is necessary for the emergence of a new order. That is, **the process of evolution occurs according to the law of unity and struggle of opposites "chaos" and "order"**.

The measure of the transformation of "order" energy into energy of "chaos" is characterized by D-entropy. That is, D-entropy plays the role of a measure of increasing "chaos" [30]. The violation of the symmetry of time is also associated with the transformation of the motion energy into internal energy. This allows us proposing a measure of "evolutionary time", defined as the rate of change of D-entropy. There is no perfect "chaos" or "order" in nature and entropy does not reach an absolute maximum. This means that the body's motion energy and internal energy cannot be equal to zero [31]. That is, matter cannot be in a state of absolute motion or absolute chaos! "Chaos" and "order" can only coexist.

The "birth" of new systems is inextricably linked with the destruction of previous systems and occurs in accordance with the laws of conservation of energy and matter of both systems and the world around them. This is reflected in the principle of dualism of symmetries, according to which the evolution of the world proceeds in the unity and struggle of "chaos" and "order".

2. The unity of the world's picture and the universality of the fundamental laws of physics. The unity of nature follows from the condition of openness. This is extremely important for building a physical picture of the world. The Universe cannot be divided into independent parts that is always done for its mathematical description. This is a huge flaw in mathematical models. Dirac suggested that it can be eliminated if one knows the principles of interaction of system elements and evolution [32]. This assumption is confirmed by the existence of principles for constructing a hierarchical structure of matter "from simple to complex", as well as the fact that matter is an ONDS hierarchy. Thus, these principles are consistent with evolutionism, with the idea of unity, interconnection and interdependence of all structures of the Universe.

If the world develops in accordance with deterministic physical laws, then in nature there is nothing that does not arise from a simpler one [12]. Thus, the physics of evolution confirms the integrity and uniqueness of the picture of the world, as well as the universality of the laws of physics for the Universe.

Let us to give a historic fact. Aristotle, not knowing the concepts of energy and acceleration, found of the body's motion equation, based on observations. According to his equation, the velocity of bodies is proportional to the force. This result is fundamentally contrary to Newtonian mechanics. However, as follows from the mechanics of SP, the Aristotle motion equation is true in the limiting case, when the body's velocity reaches its maximum value due to friction. The SP motion equation contains these two, at first glance, mutually exclusive limit cases. As follows from this equation, when the role of the structure of the body in its dynamics is small, then Newtonian mechanics is valid. However, when the work of external forces goes only for increase in internal energy due to the friction, the Aristotle equation is valid [14]. It follows that the lack of unity in the existing physical picture of the world may be due to the use of restrictions in the creation of a particular theory. If these restrictions are removed, then the contradictions may disappear, and the lost unity will be restored.

Thus, the physics of evolution reinforce the principles of the unity of the world's picture and the universality of the fundamental laws of physics [14].

3. Intensive way of constructing new knowledge. The main method for the development of physics is the study of new phenomena, objects, the identification of new laws and their explanation in the framework of existing fundamental theories. But in the process of the development of science, the limitations that were used in their construction began to appear. This can be seen in the example of elementary particle physics and cosmology. For example, here theories are faced with the problems of spontaneous symmetry breaking in understanding the Heisenberg uncertainty principle. In astrophysics there is a problem of dark matter. A similar difficulty existed to explain the problem of irreversibility in the framework of theories of classical mechanics [31]. This difficulty was overcome by the expansion of formalisms of classical mechanics as a result of taking into account the role of the structure of bodies in their dynamics. Thanks to such accounting, DMI was found, which opened the way to the physics of evolution. It follows that there is a possibility of the development of physics by identifying and eliminating the limitations on which its theories were based. This demonstrated **that physical theories can go the intensive way if existing theories are improved using more realistic models**.

Newton found new laws of classical mechanics, thanks to the simplest model of the body in the form of MP, which excludes from consideration the structure of bodies. However, as it turned out, to describe the processes of evolution in nature, it is necessary to take into account the structure. This has led to the possibility of constructing evolutionary physics based on the fundamental laws of physics. In turn, the physics of evolution has opened up the possibility of studying nature during its evolution.

The construction of the physics of evolution has revealed the need to develop new approach to the construction of a mathematical apparatus that allows us to describe the universal processes of symmetry breaking in physics. The essence new approach lies in the description of the interaction of symmetry groups arising from the motion of the ONDS in inhomogeneous external force fields. The interaction of symmetry groups leads to a violation of the symmetry of time and then to DMI. These violations are associated with "evolutionary nonlinearities" that describe the transformation of energies between independent spaces of variables from different symmetry groups [21, 23].

4. Nonlinear reductionism, the principle of causality and holism. Reductionism plays an important role in the development of science. Today this principle collides with the great difficulties. These difficulties, as a rule, indicate that the further development of knowledge about nature along the path of "*primitive reductionism*" and the extension of theory based on its existing foundation, is no longer a sufficiently effective way of understanding the world, as it was in the initial stages of the development of science.

Let us call by the "primitive reductionism" or linear reductionism such reductionism for which the sum of information about the elements gives complete information about the entire system. This reductionism does not take into account a qualitative leap in information due to the transition of quantity into quality. However, the processes of evolution in nature are impossible without these leaps. Therefore, in "*primitive reductionism*" it is impossible to search for laws and principles that establish a nonlinear relationship between the properties and laws that characterize the upper hierarchical level of matter, with the properties and laws of the elements of the lower level.

DMI, which establishes the causality principle in the physics of evolution, is nonlinear. This suggests a nonlinear relationship of qualitatively new laws of system behavior based on the laws of the dynamics of their elements. Therefore, this also speaks of the nonlinearity of reductionism and the principle of causality.

In connection with the physics of evolution, **it is necessary to use as we call "*nonlinear reductionism*" to study the evolutionary process**. "*Nonlinear reductionism*" may be one of the promising ways to develop a picture of the world. This path is justified by the existence of general principles for the synthesis of knowledge about the laws of ONDS behavior, based on knowledge of physical laws that determine the dynamic and evolutionary characteristics of their elements. Using these principles, you can build a picture of the world, climbing the hierarchical ladder of matter.

A pre-existing explanation of irreversibility is based on probabilistic principles. However, it is one thing to use the concept of randomness for a statistical description of systems, and another when it is a principle that determines the evolution of the world. If the concept of randomness belonged to the fundamental principles of nature, this would mean the absence of determinism [32]. And this, in turn, would mean the absence of "nonlinear reductionism", according to which there is the possibility of knowledge developments from "simple to complex" due to the universality and self-consistency of the physics laws.

The absence of "*nonlinear reductionism*" will mean the triumph of holism, an alternative to the principle of reductionism, which has deep roots in ancient Eastern philosophy. A brief definition of holism: "the whole is greater than the sum of its parts" [20]. For example, according to holism, life has properties that do not follow from the properties of inanimate matter. These are the properties of reproduction, homeostasis, regeneration, etc. However, if the properties of the whole are not related to the properties of its parts, this means the unknowability of nature. Consequently, the question of the validity of reductionism is a question of the cognoscibility of the world and the possibility of constructing its closed, self-consistent picture. Thus, the physics of evolution has expanded the position of the principle of cognitive ability of nature due to the "*nonlinear reductionism*", since it demonstrates the possibility of constructing a hierarchical structure of matter based on the fundamental laws of physics.

Reductionism is impossible without the principle of causality. DMI, which underlies the physics of evolution, establishes the principle of causality in evolution. Indeed, DMI made it possible to connect evolution with the fundamental laws of physics, taking into account the influence of the structure of bodies on their dynamics.

According to principles of the physics of evolution, there is a causal relationship between the laws of different hierarchical steps of matter because the laws of the dynamics of elements determine qualitatively new laws of evolution of their systems. For example, the second law of thermodynamics, reflecting the irreversibility of processes in the system, follows from the reversible laws of the dynamics of the elements of system. This led to the possibility of a deterministic description of evolutionary processes [16, 30].

Without taking into account evolution, determined by the processes of organization, development and destruction of natural systems, the evolution picture of the world not only cannot be complete, it cannot be constructed in principle. The fact that the physics of evolution satisfies the causality principle opens the way to building an evolutionary picture of the world.

Thus, in the frame of the "nonlinear reductionism" it became possible to unite different fields of science in accordance with the principles of the unity of the picture of the world and the universality of the laws of nature.

5. Transition of quantity into quality. DMI was found due to the possibility of establishing the physical properties of systems based on knowledge of the properties of their elements. For example, if we build ONDS from the MP set, then it will have the irreversibility property, while the MP motion is reversible. Therefore very existence of DMI in the framework of the laws of physics indicates the deterministic way transition of quantity to quality.

DMI follows from the ONDS motion equation. According to this equation, DMI is associated with the conversion of motion energy into internal energy [14]. This transition is characterized by D-entropy. An analysis of D-entropy for systems with different number of elements moving in an inhomogeneous space showed that key statistical laws, for example, the law of fluctuations of quadratic functions [21], follow from the fundamental laws of physics.

Using methods of numerical calculations of the dynamics of systems in an inhomogeneous force field, it was found that for the number of elements in the system $N > 100$, the D-entropy can only be positive. This number characterizes the transition of the system to a new quality, in which the laws of statistics are applied. When $N > 1000$, the behavior of D-entropy ceases to depend on an increase in the number of elements. This number determines the area of validity of thermodynamics. That is, the fundamental laws of physics determine the scope of the empirical laws [11]. For example, take the Boltzmann equation [28]. Its importance for physics is difficult to overestimate. But this is an empirical equation, and therefore it has a number of limitations and even contradictions. For example, this equation contradicts the Poincare reversibility theorem, although formally, like the Poincare theorem, it is constructed based on the Hamiltonian formalism. These shortcomings are eliminated if the Boltzmann equation is directly derived from the extended Liouville equation [25]. Even when solving the problem of N-bodies, it is impossible to do without taking into account the fact that the energy of ONDS always consists of internal energy and the motion energy of its structures.

6. The unity of the micro and the macro world. The basic laws of physics, regardless of their field of application, must be closed, self-agreements and satisfy the causality principle if the world has evolved from simple to complex. This is true for objects of classical mechanics. However, in quantum mechanics, these conditions are violated due to the Heisenberg uncertainty principle. According to this principle, it is impossible to simultaneously determine the position and momentum of microparticles [19, 33]. This violates the causality principle in the micro world and therefore, excludes the possibility of constructing an evolutionary picture of the world. However, based on the conditions of the infinite divisibility of matter and the fulfillment of PDS, a deterministic explanation of the uncertainty principle can be proposed. If matter is divisible infinitely, it should be a combination of ONDS and possess internal energy. In this case, the principles of the formation macrosystems from microsystems are valid.

Using the canonical Schrödinger equation to describe their dynamics will lead to the uncertainty of their trajectory. Indeed, the canonical Schrödinger equation is obtained from the Hamilton formalism of classical mechanics, which does not take into account the role of the structure of the system in its dynamics.

This uncertainty is similar to the trajectory uncertainty that will arise when describing the dynamics of a system using the Newton equation since this equation does not take into account the influence of the body structure on its dynamics. Thus, the uncertainty in the description of dynamics based on the canonical Schrödinger equation can also be explained by the fact that it does not take into account the influence of the structure of quantum particles on their dynamics. As in classical mechanics, this uncertainty will be determined by changes in internal energy.

It is known that in quantum mechanics the internal energy of a system cannot be equal to zero. Therefore, as in classical mechanics, this will give uncertainty in the calculations of the volume of the phase space of interacting quantum systems, which is comparable with the value of the Planck constant. To eliminate this uncertainty, one needs to use the extended Schrödinger equation. This equation takes into account the role of changes in internal energy in the dynamics of quantum particles during their interaction [34]. Therefore, the Heisenberg uncertainty principle can be associated with existing methods for describing quantum systems that do not take into account the role of their structures in dynamics, but not because it is dictated by the nature of the microworld. This conclusion is confirmed by the above calculation results of the passage of a classical oscillator through a potential barrier. Only taking into account the role of internal energy in the dynamics of the system, we take into account this effect [26]. If this dependence is not taken into account, we will come to the problem of Aristotelian dichotomy between potentiality and relevance, which Heisenberg has deeply studied and which is still controversial [19, 35].

If so, the problem of substantiating the possibility of constructing physics of evolution, associated with the principle of uncertainty in quantum mechanics, is removed. This testifies in favor of the unity of the laws of physics and in favor of the existence of a "theory of everything".

5 Conclusion

The physics of evolution is based on the mechanics of structural particles. The mechanics of structural particles arose as a result of taking into account the influence of the structure of bodies on their dynamics. This mechanics is based on the motion equation of structural particles arising from the laws of classical mechanics and the principle of dualism of symmetry. From the motion equation of structural particles, it became possible to establish how non-potential dissipative forces arise from potential external forces. This allowed us to create the physics of evolution. The task of evolutionary physics is to describe the evolutionary processes of the appearance, development, and disappearance of physical systems in the framework of the fundamental laws of physics.

According to the physics of evolution, matter is infinitely divisible and represents an infinite hierarchy of open nonequilibrium dynamical systems. There are principles that allow you to climb the hierarchical levels of the structure of matter, getting the laws of the dynamics of the upper levels of matter according to the laws of the dynamics of the lower level. As a result, you can go "from simple to complex", relying on the fundamental laws of physics, without involving any probabilistic laws. This allows you to build a hierarchical picture of matter, if the laws of behavior of only one of the hierarchical levels of matter are known.

Obviously, the number of principles that determine the relationship of hierarchical levels will increase with the development of knowledge. New principles will dictate new laws, such as the laws of evolution of living beings: the laws of self-reproduction, homeostasis, the adaptation of the body and its elements to external conditions. For example, if we stand on the position of Marx, according to which consciousness is a property of matter, which reflects itself, and not a separate, independent entity, then in principle we can develop the physics of consciousness. Thus, in the process of developing knowledge, new laws of behavior of higher hierarchical levels of matter will become known. However, since new laws are built on the basis of the well-known, none of these new laws should contradict the well-known ones, which corresponds to the principle of "common foundations".

The physics of evolution reveals the essence of the law of unity and struggle of opposites, as is seen in the example of dualism "chaos" and "order". Chaos and order coexist only together. The mediator of these opposites is D-entropy, which determines the relationship of "chaos" and "order".

The physics of evolution strengthens the positions of those philosophical concepts that confirm the possibility of constructing an evolutionary picture of the world based on the fundamental laws of physics. In particular, this applies to reductionism.

The very possibility of constructing evolutionary physics testifies in favor of the existence of the causality principle within the framework of the basic laws of physics. As Einstein said: "God does not play dice." Indeed, according to the physics of evolution, the future arises from the present in a deterministic way. In general, the physics of evolution opens up the possibility of building a picture of the world within the framework of universal evolutionism, climbing the hierarchical ladder of matter from "simple to complex."

In accordance with the physics of evolution, the possibility of constructing an evolutionary picture of the world does not mean at all that someday in a very distant future, humanity will create the final picture of the world. The fact is that the number of hierarchical levels of matter is infinite, and the complexity of the process of cognition rapidly increases with the growth of the hierarchical level. Indeed, it is easy to see the history of the explanation of the second law of thermodynamics, which began about 200 years ago. However, this is only the second step in the hierarchy of matter from MP to SP! However, any natural phenomenon will eventually become known. That is, although the processes of studying nature are endless, but the limitations of existing knowledge are associated with the limitations of theories and models used, and not with the existence of the boundaries of knowledge.

Thus, taking into account the structural nature of matter and its role in dynamics has led to the possibility of describing irreversibility, the establishment of infinite divisibility of matter, the universality of the mechanism of symmetry breaking, and, ultimately, the possibility of describing evolutionary processes within the framework of the basic laws of physics. All this means the possibility of constructing a deterministic evolutionary picture of the world within the framework of the laws of physics.

References

1. G. W't. Hooft, Free will in the theory of everything. arXiv:1709.02874v1[quant-ph]8
2. S. Weinberg, *Dreams of a Final Theory* (Vintage, New York, 1992)
3. R.B. Laughlin, D. Pines, The Theory of Everything. PNAS **97**(1), 28–31 (2000); https://doi.org/10.1073/pnas.97.1.28
4. J. Horgan, The End of Science: Facing the Limits of Knowledge in the Twilight of the Scientific Age. xii + 322 pp., bibl., index (Addison-Wesley, Reading, Mass., 1996)
5. K. Mahendra, Verma Microscopic laws vs. macroscopic laws: perspectives from kinetic theory and hydrodynamics. arXiv:1904.12044v1[physics.flu-dyn]
6. H.G. Callaway, Fundamental Physics, Partial Models and Time's Arrow. Dec.2016, https://www.researchgate.net/publication/296327588
7. I. Prigogine, *From Being to Becoming* (Nauka, Moscow, 1980)

8. V.L. Ginzburg, Special session ed. Collegium of the journal, dedicated to the 90th anniversary of the Ginzburg V.L. UFN **177**(4), 345 (2007)
9. G. Nicolis, I. Prigogine, Exploring complexity (Mir, Moscow, 1990)
10. G.M. Zaslavsky, *Stochasticity of Dynamical Systems* (Nauka, Moscow, 1984)
11. V.M. Somsikov, S.N. Azarenko, Determinism in physics and cognoscibility of a picture of the world. Open J. Phil. **9**, 265 (2019)
12. V.V. Aristov, E, Mah, L. Boltzmann, A drama of ideas, a drama of people. Metaphysics **3**(21), 100 (2016)
13. V. Anry, Modern scientific worldview. UFN **1**, 3 (1929)
14. V.M. Somsikov, *To the Basics of the Physics of Evolution* (Almaty, 2016)
15. V.F. Asmus, *Ancient Philosophy* (Moscow, 1976)
16. A. Yu, Loskutov, A.S. Mikhailov, *Introduction to Synergetics* (Moscow, 1990)
17. N.S. Kryilov, Papers on substantiation of statistical physics (L. Publishing House of USSR AS, 1950)
18. E. Wigner, Symmetry and preservation laws. UFN T LXXX111 **4**, 729 (1964)
19. W. Heisenberg, Planck's opening and the basic philosophical problems of the nuclear theory. UFN LXVI, **2**, 163–175 (1958)
20. P.W. Anderson, More is different. Science, New Series **177**, 393–396 (1972)
21. V.M. Somsikov, Deterministic irreversibility mechanism and basic element of matter, in *12th CMSIC CHAOS* (Springer, Cham, 2020), pp. 245–256
22. C. Lanczos, *The Variational Principles of Mechanics* (Peace, Moscow, 1962)
23. V. Somsikov, Deterministic irreversibility and the matter structure. JAP **1**, 26 (2019)
24. YuL Klimontovich, *Statistical Theory of Open Systems* (Janus, Moscow, 1995)
25. V.M. Somsikov, The equilibration of a hard–disks sys. IJBC **14**(11), 4027–4033 (2004)
26. V.M. Somsikov, A.B. Andreev, A.I. Mokhnatkin, V.I. Kapytin, Dual phase space of a nonequilibrium system. PEOS **20**, 1 (2018)
27. V.M. Somsikov, Transition from the mechanics of material points to the mechanics of structured particles. Mod. Phys. Lett. B, **4**, 1–11 (2016)
28. YuB Rumer, MSh Rivkin, *Thermodynamics* (Stat. Phys. Kinemat., Nauka, Moscow, 1977)
29. V.M. Somsikov, Irreversibility and physics of evolution. Chaotic Modeling and Simulation (CMSIM) 1 (2018)
30. L.D. Landau, E.M. Lifshitz, *Statistical Physics* (Nauka, Moscow, 1976)
31. J. Greenstein, A. Zaionz, Quantum challenge. Modern research of the foundations of quantum mechanics (Intellect Press, Dolgoprudny, 2012)
32. P.A.M. Dirac, The relation between mathematics and physics. Proc. R. Soc. (Edinburgh) **59**, 122–129, Part II (1938–1939)
33. R.F. Werner, T. Farrelly, Uncertainty from Heisenberg to Today (2019)
34. V.M. Somsikov, Extension of the Schrodinger equation, in *EPJ Web of Conferences Baldin*, ISHEPP XXIII, Dubna, 1–7 (2017)
35. A.N. Shirazi, Heisenberg's Equality of Inequivalents Problem. arXiv:2003.06517v1[physics.hist-ph] 14 Mar 2020

The AC Driven Frenkel-Kontorova Model: From Shapiro Steps to Chaos

Jasmina Tekić, Andre Botha, Petar Mali, and Yuri M. Shukrinov

Abstract The appearance of devil's staircase and chaos have been studied in the dc+ac driven Frenkel-Kontorova model. In the overdamped limit, the devil's staircase structure arising from the complete mode-locking of an entirely nonchaotic system was observed. Even though no chaos was found, a hierarchical ordering of the Shapiro steps was made possible through the use of a previously introduced continued fraction formula. When the inertial term is included, unlike in the overdamped case, the increase of mass led to the appearance of the whole series of subharmonic steps in the staircase of the average velocity as a function of average driving force in any commensurate structure. At certain values of parameters, the subharmonic steps became separated by chaotic windows while the whole structure retained scaling similar to the original staircase.

Keywords Frenkel-Kontorova Model · Shapiro steps · Devil's Staircase · Chaos

---

J. Tekić (✉)
Laboratory for Theoretical and Condensed Matter Physics - 020, "Vinča" Institute of Nuclear Sciences, National Institute of the Republic of Serbia, University of Belgrade, 522, 11001 Belgrade, Serbia
e-mail: jasminat@vin.bg.ac.rs

A. Botha
Department of Physics, University of South Africa, Science Campus, Private Bag X6, Florida Park 1710, South Africa
e-mail: bothaae@unisa.ac.za

P. Mali
Department of Physics, Faculty of Science, University of Novi Sad, Trg Dositeja Obradovića 4, 21000 Novi Sad, Serbia
e-mail: petar.mali@df.uns.ac.rs

Y. M. Shukrinov
BLTP, JINR, Dubna, Moscow Region 141980, Russia
e-mail: shukrinv@theor.jinr.ru

Dubna State University, Dubna 141980, Russia

C. H. Skiadas et al. (eds.), *13th Chaotic Modeling and Simulation International Conference*, Springer Proceedings in Complexity,
https://doi.org/10.1007/978-3-030-70795-8_65

1 Introduction

One of the most interesting properties of nonlinear dynamical systems with competing time scales is their ability to exhibit frequency locking phenomena. One such phenomenon, that occurs in nonlinear systems under some external radiation or force are Shapiro steps, which appear as a result of dynamical mode-locking of frequencies. Due to significance for various technological applications, for years, they have been the subject of intensive theoretical and experimental studies in charge-density wave systems [1–4], vortex matter [5–7], irradiated Josephson junctions [8–10], and, more recently, even in superconducting nanowires [11, 12]. In the search for an optimum way to control the dynamical mode-locking, one should keep in mind that there is one usually unwanted but often present phenomenon in nonlinear dynamical systems, which is highly sensitive to the initial condition, and which can affect the stability of locked states, this phenomenon is the chaos. Therefore, studies of chaotic behavior are necessary in order to get a complete microscopic picture of frequency locking in nonlinear systems.

One of the models capable of capturing the essence of frequency locking, and the appearance of Shapiro steps is the Frenkel-Kontorova (FK) model under external periodic forces [13–15]. The standard FK model represents a chain of harmonically interacting identical particles subjected to the sinusoidal substrate potential [13, 14]. When the external dc and ac forces are applied locking occurs between the frequency of the particles motion over the periodic potential and the frequency of external ac force [14]. On the macroscopic scale, this effect is characterized by the appearance of a staircase of Shapiro steps in the curve for average velocity as a function of the average external driving force $\bar{v}(\bar{F})$. The steps are called harmonic if the locking appears at integer multiples of the ac frequency or subharmonic at noninteger rational multiples.

It is well known that dissipative dynamical systems with competing frequencies can be described by the circle map. Depending on the coupling strength, the circle map can develop a cubic inflection point leading to the appearance of a devil's staircase and the transition to chaos [16, 17]. When the coupling is below some critical value, the staircase is incomplete, i.e., there are quasiperiodic intervals between the frequency locked plateaus (steps) of periodic behavior. As coupling increases, the frequency locked regions start to broaden, and at some critical value, they fill up all the space. Though the quasiperiodic intervals have zero measure, and the devil's staircase is said to be complete, they have nonzero fractal dimension (scaling index) which is *universal*, i.e. the same $D = 0.87$ for all the systems (at least for those described by the circle map with a cubic inflection point), and thus often considered as a constant of nature [16]. The mechanism leading eventually to chaos is the interaction between different resonances caused by the nonlinear coupling and overlapping of resonant regions when coupling exceeds certain critical value. However, the universality of this scenario as well as the universality of the fractal dimension have been questioned in the past years, and numerous studies in the wide range of biological, chemical, and physical systems have been devoted to models showing the occurrence of the entire nonchaotic regions with complete phase locking [18–21]. Nonchaotic transition from quasiperiodicity to complete locking [18] and *deviation from the universality* with fractal dimension varying from 0.64 to 0.98 have been observed [20, 22].

In this paper we will explore the appearance of devil's staircase and chaos in the dc+ac driven Frenkel-Kontorova model. The discrete FK model is not integrable, and in general, its dynamics can be chaotic [13] since due to its nonintegrability, atomic motion is always accompanied by energy exchange between different modes leading to intrinsic chaotisation of its dynamics [13]. We will analyze both over- and underdamped models and examine how the system dynamics changes with the changing of parameters and transferring from one limit to another.

The paper is organized as follows. The model is introduced in Sect. 2, and simulation results are presented in Sects. 3 and 4. The devil's staircase structure is revealed in Sect. 3, while the chaos was examined in Sect. 4. Finally, Sect. 5 concludes the paper.

2 Model

We consider the dynamics of the standard *damped* FK model, which consists of a series of coupled harmonic oscillators u_l of mass m, subjected to the periodic substrate potential $V(u)$:

$$V(u) = \frac{K}{(2\pi)^2}[1 - \cos(2\pi u)], \tag{1}$$

where K is the pinning strength. This potential belongs to the family of nonlinear periodic deformable potentials, introduced by Remoissent and Peyrard [23] as a way to model many specific physical situations without employing perturbation methods. By changing the shape parameter r, the potential can be tuned in a very fine way, from the simple sinusoidal one for $r = 0$ to a deformable one for $0 < |r| < 1$.

The system is driven by dc and ac forces,

$$F(t) = F_{dc} + F_{ac}\cos(2\pi\nu_0 t), \tag{2}$$

where F_{ac} and ν_0 are amplitude and frequency of the ac force respectively.

If the system is *overdamped*, its dynamics is described by the following system of equations of motion:

$$\dot{u}_l = u_{l+1} + u_{l-1} - 2u_l - V'(u_l) + F(t). \tag{3}$$

where $l = -N/2, ..., N/2$.

In the *underdamped* case we will consider the following set of equations:

$$\dot{u}_l = v_l$$
$$m\dot{v}_l = u_{l+1} + u_{l-1} - 2u_l - \frac{K}{2\pi}\sin(2\pi u_l) - v_l + F(t), \tag{4}$$

where $l = 1, ..., N$ labels the particles, and the term, which comes from the substrate potential is given for the case $r = 0$. The damping is fixed by two parameters m and K, and for some constant force F, the system is overdamped for [24, 25]:

$$0 < m \leq \frac{1}{4(2+K)}. \tag{5}$$

When the system is driven by a periodic force, two frequency scales appear in the system: the frequency ν_0 of the external periodic (ac) force and the characteristic frequency of the particle motion over the periodic substrate potential driven by the average force $\bar{F} = F_{dc}$. The competition between these two frequency scales can result in the appearance of dynamical mode-locking. The solution of the system (4) is called resonant if the time average mean velocity $\bar{v}$ satisfies the relation:

$$\bar{v} := \left\langle \frac{1}{N}\sum_{i=1}^{N} v_i \right\rangle_t = \frac{i\omega + j}{s}\nu_0, \tag{6}$$

where i, j, s are integers and ω is the winding number [15], which is fixed to rational or irrational values, characterizing commensurate or incommensurate structures, respectively.

The above systems of equations (3) and (4) have been integrated for the commensurate structures $\omega = \frac{1}{2}$. using the fourth-order Runge-Kutta method with the periodic boundary conditions for the system of $N = 8$ particles. The force has been increased from zero with the very fine discretization $10^{-4} - 10^{-6}$. Unlike in the overdamped case, the behavior of the underdamped system depends on its previous history therefore, the initial condition at the each step of driving force was obtained from the last step in the integration, at its previous value.

3 Devil's Staircase in a Nonchaotic System

We will consider first the overdamped FK model described by (3). It is well known that the standard overdamped FK model with sinusoidal substrate potential for commensurate structures with integer values of ω reduces to single particle model where no subharmonic locking exists, while for rational, noninteger ω subharmonic steps do appear, however, their size is so small that they are hardly visible on the $\bar{v}(\bar{F})$ characteristics [13]. By introducing some form of deformable potential such as the one in (1) the whole series of halinteger and higher order subharmonic steps start to emerge [14].

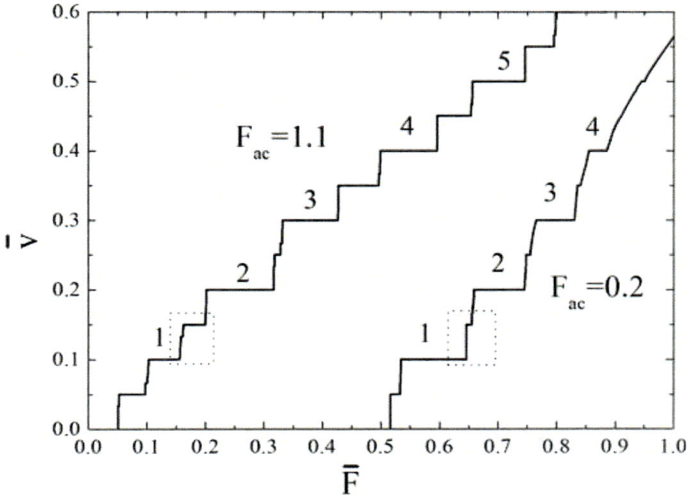

Fig. 1 The average velocity $\bar{v}$ as a function of the average driving force $\bar{F}$ for $K = 4$, $\nu_0 = 0.2$, $\omega = \frac{1}{2}$ $r = 0.5$, and different values of the ac amplitude $F_{ac} = 0.2$ and 1.1. The numbers mark harmonic steps

In Fig. 1, the average velocity as a function of the average driving force is presented for two values of the ac amplitude. The number and size of Shapiro steps, which appear on the response function are determined by the amplitude of the ac force and the extent of deformation of the potential. In Fig. 1, beside harmonic, only halfinteger steps are clearly visible, however, the high resolution analysis reveals a devil's stair case, i.e., an infinite series of higher order subharmonic steps in between them. In Fig. 2, the high resolution views of the selected areas in Fig. 1 are presented. In the devil's staircase structure, the steps appear following the continued fraction formula [9, 10], which in the case of the ac driven FK model can be written as:

$$\bar{v} = \left(i \pm \frac{1}{m \pm \frac{1}{n \pm \frac{1}{p \pm \ldots}}} \right) \omega \nu_0, \tag{7}$$

where $i, m, n, p, \ldots$ are positive integers. Harmonic steps are presented by the first-level terms, which involve only i, while the other terms involving other integers describe subharmonic or fractional steps. Terms involving i and m are called second-level terms, those with i, m, and n third-level terms, etc. In Fig. 2a, b the sequences of the third and fourth level become visible. Our calculations of fractal dimensions shows that it varies with deformation of the potential and the ac amplitude, for small deformation it changes around 0.87, while for higer deformation it decreases depending on the ac amplitude.

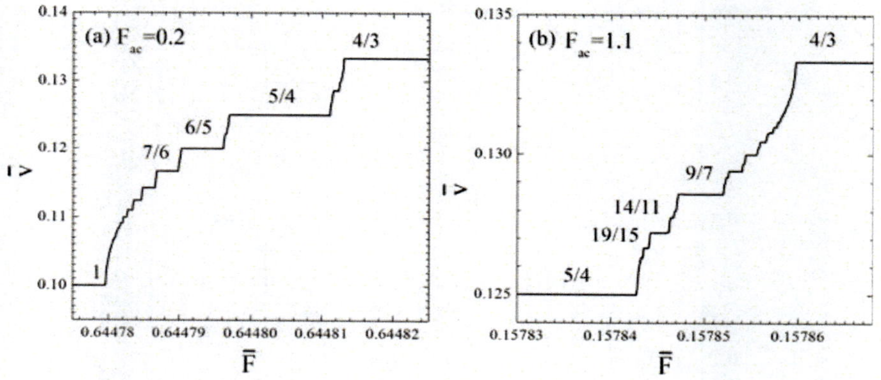

Fig. 2 The high-resolution views of the selected areas in Fig. 1 for $F_{ac} = 0.2$ in (**a**) and 1.1 in (**b**). The rest of parameters are the same as in Fig. 1

Appearance of devil's staircase in the overdamped FK model might lead to the conclusion that if it exhibits complete locking, it must also, therefore, exhibit the chaos. In our search for chaos we applied the largest Lyapunov exponent (LE) computational technique and extend our examination to a very high resolution and wider range of parameters, the ac amplitudes in particular. Regardless of system parameters, no chaos was ever observed. The overdamped Frenkel-Kontorova model remained entirely non chaotic.

The absence of chaos in the ac driven overdamped FK model can be attributed to the dissipative character of the system and the Middleton no-passing rule [26, 27]. According to this rule which applies on one-dimensional, strictly overdamped systems, the order of particles must be preserved in dynamics or, in other words, the particles cannot jump over each other while they move. In such case, there could be no overlapping of resonances which is the main cause of the chaotic behavior in frequency locking systems [10, 16, 17].

4 The Appearance of Chaos

When the inertial term is present (4) and the FK model is underdamped its behavior changes completely. One of well known inertial effects is the appearance of subharmonic mode-locking even in the commensurate structures with integer values of ω [14]. However, the increase of mass in the ac+dc driven FK model may have much more dramatic effects than just simply inducing subharmonic streps. In Fig. 3, the response function $\bar{v}(\bar{F})$ and the corresponding Lyapunov exponents (LE) λ_i are presented for two different values of mass. In order to also examine the effect of the mass on the appearance of subharmonic steps, we consider here the standard FK model with sinusoidal substrate potential ($r = 0$). As we can see in Fig. 3a since

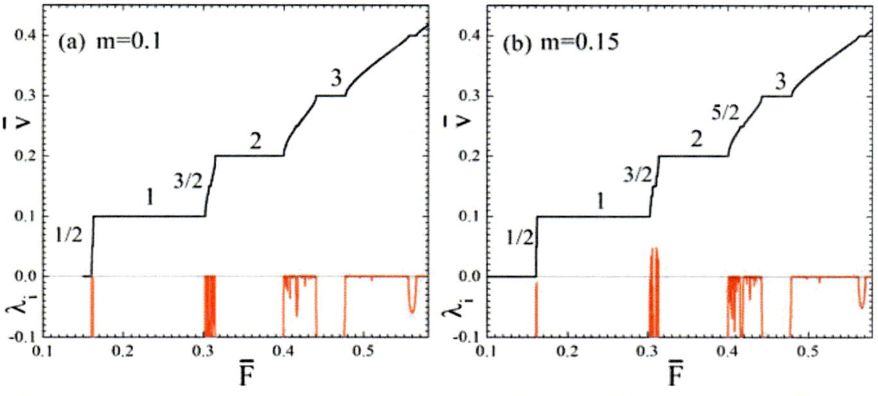

Fig. 3 The average velocity $\bar{v}$ as a function of the average driving force $\bar{F}$ and the corresponding Lyapunov exponents λ_i for $K = 4$, $r = 0$, $F_{ac} = 0.2$, $\nu_0 = 0.2$, $\omega = \frac{1}{2}$, and $m = 0.1$ and 0.15 in (a) and (b), respectively. On this scale of y axis only the largest Lyapunov exponent is visible. Numbers mark the harmonic steps

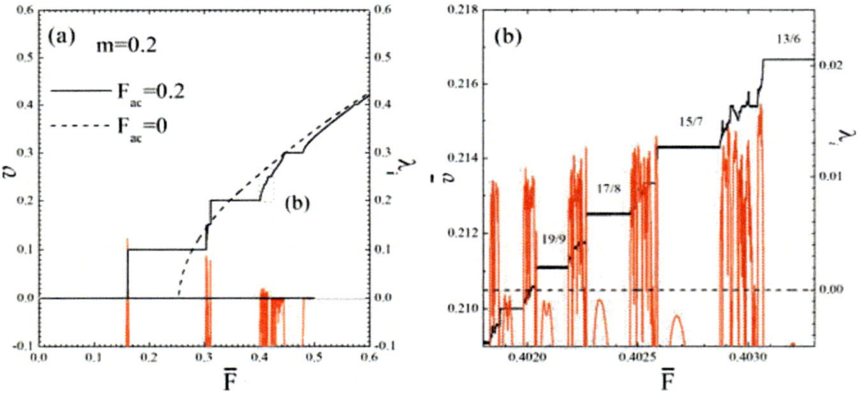

Fig. 4 **a** The average velocity $\bar{v}$ as a function of the average driving force $\bar{F}$ and the corresponding Lyapunov exponents λ_i for $K = 4$,r=0, $F_{ac} = 0.2$, $\nu_0 = 0.2$, $\omega = \frac{1}{2}$, and $m = 0.2$. Dashed line corresponds to the dc driven system $F_{ac} = 0$. **b** The high-resolution views of the selected areas in **(a)**

$m \neq 0$, subharmonic steps start to appear. As the mass increases in Fig. 3b chaotic behavior starts to appear indicated by the positive values of the LE.

Further, we will focus on the chaotic regions between the large harmonic steps, and examine in detail the onset on chaos on subharmonic steps. In Fig. 4, the staircase structure of the average velocity as a function of the average driving force $\bar{v}(\bar{F})$ and the corresponding LE for $m = 0.2$ are presented. Chaotic behavior appears only in the region of subharmonic steps as we can see in Fig. 4a. As the force increases the response of the system approaches to the that of the dc driven one. If we further examine the chaotic region, the high resolution view in Fig. 4b reveals the staircase of subharmonic steps separated by chaotic windows. Devil's staircase containing Shapiro steps separated by self-similar chaotic regions has been observed both in the single and in the one dimensional stack of Josephson junctions [10, 28]. It was shown that in the current-voltage characteristics of the junctions the staircase with chaotic intervals preserves the scaling of the original staircase with the fractal dimension close to 0.87. In our case, for the fractal dimension D in the region between the second and third harmonic step, we obtained D = 0.8759 with an uncertainty of ±0.0166.

5 Conclusion

In this work the appearance of devil's staircase and chaotic dynamics have been studied in the dc+ac driven Frenkel-Kontorova model. In the overdamped limit, though entirely non chaotic, the system exhibits the devil's staircase arising from the complete mode-locking, where the Shapiro steps appear following continued fraction formula. In the underdamped limit, on the other hand, the increase of mass leads to the appearance of the whole series of subharmonic steps in the staircase of the average velocity as a function of average driving force in any commensurate structure. At certain values of parameters, the subharmonic steps become separated by chaotic windows while the whole structure retained scaling similar to the original staircase.

This work could be important for all nonlinear physical systems with competing frequencies from physical to chemical and byological, which exhibit devil's staircase and potentially could go under the transition to chaos. Shapiro steps have been studied for years in Josephson junction systems, which posses a great potential for technological applications from device building to voltage standards and detection of Majorana fermions [29], and situations in which the parameters should be set to produce desired dynamical effects without evoking chaos are a common engineering problem [30]. In voltage standards or other applications, both quasiperiodic and chaotic behavior must be avoided; however, surprisingly, the optimum operating region is actually near the onset of chaos. Therefore, further comparative studies of the resonance phenomena in the Frenkel-Kontorova model and other physical systems, particularly experiments, would be very interesting.

Acknowledgement This work was supported by the Ministry of Education, Science and Technological Development of the Republic of Serbia, Grant no. 451-03-9/2021-14/200017 (University of Belgrade, Institute of Nuclear Sciences "Vinča") and by the Provincial Secretariat for High Education and Scientific Research of Vojvodina (Project No. APV 114-451-2201). The work was funded by the project 18-71-10095 of the Russian National Fund.

References

1. A. Zettl, G. Grüner, Phys. Rev. B **29**, 755 (1984)
2. G. Grüner, A. Zettl, Phys. Rep. **119**(3) (1985)
3. G. Grüner, Rev. Mod. Phys. **60**, 1129 (1988)
4. R.E. Thorne, J.S. Hubacek, W.G. Lyons, J.W. Lyding, J.R. Tucker, Phys. Rev. B **37**, 10055 (1988)
5. N. Kokubo, R. Besseling, V.M. Vinokur, P.H. Kes, Phys. Rev. Lett. **88** (2002)
6. C. Reichhardt, C.J. Olson Reichhardt, Phys. Rev. B **92**, 224432 (2015)
7. A.B. Kolton, D. Domínguez, N. Grønbech-Jensen, Phys. Rev. Lett. **86**, 4112 (2001)
8. S.P. Benz, M.S. Rzchowski, M. Tinkham, C.J. Lobb, Phys. Rev. Lett. **64**, 693 (1990)
9. Y.M. Shukrinov, S.Y. Medvedeva, A.E. Botha, M.R. Kolahchi, A. Irie, Phys. Rev. B **88** (2013)
10. Y.M. Shukrinov, A.E. Botha, S.Y. Medvedeva, M.R. Kolahchi, A. Irie, Chaos **24** (2014)
11. R.C. Dinsmore III, M.H. Bae, A. Bezryadin, Appl. Phys. Lett. **93** (2008)
12. M.H. Bae, R.C. Dinsmore III, T. Aref, M. Brenner, A. Bezryadin, Nano Lett. **9**, 1889 (2009)
13. O. Braun, YuS Kivshar, *The Frenkel-Kontorova Model* (Springer, Berlin, 2003)
14. J. Tekić, P. Mali, *The AC Driven Frenkel-Kontorova Model* (University of Novi Sad, Novi Sad, 2015)
15. L.M. Floría, J.J. Mazo, Adv. Phys. **45**, 505 (1996)
16. P. Bak, Phys. Today **39**, 38 (1986)
17. M.H. Jensen, P. Bak, T. Bohr, Phys. Rev. Lett. **50**, 1637 (1983)
18. P. Alstrøm, B. Christiansen, M.T. Levinsen, Phys. Rev. Lett. **61**, 1679 (1988)
19. P. Alstrøm, M. T. Levinsen, Phys. Rev. B **40**, 4609 (1989)
20. J. Maselko, H.L. Swinney, J. Chem. Phys. **85**, 6430 (1986)
21. A. Cumming, P.S. Linsay, Phys. Rev. Lett. **59**, 1633 (1987)
22. S.E. Brown, G. Mozurkewich, G. Grüner, Phys. Rev. Lett. **52**, 2277 (1984)
23. M. Remoissenet, M. Peyrard, Phys. Rev. B **29**, 3153 (1984)
24. C. Baesens, Lecture Notes in Physics, vol. 671 (Springer, Heidelberg, Berlin, 2005)
25. C. Baesens, R.S. MacCay, Nonlinearity **17**, 567 (2004)
26. A.A. Middleton, Phys. Rev. Lett. **68**, 670 (1992)
27. A.A. Middleton, D.S. Fisher, Phys. Rev. B **47**, 3530 (1993)
28. A.E. Botha, Y.M. Shukrinov, S.Y. Medvedeva, M.R. Kolahchi, J. Supercond. Nov. Magn. **28**, 349 (2015)
29. L.F. Rokhinson, X. Liu, J.K. Furdyna, Nat. Phys. **8**, 795 (2012)
30. R.I. Kauz, Rep. Prog. Phys. **59**, 935 (1996)

Switching Frequency Bifurcations in an LED Boost Driver

Elias D. Tsirbas, Frangiskos V. Topalis, and Evangelos N. Skoubris

Abstract Under specific conditions LED driver circuits can be as susceptible to chaotic bifurcations, as conventional boost converters have proven to be. A significant relationship between the switching frequency of the boost converter's transistor and the circuit's nonlinear behaviour is shown. In order to examine such transistor switching frequency effects, an open-loop configuration is employed, since a feedback control system would obscure these particular nonlinearities. A theoretical method has been devised to predict the unstable frequency regions based on certain dependence equations. There are particular nonlinear parameters which influence the circuit's behaviour, such as the reverse-recovery time of the boost diode, as well as the collective effect of the inductance and the diode's junction capacitance. The dependence equations prove a correlation between these inherent nonlinearities and the switching frequency of the boost transistor. Period doublings and transitions to chaos occur for several regions of the examined switching frequency range. The theoretical method used for the numerical analysis is based on the periodicity of certain voltage waveform peaks, probed at key points on the converter. The LED boost driver displays a wealth of nonlinear phenomena and detrimental effects on its brightness levels throughout the nonlinear frequency regions.

Keywords LED boost converter · Chaotic oscillations · Nonlinear dynamics · Bifurcations · Lighting circuits

E. D. Tsirbas (✉) · F. V. Topalis
National Technical University of Athens, Zografou, Athens, Greece
e-mail: etsirbas@central.ntua.gr

F. V. Topalis
e-mail: fvt@central.ntua.gr

E. N. Skoubris
University of West Attica, Egaleo, Athens, Greece
e-mail: eskoubris@uniwa.gr

C. H. Skiadas et al. (eds.), *13th Chaotic Modeling and Simulation International Conference*, Springer Proceedings in Complexity,
https://doi.org/10.1007/978-3-030-70795-8_66

1 Introduction

Plenty of practical applications that employ power conversion circuits would benefit from an alternative circuit model, one that takes into account the nonlinear nature of the converter's characteristics. Conventional modelling techniques utilized so far, can omit various nonlinear effects and as a result the circuit designer might miss certain operating behaviours that may significantly degrade a power converter's performance. Such unpredictable phenomena might arise in current-mode DC-DC boost converters [1] or in Power Factor Correction (PFC) power supplies in the form of slow-scale instabilities, which can degrade the power factor significantly [2]. As a consequence, the linearization approach that is still commonly used in the industry can provide inaccurate solutions to designers, since nonlinear instabilities that can affect a power converter's response, cannot be easily detected without a certain type of analysis. Such an analysis can offer an in-depth view into the behaviour of these nonlinear circuits that may embrace specific properties such as subharmonics [3] as well as several chaotic phenomena within numerous power converter configurations [4].

A diode circuit in the form of a resonator made an early introduction to electronics chaos [5]. The inductor and the diode of this simple circuit have proven to be highly nonlinear circuit elements as shown in [6] and [7] amongst others. When the nonlinear capacitance of the diode's equivalent circuit is combined with a nonlinear resistor at high frequencies chaos emerges [8].

Diode resonator circuits have also been initially used for the study of the *period-doubling* phenomenon that can lead a system towards chaos. At the course of this phenomenon, a signal waveform's period is doubled successively as a control parameter of the circuit is varied, until the time-series finally become chaotic and the signal's period becomes undetectable. The circuit diverges from its designated operation, as in the example of rectifiers that employ slow-response diodes [9]. Since LED driver circuits are essentially modified DC-DC power converters [10], it is important to explore the conditions that could cause chaos in such a system. In most published papers though, chaos appears mainly due to controller instabilities of the feedback loop in such power converters [11] or as a result of slope disturbances [12]. The possibility of chaos is examined without the feedback loop, in order to prove that chaos can be possible only due to the inherent nonlinear properties of the boost diode and inductor combined.

2 Diode Physical Characteristics

A Light Emitting Diode (LED) forms a special type of diode. A specific forward voltage is required in order to switch on an LED, which will then enable the nominal operating current through it. LEDs are nonlinear devices, in the sense that they do not possess a linear relationship between the voltage applied, and the current drawn. Once the forward voltage is reached, the current through the LED will rise exponentially, and visible radiation is emitted.

The equation shown below from the authors' published work [13], shows that the instantaneous current of the diode is linked with an exponential relationship to its reverse current,

$$i(v) = i_0\left(e^{\frac{q_e v}{kT}} - 1\right) \tag{1}$$

which is the i_o term shown above. LEDs will switch on only when they are forward-biased with a positive voltage polarity to its anode and a negative voltage to its cathode. If this voltage polarity is reversed, current should not normally flow, as the diode behaves as an insulator under reverse polarity.

Under normal current flow, a spontaneous recombination of electrons and holes takes place in the PN junction of the diode's semiconductor material, and it is through their interaction that light is generated. Under specific circumstances however, a small reverse current might flow. Such an event occurs when the temperature is varied and when minority charge carriers move inside the junction.

The power electronic circuits which drive LED devices can either operate under AC or DC voltage, although it is more common to operate LEDs under direct current.

In the case of the LED boost driver circuit, a DC-DC boost converter is utilised. A DC input voltage supplies the converter, while energy is stored in both the inductor and its output capacitor. A MOSFET transistor switch controls this energy transfer by switching on & off rapidly and the LED is supplied with this combined capacitor and inductor voltage. The combined output voltage is higher than the input voltage and it should be near the LED's nominal forward voltage.

For the purposes of the experiments of this paper, the MOSFET duty cycle was set at D = 0.5 (50% pulse on-50% pulse off) which should double the input supply voltage. For the LED boost driver, this results in an LED output voltage of $1.7 \times 2 = 3.3$ V. This is the calculated LED forward voltage in order to switch on the particular power LED module. The average LED current is also estimated in the range between 350 and 450 mA. Two main DC-DC boost converter topologies form the basis of the experiments. One open-loop boost converter with a resistive load, and one LED boost driver. The circuit in Fig. 1 shows the experimental configuration of the LED boost driver. Table 1 shows the experiment settings for the resistive boost converter.

3 Theoretical Analysis

There are certain nonlinear characteristics which influence the diode's operation. One of them is the reverse-recovery time, the time that the diode needs to recover the positions of its electrons and holes, when a reverse current goes through it.

The resonant frequency which relates the diode's reverse-recovery time is shown below.

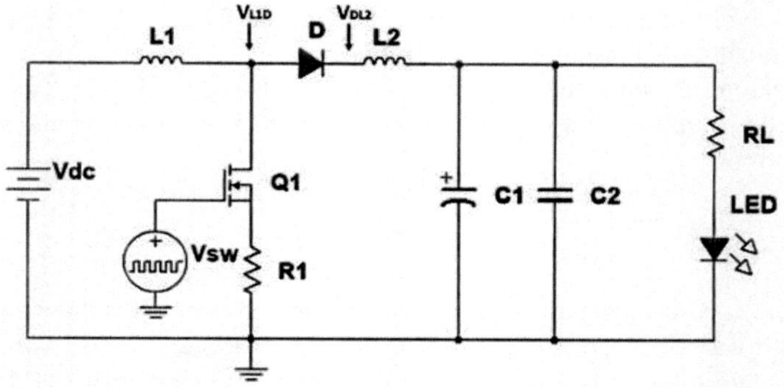

Fig. 1 LED boost driver experimental circuit schematic

Table 1 Resistive boost driver circuit configuration

Resistors	R_L = 24 kΩ (Load) R_1 = 1.2 Ω (Q1-MOSFET source)
Inductors	L_1 = 4.48 mH, L_2 = 470 μH
Boost diodes	D_2 = type BYG20J (C_j = 25 pF, τ_{RR} = 75 ns) D_5 = type 1N4007 (C_j = 10 pF, τ_{RR} = 5 μs)
Output capacitor	C = 10 μF ($C_1//C_2$)
DC Voltage source amplitude	V_{dc} = 14 V
Frequency of pulsed voltage source	f_{sw} = 1 kHz to 1 MHz

$$f_{\tau_{RR}} = \frac{1}{\tau_{RR}} \quad (2)$$

A second nonlinear parameter is the diode parasitic junction capacitance. When a reverse current flows through the diode, certain electrical charge is held at the junction, due to this capacitance. When the input voltage changes polarity again, that is from negative to positive, the inductor tries to maintain this reverse current.

The resonant frequency which relates the parasitic capacitance and the total inductance of the circuit is shown below.

$$f_{LD} = \frac{1}{2\pi\sqrt{LC_j}} \quad (3)$$

As a result, two inherent resonant frequencies emerge from these nonlinear characteristics, $f_{\tau_{RR}}$ and f_{LD}. A specific method with dependence equations has been used in order to derive the unstable switching frequency regions of this circuit. These dependence equations have been successfully utilized in resonator circuits in the past [14]. This theoretical method is supported by both simulation and laboratory experiments, whereby a circuit model has been designed for each experiment type. In other published works of Hamill [15] or Dobson [16], a different model type predicts the behaviour of generic boost converters. However, the model of this paper aims to define the physical principles behind the formation of chaos in such boost converter circuits. When these resonant frequencies are synchronised with the switching frequency of the transistor, chaos occurs at multiples of these resonant frequencies. The dependence equations shown here, were used to support this theoretical analysis.

$$f_{sw} \approx f_{LD} \tag{4}$$

$$f_{sw} \approx f_{\tau_{RR}} \tag{5}$$

In order to examine the reverse recovery effects on the diode's operation, a resonator circuit was initially simulated, built and examined. This original diode resonator includes an AC source, an inductor, a diode and a resistor load [5]. When the AC voltage changes direction, chaotic oscillations are formed at specific frequencies, right at the load's voltage due to the aforementioned reverse current effects.

The next experiment involves a pulsed input as the excitation source of the diode resonator. A similar nonlinear response to the original diode circuit was observed, which proves that chaotic and resonant oscillations under an alternating or pulsed signal input are indeed feasible. At a later stage, the resistive boost converter was used to test the theoretical model. Finally, the LED boost driver was utilised in order to verify these findings in a lighting application circuit.

For the theoretical analysis, specific voltages on the LED driver were selected, and their waveforms have been examined. This theoretical analysis follows an iteration sample procedure which examines the periodicity of the waveform peaks. The voltage waveforms have been sampled at the same switching frequency of the transistor, and if the peaks repeat at the same amplitude and shape periodically, then the system is in a stable and linear state. If the peaks show period doubling, the waveform peaks appear twice, before the waveforms' period starts again. If the peaks do not show a regular periodic pattern, chaos has been reached in the system. The suggested iteration procedure is described below. The reader may select the two voltage variables of the system, that is the MOSFET drain voltage V_{L1D} and the diode voltage V_{DL2}. The suggested peak analysis method involves an iteration algorithm for the two voltages as shown below

$$V_{L1D,(n+1)T} = f(V_{L1D,nT}) \tag{6}$$

$$V_{DL2,(n+1)T} = f(V_{DL2,nT}) \tag{7}$$

where T is the switching period and n is the iteration number ($n = 1, 2, \ldots k$), which is an independent variable. Typical switching frequencies for most boost converters

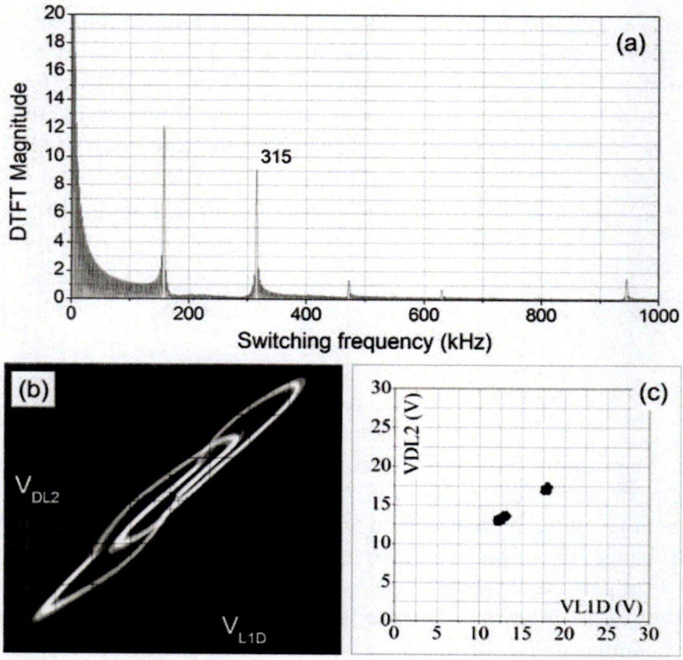

Fig. 2 (period-2, $f_{sw} = 315$ kHz): **a** VDL2 voltage Fourier spectra, **b** XY plot, **c** Poincaré plot

range from tens to hundreds of kHz, but modern LED drivers have reached 2–3 MHz. In order to observe such chaotic phenomena in the laboratory, Poincaré sections, Fourier transform spectra and phase space XY plots have been calculated and displayed as follows (Figs. 2 and 3).

For the initial experiments with the resistive boost converter, a bifurcation diagram has been generated for the entire examined frequency range. Period-doublings have been witnessed at 315 kHz and chaos at 214 kHz. Both of these frequencies were multiples of the resonant frequency that the theoretical model predicts for this particular diode type. These results are plotted in phase space plots, frequency spectra and Poincare sections.

For the transistor switching frequency range from 15 kHz up to 350 kHz the experimental voltage waveform data was analysed, and the bifurcation diagram of Fig. 4 was generated. Linear regions are indicated with single solid lines, and the sparse regions indicate strong nonlinear regions with wide voltage differences. It should be stressed that in some cases (e.g. near 25 kHz), the inspected diode voltage reached almost 80 V, with only 14 V of input voltage.

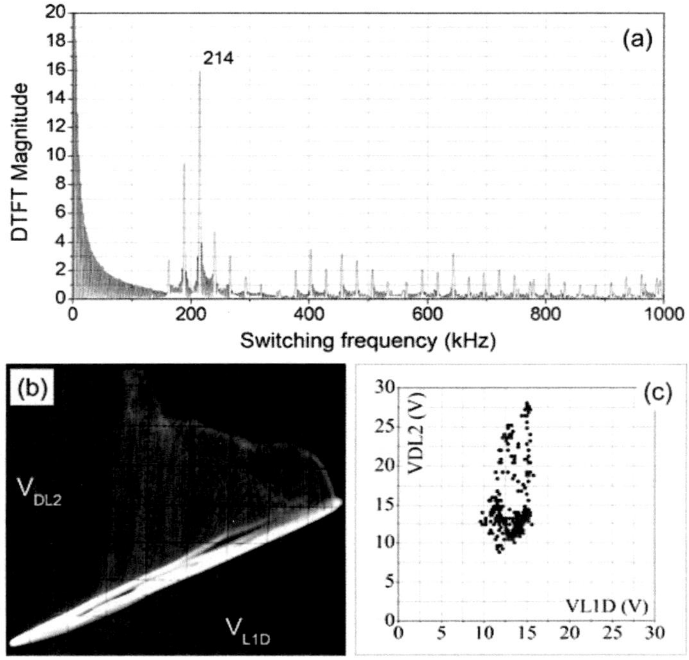

Fig. 3 (chaos, $f_{sw} = 214$ kHz): **a** VDL2 voltage Fourier spectra **b** XY plot, **c** Poincaré plot

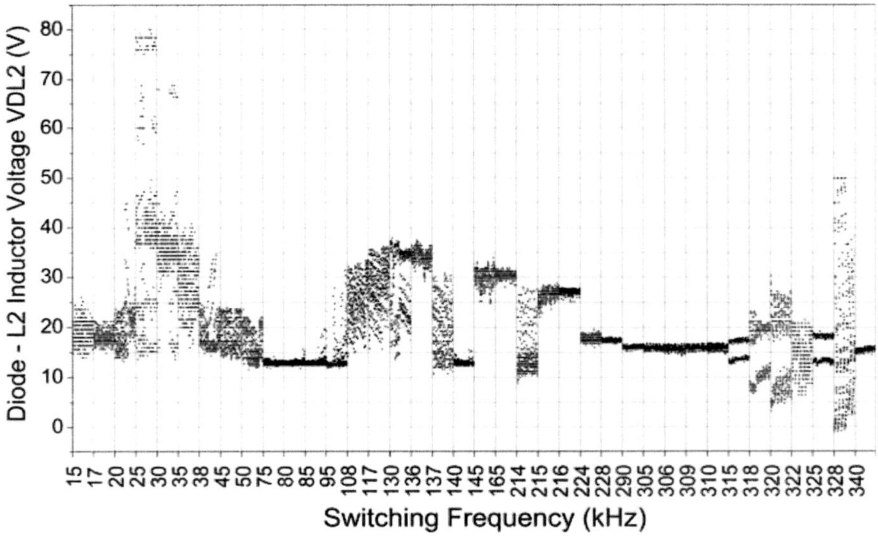

Fig. 4 Bifurcation diagram of V_{DL2} voltage-as a function of the switching frequency

4 Illuminance Experiments

Following the experiments with the resistive boost converter, the same switching frequency variation procedure was performed, in order to investigate the LED boost driver behaviour. For the LED boost driver the following circuit configuration has been used (Table 2).

Boost diode-1N4007 (slow-response): Throughout the different switching frequencies, periodic behaviour at 50 kHz was recorded, along with relatively high brightness levels (230 lx) (Figs. 5 and 6). At much lower switching frequencies the LED illuminance peaks at 460 lx. With the same diode type, a strange attractor appears at 80 and 200 kHz (Figs. 7 and 8), in a similar fashion to the resistive boost

Table 2 LED boost driver circuit configuration

Resistors	$R_L = 10\ \Omega$ (2 W) (Load) $R_1 = 1.2\ \Omega$ (Q1-MOSFET source)
Inductors	$L_1 = 4.48$ mH, $L_2 = 470\ \mu$H
Boost diodes	D_2 = type BYG20J ($C_j = 25$ pF, $\tau_{RR} = 75$ ns) D_5 = type 1N4007 ($C_j = 10$ pF, $\tau_{RR} = 5\ \mu$s)
Output capacitor	$C = 10\ \mu$F ($C_1//C_2$)
DC voltage source amplitude	$V_{dc} = 1.7$ V
Frequency of pulsed voltage source	$f_{sw} = 1$ kHz to 1 MHz

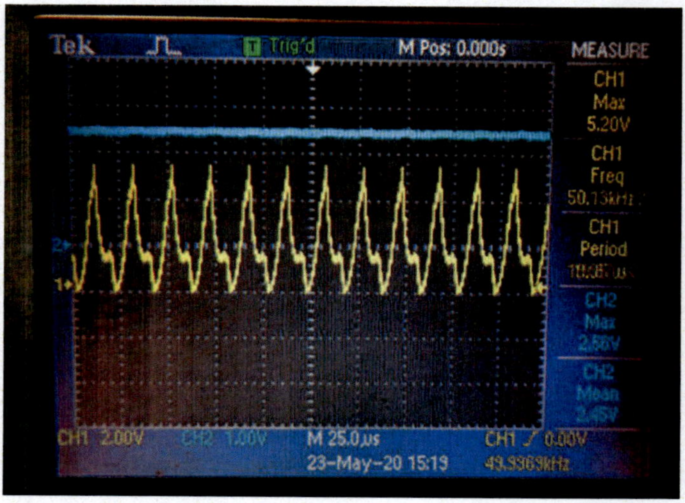

Fig. 5 LED Driver period-1 response, 50 kHz, Time-domain waveform

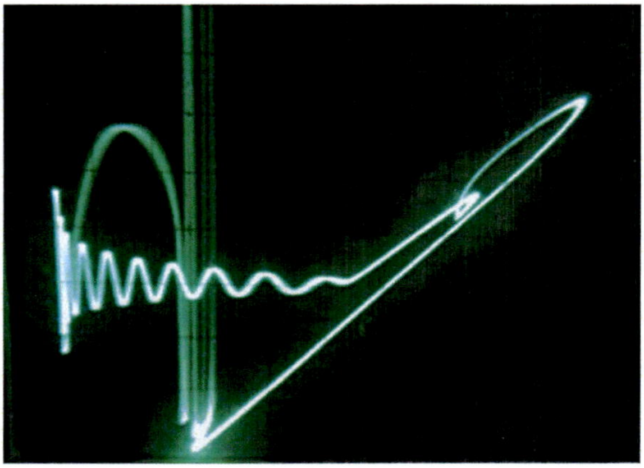

Fig. 6 LED Driver period-1 response, 50 kHz, Phase plot

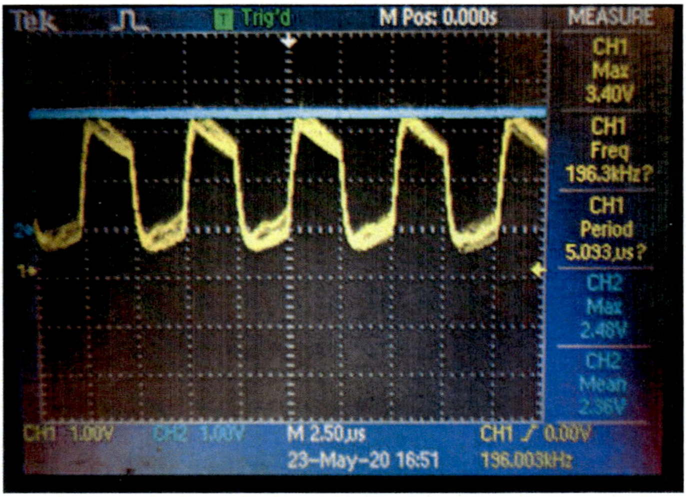

Fig. 7 LED Driver chaotic response, 200 kHz, Time-domain waveform

converter. The illuminance levels drop considerably at these frequencies, due to the high nonlinearity of the system.

The illuminance data has been processed in order to plot them against the switching frequency. The illuminance graph for the slow-response 1N4007 diode is shown in Fig. 9.

High illuminance is recorded in periodic or relatively stable period-4 or period-5 attractors. In the frequencies that the system reached chaos, extremely small illuminance levels have been recorded.

Boost diode-BYG20J (fast-response): With the ultra-fast diode, larger areas of linear behaviour have been observed, as this diode's reverse recovery time is

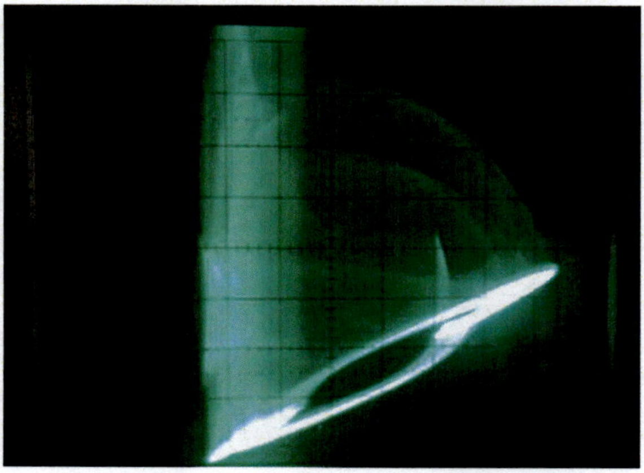

Fig. 8 LED Driver chaotic response, 200 kHz

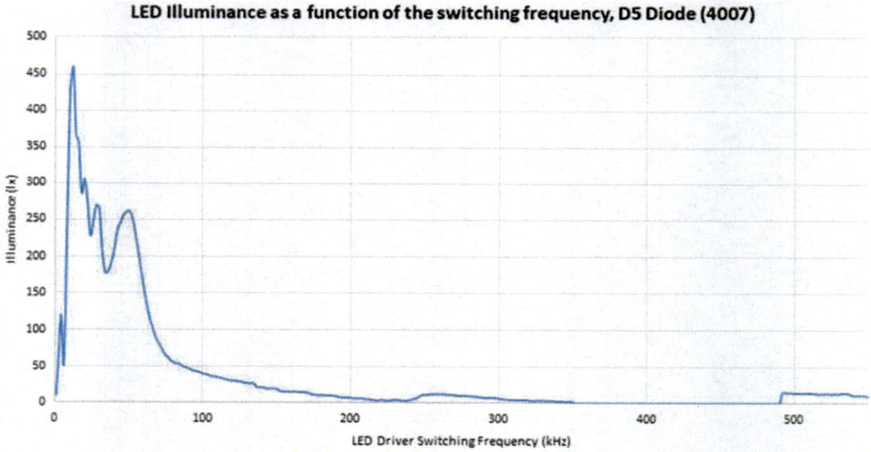

Fig. 9 Illuminance as a function of the MOSFET switching frequency (slow-response diode)

considerably faster than the 1N4007 diode. A very high illuminance was recorded in the periodic regions reaching a maximum of 1353 lx, whilst the illuminance falls to just 17 lx in the chaotic attractor regions (Fig. 10). The theoretical method predicts these nonlinear resonant frequencies where chaos dominates the system.

The ultra-fast diode exhibits more illuminance peaks at higher frequencies, where the slow diode was going through a very unstable region. Still, chaos is observed at certain frequencies at the lower peaks of the graph, e.g. at 80, 130 or 200 kHz.

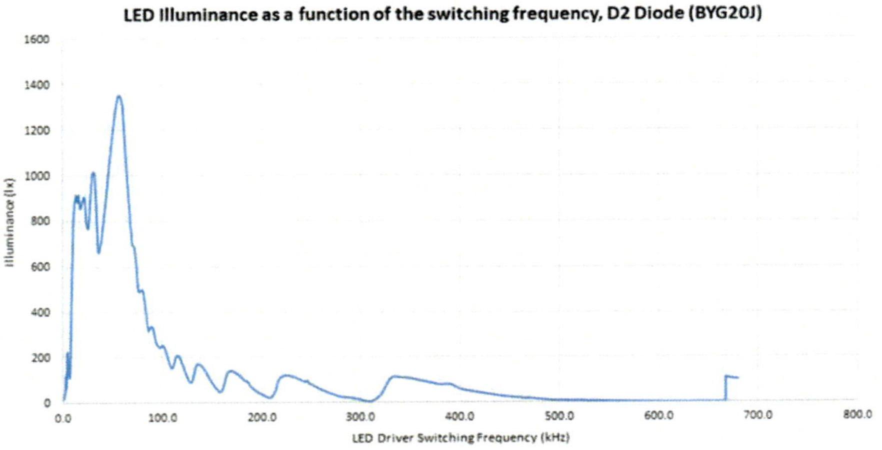

Fig. 10 Illuminance as a function of the MOSFET switching frequency (fast-response diode)

5 High Sensitivity to Initial Conditions

There is a particular type of phenomenon which occurs often in such nonlinear systems. Even with the BYG20J ultra-fast diode, a sensitive dependence on the initial conditions of the system has been recorded. With only a 100 Hz of difference between the switching frequencies of 666.6 kHz and 666.7 kHz, the LED driver circuit moves suddenly from an unstable region of almost zero illuminance, to a stable region of considerable illuminance (Fig. 11).

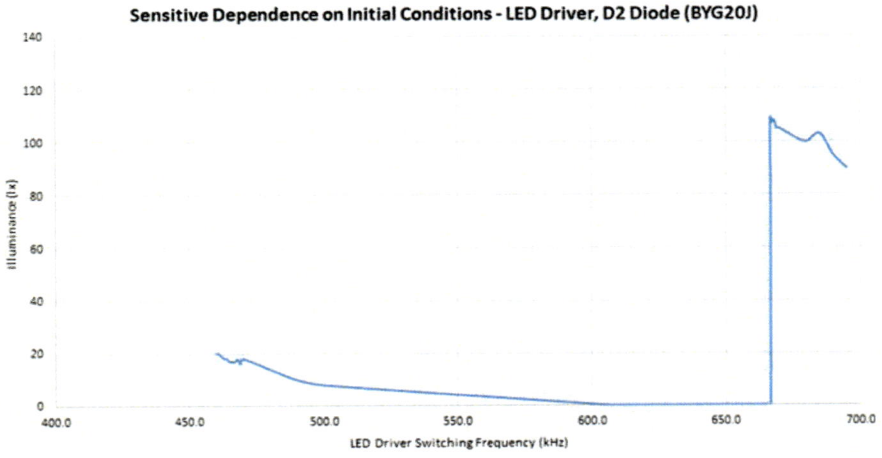

Fig. 11 Illuminance as a function of the MOSFET switching frequency

6 Conclusions

Two nonlinear resonant frequencies of an LED driver circuit have been detected, which influence harmfully its performance. This behaviour manifests itself only in the power section of the boost driver, and unlike previous literature, chaos occurs without a feedback loop. The synchronisation of these resonant frequencies with the transistor switching frequency, steers the LED converter towards chaos. The negative effects include a significant degradation in illuminance performance, especially at the chaotic regions. The suggested nonlinear analysis includes a bifurcation peak-to-peak method, in order to guide the interested researcher to avoid such unstable regions. A number of research suggestions for further steps of this work can be considered. In this context, the resonant response of even faster diodes can be investigated. The interested reader can also explore the low-frequency chaos that has been found in such fast diodes. Finally, some abrupt transitions to chaos that have been recorded at specific frequencies, could be investigated as forms of interior crisis within this nonlinear system.

References

1. D. Giaouris, S. Banerjee, O. Imrayed, K. Mandal, B. Zahawi, V. Pickert, Complex interaction between tori and onset of three-frequency quasi-periodicity in a current mode controlled boost converter. IEEE Trans. Circuits Sys. I, Reg. Papers **59**, 1, 207–214 (2012)
2. D. Dai, S. Li, X. Ma, C.K. Tse, Slow-scale instability of single-stage power-factor-correction power supplies. IEEE Trans. Circuits Sys. I, Reg. Papers **54**, 8, 1724–1735 (2007)
3. K. Mandal, A. Abusorrah, M.M. Al-Hindawi, Y. Al-Turki, D. Giaouris, S. Banerjee, Dynamical analysis of single-inductor dual output DC-DC converters, in *IEEE International Symposium on Circuits System (ISCAS)* (2013), pp. 2755–2758
4. S. Banerjee, G. Verghese, *Nonlinear Phenomena in Power Electronics* (Wiley, 2001)
5. T. Caroll, L. Pecora, *Nonlinear Dynamics in Circuits* (World Scientific Publishing, 1995)
6. L.O. Chua, R.N. Madan, Sights and sounds of chaos. IEEE Circ. Dev. Mag. **4**, 1, 3–13 (1988)
7. J.-M. Collantes, A. Suárez, Period-Doubling Analysis and Chaos Detection Using Commercial Harmonic Balance Simulators, IEEE Trans. Microw. Theory Tech., 48, 4, 574{581, 2000
8. T. Matsumoto, Chaos in electronic circuits. IEEE Proc. **75**, 8, 1033–1057 (1987)
9. L.V. Karadzinov, D.C. Hamill, Influence of diode reverse recovery on the operation and design of high-frequency rectifiers, in *IEEE Applied Power Electronics Conference Proceedings, APEC, Thirteenth Annual*, vol. 1 (1998), pp. 447–453
10. ON Semiconductors Application Note, NCP5007 Compact Backlight LED Boost Driver, Rev. 5, 1–21 (2014)
11. B. Basak, S. Parui, Exploration of bifurcation and chaos in buck converter supplied from a rectifier. IEEE Trans. Power Electron. **25**, 6, 1556–1564 (2010)
12. H. Wu, V. Pickert, D. Giaouris, B. Ji, Nonlinear analysis and control of interleaved boost converter using real-time cycle to cycle variable slope compensation. IEEE Trans. Power Electron. **32**, 9, 7256–7270 (2017)
13. E. Gluskin, E. Tsirbas, I. Kateri, F.V. Topalis, Use of logarithmic sensitivity in power system analysis: the example of lighting circuits (hot filament, LED and fluorescent lamp circuits). IET Sci. Meas. Technol **7**, 6, 297–305 (2013)
14. R.M. de Moraes, S.M. Anlage, Unified model and reverse-recovery nonlinearities of the driven diode resonator. Phys. Rev. E. 026 201, **68**, 1–10 (2003)

15. D.C. Hamill, H.B.J. Deane, instability, subharmonics, and chaos in power electronic systems. IEEE Trans. Power Electron. **5**, 3, 260–268 (1990)
16. I. Dobson, Stability of ideal thyristor and diode switching circuits. IEEE Trans. Circuits Sys. I, Reg. Papers **42**, 9, 517–529 (1995)

Halo Dynamics: From Rainbows to Black Holes

Alberto Tufaile and Adriana Pedrosa Biscaia Tufaile

Abstract Here we applied some of concepts of dynamical systems in an experiment involving a laser beam injected in a glass cylinder, recording the light patterns from the scattering of light from a finite cylinder. We have studied some aspects of the representation of dynamical systems in this experiment, along with the observation of the existence of a sequence of numbers which characterizes this dynamics, known as Farey sequence, due to its connection with trajectories following star polygons. We also report the observation of arcs with folds in these light patterns. We studied the case when the cylinder change its shape into the case of a foot of a wine glass, and compared some solutions of the cylindrical lens with epicycloid dynamics and halo formation, for the case of patterns formation based in the observation of relativistic effects.

Keywords Relativity · Farey sequence · Billiards

1 Introduction

The word halo brings to mind a picture of an optical phenomenon by light interaction with ice crystals suspended in the atmosphere, such as the circular halo, which is a representative of a family of luminous patterns whose main effect is a large ring of light around a point. Here, we consider a halo any distribution of a light pattern around a point. The essence of these phenomena is somehow related to any property that is capable of bending light from the source as the light travels towards the observer. The presence of a halo can therefore reveal the extent of the optical properties of the medium where light spreads. For example, luminous arcs are present in rainbows (Fig. 1a), 22 degrees halo around the Sun (Fig. 1b) or in relativistic effects such

A. Tufaile (✉) · A. P. B. Tufaile
Soft Matter Lab, Escola de Artes, Ciências E Humanidades, University of São Paulo, São Paulo 03828-000, Brazil
e-mail: tufaile@usp.br

A. P. B. Tufaile
e-mail: atufaile@usp.br

C. H. Skiadas et al. (eds.), *13th Chaotic Modeling and Simulation International Conference*, Springer Proceedings in Complexity,
https://doi.org/10.1007/978-3-030-70795-8_67

Fig. 1 Some halos and luminous arcs found in nature. In **a** rainbow observed at Newark, California, USA, in **b** the 22 degrees halo around the sun formed by the interaction of the sun light and ice crystals suspended in the atmosphere, and in **c** the gravitational mirage known as Einstein ring obtained with a simulation of a gravitational lens

as Einstein rings (Fig. 1c), with the presence of partial or complete circumferences. From these examples, we see that there are different types of systems that can present formation of luminous halos, based on quite different physical environments. In the previous examples, the halos are related to the presence of ice crystals or drops in space, deflecting light, in other cases the very constitution of the structure of the space-time creates the effect of the curvature of light path, due to the presence of massive objects such as quasars, galaxies or black holes.

Due to the visual appeal of these phenomena in the mankind, this subject has been observed and studied intensely by centuries with a significative impact on history of science. The study of rainbows, glories and halos helped the development of optics and mathematics [1], and many authors have described the aspects of the theory of rainbows and its applications [2], the elementary physical features of halos [3], and advanced aspects such as considering a rainbow as diffraction catastrophe [4]. The existence of multiple rainbows was done using glass rods with normally incident light and for diagonal incidence [5, 6]. Interesting patterns obtained with cylindrical symmetry in optical systems is also related with some studies of dynamical systems and solitons [7]. In addition to rainbow halos, the existence of halos in Einstein rings was predicted by the lensing effect by Albert Einstein over 70 years ago [8].

Stimulated by the observation of some patterns obtained in our experiments, we have explored the use of the concepts present in dynamical systems and topology to give a new perspective in the comprehension of halo formation. For this reason, it is appropriate to discuss the light beams like particles travelling in an open billiard, exploring these motions from the point of view of dynamical systems, and this is the aim of this paper. The description of light rays in open billiards creating light patterns forming circles was explored in our previous works [9–13], as well as the role of the hyperbolic geometry in some light pattern formation [15]. In this paper we explore the formation of some of these halos and arcs from the point of view of dynamical systems based on some concepts of optics using two lenses, a cylindrical one and another with the shape of a pseudosphere. We start studying the scattering of rays in a cylinder and present a direct analysis of the observed pattern based on the conceptual structure of the dynamical systems. After that, we propose a topological transformation of the cylinder to a pseudosphere and analyze some properties of gravitational lenses.

2 The Cylinder as an Open Billiard

Our first experiment consists of a laser beam hiting a glas rod, as it is shown in Figs. 2a, b. The theory of rainbows here is the following. For the case of circular section of Fig. 2c, light reaching the cylinder with angle $\alpha_i = 0$, creates rainbows for specific angles, due to the formation of caustics. Usually, caustics can be defined as the envelope of rays that describes the flow of energy, and this energy flow increases significantly on caustics compared to the adjacent space, forming more intense light patterns. Lenses with circular sections exhibit the formation of caustics similar to those observed in the case of the rainbow, and the main ray of the caustics is called Cartesian ray, which can be obtained considering the total deviation of a ray for the general case of k-internal reflections giving by [16]:

$$\theta_{TRT} = 180^\circ\, k + 2\theta_i - 2\gamma(k+1), \tag{1}$$

and the incident angle θ_i corresponding to the minimum deviation after k internal reflection is the Cartesian ray angle θ_{Cr}:

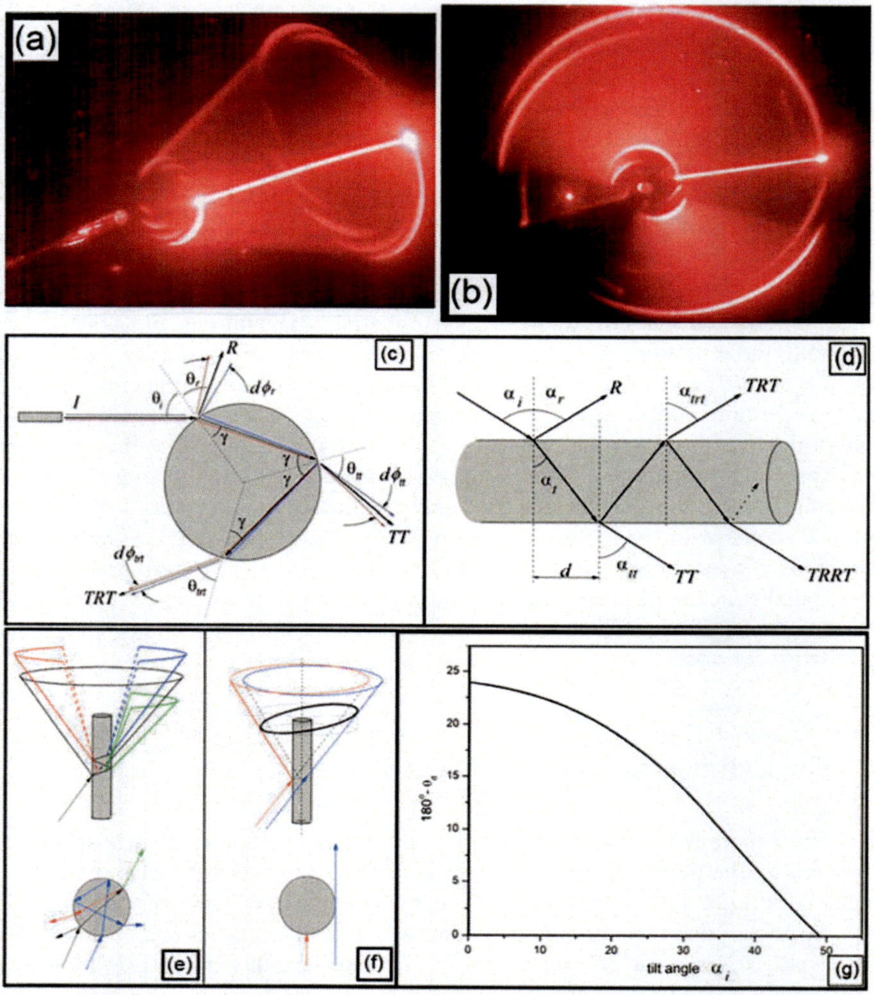

Fig. 2 Two views of a laser hitting a cinder in a tilting angle, with the emerging light traveling in space. In **a** we can see the glass cylinder at the left side and the scattered light forming a cone on the right side. In **b**, we have a frontal perspective of this experiment. Diagrams of light rays in a cylinder: **c** normal incidence and **d** oblique incidence. Diagrams of conical projections in **e** and **f**, and the angle of Cartesian ray in function of the tilt angle α_i for a glass cylinder with refraction index $n = 1.50$ in **(g)**

$$\theta_{cr} = \min(\theta_{TRT}) = \arccos\sqrt{\frac{n^2-1}{k(k+2)}}. \tag{2}$$

For the case of values of α_i different for zero of Fig. 2d and $\theta_i = 0$, the laser beam hitting the cylinder at distance *dsc* of the screen will give an image at height h_k from the axis of the cylinder for the *k*th point is giving by:

$$h_k = \{[dsc + d(1-k)(tg\alpha_r)[tg(90^\circ - \alpha_i)]\}, \tag{3}$$

in which when *k* is even the projection of this point is at upper part of the screen, and when *k* is odd the point is at the down part of the screen, considering the vertical axis perpendicular to the cylinder axis.

However, the light rays are not always scattered about a cone whose axis is in the cylinder axis [17], because the laser hits the cylinder obliquely, as it is shown in the diagrams of Fig. 2e–f. For example, one ray that hits the axis of the cylinder is associated with one cone, while another ray grazing the cylinder surface is associated with another rays' cone with intermediary impact parameters generate intermediary cone of Fig. 2f. Basically, this effect of the laser hitting the cylinder obliquely causes an optical flattening of the circular cross section of the cylinder, creating similar effects of a cylinder with an elliptical section, and coupling the radial and longitudinal modes.

3 Dynamical System Optics

Consider a light ray of a laser beam traveling inside a glass plate undergoing a sequence of internal reflections on the upper (1) and down sides (2) of this plate, like a light ray trapped in a kaleidoscope formed by two parallel mirrors. This sequence of reflections can be represented in the sequence of events with the reflections "1" evolving in space at the top of this plot and reflections "2" at the bottom part of this plate. Considering the sequence of spatial events, we can associated a temporal evolution for the trajectory of the light ray for each reflection, so that the first reflection in time is closer to the laser source, while the most distant ones occurred later. The pattern associated with this dynamics in a cylinder is shown by the caustics of Fig. 3a, and the external pattern of Fig. 3d for an angle of incidence $\theta_i = 0$. Changing this angle, we obtain the patterns of Figs. 3e–f, associated with the caustics of Fig. 3b, c.

The simplest dynamical system related to the ray tracing described previously is the one-dimensional circle map [14]:

$$f(\theta) = \theta + \Omega - {}^{K}/_{2\pi}\sin(2\pi\theta) \pmod 1, \tag{4}$$

which describes a motion on a three-dimensional torus, characterized by two frequencies ω_1 and ω_2 given by $\Omega = \omega_2/\omega_1$. If the non-linear term *K* vanishes in (4), the ratio

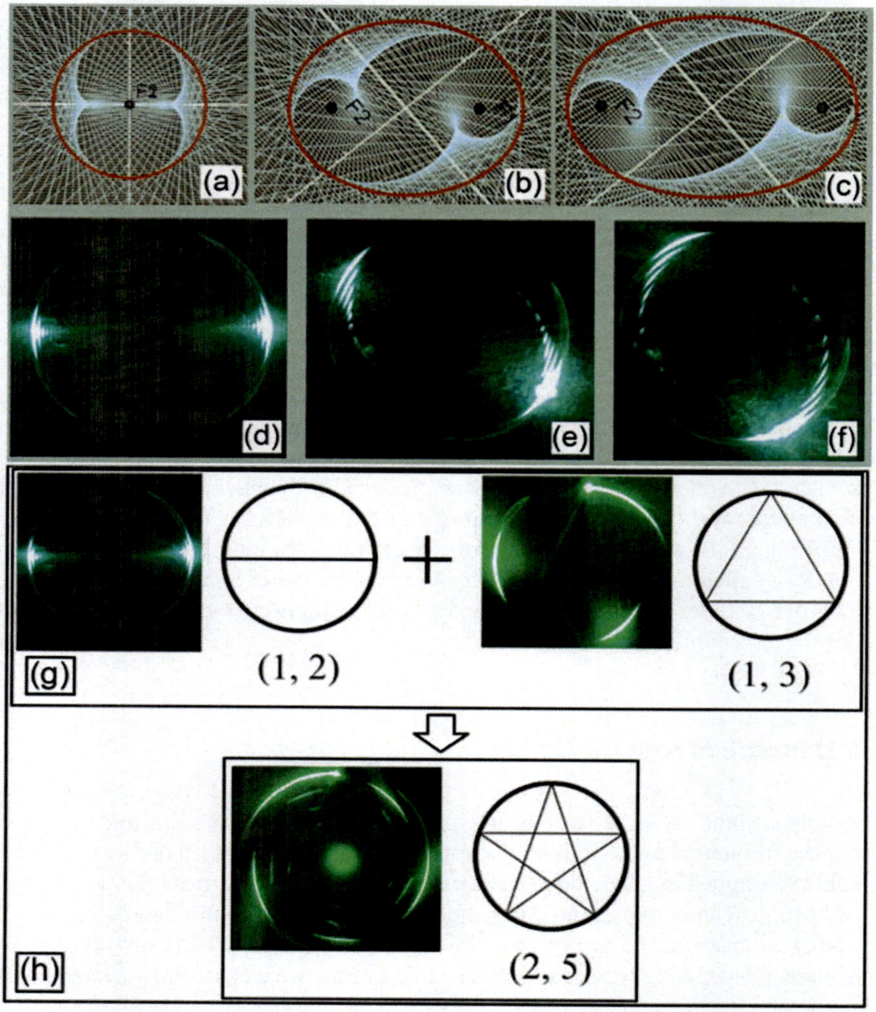

Fig. 3 Observing the flattening of the section of the cylindrical lens due to the coupling between the azimuthal and longitudinal modes. A similar behavior of flattening can be seen in the caustics from a circle in **(a)** to an ellipse in **(b)**, to another ellipse in **(c)**. In our experiment, the equivalent of a circle is shown in **(d)** ($\theta = 0°$, $\alpha = 30°$), with the elliptical sections in **(e)** ($\theta = 7°$, $\alpha = 30°$) and **(f)** ($\theta = 12°$, $\alpha = 30°$). The dynamics of Farey mediant in **(g)** and luminous pattern formation from a cylinder due the combination of different modes in **(h)**

of the two frequencies $\omega_2/\omega_1 = p/q$ is a rational number, and the trajectory is closed after q internal reflections and the motion is periodic. Irrational numbers Ω lead to quasi-periodic internal reflections which creates spiral patterns. The case when the ratio of the two frequencies gives a rational number is related to a structure in the space of control parameters known as Arnold tongues.

In Fig. 3g we can see how these patterns are associated with the dynamics of a particle bouncing in a billiard for a period-2 (1, 2) and for a particle for the case of a period-3 (1, 3), with its respective patterns. From the dynamics obtained from (4), there is the existence of a closed orbit with the shape of a pentagram of Fig. 3h, which is defined as the Farey mediant between the period-2 and period-3, because this period-5 orbit is part of a family of star polygons, linked to a special array of rational numbers in the unit interval, which is defined by the Farey mediant [14], given the three consecutive fractions p_1/q_1, p_2/q_2, p_3/q_3:

$$q_1 p_2 - q_2 p_1 = 1,$$
$$\frac{p_2}{q_2} = \frac{p_1}{q_1} \oplus \frac{p3}{q3} \equiv \frac{p_1 + p_3}{q_1 + q_3}, \tag{5}$$

which is a special addition of fractions. This idea is valid for the flattened case of the ellipsis of Fig. 4a, b. In Fig. 4c we present the Farey diagram associated with the circle map, along with some orbits obtained in our experiment.

4 Caustics-like Patterns

In our experiment, we have observed the formation of arcs with fold. Considering that the laser beam maintains its shape inside the glass cylinder, for some of the rays that escape from its surface, in addition to the divergence projected in the screen there is a fold property, forming these arcs with two branches represented in the plot of Fig. 5a. Looking for a possible explanation for these arcs with folds in the literature of optics, we have found an analogy with the mechanism of caustics formation and folded wavefront of Fig. 5b caused by the involutes of Fig. 5c, d for each branch of the caustics.

Considering the incident light as a plane wave there is a folding of this wave in the cylinder and formation of a caustic [2] as it is shown in Fig. 5. This caustic is the involute of the plane wave represented in Fig. 5c, d. The projection of these involutes in the screen forms different patterns for different values of θ_i and α_i, and for some projections is possible to observe two arcs of Fig. 5a. According to catastrophe theory each side of this caustic forms an arc parametrically modeled by a constant Ct in the following function:

$$y_n = Ct(x^{2/3}). \tag{6}$$

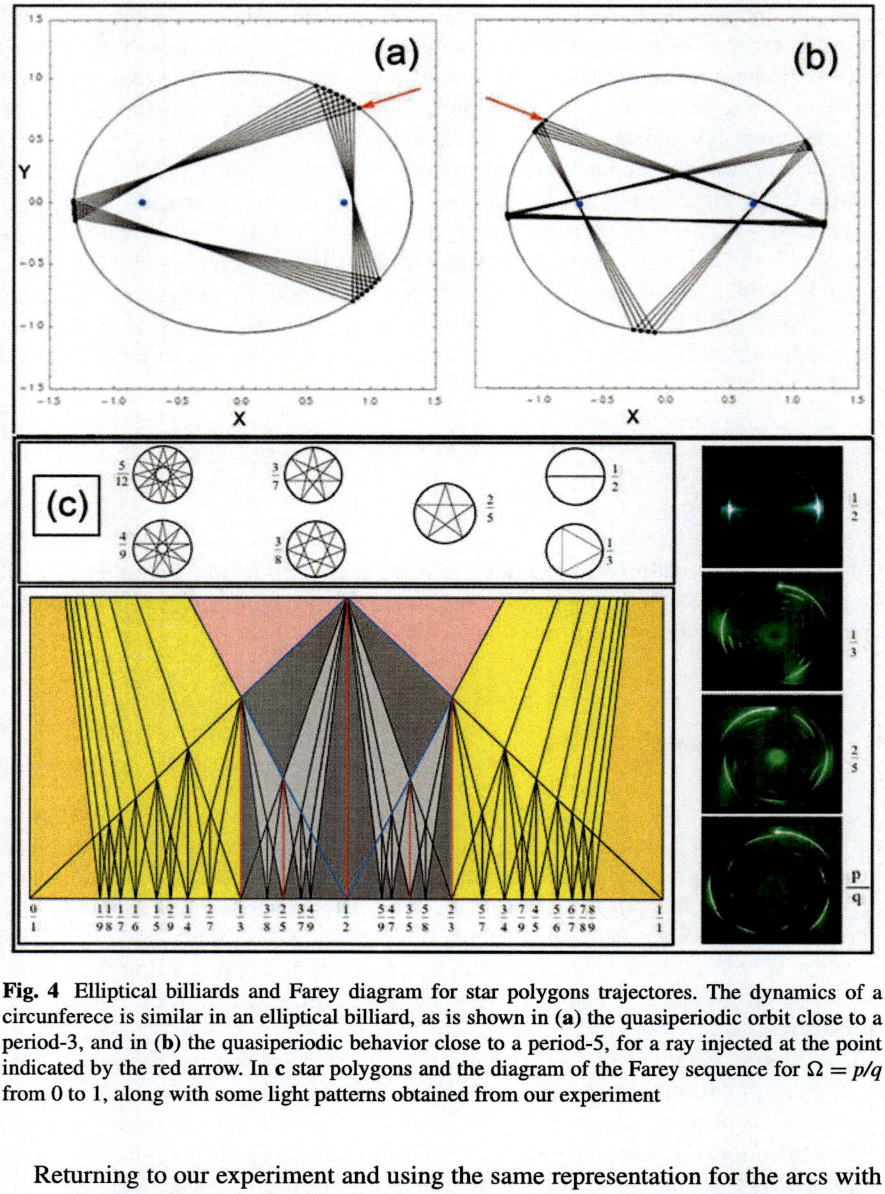

Fig. 4 Elliptical billiards and Farey diagram for star polygons trajectores. The dynamics of a circunferece is similar in an elliptical billiard, as is shown in (**a**) the quasiperiodic orbit close to a period-3, and in (**b**) the quasiperiodic behavior close to a period-5, for a ray injected at the point indicated by the red arrow. In **c** star polygons and the diagram of the Farey sequence for $\Omega = p/q$ from 0 to 1, along with some light patterns obtained from our experiment

Returning to our experiment and using the same representation for the arcs with folds discussed above, we can see some arcs with folds in the pattern of Fig. 5e, analogous to the case of the Fig. 5a.

The existence of non-closed orbits is related to quasi-periodic dynamics of the laser inside the billiard, such as the spirals shown in Fig. 5f, g. The rotation orientation of these spirals shows which side of the Farey diagram are the orbits in relation to the center of the diagram for the value of $\Omega = ½$. The green pattern of Fig. 5f rotates

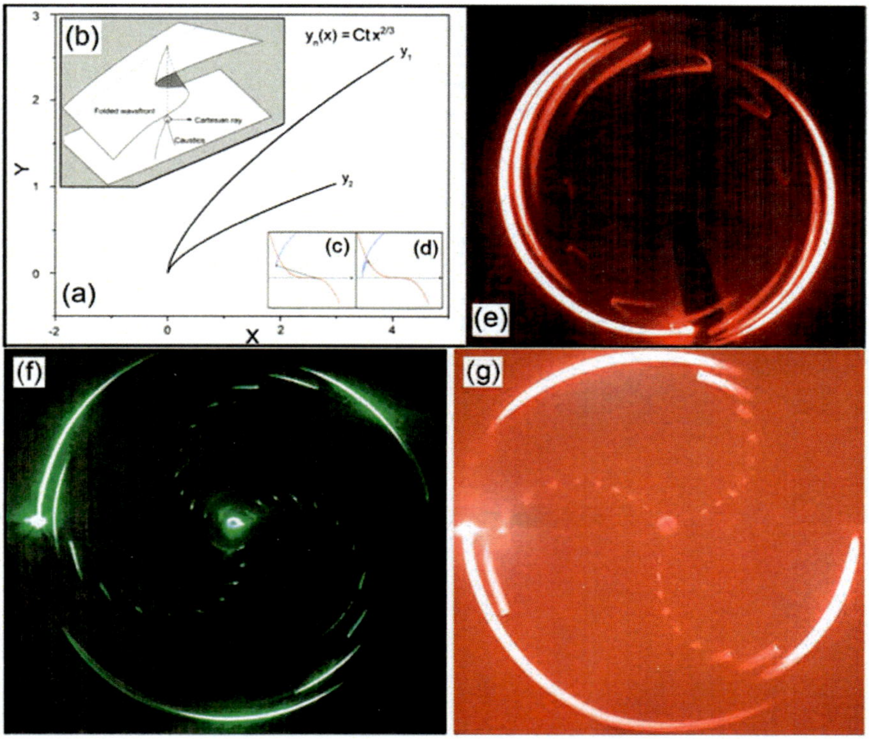

Fig. 5 Light arcs and folding light due to caustic formation. In **a** the plot of the caustic obtained from the folded wavefront **b** due to the glass cylinder. Each side of this caustic (y1 and y2) is associated with the involutes shown in (**c**) and (**d**). The image of the multiple caustics is projected in a screen (**e**) ($\theta = 27°$, $\alpha = 33°$). The pattern obtained with green laser ($\theta = 18°$, $\alpha = 13°$) rotates counterclockwise in (**f**), while the pattern with red laser in (**g**) rotates clockwise ($\theta = -18°$, $\alpha = 13°$). The longest arc of light is always the first reflected ray R of the diagram of Fig. 2a

in clockwise direction for positive values of θ_i, while using a red laser in Fig. 5g, with a negative value of θ_i, these arclets rotate in counterclockwise direction. The patterns of Figs. 5f, g have up to 30 rays each one. The longest arc of light is always the first reflected ray *R* of the diagram of Fig. 2a, while the brightest point is the laser hitting the screen directly.

5 From Cylinder to Gravitational Lens

The next step in this work of halo dynamics is to remark the association of this pattern formation with images like epicycloids, observed in another optical system related with gravitational lensing, in which there is the formation of the Einstein's ring, which here we compare with halos.

First, we can transform a cylinder to a foot of a wine glass with the transformation of Fig. 6a. The realization of this optical lens is shown in Fig. 6b. Using the terms of hyperbolic geometry, this foot of a wine glass shape is related to the surface of a tractrix curve and the pseudosphere, in our case a half pseudosphere. Second, placing this lens of three horizontal parallel lines, we have obtained the pattern of Fig. 6c, which are related with the Möbius transformations [15].

The transformations associated with this type of geometry is analogous to the path traced by a point P on the edge of a circle of radius b rolling on the outside of a circle of radius a of the period-2 of Fig. 3, and it is related with the caustic of epicycloids of Fig. 3a, which are given by the parametric equations [18]:

$$x = (a + b)\cos\phi - b\cos\left[\frac{(a+b)\phi}{b}\right], \tag{7a}$$

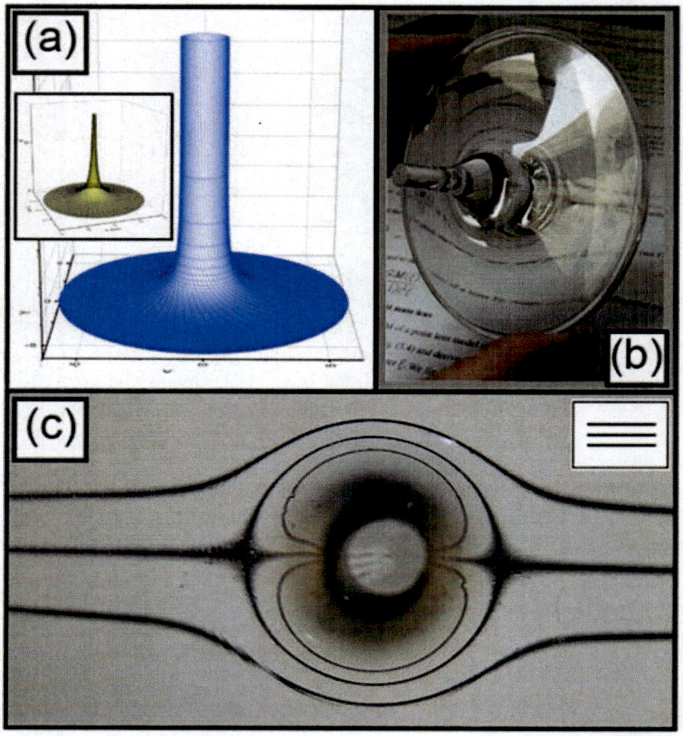

Fig. 6 Transforming a cylinder in (**a**) in a foot of wine glass (half pseudosphere) shown with yellow surface in the inset. The realization of this lens is shown in (**b**). In **c** we can see the image obtained with the lens with the shape of a pseudosphere of three parallel lines, shown in the inset at the top of this image

$$y = (a + b)\sin\phi - b\sin\left[\frac{(a+b)\phi}{b}\right], \tag{7b}$$

in which a, b are the parameters of the epicycloids. In this way, the dynamics of the caustics of period-2 of the cylindrical lens is still present in the foot of a wine glass lens, considering the profiles of caustics in a cylinder Fig. 3a and the Möbius transformation of Fig. 6c.

Now, we explain why this lens can be associated with a black hole. To obtain this lens, we start with the Einstein's gravity field equations [19]:

$$R_{\mu\upsilon} - {}^{1}\!/_{2}Rg_{\mu\upsilon} + \Lambda g_{\mu\upsilon} = \frac{8\pi G}{c^4}T_{\mu\upsilon}. \tag{8}$$

The two first terms are related with the space time-curvature, the third term is the stress from empty space-time itself, and the last term is the stress from an object in space-time.

One possible solution of these equations is the pseudosphere lens, because they can be broken into simpler equations, which are more appropriate to understand the gravitational lensing, considering the case of the lens with the shape of a foot of a wine glass for a point mass model, and the angle which light is deflected by this poit mass is [19, 20]:

$$\hat{\alpha} = \frac{4GM}{c^2 b}, \tag{9}$$

where G is the universal gravitational constant, M is the mass and b is the impact parameter. This point mass could be the black hole.

The images obtained for the solution in (9) are comparable to the period-2 observed of Fig. 7a with the cylindrical lens and are related with the parametric equations of the epicycloid of (7a), (7b) due to Möbius transformations of space-time of Fig. 7b. The diagrams of Fig. 7c, d describes how (9) is related with the image perceived by the observer in the point O, from a star source of the light at the distance of some gigaparsecs $D_S$from this observer, when a massive object L with the size of some megaparsecs is at distance D_L placed betweem them. When the star source S, the massive object L and the observer O are perfectly aligned, there is the possibilty of the formation of the ring (halo) [20], known as Einstein's ring. Like the case of the cylinder acting as a lens, the control parameter here is the alignement of the system S-L-O, with the system triggered in the period-2 due to the lens configuration. This halo pattern is present in a system which is analogous to the case of gravitational lens observed in Fig. 1c, and we observed that it is similar to the dynamics involving Farey diagram of Fig. 4, for the case of frequency ½.

One important thing to note for the case of the gravitational lensing is that the formation of both arcs occurs simultaniously, suggesting that the dynamics of halo formation here is related to a supercritical pitchfork bifurcation [14], as it is shown from the images obtained from a point source of Fig. 8 for different values of the control parameter b. Multiple images can be present in gravitational lensing [19–21],

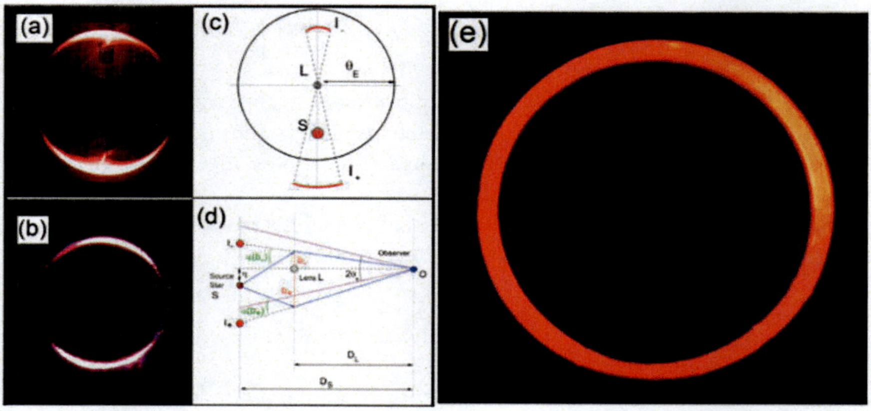

Fig. 7 The period-2 compared with the gravitational lens model. In **a** the period-2 obtained with the cylinfrical lens. The luminous arcs obtained with the foot glass. In **c** diagram of the image formation of the gravitational lens. The lateral view of the same diagram in (**d**). Simulation of the Einstein's ring (**e**) obtained from our experiment is another example of halo dynamics

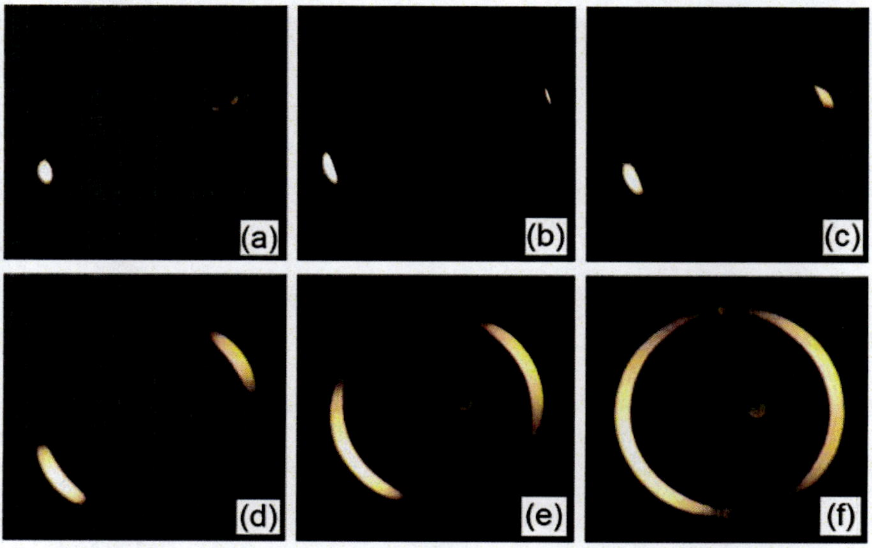

Fig. 8 Evolution of arcs using the foot of wine glass lens for a period-2, like a supercritical pitchfork bifurcation in a surface of a cylinder for the parameter b from 1.1 in (**a**), 1.0 in (**b**), 0.8 in (**c**), 0.6 in (**d**), 0.4 in (**e**), and 0.2 in (**f**)

involving a more complex analysis of the halo dynamics, due the presence of multiple lens planes, resmbling the previous case of the cylinder with oblique incidence.

6 Conclusions

We have explored the halo formation based in the concepts of rainbow ray formation and caustics in parallel with the studies of dynamical systems, such as particles bouncing in an open biliard. We have observed the existence of pattern formation following the dynamics observed in the Farey Diagram, using an approach based in the experiment of a laser beam hitting obliquelly a glass cylinder. After that, we investigated the pattern formation when this cyclinder suffers deformations until it has the shape of a foot of a wine glass lens, used to study patterns in systems affected by relativistic effects. In our explorations, we have found patterns for this system as two light spots, which converges to a halo when the light source, the lens and the observer are aligned.

Acknowledgements The authors would like to thank to Lan Ma for picture of the rainbow and Tony Klein for picture of the solar halo in Fig. 1.This work was partially supported by Conselho Nacional de Desenvolvimento Científico e Tecnológico (CNPq), Instituto Nacional de Ciência e Tecnologia de Fluidos Complexos (INCT-FCx), and by Fundação de Amparo à Pesquisa do Estado de São Paulo (FAPESP), FAPESP/CNPq#573560/2008-0.

References

1. J.A. Adams, The mathematical physics of rainbows and glories. Phys. Reports **356**, 229–365 (2002)
2. M.H. Nussenzweig, The theory of rainbow. Sci. Am. **April**, 116–127 (1977)
3. R. Greenler, *Rainbows, Halos and Glories* (Cambridge University Press, 1980)
4. M.V. Berry, C. Upstill, Catastrophe optics: Morphologies of caustics and their diffraction patterns, in *Progress in Optics*, ed. by E. Wolf, vol. XVIII, Chap. 4 (North-Holland, 1980)
5. C.L. Adler, J.A. Lock, B.R. Stone, Rainbow Scattering by a cylinder with a nearly elliptical cross section. Appl. Opt. **37**, 1540–1550 (1988)
6. P.L. Marston, Descartes glare points in scattering by icicles. Appl. Opt. **37**, 1551–1556 (1998)
7. E. Garza, S. Lopes-Aguayo, J.C. Gutierrez-Veja, Soliton dynamics in finite non local media with cylindrical symmetry. Phys. Rev. A **99**, (2019)
8. R. Nemiroff, J. Bonnell, A Horseshoe Einstein Ring from Hubble. ESA/Hubble & NASA. https://apod.nasa.gov/apod/ap111221.html
9. A. Tufaile, A.P.B. Tufaile, Rainbows, Billiards and Chaos, in *Proceedings of 11th Chaotic Modeling and Simulation International Conference. CHAOS 2018*, ed. by C. Skiadas, I. Lubashevsky. Springer Proceedings in Complexity (2019), pp. 289–301
10. A.P.B. Tufaile, J.D. da Silva, A. Tufaile, Optical elements based in dynamical systems, in *Frontiers in Optics + Laser Science APS/DLS*, OSA Technical Digest, paper JW4A.38 (Optical Society of America, 2019)
11. A. Tufaile, A.P.B. Tufaile, The dynamics of diffracted rays in foams. Phys. Lett. A **379**, 3059–3068 (2015)

12. A. Tufaile, A.P.B. Tufaile, Parhelic-like circle from light scattering in Plateau borders. Phys. Lett. A **379**, 529–534 (2015)
13. R.J. Fitzgerald, Soap Halos. Phys. Today **72**(7), 68 (2019)
14. J. Argyris, G. Faust, M. Haase, *An Exploration of Chaos, An Introduction for Natural Scientists and Engineers* (North-Holland, 1994)
15. A. Tufaile, A.P.B. Tufaile, Hyperbolic prisms and foams in Hele-Shaw cells. Phys. Lett. A **375**, 3693–3698 (2011)
16. J.D. Walker, Multiple rainbows from single drops of water and other liquids. Am. J. Phys. **44**, 421–433 (1976)
17. L. Mees, K.F. Ren, G. Gréhan, G. Gouesbet, Scattering of a Gaussian beam by an infinite cylinder with arbitrary location and arbitrary orientation: numerical results. Appl. Opt. **38**, 1867–1876 (1999)
18. E. Weistein, Epicycloid, http://mathworld.wolfram.com/Epicycloid.html
19. J. Wambsganss, *Gravitational Microlensing*, in *Gravitational Lensing: Strong, Weak and Micro*, ed. by G. Meylan, P. Jetzer, P. North (Springer, 2006)
20. M. Dominik, The physics and mathematics of gravitational lensing, http://www.artemis-uk.org/Microlensing_physmath.html
21. A.O. Petters, M.C. Werner, Mathematics of Gravitational Lensing: multiple imaging and magnification. Gen. Relativ. Gravit. **42**, 211–246 (2009)

Hysteresis Loops, Dynamical Systems and Magneto-Optics

Adriana Pedrosa Biscaia Tufaile and Alberto Tufaile

Abstract The interest in hysteresis and magnetism is shared by scientists with an impressive variety of backgrounds, such as mechanics, thermodynamics, electromagnetism, catastrophe theory, mathematics and dynamical systems, because hysteresis loop is a concept at the core of non-linear systems in which the dependence of the evolution of the states of these systems are related with their history. In this work we present a connection between dynamical systems and hysteresis loops and after that, we present some interesting hysteresis loops obtained using the Transverse Magneto-Optical Kerr effect (TMOKE), of thin films of sperimagnetic amorphous alloys with rare-earth and transition metal. The samples present first and second order transitions. The first order transition occurs at the compensation temperature when the total magnetic moment or magnetization is minimal. The second one occurs at the transition magnetic field when the behavior of the derivative of the signal changes, and it can be spin-reorientation or spin-inversion.

Keywords Bistable system · Hysteresis loop · Sperimagnetism

1 Introduction

Magnetism is a property of matter, and we can observe their magnetization **M** under several specific conditions of measurement or geometry when some materials are subjected to an external magnetic field **H**, as it is shown in Fig. 1 [1].

These curves give us some interesting ideas of the properties of the materials, helping to categorize them paramagnetic, diamagnetic, ferromagnetic, and so on. The blue curve representing the ferromagnetism presents the interesting feature known as hysteresis curve, in which the response of the system depends on its history.

A. P. B. Tufaile (✉) · A. Tufaile
Soft Matter Lab, Escola de Artes, Ciências e Humanidades, Universidade de São Paulo, São Paulo, Brazil
e-mail: atufaile@usp.br

C. H. Skiadas et al. (eds.), *13th Chaotic Modeling and Simulation International Conference*, Springer Proceedings in Complexity,
https://doi.org/10.1007/978-3-030-70795-8_68

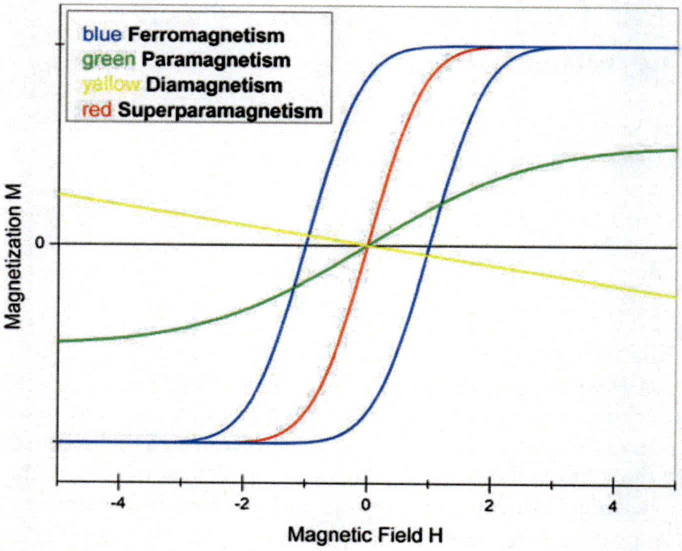

Fig. 1 Magnetic curves for different magnetic materials

Ferromagnetism here represents several kinds of materials with ferro and antiferromagnetic materials that have exchange interactions that align the atomic magnetic moment without external applied field. The sperimagnetism is included.

Systems with hysteresis are in general nonlinear systems, and they are common in nature, and the word is used in different contexts to express some kind of state of a system that is dependent of its history, with some kind of lag between the input or control parameter and the output or state variable, enabling to make memory devices, such as magnetic tapes or hard-disks [1].

The magnetic field H is the input or control variable and the magnetization or total magnetic moment M is the output or state variable. H and M are conjugate work variables meaning that their product is work. G is the Gibbs free energy of the system and the thermodynamic potential controlling spontaneous transformations. G_L is the Landau free energy, after L. D. Landau phenomenological theory of phase transitions. G is different of G_L, but after symmetry arguments and proper approximations, we can consider that Gibbs is equal Landau free energy, and search the local minima using the derivatives of G_L. The lowercases represents some convenient dimensionless form [1].

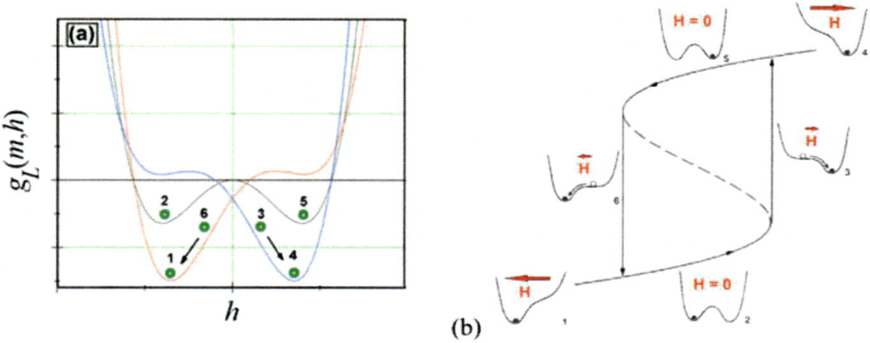

Fig. 2 The system is in the state *m* of the green dots for the temporal evolution (1, 2, 3, 4, 5 and 6) of the potential $g_L(m, h)$ controlled by the input *h* in (**a**). The three different functions of g_L are represented by red, black, and blue curves. In **b**, we have the hysteresis loop associated with this bistable system [1]

First, we will concentrate on bistable systems [2] and rate independent hysteresis, as it is shown in Fig. 2.

The potential represented with the black plot in Fig. 2a is the free energy of the bistable system for $h = 0$ for the expression [1]:

$$g_L(m, h) = m^4 - 2am^2 - hm \tag{1}$$

where *a* is a positive parameter.

The hysteresis loop of this bistable system is shown in Fig. 2b changing the control parameter *h*. This system always moves towards a minimum value, when *h* is increased from the saturated state 1, to state 2, jumps abruptly in state 3 by a Barkhausen jump, reaching another saturated state 4. Reversing the control parameter, the system evolves from state 4 to state 5, a different minimum of the bistable potential. After that, there is another Barkhausen jump in state 6 the initial saturated state 1. This behavior exemplifies how the system can have different values of one variable depending on the direction of change of the control parameter.

2 Dynamical Systems

We can observe that the hysteresis is related to a supercritical pitchfork bifurcation, with this example. Related with dynamical systems by look the derivatives, considering the state variable *x*, representing the magnetization, that will change in time.

Stable points have first derivative equal to zero, and a positive second derivative. Unstable points have first derivative equal to zero, but a negative second derivative. The first derivative is

$$\mathrm{f}(h, x) = \mathrm{d}g_\mathrm{L}/\mathrm{d}x = 4x^3 - 4\mathrm{a}x - h. \tag{2}$$

Changing the magnetic field will change the first derivative f(h,x) and one can construct a hysteresis loop. In Fig. 3a, the system evolves from only one stable fixed point to a saddle-node bifurcation, with two new stable fixed points, as it is shown in Fig. 3b. Then, in Fig. 3c, the hysteresis is seen when the system goes back and forth, and just one stable fixed point is visited each time, before the Barkhausen jump. The intermediate region between the two stable fixed points is unstable, in contrast to the case when the parameter $h < 0$. The pitchfork bifurcation of each branch of the hysteresis cycle is shown in Fig. 3d.

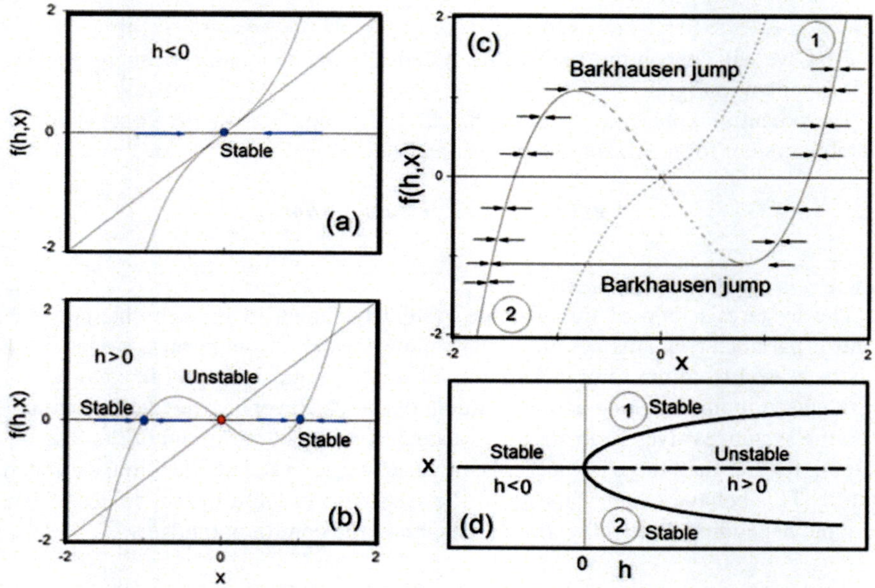

Fig. 3 Relation between hysteresis loop and a supercritical pitchfork bifurcation. In **a**, the system evolves to only one stable fixed point for a control parameter $h < 0$. A saddle-node bifurcation is shown in (**b**) as we change the control parameter h. The complete hysteresis cycle depends on the history of the initial conditions. The pitchfork bifurcation of each branch of the hysteresis cycle is shown in Fig. 3d

3 Sperimagnetism

The samples are thin films made of amorphous alloys contain rare-earth and transition metal. They have transition metal cobalt Co and a rare-earth metal as gadolinium Gd or holmium Ho. The films were deposited by magnetron sputtering in glass and silicon Si substrates cooled to free the atoms and make amorphous alloys. We took the loops of the films in glass substrates; the Si ones were used the measure the thickness of the layers. In Fig. 4, we represent the profile of the films.

The composition and thickness of the samples are shown in the Table 1. They were measured by analyzing Rutherford Backscattering Spectrometry (RBS) spectra obtained with normal beam incidence of $4He^{+}$ de 2.4 MeV. We used the films with silicon substrates for this analysis.

It is important to note that, in amorphous materials, atoms are frozen in random positions and orientations, unlike a crystalline structure, where the positions and orientations of atoms, or ions, are periodically ordered. So, in amorphous alloys, the direction of anisotropy for each ion is random. The interaction responsible for the ordering of spins in the material, and therefore, for the existence of spontaneous magnetization, was recognized by Heisenberg, in 1926, and called exchange interaction. The type of exchange interaction that occurs between two atoms depends on the electronic structure of the interacting atoms.

The antiferromagnetic exchange interaction between gadolinium or holmium and cobalt make the magnetization of these two sublattices points to opposites sides. This gives rise to a compensation temperature $\mathbf{T_{comp}}$, in which the total magnetization of the alloy is minimal.

The sperimagnetism [3] occurs because of two aspects, local or ion anisotropy and interaction between the alloy components. The intensity of exchange interaction

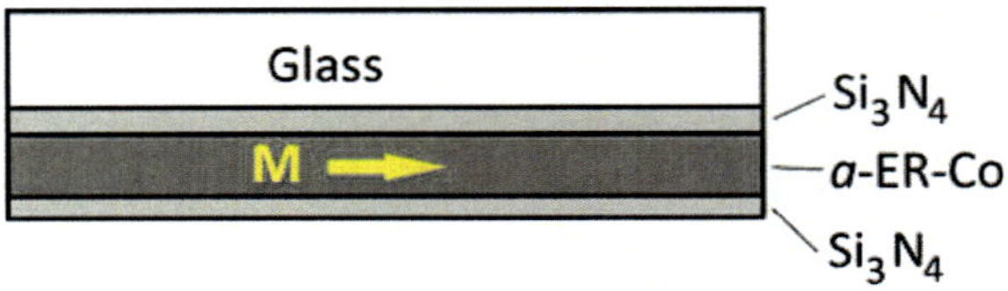

Fig. 4 The structure of the samples shown in cross section. The Si_3N_4 are antireflexive and protective layers. The axis of easy magnetization (arrow) is on the plane of the film

Table 1 Composition and thickness of the samples

Composition of the metallic layer	Layer thickness (nm)		
	Metallic layer	1st Si_3N_4	2nd Si_3N_4
$Ho33Co_{67}$	108	32	31
$Ho36Co_{64}$	116	32	31
$Gd_{20}Co_{80}$	153	55	55

between cobalt and cobalt is one order of magnitude greater than the exchange interaction between cobalt and rare-earth, which in turn is one order of magnitude greater than the interaction between rare-earth ions. So, the cobalt sublattice is well aligned, but the alignment of the rare-earth sublattice it is weaker in the struggle against temperature.

The rare-earth elements have the electronic structure of the most energetic layers represented by: $4f^N\ 5s^2\ 5p^6\ 5d^1\ 6s^2$, where N = 0 to 14, corresponding to the elements from La to Lu. For Ho, N = 10, because its atomic number is 66, for Gd N = 7 (no orbital angular momentum, L = 0). The rare-earth magnetism comes from the unpaired electrons of layer 4f, and in an alloy they are usually in the form of a 3+ ion, that is, the electrons of layers 5d and 6s are conduction electrons. Layer 4f is said to be deep, as it has an average radius of approximately 0.3 Å, while the ionic radius is around 1.8 Å. Thus, the superposition of layer 4f of neighboring ions is negligible, and therefore, there is no direct exchange interaction between these ions. Its alignment is due to a low intensity indirect exchange interaction. Gadolinium ion is spherical, much more symmetric than holmium, so the local ion anisotropy is stronger in holmium. The exchange interaction between ions of transition metal as Co happens with electrons of the 3d that have large volume of superposition, therefor a large exchange integral [3].

The sperimagnetism that appears in amorphous rare-earth (RE) alloys and transition metal (TM) is a special arrangement of spins, in which the directions of the spins corresponding to the rare-earth ions are randomly distributed inside a cone or a semi-sphere, while the spins of the transition metal are all aligned with the axis of the rare-earth spin cone. This type of arrangement is represented in Fig. 5a, the dark circles represent the rare-earth ions, and the light circles, the TM ions. In Fig. 5b, we have a schematic representation of the arrangement, showing the random distribution of the RE spins and the TM alignment. This representation corresponds to the fundamental state, that is, temperature equal to zero Kelvin.

During the hysteresis loops, this type of alloy suffers a second order phase transition, when the applied magnetic field can turn the direction of the magnetization of the sublattices, the transition magnetic field $\mathbf{H_t}$. And it can change with the temperature.

The phenomenon called spin-reorientation is the change in direction of the average magnetic moments of each sublattice, in relation to the direction of the applied field [4]. When the applied field is weak, we have sperimagnetic arrangements before and after the compensation temperature, this phase transition is first order. But if the applied magnetic field is larger than the critical field, H_{cri}, for temperatures close to $\mathbf{T_{comp}}$, then the magnetic moments of the two sublattices start to have a projection in the direction of the applied field. With the hysteresis loops, we identify the $\mathbf{H_t}$.

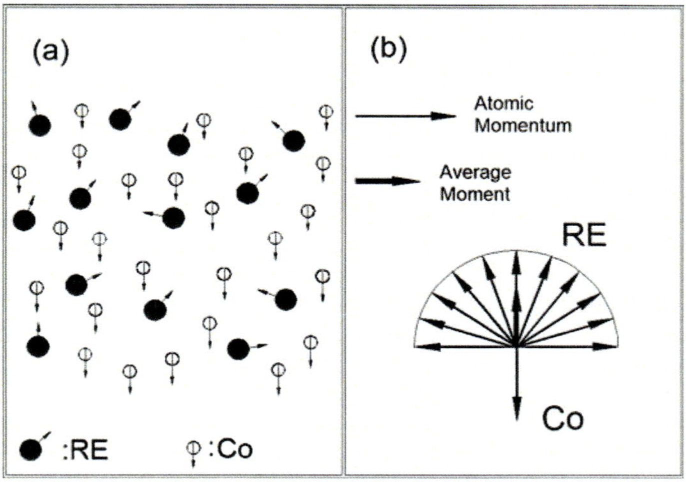

Fig. 5 T = 0. In **a**, we have a spatial representation of the amorphous RE-TM alloy (rare-earth/transition metal), where we see that the positions and orientations of the ions are random, and for RE the local anisotropy is dominant, so the orientations of its spins are random. However, for TM the dominant interaction is the exchange interaction, which promotes the perfect alignment of the spins of these ions. In **b**, we have a schematic representation of the directions of the spins, where we see that the average magnetic moment of RE is in the opposite direction to that of Co, since there is an anti-ferromagnetic coupling [3]

This phase transition is second order. In Fig. 6, before the spin reorientation, we have small shaded regions that correspond to the phases where the transition metal also has a random distribution of the spin directions. The dashed lines above H_{cri} represent the behavior of coercivity. The dashed lines below H_{cri} correspond to the loss of stability of the collinear phases. Zvezdin [4] used films with the axis of easy magnetization perpendicular to the plane of the film, our samples have it in the plane of the film.

4 Thermal Behavior of the Hysteresis Plots

The magnetic signal was measure by vibrating sample magnetometer (VSM) at room temperature or SQUID: Superconducting Quantum Interference Device, at $5\ K < T < 300\ K$. A magnetic hysteresis loop is shown in Fig. 7, where we can see the transition magnetic field.

The magneto-optical signal is the Transverse Magneto-Optical Kerr Effect (TMOKE) measured using laser diode. $\lambda = 670$ nm, at $7\ K < T < 300\ K$, angle of incidence: 45° [5, 6]. The Magneto-Optical Kerr Effects are shown in Fig. 8.

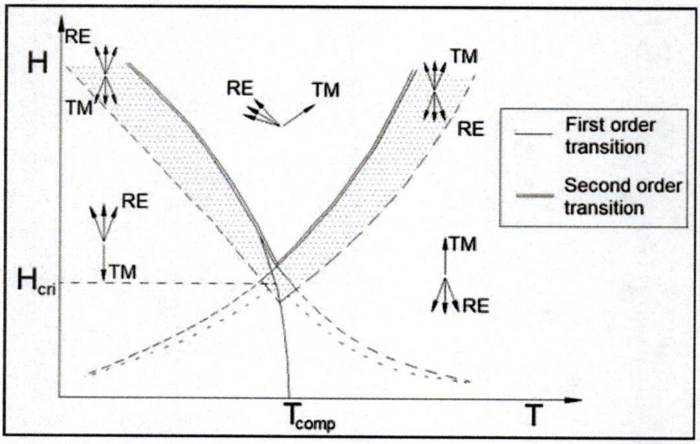

Fig. 6 Phase diagram for amorphous films of RE-TM, applied magnetic field H versus temperature T. The axis of easy magnetization is normal to the film plane. In our case, the axis of easy magnetization is in the plane of the film [4]

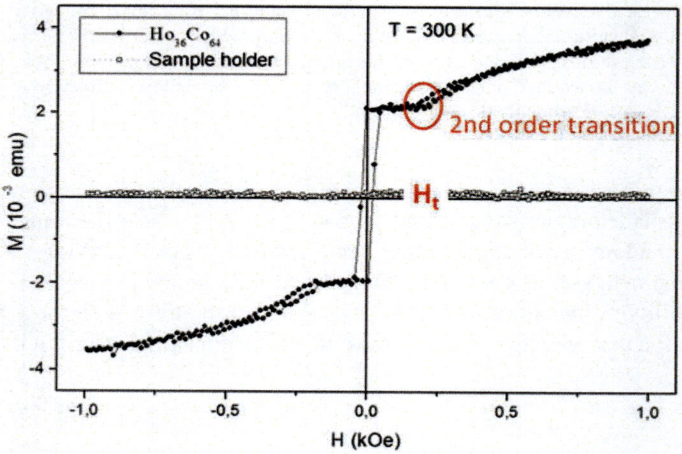

Fig. 7 Magnetic hysteresis loop (VSM), where we can see the transition magnetic field, H_t. It is a second order phase transition

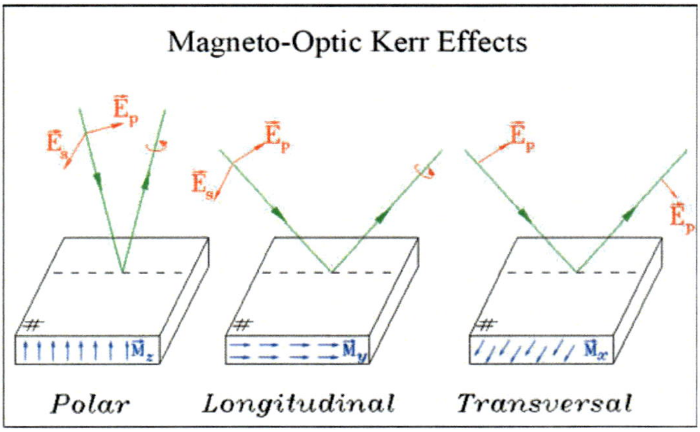

Fig. 8 The three configurations of the magneto-optical Kerr effects

Representing the light by its electric field components with respect to the plane of incidence, the material surface is represented by the R matrix:

$$\begin{pmatrix} E_p \\ E_s \end{pmatrix}^r = \mathbf{R}\begin{pmatrix} E_p \\ E_s \end{pmatrix}^i \text{ where } \mathbf{R} = \begin{pmatrix} r_{pp} & r_{ps} \\ r_{sp} & r_{ss} \end{pmatrix} \tag{3}$$

The reflectivity of the component in the plane of incidence (p) is

$$|R_p = |r_{pp}|^2 = r_{pp}.r_{pp}^* \tag{4}$$

The TMOKE signal is the relative variation in reflectivity in p component. For visible light, TMOKE is sensitive to the component of the magnetization of the transition metal sublattice parallel to the applied magnetic field:

$$\frac{\Delta R}{R} \propto M_x^{Co} \tag{5}$$

The two types of phase transition happen due to the competition between the magnetization of the sublattices. This behavior creates variety of hysteresis loops, more complex than the bistable system of Fig. 2.

In the Fig. 9, there are hysteresis loops for six different temperatures that we obtained with magneto-optical effect for another composition of amorphous holmium-cobalt alloy. As you can see, the loops are more complicated that a bistable system. But it is possible to find out what is happening. Before the compensation temperature, the magneto-optical loop is inverted, because this signal is sensitive to cobalt, which has less magnetization than holmium. The compensation temperature is around 218 K and the main thermal behavior is in the signal of saturation. The magneto-optical saturation signal increases with increasing temperature. We did not notice thermal variation of the transition magnetic field [5].

Here inverted hysteresis loops have different meaning of those ones observed by Ghising, Samantaray, and Hossain [7] (Fig. 9).

We have a superposition of the magnetic and magneto-optical loops taken of the sample *a*-Ho33Co_{67} in Fig. 10.

The comparison of magnetic and the magneto-optical loops for a film of *a*-$Gd_{20}Co_{80}$ is in Fig. 11. The compensation temperature is around 92 K, and we can see that the transition field increased with temperature. The magneto-optical saturation signal is constant, but further analyzes show that the transition magnetic field increases exponentially with temperature.

In Fig. 12, there is the schematic explanation for the phenomena we observed in the hysteresis loops.

For gadolinium-cobalt alloy, the magneto-optical signal has the same intensity before and after the transition field, indicating that cobalt sublattice has the same

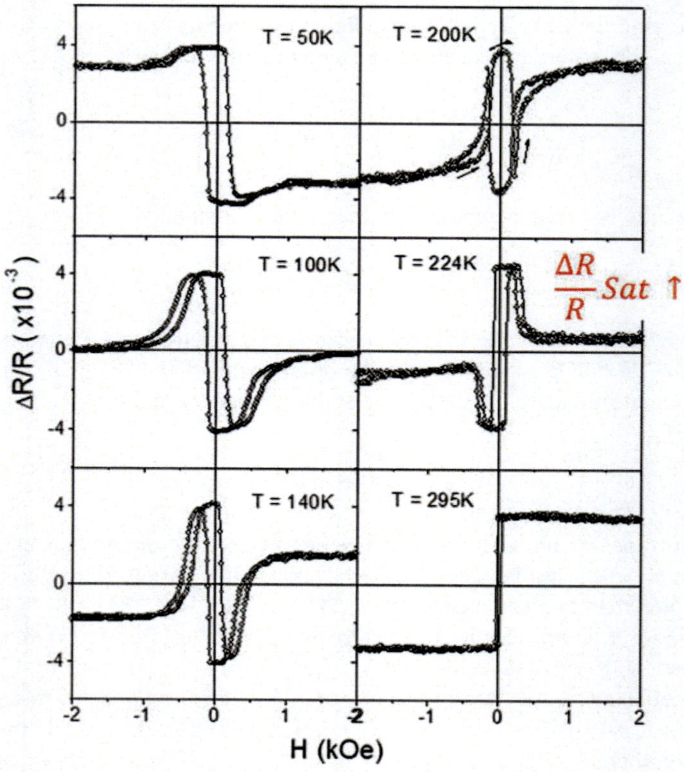

Fig. 9 Magneto-optical hysteresis loops for a thin sperimagnetic film of *a*-Ho33Co_{67}. The T_{comp} ~ 218 K, and H_t is approximately constant

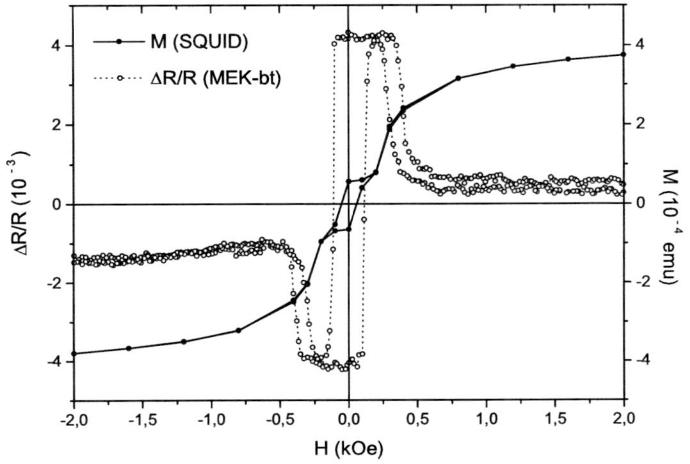

Fig. 10 Magnetic and magneto-optical hysteresis loops for the sample of a-Ho33Co_{67}, at 224 K. We can see that the effect of the second order phase transition occurs in the same value of the applied field

component of magnetization in the direction of the field, but in the opposite sense. Therefore, we conclude that the second order transition is not a spin-reorientation. We call it a spin-inversion because there is no non-collinear sublattice situation.

As gadolinium ion is spherical, so the local anisotropy is less than in the case of holmium, which has strong local anisotropy because it has a large orbital angular moment. So, it is easier for the ionic magnetic moment (spin) of Gd to turn in the direction of the magnetic field than spin of Ho.

5 Conclusions

We started showing the connections between hysteresis loops and dynamical systems. We have presented the complexity of hysteresis loops present in some sperimagnetic thin films of Gd-Co and Ho-Co, by looking magnetic and magneto-optical signal. Both samples presented a first order phase transition and a second order phase transition. The difference in local anisotropy between Gd and Ho ions led to different types of second-order phase transition. Ho-Co film has a spin-reorientation phase transition (collinear phase to non collinear phase). Gd-Co film has a spin-inversion phase transition (collinear phase to opposite collinear phase). Despite the complexity, we know the characteristics of the transitions, these types of hysteresis loops deserve a more complex bistable system model.

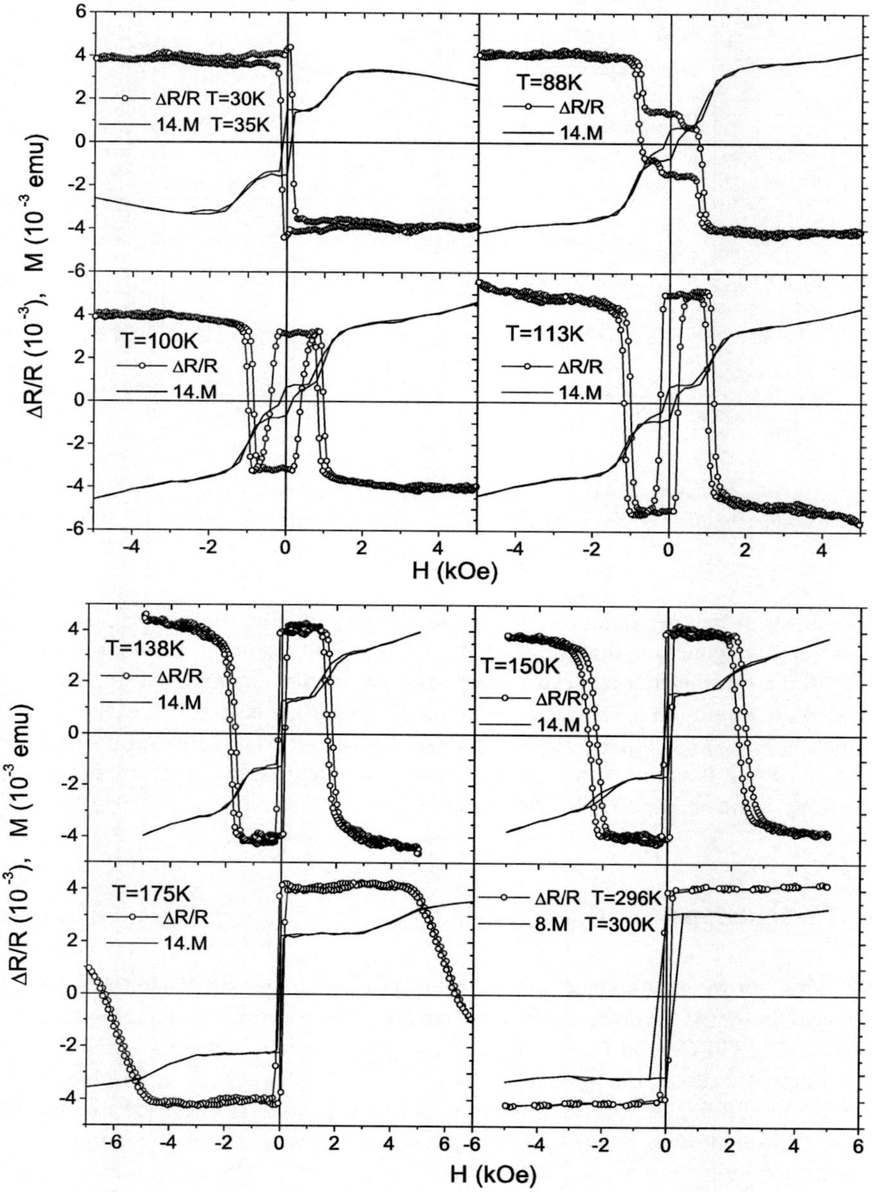

Fig. 11 Superposition of magnetic and magneto-optical hysteresis loops for the sample of *a*-$Gd_{20}Co_{80}$. The compensation temperature is around 92 K, the transition field increased with temperature [5]

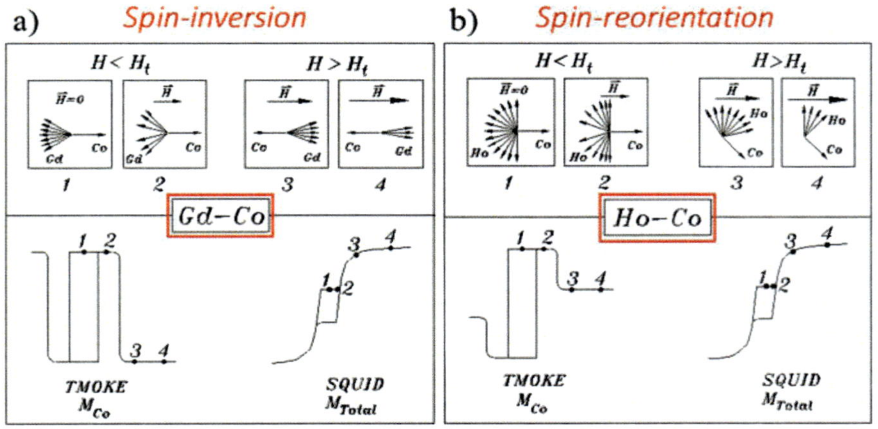

Fig. 12 The analysis of magnetic and magneto-optical hysteresis loops for the sample of *a*-$Gd_{20}Co_{80}$ indicates that it has a spin-inversion, and the sample of *a*-Ho33Co_{67} has spin-reorientation

Acknowledgements Thanks to A. D. dos Santos (LMM) and M. H. Tabacniks (LAMFI), Instituto de Física, Universidade São Paulo. This work was supported by FAPESP (Proc. 96/04507-5).

References

1. G. Bertotti, *Hysteresis in Magnetism, for Physicists, Material Scientists, and Engineers* (Academic Press, San Diego, 1998).
2. E. Colli, V.S.M. Piassi, A. Tufaile, J.C. Sartorelli, Bistability in bubble formation. Phys. Rev. E **70**, 066215 (2004)
3. R. Arrese-Boggiano, J. Chappert, J.M.D. Coey, A. Liénard, J.P. Rebouillat, Mössbauer spectra of the amorphous alloys DyM_3 (M = Fe Co, Ni). J. Physique **37**, C6-771 (1976)
4. A.K. Zvezdin, Field induced phase transitions in ferrimagnets, in *Handbook of Magnetic Materials*, ed. by K.H.J. Buschow, vol. 9 (Elsevier, London, 1995)
5. A.P.B. Tufaile, Efeito Magnetoóptico espectral e Sperimagnetism de filmes amorfos de terra rara-Co. PhD Thesis, Universidade de São Paulo (2001), https://www.teses.usp.br/teses/disponiveis/43/43134/tde-05072012-171055/pt-br.php
6. P. Ghising, B. Samantaray, Z. Hossain, Spin inhomogeneities at the interface and inverted hysteresis loop in La0.7Sr0.3Mno3/SrTiO3. Phys. Rev. B **101**, 024408 (2020)
7. A.P.B. Tufaile, A.D. Santos, Low temperature magnetic characterization of sperimagnetic films by TMOKE. Mater. Sci. Forum **302–303**, 120 (1999)

Investigating Dynamical Systems Using Optic-Fluidics

Alberto Tufaile, Michael Snyder, Timm A. Vanderelli, and Adriana Pedrosa Biscaia Tufaile

Abstract We are presenting experimental results and simulations of dynamical systems using magneto-optics. These light patterns are obtained by the observation of a thin film of ferrofluid in the presence of a magnetic in the presence of a magnetic field.

Keywords Magnetism · Isoclines · Chiral

1 Introduction

In our previous work, we have considered the analogy between the general properties of vector fields of the phase space of dynamical systems with the properties of potential of magnetic charges using magneto-optics [1], as it is shown in Fig. 1.

We have proposed this representation because the representation of both fields is comparable, the existence of two different types of "charges" enable us to obtain elliptic points and saddles. Basically, the colored lines observed from our magneto-optical system are obtained from the light diffraction of light sources in micro-needles aligned with magnetic field. In Figs. 1 and 2, we can see that the isopotentials are perpendicular to the lines of the magnetic field.

A. Tufaile (✉) · A. P. B. Tufaile
Soft Matter Lab, Escola de Artes, Ciências e Humanidades, Universidade de São Paulo, São Paulo, Brazil
e-mail: tufaile@usp.br

A. P. B. Tufaile
e-mail: atufaile@usp.br

M. Snyder
Technical Space Science Center, Morehead State University, Martindale Drive, Morehead, KY 235, USA

T. A. Vanderelli
Ferrocell USA, 739 Route 259, Ligonier, PA 15658, USA
e-mail: tvan@ferrocell.us

C. H. Skiadas et al. (eds.), *13th Chaotic Modeling and Simulation International Conference*, Springer Proceedings in Complexity,
https://doi.org/10.1007/978-3-030-70795-8_69

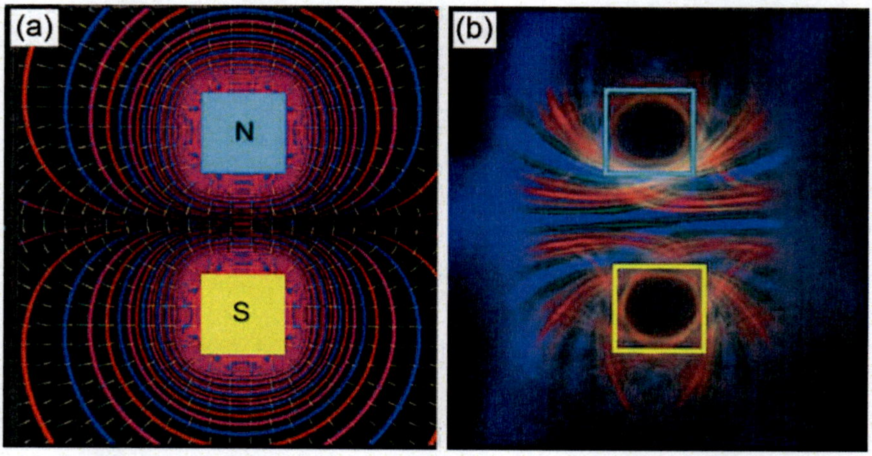

Fig. 1 **a** Two magnetic charges with the field lines represented by the arrows and isopotentials represented by the colored lines and its magneto-optical counterpart in (**b**)

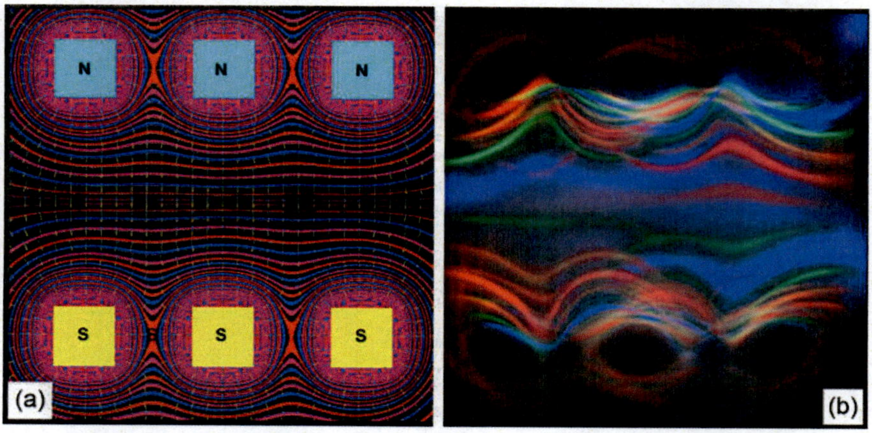

Fig. 2 **a** Three magnetic charges with the field lines represented by the arrows and isopotentials represented by the colored lines and its magneto-optical counterpart in (**b**)

It is important to note that the representation of magnetic charges is a valid way to represent the magnetic field and this is not incompatible with the idea of the Lorentz force, as the same way that a phase space represents states of motion, not the motion itself.

However, there is an apparent contradiction in this analogy, because the representation of isopotentials and the colored lines of our magneto–optical is not perfect. A close observation of the light patterns of the experiment shows the existence of crossing lines, which could imply in indeterminacy in a dynamical system, violating the classical representation of dynamical systems. The light patterns mimic the isopotentials, because the light patterns are a combination of the magnetic field and the position of the light source. For different positions of the light source, we have different diffracted lines, which eventually will cross each other. In this way, metaphorically speaking, these luminous patterns linked to the isopotentials are equivalent to the representation of the nature by the impressionist painters, with emphasis in depiction of light in its changing qualities with unusual visual angles.

Consider now the Hénon conservative map given by:

$$\begin{aligned} x' &= x\cos\alpha + \left(y - x^2\right)\sin\alpha, \\ y' &= x\sin\alpha + \left(y - x^2\right)\cos\alpha. \end{aligned} \tag{1}$$

For the case of Hamiltonian systems, we can observe the existence of chaos for perturbations close to the separatrix of the system, as it is shown in Fig. 3a, which

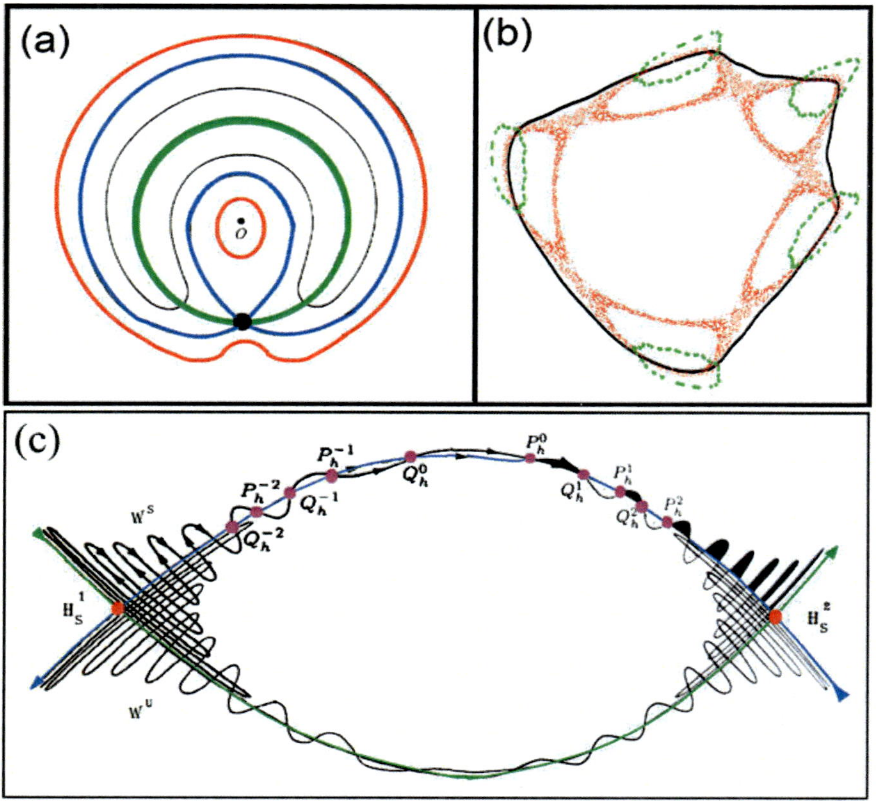

Fig. 3 In **a** diagram of separatrix chaos. In **b** the diagram of a Henon-Heyles map with chaotic oscillations. In **c** the concept of chaos in conservative systems close of hyperbolic points

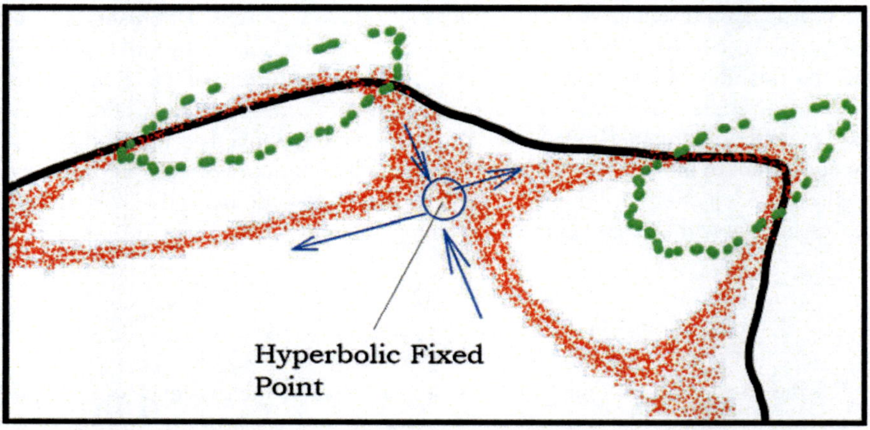

Fig. 4 Chaos close of a hyperbolic point in the Hénon map

shows the conditions of the nonlinear resonance on the phase space, in which the green line is the unperturbed trajectory. The blue line is the new separatrix of the phase oscillations. The classical plot of chaotic behavior can be obtained for the Hénon conservative map of the Fig. 3b from (1), with chaotic behavior given by red region ($\alpha = \pi/2 - 0.228$), quasiperiodic behavior in green color ($\alpha = \pi/2 - 0.200$), and another chaotic region in black ($\alpha = \pi/2 - 0.250$).

The idea o f chaos in this case can be understood if we follow the stable (W^s) and unstable manifolds (W^u) of Fig. 3c, until their intersection points in red, called homoclinic points $H_S{}^1$ and HS^2. Applying the perturbation repeatedly to $P_h{}^0$, we have the sequence of image points $P_h{}^k$ converging towards hyperbolic point for k tending to infinite, and consequently W^u and W^s can only intersect after an infinite sequence, and the same is valid for the reversing points $Q_h{}^k$. The result is an extraordinary complex view of intersecting invariant manifolds. One example of this behavior is shown in Fig. 4 using the (1), for the case of α equals to ($\pi/2$–0.228) (see [2]).

For the case of the magneto-optics in our experiment, we observed that the light patterns are oriented by the vectorial product of Fig. 5 [1, 3–6].

This approach is well known in celestial mechanics, demonstration of chaotic pendulum, or in electromagnetism, where physicists look for the dynamics of particles in magnetic fields. We are investigating in this paper the equivalent case of chaotic scenario in magnetostatics interacting with light from our experiment involving magneto-optics. We were inspired by the direct observation of luminous patterns and properties of magnetic fields.

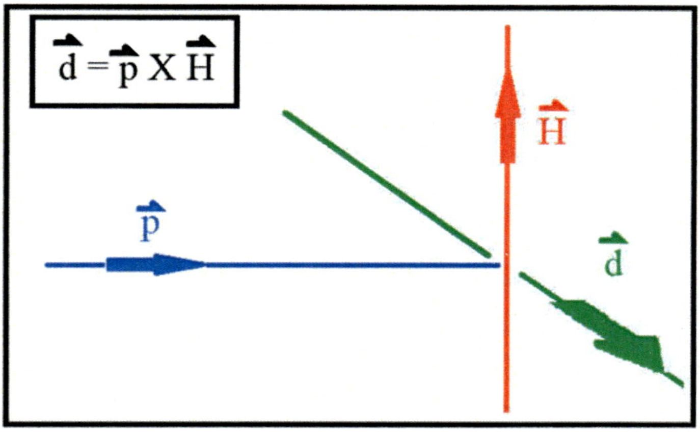

Fig. 5 The vector **d** is the direction of the tangent line of the diffracted lines, which is perpendicular to the direction of propagation of the light ray **p** and the orientation of the magnetic field **H**

2 Experimental Apparatus and Modeling Isopotentials

In Fig. 6 we present the experimental apparatus of this system. The luminous patterns observed in the thin film of ferrofluid is a direct effect of the magnetic field with the

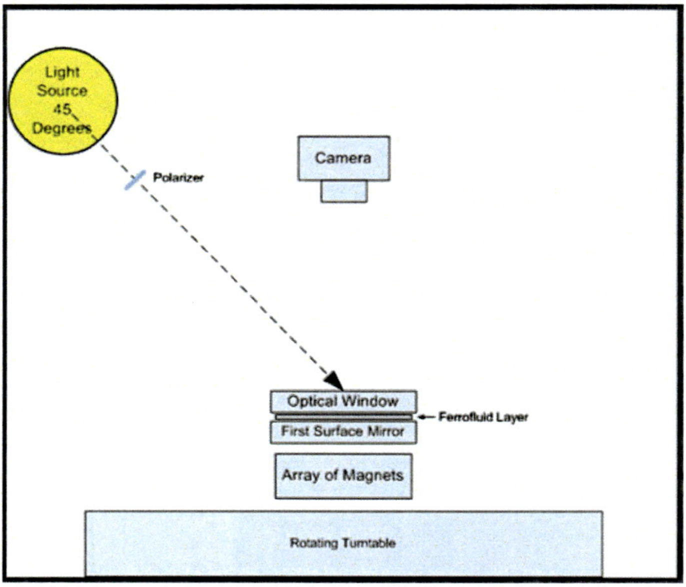

Fig. 6 Diagram of the ferro-mirror experiment setup

iron particles, which take a shape that scatters light in a certain shape for the viewer. In this way we have to use an array of magnets of Fig. 7, above the magnets we have a mirror. The device known as Ferrolens of Fig. 8, the Hele-Shaw cell containing the ferrofluid, is placed above this assembly. We use different light arrays above

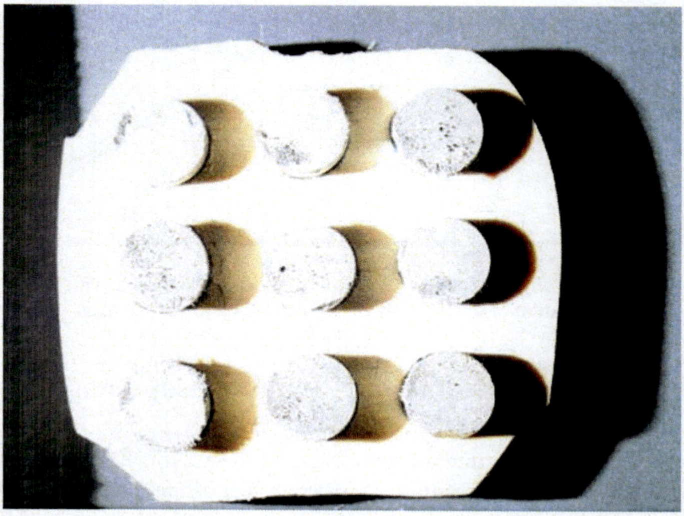

Fig. 7 Magnets fixed in a base

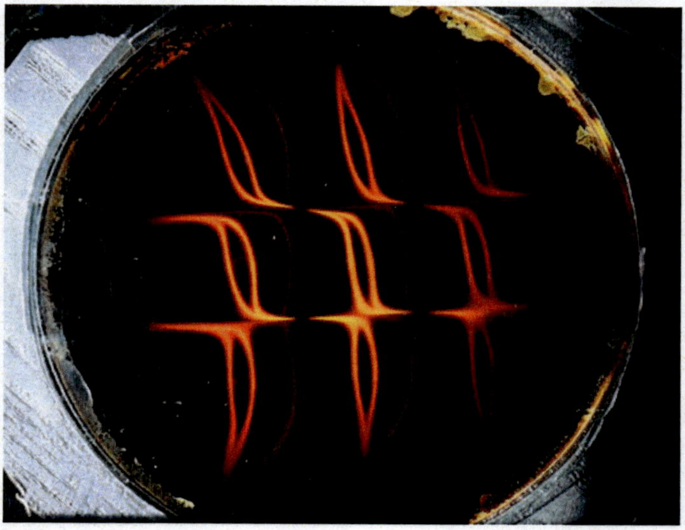

Fig. 8 Pattern obtaine with the Ferrolens

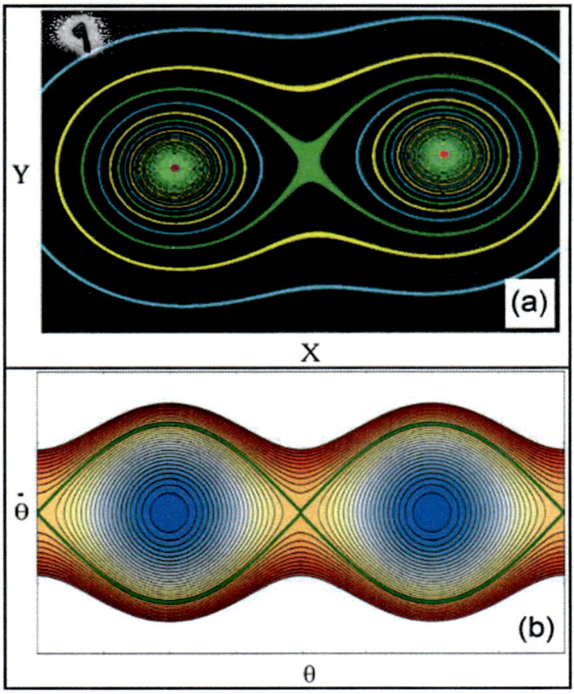

Fig. 9 Comparison between isopotentials of magnetic field and the phase space of a pendulum. The green curves represents the separatrix

this setup, which represent the ferro-mirror experiment. The ferrofluid is a stable colloidal dispersion using light mineral oil. The nanoparticles are spheres of the order of 10 nm in diameter. The magneto-optic effect results in the change of some optical parameters of the ferrofluid, forming images. For more details see [1, 3, 7, 5].

The software Pic2Mag [7] simulates some aspects of magnetic field arrangement of a magnet array, such as vector field and isopotentials, like the case of two magnets of Fig. 9a, wich can be compared to a phase space like the one in Fig. 9b, which represents the phase space of a pendulum.

3 The Patterns of Eyes and Chirality

One interesting phenomenon observed is a magneto-chiral pattern of Fig. 10, which there is three magnets to create this pattern, in a three-pole configuration formed by south-north–south poles. The pattern resembles three eyes arranged in a column-like alignment. Figure 10c is the top view of the light pattern, Figs. 10a, b is from the same system observed from the right side, and Figs. 10d, e are perspective obtained from the left. We can see that the pattern suffers distortions. However, these patterns

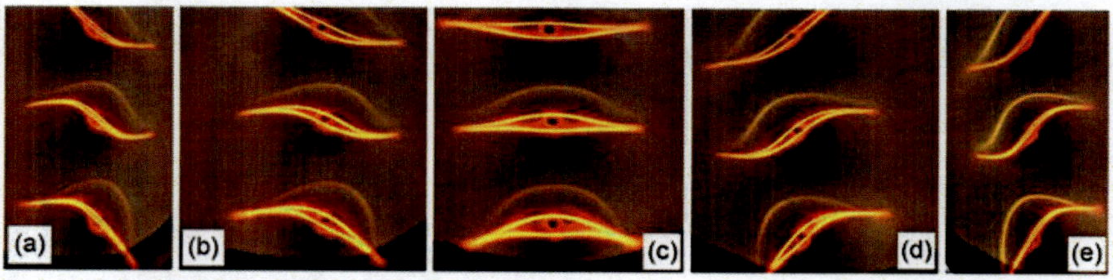

Fig. 10 Magneto chiral effect

cannot be overlapped with other in order to be reproduced. One image is a reflection of the other, in such way that there is a chiral effect. We can consider that the assembly of nanoparticles is somehow affecting these light patterns, because nanoscale particles could self-assemble into helical-like structures due to the interplay of magnetic dipoles and van der Walls interactions [3, 4, 6]. The consequence of this anisotropy is the emergence of optical chiral structures.

4 Hyperbolic Points

For the case of Hamiltonian systems, the existence of saddle point is the important key to observe the existence of chaos. The stable and unstable manifolds are called separatrices, and when a weak perturbation is added, the separatrix are destroyed and replaced by a separatrix chaotic layer. The same way as the separatrix is obtained numerically by integration of the equations with a set of initial conditions in the vicinity of the separatrix, we can explore in our system what is happening around the saddle points in our experiment. Let's consider the case of Fig. 11 with a configuration of isopotentials equivalent of a torus. Observing the detail of the central area of this image in Fig. 12a with the experiment in Fig. 12b, we can see what is happening with the four saddle points around the center of the light pattern observed with experiment. The colored lines converge to the saddle point and vanish. In contrast, the center point of the image, which represents the center point of a dynamical system, the colored lines swirls around it, and the central region is dark.

Figure 13 was obtained by placing the pattern obtained experimentally on the simulation. With this picture, we can observe that the saddle points of the simulation is slightly different from the experiment, for example the green cross at the right side, at the top of Fig. 14, in which the experiment is the red circle A, and simulation is the green circle B.

Fig. 11 Simulation of isopotentials in a a torus

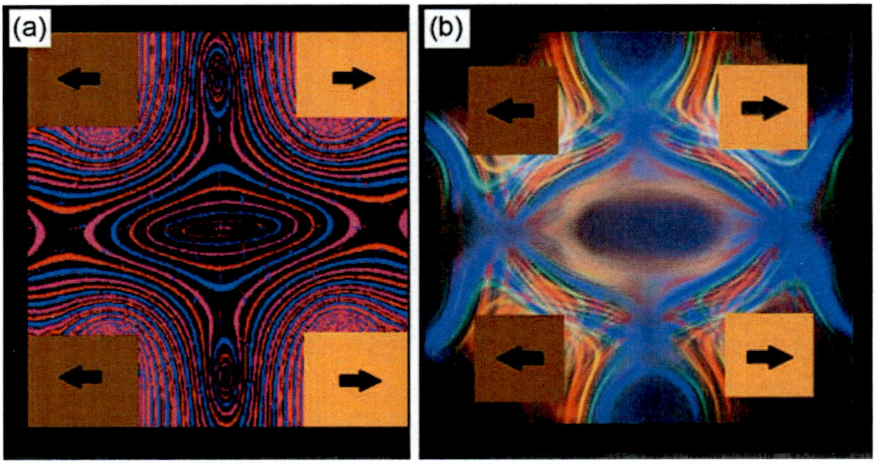

Fig. 12 Exploring Center and saddle points: simulation and experiment

Fig. 13 Another array of magnets superposed on the simulation of the magnetic field and isopotentials

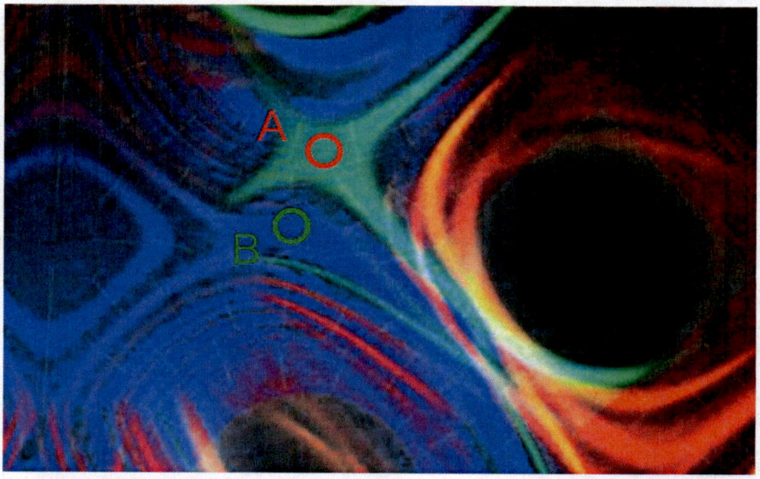

Fig. 14 Observing a hyperbolic magnetic point from the previous figure: the experiment is the red circle A, and simulation is the green circle B

5 Conclusions

We have explored some aspects of the analogy between dynamical systems and the magneto-optical system formed by a thin film of ferrofluid. Magnetic static fields have some general properties of Hamiltonian systems, and using different magnetic fields configurations, we look for hyperbolic points and observed how the experiment behaves around these regions. The light patterns are related to the vectorial product between the ray light **p** and the orientation of the magnetic field **H**, given locally the tangent vector **d**. During theses explorations of this magneto-optical system, we have found some evidences of chiral effects and we suggested that this effect is related to anysitropic properties of magnetic nanoparticles.

We have observed that the presence of the thin film of ferrofluid affects the magnetic field, and the formation of patterns can show the differences between the values of the patterns observed experimentally and the computed values.

References

1. A. Tufaile, T.A. Vanderelli, A.P.B. Tufaile, Observing jumping laser dogs. J. Appl. Math. Phys. **4**, 1977–1988 (2016)
2. J. Argyris, G. Faust, M. Haase,An Exploration of Chaos (Elsevier, Amsterdam, 1994), p. 108
3. A. Tufaile, T.A. Vanderelli, A.P.B. Tufaile,Light polarization using ferrofluids and magnetic fields. J. Adv. Condens. Matter Phys. **2583717** (2017)
4. A. Tufaile, T.A. Vanderelli, M. Snyder, A.P.B. Tufaile,Observing dynamical systems using magneto-controlled diffraction. Condens. Matter **4**(2), 35 (2019)
5. A. Tufaile, M. Snyder, T.A. Vanderelli, A.P.B. Tufaile, Non-linear stability observation using magneto-controled diffraction with optic-fluidics, in *11th Chaotic Modeling and Simulation International Conference*, ed. by C.H. Skiadas, I. Lubashevsky, Springer Proceedings in Complexity (2019), p. 275
6. M. Snyder, A. Tufaile, A.P.B. Tufaile, T.A. Vanderelli, Controlling light diffraction with nanostructures. TechConnect Briefs 2019, June 17 (2019), pp. 369–372
7. Program for simulation of magnetic field, www.pic2mag.com

Inversive Generators of Second Order

Sergey Varbanets and Yakov Vorobyov

Abstract Inversive congruential method for generating the uniform pseudorandom numbers is a particulary attractive alternative to linear congruential generators, which show many undesirable regularities. In present paper, we investigate the equidistribution of sequences produced by inversive congruential generator of second order by using the discrepancy bounds of such sequences of pseudorandom numbers (PRN's). Also there are obtained the estimates of special exponential sums of these sequences.

1 Introduction

The sequences of random numbers have the various applications in the numerical analysis and cryptography. But in practice instead the sequences of random numbers we use the pseudorandom sequences, i.e. the sequences which pass appropriate statistical test on randomness. An assortment of statistical tests depends on the type solved problem.

Our main point here is to elucidate the motivation for construction the sequences of PRN's with some specific properties that foster their applications in Quasi-Monte Carlo methods and cryptography. For the numerical analysis there are tests for the uniform distribution in certain region. Such tests go with success the sequences of linear congruential pseudorandom numbers (abbr., LCPRN's) (see, [4]). For cryptographic applications it is essential that except "equidistribution" it performs yet a requirement of "unpredictability". But to this requirement the sequence of PRN's produced by linear congruential generator does not satisfy.

S. Varbanets (✉)
Department of Computer Algebra and Discrete Mathematics, Odessa I.I. Mechnikov National University, Dvoryanskaya street 2, Odessa, Ukraine
e-mail: varb@sana.od.ua

Y. Vorobyov
Department of Mathematics, Informatics and Information Activities, Izmail State University of Humanities, Repina street 12, Odessa, Ukraine
e-mail: yashavo@mail.ru

C. H. Skiadas et al. (eds.), *13th Chaotic Modeling and Simulation International Conference*, Springer Proceedings in Complexity,
https://doi.org/10.1007/978-3-030-70795-8_70

Knuth [3] and Marsaglia [5] proved that LCPRN's are predictable. This motivated the creation of the nonlinear congruential pseudorandom sequences having the unpredictable property.

The inversive congruential generator initiated by Eichenauer and Lehn [2] in 1986 make sure the "equidistribution" and "unpredictability" of the sequence of PRN's produced by the congruential recursion

$$y_{n+1} \equiv a y_n^{-1} + b \pmod{p},$$

where $a, b \in \mathbb{Z}$, p be a prime number, y^{-1} denotes a multiplicative inverse of y modulo p, y_0 be initial value.

In present paper, we study the sequence of PRN's $\{y_n\}$ defined by the congruential recursion

$$y_{n+1} = a y_{n-1}^{-1} y_n^{-1} + b \pmod{p^m}. \tag{1}$$

This generator of $\{y_n\}$ we call the inversive congruential generator of second order. We distinguish two type of respective sequences

$$\textbf{(I) type}: \nu_p(a) = 1,\ \nu_p(b) = \nu \geq 1;$$
$$\textbf{(II) type}: \nu_p(a) = \alpha,\ \nu_p(b) = 1.$$

In case of inversive generator of first order (which was studied by Eichenauer, Lehn, Niederreiter and others) the sequence $\{y_n\}$ of type (II) steadies starting from a certain index n, and therefore it will not be the uniformly distributed sequence.

The sequences produced by the congruential generator of kth order

$$y_{n+1} \equiv f(y_n, y_{n-1}, \ldots, y_{n-k+1}) \pmod{M}$$

have the least period length $\tau \leq M^k$.

The function $f(y_n, y_{n-1}, \ldots, y_{n-k+1})$ ought to choose so as a period $\{y_n\}$ was as large as possible.

Some results about the period and distribution of inversive congruential sequences of first order can be found in the survey paper [1].

The inversive generators of first order generate the sequences of PRN's which pass the test on "equidistribution" and "unpredictability" but they do not satisfy to requirement of "security". Indeed, if we know more than one sequential values $y_n, y_{n+1}, \ldots, y_{n+k}$, we easy can obtain the parameters a and b of the inversive generator of first order. However, the "security" of sequence of PRN's can be remained if the parameter of shift c change to $c + n + p^{\mu} F(n)$, where $\mu > \nu$ and $F(n)$ is some random polynomial from $\mathbb{Z}[n]$ (see, [8, 9]).

In this paper we continue investigation of the inversive generator of second order of two types.

Our purpose in this work is to show the passing of test on equidistribution and statistical independence for the sequence $\{x_n\}$, $x_n = \frac{y_n}{p^m}$, and hence, the main point is to show the possibility for such sequences to be used in the problem of real processes modeling and in the cryptography.

In the sequel we will use the following notations.

Variables of summation automatically range over all integers satisfying the indicated condition. The letter p denotes a prime number, $p \geq 3$. For $m \in \mathbb{N}$ the notation $\mathbb{Z}_{p^m}$ (respectively, $\mathbb{Z}^*_{p^m}$) denotes the complete (respectively, reduced) system of residues modulo p^m. For $z \in \mathbb{Z}$, $(z, p) = 1$ let z^{-1} be the multiplicative inverse of z modulo p^m; instead of $\frac{a}{b} \pmod{p^m}$ we will write $a \cdot b^{-1}$. We write $\nu_p(A) = \alpha$ if $p^\alpha | A$, $p^{\alpha+1} \nmid A$ for $A \in \mathbb{Z}$. For integer t, the abbreviation $e_q(t) = e^{\frac{2\pi i t}{q}}$ is used.

Let $f(x)$ be a periodic function with a period τ. For any $N \in \mathbb{N}$, $1 \leq N \leq \tau$, we denote

$$S_N(f) := \sum_{x=1}^{N} e^{2\pi i f(x)}$$

2 Auxiliary Results

Let $a, b \in \mathbb{Z}$, $p \geq 3$ be a prime, and let $m > 1$ be a positive integer. Let us consider the transformation Ψ defined on $\mathbb{Z}^*_{p^m}$ by

$$\begin{aligned} &\Psi(y_n, y_{n-1}, \ldots, y_{n-s+1}) = a y_n^{-1} \cdots y_{n-s+1}^{-1} + b, \\ &(y_0, p) = (y_1, p) = 1, \ \ 0 \leq y_n < p^m. \end{aligned} \tag{2}$$

We put $x_n = \frac{y_n}{p^m}$, $n = 0, 1, \ldots$. And then the transformation Ψ we call the inversive congruential generator of second order of the sequence of pseudo-random numbers (abbr., PRN's).

In order that the sequence $\{y_n\}$ exists for any $n \in \mathbb{Z}$ it is sufficient that $(a, p) = 1$, $b \equiv 0 \pmod{p}$ or $a \equiv 0 \pmod{p}$, $(b, p) = 1$. These conditions generate various of sequences of PRN's. Henceforth, we call that sequences as inversive congruential sequences of type I or II.

For example, the inversive congruential sequence $\{y_n\}$ of first order with $a \equiv 0 \pmod{p}$, $(b, p) = 1$ has a period $\tau = 1$, but in the case $(a, p) = 1$, $\nu_p(b) = \nu \geq 1$, the relevant sequence of PRN's can be a period with $\tau = 2p^{m-\nu}$.

In our paper we study the inversive congruential generator of second order, i.e. $s = 2$. We will illustrate that the least period length of sequence $\{y_n\}$ can be equal to $3p^{m-\mu}$ or $p^{m-\nu}$.

We need the following lemmas.

Lemma 1 *Let $f(x)$ be a periodic function. For any $N \in \mathbb{N}$, $1 \leq N \leq \tau$ the following estimate*

$$|S_N(f)| \leq \left(\max_{1 \leq n \leq \tau} \left| \sum_{x=1}^{\tau} e^{2\pi i \left(f(x) + \frac{nx}{\tau}\right)} \right| \right) \log 2\tau$$

holds.

This statement can be derived by inequalities for complete exponential sums on a usual way.

Lemma 2 *Let h_1, h_2, k, ℓ be positive integers and let $\nu_p(h_1+h_2)=\alpha$, $\nu_p(h_1k+h_2\ell)=\beta$, $\delta=\min(\alpha,\beta)$. Then for every $j=2,3,\dots$ we have*

$$\nu_p(h_1k^{j-1}+h_2\ell^{j-1})\geq\delta.$$

Moreover, for every polynomial $G(u)=A_1u+A_2u^2+p^tG_1(u)\in\mathbb{Z}[u]$ we have

$$h_1G(k)+h_2G(\ell)=A_1(h_1k+h_2\ell)+A_2(h_1k^2+h_2\ell^2)+p^{t+s}G_2(k,\ell),$$

where $s\geq\min(\nu_p(h_1+h_2),\nu_p(h_1k+h_2\ell))$, $h_1,h_2,k,\ell\in\mathbb{Z}$, $G_2(u,v)\in\mathbb{Z}[u,v]$.

Proof By the equality

$$h_1k^j+h_2\ell^j=(h_1k^{j-1}+h_2\ell^{j-1})(k+\ell)-k\ell(h_1k^{j-2}+h_2\ell^{j-2}),$$

applying the method of mathematical induction, we obtain at once $\nu_p(h_1k^j+h_2\ell^j)\geq\delta$, $j=2,3,\dots$ □

Lemma 3 *Let $p>2$ be a prime number, $m\geq 2$ be a positive integer, $m_0=\left[\frac{m}{2}\right]$, $f(x)$, $g(x)$, $h(x)$ be polynomials over $\mathbb{Z}$*

$$f(x)=A_1x+A_2x^2+\cdots,\quad g(x)=B_1x+B_2x^2+\cdots,$$
$$h(x)=C_\ell x+C_{\ell+1}x^{\ell+1}+\cdots,\quad \ell\geq 1,$$

$$\nu_p(A_j)=\lambda_j,\quad \nu_p(B_j)=\mu_j,\quad \nu_p(C_j)=\nu_j,$$

and, moreover,

$$k=\lambda_2<\lambda_3\leq\cdots,\quad 0=\mu_1<\mu_2<\mu_3\leq\cdots,$$
$$\nu_p(C_\ell)=0,\quad \nu_p(C_j)>0,\quad j\geq\ell+1.$$

Then the following bounds occur

$$\left|\sum\nolimits_1\right|:=\left|\sum_{x\in\mathbb{Z}_{p^m}}e_m(f(x))\right|\leq\begin{cases}2p^{\frac{m+k}{2}} & if\ \nu_p(A_1)\geq k,\\ 0 & if\ \nu_p(A_1)<k;\end{cases}$$

$$\left|\sum\nolimits_2\right|:=\left|\sum_{x\in\mathbb{Z}^*_{p^m}}e_m(f(x)+g(x^{-1}))\right|\leq I(p^{m-m_0})p^{\frac{m}{2}}$$

$$\left|\sum_3\right| := \left|\sum_{x\in\mathbb{Z}^*_{p^m}} e_m(h(x))\right| \leq \begin{cases} 1 \; if \; \ell = 1, \\ 0 \; if \; \ell > 1, \end{cases}$$

where $I(p^{m-m_0})$ is a number of solutions of the congruence

$$y \cdot f'(y) \equiv g'(y^{-1}) \cdot y^{-1} \pmod{p^{m-m_0}}, \;\; y \in \mathbb{Z}^*_{p^{m-m_0}}.$$

Proposition 1 *Let the sequence $\{y_n\}$ be produced by the recursion (2) with $(a, p) = (y_0, p) = (y_1, p) = 1$, $\nu_p(b) = \nu_0 > 0$, $\nu_p(c) = \mu_0 > 2\nu_0$. There exist the polynomials $F_0(x)$, $F_1(x)$, $F_2(x) \in \mathbb{Z}[x]$ with the coefficient depending on y_0, y_1, such that*

$$y_{3k} = A_0 + A_1k + A_2k^2 + p^{\mu}G_0(k, y_0, y_0^{-1}, y_1, y_1^{-1}), \tag{3}$$

$$y_{3k+1} = B_0 + B_1k + B_2k^2 + p^{\mu}G_1(k, y_0, y_0^{-1}, y_1, y_1^{-1}), \tag{4}$$

$$y_{3k+2} = C_0 + C_1k + C_2k^2 + C_3k^3 + C_4k^4 + p^{\mu}G_2(k, y_0, y_0^{-1}, y_1, y_1^{-1}), \tag{5}$$

where

$$\begin{aligned}
A_1 &\equiv b + a^{-1}b^2y_0y_1 - \frac{1}{2}a^{-1}b^2y_0^2 - a^{-1}by_0^2y_1 - 2b^2y_0^3y_1^2 - \frac{1}{2}a^{-1}b^2y_0y_1,\\
A_2 &\equiv -a^{-1}b^2y_0y_1 - \frac{1}{2}a^{-1}b^2y_0^2 + b^2y_0^3y_1^2 + \frac{1}{2}a^{-1}b^2y_0y_1,\\
B_1 &\equiv b\left(\frac{1}{2}b\left(y_0^{-1} - a^{-1}y_1^2\right) + 1 - y_0^{-1}y_1\right),\\
B_2 &\equiv b^2\frac{1}{2}\left(-y_0^{-1} + a^{-1}b^2y_1^2\right),\\
C_1 &\equiv b\left(\left(-ay_0^{-2}y_1^{-1} + 1\right) - \frac{1}{2}by_1^{-1}\left(ay_0^{-2}y_1^{-1} - 1\right)\right),\\
C_2 &\equiv b^2\frac{1}{2}y_0^{-1}\left(-1 + a^{-1}y_0y_1^2\right),\\
\mu &= \min\left(\nu_0 + \mu_0, 3\nu_0\right).
\end{aligned}$$

(see, [7])

Corollary 1 *Let $\nu_p(y_0 - ay_1^{-2}) = \alpha \leq \nu_0$ and let τ be a period length of the sequence $\{y_n\}$ generated by recursion (2) of type (I) with initial values y_0, y_1. Then we have*

$$\tau = 3p^{m-\nu_0-\alpha},$$

and $\tau \leq 3p^{m-\nu_0}$ on all occasions.

Corollary 2 *For $k = 3, 4, \ldots$, we have modulo p^{μ}, $\mu = \min(2\nu_0, \mu_0)$*

$$y_{3k} = \left(1 + b^2 \left(a^{-1}y_1 + \frac{1}{2}a^{-1}ky_1 - a^{-1}k^2y_1 + \frac{1}{2}a^{-1}k^2y_1\right)\right) y_0 +$$
$$+ \left(a^{-1}by_1 - \frac{1}{2}a^{-1}kb^2 - \frac{1}{2}a^{-1}b^2k^2\right) y_0^2 + \left(-2ky_1^2b^2 + k^2b^2y_1^2\right) y_0^3,$$
$$y_{3k+1} = \left(1 - kby_0^{-1}\right) y_1 + \left(-a^{-1}b^2 - a^{-1}bk + a^{-1}k^2b^2\right) y_1^2,$$
$$y_{3k+2} = ay_0^{-1}y_1^{-1} + \left(\frac{1}{2}kb^2y_0^{-1} + \frac{1}{2}ab^2ky_0^{-2} + \frac{1}{2}b^2k^2\right) y_1^{-1} -$$
$$- \frac{1}{2}k^2b^2y_0^{-1} + a^{-1}b^2k^2y_0y_1^2.$$

Proposition 1' *Let $\{y_n\}$ be a sequence of PRN's generated by the recursion (2) of type (II), and let $\nu_p(a) = \alpha > 1$, $\nu_p(b) = 0$. Then for $n = 9, \ldots$ the following representation*

$$y_{n+1} = \frac{A_0^{(n+1)} + A_1^{(n+1)}y_0 + A_2^{(n+1)}y_0y_1}{B_0^{(n+1)} + B_1^{(n+1)}y_0 + B_2^{(n+1)}y_0y_1} \tag{6}$$

holds,
where

$$A_0^{(n)} = (n-4)a^2b^{n-5} + ab^{(n-2)}, \qquad B_0^{(n)} = (n-5)a^2b^{n-6} + ab^{n-3},$$
$$A_1^{(n)} = (n-5)a^2b^{n-6} + ab^{(n-3)}, \qquad B_1^{(n)} = (n-6)a^2b^{n-2} + ab^{n-4},$$
$$A_2^{(n)} = \frac{(n-5)(n-4)}{2}a^2b^{n-3} + \qquad B_2^{(n)} = \frac{(n-7)(n-6)}{2}a^2b^{n-8} +$$
$$+ (n-3)ab^{(n-4)} + b^{(n-1)}, \qquad + (n-4)ab^{n-5} + b^{n-2},$$

Proof The straightforward computations on congruent recursion (2) allow to obtain the representations for y_7 and y_8 modulo $p^{3\alpha}$:

$$y_7 = \frac{3a^2b^2 + ab^5 + (2a^2b + ab^4)y_0 + (a^2 + 4ab^3 + b^6)y_0y_1}{2a^2b + ab^2 + (a^2 + ab^3)y_0 + (3ab^2 + b^5)y_0y_1},$$

$$y_8 = \frac{4a^2b^3 + ab^6 + (3a^2b^2 + ab^5)y_0 + (3a^2b + 5ab^4 + b^7)y_0y_1}{3a^2b^2 + ab^5 + (2a^2b + ab^4)y_0 + (a + 4ab^3 + b^6)y_0y_1}$$

And now by a mathematical induction we at once have the assertion of Proposition 1'. □

Corollary 1' *The elements of sequence $\{y_n\}$ generated by (1) with $\nu_p(a) = \alpha > 1$, $\nu_p(b) = 0$ can be represented as the following polynomials on n modulo $p^{3\alpha}$*

$$y_{n+1} = A_0(y_0, y_1) + nA_1(y_0, y_1) + n^2A_2(y_0, y_1), \tag{7}$$

where

$$\begin{aligned}
A_0 &= by_0y_1 + a(b^{-4}y_0 - 3b^{-2}y_0y_1) + a^2A_0'(y_0, y_1, y_0^{-1}, y_1^{-1}),\\
A_1 &= -ab^{-2}y_0y_1 + a^2A_1'(y_0, y_1, y_0^{-1}, y_1^{-1}),\\
A_2 &= -a^2b^{-1}(b^4 - y_0y_1) + a^3A_2'(y_0, y_1, y_0^{-1}, y_1^{-1}),\\
A_i'&(y_0, y_1, y_0^{-1}, y_1^{-1}) \in \mathbb{Z}[y_0, y_1, y_0^{-1}, y_1^{-1}],\ y_0y_0^{-1} \equiv y_1y_1^{-1} \equiv 1 \pmod{p^m}.
\end{aligned}$$

Proof Indeed, (6) shows that all summands in denominator of representation y_{n+1}, except y_0, y_1, are congruent to zero modulo p^α. Thus, using a congruence $\frac{1}{c+p^\alpha d} = c^{-1}(1 - p^\alpha c^{-1}d + p^{2\alpha}(c^{-1}d)^2) \bmod p^{2\alpha}$ with $(c, p) = 1$, we obtain (4) at once. □

Corollary 2' *For every sequence of PRN's produced by (2) with $\nu_p(a) = \alpha \geq 1$, $\nu_p(b) = 0$, the least period length is equal to $p^{m-\alpha}$.*

Actually, we have modulo p^m

$$y_{n+\ell} - y_n = -2ab^{-2}y - 0y - 1(1 + aF_0(n\ell)),\quad F_0(n, \ell) \in \mathbb{Z}[n, \ell].$$

So, $y_{n+\ell} \equiv y_n \pmod{p^m}$ if only $n \geq 8$ and $\ell \equiv 0 \pmod{p^{m-\alpha}}$.

The following lemmas need to study the exponential sum of special type on the sequences of PRN's.

Lemma 4 *Let $p > 2$ be a prime number, $b_0, b_1 \in \mathbb{Z}$, $(b_0, p) = (b_1, p) = 1$. We have for $k < m$*

$$S_1 := \sum_{x\in\mathbb{Z}_{p^m}} e_{p^m}\left(\frac{a_0 + a_1p^kx}{b_0 + b_1p^kx}\right) = \begin{cases} 0 & if\ a_0b_1 \not\equiv a_1b_0 \pmod{p^{m-k}}\\ p^m & if\ a_0b_1 \equiv a_1b_0 \pmod{p^{m-k}}. \end{cases} \tag{8}$$

Proof In view of $\frac{1}{b_0+b_1p^k} \equiv b_0^{-1}(1 - b_0^{-1}b_1p^kx + (b_0^{-1}b_1)^2p^{2k}x^2 + \cdots) \pmod{p^m}$ we get by Lemma (3)

$$\begin{aligned}
S_1 &= \sum_{x\in\mathbb{Z}_{p^m}} e_{p^m}(b_0^{-1}(a_0 + a_1p^kx)(1 - b_0^{-1}b_1p^kx + (b_0^{-1}b_1)^2p^{2k}x^2 - \cdots)) =\\
&= \sum_{x\in\mathbb{Z}_{p^m}} (b_0^{-1} + (a_1 - a_0b_0^{-1}b_1)p^kx + b_0^{-1}(a_0b_0^{-2}b_1^2 - a_1b_1b_0^{-1})p^{2k}x^2 + \cdots) =\\
&= \begin{cases} 0 & if\ a_0b_1 \not\equiv a_1b_0 \pmod{p^{m-k}},\\ p^m & if\ a_0b_1 \equiv a_1b_0 \pmod{p^{m-k}}. \end{cases}
\end{aligned}$$

□

Lemma 5 *Let $a_i, b_i \in \mathbb{Z}$, $(a_i, p) = (b_i, p) = 1$, $i = 0, 1$; $p > 2$ be a prime number, and m, k be positive integers, $m \geq 3k$. Then*

$$S_2 := \sum_{x\in\mathbb{Z}_{p^m}} e_{p^m}\left(\frac{a_0 + p^k a_1 x + p^{2k} a_2 x^2}{b_0 + p^k b_1 x + p^{2k} b_2 x^2}\right) \ll p^{\frac{m}{2}+k} \tag{9}$$

with the absolute constant in the symbol "$\ll$".

Proof First we assume $m = 2m_0$, $m_0 \in \mathbb{N}$. Let $x = y(1 + p^{m_0} z)$. We obtain

$$\sum_{x\in\mathbb{Z}_{p^m}} e_{p^m}\left(\frac{a_0 + p^k a_1 x + p^{2k} a_2 x^2}{b_0 + p^k b_1 x + p^{2k} b_2 x^2}\right) =$$

$$= \sum_{y\in\mathbb{Z}_{p^{m_0}}} \sum_{z\in\mathbb{Z}_{p^{m_0}}} e_{p^m}\left(\frac{A_0 + p^{k+m_0}(a_1 y + p^k a_2 y^2) z}{B_0 + p^{k+m_0}(b_1 y + p^k b_2 y^2) z}\right),$$

where

$$A_0 = a_0 + p^k y + p^{2k} a_2 y^2,$$
$$B_0 = b_0 + p^k y + p^{2k} b_2 y^2.$$

It follows

$$\begin{aligned}
S_2 &= \sum_{y\in\mathbb{Z}_{p^{m_0}}} e_{p^m}(A_0 B_0^{-1}) \sum_{z\in\mathbb{Z}_{p^{m_0}}} e_{p^m}(-A_0 B_0^{-1}(b_1 y + p^k b_2 y^2) z p^{k+m_1}) = \\
&= p^m \sum_{y\in\mathbb{Z}_{p^{m_0}}} e_{p^m}(A_0 B_0^{-1}) \sum_{z\in\mathbb{Z}_{p^{m_1}}} e_{p^{m_1-k}}(-A_0 B_0^{-1}(b_1 y + p^k b_2 y^2) z) = \\
&= p^{m_1} \sum_{\substack{y\in\mathbb{Z}_1 \\ y\equiv 0 \pmod{p^{m_0-k}}}} e_{p^m}(A_0 B_0^{-1}) = \\
&= p^{m_0} \sum_{y_0\in\mathbb{Z}_{p^k}} e_{p^m}((1 - p^{m_0} b_0^{-1} b_1 y_0)(a_0 + p^{m_0} a_1)) = \\
&= p^{m_1} e^{2\pi i \frac{a_0}{p^m}} \sum_{y_0 \pmod{p^k}} e_{p^m}(a_0 b_0^{-1} b_1 y_0) \ll p^{m_0+k}.
\end{aligned}$$

For $m = 2m_0 + 1$ we infer by a similar way

$$S_2 \ll p^{\frac{m}{2}} + k$$

□

3 Evaluation of Exponential Sums over the Sequences of PRN's

Let $\{y_n\}$ be the sequence of PRN's produced by the inversive generator of second order of the first or second type. In the Propositions 1 and 1' it was received a description of elements y_n as polynomials at n that essentially make easier the construction of estimates for exponential sums. We will consider the following exponential sums over the sequence $\{y_n\}$ of PRN's generated by the recursion (1) of type (I) or (II) with the least period τ.

$$\sigma_{k\ell}(h_1, h_2) := \sum_{y_0, y_1 \in \mathbb{Z}^*_{p^m}} e_{p^m}(h_1 y_k + h_2 y\ell), \;\; h_1, h_2 \in \mathbb{Z};$$

$$S_N(h, y_0) := \sum_{n=0}^{N-1} e_{p^m}(h y_n), \;\; h \in \mathbb{N}, \;\; 0 < N \le \tau;$$

$$K(h_1, h_2; p^m) := \sum_{n=1}^{\tau} e_{p^m}(h_1 y_n + h_2 y_n^{-1}), \;\; (h_1, h_2 \in \mathbb{Z});$$

$$G(h, p^m) := \sum_{n=1}^{\tau} e_{p^m}(h y_n^2), \;\; (h \in \mathbb{Z}).$$

These sums are called σ-sum, S-sum, Kloosterman sum and Gauss sum, respectively.

Theorem 1 *Let the sequence $\{\omega_k\}$ has the maximal period τ, $\tau = 2p^{n-\beta}$. Then the following bound*

$$|S_\tau(h, \omega)| := \left| \sum_{k=0}^{\tau-1} e_{p^n}(h\omega_k) \right| \le \begin{cases} 0 & if \;\; \beta + \delta < n, \\ \tau & if \;\; \beta + \delta \ge n, \end{cases}$$

holds.

(see, [7], Th. 3)

Theorem 1' *Let $h_1, h_2 \in \mathbb{Z}$, $\nu_p(h_1 + h_2) = p^s$, $0 \le s \le m$. Then for the sequence $\{y_n\}$ of type (II) we have*

$$\sigma_{k,\ell}(h_1,h_2)=\begin{cases}0 & if \quad k\not\equiv\ell \pmod{p^m}\\ & and \quad \nu_p(h_1+h_2)<m-1;\\ p^{2(m-2)}(p-1)^2 & if \quad k\not\equiv\ell \pmod{p^m}\\ & and \quad \nu_p(h_1+h_2)=m-1;\\ p^{2(m-1)}(p-1)^2 & if \quad k\overset{=}{\not\equiv}\ell \pmod{p^m}\\ & and \quad \nu_p(h_1+h_2)=m.\end{cases} \tag{10}$$

Proof By Corollary 1' we can write

$$h_1y_k+h_2y_\ell=b(h-1+h_2)y_0y_1+(h_1+h_2)p^\alpha F(y_0,y_0^{-1},y_1,y_1^{-1}), \tag{11}$$

where $F(y_0,y_0^{-1},y_1,y_1^{-1})$ is a polynomial with the integer coefficients.
Now we obtain

$$\sigma_{k,\ell}(h_1,h_2)=\sum_{y_0,y_1\in\mathbb{Z}^*_{p^m}} e_{p^m}((h_1+h_2)b\ell_1\ell_2+p^k(h_1+h_2)F(y_0,y_0^{-1},y_1,y_1^{-1}))=$$
$$=\sum_{y_0\in\mathbb{Z}^*_{p^m}}\sum_{y_1\in\mathbb{Z}^*_{p^m}} e_{p^m}((h_1+h_2)by_0y_1+p^k(h_1+h_2)F_1(y_1,y_1^{-1}))=$$
$$=\sum_{y_1\in\mathbb{Z}^*_{p^m}}\begin{cases}0 & if\ \nu_p(h_1+h_2)<m-1,\\ p^{2(m-2)}(p-1)^2 & if\ \nu_p(h_1+h_2)=m-1,\\ p^{2m(m-1)}(p-1)^2 & if\ \nu_p(h_1+h_2)=m.\end{cases}$$

Here we took into account that for $(a,p)=1$

$$\sum_{x\in\mathbb{Z}^*_{p^m}} e^{2\pi i\frac{ax}{p^m}}=\begin{cases}-1 & if\ m=1,\\ 0 & if\ m\geq 1.\end{cases}$$

□

In Theorem 1' the initial values y_0,y_1 run the set $\mathbb{Z}^*_{p^m}$ independently of each other. Now we shall assume that $y_0=y_1$.

Theorem 1'' *Under conditions of Theorem 1' and the proposal* $y_0=y_1$ *we have*

$$|\sigma_{k,\ell}(h_1,h_2)|=\begin{cases}p^{\frac{m+s}{2}} & if\ s<m,\\ p^{m-1}(p-1) & if\ s=m.\end{cases} \tag{12}$$

Proof By (11) and $y_0\equiv y_1 \pmod{p^m}$ we obtain modulo p^m

$$h_1y_k+h_2y_\ell=b(h_1+h_2)y_0^2+p^\alpha(h_1+h_2)F_1(y_0,y_0^{-1}).$$

Thus Lemma 3 gives

$$|\sigma_{k,\ell}(h_1,h_2)| = \left|\sum_{y_0\in\mathbb{Z}^*_{p^m}} e_{p^m}((h_1+h_2)by_0^2 + p^\alpha(h_1+h_2)F_1(y_0,y_0^{-1}))\right| =$$

$$= \begin{cases} p^{\frac{m+2}{2}} & if\ s<m, \\ p^m(p-1) & if\ s=m. \end{cases}$$

□

Now we will construct the estimates for G- and K-sums.

Theorem 2 *Let the sequence of PRN's is generated by recursion (1) with $\nu_p(a)=0$, $1\le\nu_p(b)=\nu\le\frac{m}{3}$. Then for G-sum the following estimate*

$$G(h,p^m) \ll p^{\frac{m+s}{2}},\ \ s=\nu_p(h,p^m)$$

holds.

Proof By Proposition 1 we easy obtain

$$y_{3k}^2 = \frac{(y_0+A_0'p^\alpha)+2k(by_0+A_1'p^{2\alpha})+k^2(b^2(1+a^{-1}y_0y_1)+A_2'p^{3\alpha})}{(1+B_0'p^\alpha)+k(by_0y_1+B_1'p^{2\alpha})+k^2(b^2a^{-2}y_0^2y_1^2+B_2'p^{3\alpha})},$$

$$y_{3k+1}^2 = \frac{(y_0^2y_1^2+A_0'p^\alpha)+2k(by_0^2y_1+A_1'p^{2\alpha})+k^2(b^2y_0^2y_1^2+A_2'p^{3\alpha})}{(y_0^2+B_0'p^\alpha)+2k(by_0+B_1'p^{2\alpha})+k^2(b^2(1+a_0^{-1}y_0y_1)+B_2'p^{3\alpha})},$$

$$y_{3k+2}^2 = \frac{(1+A_0'p^\alpha)+2k(by_0y_1+A_1'p^{2\alpha})+k^2(b^2(ay_0+y_0^2y_1^2)+A_2'p^{3\alpha})}{(y_0^2y_1^2+B_0'p^\alpha)+2k(by_0y_1+B_1'p^{2\alpha})+k^2(b^2y_0^2+B_2'p^{3\alpha})},$$

where A_j', B_j' are some polynomials from $\mathbb{Z}[y_0,y_1,k]$.

Next, we have

$$G(h,p^m) = \sum_{n=1}^{3p^{m-\nu}} e_{p^m}(hy_n^2) =$$

$$= \sum_{k=1}^{p^{m-\nu}} e_{p^m}(hy_{3k}^2) + \sum_{k=1}^{p^{m-\nu}} e_{p^m}(hy_{3k+1}^2) + \sum_{k=1}^{p^{m-\nu}} e_{p^m}(hy_{3k+2}^2) \ll$$

$$\ll p^{s-\nu}\sum_{j=0}^{s}\left|\sum_{k=1}^{p^{m-s}} e_{p^{m-s}}(hy_{3k+j})\right| \ll p^{s-\nu}p^{\frac{m-s}{2}+\nu} = p^{\frac{m+s}{2}}.$$

□

It is similarly investigating the inversive congruential sequence of PRN's of second order type (II).

Theorem 2' *The G-sum for the inversive congruential sequence of type (II) with $\nu_p(a) = \alpha$ estimates by*

$$G(h, p^m) \ll p^{\frac{m+s}{2}}, \ s = \nu_p(h, p^m).$$

Lemma 5 makes possible to prove the estimates for K-sums.

Theorem 2'' *Let $\{y_n\}$ be the sequence produced by recursion (1) of type (II) with $\nu_p(a) = \alpha$, $\nu_p(b) = 0$, and let $A, B \in \mathbb{Z}$. Then we have*

$$K(A, B; p^m) = \sum_{n=1}^{p^{m-\alpha}} e^2_{p^m}(Ay_n + By_n^{-1}) \ll \begin{cases} 0 & if\ 2\alpha \geq m,\ (Ab^2 + B) = 1, \\ p^{\frac{m+\alpha}{2}} & if\ 2\alpha < m,\ (Ab^2 + B) = 1. \end{cases}$$

Proof Using Proposition 1', we after a simple calculations can obtain

$$Ay_n + By_n^{-1} = \frac{\widetilde{A}_0 + \widetilde{A}_1 n + \widetilde{A}_2 n^2}{\widetilde{B}_0 + \widetilde{B}_1 n + \widetilde{B}_2 n^2},$$

where

$$\begin{array}{ll} \widetilde{A}_0 = (Ab^2 + B)ab^5(1 + \widetilde{A}'_0 p^\alpha)y_0^2 y_1^2; & \widetilde{B}_0 = b^3(1 + \widetilde{B}'_0 p^\alpha)y_0^2 y_1^2; \\ \widetilde{A}_1 = p^\alpha 2a_1 b^3(Ab + B)(1 + \widetilde{A}'_1 p^\alpha)y_0^2 y_1^2; & \widetilde{B}_1 = p^\alpha 2a_1 b^3(1 + \widetilde{B}'_1 p^\alpha)y_0^2 y_1^2; \\ \widetilde{A}_2 = p^{2\alpha} a_1^2 b^{-1}(Ab^2 + B)(1 + \widetilde{A}'_2 p^\alpha)y_0^2 y_1^2; & \widetilde{B}_2 = p^{2\alpha} a_1(1 + \widetilde{B}'_2 p^\alpha)y_0^2 y_1^2; \end{array}$$

$\widetilde{A}'_j$, $\widetilde{B}'_j$ are some polynomials from $\mathbb{Z}[y_0, y_1, n]$.

It follows thence the assertion of theorem. □

Similarly, the result is true for the K-sum on the sequence $\{y_n\}$ produced by inversive congruential generator of second order of type (I) with $\nu_p(a) = 0$, $\nu_p(b) = \nu \geq 1$.

4 Discrepancy Bound

To analyze the equidistribution and statistical independence properties of the investigated sequences of PRN's $\{x_n\}$, $x_n = \frac{y_n}{p^m}$, $n = 1, 2, \ldots$ we use the discrepancy of points $x_0, x_1, \ldots, x_{N-1}, \ldots$ and overlapping points $X_N^{(s)} = \{\, x_n, x_{n+1}, \ldots, x_{n+s-1} \,\}$, $n = 0, 1, \ldots,$ with fixed s. For given N points $X_n^{(s)}$, the discrepancy $D_N^{(s)}$ $(X_0^{(s)}, \ldots, X_{N-1}^{(s)})$ is defined as

$$D_N^{(s)}(X_0^{(s)}, \ldots, X_{N-1}^{(s)}) = \sup_{\Delta} \left| \frac{A_N(\Delta)}{N} - vol(\Delta) \right|,$$

where the supremum is extended over all subrectangles $\Delta \subset [0, 1)^s$, $A_N(\Delta)$ is the number of points among $X_0^{(s)} \ldots, X_{N-1}^{(s)}$ falling into $[0, 1)^s$, and $vol(\Delta)$ is the area of Δ.

If for every $s = 1, 2, \ldots, S$, we have $D_N^{(s)}(X_0^{(s)}, \ldots, X_{N-1}^{(s)}) \to 0$ with a rise of N, we will say that the sequence of PRN's passes s-dimensional test on the pseudorandomness. In cryptographical applications a penetrations of s-serial test ($s \geq 2$) means that the sequence $\{x_n\}$ is unpredictable.

Beside discrepancy there exists other important criteria for the uniformity and the independence of PRN's. We shall restrict our attention to the discrepancy, since it is the most important measure of uniformity and independence related to PRN's. For upper estimate of the discrepancy of points we will use the following inequality from [6].

Lemma 6 *Let $q > 1$ and s be natural numbers and let $\{Y_n\}$, $Y_n \in \{0, 1, \ldots, q-1\}^s$, be a purely periodic sequence with a period τ. Then the points $X_n = \frac{Y_n}{q} \in [0, 1)^s$, $n \in \{0, 1, \ldots, N-1\}$, $N \geq \tau$, have discrepancy*

$$D_N^{(s)}(X_0, X_1, \ldots, X_{N-1}) \leq \frac{s}{q} + \frac{1}{N} \sum_{h_0, h_1, \ldots, h_s} \frac{1}{\overline{h}_0 \overline{h}_1 \cdots \overline{h}_s} |S|, \tag{13}$$

where the summation runs over all integers $h_0, h_1, \ldots, h_s$ for which $h_0 \in \left(\frac{-\tau}{2}, \frac{\tau}{2}\right]$, $h_i \in \left(-\frac{q}{2}, \frac{q}{2}\right]$, $(i = 1, \ldots, s)$, $(h_1, \ldots, h_s) \neq (0, \ldots, 0)$, $\overline{h}_i = \max(1, |h_i|)$, and

$$S := \sum_{n=0}^{\tau-1} e\left(h \cdot X_n + \frac{nh_0}{\tau}\right),$$

where $h \cdot X_n = \sum_{i=1}^{s} h_i x_i^{(n)}$ stands for the inner product of h and X_n in $\mathbb{Z}^s$.

The following lemma is a special version of Niederreiter's result [6].

Lemma 7 *The discrepancy of N arbitrary points $\mathbf{t}_0, \mathbf{t}_1, \ldots, \mathbf{t}_{N-1} \in [0, 1)^2$ satisfies*

$$D_N^{(2)}(\mathbf{t}_0, \mathbf{t}_1, \ldots, \mathbf{t}_{N-1}) \geq \frac{1}{2(\pi + 2)|h_1 h_2| N} \cdot \left| \sum_{k=0}^{N-1} e(\mathbf{h} \cdot \mathbf{t}_k) \right| \tag{14}$$

for any lattice point $\mathbf{h} = (h_1, h-2) \in \mathbb{Z}^2$ with $h_1 h_2 \neq 0$.

Going to the estimates of a discrepancy for the sequence of PRN's produced by the generators of second order of type (I) or (II) let us remark here that required estimates of the exponential sums

$$\left| \sum_{n=0}^{N-1} e_{p^m}(hy_n) \right| \leq \begin{cases} O(m) & if\ N = \tau; \\ 4p^{\frac{m+\delta+\nu}{2}} & if\ N \leq \tau,\ \delta + \nu \leq m,\ \nu = \nu_p(h); \\ N & otherwise. \end{cases} \tag{15}$$

we can infer at once from the Corollaries 1 or 1' and Lemma 3.

We need also some supporting data on upper and lower boundaries for the discrepancy of points $t_k = \frac{y_n}{q}$, $y_n \in \mathbb{Z}_q^s$, $n = 0, 1, \ldots, N-1$; $q \in \mathbb{N}$.

Lemma 8 *Let $C_s(q)$ be set of all nonzero point $\mathbf{h} = (h_1, \ldots, h_s) \in \mathbb{Z}^s$, $-\frac{q}{2} < h_j \leq \frac{q}{2}$, $1 \leq j \leq s$. For $\mathbf{h} \in C_s(q)$, $\mathbf{y}_0, \mathbf{y}_1, \ldots, \mathbf{y}_{N-1} \in \mathbb{Z}_q^s$, $\mathbf{y} \in [0, q)^s$, $n = 0, 1, \ldots, N-1$, we have*

$$D_N^{(s)}(\mathbf{y}_0, \mathbf{y}_1, \ldots, \mathbf{y}_{N-1}) \leq \frac{s}{q} + \frac{1}{N} \sum_{\mathbf{h}\in C_s(q)} \sum_{h_0 \in (-\frac{\tau}{2}, \frac{\tau}{2}]} \frac{1}{r(\mathbf{h}, q)} \left| \sum_{n=0}^{N-1} e_q(\mathbf{h} \cdot \mathbf{y}_n) \right|,$$

where $r(\mathbf{h}, q) = \prod_{j=1}^{s} r(h_i, q)$, $r(h, q) = \begin{cases} 1 & if\ h = 0, \\ q \sin(\pi \frac{|h|}{q}) & if\ h \neq 0, \end{cases}$

(see, [6])

Lemma 9 *The discrepancy of N arbitrary points $\mathbf{y}_0, \mathbf{y}_1, \ldots, \mathbf{y}_{N-1} \in [0, 1)^s$ suffice to inequality*

$$D_N(\mathbf{y}_0, \mathbf{y}_1, \ldots, \mathbf{y}_{N-1}) \geq \frac{1}{2^{s-1}(\pi + 2)|h_1 \cdots h_s|} \cdot \frac{1}{N} \left| \sum_{n=0}^{N-1} e(\mathbf{h} \cdot \mathbf{y}_n) \right|$$

for any point $\mathbf{h} = (h_1, \ldots, h_s) \in \mathbb{Z}^s$ under condition $h_1 \cdot \cdots \cdot h_2 \neq 0$.

(see, [6])

For the sequence $\{y_n\}$ produced by the recursion (2) we easy infer (with help Lemma 8).

Theorem 3 *Let $\{y_n\}$, $n = 0, 1, \ldots$, be the sequence of PRN's of second order type (I) with the maximal period $\tau = 3p^{m-\nu}$. Then for discrepancy $D_N(x_0, \ldots, x_{N-1})$ of the sequence $\{x_n\}$, $x_n = \frac{y_n}{p^m} \in [0, 1)^s$, the following bound for $N \leq \tau$*

$$D_N^{(1)}(x_0, x_1, \ldots, x_{N-1}) \leq \frac{1}{p^m} + \frac{3p^{\frac{m}{2}}}{N} \left(\frac{1}{p} \left(\frac{2}{\pi} \log p^m + \frac{7}{5} \right)^2 + 1 \right)$$

holds.

Proof Since $\{y_n\}$ has a maximal period, we have $\tau = 3p^{m-\nu}$. Hence by Lemma (8)

$$D_N^{(1)}(x_0, x_1, \ldots, x_{N-1}) \leq$$

$$\leq \frac{1}{p^m} + \frac{1}{N} \sum_{|h| \leq \frac{1}{2}p^m} \sum_{|h_0| \leq \frac{1}{2}\tau} (r(h, \frac{1}{2}p^m) r(h_0, \tau))^{-1} \left| \sum_{n=0}^{\tau-1} e_{p^m} \left(hx_n + \frac{hx_0 p^m}{\tau} \right) \right| \leq$$

$$\leq \frac{1}{p^m} + \frac{1}{N} \sum_h \sum_{h_0} \left(r\left(h, \frac{1}{2}p^m\right) r\left(h_0, \frac{1}{2}\tau\right) \right)^{-1} \cdot \left(\left|\sum\nolimits_1\right| + \left|\sum\nolimits_2\right| + \left|\sum\nolimits_3\right| \right)$$

where

$$\sum_j = \sum_{k=0}^{p^{m-\nu}-1} e_{p^m}\left(hx_{3k+j} + nh_0p^\nu + n^2hp^{2\nu} + \cdots\right), \quad j = 0, 1, 2.$$

Now in view of the representations of x_{3k+j}, $j = 0, 1, 2$, and Lemma (3), we obtain

$$D_N^1(x_0, x_1, \ldots, x_{N-1}) \leq \frac{1}{p^m} + \frac{3p^{\frac{m}{2}}}{N}\left(\frac{1}{p}\left(\frac{2}{\pi}\log p^m + \frac{7}{5}\right)^2 + 1\right).$$

□

Remark 1 For the case s, $2 \leq s \leq 4$, we have similarly

$$D_N^s := D_N^{(s)}(x_0^{(s)}, x_1^{(s)}, \ldots, x_{N-1}^{(s)}) \leq \frac{s}{p^m} + \frac{1}{p^{\frac{m}{2}-\nu}}\left(1 + \frac{1}{p^\nu}\left(\frac{2}{\pi}\log p^m + \frac{7}{5}\right)^s\right).$$

Theorem 4 *For every $s \in \{1, 2, \ldots, p-1\}$ we have for the sequence of PRN's produced by the inversive generator type (II) of second order the following estimates*

$$D_\tau^{(s)}(x_0^{(s)}, x_1^{(s)}, \ldots, x_{\tau-1}^{(s)}) \leq \frac{s}{p^m} + \frac{3}{p^{\frac{m}{2}-\nu}}\left(1 + \frac{1}{p^\nu}\left(\frac{s}{\pi}\log p^m + \frac{7}{5}\right)^s\right)$$

hold.

This assertion is a corollary of the representation (7) and Lemmas 3 and 7.

From Theorems 3 and 4 it follows that the sequences of PRN's $\{x_n\}$ produced by generator (2) pass the s-dimensional test ($s = 1, 2, 3, 4$) on the uniform distribution and statistical independency (unpredictability). Moreover, if we add the constant shift b to the variable shift $b(n) = b + cn + dF(n)$ with $\nu_p(c) \geq \max(\nu_p(a), \nu_p(b))$ and $\nu_p(d) \geq 2\nu_p(c)$, then all assertions of Theorems 1–4 will true.

Theorem 5 *For the sequences of PRN's produced by recursion (2) of type (I) or (II) we have*

$$\left|\overline{S}_N(h)\right| := \left|\sum_{(y_0,y_1)\in \mathbb{Z}_{p^m}^{*\,2}} \sum_{n=0}^{N-1} e_{p^m}(hy_n)\right| \leq 12N^{\frac{1}{2}} + 12Np^{-\frac{m-\delta}{2}},$$

where

$$(h, p) = 1, \quad \delta = \begin{cases} \nu_p(b) \; if \; \nu_p(a) = 0, \; \nu_p(b) = \nu, \\ \nu_p(a) \; if \; \nu_p(a) = \alpha, \; \nu_p(b) = 0. \end{cases}$$

Theorem 6 *Let the sequence $\{y_n\}$ be produced by (2) with parameters a, b, y_0, y_1, $(a, p) = (y_0y_1, p) = 1$, $\nu_p(b) = p^{\nu_0}$, $\nu_0 \geq 1$. Then for every $h \in \mathbb{Z}$, $(h, p^m) = \mu \leq$*

m, we have

$$\overline{S}_N(h) = \frac{1}{(\varphi(p^m))^2} \sum_{y_0, y_1 \in \mathbb{Z}^*_{p^m}} |S_N(h, y_0, y_1)| \leq 12N^{\frac{1}{2}} + 12Np^{-\frac{m-\nu_0}{2}}.$$

Proof Let $\nu_p(h) = 0$, i.e. $(h, p) = 1$. By the Cauchy-Schwarz inequality we get

$$\begin{aligned}
\left|\overline{S}_N(h)\right|^2 &= \frac{1}{(\varphi(p^m))^2} \left| \sum_{y_0, y_1 \in \mathbb{Z}^*_{p^m}} \sum_{n=0}^{N-1} e_m(hy_n) \right|^2 = \\
&= \frac{1}{(\varphi(p^m))^2} \sum_{y_0, y_1 \in \mathbb{Z}^*_{p^m}} \sum_{k,\ell=0}^{N-1} e_m(h(y_k - y_\ell)) \leq \\
&\leq \frac{1}{(\varphi(p^m))^2} \sum_{k,\ell=0}^{N-1} |\sigma_{k,\ell}(h, -h)| = \frac{1}{(\varphi(p^m))^2} \sum_{r=0}^{\infty} \sum_{\substack{k,\ell=0 \\ \nu_p(k-\ell)=r}}^{N-1} |\sigma_{k,\ell}(h, -h)| = \\
&= \frac{1}{(\varphi(p^m))^2} \sum_{t=0}^{m-1} \sum_{\substack{k,\ell=0 \\ \nu_p(k-\ell)=t}}^{N-1} |\sigma_{k,\ell}(h, -h)| + \frac{1}{(\varphi(p^m))^2} \sum_{k=0}^{N-1} |\sigma_{k,k}(h, -h)| = \\
&= N + \frac{1}{(\varphi(p^m))^2} \sum_{t=0}^{m-1} \sum_{\substack{k,\ell=0 \\ \nu_p(k-\ell)=t}}^{N-1} |\sigma_{k,\ell}(h, -h)|.
\end{aligned}$$

Using Theorem 1, we obtain

$$
\begin{aligned}
\left|\overline{S}_N(h)\right|^2 &\le N + \frac{1}{(\varphi(p^m))^2}\times \\
&\times \sum_{r=0}^{m-1}\left(\sum_{\substack{k,\ell=0\\ k\not\equiv \ell \pmod 3\\ \nu_p(k-\ell)=r}}^{N-1} |\sigma_{k,\ell}(h,-h)| + \sum_{\substack{k,\ell=0\\ k\equiv \ell \pmod 3\\ \nu_p(k-\ell)=r}}^{N-1} |\sigma_{k,k}(h,-h)| \right) \le \\
&\le N + \frac{1}{(\varphi(p^m))^2}\times \\
&\times \left[4p^m \sum_{r=0}^{m-1} \frac{N^2}{p^r} + \left(\sum_{r<m-\nu_0} + \sum_{m-\nu_0\le r\le m-1} \right) \sum_{\substack{k,\ell=0\\ k\equiv \ell \pmod 3}}^{N-1} |\sigma_{k,\ell}(h,-h)| \right] \le \\
&\le N + \frac{N}{(\varphi(p^{(m)}))^2}\times \\
&\times \left(4Np^m + \sum_{r<m-\nu_0} \frac{N}{p^r} p^{m+\nu_0+r} + p^m \sum_{r\ge m-\nu_0} \frac{N}{p^r} \right) \le \\
&\le N + N^2 p^{-m}\cdot 11 p^{\nu_0}(m-\nu_0).
\end{aligned}
$$

Hence, for $(h, p) = 1$ we obtain

$$
\left|\overline{S}_N(h)\right| \le N^{\frac{1}{2}} + 12Np^{-\frac{m-\nu_0}{2}}.
$$

□

5 Conclusion

Although the considered sequences produced by inversive congruential generators of second order do not reach the maximal period length $T = \left(p^{m-1}(p-1)\right)^2$, but due to simplicity of construction and cryptographic applicability they merit attention and further generalization.

References

1. W.-S. Chou, The period lengths of inversive congruential recursions. Acta Arith. **73**(4), 325–341 (1995)
2. J. Eichenauer, J. Lehn, A non-linear congruential pseudorandom number generator. Statist. Hefte **27**, 315–326 (1986)

3. D.E. Knuth, *The Art of Computer Programming II: Seminumerical Algorithms* (Addison-Wesley, 1998)
4. D. Lehmer, Mathematical methods in large-scale computing units, in *Proceedings of the 2nd Symposium on Large Scale Digital Calculating Machinery,* (Harvard University Press, Cambridge, Mass, 1951) pp. 141–146
5. G. Marsaglia, Random numbers fall mainly in the planes. Proc. Nat. Acad. Sci. **61**, 25–28 (1968)
6. H. Niederreiter, *Random Number Generation and Quasi-Monte Carlo Methods* (SIAM, Philadelphia, 1992)
7. P. Varbanets, S. Varbanets. Exponential sums on the sequences of inversive congruential pseudorandom numbers with prime-power modulus, in *VoronoI's Impact on modern science, Proceedings of the 4th International Conference on Analytic Number Theory and Spatial Tessellations,* Book 4, Vol. 1 (Kyiv, Ukraine, 2008 September 22–28), pp. 112–130
8. S. Varbanets, Inversive generator of the second order for the sequence of PRN's, in *8th CHAOS Conference Proceedings, 26–29 May 2015* (Henri Poicare Institute, Paris, France, 2016), pp. 877–889
9. P. Varbanets, S. Varbanets, Inversive generator of the second order with a variable shift for the sequence of PRN's. *Annales Univ. Sci. Budapest., Sect. Comp.* **46**, 255–273 (2017)

Variations on the Fermi-Pasta-Ulam Chain, a Survey

Ferdinand Verhulst

Abstract We will present a survey of low energy periodic Fermi-Pasta-Ulam chains with leading idea the "breaking of symmetry". The classical periodic FPU-chain (equal masses for all particles) was analysed by Rink in 2001 with main conclusions that the normal form of the beta-chain is always integrable and that in many cases this also holds for the alfa-chain. The implication is that the KAM-theorem applies to the classical chain so that at low energy most orbits are located on invariant tori and display quasi-periodic behaviour. Most of the reasoning also applies to the FPU-chain with fixed endpoints. The FPU-chain with alternating masses already shows a certain breaking of symmetry. Three exact families of periodic solutions can be identified and a few exact invariant manifolds which are related to the results of Chechin et al. (1998–2005) on bushes of periodic solutions. An alternating chain of 2n particles is present as submanifold in chains with k 2n particles, k = 2, 3, ... The normal forms are strongly dependent on the alternating masses 1, m, 1, m, ... If m is not equal to 2 or 4/3 the cubic normal form of the Hamiltonian vanishes. For alfa-chains there are some open questions regarding the integrability of the normal forms if m = 2 or 4/3. Interaction between the optical and acoustical group in the case of large mass m is demonstrated. The part played by resonance suggests the role of the mass ratios. It turns out that in the case of 4 particles there are 3 first order resonances and 10 s order ones; the 1:1:1:...:1 resonance does not arise for any number of particles and mass ratios. An interesting case is the 1:2:3 resonance that produces after a Hamilton-Hopf bifurcation and breaking symmetry chaotic behaviour in the sense of Shilnikov-Devaney. Another interesting case is the 1:2:4 resonance. As expected the analysis of various cases has a significant impact on recurrence phenomena; this will be illustrated by numerical results.

Keywords FPU-chain · Resonance · Periodic solutions · Normalisation · Chaos · Hamilton-Hopf bifurcation

F. Verhulst (✉)
Department of Mathematics, University of Utrecht, PO Box 80010, 3508 TA Utrecht
, The Netherlands
e-mail: f.verhulst@uu.nl

C. H. Skiadas et al. (eds.), *13th Chaotic Modeling and Simulation International Conference*, Springer Proceedings in Complexity,
https://doi.org/10.1007/978-3-030-70795-8_71

1 Introduction

Chains of oscillators arise naturally in systems of coupled oscillators and by discretisation of vibration problems of structures. In physics studying the Fermi-Pasta-Ulam (FPU) chain has been very influential for a different reason. The FPU-chain models a one-dimensional chain of oscillators with nearest-neighbour interaction only; see Fig. 1. It was formulated to show the thermalisation of interacting particles by starting with exciting one mode with the expectation that after some time the energy would spread out over all the modes. This is one of the basic ideas of statistical mechanics. In the first numerical experiment in 1955, 32 oscillators were used with the spectacular outcome that the dynamics was recurrent as after some time most of the energy returned to the chosen initial state. For the original report see Fermi et al. [14] and a review by Ford [15], recent references can be found in Christodoulidi et al. [10] or Bountis and Skokos [1]. Discussions can be found in Jackson [22], Campbell et al. [6] and Galavotti [16]. Note that although studies of FPU-chains are of great interest, as models for statistical mechanics problems they are too restrictive.

1.1 Formulation

The original FPU-chain was designed with fixed endpoints and choosing the initial energy small. Later research showed the presence of periodic solutions and wave phenomena, also larger values of the energy were considered. Another version of the FPU-chain is the spatially periodic chain where particle 1 is connected with the last one. In this survey we will focus mainly on the periodic chain with small initial values of the energy. The Hamiltonian $H(p, q)$ for N particles is of the form:

$$H(p, q) = \sum_{j=1}^{N} \left(\frac{1}{2m_j} p_j^2 + V(q_{j+1} - q_j) \right), \tag{1}$$

where particle 1 is connected with particle N. The coordinate system has been chosen so that $q = p = 0$ is a stable equilibrium. For FPU-chains one considers usually potentials $V(z)$ that contain quadratic, cubic and quartic terms. Explicity

$$V(z) = \frac{1}{2}z^2 + \frac{\alpha}{3}z^3 + \frac{\beta}{4}z^4.$$

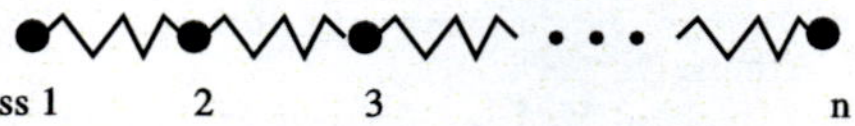

Fig. 1 A Fermi-Pasta-Ulam chain with fixed endpoints

If $\beta = 0$ we call the FPU-chain an α-chain, if $\alpha = 0$ a β-chain. Physically the 2 chains are different, for an α-chain the forces on each particle are asymmetric, for a β-chain they are symmetric.

The spatially periodic chain has a second integral of motion, the momentum integral:

$$m_1\dot{q}_1 + m_2\dot{q}_2 + \cdots + m_N\dot{q}_N = \text{constant}. \tag{2}$$

The momentum integral (2) enables us to reduce the N dof system to a $N-1$ dof Hamiltonian system by a symplectic transformation.

For low energy orbits near stable equilibrium one usually rescales $p \mapsto \varepsilon\bar{p}$, $q \mapsto \varepsilon\bar{q}$, divides the Hamiltonian by ε^2 and drops the bars. For the linearised system near stable equilibrium we find:

$$\begin{cases} m_1\ddot{q}_1 + 2q_1 - q_2 - q_N & = 0, \\ m_2\ddot{q}_2 + 2q_2 - q_3 - q_1 & = 0, \\ m_3\ddot{q}_3 + 2q_3 - q_4 - q_2 & = 0, \\ \ldots & = 0, \\ m_N\ddot{q}_N + 2q_N - q_1 - q_{N-1} & = 0. \end{cases} \tag{3}$$

The quadratic nonlinearities start with ε, the cubic ones with ε^2. The spectrum of the linear operator (the eigenvalues near stable equilibrium) determines the resonances and the nonlinear dynamics near stable equilibrium. Our survey is based on papers that make extensive use of normalisation-averaging techniques, see Sanders et al. [28], Chaps. 2 and 10. This involves near-identity transformations to simplify the equations of motion or the Hamiltonian itself if one studies such a system. A quadratic Hamiltonian indicated by H_2 corresponds with a linear system of differential equations; for a Hamiltonian with cubic terms near-identity transformation removes the non-resonant terms to higher order. Omitting the higher order terms the resulting normalised Hamiltonian $\bar{H} = H_2 + \bar{H}_3$ contains only the resonant terms $\bar{H}_3$ of the cubic H_3 (the index indicates the power of the polynomials). One can go on with the normalisation process by using a near-identity transformation to remove the non-resonant terms from H_4, etc.

In general the normalised (averaged) equations that are truncated at some level of normalisation will not be integrable, although there are many exceptions. For the FPU-Hamiltonian in homogeneous polynomials we have the notation:

$$H = H_2 + \varepsilon H_3 + \varepsilon^2 H_4, \text{ and } \bar{H} = H_2 + \varepsilon\bar{H}_3 + \varepsilon^2\bar{H}_4.$$

We will describe a number of prominent cases that show different dynamics for different choices of the masses. In the original (classical) FPU problem all masses are equal which seems a natural choice. A second natural choice is to alternate the masses $m, M, m, M, \ldots, m, M$; it is no restriction to assume $0 < m \leq M$. A quite different approach is to look for mass ratio's that produce interesting resonances and dynamics. We aim at summarising all these approaches for low energy chains. Of special interest in the analysis are integrals corresponding with approximate invariant manifolds of the averaged systems, periodic solutions, bifurcations and chaos.

An important conclusion will be that the classical FPU-chain contains so many symmetries that by symmetry breaking it is structurally unstable.

1.2 Theoretical Background

There exist an enormous amount of papers on the original FPU-chain of Fermi et al. [14]. A large number of the papers consist of numerical explorations; they are often inspiring but not always satisfactorily explaining the phenomena. Apart from normalisation-averaging, symmetry considerations are important for the qualitative results. This involves the theory of Hamiltonian systems, see for an introduction Verhulst [29] and for the more general dynamical systems context Broer and Takens [2]. New results on Hamiltonian systems and symmetry are found in Bountis and Skokos [1], Efstathiou [13] and Hanßmann [18]. Basic understanding of recurrence as formulated by Poincaré [24] vol. 3, Chap. 26 is essential.

A systematic study of dynamical systems with discrete symmetry was started by Chechin and Sakhnenko [7]. The authors introduce the notion of *bushes* with a bush comprising all modes singling out an active symmetry group in the system. A bush corresponds with a lower dimensional invariant manifold (or approximate invariant manifold in the sense of normalisation) giving insight in the various dynamical parts that compose the system. The theory is quite general, it was applied to FPU chains by Chechin et al. in [8, 9].

Independently the ideas of utilising symmetries were also developed by Rink [26] and by Bruggeman and Verhulst in [4, 5].

2 The Classical Periodic FPU-chain

In the original FPU problem one considered the so-called mono-atomic case, i.e. all masses equal; we call this the classical FPU-chain and put $m_1 = m_2 = \cdots = m_N = 1$. The recurrence of the classical FPU-chain signalled by Fermi et al. [14] was surprising at the time as this was before the time of publication of the KAM theorem (see below).

The linearised system (3) has the frequencies ω_j of the corresponding harmonic equations:

$$\omega_j = 2\sin\left(\frac{j\pi}{N}\right),\ j = 1, \ldots, N. \tag{4}$$

The implication is that we have many $1:1$ resonances, $N/2$ if N is even and $(N-1)/2$ if N is odd. Also there exist accidental other resonances like $1:2:1$. A natural first step is to reduce the system using integral (2) to $N-1$ dof.

An interesting attempt to solve the recurrence problem was made by Nishida [23] by proposing to use the KAM theorem; this theorem guarantees under the right conditions the existence of an infinite number of $(N-1)$-tori containing quasi-periodic solutions near stable equilibrium. This would solve the recurrence problem, but unfortunately the spectrum is resonant and the KAM theorem can not be applied in a simple way.

The problem was for most cases solved for the spatially periodic FPU chain by Rink in [26]; his results can also be applied to the chain with endpoints fixed. We summarise the reasoning. First the system with cubic and quartic terms in the Hamiltonian is transformed by symplectic normalisation (also called Birkhoff-Gustavson normalisation) to a simpler form. If the resulting normalised Hamiltonian $\bar{H}$ is nondegenerate in the sense of the KAM theorem and if it is integrable i.e. containing, in addition to integral (2), $N-1$ functionally independent integrals that are in involution, then the KAM theorem applies to the original Hamiltonian H. By the transformation the nonresonant terms of the cubic and quartic part are shifted to higher order. The original system contains various discrete symmetry groups, a rotation symmetry and a reflection symmetry. These symmetries carry over to the normalised Hamiltonian system with the surprising result that the cubic terms in $\bar{H}$ vanish! From theorem 8.2 of Rink [26] we have for the classical periodic FPU chain derived from Hamiltonian (1) containing cubic and quartic terms:

$$\bar{H}_3 = 0. \tag{5}$$

The analysis in Rink [26] of $\bar{H}$ produces furthermore:

1. Assume $\alpha \neq 0$ and N is odd, then $H_2 + \varepsilon^2 \bar{H}_4$ is integrable and nondegenerate in the sense of the KAM theorem.
2. Assume $\alpha \neq 0$ and N is even, then $H_2 + \varepsilon^2 \bar{H}_4$ hast at least $(3N-4)/4$ quadratic integrals (if 4 divides N) or $(3N-2)/4$ quadratic integrals (if 4 does not divides N).
3. The normalised β-chain ($\alpha = 0$) is integrable and nondegenerate in the sense of the KAM theorem. Almost all low-energy orbits are periodic or quasi-periodic and move on invariant tori near stable equilibrium.
4. Similar results can be obtained for the classical FPU-chain with fixed endpoints.

The remaining problem is the integrability of $H_2 + \varepsilon^2 \bar{H}_4$ in the case of the even α-chain. To check this one has to carry out the normalisation to quartic terms which is quite a lot of work if N is large. We will discuss an example with $\alpha = 1, \beta = -1$.

Example 1 Consider a periodic Fermi-Pasta-Ulam chain consisting of four particles of equal mass m (=1) with quadratic and cubic nearest-neighbor interaction. Periodic means that we connect the first with the fourth particle. The Hamiltonian is in this case:

$$H(p,q) = \sum_{j=1}^{4} \left(\frac{1}{2} p_j^2 + V(q_{j+1} - q_j) \right), \tag{6}$$

with

$$V(z) = \frac{1}{2} z^2 + \frac{1}{3} z^3 - \frac{1}{4} z^4.$$

The corresponding equations of motion were studied Rink and Verhulst [25].

The equations induced by Hamiltonian (6) have a second integral of motion, the momentum integral $\sum_{j=1}^{4} p_j =$ constant. This enables us to reduce the 4 dof equations of motion to 3 dof by a canonical (symplectic) transformation. From Rink and Verhulst [25] we have the reduced system:

$$\begin{cases} \ddot{x}_1 + 4x_1 &= 4x_2 x_3 + 4x_1^3 + 6x_1(x_2^2 + x_3^2), \\ \ddot{x}_2 + 2x_2 &= 4x_1 x_3 + x_2^3 + 3x_2(x_3^2 + 2x_1^2), \\ \ddot{x}_3 + 2x_3 &= 4x_1 x_2 + x_3^3 + 3x_3(x_2^2 + 2x_1^2). \end{cases} \tag{7}$$

We can identify 3 families of periodic solutions, the 3 normal modes in the coordinate planes. Consider the x_2 normal mode that satisfies the equation:

$$\ddot{x}_2 + 2x_2 = x_2^3.$$

In general, solutions far from stable equilibrium become chaotic, so we restrict ourselves to a neighbourhood of the origin by rescaling $x_1 = \varepsilon \bar{x}_1$, $x_2 = \varepsilon \bar{x}_2$, $x_3 = \varepsilon \bar{x}_3$ and then omitting the bars. Rescale also $\sqrt{2}t = s$. System (7) becomes:

$$\begin{cases} \frac{d^2 x_1}{ds^2} + 2x_1 &= 2\varepsilon x_2 x_3 + 2\varepsilon^2 x_1^3 + 3\varepsilon^2 x_1(x_2^2 + x_3^2), \\ \frac{d^2 x_2}{ds^2} + x_2 &= 2\varepsilon x_1 x_3 + \frac{1}{2}\varepsilon^2 x_2^3 + \frac{3}{2}\varepsilon^2 x_2(x_3^2 + 2x_1^2), \\ \frac{d^2 x_3}{ds^2} + x_3 &= 2\varepsilon x_1 x_2 + \frac{1}{2}\varepsilon^2 x_3^3 + \frac{3}{2}\varepsilon^2 x_3(x_2^2 + 2x_1^2). \end{cases} \tag{8}$$

The equation for the x_2 normal mode was studied in many introductions to the averaging method, where with initial values $x_2(0) = a$, $dx_2(0)/ds = 0$ we obtain the approximation:

$$\phi(s) = a \cos(s - \varepsilon^2 \frac{3}{16} a^2 s).$$

We transform $x_1 = y_1$, $x_2 = \phi(s) + y_2$, $x_3 = y_3$ in system (8) and linearising we find:

$$\begin{cases} \frac{d^2 y_1}{ds^2} + 2y_1 &= 2\varepsilon \phi(s) y_3 + 3\varepsilon^2 y_1 \phi^2(s), \\ \frac{d^2 y_2}{ds^2} + y_2 &= 0, \\ \frac{d^2 y_3}{ds^2} + y_3 &= 2\varepsilon y_1 \phi(s) + \frac{3}{2}\varepsilon^2 y_3 \phi^2(s). \end{cases} \tag{9}$$

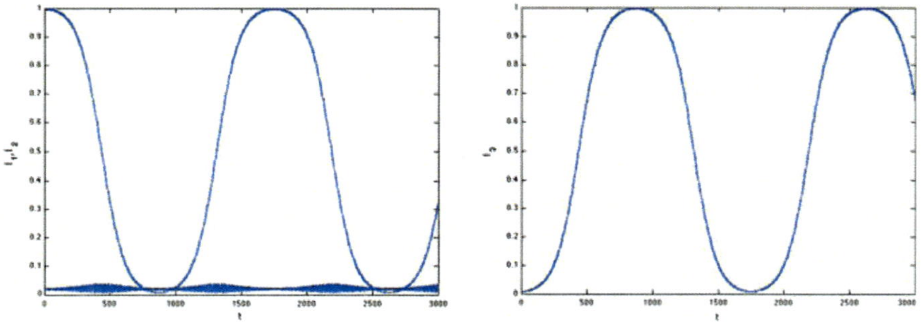

Fig. 2 The actions for 3000 timesteps near the unstable x_2 normal mode of system (7) with $\varepsilon = 0.1$, initial conditions $x_1(0) = x_3(0) = 0.1, x_2(0) = 1$ and initial velocities zero. Left the action $I_2(t) = \frac{1}{2}(\dot{x}_2^2 + 2x_2^2)$ starting near zero and increasing to values near 1; also the nonresonant $I_1(t) = \frac{1}{2}(\dot{x}_1^2 + 4x_1^2)$. Right the resonant action $I_3(t) = \frac{1}{2}(\dot{x}_3^2 + 2x_3^2)$ that exchanges energy with the x_2 mode (pictures from [31])

The first and third equations are coupled but there is no resonance because of the basic frequencies $\sqrt{2}$ and 1; we conclude that the solutions of system (9) are stable. Interestingly, it was proved in [25] that near stable equilibrium the stability in linear approximation is destroyed by the nonlinearities. See Fig. 2 for an illustration.

Example 2 (*Other examples*) The relatively simple case of 3 particles was discussed by Ford [15]; the system is identified with the Hénon-Heiles system, a 2 dof Hamiltonian system in $1:1$ resonance; for a survey see Rod and Churchill [27]. This is interesting as this system has an integrable normal form for low energy values. Between the invariant tori there exists chaos but of exponentially small measure. If the energy is increased the amount of chaos increases, destroying more and more tori until the system looks fully chaotic at higher energy. Proofs are available for this behaviour, see Holmes et al. [20], except that we do not know whether at "full chaotic behaviour" there are no tiny sets of tori left, undetected by numerics.

In Rink and Verhulst [25] the classical system with 4, 5 and 6 particles was analysed in the cases of α- and β-chains, also for mixed cubic and quartic terms. In these examples the normal forms are integrable.

3 The FPU-chain with Alternating Masses

Alternating the masses of a FPU-chain produces already a certain symmetry breaking. It is no restriction to rescale the smallest mass to 1 and have largest mass $m \geq 1$.

So we consider the periodic FPU-chain with N (even) masses that alternate: $1, m, 1, m, \ldots, 1, m$ (the case $0 < m \leq 1$ follows from symmetry considerations). The chain is related to the formulation in Galgani et al. [17] that analyses the chain and explores numerical aspects if N is large. In Bruggeman and Verhulst [5] a general analysis was started, but there are still many open questions; we summarise a number of results of this paper.

The eigenvalues $\lambda_j, j = 1, \ldots N$ of system (3) are with $a = 1/m$ in the case of alternating masses:

$$\lambda_j = 1 + a \pm \sqrt{1 + 2a\cos(2\pi j/N) + a^2}, \; j = 1, \ldots N. \tag{10}$$

Several observations can be made:

1. One eigenvalue equals 0 corresponding with the existence of the momentum integral (2).
2. If N is a multiple of 4 we have among the eigenvalues the numbers $2(a+1), 2, 2a$.
3. For large masses m ($a \to 0$) the eigenvalue spectrum consists of 2 groups, one with size $2 + O(a)$ (the so-called optical group) and one with size $O(a)$ (the so-called acoustical group). The symplectic transformation to $N-1$ dof mixes the modes because of the nearest-neighbour interactions, present already in the linearised system (3). So we cannot simply identify the dynamics of the optical group with the dynamics of the large masses.

A few qualitative and quantitative results were obtained by Bruggeman and Verhulst [5]:

1. We can identify *three explicit families of periodic solutions* characterised by the frequencies $\sqrt{2}, \sqrt{2a}, \sqrt{2(1+a)}$. The solutions are either harmonic or elliptic functions.
2. In the spirit of Chechin and Sakhenko [7] we can identify *bushes* of solutions in the following sense: the dynamics of a system with N particles will be found as a submanifold in systems with kN particles ($k = 2, 3, \ldots$). This increases the importance of studying chains with a small number of particles enormously. Note that the result is valid for large values of N, it also holds in the classical case $m = 1$.
3. First order averaging-normalisation ($m \neq 1$) produces for the α-chain only non-trivial results if $m = 2$ and $m = 4/3$. From the point of view of normalisation the case of large m ($a \to 0$) has to be treated separately.
4. An interesting discussion by Zaslavsky [32] deals with the phenomenon of delay of recurrence in Hamiltonian systems by *quasi-trapping*. This phenomenon arises for 3 and more dof if resonance manifolds, acting as subsets of the energy manifold, contain periodic solutions surrounded by invariant tori. The orbits entering such resonance manifolds may be delayed passage by staying for a number of revolutions near these tori.
   In Bruggeman and Verhulst [5] an explicit analysis and numerics of quasi-trapping is given for a number of cases with 8 particles. In the case of large mass m a second order normalisation is necessary; the recurrence is sensitive to the initial conditions.
5. For the alternating mass m large (small a) we expect different dynamics for the optical group (eigenvalues near 2) and the acoustical group (eigenvalues $O(a)$),

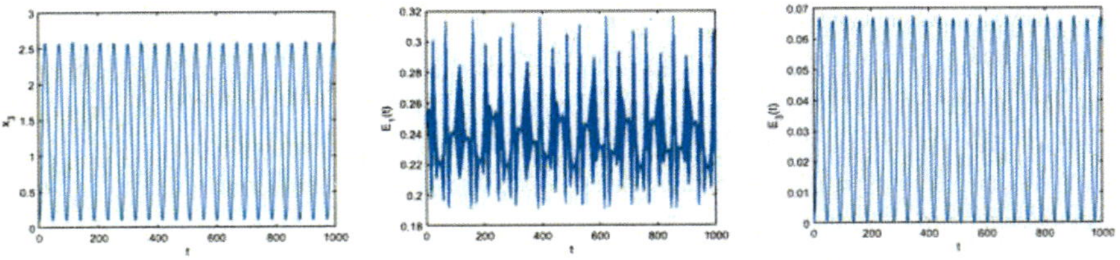

Fig. 3 Interaction between optical and acoustical group in invariant manifold M corresponding with 4 particles. The modes x_1, x_2 are near 1 : 1 resonance. We have in system (11) $a = 0.01$, $x_1(0) = x_2(0) = 0.5$, $x_3(0) = 0$ and initial velocities zero. The instability of the solution in the optical group is indicated by the action $E_1(t) = 0.5(\dot{x}_1^2 + 2(1+a)x_1^2)$ (middle). Although far from resonance, the low-frequency mode x_3 is excited; figs left $x_3(t)$ and right $E_3(t) = 0.5(\dot{x}_3^2 + 2ax_3^2)$

see Galgani et al. [17]. This raises an old question: can high frequency modes transfer energy to low frequency modes and vice versa? The answer is affirmative, see the discussion below and Fig. 3.

We summarise results for the cases $N = 4n$ and $N = 8n$.

3.1 *Chain with 4n Particles, n = 1, 2, 3, . . ., [3]*

A system with 4 particles is imbedded as an invariant manifold in a system with 4n particles. The momentum integral (2) enables reduction to 3 dof with frequencies $\sqrt{2}, \sqrt{2a}, \sqrt{2(1+a)}$. We find no 3 dof first order resonances in a system with 4 particles. The normal modes are exact periodic solutions both for the α- and the β-chain. The normal forms are in both cases integrable to second order. The recurrence of the orbits on an energy manifold depends on the initial conditions, starting near an unstable periodic orbit lengthens the recurrence times.

For the case large mass m (a small) see below.

3.2 *Chain with 8n Particles, n = 1, 2, 3, . . ., [5]*

A system with 8 particles is imbedded as an invariant manifold in a system with 8n particles. Using integral (2) produces reduction to 7 dof with frequencies:

$$\sqrt{2}, \sqrt{2a}, \sqrt{2(1+a)}, 1 + a + \sqrt{1+a^2}\ \text{(twice)}, 1 + a - \sqrt{1+a^2}\ \text{(twice)}.$$

The normal forms become much more complex (H_4 contains 49 terms) so we restrict the analysis to α-chains. As expected we recover the invariant manifold associated with the first 3 eigenvalues (or frequencies) for the system *before* normalisation; we find two more 6-dimensional invariant manifolds of the exact equations. The 3 invariant manifolds have the normal mode periodic solution associated with the frequency $2a$ in common. This mode plays a pivotal part in the dynamics.

Normalisation produces $\bar{H}_3 = 0$ except if $a = 0.5, 0.75$ and if a is close to zero (large mass). The normal form flow in the 3 invariant manifolds is integrable. In the case $a = 0.75$ we find instability of the invariant manifolds, the stability in the other cases can not be decided as the eigenvalues are purely imaginary (this is a basic stability problem of Hamiltonian systems with more than 2 dof).

A conclusion is that the presence of nested invariant manifolds (bushes) makes the equipartition of energy rather improbable.

3.3 *Interactions Between Optical and Acoustical Group*

The eigenvalues and frequencies obtained from (10) suggest that for mass m large we have two groups of oscillators, one with frequency size close to $\sqrt{2}$ and one with size $O(\sqrt{a})$. There are indications in Bruggeman and Verhulst [3] that in the case of a chain with 4 particles there exists significant interactions between the 2 groups. It turns out that in α-chains the acoustical group can be strongly excited by the optical group.

We will clarify this interaction phenomenon in the case of 4n particles using the 4 particles invariant manifold M that consists of the modes with frequencies $\sqrt{1+a}$, $\sqrt{2}$, $\sqrt{2a}$. This submanifold corresponds with the 4 particles system described above.

As $0 < a \ll 1$ there is actually no need for a scaling by small parameter ε in this case. The corresponding equations of motion are (see Bruggeman and Verhulst [5]):

$$\begin{cases} \ddot{x}_1 + 2(1+a)x_1 &= 2\sqrt{a(1+a)}x_2x_3, \\ \ddot{x}_2 + 2x_2 &= 2\sqrt{a(1+a)}x_1x_3, \\ \ddot{x}_3 + 2ax_3 &= 2\sqrt{a(1+a)}x_1x_2. \end{cases} \tag{11}$$

The modes x_1 and x_2 are in a detuned $1:1$ resonance when choosing $0 < a \ll 1$. Consider the general position periodic solution of the $1:1$ resonance of the x_1, x_2 modes, described in Bruggeman and Verhulst [3]. A normal form approximation is $x_1(t) = r_0 \cos(\sqrt{2}t + \psi_0)$, $x_1(t) = \pm x_2(t)$; the approximation is based on the equations for these modes to order $O(a)$:

$$\ddot{x}_1 + 2x_1 = 2\sqrt{a}x_2x_3 + a\ldots, \; \ddot{x}_2 + 2x_2 = 2\sqrt{a}x_1x_3 + a\ldots$$

with x_3 varying on a long timescale. The asymptotic approximation with $x_1 = x_2$ leads to a forced, linear equation for $x_3(t)$:

$$\ddot{x}_3 + 2ax_3 = 2\sqrt{a}r_0^2\cos^2(\sqrt{2}t + \psi_0), \tag{12}$$

with particular solution:

$$x_3(t) = \frac{r_0^2}{2\sqrt{a}} - \frac{r_0^2}{8r_0^2 - 2\sqrt{a}} \cos(2\sqrt{2}t + 2\psi_0). \tag{13}$$

To this expression we have to add the homogeneous solution consisting of $\cos(\sqrt{2at})$ and $\sin(\sqrt{2at})$. It is remarkable that the particular solution has a large amplitude, $O(1/\sqrt{a})$, and period $\pi/\sqrt{2}$. The homogeneous solution has long period $\pi\sqrt{2}/\sqrt{a}$. We find that the "acoustical mode" x_3 is strongly excited; x_1 and x_3 are shown in Fig. 3 in the case of large mass 100.

4 Resonances Induced by Other Mass Ratio's

The classical FPU-chain and the chain with alternating masses are natural models of physical chains. It is clear from dynamical systems theory that resonances and symmetries play a fundamental part in all these model chains; see for instance Poincaré [24] or Sanders et al. [28].

Take for instance the classical FPU-chain with $N = 6$; the 6 harmonic frequencies of system (3) are $1, \sqrt{3}, 2, \sqrt{3}, 1, 0$. As we know, both for α- and β-chains $\bar{H}_3 = 0$ so the $1 : 2 : 1$ first order resonance is not effective because of symmetry; it might appear as a $2 : 4 : 2$ resonance at higher order. The $\sqrt{3} : \sqrt{3} = 1 : 1$ resonance plays a part for β-chains.

A different choice of masses that would make the frequency spectrum of system (3) non-resonant would always have near-resonances as the rationals are dense in the set of real numbers. This would produce detuned resonances with behaviour related to exact resonance, so even in this case the analysis of Nishida [23] would not apply although his idea turns out to be correct. Thus it makes sense to explore systematically the kind of resonances that may arise in FPU-chains. As we shall see this leads to various applications.

The exploration of possible resonances was done by Bruggeman and Verhulst [4] for the case of 4 particles leading to chains described by 3 dof. In Sanders et al. [28], Chap. 10 a list of prominent Hamiltonian resonances in 3 dof is given for general Hamiltonians. In *general* for 3 dof we have 4 first order resonances (active at H_3) and 12 second order resonances (active at $H_3 + H_4$). Considering system (3) for the special case of the FPU chains with arbitrary positive masses, we find that the first order resonance $1 : 2 : 2$ does not arise, of the 12 seond order resonances $1 : 1 : 1$ and $1 : 3 : 3$ are missing. The importance of the resonances that do arise is partly determined by the size of sets in the parameter space of masses. We present the results from Bruggeman and Verhulst [4] where the sets in 3d-parameter space, the mass ratios of (m_1, m_2, m_3, m_4), with active resonance are indicated between brackets:

First order resonance
1 : 2 : 1 (4 points); 1 : 2 : 3 (4 open curves); 1 : 2 : 4 (12 open curves).

Second order resonances

1 : 1 : 3 (4 points)	1 : 2 : 5 (12 open curves)
1 : 2 : 6 (12 open curves)	1 : 3 : 4 (4 open curves)
1 : 3 : 5 (4 open curves)	1 : 3 : 6 (12 open curves)
1 : 3 : 7 (12 open curves)	1 : 3 : 9 (12 open curves)
2 : 3 : 4 (2 compact curves);	2 : 3 : 6 (2 compact curves)

To determine the possible resonances for FPU chains with more than 4 particles is a formidable linear algebra and algebraic problem that has not been solved in generality. A general result from Bruggeman and Verhulst [4] is that for $N \geq 4$ no mass distribution will produce the N dof $1 : 1 : \ldots : 1$ resonance. We will discuss some results that are known for the 1 : 2 : 3 and 1 : 2 : 4 resonances with 4 particles. The second order resonances are largely unexplored for FPU-chains.

4.1 The 1 : 2 : 3 Resonance

This resonance is of special interest as in this case for the general Hamiltonian chaos does not become exponentially small near stable equilibrium as $\varepsilon \to 0$ (see Hoveijn and Verhulst [21]). In general the normal form of the 1 : 2 : 3 resonance is not integrable; see Christov [11]. However, symmetries may change the dynamics as is shown in systems with 4 particles, see Bruggeman and Verhulst [4] and below.

The symmetric case of 4 particles α-chains, $m_1 = m_3$
Using integral (2) and symplectic transformation we find the Hamiltonian:

$$
\begin{aligned}
H(p,q) &= \frac{1}{2}\sum_{j=1}^{3}(p_j^2 + \omega_j^2 q_j^2) + \varepsilon(d_3 q_1^2 + d_{10} q_2^2 + d_6 q_3^2)q_3, \\
(\omega_1, \omega_2, \omega_3) &= (3, 2, 1),
\end{aligned} \tag{14}
$$

with coefficients $d_3, d_6, d_{10} \neq 0$. The (p_1, q_1) and the (p_2, q_2) normal modes are exact periodic solutions in the 2 coordinate planes. Averaging-normalisation produces in addition the (p_3, q_3) normal mode periodic solution. We find 3 integrals of motion of the normalised system so the normal form dynamics is integrable. The normal form system contains only one combination angle $\chi = \psi_1 - \psi_2 - \psi_3$ producing for fixed energy families of periodic solutions (tori) in general position. This is a degeneration in the sense described by Poincaré [24] vol. 1, Chap. 4.

The stability of the normal modes is indicated in Fig. 4, left; the normal 2nd and 3rd modes ($\omega = 2, 1$) are stable with purely imaginary eigenvalues, the eigenvalues are coincident for the 2nd normal mode (Krein collision of eigenvalues). The first

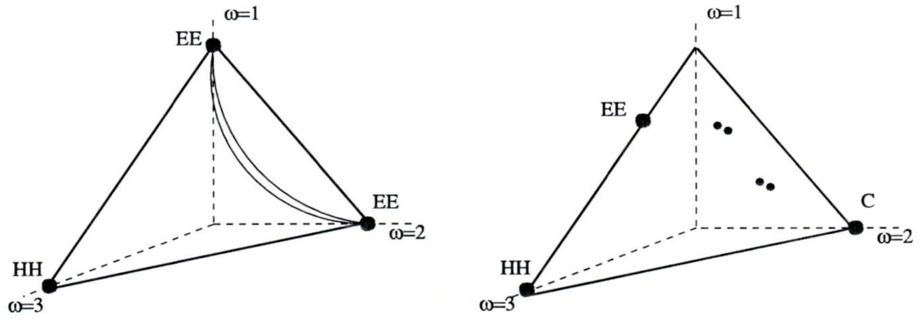

Fig. 4 The $1:2:3$ resonance with action simplex of the symmetric case $m_1 = m_3$ (left) and right a typical case with all masses different. Along the axes the actions form a triangle for fixed values of H_2 which is an integral of the normal forms. The frequencies 1, 2, 3 indicate the 3 normal mode positions at the vertices. The black dots indicate periodic solutions, the indicated stability types are HH (hyperbolic-hyperbolic), EE (elliptic-elliptic) and C (complex with real and imaginary parts nonzero). The two (roughly sketched) curves connecting the 2 normal modes in the left simplex correspond for fixed energy with two tori consisting of periodic solutions, respectively with combination angle $\chi = 0$ and π. The tori break up into 4 general position periodic solutions if all masses are different

mode ($\omega = 3$) is unstable with real eigenvalues; In Bruggeman and Verhulst [4] a detailed description is given of the motion of the orbits starting near the unstable normal mode ($\omega = 3$).

We will see that the case $m_1 = m_3$ is structurally unstable, the dynamics changes drastically if all masses are different.

The case of 4 particles α-chains, all masses different

This case presents striking differences from the case with 2 masses equal, the symmetry is broken. We summarise:

1. The 3rd normal mode ($\omega = 1$) vanishes, the periodic solution shifts to the 2 dof subspace formed by the first and 3rd mode; stability EE.
2. The second normal mode becomes complex unstable (C) by a Hamiltonian-Hopf bifurcation. In this case two pairs of coincident imaginary eigenvalues (the case $m_1 = m_3$) move into the complex plane.
3. The presence of a complex unstable periodic solution fits in the Shilnikov-Devaney scenario leading to chaotic dynamics in the normal form, see Devaney [12] and Hoveijn and Verhulst [21]; the normalised Hamiltonian is not integrable in this case. Establishing chaos involves the presence of a horseshoe map. As this map is structurally stable, finding chaos in the normal form, this chaos will persist in the original system.
4. The tori consisting of periodic solutions in the case $m_1 = m_3$ break up into 4 periodic solutions at fixed energy.

4.2 The 1 : 2 : 4 Resonance

Work in progress for the FPU-chain with 4 masses in 1 : 2 : 4 resonance can be found in Hanßmann et al. [19]; this analysis includes detuning. We mention some of the results in the case of opposing masses equal, $m_1 = m_3$.

1. The case of 2 opposing masses equal induces a $\mathbb{Z}_2$ symmetry with as consequence that for both α- and β-chains we have $\bar{H}_3 = 0$.
2. The normal form $H_2 + \bar{H}_4$ for the α- and β-chains has 3 normal mode periodic solutions and is integrable.
3. Normalisation to H_6 breaks the symmetry, only 2 integrals of the normalised Hamiltonian could be found.

Interestingly the case of 2 adjacent equal masses produces different results; in this case $\bar{H}_3 \neq 0$, the symmetry mentioned above is broken.

The case of all masses different will be studied in a forthcoming paper.

4.3 An Application to Cell-Chains

One can use low-dimensional FPU-chains as cells to form a new type of chain, see Fig. 5. This is quite natural when thinking of interactions of molecules (a small group of connected oscillators) instead of atoms leading to a chain of connected near-neighbour interacting oscillators. A few examples of such cell-chains are discussed in Verhulst [30].

Consider cells consisting of a FPU-chain with 4 particles. As we have seen before the dynamics within each cell will strongly depend on the choice of the 4 masses. A second important aspect is how the cells are linked. Connecting cells by particles where stable periodic solutions dominate is expected to produce less transfer of energy than connecting by particles with more unstable periodic solutions and more active dynamics. Also the linking of cells will detune the resonances; this effect can be stronger if the FPU-chain is structurally unstable. We will show a few examples of transfer of energy for the simplest case of two connected cells. As the systems are Hamiltonian the phase-flow will always be recurrent but if the recurrence takes a long time this will indicate active but small transfer of energy between the cells with delayed recurrence.

Fig. 5 A FPU cell-chain with 3 cells

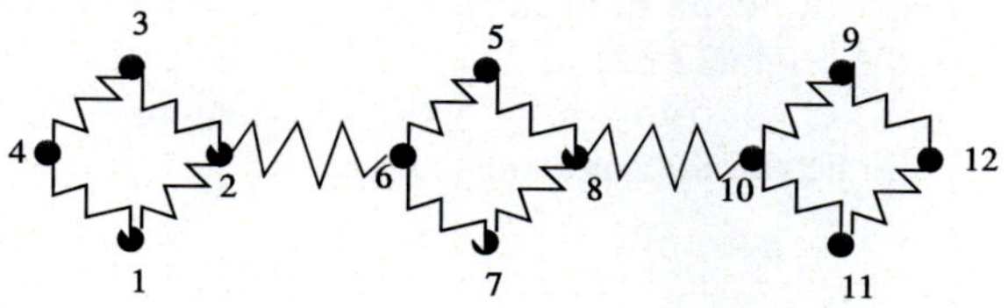

Hamiltonian (15) describes the interaction of 2 cells if $c_1 \neq 0$.

$$
\begin{aligned}
H(p,q) &= \sum_{j=1}^{4}\left(\frac{m_j}{2}p_j^2 + \frac{1}{2}(q_{j+1}-q_))^2)\right) \\
&+ \sum_{j=5}^{8}\left(\frac{m_{j-4}}{2}p_j^2 + \frac{1}{2}(q_{j+1}-q_j)^2\right) + \frac{\varepsilon}{2}c_1(q_2-q_6)^2 + H_3,
\end{aligned}
\tag{15}
$$

with

$$
H_3 = \sum_{j=1}^{8}\frac{\varepsilon}{3}(q_{j+1}-q_j)^3.
$$

In the experiments we start with zero initial values in the 2nd cell, $q_j(0) = v_j(0) = 0,\ j = 5, \ldots, 8$. If $c_1 = 0$ we have non-trivial dynamics and corresponding distance $d(t)$ to the initial values only in the first cell. Explicitly:

$$
d(t) = \sqrt{\sum_{j=1}^{8}[(q_j(t)-q_j(0))^2 + (v_j(t)-v_j(0))^2]}. \tag{16}
$$

The distance $d(t)$ can be used to consider recurrence to a δ-neighbourhood of the initial values. An upper bound L for the recurrence time has been given in Verhulst [30]. Suppose we consider a bounded Hamiltonian energy manifold with N dof, energy value E_0 and Euclidean distance $d(t)$ of an orbit to the initial conditions, than we have for the recurrence time T_r to return in a δ-neighbourhood of the initial conditions an upper bound L with:

$$
L = O\left(\frac{E_0^{N-1/2}}{\delta^{2N-1}}\right). \tag{17}
$$

For one FPU-cell we have with reduction to 3 dof $L_1 = E_0^{5/2}/\delta^5$ and for 2 linked FPU-cells $L_2 = E_0^{13/2}/\delta^{13}$. Of course, starting near a stable periodic solution or if there exist extra first integrals will reduce the recurrence time enormously.

Numerical experiments

We present numerical results for 3 cases with cells consisting of 4 masses: the classical FPU-chain with equal masses in Fig. 6 ($m = 0.1$ to have comparable timescales), the $1:2:3$ resonance case with symmetry induced by the choice $m_1 = m_3$ in Fig. 7 and the less-balanced case of the $1:2:3$ resonance where the dynamics is chaotic, Fig. 8. In each of the 3 cell-chains we have initial values $q_1(0) = 0.05$, $q_2(0) = 0.2$, $q_3(0) = 0.05$, $q_4(0) = 0.05$, $q_5(0) = q_6(0) = q_7(0) = q_8(0) = 0.0$, initial velocities are all zero. So we start in the first cell near the second normal mode plane.

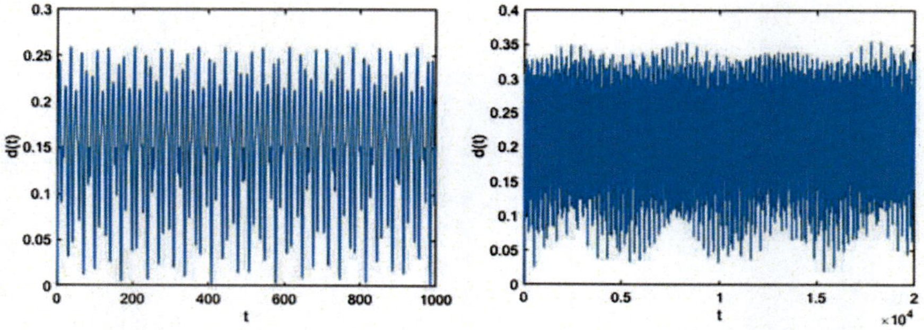

Fig. 6 The classical case $m = 0.1$ with strong recurrence for 1 cell (roughly 100 timesteps if $\delta = 0.05$) and delayed recurrence for 2 cells (roughly 5000 timesteps)

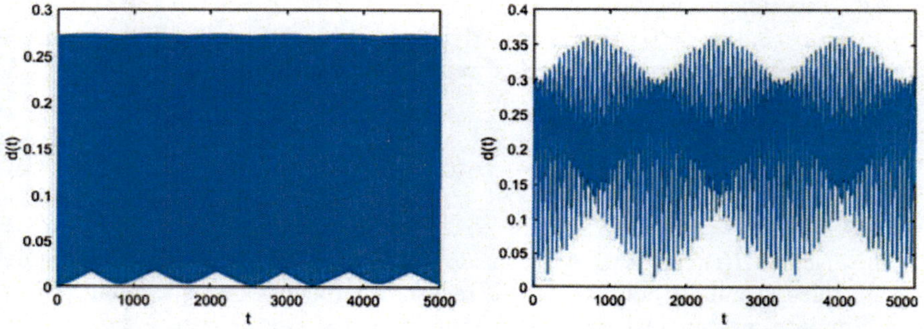

Fig. 7 The symmetric 1 : 2 : 3 case $m_1 = m_3$; the normal form is integrable, we have strong recurrence. Left one cell, with $\delta = 0.01$ roughly 800 timestep; right 2 cells with $\delta = 0.05$ roughly 1600 timesteps

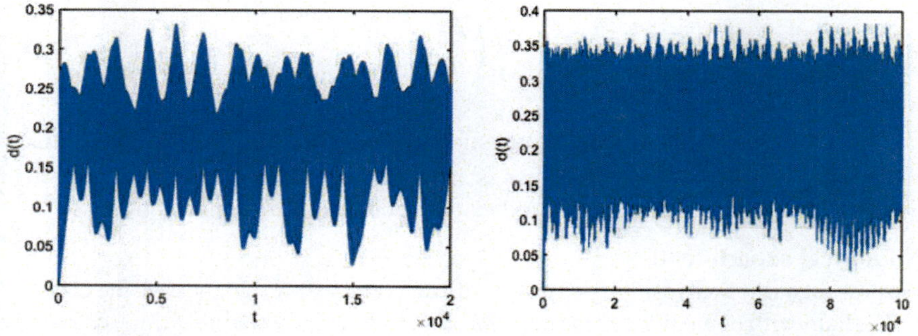

Fig. 8 The chaotic 1 : 2 : 3 cell-chain with already delayed recurrence in one cell; with $\delta = 0.05$ left 15,000 timesteps; right for 2 cells we have to integrate nearly 90,000 timesteps

As expected the recurrence times increase when adding one cell but most dramatically in the chaotic case. The inverse masses for Fig. 7 are $a_1 = 0.0357143$, $a_2 = 0.126804$, a_3=0.0357143, a_4=0.301767 (symmetric 1 : 2 : 3 resonance with $m_1 = m_3$) and for Fig. 8 a_1=0.00510292, a_2=0.117265, $a_3 = 0.0854008$, $a_4 = 0.292231$ (chaotic 1 : 2 : 3 resonance).

In all these recurrence experiments with for instance $\delta = 0.1$ or $\delta = 0.05$ the recurrence times are definitely lower than the corresponding upper bound L given by (17).

The numerics used MATLAB ode 78 with abs and rel error e^{-15}.

Acknowledgements Comments on earlier versions of this paper by Tassos Bountis and Roelof Bruggeman are gratefully acknowledged.

References

1. T. Bountis, H. Skokos, *Complex Hamiltonian Dynamics* (Springer, 2012)
2. H.W. Broer, F. Takens, *Dynamical Systems and Chaos*. Applied Mathematical Sciences, vol. 172 (Springer, 2011)
3. R.W. Bruggeman, F. Verhulst, Dynamics of a chain with four particles and nearest-neighbor interaction, in *Recent Trends in Applied Nonlinear Mechanics and Physics*, ed. by M. Belhaq, CSNDD, 2016 (Springer, 2018), pp. 103–120. https://doi.org/10.1007/978-3-319-63937-6-6
4. R. Bruggeman, F. Verhulst, The inhomogenous Fermi-Pasta-Ulam chain. Acta Appl. Math. **152**, 111–145 (2017)
5. R. Bruggeman, F. Verhulst, Near-integrability and recurrence in FPU chains with alternating masses. J. Nonlinear Sci. **29**, 183–206 (2019). https://doi.org/10.1007/s00332-018-9482-x
6. D.K. Campbell, P. Rosenau, G.M. Zaslavsky (eds.), The Fermi-Pasta-Ulam problem. The first 50 years. Chaos, Focus **15** (2005)
7. G.M. Chechin, V.P. Sakhnenko, Interaction between normal modes in nonlinear dynamical systems with discrete symmetry. Exact results. Phys. D: Nonlinear Phenom. **117**, 43–76 (1998)
8. G.M. Chechin, N.V. Novikova, A.A. Abramenko, Bushes of vibrational normal modes for Fermi-Pasta-Ulam chains. Phys. D: Nonlinear Phenom. **166**, 208–238 (2002)
9. G.M. Chechin, D.S. Ryabov, K.G. Zhukov, Stability of low-dimensional bushes of vibrational modes in the Fermi-Pasta-Ulam chains. Phys. D: Nonlinear Phenom. **203**, 121–166 (2005)
10. H. Christodoulidi, Ch. Efthymiopoulos, T. Bountis, Energy localization on q-tori, long-term stability, and the interpretation of Fermi-Pasta-Ulam recurrences. Phys. Rev. E **81**, 6210 (2010)
11. O. Christov, Non-integrability of first order resonances in Hamiltonian systems in three degrees of freedom. Celest. Mech. Dyn. Astr. **112**, 149–167 (2012)
12. R.L. Devaney, Homoclinic orbits in Hamiltonian systems. J. Diff. Eqs. **21**, 431–438 (1976)
13. K. Efstathiou, *Metamorphoses of Hamiltonian Systems with Symmetries*. Lecture Notes in Mathematics, vol. 1864 (Springer, 2005)
14. E. Fermi, J. Pasta, S. Ulam, Los Alamos report LA-1940, in *Collected Papers of Enrico Fermi*, vol. 2 (1955), pp. 977–988
15. J. Ford, The Fermi-Pasta-Ulam problem: paradox turns discovery. Phys. Rep. **213**, 271–310 (1992)
16. G. Galavotti (ed.), *The Fermi-Pasta-Ulam Problem: A Status Report*. Lecture Notes in Physics (Springer, 2008)
17. L. Galgani, A. Giorgilli, A. Martinoli, S. Vanzini, On the problem of energy partition for large systems of the Fermi-Pasta-Ulam type: analytical and numerical estimates. Phys. D: Nonlinear Phenom. **59**, 334–348 (1992)

18. H. Hanßmann, *Local and Semi-local Bifurcations in Hamiltonian Dynamical Systems*. Lecture Notes in Mathematics, vol. 1893 (Springer, 2007)
19. H. Hanßmann, R. Mazrooei-Sebdani, F. Verhulst, The 1 : 2 : 4 resonance in a particle chain. Indag. Math. (2020). https://doi.org/10.1016/j.indag.2020.06.003
20. P.J. Holmes, J.E. Marsden, J. Scheurle, Exponentially small splittings of separatrices with application to KAM theory and degenerate bifurcations. Contemp. Math. **81**, 213–244 (1988)
21. I. Hoveijn, F. Verhulst, Chaos in the 1 : 2 : 3-Hamiltonian normal form. Phys. D: Nonlinear Phenom. **44**, 397–406 (1990)
22. E.A. Jackson, *Perspectives of Nonlinear Dynamics*, 2 vols. (Cambridge University Press, Cambridge, 1991)
23. T. Nishida, A note on an existence of conditionally periodic oscillation in a one-dimensional anharmonic lattice. Mem. Fac. Eng. Univ. Kyoto **33**, 27–34 (1971)
24. H. Poincaré, *Les Méthodes Nouvelles de la Mécanique Céleste*, 3 vols. (Gauthier-Villars, Paris, 1892, 1893, 1899)
25. B. Rink, F. Verhulst, Near-integrability of periodic FPU-chains. Phys. A: Stat. Mech. Appl. **285**, 467–482 (2000)
26. B. Rink, Symmetry and resonance in periodic FPU-chains. Comm. Math. Phys. **218**, 665–685 (2001)
27. D.L. Rod, R.C. Churchill, A guide to the Hénon-Heiles Hamiltonian, in *Progress in Singularities and Dynamical Systems*, ed. by S.N. Pnevmatikos (Elsevier, 1985), pp. 385–395
28. J.A. Sanders, F. Verhulst, J. Murdock, *Averaging Methods in Nonlinear Dynamical Systems*, 2nd ed., Applied Mathematical Sciences, vol. 59 (Springer, New York, etc., 2007)
29. F. Verhulst, *Nonlinear Differential Equations and Dynamical Systems*, 2nd edn. (Springer, New York, etc., 2000)
30. F. Verhulst, Near-integrability and recurrence in FPU cells. Int. J. Bif. Chaos **26**(14) (2016). https://doi.org/10.1142/S0218127416502308
31. F. Verhulst, Linear versus nonlinear stability in Hamiltonian systems, in *Recent Trends in Applied Nonlinear Mechanics and Physics, Proceedings in Physics 199*, ed. by M. Belhaq (Springer, 2018). https://doi.org/10.1007/978-3-319-63937-6-6
32. G.M. Zaslavsky, *The Physics of Chaos in Hamiltonian Systems*, 2nd extended ed. (Imperial College Press, 2007)

Chaotic Mixing Experiments at High Temperature: Towards Unravelling a Large Magmatic Province

Caio M. Vicentini, Cristina P. de Campos, Werner Ertel-Ingrisch, Diego Perugini, Leila S. Marques, and Donald B. Dingwell

Abstract The Paraná-Etendeka Magmatic Province (PEMP) is the largest outpour of magma on the Earth surface during lower Cretaceous times (~133 Ma). Basalts (SiO_2~50%, both high-Ti and low-Ti members) predominate over an estimated volume of $7 \cdot 10^5$ km^3. However, *ca.* 2.5% of the volcanic products, are chemically more evolved (SiO_2>63%). Their genesis is still under debate. This work aims a first attempt to experimentally reproduce the impact of underplating basaltic melt into a pre-existing continental crust. Mixing dynamics is thought to greatly influence the formation/contamination conditions of the high-Ti acid member (Chapecó-type). We used a chaotic mixing protocol (Journal Bearing System) at 1,350 °C and following end-members: KS-700 basalt (high-Ti Pitanga-type from PEMP; $\eta_{1350} = 8.78$ Pa.s; $\rho_{1350} = 2.47$ g/cm^3) and LMC-027 granite (syenogranite from Capão Bonito Stock; $\eta_{1350} = 1.22 \cdot 10^5$ Pa s; $\rho_{1350} = 2.29$ g/cm^3). Homogenized glasses from the basalt and the granitic basement were used in an 80/20 proportion, respectively. The experiment was performed during 212 min in total, i.e., two periods of: (i) two clockwise rotations of outer cylinder (35 min); (ii) six anticlockwise rotations of inner cylinder (18 min). The independent and non-simultaneous movements of the two cylinders guaranteed the chaotic flow. The obtained mixed glass was cut in slices of 3 mm perpendicular to the rotation axes and two of the sections reproduced Poincaré patterns, which are theoretically expected to be resulted for this kind of dynamic mixing. With the development of stretching and folding processes, chaotic trajectories are distributed off-centered as lamellar lens-like structures in the mixed system. Vortex structures are comparable to those produced by mixing Fe-free silicate melts, however with much higher fractal dimension (D_f). These sections were preliminarily analyzed for

C. M. Vicentini (✉) · C. P. de Campos · W. Ertel-Ingrisch · D. B. Dingwell
Department of Earth and Environmental Sciences, University of Munich—DEES/LMU, Munich, Germany
e-mail: caio.vicentini@usp.br

C. M. Vicentini · L. S. Marques
Instituto de Astronomia, Geofísica e Ciências Atmosféricas—IAG/USP, São Paulo, Brazil

D. Perugini
Department of Physics and Geology, University of Perugia, Perugia, Italy

C. H. Skiadas et al. (eds.), *13th Chaotic Modeling and Simulation International Conference*, Springer Proceedings in Complexity,
https://doi.org/10.1007/978-3-030-70795-8_72

the changes in morphology by comparing the D_f using binary images (ImageJ software). Obtained D_f's (≈1.79) are close to those from similar experiments with natural melts, although widespread orbicular structures found along all basaltic morphological domains are thought to enhance the complexity. Further experiments varying the granitic end-member are planned. Raman, microprobe and LA-ICP-MS investigations will be performed to compare chemical behavior of major/minor oxides and trace elements. Furthermore, numerical simulations will follow.

Keywords Magma mixing · Chaotic dynamics · Journal bearing system at high-T · Paraná-Etendeka Magmatic Province · Chapecó-acidic type

1 Introduction

The Paraná-Etendeka Magmatic Province (PEMP—Fig. 1) is the second largest LIP (large igneous province) that occurs on the Earth's surface with over a $1 \cdot 10^6$ km^2 of area and 800,000 km^3 (Frank et al. [1]) of expelled material. It took place about 133 My ago (Piccirillo and Melfi [2]; Janasi et al. [3]; Marzoli et al. [4]). Basalts (rocks with $SiO_2 \approx 50$ wt % in their composition) are the most common members found along the province and they are mainly divided in two groups: one with low contents of titanium ($TiO_2 < 2$ wt%; LTi), and other one with high contents ($TiO_2 \geq 2$ wt%; HTi). HTi basalts also differ from LTi due to the higher contents of P_2O_5, large-ion lithophile elements (such as Sr, Zr, Ba) and rare earth elements (as La, Ce and Lu) in general (Piccirillo and Melfi [2]; Marques et al. [5]; Peate et al. [6]). However, approximately 2.5% of the volcanic products are chemically more evolved ($SiO_2 \geq 63$ wt%), named rhyolites, and they also can be divided in two further-subgroups, according to the amount of TiO_2. Each subgroup also presents a geochemical behaviour similar to those found in the basalts (Piccirillo and Melfi [2]; Nardy et al. [7]). Despite the genesis of the rhyolites from PEMP being still under debate, the most accepted ideas consider these evolved members as products of basaltic evolution and interaction with other crustal rocks. Therefore, each group of basaltic rocks (LTi and HTi) would have originated rhyolites with respective low and high contents of titanium.

Figure 2 shows the underplating model used for Piccirillo and Melfi [2] to explain the genesis of Chapecó-type, where a HTi basaltic body is trapped under the crust, which is mainly composed by granitic rocks. At some point, the basalt has been remelted and interacted with the granitic basement around. Petrogenetic models tend to consider the magma chambers as static, so that the evolution of magmas is basically linear in time. However, in the last years, scientists began to compare the structures observed in volcanic environmental with the fractal geometry (De Rosa et al. [9], Perugini et al. [10], Perugini and Poli [11]), which is a characteristic of chaotic dynamic processes, and the complex dynamics inside the magma chamber started to be considered. Since then, fractal analysis and numerical simulations have been used as a tool to quantify such processes. Therefore, this work aims to an experimental

Fig. 1 Map from Paraná-Etendeka Magmatic Province—PEMP. Africa is rotated for present position of South America. Insert: TC = Tristan da Cunha island. Modified after Nardy et al. [7], Machado et al. [8] and references therein

approach of the genesis of high-Ti acid member of PEMP (Chapecó-type). It is a first attempt to reproduce the impact of underplating basaltic melt into a pre-existing continental crust and investigate the influence of chaotic mixing dynamics in their formation/contamination processes.

In order to investigate the genesis of Chapecó-acidic type, we chose an experimental approach, which is based on previously obtained results from chaotic mixing experiments with geological materials at high temperatures. Therefore, the development and construction of the first experimental apparatus succeeded, being able to reproduce the chaotic dynamics according a specific protocol of motion at high temperature (De Campos et al. [12]). It mimics the processes that would govern the magma chambers, in where the mixing process occurs. Results with synthetic

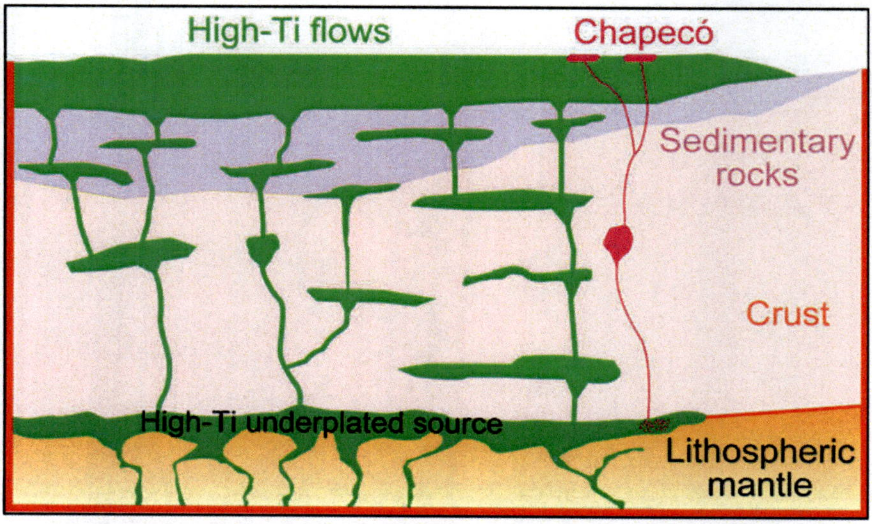

Fig. 2 Model for the underplating scenario of Chapecó-type genesis

end-members showed similar patterns comparing to those theoretically expected (e.g.: Poincaré sections). Using the same device, Perugini et al. [13] generated a hybrid sample from synthetic end-members and correlated the chemical analysis with numerical simulations considering that the particles have been submitted to a chaotic field motion. The comparison demonstrated that the patterns fit, meaning that a chaotic mixing process could explain what is observed in nature. Finally, Morgavi et al. [14] performed the same experiment using natural end-members and finding different geometrical patterns compared to those obtained from synthetic end-members. Furthermore, the chemical analysis showed that the elementary mobility is different to the linear behaviour expected.

2 Methods

The chaotic mixing device is based on the Journal Bearing System (JBS; Swanson and Ottino [15]) that is able to generate a chaotic flow using a certain protocol of motion. Our apparatus scheme is shown in Fig. 3. It is basically consisted by two independent motors. The lower motor moves a base where an Al_2O_3-rod (alumina: AL23) is fixed and allows positioning the crucible inside the hot spot of a high temperature furnace. The upper motor fixes an off-centered spindle also made from alumina and sheathed by a Pt foil in its extremity that will be positioned inside the sample. The geometry of JBS is determined basically by: (i) the ratio of the radii of the two cylinders, $r = R_{in}/R_{out} = 1/3$; (ii) the eccentricity ratio to the outer cylinder

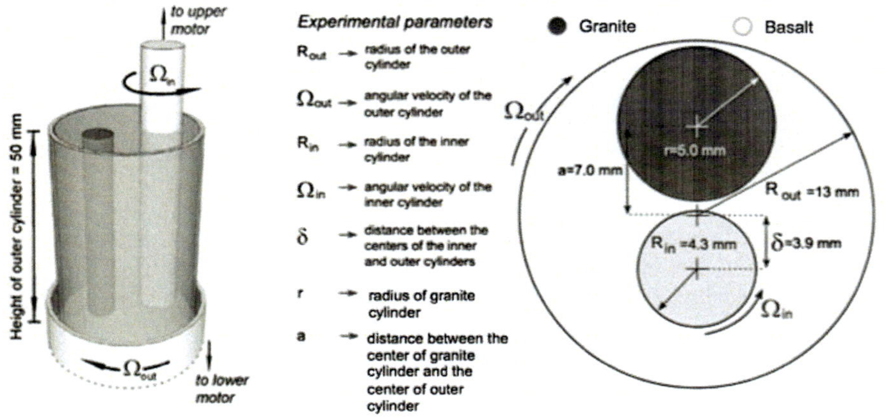

Fig. 3 Schematics of the experimental apparatus. Adapted from De Campos et al. [12] and Morgavi et al. [14]

$\varepsilon = \delta/R_{out} = 0.3$, where δ is the distance between the centers of the inner and outer cylinders (R_{in} and R_{out}). The system was designed to enable independent rotations of inner and outer cylinders at variable speeds, directions and stirring protocols and to place the apparatus in the hot spot region of a well-calibrated high-temperature oven. The motors for both (inner and outer) cylinder movements are controlled by a central mixing system, which enables independent control of rotation direction, rotation speed and number of rotations (De Campos et al. [12]).

The protocol used to generate the chaotic streamlines is: (i) two clockwise rotations of outer cylinder (35 min); (ii) six anticlockwise rotations of inner cylinder (18 min); (iii) two clockwise rotations of outer cylinder (35 min); (iv) six anticlockwise rotations of inner cylinder (18 min). It totalizes 212 min of non-simultaneous and independent movements, which guarantees the chaotic flow. This choice was made based on Morgavi et al. [14] that performed the experiment during ½, 1 and 2 protocols and demonstrated that the mixing efficiency at this time is sufficient to preserve the structures.

The end-members used in our experiment were: (i) a HTi *Pitanga*-type basalt from PEMP (mafic); (ii) a basement syenogranite from *Capão Bonito* Stock (felsic). The experimental temperature was 1,350 °C and it was chosen taking in to account the lowest value that still allows the interaction between the melts, once at low temperatures the viscosity increases and in some point no interaction would be possible. Estimated temperatures in nature where this interaction happens are lower than that, however there are other factors not considered in the experiment such as volatiles content and pressure that may greatly change the viscosity. In our case, melts are free of volatiles. At 1,350 °C, measured viscosity and calculated density for basalt was $\eta_{1350} = 8.78$ Pa.s; $\rho_{1350} = 2.47$ g/cm^3, while to rhyolite (melted granite) was $\eta_{1350} = 1.22{\cdot}10^5$ Pa s; $\rho_{1350} = 2.29$ g/cm^3, which leads to a viscosity ratio of approximately $1.4{\cdot}10^4$. For the preparation, the end-members were pulverized and heated to produce a glass. Melts have been homogenized with a viscometer following the procedure described by Morgavi et al. [14]. The dry-bubble-free glasses were inserted into a Pt_{80}–Rh_{20} crucible with 25 mm of diameter, 50 mm of length and 1 mm thick in a proportion of 80% of basalt and 20% of rhyolite. After positioning the crucible in the device, the oven is turned on and it heats up during approximately 40 min to all parts accommodate. During this time the temperature increases gradually until 1,350 °C is reached and then the mixing protocol starts. At the end, the chaotic mixing device is turned off and hybrid glass is cooled at room temperature. After cooling, the sample was drilled out. The obtained core was cut perpendicular to the rotation axis into several pieces with approximately 3 mm length for further analysis. It is important to notice that De Campos et al. [12], Perugini et al. [13] and Morgavi et al. [14] used the inverse proportion (i.e., 80% of felsic end-member with 20% of mafic inside) due to the fact the higher viscous material would stabilize the less viscous one at high temperatures. In principal, it would avoid any movement of the small cylinder before the protocol starts, although during the experiment the interaction expected between the end-members and the resultant patterns would not have been affected.

3 Results and Discussion

The initial cooling rate observed was approximately 86.5 °C/min, which is concordant with previous reports (De Campos et al. [12]; Morgavi et al. [14]). Two sections of the sample core presented morphological aspects theoretically expected and similar to Poincaré sections. These have been polished for microscopic further geochemical analysis as shown in Fig. 4.

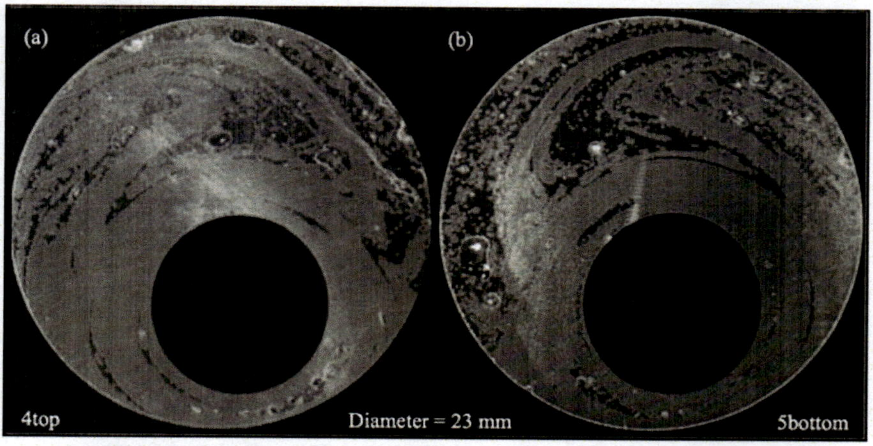

Fig. 4 Representative sections of chaotic mixing patterns after one protocol. Dark grey = felsic end-member; light grey = mafic end-member

The sections exhibit complex patterns of non-centered lamellar structures consisting of alternation of lens-like filaments in the mixed system. Around the inner cylinder these filaments are more or less concentric. At the top part it is observed a complex morphology composed by filaments strongly stretched and folded defining a lobate and banded structure pointing to the right side (4 top; 5 bottom is an adjacent face, therefore it points to opposite direction). The left side shows the felsic end-member thinning and forming a tail that almost connects to the concentric part described before. Comparing the morphology with those obtained for synthetic sample experiments the same vortex structures are observed (Fig. 5).

It is important to notice that synthetic samples prepared by De Campos et al. [12] and Perugini et al. [13] are iron free (composition: SiO_2, Al_2O_3, MgO, CaO, Na_2O and K_2O) and their experiments were executed at 1,400 °C, which facilitated the mixing process. At this temperature the authors calculated a viscosity of $1.55 \cdot 10^3$

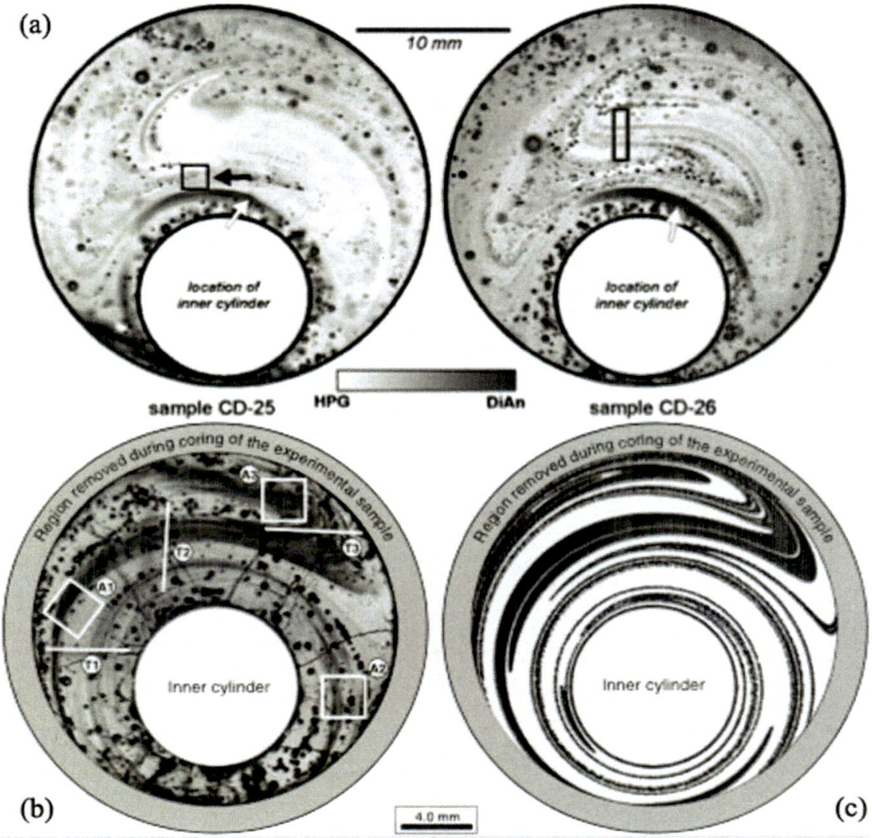

Fig. 5 Previous results by: **a** De Campos et al. [12]; (**b**, **c**) Perugini et al. [13]. Dark grey = mafic end-member; light grey = felsic end-member

Pa.s to felsic and 1.4 Pa.s to mafic end-member, corresponding to a viscosity ratio of ca. $1.1 \cdot 10^3$, and densities of 2.26 and 2.52 g/cm^3 respectively. The same morphological response to mixtures with different chemical and physical properties could indicate that this mechanism is sufficiently robust. Perugini et al. [13] also numerically simulated the trajectory of the particles (Fig. 5c) and, despite the differences due to the initial proportion (felsic = 65%; mafic = 35%), similar structures appears in the simulation such as the off-centered lamellar structures, the lobate and banded portion at the top and the concentric filaments around the inner cylinder. The agreement between experiment and numerical results corroborates to the importance of this mechanism on the genesis of this rocks.

Nonetheless, the comparison with natural material from Morgavi et al. [14] permits to observe the discrepant patterns (Fig. 6). It seems the principal mafic body dropped inside the felsic one at the right part, while the upper region suffered more influence of stretching and folding processes. Measured viscosity of the rhyolite was $5.6 \cdot 10^4$ Pa.s and of basalt was 7.2 Pa.s, corresponding to a viscosity ratio of ca. $7.8 \cdot 10^3$, and the calculated densities were 2.33 and 2.98 g/cm^3 respectively. The discrepancies between the final morphology of natural and synthetic materials were initially thought to appear due to the presence of iron, that plays an important role in the melt structure, especially because of its two possible valences (Fe^{2+} and Fe^{3+}). Nevertheless, there are other factors that apparently influence the melt interaction taking into account the physical parameters of two experiments using natural samples show no relevant differences.

One of the most interesting features observed in the mixed glass presented in the Fig. 7 are the orbicular structures along the two end-members. They are more evident on the felsic portion though. It is notice such feature was not reported on previous

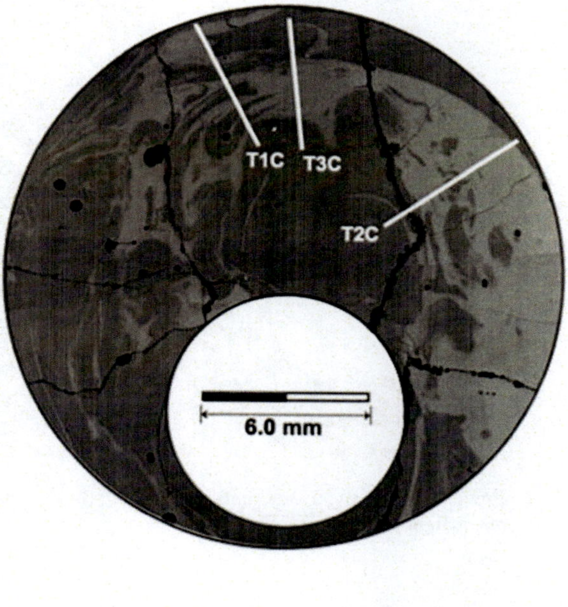

Fig. 6 Previous result by Morgavi et al. [14]. Dark grey = rhyolite; light grey = basalt

Fig. 7 4 top: 8 × zoom from Carl Zeiss Discovery V8.0 microscope. Object field = 2.9 mm. Dark grey = felsic end-member; light grey = mafic end-member

works using synthetic or natural materials (De Campos et al. [12]; Perugini et al. [13]; Morgavi et al. [14]).

In order to estimate the fractal dimension of both sections its photos were transformed in binary images using the *ImageJ* software developed by Rasband [16] (Fig. 8) and the fractal coefficient D_f was calculate using the box-counting method, developed by Mandelbrot [17]. D_f values are substantially sensitive to the input (i.e., the image) once even some blurred portions and anomalous pixels can be interpreted as interest features by the software. Therefore, a simple protocol was adopted to provide the best estimative. It implicated in changing its brightness to highlight the contrast between the end-members, transforming in binary images and using tools (*Erode* and *Dilate*) to remove some anomalous pixels and to highlight the relevant ones.

After this procedure the calculated D_f 's were 1.78 (4top) and 1.80 (5bottom). The values are consistently higher than $D_f = 0.91$ obtained by De Campos et al. [12], for a mixture composed by synthetic samples, and similar to other results obtained to natural occurrences that takes place in the intervals: $1.01 < D_f < 1.84$ (De Rosa et al. [9]); $1.67 < D_f < 1.92$ (Perugini et al. [10]); $1.39 < D_f < 1.62$ (Guimarães et al. [18]). D_f's reported are consider an initial estimative indicating the experiment reproduced the patterns expected by a natural chaotic dynamic process. More precise calculations will be further obtained improving the quality of digitalized sections.

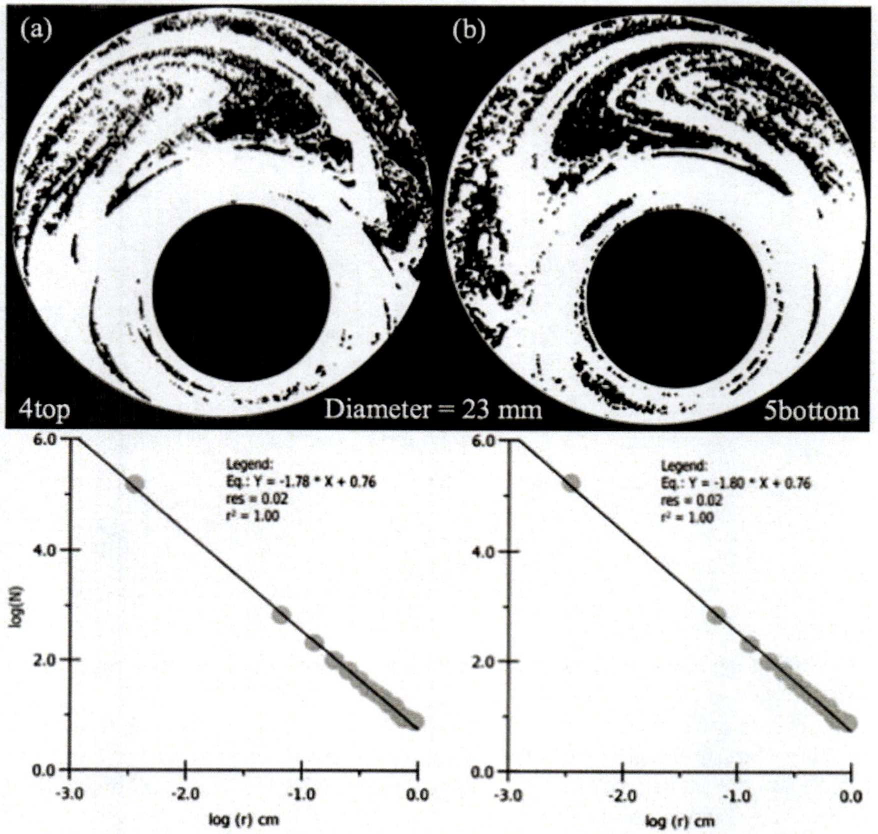

Fig. 8 Binary images used to calculate D_f (*ImageJ*) and respective curves built by box-counting method

4 Conclusions

Chaotic mixing processes are thought to be a sophisticated explanation for natural melts interaction that have to be approached carefully. Therefore, at this stage is possible to summarize some conclusions:

1. The experiment showed very good primarily results. Compared to the morphological features previously obtained by Morgavi et al. [14], results from this study are better than expected, with the generation of Poincaré sections. It is evident this line of investigation should be maintained;
2. According to the results, inserting a cylinder of a more viscous material (rhyolite) inside a less viscous material (basalt) did not affect the experimental stability at high temperatures (1,350 °C). The heating procedure adopted is thought to be vital to this fact and allows to invert the end-members proportion;

3. The patterns produced by chaotic dynamics are in concordance with those theoretically expected. This was noticed on the appearance of similar morphological elements in comparison with numerical simulations performed under similar conditions (Perugini et al. [13]);
4. Orbicular structures were identified for the first time on this type of experiment;
5. The fractal coefficients ($D_f \approx 1.79 \pm 0.01$) are in agreement with those values reported. Based on geological cases previous reported (e.g.: De Rosa et al. [9], Perugini et al. [10] and Guimarães et al. [18]) this first estimative could stand that the chaotic mixing has a role on petrogenesis of Chapecó-acidic type of PEMP.

Following investigation steps are still under development:

- Geochemical behavior of major/minor oxides and trace elements (microprobe and Laser Ablation-ICP-MS analysis);
- Numerical simulations to compare the theoretical response of the system;
- Raman analysis of the orbicular structures to study the nature and changes of the glass structure;
- Further chaotic mixing experiments varying the rhyolitic end-member in order to test other candidates as contaminants for the Chapecó-type genesis. As a consequence, the impact of chemical and physical variations on morphological aspects under the same experimental conditions can be evaluated (i.e., temperature and duration).

Acknowledgements This study was financed in part by the Coordenação de Aperfeiçoamento de Pessoal de Nível Superior—Brasil (CAPES)—process: 88887.363370/2019-00 (CAPES-DAAD project 57389574 "Large Igneous Province Paraná-Etendeka").

References

1. H.T. Frank, M.E.B. Gomes, M.L.L. Formoso, Review of the areal extent and the volume of the Serra Geral Formation—Paraná Basin South. Pesquisas Em Geociências **36**(1), 49–57 (2009)
2. E.M. Piccirillo, A.J. Melfi, in *The Mesozoic Flood Volcanism of the Paraná Basin: Petrogenetic and Geophysical Aspects*, ed. by University of São Paulo (IAG, São Paulo, Brazil, 1988), 600p
3. V.A. Janasi, V.A. Freitas, L.H. Heaman, The onset of flood basalt volcanism, Northern Paraná Basin, Brazil: A precise U-Pb baddeleyite/zircon age for a Chapecó-type dacite. Earth. Plan. Sci. Letters **302**, 147–153 (2011)
4. A. Marzoli, L. Melluso, V. Morra, P.R. Renne, I. Sgrosso, M. D'Antonio, L. Duarte Morais, E.A.A. Morais, G. Ricci, Geochronology and petrology of Cretaceous basaltic magmatism in the Kwanza basin (western Angola), and relationships with the Paraná-Etendeka continental flood basalt province. J. Geody. **28**, 341–356 (1999)
5. L.S. Marques, E.M. Piccirilo, A.J. Melfi, P. Comin-Chiaramonti, G. Bellieni, Distribuição de terras raras e outros elementos traços em basaltos da Bacia do Paraná. Geochim. Brasil. **3**, 33–50 (1989)
6. D.W. Peate, C.J. Hawkesworth, M.S.M. Mantovani, Chemical stratigraphy of the Paraná lavas, South America: classification of magma types and their spatial distribution. Bull. Volcanol. **55**, 119–139 (1992)

7. A.J.R. Nardy, F.B. Machado, M.A.F. Oliveira, As rochas vulcânicas mesozoicas ácidas da Bacia do Paraná: litoestratigrafia e considerações geoquímico-estratigráficas. Rev. Bras. De Geoc. **38**(1), 178–195 (2008)
8. F.B. Machado, E.R.V. Rocha-Júnior, L.S. Marques, A.J.R. Nardy, L.V. Zezzo, N.S. Marteleto, Geochemistry of the Northern Paraná Continental Flood Basalt (PCFB) Province: implications for regional chemostratigraphy. Braz. J. Geol. **48**, 2 (2018)
9. R. De Rosa, P. Donato, G. Ventura, Fractal analysis of mingled/mixed magmas: an example from the Upper Pollara eruption (Salina Island, southern Tyrrhenian Sea, Italy). Lithos **65**, 299–311 (2002)
10. D. Perugini, G. Poli, R. Mazzuoli, Chaotic advection, fractals and diffusion during mixing of magmas: evidence from lava flows. J. Volcanol. Geotherm. Res. **124**, 255–279 (2003)
11. D. Perugini, G. Poli, Analysis and numerical simulation of chaotic advection and chemical diffusion during magma mixing: petrological implications. Lithos **78**, 43–66 (2004)
12. C.P. de Campos, D. Perugini, W. Ertel-Ingrisch, D.B. Dingwell, G. Poli, Enhancement of magma mixing efficiency by chaotic dynamics: an experimental study. Contrib. to Mineral. Petrol. **161**, 863–881 (2011)
13. D. Perugini, C.P. de Campos, W. Ertel-Ingrisch, D.B. Dingwell, The space and time complexity of chaotic mixing of silicate melts: implications for igneous petrology. Lithos **155**, 326–340 (2012)
14. D. Morgavi, D. Perugini, C.P. de Campos, W. Ertel-Ingrisch, D. Dingwell, Time evolution of chemical exchanges during mixing of rhyolitic and basaltic melts. Contrib. Miner. Petrol. **166**, 615–638 (2013)
15. P.D. Swanson, J.M. Ottino, A comparative computational and experimental study of chaotic mixing of viscous fluids. J. Fluid. Mech. **213**, 227–249 (1990)
16. W.S. Rasband, ImageJ. U. S. National Institutes of Health, 2016. Software free available on https://rsb.info.nih.gov/nih-image
17. B.B. Mandelbrot, *The Fractal Geometry of Nature* (W.H. Freeman, New York, 1982).
18. L.F. Guimarães, C.P. de Campos, V.A. Janasi, E.F. Lima, D.B. Dingwell, Flow and fragmentation patterns in the silicic feeder system and related deposits in the Paraná-Etendeka Magmatic Province, São Marcos, South Brazil. J. Volcanol. Geoth. Res. **358**, 149–164 (2018)

Self-similar Chaotic Processes in Dynamical Systems of Nonlinear Stochastic Maps

George Vostrov, Andrii Khrinenko, and Roman Opiata

Abstract The subject of research is cyclic processes that arise in recursive nonlinear maps under the influence of external probabilistic or stochastic factors. A wide range of nonlinear maps are considered, both continuous and containing discontinuity points. It is proved that any cyclic trajectory of considerable length always contains fragments with chaos properties, which are defined as properties of maps as well as properties of numbers from the domain of their definition. The laws are established for the transition of a dynamic system from one chaotic state to another chaotic state, depending on the properties of nonlinear maps.

Keywords Nonlinear maps · Dynamical systems · Self-similarity

1 Introduction

Emergence of a book by famous authors with the name "Many-sided Chaos" is not accidental [1]. There is no doubt that the concept of "chaos" is associated with dynamic processes taking place in a certain space. In the literature [2], deterministic and stochastic chaos is defined as two general forms. However, the systematic use of the term "chaos" often leads to an incorrect analysis of its properties [3]. This fact is due to the fact that the nature of all chaotic processes is diverse and in some sense has many faces. Despite this, chaotic processes of completely different nature are often subject to general laws [4]. The study of such universal, in a certain sense, laws of chaotic processes of different nature is an actual and interesting problem of modern concept of chaos. Despite the fact that the deterministic and stochastic forms of chaos are very different from the point of view on mechanisms of its origin, from the point of view of their mathematical models it can be stated about certain patterns of interweaving of their properties in one frame.

G. Vostrov · A. Khrinenko (✉) · R. Opiata
Department of Applied Mathematics and Information Technologies, Odessa National Polytechnic University, Odessa, Ukraine
e-mail: khrinenko@stud.opu.ua

C. H. Skiadas et al. (eds.), *13th Chaotic Modeling and Simulation International Conference*, Springer Proceedings in Complexity,
https://doi.org/10.1007/978-3-030-70795-8_73

The stochastic model of chaos is certainly based on the concept of randomness. Any random variable ξ, in the simplest case with a uniform probability distribution law over a finite interval (a, b), with a large number of tests, potentially allows to get a chaotic sequence of arbitrarily large length. Unfortunately, this is only a theoretical possibility. At present moment, random generators with a uniform distribution law are absent, and, consequently, with any other probability distribution law. Moreover, as shown in [5], none of the mathematical models of algorithmic computability makes it possible to obtain a binary sequence that could be called random in the exact mathematical sense. The same result was obtained in [6] on the basis of another mathematical analysis of algorithms for constructing binary random sequences. It follows that absolute non-deterministic chaos, i.e. stochastic chaos at this stage in the development of research cannot be obtained by physical methods.

Assuming that this is "potentially" possible, the question arises as to what conditions a numerical sequence must satisfy in order to be considered as random. This question also applies to its binary representation. It is proved that absolutely random, and therefore chaotic sequence should not have any internal predicted structures. When solving problems of dynamical systems modeling [7], analyzing and processing data, and especially in the case of Big Data, when analyzing random processes [8], non-stationary time series and many other areas of pure and applied mathematics, random number generators with a given probability distribution law are needed. It follows that algorithmic models of random number generators are needed based on the theory of recursive functions, which allow us to model "absolute chaos" with maximum approximation to a given probability distribution law.

At present, it is not possible to construct random number generators based on the theory of recursive functions that could simultaneously simulate "absolute chaos" with a given distribution law. But the condition on the distribution law already violates the requirement that there are no internal properties in the generated sequences that reduce the level of randomness to some extent. It is necessary to construct pseudo-random number generators [9], but at the same time one has to admit the existence in the generated sequences of numbers of internal controlled patterns that minimize distortion of the results of modeling limited chaos. This means that we have to move to deterministic chaos [10]. The basis of such mathematical constructions is the theory of recursive functions and the theory of effective computability associated with it [10].

Such class of recursive functions is very extensive. It includes features of high computational complexity. Uncontrolled use of the theory of recursive functions can lead to a significant limitation on the speed of obtaining pseudorandom numbers. Therefore, we have to choose such classes of recursive functions that play a dual role. On the one hand, they are models of dynamical systems of a certain class, and on the other hand, their cyclic fixe points make it possible to generate their trajectories in such a way that they can be used to construct pseudorandom number generators. To solve this problem, it is could be promising to use maps on the set of integers or rational numbers, which on the one hand are models of deterministic dynamical systems, and on the other hand, the recursive sequences generate with these maps can be used to construct pseudorandom number generators.

In this regard, it seems perspective to start such studies with maps of the "tent" class in symmetric and asymmetric form, logistic map and algebraic maps that are based on the theory of residues modulo prime. The use of primes is interesting since in modern state of mathematical science there are still major blind spots in knowledge of its nature, although their logarithmic distribution law is established.

$$\pi(x) = \int_{2}^{x} \frac{dt}{\ln t} + O\left(xe^{-c\sqrt{\ln x}}\right),$$

where $\pi(x)$ is the number of primes $p \le x$, however, the exact behavior of the function $f(x) = \pi(x) - Li(x)$, where $Li(x) = \int_{2}^{x} \frac{dt}{\ln t}$ not studied. There is an assumption that the function $f(x)$ has a fractal structure. Until now the dynamics of changes in the distance between primes is unknown in number theory. Estimation of increase in the distance between primes based on the following expression:

$$p_{n+1} - p_n \ge \frac{\ln x \ln \ln x \ln \ln \ln \ln x}{\ln \ln \ln x},$$

shows that the distance between primes is continuously increasing. However, this expression does not at all follow how the properties of adjacent primes differ. Despite their simplicity, the properties of $p_{n+1} - 1$ and $p_n - 1$ can be very different, and this must be taken into account in the modern theory of the discrete logarithm, modern cryptography and in the modern theory of pseudorandom number generators.

The answers to these questions can be obtained to a certain extent by studying dynamic chaotic processes modeled by simple maps abovementioned classes at their cyclic fixed points determined by primes $p \in \mathcal{P}$. At the same time, it is important to choose such primes from the set of all primes $\mathcal{P}$ that allow obtaining of cyclic trajectories of large length and complex structure. At points given by primes $p \in \mathcal{P}$, rational numbers of the form q/p where p and q are primes allow to study the properties of dynamical systems determined on the given maps.

2 Self-similar Processes Inside the Trajectories of Nonlinear Chaotic Maps

Maps of the set of natural, integer, rational, real, complex numbers onto or into itself by means of nonlinear maps are always associated with the representation of such maps using recursive, primitively recursive, and therefore computable functions. Any such map defines a dynamic system, which is usually associated with some dynamic processes of a very different nature. As proven by well-known authors mathematical models of such processes can be built in arithmetic systems of various levels of complexity. The study of such mathematical models always includes an analysis of the trajectories of cyclic fixed points. As proved by Sharkovsky, the existence of a cycle of length three implies the existence of an iterative cycle in a dynamical system with a trajectory of any length, i.e. while the relation:

$$3 > 5 > 7 > 9 > \ldots > 2 \cdot 3 > 2 \cdot 5 > \ldots > 1.$$

It follows that there is an iterative cycle of a fixed point of any length. The existence of long cycles leads to the need to analyze their properties, since these properties are properties of a dynamical system. Information about the properties of iterable maps is important for control of dynamical systems, analysis of their properties in order to make decisions aimed at studying the behavior of systems under various conditions. As proven in many monographs, scientific articles [10, 12] studies of the properties of dynamical systems based on the analysis of their trajectories are associated with significant difficulties due to the chaotic properties of the trajectories. The term "chaos" is interpreted by different authors in very different ways. This fact is not paradoxical since the nature of dynamic systems differs significantly from dynamic processes in biology, medicine, and economics. However, it can be assumed that there are some properties that are common to any models of dynamical systems. These properties are related to the mathematical form of the system representation and the properties of numbers, which determine the conditions for the occurrence of a cyclic trajectory of a fixed point.

Let examine the class of algebraic dynamical systems, which is surprisingly connected with nonlinear dynamical systems and represented by very simple nonlinear maps. Consider the set of all primes $\mathcal{P}$ and the residue group (Z/pZ) which is a cyclic group associated with each prime $p \in \mathcal{P}$. Natural numbers a are usually called the primitive root of a prime p and, therefore, the group $(Z/pZ)^*$ if the next condition is satisfied:

$$\begin{cases} a^{p-1} \equiv 1(mod\ p) \\ a^{\frac{p-k}{k}} \not\equiv 1(mod\ p) \end{cases}, \tag{1}$$

for any divisor $k > 1$ of numbers $p - 1 = \prod_{i=1}^{n} p_i^{\alpha_i}$. These are necessary and sufficient conditions for verifying that a is the primitive root of p and it follows from Fermat's little theorem. The validation of these conditions is based on the calculation of a recursive sequence:

$$x_0 = 1, x_{n+1} = ax_n(mod\ p) \tag{2}$$

to those. until the condition $x_m = ax_{m-1}(mod\ p) = 1$ is satisfied at the m-th step of the iteration. If $m = p - 1$, then a is the antiderivative root of p. Otherwise, the number a is a generating element of the number p of some subgroup of the group (Z/pZ). And any number a is a classifier of the set of all primes $\mathcal{P}$ into classes:

$$\mathcal{P}(a, 1), \mathcal{P}(a, 2), \ldots, \mathcal{P}(a, i), \tag{3}$$

According to condition:

$$\mathcal{P}(a, i) = \{p|(p-1)card_a(p) = ind_a(p) = i\},$$

where $card_a(p)$ is the minimum recursion length (2) at which 1 modulo p is achieved, since $card_a(p)$ always divides $p-1$, then $ind_a(p)$ determines the number of adjacency classes of the subgroup of the group $(Z/pZ)^*$. Obviously, if $card_a(p) = p-1$, then a is the primitive root of the group $(Z/pZ)^*$, and if $card_a(p) < p-1$, then a is a generating element of the cyclic subgroup of this group and its order is $card_a(p)$, and the number of adjacency classes is $ind_a(p)$. Thus, the set $\mathcal{P}(a, 1)$ contains all primes p for which a is a primitive root, $\mathcal{P}(a, i)$ contains all primes p for which a is a generating element with index $ind_a(p) = i$. The infinity of the set of all primes $\mathcal{P}$ determines the existence of cyclic recursion (2) of arbitrarily large length for all classes and especially for the class $\mathcal{P}(a, 1)$. In this case, four classes of problems arise:

1. $x \equiv a^m (mod\ p)$—defined a, p, m, find x;
2. $c \equiv a^x (mod\ p)$—defined a, p, c, find x;
3. $c \equiv x^m (mod\ p)$—defined c, p, m, find x;
4. $c \equiv a^m (mod\ x)$—defined a, c, m, find $x \in \mathcal{P}$.

The first problem is solved relatively simply by the method of repeated squaring. The second problem is the discrete logarithm problem, which belongs to the class of problems of high complexity, and it is possible that to the class of algorithmically unsolvable problems. The third and fourth task is much more complicated since the literature does not describe attempts to solve them. Closely related to the solution of the second problem is modern cryptography and modern methods for constructing pseudorandom number generators [8]. And also along with the first problem these problems has relation to the problem of chaos analysis. For convenience, we consider the case when a is a primitive root $p \in \mathcal{P}$ and the number p has a larger order, for example, $p > 10^{200}$. Recursion (2) for any such type is a permutation of the set of numbers $\{2, 3, \ldots, p-2\}$, where the number $p-1$ is not included since $a^{p-1} \equiv 1(modp)$.

The prime p has $\varphi(p-1) = \prod_{i=1}^{m} p_i^{\alpha_i - 1}(p_i - 1)$ (Euler function) of primitive roots $(a_1, a_2, \ldots, a_{\varphi(p-1)}\}$ and for each of which there exists a permutation in which the order of numbers from 2 to $p-2$ differs from the order of any other primitive root. In order to solve all problems associated with the discrete logarithm, it is necessary to possess information at least some degree on the order of the placement of numbers in chaos that is generated by recursion (2) for each a_i from the set of all primitive roots.

Definition Let $p \in \mathcal{P}$ be a prime such that $p - 1 = \prod_{i=1}^{k} p_i^k$ and a is its primitive root, then in the trajectory of the recursive function (2) $x_{n+1} = ax_n(modp)$ and the index $n = kp_i$ for all $p_1, \ldots, p_k$ will be called basic indices. With respect to the set of indices of trajectories of recursion on the base a for a prime number $p \in \mathcal{P}$ and set of its primitive roots $\{a_1, \ldots, a_{q(p-1)}\}$ next theorem is true.

Theorem For any prime number $p \in \mathcal{P}$ such that $p - 1 = \prod_{i=1}^{k} p_i^{\alpha_i}$ and any primitive root of it, the set of all primitive roots $\{a_1, \ldots, a_{q(p-1)}\}$ at the base points of the trajectories (2) there cannot be numbers from the set of all its primitive roots.

The validity of the theorem follows from the fact that during operations at base points of primitive roots there cannot be created a conflict situation among them in accordance with Fermat's little theorem and set theory.

Thus, the structure of the trajectories is constructed to a certain extent, taking into account the self-similarity of the trajectories of all primitive roots of a given prime number. This theorem confirms that in deterministic chaos, the properties of the map functions and the properties of a number from the region of the trajectories of recursive fixed points to some extent affect the structure of chaos.

Consider the case of dynamical system maps that are defined by simple functions of the "tent" class [10]. Despite their simplicity, as shown in [12], their behavior, both of dynamical systems and of sources of chaos formation, is far from simple. Consider two types of maps of this class:

$$x_{n+1} = \left\{ 2x_n, \middle| x_n < \frac{1}{4}; \right. \tag{4}$$

$$x_{n+1} = \left\{ 1 - 2x_n, \middle| \frac{1}{2} > x_n \geq \frac{1}{4}. \right. \tag{5}$$

The graphic representation of these maps is elementary. Consider the behavior of these maps on a set of numbers of the form $x = 1/p$ under the condition $p \in \mathcal{P}$. This choice of the set of initial conditions is chosen due to two considerations. Firstly, the analysis of dynamic processes and the chaos accompanying them can be compared with the algebraic map of residues modulo prime (2), and secondly, the study of the properties of dynamic processes by elements of such a set of initial values allows us to study their dependence on the properties of primes without taking into account decomposition $p - 1 = \prod_{i=1}^{k} p_i^{\alpha_i}$ and the properties resulting from it. In addition, since such a rational number $0 < m/n < 1$ is expressed through decomposition into simple factors, knowing the laws of the influence of the $1/p$ properties on the dynamics and chaos features, we can switch to composite numbers of the form m/n.

Maps (4) and (5) were studied in papers where the study of map (2) led to the construction of a generalized Artin hypothesis and its solution [8, 9, 11]. Based on the results of these studies, interesting conclusions can be drawn. Note that the logistic map:

$$x_{n+1} = 4x_n(1 - x_n) \tag{6}$$

As an object of numerous studies [10] on a set of numbers of the form $1/p$, it behaves in a certain sense "similarly" to maps (4) and (5). By analogy it is ment the congruence of these maps, which was studied in [13] for maps (4) and (6). The congruence of the two mappings $f(x)$ and $g(x)$ on the set [0, 1] suggests that there is a one-to-one correspondence between their cyclic fixed points for which the lengths of the cyclic trajectories of the congruent points coincide, but the topological structure is different. The author of the congruence method in [13] described a method for proving the congruence of maps on a distinguished set from the domain of definition of maps.

Based on the results of modeling the processes of formation of classes $\mathcal{P}(a, i)$ for any $a \neq i\& \pm 1$ and $i \in \{1, 2, \ldots, n, ..\}$, theorems on congruence of maps, analysis of trajectories of maps (2), (4), (5) it is not difficult to prove the validity of the following statements.

Statement 1 The map (4) is congruent to the map (2) for $a = 4$ on the set of all primes $p \in \mathcal{P}$, but for any prime number their trajectories in the chaos structure do not coincide.

This statement means that, on the basis of map (4), we obtain the same system of primes in the generalized Artin hypothesis as on the basis of map (2), although map (2) is discrete and map (2) is continuous. As shown in [12], the display paths (2) represent a sequence of self-similar successive fragments of the trajectory located at a regular distance from each other and interconnected by successive fragments of chaotic behavior. A fragment of such a structure of cyclic trajectories of fixed points $1/p$ is shown in Fig. 1 where solid line represents sequence fragments with high degree of similarity and dashed line represent chaotic behavior of the map.

It should be noted that a completely non-algebraic function of the "tent" type allows one to solve a problem from number theory related to the discrete logarithm problem, modern cryptography and problems of constructing pseudorandom number generators, and, on the other hand, stimulates a deeper analysis of the trajectories of cyclic fixed points of dynamical systems in order to identify areas where chaotic

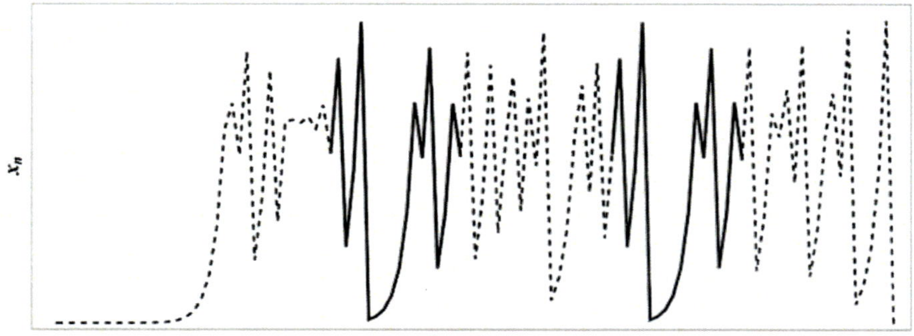

Fig. 1 Sequence with self-similar successive fragments

behavior approaches its non-deterministic models. A similar analysis of the map (5) allows us to prove the following statement.

Statement 2 The map (5) is congruent to the map (2) with $a = 2$ on the set of all primes $p \in \mathcal{P}$, but their trajectories do not coincide in structure for any p.

From this statement it follows that map (6) forms a system of classes of primes of the form:

$$\mathcal{P}(2, 1), \mathcal{P}(2, 2), \ldots, \mathcal{P}(2, i), \ldots$$

This means that $\mathcal{P}(2, 1) = \{p \in \mathcal{P} | (p - 1)/card_2(p) = ind_2(p) = 1\}$, i.e. these are all prime numbers for which the number $a = 2$ is their primitive root. The class $\mathcal{P}(2, i)$ consists of the set of all primes p for which in the group $(Z/pZ)^*$ a subgroup with index i has the number $a = 2$ by its generating element. Thus, maps of the "tent" type with an appropriate choice of parameter allow to solve problems of modern number theory. In addition, it follows from statements 1 and 2 that the properties of dynamical systems fundamentally depend on the properties of the functions that determine them and on the properties of the number of some fundamental sequences from the domain of their definition.

3 Conclusions

An analysis of the chaos structure in the cyclic trajectories of fixed points of dynamical systems reveals formation patterns of their trajectories based on the properties of the maps that determine the dynamical system. The parameters of the maps determine the classification of numbers from the domain of their definition whose properties are functions of the properties of parameters. Various studies in this direction allow us to deepen an understanding of the mechanisms of chaos formation and its dynamical structure.

References

1. E. Mishchenko, V. Sadovnichy, Many-Sided Chaos (Moscow, 2012)
2. C. Skiadas, I. Dimotikalis, *Chaotic Systems: Theory and Applications* (World Scientific, 2010)
3. C. Skiadas, C. Skiadas, *Handbook of Applications of Chaos Theory* (Chapman & Hall/CRC, 2016)
4. P. Cvitanović, R. Artuso, R. Mainieri, G. Tanner, G. Vattay, *Chaos: Classical and Quantum* (Niels Bohr Institute, Copenhagen, 2016)
5. E. Schöll, H.G. Schuster, *Handbook of Chaos Control* (Wiley-VCH, 2014)
6. H. Rogers, *Theory of Recursive Functions and Effective Computability* (MIT Press, Cambridge, Mass, 2011)
7. V. Uspenskii, A. Semenov, A. Shen, Can an individual sequence of zeros and ones be random? Russian Math. Surveys **45**(1), 121–189 pp (1990)

8. R. Crandall, C. Pomerance, *Prime Numbers* (Springer Science + Business Media Inc, New York, NY, 2005)
9. Y. Manin, A. Panchishkin, *Introduction to Modern Number Theory* (Springer, Berlin Heidelberg, Berlin, Heidelberg, 2005)
10. B. Hasselblatt, A. Katok, *A First Course in Dynamics: with a Panorama of Recent Developments* (Cambridge University Press, Cambridge, 2003)
11. *Computer Data Analysis and Modeling: Stochastics and Data Science: Proc. of the Twelfth Intern. Conf. Minsk: BSU* (2019). 319–325 pp
12. G. Vostrov, A. Khrinenko, Sequence internal structure formation during pseudorandom generation. Electr. Comput. Syst. **29**(105), 164–168 (2018)
13. J. Rauch, Conjugation of Logistic and Tent Maps (2014)

Stabilization in the Instability Region Around the Triangular Libration Points for the Restricted Three-Body Problem

Asher Yahalom and Natalia Puzanov

Abstract Stabilized restricted three-body problem in which the motion of third body is planar and circular is presented. The instability region of triangular libration points is stabilized by feedback control in integral form.

1 Introduction

We consider the plane elliptic restricted three-body problem. The differential equations of this problem in the Nechville coordinates (ξ, η) have the form [1, 3, 4]:

$$\begin{cases} \xi'' - 2\eta' = \rho(\xi - \mu + \frac{\mu-1}{(\xi^2+\eta^2)^{\frac{3}{2}}}\xi - \frac{\mu}{[(\xi-1)^2+\eta^2]^{\frac{3}{2}}}(\xi - 1)) \\ \eta'' + 2\eta' = \rho(\eta + \frac{\mu-1}{(\xi^2+\eta^2)^{\frac{3}{2}}}\eta - \frac{\mu}{[(\xi-1)^2+\eta^2]^{\frac{3}{2}}}\eta) \end{cases} \tag{1}$$

where

$$\rho = \frac{1}{1 + \epsilon \cos t}, \; \mu = \frac{m_1}{m_0 + m_1},$$

ϵ is the eccentricity of the Keplerian orbit ($0 \leq \epsilon < 1$), t is the true anomaly, m_0 and m_1 are the masses of actively gravitating bodies; thus $0 < \mu < 1$. Those equations are derived in the appendix.

A. Yahalom (✉) · N. Puzanov
Ariel University, Ariel 40700, Israel
e-mail: asya@ariel.ac.il

N. Puzanov
e-mail: puzanovn@ariel.ac.il

N. Puzanov
Princeton University, Princeton, NJ 08543, USA

C. H. Skiadas et al. (eds.), *13th Chaotic Modeling and Simulation International Conference*, Springer Proceedings in Complexity,
https://doi.org/10.1007/978-3-030-70795-8_74

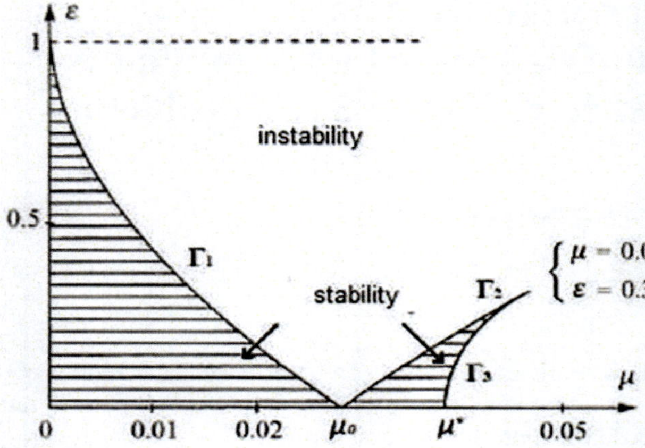

Fig. 1 The stability region of triangular libration points

The system (1) has five constant solutions—libration points: straight-line L_1, L_2, and triangular L_4 and L_5. In the plane of the variables (ξ, η) of the system (1), the straight-line libration points lie on the line $\eta = 0$, and the triangular libration points have the coordinates:

$$L_4\left(\frac{1}{2}, \frac{\sqrt{3}}{2}\right), \; L_5\left(\frac{1}{2}, -\frac{\sqrt{3}}{2}\right). \tag{2}$$

We will be interested in questions related to the stability of triangular libration points.

Figure 1 [6] describes the regions of stability and instability of the system (1) for small values of μ. The shaded area corresponds to stability region in parameter space.

Here

$$\mu^* = \frac{1}{2} - \frac{\sqrt{69}}{18} = 0.038520....,\quad \mu_0 = \frac{1}{2} - \frac{\sqrt{2}}{3} = 0.028595.... \tag{3}$$

Numerous studies addressing the regions of stability of the system (1) and the behavior of the solutions of this system near the boundaries of these regions of stability are due to the importance of these questions for celestial mechanics. However, so far no attempts have been made to stabilize system (1) in the region of instability of the libration point.

We will pass from system (1) to equivalent normal system by introduction of new variables $u_1 = \xi$, $u_2 = \eta$, $u_3 = \xi'$, $u_4 = \eta'$

$$\begin{cases} u_1' = u_3, \\ u_2' = u_4, \\ u_3' = 2u_4 + \rho(u_1 - \mu + \frac{\mu-1}{(u_1^2+u_2^2)^{\frac{3}{2}}}u_1 - \frac{\mu}{[(u_1-1)^2+u_2^2]^{\frac{3}{2}}}(u_1 - 1)) \\ u_4' = -2u_3 + \rho(u_2 + \frac{\mu-1}{(u_1^2+u_2^2)^{\frac{3}{2}}}u_2 - \frac{\mu}{[(u_1-1)^2+u_2^2]^{\frac{3}{2}}}u_2) \end{cases} \quad (4)$$

We thus have a system of the form:

$$u' = F(u, \epsilon, \mu, t), \quad u \in R^4 \quad (5)$$

$F(u, \epsilon, \mu, t)$ is the vector function defined by the right part of system (4). Libration points of system (1) correspond to constant solutions of system (5). In particular triangular libration points L_4 and L_5 correspond to following constant solutions of system (5)

$$v_4 = \begin{bmatrix} \frac{1}{2} \\ \frac{\sqrt{3}}{2} \\ 0 \\ 0 \end{bmatrix} \quad v_5 = \begin{bmatrix} \frac{1}{2} \\ -\frac{\sqrt{3}}{2} \\ 0 \\ 0 \end{bmatrix} \quad (6)$$

The behaviour of system (5) is the same in neighborhoods of libration points v_4 and v_5. For definiteness we will study behaviour of system (5) in neighborhood of libration point v_4.

Carrying out in (4) the substitution $X = u - v_4$, we arrive at the equivalent system.

$$\begin{cases} X'(t) = Z(t), \\ Y'(t) = W(t), \\ Z'(t) = 2W(t) + \rho\left[\left(X(t) + \frac{1}{2}\right) - \mu + \frac{\mu-1}{\left[\left(X(t)+\frac{1}{2}\right)^2+\left(Y(t)+\frac{\sqrt{3}}{2}\right)^2\right]^{\frac{3}{2}}}\left(X(t) + \frac{1}{2}\right) - \frac{\mu}{\left[\left(X(t)-\frac{1}{2}\right)^2+\left(Y(t)+\frac{\sqrt{3}}{2}\right)^2\right]^{\frac{3}{2}}}\left(X(t) - \frac{1}{2}\right)\right] \\ W'(t) = -2Z(t) + \rho\left[\left(Y(t) + \frac{\sqrt{3}}{2}\right) + \frac{\mu-1}{\left[\left(X(t)+\frac{1}{2}\right)^2+\left(Y(t)+\frac{\sqrt{3}}{2}\right)^2\right]^{\frac{3}{2}}}\left(Y(t) + \frac{\sqrt{3}}{2}\right) - \frac{\mu}{\left[\left(X(t)-\frac{1}{2}\right)^2+\left(Y(t)+\frac{\sqrt{3}}{2}\right)^2\right]^{\frac{3}{2}}}\left(Y(t) + \frac{\sqrt{3}}{2}\right)\right] \end{cases} \quad (7)$$

The libration point v_4 of the system (5) corresponds to an equilibrium point $X = 0$ of the system (7). System (7) can be represented as

$$X' = A(\epsilon, \mu, t)X + a(X, \epsilon, \mu, t), \qquad X \in R^4 \tag{8}$$

in which $A(\epsilon, \mu, t) = F'_X(0, \epsilon, \mu, t)$ is the Jacobi matrix of the vector function $F(X, \epsilon, \mu, t)$ calculated at the point $X = 0$, and $a(X, \epsilon, \mu, t)$ is a nonlinearity which begins with terms quadratic in X. At $\epsilon = 0$ (a circular case) the system (7) is autonomous.

The matrix $A(\epsilon, \mu, t)$ is equal to

$$A(\epsilon, \mu, t) = \begin{bmatrix} 0 & 0 & 1 & 0 \\ 0 & 0 & 0 & 1 \\ \frac{3}{4}\rho & 3\frac{\sqrt{3}}{4}(1-2\mu)\rho & 0 & 1 \\ 3\frac{\sqrt{3}}{4}(1-2\mu)\rho & \frac{9}{4}\rho & -2 & 0 \end{bmatrix} \tag{9}$$

To investigate our problem it suffices to consider the linear equation:

$$X' = A(\epsilon, \mu, t)X. \tag{10}$$

Since for $0 \le \epsilon < 1$ the equality

$$\rho = \frac{1}{1 + \epsilon \cos t} = 1 - \epsilon \cos t + \epsilon^2 cos^2 t - \epsilon^3 cos^3 t + ... \tag{11}$$

is correct then the matrix (9) can be represented in the form:

$$A(\epsilon, \mu, t) = A_0(\mu) + (-\epsilon \cos t)A_1(\mu) + ... \tag{12}$$

where

$$A_0(\mu) = \begin{bmatrix} 0 & 0 & 1 & 0 \\ 0 & 0 & 0 & 1 \\ \frac{3}{4} & 3\frac{\sqrt{3}}{4}(1-2\mu) & 0 & 1 \\ 3\frac{\sqrt{3}}{4}(1-2\mu) & \frac{9}{4} & -2 & 0 \end{bmatrix} \tag{13}$$

and for $\mu = \mu_0 = \frac{1}{2} - \frac{\sqrt{2}}{3}$

$$A_0 = A_0(\mu_0) = \begin{bmatrix} 0 & 0 & 1 & 0 \\ 0 & 0 & 0 & 1 \\ \frac{3}{4} & \frac{\sqrt{6}}{2} & 0 & 2 \\ \frac{\sqrt{6}}{2} & \frac{9}{4} & -2 & 0 \end{bmatrix}, \tag{14}$$

The linear approximation (10) for the planar case is:

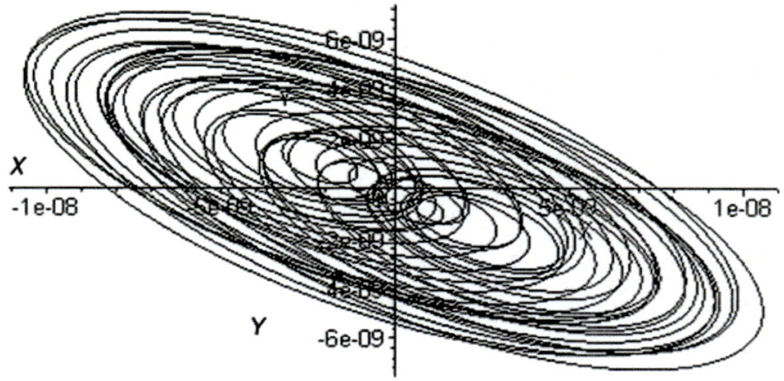

Fig. 2 The phase portrait of the solution of the system (7) for $\rho = 0$ in the (X, Y) plane with initial conditions: $X(0) = Y(0) = Z(0) = W(0) = 0$ for interval $200 \leq t \leq 500$

$$X' = A_0(\mu_0)X. \tag{15}$$

has the following simple eigenvalues:

$$\lambda_{1,2} = \pm\frac{1}{2}, \qquad \lambda_{3,4} = \pm\frac{\sqrt{3}}{2} \tag{16}$$

In accordance with the Fig. 1, system (7) is not stable at $\mu = \mu_0$, $\epsilon = 0$ $(\rho = 1)$ (Fig. 2).

2 Exponential Stabilization of Triangular Libration Points in Instability Region by Feedback Delay Control in Integral Form

We introduce the feedback delay control $V(t)$ in the form, in which all the history of the process W(t) is taken into account [5].

$$V(t) = \int_0^t e^{-\beta(t-s)} W(s)ds, \tag{17}$$

We apply stabilization by the feedback delay control signal to the system (7) assuming that the control signal $V(t)$ acts only in the first equation.

$$\begin{cases} X'(t) = Z(t) - \alpha \int\limits_0^t e^{-\beta(t-s)} W(s) ds, \\ Y'(t) = W(t), \\ Z'(t) = 2W(t) + \rho \left[\left(X(t) + \frac{1}{2}\right) - \mu + \frac{\mu - 1}{\left[\left(X(t)+\frac{1}{2}\right)^2 + \left(Y(t)+\frac{\sqrt{3}}{2}\right)^2\right]^{\frac{3}{2}}} \left(X(t) + \frac{1}{2}\right) - \right. \\ \qquad \left. \frac{\mu}{\left[\left(X(t)-\frac{1}{2}\right)^2 + \left(Y(t)+\frac{\sqrt{3}}{2}\right)^2\right]^{\frac{3}{2}}} \left(X(t) - \frac{1}{2}\right) \right], \\ W'(t) = -2Z(t) + \rho \left[\left(Y(t) + \frac{\sqrt{3}}{2}\right) + \frac{\mu - 1}{\left[\left(X(t)+\frac{1}{2}\right)^2 + \left(Y(t)+\frac{\sqrt{3}}{2}\right)^2\right]^{\frac{3}{2}}} \left(Y(t) + \frac{\sqrt{3}}{2}\right) - \right. \\ \qquad \left. \frac{\mu}{\left[\left(X(t)-\frac{1}{2}\right)^2 + \left(Y(t)+\frac{\sqrt{3}}{2}\right)^2\right]^{\frac{3}{2}}} \left(Y(t) + \frac{\sqrt{3}}{2}\right) \right], \end{cases} \tag{18}$$

where α and β the parameters needed to achieve the point $(X, Y, Z, W) = (0, 0, 0, 0)$ at which the system becomes exponentially stable. In accordance with the Leibnitz rule for differentiation under the integral sign, of the form $\frac{d}{dy} \int_{a(y)}^{b(y)} f(x, y) dt$ we get the expression for $V'(t)$. Thus we can rewrite the system (18) in a form of the system of ordinary differential equations [2]:

$$\begin{cases} X'(t) = Z(t) - \alpha V(t), \\ Y'(t) = W(t), \\ Z'(t) = 2W(t) + \rho \left[\left(X(t) + \frac{1}{2}\right) - \mu + \frac{\mu - 1}{\left[\left(X(t)+\frac{1}{2}\right)^2 + \left(Y(t)+\frac{\sqrt{3}}{2}\right)^2\right]^{\frac{3}{2}}} \left(X(t) + \frac{1}{2}\right) - \right. \\ \qquad \left. \frac{\mu}{\left[\left(X(t)-\frac{1}{2}\right)^2 + \left(Y(t)+\frac{\sqrt{3}}{2}\right)^2\right]^{\frac{3}{2}}} \left(X(t) - \frac{1}{2}\right) \right], \\ W'(t) = -2Z(t) + \rho \left[\left(Y(t) + \frac{\sqrt{3}}{2}\right) + \frac{\mu - 1}{\left[\left(X(t)+\frac{1}{2}\right)^2 + \left(Y(t)+\frac{\sqrt{3}}{2}\right)^2\right]^{\frac{3}{2}}} \left(Y(t) + \frac{\sqrt{3}}{2}\right) - \right. \\ \qquad \left. \frac{\mu}{\left[\left(X(t)-\frac{1}{2}\right)^2 + \left(Y(t)+\frac{\sqrt{3}}{2}\right)^2\right]^{\frac{3}{2}}} \left(Y(t) + \frac{\sqrt{3}}{2}\right) \right], \\ V'(t) = W(t) - \beta V(t). \end{cases} \tag{19}$$

In the linear approximation, system (19) can be represented as

$$X' = JX, \tag{20}$$

where J is Jacobi matrix for the right side of the system (19)

$$J = \begin{bmatrix} 0 & 0 & 1 & 0 & -\alpha \\ 0 & 0 & 0 & 1 & 0 \\ \frac{3}{4} & \frac{\sqrt{6}}{2} & 0 & 2 & 0 \\ \frac{\sqrt{6}}{2} & \frac{9}{4} & -2 & 0 & 0 \\ 0 & 0 & 0 & 1 & -\beta \end{bmatrix} \tag{21}$$

$$det(J - \lambda I) = \begin{vmatrix} -\lambda & 0 & 1 & 0 & -\alpha \\ 0 & -\lambda & 0 & 1 & 0 \\ \frac{3}{4} & \frac{\sqrt{6}}{2} & -\lambda & 2 & 0 \\ \frac{\sqrt{6}}{2} & \frac{9}{4} & -2 & -\lambda & 0 \\ 0 & 0 & 0 & 1 & -\lambda - \beta \end{vmatrix} \tag{22}$$

Characteristic equation is

$$-\left(\lambda^5 + \beta\lambda^4 + \lambda^3 + \left(\beta + \frac{\sqrt{6}}{2}\alpha\right)\lambda^2 + \left(-\frac{3}{2}\alpha + \frac{3}{16}\right)\lambda + \frac{3}{16}\beta\right) = 0 \tag{23}$$

According to the Hurwitz criterion for a 6-th order system all roots of the characteristic equation (23) have negative real parts if and only if

$$\begin{aligned} &a_1 = \beta > 0 \\ &a_2 = 1 > 0, \\ &a_3 = \beta + \frac{\sqrt{6}}{2}\alpha > 0, \\ &a_4 = -\frac{3}{2}\alpha + \frac{3}{16} > 0, \\ &a_5 = \frac{3}{16}\beta > 0, \\ &a_1a_2 - a_3 = -\frac{\sqrt{6}}{2}\alpha > 0, \\ &(a_1a_2 - a_3)(a_3a_4 - a_2a_5) - (a_1a_4 - a_5)^2 = \\ &\qquad \frac{3}{32}\alpha^2\left(-24\beta^2 + 8\sqrt{6} + 24\alpha - 3\right) > 0. \end{aligned} \tag{24}$$

From (24) we obtain:

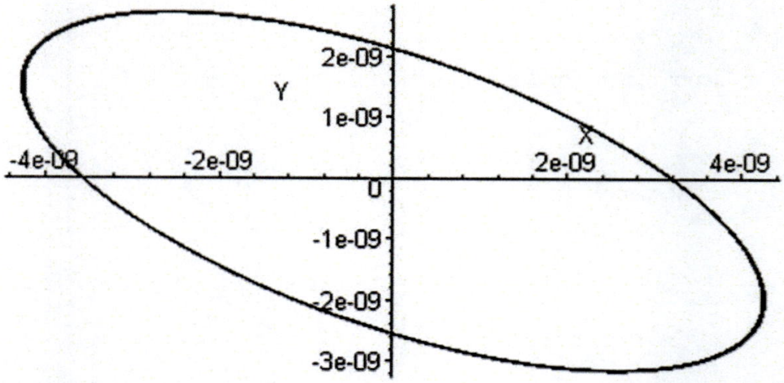

Fig. 3 The phase portrait of the solution of the system (19) for $\rho = 0$ in the (X, Y) plane with initial conditions: $X(0) = Y(0) = Z(0) = W(0) = V(0) = 0$ for interval $200 \le t \le 500$ with control parameters: $\alpha = -0.029$, $\beta = 0.5$

$$
\begin{aligned}
&-\frac{1}{24} < \alpha < 0, \\
&\frac{\sqrt{6}}{6} - \frac{\sqrt{6+144\alpha}}{6} < \beta < \frac{\sqrt{6}}{6} + \frac{\sqrt{6+144\alpha}}{6}
\end{aligned}
\tag{25}
$$

Numerical integration of the system (19) illustrated in Fig. 3.

3 Conclusion

Comparison of (2) and (3) show quite clearly the effect of stabilization on a planetary orbit. The dramatic change from a chaotic orbit to a regular elliptic orbit is quite apparent. This is achieved by adding an additional degree of freedom which stabilizes the orbit (see (19)), this approach seems quite general and may be used for many chaotic systems.

Appendix: Derivation of Nechville Differential Equations for Bounded, Restricted Three-Body Problem

We define our problem as follows: Two bodies (M_0 and M_1) revolve around their center of mass G under the influence of their mutual gravitational attraction and a third body M_2 which attracted by the previous two but not influence in their motion, movies on the plane defined by the two revolving bodies (Fig. 4). The restricted problem of three bodies is to describe the motion of this third body (M_2).

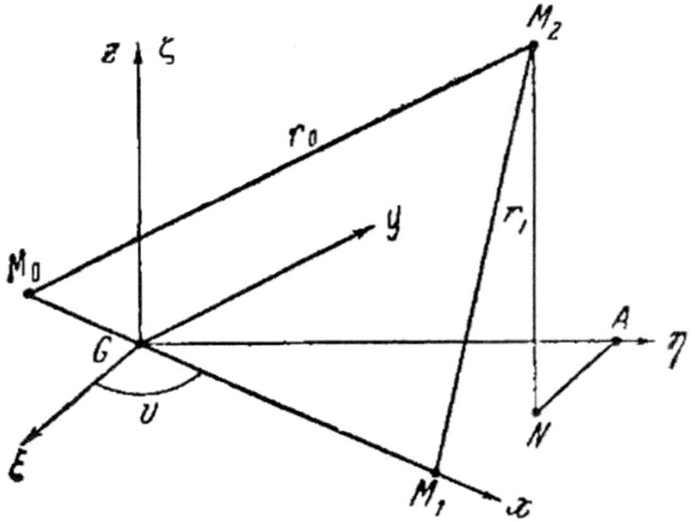

Fig. 4 To the derivation of the equations of motion presented below

The masses of bodies M_0 and M_1 are arbitrary but these bodies have such internal mass distributions that they may be considered point masses. The mass of third body M_2 does not influence the motion of M_0 and M_1.

Let $G\xi\eta\zeta$ is a coordinate system, in the plane $(\xi\eta)$ in which the point M_1 moves.

The equation of motion of M_2 ("zero" mass) in inertial (fixed) rectangular coordinate system $(\xi\eta\zeta)$ are

$$\ddot{\xi} = -\frac{\partial U}{\partial \xi}, \quad \ddot{\eta} = -\frac{\partial U}{\partial \eta}, \quad \ddot{\zeta} = -\frac{\partial U}{\partial \zeta}, \tag{26}$$

where

$$U = f\left(\frac{m_0}{r_0} + \frac{m_1}{r_1}\right), \tag{27}$$

f is constant of gravitation; m_0 and m_1 are masses of bodies M_0 and M_1, and mutual distances are determined by formulas:

$$\begin{cases} r_0^2 = (\xi - \xi_0)^2 + (\eta - \eta_0)^2 + \zeta^2 \\ r_1^2 = (\xi - \xi_1)^2 + (\eta - \eta_1)^2 + \zeta^2 \\ \zeta_0 = \zeta_1 = 0, \end{cases} \tag{28}$$

where ξ_0, η_0 and ξ_1, η_1 are coordinates of points M_0 and M_1 in the system $G\xi\eta\zeta$.

These coordinates are determined by obvious formulas.

$$\begin{cases} (m_0 + m_1)\xi_0 = -m_1 r \cos v, & (m_0 + m_1)\eta_0 = -m_1 r \sin v, \\ (m_0 + m_1)\xi_1 = +m_0 r \cos v, & (m_0 + m_1)\eta_1 = +m_0 r \sin v. \end{cases} \tag{29}$$

Values of r and v are known functions of time defined by the Keplerian motion formulas. $r = \overline{M_0 M_1}$; v is the angle of vector r with the positive direction of the axis $G\xi$ (the true anomaly).

Orbit of point M_1 in the plane $G\xi\eta$ is ellipse with focus at the point M_0. It is determined by the equation.

$$r = \frac{p}{1 + \epsilon \cos v} \tag{30}$$

where p is a focal parameter, ϵ is the eccentricity of the Keplerian orbit ($0 \le \epsilon < 1$), v is the true anomaly.

The expression for the kinetic energy of the point M_2 is

$$T = \frac{1}{2} m_2 (\dot{\xi}^2 + \dot{\eta}^2 + \dot{\zeta}^2) \tag{31}$$

Let us move in (26) from the fixed axis system $G\xi\eta\zeta$ to the rotating one around $G\zeta$ axis, so that the new abscissa passes through the points M_0 and M_1. Denoting the coordinates of the point M_2 in the new coordinate system: x, y, z we have

$$\begin{cases} \xi = x \cos v - y \sin v, \\ \eta = x \sin v + y \cos v \\ \zeta = z, \end{cases} \tag{32}$$

where v is the same angle as in formula (29), i.e. true anomaly of the Keplerian movement. Whence by means of differentiation on time one find derivatives

$$\begin{cases} \dot{\xi} = \dot{x} \cos v - \dot{y} \sin v - \dot{v}(x \sin v - y \cos v) \\ \dot{\eta} = \dot{x} \sin v + \dot{y} \cos v + \dot{v}(x \cos v - y \sin v) \\ \dot{\zeta} = \dot{z}. \end{cases} \tag{33}$$

The expression for the kinetic energy (31) in the new coordinates will take the form

$$T = \frac{1}{2} m_2 (\dot{x}^2 + \dot{y}^2 + \dot{z}^2 + (2\dot{v}(x\dot{y} - y\dot{x}) + \dot{v}^2(x^2 + y^2). \tag{34}$$

Expressions for partial derivatives of kinetic energy take the form

$$\begin{cases} \dfrac{\partial T}{\partial \dot{x}} = m_2(\dot{x} - \dot{v}y), & \dfrac{\partial T}{\partial x} = m_2\dot{v}(\dot{y} + \dot{v}x), \\ \dfrac{\partial T}{\partial \dot{y}} = m_2(\dot{y} + \dot{v}x), & \dfrac{\partial T}{\partial y} = m_2\dot{v}(-\dot{x} + \dot{v}y), \\ \dfrac{\partial T}{\partial \dot{z}} = m_2\dot{z}, & \dfrac{\partial T}{\partial z} = 0. \end{cases} \tag{35}$$

Substituting these expressions into the Lagrange equation

$$\frac{d}{dt}\left(\frac{\partial T}{\partial \dot{q}_j}\right) - \frac{\partial T}{\partial q_j} = \frac{\partial U}{\partial q_j}, \quad (j = 1, 2, 3). \tag{36}$$

We will get the equations of motion of a point M_2 in the rotating axes

$$\begin{cases} \ddot{x} - 2\dot{v}\dot{y} - \dot{v}^2 x - \ddot{v}y = \dfrac{\partial U}{\partial x} \\ \ddot{y} + 2\dot{v}\dot{x} - \dot{v}^2 y + \ddot{v}x = \dfrac{\partial U}{\partial y} \\ \ddot{z} = \dfrac{\partial U}{\partial z} \end{cases} \tag{37}$$

where U is defined by (27) but the distances r_0 and r_1 taking into account (32) are given by the formulas

$$\begin{cases} r_0^2 = (x - x_0)^2 + y^2 + z^2 \\ r_1^2 = (x - x_1)^2 + y^2 + z^2 \\ x_0 = -\dfrac{m_1 r}{m_0 + m_1} \\ x_1 = -\dfrac{m_0 r}{m_0 + m_1}. \end{cases} \tag{38}$$

Further since

$$r^2\dot{v} = c = const \tag{39}$$

where c is area integral in the orbit plane, we have

$$\dot{v} = \frac{c}{p^2}(1 + e\cos v)^2; \quad \ddot{v} = -\frac{2c^2 e}{p^4}(1 + e\cos v)^4 \tag{40}$$

In (37), we make a transition to a pulsating coordinate of system $G\tilde{\xi}\tilde{\eta}\tilde{\zeta}$ using formulas

$$x = \rho\tilde{\xi}, \;\; y = \rho\tilde{\eta}, \;\; z = \rho\tilde{\zeta} \tag{41}$$

where

$$\rho = \frac{r}{p} = \frac{1}{1 + e\cos v}, \tag{42}$$

and following Nechville [5] we input a new independent variable, namely the true anomaly v of the body M_1.

Then, as is simply to verify, we have

$$\begin{cases} \dot{x} = (\rho'\xi + \rho\tilde{\xi}')\dot{v} \\ \ddot{x} = (\rho''\xi + 2\rho'\xi' + \rho\xi'')\dot{v}^2 + (\rho'\xi + \rho\xi')\dot{v}\dot{v}' \end{cases} \tag{43}$$

(hatchs denote differentiation by variable v). Similarly, we can obtain formulas for two other coordinates.

Substituting the expressions for the old coordinates and their derivatives and the expressions for $\dot{v}$ and $\ddot{v}$ from (40) into (37), we obtain as a result, instead of system (37), the following

$$\begin{cases} \tilde{\xi}'' - 2\tilde{\eta} = \dfrac{\partial \Omega}{\partial \tilde{\xi}} \\ \tilde{\eta}'' - 2\tilde{\xi} = \dfrac{\partial \Omega}{\partial \tilde{\eta}} \\ \tilde{\zeta}'' = \dfrac{\partial \Omega}{\partial \tilde{\zeta}} \end{cases} \tag{44}$$

where

$$\begin{gathered} \Omega = \rho\left[\tfrac{1}{2}(\tilde{\xi}^2 + \tilde{\eta}^2 + e\tilde{\zeta}^2\cos v) + p^3\left(\tfrac{1-\mu}{\tilde{r}_1} + \tfrac{\mu}{\tilde{r}_2}\right)\right], \\ \tilde{r}_1 = \sqrt{(\tilde{\xi} - \tilde{\xi}_1)^2 + \tilde{\eta}^2 + \tilde{\zeta}^2},\ \tilde{r}_2 = \sqrt{(\tilde{\xi} - \tilde{\xi}_2)^2 + \tilde{\eta}^2 + \tilde{\zeta}^2}, \\ \tilde{\xi}_1 = -p\mu,\ \tilde{\xi}_2 = -p(1-\mu), \\ \mu = m_1/(m_0 + m_1),\ 1 - \mu = m_0/(m_0 + m_1). \end{gathered} \tag{45}$$

References

1. G.N. Duboshin, *Celestial Mechanics: Analytical and Qualitative Methods* (Nauka, Moscow, 1978) [in Russian]
2. Y. Goltser, A. Domoshnitsky, Bifurcation and stability of integro-differential equations. Nonlinear Anal. Theory, Methods Appl. **47**, 953–967 (2001)
3. C. Marchal, *The Three-Body Problem* (Elsevier, Amsterdam, 1990)
4. A.P. Markeev, *Libration Points in Celestial Mechanics and Astrodynamics* (Nauka, Moscow, 1978) [in Russian]
5. V. Nechville, Sur une forme nouvelle des equations differentielles due probleme restreint elliptique. Compt. Rend. **182** (1926)
6. M.G. Yumagulov, O.N. Belikova, N.R. Isanbaeva, Bifurcation near boundaries of regions of stability of libration points in the three-body problem. Astron. Rep. **62**(2), 144–153 (2018)

Influence of M-Current on Dopamine Modulation of Weak PING Gamma Rhythm

Denis Zakharov and Boris Gutkin

Abstract The human brain demonstrates electrical oscillations of various frequency ranges that are associated with a number of cognitive tasks. Here we will focus on the so-called weak (clustered) gamma rhythm (20–80 Hz). Typically, in the cortex, gamma oscillations appear in neuronal networks consisting of excitatory pyramidal cells and inhibitory interneurons. This is the Pyramidal INterneuronal Gamma (PING) rhytm. The weak clustered gamma oscillations are a specific case of PING when the pyramidal cells fire in several internally synchronous clusters producing a "collective" rhythm by alternating the cluster firing. We will analyse how characteristics of the cluster states (mainly the number of clusters) depend on the intrinsic ionic currents of the PY cells (AHP- and m-currents). Since different number of clusters mean different level of PING oscillations coherence, our work links the intrinsic cellular properties of the constitutent neurons to the coherence of the gamma rhythm.

Keywords PING gamma rhythm · Weak gamma rhythm · Cluster synchronization · Spike frequency adaptation · M-current

---

D. Zakharov (✉) · B. Gutkin
Centre for Cognition and Decision Making, National Research University Higher School of Economics, Myasnitskaya St. 20, Moscow, Russia
e-mail: dgzakharov@hse.ru

B. Gutkin
e-mail: boris.gutkin@ens.fr

D. Zakharov
Institute of Biology and Biomedicine, Lobachevsky State University of Nizhni Novgorod, Gagarin Ave. 23, Nizhny Novogorod, Russia

B. Gutkin
Group of Neural Theory, LNC INSERM U960, École Normale Supérieure PSL*, University, 29 Rue D'Ulm, Paris, France

C. H. Skiadas et al. (eds.), *13th Chaotic Modeling and Simulation International Conference*, Springer Proceedings in Complexity,
https://doi.org/10.1007/978-3-030-70795-8_75

1 Introduction

During performance of cognitive tasks, the brain demonstrates electrical activity of various frequency ranges [1]. One such prominent oscillation is the gamma rhythm (20–80 Hz). It is observed during a wide variety of cognitive tasks, such as working memory, coding and information processing [2]. In comparison with the other oscillations, gamma has a higher frequency, is comparatively much more irregular, sparse and locally distributed [3]. It is well known that typically this rhythm is generated in the networks consisting of two neuronal populations of interneurons (IN) and pyramidal (PY) cells [4]. There are two basic mechanisms of gamma oscillations generation in such networks. The first is INterneuronal Gamma (ING) rhythm which are produced in the IN population [5]. In this case the PY population has negligible influence on the IN neurons and, in fact, mirrors the IN population activity. In the second mechanism, the gamma oscillations are a result of recurrent interactions between both populations. This is so called Pyramidal-INterneuronal Gamma (PING) gamma rhythm [6]. In addition, it is possible to to construct either strong or weak gamma rhythms. In the first case, both populations have almost the same firing rate account the experimental features of the gamma and the observed large difference between intrinsic frequencies of the PY cells and IN neurons, one potentially likely mechanism is that alternatively firing of synchronous PY clusters generates the gamma oscillations (see, for example, Börgers and Kopell [7], Kilpatrick and Ermetrout [8] and Krupa et al. [9]). This is a weak clustered gamma rhythm. Each PY cell, in this case, fires with the frequency, which is lower than gamma frequency, whereas the gamma oscillations are formed by collective activity of the PY clusters. The number of clusters in a cluster state and the number of PY cells in the each cluster determine and regularly fire together with a small time lag. To take into coherence of the gamma oscillations while making them sparse and irregular. In Krupa et al. [9] it was shown that the maximum number of clusters dramatically depends on the intrinsic neuronal parameters and coupling strength of interpopulation coupling, especially inhibitory ones. In particular, Krupa et al. found that the spike frequency adaptation of PY cells has a significant influence on the cluster formation process. This was confirmed by parametric modulation studies of the network parameters on the changes in cluster states (Zakharov et al. [10, 11]). For instance, it was shown that modulation of the spike frequency adaptation is the most effective in increasing of cluster number under an increase of the spike frequency adaptation parameters and a decrease, for a negative modulation of the spike frequency adaptation. The strength of the inhibitory interpopulational coupling, in turn, also substantially influenced on the cluster modulation. Higher connection strengths stabilized the cluster states with a lower number of clusters.

In this paper, we expand on Krupa et al. [9] and focus in the influence of the intrinsic parameters of the PY cells on the cluster formation process in a PING network and therefore, on the coherence of the oscillations. In particular, we take into account both spike frequency adaptation (AHP-current) and the M-current. The former is a slow spike-dependent hyperpolarizing current (biophysically speaking it is a calcium-dependent slow potassium current), the latter is voltage a voltage-dependent slow hyperpolarizing K-current. The M-current has a significant impact on the dynamics of the PY cells, controls the intrinsic cellular excitability and synaptic responses of the pyramidal neurons [see, for example, Marrion [12] and Peng et al. [13]. This current is slowly activated when the membrane potential is depolarized towards voltage levels where the spike producing currents activate, and repolarizes the neuron back to the rest state reducing neuronal excitability In addition, the M-current has no inactivation and may play a critical role for neuronal excitability, especially near the rest state. It was previously shown that the addition of this current even in the canonical theta-neuron model with adaptation results in a change of excitability type (from 1st type to 2nd one) Gutkin et al. [14]. For the hippocampal PY cells it was shown experimentally and theoretically (within the framework of Morris-Lecar equations) that the M-current, as well as shunting inhibition, can lead to the same change of excitability type and thereby significantly change dynamics of the PY cells [14–16].

2 Influence of AHP- and M-Currents on PY Cell Activity

To describe the PY neuron activity, we use the modified the Miles-Traub equations with adaptation from Krupa et al. [9] with the addition of the M-current:

$$\begin{aligned}
\frac{dv_e}{dt} =& I_{app} - g_L(v_e - E_L) - g_K n^4(v_e - E_K); \\
& - g_{Na} m_{inf,Na}(v_e)^3 F_h(n)(v_e - E_{Na}) \\
& - g_{Ca} m_{inf,Ca}(v_e)(v_e - E_{Ca}) \\
& - g_{AHP}\frac{[Ca]}{[Ca]+1}(v_e - E_K) - g_M w_e(v_e - E_K),
\end{aligned}$$

$$\frac{d[Ca]}{dt} = -\varepsilon_{Ca} I_{Ca} - \frac{[Ca]}{\tau_{Ca}},$$

$$\frac{dn}{dt} = \alpha_n(v_e)(1-n) - \beta_n(v_e)n, \tag{1}$$

$$w_e = \varepsilon\big(w_{inf}(v_e) - w_e\big)/\tau_M(v_e)$$

$$m_{inf,Na}(v_e) = \frac{\alpha_m(v_e)}{\alpha_m(vv_e) + \beta_m(v_e)},$$

$$m_{inf,Ca}(v_e) = \frac{1}{1 + e^{-(v_e+25)/2.5}},$$

where the first equation is the current balance equation giving the dynamics of v_e the membrane potential of a PY cell, [Ca] is the calcium concentration, n, $m_{inf,Na}$, $m_{inf,Ca}$ and w_e are the voltage-dependent gating variables for the various cross-membrane currents; E_x are the reversal potentials of the various ionic species. This is a biophysical neuronal model containing on the right hand side of the first equation a combination of the ionic currents: an applied current I_{app}, a leak current I_L, a fast potassium current I_K, a sodium current I_{Na}, calcium current I_{Ca}, a slow after-hyperpolarization (AHP) current I_{AHP}, and M-current I_M respectively. The AHP current is a calcium-activated slow potassium current that effectively results in the spike frequency adaptation of a PY cell. Gating functions of the currents describes by the following equations:

$$\alpha_n(v_e) = 0.32\frac{v+52}{1-e^{-\frac{v+52}{5}}},\ \beta_n(v_e) = 0.50.28e^{-\frac{v_e+57}{40}};$$

$$\alpha_m(v_e) = 0.32\frac{v_e+54}{1-e^{-\frac{v_e+54}{4}}},\ \beta_m(v_e) = 0.28\frac{v_e+27}{e^{-\frac{v_e+27}{5}}-1};$$

$$w_{inf}(v_e) = \frac{1}{1+e^{--\frac{v_e+35}{10}}},\ \tau_m(v_e) = \frac{400}{3.3e^{\frac{v_e+35}{20}}+e^{-\frac{v_e+35}{20}}} \tag{2}$$

$$F_h(n) = \max\{1-1.25n, 0\}$$

We choose the g_{AHP} and g_M as control parameters and fix the other parameters of the PY model in the following way:

$$I_{app} = 4\,\mu A,\ g_{Na} = 100\,\text{ms/cm}^2,\ g_K = 80\,\text{ms/cm}^2,\ g_{Ca} = 1\,\text{ms/cm}^2,\ E_{Na} = 50\,\text{mV},$$
$$E_K = -100\,\text{mV},\ E_{Ca} = 120\,\text{mV},\ E_L = 67\,\text{mV},\ \tau_{Ca} = 80\,\text{ms},$$
$$\varepsilon_{Ca} = 0.01\,\text{cm}^2/(\text{ms}\,\mu A)$$

Our analysis shows that both AHP- and M-currents can effectively change the frequency of the PY cell firing but there is a qualitative difference between their influence. The M-current changes the bifurcation scenario that govern the onset of the repetitive firing (Fig. 1). If the conductance of this current is zero, the transition from the rest state (stable equilibrium) to the active one (stable limit cycle) is through the bifurcation of saddle-node in invariant circle (SNIC, Fig. 1a), whereas for sufficiently large positive values of the conductance the limit cycle appears through the saddle-node bifurcation of limit cycles (Fig. 1b). In the first case, the limit cycle has zero frequency at the bifurcation point. It means that the PY cell has the 1st type of excitability and is able to generate spike trains with arbitrary small frequencies. In the second case, it is born with a finite frequency and, thus, the neuron has a minimum frequency and can fire only in a certain frequency band. It corresponds to the 2nd type of excitability. This is confirmed by the Infinitesimal phase response curves (iPRC) which were plotted for both cases in Fig. 2. The AHP-current decreases the frequency of the PY cell but keeps iPRC positive (Fig. 2a). The M-current also increases the

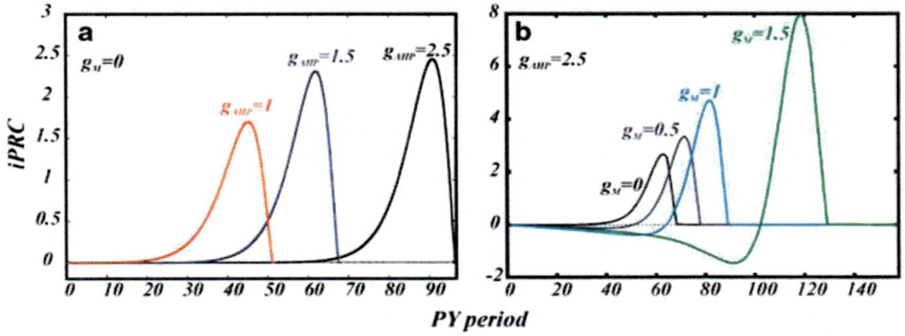

Fig. 1 Infinitesimal phase response curves (iPRC) for the different values of g_{AHP} (**a**, $g_M = 0$) and for the different values of g_M (**b**, $g_{AHP} = 2.5$). The frequency spike adaptation current keeps the 1st type of excitability ($g_M = 0$), whereas the M-current ($g_M > 0$) changes it to 2nd type

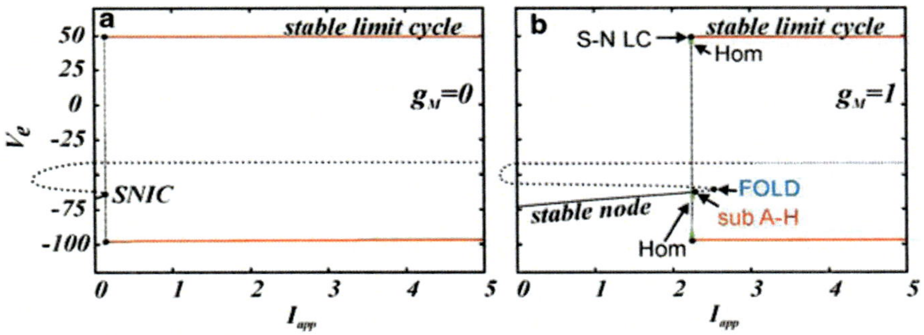

Fig. 2 One parametric bifurcation diagrams for $g_M = 0$ (**a**) and $g_M = 1$ (**b**) ($g_{AHP} = 2.5$). If $g_M = 0$ the stable equilibrium, corresponding the rest state of the neuron (the solid black curve), disappears through the SNIC bifurcation. In addition, it results in the birth of the stable limit cycle corresponding to the active state of the neuron. For values of the parameter $g_M > 0$ the stable equilibrium loses stability through the subcritical Andronov–Hopf bifurcation (A–H). An unstable limit cycle, which was also born due to the bifurcation, disappears flipping to the saddle separatrix loop (Hom, the homoclinic bifurcation) and appears once more due to another homoclinic bifurcation for the smaller value of I. In contrast to the previous case, the stable limit cycle is born by the saddle-node bifurcation of limit cycles

frequency but at the same time changes the iPRC to the 2nd type. The change of excitability type can significantly change the neurocomputational properties of the PY cells and has influence on their synchronization properties.

Both conductances of the AHP and M-currents decrease the firing rate of the PY cell the diagram showing the dependence of the frequency on both parameters has a well-expressed diagonal structure (Fig. 3). The lowest firing rate is observed in the upper right corner. We would like to note that the increase of spike frequency adaptation by the AHP does not lead to activity suppression for biologically relevant values, making this parameter to be effective for firing rate control of the PY cells.

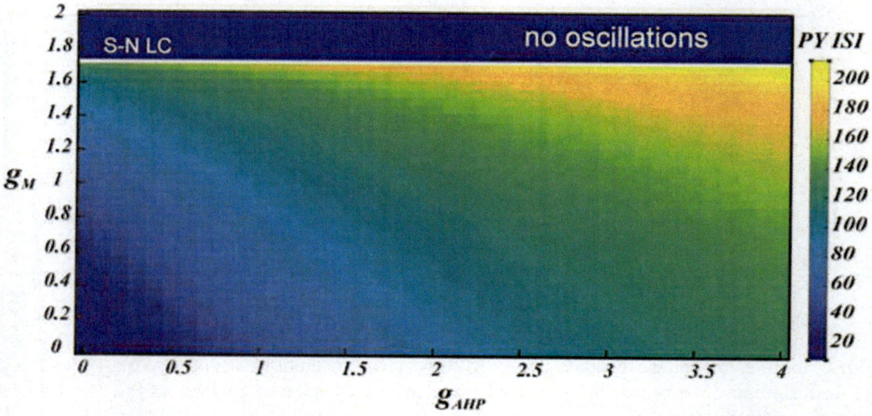

Fig. 3 Dependence of the PY cell period on the control parameter plane (g_{AHP}, g_M) ($I_{app} = 4$). The period increases with growth of both control parameters. Increase of g_M leads to the disappearance of the stable limit cycle through the saddle-node bifurcation (S–N LC)

In contrast, the PY cells can demonstrate activity only for the certain interval of the conductance of the M-current, its increase leads to the disappearance of the stable limit cycle through the saddle-node bifurcations of limit cycles.

3 Influence of AHP- and M-Currents on Cluster Formation

To describe the generation of the PING rhythm, the network model should consist of two interacting populations of neurons of the PY cells and IN neurons. Since we do not focus on the intrinsic properties of the interneurons, following the approach in Krupa et al. [9], we may describe the population of the IN neurons we need a minimal yet relevant model of a spiking neuron. Thus, we chose the quadratic Integrate-and-fire (QIF) model:

$$\frac{dv_i}{dt} = I_{\text{int}} - 2v_i(v_i - 1), \tag{3}$$

$$\text{reset}: \; if v_i \geq 1, v_i \rightarrow 0$$

where vi is a membrane potential of a IN neuron, Iint $= 0.52$ is a parameter determining the excitability of the IN neurons. We note that the QIF neuron is the canonical model for type I excitability and spike generation (SNIC driven spiking).

In the IN population, the neurons interact with each other and with the PY cells via inhibitory synapses (GABA synapse, gamma-aminobutyric acid). In contrast, the PY cells have projections only to the IN neurons via fast excitatory synapses (AMPA-synapses, α-amino-3-hydroxy-5-methyl-4-isoxazolepropionic acid). All existing couplings have an all-to-all topology. Thus, the network model has the following form:

$$\frac{dv_{e,j}}{dt} = \sum_j I_{ionic,j} - g_{ei}\left(\frac{1}{N_i}\sum_{k=1}^{N_i} s_{i,k}\right)\left(v_{e,j} - E^{ei}_{rev}\right),\ j = \overline{1, N_e}, \tag{4}$$

$$\frac{dv_{i,l}}{dt} = f\left(2v_{i,l}, I_{int}\right) - g_{ie}\left(\frac{1}{N_e}\sum_{k=1}^{N_e} s_{e,k}\right)\left(v_{i,l} - E^{ie}_{rev}\right)$$

$$- g_{ii}\left(\frac{1}{N_i}\sum_{k=1}^{N_i} s_{i,k}\right)\left(v_{i,l} - E^{ii}_{rev}\right),$$

where $v_{i,l}$ and $v_{i,j}$ are the membrane potential of the l-th IN neuron and j-th PY cell respectively. The neurons interact through chemical synapses: g_{ie} is the conductance of the inhibitory synapses located on the PY cells, g_{ei} and g_{ii} are the conductances of the excitatory and inhibitory synapses located on the IN neurons. Parameters E^{ei}_{rev}, E^{ie}_{rev} and E^{ii}_{rev} determine the reversal potentials of the synapses. The synaptic variables $s_{i,k}$ and $s_{e,j}$ have values between 0 and 1. They are set to 1 after each spike of the kth IN neuron and the j^{ht} PY cell E-cell and decay exponentially with time constant τ_i and τ_e. The parameter g_l, that controls intrinsic frequency of the PY cells, was uniformly distributed in the interval [0.075, 0.125]. As in Krupa et al. [9] the IN population contains 20 IN neurons, the PY population has 200 neurons.The coupling parameters were set in the following way: g_{ei}= 0.2, $g_{ii} = 10$, $\tau_i = 9$, $\tau_e = 2$, $E^{ei}_{rev} = -80$ mv, $E^{ie}_{rev} = 6.5$ and $E^{ii}_{rev} = -0.25$.

In agreement with the frequency distribution on Fig. 3, the diagrams on Fig. 4 have almost the same diagonal structure of the cluster states with different number of clusters (Fig. 4). Typically, simultaneous growth of the M-current conductance and the spike frequency adaptation AHP parameter leads to greater number of clusters. Because of much higher frequency of IN neurons, the PY cells split into several alternatively firing clusters and form a cluster state. Examples of the different cluster states are presented in Fig. 5. I may be one (Fig. 5a), two (Fig. 5b), three (Fig. 5c), four (Fig. 5d) and five cluster states, cluster states with "skipping" PY cells activity each three periods of IN neurons (Fig. 5e). It is important to note that each point at each diagram in the Fig. 4 is a result of network evolution from the randomly generated initial conditions. The network is multistable, thus, for the same parameter set but for different initial conditions, it is possible to get various cluster state with either the same or the different number of clusters. For instance, in the lower right corners of the diagrams in Fig. 4c, d there is a 2-cluster state (Fig. 5f) that coexists with some 2- and 3-cluster states. This is a result of the PING mechanism of gamma rhythm generation for which the interaction between the network populations can make their frequency multiples and form a state with a number of clusters determined by the

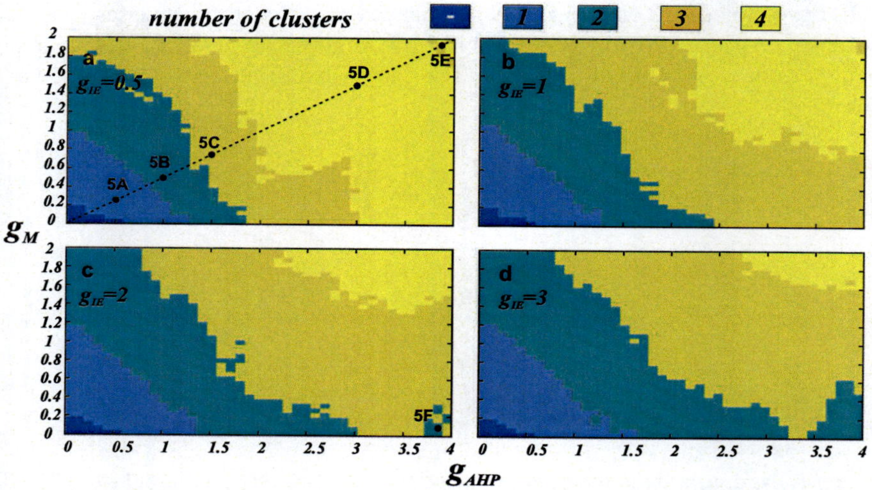

Fig. 4 Number of clusters in the cluster states for the different strengths of the interpopulational inhibitory connections. Each point on these diagrams corresponds to the number of clusters in a state to which the system has evaluated from randomly generated initial conditions. The diagrams demonstrate the well-expressed diagonal layered structure according with the changes of the PY cells period (Fig. 3). Simultaneous growth of m-current and spike frequency adaptation leads to greater number of clusters. In contrast, increasing inhibition results in smaller number of clusters. The black circles in the diagrams correspond to the rasterplots in Fig. 5. We draw your attention that here there is a multistability between different cluster states with either the same or the different numbers of clusters. For example in the region labeled 5F it is possible to get (depending on initial conditions) either 2, 3 or 4 clusters in the cluster state

ratio between the frequencies of IN and PY populations.

$$g_{ei} = 0.2,\ g_{ie} = 2,\ g_{ii} = 10,\ \tau_i = 9,\ \tau_e = 2,\ E^{eie}_{rev} = 6.5,$$
$$E^{ii}_{rev} = -0.25,\ E^{ei}_{rev} = -80\,\text{mV},\ \ E^{ee}_{rev} = 50\,\text{mV}$$

Interestingly we note that changing of excitability type of the PY cells by the M-current from type I to type II, leads to an increase in region of activity of the PY population. At first we note, that an individual PY cell with $g_L = 0.1$ does not fire for the $g_M > 1.7$ (see Fig. 3). In the PY population in our network, the parameter g_L is distributed in the interval [0.05, 0.15]. Thus, it is possible to suggest that some PY cells stop to fire before this critical value of g_M, the others above it. In contrast, all PY cells fire at least for g_M which is approximately higher two (Fig. 4). Taking into account the absence of excitatory connections between the PY cells, it is possible to conclude that such behavior (that cannot be predicted directly from single cell analysis) is likely due to inhibitory synaptic influence of the IN neurons and a generation of rebound spikes.

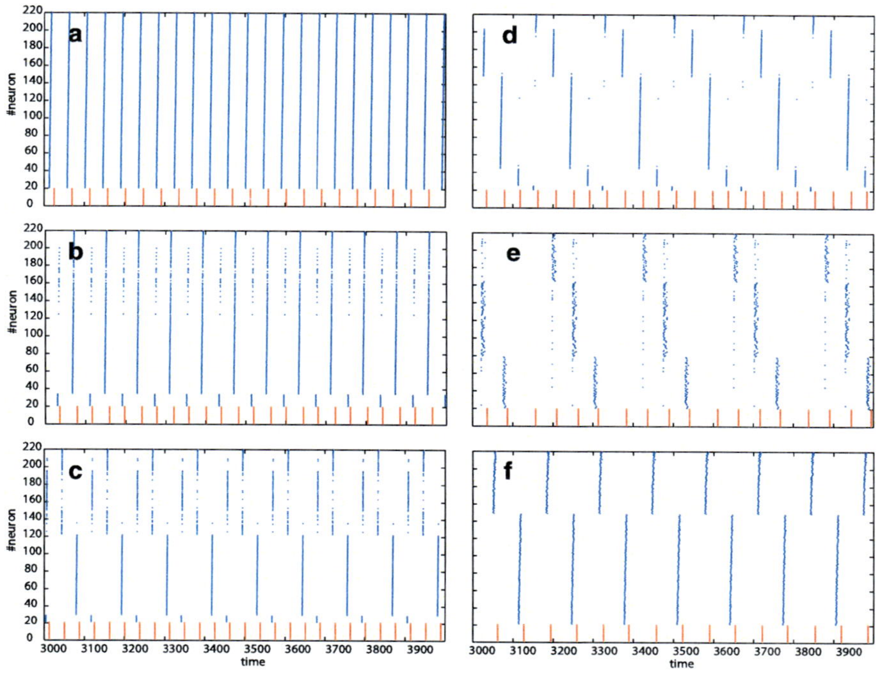

Fig. 5 Rasterplots showing different possible cluster states in the PY population for the different values of conductances of AHP and M-currents: one cluster for $g_{AHP} = 0.5$, $g_M = 0.25$ (**a**), two clusters for $g_{AHP} = 1$, $g_M = 0.5$ (**b**), three clusters for $g_{AHP} = 1.5$, $g_M = 0.75$ (**c**), four clusters for $g_{AHP} = 3$, $g_M = 1.5$ (**d**), three cluster state with "skipping" PY cells activity each three periods of IN neurons for $g_{AHP} = 3.9$, $g_M = 1.95$ (**e**) and two clusters for $g_{AHP} = 3.9$, $g_M = 0.1$ (**f**). All panels labels coincide with letters titled the points on the diagrams in the Fig. 4a, b

The inhibitory couplings from the IN neurons to the pyramidal cells have a stabilizing effect on the network. Stronger inhibition leads to lower numbers of clusters in the cluster states and, thus, increase coherence of the gamma. In particular, the size of three and especially four clusters regions decreases. It is important for investigation of modulation of such cluster states by endogenous neuromodulators (dopamine, acetylcholine and others). For example, the positive dopamine modulation can increase the PY cells inhibition and decrease the spike frequency adaptation and, vice versa, the negative dopamine modulation decreases the inhibitory connections and increases the spike frequency adaptation (see, for example, Zakharov et al. [10, 11] and cited papers). Depending on initial point in the diagrams in Fig. 4, it is possible to control the efficiency of the positive and negative dopamine modulation.

4 Conclusions

In this paper, we consider process of cluster formation in a PING network producing the weak (clustered) gamma rhythm. We have shown that the M-current significantly changes the dynamical properties of the PY cells. The M-current changes excitability type of the PY cells and decreases the region of their activity in the parameter space. In addition, the M-current, as well as AHP-current, effectively changes the frequency of the neuron and can effectively affect the cluster formation. In particular, due to the frequency drop with the increase of these currents, the color plots in the Fig. 5, showing the dependence of number of clusters in the cluster states, have the well-expressed diagonal layered structure. By changing PY cell excitability to 2nd type, the M-current also promotes the population activity of the PY cells within the PING network. In comparison with papers of Prescott et al. [14–16] we used more realistic model of the PY cell that also takes into account transport of Ca^{2+} ions. It allows us to describe spike frequency adaptation more accurately and use this model for simulation of action of endogenous neuromodulators in the brain.

In our previous studies, we have shown that DA modulation of the AHP-current can effectively change the cluster number in the cluster states of the weak PING networks and thus the coherence of their collective activity that can significantly affect information processing and decision-making. Thus, we expect that a joint modulation of AHP- and M-currents will be able to do it more efficiently. From the biological point of view, it can happen in the cases of simultaneous action of dopamine, which can affect the AHP-current [17, 18], and acetylcholine, which can decrease the M-current by, for example, muscarinic receptors [12], or due to acetylcholine modulation, which can influence both currents [19].

Acknowledgements This work was supported by the HSE Basic Research Program and the Russian Academic Excellence Project '5-100'. D.Z. acknowledges support of Russian Science Foundation Grant No. 18-11-00294 (numerical simulations). BSG acknowledges partial support from ANR project "ERMUNDY".

References

1. G. Buzsáki, *Rhythms of the Brai* (Oxford University Press, New York, 2006)
2. T. Akam, D.M. Kullmann, Oscillatory multiplexing of population codes for selective communication in the mammalian brain. Nat. Rev. Neurosci. **15**, 111–122 (2014)
3. C.T. Dickson, G. Biella, M. de Curtis, Evidence for spatial modules mediated by temporal synchronization of carbachol-induced gamma rhythm in medial entorhinal cortex. J. Neurosci. **20**, 7846–7854 (2000)
4. M. Bartos, I. Vida, P. Jonas, Synaptic mechanisms of synchronized gamma oscillations in inhibitory interneuron networks. Nat. Rev. Neurosci. **8**(1), 45–56 (2007)
5. B. Ermentrout, E.H. Buhl, Inhibition-based rhythms: experimental and mathematical observations on network dynamics. Int. J. Psychophysiol. **38**, 315–336 (2000)

6. M.A. Whittington, R.D. Traub, N. Kopell, B. Ermentrout, E.H. Buhl, Inhibition-based rhythms: experimental and mathematical observations on network dynamics. Int. J. Psychophysiol. **38**, 315–336 (2000)
7. C. Börgers, N. Kopell, Synchronization in networks of excitatory and inhibitory neurons with sparse. Random Connect., Neural Comput. **15**, 509–538 (2003)
8. Z.P. Kilpatrick, B. Ermetrout, Sparse gamma rhythms arising through clustering in adapting neuronal networks. PLoS Comput. Biol. **7**, (2011)
9. M. Krupa, S. Gielen, B. Gutkin, Adaptation and shunting inhibition leads to pyramidal/interneuron gamma with sparse firing of interneurons. J. Comput. Neurosci. **37**, 357–376 (2014)
10. D. Zakharov, V. Tyutin, M. Krupa, B. Gutkin, Modulation of synchronous gamma rhythm clusters. Cybern. Phys. **8**, 185–188 (2019)
11. D. Zakharov, M. Krupa, B. Gutkin, Modelling dopaminergic modulation of clustered gamma rhythms. Commun. Nonlinear Sci. Numer. Simul. **82**, (2020)
12. N.V. Marrion, Control of m-current. Annu. Rev. Physiol. **59**, 483–504 (1997)
13. H. Peng, X.-L. Bian, F.-C. Ma, K.-W. Wang, Pharmacological modulation of the voltage-gated neuronal Kv7/KCNQ/M-channel alters the intrinsic excitability and synaptic responses of pyramidal neurons in rat prefrontal cortex slices. Acta Pharmacol. Sin. **38**, 1248–1256 (2017)
14. B.S. Gutkin, G.B. Ermentrout, A.D. Reyes, Phase-response curves give the responses of neurons to transient inputs. J. Neurophysiol. **94**, 1623–1635 (2005)
15. S.A. Prescott, Nonlinear interaction between shunting and adaptation controls a switch between integration and coincidence detection in pyramidal neurons. J. Neurosci. **26**, 9084–9097 (2006)
16. S.A. Prescott, S. Ratté, Y. De Koninck, T.J. Sejnowski, Pyramidal neurons switch from integrators in vitro to resonators under in vivo-like conditions. J. Neurophysiol. **100**, 3030–3042 (2008)
17. S.A. Prescott, Y. De Koninck, T.J. Sejnowski, Biophysical basis for three distinct dynamical mechanisms of action potential initiation. PLoS Comput. Biol. **4**, (2008)
18. P. Pedarzani, J.F. Storm, Dopamine modulates the slow Ca^{2+} -activated K^{+} -current IAHP via cyclic AMP-dependent protein kinase in hippocampal neurons. J. Neurophysiol. **74**, 2749–2753 (1995)
19. R.A. Nicoll, The coupling of neurotransmitter receptors to ion channels in the brain. Science **241**, 545–551 (1988)

Author Index

C. H. Skiadas et al. (eds.), *13th Chaotic Modeling and Simulation International Conference*, Springer Proceedings in Complexity,
https://doi.org/10.1007/978-3-030-70795-8